Book of Abstracts of the 73rd Annual Meeting of the European Federation of Animal Science

The European Federation of Animal Science wishes to express its appreciation to the
Ministero delle Politiche Agricole Alimentari e Forestali (Italy) and the
Associazione Italiana Allevatori (Italy)
for their valuable support of its activities.

Book of Abstracts of the 73rd Annual Meeting of the European Federation of Animal Science

Porto, Portugal, 5th – 9th September, 2022

EAAP Scientific Committee:

F. Miglior
L. Pinotti
L. Boyle
D. Kenny
M. Lee
M. De Marchi
V.A.P. Cadavez
S. Millet
R. Evans
T. Veldkamp
M. Pastell
G. Pollott
H. Spoolder (chair)

EAN: 9789086863853
e-EAN: 9789086869374
ISBN: 978-90-8686-385-3
e-ISBN: 978-90-8686-937-4
DOI: 10.3920/978-90-8686-937-4

ISSN 1382-6077

First published, 2022

The individual contributions in this publication and any liabilities arising from them remain the responsibility of the authors.

The designations employed and the presentation of material in this publication do not imply the expression of any opinion whatsoever on the part of the European Federation of Animal Science concerning the legal status of any country, territory, city or area or of its authorities, or concerning the delimitation of its frontiers or boundaries.

The publisher is not responsible for possible damages, which could be a result of content derived from this publication.

Welcome to the EAAP 2022 in Porto

On behalf of the Portuguese Organizing Committee, we are honored and delighted to welcome you at the 73rd EAAP Annual Meeting being held at the wonderful world heritage city of Porto, in Portugal. The last EAAP meeting held in Portugal was in 1987. 35 years and one pandemic later, Portugal has the privilege to finally again host the annual meeting of EAAP.

The years we are living show us that our sector never stops, that animal production continues to put food in people's houses, and that we are an essential part of society. This year, recent war events at our door have put the society under high economic and societal changes. To add up we are faced with the undergoing climate urgency and still adapting to the post pandemic crisis. This conjuncture increases the challenges of Animal Science making them even more relevant than ever, with a consequent higher engagement and responsibility from the scientific community.

The program will cover various areas of knowledge, such as nutrition, genetics, physiology, animal health and welfare, livestock farming systems, precision livestock farming, insect production and use, cattle, horse, pig, sheep and goat production. These topics will be filled with innovation and recent scientific results leading animal production in the right path.

The European Federation of Animal Science (EAAP) Annual Meeting gives an opportunity for the application of new ideas in practice through many parallel sessions, poster presentations, and discussions about scientific achievements in livestock production all around the world. The Plenary Session,under the topic "The coexistence of wildlife and livestock" is a must of 2022 Porto Meeting.

Moreover, as we know, this Meeting is a privileged discussion forum where the research community meets with the industry, to discuss and plan for and how to address the multiple challenges that the animal science sector has to cope with in the upcoming years. All these activities make the EAAP Annual Meeting one of the largest animal science congresses in the world.

Of course our unforgettable social program throughout the week promotes all this scientific activities and networking even more. Starting with the welcome ceremony the programme follows with a typical Portuguese night, a gala dinner and finishes with remarkable technical tours. In parallel an exquisite accompanying persons program is available.

We hope that the 73rd Annual Meeting of EAAP: EAAP 2022, is a unique opportunity to add work with pleasure. We wish you a very pleasant stay in our beautiful city and country!

Ana Sofia Santos and Olga Moreira
Chairmen of the Portuguese Organizing Committee

National Organisers of the 73rd EAAP Annual Meeting

Porto, Portugal

5 – 9 September 2022

Members:

Portuguese Steering Committee

Presidents
Ana Sofia Santos (APEZ-FeedInov CoLab)
Olga Moreira (INIAV)
Executive Secretaries
Elisabete Mena (APEZ-UTAD)
Mariana Almeida (APEZ-UTAD)
Pedro Santos Vaz (APEZ)
Telma G. Pinto (APEZ)

Members
Amélia Ramos (APEZ-ESAC)
Ana Geraldo (APEZ)
André Almeida (APEZ-ISA-ULisboa)
Ângela Martins (APEZ-UTAD)
Divanildo Monteiro (APEZ-UTAD)
Jorge Oliveira (APEZ-ESA-IPV)
Rui Charneca (APEZ-UÉvora)
Vasco Cadavez (IPB)

Portuguese Scientific Committee

Presidents
Ana Sofia Santos (FeedInov CoLab)
Olga Moreira (INIAV)
Executive members
José Pedro Araújo (ESA-IPVC)
Divanildo Monteiro (UTAD-APEZ)
Jorge Oliveira (ESA-IPV-APEZ)
Luis Ferreira (UTAD-APEZ)
Animal Genetics
Jorge Oliveira (ESA–IPV)
Júlio Carvalheira (CIBIO-UPorto)
Luís Telo da Gama (FMV-ULisboa)
Nuno Carolino (INIAV)
Animal Health and Welfare
Joaquim Cerqueira (ESA-IPVC)
Maria Antónia Conceição (ESA-IPC)
Animal Nutrition
André Almeida (ISA-ULisboa)
Olga Moreira (INIAV)
Rui Bessa (FMV-ULisboa)
Animal Physiology
Graça Ferreira Dias (FMV-ULisboa)
Rita Payan (UÉvora)
Rosa Lino Neto (INIAV)
Livestock Farming Systems
Alfredo Teixeira (ESA-IPB)
Luís Ferreira (UTAD)
Rui Caldeira (FMV-ULisboa)

Cattle Production
Alfredo Borba (UAçores)
Ana Geraldo (UÉvora)
António Moitinho Rogrigues (ESA-IPCB)
Horse Production
António Rocha (ICBAS-UP)
António Vicente (ESA-IPS)
Maria Fradinho (FMV-ULisboa)
Pig Production
João Bengala-Freira (ISA-ULisboa)
Divanildo Monteiro (UTAD)
Rui Charneca (UÉvora)
Sheep and Goat Production
Ana Teresa Belo (INIAV)
Jorge Azevedo (UTAD)
Rosário Marques (INIAV)
Vasco Cadavez (IPB)
Insect Production
Daniel Murta (Ingredient Odyssey)
Miguel Rodrigues (UTAD)
Precision Livestock Farming
Luís Alcino (ESAE-IPP)
Pedro Carvalho (INESC TEC)
Severiano Silva (UTAD)
Mediterranean Working Group
José Santos Silva (INIAV)
Nuno Brito (ESA – IPVC)
Aquaculture Working Group
Luisa Valente (ICBAS-UP)
Paulo Rema (UTAD)

Conference website: www.eaap2022.org

European Federation of Animal Science (EAAP)

President:	Isabel Casasús
Secretary General:	Andrea Rosati
Address:	Via G. Tomassetti 3, A/I I-00161 Rome, Italy
Phone/Fax:	+39 06 4420 2639
E-mail:	eaap@eaap.org
Web:	www.eaap.org

Council Members

President	Isabel Casasús (Spain)
Secretary General	Andrea Rosati
Vice-Presidents	Hans Spoolder (the Netherlands)
	John Carty (Ireland)
Members	Ilan Halachmi (Israel)
	Stéphane Ingrand (France)
	Denis Kučevič (Serbia)
	Martin Lidauer (Finland)
	Olga Moreira (Portugal)
	Bruno Ronchi (Italy)
	Peter Sanftleben (Germany)
	Klemen Potocnik (Slovenia)
Auditors	Zdravko Barac (Croatia)
	Gerry Greally (Ireland)
Alternate Auditor	Jeanne Bormann (Luxembourg)
FAO Representative	Badi Besbes

Friends of EAAP

By creating the 'Friends of EAAP', EAAP offers the opportunity to industries to receive services from EAAP in change of a fixed sponsoring amount of support every year.

- The group of supporting industries are layered in three categories: 'silver', 'gold' and 'diamond' level.
- It is offered an important discount (one year free of charge) if the sponsoring industry will agree for a four years period.
- EAAP will offer the service to create a scientific network (with Research Institutes and Scientists) around Europe.
- Creation of a permanent Board of Industries within EAAP with the objective to inform, influence the scientific and organizational actions of EAAP, like proposing choices of the scientific sessions and invited speakers and to propose industry representatives for the Study Commissions.
- Organization of targeted workshops, proposed by industries.
- EAAP can represent and facilitate activities of the supporting industries toward international legislative and regulatory organizations.
- EAAP can facilitate the supporting industries to enter in consortia dealing with internationally supported research projects.

Furthermore EAAP offers, depending to the level of support (details on our website: www.eaap.org):

- Free entrances to the EAAP annual meeting and Gala dinner invitation.
- Free registration to journal *animal*.
- Inclusion of industry advertisement in the EAAP Newsletter, in the banner of the EAAP website, in the Book of Abstract and in the Programme Booklet of the EAAP annual meeting.
- Inclusion of industry leaflets in the annual meeting package.
- Presence of industry advertisements on the slides between presentations at selected standard sessions.
- Presence of industry logos and advertisements on the slides between presentations at the Plenary Sessions.
- Public Recognition by the EAAP President at the Plenary Opening Session of the annual meeting.
- Discounted stands at the EAAP annual meeting.
- Invitation to meetings (at every annual meeting) to discuss joint strategy EAAP/Industries with the EAAP President, Vice-President for Scientific affair, Secretary General and other selected members of the Council and of the Scientific Committee.

Contact and further information

If the industry you represent is interested to become 'Friend of EAAP' or want to have further information please contact jean-marc.perez0000@orange.fr or EAAP secretariat (eaap@eaap.org, phone : +39 06 44202639).

The Organization

EAAP (The European Federation of Animal Science) organises every year an international meeting which attracts between 900 and 1,500 people. The main aims of EAAP are to promote, by means of active co-operation between its members and other relevant international and national organisations, the advancement of scientific research, sustainable development and systems of production; experimentation, application and extension; to improve the technical and economic conditions of the livestock sector; to promote the welfare of farm animals and the conservation of the rural environment; to control and optimise the use of natural resources in general and animal genetic resources in particular; to encourage the involvement of young scientists and technicians. More information on the organisation and its activities can be found at www.eaap.org

What is the YoungEAAP?

YoungEAAP is a group of young scientists organized under the EAAP umbrella. It aims to create a platform where scientists during their early career get the opportunity to meet and share their experiences, expectations and aspirations. This is done through activities at the Annual EAAP Meetings and social media. The large constituency and diversity of the EAAP member countries, commissions and delegates create a very important platform to stay up-to-date, close the gap between our training and the future employer expectations, while fine-tuning our skills and providing young scientists applied and industry-relevant research ideas.

Committee Members at a glance

- Torun Wallgren (President)
- Ines Adriaens (Vice President)
- Marcin Pszczola (secretary)

YoungEAAP promotes Young and Early Career Scientists to:

- Stay up-to-date (i.e. EAAP activities, social media);
- Close the gap between our training and the future employer expectations;
- Fine-tune our skills through EAAP meetings, expand the special young scientists' sessions, and/ or start online webinars/trainings with industry and academic leaders;
- Meet to network and share our graduate school or early employment experiences;
- Develop research ideas, projects and proposals.

Who can be a Member of YoungEAAP?

All individual members of EAAP can join the YoungEAAP if they meet one of the following criteria: Researchers under 35 years of age OR within 10 years after PhD-graduation

Just request your membership form (torun.wallgren@slu.se) and become member of this network!

74th Annual Meeting of the European Federation of Animal Science

Lyon, France, August 26th to September 1st, 2023

Organizing Committees

The 74th EAAP annual meeting is locally co-organized by the French National Research Institute for Agriculture, Food, and Environment (INRAE) and the French Association for Animal Production (AFZ). It will be organised jointly with the scientific meetings of the World Association for Animal Production (WAAP) and of Interbull.

The organisation of this event is also in close collaboration with other French institutions, including France Génétique Elevage (FGE), the French Livestock institute (IDELE), VetAgro Sup and ISARA which are both institutions of higher education and research in agriculture and food sciences, and finally CIRAD, the French agricultural research and international cooperation organization working for the sustainable development of tropical and Mediterranean regions.

French Committees

President
Jean-François Hocquette (INRAE, AFZ)

Local organizing committee
Jean-François Hocquette (INRAE, AFZ), Adeline Dubost (INRAE), Sabrina Gasser (INRAE), Valérie Heuzé (AFZ), Jérôme Normand, Emmanuelle Caramelle-Holtz (IDELE), Charlotte Chêne (VetAgro Sup), Karima Latti (ISARA),

Steering committee
Patrick Chapoutot, Jean-Louis Peyraud, Rene Baumont, Sabrina Gasser, Philippe Chemineau, Emmanuelle Gilot-Fromont, Fabienne Blanc, Karine Chalvet-Monfray, Marie-Pierre Ellies, Latifa Najar

Scientific committee

Precision Livestock Farming	Nathalie Hostiou, Amélie Fischer
Genetics	Laurent Journaux, Jean-Pierre Bidanel
Nutrition	Latifa Najar, Jaap Van Milgen
Health and Welfare	Christian Ducrot, Alice De Boyer-des-Roches
Animal Physiology	Isabelle Louveau, Xavier Druart
Livestock Farming Systems	Vincent Thenard, Marie-Odile Nozieres-Petit
Cattle Production	René Baumont, Christophe Denoyelle
Sheep and Goat Production	Jérémie Jost
Pig Production	Ludovic Brossard, Christine Roguet
Horse Production	Marion Cressent, Léa Lansade
Insects	Franck Pierre, Anna Zaidman

Conference website: www.eaap2023.org

Commission on Animal Genetics

Filippo Miglior	President Canada	Guelph University fmiglior@uoguelph.ca
Marcin Pszczola	Vice-President Poland	Poznan University of Life Sciences marcin.pszczola@gmail.com
Morten Kargo	Vice-President Denmark	Aarhus University morten.kargo@qgg.au.dk
Alessio Cecchinato	Secretary Italy	Padova University alessio.cecchinato@unipd.it
Christa Egger-Danner	Industry rep. Austria	Zuchtdata egger-danner@zuchtdata.at
Xiao Wang	Young Club Denmark	Technical University of Denmark xiwa@dtu.dk
Ewa Sell-Kubiak	Young Club Poland	Poznan University of Life Sciences ewa.sell-kubiak@puls.edu.pl

Commission on Animal Nutrition

Luciano Pinotti	President Italy	University of Milan luciano.pinotti@unimi.it
Sam de Campaneere	Vice-President Belgium	ILVO sam.decampeneere@ilvo.vlaanderen.be
Maria José Ranilla García	Secretary Spain	Universidad de León mjrang@unileon.es
Roselinde Goselink	Secretary The Netherlands	Wageningen Livestock Research roselinde.goselink@wur.nl
Latifa Abdenneby -Najar	Secretary France	IDELE latifa.najar@idele.fr
Javier Alvarez Rodriguez	Secretary Spain	University of Lleida javier.alvarez@udl.cat
Daniele Bonvicini	Industry Rep Italy	Prosol S.p.a d.bonvicini@prosol-spa.it
Geert Bruggeman	Industry Rep Belgium	Nusciencegroup geert.bruggeman@nusciencegroup.com
Susanne Kreuzer-Redmer	Young Club Austria	Vetmed Vienna susanne.kreuzer-redmer@vetmeduni.ac.at

Commission on Health and Welfare

Laura Boyle	Acting President Ireland	Teagasc laura.boyle@teagasc.ie
Flaviana Gottardo	Vice-President Italy	University of Padova flaviana.gottardo@unipd.it
Giulietta Minozzi	Vice-President Italy	University of Milan giulietta.minozzi@unimi.it
Stefanie Ammer	Secretary Germany	Goettingen University stefanie.ammer@uni-goettingen.de
Olivier Espeisse	Industry Rep France	Ceva Santé Animale olivier.espeisse@ceva.com
Delphine Gardan	Industry Rep France	Groupe CCPA dgardan-salmon@ccpa.com
Mariana Dantas de Brito Almeida	Young Club Portugal	University of Tras-os-Montes and Alto Douro mdantas@utad.pt

Commission on Animal Physiology

David Kenny	President Ireland	Teagasc David.Kenny@teagasc.ie
Isabelle Louveau	Vice president France	INRA isabelle.louveau@inra.fr
Alan Kelly	Secretary Ireland	University College Dublin alan.kelly@ucd.ie
Yuri Montanholi	Secretary USA	North Dakota State University yuri.montanholi@ndsu.edu
Federico Randi	Industry rep. France	Ceva Sante Animale federico.randi@ceva.com
Olaia Urrutia	Young Club Spain	University of Navarra olaia.urrutia@unavarra.es
Kate Keogh	Young Club Ireland	Teagasc kate.keogh@teagasc.ie

Commission on Livestock Farming Systems

Michael Lee	President United Kingdom	Harper Adams University MRFLee@harper-adams.ac.uk
Monika Zehetmeier	Vice President Germany	Institute Agricultural Economics and Farm Management monika.zehetmeier@lfl.bayern.de
Tommy Boland	Secretary Ireland	University College Dublin tommy.boland@ucd.ie
Ioanna Poulopoulou	Secretary Italy	Free University of Bozen Ioanna.Poulopoulou@unibz.it
Vincent Thenard	Secretary France	INRA vincent.thenard@inra.fr
Alfredo J. Escribano	Industry rep. Spain	Orffa/Independent researcher and consultant ajescc@gmail.com
Tiago T. da Silva Siqueira	Young Club France	INRA tiago.teixeira.dasilva.siqueira@gmail.com
Maria-Anastasia Karatzia	Young Club Greece	Hellenic Agricultural Organization karatzia@rias.gr

Commission on Cattle Production

Massimo De Marchi	President Italy	Padova University massimo.demarchi@unipd.it
Paul Galama	Vice-President Netherlands	Wageningen Livestock Research paul.galama@wur.nl
Joel Berard	Vice-President Switzerland	Agroscope joel.berard@agroscope.admin.ch
Jean François Hocquette	Secretary France	INRA jean-francois.hocquette@inra.fr
Ray Keatinge	Industry rep United Kingdom	Agriculture & Horticulture Development Board ray.keatinge@ahdb.org.uk
Karsten Maier	Industry rep. Belgium	UECBV info@uecbv.eu
Cătălin Necula	Industry rep. Ireland	Alltech cnecula@alltech.com
Barbara Kosinska-Selvi	Young Club Poland	Wroclaw University of Environmental and Life Sciences barbara.kosinska@upwr.edu.pl

Commission on Sheep and Goat Production

Vasco Augusto Pilão Cadavez	President Portugal	CIMO - Mountain Research Centre vcadavez@ipb.pt
Lorenzo E Hernandez Castellano	Vice-President Spain	Universidad de Las Palmas de Gran Canaria lorenzo.hernandez@ulpgc.es
Antonello Cannas	Secretary Italy	University of Sassari cannas@uniss.it
Georgia Hadjipavlou	Secretary Cyprus	Agricultural Research Institute georgiah@ari.gov.cy
Neil Keane	Industry rep. Ireland	Alltech nkeane@alltech.com
Christos Dadousis	Young Club Italy	University of Parma christos.dadousis@unipr.it

Commission on Pig Production

Sam Millet	President Belgium	ILVO sam.millet@ilvo.vlaanderen.be
Paolo Trevisi	Vice president Italy	Bologna University paolo.trevisi@unibo.it
Giuseppe Bee	Vice president Switzerland	Agroscope Liebefeld-Posieux ALP giuseppe.bee@alp.admin.ch
Katja Nilsson	Secretary Sweden	Swedish University of Agricultural Science katja.nilsson@slu.se
Katarzyna Stadnicka	Secretary Poland	UTP University of Science and Technology in Bydgoszcz katarzynakasperczyk@utp.edu.pl
Egbert Knol	Industry rep. The Netherlands	TOPIGS egbert.knol@topigs.com
Stafford Vigors	Young Club Ireland	University College Dublin staffordvigors1@ucd.ie

Commission on Horse Production

Rhys Evans	President Norway	Norwegian University College of Green Development rhys@hgut.no
Klemen Potočnik	Vice president Slovenija	University of Ljubljana klemen.potocnik@bf.uni-lj.si
Roberto Mantovani	Vice president Italy	University of Padua- DAFNAE roberto.mantovani@unipd.it
Isabel Cervantes Navarro	Secretary Spain	Complutense University of Madrid icervantes@vet.ucm.es
Pasquale De Palo	Secretary Italy	University of Bari pasquale.depalo@uniba.it
Jackie Tapprest	Secretary France	Animal health laboratory (ANSES) jackie.tapprest@anses.fr
Melissa Cox	Industry rep. Germany	Generatio GmbH – Center for Animal Genetics melissa.cox@centerforanimalgenetics.de
Juliette Auclair-Ranzaud	Young Club France	Institut français du cheval et de l'équitation juliette.auclair-ronzaud@ifce.fr
Kirsty Tan	Young Club Germany	Christian-Albrechts-Universität zu Kiel kirsty.tan89@gmail.com

Commission on Insects

Teun Veldkamp	President The Netherlands	Wageningen Livestock Research teun.veldkamp@wur.nl
Laura Gasco	Vice president Italy	University of Turin laura.gasco@unito.it
Christoph Sandrock	Secretary Switzerland	Research Institute of Organic Agriculture FiBL christoph.sandrock@fibl.org
David Deruytter	Secretary Belgium	INAGRO david.deruytter@inagro.be
Marian Peters	Industry rep. The Netherlands	IIC (International Insect Centre) marianpeters@ngn.co.nl
Daniel Murta	Industry rep. Portugal	Ingredient Odyssey – EntoGreen daniel.murta@entogreen.com
Thomas Farrugia	Industry rep. UK	Beta Bugs thomas@betabugs.uk
Matteo Ottoboni	Young Club Italy	Univerity of Milan matteo.ottoboni@unimi.it
Marwa Shumo	Young Club Germany	University of Bonn mshummo@hotmail.com

Commission on Precision Livestock Farming

Matti Pastell	President Finland	Natural Resources Institute Finland (Luke) matti.pastell@luke.fi
Jarissa Maselyne	Vice president Belgium	ILVO jarissa.maselyne@ilvo.vlaanderen.be
Francisco Maroto Molina	Vice president Spain	University of Cordoba g02mamof@uco.es
Claire Morgan -Davies	Vice president United Kingdom	Scotland's Rural College (SRUC) claire.morgan-davies@sruc.ac.uk
Ines Adriaens	Vice president Belgium	KU Leuven ines.adriaens@kuleuven.be
Jean-Marc Gautier	Vice president France	IDELE jean-marc.gautier@idele.fr
Shelly Druyan	Secretary Israel	ARO, The Volcani Center shelly.druyan@mail.huji.ac.il
Radovan Kasarda	Secretary Slovakia	Slovak University of Agriculture in Nitra radovan.kasarda@uniag.sk
Hiemke Knijn	Industry rep. The Netherlands	CRV hiemke.knijn@crv4all.com
Victor Bloch	Young Club Finland	Luke victor.bloch@luke.fi

Platinum sponsor

Gold sponsors

Silver sponsors

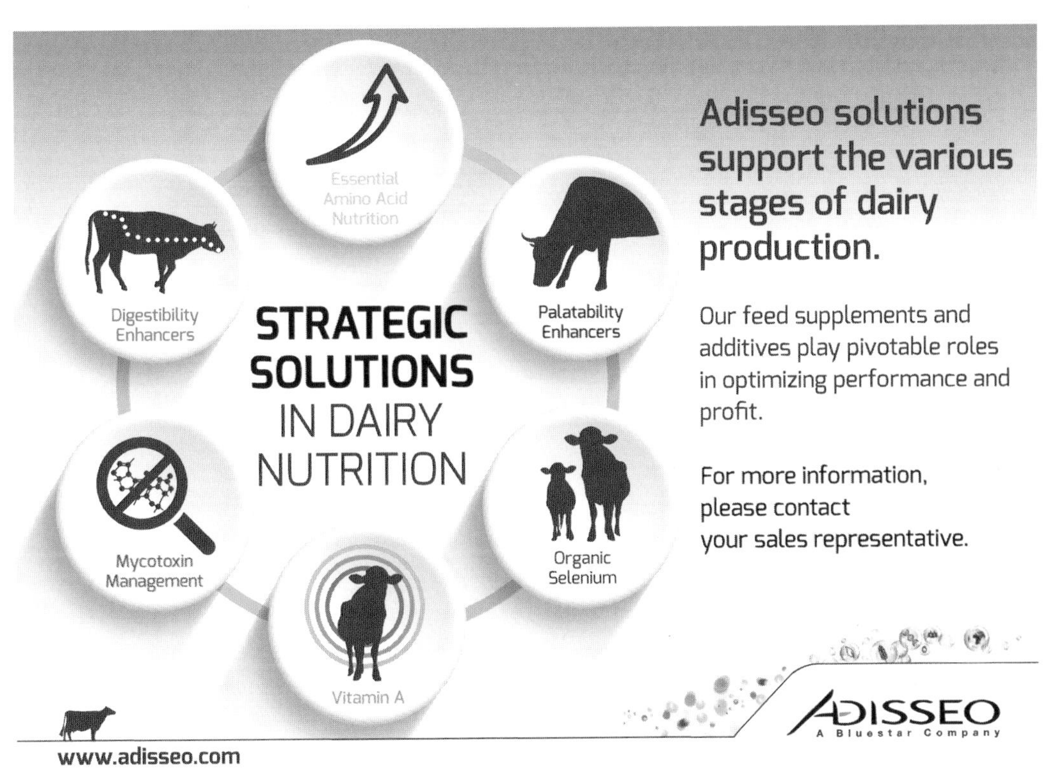

Other sponsors

Acknowledgements

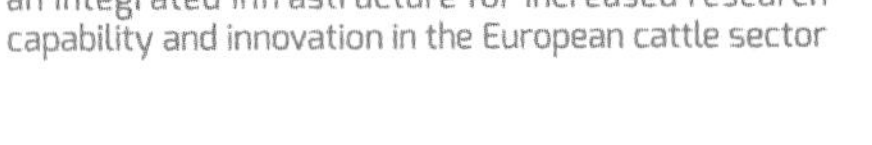

Friends of EAAP

Innovating from
what nature provides
At METEX NØØVISTAGO we produce essential amino acids
for better animal nutrition and a more sustainable world.
We provide the whole livestock value chain with innovative
amino-acids solutions at the heart of a virtuous
circular economy. More than 40 years of expertise
and high-quality R&D make us a reliable partner for
our 4000 clients all over the world.
www.metex-noovistago.com
NØØVISTA
GO
Tailored for Animals,
Inspired by Nature.

Thank you
to the 73rd EAAP Annual Conference Sponsors and Friends

Platinum sponsor

Gold sponsors

Silver sponsors

Other sponsors

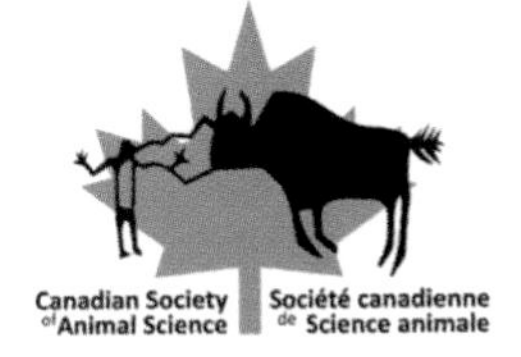

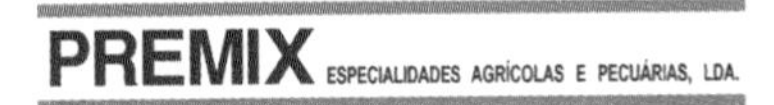

Scientific programme EAAP 2022

Monday 5 September 8.30 – 12.30	Monday 5 September 14.00 – 18.00
Session 01 Coordination of local and transboundary breed conservation; the role of *in situ* and *ex situ* strategies Chair: Bojkovski / Lino Neto Pereira	**Session 12** Social license to operate Chair: Evans
Session 02 Livestock emissions and the COP26 targets Chair: O'Mara / Lee	**Session 13** Livestock emissions and the COP26 targets Chair: Lee / O'Mara
Session 03 Novelties in genomics research and their impact on genetic selection Chair: Baes / Miglior	**Session 14** Novelties in genomics research and their impact on genetic selection Chair: Baes / Cecchinato
Session 04 Multifunctional grasslands systems for efficient cattle production promoting ecosystems services Chair: Berard / Probo	**Session 15** Young Train session: dairy research and extension Chair: Kotlarz / Kuipers
Session 05 Dairy herd reproduction management strategies for improved efficiency Chair: Chagunda / Montanholi	**Session 16** Integrated approach for sheep and goats Mediterranean farming systems Chair: Hadjipavlou / Marques
Session 06 Nutrition regulation of rumen and/or gut microbiota Chair: Goselink / Abdenneby-Najar	**Session 17** Role of nutrition in animal well-being Chair: Alvarez-Rodriguez / Kreuzer-Redmer
Session 07 Insects as food, feed, and technical applications Chair: Gasco / Ottoboni	**Session 18** Insect production and processing Chair: Deruytter / Murta
Session 08 Animal health and welfare free communications Chair: Ammer / Gardon-Salmon	**Session 19** Pig health and welfare free communications Chair: Gottardo / Trevisi
Session 09 Nutrition and fertility of cattle Chair: Lamb / Kreuzer Redmer	**Session 20** Influence of nutrition on reproductive performance Chair: Lamb / Sanz
Session 10 Early career competition 'Innovative approaches to pig and poultry production', supported by Wageningen Academic Publishers Chair: Nilsson / Vigors	**Session 21** Early career competition 'Innovative approaches to pig and poultry production', supported by Wageningen Academic Publishers Chair: Nilsson / Vigors
Session 11 Limits to productive performance and welfare in livestock: Physiological evidence and sensing strategies Chair: Montanholi / Druyan	

Tuesday 6 September 9.00 – 12.30	Tuesday 6 September 14.00 – 18.00
Session 22 **Plenary session and Leroy Lecture** Chair: Casasús	**Session 23** Young EAAP Chair: Wallgren **Session 24** Balancing metrics of impacts and services to assess sustainability Chair: da Silva / Thenard **Session 25** ClearFarm: A platform to control animal welfare in pig and cattle farming Chair: Pastell / Llonch **Session 26** >>> Climate care dairy farming Chair: Galama / Hristov **Session 27** Breeding for health, welfare and longevity Chair: Egger-Danner / Cecchinato **Session 28** SmartCow: Methods and technologies for research and smart nutrition management in dairy and beef cattle Chair: Baumont / Abdenneby-Najar **Session 29** Genetics, physiology, and behaviour of insects Chair: Sandrock / Lefebvre **Session 30** Improving animal welfare to increase animal health, performance and reduce the use of medication Chair: Boyle / Costa **Session 31** Development of novel sheep and goat products for human consumption Chair: Lorenzo / Hernández **Session 32** Role of pork in circular food production Chair: Bee / Millet **Session 33** Breeding for performance Chair: Cox

Wednesday 7 September 8.30 – 12.30	Wednesday 7 September 14.00 – 18.00
Session 34 Animal farming 4.0: the role of Big Data in animal breeding Chair: Granados / Kaya	**Session 45** Breeding for environmentally important traits (e.g. methane emission, feed efficiency etc.) Chair: Pszczola / Manzanilla
Session 35 Reimagining pasture based production systems Chair: Boland	**Session 46** The role of local (plant and animal) resources in the resilience of livestock farming systems Chair: Poulopoulou / Sturaro
Session 36 Sensing animals and fields to improve pasture and rangeland management Chair: Maroto Molina	**Session 47** Phenotyping vs monitoring: better exploitation of measurement data for breeding and management Chair: Knijn / Kargo
<<< **Session 37** Climate care dairy farming Chair: Edouard / Kuipers	**Session 48** Climate care dairy farming Chair: Keatinge / Ruska
Session 38 High quality and sustainable milk and meat products from cattle Chair: Berard / Giller	**Session 49** High quality and sustainable milk and meat products from cattle Chair: De Marchi / Visentin
Session 39 New/alternative/innovative/revised feeds in farm animals and farmed fish Chair: Ranilla García / Pinotti	**Session 50** Nutrition and environment Chair: Bruggeman / Goselink
Session 40 INSECTDOCTORS/INSECTFEED Chair: Veldkamp / Van Oers / Dicke	**Session 51** Workshop on multi-stakeholder views on using insects as feed Chair: Dicke / Veldkamp
Session 41 Development and external validation of PLF tools for animal behaviour, health and welfare: pigs, sheep, beef and poultry Chair: Maselyne / Minozzi	**Session 52** Development and external validation of PLF tools for animal behaviour, health and welfare: dairy, in collaboration with ICAR Chair: Minozzi / Maselyne
Session 42 Metabolic inflammation and insulin resistance in livestock Chair: Huber / Kelly	**Session 53** Heat stress and other adaptations to implications of climate change Chair: Montanholi / Mateescu
Session 43 Challenges in poultry nutrition, health and welfare (1) Chair: Stadnicka / Daş	**Session 54** Challenges in poultry nutrition, health and welfare (2) Chair: Stadnicka / Daş
Session 44 Sheep and goat production systems in mountain areas – where are we and where are we going? Chair: Gauly / Cannas	**Session 55** Innovation in equine production Chair: Tapprest / de Palo / Auclair-Ronzaud / Via

Thursday 8 September 8.30 – 12.30	Thursday 8 September 14.00 – 18.00
Session 56 Animal breeding Chair: Jahnel / Miglior	**Session 65** Genetics and product quality in local breeds Chair: Knol / Nilsson
Session 57 Livestock farming and value chain initiatives contributing to societal/ecological transition Chair: Thenard / da Silva	**Session 66** Digitalisation, Data integration, Detection and Decision support in PLF applications – collaboration with D4Dairy Chair: Egger-Danner / Adriaens
Session 58 PLF for social, environmental and ethical impact Chair: Gautier / Morgan-Davies	**Session 67** Dairy management to enhance farm sustainability Chair: Foskolos / Righi
Session 59 Flexitarianism, plant-based products, cultured meat: pros and cons of these new solutions Chair: Ellies-Oury / Hoquette	**Session 68** The multiple facets of adaptive capacities in sheep and goat farms: how physiological, genetics and management strategies can be used and combined to improve resilience? Chair: Puillet / Cadavez
Session 60 LIFE Green Sheep project / Meta-analysis applied to animal science Chair: Dolle / Tzamaloukas	**Session 69** Increasing animal resilience via nutrition (organized in collaboration with FEFANA) Chair: Pinotti / Goselink
Session 61 Safety and sustainability in the insect chain Chair: Gasco / Sandrock	**Session 70** Genome informed breeding Chair: Sell-Kubiak / Pocrnic
Session 62 Early career competition on measuring animal behaviour Chair: Almeida / Maigrot	**Session 71** Integrative physiology to understand whole-tissue or whole-body functions and their application to animal production, health or disease Chair: Bonnet / Keogh
Session 63 Role of antioxidants in the physiology of several tissues in livestock Chair: Boutinaud / Louveau	**Session 72** Sustainable pig production Chair: Aluwé / Bee
Session 64 Feeding sows from weaning to farrowing Chair: Trevisi / Bee	**Session 73** Equine nutrition – joint session with Animal Nutrition Chair: Pinotti / Evans

11.30 – 12.30

Commission meeting

- Cattle
- Genetics
- Health and Welfare
- Horse
- Insects
- LFS
- Nutrition
- Physiology
- Pig
- PLF
- Sheep and Goats

Scientific programme

Session 01. Coordination of local and transboundary breed conservation; the role of in situ and ex situ strategies

Date: Monday 5 September 2022; 8.30 – 12.30
Chair: Bojkovski / Lino Neto Pereira

Theatre Session 01

Poster Session 01

Session 02. Livestock emissions and the COP26 targets

Date: Monday 5 September 2022; 8.30 – 12.30
Chair: O'Mara / Lee

Theatre Session 02

Poster Session 02

Session 03. Novelties in genomics research and their impact on genetic selection

Date: Monday 5 September 2022; 8.30 – 12.30
Chair: Baes / Miglior

Theatre Session 03

Poster Session 03

Session 04. Multifunctional grasslands systems for efficient cattle production promoting ecosystems services

Date: Monday 5 September 2022; 8.30 – 12.30
Chair: Berard / Probo

Theatre Session 04

Poster Session 04

Session 05. Dairy herd reproduction management strategies for improved efficiency

Date: Monday 5 September 2022; 8.30 – 12.30
Chair: Chagunda / Montanholi

Theatre Session 05

Poster Session 05

Session 06. Nutrition regulation of rumen and/or gut microbiota

Date: Monday 5 September 2022; 8.30 – 12.30
Chair: Goselink / Abdenneby-Najar

Theatre Session 06

Poster Session 06

Session 07. Insects as food, feed, and technical applications

Date: Monday 5 September 2022; 8.30 – 12.30
Chair: Gasco / Ottoboni

Theatre Session 07

Poster Session 07

Session 08. Animal health and welfare free communications

Date: Monday 5 September 2022; 8.30 – 12.30
Chair: Ammer / Gardon-Salmon

Theatre Session 08

Poster Session 08

Session 09. Nutrition and fertility of cattle

Date: Monday 5 September 2022; 8.30 – 12.30
Chair: Lamb / Kreuzer Redmer

Theatre Session 09

Poster Session 09

Session 10. Early career competition 'Innovative approaches to pig and poultry production', supported by Wageningen Academic Publishers

Date: Monday 5 September 2022; 8.30 – 12.30
Chair: Nilsson / Vigors

Theatre Session 10

Poster Session 10

Session 11. Limits to productive performance and welfare in livestock: physiological evidence and sensing strategies

Date: Monday 5 September 2022; 8.30 – 12.30
Chair: Montanholi / Druyan

Theatre Session 11

Poster Session 11

Session 12. Social license to operate

Date: Monday 5 September 2022; 14.00 – 18.00
Chair: Evans

Theatre Session 12

Poster Session 12

Session 13. Livestock emissions and the COP26 targets

Date: Monday 5 September 2022; 14.00 – 18.00
Chair: Lee / O'Mara

Theatre Session 13

Session 14. Novelties in genomics research and their impact on genetic selection

Date: Monday 5 September 2022; 14.00 – 18.00
Chair: Baes / Cecchinato

Theatre Session 14

Poster Session 14

Session 15. Young Train session: dairy research and extension

Date: Monday 5 September 2022; 14.00 – 18.00
Chair: Kotlarz / Kuipers

Theatre Session 15

Poster Session 15

Session 16. Integrated approach for sheep and goats Mediterranean farming systems

Date: Monday 5 September 2022; 14.00 – 18.00
Chair: Hadjipavlou / Marques

Theatre Session 16

Poster Session 16

Session 17. Role of nutrition in animal well-being

Date: Monday 5 September 2022; 14.00 – 18.00
Chair: Alvarez-Rodriguez / Kreuzer-Redmer

Theatre Session 17

Poster Session 17

Session 18. Insect production and processing

Date: Monday 5 September 2022; 14.00 – 18.00
Chair: Deruytter / Murta

Theatre Session 18

Poster Session 18

Session 19. Pig health and welfare free communications

Date: Monday 5 September 2022; 14.00 – 18.00
Chair: Gottardo / Trevisi

Theatre Session 19

Session 20. Influence of nutrition on reproductive performance

Date: Monday 5 September 2022; 14.00 – 18.00
Chair: Lamb / Sanz

Theatre Session 20

Poster Session 20

Session 21. Early career competition 'Innovative approaches to pig and poultry production', supported by Wageningen Academic Publishers

Date: Monday 5 September 2022; 14.00 – 18.00
Chair: Nilsson / Vigors

Theatre Session 21

Poster Session 21

Session 22. Plenary session

Date: Tuesday 6 September 2022; 9.00 – 12.30
Chair: Casasús

Theatre Session 22

Session 23. Young EAAP

Date: Tuesday 6 September 2022; 14.00 – 18.00
Chair: Wallgren

Theatre Session 23

Session 24. Balancing metrics of impacts and services to assess sustainability

Date: Tuesday 6 September 2022; 14.00 – 18.00
Chair: da Silva / Thenard

Theatre Session 24

Session 25. ClearFarm: A platform to control animal welfare in pig and cattle farming

Date: Tuesday 6 September 2022; 14.00 – 18.00
Chair: Pastell / Llonch

Theatre Session 25

Poster Session 25

Session 26. Climate care dairy farming

Date: Tuesday 6 September 2022; 14.00 – 18.00
Chair: Galama / Hristov

Theatre Session 26

Poster Session 26

Session 27. Breeding for health, welfare and longevity

Date: Tuesday 6 September 2022; 14.00 – 18.00
Chair: Egger-Danner / Cecchinato

Theatre Session 27

Poster Session 27

Session 28. SmartCow: Methods and technologies for research and smart nutrition management in dairy and beef cattle

Date: Tuesday 6 September 2022; 14.00 – 18.00
Chair: Baumont / Abdenneby-Najar

Theatre Session 28

Poster Session 28

Session 29. Genetics, physiology, and behaviour of insects

Date: Tuesday 6 September 2022; 14.00 – 18.00
Chair: Sandrock / Lefebvre

Theatre Session 29

Poster Session 29

Session 30. Improving animal welfare to increase animal health, performance and reduce the use of medication

Date: Tuesday 6 September 2022; 14.00 – 18.00
Chair: Boyle / Costa

Theatre Session 30

Poster Session 30

Session 31. Development of novel sheep and goat products for human consumption

Date: Tuesday 6 September 2022; 14.00 – 18.00
Chair: Lorenzo / Hernández

Theatre Session 31

Poster Session 31

Session 32. Role of pork in circular food production

Date: Tuesday 6 September 2022; 14.00 – 18.00
Chair: Bee / Millet

Theatre Session 32

Poster Session 32

Session 33. Breeding for performance

Date: Tuesday 6 September 2022; 14.00 – 18.00
Chair: Cox

Theatre Session 33

Poster Session 33

Session 34. Animal farming 4.0: the role of Big Data in animal breeding

Date: Wednesday 7 September 2022; 8.30 – 12.30
Chair: Granados / Kaya

Theatre Session 34

Session 35. Reimagining pasture based production systems

Date: Wednesday 7 September 2022; 8.30 – 12.30
Chair: Boland

Theatre Session 35

Poster Session 35

Session 36. Sensing animals and fields to improve pasture and rangeland management

Date: Wednesday 7 September 2022; 8.30 – 12.30
Chair: Maroto Molina

Theatre Session 36

Poster Session 36

Session 37. Climate care dairy farming

Date: Wednesday 7 September 2022; 8.30 – 12.30
Chair: Edouard / Kuipers

Theatre Session 37

Session 38. High quality and sustainable milk and meat products from cattle

Date: Wednesday 7 September 2022; 8.30 – 12.30
Chair: Berard / Giller

Theatre Session 38

Poster Session 38

Session 39. New/alternative/innovative/revised feeds in farm animals and farmed fish

Date: Wednesday 7 September 2022; 8.30 – 12.30
Chair: Ranilla García / Pinotti

Theatre Session 39

Poster Session 39

Session 40. INSECTDOCTORS/INSECTFEED

Date: Wednesday 7 September 2022; 8.30 – 12.30
Chair: Veldkamp / Van Oers / Dicke

Theatre Session 40

Poster Session 40

Session 41. Development and external validation of PLF tools for animal behaviour, health and welfare: pigs, sheep, beef and poultry

Date: Wednesday 7 September 2022; 8.30 – 12.30
Chair: Maselyne / Minozzi

Theatre Session 41

Poster Session 41

Session 42. Metabolic inflammation and insulin resistance in livestock

Date: Wednesday 7 September 2022; 8.30 – 12.30
Chair: Huber / Kelly

Theatre Session 42

Poster Session 42

Session 43. Challenges in poultry nutrition, health and welfare (1)

Date: Wednesday 7 September 2022; 8.30 – 12.30
Chair: Stadnicka / Daş

Theatre Session 43

Session 44. Sheep and goat production systems in mountain areas – where are we and where are we going?

Date: Wednesday 7 September 2022; 8.30 – 12.30
Chair: Gauly / Cannas

Theatre Session 44

Poster Session 44

Session 45. Breeding for environmentally important traits (e.g. methane emission, feed efficiency etc.)

Date: Wednesday 7 September 2022; 14.00 – 18.00
Chair: Pszczola / Manzanilla

Theatre Session 45

Poster Session 45

Session 46. The role of local (plant and animal) resources in the resilience of livestock farming systems

Date: Wednesday 7 September 2022; 14.00 – 18.00
Chair: Poulopoulou / Sturaro

Theatre Session 46

Poster Session 46

Session 47. Phenotyping vs monitoring: better exploitation of measurement data for breeding and management

Date: Wednesday 7 September 2022; 14.00 – 18.00
Chair: Knijn / Kargo

Theatre Session 47

Poster Session 47

Session 48. Climate care dairy farming

Date: Wednesday 7 September 2022; 14.00 – 18.00
Chair: Keatinge / Ruska

Theatre Session 48

Session 49. High quality and sustainable milk and meat products from cattle

Date: Wednesday 7 September 2022; 14.00 – 18.00
Chair: De Marchi / Visentin

Theatre Session 49

Poster Session 49

Session 50. Nutrition and environment

Date: Wednesday 7 September 2022; 14.00 – 18.00
Chair: Bruggeman / Goselink

Theatre Session 50

Poster Session 50

Session 51. Workshop on multi-stakeholder views on using insects as feed

Date: Wednesday 7 September 2022; 14.00 – 18.00
Chair: Dicke / Veldkamp

Theatre Session 51

Session 52. Development and external validation of PLF tools for animal behaviour, health and welfare: dairy, in collaboration with ICAR

Date: Wednesday 7 September 2022; 14.00 – 18.00
Chair: Minozzi / Maselyne

Theatre Session 52

Poster Session 52

Session 53. Heat stress and other adaptations to implications of climate change

Date: Wednesday 7 September 2022; 14.00 – 18.00
Chair: Montanholi / Mateescu

Theatre Session 53

Poster Session 53

Session 54. Challenges in poultry nutrition, health and welfare (2)

Date: Wednesday 7 September 2022; 14.00 – 18.00
Chair: Stadnicka / Daş

Theatre Session 54

Poster Session 54

Session 55. Innovation in equine production

Date: Wednesday 7 September 2022; 14.00 – 18.00
Chair: Tapprest / de Palo / Auclair-Ronzaud / Vial

Theatre Session 55

Poster Session 55

Session 56. Animal breeding

Date: Thursday 8 September 2022; 8.30 – 12.30
Chair: Jahnel / Miglior

Theatre Session 56

Poster Session 56

Session 57. Livestock farming and value chain initiatives contributing to societal/ecological transition

Date: Thursday 8 September 2022; 8.30 – 12.30
Chair: Thenard / da Silva

Theatre Session 57

Poster Session 57

Session 58. PLF for social, environmental and ethical impact

Date: Thursday 8 September 2022; 8.30 – 12.30
Chair: Gautier / Morgan-Davies

Theatre Session 58

Session 59. Flexitarianism, plant-based products, cultured meat: pros and cons of these new solutions

Date: Thursday 8 September 2022; 8.30 – 12.30
Chair: Ellies-Oury / Hoquette

Theatre Session 59

Session 60. LIFE Green Sheep project / Meta-analysis applied to animal science

Date: Thursday 8 September 2022; 8.30 – 12.30
Chair: Dolle / Tzamaloukas

Theatre Session 60

Poster Session 60

Session 61. Safety and sustainability in the insect chain

Date: Thursday 8 September 2022; 8.30 – 12.30
Chair: Gasco / Sandrock

Theatre Session 61

Poster Session 61

Session 62. Early career competition on measuring animal behaviour

Date: Thursday 8 September 2022; 8.30 – 12.30
Chair: Almeida / Maigrot

Theatre Session 62

Session 63. Role of antioxidants in the physiology of several tissues in livestock

Date: Thursday 8 September 2022; 8.30 – 12.30
Chair: Boutinaud / Louveau

Theatre Session 63

Poster Session 63

Session 64. Feeding sows from weaning to farrowing

Date: Thursday 8 September 2022; 8.30 – 12.30
Chair: Trevisi / Bee

Theatre Session 64

Poster Session 64

Session 64. Feeding sows from weaning to farrowing

Date: Thursday 8 September 2022; 8.30 – 12.30
Chair: Trevisi / Bee

Theatre Session 65

Poster Session 65

Session 66. Digitalisation, Data integration, Detection and Decision support in PLF applications – collaboration with D4Dairy

Date: Thursday 8 September 2022; 14.00 – 18.00
Chair: Egger-Danner / Adriaens

Theatre Session 66

Poster Session 66

Session 67. Dairy management to enhance farm sustainability

Date: Thursday 8 September 2022; 14.00 – 18.00
Chair: Foskolos / Righi

Theatre Session 67

Poster Session 67

Session 68. The multiple facets of adaptive capacities in sheep and goat farms: how physiological, genetics and management strategies can be used and combined to improve resilience?

Date: Thursday 8 September 2022; 14.00 – 18.00
Chair: Puillet / Cadavez

Theatre Session 68

Poster Session 68

Session 69. Increasing animal resilience via nutrition (organized in collaboration with FEFANA)

Date: Thursday 8 September 2022; 14.00 – 18.00
Chair: Pinotti / Goselink

Theatre Session 69

Session 70. Genome informed breeding

Date: Thursday 8 September 2022; 14.00 – 18.00
Chair: Sell-Kubiak / Pocrnic

Theatre Session 70

Poster Session 70

Session 71. Integrative physiology to understand whole-tissue or whole-body functions and their application to animal production, health or disease

Date: Thursday 8 September 2022; 14.00 – 18.00
Chair: Bonnet / Keogh

Theatre Session 71

Poster Session 71

Session 72. Sustainable pig production

Date: Thursday 8 September 2022; 14.00 – 18.00
Chair: Aluwé / Bee

Theatre Session 72

Poster Session 72

Session 73. Equine nutrition – joint session with Animal Nutrition

Date: Thursday 8 September 2022; 14.00 – 18.00
Chair: Pinotti / Evans

Theatre Session 73

Poster Session 73

Session 01 Theatre 1

Strategic transboundary collaboration to conserve Dutch and Belgian draught horse populations

M.A. Schoon[1,2], L. Chapard[3], A.C. Bouwman[1], B.J. Ducro[1], S. Jansenss[3], S.J. Hiemstra[2] and N. Buys[3]
[1]Animal Breeding and Genomics, Wageningen University & Research, P.O. Box 338, 6700 AH Wageningen, the Netherlands, [2]Centre for Genetic Resources, the Netherlands, Wageningen University & Research, P.O. Box 338, 6700 AH Wageningen, the Netherlands, [3]KU Leuven, Center for Animal Breeding and Genetics, Department of Biosystems, Kasteelpark Arenberg 30, Box 2472, 3001 Leuven, Belgium; mira.schoon@wur.nl

Most of the native Draught horse populations in Western Europe are 'at risk'. The *in-situ* effective population size of the Belgium draught horse population was found to be 40 in 2021, for the Dutch population it was even less (N_e=35). Another indicator of population trend is the number of foals born per year. In the Netherlands the number of foals born reduced from >600 to <300 between 2000 and 2020. The same accounts for the Belgian draught horse population where in 2000 still over 900 foals were born, in 2020 this was half as much with 456. Research at national level showed that rate of inbreeding per generation is far beyond the save limits set by the FAO (<0.5% ΔF), 1.3% ΔF for the Dutch population and 1.25% ΔF for the Belgian population. Both studies suggested to increase the exchange of (breeding) animals across national borders. Even though breeders from the Netherlands and Belgium are known to exchange (breeding) animals, a more strategic collaboration between breeding organizations and countries would be beneficial. Inconsistencies in pedigree data, changes in names, (birth) dates and even unique identification numbers hamper implementation of a strategic breeding plan for the two populations as one, transboundary, breed. Therefore, we started a project to reconstruct the pedigree data that will allow a combined population analysis. Firstly, uniqueness of genetic diversity (Retriever software) will be studied for both populations as well as the combined level of inbreeding. Secondly, exchange scenario's (Pointer software) will be simulated, including the set-up and use of *ex-situ* genebank collection to support (future) breeding programmes. We hypothesize that optimizing the exchange between the two 'at risk' populations contributes to the conservation of both national populations as well as of the transboundary breed as a whole.

Session 01 Theatre 2

Developing, validating and improving low density multi-species SNP arrays in farm animals

R.P.M.A. Crooijmans[1], R. Gonzalez-Prendes[1] and M. Tixier-Boichard[2]
[1]Wageningen University & Research, Animal Breeding and Genomics, Droevendaalsesteeg 1, 6708 PG Wageningen, the Netherlands, [2]University Paris-Saclay, INRAE, AgroParisTech, GABI, CRJJ, 78350 Jouy-en-Josas, France; richard.crooijmans@wur.nl

Genomic variation in farm animal species is predominantly detected in commercial populations. Most SNP arrays developed so far are, therefore, mostly biased towards these commercial populations. Moreover, SNP sets on the SNP arrays changes in time and results in poor overlap between the different versions and platforms. Furthermore, in several cases, arrays are not freely available without the permission of the consortia or industry involved in their development. We created a SNP pool by SNPs detected by sequencing individuals of different species within the IMAGE -project supplemented by publicly and private available SNP sets. We developed and tested two Affymetrix multi species arrays, which are publicly available. The IMAGE001 multi species array contains 58,839K SNPs for cattle, pigs, chicken, goat, sheep and horse species. The IMAGE002 multi species array contains around 50K SNPs but now for the waterbuffalo, quail, pigeon, rabbit, duck and bees species. For both arrays, between 7 and 10K SNPs per species were selected based on overlap with public available species specific SNP arrays and supplemented by specific trait, ancestral, MHC, sex chromosome and inheritance check SNPs (when available)). This low cost SNP array will make possible to genotype a large number of samples of the different species collected for *in situ* and *ex situ* population characterization. The IMAGE project received funding from the EU H2020 Research and Innovation Programme under the grant agreement n° 677353.

Genomic diversity and relatedness between endangered German Black Pied cattle and 68 cattle breeds

G.B. Neumann[1], P. Korkuć[1], D. Arends[1], M.J. Wolf[2], K. May[2], S. König[2] and G.A. Brockmann[1]
[1]Humboldt University of Berlin, Albrecht Daniel Thaer-Institute for Agricultural and Horticultural Sciences, Unter den Linden 6, 10117 Berlin, Germany, [2]Justus-Liebig-University, Institute of Animal Breeding and Genetics, Ludwigstraße 21B, 35390 Gießen, Germany; guilherme.neumann@hu-berlin.de

According to the Food and Agriculture Organization of the United Nations, 84% of all local breeds in Europe were considered under risk in 2021. This trend is also observed for the endangered dual-purpose German Black Pied cattle population ('Deutsches Schwarzbuntes Niederungsrind', DSN). The small population size of ~2,500 DSN cattle is a consequence of a high replacement rate of DSN by high-yielding Holstein Friesian (HF) cattle. In this study, we characterized DSN's genomic diversity compared to 68 other taurine breeds and estimated their relatedness defining closely related cattle breeds to DSN. A total of 79,019,242 biallelic variants of 302 DSN animals from whole-genome sequencing data and 1,388 animals of 68 other taurine breeds from the 1000 Bull Genomes Project were used. Genomic diversity between DSN and the other breeds was estimated as nucleotide diversity and relatedness as pairwise fixation indices (FST). Further, a phylogenetic tree was constructed to visualize the relationship between all investigated breeds. Phylogenetic analysis detected four main clusters that could be assigned to distinct geographical origin: Northern Europe, Central Europe, Jersey and Guernsey islands, and an area comprising Eastern Europe, Central Italy, and Asia. DSN clustered closely to breeds from the Netherlands, Northern Germany, and Belgium, such as Dutch Friesian Red, Eastern Belgian Red White, and HF. Those breeds were closest to DSN also in terms of average FST values. Analysis of FST values across the genome showed differences between breeds pointing to potential signatures of selection. Nucleotide diversity was medium in DSN while it was highest in breeds from Asia and Eastern Europe. Nevertheless, nucleotide diversity in DSN was not significantly different from HF and other closely related breeds. The definition and characterization of breeds closely related to DSN provide an opportunity for a joint conservation plan, overcoming the limitation of the small herd sizes of those breeds.

Population structure and genetic diversity of sheep breeds in Slovenia

M. Simčič[1], N. Pogorevc[1], D. Bojkovski[1] and I. Medugorac[2]
[1]University of Ljubljana, Biotechnical Faculty, Jamnikarjeva 101, 1000 Ljubljana, Slovenia, [2]LMU Munich, Faculty of Veterinary Medicine, Lena-Christ-Straße 48, 82152 Martinsried/Planegg, Germany; mojca.simcic@bf.uni-lj.si

The aim of this study was to estimate genetic diversity parameters and population structure in local sheep breeds in Slovenia. Bovec and Jezersko-Solčava sheep could be the direct descendants of the Small White sheep, which was widely distributed in the Alps centuries ago. Istrian and Bela Krajina Pramenka, on the other hand, belong to the Zackel group from the Balkan Peninsula. In the 1970s, part of the Jezersko-Solčava population was improved with the Romanov resulting in a new breed called 'Improved Jezersko-Solčava sheep'. In the 1980s, part of the Bovec population was improved with the East Frisian,which developed a new breed called 'Improved Bovec sheep'. Genetic analyses were performed on the genome-wide Single Nucleotide Polymorphisms (SNP) of Illumina Ovine SNP50 array data from 256 animals of Bovec sheep (47), Jezersko-Solčava sheep (48), Istrian Pramenka (46), Bela Krajina Pramenka (42), Improved Jezersko-Solčava sheep (40), and Improved Bovec sheep (33), and 1.012 animals from 41 reference populations representing neighbouring and cosmopolitan breeds. Similarly, 116 animals from 5 populations of wild relatives were added. To obtain unbiased estimates we used short haplotypes spanning 4 markers instead of single SNPs to avoid an ascertainment bias of the OvineSNP50 array. Phylogenetic analyses and the genetic distance matrix presented by Neighbour-Net revealed the unique genetic identity and origin of Bovec and Jezersko-Solčava sheep, as well as Istrian and Bela Krajina Pramenka with high number of private alleles. The origin of both improved breeds was also confirmed by genetic analyses. Unsupervised clustering by the Admixture analysis among individuals suggested that all six breeds are distinct. Unfortunately, introgression from other breeds was detected in some animals. Phenotypic selection performed over several centuries resulted in the preservation of 4 unique sheep breeds in Slovenia. Despite the influence of cosmopolitan breeds, the populations of Improved Jezersko-Solčava and Improved Bovec sheep have been consolidated in recent decades and the animals have adapted to environment conditions in Slovenia.

Analysis of genomic inbreeding, diversity and population structure of the Angler Saddleback pig

A. Meder, A. Olschewsky and D. Hinrichs
University of Kassel, Organic Agriculture, Animal Breeding Section, Nordbahnhofstraße 1a, 37213 Witzenhausen, Germany; alicia.meder@uni-kassel.de

The Angler Saddleback pig originated from northern Germany. With 94 and 26 herdbook sows and boars, respectively, the breed is classified as endangered and hardly been studied at the genomic level. Hence, the objective of this study was to analyse genomic inbreeding and diversity of the breed. Furthermore, population structure including the assignment of individuals to their breeders was investigated. Therefore, genomic data of 210 Angler Saddleback pigs were analysed using PLINK 1.9. After quality control (Samples call rate ≤0.90; SNPs call rate≤0.90; MAF≤0.05; hwe < P-value 0.0001) 196 animals and 45,809 SNPs remained. The inbreeding coefficient F_{ROH} was calculated with three minimum ROH sizes (>4 Mb, >8 Mb and >16 Mb). For assessing genomic diversity observed heterozygosity HO and the relatedness coefficient g_{Rel} were computed. The population structure was studied by a principal component analysis (PCA) based on genome-wide pairwise identity-by-state distances between the individuals. Additionally, a network analysis was conducted using NetView. The results of F_{ROH} (>4 Mb=0.113±0.042; >8 Mb=0.089±0.041; >16 Mb=0.057±0.035) indicate that inbreeding occurred mainly based on old inbreeding events. Despite the extremely low population size, observed heterozygosity (HO=0.354) and relatedness (g_{Rel}=0.728) indicate a comparable high genomic diversity. PCA and network analysis show clear clusters for a part of individuals associated to their breeders. Nevertheless, no consistent structure can be derived for another part of the data. In summary, the analysed Angler Saddleback pig population exhibit moderate inbreeding and relatedness levels whilst genomic diversity in the breed is high relative to its population size. The revealed population structure with partly identified cluster of breeders provide a base to develop sustainable strategies for the breeding management in the future.

First insights into mitogenome variability of indigenous Croatian sheep breeds

V. Brajkovic[1], F. Diab[1], N. Valjak[1], V. Cubric-Curik[1], D. Novosel[2], F. Oštarić[3] and I. Curik[1]
[1]University of Zagreb Faculty of Agriculture, Department of Animal Science, Svetosimunska cesta 25, 10000 Zagreb, Croatia, [2]Croatian Veterinary Institute, Savska cesta 143, 10000 Zagreb, Croatia, [3]University of Zagreb Faculty of Agriculture, Department of Dairy Science, Svetosimunska cesta 25, 10000 Zagreb, Croatia; vbrajkovic@agr.hr

Sheep rearing is one of the most important types of extensive livestock in Croatia, occurring mainly in coastal and mountainous areas along the eastern Adriatic. We retrieved mitogenome from the whole genome (12× and 30×) of 23 individuals representing eight indigenous sheep breeds (Cres Island Sheep, Dalmatian Pramenka, Dubrovnik Ruda, Istria Sheep, Krk Island Sheep, Lika Sheep, Pag Island Sheep and Rab Island Sheep) and two mouflons. The average coverage of the mitogenome from the 12-fold sequenced genome of 17 samples was 1,627 (min 322; max 9,588) and from the 30-fold sequenced genome of eight samples was 1,851 (min 865; max 2,841). Our 16,624 bp fasta set consisted of 112 parsim-info sites and 247 singleton sites while obtained 24 haplotypes with a haplotype diversity of 0.997. When compared with the mitogenome variability of 347 sheep (86 breeds) and three mouflon (Ovis aries musimon) available in GenBank, the median-joining network showed that 22 haplotypes belonged to haplogroup B and, interestingly, one belonged to haplogroup A and one to haplogroup C. Bayesian analyses also confirmed a high frequency of haplogroup A in the breeds from Central and East Asia, and a high frequency of haplogroup B in Europe and Anatolia. The rare haplogroups C occur in the Iberian Peninsula area, the Middle East, the Caspian Sea region, China, India, and Nepal, while haplogroups D and E may be found in the Middle East. The haplotypes of our mouflons were assigned to haplogroup B along with those of the other European mouflons. The presence of two haplotypes assigned to haplogroups A and C suggests migratory contacts between the Balkans and the Middle East that need to be further explored. In order to clarify the timing of the appearance of A and C haplotypes in the eastern Adriatic, our future analyses will refer to chronological analyses of ancient sheep bones.

Session 01 — Theatre 7

Comparative analyses of inbreeding on the X-chromosome in Croatian sheep breeds

M. Shihabi[1], V. Cubric-Curik[1], I. Drzaic[1], L. Vostry[2] and I. Curik[1]
[1]University of Zagreb Faculty of Agriculture, Department of Animal Science, Svetosimunska 25, 10000 Zagreb, Croatia, [2]Czech University of Life Sciences Prague, Department of Animal Genetics and Breeding, Kamycka 129, 16500 Prague, Croatia; vcubric@agr.hr

Inbreeding is the mating of related individuals that share common ancestors, resulting in increased homozygosity. Inbreeding depression and increased prevalence of genetic defects are known negative consequences of inbreeding. Therefore, the estimation of inbreeding coefficients is important for the survival of genetically small populations. The main objective of this study was to estimate the extent of genomic X-chromosome inbreeding (FXsex), a rarely estimated parameter, and its relationship with autosomal inbreeding (FX) in seven Croatian sheep breeds (99 individuals) genotyped with a high-density SNP array (606,006 SNPs). All genomic inbreeding coefficients were calculated with respect to the ROH>4 Mb. In all 99 individuals (East Adriatic metapopulation), the mean FXsex (0.054) was 2.26 times higher than FX(0.024) and a similar trend was observed in variation, as the standard deviation of FXsex (0.116) was 2.54 times higher than that of FX (0.046). The highest observed FXsex was 0.572, while 10% of animals had an FXsex>0.250. In contrast, the highest observed FX was 0.301, with only one individual having an FX>0.2500. The highest mean X-chromosome and autosomal inbreeding coefficient (FXsex=0.079 and FX=0.042) was observed in Rab Island Sheep, while the smallest inbreeding coefficients were found in Pag Island Sheep (FXsex=0.024 and FX=0.012). Our results are preliminary because both the number of breeds (7) and the number of individuals representing each breed (13 to 20) were small. Nevertheless, the higher inbreeding estimates observed for the X-chromosome in sheep demonstrate the importance of analysing possible negative inbreeding effects of the X-chromosome.

Session 01 — Theatre 8

Regional approach to strengthen the documentation of *ex situ* in vitro collections in Europe

Z.I. Duchev[1], F. Tejerina[2], A. Stella[3] and S.J. Hiemstra[4]
[1]Institute of Animal Science – Kostinbrod, Agricultural Academy, Pochivka stn, 2232 Kostinbrod, Bulgaria, [2]Ministerio de Agricultura, Pesca y Alimentación, C/ Almagro, 33, 28071 Madrid, Spain, [3]National Research Council of Italy, via Bassini 15, 20133 Milan, Italy, [4]Wageningen University & Research, Centre for Genetic Resources, the Netherlands (CGN), P.O. Box 338, 6700 AH Wageningen, the Netherlands; zhivko.duchev@agriacad.bg

The European Regional Focal Point for Animal Genetic Resources (ERFP) has developed the European Genebank Network for Animal Genetic Resources (EUGENA) to improve the conservation and sustainable use in Europe and the monitoring of the AnGR kept in the *ex situ* collections. The EUGENA Portal collects data about the *ex situ* in vitro collections of its member gene banks. In the beginning of year 2022, the portal contained data about almost a million samples of semen, embryos, hair, blood, DNA, and somatic cells from 11 gene banks. To facilitate the optimization of complementary *ex situ* and *in situ* conservation strategies for livestock breeds, EUGENA shares data with the European Farm Animal Biodiversity Information System (EFABIS), the Domestic Animal Diversity Information System (FAO DAD-IS), and the IMAGE Data Portal. ERFP is constantly working to improve the gene banks' documentation in order to achieve correct, uniform, and interoperable data about their collections. Such data could be used in many ways, from feeding DAD-IS with information on animal genetic resources being secured in either medium or long term conservation facilities as needed for the SDG indicator 2.5.1, to coordination of transboundary breeds conservation programmes based on genomic data. A step in the direction of improved interoperability of databases was also taken by the IMAGE Data Portal, which integrates standardized gene banks/collections metadata with genomics data, geographical information systems and breeding and diversity database resources. In several ERFP Ad hoc actions, an inventory of gene banks in Europe was compiled; the status of documentation of the collections reviewed and various technical solutions for improving the situation identified.

Session 01 Theatre 9

Dutch genebank for breeding programme support and long term conservation of genetic diversity

M. Van der Sluis[1], M.A. Schoon[1,2], J.J. Windig[1,2], S.J. Hiemstra[2] and M. Neuteboom[1,2]
[1]Wageningen University & Research, Animal Breeding and Genomics, P.O. Box 338, 6700 AH Wageningen, the Netherlands, [2]Wageningen University & Research, Centre for Genetic Resources, the Netherlands, P.O. Box 338, 6700 AH Wageningen, the Netherlands; mira.schoon@wur.nl

The Dutch national genebank for animal genetic resources is one of the largest in Europe, with a collection of genetic material of more than 350,000 samples (mainly semen), from almost 9,000 unique donor animals and over 135 different livestock breeds. It is still expanding, the targeted annual growth in number of donor animals is 2% (across species). In 2021, 11,800 sperm samples from 259 donors were added (representing a growth of around 2.5%). Further growth should be in agreement with the aims of the genebank. The main aims are: (1) establishing sufficiently large collections for breeds at risk, to be able to re-establish breeds in case of calamities ('core-collection'); and (2) support live populations by using genetic material in breeding programmes today. With regard to the first aim, data in EFABIS indicate that, of the breeds with data available, roughly 25% of the Dutch breeds stored in the genebank have sufficient material stored for long term conservation. To better coordinate planning of targeted material collection in the next five years, detailed calculations are performed by species and breed to determine how much genetic material is needed to be able to re-establish a breed with a sufficient effective population size. These calculations take among other parameters litter size, life production, pregnancy rate, and quality and type (e.g. semen, embryos, ovarian tissue and primordial germ cells) of stored material into account. Next to this 'core-collection', genetic material can be requested from the genebank for use in breeding programmes, e.g. to maintain a sufficiently large effective population size. This requires close collaboration with breeding organisations. First results indicate that currently the amount of stored semen is relatively large for the native cattle breeds and commercial pig breeding lines, but requires expansion for other species. Regular use of genetic material in the live population is currently mostly confined to cattle breeds.

Session 01 Theatre 10

Gene bank material as the support for *in situ* conservation

D. Bojkovski and T. Flisar
University of Ljubljana, Biotechnical faculty, Jamnikarjeva 101, 1000, Slovenia; danijela.bojkovski@bf.uni-lj.si

Long-term conservation of genetic diversity in AnGR cannot be ensured only through *in situ* conservation approaches. Animals evolve in their traditional environments and farming systems, and focusing entirely on *in situ* conservation could lead to loss of genetic diversity due to selective breeding and/or genetic drift. The importance of *in situ* and *ex situ* conservation as complementary activities is therefore emphasized in many global policies, such as the Convention on Biological Diversity (CBD), which Slovenia has ratified in 1996. Many other factors, such as cross-breeding, changes in production systems, natural disasters and diseases can also threaten the survival of breed and in such circumstances *ex situ* conservation is crucial. For this reason, *in situ* and *ex situ* conservation should be used as complementary methods, which is also recommended by CBD and FAO. European countries are aware of the need for conservation and sustainable use of AnGR and have national programs in place, but there is a need for better prioritization and coordination of actions, as well as complementary *in situ* and *ex situ* activities. Combining the two approaches provides an effective and reliable conservation approach. While in Slovenia conservation of AnGR is included in the various sectoral strategies, plans and programs at the national level, the combination of the two conservation methods is still under development. *Ex situ* in vivo conservation activities are carried out in the Slovenian Ark Network, *ex situ* in vitro is managed at the three different sites. The *in situ* activities are carried out by farmers and supported by the Rural Development Plan and the AnGR Resources Conservation Program. Currently, the *ex situ* collection is in the establishment phase and is only used based on a specific decision by the Minister of Agriculture and food and official body responsible for the management and conservation of AnGR. This decision should be reconsidered and changed in the future when sufficient genetic material is stored for each local breed.

Session 01 Theatre 11

Impact of CAP payments on the increase in the number of local livestock populations

G.M. Polak

National Research Institute of Animal Production, Unit for the Conservation of Animals Genetic Resources, Krakowska 1, Str., 31-047 Balice, Poland; grazyna.polak@iz.edu.pl

Poland participate in activities aimed at the protection of genetic resources of native and locally adapted breeds in Poland. In 1999 work began on the National Program for the Conservation of Animal Genetic Resources, under which programs for the conservation of genetic resources for individual populations were developed. During this time, the number of native breeds increased to 83 populations. Currently the protection covers the following species: cattle, horses, pigs, goats, sheep, poultry (chickens, geese, ducks), fur animals (foxes, coypu, rabbits, chinchillas, ferrets), bees; in total, almost 114 thousand females and over 1.8 thousand bee families. An intensive increase in the part of the populations covered by protection is noticeable, unfortunately not all of them. Taking into account the data contained in the world DAD IS database, some of them stagnate or even decline. The aim of the research was analyse the influence of payments received in frame of CAP on the enlargement of the number of conserved populations.

Session 01 Theatre 12

Trends in inbreeding and challenges in Norwegian cattle breeds

P. Berg[1], A. Holene[2] and N. Svartedal[2]

[1]Norwegian University of Life Sciences, Animal and Aquacultural Sciences, IHA / NMBU, Arboretveien 5, 1432, Norway, [2]Norwegian Genetic Resource Center NIBIO, P.O. Box 115, 1431 Ås, Norway; peer.berg@nmbu.no

A systematic effort to collect pedigree records of the six local Norwegian dairy cattle breeds was initiated in 1990. These breeds are to varying degrees in traditional use in dairy herds, but increasingly used as suckler cows in meat production. Here we report changes in census and effective population size, rates of inbreeding and pedigree quality and discuss the challenges for these small breeds in different production systems. Pedigree quality is relatively low, with a pedigree completeness index for 5 generations varying between 0.56 and 0.91. This is largely due to incomplete pedigree on older animals and partly due to the use of natural service sires. Both census and effective population sizes has increased in all six populations, except for a small decrease in effective size of the largest breed STN in the last decade. Effective population sizes are above 50 in the period from 2001 to 2020 for four of the six breeds. Rates of inbreeding has decreased but are still above 1% per generation in some of the breeds. The positive trends thus need to continue to further decrease rate of inbreeding. The decreasing rates of inbreeding has been possible in part by the efforts to record pedigree on all animals, use of more sires but also by use of older sires from the gene bank as well as recruiting new AI sires with low relatedness to the population. In contrast to the increasing number of cows in the local Norwegian breeds, the number of cows from these breeds in milk recording has decreased from 1,384 to 1,292 (-6%) in the period 2012 to 2021, whereas the number of cows in suckler production increased from 685 to 3,197 (+386%) in the same period. In 2021 approximately 29% of the cows in the local breeds are in milk recording herds. Implications of a decreasing proportion of cows in milk recording for future management of the breeds is discussed.

Conservation of animal genetic resources in Poland and Ukraine

E. Sosin[1] and N. Reznikova[2]
[1]National Research Institute of Animal Production, Department of Animal Nutrition and Feed Science, Krakowska St. 1, 32-083 Balice, Poland, [2]M.V. Zubets Intitute of Animal Breeding and Genetics, Laboratory of Red breeds selection, L. Ukrainian str.17, apt.14, Shchaslyve vil. Boryspil district, 08-321 Kiev, Ukraine; ewa.sosin@iz.edu.pl

In Poland a total of 87 breeds/lines of farm animals, including cattle, horses, sheep, goats, pigs, fur animals, poultry and bees are incorporated in Programme of Animal Genetic Resources Conservation. In Ukraine the number of population covered with Conservation Programme is 17 breeds, but some of them are now officially extinct or practically are at the edge of extinction. In recent decades the main emphasis has been placed on the *in situ* conservation but full protection of the existing genetic variability is possible by supporting *in situ* activities with *ex situ* conservation methods. Since 2002 in Ukraine and 2014 in Poland there are functioning National Genebanks of Biological Material. The aim of those National Genebanks is to create a collection of biological material from the basic species of farm animals (cattle, pigs, horses, sheep and goats). The main collection of the genebanks is bulls' semen, but there are some cattle embryos and oocytes as well. In Poland, a collection of pig embryos has been created for several years. In Poland *ex situ* conservation programme is implemented mainly in cattle and in some extent in pigs. In other species *ex situ* programme encounter difficulties because of technical problems or lack of infrastructure that is required by veterinary regulations In Ukraine, *ex-situ* collection of semen is given to herds at the decision of scientific council of M.V. Zubets Institute of Animal Breeding and Genetics, NAAS for the recovery of populations and to avoid in-breeding. In such way was freshed the population of Whiteheaded Ukrainian, Grey Ukrainian and Lebedin cattle. The works on the recovery of Mirgorog pig population is leaded now. In Genebank there is as ejaculated, so epidedymal semen. In Poland the programme also aims to create the so-called cattle breeding cryoarchive, which is in fact a collection of cyclical material collected from commercial populations. The collection of reproductive material derived from breeding animals is extremely valuable, especially now and can contributes to the reconstruction of native breeds.

Genetic variations of mtDNA in Lipican horses

N. Moravčíková[1], R. Kasarda[1], J. Candrák[1], H. Vostrá-Vydrová[2] and L. Vostrý[2]
[1]Slovak University of Agriculture in Nitra, Tr. A. Hlinku 2, 94976 Nitra, Slovak Republic, [2]Czech University of Life Science Prague, Kamycka 129, 16500 Prague, Czech Republic; nina.moravcikova@uniag.sk

Mitochondrial DNA (mtDNA) provide an important indicator of population history, reporting mainly evidence of adaptive evolution of populations in dynamically changing environments. In horses, the mtDNA studies are mostly based on sequencing data with various coverage, but there is also another possibility to use information about single nucleotide polymorphisms (SNPs). Therefore, this study aimed to estimate intrapopulation genetic variation and haplotypes diversity by analysing SNPs localised in the mitochondrial genome of the Slovak population of Lipican horses. Overall, 52 animals bred in National Stud Farm Topolčianky have been genotyped by GGP Equine 70k chip that includes 1,187 SNPs spread in mtDNA. Before analysis, SNP pruning was performed to select only animals and markers with a call rate higher than 95%. The genetic variation of the population gene pool was then expressed by calculating gene diversity and the proportion of homozygous genotypes. Due to the fact that the European Lipican population is closed and the gene pool of the Slovak population can be affected only by the exchange of animals between European stud farms, we expected a relatively low level of intrapopulation mtDNA variation. The observed level of gene diversity (0.097±0.003) confirmed this assumption. The average observed heterozygosity (0.022±0.001) revealed that a major part of analysed mtDNA SNPs was homozygous. The average minor allele frequency (0.056±0.068) corresponds to the level of observed heterozygosity. Even if the mtDNA intrapopulation genetic variation was low, we identified several genetic patterns represented by different haplotypes across individuals. Some of them were shared by several individuals, but 13 were unique (the most frequent haplotype was found in 20 animals). As far as different haplotypes have been identified, these haplotypes could be used to assess mechanisms determining the intrapopulation mtDNA diversity of Lipican horses and also to evaluate the possible role of mtDNA background in performance. The study was supported by the projects APVV-20-0161, APVV-17-0060 and QK1910156.

Session 01 — Poster 15

Analysis of whole-genome sequence data on Colombian Creole pig breeds

R. Suárez Mesa[1], R. Ros-Freixedes[1], R.N. Pena[1], B. Hernández[2], I. Rondón-Barragán[3] and J. Estany[1]
[1]University of Lleida-Agrotecnio-CERCA Center, Rovira Roure 191, 25198 Lleida, Spain, [2]Agrosavia, Km 14 Bogotá-Mosquera, 250047 Bogota, Colombia, [3]University of Tolima, Santa Helena, 730006 Ibagué, Colombia; rafael.suarez@udl.cat

There is a growing interest in conserving local breeds in favour of biodiversity, particularly in a context of climate change and the challenges it poses to livestock productive systems. The three recognized Colombian Creole (CR) pig breeds (ZU: Zungo; CM: Casco de Mula; and SP: San Pedreño) are an example of environmental adaptation. Whole-genome sequencing allows for a massive characterization of genetic variation and offers new opportunities for genetic management. Here, we present the first study that characterizes sequence variation at the whole-genome level in CR pigs. Seven pigs of each CR breed were sequenced together with seven Iberian pigs, which are considered the ancestors of CR breeds, and seven pigs from each of the following cosmopolitan genetic types (Duroc; Pietrain; and Large White×Landrace), as representatives of current commercial pigs. In total, we were able to identify 13,187,918 variants across all genetic types, 6,451,218 of which were common to all of them. Molecular variability of CR breeds (6,451,218 variants) was in line with that found in cosmopolitan genetic types (6,575,953 variants). Within each CR breed, the number of variants ranged from 3,919,242, in SP, to 4,648,069, in CM, being greater than in Iberian pigs (3,346,025). We used these data to investigate the sequence variation in 34 genes that harbour at least one variant with reported effects on morphological, reproductive and productive traits. The differential and, in some cases, unique distribution of allele frequencies in these genes across breeds provided a first evidence to understand the genetic makeup of CR breeds for skin colour, ear and body size, boar taint and fatness. Thus, the discovery of a novel haplotype in the leptin receptor gene that segregates in CM and ZU but not in SP may give new clues on the role of this gene on how CR breeds were formed and then locally adapted. Our findings emphasize the need for phenotypic characterization of CR breeds in order to validate in silico predictions as well as the contribution of each variant to breed characteristics.

Session 01 — Poster 16

Genetic variability in a captive Mhorr gazelle (Nanger dama mhorr) population via pedigree analysis

S. Domínguez[1], I. Cervantes[2], J.P. Gutiérrez[2] and E. Moreno[1]
[1]Estación Experimental de Zonas Áridas-CSIC, Ctra. De Sacramento s/n, 04120 La Cañada de San Urbano, Almería, Spain, [2]Departamento de Producción Animal, Facultad de Veterinaria, UCM, Avda. Puerta de Hierro s/n, 28040 Madrid, Spain; icervantes@vet.ucm.es

The dama gazelle (Nanger dama) is a North African ungulate listed as critically endangered by the International Union for Conservation of Nature (IUCN). The mhorr subspecies is extinct in the wild and currently survives thanks to the creation in 1971 of an *ex situ* breeding program in the city of Almeria (Spain). Subsequently, this original population led to multiple subpopulations in Europe, North America, Africa and the Arabian Peninsula. It is essential to highlight the importance of breeding programs as a tool for the recovery of threatened populations through optimal genetic management and mating programs. The objective of this study was to analyse the pedigree information of the Mhorr gazelle (N. dama mhorr) to assess the evolution of genetic variability and the genetic management carried out during 50 years of captive breeding. The pedigree data included a total of 2,739 individuals. The alive population in Almeria was used as reference population (106 individuals). The probability of gen origin, inbreeding coefficients, Wright's F statistics and effective population sizes were computed using ENDOG program. The number of founders, number of ancestors, effective number of founders and effective number of ancestors were 4, 4, 3 and 3, respectively. The genetic contribution of the single founder male in the alive population is considerably higher (82.4%) than the contribution of the other three founder females (9.7%, 4.0% and 3.8%). The average values of inbreeding for the alive population were 28.3%. F-statistics results reported small differences between the population of Almeria and the other living subpopulations in Europe (FST=0.003281) due to their common origin. The effective population size by individual increase in coancestry (N_e=12.1) and by individual increase in inbreeding (N_e=13.2) were considerably low, but the higher value of that latter showed the successful mating system that preserved this population despite the low number of founders, without significant new bottlenecks having occurred.

Session 01 Poster 17

Social aspects of the protection of genetic resources of primitive and warmblooded horse bree

I. Tomczyk-Wrona and A. Chełmińska
National Research Institute of Animal Production, Department of Horse Breeding, Sarego 2, 31-047, Krakow, Poland; iwona.wrona@iz.edu.pl

In situ conservation is considered the preferred method of conserving the biodiversity of the equine population in traditional production systems. It allows the maintenance and adaptive use of animal genetic resources in the productive landscapes while preserving their cultural values. Breeding of native horse breeds may support the creation of market niches, for example in the field of agritourism services that are so fashionable nowadays. For example, the presence of recreational horses on such a farms makes the offer more attractive. The development of horse tourism offers a wide range of opportunities to use a large number of horses significantly. Promotion in this respect brings significant benefits, both in keeping the native breeds of horses and in creating new labour markets. In order to achieve these goals, however, it is necessary to constantly maintain and increase social interest and understanding for the activities for the protection of genetic resources and the importance of native horse breeds. The greatest importance in the breeding of conservative horse breeds are individual breeders, whose traditions, passion and consistency in action give opportunities to obtain valuable material. Detailed characteristics of the number and area of occurrence allow the assessment of the social status of native breeds. In 2021, a total of 1,027 herds and 5,137 mares were protected, including 498 herds of primitive breeds with 3,134 mares and 529 herds of warm-blooded horses with 2,003 mares. Numbers of primitive mares covered by protection were more than 1.5 times more than warmblooded mares. This trend has been going on for several years.

Session 01 Poster 18

Morphometric variability in the Apulian native Murgese horse

G.B. Bramante[1], E.P. Pieragostini[2] and E.C. Ciani[3]
[1]National Association of Horse Breeders of the Murge and Donkey of Martina Franca, C/da Ortolini Zona L,14, 74015 Martina Franca, Italy, [2]UNIBA, DETO, Strada provinciale 62 per Casamassima Km 3, Valenzano, Italy, [3]UNIBA, DBBB, Via Amendola 165/A, 70126 Bari, Italy; knowout@gmail.com

Morphometric variability and genetic parameters for linear type traits were estimated in the Murgese horse, a mesomorph breed from Apulia, Southern Italy. Previously selected for the saddle and harness, the breed is appreciated today, in equestrian tourism for its rusticity and good temperament, with increasing interest also in equestrian sports, mainly dressage. Hereditability for morphometric traits were estimated using multiple trait animal model, the fixed effects considered were sex and year of birth, while the herd was considered as a random effect. The dataset is composed of 3,342 subjects registered in the official stud-book and approved as breeding animals in the period from 1987 to 2017, for which morphometric records were available, namely: height at withers, thoracic circumference and cannon bone circumference. Summary statistics (average, standard deviation) were estimated from the data in order to assess the phenotypic variability of the Murgese horse population for the considered morphometric parameters. The estimates of hereditability for the three linear traits were 0.40 for cannon bone circumference, 0.45 for thoracic circumference, and 0.51 for height at withers. The estimated genetic correlations ranged from 0.81 to 0.87. The highest positive correlation was found among thoracic circumference and cannon bone circumference. Current results are discussed on the light of the genetic conservation and management strategies for this local horse breed, as well as its recent attitude towards different equestrian disciplines.

Joint *in situ* and *ex situ* efforts for the active conservation of the Gentile di Puglia sheep breed

L. Temerario[1], V. Landi[1], D. Monaco[1], C. Carrino[2], G. Bramante[2], F. D'Innocenzio[2], S. Grande[3], G. Donnini[4], F.M. Sarti[5], M. Ragni[1], R. Sardaro[6], G.M. Lacalandra[1], M. Albenzio[6], P. De Palo[1], F. Pilla[7], M.E. Dell'aquila[1] and E. Ciani[1]
[1]University of Bari, Bari, Italy, [2]Freelance professional, Apulia, Italy, [3]ASSO.NA.PA. (National Association of Pastoralists), Roma, Italy, [4]A.R.A. Puglia (Regional Breeder Association), Apulia, Bari, Italy, [5]University of Perugia, Perugia, Italy, [6]University of Foggia, Foggia, Italy, [7]University of Molise, Campobasso, Italy; elena.ciani@uniba.it

Gentile di Puglia is an ancient Merino-type sheep breed from Apulia (Southern Italy). Its origin is still debated. While being generally considered as issued from crossbreeding Spanish Merino rams with local ewes under the Aragonese domination in Apulia (XV century CE), it cannot be excluded that, as supported by classical Latin writers, it (also or instead) represented the sheep genetic stock imported from Apulia to the Iberian peninsula in order to improve wool traits of the early Merino ancestors. Traditionally reared in Apulia under transhumant systems, Gentile di Puglia is currently reared under sedentary semi-extensive farming systems. The census has remarkably declined over the last decades, and while it still represent a valuable genetic material, coordinated *in situ* and *ex situ* efforts are necessary for its preservation and valorisation. Here we describe a set of initiatives focusing on the active conservation of this breed that, in the last years, have been established at local and/or international level. Better coordination of future initiatives by different key stakeholders will significantly contribute in improving effectiveness and long-term sustainability of the Gentile di Puglia conservation strategy.

CAMEL-SHIELD project: a further step toward understanding the *Camelus dromedarius* genetic diversity

E. Ciani
University of Bari, Italy; elena.ciani@uniba.it

The EU-funded CAMEL-SHIELD project, focusing on target areas in Algeria and Morocco, will contribute to further extend current knowledge about genetic variability patterns in the Camelus dromedarius species, that has been addressed so far using microsatellite markers, mtDNA, aDNA, WGS and DD-RAD sequencing technologies. In the frame of the CAMEL-SHIELD project, the HD Illumina(R) SNP chip, that is going to be manufactured as an outcome of the eleventh Agricultural Greater Good Initiative, will be adopted to infer genetic relationships among sampled animals and possibly detect signals of environmental adaptations.

Session 01 Poster 21

Population structure of Lithuanian riding horses

A. Račkauskaitė, R. Šveistienė and Š. Marašinskienė
Lithuanian University of Health Sciences, Animal Science Institute, R. Žebenkos str. 12, Baisogala, Radviliškio distr., 82317, Lithuania; ruta.sveistiene@lsmuni.lt

In the context of breeds industrialisation, the conservation of biodiversity remains a major challenge and priority. Estimation of population structure parameters, such as rate of inbreeding (F) and effective population size (Ne), is important tool for the genetic diversity characterisation. The aim of this study was to determine and compare the effective population size and inbreeding coefficient of three Lithuanian horse populations: Baltic Hanoverian (BH), Lithuanian Saddle Horse (LSH) and Trakehner (T). BH and LSH are bred as an open populations since 2000. T is partially closed population, in which only Thoroughbred and Arab horses are seldom used for breeding. In 2010 the supervision of T horse breeding was laid under the National Conservation Programme. The software system POPREPORT was used to analyse data of live populations in 2020. In total 1,892 pedigree records of BH (dates of birth from 1982), 2,421 of LSH (dates of birth from 1968) and 2,888 of T (dates of birth from 1971) were accepted. The average pedigree completeness of 100% in one generation deep was observed in all populations in 2020, the highest pedigree completeness in six generation deep was determined in T horses (42.1%), lower in BH (31.5%) ant the lowest in LSH (29.6). The average inbreeding coefficient for horses born within last 10 years varied from 0.000 to 0.006 in BH, from 0.000 to 0.008 in LSH and from 0.149 to 0.180 in T population. The effective population size (Ne) based on inbreeding were 292 in BH, 238 - LSH and 11 - T, based on number of parents it was 599, 341, 191 respectively. Open populations of BH and LSH had wide genealogy which allowed to avoid inbreeding. In T horse population the inbreeding coefficient was higher, but due to applied breeding program the coefficient was stable. However, the low effective population size raised the need to improve the conservation program to keep population on the safe state in the future.

Session 01 Poster 22

Inbreeding level of X-chromosome in Old Kladrub Horse

L. Vostry[1], M. Shihabi[2], H. Vostra-Vydrova[1,3], B. Hofmanova[1] and I. Curik[2]
[1]CZU, Kamycka 129, Prague, Czech Republic, [2]University of Zagreb, Faculty of Agriculture, Svetosimunska 25, Zagreb, Croatia, [3]IAS, Pratelstvi 815, Prague, Czech Republic; vostry@af.czu.cz

Inbreeding is usually analysed in terms of autosomes. In contrast, inbreeding on the X-chromosome is rarely studied, although it could be quite different because genes on X-chromosome loci have different genomic population dynamics than autosomal loci. Therefore, inbreeding on the X chromosome should be analysed because it could contribute significantly to inbreeding depression in some traits, such as reproduction. The main objective of this study was to compare inbreeding coefficients on the autosomal (FX) and X-chromosome (FXsex) based on pedigree and genomic data (F_{ROH} and F_{ROHsex}) in the Old Kladruber horse. We estimated autosomal and X-chromosomal inbreeding coefficients using pedigree (9,173 individuals) and genomic (180 individuals) data. The mean of complete generation equivalents of genotyped animals was 15.5. Genomic inbreeding coefficients for the X-chromosome (3,154 SNPs) and autosomes were estimated for ROH>4Mb and ROH>8Mb, respectively. The mean inbreeding coefficients for F_{Xsex} and F_X in the pedigree were 0.18 (range of 0.01 to 0.40) and 0.13 (range of 0.03 to 0.23), respectively, with a ratio of F_{Xsex}/F_X of 1.3. The mean genomic inbreeding coefficients $F_{ROHsex>4Mb}$ (0.17, range from 0.00 to 0.72) and $F_{ROHsex>8Mb}$ (0.12, range from 0.00 to 0.65) were higher than the estimates from the autosomes of $F_{ROH>4Mb}$ (0.11, range from 0.00 to 0.22) and $F_{ROHsex>8Mb}$ (0.07, range from 0.00 to 0.10), respectively. More specifically, the $F_{ROHsex>4Mb}/F_{ROH>4Mb}$ ratio was 1.6 and the $F_{ROH-sex>8Mb}/F_{ROH>8Mb}$ ratio was 1.7, which was higher than the 1.3 observed between pedigree estimates. Correlations between autosomal and X-chromosomes were highest for pedigree-based estimates (0.65) compared to genomic inbreeding coefficients for distant, ROH>4Mb (0.21) and close ROH>8Mb (0.19) inbreeding. The inbreeding estimates for the X-chromosome showed higher variability, most likely influenced by higher genetic drift. This analysis showed that the X-chromosome, although often ignored, can make an important contribution to the understanding of inbreeding and its negative consequences. Supported by projects QK1910156, MZE-RO0719, and ANAGRAMS-IP -2018-01-8708.

Session 02 Theatre 1

Livestock emissions and the COP26 targets: welcome and introduction

F. O'Mara[1] and M. Lee[2]
[1]Teagasc / Animal Task Force, Oak Park, Co. Carlow, Ireland, [2]Harper Adams University, Newport, Shropshire, TF10 8NB, United Kingdom; frank.omara@teagasc.ie

Greenhouse gas (GHG) emissions are among the greatest challenges faced by livestock farming that has become an issue of the greatest importance for consumers and policy makers. At the same time, livestock has a strong mitigation potential: 'improved management practices could reduce emissions from livestock systems by about 30%', says FAO. Several reports are under writing among international organisations (IPCC, FAO, DG Agri, etc.) that may lead to decision making in the coming months with impacts on the livestock sector. The ATF would like to draw a state of the art on GHG emissions from livestock, the role of methane, the different metrics to measure emissions and mitigation levers at various scales, inviting international organisations, research organisations, farmers, industries, NGOs and policy makers.

Session 02 Theatre 2

IPCC AR6 Working Group III report: overview of agricultural sector GHG emissions

H. Clark
NZAGRC, Grasslands Research Centre, Tennent Drive, Private Bag 11008, Palmerston North 4442, New Zealand; harry.clark@nzagrc.org.nz

Agricultural CH_4 and N_2O emissions averaged 156 (±46.8) Mt CH_4/yr and 6.0 (±3.6) MtN_2O/yr 6.8 (±2.3) GtCO2-eq/yr (using IPCC AR6 GWP_{100} values for CH_4 and N_2O) respectively between 2010 and 2018. CH4 emissions continued to increase, with enteric fermentation forming the main source and increasing ruminant animal numbers a key driver. N2O emissions also increased, notably from manure application, nitrogen deposition, and nitrogen fertiliser use. Modelling studies indicate that between 2020 and 2050, agricultural mitigation measures could potentially reduce emissions by 5.3±0.2 GtCO_2-eq/yr (assuming carbon prices up to USD100/tCO_2-eq) from cropland and grassland soil carbon management, agroforestry, use of biochar, altered rice cultivation, and livestock and nutrient management. Demand-side measures including shifting to healthier diets in some regions, reducing food waste, building with wood and biochemicals and bio textiles could potentially reduce emissions by a further 1.9 GtCO_2-eq/yr. Considering trends in population growth, income, consumption of animal-sourced food, fertiliser use and disturbances from climate change, effective policy interventions and financing will be required for agriculture to achieve the identified mitigation potential. Realisation of mitigation potential can only be achieved when stimulated by concerted, rapid and sustained effort by all stakeholders, from policy makers and investors to landowners and managers.

Session 02 Theatre 3

Technical advisory group on methane emissions from the livestock systems & Working group on metrics

J. Lynch[1] and C. De Camillis[2]

[1]University of Oxford, Department of Physics, Parks Road, Oxford, OX1 3PU, United Kingdom, [2]FAO, Viale delle Terme di Caracalla, 00153 Rome, Italy; john.lynch@physics.ox.ac.uk

Globally, livestock systems are important contributors to nutrition and improved livelihoods, especially among developing countries. With the increasing occurrence of extreme weather events and rising global temperature due to climate change, focus has shifted to the livestock's contribution to the greenhouse gas (GHG) emissions such as methane, nitrous oxide and carbon dioxide. According to the Food and Agriculture Organization of the United Nations (FAO), livestock contribute around 14.5% to global anthropogenic GHG emissions, mainly in form of methane, which originates from rumen fermentation and manure management systems. Globally, methane emissions from the livestock sector represent 32% of the total anthropogenic methane sources. More focus has been placed on methane, a short-lived climate pollutant, to reduce the impact of climate change by 2030. Methane is central at the forthcoming UN Climate Change Conference (COP27) discussions and many countries have joined the Global Methane Pledge, led by the United States of America and the European Union, to slash methane emissions by 30% from 2020 levels by 2030. The Livestock Environmental Assessment and Performance Partnership (FAO LEAP Partnership) has recently released a technical document methane reviewing methane sources, mitigation strategies and climate change metrics. Besides methane, the LEAP Partnership has developed guidelines to assess the environmental impacts of livestock systems on climate change, biodiversity, water use, soil carbon, and those tied to nitrogen and phosphorus use. The FAO LEAP Partnership is a multi-stakeholder initiative, composed of Countries, Private Sector, Civil Society and Non-Governmental Organizations, that seeks to improve the environmental sustainability of the livestock systems through harmonized methods, metrics and data.

Session 02 Theatre 4

EU policy tools to decrease emissions from the livestock sector

V. Forlin[1] and A. Pilzecker[2]

[1]European Commission, DG CLIMA, Avenue de Beaulieu 31, 1160 Brussels, Belgium, [2]European Commission, DG AGRI, Rue de la Loi 130, 1000 Brussels, Belgium; valeria.forlin@ec.europa.eu

EU climate policies for the land sector address the Member State level by means of national targets. Until 2030, under the current framework, there is no separate mitigation target for the agricultural non-CO_2 sector; together with other sectors (buildings, transport, waste) national GHG emission reduction targets under the Effort Sharing Regulation point to emission reduction objectives. In the context of the 'Fit for 55' revision, the Commission has proposed more ambitious national targets for the LULUCF sector and an EU target to achieve a climate-neutral land sector (then combining land use, forestry and agriculture) by 2035, implying a balancing of land removals and agricultural emissions. In addition, the Commission published the Communication on 'Sustainable Carbon Cycles' (COM(2021) 800 final, 'Sustainable carbon cycles', https://europa.eu/!6pdNpk) to promote the development and deployment of natural and technological carbon removal solutions at scales conducive to the EU objective of climate neutrality. In view of delivering on the EU commitments, and in accordance with its Farm to Fork Strategy (COM/2020/381 final) and Methane Strategy (COM/2020/663 final), the Commission implements actions, prepares new and revises existing legislation to enable the livestock sector to reduce its methane emissions. Within the new Common Agricultural Policy, the Commission will ensure that support from the European Agricultural Guarantee Fund and European Agricultural Fund for Rural Development contributes to climate change mitigation. A specific objective is indeed dedicated to climate and includes the reduction of greenhouse gas emissions. A result indicator will reflect the share of livestock units supported to reduce emissions, and several relevant instruments can be used for the purpose of reducing methane emissions, like eco-schemes, agri-environmental management commitments and, investments support (Regulation (EU) 2021/2115).

Global research alliance on agricultural green house gases – GRA-GHG

H. Montgomery
Global Research Alliance on Agricultural Greenhouse Gases, Ministry for Primary Industries, Charles Fergusson Building 34-38 Bowen St Pipitea, Wellington, New Zealand; hayden.montgomery@globalresearchalliance.org

Potentials and set of development of levers and in R&I to mitigate GHG emissions. Trade-offs and levers between GHG mitigation options and animal welfare, biodiversity, protein autonomy, use of pesticides.

Session 02 Theatre 6

Production profiles and GHG emissions – pilot study on farm practices towards circular bioeconomy

M. Záhradník, O. Pastierik, J. Huba, D. Peškovičová and I. Pavlík
National Agricultural and Food Centre, Research Institute for Animal Production Nitra, Hlohovecká 2, 95141 Lužianky, Slovak Republic; dana.peskovicova@nppc.sk

The key objectives of the study were to provide information for more accurate evaluation of farm emission profile of GHG emissions and to map current practices of mitigation and application of circular bioeconomy principles. Pilot survey among 36 Simmental cattle breeders in Slovakia was performed by an expert via face-to-face meetings and data collection on farms. 11,012 cows in the study represent 31% of the total Simmental population in the national milk recording scheme. Top quartile reported average annual milk production more than 8,237.5 kg. The lowest quartile reported mean annual milk production less than 7,112.25 kg. Most of the farms (29; 80.5%) produced from 7,343 to 9,070 kg of milk. There were 22 farms with intensive system, 2 with extensive and 12 semi-intensive farms. Farms that applied grazing (12) covered 23.7% (5,257 animals) of all cows and heifers. In addition to the quantitative parameters defining the production system itself, the animal waste management systems and pilot mapping of mitigation techniques and innovations were studied. Most of the farms house animals in roof-insulated stalls with air ventilation (22) and straw for bedding (30). Other types of bedding were reported, such as solids from slurry separation (5) and sawdust (1). Majority of farms produce solid manure (30; 83.3%) coming from animal waste management systems utilizing solid manure from producing cows and deep bedding used in housing for dry cows. Almost half of the farms (48%) were equipped with roof-insulated stalls and air ventilation or slated floor, while 19% of farms used combination of both. The use of these technologies has led in some cases to a slight reduction in nitrous oxide and ammonia. Preliminary findings showed that an ammonia reduction of more than 10% was achieved on 4 farms (maximum 11.97%). Our study contributes to a better insight into farm practices and provides other relevant information aimed at level of understanding of the bioeconomy principles and the use of innovation and new technologies by livestock farmers.

Session 02 Theatre 7

Territorial-scale trade-offs of livestock performance: cattle diet composition perspective

R. Wang[1], F. Accatino[1], C. Pinsard[1], L. Puillet[2] and P. Lescoat[1]
[1]Université Paris-Saclay, INRAE, AgroParisTech, UMR SADAPT, 16 Rue Claude Bernard, 75005 Paris, France, [2]Université Paris-Saclay, INRAE, AgroParisTech, UMR MoSAR, 16 Rue Claude Bernard, 75005 Paris, France; ruizhen.wang@inrae.fr

For promoting food system's sustainability, it is paramount to study the trade-offs related to livestock system, considering food provision, emissions and land use, with a systemic approach. Livestock diet connects livestock performance with land use. We focus on cattle dietary composition (i.e. dry matter content, digestibility and gross energy content) and develop a model to explore the trade-offs among animal production (measured as total weight gain), methane emissions and land use impact indicators. Body growth is modelled through a metabolism approach based on energy balance. Methane emission is estimated using the IPCC tier 2 approach. This building block is integrated into a territorial-scale model in order to calculate a feed-food competition indicator and feed import needs, as indicators of impacts on land use. We simulated two scenarios in a grassland-dominated case study region, Bocage Bourbonnais (France), considering two events that might affect cattle diet characteristics: (1) drought and (2) pasture quality improvement. (1) is simulated by decreasing average yield (-10%) and dry matter content (-20%) of crops and grass, the grazing time on pasture (-20%) and cattle grass intake (-20%). To reach the same daily energy intake from the diet, the ratio of concentrates was increased. The results showed an increase in total weight gain (+6.3%), however with an increase in methane emission (+4.1%), feed-food competition (+1.7%) and considerable feed importation (+54.8%) compared to baseline. (2) was simulated by increasing both grass digestibility and gross energy content by 2%. Results showed that total weight gain increased by 2.6% while methane emission decreased by 1.9% and feed-food competition decreased by 2.5%, with no additional impact on feed importation. Our findings indicate that trade-offs exist in livestock system and diet composition is a lever to handle trade-offs related to livestock performance and land use. Relevant practices on pasture can provide a win-win opportunity on integrated cattle management of production, emission and land use.

Session 02 Theatre 8

State of the art in research and innovation – nutrition and supplements

J.R. Newbold and C.J. Newbold
SRUC, Peter Wilson Building, West Mains Road, Edinburgh EH9 3JG, United Kingdom; john.newbold@sruc.ac.uk

Why isn't nutrition making a bigger contribution to the reduction in greenhouse gas emissions from ruminants? One answer is that it might be, but effects of nutrition are not represented adequately in current accounting systems (e.g. Tier II methodologies). The effect of macro-nutrition (e.g. level and type of oil) on methane emissions is predictable (and can be modelled in feed formulation systems), while a small number of feed supplements (3-nitroxypropanol, nitrate, garlic and essential oils) are now on, or very close to, market. Their adoption will require incentives, such as payments to the food chain from food retailers or emerging carbon trading systems. Other feed supplements hold promise but currently lack sufficient evidence of efficacy, and/or regulatory approval and/or a viable supply chain. Identification of new candidate mitigators has been over-reliant on empiricism, for example through the blanket screening of plant extracts without complete chemical characterisation of what is being tested and clarity on what part of microbial metabolism is being targeted. However, knowledge on rumen microbiology and enzymology is advancing, for example the recognition that host genetics and nutrition differentially affect methylotrophic and hydrogenotrophic archaea. Such deeper understanding holds promise as a source of much-needed new ideas for methane mitigation.

Research and innovation for climate change mitigation from an animal breeding perspective

O. González-Recio
INIA-CSIC, Dpto. Mejora Genética Animal, Ctra La Coruña km 7.5, 28040 Madrid, Spain; gonzalez.oscar@inia.es

Livestock will face an important challenge within the next decade to cope with the objective of cut by 30% methane emissions, as agreed in the COP26. Selective breeding can contribute to reducing methane emissions from ruminants. The situation of concern about climate change makes necessary to implement direct selective breeding in order to reduce the carbon footprint of livestock and improve their adaptation to global temperatures that are expected to increase. Historically, the breeding objectives in livestock focused on maximising the profitability, balancing high productive levels and functionality while guaranteeing optimum animal welfare. More recently, and because of a raising societal concern about sustainability, a constructive debate arose on how to include sustainability, mitigation and adaptation in the breeding objectives of livestock. The impact on sustainability due to selective breeding on the traditional traits needs to be evaluated. New traits must be included in the breeding objective to achieve faster genetic gain. Some of these traits are not routinely recorded in farms, and large efforts need to be done in phenotyping these traits such as methane emissions or nitrogen accumulation. However, selective breeding faces unclear policies that pose uncertainty on designing breeding objectives in the long term. This uncertainty precludes more direct strategies to include direct mitigation strategies in the breeding goal.

State of the art in research and innovation – manure management

H. Trindade[1] and D. Fangueiro[2]
[1]CITAB, Universidade de Trás-os-Montes e Alto Douro, Quinta de Prados, 5000-801 Vila Real, Portugal, [2]Instituto Superior de Agronomia, Universidade de Lisboa, Tapada da Ajuda, 1349-017 Lisboa, Portugal; htrindad@utad.pt

Animal manures are an important source of ammonia (NH_3) and greenhouse gases (GHG) emissions. Emission of methane (CH_4) and nitrous oxide (N_2O) occur in all manure management stages – barn, storage, processing and recycling as crop fertilizer after field application. At the barn level, rapid separation of the urine and faeces by using floors with built in urine channels and the frequent removal of manure using scrappers and cleaning robots are important direct measures to reduce emissions. Indirect measures as indoor climate and ventilation control may also strongly contribute to emissions reduction. During storage, treatments like acidification, storage cover, solid-liquid separation and addition of biochar show great potential for mitigation of both GHG and NH_3 emissions. Acidification is a technique that allows the reduction of NH_3 volatilization in different stages of effluent management, also producing a reduction in CH_4 emission during the storage of animal slurries, but with inconsistent effects on N_2O emissions. Covering slurry storage tanks effectively reduces NH_3 emission and total GHG emission by reducing CH_4 emission, but with little noticeable effect on N_2O. Anaerobic digestion of slurries previous to their storage also contributes to the reduction of GHG emissions but can significantly increase NH_3 volatilization. At the field level, manure application by shallow injection and the use of nitrification inhibitors are the most promising GHG and ammonia abatement options. Properly evaluation of the effects of manure treatments or the combination of distinct treatments on the mitigation of different emissions processes should change from a single stage approach to include farm-scale and/or pilot-scale studies with a manure whole-life-cycle scale to avoid pollution swapping and simultaneously to determine the most efficient solution for manure energy and nutrients recovery.

Panel discussion (morning session)

A.S. Santos[1], A. Granados[2] and T. Boland[3]
[1]FeedInov CoLab, Estação Zootecnica Nacional – Qta da Fonte Boa, 2005-048, Santarém, Portugal, [2]EFFAB, Rue de Trèves 61, 1040 Brussels, Belgium, [3]University College Dublin, Agriculture and Food Science Centre, Belfield, Dublin 4, Ireland; assantos@utad.pt

The morning session will end with a panel discussion moderated by the Vice-Presidents of the Animal Task Force and the Secretary of the EAAP Commission on Livestock Farming Session. The panellists will be some of the speakers of the morning session + David Yañez-Ruiz – CSIC expert on mitigation nutrition and additives, a representative of the 12th World Congress on Genetics Applied to Livestock Production, a representative of the British Society of Animal Science Conference, and Nicolas DiLorenzo, University of Florida, as representative of the 8th International Greenhouse Gas and Animal Agriculture Conference.

Reducing carbon footprint in the Azorean meat industry

A.M. Pereira[1], J.L. Ramos-Suárez[2], C. Vouzela[1], J. Madruga[1] and A.E.S. Borba[1]
[1]Faculdade de Ciências Agrárias e do Ambiente, Universidade dos Açores, Instituto de Investigação em Tecnologias Agrárias e do Ambiente, rua Capitão João d'Ávila, 9700-042 Angra do Heroísmo, Portugal, [2]Universidad de La Laguna, Sección de Ingeniería Agraria de la Escuela Politécnica Superior de Ingeniería, Departamento de Ingeniería Agraria, Náutica, Civil y Marítima, Carretera de Geneto 2, 38071 La Laguna, Spain; ana.mb.pereira@uac.pt

In addition to the much-discussed enteric methane emissions, animal slaughter constitutes an important source of greenhouse gas emissions from the meat industry. Therefore, strategies to mitigate its impact must include a valorisation of the slaughterhouse waste to reduce the carbon footprint. This study addresses renewable energy production from an alternative treatment of slaughterhouse waste, namely anaerobic digestion, and the suitability of its implementation in each of the 9 Azorean islands. In the last 10 years, Azorean pork and beef production increased 12% and 25%, respectively. São Miguel and Terceira produced the largest share (>85%), whereas Corvo and Flores, the least (<2%). In 2021, 71,677 bovines and 75,560 swine were slaughtered originating ≈5,284 t of waste. This might represent a biogas self-production of 324,842 Nm^3/year (methane of 189,383 Nm^3/year), which converted into thermal (1,886 MWh) or electrical (660 MWh) energy, could prevent the emission of 557 t CO_2 equivalents by replacing the use of fossil fuels. Money savings are expected from self-consuming energy and revenue from selling energy. Moreover, the biogas digestate might be applied in an area of ≈93 ha, contributing to a lower use of artificial fertilizers. The variable amount of waste generated in smaller islands challenges designing biogas plants adjacent to each slaughtering facility; however anaerobic digestion is an economical and environmentally sustainable solution to treat slaughterhouse waste in the Azores justifying pilot studies. Funding: This work was financed by the AD4MAC Project, funded by Interreg through Program MAC 2014-2020 (MAC2/1.1/350). Further information: https://ad4mac.org/en/. AMP received funding from FCT (UIDP/00153/2020). Acknowledgments: The authors wish to thank IAMA for providing data.

Session 02 Poster 13

Optimising the integration of crop and dairy production systems in the Netherlands

L.M. Alderkamp[1], A. Van Der Linden[1], C.E. Van Middelaar[1], C.W. Klootwijk[2], A. Poyda[3] and F. Taube[3]
[1]Wageningen University & Research, Animal Production Systems group, De Elst 1, 6708 WD Wageningen, the Netherlands, [2]Wageningen Livestock Research, De Elst 1 Wageningen, 6708 WD, the Netherlands, [3]Christian Albrechts University, Grass and Forage Science/Organic Agriculture, Hermann-Rodewald Str. 9, 24118 Kiel, Germany; lianne.alderkamp@wur.nl

Renewed coupling of crop and dairy production systems is seen as a potential pathway towards more environmentally sustainable agricultural production, while it might also be an economically viable strategy. The integration of crop and dairy production can occur at the farm, but also at regional level. In the latter case specialised crop and dairy production systems cooperate to improve, for example, the cycling of nutrients across farms by exchanging land, feed and fertilizers. This can mitigate the use of external inputs and promoting the self-sufficiency of farms and regions. Integrated crop-dairy systems could contribute to the diversification of cropping systems with positive effects on various ecosystem services. Most studies analysed the environmental and economic impacts of integrated crop-dairy systems at field level, i.e. ley-arable systems, and did not address impacts at higher spatial scales. There is limited insight if the integrated crop-dairy systems can contribute to more human edible food with a net reduction of environmental impacts. The aim was to quantify and evaluate the environmental and economic benefits of integrated crop-dairy systems on typical sandy soils in the Netherlands, and to quantify the net contribution to food production. We developed an optimisation model which integrates a crop and dairy production system at regional level. The model will be used first to optimise both production systems separately. These outcomes are evaluated and used as a reference scenario. Different integration scenarios are tested, e.g. with and without land exchange or the use of only leftover streams for dairy cows. The impact on environmental (e.g. nitrogen and carbon footprint), economic (e.g. labour income) and food system indicators (e.g. human-edible protein conversion ratio) and farm configuration will be assessed. This study will show that the potential of integrated crop-dairy systems depends on the objective of optimisation and integration strategy.

Session 02 Poster 14

Carbon farming as an element of two meat quality production systems in Poland

J. Walczak[1], W. Krawczyk[1] and M. Skowrońska[2]
[1]National Research Institute of Animal Production, Department Production Systems and Environment, 1 Krakowska Street, 32-083 Balice, Poland, [2]University of Life Sciences in Lublin, Department Agricultural and Environmental Chemistry, 3 Akademicka Street, 20-950 Lublin, Poland; jacek.walczak@iz.edu.pl

The reduction in GHG emissions in livestock production is not only an urgent need from the point of view of climate change, but also consumer preferences for high-quality food systems. Hence, two of them, QMP and QAFP, implemented mitigation and sequestration methods in a special innovative project financed by the Polish Agency for the Restructuring and Modernization of Agriculture (ARMiR). A catalogue of 38 methods and a simple tool on the website for calculating the final effect (FE) of CO_{2e} emission reduction and sequestration have been developed. The algorithms used by the National Center for Emissions Management (KOBiZE) in the area of agriculture and LULUCF estimation were used to calculate Initial State (ES) and FE. Such integration allows for the direct inclusion of the farms mitigation effect in annual reports for the IPCC. The highest CO_{2e} sequestration effect for the fodder base was obtained for forest-pasture systems (5.3 t/ha/year), catch crops (1.17 t/ha/ year) or reduced tillage (3.5 t/ha/year). The best results of GHG mitigation in beef cattle breeding were achieved by reducing the level of crud feed protein (-17% N_2O) and increasing feed digestibility (-10% CH_4). For the pig rearing, the introduction of semi-slatted floor (-19% N_2O) and 5-phase feeding (-21% N_2O) were the most advantageous measures. Only methods that were cost-free or generated additional profit for the breeder, such as optimization of nutrition, were used in the research. The combination of rearing mitigation methods and simultaneous CO_{2e} sequestration (carbon farming) in the area of own feed base was the most effective method of reducing the animal production impact on climate change. Both quality systems have introduced a requirement for their members / farms to achieve a minimum of 20% CO_{2e} reduction.

Session 02 Poster 15

The use of ammonia mitigation methods in animal production for the official Polish emission inventor

J. Walczak[1], W. Krawczyk[1] and M. Skowrońska[2]
[1]National Research Institute of Animal Production, Department of Production Systems and Environment, 1 Krakowska Street, 32-083 Balice, Poland, [2]University of Life Sciences in Lublin, Department of Agricultural and Environmental Chemistry, 3 Akademicka Street, 20-950 Lublin, Poland; jacek.walczak@iz.edu.pl

Ammonia emissions have a significant impact on the quality of the natural environment and the health of societies. Due to the potential for particle formation, this pollutant can be transported thousands of kilometres and deposited beyond its formation areas. The NEC Directive requires Poland to reduce ammonia emissions by 1% annually by 2029 and 17% annually from 2023. Domestic agriculture is responsible for over 95% of these gas emissions. In 2019 the Ministry of Agriculture and Rural Development published the Advisory Code of Good Agricultural Practice on Limiting Ammonia Emissions. The management practices collected there included both soil fertilization and animal husbandry. Unfortunately, the question arose regarding their actual effect, since they are voluntary for farmers. Hence, for the needs of the National Center for Emissions Management (KOBiZE), using the results of the National Institute of Animal Production NRIAP, data of the Statistics Poland Office and the Polish Agency for the Restructuring and Modernization of Agriculture (ARMiR), the mitigation of ammonia emissions in the national livestock production was determined. The selection of methods was made on the basis of the confirmed effects described in the EMEP/EEA and IIASA methodologies, according to of the modified KOBIZE method based on Tier 2. Multiphase feeding of poultry and pigs (reduction by 20.0 and 29.8% respectively), feeding of cattle with a higher proportion of enteric protein for highly intensive dairy herds (-15.21%), extension of the pasture period for cattle (-20%), partly slated floor systems for the maintenance of pigs and cattle (-20%), poultry manure drying systems (-32.01%), covering of slurry tanks and manure plates (up to -80%) – mandatory requirement for animal husbandry, were included. The total reduction effect was 11.83% in 2019 compared to 2005 with absolute emissions from livestock farming at 137.26 Gg NH_3 and total emissions from agriculture at 300.58 Gg.

Session 03 Theatre 1

Large scale genome-wide association studies for seven production traits in Italian heavy pigs

S. Bovo[1], G. Schiavo[1], A. Ribani[1], M. Cappelloni[2], M. Gallo[2] and L. Fontanesi[1]
[1]University of Bologna, Department of Agricultural and Food Sciences, Viale Giuseppe Fanin 46, 40127, Bologna, Italy, [2]Associazione Nazionale Allevatori Suini, Via Nizza 53, 00198, Roma, Italy; giuseppina.schiavo2@unibo.it

The Italian heavy pig selection program is mainly designed to obtain suitable raw materials (i.e. green legs) for the production of Protected Designation of Origin (PDO) dry-cured hams, which overall represent an economic value of more than 3 billion € per year. Several traits are collected on pigs that are evaluated in the sib testing scheme: (1) indicators of the overall fat/lean content of the animals, including backfat thickness, ham weight and lean cuts; (2) traits related to performance and efficiency, including average daily gain and feed/gain ratio; (3) peculiar traits for dry-cured ham production, including visible intermuscular fat and ham weight loss at first salting. In this study, we carried out genome-wide association studies for these traits in the three traditional Italian breeds included in the heavy pig selection program: Italian Large White, Italian Landrace and Italian Duroc breeds. A total of more than 10,000 pigs have been genotyped with two single nucleotide polymorphism (SNP) arrays: Illumina PorcineSNP60 BeadChip and GeneSeek/Neogen GGP Porcine HD arrays. Genome scans (single-SNP and haplotype-based) were performed with GEMMA v.0.98. Several QTL regions affecting these traits were obtained. Results were poorly overlapping between traits and breeds, confirming what was obtained in our previous studies based on a smaller sample size. Overall, this study further characterized the genetic architecture of these production traits and provided additional information for improving those genomic selection plans currently running in these breeds.

Identification of gene expression regulators from pig muscle transcriptome

M. Passols[1], C. Sebastià[1,2], L. Criado-Mesas[1,2], J. Estellé[3], D. Crespo-Piazuelo[4], A. Castelló[1,2], J. Valdés-Hernández[1,2], Y. Ramayo-Caldas[4], R. González-Prendes[5], A. Sánchez[1] and J.M. Folch[1]
[1]Centre de Recerca Agrigenòmica (CRAG), Plant and Animal Genomics, Consorcio CSIC-IRTA-UAB-UB, Campus UAB, Bellaterra, Spain, [2]Facultad de Veterinaria, Departamento de Ciencia Animal y de los Alimentos, Universidad Autónoma de Barcelona (UAB), Bellaterra, Spain, [3]Université Paris-Saclay, INRAE, AgroParisTech, GABI, 78350, Jouy-en-Josas, France, [4]Institut de Recerca i Tecnologia Agroalimentària (IRTA), Genética y Mejora Animal, Torre Marimon, Caldes de Montbui, Spain, [5]Wageningen University & Research, Animal Breeding and Genomics, 6708 PB Wageningen, the Netherlands; magi.passols@cragenomica.es

Intramuscular fat content and fatty acid composition are important factors determining meat quality traits, including its organoleptic properties and nutritional value. The aim of this work was to study the Longissimus Dorsi transcriptomic profile of 129 Iberian × Duroc crossbred pigs obtained by RNA-Seq to identify potential muscle gene-expression regulators. Expression genome-wide association studies using GEMMA software were conducted between 55,445 SNPs (Axiom Porcine genotyping Array 650K, Affymetrix) distributed along the genome of the individuals and 11,055 genes expressed in muscle. A double false discovery rate filtering, by the number of SNPs and the number of genes, was used, and only the genomic regions with a minimum of 3 significant SNPs, located at less than 5 Mb, were kept. There were 1,283 eQTL regions (expression quantitative gene trait loci) significantly associated with the expression of 1,083 different genes. Respect to the associated genes, 690 eQTLs were located less than 1 Mb from the gene (cis-eQTLs) and the remaining 593 were in distal genomic regions (trans-eQTLs). Out of the total number of identified eQTLs, 341 were related to lipid metabolism (178 cis-eQTLs). Additionally, *Sus scrofa* chromosomes (SSC) SSC2, SSC7, and SSC14 showed a high number of associated SNPs with the expression of lipid metabolism genes. Moreover, 12 hotspot regions regulating the expression of 3 or more lipid metabolism genes in trans were identified. Our results increase the knowledge of the genetic basis of gene expression regulation in pig muscle.

Improved imputation of low coverage sequence data by utilization of pre-phased reads and HBimpute

T. Pook and J. Geibel
University of Goettingen, Animal Sciences, Albrecht Thaer Weg 3, 37075, Germany; torsten.pook@uni-goettingen.de

The generation of high-quality genomic data for large populations is an incremental part of animal breeding. In this work, we propose an extension for the imputation pipeline HBimpute, which was originally developed for low coverage sequence data from inbred lines in plant breeding. The suggested extension makes the method applicable for animal breeding when high-quality sequence data of one of the parents (usually the sire) of an animal is available. The key idea of our extension is to perform a phasing step before the variant calling by identifying which reads stem from the sire and which from the dam. Initially, a Hidden Markov Model is used to detect the inherited sire haplotype. Subsequently, reads are phased based on their similarity to the sire haplotype. This results in two 'pseudo'-inbred animals and henceforth allows for the application of the standard HBimpute pipeline. In particular, for breeding programs with a limited number of sires, the suggested pipeline can be a cost-effective alternative to genotyping arrays with the additional benefits of a higher number of markers. The performance of the pipeline was showcased with a simulated crossbreeding scheme in chicken providing imputation accuracies of 0.989 and thus outperforming the state-of-the-art softwares STITCH (0.966) and especially BEAGLE (0.669). While STITCH's performance was similar on common variants, error rates for variants with a frequency below 0.05 were five times as high as with our suggested HBimpute pipeline, underlining the high potential of our method for the imputation of rare variants. When using our suggested pre-phasing, imputation accuracies by STITCH / BEAGLE increased to 0.987 / 0.982. Of particular note here is that no parameter adaptation is required in HBimpute, while STITCH is very sensitive to parameter settings. For example, using K=5 instead of K=10 in STITCH resulted in a drop in accuracy to 0.829.

Genetic components of serotonergic system and its association with tail biting behaviour in pigs

G.S. Plastow[1], I. Reimert[2], W. Ursinus[3], L.E. Van Der Zande[2], E.F. Knol[4] and E. Dervishi[1]
[1]University of Alberta, AFNS, 116 St and 85 Ave, T6G 2R3, Canada, [2]Wageningen University and Research, Animal Sciences, P.O. Box 338, 6700 AH Wageningen, the Netherlands, [3]Netherlands Food and Consumer Product Authority (NVWA), Risk Assessment and Research (BuRO), Catharijnesingel 59, 3511 GG Utrecht, the Netherlands, [4]Topigs Norsvin Research Center, Schoenaker 6, 6641 SZ Beuningen, the Netherlands; dervishi@ualberta.ca

Tail biting behaviour in pigs continues to be important to the public, consumers, researchers and industry. In this study we describe association analysis between tail biting behaviour and metabolites, vitamins, catecholamines and components of the serotonergic system. As the serotonergic system has been implicated in behavioural disorders, we estimated heritabilities and genetic correlations between its components. A weighted single-step genome-wide association study (wssGWAS) was carried out to identify regions of the genome influencing the concentration of serotonin in whole blood, serum and in platelet rich plasma. The results showed that vitamins B2 and B7, and platelet number in platelet rich plasma are positively associated with tail biting behaviour ($P<0.05$). Only a suggestive association was observed between serum serotonin and the total number of biting behaviours. The heritability of the serotonergic system components was moderate to high ranging from 0.23-0.47. Monoamine oxidase activity showed the highest heritability estimate (0.47±0.05) followed by serotonin concentration in platelet rich plasma 0.35±0.06. WssGWAS results showed that the top window on chromosome 12 explained 6.4 and 4.3% of serum and whole blood serotonin genetic variance. In this region we detected *the solute carrier family 6 member 4 (SLC6A4)* gene which has been associated with serotonin transport and behaviour. A window on chromosome 14 including *the solute carrier family 18 member A2 (SLC18A2)* was observed for the platelet count in whole blood, explaining 2.1% of the total variance. In conclusion, *SLC6A4* and *SLC18A2* were identified as candidate genes associated with serotonergic system components. Our results further advance our understanding of the genetic architecture of tail biting behaviour and the serotonergic system in pigs.

Mining the K-mer patterns in whole-genome sequences of Holstein-Friesian cows

J. Szyda[1,2] and M. Mielczarek[1,2]
[1]National Research Institute of Animal Production, Krakowska 1, 32-083 Balice, Poland, [2]Wroclaw University of Environmental and Life Sciences, Biostatistics Group, Department of Genetics, Kozuchowska 7, 51-631 Wroclaw, Poland; joanna.szyda@upwr.edu.pl

K-mers, i.e. short sequences of a few and up to several bases, have very recently gained importance in the analysis of genomes. Previously, mainly utilised to assemble genomes in a de-novo mode are now much more broadly used for genome-wide association studies and evolutionary analyses. The utilisation of K-mer information in livestock genetics is still at the very initial stage. Therefore, the goal of the present study was to explore the K-mer structure of the bovine genome based on the whole genome sequences of cows representing the Holstein-Friesian breed. In particular, we explored the variation in K-mer distribution among the cows for the focus on K-mers missing in their genomes as well as on the most frequently observed K-mers. The material comprises 32 cows sequenced using the Illumina HiSeq 2000 in the paired-end mode. The average genome coverage varied between 2.7 and 13.8. In the first step SNP and InDels were identified using the pipeline consisting of: (1) raw data quality control; and (2) filtering; (3) alignment to the reference genome using the ARS-UCD1.2 assembly; (4) post-alignment processing; (5) variant calling; and (6) variant filtering. The filtered numbers of SNPs varied between 79,001 and 5,528,690. The individual with the lowest coverage and resulting lowest number of SNPs was removed from further consideration. For all the remaining cows FASTA files with individually identified SNPs and InDels pre-imposed onto the reference assembly were constructed. For each of the individual FASTA files (i.e. separately for each cow) counts of K-mers of length 10 were calculated. The numbers of unobserved 10-mers within an individual genome were similar and varied between 2,653 and 2,819 but one particular sequence: 'CGCGTCGATA' was missing in each cow. Interestingly, this sequence is observed in only 2 mammalian species. Among the 20 most commonly occurring 10-mers, overlapping among the analysed genomes, most represent repeated, microsatellite-like sequences as well as 'A' and 'T' monomers, while 8 sequences have a complex nucleotide pattern.

The GENE-SWitCH H2020 project: follows up and perspectives of functional genome annotations

E. Giuffra[1], H. Acloque[1], A.L. Archibald[2], M.C.A.M. Bink[3], M.P.L. Calus[4], P.W. Harrison[5], C. Kaya[6], W. Lackal[7], F. Martin[5], A. Rosati[8], M. Watson[2] and J.M. Wells[4]
[1]Paris-Saclay University, INRAE, AgroParisTech, GABI, Anima Genetics, Domaine de Vilvert, 78350 Jouy-en-Josas, France, [2]The Roslin Institute and R(D)SVS, University of Edinburgh, Easter Bush, EH25 9RG, U.K, United Kingdom, [3]Hendrix Genetics BV, Research and Technology Center (RTC), Villa 'de Körver', Spoorstraat 69, 5831 CK Boxmeer, the Netherlands, [4]Wageningen University & Research, Animal Breeding and Genomics, P.O. Box 338, 6700 AH Wageningen, the Netherlands, [5]EMBL-European Bioinformatics Institute, Wellcome Genome Campus, Hinxton, CB10 1SD, United Kingdom, [6]EFFAB, Rue de Trèves 61, 1040 Brussels, Belgium, [7]Epigenetics R&D, Diagenode S.A., Liège Science Park, Ru du Bois Saint-Jean 3, 4102 Liège, Belgium, [8]EAAP, Via G. Tomassetti 3/A, 00161 Roma, Italy; elisabetta.giuffra@inrae.fr

Since July 2019, the GENE-SWitCH (the regulatory GENomE of SWine and CHicken: functional annotation during development) consortium has been working to deliver underpinning knowledge on the pig and chicken genomes and to enable its translation to the pig and poultry sectors. Extensive functional genomics information has been produced for a panel of important tissues at early embryonic/foetal timepoints for both species. This information is being used to characterize the transcriptional and regulatory dynamics of the genome across development and is being integrated with existing data to deliver comprehensive genome annotation maps. Targeted studies aim to gain insights into the value of functional annotations for precision breeding, specifically by: (1) developing improved models for genomic prediction that use functional annotation combined with genotypic and phenotypic data to enhance the prediction accuracy of breeding values in commercial populations of both species; and (2) assessing if and how varying fibre content in the maternal diet affects the epigenetic patterns of the pig foetus, and the possible persistence of epigenetic marks through weaning stage. Examples of current results will be illustrated along with the further needs and future strategies they highlighted. GENE-SWitCH (https://eurofaang.eu/projects/geneswitch) has received funding from the European Union's Horizon 2020 Research and Innovation Programme under the grant agreement n° 817998.

Session 03 Theatre 7

A genome-wide association study for diarrhoea sensitivity in newborn rabbits

A. Ribani[1], S. Bovo[1], G. Schiavo[1], D. Fornasini[2], A. Frabetti[2] and L. Fontanesi[1]
[1]University of Bologna, Department of Agricultural and Food Sciences, Viale Giuseppe Fanin 46, 40127 Bologna, Italy, [2]Gruppo Martini spa, Via Emilia n. 2614, 47020 Budrio di Longiano (FC), Italy; giuseppina.schiavo2@unibo.it

Breeding and selection programs in farmed animals that aim to increase disease resistance can provide sustainable approaches to benefit animal welfare, reduce the use of antibiotics, reduce breeding production costs and improve farmers' profitability. Diarrhoea of newborn rabbits is one of the main causes of pre-weaning death in commercial rabbitries. In this study, we carried out a genome-wide association study to identify genomic regions that encompass variability involved in determining sensitivity to pre-weaning diarrhoea in a commercial rabbit population. The study was based on a case (rabbits affected by diarrhoea) and control (rabbits without any diarrhoea) experimental design within litters produced from 7 bucks and 45 does. A total of 182 affected and 149 healthy newborn rabbits selected from litters that had at least one case and one control were genotyped with the Affymetrix Axiom OrcunSNP Array (199,692 DNA markers). Association analysis was carried out via linear mixed models, as implemented in GEMMA v.0.98. A main peak of associated single nucleotide polymorphisms (SNPs) and other suggestively associated SNPs were identified in a few rabbit chromosomes. Additional cases and controls from another cohort of the same rabbit population were subsequently used to genotype the most significant SNP and validate the obtained results. Whole genome resequencing of three cases and three controls identified a few candidate causative mutations. These results are useful to implement a marker assisted selection program with the aim to improve resistance against pre-weaning diarrhoea in newborn rabbits.

Assessing workflows to call structural variants in chickens from Illumina, PacBio and Nanopore data

J. Geibel[1,2], J. Schauer[3], A. Weigend[3], C. Reimer[1,2,3], D.-J. De Koning[4], H. Simianer[1,2] and S. Weigend[1,3]
[1]University of Goettingen, Center for Integrated Breeding Research, Carl-Sprengel-Weg 1, 37075 Göttingen, Germany, [2]University of Goettingen, Animal Breeding and Genetics Group, Albrecht-Thaer-Weg 3, 37075 Göttingen, Germany, [3]Friedrich-Loeffler-Institut, Institute of Farm Animal Genetics, Höltystraße 10, 314535 Neustadt, Germany, [4]Swedish University of Agricultural Sciences, Department of Animal Breeding and Genetics, Box 7023, 750 07 Uppsala, Sweden; johannes.geibel@uni-goettingen.de

Structural variants (SV) underlie distinctive phenotypic traits of livestock, e.g. the comb mor-phology of chickens or the coat colour of pigs. In addition, they often represent deleterious mutations or influence production traits through changes in gene expression. Qualitative and quantitative knowledge about SV and their effects in breeding populations is therefore of great interest, especially since SNP arrays do not necessarily capture SV effects. At the same time, accurate determination and genotyping of SV using SNP arrays and short-read sequencing data is challenging and suffers from both, high false positive rates and low specificity. High hopes are currently pinned on the use of long-read sequencing to overcome these limitations. The two available technologies from Oxford Nanopore Technologies and Pacific Biosciences have fundamental differences in terms of read lengths and accuracy. Since the use of long reads has evolved considerably in recent years, the associated software tools and pipelines are still custom-made without a common standard and were only benchmarked in the framework of human genetics. They differ conceptually in the steps of read mapping, SV detection, across-sample SV merging, and SV genotyping. In this context, we evaluate different state-of-the-art SV calling pipelines for short- and long-read data using five chicken trios sequenced with Illumina short reads and long reads from Oxford Nanopore Technologies and Pacific Biosciences.

Exploring the link between gut microbiota, faecal SCFAs, pig muscle and backfat fatty acid profile

C. Sebastià[1,2], D. Crespo-Piazuelo[3], M. Ballester[3], J. Estellé[4], M. Passols[2], M. Muñoz[5], J.M. García-Casco[6], A.I. Fernández[5], A. Castelló[1,2], A. Sánchez[1,2] and J.M. Folch[1,2]
[1]UAB, Ciència Animal i dels Aliments, Edifici V, Travessera dels Turons, 08193 Bellaterra, Spain, [2]CRAG, Plant and Animal Genomics, Carrer de la Vall Moronta, 08193, Bellaterra, Spain, [3]IRTA, Genètica i Millora Animal, Torre Marimon, 08140 Caldes de Montbui, Spain, [4]INRAE, Génétique Animale et Biologie Integrative, Domaine de Vilvert, 78350 Jouy-en-Josas, France, [5]INIA-CSIC, Mejora Genética Animal, Ctra. de La Coruña, km 7,5, 28040 Madrid, Spain, [6]INIA-CSIC, Centro I+D en Cerdo Ibérico, Ctra Zafra-Los Santos de Maimona EX101, km 4,7, 06300 Zafra, Spain; cristina.sebastia@cragenomica.es

The gut microbiota plays an important role in energy production and in degradation of complex dietary fibres. Hence, this study aimed to assess the relationships between the gut microbiota, the faecal profile of short chain fatty acids (SCFAs) and the fatty acid (FA) composition of the host adipose tissue and muscle. Samples from muscle, backfat and the rectal content of 285 crossbred Iberian × Duroc pigs were collected at slaughter. SCFAs and FA composition was determined by gas chromatography. Faecal DNA was extracted and the V3-V4 region of the 16S rRNA gene was sequenced. Taxonomy assignment was performed with QIIME2 using the SILVA database, whereas mixOmics R package was used for exploring the relations between datasets through the regularized Canonical Correlation Analysis (rCCA) method. *Prevotella* spp. had a high relative abundance and were positively correlated with n-butyric acid content and negatively with acetic acid in faeces. Conversely, the relative abundances of *Akkermansia* spp. and *Cerasicoccus* spp. were negatively correlated with n-butyric acid and positively with acetic acid. The relative abundance of *Akkermansia* spp. was negatively correlated with oleic acid levels in backfat, while *Cerasicoccus* spp. had a positive correlation with oleic acid and a negative correlation with palmitic acid. In muscle, *Rikenellaceae RC9* spp. were positively correlated with palmitic acid and negatively with oleic acid. In summary, our results are indicative of the possible role that these genera may have in the modulation of backfat and muscle FA profile in pigs through the variation of SCFAs content in the digestive tract.

Whole genome resequencing provides information for sustainable conservation of Reggiana cattle breed

G. Schiavo[1], S. Bovo[1], A. Ribani[1], M. Bonacini[2], S. Dall'olio[1] and L. Fontanesi[1]
[1]University of Bologna, Department of Agricultural and Food Sciences, Division of Animal Sciences, Viale Giuseppe Fanin 46, 40127 Bologna, Italy, [2]Associazione Nazionale Allevatori Bovini di Razza Reggiana (ANABORARE), Via Masaccio 11, 42124 Reggio Emilia, Italy; giuseppina.schiavo2@unibo.it

Reggiana is an autochthonous cattle breed mainly reared in the province of Reggio Emilia, located in Emilia Romagna region, in the North of Italy. This geographical area corresponds to the Parmigiano Reggiano production area. Reggiana cattle are linked to the production of a unique mono-breed branded Parmigiano-Reggiano cheese, which provides the economic income to the farmers that is needed for the sustainable conservation of this autochthonous cattle genetic resource. In this study, the genome of 40 Reggiana bulls (almost all currently active sires) and 10 Reggiana cows have been sequenced at a coverage depth of ~23X and aligned against the ARS-UCD1.2 cattle genome version. About 11 million of variants were identified on average per each sequenced cattle. Comparative analyses against other ~500 resequenced genomes of several other cosmopolitan and autochthonous cattle breeds publicly available identified novel variants in candidate genes for important cheesemaking properties, disease resistance and fertility traits. Other mutations could explain the QTL that we recently identified in an extensive genome wide association for several exterior traits that we carried out in this breed. A few potentially deleterious mutations were also identified. These results will be useful to design breeding plans and conservation strategies of this local cattle genetic breed.

Perceptions of genome editing in farm animals by livestock stakeholders

E. Delanoue[1], R. Duclos[2,3], L. Journaux[4], D. Guémené[2], M. Sourdioux[2], A.-C. Dockès[5], J.-P. Bidanel[3] and R. Baumont[6]
[1]IDELE, Ifip Itavi, Monvoisin, 35652 Le Rheu, France, [2]SYSAAF, UMR BOA, INRAE, 37380 Nouzilly, France, [3]INRAE, U. Paris-Saclay, 78350 Jouy-en-Josas, France, [4]France Génétique Élevage, 149 rue de Bercy, 75012 Paris, France, [5]Institut de l'Elevage, 149 rue de Bercy, 75012 Paris, France, [6]INRAE, UMR Herbivores, 63122 Saint-Genès-Champanelle, France; elsa.delanoue@idele.fr

Since the development of the Crispr-Cas9 system, genome editing (GE) tools have gained a major place in biological research and the prospects for applications are numerous: in medicine for the treatment of many diseases, medical diagnosis or eradication of disease vectors, in bioindustry (biofuels) or in agriculture. The application of GE in agriculture is nevertheless controversial. The objective of this study was to better understand the positions of the actors of the livestock sector with respect to the use of GE techniques. A qualitative survey by 48 semi-directive interviews was conducted with different actors of the livestock sector (ruminants, pigs, poultry and aquaculture). The analysis of the surveys quickly revealed that knowledge of the subject is still limited to a very narrow sphere (mainly research and breeding sectors), who are familiar with the subject (specialists). The controversy is built around uncertainties. They are linked to the technology, its impacts on the environment and/or biodiversity, on the animal and its socio-political impact. The uncertainties are linked to the major cross-cutting issues raised almost systematically by the respondents, namely: (1) the type of agricultural model in which GE tools is used; (2) impacts of their use, particularly their irreversible nature; (3) ethical issues linked to living organisms, man's right to modify nature and ownership of living organisms. It allowed to establish a typology of stakeholders in five categories (opponent, sceptical, cautious, enthusiastic, convinced) and to characterize the main arguments associated with the position of the different stakeholders. Beyond the controversy over GE, the question of the farming system in which these techniques are used is a central issue for many stakeholders. This is often equated with highly intensive and unsustainable systems. The question of their use in more agro-ecological and sustainable systems remains highly debated.

Gene organization of the TRG locus in *Equus caballus* as deduced from the genomic assembly

A. Caputi Jambrenghi[1], G. Linguiti[2], F. Giannico[3], R. Antonacci[4] and S. Massari[5]

[1]University of Bari – DiSAAT, Via G. Amendola, 165/A, 70126, Italy, [2]University of Bari, Department of Biology, Via E. Orabona, 4, 70126, Italy, [3]University of Bari – DiMeV, S.P. 62 per Casamassima Km 3, 70010, Italy, [4]University of Bari, Department of Biology, Via E. Orabona 4, 70126, Italy, [5]University of Salento, Department of Biological and Environmental Science and Technologies, Via per Monteroni, 73100, Italy; francesco.giannico@uniba.it

The domestic horse, *Equus caballus*, belongs to the Perissodactyla order, Equidae family, and Equus genus, which also includes zebras, and African and Asian asses. Since its domestication, humans have selectively bred horses for performance traits causing a drastic reduction of the wild population and the develop of multiple breeds with a wide range of phenotypic peculiarities. As a result, most horse breeds today are closed populations with high phenotypic and genetic uniformity of individuals within the breed, but with great variation between breeds. Many of the phenotypic traits found in breeds have been successfully mapped into genomic regions, facilitated by recent methodological advances in horse genomics. Taking advantage of the latest release of the genome assembly, we studied, for the first time in an organism belonging to Perissodactyla order, the horse TRG locus encoding the gamma chain of the γδ T cell receptor. The horse TRG locus spans about 1,130 kb, from AMPH and STARD3NL genes that flank this locus in all mammalian species studied so far. It contains 38 Variable (V), 23 Joining (J) and 16 Constant (C) genes organized in 16 V-J-C gene clusters, in tandem aligned in the same transcriptional orientation. The horse TRG results the locus with the greatest extension and with a significantly higher number of genes than the orthologous locus of the other mammalian species. However, despite the dynamic evolution of the horse TRG genomic region, our phylogenetic analyses show a tight relationship of the genes encoding the variable domain with those of other mammalian species, indicating that they have been preserved as a result of a strong functional constrain. The genomic organization of the horse TRG confirms the great evolutionary plasticity of the TRG locus among the different species and the important role of the gamma chain in the adaptive immune response.

Genomic predictions including unknown parent groups for milk traits in Portuguese Holstein cattle

A.A. Silva[1], D.A. Silva[1], P.S. Lopes[2], H.T. Silva[2], R. Veroneze[2], G. Thompson[3,4], J. Carvalheira[3,4] and C.N. Costa[5]

[1]FCAV-UNESP, R. Prof. P.D. Castellane, 14884-900 Jaboticabal SP, Brazil, [2]U. F. Viçosa, Av. P. H. Rolfs, 36570-000 Viçosa MG, Brazil, [3]ICBAS, U.Porto, R. J.V. Ferreira, 4050-313 Porto, Portugal, [4]BIOPOLIS-CIBIO, U. Porto, R. Padre A. Quintas, 4485-661, Vairão, Portugal, [5]Embrapa Gado de Leite, Rua Eugenio Nascimento, 36038-330 Juiz de Fora MG, Brazil; claudio.napolis@embrapa.br

In Portugal, the autoregressive test-day (TD) model for multiple lactations has been successfully applied to predict genomic breeding values (GEBV) in the national genetic evaluation. Missing pedigree information is a recurrent issue and the objective of this study was to compare the reliability and bias of GEBV predicted with ssGBLUP with and without unknown parent groups (UPG). The Portuguese Dairy Cattle Breeders Association provided a total of 12,982,057 TD records from the first 3 lactations of milk, fat, and protein yields, and somatic cell score (SCS) of Portuguese Holstein cattle. Data from 4,485 genotyped animals were used in ssGBLUP including a total of 35,970 SNPs. To evaluate the impact of UPG on the prediction ability, the validation population included more than 860 cows and 160 bulls (>19 daughters). The bulls and cows' pseudo-phenotypes (DYD and YD) predicted in the full data set (records up to 2021) were regressed on the GEBV predicted in the reduced data set (with records truncated at 2017) for all traits. The use of UPG in the pedigree relationship led to better regression coefficients, therefore, less bias. For bulls, the b1 values with UPG (without UPG) were 0.83 (0.73) for milk yield, 0.7 (0.49) for fat yield, 0.86 (0.65) for protein yield, and 1.07 (1.02) for SCS. For cows, these values were higher than 0.85 in both approaches, probably due to the larger amount of information available for them. For bulls, the validation reliability values were 0.44 (0.38) for milk yield, 0.30 (0.20) for fat yield, 0.40 (0.30) for protein yield, and 0.42 (0.41) for SCS, respectively. The increases up to 0.10 in bull reliability achieved by fitting UPG shows the importance of including this step in ssGBLUP. In conclusion, genomic predictions for milk related traits in Portuguese Holstein cattle using ssGBLUP were more reliable and nearly unbiased when using UPG.

Session 03 Poster 14

Comparison of *longissimus lumborum* muscle transcriptomes in Romanov and Polish Merino sheep

E. Grochowska[1], Z. Cai[2] and M. Grguła-Kania[3]
[1]Bydgoszcz University of Science and Technology, Department of Animal Biotechnology and Genetics, Mazowiecka 28, 85-084, Poland, [2]Aarhus University, Center for Quantitative Genetics and Genomics, Blichers Allé 20, Postboks 50, 8830 Tjele, Denmark, [3]University of Life Sciences in Lublin, Institute of Animal Breeding and Biodiversity Conservation, Akademicka 13, 20-950 Lublin, Poland; grochowska@pbs.edu.pl

Meat quality is a complex trait, which is difficult to improve by applying only traditional breeding methods. Analysis of muscle tissue transcriptomes can enhance our understanding of the genes controlling meat quality in different sheep breeds. The study aimed to investigate the *longissimus lumborum* (LL) muscle transcriptomes in Romanov (ROM) and Polish Merino (PM) sheep to reveal differences in the expression of genes, including those associated with meat quality. The experiment was conducted on 3 ROM and 3 PM female lambs. Samples of LL muscle were collected immediately after commercial slaughter. Differential expression analysis was conducted using DESeq2 v.1.34.0 package. ClusterProfiler v4.2.2 software was applied for enrichment analysis of differentially expressed genes (DEG). A total of 381 mRNA transcripts, including 237 up- and 144 downregulated transcripts, were differentially expressed in ROM relative to PM lambs. The significant GO terms (Padj<0.05) for all DEGs included 9 GO terms in the category of a cellular component. Regarding the up-regulated genes dataset, a total of 7 GO terms were significantly enriched (Padj<0.05), including 5 and 2 GO terms in the categories of cellular component, and molecular function, respectively. The further enrichment analysis among the 381 DEGs revealed 23 significantly enriched (Padj<0.05) KEGG pathways. These results provide a foundation for identifying candidate genes and further develop the theoretical basis for new breeding strategies to optimize the production performance and meat quality of the two sheep breeds. Our outcomes provide also the foundation for further use of these breeds for crossbreeding. This work was supported by the Ministry of Science and Higher Education from the 'Innovation Incubator 2.0' programme (grant no. 8/1/2019/UTP) and by the Polish National Agency for Academic Exchange (grant no. PPI/APM/2019/1/00003).

Session 03 Poster 15

SNP substitution effects for SCS changes across environmental gradients in Portuguese dairy cattle

A.A. Silva[1], D.A. Silva[1], P.S. Lopes[2], H.T. Silva[2], R. Veroneze[2], G. Thompson[3,4], C.N. Costa[5] and J. Carvalheira[3,4]
[1]FCAV-UNESP, Animal Science, R. Prof. P.D. Castellane, 14884-900 Jaboticabal, Brazil, [2]U.F.Viçosa, Animal Science, UFV, 36570-000 Viçosa, Brazil, [3]ICBAS-U.Porto, R. J.V. Ferreira, 228, 4050-313 Porto, Portugal, [4]BIOPOLIS-CIBIO-U. Porto, R. Padre A. Quintas, 4485-661, Vairão, Portugal, [5]Embrapa Gado de Leite, Embrapa, 36038-330 Juiz de Fora, Brazil; jgc3@cibio.up.pt

Several GWAS reports for milk related traits have been performed, however, the use of this approach in a context of G by E, is yet scarce. We aimed at identifying putative candidate genes associated with milk yield (MY) and somatic cell score (SCS) according to an environmental gradient (EG). A total of 4.6 million test-day records of MY and SCS from Portuguese Holstein cows were analysed. The data included 1,537 genotyped animals for 38,615 SNPs. First, the herd-test-day (HTD) effects were estimated using an autoregressive test-day model. Then, the solutions of HTD effect were used as an EG in a reaction norm model under ssGLBUP approach. The SNP effects for MY and SCS were calculated by back solving the genomic breeding values for each EG and the proportion of genetic variance explained by 100 kb SNP windows was also computed. For MY, the SNP effects were almost constant across EG, indicating the absence of SNP by environment interaction. Nevertheless, for SCS, higher changes in the magnitude of SNP effects were observed among EG. Based on significant SNP windows for MY and SCS between the extreme EG, candidate genes were mapped. For MY (SCS), we identified 25 (33) candidate genes unique for less favourable and 11 (10) candidate genes for more favourable EG. For SCS, candidate genes SOX14, HGF, and GPRIN1 were identified only in less favourable EG while NAALADL2, ELMO2, and ZNF830 were identified exclusively in more favourable EG. NAALADL2 gene for example, is related to response to bacterium and HGF is related to regulation of MAPK cascade, important in stress responses under pathological conditions. Meaningful differences in the genetic architecture of SCS across EG were revealed in this study and may contribute to a better understanding of the SCS genetic control, permitting the design of better strategies for genomic selection in specific dairy herd environments in Portugal.

Genomic patterns of homozygosity around casein gene cluster in Italian Holstein cattle

B. Lukic[1], J.B.C.H.M. Van Kaam[2], R. Finocchiaro[2], I. Curik[3], V. Cubric-Curik[3] and M. Cassandro[2,4]
[1]Faculty of Agrobiotechnical Sciences, J.J. Strossmayer University of Osijek, Department for Animal Production and Biotechnology, Vladimira Preloga 1, 31000 Osijek, Croatia, [2]National Association of Holstein, Brown and Jersey Breeders (ANAFIBJ), R&D office, Via Bergamo 292, 26100 Cremona, Italy, [3]Faculty of Agriculture, University of Zagreb, Department of Animal Science, Svetošimunska cesta 25, 10000 Zagreb, Croatia, [4]University of Padova, Department of Agronomy, Food, Natural Resources, Animals and Environment, Viale dell'Università 16, 35020 Legnaro, Italy; blukic@fazos.hr

Most of the milk produced in Italy is processed into various types of cheese. Therefore, the insight into the milk processing properties is highly important. Caseins are milk proteins that are considered the most important in the context of cheese production. 54 SNP polymorphisms have been detected in the casein gene cluster on chromosome 6 in Italian Holsteins. With the increasing availability of SNP arrays in livestock species, a lot of genomic information has been generated, enabling a range of genomic analyses, including breeding values, inbreeding, population structure, genome-wide association studies, mapping of positive selection, and others. Runs of Homozygosity and Extended Haplotype Homozygosity are approaches that reveal precise patterns of homozygosity and provide valuable insights into past or ongoing selection events. In this study, 160,000 Italian Holstein cattle genotyped with a 50k SNP array were used to analyse the casein gene cluster on chromosome 6 for extreme Runs of Homozygosity islands and integrated Haplotype Scores. The results showed that a certain number of animals have high levels of autozygosity on chromosome 6, which is also confirmed for homozygosity at the haplotype level based on high iHS scores. These analyses confirm the effects of selection on certain casein gene variants, although they were not included in the selection scheme. Since the casein gene cluster is important for the Italian Holstein population, the results of the selection signature analysis should be evaluated and hopefully contribute to the selection plans.

Genetic parameters and GWAS analysis for lactation persistency in Brazilian Murrah buffaloes

A.A. Silva, D.A. Silva, K.R. Silveira and H. Tonhati
UNESP, Zootecnia, Rua Prof. Marcos A. Giannoni 167, 14882-225, Brazil; humberto.tonhati@unesp.br

Lactation persistency is one of the main factors influencing the total milk production over a lactation period. However, few studies have evaluated lactation persistency for dairy buffaloes. The aim with this study was to estimate genetic parameters, and GWAS analysis for different lactation persistency (LP) measures in Murrah buffaloes. A total of 323,142 test-day records of milk yield from first lactation of 4,588 Murrah buffaloes (FCAV-Unesp) were analysed. Data from 823 genotyped animals including a total of 45,376 SNP were also available. Single-trait random regression model was implemented using a single step GBLUP approach. Three LP measures were analysed: LP1- The average of estimated genomic breeding values (GEBVs) for test day milk yield from day 250 to day 305 as a deviation from the average of GEBVs from day 50 to day 70; LP2- A summation of contribution for each day from day 50 to day 250 as a deviation from day 250; and LP3- The difference between GEBVs for day 250 and day 50. In GWAS analysis, the SNP effects were calculated by back solving of the GEBV and the proportion of genetic variance explained by 200 kb SNP windows was also computed. Based on significant SNPs, candidate genes were mapped for each LP measure. Candidate genes shared between all LP measures were considered important regions. Heritability and (credibility intervals – CI) estimates obtained for LP1, LP2, and LP3 were 0.35 (0.28,0.42), 0.14 (0.1,0.18), and 0.24 (0.2,0.29), respectively. The genetic correlation and (CI) estimates between lactation persistency and milk yield were of -0.46 (-0.6,-0.3) for LP1, 0.43 (0.26,0.6) for LP2, and -0.29 (-0.46, -0.12) for LP3. In addition, 99 genes were identified for LP, e.g. *PRKG1*, *SIK1*, *RRP1B* (related to cellular response to virus), *CPNE4* (related to cellular response to calcium ion). In conclusion, the LP2 measure could be regarded as the selection criterion, due it provides more favourable low genetic correlations and moderate heritability. In addition, important candidate genes were highlighted. It contributes to a better understanding of the lactation persistency genetic control in Murrah dairy buffaloes.

Session 04 Theatre 1

Multifunctionality assessments of grazing systems: opportunity or curse?

L. Merbold[1], V. Klaus[2], A. Edlinger[1] and O. Huguenin-Elie[2]
[1]Agroscope, Integrative Agroecology, Reckenholzstrasse 191, 8046 Zurich, Switzerland, [2]Agroscope, Forage Production and Grassland Systems, Reckenholzstrasse 191, 8046 Zurich, Switzerland; lutz.merbold@agroscope.admin.ch

Permanent pastures and grasslands cover 25% of the world's terrestrial surface, and associated agricultural production systems are diverse. There is a wide range of goods and services provided by these ecosystems, comprising both private (market) and public (non-market) ecosystem services (ES). Evidence based knowledge on the multiple services grasslands provide to humans–referred to as ecosystem multifunctionality–are key for sustainable policy and land management decisions. Yet, decision making can only be achieved when information on multiple ES is provided at a resolution that is relevant to ES supply and demand. Depending on the available policy tools and the respective land-use, the spatial scale of multifunctionality assessments can vary, from farm to and/or the landscape scale. Here, we provide an overview of the current knowledge on the multifunctionality of grazing systems in the European Alps with a specific focus on synergies and trade-offs across multiple environmental and ecological dimensions. We link these considerations to the stakeholders' perspective as well as to economic and societal system characteristics. We find that in the context of European grazing systems, inevitable trade-offs among single ES require a farm-scale or landscape-scale approach with an optimization of different sets of services on co-existing types of grasslands. This is particularly relevant regarding high- versus low-intensity management. Yet, to optimize ecosystem service delivery, more applied knowledge on how differentiated grassland management can best support the delivery of multiple ecosystem services is needed. In a second step, we will bridge towards grazing systems in the Global South by highlighting the need for additionally locally tailored information in order to achieve targeted multifunctionality assessments for such systems. We suggest ways forward to improve the sustainability of these grazing systems in the future. By comparing the two geographically as well as societally distinct systems described in this talk, we aim at facilitating a regionally adapted view on the assessment and evaluation of grazing system multifunctionality.

Session 04 Theatre 2

Using crowdsourced data to assess the cultural ecosystem services provided by grasslands

A. Chai-Allah[1], S. Bimonte[2], G. Brunschwig[1] and F. Joly[1]
[1]Université Clermont Auvergne, INRAE, VetAgro Sup, UMR Herbivores, 63122 St Genes Champanelle, France, [2]University of Clermont Auvergne, INRAE, TSCF, 9 avenue Blaise Pascal, 63178 Aubière, France; frederic.joly@inrae.fr

Livestock farming systems (LFS) provide multiple services to people including cultural ecosystem services (CES). It is assumed that LFS improves the recreational attractiveness of landscape by maintaining grasslands that offer wide open views, and diverse environments. Recently, crowdsourced data derived from social media and mobile applications emerged as an important component in CES studies, by offering large datasets for broad spatial and temporal assessments. In this study, we examine the level of recreational use of each type of land cover across the mountainous region of Auvergne (France). In this area below 2,000 m of altitude (i.e. below alpine conditions), grasslands mostly result from sheep and cattle livestock farming. We used ~11,500 geo-located trails from three social media platforms (Flickr, NaturaList, and Wikiloc), to assess the characteristics of the areas visited by users. We analysed quantitatively the pattern of CES potential supply (% of each land cover in Auvergne adjusted by criteria of accessibility) and use (% of land cover in areas visited by users along daily trails). By comparing the potential supply and the use, we assessed the level of CES demand for each type of land cover. We also studied the diversity of habitats used by comparing the Shannon diversity index of land covers within trails, with the index calculated on squares of similar areas randomly distributed across Auvergne. Our results show that depending on platforms: (1) the percentage of grasslands within visitors' trails is 3 to 15% higher than the potential supply, whereas the use for the agricultural land is 3 to 16% lower (2) The Shannon index in trails is 2 to 104 times higher than in the random squares of similar areas. Our result suggests that the demand for grasslands is important across users of the three platforms we studied, and that these users appreciate diverse landscapes during their outdoor activities. Our results thus stress the importance of extensive livestock farming in shaping a diverse landscape that would be otherwise made of forests and agricultural lands.

Session 04 Theatre 3

Response of Maremmana cattle to virtual fencing for herd management during spring-summer grazing

A. Confessore[1], C. Aquilani[1], M.C. Fabbri[1], L. Nannucci[1], A. Mantino[2], F. Vichi[2], E. Gasparoni[3], G. Argenti[1], C. Dibari[1], M. Mele[3] and C. Pugliese[1]

[1]Universita' di Firenze, DAGRI, Piazzale delle Cascine 18, 50144, Italy, [2]Tellus srl, P.za S.Antonio 4, Pisa, 56125, Italy, [3]Centro di Ricerche Agro-Ambientali Enrico Avanzi, Via Vecchia di Marina, 6, Pisa, 56122, Italy; andrea.confessore@unifi.it

Mediterranean agrosilvopastoral systems can provide fodder and alternative feed resources throughout the year. However, to fully exploit their potential, a rational management of grazing is impelling. The aim of the study was to test the feasibility of using Virtual Fencing (VF) to manage Maremmana cattle at pasture in an agro-forestry system. VF is a new technology based on GPS collars to manage animals by setting virtual boundaries. When the animals approach the fences, they receive an audio cue followed by a low electrical pulse, if animals cross over the fences. VF was tested in a pilot farm in Maremma (southern Tuscany, Italy) on 30 hectares composed of different feed resources (meadows, wood, corn crops after harvesting). The trial covered the spring and summer of 2021 (from May to September); the herd was led through 12 different VF areas (named from VF1 to VF12) according to seasonal biomass availability. From 26th June to 16th July animals were kept inside a physical fence due to operational needs, so the records related to VF6 and VF7 were discarded from the dataset. The number of audio cues (S), electrical pulses (Z) and their ratio (Z/S) were analysed to investigate the capacity of the animals to interact with the VF system correctly and safely in a real grazing management framework. The Z/S ratio was under 1 for every VFs, with the lowest value registered in VF10 (0.05) and the highest in VF12 (0.25). The maximum Z and S individually received were 21 Z in VF7 and 120 S in VF12, while the minimum was 0 for all the VFs, denoting individual variability in number and type of interaction within the herd. Z was significantly lower in VF4 than in VF1 (1.40vs0.61), while S remained similar. This showed that animals progressively learned to identify the exact location of the VF borders only relying on the sounds. The GPS tracking showed that animals remained within the assigned areas, few escapes were registered mainly in the hours immediately after VF changes and along the VF in wood.

Session 04 Theatre 4

Links between the sum of temperatures and forage yield and quality in intensive permanent grasslands

P. Mariotte[1], E. Perotti[1], O. Huguenin-Elie[2], P. Calanca[2], D. Frund[1] and M. Probo[1]

[1]Agroscope, Route de la Tioleyre 4, 1725 Posieux, Switzerland, [2]Agroscope, Reckenholzstrasse 191, 8046 Zurich, Switzerland; massimiliano.probo@agroscope.admin.ch

Forage production is the one of the main ecosystem services provided by grasslands. Maximizing the yield and quality of intensively managed grasslands is becoming an important challenge, especially in the face of Climate Change. Since temperature is the most important factor affecting forage yield and quality during the first growth cycle, the sum of temperatures has been used as a proxy for estimating forage yield. However, the validity of this index, which is freely and real-time available on meteorological websites, remains unclear. We conducted this experiment during 3 years in 23 intensively managed Swiss permanent grasslands. We calculated the cumulated sum of temperature (growing degree days, GDD) following the method developed at INRA. At each site, 4 replicated plots of 1×5 m were established in 2017 and managed using common intensive cutting practices (i.e. 4-6 harvests/year). Harvested forages (4 successive harvests in different subplots per plot) were weighed and analysed for protein content and fibres, which then allowed estimating forage digestibility, net energy for lactation (NEL) and absorbable protein in the intestine (PAIN). We found a strong positive relationship between forage yield and degrees day and strong negative relationship between forage digestibility, protein content, NEL and PAIN and degrees day, independently of grassland functional types. Mowing between 650 and 750 GDD provided the best trade-off between forage yield and quality, i.e. 3.6 to 4.3 t DM/ha for forage yield, 76 to 74% for forage digestibility, 147 to 128 g/kg for protein content, 6.2 and 6.0 MJ/kg DM for NEL, and 97 to 85 g/kg DM for PAIN. Mowing earlier would increase forage quality whereas mowing later would increase forage yield. Findings from our study show that farmers could optimize their forage production by adjusting the cutting time according to specific sum of temperatures.

Does stocking method affect enteric methane emissions and performance of dairy × beef steers?

P. Meo-Filho[1], S. Morgan[2], P. Nightingale[1], A. Cooke[1], M. Lee[2], A. Berndt[3] and M. Jordana Rivero[1]
[1]Rothamsted Research, North Wyke, Okehampton, Devon, EX20 2SB, United Kingdom, [2]Harper Adams University, Edgmond, Newport, TF10 8NB, United Kingdom, [3]Embrapa Southeast Livestock, Sao Carlos, Sao Paulo, 13560-970, Brazil; paulo.de-meo-filho@rothamsted.ac.uk

Methane (CH_4) emissions from agriculture account for 40% of those produced anthropogenic activities, being the ruminants' enteric methane emission (EME) the single largest source (25%). Previous works show that grazing management practices can increase animal productivity and decrease EME per kg of animal product. Our aim was to measure the EME and performance of dairy × beef steers under two stocking methods, set-stocking (SS, three 1.75-ha paddocks) and cell grazing (CG, three 1-ha paddocks). The experiment took place at Rothamsted Research, North Wyke (Devon, UK) where 24 castrated steers (Fleckvieh or Hereford, × dairy breed) with initial average liveweight (LW) 503.7±37.8 kg were assessed during the grazing season (April-October 2021). The CG cells were grazed for 1 day, followed by 56 days (May/June) and down to 21 days (October) rest period, while the SS paddocks were grazed continuously. Pastures comprised mainly perennial ryegrass receiving 100 kg/N/ha/year. EME were measured using the SF_6 tracer gas technique and animals were weighed monthly to assess LW gain (LWG). The data were analysed using the SAS 9.4 statistical software, and the separation of means was performed using the F-test. The treatment effect was considered significant at $P<0.05$. Average daily LW gain (ADG) and final LW were greater ($P<0.05$) for steers grazing the SS system compared to CG; 0.82 vs 0.63 kg/day and 657.5 vs 594.7 kg, respectively, whereas LWG/ha was greater in the CG; 42.0 vs 51.4 kg/ha. EME in g/day, and in relation to ADG (g/kg/day) and LW (g/kg) did not vary ($P>0.05$) between treatments. An interaction effect was observed ($P<0.05$) for EME divided by the LWG/ha; EME was greater in the Autumn in the SS compared with CG; 9.05 vs 6.35 g CH_4/kg LWG/ha, respectively. In conclusion, the stocking method can affect animal performance (ADG and final LW) and EME regarding the LWG per area. The SS provides better individual animal performance whereas CG method provides better animal productivity per ha and thus less EME per kg gained per hectare.

Session 04 Theatre 6

Silage or hay: does feeding regime affect health related traits of primiparous dairy cows?

B. Fuerst-Waltl[1], S. Ivemeyer[2], M. Coppa[3], C. Fuerst[4], M. Klopčič[5], U. Knierim[2], B. Martin[3], M. Musati[3,6] and C. Winckler[1]
[1]University of Natural Resources and Life Sciences, Vienna, Gregor-Mendel-Str. 33, 1180 Vienna, Austria, [2]University of Kassel, Faculty of Organic Agriculture, 37213 Witzenhausen, Germany, [3]INRAE, UMR Herbivores, 63122 Saint-Genès-Champanelle, France, [4]ZuchtData EDV-Dienstleistungen GmbH, Dresdner Str. 89/B1/18, 1200 Vienna, Austria, [5]Slovenian Holstein Association, Groblje 3, 1230 Domžale, Slovenia, [6]University of Catania, Department Di3A, 95123 Catania, Italy; birgit.fuerst-waltl@boku.ac.at

In this study, we aimed to assess health related traits of first lactating cows that were kept in farms, mostly organic or low-input, with or without silage feeding of heifers in Austria (AT; 392 farms, 10,736 Fleckvieh), France (FR; 20 farms, 2,360 Holstein and Montbéliarde), Germany (DE; 34 farms, 1,566 Brown Swiss and Fleckvieh) and Slovenia (SI; 18 farms, 1,269 Brown Swiss, Fleckvieh, Holstein and crosses). Cows calved between 2010 and 2019. Data originated from routine performance recording and on-site data collection. As data structure differed, analyses were performed within country. Traits analysed using the logistic regression methodology were the percentage of elevated cell counts (>100,000 cells/ml; SCC100P) during the cow's 305-day standard lactation, fat-protein-ratio (FPR), defined as binary trait whether it was or was not >1.4 or <1.1 at least once during the cow's first 120 days in milk, and mastitis, metabolic and fertility disorders. SCC100P was higher in DE silage feeding farms. In AT no differences were found, while in FR and SI results significantly differed between breeds. In AT and DE silage feeding farms, FPR>1.4 during the first 120 days in milk was higher, while no significant differences between feeding regimes were found in FR and SI. However, FPR<1.1 was significantly lower in AT silage feeding farms; no differences were found in the other countries. Disease frequencies were generally low; no significant differences could be found between silage and non-silage feeding farms in any of the countries involved. The rearing and feeding regime of replacement stock is complex and influenced by numerous factors. Differences between countries with regard to geographical location and management could have led to the partly contradicting results.

The role of pasture use intensity for ecosystem services in alpine pastures

C.M. Pauler[1], H. Homburger[1,2], A. Lüscher[1], M. Scherer-Lorenzen[2] and M.K. Schneider[1]
[1]Agroscope, Forage Production and Grassland Systems, Reckenholzstrasse 191, 8046 Zürich, Switzerland, [2]University of Freiburg, Faculty of Biology, Schaenzlestr. 1, 79104 Freiburg, Germany; caren.pauler@agroscope.admin.ch

Alpine pastures provide multiple services for society such as forage for grazing animals, carbon sequestration and unique biodiversity. Adequate management can have a positive impact on these ecosystem services. However, little is known about the underlying mechanisms influencing ecosystem services. Therefore, we aimed at disentangling the interactions of soil, animals and vegetation. The study was conducted in six alpine grazing areas in the Swiss Alps. We quantified four ecosystem services (forage quantity, forage quality, carbon storage and plant species diversity). As the primary management instrument at hand, we analysed pasture use intensity of cows by GPS tracking. These were related to topography and soil properties. We found that the covariates explained 54% of the variation of forage quantity, 48% of forage quality, 41% of carbon storage, and 44% of floral diversity. Each ecosystem service was determined by a specific subset of covariates. Forage quantity was the ecosystem service explained best by pasture use intensity, followed by forage quality, plant species diversity, and carbon storage in decreasing order. The higher the forage quantity, the higher was the forage quality and carbon storage. Negative non-linear relationships were found between plant species diversity and forage quantity as well as forage quality. Our findings indicate that there are trade-offs between different ecosystem services. Consequently, these services can be maintained only by differentiated pasture use intensities. For example, highly used zones provide high forage quantity and quality but contribute little to plant species diversity. Different pasture use intensities provide an adequate tool to maintain a plethora of services provided by alpine pasture ecosystems.

Generic relationships between field uses and geographical characteristics in mountain-area farms

G. Brunschwig and C. Sibra
VetAgro Sup, 89 avenue de l'Europe, 63370 Lempdes, France; gilles.brunschwig@vetagro-sup.fr

In mountain farms, challenges posed by the degree of land slope, altitude and harsh climate further compound multiple other possible constraints, particularly in relation to the distance of the farm from the farmstead. This study focused on how mountain-area dairy farmers factor the geographical characteristics of their fields into their field-use decisions. To that end, we surveyed 72 farmers who farm the traditional Salers breed of cattle and 28 specialised dairy system farmers in the central Massif region, France. Information was collected on the uses and geographical characteristics of all grassland fields (n=2,341) throughout the entire outdoor grazing season, without identifying farmers' rationales for their field-use decisions. Field-use classes were constructed for the traditional Salers system per group of fields (grazed-only, cut-only, grazed-and-cut) and then used to classify fields in the specialized dairy system. The geographical characteristics, which were associated afterwards, were significantly different between the field groups and between field-use classes. Grazed-only fields were found to be more sloping and cut-only fields were smaller and further from the farmstead. Distance/area combinations were different according to field use (animal category, earliness of first cut, grazing and cutting sequence) and were decisive for all field-use classes. This study allowed the identification of generic relationships between field uses and their geographical characteristics in mountain-area dairy cattle farms. The exceptional size of the database used gives a real scientific solidity to the results obtained. Graphical approaches are used to document the differences found.

Optimizing age at slaughter and eco-efficiency for four cattle breeds in Alentejo, Portugal

M.P. Dos Santos, T.G. Morais, T. Domingos and R.F.M. Teixeira
Instituto Superior Técnico, Universidade de Lisboa, MARETEC – Marine, Environment and Technology Centre, LARSyS, Avenida Rovisco Pais no. 1, 1049-001, Portugal; manueldossantos@tecnico.ulisboa.pt

There is an increasing need to optimize cattle production towards the highest possible environmental efficiency, taking into account the economic performance and viability of the farms. Here we developed a new version of the BalSim a carbon (C) and nitrogen (N) balance model for extensive beef production in Alentejo, for different ages at slaughter, taking in account the economic performance of production for each age. We assumed that in the typical production system, during the first 12 months calves remain in the pasture (with roughage supplementation). After 12 months, calves are confined and fattened using a mix of silage and concentrate feed. We considered average regional data for semi-natural pastures (SNP) and sown biodiverse permanent pastures rich in legumes (SBP). We assumed potential ages at slaughter between 9 and 24 months, for the Alentejana, Angus, Limousine and Charolais breeds. An extended version of the C and N mass-balance pasture model 'BalSim' (that includes C and N flows during confinement) was applied, together with an extension of a simple economic balance, based on previous experiments and agricultural regional statistics. Results signal the best age for slaughter in environmental terms for each breed and both types of pastures. The minimum possible greenhouse gas emissions (GHG) are 15.0 kg CO2e/ 100 g of protein for SBP and 18.4 kg CO2e/ 100 g of protein for SNP, both for a limousine calf with 12 months. The minima for the other breeds in terms of kg CO2e/100 g of protein are: Angus, SBP, 16.8 (12 months); Angus, SNP, 20.6 (12 months); Charolais, SBP, 19.1 (15 months); Charolais, SNP, 22,3 (17 months); Alentejana, SBP, 19.9 (16 months); Alentejana, SNP, 23,3 (19 months). The maximum eco-efficiency was achieved in SBP for a Limousine calf at 14 months, and in SNP for a Limousine calf at 17 months. The same ratio was calculated for the other breeds, and shows that across the same production system, each breed as its environmental and economic optimum that should be taken in account when determining an eco-efficient age at slaughter.

Effects of field peas on feed intake, milk production and metabolism in grazing dairy cows

R.G. Pulido[1], I.E. Beltran[2], J.A. Aleixo[1], A. Morales[1], M. Gutierrez[3] and P. Melendez[4]
[1]Universidad Austral de Chile, Independencia 641, 5110566, Chile, [2]Instituto de Investigaciones Agropecuarias, Remehue, Osorno, 5290000, Chile, [3]ALISUR S.A., San Pablo, Los Lagos, 5350000, Chile, [4]School of Veterinary Medicine, Texas Tech University, 7671 Evans Drive, Amarillo, TX 79106, USA; rpulido@uach.cl

raw peas on dairy performance and metabolism in grazing dairy cow. Twelve multiparous lactating Holstein cows were assigned to a replicated 3×3 Latin square study design. Cows were matched by milk yield, body weight and days in milk, and then randomly allocated to one of three treatments: (1) Control: 6 kg DM of fresh pasture (*Lolium perenne*), 7.2 kg DM of grass silage, 7 kg concentrate (Pea-0; 100% concentrate with 0% of pea) and 0.21 kg of mineral salts; (2) Pea30: 6 kg DM of fresh pasture, 7.2 kg DM of grass silage, 7 kg concentrate (Pea-30; 30% pea in the concentrate) and 0.21 kg of mineral salts; (3) Pea60: 6 kg DM of fresh pasture, 7.2 kg DM of grass silage, 7 kg concentrate (Pea-60; 60% pea in the concentrate) and 0.21 kg of mineral salts. The effect of treatments on milk production, composition, body weight, rumen metabolites and blood parameters were analysed using a linear-mixed model. Intake of pasture, grass silage and concentrate were not modified by treatments. Crude protein intake was lower for Control compared with other treatments, while ME intake was increased as pea was included in the concentrate. Milk production and composition and milk urea nitrogen did not differ among treatments. Blood parameters (urea, albumin, Ca, P, Mg and AST) were not modified by treatments, with exception of plasma BHB, being lower for Pea-60 compared to other treatments. Ruminal parameters (pH and volatile fatty acids) did not differ among treatments; however, it was observed that concentration of butyrate, propionate and valeric acid were greater in the afternoon compared to the morning sampling. In conclusion, pea grains could be used to replace corn grain and soybean meal in the concentrate, allowing to maintain a similar milk production and its composition of grazing dairy cows, without alteration in the intake and ruminal and blood parameters.

Genome-wide CNV discovery in the autochthonous Aosta Breeds

M.G. Strillacci[1], F. Berenini[1], C. Punturiero[1], R. Milanesi[1], M. Vevey[2], V. Blanchet[2] and A. Bagnato[1]
[1]Università degli Studi di Milano, Department of Veterinary Medicine and Animal Sciences, Via dell'Università 6, 26900 Lodi, Italy, [2]ANABORAVA, Fraz. Favret 5, 11020 Gressan (AO), Italy; maria.strillacci@unimi.it

Copy Number Variants (CNVs) are structural variants affecting genetic diversity and phenotypic variation of populations. Different authors underlined the relevance of CNVs in relation to the adaptation to environmental conditions (e.g. altitude, harsh farming environment). Valdostana cattle (Valdostana Red Pie – VRP; Valdostana Black Pied/Chestnut – VBC and Mixed Chestnut-Herèn – CH) farmed in the Aosta Valley, are autochthonous dual-purpose breeds well-adapted to be reared in the Alps mountain area. The aim of this study is to characterize these three breeds using CNV as markers and possibly relate them to the adaptative selection to the mountain farming system. ANABORAVA provided the LogR Ratio (LRR) and the B allele frequency of 2,254 females (VRP=1,537; VBC=622; CH=95) obtained with the NEOGEN's GGP Bovine100K. The number of sampled cows is proportional to the breed size. After the quality check, the CNV were called on autosomes using the SVS 8.9 software, considering: (1) the overall distribution of Derivative Log Ratio Spread (DLRS) values; (2) the GC content screening, correlated to a long-range waviness of LRR values. CNVs were then aggregated into CNV regions (CNVRs) based on at least 1 bp overlap, and only CNVRs identified in at least 1% of the cows were considered to infer statistics at population level. A total of 62,634 CNVs and 2,486 CNVRs were obtained. The number of identified CNVs is consistent with the number of cows per breed (correlation=99%). In the PCA performed using the CNVs, the three breeds appear overlapping, without any defined cluster, showing that these markers do not segregate for the selection occurring intra-breed, but mark the adaptive selection common to all the three breeds. In fact, about 46% of the CNVRs identified in at least of 1% of samples resulted common among all the three populations. Instead, only 17% (VRP), 18% (VBC) and 8% (CH) CNVRs were proper of each population. Annotated genes and QTL overlapping the CNVRs are functionally associated with productive, functional and health traits. Funded by PSRN DUAL BREEDING_Fase_2.

Training Maremmana heifers to virtual fencing management: preliminary results

C. Aquilani[1], A. Confessore[1], M.C. Fabbri[1], L. Nannucci[1], E. Gasparoni[2], F. Vichi[3], A. Mantino[3], R. Bozzi[1], C. Dibari[1], M. Mele[2] and C. Pugliese[1]
[1]Universita' di Firenze, DAGRI, Piazzale delle Cascine 18, 50144, Italy, [2]Centro di Ricerche Agro-Ambientali Enrico Avanzi, Via Vecchia di Marina, 6, Pisa PI, 56122, Italy, [3]Tellus srl, P.za S.Antonio 4, Pisa, 56125, Italy; andrea.confessore@unifi.it

Virtual Fencing (VF) is an innovative technology based on GPS collars. VF allows to set the virtual grazing area limits and when the animals approach the fences, they receive an audio cue followed by a low electrical pulse, if animals cross over the fences. The study aimed to test the feasibility of using VF in an agrosilvopastoral system to facilitate the management of 22 Maremmana heifers reared in a pilot farm in Maremma (southern Tuscany, Italy). To this, a training trial was carried out to test the reaction of heifers to this innovative managing tool. The testing activity was carried out within a 1.5 ha physically fenced paddock. It lasted 60 days in total and it was divided into 4 phases, during which the data recorded by collars were collected (S=number of sounds, Z=number of electrical pulses, Z/S=ratio, and animal position) every 15 minutes. For 5 days (T0), virtual boundaries were set coincident to the physical ones as to allow the animals to adapt in wearing the VF collars, then 1/3 (T1, 14 days) and 1/2 (T2, 14 days) were excluded of the paddock by reshaping the virtual boarder. After 14 days the area available for grazing was shift to the other half (T3), and animal were excluded from the previous one. Data were analysed by ANOVA using SAS's GLM procedure. S was significantly lower in T1 (14.46) than in T2 (27.16). Z did not show significant differences between the training phases, only a decreasing trend was identified from T1 (3.30) to T3 (2.60). The Z/S ratio, instead, decreased in T2 and T3 with respect to T1, as the S increased and Z decreased in the last phases. Escapes recordings were limited to few animals in the herd during the first hours of each VF change. Aggregate animals' positions for the T1, T2, and T3 showed that every time animals used all the area available, even some hours from the beginning of each phase were needed to start exploring the zones previously excluded.

Session 04 Poster 13

Hydroxychloride trace minerals improve nutrient digestibility in beef cattle compared to inorganics

S. Van Kuijk[1], P. Swiegers[2], Y. Han[1] and D. Brito De Araujo[3]

[1]Trouw Nutrition, Global R&D Department, Stationsstraat 7, 3811 MH, Amersfoort, Utrecht, the Netherlands, [2]Rumen-8 (PTY) Ltd., 41 La Provence Road, 9701 Bethlehem, South Africa, [3]Nutreco Nederland BV, Global Selko Feed Additives, Stationsstraat 7, 3811 MH, Amersfoort, Utrecht, the Netherlands; davi.araujo@trouwnutrition.com

The objective of this study was to answer the question 'Does the complete substitution of sulphate trace mineral (STM) by hydroxychloride (HTM) sources of Cu, Zn and Mn improve nutrient digestibility in beef cattle diets?'. Eight 12 month-old Bonsmara beef heifers were housed individually and fed according to a duplicated 4×4 Latin square design. Two levels of protein supplementation, being 12.8% CP in the low protein concentrate and 30.2% CP in the high protein concentrate, were combined with two sources of trace minerals, being HTM and STM. All four diets contained 15 ppm added Cu, 50 ppm added Zn and 33 ppm added Mn. The Eragrostis tef hay roughage was fed *ad libitum*. The supplements including the minerals were fed separately daily at a fixed rate of 1.4 kg/h/day. Each period was 24 d starting with 18 d adaptation period followed by 6 d sampling period. The body weight and intake were measured during each period, to calculate performance. Feed and fecal samples were collected during sampling period and analysed for nutrient digestibility, while acid insoluble ash was used as indigestible marker. On the last day of each period, rumen fluid was collected for rumen pH and volatile fatty acid measurements. Results showed that HTM increased (P<0.05) DM (+1.78%, P=0.042), OM (+1.81%, P=0.039), NDF (+1.80%, P=0.020) and ADF (+2.41%, P=0.019) digestibility in Bonsmara beef cattle compared to STM, regardless of protein supplementation. Feeding higher protein level increased CP digestibility compared to low protein supplementation (+7.92%, P<0.001), while overall the mineral sources only tended to improve CP digestibility when HTM were fed (+3.18%, P=0.052). High CP supplementation resulted in a higher acetate (+0.638 g/l, P<0.001), and butyrate (+ 0.196 g/l, P=0.010) production in the rumen compared to low CP supplementation, regardless of mineral source. Propionate production tended to be higher in the rumen of protein supplemented heifers (+0.114 g/l, P=0.060), which tended to be more evident in heifers fed HTM.

Session 04 Poster 14

Effect of TM source on nutrient digestibility, mineral & antioxidant status of crossbred Indian cows

A. Khare[1], S. Nayak[1], V.V. Reddy[1], S. Vardhan[2], L. Pineda[3] and D. Brito De Araujo[4]

[1]Nanaji Deshmukh Veterinary Science University, College of Veterinary Science & Animal Husbandry, 736 Indira Gandhi Marg., 482001 Jabalpur, Madhya Pradesh, India, [2]Trouw Nutrition India Pvt Ltd, Ruminant Technical Department, 16, Jayabheri Enclave, 500032 Hyderabad, Telangana, India, [3]Trouw Nutrition, Global R&D Department, Stationsstraat 7, 3811 MH Amersfoort, the Netherlands, [4]Nutreco Nederland BV, Global Selko Feed Additives, Stationsstraat 7, 3811 MH Amersfoort, the Netherlands; davi.araujo@trouwnutrition.com

The effect 3 sources of trace minerals on mineral retention, nutrient digestibility, antioxidant potential and performance were tested in low milk yield crossbred Indian cows. Crossbred Indian lactating cows (n=45; BW=479.28; MY=8.49 kg/d; DIM=15 d) were assigned into 1 of 3 treatments (15 cows/trt), based on MY, parity, BW and DIM. Animals in all trt were fed a similar TMR basal diet, comprising wheat straw, green fodder, and concentrate mix for 90 d. TMR was given to meet the nutrient requirement for maintenance and milk production as per ICAR (2013) feeding standard. All 3 diets contained 10 ppm added Cu, 40 ppm added Zn and 27.5 ppm added Mn. (1) ITM: 10 ppm Cu from Cu-sulphate, 40 ppm Zn from Zn-sulphate, 27.5 ppm Mn from Mn-sulphate; (2) OTM: 5 ppm Cu from Cu-sulphate + 5 ppm from Cu-proteinate, 20 ppm Zn from Zn-sulphate + 20 ppm Zn from Zn-proteinate, 13.75 ppm Mn from Mn-sulphate + 13.75 ppm Mn from Mn-proteinate; (3) HTM: 10 ppm Cu from Cu-hydroxychloride, 40 ppm Zn from Zn-hydroxychloride, 27.5 ppm Mn from Mn-hydroxychloride. A metabolic trial of 7 days was conducted at the end of the feeding trial to evaluate digestibility of nutrients and retention of minerals. Results showed that HTM had a positive effect on total antioxidant capacity (TAC, mM/l: 0.93 vs 0.90; P<0.02 and ALP, U/l: 41.71 vs 40.20; P<0.09) and mineral retention (Cu, ppm: 4.93 vs 3.79; P<0.05 and Mn, ppm: 11.35 vs 9.70; P<0.09) compared to STM. Also compared to STM, HTM cows increased digestibility of DM (+1.29%, P<0.02), EE (+1.31%, P<0.07), OM (+1.16%, P<0.02), NDF (+2.23%, P<0.05) and CP (+1.08%, P<0.04).

Session 04 Poster 15

Mineral nutrition in suckler beef cows during the grazing season

H. Scholz[1] and G. Heckenberger[2]
[1]Anhalt University of Applied Sciences, Faculty LOEL, Strenzfelder Allee 28, 06406 Bernburg, Germany, [2]State Institute for Agriculture and Horticulture Saxony-Anhalt, Lindesntraße 18, 39606 Iden, Germany; heiko.scholz@hs-anhalt.de

Beef cattle require a number of minerals for optimal growth and reproduction. Selecting the correct mineral supplement is important for maintaining healthy animals, and optimal growth and reproduction. Since high-quality forages can furnish a large portion of the required minerals. However, we do not always know the quality of the pasture forage or we cannot estimate the degree of selection of the grazing animals. The present study investigated the mineral content of gras and the amount of daily intake of supplemented minerals (different confections) in 10 herds of suckler beef cows, which can response to three different intensities of extensivation of the grass land (in contrast of N fertilizer: (1) more than 100 kg N/ha; (2) 60-100 kg N/ha; (3) 0 kg N/ha). It was carried out over 2 grazing seasons from May to September in Germany. During each farm (10×), 10 suckler cows were used as a indicator animal for the whole herd (30-50 cow-calf-pairs). Pasture allocated to each herd were sampled monthly (DM, CA, CP, e.g. minerals). Supplemented minerals were free-access feeding during the grazing period. Statistical analyses were carried out with the software SPSS (26.0). A significance level of $P \leq 0.05$ was chosen throughout. Content of Copper was higher ($P \leq 0.05$) in: (1) with 7.9±1.5 mg/kg DM; than (2) (6.1±1.7 mg); or (3) (6.1±1.9 mg). In contrast was Manganese significantly higher in the extensive production system. The grazing season average consumption of supplemented mineral was 66±24 g per cow and day. No significant differences between the extensivation of grass land ($P=0.388$) or year ($P=0.986$) were observed. Average daily intakes were 60 g/d (1) and 73 g/d (2) and 55 g/d (3). The content of a ration can be calculated from intake of grass are linked to the consumed amounts of supplemented mineral. Content of Copper in the calculated ration were higher ($P \leq 0.001$) in intensive grassland (1) with average 12±1 mg/kg DM than (2) by 10±4 mg/kg DM than (3) 8±3 mg/kg DM. The results obtained show that up a decrease in minerals with an increase of extensivation in grass land under German conditions.

Session 04 Poster 16

Ruminal protein degradation of alfalfa hay by *in situ* measurement and nutritional system prediction

A. Vigh and C. Gerard
ADM Animal Nutrition, Neovia Talhouet, 56250, France; antal.vigh@adm.com

Dairy rations containing significant amounts of alfalfa hay are often difficult to balance, and the actual milk production is regularly much lower than the predicted one. We hypothesis that this discrepancy could be related to an overestimation of the rumen degraded protein of alfalfa hay by the prediction equation proposed by INRA2018 nutritional system, based only on crude protein (CP) level, leading to diets deficient in fermentable proteins. Hence, the aim of this study was to compare the protein effective degradability (ED_N) of alfalfa hay measured via the standard *in situ* method, with the predicted values using the INRA2018 equations, and, if relevant, to propose some adjustments. Alfalfa hay was sampled from 40 commercial dairy farms from different European countries and analysed for Dry matter (DM), cellulose, NDF, ADF, CP and enzymatic degradability (DE1). Among the initial samples, 10 were selected based on their high variability of DE1 (sample variance 39.4) and low variability of CP (sample variance 6.9) contents, and incubated between 0-72 h (7 time points) in the rumen of 3 dry cows. The residues were analysed for DM, CP and NDF. The CP degradation kinetics were then adjusted using the Orskov model. Based on the initial CP content (mean 170.0±9.2 g/kg DM) of the incubated samples, the ED_N was also estimated using the INRA2018 prediction equation. There was a significant difference ($P<0.001$) between the ED_N measured *in situ* and the estimated value (resp. 56.1±2.1% and 70.9±0.5%). NDF content and DE1 were identified as factors significantly correlated to this difference (mean Δ=14.7±1.8%). By including these parameters in an adjustment factor (Δ) to refine the prediction equation ($ED\_Nadjusted = ED\_N_{INRA2018} - \Delta$), a good correlation ($R^2=0.95$, $P<0.001$) and similar absolute values were found between measured and estimated values (resp. 56.1±2.1% and 56.2±1.9%). Although these results obtained on this first dataset have to be confirmed on a larger one, the initial hypothesis of an overestimation of the ruminal protein degradability seems to be confirmed, and shows that for dairy cow rations that include alfalfa hay, ruminal protein availability could be the limiting factor for optimal diet valorisation.

Session 04 Poster 17

Residual feed intake calculated after weaning and finishing tests in Nellore cattle

S.F.M. Bonilha, B.R. Amâncio, J.N.S.G. Cyrillo, R.H. Branco, R.C. Canesin and M.E.Z. Mercadante
Instituto de Zootecnia, Rodovia Carlos Tonani, km 94, 14.174-000 Sertãozinho, SP, Brazil; sarah.bonilha@sp.gov.br

Residual feed intake (RFI) is a measure of feed efficiency related to animal metabolism, being influenced by various physiological processes at different stages of animal growth. This study aimed to evaluate the reclassification of 155 uncastrated Nellore males, based on RFI calculated after feed efficiency tests conducted in two different periods, post-weaning with diet formulated for growth, and at finishing with diet formulated for carcass fattening. Based on individual information of dry matter intake (DMI) and average daily weight gain (ADG), RFI was calculated for the same animals after each feed efficiency test (post-weaning and finishing) as the residual of the regression equation of DMI as a function of ADG and metabolic average body weight. Animals were classified as negative (RFI<0; efficient) or positive (RFI>0; not efficient) RFI after each test. Spearman rank correlations were estimated between DMI, ADG and RFI values in both tests. Post-weaning RFI values were plotted against finishing RFI values in order to determine the change in ranking. Rank correlation coefficients found between DMI, ADG and RFI measured after post-weaning and finishing tests were 0.17 (P<0.05), -0.08 (P>0.05), and -0.23 (P<0.01), respectively. From the 155 animals evaluated, 67 remained in the same RFI class in both tests (post-weaning and finishing) and 88 changed classes. In other words, 56.7% of the animals were reclassified and 43.3% of the animals maintained the same classification when comparing both efficiency tests. From the 88 animals that changed classification, 44 changed from negative to positive RFI class and 44 changed from positive to negative RFI class, meaning that 28.4% of the animals changed from efficient to not efficient and another 28.4% of the animals changed from not efficient to efficient in feed use. In conclusion, there is reclassification of animals when the age of assessment and/or the type of diet used in the test is modified.

Session 05 Theatre 1

Dairy herd reproduction management strategies for improved efficiency

J.M.E. Statham, M.W. Spilman and K.L. Burton
RAFT Solutions Ltd, RAFT Solutions Ltd, The Farm Yard, Sunley Raynes Farm, Galphay Road, HG4 3AJ, United Kingdom; jonathan@raftsolutions.co.uk

Optimum dairy herd economic performance is generally achieved by maintaining a calving interval of 365 days. Mitigating the environmental impact of dairy production requires efficient reproductive performance. Statham *et al.* modelled reduction in greenhouse gas emissions intensity when reproductive management strategies such as fixed-time insemination and sensor technologies were deployed. modelled the effects of fertility on emissions linking changes in fertility to herd structure, number of replacements, milk yield, nutrient requirements and gas emissions. Improving submission rate from 50 to 70% could reduce emissions of methane by up to 24% by reducing the number of heifer replacements required to maintain herd size. A range of strategies are available to boost submission rates, however adequate conception risk requires effective management of the root causes of poor reproductive efficiency. Oestrus synchronisation techniques offer an effective approach to increased submission rate, but require intensive hormonal intervention such as 'Ovsynch' and variations Co-synch, Heat-synch, Select-synch and Pre-Synch. Inclusion of progesterone releasing devices has offered an alternative approach. Automated heat detection systems utilising sensors based on cow activity, positioning or temperature sensors can accurately predict oestrus as can low milk progesterone from in-line milking parlour systems. Root causes of poor reproductive performance require a holistic herd-based approach and reproductive challenges such as early and late embryonic death may still have a significant effect on reproductive outcome and herd performance. Both female and male factors should be considered at herd level and include nutrition, infectious disease & environmental factors such as heat stress. Screening semen quality using objective multimodal systems offers an important opportunity to address variations in male factors. Perturbations in time-series data from sensors such as milk meters and activity meters offer emerging opportunities for managing the resilience and efficiency of dairy herds, ranking cows on those parameters to manage herd status over time.

Randomized field study of extended lactation in 48 Danish dairy herds

V.M. Thorup[1], G. Simpson[1], A.M.H. Kjeldsen[2], L.A.H. Nielsen[2] and S. Østergaard[1]
[1]Aarhus University, Department of Animal Science, Blichers Alle 20, 8830 Tjele, Denmark, [2]SEGES Innovation P/S, Livestock, Agro Food Park 15, 8200 Aarhus N, Denmark; vivim.thorup@anis.au.dk

Deliberately delaying the first insemination is a reproduction management strategy, which potentially benefits farmer economy, greenhouse gas emission and animal health. Delaying pregnancy improves milk production persistency and reproductive performance. Cows with a high milk yield throughout an extended lactation have a high yield in previous lactation and early in the present lactation. Our study aims to develop a more detailed model for selecting cows suitable for extended lactation. We recruited 48 Danish dairy herds; hereof 37 Holstein and 11 Jersey; 40 conventional and 8 organic; 13 robotic and 35 parlour milking herds. Within each herd we randomly selected 20 cows for extended lactation and 20 cows as control, in total 1,920 cows. We defined extended lactation as current voluntary waiting period before first insemination in a given herd plus 100 days. No changes to replacement strategy were requested. We intended to follow cows until 150 days into the lactation following the extended lactation. Data collection started August 2020 with expected end in March 2023. In February 2022, 14 cows have completed the study, 553 left prematurely, and 1,353 cows are ongoing. Of the ongoing, 81% are undergoing the extended lactation or serving as control. Extended cows are inseminated on average 4 days (±38 days) earlier and control cows 29 days (±35 days) later than expected. Preliminary results show that pregnancy rate after first insemination in the extended lactation was 52.3% for extended cows and 46.6% for control cows. Additional production variables will be analysed. In conclusion, reproductive performance was improved in cows undergoing extended lactation. Future work will involve further investigation of reproduction and production traits, as well as early lactation prediction of late lactation milk yield. We thank the Danish Milk Levy Foundation for supporting our study.

Consequences of an extended lactation for metabolism of dairy cows in different stages of lactation

E.E.A. Burgers[1], R.M.A. Goselink[1], R.M. Bruckmaier[2], J.J. Gross[2], A. Kok[1] and A.T.M. Van Knegsel[1]
[1]Wageningen University, De Elst 1, 6700 AH Wageningen, the Netherlands, [2]Veterinary Physiology, Vetsuisse Faculty, University of Bern, 3012 Bern, Switzerland; ariette.vanknegsel@wur.nl

An extended calving interval by extending the voluntary waiting period (VWP) reduces the calving frequency. An extended VWP could be associated with altered metabolism, due to delayed gestation and more time in late lactation with lower milk production. This study evaluated the metabolism of cows with different VWP during different stages of lactation. Moreover, relations between cow characteristics before insemination and lactation performance were analysed, to identify cows with lower risk for low milk production and increased body condition end lactation. Holstein-Friesian cows (n=153) were blocked and randomly assigned to a VWP of 50, 125, or 200 days (VWP50, VWP125, or VWP200), and followed from calving until 6 wk after next calving. Weekly, from 2 wk before until 6 wk after both calvings, plasma samples were analysed for non-esterified fatty acids (NEFA), β-hydroxybutyrate, glucose, insulin, and IGF-1. During lactation, plasma samples were analysed for insulin and IGF-1 every 2 wk. Fat-and-protein-corrected milk (FPCM) and body weight (BW) gain were calculated weekly. Cows were divided in two parity classes (primiparous and multiparous). During gestation, multiparous cows in VWP200 had greater plasma insulin and IGF-1 concentrations and lower FPCM compared with cows in VWP50 or VWP125, and greater BW gain compared with cows in VWP50 (3.6 vs 2.5 kg/wk). During gestation, primiparous cows in VWP125 had greater plasma insulin concentration compared with cows in VWP50, but the VWP did not affect body condition or FPCM. During the first 6 wk of the next lactation, multiparous cows in VWP200 had greater plasma NEFA concentration compared with cows in VWP125 or VWP50 (0.41 vs 0.30 or 0.26 mmol/l). For primiparous cows, the VWP did not affect metabolism, BW, or FPCM during the next lactation. Independent of VWP, higher milk production and lower body condition before insemination were associated with higher FPCM and lower body condition end of the lactation. Variation in these individual cow characteristics could call for an individual approach for an extended VWP.

Milk yield during second lactation after an extended voluntary waiting period in primiparous cows

A. Edvardsson Rasmussen[1], C. Kronqvist[1], E. Strandberg[2] and K. Holtenius[1]
[1]Swedish University of Agricultural Sciences, Department of Animal Nutrition and Management, Box 7024, 750 07 Uppsala, Sweden, [2]Swedish University of Agricultural Sciences, Department of Animal Breeding and Genetics, Box 7023, 750 07 Uppsala, Sweden; anna.edvardsson.rasmussen@slu.se

A prolonged voluntary waiting period (VWP) may be beneficial for primiparous cows in terms of better fertility without compromising average milk yield per day in the calving interval (CI). A few studies have also reported increased milk yield during the beginning of the subsequent lactation. The aim of this study was to investigate the effect of a prolonged first lactation VWP on milk yield during a complete second lactation. This was done as a part of a randomized controlled trial with 631 primiparous cows in 16 commercial farms in the south of Sweden. Data including calvings, inseminations, and test-day yields was collected during two lactations from the Swedish milk recording system. Cows that followed the planned VWP during their first lactation (25-95 and 145-215 days for the traditional and extended VWP group, respectively) and completed their second lactation with a third calving were included in the current results. No intervention was performed for the VWP during the second lactation. The preliminary results show that the mean 305-day lactation yield was higher in the extended VWP group (12,069 kg ECM, n=115) than in the traditional VWP group (11,392 kg ECM, n=157) during the second lactation ($P<0.01$). Moreover, the average milk yield per day during the second CI was higher for the cows with an extended VWP (34.6 kg ECM/CI day) than for the cows with a traditional VWP (32.5 kg ECM/CI day, $P<0.01$). In conclusion, these results suggest that cows that receive an extended VWP during their first lactation have higher milk yield during their second lactation than cows receiving a traditional first lactation VWP.

Fertility in high yielding dairy cows with extended second lactation – a randomized controlled trial

A. Hansson[1], C. Kronqvist[2], R. Båge[3] and K. Holtenius[2]
[1]Växa, P.O. Box 30204, 104 35 Stockholm, Sweden, [2]Swedish University of Agricultural Sciences, Department of Animal Nutrition and Management, P.O. Box 7024, 75007 Uppsala, Sweden, [3]Swedish University of Agricultural Sciences, Department of Clinical Sciences, P.O. Box 7054, 75007 Uppsala, Sweden; annica.hansson@vxa.se

The present recommendation is that Swedish dairy herds should aim for 12.5 months calving interval (CI). In practice, farmers are advised to strive for a voluntary waiting period (VWP) of 50 days. However with extended VWP the high yielding cow might be in a better energy balance and better prepared for a new pregnancy. High yielding cows with persistent lactation might improve production result with a longer CI. The metabolic stress during the transition at dry off might be milder because milk yield likely would be lower. The period around calving when the cows are at highest risk for diseases would occur less frequently in the cow's life. Also the length of the milk producing life could increase with a longer CI. The aim of this study was to evaluate the effect on fertility from prolonged VWP in potentially high yielding dairy cows in their second lactation. Two herds connected to the Swedish milk recording scheme volunteered for a randomized controlled study. Cows having their second calf from December 2019 until December 2020 were randomized to 50 days VWP (VWP50) or 140 days VWP (VWP140). Conception rate was analysed using a chi^2 test. Average herd milk production exceeded 11,600 kg milk. In VWP50, seven cows got a culling decision before insemination yielding a final dataset of n=97 inseminated cows. In VWP140, 12 cows got a culling decision before insemination yielding a final dataset VWP140 n=82 inseminated cows. Preliminary results show that first service conception rate was higher (68.3%) with VWP140 compared to VWP50 (51.5%) ($P<0.05$). There was no significant difference in proportion of cows finally getting pregnant, VWP 50 90.7% and VWP 140 91.5%, ($P=1.00$). In conclusion, these preliminary results suggest that longer VWP can improve conception rate at first insemination compared with traditional VWP without altering culling rate due to poor fertility for high yielding dairy cows in their second lactation.

Extended lactation as a potential strategy to effectively manage calf numbers in dairy farming

J. Gresham, C. Reiber and M.G. Chagunda
Universtiy of Hohenheim, Department of Animal Breeding and Husbandry in the Tropics and Subtropics (49⁰h), Garbenstrasse 17, 70599 Stuttgart, Germany; josephine.gresham@uni-hohenheim.de

Conventional and organic dairy production rely predominantly on a few selected high-yielding breeds. The focus on milk production results into male offspring that have low fattening abilities and hence little to no economic value in the beef industry. This leads into a conflict between production goals and ethical considerations. Delaying insemination and the dry-off time extends the lactation and thereby prolongs calving interval, which could naturally reduce the number of calves born in a cow's lifetime. The current study aimed to: (1) examine the current application of extended lactation (EL) on dairy farms in Southern Germany; (2) identify the assessed adoption potential and the biological potential of EL; and (3) assess the potential of EL in managing calf numbers using the assessed adoption potential and biological potential. Qualitative and quantitative data were collected from 310 farms, who indicated their interest in EL and their perceived feasibility in its use on a Likert scale. The combination of interest and feasibility was used to assess the adoption potential. From the study, 46.8% of farmers used EL. Through correlation analysis significant relationships between use of EL and breed as well as milk yield/cow were found. Further, significant relationship was found between current users and interest in use of EL. In the survey, 39.8% of farmers perceived the feasibility of EL as 'high' to 'very high'. Of the farmers currently not using extended lactation, 17 (12.8%) were identified as potential adopters. If all potential adopters were to apply EL by three or six months, it could result in a reduction of 1.8 or 3.5% in calve numbers respectively. Biological and assessed potential indicate that extended lactation can be used as a strategy to manage calf numbers.

Individual approach of extended lactation period in dairy cows to reduce the use of antibiotics

P. Sanftleben, T. Kuhlow and A. Römer
State Research Centre of Agriculture and Fisheries Mecklenburg-Vorpommern (LFA), Wilhelm-Stahl-Allee 2, 18196 Dummerstorf, Germany; p.sanftleben@lfa.mvnet.de

The average milk yield of cows in Germany increased by about 2% per year, which in review leads to a doubling of milk yield in relation to the last 50 years. This is associated with an increase in management requirements. Despite slight modifications of breeding values, longevity and fertility are still below the genetic potential. Although first luteal activity is delayed as milk yield increases, calving interval (CI) limit recommendations of 365 to 400 days are maintained. This aim makes early insemination necessary. However, this leads to still high milk yields at the time of dry-off with a correspondingly increasing risk of mastitis. Increasingly, detection of mastitis- and environmentally associated bacteria with an extended resistance spectrum is becoming more common. However, it has been proven that good udder health can be achieved with a minimized use of antibiotics, which is also in the interest of consumers. By extended lactation, a reduction in antibiotic use considered over time is possible. The reason for this is a lower frequency of passing through critical phases for udder health. Furthermore, fertility and milk yield can be improved. Thus, the results showed an increase in 305-day milk yield by 1,048 kg milk per cow on average for extended CI from over 460 days compared with a CI of less than 370 days. Due to heterogeneity of cows in a herd, an individual animal approach is required to facilitate a tailored lactation pattern based on current performance parameters. For this purpose, a calculator was developed as a matrix for the decision of a extended voluntary waiting period (VWP). This calculator will be available to farms as a Shiny web app in future. The underlying algorithm is a regression model (R^2_{adj}=0.96) and gives an estimate of a least VWP based on milk yield (average of the last 7 days) and current days in milk. Additionally, a differentiation into primiparous and multiparous cows is made based on their persistence of lactation. In order to train and validate the model, applicability, practicability and robustness of the prototype will be tested on 11 dairy farms in Germany and newly acquired data will be implemented.

What is an optimal body condition profile for reproduction in dairy cows?

C. Dezetter[1], F. Bidan[2], L. Delaby[3], S. Fréret[4] and N. Bédère[3]
[1]Ecole Supérieure d'Agricultures, USC 1481 URSE, INRAE, 49007 Angers, France, [2]Institut de l'élevage, 149 rue de Bercy, 75595 Paris, France, [3]PEGASE INRAE, Institut Agro, 35590 Saint Gilles, France, [4]CNRS, IFCE, INRAE, Université de Tours, PRC, 37380 Nouzilly, France; c.dezetter@groupe-esa.com

The relationships between body condition score (BCS) and reproductive performance of dairy cows exist. A BCS of 3.0 at calving and a maximum loss of 0.5 points at the beginning of lactation are recommended. However, individual profiles of BCS vary between cows. This work aimed to study the relationships between reproductive performance and different BCS profiles. Data from 6 French experimental farms with Holstein or Normande cows were used. Only lactations of cows inseminated at least once and with enough BCS records to estimate a BCS profile were used, i.e. 1,685 lactations of Holstein cows and 482 lactations of Normande cows. Artificial insemination (AI) and successive calving dates, total and maximum milk yield, average fat plus protein, and sanitary events related to reproduction (metritis, calving difficulty, etc.) were recorded. For 721 lactations, ovarian cyclicity status, determined using milk progesterone profile were available. BCS profiles were established by breed using a PCA on 9 variables: 5 BCS variables at fixed stages (calving, then 28 d, 56 d, 98 d, and 210 d postpartum) and 4 variables of BCS variation between these stages. Then, variance analyses were performed to identify differences between BCS profiles for reproductive performance. Total milk yield averaged 7,500 kg (±1,816 kg) and the total calving rate averaged 73% (from 55% to 88%). Four BCS profiles were identified in the Holstein breed and 3 in the Normande breed. In the Holstein breed, a 'lean' profile was identified with a total calving rate lower than for the other three profiles. In the Normande breed, a 'fat' profile was identified with a lower calving rate at first AI than the other two profiles. In conclusion, the profiles identified were consistent with the literature. Cows that were too thin or too fat were more likely to fail at breeding. However, the differences in calving rate at successive AI, days open, and ovarian cyclicity were small between BCS profiles and showed that reproductive failure is multifactorial and difficult to attribute solely to one BCS profile.

Increasing longevity of dairy cows reduces methane emissions and improves economy in dairy herds

J.B. Clasen[1], W.F. Fikse[2], M. Ramin[3] and M. Lindberg[4]
[1]Aarhus University, Center for Quantitative Genetics and Genomics, Blichers Allé, Tjele, Denmark, [2]Växa, Ulls väg, Uppsala, Sweden, [3]Swedish University of Agricultural Sciences, Agricultural Research for Northern Sweden, Verkstadsgatan, Umeå, Sweden, [4]Swedish University of Agricultural Sciences, Animal Nutrition and Management, Ulls väg, Uppsala, Sweden; julie.clasen@qgg.au.dk

Animal welfare and climate impact are important factors in dairy production. Increased cow longevity is highly associated with better animal welfare, and methane (CH_4) emissions per kg milk and meat are expected to decrease as productive life of the cow increases. The aim of this study was to investigate different herd management strategies to improve cow longevity and the climate impact of milk and meat output. A simulated base scenario illustrated an average Swedish dairy herd. Economic calculations included costs and income associated with rearing dairy bulls and beef × dairy calves until slaughter. Cows in the base herd had 2.8 years of productive life and emitted 19.4 g CH_4/kg energy corrected milk (ECM) produced during their lifetime. Adding meat from the culled cow and her offspring resulted in emissions of 211 g CH_4/kg meat (carcass weight), and a yearly profit of €27.4 per cow. Improving fertility increased productive years per cow by 1 year, an increase in meat output, and a reduction of replacement heifers. Lifetime emissions decreased to 197 g CH_4/kg meat, and 17.6 g CH_4/kg ECM, and the profit was €38.4 more per cow-year than in the base. Halving the risk of diseases or cow mortality increased productive life with 0.1 and 0.3 years, respectively, and had similar CH_4 emissions as the base herd, but only the latter would earn substantially more (+€36.3) because of a reduction in replacement heifers. Keeping all heifers for replacement reduced productive life by 0.8 years, and increased milk, meat and number of replacement heifers per year. CH_4 emissions increased to 240 g CH_4/kg meat and 21.6 g CH_4/kg ECM produced, with a loss of €13.2 per cow-year relative to the base. In conclusion: management strategies aiming to improve longevity and reduce the number of replacement heifers in favour of more offspring to meat production was found to be beneficial from both an economic and environmental point of view.

Session 05 Theatre 10

Evolution in performances of French dairy cattle herds transitioning towards 3-breed crossbreeding

J. Quénon[1], S. Ingrand[2] and M.-A. Magne[1,3]
[1]INRAE, AGIR, 24 chemin de Borde Rouge, Auzeville CS 52627, 31326 Castanet Tolosan Cedex, France, [2]INRAE, UMR Territoires, 9 avenue Blaise Pascal CS 20085, 63178 Aubière Cedex, France, [3]ENSFEA, BP 22687 2 route de Narbonne, 31326 Castanet Tolosan Cedex, France; julien.quenon@inrae.fr

There is a lack of references on the evolution of the performance of dairy cattle herds from commercial farms, in which dairy crossbreeding is implemented in various initial breeding situations, modalities and evolution dynamics. The objective of this study was therefore to analyse the changes in zootechnical performance at the herd level (milk production, useful matter rate, fertility and udder health) induced by the introduction of crossbreeding and to evaluate the factors explaining these changes. We relied on a sample of 13 French dairy herds, from which we extracted breed and 6,628 lactation performances data from 2009 to 2017. We modelled the evolution of performances over this period by linear regression. We considered several explanatory variables: the initial state of zootechnical performance, the initial state of farm structure and management and their evolution, modelled by linear regression. In order to assess the factors that explain evolution in performances among the explanatory variables, we performed several partial least squares (PLS) regressions. Our results show that crossbreeding has a significant effect on the evolution of the zootechnical performance of the herds, but as one explanatory factor among others, such as the systemic changes implemented or the stage of progress in the practice of crossbreeding. The implementation of dairy crossbreeding in purebred herds contributes to improving their fertility performance at the expense of milk productivity per cow, which the vast majority of farmers compensate by increasing the size of the herd. The effects of crossbreeding on milk quality, udder health and longevity are more contrasted. These results enhance our knowledge of dairy crossbreeding by taking into account the biological and decision-making dimensions that, in the long term, build the multi-performance of herds and the genetic diversification induced in herds by the use of crossbreeding.

Session 05 Theatre 11

Effect of pre & postpartum DCAD levels on productive and reproductive performance of female buffaloe

H. Metwally[1] and S. Elmashed[2]
[1]Ain Shams University, Faculty of Agriculture, Animal Production, Shubra ElKhaime, 11241, Egypt, [2]Animal Production Research Institute, Animal Production, Dokki, 12681, Egypt; sh.elmashed@hotmail.com

A total of 96 female multiparus (2, 3 and 4 seasons) buffalo cows were used to investigate the effect of pre and postpartum DCAD levels on productive and reproductive performance. Animals were divided into two main groups three weeks before expected parturition (48 each). Each group received one of two tested DCAD level Group A: (0 DCAD level mEq/kg DM) Group B: (- 100 DCAD level mEq/kg DM) Each group of prepartum treatments was divided into two subgroups postpartum DCAD levels. Group A: (0 DCAD level) 1 (+150 DCAD level) 2(+250 DCAD level) Group B: (-100 DCAD) 3(+ 150 DCAD level) 4(+ 250 DCAD level) Treatments started on day (0) postpartum and lasted for 150 days of milk (DIM). Blood &urine samples were collected from four animals in each group every week pre& postpartum. Results obtained indicated that: 1. Incidence of subclinical hypocalcaemia was noticed in 16 females 3 days postpartum in group A. 2. Serum Ca conc. 3d. after parturition was the least for group A(1,2) (6.1,6.2 mg/dl) while, it was higher for group B(3,4) (7.9,8.1 mg/dl) 3. Serum Mg& P Conc. Were not affected by treatment ranged between (1.8-2.1) and (4.9,5.3 mg/dl) respectively. 4. Blood parameters showed that serum urea, Creatinine, AST & ALT Averaged (30.2,1.3 mg/dl, 44.5, 21.0 Iu/l) respectively. 5. Group B (3)showed the highest ($P < 0.5$) milk yield 3,660 kg/305 d., fat percent 6.9% and fat yield 252.54 kg/305 d. 6. No. of cervices per conception was lower for groups A(2), B(3) and B(4) 1.1,1.2&1.3 respectively, compared with group A(1) (2.1) and calving intervals decreased for the same groups. 7. Serum Progestrone levels decreased from 1.6 prepartum to 0.35 postpartum (µg/dl), while treatments didn't affect progesterone conc. 1d. postpartum (0.71 overall mean µg/dl). It was recommended that (-100 prepartum with +150 or + 250 postpartum prevented subclinical hypocalcaemia and improved productive and reproductive buffalo performance.

System dynamics for complexity understanding in animal nutrition and farm management

A.S. Atzori[1,2], B. Atamer-Balkan[1,3] and A. Gallo[1,3]
[1]System Dynamics Italian Chapter, Via Marconi 19, 00146 Rome, Italy, [2]Dipartimento di Agraria, University of Sassari, Italy, V. Italia39, 07100 Sassari, Italy, [3]Catholic University of Sacro Cuore of Piacenza, Via E. Parmense 84, 29121 Piacenza, Italy; asatzori@uniss.it

System Dynamic (SD) is a modelling technique to improve complexity understanding and forecast future trends for managerial purposes. Qualitative SD modelling include the causal system description with links, feedbacks and connections among variables drawing loop diagrams (CLD) of circular interactions. It represent the endogenous structure driving the system behaviour over time. Quantitative modelling translate it in Stocks and Flows diagrams (S&F) were connections are explicit as differential equations. Feedback loops determine the system behaviour, positive and negative loops generate exponential and asymptotic behaviour over time, respectively. The modelling process is object-oriented particularly adapt to field experts, not requiring high mathematical skills. This work aimed to present SD application in nutrition (rumen dynamics) and farm management (herd model) to show simple SD structures and behaviours. A more complex model shows powerful SD applications in dairy management. The modelling step included: (1) a survey to gather farm data from 2015 to 2019; (2) the development of a S&F model on Vensim® (Ventana, Inc) describing heifer stock raising and the reproduction loop of cows, described with 4 basic stocks (early lactating, open, pregnant and dry cows). Calibration was performed on historical records of cattle categories, reproduction parameters and culling rates. Average farm characteristics in the same period consisted of 1,154±58 milking cows and 1,375±45.2 heifers whereas dry matter intake (DMI) and milk yield were on average equal to 24.2±2.4 and 34.0±3.4 kg/d per head, respectively. From model input based on farm data in Jan 2015 the model predicted with good accuracy the oscillating seasonal pattern of historical farm records of cattle consistency and milk deliveries until 2019. Cow consistency and milk deliveries were predicted with a RMSPE of 3.5% and 7.2% of observed values and good precision (r^2=0.68 and 0.72) respectively. Economic and environmental submodels are under development. SD can have broad applications in any sector of the animal science.

Relations between animal, carcass and meat characteristics across 15 European breeds

J. Albechaalany[1,2], M.P. Ellies-Oury[1,2], J. Saracco[3], M.M. Campo[4], I. Richardson[5], P. Ertbjerg[6], M. Christensen[7], B. Panea[4], S. Failla[8], J.L. Williams[9,10] and J.F. Hocquette[1]
[1]INRAE, UMR1213, 63122 Theix, France, [2]Bordeaux Sciences Agro, CS 40201, 33175 Gradignan, France, [3]Univ.de Bordeaux,INRIA, Av. de la Vieille Tour, 33400 Talence, France, [4]Univ.of Zaragoza – CITA, Av.Montañana, 50013 Zaragoza, Spain, [5]Univ.of Bristol, Langford, BS81TH Bristol, United Kingdom, [6]Univ.of Helsinki, Koetilankuja, 00014 Helsinki, Finland, [7]Univ.of Copenhagen, Frontmatec-Smoerum A/S, 2765 Smoerum, Denmark, [8]CREA, Via Salaria, 00015 Roma, Italy, [9]Univ. Cattolica del Sacro Cuore, Via Emilia Parmense, 29122 Piacenza, Italy, [10]Univ. of Adelaide, Adelaide, SA 5005, Australia; john.albechaalany@doctorant.uca.fr

In this study, 436 young bulls from 15 cattle breeds were reared in five different countries following the same experimental protocol. In all stations, the energy density ratio and protein content of the diet were similar. Animals were slaughtered at 15 months of age, and samples were collected 24 h post-mortem from the Longissimus thoracis muscle between the 6th and 13th left ribs. Each animal was characterized by a total of 51 variables representing animal performances, muscle biochemistry as well as sensorial and nutritional characteristics of beef quality. A statistical approach based on clustering of variables has been elaborated using the ClustOfVar package (R software 3.6.1). Using the parturition stability method, nine clusters were retained. The relationships between the variables in each cluster were studied using the principal component analysis (PCA) approach. The 1st dimension of eight clusters were considered representative, showing the highest variations between variables except from one cluster where two dimensions were considered. Each group of variables in each cluster emphasizes different characteristics: physiological traits such as the muscle mass (clusters 1 and 2), fat mass (cluster 2), maturity (cluster 3); muscle biochemistry such as levels of enzymes involved in ageing (cluster 4); sensorial traits such as tenderness and juiciness (cluster 5), the muscle oxidative type (cluster 6); meat colour (clusters 7 and 9) and lipids (cluster 8). As a conclusion, the characteristics related to body composition, muscle biochemistry, sensorial traits and meat colour show an independent relationship across breeds.

Multivariate analysis of beef carcass traits processed by a slaughterhouse in Brazil

N.S.R. Mendes[1,2], S. Chriki[3], M.P. Ellies-Oury[1,4], V. Payet[3], T.F. Oliveira[2] and J.F. Hocquette[1]
[1]INRAE, UMR1213, Theix, 63122, France, [2]School of Agronomy, Federal University of Goiás, Campus Samambaia, Rodovia Goiânia-Nova Veneza Km-0, Caixa Postal 131, 74690-900, Brazil, [3]Isara – Agro School for Life, 23 rue Jean Baldassini, CEDEX 07, Lyon, France, 69364, France, [4]Bordeaux Sciences Agro, CS 40201, Gradignan, 33175, France; nathaliasrm@gmail.com

Thanks to the increasing demand in foreign markets for Brazilian beef and to the need to produce high-quality meat, we investigated the effects of pre- and post-slaughter factors on beef carcass traits. In particular, we hypothesized, according to the scientific literature, that pre-slaughter transport of bovines over distances would cause stress to the animals. This would result in a carcass with reduced quality, due to a higher ultimate pH resulting from stress induced by long distance transportation. The data used were from 30,232 Nelore carcasses, supplied by a commercial slaughterhouse in Inhumas, Brazil. All carcasses were evaluated at 24 h post mortem. Principal component analysis was performed using R software (version: 4.1.2). The results show a correlation between animal maturity (dentition, relative to the age of the animals) and weight (axis 1). Then, a linear regression model indicates that distance does not have any significant effect on pH ($P=0.51$), while carcass weight significantly affects pH ($P<2\times10^{-6}$ ***) with a low coefficient of 2.67×10^{-4}, thus with a very small effect. Maturity negatively affects pH as well ($P=1.86\times10^{-4}$ ***) but also with a very small effect (-7.87×10^{-4}). In brief, $pH=5.67 + 2.3\times10^{-6}$ distance $+ 2.67\times10^{-4}$ weight -7.87×10^{-4} maturity. In conclusion, beef carcass characteristics can be affected by animal weight and maturity, but only to a moderate extent. The hypothesis that long distances would significantly interfere with beef quality was not confirmed within the Nelore breed, probably because of a low stress sensitivity during transportation with the studied animals.

Comparison of culling reasons of 30 Swiss dairy farms with low versus high productive lifespans

R.C. Eppenstein and M. Walkenhorst
Research Institute for organic agriculture FiBL, Animal Science, Ackerstrasse 113, 5070 Frick, Switzerland; rennie.eppenstein@gmail.com

Increasing the productive lifespan of dairy cows is an important means to lowering the environmental impact of dairy production. Since the culling rate is directly linked to the productive lifespan, understanding differences in culling rates and culling reasons across farms with differing productive lifespans can indicate routes to improve herd management strategies. Previous works have shown that reproductive problems, poor udder health, feet and claw disorders, metabolic and digestive disorders as well as low production levels are the major culling reasons among dairy cow herds across Europe. However, few studies have explored to which extend culling reasons differ among dairy farms with high productive lifespans, versus farms with low productive lifespans of their dairy herd. Within the framework of a larger Swiss research project, we conducted interviews with 30 Swiss dairy farms and collected detailed information on each culling decision 5 years retrospectively. The interviewed farms were preselected as matched pairs, such as to be similar in their characteristics (cow breed, production zone, production type, productivity level, herd management strategies), but to differ with regards to the average productive lifespan (APL) of their dairy herds. The characteristics of the 15 pairs were furthermore selected to be representative of the Swiss dairy sector. On average, farms with a low APL culled 43% of their herd per year, while farms with a high APL culled 29% per year. Reproductive problems were cited with a similar frequency among both low and high APL farms. Farms with low APL culled more frequently due to problems with udder health, metabolic disorders and low production levels, while farms with high APL culled more frequently due to feet and claw disorders, 'old age' and sales for dairy purposes. The present study sheds light onto culling reasons that most strongly reduce the productive lifespan of dairy cows, thus showing a route to improved dairy herd management.

Genetic evaluation of reproductive efficiency traits in Portuguese Holstein Cattle

A. Rocha[1,2], H.T. Silva[3], C.N. Costa[4] and J. Carvalheira[2,5]
[1]CECA-ICETA, P. Gomes Teixeira, Apartado 55142, 4051-401 Porto, Portugal, [2]ICBAS, U.Porto, Rua J.V. Ferreira, 228, 4050-313 Porto, Portugal, [3]U.F.Viçosa, Dep. Animal Science, U.F.V., 36570-000 Viçosa, Brazil, [4]Embrapa Gado de Leite, Embrapa, 36038-330 Juiz de Fora, Brazil, [5]BIOPOLIS–CIBIO, U.Porto, R. Padre A. Quintas, 4485-661 Vairão, Portugal; jgc3@cibio.up.pt

Increased milk production has been the main objective of most breeding programs for the Holstein breed. In Portugal for the last 50 years, selection of dairy cattle coupled with management improvements resulted in a production increase of almost 6,000 kg of milk/cow/year. However, reproductive efficiency has decreased over the years with a steady increase in the number of artificial inseminations per pregnancy, on calving to 1st AI interval (CAI), on days open (DO), and on calving intervals (CI). Initially, the program of genetic improvement took into account only production parameters (milk and milk components), functional characteristics (somatic cells count), and conformation, to construct the Total Merit Index (M€T). Routine genetic analysis for those reproductive traits including daughter pregnancy rate (DPR) defined as the percentage of oestrus cycles of 21 days, needed to get a pregnancy after 42 days of voluntary waiting period, are now established with 2 official genetic evaluations per year. Phenotypic mean ± SD for 2021 were 427±84 d for CI, 88±32 d for CAI, 124±53 d for DO and 28±17% for DPR. Heritability for these traits were low and varied between 0.05 and 0.09 with negative genetic correlation with production traits (CI-milk: -0.04, CAI-milk: -0.16, DO-milk: -0.08). Larger CI, CAI, DO represent lower reproductive efficiency, and inversely, higher DPR indicates increased reproductive efficiency making it more ease to understand in a selection index context. CI, CAI, and DO had an almost coincident pattern for the genetic progress curve over the years, while the curve for DPR is almost the inverse. Genetic progress for DPR have been improving in recent years, with positive trend of 0.4%/year in males and 0.3%/year in females. This improvement may be the result of indirect selection due to the use of imported semen from sires selected for high fertility. Since 2022, reproductive traits are also included in IPT, a new total merit index for dairy cattle in Portugal.

Effect of cow resting time during dry period on colostrum production and quality

G. Gislon, A. Sandrucci, A. Tamburini, S. Mondini, M. Zucali, S. Bonizzi and L. Bava
Università degli Studi di Milano, Scienze Agrarie e Ambientali, via Celoria 2, area Zootecnia, 20133, Italy; giulia.gislon@unimi.it

Colostrum quality is crucial for adequate transfer of passive immunity to the new-born calves, and it may be influenced by several animal and management factors: among others it may be affected also by cows resting time, during the dry period. The aim of the study was to investigate the effect of dry period length, parity number and resting time during cow dry period, on colostrum production and quality. In addition, the effect of colostrum production on the quality was investigated. Data were collected in two dairy farms, located in Lombardy (Italy), where dry cows were reared in a free-stall pen, either with cubicle housing system or deep litter system. Activity of 37 cows were automatically recorded by individual 3-axes acceleration loggers (HOBO Pendant G Data Loggers, Pocasset, MA), throughout the duration of dry period. Colostrum quality was measured both by electrophoresis, as immunoglobulins concentration (IgG g/l), and by digital refractometer as Brix %. Data were analysed by Proc GLM, using SAS 9.4. The results highlight that colostrum quality, in terms of IgG concentration, was significantly affected by resting time (P=0.0597), number of parity (P=0.0028) and colostrum production (P=0.0118). In particular, a higher IgG concentration was associated with a shorter resting time (67.6 vs 47.1 g/l, for resting time <767 and >767 minutes/day, respectively), higher number of parity (74.3 vs 40.3 g/l, for multiparous and secondiparous cows, respectively) and lower colostrum production (71.3 vs 43.4 g/l, for <8 and >8 l of colostrum, respectively). No effect of dry period length on IgG concentration was detected. However, colostrum quality, as Brix%, was significantly affected only by colostrum production (P=0.0047), with higher Brix percentage in the presence of lower colostrum production (26.6 and 21.7% Brix, for <8 and >8 l of colostrum, respectively). Colostrum production was not significantly affected by animals resting time during the dry period, nor by the other parameters tested. Acknowledgements: The study was supported by Progetto MAGA Regione Lombardia, OPERAZIONE 16.1.01 – Gruppi Operativi PEI.

Session 05 Poster 18

Gestation length in Swiss Holstein cows

A. Burren and H. Joerg
Bern University of Applied Sciences, School of Agricultural, Forest and Food Sciences HAFL, Länggasse 85, 3052, Switzerland; hannes.joerg@bfh.ch

This study examines the impact of the sire breed on the gestation length in Swiss Holstein (HO) cows. Data from a total of 501,709 HO cows from the years 2000-2021 with a total of 19 different sire breeds were provided by the breeding association Swissherdbook. A random sample of 99,921 observations was taken from this data set, with the various sire breeds represented as follows: Purebred Simmental Cattle descending from Swiss genetics (SI60) 1,640, Purebred Simmental Cattle descending from international genetics (SI70) 143, Angus (AN) 1,308, Belgian Blue (BB) 1,261, Blonde d'Aquitaine (BD) 389, Swiss Brown (BV) 211, Charolais (CH) 433, Holstein (HO) 69,603, Eringer (HR) 173, Jersey (JE) 177, Limousin (LM) 10,997, Montbéliard (MO) 626, Normande (NO) 246, Original Swiss Brown (OB) 57, Brown Swiss (BS) 148, Pinzgauer (PI) 373, Red factor carrier (RF) 1,299, Red Holstein (RH) 2,413 and Swiss Fleckvieh (SF) 8,424. The relation between the gestation length and sire breed has been investigated by a linear mixed model. The model included the fixed effects month of calving, milk linear and squared, calf age linear and squared, number of calves born by gender (litter size), sire breed, as well as sire as random effects and residual effect. The longest gestation length occurred with the sire breed Blonde d'Aquitaine of 289.0 days, followed by the sire breeds PI, LM, OB and SI at 288.9, 288.6, 288.4 and 288.0 days, respectively. The shortest gestation length resulted from the sire breed Red factor carrier at 282.5 days, followed by the breeds JE, RH, HO and AN, at 282.6, 282.6, 283.0 and 283.0 days, respectively. The remaining breeds had gestation lengths of between 283.2 and 286.8 days. Both the gender and the number of calves born have an impact on the gestation length. The gestation length of male calves is approximately a day longer than of female calves (f=284.7 days; m=286.0 days). In the case of twins, the gestation length of female twins and male and female twins was 280.5 and 281.1 days, respectively. The gestation length of male twins was 281.3 days. The results show that breeders can influence the gestation length by their selection of sire breed.

Session 05 Poster 19

Theobromine promotes short-term and long-term sperm survival in a dose-dependent manner

M. Ďuračka, J. Kováč, F. Benko, N. Lukáč and E. Tvrdá
Institute of Applied Biology, Faculty of Biotechnology and Food Sciences, SUA in Nitra, Tr. A. Hlinku 2, 949 76 Nitra, Slovak Republic; michaelduracka@gmail.com

Theobromine (TBR) was found as a powerful antioxidant and antimicrobial substance in tea, coffee and cocoa. Bioactive substances are popularly used as effective supplements for bovine semen preservation media. So far, there are no published data about the effects of TBR on the sperm quality. The aim of this study was to analyse effect of different TBR concentrations (0 – Control group, 5, 10, 50, 100 and 200 μmol/l) during incubation times of 0, 2 and 24 h at room temperature. For this purpose, computer-aided semen analysis (CASA) was performed to analyse the sperm motility. The sperm membrane and acrosome integrity were determined using the carboxyfluorescein- and fluorescein isothiocyanate-peanut agglutinin-based methods. Mitochondrial succinate dehydrogenase activity was evaluated using the mitochondrial toxicity test. Reactive oxygen species (ROS) were chemiluminescently determined using luminol as a probe. One-way ANOVA and the Dunnett's comparison test were performed within the statistical evaluation. Statistical significance was set to $P<0.05$, $P<0.01$ and $P<0.001$. The CASA analysis showed a significantly higher motility to control group (Ctrl; $P<0.001$ in case of 200 μmol/l, and $P<0.01$ in case of 50-100 μmol/l) when treated by TBR concentrations ≥50 μmol/l after 2 h. Long-term incubation showed a significantly higher motility when treated with 50 and 100 μmol/l ($P<0.05$). A short-term horizon showed TBR concentrations of 100-200 μmol/l TBR to be more effective in preserving membrane and acrosome integrity, while concentrations of 50-100 μmol/l were more effective in a long-term range. Similar results were observed analysing the mitochondrial activity. ROS were significantly decreased when treated with ≥50 μmol/l TBR after 2 h, while only concentrations of 50-100 μmol/l were significantly effective in ROS scavenging ($P<0.05$) after 24 h. In conclusion, TBR provides beneficial effects to stored spermatozoa, while concentrations of 100-200 μmol/l showed the most effectivity during short-term storage. Concentrations of 50-100 μmol/l were beneficial in a long-term horizon. This study was supported by APVV-15-0544, VEGA 1/0239/20, KEGA 008SPU-4/2021.

Metabolic and endocrine profiles in Holstein cows with different managements during early lactation

A.L. Astessiano[1], M.N. Viera[2], E.J. Smeding[1] and A.I. Trujillo[1]
[1]Facultad de Agronomia, Garzon 780, 12900, Uruguay, [2]Facultad de Veterinaria, Ruta 8, km 18, 13000, Uruguay; lauaste@gmail.com

During the first 90 days postpartum (DPP), multiparous cows were used in a randomized block design to study the effects of nutrition and cow management on endocrine and metabolic profiles. Cows (n=36, 667.5±9.5 kg BW, 2.77±0.02 BCS) were assigned to 3 treatments (TREAT): total mixed ration and confinement (28 kg DM/d offered; 54% forage, 46% concentrate; TMR); grazing and confinement (pasture allowance=20 kg DM/d + 8 h confinement + 50% TMR; G1) and grazing (pasture allowance=20 kg DM/d + 50% TMR offered; G2). Milk yield was determined daily, BCS every 15 days and blood samples were taken at -15, 15, 45, 60 and 90 DPP to quantify endocrine and metabolic profiles. Means from a repeated measures analysis differed when P<0.05. Milk yield was greater (P<0.01) in TMR than in G1 and G2 cows (43.3, 33.8 and 34.6±1.2 l, respectively) and BCS did not differ among TREAT, but decreased (P<0.01) from -15 to 15 DPP and then remained stable. Insulin concentration decreased (P<0.01) around calving and tended to be recovered (P=0.10) at 90 DPP (10.3, 9.2 and 10.8±0.7 uIU/ml for TMR, G1 and G2, respectively) while glucose concentration increased (P<0.01) from pre to postpartum (60.7, 59.4 and 58.8±2.2 mg/dl for TMR, G1 and G2, respectively) in all cows. Non-esterified fatty acids (NEFA) and beta-hydroxybutyrate (BHB) concentration increased (P<0.01) from pre to postpartum, NEFA from -15 to 15 DPP while BHB from -15 to 45 DPP. The physiological imbalance index increased (P<0.01) and RQUICKI decreased from -15 to 15 DPP in all cows. Interestingly, RQUICKI index was affected by interaction (P=0.03), reaching a lower value at 15 DPP increasing then at 60 DPP and reaching greater values in G2 than TMR and G1 cows. These results reflected the physiological imbalance characteristics of the transition period. Although we found non-differences in metabolic and endocrine profiles among TREAT, G2 cows had a lower sensitivity to insulin consistent with a better metabolic adaptation to lower milk production than TMR cows.

Monitoring reproductive and genetic health of F1 dairy × beef cows through cytogenetic investigation

I. Nicolae, A. Sipos, B. Groseanu and D. Gavojdian
Research and Development Institute for Bovine, Genetics, sos. Bucuresti-Ploiesti, km 21, Balotesti, Ilfov, 077015, Romania; ioana_nicolae2002@yahoo.com

During the last years, a research project concerning the crossbreeding of Romanian Black and White Spotted (RBS) dairy breed with meat specialized breeds such as Aberdeen Angus (AA), Charolaise (Cha), Limousine (Lim) and Belgian Blue (BB) was implemented. In order to evaluate the reproductive efficiency and genetic health of the F_1 crossbreed cows, a cytogenetic investigation was carried out. Karyotype analyses were performed on a group of 10 F_1 crossbreed cows: 5 females F_1 RBSxAA, 2 females F_1 RBSxCha, 1 female F_1 RBSxLim and 2 females F_1 RBSxBB. In the current study we identified 4 F_1 animals with abnormal chromosomal configuration, represented by a large number of mono-and bi-chromatidic breakages on autosomes and heterosomes, loss of chromosome fragments and gaps. Our investigation continued through SCEs-test, and for animals with numerous chromosomal breakages, the number of sister chromatid exchanges (SCEs) was very high (11-14 SCEs/cell) compared to the normal animals. Although the carriers found had normal phenotypes, the analysis of their reproductive activity revealed reproductive disturbances characterized by repeated inseminations and/or lack of oestrus (2 females F_1 RBSxBB and 1 female F_1 RBSxLim) and abortions (1 female F_1 RBSxAA). In the case of the F_1 RBSxCha crossbreeds, the karyotype was normal, with repeated inseminations found for a single cow, which was caused by a congenital vaginal stricture. According to current results, the chromosomal instability identified in 4 of the 10 investigated F_1 crossbreed cows, it is demonstrated once again the role of the chromosomal structural defects in the ethiology of different levels of infertility.

Defoliation intensity: milk production and composition of late lactating Holstein cows

O. Fast[1], G. Menegazzi[2], M. Oborsky[2], P. Chilibroste[2] and D. Mattiauda[2]
[1]Colonia El Ombú, Ruta 3 km 284, Fray Bentos, 65100, Uruguay, [2]Universidad de la República Uruguay, Facultad de Agronomía, Animal production and pasture-EEMAC, Ruta 3, km 636. Paysandú, 60000, Uruguay; dma@fagro.edu.uy

An experiment was carried out to study the effect of two contrasting defoliation intensities of a tall fescue based pasture on milk production and composition of 24 late lactating multiparous cows during summer. The animals were blocked according to parity (3.1±0.82), body weight (626±43 kg), body condition score (2.7±0.18) and days in milk (258±13 d), and were randomly allocated to the following treatments: 15 cm (TL) or 9 cm (TC) post grazing sward height, with four spatial replicates of 0.3 ha each. It was used a second-year pasture of *Festuca arundinacea* and *Lotus corniculatus* with an initial herbage mass of 3,710±106 kg DM/ha and 16.5±0.38 cm of sward height. Each plot was grazed by three cows. The criteria to start grazing was when the pasture reached the three-leaf physiologic stage. Animals had 16 h access to pasture, without supplementation, and were only removed twice a day for milking (05:00 and 16:00 h). Milk production was determined at the beginning (days 1 and 1+2), middle (days 2 and 3+4) and the end (days 3 and 5+6) of the occupation period (moment) for TL (3d) and TC (6d); respectively. Milk composition was sampled in the last day of the grazing period. A mix model was used (Glimmix procedure, SAS 9.2) for milk yield with treatment, moment and their interaction as fixed effects and block as random effect. Average energy corrected milk (ECM) yield was greater for TL than TC (13.6 vs 16.6±0.83 kg; for TC and TL respectively). Treatment by moment interaction showed that TC decreased ECM yield at the end, while TL remained unchanged from the beginning until the end of the occupation period. Fat concentration differed between treatments, being greater in TL than TC (4.6 vs 4.1±0.23%). There was no difference between treatment for protein and lactose content (3.34±0.068% and 4.59±0.051%; respectively). As result, milk component yields were greater in all cases in TL than TC. Pasture management through defoliation intensities has the potential to allow stable milk production throughout the grazing process in cows with relatively low milk yield potential.

Prediction of live weight from linear conformation traits in dairy cattle

A. Burren and S. Probst
Bern University of Applied Sciences, School of Agricultural, Forest and Food Sciences, Länggasse 85, 3052 Zollikofen, Switzerland; alexander.burren@bfh.ch

The live weight of dairy cattle plays an important role in the calculation of greenhouse gas emissions. However, many dairy farms do not know the live weights of their cattle. The objective of this study was to predict the live weight of 309 dairy cattle (121 Holstein/Red Holstein, 110 Swiss Fleckvieh and 78 Braunvieh) based on linear conformation traits. At the time of the linear assessment, four animals were in milk for the second time and all others for the first time. The correlation between live weight and linear conformation traits was investigated with 11 different multiple linear models. Depending on the model, the following factors and covariates were used: the fixed effects stature, chest width, body depth (numerical or factor), rump angle, breed, month of weight measurement (numerical or factor), days in milk (linear and quadratic), gestation length (linear and quadratic), age at weight measurement (linear, quadratic or cubic), calving month, birthing month and heart girth. R^2 ranged between 66.5-79.5%. An R^2 of 79.5% and a RSE of 37.9 kg was only achieved when heart girth was considered. However, a model using heart girth is not meaningful as data on the trait heart girth is no longer collected within the framework of the linear assessment. The baseline model, which achieved an R^2 of 66.5% and a residual standard error (RSE) of 44.5 kg, consisted of the traits stature, chest width, body depth (numerical), rump angle, breed, month of weight measurement (numerical), linear days in milk, linear gestation length and linear age. Considering the month of weight measurement as a factor and the quadratic age effect on weight measurement resulted in an increase in R^2 to 74.2% and the RSE dropped to 39.1 kg. R^2 increased to 75.8% (RSE=37.9 kg) when a quadratic effect for days in milk and the interaction between breed and month of weight measurement were also considered. It is important to note that the breed only explains a very small proportion of variance in live weight (approx. 1% R^2). Considering body depth as a factor and the quadratic effect of gestation length only resulted in a small improvement to the model (R^2=76.6%; RSE=37.0 kg). In conclusion, linear conformation traits can be used to estimate live weight of dairy cattle herds.

Session 06 Theatre 1

Trace minerals and rumen function

J.W. Spears
North Carolina State University, Department of Animal Science, Polk Hall 319B, Raleigh, NC 27695, USA; jerry_spears@ncsu.edu

This presentation will discuss trace mineral requirements of ruminal microorganisms, and the effect of trace mineral source on ruminal fermentation. A number of trace minerals are required in low concentrations by ruminal microorganisms. With the exception of cobalt (Co), minimal dietary trace mineral requirements of the host ruminant appear to be considerably greater than that needed for rumen microbial requirements. It is well known that certain bacteria can synthesize vitamin B_{12} from inorganic Co. Some bacteria species require vitamin B_{12} as a growth factor, and adequate dietary Co is needed to allow sufficient ruminal B_{12} synthesis to meet their requirement. Vitamin B_{12} is needed as a cofactor for ruminal microorganisms to convert succinate to propionate. Dietary Co deficiency results in decreased ruminal propionate in ruminants fed high concentrate diets and decreased fibre digestion in ruminants fed high fibre diets. Attempts have been made to use high concentrations of certain trace minerals to favourably manipulate ruminal fermentation. For example, attempts have been made to increase rumen protein bypass by feeding high dietary zinc (Zn). However, studies have indicated that high concentrations of copper (Cu), Zn, and iron reduce cellulose digestion *in vitro*. Recent studies have indicated lower fibre digestibility in cattle supplemented with sulphate sources of Cu, Zn, and manganese compared with those fed similar concentrations from hydroxy or certain organic sources. Research has clearly indicated that hydroxy sources of Zn and Cu are metabolized differently in the rumen than sulphate sources and limited research suggests that trace mineral source may affect the rumen microbiome. Additional research is needed to elucidate the mechanism(s) whereby trace mineral sources affect fibre digestibility differently.

Session 06 Theatre 2

Interaction between faecal microbiota and plasma metabolome of calves at two different weaning times

J. Seifert[1], N. Amin[1], S. Schwarzkopf[1], A. Camarinha-Silva[1], S. Dänicke[2], K. Huber[1] and J. Frahm[2]
[1]University of Hohenheim, Animal Science, Emil-Wolff-Str. 6-10, 70599 Stuttgart, Germany, [2]Friedrich-Loeffler-Institute, Animal Nutrition, Bundesallee 37, 38116 Braunschweig, Germany; jseifert@uni-hohenheim.de

Weaning is a stressful transition for dairy calves, influencing their feed intake, growth, health and production performance. An appropriate determination of weaning age is important to lessen such effects. This study investigated the impact of early vs late-weaning on faecal microbiome, plasma metabolome and host-microbiome interactions in calves. 59 female Holstein calves (8±1.9 days of age) were randomly allocated in equal quantities to either an early-weaning (earlyC, 7 weeks of age) or late-weaning (lateC, 17 weeks of age) groups. Weaning was done by gradually reducing the milk replacer amount in a 14 days step-down approach, followed by the dietary addition of hay (ad lib) and a total mixed ration. Faecal and plasma samples were collected from each calf at days 1, 28, 42, 70, 98, 112, and 140 of the trial and processed for faecal microbiome analysis by 16S rRNA amplicon sequencing and targeted plasma metabolome analysis using AbsoluteIDQ®p180Kit (Biocrates Life Science, Austria). Both showed a clear influence by weaning time, specifically, during days 42-112. EarlyC had significantly lower relative abundance of phylum Firmicutes and higher Bacteroidetes compared to the same-day-old lateC group (days 42-98). At genus-level, early-weaning increased the abundances of fibre-degraders and decreased lactose- and starch-degraders as well as butyrate-producing bacteria (days 42-98). Likewise, the plasma metabolic profiles of earlyC groups showed significant lower concentrations of most of the amino acids, biogenic amines and sphingomyelins compared to the lateC groups (days 42-112). Strong host-microbiome associations were detected during days 42-98. No significant differences between microbial composition of weaning groups were observed during days 112 and 140, indicating a more mature GIT at weaning enabling rapid adaptation to complete milk removal. In conclusion, weaning related-dietary changes were found to have more abrupt but less-persistent impact on microbiome compared to the host metabolism.

Influence of agroecological diets on the rumen microbiota of three dairy cattle breeds

S. Roques[1], L. Koning[1], J. Van Riel[1], A. Bossers[2], D. Schokker[1], S.K. Kar[1] and L. Sebek[1]
[1]Wageningen Livestock Research, De Elst 1, 6708 WD Wageningen, the Netherlands, [2]Wageningen Bioveterinary Research, Houtribweg, 8221 RA Lelystad, the Netherlands; simon.roques@wur.nl

The sustainability of cattle farming is questioned due to the loss of biological diversity in farm environment and because of the emission of greenhouse gases such as methane. Alternative farming systems, such as agroecology farming, aim to promote the use of natural resource and ecosystem services for food production. A core concept of agroecology is the enhancement of diversity in feeding resources or animals, to improve farm sustainability on the long term. Rumen microbiota has been associated with productive traits and methane emission and highly depends on diet and breed. Thus, the objective of this study was to assess the modulations of rumen microbiota in three dairy cattle breeds (Holstein Friesian, Groninger Blaarkop and Jersey) fed agroecological diets based on silage from grassland with high botanical diversity and from late-mowed grassland compared to a reference diet based on silage from monoculture-specie grassland. Both the rumen microbial diversity and composition were analysed depending on diets and cattle breed. In addition, methane emission were measured for each group of breed and diet. The richness and Shannon's index of the rumen microbiota were significantly improved by the interaction between diets and breed and by the diet respectively. Similarly, the composition of the microbiota was affected mostly by diet. The relative abundance of taxa associated with production of succinate and propionate were significantly reduced in cows fed the agroecological diets. Besides, rumen microbiota of cows associated with higher methane emission intensity presented significantly higher relative abundance of *Methanobrevibacter* taxa. In conclusion, these results suggest that agroecological diets modulate the rumen microbiota and may have consequences either on the cow metabolism or methane production. Further investigations should assess if these modulations might also impact other aspects of agroecological farming such as farm resilience.

Effect of a *Bacillus* probiotic on rumen microbiota, fluid composition and performance in dairy cows

A. Klop[1], S. Kar[1], R.M.A. Goselink[1] and T. Marubashi[2]
[1]Wageningen Livestock Research, De Elst 1, 6708 WD Wageningen, the Netherlands, [2]Calpis America Inc, 455 Dividend Drive, Peachtree City 30269, USA; roselinde.goselink@wur.nl

The objective of this study was to evaluate the efficacy of a *Bacillus* probiotic in dairy cows, on rumen microbiota, VFA concentrations, feed digestibility, intake and milk production. A total of 64 multiparous Holstein Friesian dairy cows were enrolled in a study with a 15 day pre-period and a 105 day study period. After the pre-period cows were 60-160 days in lactation and allocated to either a control group (T0) or a treatment group receiving 3×10^9 cfu/cow/day (T1). Groups were balanced for parity, days in lactation and milk yield and received an identical PMR diet, with or without the treatment. Additional concentrates were fed in concentrate feeders including 1.6 kg test concentrates with or without the additive for treatment T0 and T1 respectively. Individual feed intake and milk production were registered daily, while milk composition was determined weekly. Rumen and faecal samples were taken from 10 cows per treatment group at day-1, day 41 and day 104 of the study period. Faecal digestibility was estimated by using an inert marker supplemented via the diet. Total dry matter intake, milk yield and composition were not affected by treatment (P>0.100). Rumen VFA concentrations significantly differed: cows in T1 had relatively more (P=0.008) acetic acid (+2.1%), less (P=0.011) propionic (-2.7%) and less (P=0.022) valeric acid (-0.3%) compared to cows in T0. The faecal digestibility of NDF was higher (P=0.004) for T1 (72.2%) compared to T0 (67.0%), also resulting in a higher digestibility of ADF, ADL and (hemi)cellulose. Listing of dominant bacteria and archaea at phylum, family and genus levels in samples taken at three time points were influenced by the treatments applied. Based on the taxonomic abundance profiles of the bacterial-dominated microbiome (16S v5v6 region) of the different samples, rumen fluid samples showed significant difference in bacterial community composition for time and treatment. To conclude, this Bacillus probiotic improved fibre digestibility, and substantial differences occurred in rumen VFA and microbiota profile. In this trial with mid-lactating dairy cattle the product did not positively nor negatively affect performance.

Digestibility of α-, γ-tocopherol, and α-tocotrienol in the digestive tract of dairy cows

S. Lashkari, F.M. Panah, M.R. Weisbjerg and S.K. Jensen
Aarhus University, AU Foulum, Department of Animal Science, Blichers allé 20, 8830 Tjele, Denmark; saman.l@anis.au.dk

The current study examined the impact of toasting and decortication on ruminal degradation and intestinal digestibility of tocopherols in dairy cows using four ruminal and intestinal cannulated Danish Holstein cows. Cows were fed experimental diets *ad libitum*. Treatments were one of four different forms of rolled oat arranged 2×2 factorial: whole oat, decorticated oat, toasted oat, and decorticated toasted oat (all as 21% of diet DM). Toasting reduced RRR-α-tocopherol by 3.4 mg/kg of DM (as the difference between toasted and non toasted). Decortication, toasting, and their interaction had no effect on ruminal digested amount and digestibility of α-tocopherols. Average across treatments showed a ruminal disappearance of total α-tocopherol (102 mg/d), synthetic α-tocopherol (279 mg/d), synthetic 2R-α-tocopherol (133 mg/d), and 2S-α-tocopherol (190 mg/d), while RRR-α-tocopherol increased in the rumen (221 mg/d). The average across treatments showed that small intestinal digestibility of tocopherols ranked in the following order: natural α-tocopherol > synthetic α-tocopherols > γ-tocopherol > α-tocotrienol. The average across treatments for small intestinal and feed-ileum digested amount ranked in the following order: RRR > synthetic 2R > 2S-α-tocopherol with no effect of decortication and toasting. In conclusion, the results indicate RRR-α-tocopherol synthesis in rumen. In addition, the results showed that small intestinal digestion discriminates against the synthetic 2R and 2S-α-tocopherol.

Nutrients total-tract digestibility results regarding parity in dairy cows during transition period

J. De Matos Vettori[1], D. Cavallini[2], L. Lanzoni[1], G. Vignola[1], L. Lomellini[1], A. Formigoni[2] and I. Fusaro[1]
[1]University of Teramo, Faculty of Veterinary Medicine, Piano D'Accio, 64100 Teramo, Italy, [2]University of Bologna, Veterinary Medical Sciences, Via Tolara di Sopra 50, 40064 Ozzano Emilia, Italy; jdematosvettori@unite.it

Proper nutrition during the dairy cow's transition period is really important, but there are some differences between primiparous (PP) and multiparous (MP) that must be observed. This study aims to evaluate the differences in nutrients total-tract digestibility (nTTD) between PP and MP through the transition period. Diet and faeces from 25 Holstein cows (11 PP and 14 MP) were collected on days -30, -7, 0 (calving), 7, 14, and 30, as well as daily rumination time through activometers. nTTD (%) was obtained indirectly through calculation. A linear mixed model was applied using the time and the parity as a fixed effect. The results obtained between PP and MP show that average nTTD were different ($P \leq 0.02$) for aNDFom and pdNDF240 (52.5 vs 54.0 and 78.8 vs 81.3, respectively), while no differences were found regarding pdNDF24 and starch (88.5 vs 88.6 and 95.1 vs 96.1, respectively). No differences in daily rumination time ($P=0.92$) were found between PP and MP. Digestibility is mostly driven by feed quality and passage rate (kp); as the quality of feedstuffs was the same for all groups, differences in kp could be the reason for the results mentioned. It is also true that since PP cows are still growing, the gastrointestinal tract (mainly rumen-reticulum) is still in development, then it could still not have achieved the maximum potential of digestion. Besides gastrointestinal tissue development, the rumen microbiome might play a fundamental role in explaining these results since the rumen microbiome can be in development and adaption until 2 years of age in dairy cattle. Furthermore, the rumen microbiome changes a lot depending on diet, hence it could be more challenging to PP because heifers' diets are much more different than a tail-end cow diet. No differences were found regarding TTStarchD and TTpdNDF24D probably because of post-ruminal digestion. Further research needs to be done in order to clarify the reasons for differences in nutrients total-tract digestibility between PP and MP.

Towards the integration of microbial genomic data into a mechanistic model of the rumen microbiome

M. Davoudkhani[1], F. Rubino[2], C.J. Creevey[2] and R. Muñoz-Tamayo[1]
[1]Université Paris-Saclay, INRAE, AgroParisTech, UMR Modélisation Systémique Appliquée aux Ruminants, 75005 Paris, France, [2]Institute of Global Food Security, School of Biological Sciences, Queen's University Belfast, BT9 5DL, Northern Ireland, United Kingdom; rafael.munoz-tamayo@inrae.fr

The objective of this work was to develop a mathematical framework enabling the integration of microbial genomic information (i.e. 16S rDNA) into mechanistic models of the rumen microbiome. Mechanistic modelling provides promising tools to enhance our system-level understanding of the rumen ecosystem. Existing mechanistic models of the rumen fermentation consider an aggregated representation of the rumen microbiota and its metabolic function. However, none of these models integrate microbial genomic knowledge and are thus limited to capitalize on the rich information of microbial genomic sequencing. In this work, we investigated theoretically how microbial time series of the rumen microbiota determined by 16S rDNA can be integrated into a mechanistic fermentation model of the rumen microbiome. Our study used a previously developed model of *in vitro* rumen fermentation that represents the microbiota by three microbial functional groups, namely sugar utilisers, amino acid utilisers, and methanogens. The model was extended to represent the fermentation under continuous operation (i.e. chemostat). We considered a hypothetical scenario where 16S rumen microbial time series are available. Our rationale is that the microbial species can be categorized into the macroscopic functions of the mechanistic model by using tools such as CowPI. In a simulation case study, we used the theory of state observers to integrate the mechanistic model and the microbial data to provide estimations of the dynamics of the volatile fatty acids (VFA) acetate, butyrate, and propionate. The estimated VFA converge towards the real VFA dynamics demonstrating the promising potential of our approach. The next step we are currently working is the validation of our approach using *in vitro* experimental data.

Ellagic acid and gallic acid mitigate methane production in an *in vitro* model of rumen fermentation

M. Manoni[1], M. Terranova[2], S. Amelchanka[2], L. Pinotti[1,3], P. Silacci[4] and M. Tretola[1,4]
[1]University of Milan, Department of Veterinary medicine and animal sciences, Via dell'Università 6, 26900 Lodi, Italy, [2]ETH Zurich, AgroVet-Strickhof, Eschikon 27, 8315 Lindau, Switzerland, [3]CRC I-WE, Coordinating Research Centre: Innovation for Well-Being and Environment, University of Milan, Via Festa del Perdono 7, 20134 Milan, Italy, [4]Agroscope, Institute for Livestock Sciences, Rte de la Tioleyre 4, 1725 Posieux, Switzerland; michele.manoni@unimi.it

Ruminant production accounts for 81% of greenhouse gas emissions in the livestock sector. Of this quantity, enteric fermentation methane accounts for 90%. Dietary supplementation with tannins is known to mitigate methane production, but also affect feed digestibility. The aim of this study was to investigate the effect of ellagic acid (EA) and gallic acid (GA), alone or in combination, on rumen fermentation in a short-term *in vitro* experiment using Hohenheim Gas Test. EA and GA were supplemented to a control diet (hay, 200 mg DM). Five different conditions were applied to this study (% of DM): (1) EA 7.5; (2) EA 15; (3) GA 7.5; (4) GA 15; and (5) EA 7.5 + GA 7.5. After an incubation of 24 h at 39 °C, pH, ammonia formation, gas production, short-chain fatty acids (SCFA), *in vitro* organic matter digestibility (IVOMD) and the microbial count were determined. The treatments did not alter microbial count and pH. Total SCFA production decreased by approximately 10% after all treatments ($P<0.001$). In the ruminal SCFA profile, the differences were still significant (except for valeric acid) but less evident than total SCFA. Total gas production slightly decreased (-10%) after all treatments, except for GA 7.5. EA 15 and EA+GA treatments decreased methane production per g DM by 20 and 25%, respectively. These two treatments decreased by 10% the production of CO_2 per g DM and the production of ammonia on DM basis, respectively by 13 and 20%. All the treatments caused a slight but significant 10% reduction of IVOMD. In conclusion, diet supplementation with EA and GA, alone or in combination, may be a promising dietary strategy to mitigate methane production in ruminants. Further long-term *in vitro* ruminal fermentation studies are needed to validate these results, before assessing these treatments *in vivo*.

Session 06 Theatre 9

Faecal microbial transplant to boost gut microbial development after antibiotic therapy in goat kids

A. Belanche[1,2], A.I. Martín-García[1] and D.R. Yáñez-Ruiz[1]
[1]Estación Experimental del Zaidín (CSIC), Profesor Albareda 1, 18008, Granada, Spain, [2]University of Zaragoza, Animal Production and Food Sciences, IA2, Miguel Servet 177, 50013 Zaragoza, Spain; belanche@unizar.es

This study explored the concept of faecal microbial transplant from adult to young ruminants in order to accelerate the gastro-intestinal microbial development as a strategy to prevent and/or facilitate the recovery after antibiotic treatment. A total of 44 newborn goat kids were randomly distributed into 4 groups according a 2×2 factorial design: half of the animals were inoculated on day 1 and 8 with faeces from healthy adult goats which were mixed with the colostrum (INO), whereas the rest received colostrum alone (CTL). From day 3 to 7 of age, half of the animals (+) were daily dosed with an oral antibiotic (oxytetracycline) and the rest remained antibiotic-free (-), resulting in 4 experimental groups (CTL-, CTL+, INO- and INO+). Faecal samples were collected on days 7 and 14 and animals were slaughtered at weaning (week 7). Gut microbiota in faeces, rumen, ileum and colon was studied by qPCR and 16S amplicon sequencing. Faecal microbial transplant prior antibiotic treatment led to higher faecal bacterial diversity (179 vs 130 OTUs for INO- and CTL-) and allowed to prevent a substantial decrease in the diversity (136 and 91 OTUs for INI+ and CTL+). A second faecal microbial transplant after antibiotic treatment favoured the recovery of the bacterial diversity (158, 126, 188 and 177 OTUs for CTL-, CTL+, INO- and INO+). The persistency of these effects was also detected at weaning since INO animals had higher rumen concentration of protozoa (P=0.025) and treatment CTL+ showed the lowest methanogens diversity in the rumen (-16% OTUs), ileum (-27%) and colon (-24%) across treatments. As a result, treatment CTL+ showed the lowest ADG from birth to weaning (-9.2%) across treatments. These findings suggest that faecal microbial transplant could represent a strategy to minimize the negative effects derived from antibiotic therapy on gut microbial diversity and productive outcomes.

Session 06 Theatre 10

Effect of a combination of three yeasts on growth performance and faecal microbiota of weaning piglet

S. Sandrini[1], V. Perricone[1], P. Cremonesi[2], B. Castiglioni[2], F. Biscarini[2], E.R. Parra Titos[3], G. Savoini[1] and A. Agazzi[1]
[1]University of Milan, Department of Veterinary Medicine and Animal Science (DIVAS), Via dell'Università 6, 26900 Lodi, Italy, [2]Institute of Agricultural Biology and Biotechnology (IBBA-CNR), Via Einstein, 26900 Lodi, Italy, [3]Vétoquinol Italia S.r.l., Via Piana 265, 47032 Bertinoro, Italy; silvia.sandrini@unimi.it

Yeast supplementation has proven useful in reducing weaning stress and improving performance parameters of piglets with a modulatory effect on gut microbiota. This study evaluated the effect of a yeast mixture (YM, Levustim B0399, Vetoquinol; *Kluyveromyces marxianus fragilis B0399*, *Pichia guilliermondii*, and *Saccharomyces cerevisiae*, intact inactivated cells) on growth performance and gut microbiota of post-weaning piglets. Forty-eight male weaned piglets (27±1.7 d, 7.19±0.54 kg) were randomly allocated to two homogeneous experimental groups and enrolled in a 28-days trial. Both groups received a basal diet with (T) or without (C) inclusion of 0.8% YM during weeks 1 and 2, and 0.6% during weeks 3 and 4. Bodyweight (BW) was individually evaluated on days 0, 14, and 28, and faecal score was determined daily. Faecal samples were collected on days 4, 14, 21, and 28 for microbiota analysis. Growth performance and faecal score were analysed by means of ANOVA for repeated measures, accounting for the effect of the treatment, time, and their interaction. No significant differences were observed on BW, average daily gain, feed intake, and F:G ratio, while increased faecal consistency (P<0.01) was outlined in T compared to C. Sequencing the V3-V4 regions of the bacterial 16S rRNA gene produced a total of 31.573.670 reads. After quality filtering, 23.794.504 sequences were left for analysis. Firmicutes and Bacteroidetes were the most abundant phyla in both groups at all sampling points, representing approximately 70% and 20% of the total faecal microbiota, respectively. Differences between groups have been observed at the genus level. YM resulted in differentially abundant taxa compared to C, among which *Bifidobacterium* (P=0.0057), *Coprococcus* 2 (P=0.0145), and *Clostridium sensu stricto* 1 (P=0.0185). In conclusion, YM administration improved faecal consistency and modified the faecal microbiota, without affecting growth performance.

Arginine supply to lactating sows influence piglet's performance and gut bacterial and viral profile

P. Trevisi[1], D. Luise[1], T. Chalvon-Demersay[2], B. Colitti[3] and L. Bertolotti[3]
[1]University of Bologna, Distal, Bologna, 40127, Italy, [2]Metex Noovistago, rue Guersant, Paris, 75017, France, [3]University of Turin, Dep. Vet. Sci, Turin, 10124, Italy; diana.luise@libero.it

Early colonization of gut microbiome during the neonatal stage plays a key role in the development of the gut physiology, immunity and growth of the host. The early development of gut microbiome is affected by several factors including mother microbiome, which can be manipulated by the diet. This study aims to investigate if dietary supplementation of arginine (Arg) to lactating sows can influence the performance and gut microbiome of sows and their litters. 16 sows were divided into 2 groups balanced for parity and body weight: (1) control (CO) (fed corn-based diet); (2) CO + 22.5 g/d/sow of on-top Arg (ARG). Diets were fed from 4 days before farrowing (d-4) to weaning (d27). Piglets were weighed at d0, d7, d14, d27, d34 and d41. Colostrum and milk were sampled at farrowing, d10 and d20 to analyse immunoglobulins (Igs). Faecal and caecal samples were collected at d27 from all sows and piglets (8 piglets/group) respectively. Library preparation and amplicon sequencing of the bacterial V3-V4 regions of the 16S rRNA gene was performed using the Illumina MiSeq technology. Virus particles were purified and then extracted using AllPrep Power Viral Kit. Illumina libraries of viral DNA and cDNA were prepared using Nextera XT DNA and sequenced on MiSeq platform. Data were fitted using an Anova linear model with diet as factor. ARG sows had a higher IgA in milk at d20 (P=0.004). At d41, piglets from ARG sows tended to be heavier (P=0.06) and to have a higher average daily gain from d0 to d41 (P=0.08). The gut microbiome of sows and piglets clearly differ in terms of bacterial and viral community (alpha and beta diversity, P<0.05). Interestingly, bacteriophages viral species changed drastically between sows and piglets. Piglets had a higher frequency of *Caudovirales* (P<0.01), and sows of *Petitvirales* (P<0.01). Sow's faecal microbiome was not affected by the diet, while piglet's cecum viral profile was affected by the sows' diet (higher alpha diversity in ARG group, P=0.01). In conclusion, Arg supplementation to lactating sows can improve the performance and modulate the gut microbiome of their litters.

Protipig: relationship between the intestinal microbiota and nitrogen utilization efficiency in pigs

N. Sarpong, A. Kurz, D. Berghaus, R. Weishaar, E. Haese, J. Bennewitz, M. Rodehutscord, J. Seifert and A. Camarinha-Silva
Institute of Animal Science, University of Hohenheim, Emil-Wolff-Str. 10, 70599 Stuttgart, Germany; amelia.silva@uni-hohenheim.de

Improving nitrogen utilization efficiency (NUE) is crucial for sustainable pig production. The gut microbiota plays a vital role in feed digestion, but the correlations with NUE are still not clear. This study aimed to describe the interaction between gut microbiota and NUE. A total of 48 barrows (Pietrain × German Landrace) were fed with a marginal lysine deficient diet (90% of supply recommendation) in a two-phase feeding system. Faecal samples were collected in both phases, nitrogen retention estimated, and NUE calculated. Pigs were euthanized (average BW 97.1 kg), and digesta and mucosa samples were collected from the stomach, duodenum, jejunum, ileum, caecum and colon. Volatile fatty acids (VFA) were measured in gut content. Gut microbiota was characterized with target amplicon sequencing of the 16S rRNA gene. Pigs were divided into two groups based on <25%- and >75%-quantile of NUE and correlated with gut microbiota. The mucosal microbial community diverged from digesta (P=0.001). *Bifidobacterium* abundance was higher in digesta of the stomach (5.5%) than the mucosa of pars pylorica (2%) (P<0.05). *Ligilactobacillus* was more abundant in pars nonglandularis than pars pylorica (P<0.05). *Ralstonia* had the highest abundance in the duodenum mucosa (22.8%) and decreased in the mucosa of other small intestine sections (<9%) (P<0.05). In caecum mucosa samples, *Alloprevotella* was more abundant (7.2%) compared to the colon mucosa (4.4%) (P<0.05). PERMANOVA revealed for the NUE in each phase a relationship to the overall microbiota of the digestive tract (P<0.05). Pigs characterised by high and low NUE in phase 2, showed differences in the jejunum digesta and the pars nonlandularis mucosa (P<0.05). VFA concentration differed between the parts of the gut. The concentration of acetic acids was highest in all sections and showed differences between the segments (P<0.05). Finally, microbial and VFA composition varies through the pig's gastrointestinal tract, and an association between the NUE and the microbiota composition of jejunum and pars nonglandularis was detected. Funding by H. Wilhelm Schaumann Stiftung is gratefully acknowledged.

Session 06 Poster 13

In vitro evaluation of Thymus capitatus essential oil (and compounds) effects on rumen microbiota

C. Gini[1], F. Biscarini[2], S. Andrés[3], L. Abdennebi-Najar[4], I. Hamrouni[5], M.J. Ranilla[3], P. Cremonesi[2], B. Castiglioni[2] and F. Ceciliani[1]

[1]UNIMI, Via dell'Università 6, 26900 Lodi, Italy, [2]CNR, Via Bassini 15, 20133 Milan, Italy, [3]Instituto de Ganadería de Montaña (CSIC-Universidad de León), Finca Marzanas s/n, 24346, Grulleros, Spain, [4]IDELE, 149 R.de Bercy, 75012 Paris, France, [5]CBBC, Technopole de Borj-Cédria, 901 Hammam-Lif, Tunisia; chiara.gini@unimi.it

Over the last few decades, efforts have been made in the research of alternative to antibiotics: the MILKQUA project focuses on exploring the use of essential oils (EOs) as replacement in dairy cows farming. The present trial was designed to study *in vitro* the effects of a *Thymus capitatus* natural essential oil (NEO), its corresponding synthetic formula (SEO), and those of the main pure compounds (carvacrol, p-cymene, γ-terpinene) on the ruminal microbiome. Samples (500 mg) of a 70:30 F:C diet were incubated *in vitro* with buffered rumen fluid (50 ml) from cows, without (control) or with EOs at 75 mg carvacrol/l or bioactive compounds added to the vials according to the proportions present in NEO at such rate (carvacrol 70.62%; p-cymene 7.06%; γ-terpinene 7.58%). Four replicates (inoculated from 4 cows) per treatment were incubated anaerobically for 24 hours at 39 °C. Fermentation was then stopped and total digesta samples were obtained for DNA extraction. V3-V4 regions of the 16S rRNA-gene were sequenced using Illumina Sequencer. Data were analysed with the QIIME 1.9 software. Alpha and beta diversity analysis showed a moderate effect of carvacrol alone and SEO on the rumen microbiota as compared to untreated controls: for two and three, respectively, out of 8 alpha diversity indices, the P-values ranged between 0.05-0.10; weighted UniFrac distances clustered perfectly per cow, except for 4 outliers belonging to the carvacrol (3) and *T. capitatus* SEO (1) treatments. The analysis of taxonomies revealed 39 OTUs significantly different between treatments and controls. In particular, 3 genera had a P<0.01: *Bhargavaea* (increased in γ-terpinene), *Atopobium* (reduced in carvacrol, p-cymene and γ-terpinene) and *Lysinibacillus* (reduced in all treatments as compared to controls). In conclusion, our results indicate only small effects of *T. capitatus* EOs on modificating rumen microbiota, except for carvacrol and *T. capitatus* SEO.

Session 06 Poster 14

EU-CIRCLES project: microbial and health evolution of pigs reared in high and low sanitary condition

P. Trevisi[1], G. Palladino[2], D. Luise[1], S. Turroni[2], F. Correa[1], P. Brigidi[3], P. Bosi[1], D. Scicchitano[2], G. Babbi[2], S. Rampelli[2], M. Candela[2] and P.L. Martelli[2]

[1]University of Bologna, Department of Agro-Food Sciences and Technologies, Bologna, 40127, Italy, [2]University of Bologna, Department of Pharmacy and Biotechnology, Bologna, 40126, Italy, [3]University of Bologna, Department of Medical and Surgical Sciences, Bologna, 40138, Italy; paolo.trevisi@unibo.it

The aim of the EU CIRCLES project is to investigate the evolution of pig's microbiome from early colonization to slaughtering and to capture the microbial features associated with piglet's health status in low and high sanitary conditions to educate microbial targeted intervention. The first step was an observational study where at 21 days (d) of age, 96 piglets from 22 litters were divided into 2 groups balanced for the litter of origin and assigned to different weaning and fattening farms: H (high sanitary condition) and L (low sanitary condition) and followed until slaughter. Per each phase, pigs were fed the same diets whatever was farm. Samples of faeces and blood were collected at d21 of age (farrowing unit, T1); 42 and 80 d of age (weaning unit, T2 and T3); 98 and 278 of age (fattening unit, T4 and T5). Microbial profile (V3-V4 regions of 16S rRNA), blood formula and immunoglobulins (Igs) were analysed. The evolution of faecal microbiota from piglets to adult pigs was significantly highlighted by the alpha and beta diversity indices (P<0.001). The sanitary status of the farm significantly affected the faecal microbiome at T2, T3 and T4. At T2, H pigs had a lower alpha diversity (P=0.002) and were richer in *Prevotella*-related bacteria compared with L pigs which were richer in *Lactobacillus* and *Collinsella*. At T3, H pigs had a higher alpha index (P<0.001) and were richer in *Bacteriodales* and *Treponema*, L pigs were richer in *Terrisporobacter*, *Veillonellaceae* and *Catenibacterium*. In the weaning phase for several blood contents (red blood cells, haemoglobin, haematocrit, neutrophils, lymphocytes, basophils) an interaction between the sanitary condition and time was seen. Blood formula and Igs clearly clustered according to time (3 clusters: T1, T2-T4 and T5). Overall, the project shows a notable impact of the sanitary condition and animal maturation on the gut microbiota of pigs.

Ruminal fungi and solid-associated bacteria as affected by *in vitro* inclusion of winery by-products

R. Khiaosa-Ard[1], C. Pacífico[1], M. Mahmood[2], E. Mickdam[3] and Q. Zebeli[1]
[1]University of Veterinary Medicine Vienna, Veterinaerplatz 1, 1210 Vienna, Austria, [2]University of Veterinary and Animal Sciences, Subcampus Jhang, 12 km Chiniot Road, 35200 Lahore, Pakistan, [3]South Valley University, Kilo 6 Safaga Road, 83523 Qena, Egypt; ratchaneewan.khiaosa-ard@vetmeduni.ac.at

Our study explored the effect of wine industry by-products, which are fibre- and polyphenol-rich feedstuffs, on bacterial and fungal composition and the fibre degradation activities. We tested six diets: a negative control (CON) (70% hay + 30% concentrate on dry matter (DM) basis), grapeseed (GS)-low (70% hay + 25% concentrate + 5% GS), GS-high (65% hay + 25% concentrate + 10% GS), grape pomace (GP)-low (65% hay + 25% concentrate + 10% GP), GP-high (56% hay + 24% concentrate + 20% GP), and EXT (control diet top-dressed with grapeseed extract at 3.7% of diet DM) as positive control. Dietary total phenol contents were 2.9, 3.6, 4.1, 4.0, 5.0, and 6.4% of diet DM, respectively. Three rumen simulation technique trials were performed (n=6 per diet). The 24-h incubated feed bags taken on the last day of trial (day 10) were sequenced for their microbial composition (16S rRNA of bacteria and archaea using V4 region and anaerobic fungi using ITS2 region). Beta-diversity of fungi tended to be affected by diet (Aitchison distance, PERMANOVA P=0.07). *Neocallismatigomycota* and *Ascomycota* were the major fungal phyla identified. At lower taxonomic levels, GS-high and EXT consistently affected fungal composition. GS-high increased the read abundance of *Saturnispora, Mucor, Microdochium, Lachancea, Pichia, Zygosaccharomyces, Candida* and *Penicillium* compared to CON (q-value<0.05), while EXT increased *Microdochium, Cyberlindnera* and *Hanseniaspora* but decreased *Sporobolomyces* and an unidentified genus of *Neocallimastigaceae* (q-value<0.05). Some bacterial genera (*Prevotellaceae Ga6A1 group, Lachnospiraceae* FCS020 group, possible genus SK018, and an uncultured bacterium of *Muribaculaceae*) were negatively affected by EXT compared to CON (q-value<0.05). EXT and GP-low decreased apparent degradation of DM and acid detergent fibre. Our data suggests that solid-associated microbiota, especially anaerobic fungi, can tolerate grape's phenols at least up to 3.4% of diet DM.

Session 06 Poster 16

Integrated analysis of faecal microbiota and host transcriptome of Dutch traditional cattle breeds

R. Gonzalez-Prendes[1], R. Gomez Exposito[2], T. Reilas[3], M. Makgahlela[4], J. Kantanen[3], C. Ginja[5], D.R. Kugonza[6], N. Ghanem[7], H. Smidt[2] and R.P.M.A. Crooijmans[1]
[1]Wageningen University & Research, Animal Breeding and Genomics, Droevendaalsesteeg 4, 6708 PB Wageningen, the Netherlands, [2]Wageningen University & Research, Laboratory of Micriobiology, Droevendaalsesteeg 4, 6708 PB, the Netherlands, [3]Natural Resources Institute Finland (Luke), Humppilantie 14, 31600, Finland, [4]Agricultural Research Council-Animal Production Institute, Old Olifantsfontein Road, 0002, South Africa, [5]BIOPOLIS-CIBIO-InBIO, Centro de Investigação em Biodiversidade e Recursos Genéticos, R. Padre Armando Quintas 7, Portugal, [6]School of Agricultural Sciences, College of Agricultural and Environmental Sciences. Makerere Univ., Department of Agricultural Production, Kampala, Uganda, [7]Faculty of Agriculture, Cairo University, Animal Production Department, Giza, Egypt; rayner.prendes@gmail.com

Molecular characterization of well-adapted native cattle breeds is essential for improving the adaptation of less diverse commercial breeds to their ecosystem. In this study, we investigated host gene expression and faecal microbiota composition in local Dutch cattle breeds to identify significant links between breed-specific microbiota and host genes functionally relevant for the adaptation. Faecal microbiota and host transcriptome were evaluated in 46 animals from five traditional cattle breeds, namely: the Deep Red (11), Groningen White Headed (8), Meuse-Rhine-Yssel (9), Dutch Belted (11), Dutch Friesian (7); and three commercial Holstein Friesian animals sampled in The Netherlands. A total of 1,891 16S rRNA gene amplicon sequence variants (ASV) and 19,368 expressed genes were identified in the dataset. Significant differences (P<0.05) were observed in microbial composition between breeds at ASV, genus and phylum levels. The microbial relative abundance was affected (P<0.05) by breeds and environmental factors such as seasonal feeding. In contrast, the cattle transcriptome was mainly influenced by the type of feed. Transcript-microbiota correlation was modest, but we identified potential interactions between faecal microbes and host genes that are breed specific, involving host genes implicated in inflammation and energy metabolism. Our results provide insight into beneficial microbes and functional genes in local cattle breeds.

Session 06 Poster 17

The effect of feed processing and nutrient density on growth parameters of feedlot lambs

J.H.C. Van Zyl, P.H.S. Uys and C.W. Cruywagen
Stellenbosch University, Animal Sciences, Private Bag X1, 7602 Stellenbosch, South Africa; cwc@sun.ac.za

The objective of the trial was to evaluate the effects of feed pelleting and nutrient density on the production performance and profitability of finishing lambs. Sixty-four male SA Mutton Merino lambs were randomly assigned to a low (L) or high (H) nutrient density diet that was either in a mash or pelleted form. Pellets were 4 mm, 6 mm, or 8 mm in diameter. Treatments were thus L0, H0 (mashed diets), L4, H4, L6, H6, L8, and H8, resulting in a 2×4 factorial arrangement. There were eight repetitions per treatment and the experimental period was eight weeks. Treatment effects on dry matter intake (DMI), growth rate, feed efficiency and feedlot economy were determined. The mean DMI of the H4 group (1.11 kg) was lower (P<0.05) than that of the other groups (1.24 kg), except for the L0 group. Lambs that received the L0 diet had the lowest ADG (0.20 kg). The H6 diet resulted in the highest ADG (0.31 kg), although it was only significantly higher (P<0.05) than that of the lambs on the H4 diet. Pelleting increased ADG (P<0.05) in the low-density diets, but not in the high-density diets. The worst feed conversion rate (FCR) was observed in the L0 treatment (6.9 kg DM/kg gain). The FCR of the high-density treatments (4.91 kg DM/kg gain) was better (P<0.05) than that of the low-density treatments (5.69 kg DM/kg gain). Across densities, an improvement of FCR was observed as pellet size increased. Dressing percentage of the HD lambs (47.7%) was higher (P<0.05) than that of the LD lambs (44.5%). The highest mean gross profit (GP) was obtained in the H8 and the H0 treatments. Pelleting of the LD TMR improved growth and economic performance parameters of fattening lambs due to an increase in DMI and improved FCR, while pelleting of HD diets did not show any significant effects. Larger diameter pellets in HD diets appear to be more beneficial in lamb performance than the use of smaller pellets.

Session 06 Poster 18

Effect of faecal microbial transplant on animal performance and health in young goats

A. Belanche[1,2], A.I. Martín-García[1] and D.R. Yáñez-Ruiz[1]
[1]Estación Experimental del Zaidín (CSIC), Profesor Albareda 1, 18008 Granada, Spain, [2]University of Zaragoza, Animal Production and Food Sciences, IA2, Miguel Servet 177, 50013 Zaragoza, Spain; belanche@unizar.es

Therapeutic antibiotic treatments can lead to low performances in ruminants with insufficient gut microbial development. It was hypostatized that faecal microbial transplant from adult to young ruminants could minimize this problem. A total of 44 newborn goat kids were randomly distributed into 4 groups according a 2×2 factorial: half of the animals were inoculated on the first day of life with faeces from healthy goats mixed with the colostrum (INO), whereas the rest received colostrum alone (CTL). From days 3 to 7 of age, half of the animals (+) were daily dosed with an oral antibiotic (oxytetracycline) whereas the rest remained antibiotic-free (-). On day 8 the faecal microbial inoculation was repeated resulting in 4 groups (CTL-, CTL+, INO- and INO+). Health-related metabolites in blood were measured on weeks 1, 2 and 7 and animals were slaughtered at weaning (week 7). Results indicated that the faecal transplant prior antibiotic treatment prevented the increase of aspartate aminotransferase (AST), alanine aminotransferase (ALT) and lactate dehydrogenase (LDH) as indicators of tissue damage (P<0.05). At weaning, inoculated animals showed higher levels of blood glucose (P=0.05) and urea (P<0.001), whereas CTL animals had higher levels of creatinine (P=0.01) and total proteins and globulins (P<0.001). Faecal microbial transplant also led to higher butyrate molar proportion (P=0.038) in the hindgut as indicator of intestinal health. On the contrary, antibiotic treatment led to a lower length-to-width ratio of the rumen papillae and lower acetate molar proportion in the rumen (P=0.044). Treatment CTL+ showed a lower ADG until weaning in comparison to CTL- (-7.5%) and INO treatments (-10%). These findings suggest that faecal microbial transplant could help to prevent potential metabolic disorders derived from antibiotic therapy.

Effect of a methionine supplementation on the response of goats infected by *H. contortus*

L. Montout, H. Archimède, M. Jean-Bart, D. Feuillet, Y. Félicité and J.-C. Bambou
INRAE (National Research Institute for Agriculture, Food and Environment), GA (Animal Genetic), INRAE Antilles-Guyane, domaine de Duclos, 97170, Petit-Bourg (Guadeloupe), France; laura.montout@inrae.fr

Gastrointestinal nematode (GIN) infections are considered as one of the major constraints in small ruminant production at pasture. The negative impact of the drugs classically used to control these infections, on soil biodiversity coupled with concerns over the presence of residues in animal products led some states in the world to advocate a significant reduction of the use of chemical molecules in animal production. The nutritional manipulation of small ruminants has long been considered as a tool for the control of GIN infections. The aim of this study was to measure the impact of a rumen-protected methionine supplementation on the response of Creole goat kids experimentally infected with *H. contortus.* The animals were divided in three groups and infested by 10,000 larvae of *H. contortus*: a control group with no methionine supplementation (C: basal diet), a low methionine group (LM: basal diet + 4.5 g/head of rumen-protected methionine) and a high methionine group (HM: basal diet + 10 g/ head rumen-protected of methionine). Parasitological, haematological and immunological data were collected each week during 42 day of experiment. The results showed that the level of parasitism, measured through the faecal eggs counts was significantly lower in the HM group. The first immunological and haematological parameters available by now, are correlated with the parasitological ones. Further analysis under course (RNAseq) will provide more information on the underlying mechanisms. In conclusion, the supplementation with rumen-protected methionine appear as a promising strategy for GIN control in goats.

Does diet starch and fibre content affect variation in feed intake in 6 and 8.5 months bull calves

M. Vestergaard[1], M. Bjerring[1] and A.M. Kjeldsen[2]
[1]Aarhus University, Department of Animal Science, 8830 Tjele, Denmark, [2]SEGES, Livestock Innovation, 8200 Aarhus N, Denmark; mogens.vestergaard@anis.au.dk

It has been presumed that there is a larger day to day (D-D) variation in feed intake of fattening calves fed high-starch, low-fibre diets compared with moderate-starch and high-fibre diets. A high D-D variation expects to give larger fluctuations in rumen pH and cause subacute rumen acidosis that might lead to liver abscesses. Little is known about the actual size of the D-D variation in feed intake among rosé veal calves usually fed high-energy diets. The objective was to analyse how various starch contents in pelleted concentrates (CONC) (9 types of CONC) as well as various fibre contents in total mixed rations (TMR) (10 types of TMR) affected D-D variation in Net Energy Intake (NEI, Scandinavian Feeding Units/d) and Feeding Time (FT, min/d). Data from 9 experiments and a total of 768 calves were used. The CONC types were grouped according to starch content (g/kg DM) into low (275), medium (403), and high (448), and the TMR types were grouped based on energy concentration into three levels of digestible cell wall content (g/kg DM) into low (209), medium (245), and high (304). Data included individual daily NEI and FT for a 30-day period from 5.5 to 6.5 and from 8 to 9 months of age; two periods in which such calves typically attain daily gains of 1.5-1.8 kg/d. The std. dev. (SD) of the 30 d per calf was used to describe the variation in NEI and FT. Results for CONC showed similar NEI at both ages, but lower SD for NEI with high compared with low starch content at 5.5 to 6.5 ($P<0.05$) but not at 8 to 9 months. FT was numerically lower for high compared with low starch, but SD for FT was similar at both ages. For TMR, NEI decreased ($P<0.05$) with high compared with low fibre content, especially for younger calves, without affecting SD for NEI. FT was increased ($P<0.05$) with high compared with low fibre content of TMR at both ages and the same effect was found for SD of FT ($P<0.05$) for younger calves. Overall, these results suggest that for concentrate pellet based-diets, variation in NEI and eating time do not increase with high starch content. For TMR diets, a high cell wall content will increase eating time and its variation and reduce NEI but not its variation.

Session 06 Poster 21

Dietary administration of essential oil to newborn calves and long term effects on immunity

S. Andrés[1], N. Arteche-Villasol[1], D. Gutiérrez-Expósito[1], A. Martin[1], L. Abdennebi-Najar[2], F. Ceciliani[3] and F.J. Giráldez[1]

[1]Instituto de Ganadería de Montaña (CSIC-Universidad de León), Finca Marzanas s/n, 24346, Grulleros (León), Spain, [2]IDELE Institute, Quality and Health Department, 149 rue de Bercy, 75595 Paris Cedex 12, France, [3]Università degli Studi di Milano, Department of Veterinary and Animal Sciences, Via dell'Università 6, 26900 Lodi, Italy; sonia.andres@eae.csic.es

Essential oils (EO) can modulate the microbiome of the gut when included in the milk replacer (MR) of newborn dairy calves causing long-term effects on health status and feed efficiency. To test this hypothesis 16 newborn Holstein calves 3-days aged were assigned to control (n=8) and EO (n=8) groups. EO group was supplied daily with 0.23 ml of oregano EO accounting for 200 mg of carvacrol (Zane Hellas 100%) diluted in the MR during 45 days. After being weaned, calves were managed in a feedlot being fed *ad libitum* a total mixed ration to cover their nutritional requirements. Voluntary feed intake was recorded individually during 70 days to measure feed efficiency during the replacement phase (residual feed intake, RFI). In addition, blood samples were collected three times (day 3, day 45 and day 370) to: (1) measure biochemical parameters (glucose, cholesterol, insulin, etc.); and (2) characterize peripheral blood mononuclear cells (PBMCs) by flow cytometry (CD4+, CD8+, CD14+, CD21+ and WC1+). Results revealed that whereas positive effects on daily weight gain were observed during the period of EO administration (101 vs 158 g/day for control and EO group, respectively; P=0.001), no long-term effects on feed efficiency (RFI -0.165 vs +0.165 for control and EO group, respectively; P=0.124) were detected later on along the replacement phase. Regarding PBMCs, mean values of the CD14+ population (monocytes) along the whole experiment were higher in the EO group (29.0 vs 35.1% of positive cells; P=0.006), whereas the decrease of CD21+ (B lymphocytes) along the time was lower in the EO group. Finally, no changes were detected in the biochemical profile along the study. Further analyses are in progress to verify the existence of long-term effects on the microbiome of the gut and the metabolomic profile caused by EO administration during the suckling period of newborn dairy calves.

Session 06 Poster 22

Dietary administration of L-carnitine during the fattening period of early feed restricted lambs

A. Martín[1], F.J. Giráldez[1], J. Mateo[2], I. Caro[3] and S. Andrés[1]

[1]Instituto de Ganadería de Montaña (CSIC-Universidad de León), Finca Marzanas s/n, 24346 Grulleros (León), Spain, [2]Universidad de León, Departamento de Higiene y Tecnología de los Alimentos, Campus Vegazana s/n, 24071 León, Spain, [3]Universidad de Valladolid, Área de Nutrición y Bromatología, Avda. Ramón y Cajal, 47005 Valladolid, Spain; sonia.andres@eae.csic.es

Early feed restriction of lambs causes a permanent mitochondrial dysfunction and a consequent increase of fat accumulation. The main function of L-carnitine is the activation and transportation of long-chain fatty acids into the mitochondria, where fatty acid oxidation takes place. The aim of the present study was to clarify the effects of dietary L-carnitine on the fatty acid profile of meat from early feed restricted lambs. Twenty-two new-born Merino lambs were separated from the dams for 9 h daily to allow early feed restriction. During the fattening period the lambs were divided in two groups: the control group (CTRL, n=11) and the L-carnitine group (CARN, n=11). Both groups received a complete pelleted diet *ad libitum*, being the CARN group supplemented with 3 g of L-carnitine/ kg diet. The samples for the analysis of fatty acids were taken from the M. longissimus dorsi. The statistical analysis was carried out by one-way ANOVA being the dietary treatment (CARN vs CTRL) the only source of variation. The numerical values for all the long-chain polyunsaturated fatty acids (PUFA) were lower in the meat of the CARN lambs, with significant differences for 20:4n–6 (arachidonic acid, ARA; 93.18 vs 78.45 mg/ 100 g meat; P=0.012) and 22:6n–3 (docosa-hexaenoic-acid, DHA; 2.57 vs 1.76 mg/ 100 g meat; P=0.023). Consequently, the ratio PUFA/SFA (0.42 vs 0.33; P=0.023) was lower in the CARN group. According to these results, the administration of 3 g L-carnitine/ kg diet during the fattening period of early feed restricted Merino lambs seems to improve the transport of long-chain PUFA into the mitochondria, with the consequent increase of β-oxidation of these fatty acids, thus worsening the nutritional index of the meat.

Session 06 Poster 23

Immunomodulatory properties of acid whey from Greek yogurt of different animal origin

E. Dalaka, M. Karavoulia, A. Vaggeli, G.C. Stefos, I. Palamidi, I. Politis and G. Theodorou
Agricultural University of Athens, Department of Animal Science, Iera Odos 75, 11855, Greece; elenidalaka@yahoo.gr

With the increased consumer demand for biofunctional foods, it is important to develop novel products with the concurrent utilization of by-products. Among them, acid whey (AW), the yellowish liquid by-product derived from Greek yogurt production should be of considerable interest. In recent years, Greek yogurt has gained immense popularity because of its high nutritional value, thus AW production has also increased making its disposal problematic due to its high oxygen demand (COD; BOD). The objective of this study was to evaluate *in vitro* immunomodulatory properties of AW derived from yogurts of different animal origin (cow, sheep, goat) after a static three phase simulated *in vitro* digestion mimicking oral, gastric and intestinal digestion. THP-1 cell line, a human leukaemia monocytic cell line is a suitable and reliable *in vitro* tool to study the immunomodulating effects of a food compound. Before assessing immunomodulatory capacity, cell viability and potential cytotoxicity of digested AW were evaluated. Gene expression of cytokines, inflammation-related enzymes, and transcription factors were analysed using qPCR. In conclusion, after exposure to digested AW, THP-1-derived-macrophages showed attenuated expression of some proinflammatoty and increased expression of anti-inflammatory genes, that may exert a beneficial effect on immunoregulation. In sum, this study demonstrates that AW regardless of animal origin could be used as food or feed ingredient that might have a positive effect on health through its immunomodulatory capacity. This research is co-financed by Greece and the European Union through the Operational Programme 'Human Resources Development, Education and Lifelong Learning' in the context of the project 'Strengthening Human Resources Research Potential via Doctorate Research' (MIS-5000432), implemented by the State Scholarships Foundation (IKY). This research is co-financed by the European Regional Development Fund of the European Union and Greek national funds through the Operational Program Competitiveness, Entrepreneurship and Innovation, under the call RESEARCH – CREATE – INNOVATE (project code: T2EDK-00783).

Session 06 Poster 24

Citrus extract characterisation and effect on bacterial growth to explain the consequence on animal

S. Cisse[1,2,3], M. Gautron[3], M.E.A. Benarbia[1] and D. Guilet[1,2,3]
[1]FeedInTech, R&D, 42 rue Geroges Morel, 49070 Beaucouzé, France, [2]SONAS, SFR QUASAV, Université d'Angers, R&D, 42 rue Georges Morel, 49070 Beaucouzé, France, [3]Nor-Feed, R&D, 3 rue Amédéo Avogadro, 49070 Beaucouzé, France; sekhou.cisse@norfeed.net

Citrus extracts are increasingly used in poultry and swine production, due to the benefits they provide to animals. However, there is a lack of data regarding their mode of action. Most of the studies available indicate an effect on the intestinal microbiota. The aim of this study was to evaluate the effect of a standardized citrus extract (SCE, Nor-Spice® AB, Nor-Feed SAS) on the growth of several bacterial strains of interest. The SCE has also been characterized in order to better understand the observed effect on pigs and poultry performances. The effect of SCE has been monitored using 2 strains: *E. coli* and *L. acidophilus*. Briefly, 200 µl of a 10% SCE solution were added to 10 ml of medium, which was seeded with 5×10^5 bacteria/ml. Bacterial growth was monitored by measuring the medium turbidity at 650 nm. The same experiment was performed without SCE, as a negative control. In parallel, apolar and polar compounds of SCE were characterized by GC-MS and LC-MS (dereplication) respectively. Bacterial growth showed that SCE decreased the doubling time of L. acidophilus (256 min) and increased the doubling time of *E. coli* (245 min), compared to the control group (respectively 353 and 80 min). Characterization of SCE allowed to identify pectic oligosaccharides as SCE major components and confirm the presence of 39 secondary metabolites including eriocitrin, hesperidin and naringin. These compounds have already shown beneficial effect on gut health, according to the literature. As example, it has been demonstrated that hesperidin promotes the production of short chain fatty acids in the colon, which results in pH diminution and pathogenic bacteria inhibition. According to these results, the beneficial effect of SCE on animals may be due to the modulation of the intestinal microbiota. In addition, some compounds identified on SCE are well known for their positive effect on different compartments of the gut and microbiota. Further studies will be necessary to confirm the role of these molecules.

Session 06 Poster 25

Lithothamnium calcareum as natural additive in roughage-lacking diets of Nellore young bulls

G. Balieiro Neto[1], L. Machado[2] and B.R. Amâncio[1]

[1]Instituto de Zootecnia, Brazil, Department of Agriculture and Food Supply of São Paulo State, Bandeirantes avenue 2419, 14030670 Ribeirão Preto, Brazil, [2]Twenty5[0]Now Brasil LTDA, Rua Peixoto Gomide, 1186, 01409-000, São Paulo, Brazil; gebalieiro@sp.gov.br

In order to improve animal performance, facilitate cattle management and have cost-effective production, there has been an increase in the nutritional values of the administered feedlot diets. However, feeding high-grain diets can induce laminitis. The inclusion of monensin in the diet has been commonly used as a strategy to achieve higher daily weight gains and control the laminitis. However, the use of antibiotics in animal feed might result in the selection of resistant microorganisms and certain bacterial strains isolated from cattle are resistant to multiple antibiotics and could be human health hazards. Twenty-four Nellore young bulls fed with high-grain diet containing *Brachiaria brizantha* hay (15% dry matter) were divided into two groups to evaluate the marine algae *Lithothamnium calcareum* as an antibiotic alternative. The first group was fed a diet containing sodium monensin (120 mg/animal/day) and the second group was fed a diet without antibiotics, containing 120 g/animal/day of *L. calcareum*. Daily dry matter intake, early detection of lameness, laminitis and feeding behaviour traits (frequency and duration of bunk visit events, head-down duration, variance of nonfeeding intervals, and time to approach feed bunk following feed-truck delivery) were measured 13 d, preceded by 26 d of adaptation to the diet and facilities with a GrowSafe system. A completely randomized design was used. There was no detection of lameness or laminitis in any groups and there were no changes in dry matter intake (11,6 vs 11,4) or feeding behaviour traits. We concluded that marine algae *L. calcareum* could be used as an antibiotic alternative allowing the diet to be prepared with 15% roughage, that could be convenient in several commercial conditions. Appears the marine algae *L. calcareum*re covered ruminal pH and could be used as sustainable additives in ruminant diets with low roughage. Feed efficiency and methane production are topics for further research.

Session 06 Poster 26

Comparison on the microbial community structure of lamb faeces, milk ewe and cow colostrum samples

L.C. Lopes[1], T. Correia[2] and P. Baptista[3]

[1]Evora University, Apartado 94, 7006-554, Portugal, [2]Agrarian Superior School of Bragança, Department of Animal Science, Alameda de Santa Apolónia 253, 5300-252, Portugal, [3]Polytechnic Institute of Braganza, Mountain Investigation Center, Alameda de Santa Apolónia, 5300-252, Portugal; laila.lopes@uevora.pt

The sheep breed *Churra Galega Bragançana* from the northeast of Portugal is an autochthonous sheep breed, very economically important in the area. This study used cow colostrum as a supplement for lambs, as it is surplus and often discarded in intensive dairy cattle farms. In this study 18 of the Churra Galega Bragançana sheep breed were used and divided into two groups: the control and the treatment group (3 ewes and 6 lambs each). The lambs of the treatment group, from the third day of age were supplemented with 50 ml per day of colostrum from two Holstein-Friesian cows, for 4 weeks. The lambs were weighed. The bacterial community present in the faeces samples of lambs, cow colostrum and ewe milk was evaluated using non-cultivation methods using the *Illumina miseq* platform. In faeces samples, bacterial diversity was additionally evaluated by culture-dependent methods in order to evaluate the lytic enzyme production capacity of the isolates obtained. Regarding the average daily gain (ADG) of lambs, in the period of 10 days after birth, the treatment group showed a ADG (0.232 g/d) significantly higher (P=0.031) than the control group (0.127 g/d). In total, the two methods identified 93 genera in faeces and 231 genera in milk samples, with *Bacteroids* and *Acinetobacter* being the most abundant genera in the bacterial community respectively. It was found that the composition of the bacterial community of the control lambs' faeces is significantly different (R=0.48; P=0.029) from the community present in the faeces of lambs that ingested cow colostrum. From the bacterial isolates obtained from faeces, three OTUs (*Acinetobacter* sp., *Jeotgalicoccus* sp. and *Bacillus* sp.) were identified by cultivation-dependent methods that showed capacity to produce digestive enzymes such as amylase, cellulase, lipase and proteinase. In this way, these isolates present an enormous potential in animal nutrition and additionally as immunological and/or probiotic adjuvants. Although the results obtained are quite promising, further studies should be carried out.

Session 07 Theatre 1

Consumer perceptions and acceptance of insects as feed: current findings and future outlook

G. Sogari
University of Parma, Department of Food and Drug, Parco Area delle Scienze 47A, 43124, Italy; giovanni.sogari@unipr.it

Research into alternative protein sources might help to reduce environmental burden on the planet and increase animal welfare. Insect proteins used in feed production could represent an innovative solution for these environmental and ethical problems. However, the inclusion of insects as a protein source in feed production is not only related to technical, economical, and normative restrictions but is also affected consumer responses such as attitude, intention to purchase and consume, and willingness to pay for insect-fed animals. Therefore, consumer acceptance of insects as feed is crucial to determine the development of the sector of insect farming and of the foods obtained from animals fed on an insect-based diet. Despite the growing literature on consumer acceptance of edible insects as foods in Western countries few studies have examined consumers' preferences and attitudes towards the use of insects as feed. Currently, the general public is still unaware of the potential benefits of this alternative protein source for farmed animals, and the role of information may be greater for individuals who are uninformed or misinformed about the benefits of insect-based feed. Our studies showed that information about the benefits of insects as feed could improve consumers' attitudes towards animal-based products fed with insects. For instance, our results suggest that communicating about the insects as being a natural feed for several animal species (e.g. poultry and fish) and that the final meat taste is unchanged are important factors to increase acceptance. Thus, legislators should consider this lack of knowledge and help inform consumers about the benefits of insect meals. In the Global North, entomophagy has received global media attention in recent years, contributing to an increase in curiosity among consumers and providing publicity for the private sector. The same phenomenon might happen for this novel feedstuff in the near future. This presentation will contribute to the discussion on the factors influencing consumers' purchase motivations and willingness to pay for meat products from poultry, aquaculture, and pigs fed with insects.

Session 07 Theatre 2

Microbiome in the insect-based feed chain: current scenario and future perspectives

I. Biasato
University of Turin, Deparment of Agricultural, Forest and Food Sciences, Largo Paolo Braccini 2, 1005 Grugliasco (TO), Italy; ilaria.biasato@unito.it

Gut microbiome – defined as the genes and genomes of the gut microbiota, as well as their products and the host environment – has a key role in monogastric animals, as it is involved in the maintenance of gut health, which is of vital importance to animal health and performance. Diet is known to be an important driver of gut microbiome variations, especially when novel feed ingredients – such as insects – are considered. The impact of insect-based products on gut microbiome has extensively been characterized in poultry, fish, pigs and rabbits, with common findings being outlined despite the species heterogeneity. The use of *Hermetia illucens* (HI) and *Tenebrio molitor* meals (especially at low inclusion levels [5-10%]) and oils usually leads to increased microbial diversity, selection of short chain fatty acids (SCFAs)-producing bacteria (with chitinolytic activity as well [*Blautia*, *Roseburia*, *Enterococcus*]), and reduction of pathogenic bacteria (due to the HI anti-microbial properties [*Streptococcus*, *Corynebacterium*, *Campylobacter*, *Listeria*]). High inclusion levels of HI meals (15%) may, however, reduce the microbial diversity and potentially beneficial bacteria (*Ruminococcus*), as well as selecting bacteria with mucolytic activity (*Helicobacter*). Furthermore, since insects are considered as 'farmed animals', diet-microbiome interactions may have a remarkable influence on insect farming as well. The progressively increasing research about gut microbiome of insects – mainly focused on HI for its high potential for bio-conversion – revealed that, even if some members (*Providencia*, *Enterococcus*, *Morganella*) and metabolic pathways remain constant, both host genetics and feed substrate are capable of driving the diversity and metabolic potential of gut bacteria, with potential implications for larval rearing traits. Considering the heterogeneous feed substrate-related changes in HI gut microbiome, as well as the feasibility to directionally transform it by manipulating that of the rearing substrate, future research may be focused on the optimization of the HI gut microbiome through feed substrates with selected microbiome profiles, also evaluating their potential impact on gut microbiome of insect-fed animals.

Session 07 Theatre 3

Long-term evaluation of black soldier fly larvae meal on Atlantic salmon health in net pens

G. Radhakrishnan[1,2], A.J.P. Philip[2], N.S. Liland[2], M.W. Koch[1,2], E.J. Lock[1,2] and I. Belghit[2]
[1]University of Bergen, University of Bergen, P.O. Box 7800, 5020 Bergen, Norway, [2]Institute of Marine Research, Institute of Marine Research, Bergen, Norway, P.O. Box 1870 Nordnes, 5817 Bergen, Norway; Gopika.Radhakrishnan@hi.no

This study evaluated the effects of black soldier fly larvae meal (BSFM) on the performance and health parameters of Atlantic salmon (*Salmo salar*) farmed under industry relevant conditions. The trial was performed with almost 54,000 salmon in open water sea cages, exposed to natural stressors resembling a realistic commercial farm condition. Fish were fed with diets where plant-based proteins were partially replaced with BSFM, corresponding to dietary inclusion levels of 0% (control), 5% and 10% BSFM. Diet groups consisted of triplicates cages (12×12 m) with ~6,000 fish per cage from sea transfer (~0.2 kg) to harvest size (~4.5-5 kg). The study assessed fish health and monitored fish welfare parameters. Fish were monitored weekly to check for the weight, and general welfare indicators that included the gill score, opercula, deformities, emaciation, fin status, cataract, and the sea lice count. Also, the skin mucus samples were collected each month to analyse mucus lysozyme activities along with blood and plasma samples to study the biochemical conditions. Results will be presented and discussed. This trial is part of the EU-funded SUSINCHAIN project (nr. 861976).

Session 07 Theatre 4

Ability of black soldier fly proteins to decrease oxidative stress in companion animals

A. Paul[1], N.M. Tome[1], M. Dalim[1], T. Franck[2], D. Serteyn[2], K. Aarts[1] and A. Mouithys-Mickalad[2]
[1]Protix B.V., Industriestraat 3, 5107 NC, the Netherlands, [2]Centre for Oxygen, Research & Development (CORD), Bât. B6A, Allée du six Août 11, 4000 Liege, Belgium; nuria.tome@protix.eu

In Europe, black soldier fly (BSF) proteins are gaining popularity as sustainable and health promoting pet food ingredients. These ingredients contain substantial amounts of low molecular peptides which are known to supress oxidative stress in pets and other animals. In this session we will show results of our *in vitro* studies evaluating the ability of BSF proteins to supress oxidative stress derived from the excessive neutrophil response in pets. Two different studies were conducted. The first study evaluates the ability of BSF proteins to: (1) scavenge free radical species, (2) modulate the inflammation-like activity linked to neutrophil response, and (3) modulate myeloperoxidase (MPO) activity, a pro-oxidant enzyme released by activated neutrophils, using various *in vitro* models. During this study, commercial chicken meal and fishmeal were used as industrial benchmarks. Second study was planned to evaluate the efficacy of commercial pet foods containing BSF proteins to reduce oxidative stress. For this study, commercial non-insect-based pet foods were used as industrial benchmarks. Both pet foods with and without BSF proteins were evaluated using similar *in vitro* assays as mentioned in previous paragraph. To conclude, we found that BSF proteins have strong ability to reduce oxidative stress in pets as shown on free radicals, neutrophil and MPO assays. During this study, we found that chicken meal and fishmeal have pro-oxidant activity in some of the tested assays. Finally, commercial diets containing BSF proteins perform better in comparison to non-insect-based diets in terms of decreasing oxidative stress.

Effect of black soldier fly meal dietary integration on production and eggs quality of laying hens

A. Vitali[1], M. Meneguz[2], F. Grosso[2], E. Sezzi[3], R. Lorenzetti[3], D. Santori[3], N. Ferrarini[4], G. Grossi[1], E. Batistini[5] and N. Lacetera[1]

[1]Università della Tuscia, Via s. Camillo de Lellis, 01100 Viterbo, Italy, [2]Bef Biosystems S.r.l., Via Tancredi Canonico, 18/c, 10156 Torino, Italy, [3]Istituto Zooprofilattico Sperimentale del Lazio e della Toscana, Str. Bagni, 4, 01100 Viterbo, Italy, [4]Azienda Sanitaria Locale, Via Enrico Fermi, 15, 01100 Viterbo, Italy, [5]SE.CO.M. S.r.l., Via dell'Artigianato, 3, 06089 Torgiano, Italy; vitali@unitus.it

The study investigated the effect of dietary substitution of soybean meal with insect meal on production and eggs quality in laying hens. The trial lasted 56 days and involved 108 laying hens (Lohmann Brown-Classic) 18-week-old. Hens were housed into arches and fed with soybean-based meal for 4 weeks. At 22-week-old, hens were weighed, divided into 27 groups homogeneous for live (4 hens each) and housed into arches with an artificial light regime of 16L:8D. Arches were 60×90×73 cm (1,35 cm^2/hens) and enriched with nest box and perches. Groups were randomly assigned (nine replicates) to one of the three experimental diets isoenergetic, iso-proteic, balanced for amino acids and provided *ad libitum*: soybean-based meal as control (C); treated 50% (T50) and 100% (T100) where the fraction of soybean meal was replaced by *Hermetia illucens* meal at 50 or 100%, respectively. Feed consumption and eggs production were monitored daily. Eggs were sampled bi-weekly (T0, T14, T28, T42 and T56) and evaluated for quality at t0 (fresh eggs) and after 21 days storage at ambient temperature (t21), through measurement of shell thickness and calculation of Haugh, yolk and albumen indices. The integration of insect meal in the diet did not negatively impact on eggs production. On the other hand, insect meal reduced feed intake with a positive effect on the feed conversion rate. The quality analysis pointed out higher values of shell thickness in the T100 group, and higher albumen index in the T50 group. *H. illucens* meal had no adverse effects on production and eggs quality, so it may be considered an alternative protein source for developing feeding strategies to support circular economy. However, further investigation is needed to verify the effects of insect meal on a commercial scale and analyse its economic and environmental sustainability.

Effect of dietary inclusion of chitinolytic bacteria on insect meal diets for *Dicentrarchus labrax*

F. Rangel[1,2], R. Cortinhas[1,2], L. Gasco[3], F. Gai[4], A. Oliva-Teles[1,2], C.R. Serra[1] and P. Enes[1,2]

[1]CIMAR/CIIMAR Centro Interdisciplinar de Investigação Marinha e Ambiental, Terminal de Cruzeiros de Leixões. Av. General Norton de Matos s/n, 4450-208 Matosinhos, Portugal, [2]Faculdade de Ciências, Universidade do Porto, Biology department, Rua do Campo Alegre s/n, 4169-007 Porto, Portugal, [3]Forest and Food Sciences, University of Turin, Department of Agricultural, Via Verdi, 8, 10124 Turin, Italy, [4]Institute of Science of Food Production, National Research Council, Via Amendola, 122/O, 70126 Bari, Italy; fjorangel@gmail.com

Incremental worldwide fish demand promoted a rapid aquaculture intensification which, in turn, put a strain on resource management for feed production, rendering key ingredients such as fish meal (FM) economically and ecologically unviable. To respond to the increased feed demand, alternative ingredients to substitute FM are being pondered with insect meal (IM), being one of the main candidates. However, insects' exoskeleton is rich in chitin, and with higher inclusion levels of IM, it negatively interferes with fish performance and nutrient digestibility. In fact, just the removal of the cuticle of the insect positively impacts IM digestibility. In carnivorous fish species with an inexistent or low ability to degrade chitin, the inclusion of chitin-degrading bacteria could help to increase IM potential as a FM substitute. To test this hypothesis, two bacteria with the ability to degrade chitin isolated from European sea bass (ESB) gastrointestinal tract (FI645 and FI658), were incorporated individually or as a mixture in ESB diets with high defatted *Hermetia illucens* larvae meal (HM) levels. As such, four isoproteic (45%) and isolipidic (18%) diets were formulated: a control diet, containing 30% of HM (CTR); two diets like the CTR containing 2×10^9 cfu/kg of either FI645 (D645) or FI658 (D658); a diet like the CTR containing 1×10^9 cfu/kg of both FI645 and FI658 (DMIX). Preliminary results showed that DMIX significantly increased feed efficiency, although no effect was observed in growth performance. A digestibility trial, digestive enzyme activity measurements, and gut histological and microbiota evaluation are being performed to support the results. The researcher F.R. was supported by a grant from FCT, Portugal (SFRH/BD/138375/2018).

Broiler eating rate suggests preference for black soldier fly larvae over regular feed

M.M. Seyedalmoosavi, M. Mielenz, G. Daş and C.C. Metges
Research Institute for Farm Animal Biology, Institute of Nutritional Physiology, Wilhelm-Stahl-Allee 2, 18196, Dummerstorf, Germany; seyedalmoosavi@fbn-dummerstorf.de

Whole black soldier fly larvae (BSFL) can be fed to broilers not only for decreasing soybean-protein in diets but also to improve animal welfare. How much whole BSFL could be fed to broilers is however unknown. We assessed apparent interest of broilers in eating BSFL when offered with either 10% (L10), 20% (L20) or 30% (L30) of voluntary feed intake (FI) of control chickens (CON) that received no BSFL but age-specific diets. Ross 308 chicks (n=252) were allocated to one of either 4 groups (n=6 pens/group). Time spent by birds for eating their daily portion of larvae (TSL, min) was recorded. Larvae eating rate (LER, g/min) and body weight (BW) adjusted LER of the birds (LER_BW, g BSFL / kg BW and min) were calculated. Similarly, a theoretical feed eating rate (FER, g feed / light period of a day in min) with adjustment for BW (FER_BW) was calculated. Lastly, the ratio of LER:FER was calculated as fold change difference (FCD). Overall, TSL did not differ among 3 groups (P=0.982) expect for the first day (P<0.05), on which L30 spent more time (P<0.05) than did L10 and L20. The L10 had a higher LER than did the L20 and L30 bird (P<0.05). LER increased from 0.03 (d1) to 6.8 g/min (d42) by more than 200 folds. In contrast, LER_BW decreased from 8.9 to 2.9 (g/BW and min) from wk1 to wk6 (P<0.001), and did not depend on the amount of BSFL fed to birds (P=0.138). Although both FI and FER increased linearly over time (P<0.001), provision of BSFL reduced FI and FER in both L10 and L20, and did more so in L30 as compared to CON (P<0.05). FER increased from wk 1 to wk 6 by about 25 folds in CON. In contrast, FER_BW decreased linearly in all four groups over time (P<0.05). While CON had a higher FER_BW than did L30 at wk 1, 3 and 4 (P<0.05), there was no significant difference at wk 5 (P>0.05). The FCD (on average >50) was not influenced by the amount of larvae offered to the birds (P=0.195), but decreased slightly from wk 1 and wk 6 (P.018). We conclude that chickens can consume up to 30% of their voluntary FI as BSFL in just a few minutes. Apparent interest of chickens in BSFL as compared to regular feed is at least 50 times higher, implying the potential of BSFL to be used as an edible environmental enrichment tool.

InsectERA – the 57M€ plan to transform Portugal into an insect technology hub

D. Murta[1,2]
[1]CiiEM – Centro de Investigação Interdisciplinar Egas Moniz, Caparica, 2829-511, Portugal, [2]EntoGreen, Ingredient Odyssey, Santarém, Portugal, Rua Cidade de Santarém, 140, 2005-079, Portugal; daniel.murta@entogreen.com

Insects are recognized as a new nutrient source for human, animal and plants, being appointed as a bioremediation tool and a new source of raw-material for a large range of industrial applications. InsectERA is a consortium led by EntoGreen and comprises a total of 39 entities, including 17 Portuguese industrial and service companies (including 3 insect producers), 17 Non-Business Entities, 1 public administration entity (Portuguese food authority), 3 Industrial Associations and a Scientific Society. This consortium is the result of an in-depth discussion enrolling multiple actors with impact on the value chain, with discussions through collaboration with companies in both sides of the value chain, from raw-material to end-users, and also enrolling a large group of R&D centres focused on this novel field. It will contribute to upscale the production of insects, creating a strong link that closes the circular economy in the agri-food sector, by converting vegetable by-products into high value proteins, oils and organic fertilizers. It is organized in four main pillars, Insects as food, as feed, as new industry raw-materials and as a bioremediation tool, creating four sub-sectors, with specific challenges. However, each pillar has two other transversal challenges, such as raw-materials for insect production and insect production industrialization, and also results in the production of a high value product, insect frass. The final products developed under the scope of this project will be closely evaluated and submitted to market preparation, from consumer acceptance to legal framework adaptations. InsectERA will contribute to develop a novel market opportunity and to upgrade the technology associated with the insect production field. The final result will be the creation of 3 new factories, a logistical centre and 2 R&D units, that will generate new products, processes and services ready to the international markets creating a novel and competitive industry in Portugal.

Evaluation of performance and carcass characteristics of growing lambs fed *Tenebrio molitor* diet

L.E. Robles Jimenez[1], E. Cardoso Gutierrez[1], J.M. Pino-Moreno[2], S. Angeles[2], B. Fuente[2], H. Ramirez[2], I.A. Dominguez-Vara[1], E. Vargas-Bello-Perez[3] and M. Gonzalez-Ronquillo[1]
[1]Universidad Autónoma del Estado de Mexico, Instituto Literario 100, 50000, Mexico, [2]Universidad Nacional Autonoma de Mexico/, Av. Universidad #3000, 04510, Mexico, [3]University of Copenhagen, Grønnegårdsvej 3, 1870, Denmark; lizroblez@hotmail.com

Tenebrio molitor (TM) is an insect rich in proteins and lipids, it can be used as an alternative protein source in ruminant feeding. The objective of this study was to evaluate the effect of TM meal as an alternative to vegetable- (soya bean meal, SBM) and animal- (Fish meal, FM) protein sources on slaughter weight and carcass yield of lambs. This study lasted 60 days; 24 lambs were distributed randomized (8 lambs / treatment) to one of three isoprotein and isoenergetic diets (15% CP; 11 MJ ME / kg DM). The protein source was the only factor of diet variation. The control diet (SBM) included 150 g of SBM / kg DM, which was replaced by FM or TM meal in the other diets, Lambs were slaughtered (NOM-033-SAG / ZOO-2014). the slaughter body weight (SBW, kg), hot carcass weight (HCW, kg) and used to calculate the hot carcass yield (HCY%). Back fat (mm) was measured at the level of the 10th rib. After 24 hours post slaughter, cold carcass weight (CCW,%) cold carcass yield (CCY,%) pH and temperature (°C) were recorded. Meat colour (CIELAB) was evaluated on Longissimus dorsi calibrated to a standard white tile (L*) luminosity. Data were analysed for a completely randomized design, significant effects were declared at $P<0.05$. Slaughter body weight (37.7±1 kg), hot carcass yield % (45.2±0.5), cold carcass yield % (43.9±0.5), fat thickness at 10th rib (2.9±0.3), pH (6.75±0.2), L*(44.5±0.2) and temperature (11.86±0.05) were similar ($P>0.05$) between treatments. Lower carcass weights ($P<0.07$) were registered for lambs feeding TM (36.04 kg) compared to SBM (40.15 kg) diet. The lower dry matter intake (1,163 vs 1,343 g/d) of the TM diet compared with SBM diet, may have been due to the chitin content present in the exoskeleton resulting in lower HCW and CCW. TM meal is an alternative protein source that can substitute SBM or FM without compromising meat quality.

Impact of adult density and temperature on oviposition and hatchability of *Alphitobius diaperinus*

G. Baliota[1], N. Ormanoğlu[2], C. Rumbos[1] and C. Athanassiou[1]
[1]Laboratory of Entomology and Agricultural Zoology, Department of Agriculture, Crop Production and Rural Environment, Phytokou str., 38446 Volos, Magnesia, Greece, [2]Ankara University Faculty of Agriculture, Plant Protection Department, Ziraat Mah., Şehit Ömer Halisdemir Bulvarı, 06110 Dişkapi – Ankara, Turkey; mpaliota@agr.uth.gr

Knowledge of factors affecting *Alphitobius diaperinus* oviposition and hatchability is crucial in terms of industrialization of the mass production of this species for feed. In this context, the objective of the present study was to evaluate, in laboratory trials, the effect of adult density and temperature on the oviposition performance and hatchability of *A. diaperinus*. In a first series of bioassays, we assessed the impact of adult density on female fecundity and larval hatchability. Briefly, adult groups of different size (from 4 to 28 unmated adults at 1:1 sex ratio) were placed in oviposition arenas (44.1 cm^2 bottom surface area) resulting in different adult densities. Eggs laid at 30 °C and 55% relative humidity were continuously recorded every 3 days for a period of 45 days. In a second series of bioassays, the female oviposition and larval hatchability of *A. diaperinus* was assessed at different temperatures, i.e. at 20, 25, 30 and 32 °C. Overall, the results of the present study indicate that the number of eggs laid per adult was not affected by adult density, at least for the density levels tested here. Moreover, in the same bioassay larval hatchability was reduced with the increase of the observation period. Additionally, we found that temperature affected the oviposition output, as well as larval hatchability. Indicatively, significantly less eggs were laid at 20 °C compared with the rest of the temperatures tested. These data can be further utilized to optimize the production of *A. diaperinus* in mass rearing protocols. This research has been co-financed by the European Regional Development Fund of the European Union and Greek national funds through the Operational Program Competitiveness, Entrepreneurship and Innovation, under the call RESEARCH-CREATE-INNOVATE (project code: T2EDK-01528).

Session 07 — Theatre 11

Aminoacid content of *Tenebrio molitor* L. fed on different diets

F. Carvão[1], L.M. Cunha[2], G. Pereira[3], D. Monteiro[4], M. Almeida[4], L.M.M. Ferreira[5], I. Barros[5], I. Gouvinhas[5] and M.A.M. Rodrigues[5]

[1]UTAD, Department of Animal Science, Universidade de Trás-os-Montes e Alto Douro, Quinta de Prados, 5000, Átrio do Edifício Ciências Agrárias, UTAD. QTA de Prados, 5000-801, Portugal, [2]GreenUPorto, Centro de Investigação em Produção Agroalimentar Sustentável, Campus de Vairão, Edifício Ciências Agrárias, Rua da Agrária, no. 747, 4485-646 Vila do Conde, Portugal, 4485-646 Vila do Conde, Portugal, [3]Portugal Bugs, Rua do Rosmaninho 213, 4455-551 Perafita, Portugal, [4]UTAD, Veterinary and Animal Research Centre (CECAV), Átrio do Edifício Ciências Agrárias, UTAD. QTA de Prados, 5000-801, Portugal, [5]UTAD, Centre for the Research and Technology Agro-Environmental and Biological Sciences (CITAB), Quinta de Prados, 5000-801, Portugal; f.carvao95@hotmail.com

Although animal feed industry is increasingly using alternative animal protein sources, detailed information on its nutritional value is still scarce and diets normally supplied to these animals should be carefully evaluated so that possible effects on the chemical composition of larvae can be detected. This study aimed to evaluate how different levels of incorporation of wheat bran (65-95%), alfalfa (0-30%) and brewer's yeast (0-8.5%) could influence amino acid composition of mealworms according to a Central Composite Design using 10 diets. Amino acid composition of mealworms was analysed by principal component analysis (PCA). aspartate+asparagin and glutamate+glutamine were the most abundant amino acids for all the diets representing 9.1 and 12.5% (DM), respectively. Diet containing 25.6% of alfalfa, 1.5% of yeast and 72.9% of wheat bran showed the highest values of amino acids (90.3%; DM) compared to the medium value of the remaining diets (61.4%; DM). PCA analysis did not allow a clear distinction between diets indicating a tendency for higher amino acid concentration in diets containing higher levels of alfalfa. The inclusion of brewer's yeast did not affect amino acid composition of mealworms. However, an interaction response might have occurred according to the different levels of substrates in the diets. In this way, future trials should be planned testing the inclusion of one substrate at the time.

Session 07 — Poster 12

Performance of slow-growing broilers fed *Acheta domesticus* meal during the first four weeks of life

J. Nieto[1], J. Plaza[1], J.A. Abecia[2], I. Revilla[3] and C. Palacios[1]

[1]Universidad de Salamanca, Av. Filiberto Villalobos, 119, 37007, 37007, Spain, [2]IUCA. Universidad de Zaragoza, Miguel Servet, 177, 50013, Spain, [3]Universidad de Salamanca, Avenida Requejo 33, 49022, Spain; carlospalacios@usal.es

Insects are emerging as a sustainable alternative for poultry feed to the current monopoly of soybean. Therefore, the objective of this study was to assess the development of slow-growing chicks fed *Acheta domesticus* meal as a total replacement for soybean meal during the first 4 weeks of life, using isoproteic and isoenergetic diets. A total of 128 one-day-old male slow-growing chickens (RedBro), with a live weight (LW) of 36.27±1.88 g on average, were used in this trial. Once in the experimental facilities, they were homogeneously divided into two groups (n=64 chicks each): (1) the control group (C) was fed soybean meal as the main protein source; and (2) the *Acheta* group (AD) was fed *A. domesticus* meal. Each group was divided in turn into eight replicates of eight birds each. During the whole experiment, the following parameters were weekly measured: live weight (LW), feed intake (FI), water intake (WI), body weight gain (BWG), as an indicator of growth rate, and feed conversion ratio (FCR), which indicates the relationship between feed intake and weight gain of the animal. LW was significantly higher ($P<0.01$) in group C than in group AD from day 15 to day 29 (353.7±25.3 vs 291.4±21.0 g/chick). BWG showed significantly higher results ($P<0.05$) for chicks in group C from the 2nd week onwards. Mean BWG for the whole procedure was 11.3±0.9 vs 9.1±0.7 g/chick for groups C and AD, respectively. FI was significantly higher ($P<0.05$) in group C than in group AD throughout all the experiment, except for the 4th week. Total FI was significantly higher ($P<0.01$) in group C than in group AD (896.9±64.5 vs 766.3±54.9 g/chick). WI exhibited significant differences in the 2nd and 4th week ($P<0.01$), with higher values for the C group. Regarding FCR, significant higher values ($P<0.05$) were obtained for AD groups during the 1st week. However, in the 2nd and 4th, it was group C the one that showed significant higher FCR values ($P<0.01$). In conclusion, the complete replacement of soybean meal by *A. domesticus* meal in slow-growing chicks caused a decrease in their performance during the first 28 days of life.

ADVAGROMED: basis and structure of PRIMA 2021 project

L. Gasco[1], C. Athanassiou[2], S. Smetana[3], F. Montesano[4], M. Nouri[5], R. Rosa-Garcia[6] and D. Murta[7]
[1]University of Turin, largo P. Braccini 2, 10095, Italy, [2]University of Thelassy, Phytokou str., 38443, N. Ionia, Volos, Greece, [3]German Institutes of Food Technologies, Prof.-von-Klitzing-Str. 7, 49610 Quakenbrück, Germany, [4]National Research Council, Via Amendola, 122/O, 70126 Bari, Italy, [5]University of Sultan Moulay Slimane, Avenue Mohamed V, B 591, Beni Mellal, Morocco, [6]Servicio Regional de Investigación y Desarrollo Agroalimentario, Ctra. AS-267, PK 19, 33300, Villaviciosa, Spain, [7]Ingredient Odyssey Lda, R. Eng. Albertino Filipe Pisca Eugénio 140, 2005-079, Portugal; laura.gasco@unito.it

In intensive agricultural and animal production systems, loss of natural habitats by conversion to agricultural land and extensive use of chemical inputs (fertilizers and pesticides) are the major causal agents of biodiversity loss and decreased habitat heterogeneity, as well as widespread contamination of ecosystems. Integrating agroecological practices in current agricultural farming systems could offer a sustainable means to conserve and enhance the endangered farming biodiversity and increase ecosystems services. The ADVanced AGROecological approaches based on the integration of insect farming with local field practices in MEDiterranean countries project aims to develop innovative and holistic food system based on agro-ecological principles and circular economy practices, to increase the resilience of the agro livelihood systems. ADVAGROMED uses by-products from local agricultural productions for rearing insects to deliver different products: (1) insect frass to be used as bio-product to improve soil fertility, deliver plant protection effects and enhance soil microbial biodiversity, by reducing mineral fertilizers and chemical pesticides; and (2) live larvae to feed local poultry breeds ensuring optimal animal performances, health and product quality (decreasing the use of imported feeds). Biodiversity is promoted at various levels, i.e. at farm level, by promoting the genetic variability of local crops and varieties/animal breeds, but also at a regional level by minimizing the negative impact of chemical inputs on the microfauna through the exploitation of insect frass as bio-products for sustainable soil fertilization and plant protection.

SensiBug – the use of edible insects in the veterinary nutrition of companion animals

R. Gałęcki
University of Warmia and Mazury in Olsztyn, Department of Veterinary Prevention and Feed Hygiene, Faculty of Veterinary Medicine, Oczapowskiego 13, 10-719, Poland; remigiusz.galecki@uwm.edu.pl

Companion animals generate a total food requirement worth of EUR 21 billion in Europe. The number of cases of diet-dependent enteropathy in these animals is increasing. As part of the assessment of the possibility of using mealworms (*Tenebrio molitor*) in Poland, there is a need to develop dog food recipes with the use of mealworms protein and to conduct research on their impact on the health of pets. Such dog food, apart from high nutritional value, should be distinguished by its hypoallergenic nature, confirmed by reliable scientific research. The aim of the project is to develop recipes for dog food formula with the use of insect protein and to evaluate its influence on the symptoms of diet-dependent enteropathies. Design innovation is part of 3 out of 5 Schumpeter cases, including: sourcing poorly used raw materials, creating a new market, and creating a new product. The implemented formula is innovative with novel functionality. The dimension of the changes created by the implementation was assessed as incremental with breakthrough elements and may also cause strategic reorientation, potentially influencing the shaping of the insect rearing sector. The novelty of the invention was rated as creative with imitation elements. The scale of the complexity of the changes was defined as related because the possibility of using mealworms by the companies producing dog foods will have a positive impact on the development of the insect rearing industry. The pro-ecological aspect of the project is important because the planned activities will lead to the upcycling of by-products of the agri-food industry. The final result of the project will be a recipe of an insect protein-based food for specialized dog nutrition. Research supported by a project 'Lider XII' entitled 'Development of an insect protein food for companion animals with diet-dependent enteropathies'; financed by the National Centre for Research and Development (NCBiR) (LIDER/5/0029/L-12/20/NCBR/2021). Publication financially co-supported by Minister of Science and Higher Education in the range of the program entitled 'Regional initiative of Excellence' for the years 2019-2022, Project No. 010/RID/2018/19, amount of funding 12.000.000 PLN.

Artificial senses for quality evaluation of *Sparus aurata* diet containing *Hermetia illucens* meal

A.R. Di Rosa, M. Oteri, F. Accetta, D. Aliquò and B. Chiofalo
University of Messina, Department of Veterinary Sciences, Viale Palatucci, 98168 Messina, Italy; marianna.oteri@unime.it

Hermetia illucens meal (HIM) in fish diet represents a way to achieve more sustainable food production; however, its acceptability must be considered, as fish have highly developed chemical and chemosensory signalling systems due to their living in an aquatic environment. With the aim of evaluating the effects of HIM in partial substitution of fish meal (FM) on sensorial quality of diets formulated for *Sparus aurata* L. farmed offshore, a sensor-based instruments platform consisting of E-eye, E-nose with 18 MOS sensors and a potentiometric E-tongue with 7 chemical sensors was used to evaluate colour, volatile profile and taste of the HIM and of the diets, one (HIM0) with fish meal as exclusive protein source of animal origin, and the other (HIM35) containing HIM as a partial replacement (35%) for fish meal. A Principal Component Analysis (PCA) of the e-sensing data was performed. Successively, data provided by the e-senses platform was combined to improve the discrimination capability. An intermediate fusion level was adopted. The sensors and colour codes with the highest discrimination power were chosen; in particular, data of 5 E-nose sensors (LY2/G, LY2/gCTL, T30/1, P30/1, P40/2), those of 3 E-tongue sensors (AHS, CTS, NMS) and those of 8 colours extracted from E-eye (1,619, 1,636, 1,877, 1,892, 2,149, 2,165, 2,182, 2,712) were chosen. Datasets were reduced, due to the different data size, and a new PCA was performed. On the basis of Mahalanobis distance, the pattern discrimination index (PDI) between the two diets was calculated. The two PCs explained 99.99% (PC1=96.25%; PC2=3.75%) of the total variance. The PCA map showed HIM0 on the right of the graph and HIM on the left. HIM35 ranks between the two groups, closest to HIM0. The Mahalanobis distance was higher ($P<0.001$) between HIM0 and HIM35 and PDI was 54.53%, highlighting a clear effect of the HIM inclusion on the characteristics of HIM35 diet, mostly in terms of colour and taste. Artificial senses permitted distinguishing colour, odour and taste between the diets, proving to be powerful tools for assessing the different organoleptic profiles of diets. Research supported by the project 'FIFA-Feed Insects For Aquaculture', PO FEAMP 2014-2020 mis. 2.47 CUP J46C18000570006.

CELLOW-FeeP project: growth performance of poultry fed live black soldier fly larvae

C. Tognoli[1], S. Bellezza Oddon[2], M. Renna[3], I. Biasato[2], A. Schiavone[3], S. Cerolini[1], L. Gasco[2] and L. Zaniboni[1]
[1]University of Milan, Department of Veterinary Medicine and Animal Science, via dell'Università 6, 26900 Lodi, Italy, [2]University of Turin, Department of Agricultural, Forest and Food Science, largo P. Braccini 2, 10095, Italy, [3]University of Turin, Department of Veterinary Sciences, largo P. Braccini 2, 10095, Italy; laura.gasco@unito.it

The use of insects in animal feed has a great potential in poultry production and could help to mitigate negative impacts of soybean production. In the frame of the CELLOW-FeeP (Circular Economy: Live Larvae recycling Organic Waste for rural Poultry) project, black soldier (*Hermetia illucens*) live larvae (BSFL) were produced on a local organic waste-based diet, and fed to 14 days-old intermediate-growing chickens (Label Naked Neck, individually labelled) for 68 days. Experimental treatments were as follow: (1) control diet (C), commercial feed; (2) HI10, C + BSFL corresponding to 10% of the expected daily feed intake (DFI); and (3) HI20, C + BSFL corresponding to 20% of DFI. The C diets was distributed *ad libitum* in all the treatments. All treatments were randomly allocated to 4 pens (7 birds / pen) and data on feed ingestion recorded. Birds were individually weighted (BW) at the beginning of the trial, every two weeks, and at the end of the trial (82 days of age). Feed intake (FI) and feed conversion ratio (FCR) were calculated. Analysis of variance was performed using the software SAS 9.4 ($P<0.05$). Diet and age affected all the growth parameters ($P<0.001$). Moreover, FCR was affected by the diet × age interaction ($P=0.024$), with the lowest value being recorded in the HI20 treatment from 21 to 82 days of age. Diet affected the final BW ($P<0.001$), with birds fed HI20 diet showing the highest value. The CELLOW-FeeP research (Project n. 2019-1944) received the support of Fondazione Cariplo (Italy).

Session 07 Poster 17

NETA – new strategies in wastewater treatment

I. Rehan[1], M. Carvalho[2], D. Murta[3,4], I. Lopes[3,4] and O. Moreira[1,5,6]
[1]INIAV – Polo de Inovação da Fonte Boa, Fonte Boa, Vale de Santarém, 2005-048, Portugal, [2]Instituto Politécnico de Beja, Beja, 7800-000, Portugal, [3]CiiEM- Multidisciplinary Research Center of Egas Moniz, Monte de Caparica, Portugal, Monte de Caparica, 2829-511, Portugal, [4]EntoGreen, Ingredient Odyssey SA, Santarém, 2005-079, Portugal, [5]Associate Laboratory for Animal and Veterinary Sciences (AL4AnimalS), Vila Real, 5000-801, Portugal, [6]CIISA – Centre for Interdisciplinary Research in Animal Health, University of Lisbon, 1300-477, Portugal; iryna.rehan@iniav.pt

Water is the essence of life and it is a human right. However, three out of ten people do not have access to potable water, and six out of ten people do not have access to safely managed sanitation services. Agriculture (including irrigation, livestock, and aquaculture) is by far the largest water consumer, accounting for 69% of annual water withdrawals globally. Agriculture activities produce a big quantity of waste, including, wastewater that can be valorised and transformed into new products. NETA (supported by Portugal 2020 – SI IDT – National Agency of Innovation, S. A.), aims to apply a Chemical Precipitation Technique (CPT) in livestock wastewater, producing new products and promoting a circular economy. Now, CPT is successfully applied on a lab-scale and the aim is to apply TPQ to TRL-6 at the Portuguese Research Station of Animal Production (EZN). EZN is a farm with 204 ha with experimental units of pigs, cattle, sheep and goats production, slaughterhouse and different laboratories. EZN acting as an open air laboratory produces many different types of wastewaters and will be the perfect place to apply CPT. The treated water with CPT will be tested in aquaponics studies and the precipitated sludge's will be valorised agronomically after remediation with black soldier fly (BSF) larvae. The produced forage will be used in a case study with growing lambs. The reared larvae, will be directed to bio refinery. This project will contribute to close nutrient cycles in agriculture, promoting a circular economy and zero waste politics. Acknowledgments The present work was funded under the scope of NETA – New Strategies in Wastewater Treatment – POCI-01-0247-FEDER-046959.

Session 07 Poster 18

Effect of dietary yellow mealworm (*Tenebrio molitor*) in broilers performance and MAPKs activation

S. Vasilopoulos[1], I. Giannenas[1], S. Savvidou[2], E. Bonos[3], C.I. Rumbos[4], N. Panteli[5], C. Tatoudi[5], I. Voutsinou[5], E. Antonopoulou[5] and C. Athanassiou[4]
[1]Aristotle University of Thessaloniki, Veterinary Medicine, University Campus, Thessaloniki 54636, Greece, [2]Hellenic Agricultural Organisation DEMETER, Animal Science, Paralimni Giannitson, 58100 Pella, Greece, [3]University of Ioannina, School of Agriculture, Kostakioi Arta, 47100, Arta, Greece, [4]University of Thessaly, Agriculture, Crop Production and Rural Development, Phytokou str., 38443 Volos, Greece, [5]Aristotle University of Thessaloniki, Zoology, University Campus, Thessaloniki 54636, Greece; igiannenas@vet.auth.gr

The use of insects as poultry feed gained significant interest recently. Insects are regarded as a very promising development in feed production, potentially serving as a viable alternative to increasingly fluctuating traditional commodities. Mitogen-activated protein kinases (MAPK) are key components of signal transduction pathways that regulate pivotal cell processes, such as proliferation and apoptosis, induced by several extracellular stimuli including nutrient availability. The trial examined the dietary use of *Tenebrio molitor* larvae meal (LM) on performance indices and cellular responses in the breast. In a 35 day trial, 120 one-day-old mixed Ross-308 broiler chicks were randomly allocated to 3 groups with 4 replicates, housed in floor pens with wheat straw litter. Commercial breeding, management and vaccination procedures were employed. Control Group A (CL) was fed basal diet (based on maize and soybean meal) in mash form, without anticoccidials or antibiotics. The diets of Group B and C were further supplemented with 5% (LM5) and 10% (LM10) of LM respectively. Body weight (BW), feed intake (FI), average daily body weight gain (ADG) and feed conversion ratio (FCR) during the period of 1-10d, 11-24d, 25-35d and 1-35d were evaluated. Cellular responses to dietary inclusion of *T. molitor* were addressed through the phosphorylation of p38 and ERK-1/2 MAPKs using Western blot analysis. Results showed that on day 35, BW and FCR values did not differ among the groups; however, FI and ADG were lower in group B. Dietary inclusion of *T. molitor* induced significantly, in comparison to the CL, the activation of p38 MAPK in a dose-depended manner. Acknowledgements: Co-financed by GR and EU (T2EΔK-02356, InsectFeedAroma)

Session 08 Theatre 1

30 years of research on slaughter conditions of livestock in France: ex-post impact evaluation

C. Moulin, B. Ducreux, V. David and L. Clavel
Institut de L'elevage, 149 Rue de Bercy, 75595 Paris Cedex 12, France; christine.moulin@idele.fr

Understanding the impacts of research studies on the slaughter conditions of livestock in France, identifying and then promoting levers in order to improve them, are major challenges today, especially in the actual context of growing social issues about animal welfare. But, it is also an important concern for the various actors of the sector, seeking to enhance general working conditions. In this paper, we will question these essential issues through an original methodological approach: the 'a posteriori impact analysis'. The 'impact pathway' will be clarified by the 'mapping of the actors' (a representation of the actors' ecosystem) emphasised by a focus on the direct beneficiaries of the results of the various research studies and tools developed through the thirty years analysed. The 'chronology of innovation' will bring out the main facts that have marked the history of this theme. Thereby this paper aims to demonstrate how the impact analysis sheds light on the facilitating events in the development of the thematic, but also the intrinsic or suffered locks inducing difficulties to concretize the emergence of actions and tools, slowing down the deployment by the relay actors. The 'impact story' synthesized during this ex-post analysis will also explain the expected transformations affecting the various direct and indirect recipients of the research studies. Thus, this article will present impact pathways producing additional operational knowledge and partnership organizations able to promote the deployment of appropriate solutions; and therefore likely to generate the necessary and expected transformations within the livestock sector. The outputs and outcomes of this research were first useful to 'researchers' and 'advisers', who needed methods and objectified references. The results were gradually being deployed towards slaughterhouse staff, livestock transport truck drivers and also stockbreeders, thanks to the 'relay actors'.

Session 08 Theatre 2

Main animal welfare issues at transport and slaughter in Germany, according to stakeholders

R. Magner[1], C. Over[2], C. Gröner[2], J. Johns[2], A. Bergschmidt[2] and U. Schultheiß[1]
[1]Association for Technology and Structures in Agriculture (KTBL), Bartningstraße 49, 64289 Darmstadt, Germany,
[2]Thünen-Institute of Farm Economics, Bundesallee 63, 38116 Braunschweig, Germany; r.magner@ktbl.de

The debate about animal welfare during transport and slaughter increased over the last decade in Germany, but discussions are still characterised by a lack of information. As a consequence, the effects of animal welfare politics are difficult to measure, farmers have no benchmarking opportunities and society lacks an objective data source. Therefore, the project national animal welfare monitoring will suggest indicators to measure the welfare of cattle, pigs, poultry, sheep and goats as well as rainbow trouts and carp during husbandry, transport and slaughter. To include stakeholders' perceptions regarding the choice of indicators for transport and slaughter, 63 representatives of interest groups from agriculture, transport and slaughter, food industries, NGOs, veterinarians, politics, churches, and administration were interviewed in guided telephone interviews. We analysed the transcribed data by qualitative content analysis. For transport, the most frequently suggested indicator was duration of transport (n=41), followed by international transports due also to their duration and partially uncontrollable conditions for the animals (n=20). Other suggested criteria were, in decreasing frequency: climate conditions, space per animal, food and water supply, handling, suitability of the vehicle, litter, DOA (dead on arrival) and injuries. For slaughter, the most frequently mentioned indicator was the effectiveness of stunning (n=19), followed by speed of assembly line, handling and vocalisation of animals, presence of pregnant animals, presence of cameras, design of waiting areas, and waiting time before slaughter. Drivers' and slaughterhouse workers' certificate of competence regarding animal welfare was seen as an important indicator because it is positively connotated, different to e.g. DOA which is assumed to be generally perceived as negative by the public since it refers to death. Respondents stated a difficulty to cover affective states and natural behaviour during transport and slaughter.

Deck height during transport of weaners – piglet height and microclimatic conditions inside trucks

L. Foldager[1,2], M. Kaiser[2], G. Chen[3], L.D. Jensen[2], J.K. Kristensen[4], C. Kobek-Kjeldager[2], K. Thodberg[2], L. Rong[3] and M.S. Herskin[2]

[1]Aarhus University, Bioinformatics Research Centre, Universitetsbyen 81, 8000 Aarhus, Denmark, [2]Aarhus University, Dept. of Animal Science, 8830 Tjele, Denmark, [3]Aarhus University, Dept. of Civil and Architectural Engineering – Design and Construction, Inge Lehmanns Gade 10, 8000 Aarhus, Denmark, [4]Aarhus University, Dept. of Electrical and Computer Engineering, Blichers Allé 20, 8830 Tjele, Denmark; leslie@anis.au.dk

This study was part of a policy support request commissioned by the Ministry of Environment and Food of Denmark, focusing on deck height when transporting weaner pigs on long journeys. Former recommendations on deck heights have been based on an equation from Vorup and Barton-Gade (1991) relating live weight and average height based on 87 pigs (25-160 kg) and 21 sows (130-260 kg), including only 16 pigs in 26-40 kg. Part 1 of the study determined height of pigs weighing approx. 5-40 kg and to establish new or confirm validity of old equations predicting height from live body weight in this interval. Prediction equations were built using 1,435 pigs (47% castrates, 53% females) from 9 Danish herds and validated on 179 pigs (54% castrates) from a 10th herd. The old equation systematically overfitted the height of these pigs, whereas an equation developed by Condotta *et al.* (2018) using 150 pigs (age 4-20 weeks) predicted equally well as models from the present study. Part 2 of the study examined microclimatic conditions during transport of pigs weighing 20-25 kg inside two decks on the truck (70/90 cm height) and two on the trailer (60/80 cm). Sixteen drives for each of two durations (8/23 h) were run in Denmark from June 2021 to March 2022, ensuring transport under varying weather conditions. Data consisted of GPS tracking (speed/position), opening of shutters (cm), use of mechanical ventilation (0/1), weather information, temperature from two sensors outside truck and trailer, and temperature, relative humidity and CO_2 from sensors placed inside each of the 12 compartments. As per contract, results are not allowed to be revealed before handing in the advisory report to the ministry in July. Results from analyses of relations between deck height, microclimate and external factors will be presented at EAAP.

The metabolomic fingerprint of pigs' exposure to acute stress

L. Morgan[1,2], R.I.D. Birkler[2], S. Shaham-Niv[2], Y. Dong[2], L. Carmi[3], T. Wachsman[2], B. Yakobson[4], H. Cohen[5], J. Zohar[3], M. Bateson[1] and E. Gazit[2]

[1]Newcastle University, Newcastle, NE1 7RU, United Kingdom, [2]Tel Aviv University, Tel Aviv, 69978, Israel, [3]Chaim Sheba Medical Center, Ramat Gan, 52621, Israel, [4]Ministry of Agriculture and Rural Development, Rishon Lezion, 7505100, Israel, [5]Ben-Gurion University of the Negev, Beer Sheva, 84105, Israel; liat.morgan@mail.huji.ac.il

Handling of pigs pre-slaughter commonly involves a period of time when animals are transported, re-mixed with unfamiliar pen-mates and held in a new environment where they are exposed to the noise and smell of the slaughterhouse. Measuring the physiological changes before and after these procedures, represents an extreme acute stress response. Our objective was to identify the metabolomic fingerprint of the exposure to these acute stressors in pigs. Pooled saliva samples from 200 pigs were obtained non-invasively using cotton ropes provided as environmental enrichment. Samples were collected at two time points: (1) in the familiar environment, as a baseline (n=31); and (2) 24 hours after transport, regrouping and a night at the slaughterhouse (n=32). Untargeted metabolomics was utilized by ultra-high-performance liquid chromatography, high-resolution mass spectrometry. Four different analyses detected 9,327, 9,188, 2,518 and 1,165 metabolite features. Of which, 3,320, 3,271, 243, and 357 metabolite features respectively were significantly different after the stressful exposure, with at least a two-fold change difference (Adj $P<0.05$). These suggest that the exposure of pigs to acute stress change their physiology extensively, and potentially has implications on their well-being. The results highlight the potential of metabolomics as a promising future in animal welfare research. Moreover, it provides potential prophylactic candidates as treatments that may allow pigs to cope better with stress.

Inter-observer repeatability of indicators of consciousness after waterbath stunning in turkeys

A. Contreras[1], A. Varvaró-Porter[1], V. Michel[2] and A. Velarde[1]
[1]Institute of Agrifood Research and Technology (IRTA), Animal welfare program, Finca Camps i Armet s/n, 17121 Monells, Spain, [2]French Agency for Food, Environmental and Occupational Health & Safety, Direction de la stratégie et des programmes, 14 Rue Pierre et Marie Curie, 94701 Maisons-Alfort, France; alexandra.contreras@irta.cat

One of the main challenges in monitoring the state of consciousness in turkeys after waterbath stunning (WBS) is the selection of animal-based indicators (ABIs) ensuring consistency of controls. ABIs should be valid, feasible and repeatable. The validity and feasibility of ABIs for the state of consciousness in poultry have been assessed by EFSA (2013). Thus, the main goal of the study was to assess the inter-observer repeatability of the most valid and feasible ones after WBS in turkeys both before bleeding (i.e. tonic seizure, breathing, spontaneous blinking and vocalisation) and during bleeding (i.e. wing flapping, breathing, spontaneous swallowing and head shaking) and the correlation among them. This study compares the assessment of four observers in 7,877 turkeys from 24 batches of eight different slaughterhouses. Before bleeding, the most repeatable ABI was vocalisation followed by spontaneous blinking. However, both were artificially highly repeatable as hardly ever were observed. On the other hand, absence of tonic seizure is not correlated to other ABI before bleeding probably because tonic seizure occurred in some birds while the bird was still in the waterbath. Therefore, it seems difficult to rely on the absence of tonic seizure to assess consciousness. Thus, we recommend focusing on presence of breathing as indicator of consciousness. During bleeding, the most repeatable ABI was spontaneous swallowing followed by head shacking, breathing and wing flapping. However, spontaneous swallowing is artificially repeatable as was not observed in any turkey assessed. Therefore, we recommend focus on presence of breathing, head shaking and wing flapping assessment although less repeatable. Sometimes turkeys showed simultaneously more than one outcome of consciousness being breathing and head shaking and breathing and wing flapping the most observed combinations. This work will serve at proposing a refined list of ABIs so can be used to assess the consciousness of turkeys in commercial slaughterhouses.

Using water pumps as an environmental enrichment method in rainbow trout (*Oncorhynchus mykiss*)

A. Martínez[1], A. De La Llave-Propín[2], J. De La Fuente[1], C. Pérez[3], M. Villarroel[2], M.T. Díaz[1], A. Cabezas[1], E. González De Chavarri[1] and R. Bermejo-Poza[1]
[1]Universidad Complutense de Madrid, Animal Production, Avenida Puerta de Hierro s/n, 28040 Madrid, Spain, [2]Escuela Técnica Superior de Ingeniería Agronómica, Alimentaria y de Biosistemas, Universidad Politéc, Agrarian Production, Avenida Complutense 3, 28040 Madrid, Spain, [3]Universidad Complutense de Madrid, Animal fisiology, Avenida Puerta de Hierro s/n, 28040 Madrid, Spain; andrea.martinezvi@hotmail.com

Environmental enrichment is defined as a deliberate increase in rearing environmental complexity with the aim of reducing maladaptive and aberrant traits in fish reared. Therefore, it can reduce stress response produced during pre-slaughter and provides an improvement of fish welfare. This study was conducted to evaluate the effects of continuous/randomly fired water currents on stress response of rainbow trout. We used 540 rainbow trout (*Oncorhynchus mykiss*), separated into three groups with different occupational enrichment (control or no water pumps - C, and continuous -CT or aleatory -A water pumps) and two durations of pre-slaughter fasting (no fasting - 0D or five days of fasting – 5D, 45.2 degree days). After slaughter, liver/skin colour parameters and liver glycogen concentration were measured. Fish with aleatory water pumps presented a higher liver glycogen concentration than the other groups (A: 159.3±7.35 vs C: 101.2±11.5 and CT: 134.9±17.9 mg/g), suggesting a lower stress response. A higher skin luminosity was observed in C fish compared to the other groups (C: 45.3±0.98 vs A: 41.1±0.66 and CT: 41.7±1.37), possibly due to a change to a paler colour after the higher stress response comparing with A and CT fish. Pre-slaughter fasted trout presented a darker liver colour than no fasted (L*: 41.2±1.11 vs 44.3±1.30; h*: 46.9±1.39 vs 51.8±1.74°; C*: 11.0±0.47 vs 10.2±0.45), possibly due to a decrease in liver lipid concentration because of acute stress response. Based on our results, we can conclude that using water pumps as environmental enrichment method in fish seems to be a good method to reduce stress response during pre-slaughter, highlighting the use of aleatory water pumps better than continuous.

Session 08 Theatre 7

Oral stereotypes and abnormal interactions indicate animal welfare issues in fattening bulls

B. Spindler[1], N. Volkmann[2] and N. Kemper[1]
[1]Institute for Animal Hygiene, Animal Welfare and Animal Behaviour, University of Veterinary Medicine Hannover, Foundation, Bischofsholer Damm 15, 3073 Hannover, Germany, [2]WING, University of Veterinary Medicine Hannover, Foundation, Heinestraße 1, 49377 Vechta, Germany; birgit.spindler@tiho-hannover.de

In cattle, oral stereotypes such as tongue rolling and licking objects in the barn or body parts of conspecifics are described. Besides this stereotypic behaviour, the occurrence of social interactions related to dominance relationships are a common welfare problem in intensive bull fattening. In order to determine the extent to which oral behaviour and social interactions occur in intensive conditions, observations were carried out on a farm housing fattening bulls in large groups. Therefore, three groups with 22 Simmental bulls, each housed in straw-bedded pens, were video recorded at three times (7, 12 and 17 months) during their fattening period of 18 months. Behavioural observations comprised the number of oral stereotypies over 48 hours (scan sampling with an interval of two minutes) and social interactions per group and hour over 16 hours. In addition, three focus animals per group were observed continuously for 16 hours to detect the duration of stereotypic behaviour. The results showed that up to 95% of the bulls exhibited oral stereotypies with a mean of 0.83 actions/animal/hour. Tongue rolling was observed most frequently (95% of the bulls with 0.5 actions/animal/hour), followed by manipulating objects (71% of the bulls with 0.28 actions/animal/hour). Manipulating conspecifics occurred least frequently (47% of the bulls with 0.05 actions/animal/hour). The average duration of oral stereotypies was 25.4 seconds. Moreover, this behaviour occurred in almost all animals. Depending on the age of bulls, head butting was observed with a mean between 23.75 and 29.10 actions/group/hour, and mounting between 1.87 and 5.52 actions. In general, stereotypic behaviour as well as abnormal interactions increased during feeding time. There is an urgent need to analyse the causes of these behavioural disorders and to find ways to reduce their occurrence in intensive fattening bull housing to improve animal welfare.

Session 08 Theatre 8

Prevalence of BRD-related viral pathogens in the upper respiratory tract of Swiss veal calves

E. Studer[1], L. Schönecker[1,2], M. Meylan[1], D. Stucki[1], R. Dijkman[3,4,5], M. Holwerda[3,4,5,6], A. Glaus[3,4] and J. Becker[1]
[1]Vetsuisse-Faculty, University of Bern, Clinic for Ruminants, Bremgartenstrasse 109a, 3012, Switzerland, [2]Vetsuisse-Faculty, University of Bern, Institute of Veterinary Bacteriology, Länggassstrasse 122, 3012 Bern, Switzerland, [3]Vetsuisse-Faculty, University of Bern, Department of Infectious Diseases and Pathobiology, Länggassstrasse 122, 3012 Bern, Switzerland, [4]Vetsuisse-Faculty, University of Bern, Institute of Virology and Immunology, Länggassstrasse 122, 3012 Bern, Switzerland, [5]University of Bern, Institute for Infectious Diseases, Friedbühlstrasse 51, 3001 Bern, Switzerland, [6]University of Bern, Graduate School for Cellular and Biomedical Science, Mittelstrasse 43, 3012 Bern, Switzerland; jens.becker@vetsuisse.unibe.ch

The prevention of bovine respiratory disease is important, as it may lead to impaired welfare, economic losses, and considerable antimicrobial use, which can be associated with antimi- crobial resistance. The aim of this study was to describe the prevalence of respiratory viruses and to identify risk factors for their occurrence. A convenience sample of 764 deep nasopharyngeal swab samples from veal calves was screened by PCR for bovine respiratory syncytial virus (BRSV), bovine parainfluenza-3 virus (BPI3V), bovine coronavirus (BCoV), influenza D virus (IDV), and influenza C virus (ICV). The following prevalence rates were observed: BRSV, 2.1%; BPI3V, 3.3%; BCoV, 53.5%; IDV, 4.1%; ICV, 0%. Logistic mixed regression models were built for BCoV to explore associations with calf management and housing. Positive swab samples were more frequent in younger calves than older calves (>100 days; $P<0.001$). The probability of detecting BCoV increased with increasing group size in young calves. Findings from this study suggested that young calves should be fattened in small groups to limit the risk of occurrence of BCoV. Influenza D viruses were isolated from Swiss veal calves for the first time.

Comparing the impact of two dry cow therapy approaches on udder health

P. Silva Bolona and C. Clabby
Teagasc, Animal and Grassland Research and Innovation Programme, Teagasc Moorepark, P61 C996, Fermoy, Ireland; pablo.silvabolona@teagasc.ie

The increasing concern on antimicrobial resistance has led to the development of regulation by the European Union on the use veterinary medicines, which states that preventive use of antimicrobials should be avoided. Blanket dry cow therapy (treating all quarters of a cow at the end of lactation) has been advised for its role in eliminating existing infections and preventing new infections (preventive use) during the dry period. Different strategies need to be explored to reduce the use of antimicrobials and assess their impact on udder health. This trial was conducted in 3 pasture-based seasonal research herds in Ireland. Quarters of the cows were milk sampled between 1 and 7 days prior to dry-off, cultured for mastitis pathogens and quarter SCC measured. Cows with 1 infected or 1 high SCC quarter were randomly assigned to receive either: (1) antibiotic plus teat seal in the infected quarter (ABTS) while the rest received teat seal (TS); or (2) antibiotic plus teat seal in all quarters. Ninety-six cows were enrolled for the study. Quarter milk samples were collected at calving (1 and 10 DIM) and the impact of treatment on quarter SCC and quarter infection was assessed using a mixed model and logistic regression. Quarter SCC was transformed to log10 (logSCC) for analysis. We did not find bacteria in the quarter samples collected at calving. Quarters treated with ABTS had 0.26 log points lower logSCC than quarters treated with TS. Infection or high SCC (>200,000 or <200,000 cells/ml) at dry off had no impact on logSCC. Quarters treated with TS had a 0.93 higher odds of having a high SCC quarter at calving compared to quarters treated with ABTS. The raw average SCC of quarters treated with TS was 206,000 cells/ml, while for quarters treated with ABTS it was 68,000 cells/ml. Ninety percent of quarters treated with ABTS had an SCC of <120,000 cells/ml while 85% of the quarters treated with TS were below this threshold. The main bacteria present at the dry-off was Staphylococcus aureus. This research reproduced the findings of other Irish research that showed that cows treated with TS had higher SCC in the following lactation. Further research is needed to understand the use of TS only in Irish herds without an impact on udder health.

Objective gait analysis in clinically lame dairy cows

A. Leclercq[1], J. Carlander[1], S. Andersson[1], M. Söderlind[1], N. Högberg[2], M. Rhodin[1], K. Ask[1], F. Serra Braganca[3] and E. Hernlund[1]
[1]Swedish university of agricultural sciences, Department of anatomy, physiology and biochemistry, Box 7011, 750 07 Uppsala, Sweden, [2]Swedish university of agricultural sciences, Department of clinical sciences, Box 7054, 75007, Sweden, [3]Utrecht university, Department of clinical sciences, 3584 CM Utrecht, the Netherlands; anna.leclercq@slu.se

Lameness is common in dairy herds and early detection is essential for timely intervention. However, farmers tend to underestimate the prevalence of lameness in herds, and animals with mild lameness might remain unnoticed. In horses, vertical displacement asymmetry parameters of upper body landmarks are used to detect lameness. We aimed to investigate whether these landmarks are useful to detect lameness in dairy cows. A total of 18 clinically lame dairy cows (forelimb: n=9, hindlimb: n=9) were measured during straight line walk on hard surface using body-mounted inertial sensors attached to the head, withers (forelimb lame cows only) and between the tubera sacrale (TS). Data were transmitted wirelessly to a tablet. A follow-up measurement after treatment was performed for each cow 1-4 months later, when the lameness was visually reduced. Trial means of within-stride difference in vertical displacement minima and maxima (minDiff and maxDiff) were compared between lame condition and follow-up for head, withers and TS sensors. Values from cows with left limb lameness were mirrored to allow group level comparison. In the hindlimb lame cows, TS minDiff values were significantly lower at follow-up than at baseline (median in mm (IQR): 1.7 (10.8) and 12.4 (14.4), respectively, P=0.008, Wilcoxon signed rank test). In the forelimb lame cows, head maxDiff values were significantly lower at follow-up than at lame condition (median in mm (IQR): -0.7 (5.5) and 19.0 (46.2), respectively, P=0.04, Wilcoxon signed rank test). No other differences were found. Preliminary results of this study involving a limited number of animals indicate that vertical displacement asymmetry of upper body landmarks could be used to detect clinical lameness in cows. Further research is needed to investigate biomechanical adaptation strategies in lame cows in order to identify motion parameters that can be used for early lameness detection.

Hoof disinfectants for dairy cows: novel on-farm efficacy testing of footbath and spray application

M.A. Palmer[1], M.J. Garland[2] and N.E. O'Connell[1]
[1]Queen's University Belfast, 19 Chlorine Gardens, BT9 5DL, United Kingdom, [2]Kersia Group, [1]A Trench Rd, Newtownabbey BT36 4TY, United Kingdom; m.a.palmer@qub.ac.uk

Digital dermatitis (DD) is a major cause of lameness for dairy cows and is thought to have a bacterial etiology. DD control regimens often include disinfectant footbaths, however footbaths become contaminated with manure (likely to reduce their efficacy) so some farms now apply disinfectants by spraying. Efficacy tests are vital for product development but laboratory tests cannot match the complexity of organic challenges on farms, hence on-farm tests are needed. This study aimed to develop an on-farm test of hoof disinfectant antibacterial efficacy and use this to examine the efficacy of three products (compared to copper sulphate) when applied after use in a footbath or via spray. The study used two dairy farms, one to test the disinfectants after use in a footbath for 100 cows (the stated limit of use) and the other to test the sprayed products. Products were copper sulphate, P1 (acetic and peracetic acid based), P2 (amine) and P3 (lactic acid and chlorocresol) all at 5% final concentration. Before product application one rear hoof was cleaned and a 5×5 cm pre-treatment swab was taken (sterile swabs with neutraliser). The product was applied - used footbath solution was poured on the area for 2×4 s to simulate two foot dips, or product was sprayed for 5 s. After 5 min a post-treatment swab was taken. A mean of 38 swab pairs were analysed per product and application type. Total aerobic colony count was determined after 24 h incubation and log reduction after treatment was determined. Comparison of results was by one-way ANOVA (post-hoc comparisons with Bonferroni correction) in SPSS. For the used footbath solution the mean (±SE) log reduction caused by copper sulphate (1.02±0.12) was larger than P2 (0.40±0.10; $P<0.05$) but not different to P1 (1.15±0.19) or P3 (0.71±0.13). For the spray the log reduction caused by copper sulphate (0.87±0.10) was smaller than P1 (1.59±0.10; $P<0.05$) but not different to P2 (0.93±0.09) or P3 (0.92±0.09). This practical testing method can discriminate products that performed better or worse than copper sulphate under farm conditions and provide useful information on the relative performance of the products when different application methods are used.

Simplified welfare assessment method for young bulls in pens

A. Cheype[1], M. El Jabri[2], B. Mounaix[2], C. Mindus[2], A. Aupiais[2], C. Dugué[3], L.-A. Merle[4] and X. Boivin[5]
[1]Institut de l'Elevage, boulevard des Arcades, 87000 Limoges, France, [2]Institut de l'Elevage, 149 rue de Bercy, 75012 Paris, France, [3]France Limousin Sélection, Pôle de Lanaud, 87220 Boisseuil, France, [4]Ferme des Etablières, route du Moulin-Papon, 85000 la Roche-sur-Yon, France, [5]Université Clermont Auvergne, INRAE, VetAgroSup, UMR 1213 Herbivores, 63122 Saint-Genès Champanelle, France; agathe.cheype@idele.fr

For young bulls fattened in pens, welfare evaluation protocols are currently limited. One of the objectives of the BeBoP project is to design a simplified welfare assessment method for young bulls housed in pens that is accurate, practical for routine evaluation, riskless for animals and evaluators. The first step was to agree on indicators and a methodological framework based on existing scientific literature but also with the help of two focus-groups ran with professionals. This simplified method is designed with 3 types of observation: (1) overall observations of the fattening surrounding and pens to describe indicators such as overall emotion state, posture and activities; (2) individual observations in each pen to evaluate animal-human relationship, stress behaviour, body and health conditions; (3)/ indicators of performances (mortality and growth). This simplified protocol was used on 150 young bulls between 9 and 18 months of age in 2 experimental farms. The records were compared with those collected with a 'gold standard' method where the same animals were evaluated in individual condition. All measures were simultaneously recorded by two pairs of trained observers on two consecutive days. After training, moderate to good inter and intra-observer reliability (kappa coefficients K) for each welfare indicators were found for both simplified and 'gold standard' methods. Additionally, the comparison between the two protocols will be presented. This promising simplified protocol will be tested on 30 French commercial farms for practicability and prevalence for each indicator.

Session 08 Theatre 13

High diversity of meat-producing wild game species in Europe – a One Health challenge

D. Maaz[1], A. Mader[1] and M. Lahrssen-Wiederholt[2]
[1]German Federal Institute for Risk Assessment (BfR), Safety in the Food Chain, Max-Dohrn-Straße 8-10, 10589 Berlin, Germany, [2]Federal Ministry of Food and Agriculture (BMEL), Wilhelmstraße 54, 10117 Berlin, Germany; denny.maaz@bfr.bund.de

A considerable proportion of hunted wild game species serve for human consumption. Whether and to what extent these wild animals can come in contact with environmental contaminants and zoonotic pathogens in their habitat also depends on the characteristics of the animal species. To estimate the number and diversity of meat-producing game species in Europe, species that can be legally hunted in Europe were listed on the basis of hunting laws and regulations of the states, while species were identified that were mainly hunted for meat consumption. In addition, hunting bag data for the hunting year 2018/19 were collected from 30 countries, which reflect 94% of the area of Europe (excluding Russia). A total of 114 meat-producing species are legally huntable in Europe. According to bag data, animals of at least 93 of these species – 73% of them feathered game – were hunted in 2018/19, belonging to 21 families. The total number of individuals was 88.3 million meat-producing wild animals in Europe, including the most important species (>2 million individuals each) roe deer and wild boar, as well as wild rabbit, pheasant, wood pigeon, red partridge, song thrush and other thrush species, mallard and quail. The United Kingdom, France and Spain exhibited the highest hunting bags (in total 72% of Europe's bag), mainly due to the intense bird hunting. According to the respective mean weights of the species and the proportion of usable meat of normally 60% of the body weight, the equivalent of 306,000 t game meat was produced. However, the meat of the majority of species is not placed in the market but primarily consumed by hunter families or parties related to them. Due to the high weight in comparison to birds, ungulates represent 85% of the total game meat. Germany and France together accounted for 45% of the hunted ungulates in Europe. Although, the game meat chain is small, the high diversity of the species consumed by hunters elucidates the complexity regarding zoonotic pathogens and contaminants that must be taken into account in terms of One Health and food safety research.

Session 08 Poster 14

Time profiles of energy balance in dairy cows and its relation to metabolic status and disease

F. Vossebeld[1,2], A.T.M. Van Knegsel[1] and E. Saccenti[2]
[1]Adaptation Physiology group, Wageningen University, De Elst 1, 6700 AH Wageningen, the Netherlands, [2]Laboratory of Systems and Synthetic Biology, Wageningen University, Stippeneng 4, 6708 WE Wageningen, the Netherlands; ariette.vanknegsel@wur.nl

Due to a combination of a low energy intake and a high demand of energy required for milk production, dairy cows experience a negative energy balance (EB) at the start of lactation. This energy deficit causes bodyweight reduction and an increased risk for metabolic diseases. Severity and length of negative EB can differ among cows. A common approach to describe and compare EB between cows or treatments is based on averages per time period in the peripartum period. Time profiles for the EB peripartum with corresponding metabolic status and disease treatments could improve understanding the relationship between EB and metabolic status, as well as enhance identification of cows at risk for compromised metabolic status and disease. In this research we applied clustering of time series of the calculated EB and examine associated metabolic status and disease treatments of dairy cows in the peripartum period, analysing data from 3 studies. Data integration resulted in information for 419 Holstein-Friesian dairy cows consisting of the calculated EB, body weight, body condition score, milk yield, milk composition, plasma non-esterified fatty acids, β-hydroxybutyrate, glucose, insulin and IGF-1 and disease treatments. Four dairy cow clusters for time profiles of EB from week -3 until 7 relative to calving were generated by the GAK algorithm. For each cluster, mean of bodyweight prepartum was distinguishable, indicating this might be a possible on-farm biomarker for the peripartum EB profile. Moreover, cows with severe EB drop postpartum were more treated for milk fever and had high plasma NEFA and BHB concentration, and low IGF-1, insulin and glucose concentration in the first 7 weeks of lactation. Overall, this study demonstrated that cows can be clustered based on EB time profiles and that characteristics like prepartum bodyweight, and postpartum NEFA and glucose concentration are promising biomarkers to identify the time profile of EB and potentially the risk for metabolic diseases.

Veterinary advise and health management in bovine dairy farms from São Miguel Island (Azores)

I. Medeiros[1], A. Fernandez-Novo[2], J. Simões[1] and S. Astiz[3]
[1]University of Trás-os-Montes e Alto Douro, Department of Veterinary Medicine, Animal and Veterinary Research Centre (CECAV), Quinta de Prados, 5370-801 Vila Real, Portugal, [2]School of Biomedical and Health Sciences. Universidad Europea de Madrid, Department of Veterinary Medicine, Madrid, C/ Tajo s/n, 28670 Villaviciosa de Odón, Madrid, Spain, [3]National Institute of Agronomic Research (INIA), Animal Reproduction Department, Puerta de Hierro avenue s/n, 28040 Madrid, Spain; jsimoes@utad.pt

Dairy farms change the veterinary demanded services towards veterinary herd health management. In this study, veterinarians were surveyed to assess problems and veterinary strategies to deal with in S. Miguel Island dairy farms (Azores), depending on the veterinarians' sector (Cooperative vs Private Vets). All veterinarians, in the whole S. Miguel Island, providing services to dairy farms were contacted and the response rate was 67% (20/30). Individual medicine was the main service provided (56%), with 15% of practitioners implementing veterinary consultancy in, at least, one farm; 78% fulfilled health plans, more frequently by Private Vets. Herd-level breeding, fertility, lameness, postpartum, dry cow, rearing and nutrition plans were more frequently executed by Private than Coop Vets ($P<0.05$). Private Vets tend to address more commonly prevention policies such as faecal-sampling ($P=0.08$), reproductive programs ($P=0.10$) and body condition scoring ($P=0.06$); whereas Coop Vets cover predominantly individual problems (surgery; $P=0.10$, dystocia; $P=0.06$) and brucellosis or tuberculosis surveillance ($P=0.04$). Poor fertility and pneumonia are major problems, but Coop Vets report more frequently medical issues (neonatal diarrhoea, mastitis, ketosis, acidosis, pneumonia, retained placenta, paratuberculosis; $P<0.05$). Veterinarians agreed that farmers should increase biosecurity and preventive herd-level strategies. This survey helps to further point out areas to be improved and the increased veterinary herd health management demand with a maintained need for veterinarians performing individual medicine.

Session 08 Poster 16

Most common reasons of the use of emergency slaughter in dairy farms: an Italian experience

F. Mazza[1], F. Fusi[1], V. Lorenzi[1], A. Vitali[2], A. Gregori[2], G. Clemente[1], C. Montagnin[1] and L. Bertocchi[1]
[1]Istituto Zooprofilattico Sperimentale della Lombardia e dell'Emilia Romagna 'Bruno Ubertini', Italian National Reference Centre for Animal Welfare, Via A. Bianchi 9, 25124 Brescia, Italy, [2]Agenzia di Tutela della Salute di Brescia, Viale Duca degli Abruzzi, 15, 25124 Brescia, Italy; luigi.bertocchi@izsler.it

On-farm emergency slaughter (OFES) is the slaughter of an animal outside the slaughterhouse. Concerning domestic ungulates, OFES can be carried out only on otherwise healthy animals that have suffered an accident and that cannot be transported to an abattoir for animal welfare reasons. A survey was carried out to investigate the main causes of OFES in dairy farms located in the sanitary district of Brescia (Italy). This district has the 5% of all Italian dairy farms, which corresponds to the 11.7% of all Italian dairy cattle. During 2021, 3,507 cattle underwent OFES in the investigated district. Complete data were available for 2,942 OFES cases (83.9%), referred to as many animals. Of these 2,942 slaughtered cattle, 94.15% were female and 5.85% male. Cattle age ranged from 79 days to 13 years old, with a median of 3.5 years old. Hip and hind leg injuries not related to calving reasons were the most common types of injuries recorded during the *ante-mortem* inspections (49.9%), followed by injuries of the front legs or of any other anatomical site (25.4%). Calving problems (injuries or non-traumatic pathologies) were recorded in 16.4% of the *ante-mortem* inspections carried out during OFES. Uncommon reasons of OFES were digestive tract pathologies, foot pathologies and udder pathologies. Collected data confirmed that OFES was carried out mainly on animals with traumatic injuries or acute pathologies. Guidelines and on-farm decision-making protocols should be available in any farm in order to support prompt decision in case of emergency, prevent delays in action and avoid prolonging of animal suffering.

Impact of wing feather loss on muscle thickness and keel bone fracture in laying hens

R. Garant[1], B. Tobalske[2], N. Bensassi[1], N. Van Staaveren[1], D. Tulpan[1], T. Widowski[1], D. Powers[3] and A. Harlander[1]
[1]University of Guelph, N1G2W1, Canada, [2]University of Montana, Missoula, MT 59812, USA, [3]George Fox University, Newberg, 97132, USA; aharland@uoguelph.ca

Feather loss can reduce wing use and access to elevated resources. Physiologically, there can be pectoral muscle (PM) loss which may have implications for keel bone fractures (KBF); a condition reported in up to 98% of laying hens. To investigate the impact of flight feather loss on PM thickness and KBF, 120 white and brown-feathered laying hens received a wing feather treatment: full-clip (55% reduced wing area), half-clip (33% reduced wing area), or no clip (control). Feeders with RFID technology were used to assess the birds' ability to use elevated (70 cm) and ground resources. We used ultrasound to measure PM thickness and X-rays to identify KBF before the wing treatments (week 0) and six weeks (week 6) after clipping. Linear mixed models were used to analyse the effect of wing treatment, strain, and week on PM thickness and KBF prevalence. At week 0, white-feathered hens spent more time at the elevated feeders than brown-feathered hens (53.4 vs 24.0%; P=0.0218). Full clipping of white-feathered hens led to a 39.2% reduction in time on elevated feeders by week 6 (from 53 to 25%; P<0.0001). Time spent on elevated feeders did not change in half-clipped white-feathered hens and brown-feathered hens from either clipping treatment. At week 6, PM thickness decreased by ~5% in white-feathered birds (half-clipped; P=0.0165 and full-clipped; P=0.0129) which might reflect muscle atrophy due to lack of wing use. There was no effect of wing treatment on PM thickness in brown-feathered hens. The KBF prevalence was unchanged (48% affected) in all birds at weeks 0 and 6. These results indicate that wing feather loss and PM thickness are not associated with KBF in either strain of hens. However, results indicate that white-feathered birds are more sensitive to feather loss due to reduced PM thickness and access to elevated feeders seen in the clipping treatments. Future feather loss prevention strategies in laying hens should focus on strains that are more likely to experience muscle loss associated with wing feather damage.

A diagnostic chart to classify bovine rumen lesions at slaughter

A.M. Du Preez[1], L. Prozesky[2], C.J.L. Du Toit[3] and M. Faulhaber[4]
[1]Devenish, Lagan House, Belfast BT13BG, United Kingdom, [2]University of Pretoria, Dept. of Paraclinical Sciences, 100 Old Soutpansweg, Pretoria 0002, South Africa, [3]University of Pretoria, Dept. of Animal Sciences, Lynnwood road, Pretoria 0002, South Africa, [4]Cavalier Foods, Farm Tweefontein, Cullinan 1000, South Africa; amelia.may.dp@gmail.com

Feedlot cattle are fed high energy, readily fermentable diets which ensure high propionic acid production, thereby efficiently and economically providing energy to the bovine. The increase in volatile fatty acid and lactate production can cause the rumen pH to decrease to below 5.5 leading to sub-acute rumen acidosis and rumen acidosis. This acidic environment results in the erosion of rumen papillae and alteration of the rumen epithelial structure causing rumen wall ulceration and lesions. The open lesions are a significant animal welfare concern, whilst also having an impact on economy. The lesions will result in an immune response, leaky gut, liver and lung lesions, and reduce the surface area available for nutrient absorption, thereby reducing production efficiency of the animal. The aim of this paper is to create a standard, easy-to-use rumen lesion diagnostic chart which producers can use to evaluate the severity of rumen scaring at slaughter. The rumens of two batches of 200 cattle each were inspected, scored, and photographed by veterinarians and animal scientists at a commercial abattoir. Rumens were classified into six categories: Normal – no visible redness, normal rumen papillae; Mild – slight redness/one or two red areas, fewer papillae on the ventral rumen; Moderate – distinct redness/larger red areas, complete absence of papillae on the ventral rumen, older lesions starting to heal; Severe – open lesions, thickening of the rumen wall, complete absence of papillae on the ventral rumen; Normal with old lesions – no visible redness, normal papillae, old and completely healed lesions; Parakeratosis – thickening of the rumen wall and clumping of papillae. Three representative photos of each category will be selected to appear on the diagnostic chart. The diagnostic chart will help to standardise rumen scoring, assist producers to adjust feeding regimes, provide abattoir operators the opportunity to give feedback to feedlot managers, and ensure healthy, efficient animals.

Session 08 Poster 19

Effect of a phytobiotic blend to *Cryptosporidium* control in calves

M. Olvera-García, V. Cordova, L.R. Pérez, J.C. Baltazar and G. Villar-Patiño

Grupo NUTEC, Research-Innovation & Nutrition Departments, El Marqués avenue 32, Industrial Park Bernardo Quintana, Querétaro, 76246, Mexico; lrperez@gponutec.com

The *Cryptosporidium* parasite is recognized as one of the main causes of diarrhoea in young cattle causing high rates of morbidity and mortality, intestinal damage, and impact the feed conversion. Grupo NUTEC designed a phytobiotic blend based on garlic and cinnamon extracts, which has shown an anticoccidial effect in poultry, by coccidia membrane rupture and microbiota modulation. This work aimed to evaluate the effect of the phytobiotic as an alternative to the antiparasitic used for the prevention and control of cryptosporidiosis, which can be supplemented in milk during the first 30 days since calves' birth. 20 calves were divided into the treatments since the birth to 30 and 60 d. Normal feed program (avoiding extra additives) medication and management were followed in this trial. Treatments: (1) Control (milk and started feed); and (2) Control + 500 ppm BIOADD®. Productivity was documented: weight gain, size, started feed consumption, milk consumption, and diarrhoea records, at initial time of the trial, 30 and 60 d. *Cryptosporidium* was evaluated in faeces obtained at 7, 15, and 28 d. The ooquists identification was made by the sedimentation and Kinyoun stain. Quantification was based on dilutions visualized by microscopy. Viability was determined by a modification in the eosin stain. The number of infective oocysts quantified at the beginning was around 32,428 oocysts per gram of faeces (opg). At 7 d and 15 d, an increase in the number of oocysts found in faeces were observed no matter the treatment, due to the normal life cycle of *Cryptosporidium* and because a percentage of the oocysts have a thin cell wall, which excysts endogenously causing an autoinfection phenomenon, nevertheless, in the treatment 2 the counts were lower than the control in 23% (7 d) and 15% (15 d). At 28 d, a considerable reduction was observed in the number of viable oocysts found in faeces for the treatment with the phytobiotic. Phytobiotic supplementation promotes an increase in the final body weight in 4.3 kg, compared with the control. In conclusion, BIOADD® considerably decreases total oocyst and their viability reducing the infection incidence, regarding in a better performance.

Session 08 Poster 20

Effects of environmental enrichment on flesh quality in rainbow trout (*Oncorhynchus mykiss*)

A. De La Llave-Propín[1], A. Martínez-Villalba[2], R. Bermejo-Poza[2], J. De La Fuente[2], E. González De Chávarri[2], C. Pérez[3], M.T. Díaz[2], A. Cabezas[2] and M. Villarroel[1]

[1]Universidad Politécnica de Madrid, CEIGRAM, Avenida Puerta de Hierro, 2, 28040, Spain, [2]Universidad Complutense de Madrid, Departamento de Producción animal, Avenida Puerta de Hierro, 28040, Spain, [3]Universidad Complutense de Madrid, Departamento de Fisiología animal, Avenida Puerta de Hierro, 28040, Spain; alvarodelallavepropin@gmail.com

Fish welfare is assuming greater significance in the global aquaculture industry, but it must be feasible in a productive and quality manner. Environmental enrichment plays a key role in livestock industries. One of the occupational enrichment variants is the creation of water currents that allow a positive interaction between growth and exercise, as well as stress response reduction. In addition, fasting is essential to avoid contamination of fish carcasses during the gutting process. The objective of this study was to evaluate the environmental enrichment and pre-slaughter fasting effects on fish flesh quality. Tests were performed using 540 rainbow trout (*Oncorhynchus mykiss*) with an average weight of 195.0±42.4 g, separated into raceways with different environmental enrichment (without 'C', aleatory 'A' or continuous 'CT' water pumps) and two pre-slaughter fasting durations (no fasting '0D' or 5 days of fast-45.2 °C d '5D'). After slaughter, biometric (slaughter weight, standard length, condition factor, relative growth and hepato-somatic index) and flesh quality parameters (muscle colour, muscle pH and *rigor mortis*) were measured. 5D fish presented lower slaughter weight, condition factor and relative growth than 0D. Besides, A and CT fish showed higher values of those biometric parameters than C. Muscle colour parameters such as L* and h* at 0 hours *post mortem* were lower in A and CT fish than C, highlighting the lowest values in A group. On contrast, fish of group A showed the highest values of a*, b* and C*. Based on the results, the use of water pumps in rainbow trout farming seems to produce a higher fish growth, as well as a better flesh quality, with this beneficial effect being higher on trout subjected to aleatory water pumps.

Coping style: standardization of the net test

A. De La Llave-Propín[1], R. Bermejo-Poza[2], J. De La Fuente[2], E. González De Chávarri[2], C. Pérez[3], F. Torrent[4] and M. Villarroel[1]

[1]Universidad Politécnica de Madrid, CEIGRAM, Avenida Puerta de Hierro, 2, 28040, Spain, [2]Universidad Complutense de Madrid, Dept. de Producción animal, Avenida Puerta de Hierro, 28040, Spain, [3]Universidad Complutense de Madrid, Dept. de Fisiología animal, Avenida Puerta de Hierro, 28040, Spain, [4]Universidad Politécnica de Madrid, Dept. de Ingeniería Agroforestal, Ciudad Univeristaria, 28040, Spain; alvarodelallavepropin@gmail.com

The classification of individuals as proactive or reactive, according to coping style, is related to the adaptation and welfare of aquaculture species. The net test, consisted in holding fish suspended on a net out of water, facilitates the identification of reactive and proactive coping styles in fish, quantifying their escape movements during an out-net time. However, the period required for the test has not been standardized and its repeatability is unclear. To standardize that, 30 rainbow trout fingerlings (mean weight 37.5±16.5 g), identified using Pit-Tags, were systematically subjected to a net test in four trials spaced apart 14 days to assess the repeatability. Fish were distributed into three groups, each with 10 trout. An out-net time of 15 seconds was set for the 1st and 2nd groups in all trials, while the out-net time of the 3rd group was 15 seconds for the first two trials and 65 seconds for the last two trials. However, all groups were subjected to a fasting period of 4 days on the third trial. Afterwards, a video analysis was carried out using BORIS software for the recording of movements. Results showed that there were no differences between the number of movements for each fish in each test (G1: 19.1±3.3; G2: 17.5±5.0; G3: 13.2±1.6). Regarding the out-net time taken to perform the tests, it was found that the greatest concentration of information occurred in the range of 0-5 seconds (11.0±1.8), decreasing thereafter (6.6±1.6). The test preceded by a fasting period showed similar coping values (15.8±3.3) to the other tests (17.2±3.8). Regarding standardization, it is concluded that the net test helps to differentiate between individuals, remaining intact over time, without effects in terms of pre-test fasting. Finally, the study concludes that anoxia time-basis can be shortened to 5 seconds without a loss of relevant data.

Herd of origin of purchased calves may influence clinical scoring and performance until weaning

A.K. Andreassen, C. Juhl, M.B. Jensen and M. Vestergaard

Aarhus University, Department of Animal Science, 8830 Tjele, Denmark; anso@anis.au.dk

In rosé veal calf production young calves from different herds are often mixed and grouped upon arrival, which creates a risk for compromised health, disease outbreaks and reduced growth. Some calves do better than others and this may be due to better health at arrival due to herd of origin. The objective was: (1) to investigate the effect of herd of origin on performance and clinical health scores; and (2) to investigate the relation between performance and clinical health scores. Thirty-two Holstein bull calves were purchased from five dairy herds at two weeks of age (14.0±3.9 days). Calves were allocated to 4 groups of 8 calves balanced for LW, age and herd of origin. Calves were offered milk from an automated milk feeder at 8 l/d in week 3 to 6, 6 l/d in week 7 to 8, 3 l/d in week 9 to 10, and were followed until week 12 of age. Clinical scoring (CS) and rectal temperature (RT) was recorded two times a week, and calves were weighed once a week. A descriptive analysis of data was performed. Calves from herd 5 had lower average daily gain (ADG) than calves from the other four herds (0.87, 0.92, 0.93, 0.90 and 0.81 (±0.3) kg/d for herds 1, 2, 3, 4 and 5 respectively). Herd 1 had the largest variation in ADG over the weeks ranging from 1.18±0.3 kg/d in week 3 to 0.53±0.4 kg/d in week 4. Overall, CS was highest in week 3 and 4, then decreased until week 10, after which CS increased until week 12. Over the entire period herd 5 had an average CS of 6.0±1.8, compared to 4.9, 5.4, 5.4, 5.3 (±1.8) for herd 1, 2, 3, 4, respectively. There was a slightly decrease in RT over the weeks for calves from all herds. Herd 1 had a peak in RT in week 5 (39.4±0.9 C), while calves from the other herds had lower RT (38.7±0.1 C). Calves from herd 5 had a similar peak in RT the following week. In week 10, RT across herds aligned and followed the same pattern until week 12 (38.9±0.06 C). Results suggest that herd of origin contributes to the health status and growth performance when calves are raised in mixed-herd groups, and that growth rate might be influenced by clinical scoring.

Practices and attitudes for the record of veterinary treatments in beef farming

E. Royer, S. Ferrer Diaz and A. Dayonnet
Institut de l'élevage, chemin de Borde Rouge, 31321 Castanet-Tolosan, France; eric.royer@idele.fr

A prudent use of antimicrobials requires the support of precise monitoring systems. Currently, veterinary medicines are recorded in the Animal Remedies Records (ARR), but few information is available on practices for entering and using these data. A survey was conducted in France to collect practices of beef farmers on traceability of veterinary treatments as well as attitudes about digital recording. 262 complete responses were collected in December 2019 using a LimeSurvey® link, e-mail or face-to-face, mainly from cow-calf (46%), and suckling to beef systems without (38%) or with purchases of calves (10%). A majority of farmers were satisfied with their ARR, especially if they updated it frequently. However, only a quarter of the farmers (73 out of 262) used a software tool to record the veterinary treatments. However, two-thirds (181 out of 262) were equipped with livestock management software. However, this software was primarily used for recording births and animal movements (98%), then monitoring of genealogy (49%) or reproduction (45%) and finally for health management (40%). Users of digital ARR mentioned as advantage the easy and quick consultation of information (68%) in a single interface (63%). They would like tools to facilitate data entry thanks to drop-down lists for medicines (34%) and pathologies (27%), as well as pre-registered protocols (32%). The cost of the software was reported as a key-point. Half of farmers currently using a paper ARR were in favour of switching to software (96 out of 189), but the other half were reluctant (93 out of 189). The respondents who were willing to computerize the ARR asked that data entry and consultation to be intuitive and not too time-consuming (83% and 79% respectively). Respondents who did not want to computerize the ARR mainly gave personal reasons (87%), i.e. age, routine of using paper, lack of time, or lack of interest or even reluctance in computers (24%). A quarter did not have the appropriate equipment or sufficient Internet connection (26%). In conclusion, the software equipment on beef farms is satisfactory, and the proportion of farmers using a computerized ARR, even if still in the minority, could be increased thanks to automation and simplification of data entry.

Session 08 Poster 24

A study on colostrum quality and passive immunity transfer in dairy farms of Barcelos municipality

J. Cunha[1], A. Torres[1], J. Almeida[2,3], S. Silva[2,3], H. Trindade[3,4] and M. Gomes[2,3]
[1]Cooperativa Agrícola de Barcelos, R. Fernando de Magalhães, 206, 4750-290 Barcelos, Portugal, [2]CECAV, Quinta Prados, 5000-801 Vila Real, Portugal, [3]Universidade Tras-os-Montes e Alto Douro, Quinta dos Prados, Folhadela, 5000-801, Portugal, [4]CITAB, Quinta Prados, 5000-801 Vila Real, Portugal; mjmg@utad.pt

Calves are born almost agammaglobulinemic, and optimisation of colostrum management is of utmost importance to avoid failure of transfer of passive immunity (FPT). Few data currently exist about management practices in dairy calves in the municipality of Barcelos (North Portugal). The Barcelos Cooperative selected 6 herds, ranging in size from 62 to 300 lactating cows, to take part in a pilot study conducted from April to June 2020. During this study, 81 colostrum samples were evaluated using a digital Brix refractometer (ATAGO, model PAL-1). Blood samples of the 81 calves were collected c.a. 48 h post-partum by jugular venepuncture, allowed to clot at ambient temperature and a serum sample was tested. The cut-off points for identifying good quality colostrum and FPT were fixed at >22% and <8.4% Brix, respectively. The majority of calves (72%) received the colostrum within 4 h post-partum and the mean amount of colostrum fed at first feeding was 4 l. Only 3 farmers were routinely evaluated for colostrum quality. Regarding the colostrum storage practice, only 4 farms had a colostrum bank. Regarding colostrum quality, the mean colostrum %Brix was 24.3 (SD±4.4) ranging from 15.8 to 35.5%. Only 66.7% of samples met the criterion of high-quality colostrum. The mean %Brix in the serum samples was 9 (SD±0,9), with a range from 7.5 to 10.9%. Twenty animals showed values of %Brix in serum lower than 8.4, that is, the (general) rate of FPT was 24.7%. Brix% of colostrum was correlated with Brix% of serum (r=0.47; P<0.0001). Colostrum quality increase with parity (P<0.01). Although the present study is limited in size and the data presented may not be representative of the colostrum being fed to all replacement dairy heifers, the herds enrolled in the study are believed to be broadly illustrative of commercial dairy herds in this municipality of Portugal. The findings of a considerable high FPT prevalence and inferior colostrum quality indicates a need for improved awareness among dairy producers.

Long-term antibody production and viremia of American mink challenged with Aleutian

A.H. Farid, I. Hussain, P.P. Rupasinghe, J. Steohen and I. Arju
Faculty of Agriculture, Dalhousie University, Department of Animal Science and Aquaculture, 58 Sipu Road, Truro, B2N 5E3, Nova Scotia, Canada; ah.farid@dal.ca

Black American mink (*Neovison vison*) were inoculated with a local isolate of Aleutian mink disease virus (AMDV) over four years (n=1,742). The animals had been selected for tolerance to AMDV for more than 20 years (TG100) or were from herds that have been free of AMDV (TG0). The progenies of TG100 and TG0, and their crosses with 25, 50 and 75% tolerance ancestry were used. Blood was collected from each mink for up to 14 times between 35 and 1,211 days post-inoculation (dpi) and tested for viremia by PCR and for antibodies by counter-immunoelectrophoresis (CIEP). At termination, antibodies in the blood and antibody titer (n=1,217) were measured by CIEP, and viremia (n=1,217) was tested by PCR. The peak incidences of viremia (66.7%) and seropositivity (93.5%) occurred at 35 dpi. The incidence of viremia decreased whereas the incidence of seroconversion increased over time. Viremia had a negative effect and antibodies had a positive effect on productivity. Differences among tolerant ancestry were significant for every trait measured. The incidences of viremia over time, terminal viremia, seropositivity over time, and antibody titer were the highest in the susceptible groups (TG0 or TG25) and the lowest in the tolerant groups (TG100 or TG75), implying that previous selection for tolerance resulted in mink with reduced viral replication and antibody titer. Males and females had comparable incidences of seropositivity over time, but females had a significantly higher incidence of persistently seropositive, higher antibody titer, and a lower incidence of being persistently nonviremic than males.

Using infrared thermography imaging to assess cattle welfare using eye temperature as indicator

P. Valentim[1], C. Venâncio[1,2] and S. Silva[1,2]
[1]University of Trás-os-Montes e Alto Douro, Quinta de Prados, 5000-801, Vila Real, Portugal, [2]CECAV-Associate Laboratory of Animal and Veterinary Science (AL4AnimalS), Quinta de Prados, 5000-801, Vila Real, Portugal; ssilva@utad.pt

Animal welfare is a concurrent topic across all fields of animal science, and it is challenging to find valuable animal-based indicators that can adequately assess welfare. The temperature has been a studied indicator, and infrared thermography (IRT) does present some appealing advantages as an easy-to-use tool that records and measures surface temperature at a distance. Thus, the present work aims to verify the usefulness of maximum eye IRT temperature (EyeIRT) in cattle during stressful events, such as weighing or restraining. Forty-two animals of the Arouquesa cattle breed were distributed in weighing (W) and the health program (HP) groups with 10 and 32 animals, respectively. For the W group, EyeIRT were taken before, during, and after animals were weighed. For the HP group, 14 calves and 18 cows were studied. The EyeIRT were taken before and after sanitation procedures. The IRT images were captured with a FLIR F8 camera with a resolution of 320×240. To capture IRT images, the camera was set for 1 meter. The thermograms were analysed using the FLIR Tools+ software. An ellipse was adjusted to the eye, and data were exported to an Excel spreadsheet to determine EyeIRT. An ANOVA with an LSMeans Differences Student's test was performed to compare the EyeIRT in the different stages of the study. There is an increase in EyeIRT with weighing, handling, or restraining procedures for W and HP groups. For the W group, the EyeIRT increase ($P<0.05$) from 33.5 °C (before) to 35.8 °C (during) and after the weighing procedure decrease ($P<0.05$) to 34.6 °C. For the animals of the HP group, an increase ($P<0.05$) of EyeIRT was also observed with handling or restraining procedures. Therefore, it can be concluded that IRT can potentially be used as a practical and safe method to detect changes in ocular temperature under different conditions, but further research and testing are required to establish a temperature range and determine the validity of surface temperature as the welfare indicator. Acknowledgements: Operational Group PDR2020-101-031094.

Session 09 Theatre 1

Impact of nutritional changes pre and post breeding on reproduction

K.M. Epperson[1] and G.A. Perry[1,2]
[1]Texas A&M University, Department of Animal Science, College Station, TX, USA, [2]Texas A&M, AgriLife Research, Overton, TX, USA; george.perry@ag.tamu.edu

Nutrition plays a critical role in reproductive management both before and after insemination. When nutrients were restricted (≤90% NEm) for 33-36 days before insemination, embryo development was stunted (P=0.05), fewer females exhibited oestrus (P=0.02), and subsequent progesterone concentrations were decreased (P=0.02) compared to non-restricted (≥139% NEm) females. Restoring nutrient intake for 6-8 days after AI advanced embryo development (P=0.05). Despite the short time frame and small degree of restriction, embryo quality was improved when females were not restricted for 6-8 days post-AI. When nutrition was restricted post breeding, embryo stage (P<0.01), grade (P=0.02), total number of cells (P=0.03), and percentage of cells alive (P=0.01) were decreased compared to non-restricted females. Changes in nutritional intake can be the result of drought, changes in diet or movement to new locations. Specifically grazing behaviour may dictate an animal's intake and mitigate the ability of a female to become or remain pregnant. Heifers that graze forage from weaning to breeding rather than being placed in drylots retained better grazing skills and had increased average daily gains (ADG) into the subsequent summer and moving females from a drylot to pasture around breeding decreased ADG, and pregnancy success (P=0.04). While conception rates after artificial insemination (AI) and ADG were improved (P=0.02) when heifers moved to pasture after AI were supplemented compared to non-supplemented heifers, indicating a possible solution to preventing negative energy balance if heifers must be moved after insemination. These results illustrate the possibility of inducing negative energy balance by simple a change in diet or location around breeding, and the reproductive consequences that follow if measures are not taken to prevent a decrease in feed intake or quality. In summary, stress induced by diet change or decreased nutrient intake can result in embryonic loss, decreased embryo quality, stunted embryo growth, and reduced pregnancy success. Special considerations should be made prior to and during the breeding season to keep nutrient intake consistent and animals on a positive plane of nutrition.

Session 09 Theatre 2

Maternal nutrition during gestation and its consequences to beef offspring performance

P. Moriel[1], J.M.B. Vendramini[1] and R.F. Cooke[2]
[1]University of Florida, IFAS, Range Cattle Research and Education Center, Ona, FL 33865, USA, [2]Texas A&M University, Department of Animal Science, College Station, TX, USA; reinaldo.cooke@ag.tamu.edu

Multiple studies over the last decade demonstrated how offspring postnatal growth, immune response, and reproduction could be optimized by implementing maternal supplementation of protein and energy, polyunsaturated fatty acids (PUFA), trace minerals, frequency of supplementation, specific amino acids, vitamins, and increasing cow body condition score (BCS) throughout pregnancy. Specific outcomes to on offspring performance were variable and require further investigations. For instance, greater offspring growth performance following maternal supplementation of protein and energy was reported more consistently for preweaning period compared to post-weaning and finishing periods. Longer periods (entire vs half of third trimester of gestation) of frequent (daily vs 3 times weekly) supplementation of protein and energy during late gestation were required to optimize calf preweaning growth. Maternal supplementation of specific sources of PUFA and trace minerals during late gestation increased multiple growth, immune function, and carcass quality parameters. Despite the relatively recent interest on gestational nutrition of beef cattle, the complexity and obstacles required to advance our existing knowledge on foetal programming research for beef cattle system also exponentially increased. Some of these challenges include: accurate evaluation of cow milk production; unbalanced focus on *Bos taurus* vs *Bos indicus*-influenced breeds; offspring sex- and breed-specific outcomes to maternal gestational nutrition; limited knowledge on the interaction between prenatal nutrition and subsequent calf postnatal nutritional management; and lack of studies implementing immunological challenges and offspring performance through multiple generations. In summary, nutritional management of pregnant cows provide an opportunity for beef producers to enhance offspring productive performance and health, but overcoming current challenges described above are vital to expand the existing knowledge of foetal programming in beef cattle and increase beef production.

Supplementing omega-6 fatty acids to enhance early embryonic development and pregnancy establishment

R.F. Cooke
Texas A&M University, College Station, TX 77845, USA; reinaldo.cooke@ag.tamu.edu

Our research group investigated the impacts of supplementing Ca salts of soybean oil (CSSO), a source of omega-6 fatty acids (FA), on reproductive performance of beef cows. Initial studies were conducted with Nelore (*Bos indicus*) cows grazing tropical pastures. Cows were assigned to fixed-time artificial insemination (AI) and supplemented or not with 100 g/d (as-fed basis) of CSSO, and supplementation regimens ranged from d -11 to 28 relative to AI. Overall, CSSO supplementation during the 21 d after AI increased (P<0.01) pregnancy rates from 38.1% (623/1,635 as pregnant/total non-supplemented cows) to 49.0% (843/1,720 as pregnant/total CSSO-supplemented cows), and these outcomes were associated with enhanced early embryonic development when omega-6 FA were supplemented. To verify this rationale, our group compared FA incorporation in grazing Nelore cows (n=90) supplemented or not with CSSO (100 g/d; as-fed basis) beginning at fixed-time AI until slaughter at d 19 of gestation. Supplementing CSSO increased (P≤0.05) incorporation of linoleic acid and its omega-6 derivatives in plasma, endometrium, corpus luteum, and conceptus. Complementing these findings, grazing Nelore cows (n=100) were supplemented or not with CSSO (100 g/d; as-fed basis) beginning at fixed-time AI, and assigned to uterine flush on d 15 of gestation. Supplementing CSSO increased (P≤0.04) conceptus length (2.58 vs 1.15 cm) and mRNA expression of interferon-tau (4.1-fold increase) and prostaglandin E synthase 2 (2.6-fold increase). These outcomes were recently replicated in B. taurus beef cows consuming temperate forages. Pregnancy rates were greater (P=0.01) in Angus cows receiving CSSO (100 g/d; as-fed basis) for 21 d after fixed-time AI (60.2%; 226/383 as pregnant/total cows) compared with non-supplemented cows (51.7%; 193/388 as pregnant/total cows). Supplementing CSSO to Angus × Hereford cows (n=96) beginning after AI also increased (P=0.05) mRNA expression of interferon-tau in d 15 conceptuses (1.8-fold increase). Collectively, post-AI CSSO supplementation favours incorporation of omega-6 FA into maternal and embryonic tissues, which enhances interferon-tau synthesis by the conceptus and increases pregnancy rates to fixed-time AI in *B. indicus* and *B. taurus* beef cows.

Nutritional influences on reproductive performance of *Bos indicus* beef cows

J.L.M. Vasconcelos
Sao Paulo State University (UNESP), Botucatu, 18168-000, Brazil; jose.vasconcelos@unesp.br

This experiment evaluated pregnancy rates to fixed-time artificial insemination (FTAI) in *Bos indicus* beef cows according to their body condition score (BCS) at calving and subsequent change until 30 days after FTAI. Non-pregnant, suckling Nelore cows (n=593 primiparous, 461 secundiparous, and 893 multiparous) were evaluated for BCS at calving and FTAI, and at 30 days after FTAI when cow pregnancy status was verified. Cow BCS at calving was subtracted from BCS recorded at pregnancy diagnosis, and cows classified as those that lost BCS (L), maintained BCS (M), or gained BCS (G) during this period. Cows that became pregnant to the FTAI protocol had greater (P≤0.05) BCS at calving, FTAI, and at pregnancy diagnosis compared to cows that did not become pregnant. Cows that calved with BCS≥5.0 had greater (P<0.01) BCS throughout the experiment, and greater (P<0.01) pregnancy rates to FTAI compared with cows that calved with BCS<5.0 (54.8 vs 34.2%). Pregnancy rates to FTAI were greater (P<0.01) for G and M cows compared with L cows (50.0, 47.5, and 36.0%, respectively), and similar (P=0.46) between G and M cows. Moreover, pregnancy rates to FTAI in G cows that calved with BCS<5.0 were less compared with L (P=0.08) and M cows (P<0.01) that calved with BCS≥5.0 (42.2, 48.3, and 58.3%, respectively). Collectively and across parities, pregnancy rates to FTAI were greater in B. indicus cows that calved with a BCS≥5.0 regardless of post-calving BCS change, and greater in M and G cows within those that calved with BCS<5.0 or≥5.0.

Phenotypic and genetic variation in reproductive development and semen traits in young Holstein bull

S. Coen[1], K. Keogh[1], C. Byrne[1], S. Fair[2], J.M. Sanchez[3], M. McDonald[3], P. Lonergan[3] and D.A. Kenny[1]
[1]Teagasc, Animal and Bioscience Research Department, Grange, Dunsany, Co. Meath, C15 PW93, Ireland, [2]University of Limerick, Laboratory of Animal Reproduction, Department of Biological Sciences, Limerick, V94 T9PX, Ireland, [3]University College Dublin, School of Agriculture & Food Science, Belfield, Dublin 4, D04 V1W8, Ireland; scoen101@gmail.com

This study aimed to assess the variation in reproductive development and semen traits in spring born 13-month-old Holstein-Friesian (HF; n=1,117) bulls. The animals were reared under standardized regimen on 8 commercial herds and had a mean (±SD) age and bodyweight of 394 (5.28) days and 412 (47.5) kg. A breeding soundness evaluation was conducted for each animal and semen was collected using electro-ejaculation. Scrotal circumference (SC) was measured with a range of semen quality traits including total sperm number (TSN), sperm motility and kinematics using CASA. Sperm morphology was assessed on nigrosin-eosin-stained sperm via light microscopy. Heritability coefficients were calculated using 5 pedigree generations. Mean SC was 32.6 (±0.07 cm). Almost all (98.4%) bulls reached puberty based on a widely accepted SC threshold of 28 cm. Mean (±SEM) values for TSN, volume and sperm concentration 1.86 (0.06) billion, 3.42 (0.07) ml and 537.7 (14.5) million/ml. Mean percentage of normal, motile and progressively motile sperm was 81.8 (0.3), 59.52 (0.7) and 32.9 (0.5), and 4.3% of bulls were deemed to have physical abnormalities. 67.6% of bulls produced an ejaculate of >0.75 billion TSN, widely accepted by the AI industry as a threshold to produce a minimum of 50 straws/ejaculate (15 million sperm cells/straw) for elite bulls in their first season. Additionally, 88.5% of bulls produced ejaculates containing >70% morphologically normal sperm, a required standard for an ejaculate to be suitable for AI or natural service. The heritability of SC was 0.66 and semen traits ranged from 0.15 to 0.66. Results highlight the phenotypic and genetic variability in industry standard semen quality traits. These data provide a basis for future studies on genetic and genomic control of semen quality in post-pubertal bulls. This study was funded by Science Foundation Ireland (16/IA/4474).

Maternal and paternal factors driving pregnancy loss

K.G. Pohler
Texas A&M University, College Station, TX 77845, USA; ky.pohler@ag.tamu.edu

Pregnancy loss in cattle causes both management and economic challenges to a producer. Recent studies have been conducted to quantify reproductive failures that occur during fertilization, early embryonic development, and late embryonic/early foetal development periods of gestation in beef cattle. Minimizing reproductive inefficiency, specifically embryonic mortality (EM) is vital. Although fertilization rates are reportedly high in beef cattle, significant developmental failure occurs within the first 7 days of gestation. Approximately 28.4% of embryos will not develop past day 7 of gestation with most embryonic losses occurring before day 4. By the conclusion of the first month of gestation, 47.9% of cows submitted to a single insemination at day 0 will not be pregnant. Overall, late embryonic/foetal development between days 32 to 60 and 100 is 5.8% with a range of 3.2 to 42.7%. This talk will highlight some of the work our group is focusing on to determine timing, detection and causes of pregnancy loss during these pivotal periods of pregnancy loss and potential management aspects to mitigate reproductive inefficiency.

Session 09 Theatre 7

Forage planning for heifers and cows

J. Vendramini
University of Florida, Range Cattle Research and Education Center, Ona, FL 33865, USA; jv@ufl.edu

In order to implement an effective forage management in beef cattle production, there are a number of important issues to be addressed. These include: (1) what is required for plants and animals to be productive in a pasture-livestock system; (2) what management choices have the greatest impact on success or failure of a grazing system; (3) how can the nutritional requirements of the animal be matched with the ability of the forage to supply nutrients, and d) the role of conserved forage in grazing systems. The choices of forage species and nutrient management programs are the most basic and important decisions on beef cattle production. Forage production, nutritive value, and persistence under defoliation are species-specific characteristics and greatly altered by fertilization. Animal performance, which is a product of forage quality, is affected by intake (50-70%), digestibility (24-40%) and metabolism (5-15%). Increasing stocking rates tend to decrease herbage allowance and may limit herbage quantity and intake by livestock. In general, bite weight and herbage intake increase linearly with sward surface height and herbage mass, but the results of those variables in animal performance are inconsistent. Nutritive value is also an influential factor in animal performance. The difference in nutritive value of forage species is primarily the result of differences in anatomy and morphology among the plants. The decrease in leaf:stem ratio caused by the onset of reproductive stems elongation is usually one of the main factors decreasing the nutritive value of forages. It is expected that nutritive value determines the forage mass at which the animal performance plateau, with forages with greater digestibility requiring less forage quantity to reach maximum performance. Forage harvest and conservation is required to maintain the forage supply to beef cattle in several locations worldwide. Proper harvest, conservation, and feeding are crucial to meet the animal requirements and achieve the target animal performance during the seasons with shortage of grazing forage. In addition to beef cattle production, it has been identified that grazing systems can provide several ecosystem systems, and forage management has an important role on greenhouse gas mitigation and climate change.

Session 09 Poster 8

Resumption of ovarian cyclicity in *post partum* native beef cows

R.M.L.N. Pereira[1], R. Romão[2], C.C. Marques[1], F. Ferreira[1], E. Bettencourt[2], L. Capela[1], J. Pimenta[1], M.C. Abreu[2], J. Pais[3], P. Espadinha[4], N. Carolino[1] and C. Bovmais[1]
[1]Instituto Nacional de Investigação Agrária e Veterinária, Fonte Boa, 2005-048 Santarem, Portugal, [2]Universidade de Évora, Mitra, 7002-554 Évora, Portugal, [3]ACBM, Rua Diana De Liz, 7006-806 Évora, Portugal, [4]ACBRA, H Coutada Real, 7450-051 Assumar, Portugal; rosa.linoneto@iniav.pt

Resumption of ovarian cyclicity in *post partum* beef cows is determinant for reproductive efficiency and productivity of herds. Our goal was to identify the resumption of ovarian cyclicity in *post partum* Mertolenga (MT) and Alentejana (AL) beef cows raised extensively and their relationship with age, score condition, weight, season and β-hydroxybutyrate (BHB) blood concentration. This work was carried out in three herds in the Alentejo region (2018-2021). The MT (n=53) and AL (n=73) cows were evaluated between 17 and 186 days *post partum*. To identify the beginning of regular ovarian activity, gynecological examinations of the uterus and ovaries were performed by transrectal palpation and ultrasound. Cow's body condition (scale 1-5) and weight were also assessed. Blood BHB were measured using FreeStyle Precision β Ketone. Data were analysed by the PROC FREQ and PROC LOGISTICA. *Post partum* AL and MT cows had a mean condition score of 3.1±0.82 (ranging from 1.5 to 4.5) and 3.7±0.53 (ranging from 2.25 to 4.5), weighing 573.6±85.94 and 470.3±63.28 kg, respectively. Ovaries dimensions were: left (AL=29.7±8.04 and MT=24.7±6.96 mm) and right (AL=34.4±7.85 and MT=25.2±6.63 mm) sides, with developing follicles (right, AL=10.2±6.26 and MT=10.9±5.94 mm; left, AL=9.2±5.94 and MT=11.2±6.40 mm) and mature corpus luteum (right, AL=10.2±6.26 and MT=10.9±5.94 mm; left, AL=9.2±5.94 and MT=11.2±6.40 mm). In both breeds, in cyclic cows the uterus was classified more often ($P<0.05$) as involuted without content while in acyclic cows in involution with content (>7 mm). The probability for resumption of ovarian cyclicity is very dependent on the length of *post partum* interval ($P<0.001$). The present work studied different factors influencing the resumption of ovarian cyclicity in the *post partum* native beef breeds, contributing to establish management strategies to optimize the reproductive performance of beef cattle. Funded by PDR2020-101-3112 and UIDB/00276/2020.

Effects of maternal undernutrition in late gestation on uterine haemodynamics in suckler cows

L. López De Armentia[1], A. Noya[1], J. Ferrer[1], P. Gómez-Ochoa[2], I. Casasús[1] and A. Sanz[1]
[1]CITA de Aragón – IA2 (Universidad de Zaragoza), Avda Montañana 930, 50059 Zaragoza, Spain, [2]Vet Corner, C. Mosén José Bosqued 2, 50012 Zaragoza, Spain; asanz@aragon.es

Undernutrition in late pregnancy is a common scenario in extensive systems. The impact on foetal development and the physiological mechanisms involved have to be ascertained, as 75% of foetal growth occurs in this period. We examined the hemodynamic changes in the uteroplacental unit during the late third of gestation in beef cattle in an undernutrition environment. Sixteen lactating cows were synchronised and artificially inseminated. From the 7th month of gestation to calving cows were allocated to two diets (CONTROL (100% requirements) or SUBNUT (60%)), which resulted in 44 and 12 kg of total weight gain). Uterine arteries were interrogated by means of transrectal Doppler ultrasonography (EXAGO, imv-imaging, France) on days (d) 195, 221 and 250 of pregnancy. Velocity, resistance index (RI), area and blood flow of uterine arteries were measured. Data were analysed with a mixed linear model with maternal diet and day of gestation as fixed effects. All parameters were dependent on the foetus location, ipsilateral uterine artery quadrupling the blood flow compared to the contralateral one (13.836 and 3.579 ml/min; $P<0.001$). Focusing on the ipsilateral artery, the increase of systolic velocity from d 195 to d 250 was higher in SUBNUT cows (193 vs 177 cm/s, for CONTROL and SUBNUT at d 195; 215 vs 247 cm/s for CONTROL and SUBNUT at d 250). On d 250 RI of ipsilateral artery was higher in SUBNUT cows (0.46 vs 0.55 RI; $P<0.05$). The area of ipsilateral artery increased as the pregnancy progressed (1.26, 1.58 and 1.68 cm^2, on d 195, 221 and 250 of pregnancy; $P<0.001$). At d 250, the area was higher in CONTROL cows (1.78 vs 1.59 cm^2), although the difference was not significant. On d 221 of gestation, total uterine artery blood flow was lower in SUBNUT cows (21.079 vs 14.103 ml/min; $P<0.05$), this difference disappearing on d 250. In conclusion, a high significant resistance index should be considered an indicator for deficient nutrition to the foetus due to a decrement in diastolic perfusion. Therefore, maternal undernutrition may be linked to a lower uterine artery blood flow. Funded by Project PID2020-113617RR-C21 FETALNUT.

Immunocastration buffers changes in blood biochemistry markers in fallow deer fed low level nutrient

V. Ny[1,2], T. Needham[2], L. Bartoň[1], D. Bureš[1], R. Kotrba[2,3], A.S. Musa[2] and F. Ceacero[2]
[1]Institute of Animal Science, Department of Cattle Breeding, Přátelství 815, 104 00 Praha 22-Uhříněves, 10400, Czech Republic, [2]Czech University of Life Sciences Prague, Department of Animal Science and Food Processing, Faculty of Tropical AgriSciences, Kamýcká 129, 165 00 Praha-Suchdol, 16500, Czech Republic, [3]Praha 22-Uhříněves, Department of Ethology, Přátelství 815, 104 00 Praha 22-Uhříněves, 10400, Czech Republic; nyv@ftz.czu.cz

In cervids, blood biochemical markers may indicate changes in various environmental factors, especially in response to the influence of nutrient supplements. Decreasing male androgen hormone by immunocastration (IC) to ease the husbandry of male animals is currently a more acceptable method than physical castration, due to numerous welfare concerns linked to physical castration methods, but it is unexplored in fallow deer. Forty yearling male fallow deer were grouped into four treatment combinations: IC on high or low feed level supplementation, or non-castrated bucks on high or low feed level supplementation. Diet affected all the body growth parameters (slaughter weight, ADG, carcass weight, dressing percentage, and body condition score). Fallow deer increased fat and energy blood biochemistry markers at slaughtering such as plasma glucose (GLU) and triglyceride (TRIG) concentrations which might be explained by the physiological response of the deer to annual seasonal variations, as fallow deer increase fat storage for winter utilization. Higher-level supplementary feeding decreased plasma albumin (ALB) and creatinine (CREA), while increased globulin (GLOB) might be due to an imbalance of hepatic function of protein markers in this group. Diet had milder effects on blood biochemistry markers in the IC group compared to the control group. This might be because the IC animals did not produce testosterone first rutting period, thus the IC group undergoing less physiological stress related to the process of puberty and associated agitation often manifested in non-castrated males. Overall, IC can be used as an acceptable welfare tool to reduce physiological stress during the first rutting period and for better feed utilization even in low-level feed supplementation in yearling male fallow deer.

Precision feeding of lactating sows: evaluation of a decision support system in farm conditions

R. Gauthier[1,2], C. Largouët[3], D. Bussières[4], J.P. Martineau[4] and J.Y. Dourmad[2]
[1]Univ Rennes, CNRS, Inria, IRISA, UMR 6074, 35000 Rennes, France, [2]PEGASE, INRAE, Institut Agro, 35590 Saint Gilles, France, [3]Institut Agro, Univ Rennes, CNRS, INRIA, IRISA, 35000 Rennes, France, [4]Groupe Cérès inc., 845, route Marie-Victorin, Lévis, Québec G7A 3S8, Canada; raphael.gauthier@gmx.com

Precision feeding (PF) aims at providing the right amount of nutrients at the right time for each animal. Lactating sows generally receive the same diet, which either results in insufficient supply and body reserve mobilization, or excessive supply and high nutrient excretion. With the help of online measuring devices, computational methods, and smart feeders, we introduced the first decision support system (DSS) for lactating sows PF. Precision (PRE) and conventional (STD) feeding strategies were compared in commercial conditions. Each PRE sow received each day a tailored ration, computed by the DSS. This ration was obtained from the blend of a diet with high AA and mineral contents (13.00 g/kg SID Lys, 4.50 g/kg digestible P) and a diet with low contents (6.50 g/kg SID Lys, 2.90 g/kg digestible P). All STD sows received a conventional diet (10.08 g/kg SID Lys, 3.78 g/kg digestible P). Before the trial, the DSS was fitted to farm performance for the prediction of piglet average daily gain (PADG) and sow daily feed intake (DFI), with data from 1,691 and 3,712 lactations, respectively. Sow and litter performance were analysed for the effect of feeding strategy with ANOVA, considering statistical significance when $P<0.05$. The experiment involved 239 PRE and 240 STD sows. DFI was high and similar in both treatments (PRE: 6.59, STD: 6.45 kg/d; $P=0.11$). Litter growth was high (PRE: 2.96, STD: 3.06 kg/d), although slightly decreased by about 3% in PRE compared to STD ($P<0.05$). Sow body weight loss was low, although slightly higher in PRE sows (7.7 vs 2.1 kg, $P<0.001$), which might be due to insufficient AA supplies in some sows. Weaning to oestrus interval (5.6 d) did not differ. SID Lys intake (PRE: 7.7, STD: 10.0 g/kg; $P<0.001$) and digestible P intake (PRE: 3.2, STD: 3.8 g/kg; $P<0.001$) were reduced by 23% and 14% in PRE sows and feed cost by 12%. Excretion of N and P were reduced for PRE sows by 28% and 42%, respectively. According to these results, PF appears to be a very promising strategy for lactating sows.

Growth performance and gut health of low and normal birth weight piglets fed different zinc sources

C. Negrini[1], D. Luise[1], F. Correa[1], L. Amatucci[1], S. Virdis[1], A. Romeo[2], N. Manzke[2], P. Bosi[1] and P. Trevisi[1]
[1]University of Bologna, DISTAL, Viale G. Fanin 44, 40127, Bologna, Italy, [2]Animine, 10 rue Léon Rey Grange, 74960 Annecy, France; clara.negrini2@unibo.it

The study aims to compare the effect of two Zn sources on performance, microbial profile, and gut status in low (L) and normal (N) birth body weight (BW) piglets after weaning. At farrowing, 64 piglets from 13 litters were selected based on their birth BW and divided into normal BW (NBW>1 kg; 32 piglets) and low BW (LBW<1 kg; 32 piglets). At weaning (25 days of age, d0), piglets were allotted into 4 groups (8 replicates of 2 piglets/group) as follows: (1) LBW piglets fed a standard diet plus 120 ppm of Zn from $ZnSO_4$; (2) NBW, piglets fed a standard diet plus 120 ppm of Zn from $ZnSO_4$; (3) LBW fed standard diet plus 120 ppm of Zn from a potentiated Zn source; (4) NBW piglets fed a standard diet plus 120 ppm of Zn from a potentiated Zn source. Piglets were weighed at d0, d7, d14, and d21. Feed intake (FI) and faecal index were recorded daily. On days (d) 9 and 21, 1 piglet per replicate was slaughtered and colon content was collected for microbiota analysis, as well as the pH from the distal jejunum, cecum, and colon was measured. Data were analysed using a linear mixed model or a generalized linear mixed model with a Poisson distribution including treatment, class of BW and their interaction as fixed factors, and the litter as a random factor. There was an interaction between Zn source and birth BW for faecal index during d0-14 and d0-21 ($P<0.01$). Faecal index was lower for pigs fed a potentiated source of zinc compared to piglets fed $ZnSO_4$ (d0-d9, $P=0.04$; d0-d14 and d0-d21, $P<0.001$). Piglets from LBW group had consistent lower BW and FI throughout the study ($P<0.01$) compared to NBW piglets. Pigs fed potentiated Zn tended to have higher ADG from d0 to d9 ($P=0.07$) and from d9 to d14 ($P=0.08$). On d14, piglets fed potentiated Zn tended to have higher BW ($P=0.09$) than pigs fed $ZnSO_4$. G:F in the overall period was higher ($P<0.05$) for piglets fed potentiated Zn compared to piglets fed $ZnSO_4$. The pH of the jejunum in piglets fed potentiated Zn was lower than in piglets fed $ZnSO_4$ ($P=0.02$). In conclusion, potentiated Zn at the EU authorized dietary Zn levels positively affected performance and gut pH of LBW and NBW piglets.

Effect of maternal diet on slow and fast growing piglet faecal microbiota and volatile fatty acids

F. Palumbo[1,2], G. Bee[2], P. Trevisi[1], F. Correa[1], S. Dubois[2] and M. Girard[2]
[1]University of Bologna, Department of Agricultural and Food Sciences, Viale G Fanin 44, 40127 Bologna, Italy, [2]Agroscope Posieux, Route De La Tioleyre 4, 1725 Posieux, Switzerland; francesco.palumbo@agroscope.admin.ch

Specific fractions of dietary fibres (DFs), such as hemicelluloses (HCs), can affect the gut microbiota of lactating sows. In a recent study, we showed that a decreasing level of HC during lactation affected sow's faecal microbiota, increased the proportion of acetate and propionate and decreased the proportion of butyrate and valerate in faeces. We hypothesise that those changes can in turn affect the faecal microbiota of the offspring. As slow growing (SG) piglets are more prone to diseases and show greater mortality than their fast growing (FG) siblings, the aim of the study was to test whether decreasing the level of HC in sow's lactation diet differently affect the faecal microbiota and VFA profile of SG and FG piglets. From 110 days of gestation to weaning (25±0.4 day post-farrowing), 35 sows were assigned to one of four diets, formulated to contain either 13, 11, 9 or 8% of HC, the same level of DFs, to differ in the DF sources and to be isonitrogenous and isocaloric. Piglets were weighed at birth and at 16 days of lactation. According to their average daily gain (ADG), two piglets per litter were selected and divided in two categories: SG (n=35; 167±10.1 g/day) and FG (n=35; 280±10.1 g/day). Faeces were collected at 16 days of lactation. Results showed no interactions between the maternal diet and the growth rate. Regardless of the growth rate, 11, 9 and 8% of HC showed three *genera* that significantly differed (P adjusted<0.05) compared with 13% of HC: *Faecalibacterium*, *Parasutterella* and *Dialister*. Regardless of the maternal diet, FG piglets had a greater proportion of isobutyrate and isovalerate (P<0.01) in faeces compared to SG piglets. At the genus level, *Enterococcus* and *Succinovibrio* were more abundant (P adjusted<0.01) and *Olsenella* was less abundant (Padjusted <0.01) in FG than SG piglets. In conclusion, the present study confirmed an effect of the maternal diet on the microbiota of the progeny. Further studies are needed to better understand the association between the differences in ADG within the litter during the pre-weaning period and the faecal microbial composition.

Effect of dietary *Ulva lactuca* inclusion on weaned piglet small intestinal morphology and proteome

D.M. Ribeiro[1], D.F.P. Carvalho[1], C. Leclercq[2], M. Pinho[3], J. Renaut[2], J.A.M. Prates[3], J.P.B. Freire[1] and A.M. Almeida[1]
[1]Instituto Superior de Agronomia, Universidade de Lisboa, LEAF – Linking Landscape, Environment, Agriculture and Food, Tapada da Ajuda, 1349-017, Lisboa, Portugal, [2]LIST – Luxembourg Institute of Science and Technology, Green Tech Platform, Environmental Research and Innovation Department, Belvaux, 4422, Luxembourg, [3]Faculdade de Medicina Veterinária, Universidade de Lisboa, CIISA – Centro de Investigação Interdisciplinar em Sanidade Animal, Avenida da Universidade Técnica, 1300-477, Lisboa, Portugal; davidribeiro@isa.ulisboa.pt

Seaweeds can potentially replace conventional feedstuffs whilst improving environmental sustainability of pig production. *Ulva lactuca* is a green seaweed whose recalcitrant cell wall has antinutritional effects. We hypothesize that carbohydrate-active enzyme (CAZymes) supplementation can promote its degradation during digestion. A total of 44 male piglets (Large White × Duroc) were bought from a commercial farm and randomly assigned to four dietary treatments: control (maize, wheat and soybean meal-based diet), UL (7% *U. lactuca* replacing the basal diet), ULR (UL + 0.005% Rovabio Excel AP®) and ULU (UL + 0.01% ulvan lyase). The trial lasted for two weeks. At the end, piglets were slaughtered. Small intestine samples were taken for histological (duodenum, jejunum and ileum) and LC-MS proteomic (ileum) analysis. Data was analysed by ANOVA using the SAS and Progenesis software, respectively. Duodenum crypt depths were significantly higher in control (477.21 μm) compared to UL (377.80 μm). Consequently, the villus/crypt ratio was significantly higher in UL compared to control and ULR. In the ileum, control piglets had higher abundance of ribosomal (RPS7, RPS5) and proteasome (PSMC4, PSMB4) proteins compared to seaweed diets, reflecting higher protein turnover rates. Control piglets had increased abundance (vs UL) of carbamoyl-phosphate synthase, which removes excess ammonia originating from amino acid (e.g. glutamine) degradation, a major energy source for enterocytes. Our study indicates that dietary *U. lactuca* and CAZyme supplementation affects enterocyte protein and nitrogen metabolism, possibly through differential nutrient availability.

Effect of dietary carob pulp inclusion and high vitamin E on pigs carcass characteristics

D.N. Bottegal[1], L. Bernaus[1], B. Casado[1], I. Argemí-Armengol[1], S. Lobón[2], M.A. Latorre[3] and J. Álvarez Rodríguez[1]
[1]Universitat de Lleida, Av. Rovira Roure 191, 25198, Spain, [2]CITA-Aragón, Av. Montañana 930, 50059, Spain, [3]Universidad de Zaragoza, C/Miguel Servet 177, 50013, Spain; diego.bottegal@udl.cat

The aim of the present research was to evaluate the effect of dietary inclusion of carob pulp (Cp) and a high dose of vitamin E (vit E) on pigs' carcass characteristics. A total of 211 carcasses from Duroc×Landrace×Large-White crossbred pigs (94 entire males and 117 females of 170 days of age and 125.1 kg body weight (BW)) were used. In the Bonàrea Group experimental farm and for 39 days, animals were fed *ad libitum* with one of 4 iso-energetic (2,300 kcal NE/kg feed) and iso-protein diets (14.63% crude protein, 0.86% Lys) in a 2×2 factorial design with 2 Cp levels (0 vs 20%) and vit E (40 vs 300 IU/kg). Before slaughter and after a fasted period of approximately 12 h, animals were weighed. In the slaughterhouse, the hot carcasses were weighed (HCW) to calculate carcass yield (CY) and an on-line ultrasound automatic scanner was used to measure backfat thickness (BF) between 12th and 13th ribs, and fat depth at gluteus medius muscle (ham). The carcass lean (Lean%) content was estimated, and carcasses were graded based on European Union scale (S:>60% or E: 55-60% lean). On 7 loins/sex/diet, the loin meat pH was measured at 45 min and 24 h *post mortem*. No effects (P>0.1) of Cp or Vit E (nor interactions) were found on HCW (92.0±1.78 kg), CY (73.6±0.2%), Lean% (60.65±0.18%), BF (16.3±0.25 mm) and ham fat depth (11.4±0.52 mm). Nevertheless, there were differences (P<0,01) between entire males vs female on almost every parameter assessed: final BW, 130.0 vs 120.3 kg; HCW, 94.6 vs 89.8 kg; CY; 72.7 vs 74.6%; Lean%, 60.0 vs 61.6%, BF, 17.7 vs 15.0 mm, respectively. No dietary effects were found on the proportions of carcass percentage grades:72.5% for S and 26.3% for E. Female carcasses were graded more as S and less as E than males (P<0.0001, 86.2 and 13.8% vs 55.8 and 42.1%, respectively). Lastly, no effect was found on pH_{45} (6.2±0.04) or pH_{24} (5.4±0.01). We can conclude that the use of Carob pulp and a high dose of vitamin E does not modify carcass characteristics from crossbred pigs intended for improved meat quality, but females are leaner and lighter than entire males.

Dietary yeast mannan-rich fraction as a gut health solution for sustainable broiler production

S.A. Salami[1], C.A. Moran[2] and J. Taylor-Pickard[3]
[1]Alltech (UK) Ltd., Ryhall Road, Stamford, United Kingdom, [2]Alltech SARL, Rue Charles Amand, Vire, France, [3]Alltech Biotechnology Centre, Summerhill Road, Dunboyne, Ireland; saheed.salami@alltech.com

Dietary supplementation of yeast-derived mannan-rich fraction (MRF) could improve the gut health and performance of broilers, and these positive effects could result in lower environmental impacts in chicken production. In this study, a meta-analysis was conducted to quantify the retrospective effects of feeding MRF (Actigen®, Alltech Inc., USA) on the production performance of broilers. The meta-analysis database included 27 studies and comprised of 66 comparisons of basal (negative control) and antibiotic-supplemented (positive control) diets vs MRF-supplemented diets. A total of 34,596 broilers were involved in the comparisons and the average final age of the birds was 35 days. Additionally, the impact of feeding MRF on the carbon footprint (feed and total emission intensities) of chicken production was evaluated by using the meta-analysis results of broiler performance (basal vs MRF diets) to develop a scenario simulation that was analysed by a life cycle assessment (LCA) model. A database of basal and antibiotic diets vs MRF diets indicated that feeding MRF increased average daily feed intake (ADFI; +3.7%), final body weight (FBW; +3.5%) and average daily gain (ADG; +4.1%) and reduced feed conversion ratio (FCR; -1.7%) without affecting mortality. A sub-database of basal vs MRF diets indicated that dietary MRF increased ADFI (+4.5%), FBW (+4.7%) and ADG (+6.3%), whereas FCR (-2.2%) and mortality (-21.1%) were decreased. For the sub-database of antibiotic vs MRF diets, dietary MRF and antibiotics exhibited equivalent effects on broiler performance parameters, suggesting that MRF could be an effective alternative to in-feed antibiotics. Subgroup analysis revealed that different covariates (year of study, breed/strain, production challenges and feeding period) influenced the effect of dietary MRF on broiler performance. The simulated LCA indicated that feeding MRF decreased feed and total emission intensities by an average of -2.4% and -2.2%, respectively. In conclusion, dietary supplementation of MRF is an effective nutritional strategy for improving broiler performance and reducing the environmental impacts of poultry meat production.

Jejunal and ileal nutrient uptake and epithelium integrity in pigs differing in protein efficiency

M. Tretola[1,2], P. Silacci[2], E.O. Ewaoluwagbemiga[3], G. Bee[1] and C. Kasper[3]
[1]Agroscope Posieux, Swine Research Group, Route de la Tioleyre 4, 1725 Posieux, Switzerland, [2]Agroscope Posieux, Animal Biology Group, Route de la Tioleyre 4, 1725 Posieux, Switzerland, [3]Agroscope Posieux, Animal GenoPhenomics, Route de la Tioleyre 4, 1725 Posieux, Switzerland; marco.tretola@agroscope.admin.ch

Protein efficiency (PE) improvement is essential to develop sustainable pig production. The inefficient use of dietary protein by the animal results in excess excretion of unused protein that contributes to ammonia pollution. This study hypothesized that PE depends on a range of physiological processes such as intestinal amino acid uptake or the intestinal epithelial integrity in the pig. Thirty one Swiss Large White pigs (13 females and 18 castrated males) were reared until an average body weight (BW) of 22.5 kg (±1.6 kg) and subsequently allocated to a pen where they remained until slaughter (BW 106±5 kg). Pigs had *ad libitum* access to a protein-reduced grower and finisher diet and water. For the *ex vivo* evaluation, jejunum segments from the third meter distal to the pylorus were removed within 15 min after exsanguination. Tissues were stripped of outer muscle layers and mounted in Ussing chambers. Trans-epithelial difference (TEER) and short-circuit current (Isc) were continuously monitored. To evaluate L-Glutamate (L-Glut), L-Arginine (L-Arg) and D-Glucose (D-Gluc) uptake, tissues were equilibrated for 20 min before the mucosal addition of 5 mM L-Glut, followed by the addition of L-Arg, L-Meth and D-Gluc at the same concentration every 15 min. Simple linear models were run in R V 4.1.2 using lm() to investigate the influence of Isc, TEER, D-Glut, L-Arg and D-Gluc with PE, respectively. Sex did not influence any of the measured parameters. The TEER showed a negative relationship ($P=0.009$, $R^2=0.21$) with PE, while no relationships were found between PE and amino acids uptake. A significant positive relationship between PE and D-Gluc uptake was also found ($P=0.048$, $R^2=0.14$). The TEER strongly depends on the tight junction protein (TJs) expression. Further analysis are planned to investigate the intestinal TJs protein expression. Similarly, the protein expression of D-Gluc transporters will be the subject of future studies to clarify the correlation between PE and D-Gluc active transport.

Effects of Bacillus amyloliquefaciens CECT 5940 or *Bacillus subtilis* DSM 32315 on chicken PBMCs

F. Larsberg[1], M. Sprechert[1], D. Hesse[1], G. Loh[2], G.A. Brockmann[1] and S. Kreuzer-Redmer[3]
[1]Humboldt-University of Berlin, Breeding Biology and Molecular Genetics, Unter den Linden 6, 10099 Berlin, Germany, [2]Evonik Operations GmbH – Research, Development & Innovation Nutrition & Care, Kantstr. 2, 33790 Halle, Germany, [3]University of Veterinary Medicine Vienna, Nutrigenomics, Veterinärplatz 1, 1210 Vienna, Austria; filip.larsberg.1@hu-berlin.de

Feeding of the spore forming bacteria *Bacillus amyloliquefaciens* CECT 5940 and *Bacillus subtilis* DSM 32315, two commercially available probiotics for chicken, have been described to promote growth performance and health in chicken. However, the underlying mechanisms of probiotic feed additives are still elusive as they may directly influence immune cells or influence the intestinal milieu. We established a chicken *in vitro* cell culture model to explore direct interactions of chicken adaptive immune cells and probiotics. We investigated these *Bacillus* strains as an alternative dietary additive to improve animal health. To investigate direct effects of probiotics, we performed cell culture experiments with primary cultured chicken immune cells of Cobb500 broiler chicken in a co-culture with vital *B. amyloliquefaciens* or *B. subtilis*. Peripheral blood mononuclear cells (PBMCs) were treated with *B. amyloliquefaciens* or *B. subtilis* in a ratio of 1:3 (PBMCs:*Bacilli*) for 24 hours. We found a higher Δ relative cell count of CD4+ T-helper cells ($P<0.1$) as well as CD4+CD25+ activated T-helper cells ($P<0.05$) after treatment with vital *B. amyloliquefaciens*. Furthermore, we found no effect on the Δ relative cell count of CD8+ cytotoxic T-cells and Bu1a+ B-cells. After treatment with *B. subtilis*, we found a higher Δ relative cell count of CD4+ T-helper cells ($P<0.05$) and CD4+CD25+ activated T-helper cells ($P<0.01$). Furthermore, the Δ relative cell count of CD8+ cytotoxic T-cells ($P<0.05$) and CD8+CD25+ activated cytotoxic T-cells ($P<0.05$) was increased after treatment with *B. subtilis*. The Δ relative cell count of B-cells was not affected. These results suggest that T-helper cells are stimulated by *B. amyloliquefaciens* and *B. subtilis,* whereas *B. subtilis* also could activate cytotoxic T-cells. This study could provide evidence of a direct immunomodulatory effect of *B. amyloliquefaciens and B. subtilis* on adaptive immune cells *in vitro.*

Improving welfare of pigs through selection for resilience

A.T. Kavlak and P. Uimari
University of Helsinki, Agricultural Sciences, Koetilantie 5, 00014 Helsinki, Finland; alper.kavlak@helsinki.fi

Selection for high resilience in pigs will improve animal welfare and increased the sustainability of Finnish pork production. The objective of this study is to: (1) extract new phenotypes related to resilience based on fluctuations in daily feed intake (DFI, g) and time spent in feeding per day (TPD, min) using the feed intake observations; (2) estimate heritability of these traits; (3) estimate genetic correlations with production traits (PT). Data consisted of 7,347 Finnish Yorkshire, Landrace, and crossbred pigs from a research facility in Längelmäki, Finland. Four phenotypes were extracted from the individual DFI data as novel measures of resilience. The first two resilience traits were pig specific variations in the DFI and TPD that are quantified as a root mean square error (RMSE) using the difference between the observed DFI and TPD and the predicted values base on the within pig regressions of DFI and TPD on age ($RMSE_{DFI}$ and $RMSE_{TPD}$). A pig, that has a small value of RMSE than other pigs, is expected to be more resilient than other pigs. The other two traits are the rate of off-feed days classified as a proportion of days during the test period that pig's DFI and TPD are so small belong to the lowest 5% quantile (QR_{DFI} and QR_{TPD}) compared to all pigs in Längelmäki at the same age. If these resiliency traits are heritable and do not have strong negative genetic correlations with production traits, they can be used in selection to improve animal resiliency and welfare. The production traits considered were average daily gain (ADG, g), backfat thickness (BF, mm), and feed conversion rate (FCR, g/g). The heritability estimates (h^2±standard error) of $RMSE_{DFI}$, $RMSE_{TPD}$, QR_{DFI}, and QR_{TPD} were 0.11±0.03, 0.20±0.03, 0.07±0.02, and 0.16±0.02, respectively. Genetic correlations between the resilience-related traits and PT varied from very strong (e.g. between QR_{DFI} and ADG: −0.89±0.06) to weak (e.g. between $RMSE_{TPD}$ and ADG: 0.15±0.13). In conclusion, the resilience-related traits show moderate genetic variation. The most promising resilience-related trait is $RMSE_{TPD}$ which has a moderate heritability and has only a weak genetic correlation with PT, thus selection for low $RMSE_{TPD}$ would improve pig resilience without having a negative effect on PT.

Single-step genomic prediction of sperm quality traits in Pietrain pigs

M. Bauer[1], C. Pfeiffer[2] and J. Soelkner[1]
[1]University of Natural Resources and Life Sciences Vienna, Division of Livestock Sciences, Department of Sustainable Agricultural Systems, Gregor-Mendel-Straße 33, 1180 Vienna, Austria, [2]PIG Austria GmbH, Waldstraße 4, 4641 Steinhaus, Austria; martin.bauer@boku.ac.at

In the present study genetic parameter estimation and breeding value prediction were carried out for semen quality and quantity traits in Pietrain pigs. The traits inferred were total number of sperm, motility, volume and density. Trait recording was done using a CASA (*computed aided sperm analysis*) system from 2011 to 2021 and delivered data of 111,500 ejaculations from 1,795 Pietrain boars. The animals were kept at three Austrian AI stations. Genomic data was assessed using Illumina 60k and 80k SNP arrays, yielding 43,430 markers from 909 individuals after quality control. Both genetic parameters and breeding values were estimated using a multivariate single-step genomic procedure. Estimated heritabilities for total number of sperm, motility, ejaculate volume and cell density were 24.7, 13.1, 30.4 and 34.0% respectively. We found a small positive genetic correlation (0.205) between motility and sperm density, but negative genetic correlation (-0.307) between motility and ejaculate volume. Pedigree-based and single-step breeding values showed high correlations between 0.945 and 0.982. The breeding value estimation provided proof of the advantage of ssGBLUP compared to pedigree-based BLUP in this study as average reliabilities (r^2) for young boars without phenotypes were substantially higher (0.369 vs 0.217 for total number of sperm, 0.270 vs 0.161 for motility). The overall reliabilities were relatively high for pig evaluations. This implies high selection response if included in routine selection. A negative genetic trend for total number of sperm was observed over the last ten years, indicating the need of monitoring in the future.

From genotyping strategies to phenotype suitability – simulation of pig breeding schemes in MoBPS

T. Pook[1], L. Büttgen[1], L. Hanekamp[1], C. Reimer[2], H. Simianer[1], H. Henne[3] and R. Sharifi[1]
[1]University of Goettingen, Animal Sciences, Albrecht Thaer Weg 3, 37075, Germany, [2]Friedrich-Loeffler-Institut, Höltystrasse 10, 31535 Neustadt, Germany, [3]BHZP GmbH, An der Wassermühle 6, 21368 Dahlenburg, Germany; torsten.pook@uni-goettingen.de

The quantitative analysis and subsequent optimization of a pig breeding program is highly challenging. The design of pig breeding programs is characterized by a complex combination of direct and auxiliary traits for which information is available from self to progeny performances and at various ages. Complexity is added when e.g. stepwise selection is based on incomplete phenotyping, G×E needs to be accounted for or generations overlap. In this study, the breeding scheme of a sire line of the BHZP GmbH is modelled and optimized by the use of stochastic simulations. For this, the software MoBPS was extended to address the aforementioned challenges. Furthermore, the effect of genotyping, changes in the age structure of the breeding nucleus, breeding against behavioural disorders like tail biting, and changes in the index weights are analysed. As one would expect, genotyping of boars leads to a substantial increase in genetic gain (+20.9%) on the target index compared to a pedigree-based evaluation. Genotyping of a small subset of the boars (20%) only leads to a marginal increase in genetic gain (2.9%) while with an increasing proportion of genotyped boars, genetic gains increase over proportionally, as this allows for better differentiation between siblings in the selection process (e.g. 50% boars: 9.7%; 70% boars: 15.5%). Genotyping of gilts leads to increasing prediction accuracies and reliabilities in the boars for traits that were exclusively recorded in sows, while basically no gains were observed for traits measured in both sexes. Overall, genetic gains by additionally genotyping all gilts are only 3.9%. In regard to breeding against tail biting, overall success via breeding is limited, as obtained prediction accuracies (~0.35) are low and basically not influenced by genotyping. Hence, considerable genetic gains can only be obtained at the cost of substantial losses in genetic gain for the other traits. Overall this indicates that more advanced phenotyping strategies than the use of a rare binary phenotype are needed to efficiently breed against tail biting.

An open source pose estimation model for fattening pigs during weighing

C. Winters[1], W. Gorssen[2], R. Meyermans[2], S. Janssens[2], R. D'Hooge[1] and N. Buys[2]
[1]KU Leuven, Laboratory for Biological Psychology, Tiensestraat, 102 – Box 3714, 3000 Leuven, Belgium, [2]KU Leuven, Centre for Animal Breeding and Genetics, Department of Biosystems, Kasteelpark Arenberg 30 – Box 2472, 3001 Leuven, Belgium; wim.gorssen@kuleuven.be

Precision livestock farming technologies are being developed with unprecedented speed. Hereby, mainly computer vision systems are promising as they offer an easy, non-invasive method to monitor animals or collect phenotypes on a large scale. However, data processing, individual identification and availability of adequate (open source) software present the main challenges. This study aims to develop a pig pose estimation algorithm based on videos from an overhead perspective of pig weighings using *DeepLabCut*, an automated video-analytic system. The pose estimation model was in first instance developed in *DeepLabCut* based on 457 annotated frames from seven videos including white pigs of different sizes (20-120 kg). Hereafter, 150 outlier frames were selected from six novel videos with only one pig and added to the dataset to improve tracking performance. For our final model, the mean training dataset pixel error without probability cut-off was 99.2 pixels or 3.4 cm for the training dataset and 95.9 pixels or 3.3 cm for the test dataset. The performance of the automated pose estimation algorithm was validated by comparing tail-neck length, hip width and pig surface area with manual recordings. Pearson correlations between video analysis and manual observations were high for tail-neck length (r=0.94), hip width (r=0.80) and pig surface area (r=0.91). We developed a pose estimation model of individual fattening pigs using *DeepLabCut*. Body parts were estimated accurately with an average tracking error of 3.3 cm and high correlations with manual observations. Moreover, this model can be used to phenotype pigs' body dimensions and activity related traits. This body part tracking model could be valuable for breeding organizations, as it uses open source software which is transferable to other situations. Moreover, it offers a way to expand routine pig weighing with automated phenotyping of pigs' body dimensions and activity.

Poster Presentation

K. Nilsson[1] *and S. Vigors*[2]
[1]*Swedish University of Agricultural Sciences, Ulls väg 26 Uppsala, 750 07 Uppsala, Sweden,* [2]*University College Dublin, Belfield, Ireland; katja.nilsson@slu.se*

In this time slot we allocate time to view the posters.

Capsule for sampling (CapSa): a less invasive tool to sample small-intestinal content in pigs

I. García Viñado, M. Tretola, G. Bee and C. Ollagnier
Agroscope, Pig Research Unit, Route de la Tioleyre 4, 1725 Posieux, Switzerland; ines.garciavinado@agroscope.admin.ch

The spatial organization of bacterial communities along the gastrointestinal tract complicates the analysis of microbiomes. The bacterial composition varies in the different gut compartments, its composition changes as pigs grow. Unless ileal cannulated, it is up to now impossible to collect multiple intestinal content from the same pig. Nowadays, sampling microbiota is limited to faeces or can be done after slaughtering to directly collect intestinal content. A new sampling capsule (CapSa) was developed to perform non-invasive sampling of the small intestinal content on the same individual. This novel capsule is designed to resist acidic pH, therefore it should stay closed in the stomach, but it opens at basic pH allowing collection of microbiota samples in the small intestine. This study assessed the *in vitro* sampling mechanism of CapSa under various simulated gastrointestinal environments. Three *in vitro* assays were carried out to test the CapSa at both low (pH≤4.2) and neutral pH (pH=7) to simulate the gastric and the small intestine environment, respectively. In trial 1, the CapSas (n=51) were placed in water at pH2 for 30 min and then moved into water at pH7 until all CapSas had sampled. In trial 2, the CapSas (n=8) were left for 120 min at pH=2 and then moved to pH7. In trial 3, the CapSas (n=5) were kept for 2 h in a solution simulating gastric juices at pH4.2 and then moved to a buffer at pH=7. In all 3 trials, none of the CapSas started sampling under acidic conditions. However, as per design, all CapSas correctly collected samples under neutral conditions. The majority of the CapSas (72.5, 100 and 100%, in trial 1, 2 and 3, respectively) collected a sample within 60 minutes of being moved to the simulated small intestinal environment. The sampling (opening and closing) took less than 10 seconds and sampled volumes ranging from 100 to 250 µl. In conclusion, these preliminary trials revealed that the CapSas performed well in all the simulated environments tested. In the following steps, we will perform *in vivo* experiments to confirm the efficiency of CapSas to collect small intestine content. This project has received funding from the European Union's Horizon 2020 research and innovation program under grand agreement no. 955374.

Session 10 — Poster 15

Influence of body lesion score on oxidative status and gut microbiota of weaned pigs

F. Correa[1], D. Luise[1], G. Palladino[2], P. Bosi[1], D. Scicchitano[2], P. Brigidi[3], P.L. Martelli[2], G. Babbi[2], S. Turroni[2], M. Candela[2], S. Rampelli[2] and T. Paolo[1]
[1]University of Bologna, DISTAL, Viale Fanin 46, 40127, Italy, [2]University of Bologna, FaBiT, Via Zamboni 33, 40126, Italy, [3]University of Bologna, DiMeC, Via Zamboni 33, 40126, Italy; diana.luise2@unibo.it

The study aims to evaluate whether social stress can affect faecal microbiota, oxidative status, and serum profile of weaned pigs. 43 pigs were included in the study. Lesion incidence was assessed at two time points on the same subject: one-week post-weaning (28 d of age, T1) and 7 weeks post-weaning (T2). Lesions were measured on skin, tail, ear, neck, middle trunk, and hind quarters of each pig and scored according to the WQ® (2009) on a scale from 0 to 2. Based on the lesion score (LS), pigs were classified at each time point as High LS (H), when LS was higher than 1 in at least 2 of the areas considered, otherwise they were classified as Low LS (L). Based on the same scoring system, at T2, pigs were divided into 4 groups: High LS to Low LS (H-L), High LS to High LS (H-H), Low LS to Low LS (L-L) and Low LS to High LS (L-H). Blood and faecal samples were collected at T1 and T2. At T1, H group had a lower biological antioxidant potential compared with L group ($P<0.02$). At T2, the H-L group had a lower d-ROM concentration compared with H-H and H-L groups ($P<0.05$) and the H-H group had a higher concentration of IgA compared with L-L group ($P<0.05$). At T1, the microbial profile of H and L groups was significantly different ($R^2=0.04$, $P<0.01$) but no differences were observed at T2. LEfSe analysis showed that at T1 L were characterized by a higher abundance of *Blautia, Eubacterium coprostanoligenes, Faecalibacterium, Megasphaera, Subdoligranulum* (Padj <0.05). The H group was characterized by Rikenellaceae RC9, Prevotellaceae UCG-003, uncultured- Lachnospiraceae and Lachnospiraceae-Oscillospiraceae (Padj <0.05). At T2, H-H were characterized by UCG-010, H-L by *Agatobachter* and L-L by *Alloprevotella* (Padj <0.05). In conclusion, one-week post-weaning pigs with a high LS had an altered oxidative status and a different microbial profile compared with pigs with low LS; 6 weeks later, pigs that continued to have a high LS had a higher oxidative stress and a higher activation of the immune system.

Session 10 — Poster 16

Effects of a diet containing a mixture of *Chlorella* and *Spirulina* on broiler performance

C. Paiva, R. Reis, D. Carvalho, A.M. Almeida and M. Lordelo
Ins. Sup. de Agronomia, University of Lisbon, LEAF – Linking Landscape, Environment, Agriculture and Food, Tapada da Ajuda, 1349-017 Lisboa, Portugal; cristianamtpaiva@gmail.com

In a world threatened by global warming that heavily conditions animal production, microalgae offer a sustainable way to feed animals by locally producing nutritious biomass from wastewater whilst removing CO_2 from the atmosphere. One hundred and twenty, 1-day-old male broilers were assigned to 4 dietary groups with 5 replicates each: (1) a control diet; (2) a diet with 10% of *Spirulina* spp. (10S); (3) a diet with 10% *Chlorella vulgaris* (10C); (4) a diet with 5% of *Spirulina* spp. and 5% *C. vulgaris* (5S5C). Animals were fed for 36 days and body weight and feed consumption were registered weekly. Carcass traits, the length and weight of the organs and the intestinal viscosity were determined after slaughter. Final broiler weight and overall average daily gain were significantly higher in broilers fed 10C diets in comparison to the remaining treatments ($P<0.0001$), while for the same parameters broilers fed 5S5C were higher than those fed the control and 10S. In the started period (1-15 days), feed consumption by broilers fed 10C and 5S5C was significantly higher than the control and 10S groups ($P<0.0001$). No significant differences in feed consumption were found during the growing period ($P=0.1867$). Overall feed conversion ratio was lower in the groups fed 10C, followed by groups fed 5S5C in comparison to the remaining treatments ($P=0.001$). As for carcass and breast yield, birds fed the 10C treatment had the higher values ($P<0.0001$). As for the relative weight and length of the organs, the proventriculus and gizzard from birds fed 10C were smaller than the rest of the treatments. In contrast, crop, gizzard, pancreas, liver and duodenum were larger in birds that were fed 10S, in comparison to the other groups ($P<0.05$). Ileum viscosity was lower in *Spirulina* spp. fed groups ($P<0.005$) whilst there were no significant differences in duodenum and jejunum viscosity between treatment groups ($P=0.4406$). From the present results, it is clear that *C. vulgaris* may be used as a more sustainable feed ingredient while improving performance parameters of broilers. The deleterious effects on performance from adding Spirulina to the diet may be counteracted by mixing it with *Chlorella*.

Session 10 Poster 17

B-COS – key to resilient animals, plants and people

C. Guidi
Centre for Synthetic Biology / B-COS, Biotechnology, Coupure Links 653, 9000 Ghent, Belgium; chiara.guidi@ugent.be

The estimation is that by 2050, the rise in antibiotic resistance will cause 10 million deaths every year. A large part of this microbial resistance is caused by the use of antibiotics in animal feed. Hence, universities and industry are fully committed to find nutritional strategies that focus on prevention instead of cure. This search is especially challenging due to the growing world population: by 2050 the world population will increase to 10 billion people who will expect a certain quality of life. As a result, we are in need of an additional 44 million ton of animal protein by 2030. So how do we assure healthy and resilient animals without using antibiotics but still be able to feed everyone on a sustainable and responsible manner in 2050? A reliable, robust and responsible nutritional strategy is the microbial production and subsequent application of chitooligosaccharides (COS). These specialty carbohydrates function according to the lock-and-key theory. The specific structure of the COS-molecule is the key that fits a unique lock, e.g. a specific receptor in plants, animals or humans. In this research, defined COS-molecules were made using *Escherichia coli*. The developed microbial cell factories – optimized through metabolic engineering and synthetic biology tools – are able to produce defined COS-molecules and in high amounts, i.e. gram to kilogram amounts of a specific COS-molecule have been produced at bioreactor scale. After years of intensive research, the fermentation titers (15 g/l) and productivities are sufficient for application testing and subsequent valorisation. More specifically, we want to explore the valorisation for animal health using specific and defined COS-molecules as innovative antibiotic alternatives in pig and poultry production, focusing on prevention instead of cure. In the near future, these first industrial proof of concepts, i.e. *in vitro* immunity studies, will allow us to strategically approach partners and investors. The vision of the to-be founded spin-off company, B-COS, is to become key to resilient plant, animals and people. We want to do this by delivering a scalable, reliable and cost-competitive production technology for specialty carbohydrates, initially focusing on tailor-made and structurally perfectly defined COS.

Session 11 Theatre 1

Fundamentals, difficulties and pitfalls on the development of precision nutrition techniques

C. Pomar and A. Remus
Agriculture and Agri-food Canada, 2000 College St., Sherbrooke, QC, J1M 0C8, Canada; candido.pomar@agr.gc.ca

Precision livestock farming (PLF) concerns the management of livestock using the principles and technologies of process engineering. Precision nutrition (PN) is part of the PLF approach and involves the use of feeding techniques that allow the proper amount of feed with the suitable composition to be supplied in a timely manner to individual animals or groups of animals. Automatic data collection, data processing, and control actions are required activities for PN applications. Despite the benefits that PN offers to producers, few systems have been successfully implemented so far. Besides the economical and logistical challenges, there are conceptual limitations and pitfalls that threaten the widespread adoption of PN. Developers have to avoid the temptation of looking for the application of available sensors and instead concentrate on identifying the most appropriate and relevant information needed for the optimal functioning of PN applications. Efficient PN applications are obtained by controlling the nutrient requirement variations occurring between animals, or groups of animals, over time. The utilization of feedback control algorithms for the automatic determination of optimal nutrient supply is not recommended. Mathematical models are the preferred data-processing method for PN, but these models have to be designed to operate in real-time using up-to-date information. These models should therefore be structurally different from traditional nutrition or growth models which are also limited by the inaccuracy of the conceptual principles used to estimate nutrient requirements, such as the constant efficiency of nutrient utilization, body protein amino acid composition. Combining knowledge- and data-driven models using machine learning and deep learning algorithms will enhance our ability to use real-time farm data, thus opening up new opportunities for PN. To facilitate the implementation of PN in farms, different experts and stakeholders should be involved in the development of the fully integrated and automatic precision livestock farming system. Precision livestock farming and PN should not be seen as just being a question of technology, but a successful combination of knowledge and technology.

Contact voltages <0.5V in feeders and drinkers affects inflammatory and oxidative status of piglets

T. Nicolazo[1], E. Merlot[2], G. Boulbria[1], C. Clouard[2], A. Lebret[1], R. Comte[2], C. Chevance[1], J. Jeusselin[1], V. Normand[1] and C. Teixeira-Costa[1]
[1]REZOOLUTION, Pig Consulting Services, Parc d'Activités de Gohélève, 56920 Noyal-Pontivy, France, [2]PEGASE, INRAE, Institut Agro, Le Clos, 35590 Saint-Gilles, France; t.nicolazo@groupe-esa.net

The aim of our study was to describe the effect of stray voltages lower than 0.5 V on inflammation and oxidative status biomarkers in weaned piglets. The study was conducted on a nursery barn with two rooms of 12 pens of 38 28-day old piglets each, on two batches in France. In each pen, stray voltages were measured for each drinker and feeder every two weeks (9, 23, 37 and 50 days after inclusion). On the same days, two cotton ropes per pen were suspended for 30 min in order to collect oral fluid for salivary cortisol dosage. Two pigs per pen (84 pigs in total) were randomly selected at inclusion and blood sampled on days 9, 30 and 50 after weaning for the dosage of haptoglobin, hydroperoxides (HPO), and blood antioxidant potential (BAP). Pens were allocated to 4 groups based on their voltage levels in drinkers and feeders, with high (HVD, >125 mV) or low (LVD, <125 mV) voltage in drinkers and high (HVF, >50 mV) and low (LVF, <50 mV) voltage in feeders (LVD-LVF: n=24, LVD-HVF: n=15, HVD-LVF: n=26, HVD-HVF: n=19). Effects of voltage in drinkers, in feeders and their interaction on haptoglobin, HPO, BAP and salivary cortisol were analysed using a linear mixed model with pigs, pens and replicates as a random effect. Haptoglobin concentration was numerically higher in pigs exposed to HV in drinkers and feeders compared to the others, but without significance. HPO concentration tended to be affected by drinker × feeder voltage interaction (P=0.06). On day 50, HPO concentration was significantly higher in the group exposed to HV in both drinkers and in feeders compared to the 3 other groups (P=0.02). BAP and salivary cortisol concentrations were not different between groups. This is the first time that moderate consequences on oxidative status in weaned piglets of stray voltage lower than 0.5 V is reported in pigs. This suggested a possible detrimental effect for the health of pigs, but these results must be confirmed by further trials.

The ECO-PIG project: use of a new high fibre feed for outdoor finishing of intact male local pigs

J.M. Martins[1,2], R. Charneca[1,2], R. Varino[1,2], A. Albuquerque[1,2], A. Freitas[1,2], J. Neves[1,2], F. Costa[1], C. Marmelo[1], A. Ramos[1,3] and L. Martin[1,3]
[1]ECO-PIG Consortium, Z.I. Catraia, Ap. 50, 3441 S. Comba Dão, Portugal, [2]MED – Mediterranean Institute for Agriculture, Environment and Development, Un. Évora, Ap. 94, 7000 Évora, Portugal, [3]Inst. Politécnico Coimbra, S. Martinho do Bispo, 3045 Coimbra, Portugal; rmcc@uevora.pt

Thirty male pigs of the Portuguese Alentejano (AL) breed raised outdoors with *ad libitum* water and feed were used to test the effects of a new high soluble dietary fibre feed on animal performance and carcass traits. From 40 to 130 kg body weight, surgically castrated (group C) and intact pigs (groups I and IE) were fed with commercial feeds. From 130 kg until slaughter (160 kg), groups C and I ate a commercial diet, while group IE was fed the isoproteic and isoenergetic experimental diet, with the incorporation of agro-industrial by-products. Average daily gain (ADG, g/d) was different between groups, with IE pigs presenting a higher ADG than C (691±15 in IE, 649±22 in I, and 610±12 in C pigs, P=0.008). This led to fewer days on trial of IE and I pigs, when compared to C pigs (167±4 in IE, 175±2 in I, and 193±5 in C pigs, P<0.001). Feed conversion ratio was different in all groups, with the lower value in IE and the higher in C group (3.9±0.1 in IE, 4.2±0.1 in I, and 4.6±0.1 in C pigs, P<0.0001). Commercial yield (%) was higher in IE and I groups (48.9±0.3 in IE, 48.8±0.3 in I, and 46.6±05 in C pigs, P<0.001), mainly due to their higher proportion of untrimmed ham. The opposite happened with the fat cuts (%) (24.7±0.4 in IE, 25.0±0.4 in I, and 28.7±0.3 in C pigs, P<0.0001), due to a lower proportion of belly and backfat cuts in IE and I groups (P<0.001 and P=0.002 respectively). ZP fat depth and average backfat thickness were also lower in IE and I groups than in C group (P<0.0001). Overall, these data show that the experimental diet had no effect on growth and carcass traits of intact AL pigs when compared to the ones obtained in intact AL pigs consuming commercial diets. On the other hand, intact AL pigs raised outdoors reached slaughter weigh faster and produced leaner carcasses than castrated ones. Further studies will test the effect of the experimental high fibre feed on pork boar taint and meat quality of intact AL pigs raised outdoors.

Effect of a methionine supplementation on the response of goats infected by *H. contortus*

L. Montout, H. Archimède, N. Minatchy, D. Feuillet, Y. Félicité and J.-C. Bambou
INRAE (National Research Institute for Agriculture, Food and Environment), GA (Animal Genetic), INRAE Antilles-Guyane, domaine de Duclos, 97170 Petit-Bourg, Guadeloupe; laura.montout@inrae.fr

Gastrointestinal nematode (GIN) infections are considered as one of the major constraints in small ruminant production at pasture. The negative impact of the drugs classically used to control these infections, on soil biodiversity coupled with concerns over the presence of residues in animal products led some states in the world to advocate a significant reduction of the use of chemical molecules in animal production. The nutritional manipulation of small ruminants has long been considered as a tool for the control of GIN infections. The aim of this study was to measure the impact of a rumen-protected methionine supplementation on the response of Creole goat kids experimentally infected with *H. contortus*. The animals were divided in three groups and infested by 10,000 larvae of *H. contortus*: a control group with no methionine supplementation (C: basal diet), a low methionine group (LM: basal diet + 4.5 g/head of rumen-protected methionine) and a high methionine group (HM: basal diet + 10 g/ head rumen-protected of methionine). Parasitological, haematological and immunological data were collected each week during 42 day of experiment. The results showed that the level of parasitism, measured through the faecal eggs counts was significantly lower in the HM group. The first immunological and haematological parameters available by now, are correlated with the parasitological ones. Further analysis under course (RNAseq) will provide more information on the underlying mechanisms. In conclusion, the supplementation with rumen-protected methionine appear as a promising strategy for GIN control in goats.

Effect of melatonin in pregnant ewes on productive and immunological parameters of their offspring

F. Canto, S. Luis and J.A. Abecia
IUCA, Miguel Servet 177, 50013, Spain; francisco.canto@inia.cl

The objective of this study was to evaluate the effects of melatonin implants in pregnant Rasa Aragonesa ewes on birthweight (BW), liveweight (LW), average daily gain (ADG) and passive immunity transfer of female and male lambs under housed conditions. Pregnant ewes were implanted with subcutaneous melatonin (18 mg, Melovine, CEVA, Spain) at fourth month of gestation (MEL, n=11), while control ewes (CON, n=11) did not receive implants. Lamb weight was recorded in the day of birth and weekly up to weaning (42±2 d). ADG from birth to weaning was calculated. Lamb rectal temperature (°C) was measured with a digital thermometer immediately after lambing. Lamb blood samples for inmuglobulin G (IgG, mg/ml) determination were collected between 24 to 48 h of lamb´s life and serum was stored at -20 °C until analysis (Calokit-Sheep test, Zeulab, Spain). A total of 36 lambs were born, of which 21 were females (MEL, n=13; CON, n=8) and 15 males (MEL, n=5; CON, n=10). There was no significant effect of melatonin on BW (kg) of males (MEL=4.29±0.7; CON=4.02±1.3). However, there was a significant effect on BW of female lambs (MEL=3.97±0.6; CON=3.42±0.7; P<0.05). Lambs reared by melatonin-implanted ewes did not show significant differences in any of the six LW evaluations (female: MEL=8.7±1.5; CON=8.3±1.5, male: MEL=8.5±4.2; CON=9.8±1.7) and ADG (female: MEL=0.207±0.05; CON=0.208±0.05, male: MEL=0.207±0.08; CON=0.245±0.05). No significant differences were observed in the rectal temperature of those lambs lambed by ewes treated with melatonin (female=39.0±0.5, male=38.7±0.2 °C) in relation to the control group (female=38.9±0.3, male=39.1±0.5 °C). When analysing immunological aspects of offspring, implantation of ewes with melatonin not presented effects IgG concentrations in female (MEL=4.14±2.1, CON=4.66±2.1 mg/ml) and male lambs (MEL=4.49±2.0, CON=4.85±2.5 mg/ml). In conclusion, melatonin implants during advanced gestation of ewes did not affect either ADG of lambs or their plasma IgG levels, although an increased birthweight of females lambs was observed. This work was funded by the National Agency for Research and Development (ANID) / Scholarship Program / DOCTORADO BECAS CHILE/2020 – 72210031.

Rumen fill limits fractional absorption rate of volatile fatty acids

K. Dieho[1], P. Piantoni[1], G. Schroeder[1] and J. Dijkstra[2]
[1]Cargill Animal Nutrition, Veilingweg 23, 5443 LD Velddriel, the Netherlands, [2]Wageningen University, Animal Nutrition Group, De Elst 1, 6700 AH Wageningen, the Netherlands; kasper_dieho@cargill.com

The rumen environment is heterogeneous, with higher VFA concentrations and lower pH in the dorsal rumen compared with the ventral rumen and the reticulum. One explaining factor is impaired intraruminal fluid exchange within the rumen by feed particles. The objective of this experiment was to evaluate if artificial rumen fill affects the fractional fluid passage and VFA absorption kinetics in dairy cattle. Nine lactating Holstein cows were used in a 3×3 Latin square design experiment, where 45 l of rumen buffer at 39 °C and pH 6.0 containing Co-EDTA (fluid passage marker) and 120 mM of VFA [72, 30 and 18 mM of acetic (Ac), propionic (Pr) and butyric (Bu) acid, respectively] were infused into an empty, washed rumen. The presence of feed was simulated by inserting either 0 (control), 40, or 80 cellulose sponges (16×12×4 cm) into the buffer filled rumen, and represented an empty, half-full, and full rumen. After 1 h of incubation, the sponges and buffer were recovered, weighed and sampled, and the VFA and Co concentrations measured. The fractional clearance rate (/h) of Ac (0.68±0.035), Pr (0.82±0.038), and Bu (0.79±0.037) was not affected ($P>0.63$) by the presence of sponges. However, the presence of 80 sponges decreased ($P<0.05$) the fractional absorption rate of Ac (0.34 vs 0.42±0.022 /h), Pr (0.48 vs 0.57±0.025 /h), and Bu (0.44 vs 0.55±0.025 /h) compared with control. Fractional absorption rates were intermediate with 40 sponges. In contrast, the presence of 80 sponges increased ($P<0.05$) the liquid fractional passage rate (0.34 vs 0.24±0.028 /h), driven by a greater ($P<0.05$) water inflow (20.0 vs 14.0±1.22 l/h) compared with control. Rumen contractions (16±2.7 /5 min) during the buffer incubation were not affected ($P=0.80$) by the presence of sponges. These results indicate that impaired intraruminal fluid exchange may lead to a reduction of the fractional absorption rate of VFA by the rumen wall. This can help explain the occurrence of acidotic conditions in lactating cows, although clearance rate was not affected when using artificial rumen fill and buffer in the present experiment.

Combined approaches to reduce stress and improve welfare and production efficiency in beef cattle

M. Cohen-Zinder[1], Y. Ben-Meir[2], R. Agmon[1], F. Gracia[1], E. Shor-Shimoni[1], S. Kaakoosh[1], Y. Salzer[3] and A. Shabtay[1]
[1] Agricultural Research Organization, Beef Cattle Unit, Ramat Yishay, 39005, Israel, [2]Agricultural Research Organization, Department of Ruminant Science, Rishon Lezion, 50250, Israel, [3]Agricultural Research Organization, Growing, Production and Environmental Engineering Unit, Rishon Lezion, 7528809, Israel; shabtay@volcani.agri.gov.il

The ongoing global population rise alongside the increasing demand for animal-based products, impose intensification burden on the livestock industry, the outcome of which are deleterious effects on health, welfare, efficiency and product quality. In an attempt to improve these variables, our research team uses genetic, biochemical and behavioural approaches, in order to: (1) select for resilient individuals, fit for good quality production, based on the indigenous Baladi cattle; (2) detect and develop early biomarkers in biological fluids, that can predict the risk to develop diseases later in life; (3) implement preventive nutrition to reduce oxidative stress related intestinal and respiratory diseases; and (4) develop tools to estimate affiliative and agonistic traits within the social network dynamics of suckling calves reared in groups. In the current lecture, means to monitor and predict morbidity, stress and pain in ruminants will be presented. However, special emphasis is placed on Social Network Analysis (SNA), a research discipline that deals with monitoring social connections, and provides a deep understanding of the complexity of the social structure. Rearing suckling calves in groups is a 'double-edge sword'; on one hand, it apparently improves animal welfare by allowing individuals to express normal behaviour. On the other hand, it possessed higher risk of infectious diseases transmission, and may uncover negative factors of social network dynamics, which impose stress on less dominant individuals in the group, thus affecting their health status, nutritional state and development. Gaining insight into components is therefore prerequisite to successful management of calves reared in groups.

Effect of feed particle size on rumen fluid dynamics in dairy cows

K. Dieho[1], P. Piantoni[1], G. Schroeder[1], D. Braamhaar[2], A.T.M. Van Knegsel[2] and J. Dijkstra[2]
[1]Cargill Animal Nutrition, Veilingweg 23, 5443 LD Velddriel, the Netherlands, [2]Wageningen University, Department of Animal Sciences, De Elst 1, 6700 AH Wageningen, the Netherlands; kasper_Dieho@cargill.com

The rumen environment is heterogenous, with lowest pH and highest VFA concentrations in the dorsal rumen. Impairment of fluid flow by feed particles may contribute to this heterogeneity. Eight lactating rumen-cannulated Holstein cows (110±20 DIM) were used in a cross-over design experiment with the goal to evaluate the effect of feed particle size on intraruminal fluid flow. Treatments were COARSE (48% DM and 30.1% of DM≥8 mm screen) or FINE (34% DM and 15.8% of DM≥8 mm screen) total-mixed rations. Ingredient and nutrient composition (DM basis) was the same across rations (39% grass silage, 12% corn silage, 6% wheat straw, and 43% concentrates; 38.0% NDF, 15.8% CP, and 15.1% starch). To prepare the different rations, COARSE was mixed for 13 min, and FINE was mixed for an additional 35 min with added water to aid particle size reduction. Cows were adapted to each ration for 20 d before measurement of the rumen fluid dynamics on d 21 of each period. To evaluate intraruminal fluid flows, a pulse dose of Co-EDTA (25 g diluted in 0.5 L) was introduced in the dorsal rumen. Rumen fluid was sampled overtime during the first 4 h after infusion at six sites (3 in dorsal to cranial direction, 3 in dorsal to ventral direction) in the rumen to determine Co concentration. The fractional diffusion rate of Co followed a two-phase model. During phase 1 (equilibration phase), which lasted approximately 1 h, Co concentration decreased rapidly in the dorsal rumen (-2.44±0.538 /h), and increased in the cranial rumen at 7× the rate (1.64±0.371/h) of the ventral rumen (0.24±0.856 /h). During phase 2, fractional diffusion rate of Co in the dorsal and cranial rumen was (-0.27±0.080 /h) and in the ventral rumen (-0.01±0.174 /h). COARSE or FINE did not affect (P>0.12) Co fractional diffusion rates from the dorsal, cranial or ventral rumen sacs. Results indicate that there is a strong fluid flow from the dorsal towards the cranial rumen, and a limited exchange of fluid between the dorsal and ventral rumen. These flows were not affected by the rations differing in DM content and particle size.

Hypocalcaemia prevention programs in commercial dairy herds: from experimental results to practice

T. Aubineau[1,2], R. Guatteo[1] and A. Boudon[3]
[1]INRAE, Oniris, BIOEPAR, Route de Gachet, 44300 Nantes, France, [2]Innoval, Rue Eric Tabarly, 35538 Noyal-sur-Vilaine, France, [3]INRAE, L'institut agro, PEGASE, La Prise, 35590 Saint-Gilles, France; thomas.aubineau@innoval.com

Hypocalcaemia prevention programs have been widely studied in experimental setting, but constraints of their implementation in field conditions could exist. The main objective of this study was to describe, in a context of typical dairy farms in western France, if and how dairy farmers implement prevention programs to prevent the occurrence of hypocalcaemia. Seventy-nine commercial Holstein dairy farms from Brittany (France) were enrolled in a descriptive survey in 2019. We conducted in-person interviews: (1) to describe the nature, frequency, and seasonality of prevention programs reported by farmers among commercial dairy farms; and (2) to assess the quality of their implementation regarding common recommendations. A majority of farmers (80%) used at least one prevention program. Supplying a mineral-mix formulated for dry cow needs in late gestation (53%), acidifying diet in late gestation or supplying calcium at calving (37% for both) were the most frequent. The programs implemented were variable between farms depending on the stability of diet composition throughout the year. A quarter of acidified diet users didn't supply dry cows with a specific mineral-mix supposed to supply adequate amount of phosphorus, calcium, and magnesium, leading to a drop of effectiveness of acidification program. A lack of reliability of feeding practices, such as absence of feed weighing or insufficient frequency of feed delivery, was reported for 61% of surveyed farms. In addition, management practices, such as inappropriate batching practices around calving reported for 22% (cows) to 32% (heifers) of farms, may lead to unsuitable amounts of supplied phosphorus, calcium, or magnesium right before calving. In addition, almost all surveyed farmers didn't use any monitoring process to control the effectiveness of implemented programs. The barriers and reasons for such lack of compliance will be discussed.

Variability of lactose in milk and in blood during the first 36 weeks of lactation of dairy cows

C. Gaillard[1], M. Boutinaud[1], J. Sehested[2] and J. Guinard-Flament[1]
[1]Institut Agro, PEGASE, INRAE, 16 Le Clos, 35590 Saint Gilles, France, [2]ICROFS, Aarhus University, Blichers Allé 20, 8830 Tjele, Denmark; charlotte.gaillard@inrae.fr

The lactose content in dairy cows' milk (LM) varies during an energy deficit and can represent an interesting indicator for production and health management systems. One reason of its variation is the leak of lactose in the blood (LB) also revealing the permeability status of the mammary epithelium. Up to now, the variations of LM and LB have not been studied together over lactation. Therefore, the objective was to describe the variability of LM and LB over dairy cows' lactation. A total of 53 Holstein cows (including 17 primiparous) managed for a 16 months extended lactation were involved. The cows were divided into two feeding strategies, a control feeding strategy (NOR) and an individualized feeding strategy (EXP) during which the cows received a ration enriched in energy in early lactation (on average until week 7 of lactation). The LM and LB averages by animal were highly variable between animals (from 4.65 to 5.10% for LM, and from 18.7 to 104 mg/l for LB). The LM increased from calving to week 3 (from 4.69 to 4.89%), then kept stable, and decreased slowly from week 26 to 36 (from 4.88 to 4.79%). The LB decreased in early lactation (from 77.7 to 47.0 mg/l) and kept stable and low after week 3 until week 36 (on average 37.8 mg/l). Before week 3, the feeding strategy had no effect on LM, while between week 3 and 7 LM was higher for the EXP than the NOR cows, and it was the opposite after week 7. On average over the lactation, the EXP cows had a higher LB than the NOR cows (P=0.02). After week 3, LM was higher for the primiparous than the multiparous cows (P=0.01) while parity did not affected LB (P=0.22). Overall, over the weeks, LM and LB were not correlated. To conclude, LM and LB varied among cows, at the beginning of the lactation, and regarding the feeding treatment. In this study, parity only affected LM, and LB did not provide information to explain LM variations. It would be interesting to continue these observations in the last part of the extended lactation and relate potential health events with LM and LB concentrations.

Locomotor activity of lambs reared by their mothers or artificially reared, measured by actigraphy

J.A. Abecia, C. Canudo and F. Canto
IUCA-UNIZAR, FacVet, 50013, Spain; alf@unizar.es

In order to compare locomotor activity (ACT) of lambs reared or not by their mothers, during their first week of life, 16 lambs (8 twin pairs; 8 males, 8 females) were fitted with an accelerometer (ActiGraph wGT3X-BT; ActiGraph, Pensacola, USA) (46×33×15 mm in size, weight 19 g), 24 hours after birth, attached on a neck collar. Lambs were divided into two groups: group M (n=8), which were reared by their mothers, and group A, which were separated from their mothers after colostrum intake, and artificially fed reconstituted milk replacer using an automatic lamb feeder. Ewes and M lambs were housed in a collective pen (6×7 m), and A lambs were housed in a pen (5×3 m) with another 12 lambs. The sensors were programmed to collect acceleration data from day 2 to day 8 of age. Mean (±SE) hourly ACT (counts/min) of the seven days was calculated, and the effect of lamb gender, rearing mode and daytime (8:00-18:00) or night-time (19:00-7:00) on locomotor ACT was tested by the GLM procedure. The proportion of lambs presenting circadian rhythmicity of ACT was tested by the X^2 test. Significant effects of gender (P<0.001), rearing mode (P<0.001), and day/night-time (P<0.05), and their interactions (P<0.01) on ACT count were observed. A lambs presented more ACT (P<0.05) than M lambs (159±6 vs 143±5, resp.), and both groups of lambs had more ACT during daytime (175±6) than night-time (131±5) (P<0.001). Female lambs (167±6) exhibited a higher overall ACT than male lambs (135±5) (P<0.001), but only in the artificial rearing method (192±8 vs 126±7) (P<0.001). Lambs had two peaks of activity, with no differences between rearing modes, in the morning (298±19) and the evening (263±17), although rearing mode affected the time of both peaks: peak 1 was at 7.22±0.2 am in the M group and at 8:00±0.0 in A group (P<0.01), and the second peak at 19:00±0.5 am in the M group and at 17:37±0.2 in A group (P<0.05). The proportion of A lambs presenting circadian rhythmicity (7/8) was significantly higher than the M group (3/8) (P<0.05). In conclusion, both groups had a similar pattern of ACT, although with peaks at different times, but lambs reared artificially presented a higher locomotor ACT and an earlier circadianity. Female lambs were more active than male lambs in the AR group.

Litter size and reproductive response to hormonal induction in spring-lambing ewes

R. Pérez-Clariget[1], J. Bottino[2], F. Corrales[1] and R. Ungerfeld[2]
[1]Facultad de Agronomia, Universidad de la Republica, Produccion animal y pasturas, Garzón 780, 12900, Uruguay, [2]Facultad de Veterinaria, Universidad de la Republica, Departamento de Biociencias Veterinarias, Ruta 8 kilómetro 18, 13000, Uruguay; raquelperezclariget@gmail.com

An accelerated reproductive program in ewes needs that the females get pregnant during postpartum, even during the seasonal anoestrus while they are nursing. The aim was to determine the effect of nursing litter size on the reproductive responses during spring (October, SH). While 30 multiparous, Corriedale ewes were nursing 1 lamb (GS), 26 had two (GT). Other 68 ewes that were not bred on autumn, and thus did not lamb, were included as controls (GC) (Body condition score: 3.02±0.06; 3.33±0.06; 3.54±0.04, GS, GT and GC, respectively). Intravaginal sponges impregnated with 60 mg of medroxiprogeserone were inserted ~38 days after lambing, remaining *in situ* 7 d, when 350 IU of eCG were administered. Oestrus was controlled with vasectomized rams, and ewes that came into oestrus were cervically inseminated with semen of three adult Corriedale rams (108 sperm, 0.2 ml). The presence of a corpus luteum (CL) was recorded 8 d later in all GS and GT ewes and 42/68 GC ewes, and pregnancy was determined 40 d after oestrus. Data were compared using generalized linear models (PROC GENMOD, SAS OnDemand for Academics). Although the percentage of ewes that came into oestrus did not differ [GS: 73% (22/30); GT: 65% (17/26); GC: 84% (57/68); P=0.14], more GC ewes had a CL than GT ewes [78% (33/42) vs 50% (13/26)] (P=0.05), with GS not differing from the other groups [70% (21/30)]. The conception and pregnancy rates were greater in GC than in GS and GT ewes [conception rate: 60% (34/57), 18% (4/22), and 29% (5/17), pregnancy rate: 50% (34/68); 13% (4/30), and 19% (5/26), for GC, GS and GT, respectively, P≤0.05]. We concluded that, although nursing one or two lambs did not affect the oestrous response to a hormonal treatment, nursing reduced the incidence of ovulations and pregnancies without differences according to litter size.

Welfare evaluation in buffalo species by risk assessment methodology: the Classyfarm system

D. Vecchio[1], G. Cappelli[1], G. Di Vuolo[1], E. De Carlo[1], F. Fusi[2], V. Lorenzi[2], G. De Rosa[3], F. Napolitano[4] and L. Bertocchi[2]
[1]IZSM, Animal Science and Welfare – CReNBuf, Via Della Salute, 2 Portici, 80055, Italy, [2]IZSLER, CReNBa, Via Bianchi, 9, 25124, Italy, [3]Università degli Studi di Napoli Federico II, Agraria, Via della Salute, 1 Portici, 80055, Italy, [4]Università degli Studi della Basilicata, Scienze Agrarie, Forestali, Alimentari ed Ambientali, V.le dell'ateneo lucano 10, Potenza, 85100, Italy; domenico.vecchio@izsmportici.it

ClassyFarm is an integrated system for the categorization of the farms according to the risk assessment methodology (RA). It is an Italian innovation for facilitating and improving the collaboration and the dialogue between the breeders and the competent authority, in order to raise the level of safety and quality of food of animal origin. ClassyFarm collects, gather and process data referred to the following evaluation areas: biosecurity, animal welfare, health, antimicrobial usage. It can be applied to several livestock species, included buffaloes. On request of the Italian Ministry of Health (IMH), CreNbuf, in collaboration with CReNBa, has developed a checklist (CL) for buffalo welfare assessment, based on RA and included in the ClassyFarm system. The multiple-choice CL consists of 80 items. Each item is scored according to 3 categories: 'insufficient', 'acceptable' and 'excellent'. The assessment system includes non-animal based (N-ABMs) and animal-based measures (ABMs). N-ABMs are divided into 2 macro-areas: Area A (28 items) 'Management factors'; Area B (30 items) 'Housing factors'. ABMs are assessed in Area C (14 items). The CL has been tested in 341 farms, with an average of 353 head (min 10, max 2,240). The average overall welfare value was 61.27% (on a scale 0 to 100%). The values of the specific areas were: A, 60.20%; B, 41.34%; C, 71.55%. At least one potential legislative non-compliance was recorded in 39.80% of the farms. These CL represent a functional and smart tool to assign animal welfare index to each farm, and to improve farm management and housing conditions by using the data collected in each Area, giving answers to consumers and add value to farmers' good practices. The IMH is promoting the application of this system at European and international levels.

Relationship between selection for yearling weight and mature size in Nellore females

J.N.S.G. Cyrillo, J.V. Portes, L. El Faro, S.F.M. Bonilha, C.H.F. Zago, R.C. Canesin, L.T. Dias and M.E.Z. Mercadante
Instituto de Zootecnia, Centro Avançado de Pesquisa de Bovinos de Corte, Rodovia Carlos Tonani, 14 175-000, Brazil; jgcyrillo@sp.gov.br

The aim of this study was to estimate genetic parameters of body weights and hip heights of Nellore cows from 1 to 8 years old, using a Bayesian multiple-trait random regression model, and determine the best time to select females for mature size control. The dataset comprised 33,569 body weight and 31,804 height records from 3,860 and 3,487 females, respectively. For body weight, the model included the fixed effects of a contemporary groups and physiological state for pregnancy and lactation, and the age at calving (linear and quadratic) and Legendre polynomials of age with k=3 (quadratic regression) as covariates. Random effects, including the additive genetic effects and the animal permanent environment effect, were also included in the model. For hip height, the model included the fixed effects of a contemporary groups, the age at calving (linear and quadratic) and Legendre polynomials of age with k=3 (quadratic regression) as covariates and random effects of additive genetic and the animal permanent environment effects. The analysis considered polynomials of k=3 to model random effects and the residual variance was considered to be homogeneous. The heritability estimates were 0.26 to 0.59 for body weight and 0.52 to 0.86 for hip height. Positive and high genetic correlations between body weights (0.49-0.99), between hip heights (0.83-0.99) and between body weight and hip height (0.76-0.82) at different ages, suggest that selection at any age will change mature size. The use of selection indices that include body weights in different ages, including mature weight, and hip height can be effective in obtaining a appropriate mature size of the cows to production system. Grant#2017/50339-5, São Paulo Research Foundation (FAPESP)

Reproductive traits in Portuguese Minhota beef cattle

P.M. Araújo[1], J.L. Cerqueira[1,2], A. Kowalczyk[3], M. Camiña[4], J. Sobreiro[5] and J.P. Araújo[1,6]
[1]ESA, Instituto Politécnico de Viana do Castelo, Refóios, 4990-706 P Lima, Portugal, [2]Centro de Ciência Animal e Veterinária, Q. Prados, 5001-801 V. Real, Portugal, [3]Wrocław Uni. Env. Life Sci., Wrocław, 71-270 Szczeci, Poland, [4]Fac. Vet., Univ. San. Compostela, Lugo, 27002 Lugo, Spain, [5]Ass. Port. Cri. Bovinos Raça Minhota, P Lima, 4990-081 P Lima, Portugal, [6]Centro de Investigação de Montanha, IPVC, V Castelo, 4990-706 Refoios, Portugal; pedropi@esa.ipvc.pt

Minhota cattle located on the northwest of Portugal is one of the most important breed on meat aptitude mainly of veal and beef animals. The livestock production system of this breed mainly involve small family farms, using indoor systems or traditional grazing, which calves are fed with maternal suckling, grass hay and complementary concentrate-based diet. Reproductive parameters evaluation in beef cattle have an enormous impact because they are elements essential to obtain a greater animal productivity. Age at first calving (AFC), calving interval (CI) and calving distribution (CD) are mentioned traits because birth date is collected in all production systems. Reproductive data of 2812 cows enrolled in the Minhota Herdbook (Portuguese Association for the Minhota Breed Farmers – APACRA), from 106 farms of the Entre Douro and Minho region were used. Data was analysed using Stat. Package for Social Sciences (SPSS) 22. The averages for AFC and CI with their corresponding standard deviations were 27.3±3.8 months and 401.7±45.3 days, respectively. The majority of the cows (82.0%) had their first calving before 31 months of age, with 30.5% occurrence in the interval between 25 and 28 months (minimum value of 19 months and maximum of 36). There was no relationship between season of birth of females and respective AFC (R=0.005), and AFC and CI (R=0.120). 61% of the CI values were less than 14 months. There is no seasonality in the calving of cows, with 25.9% on winter, 27.5% on spring, 22.2% on summer and 24.4% on autumn, enabling the sale of animals throughout the year. On a total of 14,892 births the percentage of cows calving twins was 1.5%. We can conclude that it would be important to reduce the age at first calving, in order to reduce the cost of rearing heifers.

Session 11 Poster 16

Young cattle fattening with grass fodder and by-products from agri-food industry

B. Sepchat[1] and P. Dimon[2]
[1]INRAE, Herbipole, Theix, 63122 Saint-Genès-Champanelle, France, [2]Institut de l'Elevage, Maison Régionale de l'Agriculture, 87060 Limoges cedex2, France; bernard.sepchat@inrae.fr

The economic context of livestock farming, debates on environmental impacts and feed food competition reinforce the need to build new livestock systems. Question about capacity to fatten animals with grassland resources appears central, particularly where grass is the main resource. In this context, we set up a fattening trial using Angus breed known for its early fattening and ability to valorize grass. We tested, over 2 consecutive years, fattening performance of 3 lots of young bulls (12 Angus × Salers (AS), 12 Charolais × Salers (CS), 12 pure Salers (S)), i.e. 72 animals. They had the same diet, 65% of a mix of ensiled or wrapped grass, supplemented with by-products of agri-food industry (wheat draff and beet pulp). Individual daily records of feed intake were made. Animals were weighed every 15 days, measurements of fat cover by body condition score (BCS) and biopsies of subcutaneous tissue were performed. An economic evaluation was made by calculating gross margin and cost price. Feed intake by AS was higher: 1.9 kg dry matter intake (DMI)/100 kg live weight (LW) vs 1.6 and 1.7 for CS and S. Overall, AS intake 400 kg less dry matter because they were slaughtered earlier. AS were more efficient with 165 g of growth/kg DMI vs 150 g for other lots. AS achieved a growth of 1,490 g/d, 180 and 142 g higher than S and CS. Fattening duration of AS was 222 days, 41 and 58 less than CS and S. Lipids were deposited earlier by AS, since the middle of fattening we note differences:14 kg/100 kg PV vs 13.7 for CS and 12.9 for S. With current marketing conditions and experimental constraints, the three technical itineraries studied lead to a negative gross margin. AS are the most profitable due to a shorter fattening period (and therefore lower feed and bedding costs) and a lower purchase price of grazers (linked to a lower initial LW). An increase of 23, 31 and 47 c€ respectively for AS, S and CS allows to balance the margins. Results illustrate good valorisation of grass fodder and precocity of fattening for AS lot. It is possible to obtain lighter carcasses (350 kg) with a good fattening condition (BCS=3.5). A fair valuation of selling price, in a contractualization logic, is necessary to preserve economic interest.

Session 11 Poster 17

Effect of melatonin implants in pregnant dairy ewes on milk yield and composition

F. Canto and J.A. Abecia
IUCA. Fac.Vet., Miguel Servet 177, 50013, Spain; francisco.canto@inia.cl

The aim of this trial was to evaluate the effects of melatonin implants in late pregnant Lacaune ewes on milk yield, % milk fat (F), % milk protein (P) and % milk lactose (L) under intensive housing conditions. Twenty-three pregnant Lacaune ewes were implanted with subcutaneous melatonin (18 mg, Melovine, CEVA, Spain) 15 days before lambing (MEL, n=13), while control ewes (CON, n=10) did not receive implants. The lambs were removed from the mothers after consuming colostrum, and from the third day of lactation the ewes were milked once per day in a parallel milking parlour system. Daily milk yield was recorded weekly (up to the sixth week) using the milk-flow of the milking parlour (MM25SG, SCR DeLaval). The nutritional composition of the milk samples were analysed weekly until the sixth week of lactation using a milk analyser (Milskoscan SP+). This instrument is calibrated for ewes according to the manufacturer's instructions (Milkotronic Ltd., Bulgaria). The effects of the treatment on milk yield, F, P and L were evaluated statistically using the GLM PROC (SPSS v. 26) in a model that included melatonin treatment and the week of lactation, and their interaction. Treatment and week had no significant effect on milk yield (P=0.17, P=0.39, resp.) throughout the milking period. However, there was significant effect on F, P and L for treatment (P=0.003; P=0.03; 0.03, resp) and week (P=0.005; P=0.00; P=0.00, resp.). In relation to the treatment × week interaction for the variables analysed, no significant effects were observed. There was significant effect (P<0.05) of melatonin on F during week 3 (MEL=6.54±0.37, CON=5.30±0.22%), 4 (MEL=5.91±0.27, CON=5.27±0.22%) and 5 (MEL=6.53±0.29, CON=5.19±0.25%). Additionally, there was a significant effect (P<0.05) on P and L at week 6 of lactation (MEL=4.60±0.04, CON=4.72±0.03; MEL=4.35±0.04,CON=4.47±0.03%, resp.). In conclusion, exogenous melatonin treatment during late pregnancy in dairy ewes did not have effect on milk yield. However, there were compositional effects, increasing milk fat concentration and decreasing milk protein and lactose. This work was funded by the National Agency for Research and Development (ANID)/ Scholarship Program / DOCTORADO BECAS CHILE/2020 – 72210031.

Benchmarking of claw health and lameness in Austrian dairy cattle

C. Egger-Danner[1], M. Suntinger[1], M. Mayerhofer[1], K. Linke[1], L. Maurer[2], A. Fiedler[3], A. Hund[4], J. Duda[5] and J. Kofler[6]

[1]ZuchtData, Dresdner Straße 89/B1/18, 1200 Vienna, Austria, [2]University of Natural Ressources and Life Sciences, Gregor-Mendel-Str. 33, 1180 Vienna, Austria, [3]Praxisgemeinschaft Klauengesundheit, Heerstr. 3, 81247 Munich, Germany, [4]LAZ Baden-Württemberg, Atzenberger Weg 99, 88326 Aulendorf, Germany, [5]Landeskontrollverband Bayern, Landsberger Str. 282, 80687 Munich, Germany, [6]University of Veterinary Medicine Vienna, Veterinaerplatz 1, 1210 Vienna, Austria; egger-danner@zuchtdata.at

The aim of this contribution was to describe the establishment of a benchmarking tool for claw health in Austrian dairy cattle. The basis for documentation of claw lesions by the participating hoof trimmers was the harmonized terminology described in the ICAR Claw Health Atlas. As data basis served the electronically documented claw health data of cows of 512 dairy herds documented on each trimming visit by well-trained hoof trimmers, the recorded culling data of cows of the same 512 herds, and locomotion scorings (score 1-5) taken at monthly performance testings in 99 herds of the year 2020. The following key figures were regarded to be suitable for describing claw health within a benchmarking system: the mean, median and the 10th, 25th, 75th and 90th percentiles of the incidence rate of lameness, the incidence rate of 13 claw lesions, and of the annual culling rate due to claw and limb disorders. For calculation of these key figures only validated data sets were used. Painful so called 'alarm' lesions were found on average in 30.1% of cows (median: 25.9%), however in the farms of the 10th percentile only 6.3% of cows had 'alarm' lesions in contrast to 62.2% of cows in the farms of the 90th percentile. The presented benchmarking system allows a comparison of claw health of the own farm with a large number of other dairy farms showing a similar performance level, similar cow number and the same breed in the frame of the transnational Cattle-Data-Network. Benchmarking of claw health may further support the analysis of improvement potential for the own farm, may encourage farmers to improve animal welfare, and may be helpful for minimizing economic losses due to lame cows suffering from claw disorders.

Interplay between IFN gamma and prokineticin 1 to regulate genes in porcine uterus

S.E. Song and J. Kim

Dankook University, Animal resources science, 119, Dandae-ro, Dongnam-gu, Cheonan-si, Chungnam, Korea, 31116, Korea, South; thdtjdms@naver.com

Pig conceptus trophectoderm secret interferons (IFNs) gamma, chemokines and prokineticin 1 (PROK1) during early pregnancy. Maternal uterus in response to secreted proteins by trophectoderm provide a microenvironment required for growth and development of conceptus and receptivity of the uterus to implantation. It has been shown that both IFNG and PROK1 regulate the endometrial expression of pregnancy related gene including chemokine, cytokine and prostaglands in pig. However, it is unclear the molecular mechanism underlying the effect of IFNG and PROK1 on uterine function. The present study examined the expression, regulation, and function of IFNG and PROK1 on porcine endometrial cells. IFNG treatment in combination with PROK1 stimulated cell migration, adhesion, proliferation and activated phosphatidylinositol 3-kinase (PI3K)/AKT and mitogen-activated protein kinase (MAPK) pathway proteins. In the presence of PI3K inhibitor (Wortmannin), ERK1/2 MAPK inhibitor (U0126) or P38 MAPK inhibitor (SB203580), IFNG/PROK1-mediated uterine endometrial cells adhesion and proliferation were reduced. Expression of CXCL8, CXCL11 and CXCL12 was increased at transcript and protein level in IFNG/PROK1-treated endometrial cells, while the IFNG/PROK1-induced chemokines expression was partially inhibited in the presence of IFNG-R1, IFNG receptor antagonist. In addition, the IFNG/PROK1-induced increase CXCL8, CXCL11 and CXCL12 were inhibited by blockage of PI3K and/or MAPK pathways. These findings provide evidence that interplay between IFNG and PROK1 on PI3K and MAPK signalling pathways and up-regulates expression of chemokines is critical for the attachment/implantation during peri-implantation period.

Session 11 Poster 20

Haemonchus contortus replacement in a sheep flock as an approach to control anthelmintic resistance

A.C.S. Chagas[1], A.P. Minho[1], L.A. Anholeto[1], R.T.I. Kapritchkoff[2], L.A.L. Santos[2] and S.N. Esteves[1]
[1]Embrapa Southeastern Livestock (CPPSE), Rod. Washington Luiz, Km 234, 13560-970, São Carlos, SP, Brazil, [2]Faculty of Agriculture and Veterinary Sciences, São Paulo State University (UNESP), Rod. Prof. Paulo Donato Castellane, 14884-900, Jaboticabal, SP, Brazil; carolina.chagas@embrapa.br

The current situation of anthelmintic resistance in Brazil indicates low expectations for increased breeding of small ruminants. However, studies indicate that the efficacy of anthelmintics (AHs) can be restored by worm replacement (WR) in the hosts. This study aimed to detect the anthelmintic resistance status of a sheep flock for subsequent performance of WR with a susceptible *Haemonchus contortus* isolate, and then to monitor its effect in different sheep breeds. Blood (packed cell volume) and faecal samples (eggs per gram of faeces count (EPG); coprocultures) were collected from 180 ewes of the Santa Inês, Texel and White Dorper breeds. They were weighed and submitted to FAMACHA grading and body condition scoring. A total of 77 ewes were selected (EPG>200), randomized in groups (n=10) and submitted to the Faecal Egg Count Reduction Test (FECRT) (day zero – D0), treated with the following AHs: (1) Albendazole (5 mg/kg BW); (2) Levamisole (6.2 mg/kg BW); (3) Closantel (10 mg/kg BW); (4) Ivermectin (0.2 mg/kg BW); (5) Moxidectin (0.2 mg/kg BW); (6) Monepantel (2.5 mg/kg BW); and (7) control (C, untreated). On D14, a new faecal collection was performed for EPG, coprocultures and to estimate the efficacy of the AHs by the RESO 4.0 program. Faecal cultures indicated the predominance of *H. contortus* in the flock (95.7%), followed by *Trichostrongylus* sp. (4.3%). The efficacy values to the AHs from 1 to 6 were: 0, 81.2, 84.0, 39.8, 80.3 and 0%. Efficacy <95% means that *H. contortus* was resistant to all chemical groups. So, in this five-year project (REVERTA Project), we will be able to monitor the *H. contortus* resistance status after WR of the 180 ewes, also performing *in vitro* and molecular tests. We expect to generate new information for breeds with different degrees of resistance against *H. contortus* and to verify if WR is effective, so that it can be incorporated into integrated parasite control programs.

Session 11 Poster 21

The impact of using internal teat seal or antibiotic at dry-off on SCC in the following lactation

C. Clabby[1,2], S. McParland[2], P. Dillon[2], S. Arkins[1] and P. Silva Boloña[2]
[1]University of Limerick, Faculty of Science and Engineering, Co. Limerick, V94 C61W, Ireland, [2]Teagasc, Animal & Grassland Research and Innovation Centre, Moorepark, Fermoy, Co. Cork, P61C996, Ireland; clare.clabby@teagasc.ie

The objective of this study was to assess the impact of using internal teat seal (ITS) alone, antibiotic alone (AB), or antibiotic plus ITS (AB+ITS) at dry-off in low SCC cows on somatic cell count (SCC) in the following lactation on three research Irish dairy herds. The study was conducted from November 2017 (dry-off season) to the end of lactation of 2018. All herds were spring calving pasture-based systems. Herds mostly comprised of Holstein-Friesian × Jersey crossbreds. Cows which had every test day SCC below 200,000 cells/ml were blocked according to lactation, proportion of Holstein-Friesian genetics, average SCC and expected week of calving in the spring 2018. Cows then were sequentially assigned to receive ITS, AB or AB+ITS at dry-off. Across the three research herds, 76, 71 and 73 cows were enrolled in ITS, AB or AB+ITS treatment groups, respectively. Herds undertook weekly milk recordings during the 2018 lactation which provided information on cow SCC. Cow SCC was log transformed to log 10 SCC (LogSCC) for analysis. The effect of dry-off group on LogSCC was analysed using a mixed model with a cow random effect and fixed effects of parity (2, 3, 4 and 5+), days in milk and herd (1, 2 and 3). The effects of dry-off group on the odds of a cow exceeding an SCC of 100 or 200 × 1000 cells/ml were quantified using logistic regression model adjusted for the same fixed effects as the mixed model. The LogSCC of ITS, AB and AB+TS treatments groups was 4.7(0.04), 4.5(0.04) and 4.5(0.04) respectively, for which there was no significant difference. The odds of a cow exceeding a SCC of 100,000 cells/ml was 2.3 (CI: 1.5 to 3.6) and 2.1 (CI: 1.3 to 3.3) times higher in the ITS group and the odds of a cow exceeding 200,000 cells/ml was 2.8 (CI: 1.7 to 4.8) and 1.8 (CI: 1.1 to 3.1) times higher in the ITS group compared to the AB+ITS and AB groups respectively. This study showed that using ITS alone at dry-off increased the odds of a higher SCC compared to antimicrobials alone or antimicrobials plus ITS.

Session 11 Poster 22

Phytoparasiticides in livestock animals – the ethnobotany and ethnopharmacology approach

R. Carvalho Da Silva[1], N. Farinha[2], O. Póvoa[3] and L. Meisel[4]

[1]Faculdade de Farmácia-Lisboa, Av. Prof. Gama Pinto, 1649-003 Lisboa, Portugal, [2]Instituto Politécnico de Portalegre, Praça do Município, 7300-110 Portalegre, Portugal, [3]VALORIZA-Polytechnic Institute of Portalegre, Praça do Município, 7300-110 Portalegre, Portugal, [4]iMED.ULisboa, Av. Prof. Gama Pinto, 1649-003 Lisboa, Portugal; ritaapsc@gmail.com

Given that Portugal is a country rich in plant biodiversity, ethnobotany and ethnopharmacology represent crucial instruments for the bioprospection of its traditional valorisation. In essence, considering the progressive increase in resistance to compounds used in synthetic-ectoparasiticides, a sustainable question imposes itself, and the plant-derived products (PDPs) have shown a potential to struggle ectoparasites. Considering an ethnobotany-research previously performed by Escola Superior Agrária de Elvas, a new analysis-arrangement of interviews to obtain information on the use of various plant species was carried out to achieve the main objective: to select PDPs based on their ethnopharmacology to reach alternative and sustainable phytomedicines. The researched and analysed data was gathered through fifty-sixth semi-structured interviews in the Alentejo region and plants with ectoparasiticide potential livestock animals identified. After recognizing the ectoparasiticide potential of the different selected plants, it is supported that phytoparasiticides for livestock animals could be developed with extracts of *Juglans regia*, *Daphne gnidium* and *Ruta graveolens*. Toxicity, efficacy, stability, and safety studies should be conducted to prove the possibility of using extracts from these different species of plants. The provided case study allowed us to understand that there are plants with ectoparasiticide potential that have been identified through the traditional knowledge of the populations. An alternative approach plays a critical role in ethnobotany/ ethnopharmacology evolution.

Session 11 Poster 23

Detection of chemical residues of acaricide in Wagyu breed oocyte donor cows

V.E.G. Silva, M.A. Andreazzi, F.L.B. Cavalieri, J.E. Gonçalves, J.E.G. Santos, I.P. Emanuelli, D.A.B. Moreski, G.G. Fanhani and L.P. Mardigan

Unicesumar University and Cesumar Institute of Science and Technology, Veterinary Medicine and master's in Clean Technologies, Avenue Guedner, 1610, Maringá, Paraná, 86.057-970, Brazil; marcia.andreazzi@unicesumar.edu.br

Brazil stands out in the world scenario of beef production and an important point to be considered is the sanitary control of the herd, and a common problem in Brazilian herds, mainly of imported breeds, such as the Wagyu breed, are ectoparasites, such as the tick *Rhipicephalus (Boophilus) microplus*, which require constant control. There are several ways to reduce and control the incidence of ticks, especially chemical control. However, in practice, it is observed that in many cases, producers do not respect the rules for the use of tick pesticide and/or do not respect the grace period and, at the same time, it is verified that the grace period established by the manufacturers can not be suitable for other characteristics inherent to livestock, such as the use of reproductive biotechnologies, evidencing knowledge gaps in this area. Thus, the aim of this study was to evaluate the presence of chemical residues of tick pesticide or their metabolites in the blood and in the follicular fluid and to verify the *in vitro* production of embryos (IVP) in Wagyu cows submitted to treatment with tick pesticide. The experiment was conducted at the Biotechnology Center of school farm Unicesumar, South region of Brazil. Twenty adult female Wagyu bovine oocyte donors were used, distributed in 2 groups: G1 – animals that were not subjected to tick control and G2, animals that were subjected to tick control with a fluazuron based product. After application, all cows were subjected to oestrous synchronization at four times during the experiment, with intervals of 21 days. On days D12, D33, D54 and D75, oocytes were collected for IVP, follicular fluid aspirations were performed and blood samples were collected for analysis. Fluazuron residues were detected in the plasma of the animals up to 54 days after the application of the tick pesticide, however, metabolites of the tick pesticide were not detected in the follicular fluid. The tick treated animals had a higher number of total and viable oocytes, however the viability rate and the blastocyst rate were lower.

Induction of twin gestation in beef cows

K.C. Bazzo, M.A. Andreazzi, F.L.B. Cavalieri, A.I. Toma, K.S. Silva, I.P. Emanuelli and A.H.B. Colombo
Unicesumar University and Cesumar Institute of Science and Technology, Veterinary Medicine and Master's in Clean Technologies, Avenue Guedner, 1610, Maringá, Paraná, 86.057-970, Brazil; marcia.andreazzi@unicesumar.edu.br

Brazilian beef cattle stands out on the world stage, so ranchers have focused on the production of breeds whose meat is more valuable, such as Wagyu, recognized by international cuisine as a noble meat. However, to increase the production of Wagyu cattle, the use of reproduction technologies is essential and, among them, we point out the use of induction of twin gestation, which contributes to the increase in the number of animals born per cow and, consequently, encourages meat production. Thus, the objective of this study was to evaluate the feasibility of using the technique of inducing twin gestation, inovulating Wagyu bovine embryos in mixed-breed receptors. The experiment was conducted at the Biotechnology Center of school farm Unicesumar, South region of Brazil, from June 2021, and 294 adult cows, receptors, with no defined breed, were used, distributed in 2 groups: G1 – 224 cows that were inovulated with only one embryo and G2 – 70 cows that were inovulated with two embryos. All receptors were synchronized and, on D17 of synchronization, the cows were inovulated with 1 or 2 embryos produced *in vitro* from oocytes from donor cows of the Wagyu breed, according to the experimental group. The gestation rate at 30 and 60 days after inovulation was evaluated with the aid of an ultrasound device. The results showed a gestation rate of 37% at 30 days and 36.61% at 60 days for the group that received only one embryo. The group that was inovulated with two embryos had a gestation rate at 30 and 60 days of 54.29%. The results indicate that the technique is promising, because even if the birth of two products does not occur, double inovulation improved the gestation rate. However, more studies must be conducted in order to prove the benefits of double embryo fertilization, and thus, promote the production of quality meat and contribute to the economic sustainability of Brazilian livestock.

Session 12 Theatre 1

Introduction to the social licence to operate for animal production systems

R. Evans
Hogskulen for gron utvikling, Arne Garborgsveg 22, 4340 Bryne, Norway; rhys@hgut.no

In these days of social media, proliferating news channels and growing environmental and ethical concerns about the production and consumption of animals, animal production systems are increasingly coming under the purview of what is called the 'social licence to operate' (SLO). This paper introduces the concept, explores several academic understandings of it and offers tools which animal production systems can use to assure that they can minimize negative public perception of their operations and increase public support for their science, practitioners and operations.

Session 12 Theatre 2

Ethical dog breeding

O. Vangen
Norwegian University of Life Sciences, Animal and aquacultural sciences, P.O. Box 5003, 1432 Aas, Norway; odd.vangen@nmbu.no

In January 2022 the federal court (Oslo District Court) in Norway banned breeding of two of the most unhealthy dog breeds, the English bulldog and King Charles Cavalier spaniel. The Norwegian Kennel Club, the breeding organizations for the two breeds and 6 breeders of both breeds was taken to court by Animal Protection Norway. The court state that breeding these breeds is a violation of the Norwegian Animal Welfare Act due to the heavy load of genetic defects and man-made overbreeding on unhealthy conformation traits. The decision by the court has got a lot of international attention, and many questions and comments have been raised and discussed in many countries. The public awareness has increased and the direction of pure breeding, especially in companion dog breeds is more and more questioned. Why have we allowed the overbreeding on conformation traits to come to the extreme in many dog breeds. Why have the society accepted that many of more than 200 genetic defects/traits in dogs and development of inbreeding levels and health statuses far above anything we would have accepted in farm animals? Many scientific papers on genetic defects due to breeding for extreme conformation traits, have been published. They include breathing problems, birth problems, walking problems, eye problems and skin problems. Several other publications list dog breeds where genetic defects due to selection or extreme conformation, counts for up to 40% of all defects in dogs. Among these countries, Britain, the country of origin for many of the dog breeds in the Kennel Club societies. Animal geneticists with their expertise on breed plans, breeding goals, genetics of traits and experts on inbreeding and genetic defects should play a more mayor role as advisors for dog breed societies and Kennel Clubs. Scientific based discussions on ethical dog breeding should be included among the topics discussed among animal geneticists.

Session 12 Theatre 3

Expectations of stakeholders regarding a German animal welfare monitoring: results of a survey

H. Treu[1], R. Magner[2], U. Schultheiß[2], A. Bergschmidt[1] and J. Johns[1]
[1]Thünen Institute of Farm Economics, Bundesallee 63, 38116 Braunschweig, Germany, [2]Association for Technology and Structures in Agriculture (KTBL), Bartningstraße 49, 64289 Darmstadt, Germany; hanna.treu@thuenen.de

The welfare of farm animals is controversially discussed in Germany: Some stakeholders see it improving whereas others, e.g. animal welfare NGOs, assume deterioration. So far, there has been no regular reporting on the status quo and the development of welfare in livestock farming. Although some data are already collected, these do not provide a complete picture of animal welfare. The aim of the project 'National Animal Welfare Monitoring' is to develop the scientific basis for a regular, indicator-based animal welfare monitoring (AWM). To gather expectations of stakeholders regarding a future AWM, an online survey (convenience sample) was conducted in 2021. Citizens, scientists, veterinarians, farmers, animal welfare NGOs, etc. were asked to rate the importance of different welfare aspects as well as challenges of an AWM. 1,900 people participated in the survey. 87% of respondents find it 'important' that an AWM is implemented. Respondents ranked the following aspects as (very) important in decreasing frequency: physical health (98%), appropriate feed and water, husbandry conditions in general, welfare during slaughter and transport (with 96% each), natural behaviour (93%), water quality of fish (89%), quantity of medicaments (87%), absence of amputations (83%), emotional state (83%), animal welfare training of farmers (80%) and production performance (50%). However, substantial differences between groups such as 'citizens' and 'farmers' exist in the perceived importance of welfare aspects: While agricultural practitioners consider the animal welfare dimension 'health' to be more important than those of 'natural behaviour' and 'affective states', animal welfare NGOs rank them as equally important. If an AWM is to be implemented, most respondents wish the results to be published once a year (70%). Expectations towards an AWM are high: More than 80% of participants expect that farm animal welfare will increase as a consequence of its implementation. The results of the survey will feed the recommendations to the German Federal Ministry of Agriculture which will consider the implementation of an AWM.

The role of human-animal interaction in farm animal welfare

C. Marmelo
Rações Santiago Lda, Rua Namorados, 7500-062, Portugal; csofiamarmelo@gmail.com

The present study consists of a theoretical approach to the interactions between humans and nonhuman animals, in which the contribution of anthrozoology and animal welfare is used to understand the factors that shape them, and a compilation of some studies is set out to exemplify ways to improve them in the livestock context. In the second part, different approaches to animal ethics that can contribute to understanding the nature of some interactions are summarized and three case studies are analysed. This critical analysis, as well as the discussion of the fundamental elements that shape the interactions, allows the systematization of possible solutions to improve the role of the keeper/ operator in coexistence with animals. It is concluded that poor quality in the interaction between humans and animals makes the handling of animals difficult and contributes to the impoverishment of the quality of life of both. This work underlines the importance that the keeper's personal profile, skills and technical and personal training have for fulfilling their professional tasks with high animal welfare criteria, thus establishing a mutually beneficial interaction.

Citizens' attitudes towards an animal welfare label

V.D.L.F. Mansky[1], D. Enriquez-Hidalgo[1,2,3], M.J. Hotzel[4] and D.L. Teixeira[1,5,6]
[1]Pontificia Universidad Catolica de Chile, Departamento de Ciencias Animales, Santiago, Chile, [2]Rothamsted Research, North Wyke, Okehampton, EX20 2SB, United Kingdom, [3]University of Bristol, Bristol Veterinary School, Langford, United Kingdom, [4]Universidade Federal de Santa Catarina, Departamento de Zootecnia, Florianopolis, Brazil, [5]Hartpury University, Department of Animal and Agriculture, Gloucester, United Kingdom, [6]Universidad de O'Higgins, ICA3, San Fernando, Chile; dadaylt@hotmail.com

The aim of this study was to evaluate Chilean citizens' attitudes towards animal welfare labels in different animal products. Participants (n=600) were recruited online and asked their opinion on a hypothetic label to certify animal welfare in one among four animal products (n=150 each): cow's milk, pork, fish, or eggs. Participants were asked: (1) if they would choose the product with instead without the label (always/sometimes/never); (2) how much more they would be willing to pay for the product certified with this label compared to a conventionally produced; and (3) if they thought that this label would ensure a better life for the animals (yes/no). Data were analysed using Chi square. More participants were willing to pay more for the certified compared to the non-certified product in the case of eggs (69%), milk (65%), fish (61%), and pork (57%) ($P<0.05$) which was in the same order with participants' thoughts about whether the label would improve the quality of life of the animals (laying hens: 38%, cows: 31%, fish: 26%, pigs: 19%; $P<0.01$) and in which the participants would always choose to buy the certified product compared to the conventional one (eggs: 66%, milk: 55%, fish: 54%, pork: 42%; $P<0.01$). Participants were favourable to an animal welfare label for animal products and believed that the label could generate an improvement in animal welfare, to a greater or lesser extent depending on the species. This would condition the eligibility and willingness to pay for it.

Session 12 Theatre 6

mEATquality: a project to promote sustainable pork and broiler meat production

H.A.M. Spoolder[1], B. Kemp[2] and B. De Bruijn[1]

[1]Wageningen Livestock Research, Animal Health and Welfare, De Elst 1, 6708 WD, the Netherlands, [2]Wageningen University, Adaptation Physiology, De Elst 1, 6708 WD, the Netherlands; hans.spoolder@wur.nl

The mEATquality project aims to provide consumers with sustainably produced pork and broiler meat, by developing novel solutions that address societal demands, environmental concerns and economic needs on farm and in the chain. The 'extensiveness' of production is a key issue, and will be developed in a stepwise approach. The first step surveys extensive husbandry factors in relation to intrinsic meat quality, through data collection on conventional, free-range and organic farms, and through consumer expectation studies. The second will include controlled experiments on-farm to investigate intrinsic meat quality characteristics in relation to husbandry factors: genetics, forage, space and enrichment. It will also develop innovative techniques for automated meat quality assessment at high line speeds, and combat food fraud through authentication of the final product via 'fingerprinting techniques' and blockchain technology. The third step will check the novel farming practices against sustainability aspects: animal welfare, environmental impact and economic viability. Market acceptance of the new products and ways to communicate them to consumers will be studied. The fourth and final step will communicate and disseminate the results. Key outputs are an 'Extensive Practices' app, animated movies and EIP Practice Abstracts for farmers, educational tutorials for consumers, retailers and restaurants, and an EU Meat Database for authentication purposes. mEATquality is proposed and co-designed by organic sector representatives Ecovalia and Naturland, in collaboration with CLITRAVI, the Liaison Centre for the Meat Processing Industry in the European Union. They joined forces with academic partners (including 5 of Europe's leading meat quality laboratories), Marel (poultry, fish, meat & further processing equipment), retailer Carrefour S.A. and poultry breeder Hubbard. Finally, Plukon Food Group supports the project in-kind. Collectively these partners span the chain from Farm to Fork.

Session 12 Theatre 7

European consumers' evaluation of pork and broiler meat quality attributes

A.O. Peschel, K. Thomsen and K.G. Grunert

Aarhus University, Department of Management, Fuglesangs Allé 4 building 2622, C109, 8210 Aarhus V, Denmark; peschel@mgmt.au.dk

Consumers' evaluation of meat quality attributes is shifting to include not only sensory and health attributes, but also aspects of animal welfare and ethical attributes. The mEATquality project investigates these credence attributes (which cannot be assessed by consumers, but need to be conveyed via the product packaging), as they offer an additional means to achieve product differentiation in the market. For successful product differentiation, it is necessary to understand which of these animal welfare attributes consumers view as relevant when choosing between different meat products. As consumers differ in terms of their preferences, it is likewise important to understand how consumer groups in the market differ in their evaluation of these attributes. Moving towards a better understanding of this matter, we conduct semi-structured expert interviews with pork and broiler meat professionals in six countries (NL, DE, ES, DK, IT, PL). Based on these interviews, we identify the most relevant pork and broiler meat attributes across the dimensions of breed, forage/feed, space allocation and space quality. This information feeds into the design of a discrete choice experiment with 500 consumers in each of the aforementioned six countries. During this task, consumers make trade-offs between different attributes displayed with pork and broiler meat packaging to arrive at the choice of their most preferred product. At EAAP, we will be able to present the results of the segmentation analysis and discuss European consumers' choice behaviour for pork and broiler meat produced under different animal welfare scenarios.

Assessment of animal welfare in Germany according to stakeholders

C. Over[1], R. Magner[2], C. Gröner[1], J. Johns[1], U. Schultheiss[2] and A. Bergschmidt[1]
[1]Thünen Institut, Farm Economics, Bundesallee 63, 38116 Braunschweig, Germany, [2]Association for Technology and Structures in Agriculture (KTBL), Bartningstrasse 49, 64289 Darmstadt, Germany; caroline.over@thuenen.de

The aim of the German project 'National Animal Welfare Monitoring' is to develop a scientific basis for a regular, indicator-based animal welfare monitoring. To include the perspective of different stakeholders, 63 representatives from stakeholder associations and civil society organizations were interviewed in guided qualitative telephone interviews on their definition of animal welfare, the implementation of a national monitoring and the current situation of animal welfare in Germany. Transcribed data was analysed by qualitative content analysis using the software MaxQDA2020. This abstract describes the answers regarding the current animal welfare situation. Of the 63 Interviewees, 16 perceived the animal welfare situation in Germany as good, 12 as neither good nor bad and 32 as problematic or bad. Although more than half of the respondents have a negative perception of the current animal welfare situation in Germany, 34 mentioned an improvement in animal welfare in the last ten years. Reasons for this development were seen regarding the change in attitude towards animals in agriculture and society (n=30) as well as policy changes (n=16) (e.g. the ban of piglet castration without anaesthesia). The introduction of animal welfare labels (n=7), the scientific and technical progress (n=9) and advanced production and management systems (n=14) were seen as additional reasons for the improvement. Regardless of the overall positive assessment of the development, the respondents mentioned numerous adverse influencing factors on the animal welfare situation. Especially the economic situation of farmers (n=15), structural changes in agriculture (n=10) in conjunction with the intensification of production and increases in performance of livestock through breeding and management procedures (n=9) were seen as negative influences on animal welfare. Trade-offs with ecology and economy were perceived as major obstacles in the implementation of welfare friendly animal husbandry (n=10).

The modalities of the retirement of equidae in France

V. Deneux - Le Barh
INRAE, UMR Innovation Bat 27 2 place Pierre Viala, 34090 Montpellier, France; vanina.deneux@inrae.fr

In France, socio-economic studies carried out in the equine sector have established that over the period 2010-2018 the population of equidae over 20 years old has increased by 30%. In 2018, it is estimated that retired equines constitute 16% of the French equine population. At the same time, there has been a 46% decrease in the number of horses sent to the slaughterhouse. From these facts we can hypothesise that professionals are paying increasing attention to the retirement of horses. What are the determinants of this evolution? What are their strategies for ensuring the retirement of horses? The proposed communication presents first results from two research fields. The first one corresponds to thesis results on the living and working conditions between humans and horses. The second field study is an exploratory research, entitled EXIT, on the leaving of the work of the animals, funded by the SANBA (animal health and welfare) metaprogramme of INRAE. The methodology used in both studies is based on a qualitative analysis of over 120 semi-structured interviews. Firstly, the representations that professionals have of this problem of leaving work for their horses will be presented. In particular, their position on sending their horses to the slaughterhouse will be discussed. Then, the different strategies for retiring horses will be presented. Indeed, the strategies deployed for the retirement of horses depend on the number of horses, the sex of the animals, the number of years of collaboration, land and time availability and the economic income of the professionals. Finally, the moral tensions experienced by professionals towards their animals will be exposed. The particularity of working with domestic animals lies in the fact that humans enter into a tacit contract of gift – against gift with their animals. In other words, to the gift that animals make to humans by responding to the tasks requested, they undertake in return to provide them with protection and care. This moral commitment of humans towards their animals is exacerbated when the latter leave their work. This commitment is not without moral tensions between the ideal of 'good' equine retirement and the reality imposed by economic and land rationalities.

American consumers rate grass and grain-fed Australian lamb equally

M.T. Corlett[1,2], G.E. Gardner[1,2], L. Pannier[1,2], A.J. Garmyn[3], M.F. Miller[4] and D.W. Pethick[1,2]
[1]Murdoch University, College of Science, Health, Engineering and Education, Murdoch, WA, 6150, Australia, [2]Advanced Livestock Measurement Technologies, Armidale, NSW, 2350, Australia, [3]Michigan State University, Department is Food Science and Human Nutrition, East Lansing, Michigan, 48824, USA, [4]Texas Tech University, Department of Animal and Food Sciences, Lubbock, Texas, 79409, USA; m.corlett@murdoch.edu.au

Anecdotal reports suggest consumers in the USA perceive Australian lamb meat as more 'gamey' or 'stale' compared to USA lamb. These perceptions may be due to a combination of grass feeding and long storage time under chilled shipping conditions. This study tested American consumers for their eating quality preferences of Australian lamb (n=72) from a commercial flock, fed grass or grain, with meat samples stored for either 5, 21 or 45 days. Meat from 4 cuts (loin, rump, topside and outside) were vacuum packaged 24 hours after slaughter and stored at 1-2 °C for each storage period. Untrained consumers (n=720) at Texas Tech University assessed samples for overall liking using a scale from 1 (worst) to 100 (best). There were significant interactions ($P<0.001$) between diet, ageing period and cut. Consumers preferred grain-fed to grass-fed meat when stored for 5 days, except for outside samples which were non-significantly different. When these cuts were stored from 5 to 21 days their overall-liking scores improved by between 6.1 to 19.4 scores, with the only exception being grain-fed rumps which remained relatively unchanged. Increasing storage time from 21 to 45 days generally resulted in no further change in consumer scores. The exception to this finding was grass-fed loins in which overall liking scores increased by 8.4 units. The results suggest that when lamb is aged for only 5 days, American consumers prefer meat from grain-fed lambs. However, when stored for 21 to 45 days, as occurs under commercial shipping arrangements, American consumers will likely not discriminate between grass and grain-fed lamb from Australia. Moreover, the long storage time will also have no negative impact on lamb eating quality. For this reason, grain feeding is not warranted for lamb shipped by sea to the USA, for improving consumer sensory perception.

Has the Covid-19 pandemic influenced citizens' attitudes towards beef consumption?

D. Enriquez-Hidalgo[1,2,3], V.D.L.F. Mansky[1], M.J. Hötzel[4], D.L. Teixeira[5] and R. Larraín[1]
[1]Pontificia Universidad Católica de Chile, Facultad de Agronomia, Santiago, Chile, [2]Rothamsted Research, North Wyke, Okehampton, United Kingdom, [3]University of Bristol, Bristol Veterinary School, Langford, United Kingdom, [4]Universidade Federal de Santa Catarina, Laboratorio de Etologia Aplicada, Florianópolis, Brazil, [5]Hartpury University, Department of Animal and Agriculture, Gloucester, United Kingdom; daniel.enriquez@bristol.ac.uk

The Covid-19 pandemic caused lockdowns, food access restrictions and highlighted information regarding food production and zoonotic diseases which may have affected citizens' opinions regarding beef production and consumption. The aim of the study was to explore citizens' attitudes regarding beef production and consumption as affected by the Covid-19 pandemic in Chile. Participants were surveyed in April 2020 (n=1,141) and in December 2021 (n=1,221). They were asked: Q1: if they consume meat, Q2: what do they plan to do with their beef consumption in 3 to 5 years; Q3: what do they think the Chilean population WILL and Q4a: SHOULD do about beef consumption and Q4b: why. Then they were asked their level of agreement (0 totally disagree to 4 totally agree) with: Q5: greenhouse gases are emitted in beef production; Q6: beef production is negative for the environment and Q7: beef is bad for human health. Data were analysed using Chi square. Less participants did not eat meat in 2020 than in 2021 (21 vs 27%, $P<0.001$) and fewer planned to reduce their beef consumption in 2021 (50 vs 46%; $P<0.05$). More participants in 2021 thought that Chileans WILL reduce beef consumption (45 vs 49%; $P<0.01$), but their expectations that Chileans SHOULD do it were similar (increase 3%, maintain 18%, reduce 80%). Their answers were mainly due to environmental (38%), health (25%) and animal welfare (14%) concerns. 70% agreed with Q5, but 48% did not agreed that beef is bad for human health (Q7). Agreement level increased from 2020 to 2021 only for Q6 (51 vs 54%; $P<0.05$). During the pandemic citizens reduced their will to reduce beef consumption but increased their expectations that the population will, but not that they should reduce their beef consumption. Citizens awareness of the negative impact of beef production to the environment is high and partially increased during the pandemic.

Session 13 Theatre 1

At farm gate: feed systems, increased soil carbon sequestration and energy production

A.C. Dalcq

CEJA, Rue de la Loi 67, 7th floor, 1040 Brussels, Belgium; vp.dalcq@ceja.eu

At the farm 'de la Gobie' in the centre of Belgium, several practices are implemented to deal with carbon at the level of the farm. Grazing was kept to feed the cows and is optimized. So the permanent grasslands allow carbon sequestration. The temporary grasslands also, they are seed for minimum 3 years, which allows also to sequestrate carbon. Moreover, the mix of seeds contains an important part of legumes. This decreases the needs in nitrogen fertilization and produces a forage more concentrated in protein, decreasing our dependence to purchased protein. We sow also alfalfa to have more forage with higher amount of nitrogen. We are going to test the direct sowing of corn silage in the temporary grasslands for the first time this year. We use mix of cereals and legumes as cover crops that we harvest at Spring to feed a part of our herd. Manure, slurry and soil analyses are realised to have an optimised use of nitrogen. We calculate also the benefit of nitrogen fertilisation on the delay of the first cut. Indeed, in time of high of price of nitrogen fertilizer, it can be more interesting to wait two weeks more for the first cut.

Session 13 Theatre 2

Breeding towards efficiency in Finnish dairy and beef cattle improves environmental performance

S. Hietala[1], E. Negussie[1], A. Astaptsev[2], A.M. Leino[1] and M. Lidauer[1]

[1]Natural Resources Institute Finland (Luke), Latokartanonkaari 9, 00790 Helsinki, Finland, [2]Valio ltd., Meijeritie 6, 00370 Helsinki, Finland; sanna.hietala@luke.fi

Livestock production is acknowledged for its large footprint on environment, and it has been estimated that it is responsible for 14.5% of all anthropogenic GHG emissions. The emissions of the food system are understood to be mainly generated in primary production and agricultural processes, on farms. As a large part of livestock sector's emissions are generated in ruminant production, especially the cattle-based primary production is driven to mitigate its climate change impact. Regarding cattle production, good feed efficiency and improved production yields can be seen as major factors in also improving the environmental performance and efficiency. The key elements in improving these relies in genetic improvement. In the ongoing project A++ Cow in Finland, the main objective is to develop genomic prediction models to identify dairy cattle that inherit genetics for improved feed efficiency, lower methane emissions from enteric fermentation and thus, to improved resource efficiency and environmental impacts. This work is conducted in close collaboration with dairy cattle breeding stakeholders and with the largest Finnish dairy producer Valio ltd, utilizing vast farm data from company's suppliers. Here, the life cycle assessment and impacts of genetic selection of dairy cattle to climate change impact are presented for Finnish dairy in collaboration with Valio ltd. In preliminary results for dairy cows, the impact of breeding was seen to achieve -10% to -19% reduction to dairy and dairy beef production at farm level with scenarios for breeding to 2035 and 2050. Similarly, in ongoing Beefgeno project, the preliminary results indicate up to 15% difference in climate change impact of beef fattening phase, when assessed cattle was ranked based on the Finnish breeding scheme traits. Thus, it can be stated that efficient breeding is an effective tool to mitigate emissions to environment.

GHG emissions mitigation in practice: a Dutch farmer with monogastrics & ruminants

I. Gijsbers
Normec Foodcare, Balkweg 3, 5232 BT Den Bosch, the Netherlands; iwan@gijsbersbeheer.nl

In the Netherlands, there is a 'climate-deal' with industries and farmers. As farmers we work with the government, scientists and farmers in a programme to reduce GHG emissions that is based on an integrated approach to feed animals, manure and housing. As farmers we already produce with a very efficient carbon footprint by state of the art production systems. Now we are going to the next level: our carbon footprint becomes even more efficient by integrating smart solutions. Our ruminants eat all the feed we can't digest, our monogastrics eat all our leftover food. With smart animal- and feed management we reduce digestive GHG emissions. With smart manure removal systems we mitigate GHG emissions from animal housing. We also real-time monitor these emissions so we can be held accountable based on real emissions. In combination with manure processing we produce natural gas and organic fertilisers to replace synthetic fertilisers. With this integrated approach, we farm in a circular and sustainable way to keep feeding the world.

Danone & its project that aims to support French dairy farmers in their Carbon footprint reduction

M. Baudet
Danone, 17 boulebard Haussmann, 75009 Paris, France; maeline.baudet@danone.com

Dairy farming contributes to 6% of France's greenhouse gas emissions. More than 90% of these emissions come from animals and farm inputs. As such, there is considerable need for creating awareness and supporting milk producers to improve farming practices. The *Danone Ecosystem Fund* has joined forces with *Danone Produits Frais France*, *Les prés rient bios*, *Idele* (French Livestock Institute), and *MiiMOSA* (a crowdfunding website dedicated to agriculture and food) in order to implement *Les 2 Pieds Sur Terre*. The program consists of supporting French dairy farmers as they reduce their carbon footprints and improve soil health, while increasing their competitiveness and enhancing the image of agriculture among the general public. The program covers the following dimensions: (1) measuring the carbon footprint of milk and creating awareness among farmers by conducting carbon diagnostics (1,604 in march 2021); (2) supporting farmers in identifying concrete carbon reduction projects through further carbon diagnostics (351 in march 2021); (3) implementing a collaborative and digital crowdfunding solution to help finance those projects, then offering technical and financial support (149 farmers in march 2021). With a range of technical partners, *les 2 Pieds sur Terre* also launches pilot projects (4 existing in 2020), involving groups of farmers who try out innovative practices regarding soil preservation, feed autonomy and pesticides reduction. The goal is to formalize results & keys for success in order to share them with the professional dairy network. Beyond the significative benefits on soil health, carbon sequestration, the decrease in emissions and the better understanding farmers have of their levers of action, the program actually raises pride among farmers since 41% of interviewed producers say they are more eager to talk about their job and feel proud about it.

Session 13 Theatre 5

Panel discussion (afternoon session)

A.S. Santos[1], A. Granados[2] and T. Boland[3]
[1]FeedInov CoLab, Estação Zootecnica Nacional, Qta da Fonte Boa, 2005-048, Santarém, Portugal, [2]EFFAB, Rue de Trèves 61, 1040 Brussels, Belgium, [3]University College Dublin, Agriculture and Food Science Centre, Belfield, Dublin 4, Ireland; assantos@utad.pt

The afternoon session will end with a panel discussion moderated by the Vice-Presidents of the Animal Task Force and the Secretary of the EAAP Commission on Livestock Farming Session. The panelists will be the speakers of the afternoon session + Josselin Andurand – Idele for a state of implementation of good practices in various EU projects (Climate Farm Demo, Life Carbon Dairy, Life Beef Carbon, Life Carbon Farming, ClieNFarms).

Session 13 Theatre 6

State of implementation of good practices in various EU projects

J. Andurand
Institut de l'Elevage, Service Environnement, Site de Monvoisin, BP 85225, 35652 Le Rheu Cedex, France; josselin.andurand@idele.fr

To meet the objectives of the Paris Agreement, the EU livestock sectors have taken steps to reduce the products carbon footprint by 15 to 20%. These approaches are based on training for stakeholders involved in the sector, farms environmental and carbon assessment, development of action plans and support for livestock farmers. Thanks to LIFE CARBON DAIRY, LIFE BEEF CARBON, LIFE GREEN SHEEP, more than 20,000 mixed crops and livestock farmers are moving to low carbon production systems. Based on a farm global approach, it makes it possible to ensure consistency between all the productions of a farm and thus to optimise the carbon cycle, forage and protein autonomy, and the use of animal waste. In France a MRV (Monitoring Reporting Verification) CAP'2ER&CARBON AGRI approach has been developed to ensure the certification of carbon gains under the Low carbon standard. This approach is essential for optimising farm operations, applying low carbon practices and certifying carbon gains. Combined with a multi-criteria environmental assessment (carbon, water, air, biodiversity), it also avoids the transfer of environmental impacts. This mechanism is an operationalisation of the European CARBON FARMING initiative which refers to the management of carbon pools, flows and GHG fluxes at farm level, with the purpose of mitigating climate change. For upscaling these initiatives, 3 EU projects i.e. LIFE CARBON FARMING, CLIENFARMS, CLIMATE FARM DEMO with up to 28 countries involved are in progress. They will permit to finalize a common accounting framework at EU scale, to certify GHG reductions and to reward farmers for applying mitigation practices in accordance with the CARBON FARMING initiative supported by the commission.

Time off feed affects enteric methane and rumen archaea

S. McLoughlin[1], S.F. Kirwan[2], P.E. Smith[2], F. McGovern[3], E. O'Conor[3] and S.M. Waters[1,2]
[1]National University of Ireland Galway, Ryan Institute, University Road, H91 TK33, Ireland, [2]Teagasc, Animal Bioscience Research Centre, Grange, Dunsany, C15PW93, Ireland, [3]Teagasc, Animal & Grassland Research and Innovation Centre, Mellows Campus, Athenry, Ireland; stuart.kirwan@teagasc.ie

Enteric fermentation of feed is the primary source of methane emission in agriculture, accounting for 5% of global GHG emission. Enteric methane is a by-product of microbial fermentation that is exclusively generated by methanogenic archaea in the rumen. In addition to the microbiome, the quantity and quality of feed consumed by ruminants are important factors influencing methane emissions, with higher forage intake associated with increased methane production. However, there is a lack of studies investigating the impact of time off feed on ruminant methane emissions and rumen microbiome. In this study, Portable Accumulation Chambers (PAC) were used to quantify methane emissions from 96 Belclare ewes grazing permanent pasture. All ewes were removed from pasture in the morning and put through PACs in lots of 12, on two separate occasions, 14 days apart, for a duration of 50 min, with methane (ppm) recorded at 0, 25, 50 min. After exiting the PAC, a sample of rumen fluid from each animal was collected using a trans-oesophageal apparatus for further microbial analysis. Statistical analysis of methane production data was conducted using a repeated mixed effects regression. Taxonomic and functional analysis of the rumen microbiome was conducted using Metaphlan3 and Humann3. Statistical analysis of microbiome data was conducted using Maaslin2 and Kruskal Wallace test. BH was used to correct for multiple hypothesis testing. Methane emissions decreased as the duration off pasture increased ($P<0.001$). Time off feed did not influence the taxonomic and functional profiles of the bacterial community ($P.adj>0.05$). However, the abundance of *Methanobrevibacter* sp. *YE315* was affected by time off feed, showing increased abundance in both early and late time points (*Kruskal* $P=0.0052$). To conclude, methane emissions and the rumen microbiome are affected as the duration of time off feed increases.

Alentejo's beef farms economic and environmental performance through a regionally developed tool

M.P. Dos Santos, T.G. Morais, T. Domingos and R.F.M. Teixeira
Instituto Superior Técnico, Universidade de Lisboa, MARETEC – Marine, Environment and Technology Centre, LARSyS, Avenida Rovisco Pais no. 1, 1049-001, Portugal; manueldossantos@tecnico.ulisboa.pt

In this work we applied a decision support tool for extensive animal production, focused on ruminant production in Alentejo (Portugal). This region plays a crucial supporting role for meat production in Portugal due to its availability of grazing land. The applied tool was tailored to the region's characteristics and main production systems and consists of integrated modules for farming management, including environmental and economic aspects. It includes an innovative and specialized grazing management system based on the estimation of pastures productivity, an environmental assessment system for calculating the environmental impact of each farm, based on Life Cycle Assessment, and a module for determination of the costs of the different farming practices associated with ruminant production. This multiplicity allows to simulate different scenarios to test the effects of several practices and the consequent environmental and economic outcomes. In this work we used the tool to model 27 beef and/or sheep farms. Farm-level data was collected under the scope of Animal Future project. Results can be divided in three levels. The first refers to economic performance of the surveyed farms in terms of production and efficiency, including a social dimension analysis (as labour and land tenancy indicators). The second level of results refers to environmental performance of the farms, that are returned by the tool after a complete data insertion on the platform. The accounted environmental impacts were greenhouse gases emissions, water and energy consumptions, and use of pesticides. Finally, the third level of results relates economic and environmental performance in multiple scenarios, in order to retrieve optimal solutions and delineate farm-level recommendations. This work allows to incorporate the consequences of global climate change in terms of farm management, in a regionally adapted way.

Phytochemical and *in vitro* methane inhibition of plant extracts as influenced by extractive solvents

T.A. Ibrahim and A. Hassen
University of Pretoria, Animal Science, Agriculture building Room 10-32, Department of Animal and Wildlife Sciences, University of Pretoria, Private Bag x20, South Africa; abubeker.hassen@up.ac.za

Plant phytochemicals are an important area of study in ruminant nutrition majorly due to their antimethanogenic potentials. This study evaluated the yields and phytochemical constituents of four plant extracts as affected by aqueous-methanolic (H_2O-MeOH) extraction and their antimethanogenic properties on *in vitro* methane production. The plant extracts include Aloe vera, Jatropha curcas, Moringa oleifera, and Piper betle leaves with three levels of extractions (70, 85 and 100% MeOH). The plant extract yields were affected by extractive solvents. The extract yields increased with a decrease amount of %MeOH replaced with an equivalent amount of deionised water. The phytochemical screening revealed that aloin A, aloin B, methoxycoumaroylaloeresin B, homocitric acid (in Aloe vera); tryptophan, vitexin-7-olate, isovitexin-7-olate, quercetin rhamnoside, apigenin-O-rutinoside and apigenin-6-C arabinosyl-8-C hexoside (in *Jatropha curcas*), cinnamoylquinic acids, Dihydrocaffeic acid (in Piper betle) and most metabolites in Moringa oleifera were relatively more hydrophilic. The methane mitigating potentials of these extracts were evaluated as additives on Eragrostis curvula hay at a recommended rate of 50 mg/kg DM. Although, the plant extracts exhibited antimethanogenic properties in various degrees, reducing ($P<0.05$) *in vitro* methane production on Eragrostis curvula hay, their antimethanogenic efficacy, TGP and IVOMD were unaffected by extraction solvents and yields. This may be due to an increase in the extraction of both useful and counter-useful antimethanogenic bioactive compounds, as well as little variation in the physical and chemical properties of the extractive solvents. Alkaloids, kaempferol, quercetin, neochlorogenic acid and feruloylquinic acid were noted to exhibit methane reducing properties and directly correlated with crude extract yields. Plant extracts could be more promising and hence, further study is necessary to explore other extraction methods as well as encapsulation of extracts for improved delivery of core materials to the target sites and to enhance methane reducing properties.

Considerations from modelling UK livestock farming: how farmers can reduce emissions

C. Kamilaris[1], X. Chen[2], R.M. Rees[3], T. Takahashi[4], T. Misselbrook[4], I. Kyriazakis[5], M.J. Young[1], S. Morrison[2] and E. Magowan[2]
[1]Centre for Innovation Excellence in Livestock, Innovation Centre, York Science Park, Innovation Way, Heslington, YO10 5DG, York, United Kingdom, [2]Agri-Food and Biosciences Institute, 18a Newforge Lane, Belfast, Co Antrim, BT9 5PX, Northern Ireland, United Kingdom, [3]Scotland's Rural College (SRUC), Kings Buildings, West Mains Road, EH9 3JG, Edinburgh, United Kingdom, [4]Rothamsted Research, North Wyke, Okehampton, EX20 2SB, Devon, United Kingdom, [5]Queen's University Belfast, University Rd, BT7 1NN, Belfast, United Kingdom; harry.kamilaris@cielivestock.co.uk

A high-level guide, as well as the impact, of some key mitigations that farmers can adopt to reduce their farm carbon footprint and drive down net emissions as reported through the UK National Inventory was compiled. Using real farm case studies, the mitigations with the biggest effects in ruminant systems were, in order of impact: use of dietary methane inhibitors, increased production efficiency (through improvements in such things as fertility, health, and genetic merit) and converting land freed up by such improvements to woodland forestry. Age at first calving, adoption of anaerobic digestion and use of nitrification inhibitors were also addressed, and modelling found that they can also contribute positively. For monogastric systems, their carbon footprint was greatly influenced by source of feed ingredients. The impact of land use change associated with protein ingredients in diets had the greatest impact on carbon footprint within farm case studies (increasing it by 100% in some circumstances), while the type of protein crop had less effect when no land use change was associated with it. The impact of crude protein and the use of anaerobic digestion were also modelled. Outcomes highlighted complementary as well as contradictory scenarios between the LCA outcomes using a farm carbon calculator, in contrast to the results obtained for the national calculation of GHG emissions and carbon sequestration using the National Inventory. This report highlights that through wide scale adoption of the most impactful mitigations available, a 23% reduction in GHGs and a 15% reduction in ammonia emissions from UK agriculture could be achieved. Much more mitigation and carbon capture is therefore needed to contribute to the Net Zero.

Session 13 Theatre 11

Closing of the ATF-LFS one-day symposium

F. O'Mara[1] and M. Lee[2]
[1]Teagasc / Animal Task Force, Oak Park, Co. Carlow, Ireland, [2]Harper Adams University, Newport, Shropshire, TF10 8NB, United Kingdom; frank.omara@teagasc.ie

Frank O'Mara, ATF President, and Michael Lee, President of the EAAP Commission on Livestock Farming session, will conclude this one-day symposium on 'Livestock emissions and the COP26 targets'. The outcomes of the one-day symposium will be discussed with a large panel of European stakeholders during the 12th ATF seminar, in Brussels, on 17 November 2022.

Session 14 Theatre 1

Analysis of ROH in two divergent lines of mouse selected for homogeneity of birth weight

C. Ojeda-Marín[1], J.P. Gutiérrez[1], N. Formoso-Rafferty[2], F. Goyache[3] and I. Cervantes[1]
[1]Universidad Complutense de Madrid, Dpto.Producción Animal, Avda Puerta del Hierro, s/n, Madrid, 28040, Spain, [2]Universidad Politécnica de Madrid, Dpto. Producción Agraria, C/Senda del Rey, 18, Madrid, 28040, Spain, [3]S.E.R.I.D.A-Deva, Gijón, Asturias, 33394, Spain; candelao@ucm.es

Runs of homozygosity (ROH) are defined as long continuous homozygous stretches in the genome which are assumed to arise from a common ancestor. It has been demonstrated that divergent selection for variability in mice is possible, and that low variability in birth weight is associated with robustness. The aim of this study was to analyse ROH patterns, and genomic inbreeding based on ROH in two mice lines selected for high variability (H-Line) and low variability (L-Line) for birth weight during 26 generations with sliding windows and consecutive runs methodologies to ascertain the differences due to divergent selection between lines across generations in this population, testing two different ROH detection algorithms. Up to 844 H-Line, 855 L-line and 19 of the founder population (FP) mice were genotyped using the Affimetrix Mouse Diversity Genotyping Array. After quality filtering, 173,642 SNPs were kept. ROH were computed using consecutive runs (CR) and sliding windows (SW) approaches. Three inbreeding coefficients were calculated for each individual and generation for both selection line and FP: inbreeding based on pedigree (F_{PED}), inbreeding based on ROH calculated with sliding windows (F_{ROHSW}) and with consecutive runs (F_{ROHCR}). H-Line presented slightly higher mean genomic inbreeding than L-Line. Between-lines F_{ROH} differences were higher at the chromosome level (mainly in chr 15, chr 14 and chr 3). Correlations between F_{PED} and F_{ROH} were high, particularly for the SW method except when segments between 1-4 Mb were considered. On the contrary, correlation between F_{PED} and F_{ROHCR} was lower when only longer segments were considered. Sliding windows seemed to detect better long segments related with recent inbreeding. However, consecutive runs correlated better with short segments related with ancient inbreeding. Differences in ROH distribution between lines might be caused by divergent selection. However, further analyses should confirm that new homozygous stretches are caused by selection.

Genes are shared across short, medium and long range ROHs in Avileña-Negra Ibérica local beef cattle

A. Rubio[1], M. Ramón[1], C. Meneses[2], A. Hernández-Pumar[2], M.J. Carabaño[2], L. Varona[3] and C. Diaz[2]
[1]IRIAF, Av. del Vino s/n, 13300 Valdepeñas, Spain, [2]INIA-CSIC, Animal Breeding Department, Ctra de la Coruña Km 7.5, 28040 Madrid, Spain, [3]University of Zaragoza, Miguel Servet s/n, 50013 Zaragoza, Spain; mramon@jccm.es

Runs of homozygosity are long continuous stretchers of DNA homozygosity that appear in individuals of different populations as a result of selection and other processes. It is well accepted that length of ROHs represent timing of selection events, thus, short ROHs refer to old selection processes while long ROHS reflect recent processes. Furthermore, analysis of ROHs allows for a good description of realized inbreeding at population and individual level (FROH). The objective of the work was to evaluate patterns of genomic regions containing ROHs. To do so, 284 bulls born between 1975 and 2018 were genotyped with the HD Illumina bovine BeadChip. Those animals were selected to capture variability and represent the Avileña-Negra Ibérica local beef cattle population. ROHs were classified into three groups based on their length, small ranges (from 1 to 2.5 Mb), mid ranges (from 2.5 to 5 Mb) and long ranges (>5 Mb). After finding ROHs, genes located within those regions were extracted. The number of ROHs (16,393 in total,57.7 in average per bull), varied among animals ranging from 6 to 165. The level of FROH also ranged from 0.3 to 42% being P50 above 7%. Chromosomes 6, 7, 21 and 23 showed ROHs in all individuals in specific regions (selection footprint). The number of different genes contained in those regions varied across individuals from small ROHs to long ROHs. There was an important number of genes (4,117) in common between classes of ROHs, meaning that the size of ROHs containing those genes varied among individuals suggesting that selection has persisted in specific regions of the genome and/or that recombination rate varies among individuals.

Genotypic interaction between gametes: mate incompatibility in dairy cattle

A.A.A. Martin[1], S. Id-Lahoucine[1,2], P.A.S. Fonseca[1], C.M. Rochus[1], L. Alcantara[1], D. Tulpan[1], S. Leblanc[3], F. Miglior[1], J. Casellas[4], A. Cánovas[1], C.F. Baes[1,5] and F.S. Schenkel[1]
[1]University of Guelph, Centre of Genetic Improvement of Livestock, 50 Stone Rd E, N1G 2W1 Guelph ON, Canada, [2]Scotland's Rural College, Department of Animal and Veterinary Sciences, Easter Bush, Midlothian EH25 9RG, United Kingdom, [3]University of Guelph, Ontario Veterinary College, 50 Stone Rd E, N1G 2W1 Guelph ON, Canada, [4]Universitat Autònoma de Barcelona, Departament de Ciència Animal i dels Aliments, Travessera dels Turons, 08193 Bellaterra, Spain, [5]Vetsuisse Faculty, Institute of Genetics, Länggassstrasse 120, 3012 Bern, Switzerland; amarti52@uoguelph.ca

In the dairy industry, producers determine mate allocation by selecting future offspring with genetic potential aligning with their breeding goal. However, not all sire-dam combinations are equally fertile, influencing the mating outcome. Under Mendel's laws, alleles on the same locus should segregate independently and be equally transmitted to the next generation. However, there are known exceptions, such as gametic incompatibility, generating a biased ratio of genotypes within the offspring. This bias can be investigated through transmission ratio distortion (TRD) analysis, exploring the over- or under-representation of parental alleles in descendants. To study gametic incompatibility specifically, the classical allelic TRD model was reparameterized to capture genotypic interactions. The objective of this study was to identify genomic regions, candidate genes and underlying biological pathways associated with incompatible gametic interactions using network, overrepresentation, and guilt-by-association analyses. Using 283,817 genotyped Canadian Holstein trios, 422 regions containing 2,075 positional genes were identified. The functional analyses pointed to signalling and immunological functions, and more specifically, to processes of self-recognition. A total of 16 candidate genes were identified and categorized into three groups based on the fertilization mechanisms they influence: fertilization ability of the sperm, gametic interaction, and female immuno-acceptance. Further investigation on gametic incompatibility will provide opportunities to improve mate allocation for the dairy cattle industry.

Genomic alterations underly the identification of Mendelian errors in SNP array data

K.D. Arias[1], I. Álvarez[1], J.P. Gutiérrez[2], I. Fernandez[1], J. Menéndez[3], N.A. Menéndez-Arias[1] and F. Goyache[1]
[1]SERIDA, Camino de Rioseco 1225, 33394 Gijón, Spain, [2]Universidad Complutense de Madrid, Departamento de Producción Animal, Avda. Puerta del Hierro,s/n, 28040 Madrid, Spain, [3]ACGA, C/ Párroco José Fernández Teral [5]A., 33403 Avilés, Spain; kathyah18@gmail.com

Performance of genotyping platforms may be worse when the SNPs typed are located in genomic regions carrying copy number alterations (CNV) or short (SINE) and long (LINE) interspersed nuclear elements. This research uses a Gochu Asturcelta pig pedigree to ascertain whether family structure and genomic features (occurrence of CNV, SINE or LINE at a locus) can underly Mendelian Errors (ME). Up to 492 individuals forming 478 parent-offspring trios were genotyped using the Axiom_PigHDv1 array. No thresholds for minor allele frequency (MAF) or Hardy-Weinberg test were applied. A total of 545,364 SNPs were retained. ME were identified using the program PLINK v1.9. Eight different ME classes were created according to the trio member (with 'Trio' meaning no assignment) and the allele on which ME was identified (A: the most frequent allele; B: the less frequent allele): TrioA; TrioB; FatherA; FatherB; MotherA; MotherB; OffspringA; and OffspringB. CNV calling on each individual was performed using PennCNV. The ME set was overlapped with the *Sscrofa* genome build 11.1 and the CNV identified using the BedTools software. A total of 40,540 SNPs had at least one ME in one individual gathering 292,297 allelic mismatches. ME assigned to TrioB class occurred on 33,483 loci (82% of the total) gathering 65,682 mismatches (11.5%). TrioB ME were likely due to genomic alterations, namely CNV. Except for the TrioA and Offspring errors, ME are biased with family size. TrioA and TrioB ME were not likely to affect allelic diversity at a locus in a sample. ME assigned to Trio classes occurred at loci with balanced and non-balanced allelic frequencies, therefore being difficult to be corrected using filters for MAF. The ME assigned to Mother and Father classes can partially be explained by the presence of null alleles. Identification of ME classes can contribute to gain consistency in association analyses. Work partially funded by grants PID2019-103951RB/AEI/10.13039/501100011033 and PRE2020-092905 (KDA).

Haplotype richness drop: a new method for mapping selection signatures

M. Shihabi[1], B. Lukic[2], V. Cubric-Curik[1], V. Brajkovic[1], L. Vostry[3] and I. Curik[1]
[1]University of Zagreb Faculty of Agriculture, Department of Animal Science, Svetosimunska 25, 10000 Zagreb, Croatia, [2]J.J. Strossmayer University of Osijek, Faculty of Agrobiotechnical Sciences Osijek, Department for Animal Production and Biotechnology, Vladimira Preloga 1, 31000 Osijek, Croatia, [3]Czech University of Life Sciences Prague, Department of Animal genetics and Breeding, Kamycka 129, 16500 Prague, Czech Republic; icurik@agr.hr

Under the pressure of natural and artificial selection, the pattern of genomic landscape of domestic animals is constantly changing, leaving visible signs of selection. Here, we propose a new approach called haplotype richness drop (HRiD), which is able to identify selection signals from genomic information contained in male haplotypes. This applies to males genotyped on the X-chromosome, although extension to diploid genotypes that are accurately phased is also possible. HRiD is based on the assumption that genomic regions subject to positive selection exhibit a sudden decrease (drop) in haplotype richness. Thus, the main task of HRiD is to identify outlying genomic regions that exhibit a sudden drop in haplotype richness. Using empirical datasets, we have shown that HRiD complements other within-population approaches, such as extreme Runs of Homozygosity islands, integration Haplotype Scores, and the number of Segregating Sites by Length. In addition, we demonstrated that HRiD results can be readily combined with phylogenetic analyses of selection signals (haplotypes). This study was supported by the Croatian Science Foundation under the project ANAGRAMS, IP-2018-01-8708, and by the Operational Programme for Competitiveness and Cohesion from 2014 to 2020 under the project SirjeIN, KK.01.1.1.04.0058, while the participation of Luboš Vostry was supported by the Ministry of Agriculture of the Czech Republic (project no. QK1910156).

Single- and multi-breed ssGBLUP evaluations with sequence data for over 200k pigs

D. Lourenco[1], S. Jang[1], R. Ros-Freixedes[2], J. Hickey[3], C.Y. Chen[4], W. Herring[4] and I. Misztal[1]
[1]University of Georgia, 425 River Rd, 30602, Athens, GA, USA, [2]Universitat de Lleida, 181 Rovira Roure, 25198, Lleida, Spain, [3]The Roslin Institute and Royal (Dick) School of Veterinary Studies, The University of Edinburgh, Easter Bush Campus, Midlothian EH25 9RG, United Kingdom, [4]Genus PIC, 100 Bluegrass Commons Blvd, 37075, Hendersonville, TN, USA; danilino@uga.edu

Whole-genome sequence data harbour causative variants that may not be present in regular SNP chips. As such data are becoming available for many farm animal species, there is a question on whether this information can help increase the accuracy of genomic predictions (GP) beyond that already achieved with SNP chips. We investigated the impact of using preselected variants from sequence data in up to 105k animals in single-breed (SB) and 207k in multi-breed (MB) ssGBLUP evaluations. The number of animals imputed (sequenced) was 60k (731), 41k (760), and 104k (1,856) for breeds A, B, and C, respectively. Imputation was performed separately for each breed. We examined three genotype sets: regular SNP chip, top markers in each 40k genomic window, and significant SNP based on GWAS combined with the chip. The three terminal pig breeds were evaluated for five traits, namely average daily feed intake (ADFI), average daily gain (ADG), backfat (BF), ADG recorded in crossbred animals (ADGX), and BF recorded in crossbred animals (BFX). All three terminal breeds were combined for the MB analyses. Maximum gains of 0.03 and 0.01 in accuracy were observed for SB and MB with selected sequence variants. Regarding MB evaluations, breeds with fewer genotyped animals (A and B) observed a decrease in predictive ability when the models involved the top markers in each genomic window of 40k SNP. This might be because those top SNP selected from MB GWAS could be related to causative variants in the largest breed (C) but not in the small ones. Larger breeds can easily overwhelm GWAS and GP results when combined with smaller breeds. The benefits of using sequence data for SB and MB predictions are limited even when information on many sequenced or imputed animals is available. Further investigations will evaluate the impact of sequence data on predictions across-breed.

Application of a single-step SNP BLUP maternal-effect model to calving trait genomic evaluation

Z. Liu, H. Alkhoder, D. Segelke and R. Reents
IT Solutions for Animal Production (vit), Division of Genetic Evaluation and Biometrics, Heinrich-Schroeder-Weg 1, 27283 Verden (Aller), Germany; zengting.liu@vit.de

Calving ease (CE) and stillbirth (SB) in first three parities were treated as genetically correlated traits in the linear animal model with correlated direct and maternal genetic effects. A single-step SNP BLUP model by Liu *et al.* (2014) was applied to the national calving data. Because the MACE evaluation for the calving traits uses a single-effect model per trait, we treated the bull MACE EBV of calving traits as a correlated trait to the national calving traits in order to utilise deregressed bull MACE EBV of all the four MACE traits. The aim of this study was to test the application of the single-step SNP BLUP model to the calving traits with foreign bull data included. Phenotype data from official German evaluation in August 2021 were analysed together with deregressed bull MACE EBV from Interbull evaluation. A total of 1,003,041 genotyped animals were considered, including young candidates and culled animals. The number of animals with own phenotype data, calving cows and calves, was 31,167,053. The complete pedigree file included 38,150,805 animals and 90 phantom parent groups. The total number of effects to be estimated for the single-step model was 615,848,330. A genomic validation was conducted to simulate an evaluation four years ago. Due to the short history of female genotyping in Germany, about 75% genotyped calving cows and 66% genotyped calves were removed. All bulls included in the MACE evaluation for the four calving traits were added to the national data, if the bulls had calving daughters for maternal effects or calves born for direct effects in foreign countries. For a comparison to the single-step model, results of a genomic validation based on the current two-step genomic model were used. SNP effect estimates were compared between the two genomic models as well as between the full and truncated data sets within each model. GEBV of about 1,650 national validation bulls were analysed between the full and truncated data for checking the predictive ability of the genomic models. GEBV of three groups of genotyped animals were investigated in detail: genotyped AI bulls, genotyped female animals and genotyped male candidates. Results of this study will be shown.

Computing strategies for national dairy cattle evaluations

M. Bermann[1], A. Cesarani[2], D. Lourenco[1] and I. Misztal[1]
[1]University of Georgia, Animal and Dairy Science, 425 River Road, 30602, USA, [2]Università degli Studi di Sassari, Dipartimento di Agraria, Piazza Università, 21, 07100 Sassari, Italy; mbermann@uga.edu

The Council on Dairy Cattle Breeding (CDCB) performs the U.S. national dairy cattle evaluations. Their dataset has over 138 million records with a pedigree that exceeds 90 million animals, from which more than 5 million are genotyped. Multi-step has been used to estimate genomic EBV (GEBV) since 2009; however, the feasibility of single-step (ss) GBLUP as the official evaluation method for CDCB is under investigation. The size of the U.S. dairy datasets creates computational challenges when estimating breeding values and their accuracies in a timely way. This study presents strategies for making the U.S. dairy cattle evaluation computationally feasible under ssGBLUP. The process for estimating breeding values in ssGBLUP can be divided into two parts. In the first part, hereafter called genomic setup, the genotype file is read, and the inverse of the genomic relationship matrix (G^{-1}) is calculated using the Algorithm for Proven and Young (APY). Optimizations included reading the genotype file in a stream-unformatted format, using bit-wise storage for markers, having an efficient algorithm to compute the residual polygenic effect, and calculating G^{-1} using parallel computing and enhanced matrix algebra procedures. The second part estimates breeding values by iteration on data with a block diagonal preconditioner. Finally, the reliability of GEBV is calculated by combining the pedigree and genomic information and using a block sparse inversion algorithm accounting for the sparsity of G^{-1}. For a Holstein dataset with 30 million animals in the pedigree and 3.8 million genotypes, the genomic setup took 50 hours without and 8 hours with optimizations. Solutions for a three-trait model including milk, fat, and protein yields were obtained in 21 hours. Approximating the reliability of GEBV took less time than the optimized genomic setup. Our results show that weekly national dairy cattle evaluations are computationally feasible.

Impact of foreign phenotype data on single-step genomic prediction using test-day protein yields

H. Alkhoder, Z. Liu, D. Segelke and R. Reents
IT solutions for animal production (vit), Genetic evaluation and biometrics, Heinrich-Schroeder-Weg 1, 27283 Verden, Germany; hatem.alkhoder@vit.de

For genomic evaluation of the international dairy breed Holstein, foreign phenotype data of bulls are usually combined with national cow phenotype data to enhance the accuracy of genomic prediction for young animals. More than 242 millions of test-day records from 12.4 millions of Holstein cows were evaluated together with deregressed MACE EBV of more than 138,000 Holstein bulls. Genotype data of more than 1 million animals were analysed with a single-step SNP BLUP model that included a residual polygenic effect. Correlations and regressions of the GEBV of validation bulls between the full and truncated evaluations were compared among five scenarios: (1) using all bulls included in the MACE evaluation (ALL); (2) using only bulls linked to the national genotyped population or cows with phenotype data (LINK); (3) using only reference bulls with both own phenotype and genotype data (REF); (4) using only MACE bulls born in 1995 and later (1995); and (5) ignoring all bull MACE data by using only national cow phenotype data (NAT). In contrast to the first four scenarios using the bull MACE data, the scenario of ignoring MACE data (NAT) led to highest GEBV correlation for the validation bulls between the truncated and full evaluations and regression coefficient most closer to 1. Young genotyped animals had the highest level and variance in GEBV under this scenario NAT than all the others. However, this scenario NAT resulted in largest overestimation of GEBV for young animals when comparing their GEBV between the early truncated to the later full genomic evaluation. In comparison to the scenario of including all MACE data (ALL), equal regression coefficients b0 and b1 were found for scenario 1995 and lower for scenarios LINK and REF. All the four scenarios with MACE data had nearly equal R^2 value. Highest overestimation of GEBV in young animals was found for scenario REF, followed by LINK and 1995. Within-scenario early genomic prediction shows the scenarios ALL or 1995 led to the smallest inflation, followed by LINK and REF. For the international breed Holstein, using MACE data of bulls from all countries in all birth years helps reduce the GEBV inflation in young animals.

Impact of different sources of genomic data on predictions on the local dual-purpose Rendena breed

E. Mancin, C. Sartori, B. Tuliozi, G.G. Proto and R. Mantovani
University of Padova, DAFNAE, Viale del Università 16, Legnaro, 35020, Italy; enrico.mancin@studenti.unipd.it

Although the genomic selection is used in many livestock species, it has not yet been considered in local breeds due to the reduced population size and the less expected impact. In this study, the local Rendena dual-purpose breed was considered. This breed is selected mainly for milk and with a lesser emphasis on beef traits. In this situation, increasing accuracy for beef traits could enhance the selection efficacy due to the antagonism with milk production. The aim of this study was: (1) to evaluate the impact of genomic data on breeding values prediction; (2) to analyse the effect on EBVs accuracy due to the inclusion of a variable amount of genomic information from relatives (i.e. females). In fact, in dual-purpose breeds, it is possible to increase the number of genotyped animals by including cows' genomic information that would be normally used for genomic selection for milk. Data analysed were performance test traits, i.e. the average daily gain, *in vivo* SEUROP, and dressing percentage, recorded on 1,691 young bulls, 700 of which with genomic information. Extending to the whole population, about 1,800 animals with their own genotype were accounted for. After imputation and quality control animals had an average density of 120k SNPs. The impact of genomic information was evaluated by comparing pedigree-BLUP (PBLUP) with single-step GBLUP (ssGBLUP) using LR method and repeated 5-fold cross-validations. As a second step we evaluated the impact of including: (1) only dams' information; (2) only female with a kinship of 0.30 with at least one male; (3) only female with a kinship of 0.15 with at least one male; and, finally (4) all females with genomic information. We observed lower bias and greater accuracy values in EBVs when ssGBLUP was used, while a further substantial increase of accuracy was observed when only females with a kinship of 0.15 were included. Otherwise, no substantial increase of accuracy was observed when all females with genomic information were considered. Results suggest the importance of including both male and female genomic information in routine genetic evaluations of beef traits in local breeds.

Genetic associations between milk protein fractions and FTIR milk-spectra

H.O. Toledo-Alvarado[1], G. De Los Campos[2,3,4] and G. Bittante[5]
[1]National Autonomous University of Mexico, Department of Genetics and Biostatistics, 04510, Mexico, [2]Michigan State University, Institute for Quantitative Health Science and Engineering, 48824, USA, [3]Michigan State University, Department of Statistics and Probability, 48824, USA, [4]Michigan State University, Department of Epidemiology and Biostatistics, 48824, USA, [5]University of Padova, Department of Agronomy, Food, Natural Resources, Animals and Environment, 35020, Italy; h.toledo.a@fmvz.unam.mx

Records from 971 Brown Swiss cows (genotyped with 37,519 SNP) from 85 herds located in Trentino (Italy) were used to study the genetic relationship between major milk protein fractions (caseins: α_{S1}-CN, α_{S2}-CN, β-CN, κ-CN; whey proteins: α-LA, β-LG;), and each wavenumber (wn) of FTIR milk-spectra (1,060 wn). The heritabilities, genetic and phenotypic correlations, as well as the accuracy of an indirect selection using each wn were calculated. A bivariate (y, wn) GBLUP was used to estimate the parameters using the Multitrait() function of the BGLR R-package, where y = α_{S1}-CN, α_{S2}-CN, β-CN, κ-CN α-LA, β-LG; and wn = 1…1,060 wn. The model included the fixed effects of parity and days in milk, and the random effects of herd and animal. The accuracy of an indirect selection index for each protein fraction using a specific wn was calculated as: $Acc(g_{yi},wn_i)=r_{gy},g_{wn}\ \sqrt{(h^2_{wn})}$. The maximum absolute genetic correlations $|r_{gy},g_{wn}|$ for α_{S1}-CN, α_{S2}-CN, β-CN, κ-CN, α-LA, β-LG, were 0.48 (wn 3,694), 0.32 (wn 1,067), 0.33 (wn 3,710), 0.38 (wn 2,587), 0.38 (1,542), 0.39 (wn 990), respectively. Genetic correlations were higher than phenotypic correlations in general. The accuracies for an indirect selection index (IS) and direct selection (DS = $\sqrt{(h^2_y)}$) were: α_{S1}-CN (IS=0.28 [wn 1,252], DS=0.73), α_{S2}-CN (IS=0.16 [wn 1,094], DS=0.57), β-CN (IS=0.13 [wn 2,552], DS=0.83), κ-CN (IS=0.20 [wn 2,587], DS=0.80), α-LA (IS=0.22 [wn 1,461], DS=0.50), β-LG (IS=0.16 [wn 1,021], DS=0.71). The highest relative efficiency (IS/DS) was observed for α-LA with 0.44, while the lowest was estimated for β-CN (0.16). These results can be used to derive methodologies to improve the prediction of genetic values using the FTIR-spectra together with genotypes in order to increase the accuracy of the breeding values.

Genetic evaluations using a novel custom-made SNP chip for local cattle breeds

M.J. Wolf[1], G.B. Neumann[2], P. Korkuć[2], K. May[1], G.A. Brockmann[2] and S. König[1]
[1]Justus-Liebig-University Gießen, Institute for Animal Breeding and Genetics, Ludwigstraße 21b, 35390 Gießen, Germany, [2]Humboldt-University zu Berlin, Thaer-Institute for Agricultural and Horticultural Sciences, Animal Breeding Biology and Molecular Genetics, Invalidenstraße 42, 10115 Berlin, Germany; manuel.j.wolf@agrar.uni-giessen.de

This study evaluates the performance of a novel custom-made 200K SNP chip (DSN200K), designed for a small local dual-purpose breed German Black Pied Cattle (DSN, German: 'Deutsches Schwarzbuntes Niederungsrind'). Genotype information of 3,702 DSN cattle based on the Illumina BovineSNP50 BeadChip (50K), on whole-genome sequence (WGS) data (~16.1 M) and on the DSN200K. On such genomic databases, we performed genome-wide association studies (GWAS), and we estimated genetic parameters and breeding values for milk production, health indicator and calving traits. The accuracies of genomic estimated breeding values were determined in ten-fold cross-validations for the DSN200K chip, and for 20 randomly constructed chips based on WGS data using the same marker density as considered for the DSN200K chip. We identified significant SNP marker associations and potential candidate genes for all traits with the specifically designed DSN200K chip, which were not detected with the commercial 50K chip. The annotated genes (e.g. *MGST1* for 305-day lactation milk yield) were confirmed based on WGS data, indicating breed-specific genomic mechanisms. Heritabilities and accuracies of genomic breeding values increased slightly when using the novel DSN200K array compared with the random constructed chips of same density. The breed specific DSN200K was designed considering important chromosome segments for novel traits reflecting disease resistances such as endoparasite infections. Consequently, especially for such breed specific novel traits, we hypothesize additional gain in ongoing studies regarding the significance in genome-wide associations and accuracies in genetic evaluations when compared to the results from the commercial 50K chip.

Comparison of different validation strategies for ssGBLUP based on a simulated cattle population

J. Himmelbauer[1], H. Schwarzenbacher[1], C. Fuerst[1] and B. Fuerst-Waltl[2]
[1]ZuchtData EDV-Dienstleistungen GmbH, Dresdner Straße 89/B1/18, 1200, Austria, [2]University of Natural Resources and Life Sciences, Vienna, Department of Sustainable Agricultural Systems, Gregor-Mendel Str. 33, 1180 Vienna, Austria; himmelbauer@zuchtdata.at

The validation of estimated breeding values from single-step GBLUP (ssGBLUP) is an important topic, as more and more countries and animal populations are currently changing their genomic selection to single-step. The objective of this work is to compare different strategies to validate single-step genomic breeding values (GEBVS). The investigations were carried out using a simulation study based on the German-Austrian-Czech Fleckvieh population. In order to test the validation strategies under different conditions, several scenarios were simulated. Firstly, the validation strategies were applied on base scenarios with more or less ideal conditions. Secondly, scenarios with biases were induced by using wrong heritabilities, genotyping only the best progeny and excluding genotypes of non-selected animals. The application of the widely used Interbull GEBV test on the single-step method is only possible to a limited extent, partly because of genomic preselection, which biases conventional EBVs Alternative validation strategies considered in the study are the Linear Regression method (LR) proposed by Legarra and Reverter, the improved genomic validation including extra regressions as suggested by VanRaden and an adaption of the Interbull GEBV test using (daughter) yield deviations from ssGBLUP instead of BLUP. On the one hand, it was investigated how well the different methods are suited for the validation of single-step GEBVs and, on the other hand, how reliably bias and overdispersion caused by different factors are detected. The validation results of different methods are compared for the different simulation scenarios in order to identify the strengths and weaknesses of the different methods.

Comparison of different ROH-based genomic inbreeding coefficients in a large cohort of horses

N. Laseca[1], A. Molina[1], M. Valera[2], A. Antonini[3] and S. Demyda-Peyrás[3]
[1]University of Cordoba, Dept. of Genetics, 14071, Córdoba, Spain, [2]University of Sevilla, Dept. of Agronomie, ETSIA, 41013, Sevilla, Spain, [3]National University of La Plata, Fac. of Veterinary Sciences, La Plata, Argentina; ge2lagan@uco.es

The inbreeding value has been historically calculated in livestock using pedigree information. However, this estimate heavily depends on the depth and reliability of the dataset employed. Nowadays, the availability of high-throughput SNP genotyping in livestock has shifted inbreeding estimations towards methodologies based on genomic information, increasing its reliability and accuracy. In this study, we aimed to compare different estimates of genomic inbreeding coefficients using two array-based SNP datasets from a large cohort (n=805) of Pura Raza Español horses. Genomic inbreeding values were estimated using four runs of homozygosity (ROH) based approaches: two observational, 'consecutive runs' (F_{ROH}-DR) and 'sliding-windows' (F_{ROH}-PL), and two probabilistic methods (based on hidden Markov models) with (F_{IBC}) or without (F_{ROH}-RZ) taking into account linkage disequilibrium (LD). The genomic datasets, including 540,294 (HD) and 55,176 markers (MD), were analysed using *DetectRuns*, *RZooROH*, and *IBD-LD* packages. Results were compared with a pedigree-based inbreeding value (F_{PED}) estimated from a robust pedigree database including 400,000 records (average WG=9.46). Results showed differences between the methodology and the array density only in the HD dataset using the consecutive runs approach, which showed a lower correlation with F_{PED}. On the contrary, F_{ROH}-PL, F_{IBC}, and F_{ROH}-RZ showed similar results in HD and MD datasets, with correlations with F_{PED} close to 0.8. In addition, most of the estimates showed higher correlations between them except for HD F_{ROH}-DR which showed the lowest correlation with the rest of the coefficients. Finally, it is noteworthy that genomic inbreeding coefficients calculated using the same method but analysing HD or MD data were highly correlated (>.95 in most of the cases with a maximum of 0.995) except for the F_{ROH}-DR (0.89), suggesting an increased sensitivity to the type of genomic data employed. We concluded that the methodology employed, rather than density or LD correction, is the key factor in the evaluation of genomic inbreeding in the domestic horse.

Preliminary analysis of the additive genetic variance associated with Mendelian Errors

K.D. Arias[1], I. Álvarez[1], J.P. Gutiérrez[2], I. Fernandez[1], J. Menéndez[3], N.A. Menéndez-Arias[1] and F. Goyache[1]
[1]SERIDA, Camino de Rioseco 1225, 33394-Gijón, Spain, [2]Universidad Complutense de Madrid, Departamento de Producción Animal, Avda. Puerta del Hierro, s/n, 28040-Madrid, Spain, [3]ACGA, C/ Párroco José Fernández Teral 5A, 33403-Avilés, Spain; kathyah18@gmail.com

Mendelian Errors (ME) can reflect true genomic variation arising from *de novo* mutations. However, in case of correct pedigree information they more likely result from genotype calling errors. The objective of this work was to ascertain the existence of a possible additive genetic component influencing the occurrence of ME in SNP arrays. A total of 478 parent-offspring trios registered in the Gochu Asturcelta pig breed pedigree were genotyped using the Axiom Porcine Genotyping Array (Axiom_PigHDv1). Up to 492 individuals belonging to 61 different families and 96 litters were available to estimate genetic parameters associated with the presence of ME. ME were identified using the –me option of the program PLINK v 1.9. Using the BedTools software SNPs were mapped on the Sscrofa genome build 3.0 to ascertain if they overlapped with long (LINE) or short (SINE) interspersed nuclear elements. Up to 1,445 SNPs gathering ME on 10 or more individuals (29,932 mismatches) were used as phenotypes. Genetic parameters were estimated via a Bayesian procedure using the program TM assuming a threshold model. Model fitted for analyses included the following effects: sex of the individual (2 levels: male/female), location of the SNP within a LINE or SINE (three levels: NO/LINE/SINE), porcine autosome (18 levels), family (61 levels) as a random effect and the animal genetic effect. Marginal posterior distributions for each parameter were obtained with a period of data collection of 200,000 iterations and a burn-in period of 50,000 iterations. Estimate of heritability (h2) was 0.101±0.0068. Estimate for the family random effect was 0.031±0.009. The solutions obtained for the systematic effects (sex, autosome and interspersed nuclear elements) suggested that they did not influence data significantly. ME could be associated with a low but non-negligible heritability. Genetic basis of this additive variation remains to be elucidated. Work partially funded by grants PID2019-103951RB/AEI/10.13039/501100011033 and PRE2020-092905 (KDA).

Session 15 Theatre 1

Identifying important risk factors for lameness in partly-housed pasture-based dairy cows

N. Browne[1,2], C.D. Hudson[1], R.E. Crossley[2,3], K. Sugrue[2], E. Kennedy[2], J.N. Huxley[4] and M. Conneely[2]
[1]University of Nottingham, School of Veterinary Medicine and Science, Loughborough, LE12 5RD, United Kingdom, [2]Teagasc, Moorepark, Co. Cork, P61 P302, Ireland, [3]Wageningen University & Research, Department of Animal Sciences, Wageningen, 6700 AH, the Netherlands, [4]Massey University, School of Veterinary Science, Palmerston North, 4442, New Zealand; natasha.browne@hotmail.co.uk

Lameness in dairy cows is widely accepted as a welfare concern, which also has negative economic and environmental impacts. Most risk factor studies on lameness have focused on fully housed or fully pasture-based cows. This large scale study aimed to ascertain the most important risk factors in a system where cows are exposed to risk factors of grazing for most of the year but also to housing risk factors over winter. Cows were lameness scored (0-3 scale; 2 & 3 = lame) during grazing (99 farms) and the subsequent housing period (85 farms). Routinely recorded cow-level management data, questionnaire data and infrastructure measurements were collated as potential predictors. Two different types of regression (elastic net regression and logistic regression using modified Bayesian information criterion) were used and combined to triangulate results, with lameness status as the binary outcome. This novel statistical method enabled a robust list of risk factors to be identified from a large number of potential predictors (~200), which would usually result in overfitting if simple logistic regression modelling was performed. Median lameness prevalence across farms was 7.9% (grazing) and 9.1% (housing). Older cows and those with a positive Predicted Transmitting Ability for lameness had increased risk, as did farms with a smaller grazing platform and herd size. Cow track and parlour characteristics (stones in paddock gateways, the presence of slats at the start of the track and a short distance to turn at the milking parlour exit) were also shown to increase risk. When the farmer recognised lameness (and digital dermatitis presence) as a problem on their farm, the risk for lameness increased, indicating that farmers in this study were aware of lameness. Results from this study will help prioritise risk factors to study in future randomised controlled trials within partly-housed, pasture-based systems.

Session 15 Theatre 2

Seasonal variation of methane emissions in buffaloes captured using laser methane detector®

L. Lanzoni[1], M.G.G. Chagunda[2], M. Chincarini[1], I. Fusaro[1], M. Giammarco[1], M. Podaliri[3] and G. Vignola[1]
[1]University of Teramo, Faculty of Veterinary Medicine, Piano d'Accio, 64100 Teramo, Italy, [2]University of Hohenheim, Animal Breeding and Husbandry in the Tropics and Subtropics, Garbenstr. 17, 70599 Stuttgart, Germany, [3]Istituto Zooprofilattico Sperimentale dell'Abruzzo e del Molise, Campo Boario, 64100 Teramo, Italy; llanzoni@unite.it

Seasonal variations can act as a physiological stressor for animals, triggering a range of behavioural and biological responses. The present study aimed to assess the variability of CH_4 emissions from Italian Mediterranean Buffaloes (IMB) between the summer and winter seasons. Direct CH_4 emissions of twenty non-productive IMB, under the same feeding regimen, were monitored daily for 12 days in summer and winter using a Laser Methane Detector® (LMD). LMD is a portable, smart, non-invasive tool to measure CH_4 emissions. Each measurement lasted 4 minutes with a distance of 1 meter from the LMD to the animal. Daily temperature-humidity index (THI) and dry matter intake (DMI) were recorded during the trial. LMD output data were divided for each day and each subject into CH_4 from eructation (CH_E), breathing (CH_B), and overall average (CH_{AV}). Linear and linear mixed models were used to analyse the effect of the variables. The average CH_E, CH_B, and CH_{AV} measured during the trial period were 115.76±56.38, 18.33±7.41, and 34.16±14.35 ppm-m, respectively. Results showed a significant influence of season on CH_E, CH_B and CH_{AV} ($P<0.001$). Winter season had higher emission values than summer (CH_E=130.20±40.98, CH_B=21.60±5.07, CH_{AV}=40.62±9.83 ppm-m for winter Vs CH_E=84.37±42.62, CH_B=11.34±3.45, CH_{AV}=20.32±6.38 ppm-m for summer). There was significantly lower DMI per live weight ($P<0.001$) and higher THI ($P<0.001$) in summer than in winter (summer: %DMI=2.24, THI=74.5±1.9; winter: %DMI=2.51, THI=49.6±4.9). Linear regression confirmed that a reduction in THI was significantly ($P<0.001$) associated with an increase in CH_E (R^2=0.69), CH_B (R^2=0.83) and CH_{AV} (R^2=0.85). The variation of CH_4 between the two seasons could therefore be attributable to THI changes, which also influence the feed intake. In conclusion, our results suggest the need to account for the season of assessment in LMD protocols.

Effect of lysolecithin supplementation in milk on performance of suckling female dairy calves

M.I. Mandouh, R.A. Elbanna and H.A. Abdellatif
Cairo university, Nutrition and Clinical Nutrition, Faculty of Veterinary medicine, Giza square, Giza, 12211, Egypt; mandouh.m93@gmail.com

The early life stage of calves is very important for their future production. Nutrition plays a vital role during this stage. Lysolecithin is an emulsifier that enhances animal production. Therefore, the aim of the current study is to investigate the growth and general health-promoting effect of lysolecithin on newborn female dairy calves. In a 12 weeks experiment, a total of 30 female Holstein calves (3 days old) were blocked by body weight and allocated into three groups (n=10/ group). The first group was fed on pasteurised whole milk only (control group; CON), the second group was fed on whole pasteurised milk supplemented with lysolecithin (dissolved in milk at a dose of 6 grams/ head/ day; LECL). The third group received the lysolecithin in milk at a dose of 12 grams/ head/ day (LECH) up to weaning (d 84), calves were fed individually. Calf starter was offered free of choice to all calves in the different experimental groups starting from two weeks of age. Bodyweight and body conformation were recorded every 4 weeks and calf starter intake was recorded weekly. Blood samples were collected at the end of the experiment and used for the determination of liver and kidney function metabolites, data were statistically analysed using PROC MIX procedures of SAS 9.4. Lysolecithin at both levels of supplementation did not affect the body weight as well as body conformation (P>0.05). Calf starter intake was significantly reduced in a dose-response manner due to lysolecithin supplementation compared to the control group (P=0.03) at the third month of the experiment (w8-12). Serum levels of ALT, AST, albumin, A: G ratio, and creatinine were significantly reduced (P=0.05) in lysolecithin supplemented calves compared to the control ones. Lysolecithin did not affect calf starter intake at the first two months of age and it improved nutrient utilization afterward as seen by reduced calf starter intake, and reaching the same body weight, therefore, it could be concluded that supplementation of lysolecithin as an emulsifier to suckling dairy calves could improve their performance and general health status.

Net protein productivity to evaluate the contribution of specialized dairy farms to food security

C. Battheu-Noirfalise[1,2], A. Mertens[2], E. Froidmont[2], M. Mathot[2] and D. Stilmant[2]
[1]Université de Liège, Unité de Zootechnie, 2 Passage des déportés, 5030 Gembloux, Belgium, [2]Centre wallon de Recherches Agronomiques, Department sustainability, systems and prospectives, 100 Rue du Serpont, 6800 Libramont, Belgium; c.battheu@cra.wallonie.be

The demand in animal source food (ASF) is expected to increase as the world's population keeps expanding and the food consumption patterns are changing towards a higher incorporation of ASF in diets. This represents a threat to food security as livestock is partly in competition with human-edible food and uses high amounts of agricultural land. However, livestock can also contribute positively to food security by recycling non-edible feeds (industrial by-products or fodders) into high quality food products. If different metrics have been proposed to represent the capacity of livestock to contribute to human food supply, they often remain in a certain dichotomy by approaching: (1) the efficiency with which animal productions use human-edible feedstuffs; or (2) the efficiency with which they use agricultural land. This study approach the contribution of the animal system to food security by evaluating the net productivity, an indicator which aims to include both of these pre-cited aspects. Net productivity is the difference between human-edible animal productions and human-edible animal consumptions per unit of non-edible surface. We tested this indicator on accounting data of 115 specialized dairy farms all over the Walloon Region (Belgium), characterized by a large diversity of pedoclimatic conditions. Dairy farms with the highest net productivity are those showing simultaneously a high milk production per cow, high fodder yields, an intermediate use of maize silage, use of concentrates with intermediate protein content and an efficient herd management by reducing unproductive phases (age at first calving and number of female followers per cow). In the future, net productivity should be evaluated in other European animal systems and related to environmental and socio-economical performances of dairy farms while taking into account pedoclimatic constraints.

Greenhouse gases and ammonia emissions: a holistic approach towards dairy production sustainability

A.R.F. Rodrigues[1], M.R.G. Maia[1], A.R.J. Cabrita[1], A. Gomes[2], L. Ferreira[3], H. Trindade[4], A.J.M. Fonseca[1] and J.L. Pereira[4,5]

[1]REQUIMTE, LAQV, ICBAS, School of Medicine and Biomedical Sciences, University of Porto, R. Jorge Viterbo Ferreira 228, 4050-313 Porto, Portugal, [2]Cooperativa Agrícola de Vila do Conde CRL, R. Lapa 293, 4480-757 Vila do Conde, Portugal, [3]AGROS UCRL, R. Cidade Póvoa Varzim 55, 4490-295 Argivai, Portugal, [4]CITAB, University of Trás-os-Montes and Alto Douro, Quinta de Prados, 5000-801 Vila Real, Portugal, [5]Agrarian School of Viseu, Polytechnic Institute of Viseu, Quinta da Alagoa, 3500-606 Viseu, Portugal; anaferodrigues@gmail.com

Dairy sector is facing new paradigms and challenges to reduce greenhouse gases (GHG) and ammonia (NH_3) emissions towards the goals established by international agreements and European Directives for a more sustainable production. To achieve these goals, accurate mitigation strategies are needed. However, in Portugal GHG emissions from dairy buildings has not yet been characterized and only two studies evaluated NH_3. Thus, this work aimed to improve the sustainability of Portuguese dairy sector in a holistic approach. First, GHG and NH_3 emissions from dairy buildings of NW Portugal, the second most important Portuguese dairy region, were assessed for at least 7 days, in each season, for two consecutive years. Indoor temperature and relative humidity were also monitored. Farms differed in feeding systems, yet a maize silage-based diet was used all year round. In a second study, excreta from dairy cows of different stages of lactation and feeding systems were collected at distinct sampling times to evaluate GHG and NH_3 emissions in a lab chamber system. Moreover, faecal microbiome diversity and abundance and end-fermentation profile were determined, as diet has been reported to affect faecal microbiome. In a third study, biochar supplementation on enteric CH_4 production and fermentation pattern was evaluated *in vitro* to assess its potential use as a dietary mitigation strategy. Overall, these studies showed variability in gaseous emissions output suggesting that mitigation strategies should be tailored to on-farm management systems in order to be effective. Funding of FCT, AGROS and CAVC to ARFR (PDE/BDE/114434/2016) and of FCT to MRGM (DL57-Norma Transitória), REQUIMTE (UIDB/50006/2020) and CITAB (UIDB/04033/2020) is acknowledge.

QTL for multiple birth in Brown Swiss and Original Braunvieh cattle on chromosome 15 and 11

S. Widmer[1], F.R. Seefried[2], P. Von Rohr[2], I.M. Häfliger[1], M. Spengeler[2] and C. Drögemüller[1]

[1]Institute of Genetics, University of Bern, Bremgartenstrasse 109a, 3012 Bern, Switzerland, [2]Qualitas AG, Chamerstrasse 56, 6300 Zug, Switzerland; sarah.widmer@vetsuisse.unibe.ch

Twin and multiple births are rare in cattle and have a negative impact on the performance and health of cows and calves. To decipher the genetic architecture of this trait in the two Swiss Brown Swiss cattle populations, we performed various association analyses based on de-regressed breeding values. Genome-wide association analyses were performed using ~600 K imputed SNPs for the maternal multiple birth trait in ~3,500 Original Braunvieh and ~7,800 Brown Swiss animals. Significantly associated QTL were observed on different chromosomes for both breeds. We have identified a QTL on chromosome 11 for maternal multiple birth that explains ~6% of the total genetic variance in Original Braunvieh. For the Brown Swiss breed, we have discovered a QTL on chromosome 15 that accounts for ~4% of the total genetic variance. A haplotype-based approach was used for fine-mapping of both QTL signals. For Original Braunvieh, a 90-kb window was identified on chromosome 11 at 88 Mb, where a most likely regulatory region is located close to the *ID2* gene. The signal for Brown Swiss was fine-mapped to a 130-kb window at 75 Mb on chromosome 15. Analysis of WGS data using linkage-disequilibrium estimation revealed possible causal variants for the identified QTL. A regulatory variant in the non-coding 5′ region of the *ID2* gene was strongly associated with the haplotype for Original Braunvieh. In Brown Swiss, an intron variant in *PRDM11*, one 3' UTR variant in *SYT13* and three intergenic variants 5' upstream of *SYT13* were identified as candidate variants for the trait multiple birth maternal. In this study, we report for the first time fine-mapped QTL for the trait of multiple births in Original Braunvieh and Brown Swiss cattle. Moreover, our findings are another step towards a better understanding of the complex genetic architecture of this polygenic trait.

Predicting cheese making traits in Grana Padano PDO via milk Fourier-transform infrared spectra

A. Molle[1], G. Stocco[1], A. Summer[1], A. Ferragina[2] and C. Cipolat-Gotet[1]
[1]University of Parma, Department of Veterinary Science, via del Taglio 10, 43126 Parma (PR), Italy, [2]Teagasc Food Research Centre, Food Quality and Sensory Science Department, KN3K, Dublin, D15, Ireland; arnaudpaulj.molle@unipr.it

The aim of this study was to investigate the prediction reliability of cheese making traits [3 measures of cheese yield (%CY): fresh, solids and retained water (%CYCURD, %CYSOLIDS, %CYWATER); 4 recovery traits (%REC): milk fat, protein, solids and energy in the curd (%RECFAT, %RECPROTEIN, %RECSOLIDS and %RECENERGY)] applying Bayesian models on the Fourier-transform infrared spectroscopy (FTIR) spectra of vat milk samples used for Grana Padano PDO production. Information from 50 cheese-making days (2-3 vats per day depending on the number of batches/day; in total 139 vats) from two dairy industries were collected. For each vat, weights of the vat milk, and the cheese after 48h from cheese making, were measured and milk and whey were sampled for their composition (total solids, lactose, protein and fat). Two spectra from each milk sample were collected using a MilkoScan FT 6000 in the range between 5,011 and 925 cm^{-1} and averaged prior the data analysis. A Bayesian approach was implemented to develop calibration models by using the BGLR (Bayesian Generalized Linear Regression) package of R software. Performance of models was assessed by coefficient of determination (R2VAL) and the root mean squared error of validation (RMSEVAL). A random cross-validation (CV) was applied [80% calibration (CAL) and 20% validation (VAL) set] with 10 replicates. Results from CV showed that the most accurate predictions were obtained for %CYCURD and %CYSOLIDS, which exhibited R2VAL and RMSEVAL values of 0.55 and 0.27, and of 0.65 and 0.18, respectively. The %CYWATER showed the lowest R2VAL (0.53), being the least repeatable among cheese making traits. Considering %REC traits, promising results were obtained for the recovery of protein (RMSEVAL=0.31%). In opposite, the recovery of energy (RMSEVAL=0.81%) and fat (RMSEVAL= 1.83%) showed a less favourable result. These results demonstrate FTIR spectroscopy could be a valid method to indirect monitor cheese making efficiency at the dairy industry level for Grana Padano PDO production.

Saliva as a potential non-invasive fluid for passive immune transfer surveillance in calves

F.G. Silva[1,2], E. Lamy[1], S. Pedro[1], I. Azevedo[1], P. Caetano[3], J. Ramalho[3], L. Martins[3], J.O.L. Cerqueira[2,4], S.R. Silva[2] and C. Conceição[1]
[1]Mediterranean Institute for Agriculture Environment and Development (MED), University of Évora, Department of Zootechnic, Largo dos Colegiais 2, 7004-516 Évora, Portugal, [2]Veterinary and Animal Research Centre (CECAV), University of Trás-os-Montes e Alto Douro, Department of Animal Science, Quinta de Prados, 5000-801 Vila Real, Portugal, [3]Mediterranean Institute for Agriculture Environment and Development (MED), University of Évora, Department of Veterinary Medicine, Largo dos Colegiais 2, 7004-516 Évora, Portugal, [4]Polytechnic Institute of Viana do Castelo (IPVC), Agrarian School of Ponte de Lima, Rua D. Mendo Afonso, 147 Refóios do Lima, 4990-706 Ponte de Lima, Portugal; fsilva@uevora.pt

Passive immune transfer (PIT) is detrimental to the calf's welfare. PIT is currently assessed by immunoglobulins or total proteins concentration (TP) in blood. Blood collection has some constraints, ergo, a simple and non-invasive alternative method is of great interest. This study aimed to evaluate the potential use of saliva as a non-invasive fluid to assess PIT. A total of eighty saliva and blood samples were taken from twenty calves at 4 time points: at birth (approximately 30 min before colostrum consumption), at 24h, 48h and at day 7, for total protein and IgG levels concentration assessment. A hand-held refractometer was used for serum samples and the Bradford method for total protein determination in both saliva and serum samples. Dot-blot analysis was performed in a sub-sample to check for IgG in saliva and serum (n=5 calves). With both methods, TP in serum at 24h, 48h and at day 7 was significative higher than TP at birth ($P<0.001$). A tendency for elevated TP at 24h ($P<0.1$) was found in saliva samples. A significant Pearson's positive correlation was found between TP in saliva and in serum by refractometer at 48h ($r=0.45$; $P=0.0474$). Dot-blot analysis showed an increase in IgG from birth to following phases in blood and saliva. These results suggest that saliva can be used for PIT surveillance, highlighting the importance of further studies.

Effect of grazing on cow milk metabolites

G. Niero[1], G. Meoni[2], L. Tenori[2], C. Luchinat[2], S. Callegaro[3], M. De Marchi[1] and M. Penasa[1]
[1]University of Padova, Department of Agronomy, Food, Natural resources, Animals and Environment, Viale dell'Università 16, 35020 Legnaro, Italy, [2]Magnetic Resonance Center, Via Luigi Sacconi 6, 50019 Sesto Fiorentino, Italy, [3]Associazione Nazionale Allevatori delle Razze Bovine Charolaise e Limousine Italiane, Via Ventiquattro Maggio 44/45, 00187 Roma, Italy; g.niero@unipd.it

Traceability tools able to characterize milk from pasture are important to safeguard low-input farming systems and their surrounding environments. The aim of the present study was to investigate nuclear magnetic resonance (^{1}H NMR) spectroscopy as candidate technique to discriminate between milk produced by grazing cows and milk produced by non-grazing cows. The trial involved an alpine herd of 72 lactating cows. Individual milk samples were repeatedly collected on the same animals, before (days -3 and -1) and after (days 2, 3, 7, 10, and 14) the onset of the grazing season. One dimensional ^{1}H NMR spectra of milk extracts were collected through a Bruker spectrometer. Sources of variation of the relative abundances of each milk metabolite were investigated through a mixed model which accounted for the fixed effects of period of sampling, cow breed, stage of lactation, and parity, and for the random effect of cow nested within breed. In addition, a random forest model was applied to the data and results showed that it was highly accurate (93.1%) in discriminating between samples collected in days -3, -1, 2, and 3 (period 1) and samples collected in days 7, 10, and 14 (period 2). Univariate analysis performed on the 40 detected metabolites highlighted that milk from period 2 had lower content of 14 compounds (with fumarate being the most depressed metabolite) and greater content of 15 compounds (with methanol and hippurate being the most elevated metabolites) than milk from period 1. Results suggest that milk ^{1}H NMR spectra are useful to identify milk from grazing and non-grazing cows based on milk metabolic profile, meaning that some compounds are of potential interest as candidate traceability indicators of milk.
This research was funded by Veneto Region with the DGR n. 376/2018, Misura 16.1.1 and 16.1.2, FITOCHE project.

Impact of prolonged cow-calf contact on dairy cow production in a pasture-based system

S.E. McPherson[1,2,3], L.E. Webb[2], A. Sinnott[2,3], E.A.M. Bokkers[2] and E. Kennedy[1,3]
[1]VistaMilk SFI Research Centre, Teagasc Moorepark, Fermoy, Co. Cork, P61C996, Ireland, [2]Wageningen University & Research, Animal Production Systems group, P.O. Box 338, 6700 AH Wageningen, the Netherlands, [3]Teagasc, Teagasc Moorepark, Fermoy, Co. Cork, P61C996, Ireland; sarah.mcpherson@teagasc.ie

With consumers becoming increasingly concerned about the welfare of farm animals, two aspects of dairy farming have come under scrutiny: cow-calf separation soon after birth and access to pasture. The aim of this study was to investigate the impact of prolonged cow-calf contact (CCC), from birth to weaning, on dairy cow (n=50) production within the context of a pasture-based, seasonal calving system in Ireland. The 3 CCC rearing systems were no contact (control (C), 18 cows), full-time contact (FT), 14 cows; part-time contact (PT), 18 cows). The FT cows and their calves and C cows grazed fulltime; cows were milked twice a day (TAD). The PT cows were milked once a day (08:00), went to grass and returned indoors at 15:00 to join their calves until milking the following morning. All calves were gradually weaned from milk at 8-weeks old. At weaning PT cows switched to TAD milking. Individual parlour milk yield (MY) was recorded daily and average daily milk solid yield (MSY; kg fat + kg protein) was recorded weekly. During weeks 1-8, C, FT, and PT all had different (P<0.001) MY and MSY, during weeks 1-4 MY were 22.1, 13.4, and 9.7 kg/day, respectively, and MSY were 2.1, 1.1, and 0.8 kg/day, respectively. In line with stage of lactation, MY of C (23.9 kg) increased during weeks 5-8 (P<0.001) but decreased for CCC cows (FT: 12.0 and PT: 9.5 kg) and MSY followed the same trend (2.0, 0.9 and 0.7 kg/day respectively). Following weaning (weeks 9-12) FT and PT cows increased MY by 34 and 47%, respectively, and increased MSY by 40 and 51%, respectively, but remained lower (P<0.001) than the C cows (MY: 23.1 kg; MSY: 1.9 kg) for both. From weeks 13 to 35 MY and MSY of FT and PT cows were similar but significantly lower than C (MY: 15.6, 14.9 and 18.8 kg, respectively; MSY: 1.4, 1.3, 1.7 kg, respectively). Over the 35-week lactation C had a higher parlour recorded MY and produced more milk solids than CCC treatments. Further investigation is required to understand the divergence in MY post-weaning.

Automated, incomplete milking as energy management strategy in early lactation of dairy cows

I. Meyer, E. Haese, K.-H. Südekum, H. Sauerwein and U. Müller
University of Bonn, Institute of Animal Science, Katzenburgweg 7, 53115 Bonn, Germany; ismey@uni-bonn.de

The negative energy balance (EB) in early lactation imposes metabolic stress with an increased risk for diseases in dairy cows. Incomplete milking (IM) is one way of attenuating the negative EB; suitable technical solutions must be easy to integrate into the milking routine while maintaining udder health during the application. Our objective was to test a software module by which milking clusters are automatically removed when a pre-defined amount of milk is obtained. Forty-two Holstein cows without any signs of impaired udder health were stratified by calving date and lactation number, and allocated to either the IM group or the control (CON) group (n=21/group). The IM was applied for 2 weeks starting from 8 (±1) days in milk (DIM) by clamping the amount of milk removed to the yield of the preceding day. Thereafter the IM cows were milked in the same way as the control group (CON), i.e. clusters were removed at milk flow rates <0.3 kg milk/min. Records of daily milk yield and feed intake, of weekly feed- and milk composition, as well as of multiple cytobacteriological milk analyses for evaluating udder health were taken until 35 DIM. All data were evaluated using mixed models and t-tests (SPSS). During the software application, daily yields in IM were clamped to 30.4±4.8 kg milk/d; in CON, milk yield increased at the same time from 33.4±5.4 to 35.8±6.4 kg milk/d. Considering the treatment period, milk yield was affected by time (P<0.001) and group (P=0.029). Feed intake was not different between groups; energy corrected milk (ECM) tended to be lower in IM (P=0.072); accordingly, the calculated EB tended to be higher in the second treatment week in IM (P=0.064) than in CON. After the treatment period, milk yield increased also in the IM group and thus levelled off the differences between groups. Both ECM and EB did not differ between the groups after the IM period. In none of the groups udder health was impaired, and groups did also not differ in this respect during the entire observation period. Our results confirm the applicability of the software in the first weeks of lactation. Further investigations about the effects of IM on the individual`s metabolic health using blood analyses are ongoing.

Characterization of bovine vaginal microbiota and its relationship with host phenotypes

L. Brulin[1,2], S. Ducrocq[1], G. Even[1], M.P. Sanchez[2], S. Martel[1], C. Audebert[1], P. Croiseau[2] and J. Estellé[2]
[1]Gènes Diffusion/GDBiotech, 3595 route de Tournai, 59500 Douai, France, [2]Université de Paris-Saclay, INRAE, AgroParisTech, GABI, Domaine de Vilvert, 78350 Jouy-en-Josas, France; louise.brulin@inrae.fr

Many studies highlighted the important role of microbiota for the physiology and homeostasis of their host animals. In cattle, these studies mainly focused on the digestive tract and, more especially, on the rumen. Yet, there is a growing interest in the microbiota associated with the female reproductive tract because it could influence fertility-related traits or gestation ability, which are key components of a cow's productive life. Thus, the present study aims to better characterize the vaginal microbiota, identify possible factors influencing its composition and diversity, and explore its relationships with relevant host phenotypes. For this purpose, 1,365 vaginal swabs were sampled on 1,101 non-pregnant Holstein cows, with one to five samples per animal. These dairy cows, from 19 commercial farms located in northern France, could be either heifers or lactating cows in their first to fifth lactation. Oestrous cycle, health status, gestation, and production were monitored for all cows. The frozen vaginal swabs were sequenced for the 16S bacterial gene using the Miseq Illumina technology. Then, the data were processed using the DADA2 package to obtain the ASV tables. The existence of a vaginal core microbiota has been investigated and we have explored how animal characteristics or environmental factors shape the vaginal microbiota. Apart from the herd management and sampling conditions, the diversity of the vaginal microbiota, mainly composed of Firmicutes (46%), was significantly associated with the lactation rank and the oestrus cycle stages. In addition, we evaluated the association between the microbiota α-diversity and composition and major fertility-related traits, such as the insemination success, to estimate the predictive potential of the vaginal microbiota on the host reproductive performances. Initial results showed correlations between the variations in vaginal microbiota diversity and the number of inseminations. Overall, our study will potentially represent a step forward in the comprehension of the relationships between the vaginal microbiota composition and fertility-related traits.

Empirical test of strategies for implementing genomic prediction in small Holstein populations

A. Ule[1], G. Gorjanc[2] and M. Klopčič[1]
[1]University of Ljubljana, Biotechnical Faculty, Dept. of Animal Science, Groblje 3, 1230, Slovenia, [2]The Roslin Institute and Royal (Dick) School of Veterinary Studies, The University of Edinburgh, Easter Bush, Midlothian, Scotland, United Kingdom; anita.ule@gmail.com

This study aimed to evaluate how compositions of the reference population and incorporating non-genotyped animals through a single-step genomic BLUP (ssGBLUP) impact the accuracy of predictions of breeding values (EBV) in small Holstein populations. The most reference population consists of progeny tested bulls, sufficient to estimate the breeding value in big populations. The reference population's size and accuracy of observations are the most crucial factors for accurate genomic prediction. The accuracy of a genomic EBV (GEBV) will be higher when the number of genotyped individuals with their performance or progeny records is large. Therefore, the accuracy of genomic EBV for small reference populations can be as low as in pedigree-based evaluation. Countries with a small Holstein population have a small number of progeny-tested bulls, and hence estimated breeding values (EBV) are less reliable. Five different scenarios were defined to show the influence of different methods, sources of data, and compositions of the reference population on the reliability of breeding values. The effect of increasing the reference population on the reliability of the GEBV was examined for five traits: protein yield (kg), fat yield (kg), somatic cell count (log), overall leg score, and overall udder score. Phenotypic data comprises 3,533,390 test-day records for protein and fat yield and somatic cell count for 178,739 cows between 2000 and 2021. The associated pedigree consisting of 4 generations comprises 242,425 animals. Genotypes for 3,773 animals were available: 401 males and 3,372 females. The combined reference population helps achieve higher accuracy of estimated breeding value than the male reference population. We obtained the highest accuracy when using the ssGBLUP method, which includes phenotypes, genomic data, and pedigree in a single step and both male and female animals in the reference population.

Quality control of milk containing only A2 β-casein using locked nucleic acid-based real-time PCR

L. Jiménez-Montenegro, J.A. Mendizabal, L. Alfonso and O. Urrutia
ETSIA-ISFOOD. Public University of Navarre, Campus de Arrosadia, 31006, Spain; lucia.jimenez@unavarra.es

Bovine milk mainly contains two types of β-casein: A1 and A2 variants. In recent years, a new variety of cows' milk has emerged in dairy sector called 'A2 milk'. This novel product is characterized by the absence of A1 β-casein, which has been associated with possible gastrointestinal discomfort due to β-casomorphin-7 (BCM-7) release during gastrointestinal digestion. In this context, methods to verify A1 allele absence in A2 milk are required as a quality control in the A2 milk commercialization. Therefore, the aim of the present study was to develop a locked nucleic acid (LNA) probe-based duplex real-time PCR (qPCR) assay for A1 allele detection in A2 milk samples. Firstly, four DNA isolation methods from milk somatic cells were optimized and evaluated. Results suggests that the commercial kit *Nucleospin Tissue* was the most suitable method in terms of DNA quality and amplificability for downstream applications. Then, optimization and validation of the qPCR assay were carried out. For both A1 and A2 alleles, the absolute limits of detection of this qPCR assay were 7.3 DNA copies/reaction (2×10^{-5} ng DNA) and 30.4 DNA copies/reaction (0.1 ng DNA) at a 95% confidence level with synthetic reference DNA samples and heterozygous genotyped DNA sample, respectively. The relative limits of detection were 2% (15 copies) and 5% (152 copies) for A1 allele in A2 samples at a 95% confidence with synthetic reference and genotyped DNA samples, respectively. The qPCR assay was robust, with intra- and inter-assay variability below 4.3%, and specific, differentiating between A1 and A2 alleles with 100% genotyping accuracy. In conclusion, this cost-effective and fast method could be used to discriminate A1 allele in A2 samples and, consequently, to verify A1 allele absence in 'A2 milk' by screening commercial product on the market. The method could also be valid for genotyping individual cows, since it has been shown to be highly reliable for discriminating A1A1, A2A2 and A1A2 genotypes of the *CSN2* gene, consequently providing an alternative to sequencing techniques and other DNA-based methods used in dairy cattle genotyping.

Session 15 Poster 15

Identification and expression profiling of miRNAs linked to early life performance in dairy cattle

M.K. MacLeay[1], G. Banos[2] and F.X. Donadeu[1]
[1]The Roslin Institute and Royal (Dick) School of Veterinary Studies, University of Edinburgh, Easter Bush, Midlothian, EH25 9RG, United Kingdom, [2]Scotland's Rural College, Department of Animal and Veterinary Sciences, Roslin Institute Building, Easter Bush, Midlothian, EH25 9RG, United Kingdom; s1204766@ed.ac.uk

Early life performance traits in dairy cattle can have important influences on lifetime productivity. Poor health and fertility may cause large losses before the first lactation and early growth impacts the time to first service and subsequent milk yields. Circulating miRNAs have been linked to livestock traits such as resistance to infection, early pregnancy, and muscle development. This study aimed to identify and characterize circulating miRNAs in dairy cattle that are associated with early life performance. Plasma samples from 12 calves identified retrospectively as differing in health, growth, and early fertility were analysed using PCR arrays detecting 378 miRNAs, of which 6 significantly differed between poor growth/fertility and control calves ($P<0.05$). General(ised) linear mixed models (GLMMs) found 1 miRNA associated with average daily gain until weaning (ADG), 22 with one year bodyweight, 47 with age at first service, and 19 with number of infections recorded before first calving (85 distinct miRNAs). Of these, 9 selected miRNAs were further analysed by RT-qPCR in samples from 91 animals, including 3 age groups (calf, heifer, and cow). Levels of 8 miRNAs (miR-126-3p, miR-127, miR-142-5p, miR-154b, miR-27b, miR-30c-5p, miR-34a, miR-363) changed significantly with age, miR-30c-5p was associated with ADG, and miR-126-3p with number of infections (GLMMs). Moreover, 2 ratios between miRNAs were associated with ADG, 2 with age at first service, and 4 with number of infections. Age-related changes in miRNAs involved mainly the calf to heifer transition but no associations were found between fold-changes in miRNA levels between calves and heifers and early life traits. Finally, to further investigate potential functions of these miRNAs, comparative expression analyses were performed by RT-qPCR across 25 calf body tissues, and putative targets and associated biological pathways were identified from online databases. Our results suggest that miRNAs may have important roles in dairy cow early development.

Session 16 Theatre 1

The prospective methods to support pastoral strategies in the Mediterranean: the case of Corsica

J.-P. Dubeuf
INRAE, SELMET, 20250 Corte, France; jean-paul.dubeuf@inrae.fr

Pastoralism is a practice of animal production based on the exclusive or partial use of spontaneous resources. It is therefore likely to meet the challenges of the agro-ecological transition. As in many Mediterranean regions, Corsica has a very old pastoral transition, today questioned by the reduction in the number of farms, the closure of environments, the abandonment of grazing management practices on rangelands and a reduction of transhumance. In this context, what can be the prospects for pastoral farming in Corsica? To support the implementation of a pastoral strategy for public action, a prospective workshop based on the territorial prospective approach tried to answer this complex question. A group of thirty volunteer experts and actors have met during four one-day sessions. They problematized the question, established a shared diagnosis of the present situation and planned future prospects by identifying transformative processes, to build scenarios and challenge stakeholders. Each of four scenarios was built by considering successively, the men and women as actors of pastoralism, the animals and their valorisation, the policies and the organizations, the territories of pastoralism: (1) The memorial pastoralism scenario: Pastoralism remains the reference backdrop but development priorities go to other non-agricultural activities. (2) The scenario of a rediscovered pastoralism: We are witnessing a vast movement to live and work in the country by developing local production and a sylvo-pastoral orientation. (3) The scenario of a pastoralism redesigned by the downstream sector: Pastoral components are marginal but symbolic contributors for Corsican livestock farming with other productive ambitions. (4) The scenario of the new pastoral enterprise: In a world of radical and rapid changes, pastoralism is part of a new pluri active and innovative farm and start up economy Then, we discuss the conditions for stabilizing a pastoral strategy towards an assumed final scenario that would combine these different proposals. The conclusions of the workshop underlined the absence of a true pastoral technical model still to build, the need to invest in engineering, pluri disciplinary research, education and permanent training for breeders while limiting the perverse effects of CAP subsidies.

Breeding for resistance and robustness against gastrointestinal nematodes in German Merino lambs

J. Gürtler, M. Schmid and J. Bennewitz
Institute of Animal Science, University of Hohenheim, Garbenstraße 17, 70599 Stuttgart, Germany; johannes.guertler@uni-hohenheim.de

Sheep farmers in Germany generate a large part of their income from the management and the care of extensive grassland areas. On the pastures, the famers face challenges such as infestation with gastrointestinal nematodes (GIN). Due to the increasing resistance of GIN to anthelmintics, but also due to the treatment costs, the interest in alternative control strategies is large. The aim of this study was to improve the resistance and robustness of German Merino sheep to GIN. The field work covers the grazing periods of the years 2021 and 2022. The aim is to find a suitable sire breed for the German Merino for certain pasture forms, which can also cope with GIN pressure on pasture. For this purpose, three farms on the Swabian Alb (Southern Germany) were selected, each of them housing F1-Merinoland crosses descending from five sire breeds (i.e. Merinoland, Suffolk, Texel, German black-headed mutton and Ile de France) in two grazing systems (extensive and intensive grazing). Per year and farm, around 50 lambs per cross were generated (in total 250 lambs per year and farm). Preliminary evaluations from the dataset of the year 2021 which included the phenotypes of the traits average daily gain on pasture (ADG, n=469) and faecal worm egg count (FEC, n=485). All labs were genotyped with a 50k SNP chip. The lambs were frequently weighed and individually faecal sampled once after grazing 6 weeks on pasture. To achieve approximately normally distributed data, FEC was transformed to $\log_e$ (FEC+10). From these trait records, resistance and robustness were derived. Animals exhibiting a lower FEC compared with the population average were considered resistant and animals showing less growth depression than the population average when exposed to GIN were considered robust. Moderate heritabilities were estimated for FEC and ADG. About half of the lambs were resistant and about one third of the animals were robust against GIN. Due to the limited sample size, significant breed differences in robustness, resistance and FEC traits have not yet been observed. However, selective deworming was successful, one fourth of the lambs did not need to be treated, indicating a natural resistance and robustness in these animals.

Introducing hormone-free insemination in dairy sheep farms challenges their feeding system design

E. Laclef[1], N. Debus[1], P. Taillandier[2], P. Hassoun[1], S. Parisot[3], E. González-García[1] and A. Lurette[1]
[1]CIRAD,INRAE,L'Institut Agro, UMR SELMET, 2 Place Pierre Viala, 34000 Montpellier, France, [2]INRAE,Toulouse University, UR MIAT, 24 Chemin de Borde Rouge, 31320 Castanet Tolosan, France, [3]INRAE, La Fage, 12250 Saint Jean et Saint Paul, France; ellen.laclef@inrae.fr

Hormone-free (HF) reproduction management in dairy sheep farming is a way to address current societal demands. However, we must be prepared for the collateral impacts to be expected in the rest of the farming system components. Indeed, the efficiency of related HF practices (e.g. using male effect for oestrus induction and synchronisation) is more uncertain than that of hormonal treatment (HO). Direct consequences, on the feeding system design, will thus reflect a higher variability of physiological stages present in the flock throughout the year. This work aims to simulate the impact of a HF reproduction including artificial insemination (AI) on the distribution of productive performances and nutritional requirements of a conventional dairy sheep flock, using a dynamic agent-based model. Six reproductive management scenarios, for managing the same flock (n=597 Lacaune ewes), in the Mediterranean Roquefort Basin, were simulated and compared over one full and representative production season i.e.: HO+AI in mid-May (Early); HO+AI in July (SumLate); HO+AI in November (AutLate); as well as their HF version (i.e. HF-Early; HF-SumLate and HF-AutLate, respectively). In all HF scenarios, a reduction in the number of ewes lambing and in the annual milk production (-1 to -8%) was observed, which was accompanied by a subsequent decrease of the flock's annual nutritional requirements (-2 to -7%). The HF scenarios also resulted in a staggering of lambing occurrences which led to a shift in the peaks of milk production and, consequently, in the nutritional requirements during the production season. In conclusion, transitioning to HF reproduction management, while preserving AI, would imply an increase in the workload, a prolongation of the programmed milking period, and an essential readjustment of the feeding management strategy of the flock with regard to farm's feed resources availability. Our simulation results are used to support discussions within the dairy sheep industry looking for sustainable alternatives to using hormones.

Session 16 Theatre 4

Response of lactating dairy ewes to exogenous melatonin varies according to breed

A. Elhadi, A.A.K. Salama, X. Such and G. Caja
Universitat Autonoma de Barcelona (UAB), Animal and Food Sciences, G2R, Campus Universitari de la UAB, 08193 Bellaterra, Barcelona, Spain; gerardo.caja@uab.es

Previous research on melatonin (MEL) in lactating dairy ewes, showed unexpected differences in plasmatic MEL according to breed. With this aim, 20 lactating ewes of 2 dairy breeds (MN: Manchega, n=10, 84.9±2.3 kg; LC: Lacaune, n=10, 88.6±2.9 kg) in late lactation (MN: 0.98±0.05 kg/d; LC: 1.22±0.09 kg/d), were used to study plasmatic MEL according to breed. Lactating ewes were in pens, fed a total mixed ration *ad libitum* and milked ×2 (08:00 and 17:00). Ewes were allocated into 4 balanced groups of 5 ewes to which the treatments were applied by breed: (1) Control (CO); and (2) Melatonin (MEL), ear base implanted (18 mg/ewe; Melovine, Ceva Animal Health) before the p.m. milking. Artificial light (50 lx; 06:00-10:00 and 18:00-21:00) was also provided for management. Blood samples post implantation were taken from the jugular at d 12 (09:20), for basal MEL, and at d 20 and 21 (02:00 to 24:00), for assessing the MEL variation by breed and treatment. Red light was used during bñood sampling at night. Blood was processed for plasma and frozen until analysis. MEL concentration was analysed by ELISA sandwich kits (LDN). Data were analysed by PROC MIXED for repeated measurements of SAS (v.9.4). On average, daily MEL concentration was greater in the MEL treated than the CO ewes (75±18 vs 31±14 pg/ml; P<0.01), showing an expected circadian daily pattern, increasing after the sunset and decreasing after the dawn. Maximum MEL values were found at 02:00 (146±35 pg/ml) in both breeds and MEL treatments. MEL implants increased the plasmatic MEL in both breeds, but differences between CO and MEL ewes were only detected from 08:00 to 22:00 (P<0.10 to P<0.01). Regarding the MEL implanted ewes, greater MEL values were detected in MN than in LC ewes at 8:00 and 20:00 (P<0.05), indicating that MEL dropped later after the dawn and increased earlier at the sunset in MN than in LC ewes. In conclusion, differences in the response to exogenous MEL were confirmed according to breed in spring, the MN (lighter, mid yield) showing greater plasmatic concentrations than LC (heavier, high yield), which may be related to their weight, yield and MEL receptor gene MTNR1A polymorphisms.

Session 16 Theatre 5

Joint effects of month of kidding and stage of lactation on the chemical composition of goat milk

F. Zamuner, A.W.N. Cameron, B.J. Leury and K. Digiacomo
The University of Melbourne, Faculty of Veterinary and Agricultural Sciences, Building 184 Royal Parade, Parkville, 3010, Australia; fernanda.zamuner@unimelb.edu.au

It is well known that goat milk composition is influenced by several factors such as the stage of lactation (SL), month of kidding (MK), and environmental conditions. However, information on the effects of MK on milk composition at different SL, and the joint effects of MK and SL upon year-round bulk-milk composition in dairy goat herds managed in multiple kidding seasons per year, are still scarce in the literature. Therefore, this study aimed to characterize the joint effects of MK and SL on percentages of fat, protein, and lactose in three commercial herds in Australia. In total, 5,811 (8,467 lactations: MAR, n=792; JUN, n=713; SEP, n=1,279; and NOV, n=972)) Saanen goats were enrolled in the study. Individual milk samples were collected in early, mid, and late lactation and bulk-milk samples collected weekly from each farm from September 2017 to July 2019. We found a significant effect (P<0.001) of MK, SL, and MK Í SL on percentage milk fat, protein, and lactose fat. In early, mid and late lactation the mean percentage of the following were; fat = 3.2, 3.2, and 3.8, protein = 2.9, 3.0, 3.2, and lactose = 4.7, 4.6, 4.8, respectively.

Session 16 — Theatre 6

Environmental impact of suckling dairy lamb

M.F. Lunesu, G. Battacone, S.P.G. Rassu, G. Pulina, A. Fenu, A. Mazza and A. Nudda
University of Sassari, Dipartimento di Agraria, Viale Italia, 39, 07100 Sassari, Italy; mflunesu@uniss.it

Dairy lamb is a secondary product of dairy sheep farms. Its nutritional and organoleptic characteristics, partly due to the young age of slaughter (4-6 weeks) and in part to the quality of suckled milk obtained by ewes grazing natural pasture, make this a niche product appreciated by consumers. In European Union, the local foods should become the global standard for sustainability. However, the carbon footprint (CF) of lamb meat has received less interest and the information available in the scientific literature refer only to heavy lambs. This study aimed to estimate the environmental impact of suckling lamb. Carbon footprint of suckling lambs was estimated using data collected in a commercial dairy sheep farm located in the northwest of Sardinia (Italy). Ninety-six Sarda ewes after parturition were monitored with their lambs (n=44 females and 52 males) for a period of 28 days. Traditional (T; suckling lambs followed their mothers on pasture during the grazing time) or separated (S; suckling lambs were maintained indoor during the grazing time of mothers) farming systems were considered. The sex of lamb and the parity (single and twins) were also considered in the analyses. The CF was calculated within a cradle to farm gate system boundary considering maternal enteric CH_4 emissions and milk suckled by the lambs as the main emissions hotspots. The functional unit (FU) considered was 1 kg of body weight gain (BWG) during the suckling period. Our results evidenced that CF of suckling lamb varied between 4.56 and 7.30 kg CO_2eq/kg lamb BWG with the highest values ($P<0.05$) referring to the S farming system. Lamb BWG was the best predictor of CF ($CF=37.975e^{-0.245lambBWG}$; $R^2=0.83$). The sex of lambs affected their CF, with males having lower ($P<0.05$) values than females. The CF of twins was markedly lower than singles (4.56 vs 7.30 kg of CO_2eq/kg lamb BWG). In conclusion, the CF of the suckling lamb meat can be reduced by maintaining the traditional lamb rearing system and improving the flock prolificacy. Research funded by RESTART-UNINUORO Project 'Actions for the valorisation of agroforestry resources in central Sardinia' Regione Autonoma della Sardegna, fondi FSC 2014-2020.

Session 16 — Theatre 7

Effect of the incorporation of forage alternatives in Chilean Mediterranean sheep farming

P. Toro-Mujica
Universidad de O'Higgins, Instituto de Ciencias agroalimentarias, animales y ambientales, Ruta I-50.Km 3, 3070000, Chile; paula.toro@uoh.cl

In Chilean Mediterranean sheep farming systems, the production is based on annual pastures. Thus, the dependence of the sheep production system on pasture growth determines the total production and the productivity. Climate change has affected the rainfall pattern, shortening the rangeland growing season, decreasing total dry matter production and, hence constraining the production. In this scenario, the agricultural research institute (Instituto de Investigaciones Agropecuarias, INIA), as well as research financed by the national research and development agency (Agencia Nacional de Investigación. ANID), will be in the search for new forage alternatives that allow to extend the growing season or expand the availability of food for the livestock systems of dry areas. Among these alternatives, two species have shown good results given their productivity and resistance to drought: Medicago sativa and Triticale. Although the introduction of these new species increases forage production, it determines changes in production systems that can increase greenhouse gas emissions. These changes include the processes of soil preparation, sowing, fertilization, application of pesticides and herbicides, harvesting, processing and/or storage of forage and/or grain. The objective of this work was to determine how the inclusion of these crops affects the carbon footprint, productivity, and profitability of sheep production systems in rainfed areas through the use of a simulation model. The simulation model allows to calculate the performance of the sheep against supplementation strategies, pasture management and stocking rates throughout the production cycle.

Session 16 — Theatre 8

Sustainability in the sheep sector: a systems perspective, from good practices to policy

A.S. Atzori[1], A. Franca[2], P. Arca[3], G. Molle[4], M. Decandia[4], P. Duce[3] and E. Vagnoni[3]
[1]Dipartimento di Agraria, University of Sassari, Sassari, Italy, Vial Italia 39, 07100, Italy, [2]Institute for Animal Production System in Mediterranean Environment, National Research Council, Traversa La Crucca, 07100 Sassari, Italy, [3]Institute of BioEconomy, National Research Council, Sassari, Italy, Traversa La Crucca, 07100 Sassari, Italy, [4]AGRIS Sardegna, Loc Totubella, 07100 Sassari, Italy, Italy; asatzori@uniss.it

Three million sheep are raised in the Island of Sardinia (Italy), on 10,000 active farms in mixed systems, producing about 13% of sheep milk in the European Union (EU). Almost all delivered milk is processed to sheep cheese and destined for world trade. Oppositely, the Sardinian dairy sheep sector emits about 1,600 kt CO_2eq/year. The SheepToShip LIFE project (EU-LIFE CCM 2014-2020) represented a laboratory of territorial mitigation action. This work summarizes the methodological approach followed by SheepTo Ship to promote ecoinnovation strategy for carbon emissions mitigation under the case study of the Sardinian sheep sector. The actions carried out In Sardinia included: (1) a life cycle assessment approach featuring environmental and socioeconomic traits of the dairy sheep farms; (2) on-farm demonstration of mitigation techniques, to define the most viable and profitable practices of sheep impacts mitigation; (3) a series of focus groups, discussing beliefs and reactions of the main stakeholders to proposed practices; and (4) a group model building that developed a causal loop diagram under a systemic perspective of the sheep supply chain, in order to stimulate regional policy-making aligned with the Sustainable Development Goals. The project outcome is summarized in a causal loop diagram showing viable governmental policies to reduce emissions in Sardinia by 20% in 10 years by: (1) improving technical efficiency in those farms with highest milk delivery potential to support food provisioning; (2) enhancing ecosystem services in those farms with high potential of nonmarketable good and environmental regulation roles. The outcome included a causal loop diagram showing viable governmental policies oriented at sustaining the cheese production and profit of the supply chain also enhancing ecological benefits and social impacts.

Session 16 — Theatre 9

Whole-genome analysis of diversity and population structure in Portuguese native sheep breeds

D. Gaspar[1,2], A. Usié[2,3], C. Leão[2,3], C. Matos[4], A.M. Ramos[2,3] and C. Ginja[1]
[1]BIOPOIS CIBIO-InBIO, Centro de Investigação em Biodiversidade e Recursos Genéticos, Universidade do Porto, Vairão, Portugal, [2]CEBAL – Centro de Biotecnologia Agrícola e Agro-Alimentar do Alentejo, Campus IPBeja, Beja, Portugal, [3]MED – Mediterranean Institute for Agriculture, Environment and Development, Pólo da Mitra, Évora, Portugal, [4]ACOS – Agricultores do Sul, Rua Cidade S. Paulo, Beja, Portugal; danibgaspar@gmail.com

Since their domestication, approximately 10,500 years before present, sheep accompanied humankind in all its history. In Portugal, sheep are reared nationwide mainly in agrosilvopastoral systems. Merino Branco, Merino Preto, Campaniça and Bordaleira Serra da Estrela are among the most abundant local breeds. Merino and Campaniça are raised in the Alentejo region to produce meat and wool. Serra da Estrela is the most important dairy breed in Portugal used to produce a typical high-value cheese with a protected designation of origin certification. The purpose of this study was to estimate genomic variation in these four native sheep breeds and a population of crossed Merino, as well as to describe their genetic structure in the context of worldwide sheep and other Iberian breeds. High-throughput resequencing data was generated for 56 individuals, and a total of 31,320,380 high-quality SNPs were used in subsequent analyses. The overall levels of genomic variability were very similar across Portuguese sheep breeds ($0.30 \leq Ho \leq 0.34$; $0.30 \leq He \leq 0.35$). The Principal Components and Bayesian clustering analyses separated these breeds in two clusters: one comprising Campaniça and Serra da Estrela together with transboundary dairy breeds; and another of the well-differentiated multi-purpose Portuguese Merino sheep breeds along with Spanish Merino. Runs of homozygosity analysis yielded 1,690 ROH segments comprising an average of 2.27 Gb across the genome in all individuals. Campaniça showed the highest mean number of homozygous segments per animal (nROH=44.5) comprising on average 61.29 Mb. The results of this study are useful to develop genomic tools for genetic improvement, management and conservation of these breeds, including the traceability of certified products.

Detailed protein fraction profile of goat milk of six breeds

G. Secchi[1], N. Amalfitano[1], S. Pegolo[1], M.L. Dettori[2], M. Pazzola[2], G.M. Vacca[2] and G. Bittante[1]
[1]University of Padova, Department of Agronomy, Food, Natural resources, Animals and Environment, Viale dell'Università, 16, 35020, Legnaro (PD), Italy, [2]University of Sassari, Department of Veterinary Medicine, Via Vienna, 2, 07100, Sassari (SS), Italy; giorgia.secchi@phd.unipd.it

The protein profile of goat milk is very different from cow milk in terms of proportion of the different protein fractions and especially for its much higher genetic polymorphism. The aim of this study was to investigate breed effect on the detailed protein profile of goat milk. This research is part of the GOOD-MILK project which involved 1,272 goats from 6 different breeds reared in 35 farms located in Sardinia (Italy). The breeds are divided in two categories: the Alpine breeds (ALP: Camosciata delle Alpi and Saanen); and the Mediterranean breeds (MED: Murciano-Granadina, Maltese, Sarda, and Sarda Primitiva). A total of 1,090 individual milk samples from 23 farms (823 samples + 267 duplicates) were analysed for identification and quantification of milk protein fractions by RP-HPLC method. The data were analysed using a linear mixed model with the classes of days in milk (DIM), parity, and breed as fixed effects, while the flock effect was included as random. In a second model, the fixed effects of casein genotypes were also included. Breed affected all milk protein fractions, while DIM and parity had small effects. The main differences in protein profile among breeds were between ALP and MED types, being the former richer in whey proteins and lower in caseins than the latter; within MED, Sarda and Sarda Primitiva showed higher concentration in caseins than the other breeds, and no significant differences from each other. The inclusion of casein genotypes did not change relevantly the breed effect estimates but it decreased the residual variance, especially for α_{S1}- and β-caseins. Further investigation about the relationships between milk protein profile and milk quality and cheese-making ability in different goat breeds are envisaged.

Assessment of dairy sheep carcass composition with X-ray computed tomography

A. Argyriadou[1], M. Monziols[2], M. Patsikas[1] and G. Arsenos[1]
[1]School of Veterinary Medicine, Aristotle University, University Campus, 54124 Thessaloniki, Greece, [2]IFIP institut du porc, La Motte au Vicomte BP 35104, 35651 Le Rheu Cedex, France; argyrian@vet.auth.gr

The objective was to test a CT-acquisition protocol, with automated settings, for post-mortem assessment of dairy sheep carcasses using also two image analysis protocols (IAP). Twenty-six sheep of both sexes were used. After slaughter, hot carcass weight was recorded. Carcasses were chilled for 24h, then subjected to CT-scanning (Optima CT520, GE). The two CT-acquisition protocols (CT1, CT2) were tested on 4 carcasses. CT1 was automated and involved dose efficiency parameters, commonly used in medical practice (ranging field of view-FoV and tube current-TC, standard tube tension-TT of 120 kV). In CT2, parameters were manually set as defaults throughout scanning (Large FoV, TC: 80 mA, TT: 140 kV). Slice thickness was 0.625 mm in both CT1 and CT2. The remaining 22 carcasses were scanned only with CT1. Two IAPs were tested on CT1 images of all carcasses; different thresholds were used to allocate voxels to muscle (IAP1: 0-120 HU, IAP2: 0-200 HU) and bone tissue (IAP1: HU>120, IAP2: HU>200), whereas fat tissue thresholds were standard; -200 to -1 HU. Tissue volumes and weights were calculated based on voxel volume, number of tissue voxels and tissue densities. Bone weight was calculated by subtracting fat and muscle weight from carcass weight. Paired sample t-tests and Wilcoxon signed-rank tests were used to assess estimation differences between CT1 and CT2. Mann-Whitney U tests were used for tissue estimations of both IAPs with available dissection data from similar sheep of the same population (n=101). Differences between CT1 and CT2 were not statistically significant. Bone estimations of both IAPs and muscle content estimation of IAP1 significantly differed from dissection data (P<0.01). Both CT-protocols are suitable for post-mortem assessment of dairy sheep carcass composition; hence automated CT1 can replace CT2. IAP2 better estimates muscle tissue compared to IAP1. However, bone tissue estimations are more complex. Work was funded by GreQuM project (T1EDK-05479), co-financed by Greece and EU (EPAnEK 2014-2020, Partnership Agreement 2014-2020).

Session 16 Theatre 12

Comparative study of fat-tailed and thin-tailed sheep carcass quality

A. Argyriadou, A. Tsitsos, I. Stylianaki, S. Vouraki, T. Kallitsis, V. Economou and G. Arsenos
School of Veterinary Medicine, Aristotle University, University Campus, 54124 Thessaloniki, Greece; argyrian@vet.auth.gr

In dairy sheep production, fat-tailed sheep (FTS) are popular for their milk but not for their carcasses which are significantly undervalued compared to those of thin-tailed sheep (TTS). The latter view is dominant in Greece and hence, our objective here was to assess carcass quality of fat-tailed and thin-tailed sheep. In total, 146 fat-tailed and 97 thin-tailed dairy sheep crosses of both sexes were slaughtered at five different live weights comprising 5 slaughter groups (SGs). Wither height, carcass weight and carcass yield were recorded. Carcass pH was measured non-destructively, 1h after refrigeration. Samples of psoas major muscle, from 76 carcasses, were obtained for histomorphometry; muscle fibre minimum Feret's diameter was calculated. Samples from the quadriceps muscle (n=43) and the 13th rib (n=99) were collected 24h after slaughter; the latter were subjected to texture profile analysis (TPA), pH measurement and colorimetry. Double compression cycle tests were performed and TPA parameters were calculated (hardness 1 and 2, springiness, cohesiveness and chewiness). Meat colour lightness (L*), redness (a*) and yellowness (b*) were measured; chroma and hue angle were calculated based on a* and b*. Moisture, lipid and protein content of quadriceps samples were determined. Mann-Whitney U tests were used to compare fat-tailed with thin-tailed sheep concerning the above traits. To account for effects of sex and SGs and their interactions with sheep population, two-way ANOVA and Scheirer-Ray-Hare tests were used. Significant differences were observed regarding carcass yield and a* medians ($P<0.05$); FTS presented higher values for carcass yield and lower for a*. When accounting for sex and weight groups, effect of sheep population on the latter traits remained significant ($P<0.05$). Significant effects of interactions between sheep population and sex or SGs on meat quality traits were also observed ($P<0.05$). Results show that carcass and meat quality traits of FTS and TTS vary. Underestimated FTS carcass quality often exceeds that of TTS. Work was funded by GreQuM project (T1EDK-05479), co-financed by Greece and EU (EPAnEK 2014-2020, Partnership Agreement 2014-2020).

Session 16 Theatre 13

Loin intramuscular fat as a predictor of sheepmeat eating quality

L. Pannier, G.E. Gardner, R.A. O'reilly, F. Anderson and D.W. Pethick
Murdoch University, College of Science, Health, Engineering and Education, Murdoch, Western Australia 6150, Australia; l.pannier@murdoch.edu.au

Sensory perception of sheepmeat is a key factor influencing consumer demand. This is strongly influenced by intramuscular fat percentage (IMF%), and therefore has been included as a positive predictor of sheepmeat eating quality in the Meat Standards Australia grading system. The IMF% value is determined in the loin, and relies on low to moderate correlations for IMF% between loin and other cuts to predict their eating quality. However, IMF% of individual cuts may account for more variability in eating quality. Therefore, we hypothesised that individual cut IMF% will provide a more accurate description of that cuts eating quality than using loin IMF%. Eating quality cuts were collected from lambs (n=3,119) from the Meat and Livestock Australia Resource Flock. Loin and topside cuts were collected of all lambs, whereas the knuckle, outside and rump cuts were additionally collected from 830 carcasses. In total, 8,479 cuts were collected for grilled sensory testing by untrained consumers to assess overall liking. IMF% was measured on every loin, as well as on most topside, outside and rump cuts of the 830 carcasses. Linear mixed effects models in R (Version 4.0.3) were used to analyse consumer scores for overall liking, including fixed effects for cut, production factors, and either loin IMF% or individual cut IMF% fitted as a covariate. A 1 unit increase in loin IMF% was associated with an increase ($P<0.01$) in overall liking of 1.9 and 1.1 units in the loin and the topside cuts. However, loin IMF% had no association with overall liking in the knuckle, outside and rump cuts. Alternatively, individual cut IMF% was associated with an increase ($P<0.01$) in overall liking by 1.4 and 1.3 units per unit IMF% for the topside and outside cuts. This partly supports our hypothesis with the individual cut IMF% for the topside and outside showing stronger associations with eating quality than loin IMF%. This suggests that the development of on-line measurement technologies with the capacity to measure IMF% within multiple cuts would enhance a sheepmeat eating quality prediction system.

Risk factors of abortion in dairy Florida goats followed by a new lactation

P. Rodríguez-Hernández[1], J. Simões[2], C. Díaz-Gaona[1], M.D. López-Fariña[1], M. Sánchez-Rodríguez[1] and V. Rodríguez Estévez[1]
[1]University of Córdoba, Department of Animal Production, Campus de Rabanales, 14071 Córdoba, Spain, [2]University of Trás-os-Montes e Alto Douro, Animal and Veterinary Research Centre (CECAV), Quinta de Prados, 5370-801 Vila Real, Portugal; jsimoes@utad.pt

The stimulation of a new lactation after late abortion is a common practice among goat farmers. This study aimed to determine if the number of lactation, system production and season kidding have been risk factors of abortions followed by lactation (AFL) in 49 dairy Florida goat farms of Spain during the last two decades. The daily milk yield and lactation length differences between AFL (n=1,961) and normal kidding (NK; n=59,135) groups were tested by ANOVA. A multivariable logistic regression analysis was used to evaluate the following potential risk factors: period (2003-2010 and 2011-2019), number of lactation (1 to >6), system production (extensive and intensive) and kidding season (4 seasons). AFL goats produced less milk yield than NK group ones (1.84±0.02 vs 2.08±0.01 kg/day, respectively; ±SE; P<0.001) but the lactation length was similar (250.5±0.03 days; P=0.26). No significant effect of production system on AFL prevalence was observed (3.2%; P=0.58), but its likelihood decreased 60% in 2011-2019 period (odds ratios, OR=0.40; 95%IC: 0.36-0.43; P<0.001). Goats at 3rd lactation (as reference) have had less probability to present AFL than goats at 1st (OR=1.16; 95%IC: 1.00-1.34; P<0.05), 2nd (OR=1.24; 95%IC: 1.08-1.42; P<0.01) and 4th (OR=1.30; 95%IC: 1.10-1.53; P<0.01). Also AFL had more probability to occurs in Spring (OR=1.48; 95%IC: 1.27-1.72; P<0.001), Autumn (OR=1.52; 95%IC: 1.33-1.73; P<0.001) and Winter (OR=1.16; 95%IC: 1.02-1.32; P<0.05) than in Summer (as reference); or in Autumn (OR=1.31; 95%IC: 1.17-1.47; P<0.001) and Spring (OR=1.28; 95%IC: 1.11-1.46; P<0.001) than in Winter (as reference). Overall, the AFL prevalence remained at low levels and was similar in extensive and intensive production systems. The lactation length was similar between NK and AFL goats but these last ones produce 11.5% less of milk yield. Goats at 3rd lactation were among the less affected during the first four lactations. Goats kidding in Summer or Winter had less chances to be affected by AFL. The target proposed for AFL is ≤3%.

Factors effecting the performance of lambs from birth to weaning

F.P. Campion[1], N. McHugh[2] and M.G. Diskin[1]
[1]Teagasc, Animal & Bioscience Department, Mellows Campus, Athenry, Co. Galway, H65 R718, Ireland, [2]Teagasc, Animal & Bioscience Department, Fermoy, Cork, P61 P302, Ireland; francis.campion@teagasc.ie

The performance of lambs from birth to weaning is often extremely variable within and between sheep flocks operating similar farming systems. While the effects of genetics and nutrition have been shown to have an impact on this there are also other elements within the flock that can influence this that are sometime overlooked by producers and researchers alike. This study aimed to assess the effect of ewe mating body condition score (BCS), ewe live weight and ewe age has on the subsequent performance of the lamb crop. Over three separate production years' ewes of mixed age and breed were mated on three separate mid-season lambing, grass based sheep flocks in Ireland. Ewes had their live weight and BCS assessed at the point of mating and performance data including litter size, lamb birth weights and live weights at seven and 14 weeks post-partum were recorded on each individual lamb. These weights were then used to calculate the performance of the litter from each ewe across three production years. In total 1,613 ewes were included in the final analysis. Mating BCS had no effect on the performance of lambs from birth to weaning (P>0.05). Ewe live weight had a significant effect on lamb weight (P<0.05). As ewe live weight increased the weight of the litter also increased, however, the weight of the litter relative to the live weight of the ewe decreased, indicating a decrease in ewe efficiency (P<0.01). Age of the ewe at parturition had a significant effect on the performance of lambs (P<0.05). Twin bearing ewes that were >5 years of age had significantly lighter lambs than ewes aged 2 and 3 years of age (P<0.05) and tended to have lighter lambs than ewes 4 years of age (P<0.10). The results from this study show that the composition of the flock such as age and mature ewe weight have an important influence on the performance of the flock and needs to be considered when assessing and comparing individual flock performance.

Trans fatty acids in the fat of lambs produced in south of Portugal

E. Jerónimo[1,2], A. Silva[2], O. Guerreiro[1,2], L. Fialho[2,3], S.P. Alves[3,4], J. Santos-Silva[3,4,5] and R.J.B. Bessa[3,4]
[1]Mediterranean Institute for Agriculture, Environment and Development, MED, Beja, Portugal, [2]Centro de Biotecnologia Agrícola e Agro-Alimentar do Alentejo, IPBeja, Beja, Portugal, [3]Centro Investigação Interdisciplinar em Sanidade Animal, CIISA, Lisboa, Portugal, [4]Associate Laboratory for Animal and Veterinary Sciences, AL4AnimalS, Lisboa, Portugal, [5]Instituto Nacional de Investigação Agrária e Veterinária, Quinta da Fonte Boa, Santarém, Portugal; leticia.fialho@cebal.pt

Due to the impact of trans fatty acids (FA) on human health, the presence of these FA in ruminant-derived foods has been extensively scrutinized. Vaccenic acid (t11-18:1) has beneficial health effects and is the main precursor for the endogenous synthesis of rumenic acid (c9,t11-18:2), the major conjugated linoleic acid isomer in ruminant fat that shows diverse biological activities. Conversely, t10-18:1 is associated with detrimental health effects. The t11-18:1 is often the major trans FA, but under specific dietary conditions can occur higher production of t10-18:1 than t11-18:1 (known as t10-shift). The objective of this work was to characterize the trans FA composition of fat from lambs produced in the south of Portugal, with particular interest for t11-18:1 and t10-18:1. Samples of kidney knob channel fat from 142 lambs produced in 15 commercial farms from the Alentejo region were collected between late autumn and early winter. The t10-18:1 was the predominant trans FA in 61% of samples (shifted samples, t10/t11>1), ranging in these samples from 1.1 to 11.1 g/100 g total FA, while t11-18:1 only represented 0.22-4.61 g/100 g total FA. In remained samples (non-shifted samples, t10/t11<1), the t11-18:1 was the predominant trans FA and ranged from 1.23 to 7.35 g/100 g total FA. In non-shifted samples, the t10-18:1 varied from 0.15 to 2.75 g/100 g total FA. The c9,t11-18:2 content was lower in shifted samples than in non-shifted samples (0.36 vs 0.85 g/100 total FA, respectively). Results suggest that lamb meat produced in the south of Portugal in late autumn and early winter, when pastures are scarce, can have lower nutritional value due to their high t10-18:1 content. Funding: Project Val+Alentejo (ALT20-03-0246-FEDER-000049) – FEDER; Projects UIDB/05183/2020 (MED) and UIDB/00276/2020 (CIISA), and PhD grant of LF (2020.04456.BD) – FCT.

Session 16 Poster 17

Breed effects on lamb and ewe traits in Dorper & Namaqua Afrikaner sheep in an extensive environment

M.A. Kao[1], J.B. Van Wyk[1], A.J. Scholtz[2], J.J.E. Cloete[3] and S.W.P. Cloete[2,4]
[1]University of the Free State, Department of Animal, Wildlife and Grassland Sciences, P.O. Box 339, 9300 Bloemfontein, South Africa, [2]Directorate: Animal Sciences, Elsenburg, 7607, South Africa, [3]Cape Institute for Agricultural Training, Elsenburg, 7607, South Africa, [4]University of Stellenbosch, Department of Animal Sciences, Matieland, 7602 Elsenburg, South Africa; vanwykjb1@gmail.com

Harsh, extensive environments pose a challenge to sheep production in many developing regions of the world. There are suggestions that unimproved indigenous animals are better equipped to deal with stressors typical of such marginal environments when compared to commercial livestock. This study considered 914 to 1,530 individual lamb records and 1,068 to 1,322 ewe-year records of the indigenous, fat-tailed Namaqua Afrikaner (NA) and the largest commercial meat breed in South Africa, the Dorper. Lamb traits included early growth (birth and weaning weight) as well as fitness traits (lamb survival and log-transformed total body tick count under natural challenge). Ewe traits analysed were mating weight, number of lambs born and weaned as well as age and sex corrected total weight of lamb weaned (the reproduction traits were all expressed per ewe mated). These traits were studied in an extensive environment in an effort to understand the adaptability and robustness of these breeds to a limiting production environment. Relative to NA lambs, Dorper lambs were heavier at birth and weaning ($P<0.05$). NA lambs were more resistant to tick infestation ($P<0.05$). Although the magnitude of the breed effect varied between years, as suggested by a ewe breed × year interaction, the direction of the breed effect was consistent in all years. Lamb survival was unaffected by breed. Dorper ewes were heavier than NA ewes at mating ($P<0.05$). None of the reproduction traits were affected by breed ($P>0.05$), although NA tended to outperform Dorpers for number of lambs weaned per ewe mated ($P=0.07$). As experienced for lambs, log-transformed tick counts of NA ewes were lower than the comparable mean for Dorpers. It was contended that hardy, indigenous genetic resources like the NA should be conserved and used in further studies of ovine fitness and particularly on the genetics of resistance to ticks.

Dairy goat performances fed whole or processed cereal-legume meslin in a based ventilated hay ration

H. Caillat[1], L. Perrin[1], C. Boisseau[1] and R. Delagarde[2]
[1]INRAE, UE FERLUS, Les Verrines, 86600 Lusignan, France, [2]PEGASE, INRAE, Institut Agro, 16, le Clos, 35590 Saint Gilles, France; hugues.caillat@inrae.fr

The challenge of the French dairy goat sector is to reduce feeding costs. One of technical solutions is to increase forage intake and produce cereal-legume mixtures on farms. The aim of this trial was to evaluate the effects of the distribution of a meslin in whole or processed grain on milk performance of dairy goats fed with a ventilated hay and whole grain loss in the faeces. Two homogeneous groups of 23 Alpine goats were created based on their individual characteristics measured during a reference week: lactation number (2.6±1.5), stage of lactation (164±12 days), milk production (2.5±0.6 kg/d), milk fat concentration (33.8±4.1 g/kg), milk protein concentration (35.6±2.5 g/kg), body weight (57.5±9.7 kg). During 14 days, goats were fed in groups *ad lib* (>13% of refuses) with a ventilated hay. Each goat received individually 500 g/d of a commercial concentrate and 80 g/d of flattened sunflowers grains, fed at milking time (8:30 and 16:30). At 11:00, 400 g/d of cereal-legume mixture of whole (W) or flattened (F) grains were distributed individually. The mixture was composed of 65% of triticale and 35% of peas. The particle size more than 5 mm was 29% and 5% and those of less than 2 mm was 1 and 39% for the W and F mixture, respectively. After 7 days of adaptation, individual milk production and composition were measured during 3 days. Faeces were individually sampled twice a day during 3 last days from 15 goats of each group. Processing grains had no effect on forage intake (1.82±0.05 kg DM/goat/d). Milk production, milk protein concentration, milk urea and milk solids were not different between the groups. Milk fat concentration was significantly higher by 5.9 g/kg (P<0.01) for goats of the W group (44.2 vs 38.3 g/kg). The whole grains (mainly triticale) represented 12% of DM faeces from goats fed with whole grains. The distribution of a cereal-legume mixture in whole grains with a high value forage modify the production of milk fat concentration but the quantity of milk solids is not different between the groups. The losses of grains in faeces are limited but the efficiency of the distributed ration could be modified.

Dorper-Sarda lamb breed valorisation throughout enhancement of meat quality

M.F. Lunesu, M.R. Mellino, A. Fenu, A. Mazza, G. Battacone, S.P.G. Rassu, G. Pulina and A. Nudda
University of Sassari, Dipartimento di Agraria, Viale Italia, 39, 07100 Sassari, Italy; mflunesu@uniss.it

The aim of this study was to assess the effect of crossbreeding a local sheep dairy breed (Sarda) with a cosmopolitan sheep meat breed (Dorper) as the sire line on growth performance, carcass and meat traits and environmental impact of lambs. A total of 40 heavy lambs (20±0.28 kg; mean ± SEM) of Sarda (S; n=20; female=9, male=10) and Dorper×Sarda (D×S; n=20; female=11, male=10) breeds were used. Lambs were raised traditionally, suckling milk from their dams until slaughter at 2 months of age. During the trial, body weight (BW) and average daily gain (ADG) were recorded every two weeks. Data collected were used to estimate carbon footprint (CF) within a cradle to farm gate system boundary, considering 1 kg of body weight gain (BWG) as the functional unit (FU). At slaughter, carcass weight (CW), proximate composition, pH, colour traits were recorded from the left side of the carcass. In addition, cooking and drip loss were also measured. Data were analysed by the PROC GLM procedure of SAS to test the effect of breed, the effect of sex and their interaction. Body weight and CW did not differ between breeds. The colour variables, pH, cooking and drip loss were also not affected. The crossbreeding with Dorper enhanced meat quality increasing UFA content (51.34 vs 49.75±0.33%; P=0.017) and decreasing SFA content (49.75 vs 51.28±0.312%; P=0.016) in meat, compared to S lamb meat. In addition, D×S lamb meat showed a lower rib fat thickness (2.27 vs 4.29±0.30 mm; P=0.0002) than S lamb meat. In relation to the sex, females showed lower ADG (0.21 vs 0.25±0.009 kg/d; P=0.01) and higher carcass yield (62 vs 60±0.004%; P=0.0015) and rib fat thickness (3.72 vs 2.76±0.30; P=0.034) than males. The effect of sex was evident also for the CF which was higher in female than male lambs (25.41 vs 22.20±0.69 kg CO_2eq/kg lamb BWG; P=0.02) whereas did not differ between breeds. In conclusion, the crossbreeding of Sarda ewes with Dorper rams can be a useful tool to enhance carcass meat quality of ovine meat in local breeds without altering the environmental performance. Research funded by PSR Sardegna 2014-2020-misura 16.2 – Probovis project.

Livestock predated by wolf (*Canis lupus*) in agroforestry systems of Lazio region: first evidences

C. Tiberi[1], A. Amici[1], R. Primi[1], P. Viola[1], M.M. Madonia[2] and A. Vitali[1]
[1]University of Tuscia, DAFNE, Via San Camillo e Lellis snc, 01100 Viterbo, Italy, [2]Lazio Region, Department of agriculture, Via di Campo Romano 65, 00173 Roma, Italy; cristiano.tiberi@unitus.it

Agroforestry systems in which a livestock farming component is present are widespread and increasing in Europe. Accordingly, contact zones and overlaps with wildlife are increasing too determining an escalation of conflicts between wild carnivores and farming activities. The aim of the study was to highlight the incidence of wolf predations in livestock agroforestry systems of Lazio Region (central Italy). The following data, related to the 2016-2019 period, were shared by regional hunting office. Data were aggregated on a provincial scale recording: year of predation, predator species, prey species, predator events (n°), individuals preyed per event (n°) and economic estimated damage (€). Descriptive statistics were performed. Total predation events were 307 with different frequencies in the investigated Provinces (Rieti=51.79%, Rome=23.45%, Viterbo=21.50%, Frosinone=2.93%, Latina=0.33%). Total individuals predated were 1,310 of which 83.3% sheep, 8.6% cattle, 4.9% equine and 3.2% goat. Although the mountain areas of Rieti province could potentially appear the most exposed to predation risk, a major part of the predation events was recorded in the anthropized hilly areas of Rome and Viterbo provinces. Extensive ovine farms were the most impacted. Respect to the overall sheep's regional population (525,000 individuals) the predation incidence in the worst year (2017) was 0.068%. Total estimated economic losses were 191,053 €. Although this may seem a small loss compared to the overall value of the regional livestock (estimated average value of 56,778,750 €), at the local scale, predation can severely impact on the economic balance of marginal agroforestry systems leading to the risk of illegal defensive actions such as the use of poisons. It is necessary to plan coexistence strategies between marginal areas and the presence of wolves. To mitigate conflicts and restore a condition of socio-economic acceptability, it is essential to adopt adequate, effective and integrated prevention tools, in addition to economic compensation (e.g. indemnities and insurance).

Evaluation of butyric acid for inflammation mitigation using *in vitro* and *in vivo* approaches in pig

D. Gardan-Salmon[1], S. Molez[2], B. Saldaña[3], J.I. Ferrero[3], A. Bruyère[2] and M. Arturo-Schaan[1]
[1]Deltavit, CCPA Group, 35150, France, [2]Irset, Inserm UMR1085, Université Rennes 1, France, [3]NOVATION, CCPA Group, 42250, Spain; dgardan-salmon@ccpa.com

Butyric acid (BA) is naturally formed in animal gut. The use of protected BA in young animal diets is known as a safe alternative to AGP, with beneficial effects on villus development, microbial population, and nutrient digestibility. Recent studies show that BA is a regulator of inflammatory processes in the small intestine and could therefore improve digestive health in post-weaning piglets fed with antibiotic free diets. Our study aimed at investigating the ability of BA to modulate inflammation *in vitro* in an enterocyte model, and to evaluate the resulting benefits *in vivo* in diarrhoea incidence in post-weaning piglets. Porcine IPEC-J2 cell line was used as an *in vitro* model, with lipopolysaccharide (LPS)-induced inflammation. Cells were pre-treated, with BA (4 to 200 µg/ml) at 37 °C for 1h and then co-incubated with LPS (10 ng/ml) for 1 h. Cytokine gene expressions were measured. *In vivo*, 1,600 piglets (weaned at 21 days, housed in 32 pens) were divided into 2 groups: control fed without supplementation and BA group fed with Butirex C4® in the feed (3 kg/t from age 21 to 35 and then 2 kg/t up 63 d of age). Growth performances were recorded by pen every 2 weeks. Faeces were scored daily. In the cell line, mRNA levels for cytokines IL-8, TNFα and CCL20 were significantly increased by LPS (12, 54 and 137-fold respectively; $P<0.001$) compared to control. BA alleviated inflammation induced by LPS by significantly reducing IL-8, TNFα and CCL20 gene expression compared to LPS group (-4 to 17-fold depending on cytokines and doses; $P<0.001$). *In vivo*, from 21 to 63 d of age, ADG (+4.2%) and FCR (-3.4%) were significantly improved in piglets receiving BA feed compared to control ($P<0.05$). The incidence of diarrhoea was lower (-26.2%; $P<0.05$) in piglets fed BA diet compared to control diet. Using pig enterocytes challenged by LPS, ours results showed that BA can exert direct anti-inflammatory effects, linked to growth and sanitary status improvement observed in our piglet trial. This *in vitro* approach is a promising sustainable tool to better evaluate the biological properties of feed ingredients prior to *in vivo* evaluation.

Citrus pectin modulates chicken peripheral blood mononuclear cells proteome

G. Ávila[1], M. Bonnet[2], A. Delavaud[2], D. Viala[2], S. Déjean[3], G. Grilli[1], S. Di Mauro[1], M. Ciccola[1], C. Lecchi[1] and F. Ceciliani[1]

[1] Università Degli Studi di Milano, Department of Veterinary Medicine, Via dell'Università 6, Lodi, 26900, Lodi, Italy, [2]INRAE, Université Clermont Auvergne, Vetagro Sup, Unité Mixte de Recherches sur les Herbivores (UMRH), Rte de Theix, 63122, Saint-Genès-Champanelle, France, [3]Institut de Mathématiques de Toulouse, Université de Toulouse, CNRS, UPS, UMR 5219, 118 Rte de Narbonne, 31062 Toulouse, France; fabrizio.ceciliani@unimi.it

Citrus pectin (CP) is a dietary fibre commonly used in animal nutrition with known anti-inflammatory properties. CP supplementation in broilers has improved energy utilization and nutrient digestibility *in vivo* and has suppressed two main inflammatory functions of chicken monocytes *in vitro*: chemotaxis and phagocytosis. Limited information on the molecular mechanisms underlying these immunomodulatory effects is available so far. Thus, this study aimed to assess the effects of CP on chicken peripheral blood mononuclear cells (PBMC) proteome. Cells purified from 7 different whole blood pools of healthy chickens were incubated with 0.5 mg/ml of CP for 20 h at 41 °C. A label-free quantitative proteomics was performed using nano-LC-MS/MS. A total of 1,503 was identified, and a supervised multivariate statistical analysis (sPLS-DA) was applied to identify discriminant proteins (DP). A total of 373 DP was identified, and a Gene Ontology (GO) analysis was performed using ProteINSIDE. The main enriched GO terms in the biological process category (FDR<0.05) were related to peptide metabolic process, translation, actin cytoskeleton organization and cell migration. To further elucidate the specific role of the DP identified in CP group, 50 proteins with the highest abundance in CP were again mined with ProteINSIDE. The DP were mainly involved in actin cytoskeleton organization and negative regulation of cell migration. Among the proteins with the highest abundance in CP, MARCKSL1, a protein that restricts cell movement, was detected. MARCKS and LGALS3 showed the lowest abundance in CP, these proteins mainly being related to cell migration and phagocytosis. In conclusion, these results suggest that CP does modulate chicken PBMC proteome and support the suppressive effects we previously observed *in vitro* on chicken monocytes phagocytosis and chemotaxis.

Reducing aggressive behaviour using a liquid calming herbal extracts in finishing pigs

C. Farias-Kovac[1], D. Carrion[1], E. Mainau[2], M. Gispert[3] and E. Jiménez-Moreno[1]

[1]Cargill Animal Nutrition and Health, Poligono Industrial Riols s/n, 50170 Mequinenza, Zaragoza, Spain, [2]AWEC Advisors SL, Parc de Recerca de la Universitat Autonoma de Barcelona, 08193 Bellatera, Barcelona, Spain, [3]IRTA-Food Quality and Technology Program, Finca Camps i Armet, 17121 Monells, Girona, Spain; carloseduardo_farias@cargill.com

Handling, transportation, lairage and slaughter can affect animal welfare and carcass and meat quality. Poor pre-slaughter practices may result loss of profits due to animal losses, carcass downgrading, and poor meat quality. The aim of this study was to evaluate the effect of the inclusion of a calming herbal extract blend in liquid form (BehavePro® L) on feeding and aggressive behaviours and carcass quality for the last 8 days of fattening. Five hundred twenty-eight [Pietrain × (Landrace × Large White)] pigs, were sexed (entire males and females) and allotted randomly to two treatments (Control and Treatment group) with twenty-four conventional pens of 19 pigs each and six pens of 12 pigs each with one automatic system (Nedap ProSense, the Netherlands) per pen. BehavePro L is a nutritional solution containing natural compounds and plant extracts added to drinking water at level of 0.6%. Pigs were slaughtered at 108 kg live weight. All pigs had free access to water until transportation to the abattoir. Fasting lasted for 18 h. The transportation and lairage time in both groups was of around 3.5 h. Treatment had no impact on the number of feeder visits, average daily feed intake and time spent eating (P>0.1). The percentage of resting pigs (inactive lying) was of 9.1 and 13.1% and of active laying down pigs was of 32.6 and 35.6% for the control and treatment group, respectively (P>0.1). Hot carcass weight and lean content were not affected by the treatment (P>0.1). The incidence and severity of the carcass skin lesions were reduced with the treatment (25.2 vs 47.4% pigs with less than 2 scratches, 59.6 vs 50.6% pigs with between 2 and 10 scratches, and 15.2 and 2% pigs with more than 10 scratches for the control and treatment group, respectively, P<0.001). In conclusion, the supplementation of 0.6% of BehavePro L via water for the last 8 days of fattening reduced the skin damage at slaughter probably due to its calming effect on aggressive behaviour.

Do dietary carob pulp and vitamin E affect performance and welfare in fattening pigs?

D.N. Bottegal[1], B. Casado[1], L. Bernaus[1], S. Lobón[2], I. Argemí-Armengol[1], J. Álvarez Rodríguez[1] and M.A. Latorre[3]
[1]Universitat de Lleida, Rovira Roure 191, 25198, Spain, [2]CITA-Aragón, Av. Montañana 930, 50059, Spain, [3]Universidad de Zaragoza, C/Miguel Servet 177, 50013, Spain; diego.bottegal@udl.cat

The use of alternative feedstuffs, as Carob pulp (Cp), may prevent oxidative stress and improve animal health. The Cp is a by- product with a high level of fibre, condensed tannins, and hence good antioxidant capacity. Nevertheless, high levels of dietary tannins and fibres might affect performance or even social and eating behaviours in pigs. To assess the effect of dietary inclusion of carob pulp and vitamin E (vit E) level on the growth performance and skin lesion scores in fattening pigs, a study was performed in an experimental farm of Bonàrea Group. A total of 220 crossbred pigs [Duroc × (Landrace × Large-White)] of 130±4.5 days (d.) of age and 78.4±8.93 kg body-weight were used. Animals were randomly distributed in 44 pens (4-6 pigs/pen) and fed *ad libitum* one of four iso-energetic (2,300 kcal net energy/kg feed) and iso-protein diets (14.63% crude protein, 0.86% Lys) (5-6 pens/diet/sex) with different inclusion levels of Cp and/or vit E; 20% Cp+300 IU/kg vit E, 20% Cp+40 IU/kg vit E, 0% Cp+300 IU/kg vit E and 0% Cp+40 IU/kg vit E. At day 130, 151 and 169 of age, the animals were individually weighed, and the Welfare Quality protocol's skin lesion score (SLS) was performed in every pig. At 39 d. of the study, no effects of Cp or vit E (nor interactions) were found on final body weight (BWf, 124.7±2.16 kg), average daily gain (ADG, 1.18±0.02 kg/d), daily feed intake (ADFI, 3.08±0.05 kg/d) and Feed to gain (F:G, 2.65±0.03). Sex differences were found between entire males vs females (BWf; $P<0.01$, 129.7 vs 119,4±2.16 kg, ADG; $P<0.001$, 1.28 vs 1.09±0.02 kg BW/d, ADFI; $P<0.001$, 3.17 vs 2.99±0.05 kg/d, F:G; $P<0.001$, 2.49 vs 2.71±0.03, respectively). The day control affected the SLS ($P<0.01$); the lowest levels of SLS 1 and 2 (being SLS 2 the worst score) were found on day 169, while the highest level of SLS were showed on day 151. Besides, sex effect was found only in severe SLS (5% of females vs 1% of males, $P<0.01$). Both the inclusion of Cp and vit E seem to have no negative effect on the animal welfare parameters assessed or the pig's productive performance.

MicroRNAs in plasma and leucocytes as potential biomarkers for rumen health in a SARA cow model

O.E. Ojo[1,2], L. Hajek[1], S. Johanns[1], C. Pacifico[2], S. Ricci[2], R. Rivera-Chácon[2], A. Sener-Aydemir[2], E. Castillo-Lopez[2], N. Reisinger[3], Q. Zebeli[2] and S. Kreuzer-Redmer[1,2]
[1]Nutrigenomics, Institute of Animal Nutrition, University of Veterinary Medicine Vienna, 1220, Austria, [2]CD Lab Innovative Gut Health Concepts of Livestock, Institute of Animal Nutrition, University of Veterinary Medicine Vienna, 1220, Austria, [3]BIOMIN Research Center, Tulln an der Donau, 3430, Austria; susanne.kreuzer-redmer@vetmeduni.ac.at

MicroRNAs (miRNAs) are small non-coding RNAs which are crucial regulators of gene expression and could serve as biomarkers for several biological conditions. Sub-acute ruminal acidosis (SARA) is a metabolic disorder marked by low ruminal pH, and has been linked to the feeding of starch-rich diets. Biomarkers are urgently needed to identify cows that suffer from SARA, since there are no clear clinical signs, but subsequently a huge loss of productiveness. Four rumen-cannulated dry Holstein cows were fed a control diet of 100% forage and transitioned over one week to a 65% grain diet (HG). The occurrence of SARA was verified after diet transition using a ruminal pH threshold of <5.8 for 320 min/d. Blood samples were taken during forage feeding and after one week high grain. Total RNA was isolated and small RNA libraries were prepared with the NEXTflex Small RNA-Seq Kit and sequenced on an Illumina NovaSeq 6000 platform. Unique and shared miRNAs between diets were discovered after read counts were filtered (>10 reads) using tidyr in R (version 4.0.1), yielding a total of 520 miRNAs in plasma and 730 miRNAs in leucocytes. When cows were fed a high-grain diet, 63 circulating miRNAs were discovered in plasma that were not identified when they were fed forage and 6 potential biomarkers (bta-miR-11982, bta-miR-1388-5p, bta-miR-1306, bta-miR-12034, bta-miR2285u, bta-miR-30b-3p) were selected from these list with respect to the total read counts and from 12 differentially expressed miRNAs (10 upregulated and 2 down-regulated) in terms of their log2 fold change (FDR<0.05). Validation of expression by RT-qPCR revealed that bta-miR-30b-3p and bta-miR-2285 appear to be good candidates as biomarkers for rumen health. Our results suggested that dietary modifications affect the expression of miRNAs in systemic circulation, potentially influencing post-transcriptional gene expression in SARA-affected cattle.

Session 17 — Theatre 6

Evaluation of pre-transport feeding strategies in bull calves' performance and gut permeability

L. Pisoni[1], S. Marti[1], A.M. Bassols[2], Y. Saco[2], J. Pujols[3], N. Gómez[2] and M. Devant[1]
[1]IRTA, Ruminant Production, Caldes de Montbui, 08140, Spain, [2]UAB-Faculty of Veterinary Science, Biochemistry and Molecular Biology, Bellaterra, 08193, Spain, [3]IRTA – CReSA, Animal Health Research Centre, Bellaterra, 08193, Spain; lucia.pisoni@irta.cat

Unweaned calves in the dairy-beef production are normally subjected to long-distance transportations that compromise their health and performance due to feed restriction and fasting. Pre-transport feeding strategies able to mitigate the negative impacts of transportation on calves' performance and gut function were studied. Sixty-five Holstein bull calves were subjected to an assembly centre simulation for 3 days (from d-4 to d-1 of the study) and during this period randomly assigned to 1 of 5 treatments: Control (CTR; n=13), fed 2.5 l of milk replacer (MR) twice daily and *ad libitum* access to concentrate; MLR (n=13), fed 2.5 l of MR twice daily; MRC (n=13), fed 2.5 l of MR twice daily and *ad libitum* access to concentrate; MRA (n=13), fed 2.5 l of MR twice daily with *ad libitum* access to acidified milk; and MAC (n=13), fed 2.5 l of MR twice daily with *ad libitum* access to acidified milk and concentrate. On d -1, MLR, MRC, MRA, and MAC calves were transported during 19 h. At arrival on d 0 until d 7 all calves were fed MR and had *ad libitum* access to concentrate and straw. Serum concentration of Cr-EDTA (a biomarker of gut permeability) and BW were recorded on d -1, 0, 1, 2, and 7 and intake was recorded daily. Data were analysed using MIXED procedure of SAS. MLR calves had the lesser (P<0.01) BW by d -1 at the end of the assembly centre simulation while MAC showed the greatest BW in the same period, indicating that calves fed only MR suffered greater BW losses compared with those fed a higher plane of nutrition. Concentrate intake was greater (P<0.01) in MRA after transportation, and the greatest (P<0.01) drop in intake on d 1 was observed in MLR. Serum Cr-EDTA concentration was lesser (P<0.01) in CTR and MAC on d 0. In conclusion, calves with the greatest plane of nutrition showed greater BW before transport, lesser drops in intake, and lesser serum Cr-EDTA concentration indicating that pre-transport nutritional strategies may be effective avoiding the side effects of transportation on calf performance and gut function.

Session 17 — Theatre 7

Can real lavender essential oil reduce young bulls' stress and prevent respiratory diseases?

E. Vanbergue[1], M. Guiadeur[1], B. Mounaix[1], S. Masselin-Sylvin[1], J. Noppe[1], L. Jouet[2], S. Guedon[3] and S. Assié[4]
[1]IDELE – French Livestock Institute, 149 rue de bercy, 75012 Paris, France, [2]RIVIERESVET clinique vétérinaire, 87 Rue de la Châtaigneraie, 35600 Redon, France, [3]Ter 'élevage, La blanchardière, 44522 Mésanger, France, [4]ONIRIS, 101 Route de Gachet, 44307 Nantes, France; elise.vanbergue@idele.fr

The study was performed in young-bulls (YB) fattening units in western France. In these systems, non-castrated 6-10 month old YB are weaned, transported from cow-calf farms to sorting facilities where they are handled to be commingled with peers of similar body weights and then transported to fattening units. These many stress factors cause immune depression and promote respiratory diseases (BRD) development, traditionally controlled by prophylactic use of antibiotics. Because of its anxiolytic properties, real lavender essential oil (RL-EO) could avoid immune function impairment resulting from stress and reduce BRD incidence. Two hundred and twenty-two young bulls had been monitored in a control/case study to evaluate the interest of the use of RL-EO to support the transition to fattening. Behaviour, health and weight have been recorded and analysed by linear models. No significant effect of RL-EO has been observed on behaviour, morbidity, BRD severity, antibiotic treatments, average daily gain, carcass weight, and fattening duration. This study has failed to provide the evidence of the RL-EO anxiolytics properties for YB and consecutive expectations on BRD incidence reduction. Several factors could explain these results. First, the stress experienced by the YB is an 'acute' stress, and according to some studies in human medicine, RL-EO would be better suited to control the effects of so-called 'anticipatory' stress. Furthermore, the timing of administration was late in relation to the onset of stressful events. Protocols for the administration of other EO, prescribed for 'acute' stress before transport, would deserve to be studied. Then, the level of stress undergone by the YB could be such that the only administration of an EO would not be enough to decrease it significantly. Thus, taking into account the multiple risk factors at the farm level and at the sector organization level would be more efficient to control BRD in YB fattening units.

Fat sources in milk replacer for calves affect *ad libitum* intake behaviour and performance

J.N. Wilms[1,2], V. Van Der Nat[3], M.H. Ghaffari[4], M.A. Steele[1], H. Sauerwein[4], J. Martín-Tereso[2,3] and L.N. Leal[2]
[1]University of Guelph, Department of Animal Bioscience, 50 Stone Road East, Guelph, Canada, [2]Trouw Nutrition, Research and Development, P.O. Box 299, 3800 AG Amersfoort, the Netherlands, [3]Wageningen University, Animal Nutrition Group, P.O. Box 338, 6700 AH Wageningen, the Netherlands, [4]University of Bonn, Institute of Animal Science, Regina-Pacis-Weg 3, 53111 Bonn, Germany; juliette.wilms@trouwnutrition.com

Fat composition in milk replacers (MR) for calves differs from milk fat in multiple ways. Our objective was to investigate how common fat sources used in MR affect growth, *ad libitum* intake behaviour, and blood metabolites in 72 calves. Calves were blocked based on arrival day, and within each block, calves were randomly assigned to one of three treatments (n=21 per group): VEG was a MR with vegetable oils (20% coconut, 80% a mixture of coconut and rapeseed oils), in ANIM, the MR contained animal fats (65% lard, 35% dairy cream), and in MIX, a mixture of 80% lard and 20% coconut oil was used. The three MR contained 30% fat, 24% crude protein, and 36% lactose and were formulated with a fatty acid profile aiming to mimic that of milk fat. From arrival onwards, calves were group-housed and fed *ad libitum* (135 g/l). Weaning was gradual and occurred between wk 7 to 10, and calves were fed solids only thereafter. Starter feed, chopped straw, and water were offered *ad libitum*. Calves were weighed and blood was collected weekly until d70. Throughout the experiment, ADG was greater in ANIM (960 g/d) than VEG (858 g/d; P=0.05), whereas MIX (940 g/d) did not differ from other treatments. Preweaning MR intakes were 21% greater in ANIM than MIX (P=0.02), whereas VEG showed no differences. Consistently, the number of rewarded visits during the milk phase was 10% greater in ANIM than MIX (P=0.03), whereas VEG showed no differences. In contrast, fat sources did not affect starter feed intake. This resulted in a 14% greater preweaning ME intake in ANIM than in VEG and MIX (P=0.01). Serum cholesterol was 26% greater (P<0.01) and serum albumin was 3% lower in VEG (P<0.01) than in ANIM and MIX. Feeding ANIM improved growth compared to VEG and MR intake compared to MIX. Feeding VEG resulted in metabolic differences compared with ANIM and MIX, although their biological relevance is unclear.

Feeding concentrated colostrum ensures sufficient uptake of IgG in Holstein calves

T. Jarltoft[1,2], C.B. Jessen[2], M.B. Samarasinghe[2] and M. Vestergaard[2]
[1]SAGRO I/S, Billund, 7190, Denmark, [2]Aarhus University, Animal Science, Tjele, 8830, Denmark; tja@sagro.dk

Failure of passive transfer of IgG is one of the major reasons for increased morbidity and mortality among neonatal calves. By applying a modified processing method that concentrates IgG in medium-quality colostrum, the total volume of high-quality colostrum can be increased. The objective of this study was to investigate whether calves fed concentrated colostrum 0-4 h after birth, attain a sufficient passive immunization compared to calves fed non-processed (i.e. 50 g IgG/l) colostrum, with no negative short-term consequences to morbidity, mortality, ADG and daily milk intake. The experiment included 48 newborn Holstein calves with 24 calves randomly fed (i.e. 10% of birth BW) control colostrum (CC: 68 g IgG/l, 14.6% DM protein) or processed colostrum (PC: 54 g IgG/l, 16.3% DM protein). The calves were health screened three times a week, weighed immediately after birth and once a week for four weeks. Blood was sampled 0-4 and 24-72 h after birth and in week 1, 2, 3 and 4. Faeces were sampled 24-72 h after birth and in week 2 and 3. Calves had a birth BW of 39.2±3.4 kg and were 2±1 h old at colostrum feeding. Results are given in mean ± SD. Calves fed PC compared to CC ingested a lower volume of IgG (5.5±0.5 vs 6.9±1.9 g/kg birth BW) but had similar concentration of IgG in plasma at 50±19 h of age (25.6±7.0 vs 27.9±9.0 g/l) and a slightly higher apparent efficiency of absorption (47±14 vs 42±16%). ADG of PC compared to CC calves was 923±314 vs 963±249 g/d until day 14 and 800±101 vs 760±139 g/d until day 31. Milk intake was 7.1±0.8 l/d, with no differences between treatments. One CC calf died and one PC calf was excluded from the experiment due to illness from birth. No differences were observed between treatments in relation to health screenings, DM- and Enterit 4 PCR analysis of faeces. The findings indicate that medium-quality colostrum concentrated to a well-defined high-quality colostrum (≥50 g IgG/l) can be used as a substitute for high-quality colostrum. Thus, the total quantity of high-quality colostrum can be increased and thereby the total rate of passive immunized newborn calves can be ensured, with no short-term consequences to morbidity, mortality, and growth.

Session 17 Theatre 10

Live yeast supplementation improves performance of calves under stressful conditions

S.J. Davies[1], G. Esposito[1,2], C. Villot[3], E. Chevaux[3], L. Bailoni[4] and E. Raffrenato[1,4]
[1]Stellenbosch University, Animal Sciences, Merriman Ave, 7602 Matieland, South Africa, [2]Università di Parma, Dipartimento di Scienze Medico Veterinarie, Via del Taglio 10, 43126 Parma, Italy, [3]Lallemand SAS, 19 Rue des Briquetiers, 31702 Blagnac, France, [4]Università di Padova, Biomedicina Comparata e Alimentazione, Viale dell'Università 16, 35020 Legnaro, Italy; emiliano@sun.ac.za

The aim of this study was to determine the effects of a live yeast supplement (*Saccharomyces cerevisiae boulardii*) on calf performance in the presence or absence of a nutritional stress. Eighty Holstein heifer calves were enrolled from day 5 of age until weaning (90d) in a randomized block design and randomly assigned to one of the four treatment groups devised according to the presence or absence of a nutritional stress and the inclusion or exclusion of yeast. The daily inclusion level of the yeast supplement corresponded to 1 billion cfu/calf. The nutritional stress was implemented every other day by delaying one feeding by 2 h. All calves were fed milk replacer and a pelleted starter was offered *ad libitum*. Intake of the starter was measured daily and body weight (BW) was recorded weekly. Faecal score and health status were recorded daily. Blood samples were collected every 10d, for glucose, β-hydroxybutyrate, and total blood count. A repeated-measure (i.e. day) mixed model was used with yeast and stress and their interaction as fixed components, and calves as random component. Calves in the stress group had a lower intake of starter ($P<0.01$) at 724 g/d vs 830 g/d of the no-stress group and the inclusion of yeast tended to result in higher intake ($P=0.08$) at 799 g/d vs 754 g/d of the no-yeast group. The stress resulted in lower average BW ($P=0.03$; 66.8 vs 64.1 kg) and the yeast tended to increase BW ($P=0.07$). A physiologic leukocytosis was apparent for stressed calves ($P=0.04$), but not for the yeast-supplemented calves ($P=0.17$). The leukogram of the stressed calves not receiving the yeast was characterized by neutrophilia, lymphocytopenia and eosinopenia ($P<0.05$). This was also associated with lower faecal scores ($P<0.05$) and less health problems incidences ($P=0.01$). The inclusion of the yeast was therefore able to decrease the negative effect of the stressful condition for both performance and health of the calves.

Session 17 Theatre 11

Lying behaviour of dairy cows is influenced by the characteristics of heavily forage-based rations

A. Haselmann[1], M. Wenter[2], W. Knaus[3], K. Bauer[3], B. Fuerst-Waltl[3], Q. Zebeli[4] and C. Winckler[3]
[1]Research Institute of Organic Agriculture (FiBL), Ackerstrasse 113, 5070 Frick, Switzerland, [2]Südtiroler Rinderzuchtverband, Luigi Galvani Strasse 38, 39100 Bozen, Italy, [3]University of Natural Resources and Life Sciences, Vienna, (BOKU), Gregor Mendel Strasse 33, 1180 Vienna, Austria, [4]University of Veterinary Medicine Vienna, Veterinaerplatz 1, 1210 Vienna, Austria; katrin.bauer@boku.ac.at

The aim of the present work was to investigate whether differences in the particle size (Trial 1) or the preservation method of forages (Trial 2) alter cows' lying behaviour. In Trial 1, a forage-based (80% DM) total mixed ration (TMR) was fed to 2 feeding groups (n=10 Holsteins each). Forages included were grass-hay and clover-grass silage; either with a conventional particle size (LONG; ≈52 mm) or re-chopped (SHORT; ≈7 mm). In Trial 2, feeding groups (n=9 cows each) received the same forage; either as HAY or SILAGE, and 3.6 kg (DM) concentrate per cow and day. Trials lasted for 5 wks whereby the first 2 wks were used as an adaptation period. Cows were on average 192 DIM and yielded 27.9 kg/d, and were housed in a free-stall barn with deep straw-bedded cubicles and rubber flooring in the alleys. Lying behaviour of cows was recorded for at least 3 consecutive days per animal using HOBO Pendant® acceleration data loggers. For both trials, linear mixed models included the treatment, the placement of the loggers (left/right), the day of the experiment, and corresponding covariate data, derived from the preceding feeding period when all cows were offered a TMR. Average daily lying time in trials was 12.5 h/d and was divided into roughly 11 bouts. Cows on the SHORT ration had a longer daily lying time (+ 1.1 h/d; $P=0.003$) and more lying bouts per day (+ 3/d; $P=0.090$) than LONG cows. Cows of both groups spent more time lying on the left side (57.4%). In Trial 2, total lying time was not affected by the treatment ($P=0.608$) but cows on HAY showed about 2.5 bouts less per day ($P=0.039$) than cows on SILAGE. Cows on SILAGE again preferred lying on the left side (55.6%) whereas cows on HAY did not show any side preference. Effects could be attributed to the modified physical structure of the forages via changes in eating behaviour as well as the degree of the ruminal fill.

Season and cleaning interval affect biological water quality and drinking behaviour of dairy cows

F. Burkhardt[1,2], *J.J. Hayer*[1,2], *C. Heinemann*[2] *and J. Steinhoff-Wagner*[1]
[1]*TUM School of Life Science, Liesel-Beckmann-Straße 2, 85354 Freising-Weihenstephan, Germany,* [2]*Institute of Animal Science, University of Bonn, Katzenburgweg 7-9, 53115 Bonn, Germany; burkhardt@uni-bonn.de*

Sufficient water supply and quality are necessary for optimal feed intake, productivity, health, and animal welfare. This study aimed to investigate the effect of season (winter vs summer) and cleaning interval (cleaned daily vs not cleaned) on dairy cows' drinking behaviour. The study was conducted on a commercial dairy farm with 135 lactating cows housed in a symmetrical free-stall barn. The quality of the animals' drinking water was analysed at the beginning and end of the study and monitored daily by rapid ATP tests. Drinking events were video recorded 2 h after feeding and characterized by 12 drinking behaviour parameters. Data of continuous behavioural parameters were logarithmized to base 10 and analysed using a mixed model with cleaning interval and season as fixed factors in SAS (SAS 9.4), whereas categorical data were modelled by binary logistic regression using the package 'MASS' in R (version 4.1.1). All physico-chemical and microbiological water quality indicators were acceptable. ATP-value of the drinking water was significantly higher in summer than in winter and tendentially higher in not cleaned compared to cleaned troughs. More drinking events were observed in winter (n=4,103) than in summer (n=3,978). The cleaning interval alone failed to be mirrored in drinking behaviour. However, significant behavioural alterations were observed for the interaction of cleaning interval × season: The total drinking time ($P<0.01$) and number ($P<0.05$) and duration of active water intake ($P<0.01$) was significantly higher in summer at cleaned troughs than at not cleaned troughs. The average number of sips was lower in summer compared to winter ($P<0.0001$), with more sips being recorded at cleaned troughs compared to not cleaned troughs in summer and winter ($P<0.01$). The odds ratios (OR) for agonistic behaviours and resulting displacements at the troughs were higher in summer than in winter ($P<0.0001$). The duration of drinking events, the active water intake, and the occurrences of agonistic behaviours at troughs might be useful behavioural indicators for evaluating water provision on farm.

Evaluation of the antibacterial and anti-inflammatory effects of *Thymus capitatus* against mastitis

R. Nehme[1,2], *E. Vanbergue*[2], *S. Even*[1], *H. Falleh*[3], *R. Ksouri*[3], *S. Bouhallab*[1], *L. Rault*[1] *and L. Abdennebi-Najar*[2]
[1]*INRAE, Institut Agro, STLO, 65 Rue de Saint-Brieuc, Rennes, France,* [2] *IDELE Institute, Quality and Health Department, 149 Rue de Bercy, Paris, France,* [3]*Laboratory of Aromatic and Medicinal Plants, Biotechnology Center of Borj-Cédria, BP 901, 2050 Hamma, Tunisia; ralph.nehme@idele.fr*

Essential oils (EO) are known to inhibit the growth of bacteria responsible for mastitis in dairy cows. We recently have found that, among 10 medicinal promising Mediterranean plants, *Thymus capitatus* essential oils (TCEO) has the highest anti-microbial effects against *Staphylococcus aureus* strains isolated from mastitis dairy cows. The aim was to test whether the antibacterial effects of TCEO were accompanied by any anti-inflammatory activity in *in vivo* and *in vitro* models, by measuring gene expression and cytokine secretion of the major inflammatory components, and by studying the evolution of somatic cell counts (SCC) in milk for the *in vivo* model. *In vivo* study: 12 Holstein cows with subclinical mastitis were assigned to either control (n=6) or EO (n=6) group. Udder infected in the EO group was treated two times daily, after the milking, with TCEO diluted in milking grease for 7 days. Milk samples were collected to measure the level of IL8, SCC and for bacteriological analysis. We measured the abundance of some inflammatory genes on PBMC isolated cells. Statistical analysis was carried out by using one-way ANOVA. *In vitro* study: PBMC were purified from fresh blood derived from 3 healthy cows stimulated with LPS and with TCEO or its major components. The secretion of inflammatory cytokines in the culture supernatant was measured by using a Bovine Cytokine/Chemokine Magnetic Bead Panel. No statistical differences on SCC, IL8 secretion in the milk, bacteriological status, nor PBMC cytokine gene expression were observed between Control and TCEO groups in the *in vivo* model. These preliminary results suggest that the antibacterial effects of TCEO highlighted in the *in vitro* study were not followed by any anti-inflammatory *in vivo* effects against mastitis. Further analyses are in progress to evaluate the anti-inflammatory activity of TCEO and its major components in the *in vitro* model.

Milk production responses of transition dairy cows to supplementation of rumen protected choline

H.T. Holdorf[1], K.A. Estes[1], C. Zimmerman[2], K.E. Ritz[1], M.J. Martin[1], G.J. Combs[1], S.J. Henisz[1], S.J. Erb[1], W.E. Brown[1] and H.M. White[1]
[1]University of Wisconsin-Madison, Madison, WI, USA, [2]Balchem Corporation, New Hampton, NY, USA; czimmerman@balchem.com

Rumen protected choline (RPC) improves milk and ECM yields in transition dairy cows producing 15 to 45 kg of milk/d. The objective of this study was to evaluate the effects of dietary RPC supplementation during the transition period on the performance of high producing dairy cows. As a part of a larger study, multiparous Holstein cows (n=58) were randomly distributed into 2 groups and fed either 0 (CTL) or 12.9 g/d of choline ion of a new generation RPC from 21d pre-calving through 21d post-calving. Treatments (trt) were mixed into the TMR and cows had *ad libitum* access. Prepartum, cows were fed via Insentec feeders (Hokofarm Group; 4 feeders/trt) which allowed quantification of individual intake. Postpartum, cows were housed in pens without individual DMI measurements and trt were maintained until 21 DIM, after which all cows received a common lactating diet without RPC. Individual DMI and milk yield were recorded daily while milk components were analysed weekly. Mixed models analysing trt effects and trt × time interactions were performed in PROC MIXED, SAS 9.4. Initial BW (820 and 842 ±19 kg, CTL vs RPC) and BCS (3.7 and 3.8 ±0.2, CTL vs RPC) at enrolment were similar. Prepartum DMI was lower ($P<0.01$) in cows fed RPC. During the first 21d post-partum, milk yield (47.3 RPC vs 46.4 kg/d CTL, trt $P=0.58$; trt × time $P=0.18$) and ECM yield (56.0 RPC vs 56.8 kg/d CTL, trt $P=0.63$; trt × time $P=0.18$) did not differ. There were no trt effects on milk components in the first 21 DIM. From 21 to 100 DIM, milk yield (58.5 RPC vs 56.6 kg/d CTL, $P=0.29$) and ECM yield (58.0 RPC vs 56.1 kg/d CTL, $P=0.18$) were similar. From 22 to 100 DIM, RPC tended to increase protein yield ($P=0.09$). Supplementation with RPC also increased ($P=0.03$) the de novo proportion and reduced ($P=0.03$) the preformed proportion of total milk FA. Supplementation of a new generation RPC during the transition period resulted in a 1.9 kg advantage in milk and ECM yields consistent with past literature and meta-analyses, despite higher levels of production in the current experiment.

Gastrointestinal health and oxidative status in calves fed milk replacer differing in fat sources

J.N. Wilms[1,2], N. Kleinveld[3], G.H. Ghaffari[4], M.A. Steele[2], H. Sauerwein[4], J.M. Martín-Tereso[1,3] and L.N. Leal[1]
[1]Trouw Nutrition, Research and Development, P.O. Box 299, 3800 AG Amersfoort, the Netherlands, [2]Guelph University, Department of Animal Biosciences, 50 Stone Road East, Guelph, Canada, [3]Wageningen University, Animal Nutrition Group, P.O. Box 338, 6700 AH, Wageningen, the Netherlands, [4]Bonn University, Institute of Animal Science, Regina-Pacis-Weg 3, 53111 Bonn, Germany; juliette.wilms@trouwnutrition.com

Fat in milk replacers (MR) differs from milk fat because it consists of combinations of vegetable and animal oils. Our objective was to investigate how common fat sources used in MR affect digestibility, gut permeability, and indicators of oxidative status. Forty-five male calves arrived at the facility between 0 to 3 d of age and were assigned to blocks based on arrival day. Within each block, calves were randomly assigned to one of three experimental diets (n=15 per treatment): VEG was a MR containing vegetable oils (20% coconut, 80% of a mixture of coconut and rapeseed oils); in ANIM, the MR contained animal fats (65% lard, 35% dairy cream), and in MIX, a mixture of 80% lard and 20% coconut oil was used. The three MR had 30% fat, 24% crude protein, and 36% lactose, and were formulated with a fatty acid profile aiming to mimic that of milk fat. Calves were individually housed, and daily milk volumes were 6.0 l from d1-5, 7.0 l from d6-9, and 8.0 l from d10-35, divided into two equal meals (135 g/l). Faeces were scored daily, and calves were weighed, and blood was drawn at arrival and weekly thereafter. Urine and faeces were collected over 24h at wk 3 and 5 to assess gut permeability using indigestible markers and digestibility, respectively. Fat sources in MR did not affect growth, intakes, gut permeability, and MR digestibility. Diarrhoea prevalence was lower in VEG than in MIX in wk 2 after arrival ($P<0.01$), whereas ANIM did not differ from other treatments. Calves fed ANIM and MIX had 16% more thiobarbituric acid reactive substances in serum than VEG ($P<0.01$), whereas the ferric reducing ability of plasma was 7% greater in ANIM than MIX ($P=0.03$). Formulating the fat fraction of MR based on the fatty acid profile of milk fat resulted in marginal differences in terms of gastrointestinal health. However, specific serum redox parameters were altered by fat composition.

Session 17 Poster 16

Effect of a spice functional feed additive on behaviour and ruminal ecosystem in cows

C. Omphalius[1], J.-F. Gabarrou[2], G. Desrousseaux[2] and S. Julliand[1]
[1]Lab To Field, 26 bd Dr Petitjean, 21000 Dijon, France, [2]Phodé, Z.I. Albipôle, 81150 Terssac, France; gdesrousseaux@phode.fr

Eight cows received an individual controlled diet (in DM: 77% corn silage, 5% wheat, 14% soybean meal, 3% alfalfa and 1% mineral and vitamins, at 2.3% BW DMI) without (CTRL) or with 3.2 g/d of a spices-based feed additive (SA) mixed to the ration to evaluate the impact on behaviour and ruminal ecosystem. Cows were distributed in a Latin square design, with 3-wk experimental periods and a 1-wk wash-out between them. On d20 of each period, behaviour was recorded over 24-h. With a scan sampling technique, the number of behaviour bouts (time feeding, ruminating standing/lying, drinking, licking, auto-auscultating, walking, exploring, listening, interacting socially, contacting physically, sleeping) was determined and the mean feeding duration was calculated. Activity of the ruminal ecosystem was assessed on d21 (pre-feeding, 3h, 6h, 9h after feeding) of each period through ruminal pH, volatile fatty acids (VFA), lactate and ammonia concentrations. At time 3h, total anaerobic, cellulolytic, proteolytic and amylolytic bacteria were enumerated in ruminal content. The effects of SA, period and their interaction were evaluated using a GLIMMIX procedure (SAS) with cow as random effect. Parameters monitored in kinetics were analysed with adding time effect in the model. Pearson correlations between behavioural and microbial parameters were calculated. Rumination bouts (CTRL: 4.83; SA: 6.35 nb/d) and time ruminating (CTRL: 16.1; SA: 21.2% of observations) tended to be higher (P=0.06) with SA. Ingestion was numerically faster (CTRL: 8.0; SA: 7.7 min/kg DMI; P=0.13) and rumination started earlier after feeding (P=0.02) with SA. Rumination bouts and time ruminating over 24h were positively correlated with VFA concentrations at time 9h (P<0.08). Rumination time while standing also tended to be positively correlated to total anaerobic bacteria concentration (P=0.06). In this study, including 3.2 g/d of SA to the ration altered ingestion and rumination behaviours in cows. Our results suggest that this SA modified ruminal ecosystem microbiota and its activity over time. Ongoing analyses of ruminal bacterial structure and diversity will provide deeper insight into the relationship between feeding behaviour, ruminal microbiota, and health.

Session 17 Poster 17

Efficacy of different mycotoxin binders to adsorb mycotoxins: a review from *in vitro* studies

A. Kihal, M. Rodríguez-Prado and S. Calsamiglia
Universitat Autonoma de Barcelona, Animal welfare and nutrition, Edifici V, Travessera dels Turons, 08193 Bellaterra, Spain; abdelhacib.khl@gmail.com

The objective of this study was to determine the efficacy of different mycotoxin binders (MTB) to adsorb mycotoxins (MTx) *in vitro*. A literature search was conducted to identify *in vitro* research papers from different databases. The search was based on 7 MTB [active carbon (AC), bentonite, clinoptilolite, hydrated sodium calcium aluminosilicate (HSCAS), sepiolite, yeast cell walls (YCW) and zeolite] and 6 MTx [aflatoxin B1 (AF), deoxynivalenol (DON), fumonisin B1 (FUM), ochratoxin (OTA), T-2 toxin and zearalenone (ZEA)]. Inclusion criteria were: *in vitro* studies, description of the incubation media and pH, and percentage of MTx adsorption. Sixty-eight papers with 1,842 data were selected. The response variable (percentage MTx adsorption by MTB) was analysed with the PROC MIXED of SAS (SAS Inst. Inc., Cary, NC). The model included also pH, incubation media [water, methanol, HCl, citrate acetate phosphate (CAP) buffer, a simulation of the gastro-intestinal digestion and gastric juice (GI)] and their interactions. Difference among incubation media were only significant between CAP and GI (P<0.05), and data from GI were excluded. The MTx adsorption capacity was 62%±1.0 for bentonite (from 18 with DON to 93% with AF, P<0.05), 52%±4.3 for clinoptilolite (from 0 with DON to 75% with AF, P<0.05), 55%±1.9 for HSCAS (from 11 with DON to 83% with AF, P<0.05), 76%±3.1 for MMT (from 9 with DON to 88% with AF, P<0.05), 83%±1.0 for AC (from 53 with T-2 to 93% with AF), 44%±0.4 for YCW (from 19 with DON to 49% with AF) and 52%±9.1 for sepiolite (from 12 with DON to 95% with AF). The adsorption of AF was on average 76%±0.6 (from 49 with YCW to 95% with sepiolite, P<0.05), for DON was 35%±1.6 (from 0 with clinoptilolite to 69% with AC, P<0.05), for FUM was 50%±1.8 (from 25 with sepiolite to 86% with AC, P<0.05), for OTA was 42%±1.0 (from 17 with sepiolite to 88% with AC, P<0.05), for ZEA was 48%±1.1 (from 14 with clinoptilolite to 85% with AC, P<0.05), and for T-2 was 27%±2.8 (from 5 with zeolite to 52% with AC). The absorption of AF, DON, FUM, OTA and ZEA was affected by pH.

Session 17 Poster 18

Meta-analysis on the efficacy of mycotoxin binders to reduce aflatoxin M1 in dairy cow's milk

A. Kihal, M. Rodríguez-Prado and S. Calsamiglia
Universitat Autonoma de Barcelona, Animal welfare and nutrition, Edifici V, Travessera dels Turons, 08193 Bellaterra, Spain; abdelhacib.khl@gmail.com

The objective of this meta-analysis was to determine the efficacy of different mycotoxin binders (MTB) to reduce aflatoxin M1 (AFM1) in milk. A literature search was conducted to identify *in vivo* research papers from different databases. Inclusion criteria where: *in vivo*, dairy cows, description of the MTB used, doses of MTB, aflatoxin inclusion in the diet and concentration of AFM1 in milk. Twenty-two papers with 108 data were selected. Binders used in the studies were: hydrated sodium calcium aluminosilicate (HSCAS), yeast cell wall (YCW), bentonite (BEN), and mixes of several MTB (MIX). The response variables were: AFM1 percentage reduction in milk, total AFM1 concentration excreted in milk per day and transfer percentage of aflatoxin from feed to AFM1 in milk; and AFM1 concentration in urine and faeces. Data were analysed with GLIMMIX procedure and WEIGHT statement of SAS (SAS Inst. Inc., Cary, NC). The percentage reduction of AFM1 in milk was 54%±2.7 ($P<0.05$) for BEN, 27%±2.9 ($P<0.05$) for MIX, 18%±4.5 for YCW and 7%±3.5 for HSCAS. The excretion of AFM1 in milk (µg/d) was lower in HSCAS (14.4±0.97), YCW (16.3±1.21), and MIX (18.7±1.27) ($P<0.05$), and tended to be lower (13.9±1.44) in BEN ($P<0.08$) compared with control (22.6±1.00). The transfer of AFM1 to milk was lower in HSCAS (1.5%±0.10), YCW (0.9%±0.03) and BEN (0.7%±0.10) ($P<0.05$) compared with control (2.3%±0.11). Urine and faecal excretion were only identified in HSCAS and MIX treatments. Urine concentration of AFM1 (µg/l) tended to decrease with HSCAS (2.1±0.70, $P<0.08$) compared to control (7.6±1.25). Faecal concentration of AFM1 (µg/kg) tended to decrease with MIX (4.9±0.56, $P<0.09$) compared to control (6.8±1.05). The meta-analysis results showed that BEN has the highest capacity to reduce AFM1 from milk and YCW the lowest.

Session 17 Poster 19

Effect of dietary herbal extracts in broiler breast composition, oxidation and MAPKs activation

S. Dokou[1], I. Giannenas[1], S. Savvidou[2], N. Panteli[3], C. Tatoudi[3], I. Voutsinou[3], I. Mourtzinos[4], E. Antonopoulou[3], D. Lazari[5], J. Wang[6] and K. Grigoriadou[2]
[1]Aristotle University, Veterinary Medicine, University Campus, 54124 Thessaloniki, Greece, [2]Hellenic Agricultural Organisation-DEMETER, Animal Science & Plant Breeding and Genetic Resources, Paralimni, 58100 Giannitsa, Greece, [3]Aristotle University, Zoology, University campus, 54124 Thessaloniki, Greece, [4]Aristotle University, Food Science and Technology, University campus, 54124 Thessaloniki, Greece, [5]Aristotle University, Pharmacy, University Campus, 54124 Thessaloniki, Greece, [6]Nanjing Agricultural University, Gastrointestinal Nutrition and Animal Health, University campus, CN-210095 Nanjing, China, P.R.; igiannenas@vet.auth.gr

Mitogen-activated protein kinases (MAPKs) mediate signal transduction in response to several extracellular stimuli, including nutrient availability, in order to regulate pivotal cell processes, such as proliferation and apoptosis. This trial examined the effects of dietary use of an extract of oregano, garlic, camelina and crithmum either encapsulated in cyclodextrin or in an aqueous form, on breast meat composition, lipid and protein oxidation and cellular responses in the breast meat. Control Group A (CL) was fed basal diets (based on maize and soybean meal) in mash form. Duration of the trial was 35 days. 120 one-day-old mixed broiler chicks (Ross-308) were randomly allocated to 3 groups with 4 replicates, housed in floor pens with wheat straw litter. Commercial breeding, management and vaccination procedures were employed. Results showed that breast chemical composition was similar among the three groups. Lipid oxidation and protein carbonyls were reduced in both groups that received the herbal extracts by the encapsulated or the aqueous form. Cellular responses to dietary inclusion of plant extracts were addressed through the activation of p38 and ERK-1/2 MAPKs using Western blot analysis. Dietary inclusion of the mixture of plant extracts in aqueous form significantly reduced the activation of ERK-1/2 MAPK, In conclusion, protein and lipid metabolism and oxidation may be significantly affected by dietary mixtures of herbal extracts in broilers. Acknowledgments: Funded by EU and Greek Funds, Bilateral R&T Cooperation Greece-China, 2014-2020 (T7ΔKI-00313-GreenPro).

Session 17 Poster 20

Dietary olive cake inclusion and metabolic change of lactating Holstein, Simmental and Modicana cows

V. Lopreiato[1], A. Bionda[1,2], A. Amato[1], E. Fazio[1], P. Crepaldi[2], V. Chiofalo[1] and L. Liotta[1]
[1]Università di Messina, Dept. of Vet. Sciences, Polo Universitario dell'Annunziata, 98168 Messina, Italy, [2]Università di Milano, Via Celoria, 2, 20133 Milan, Italy; vincenzo.lopreiato@unime.it

The use of Olive Cake (OC) as feed represents a sustainable alternative to reduce the costs associated with animal nutrition while allowing a rational utilization of this by-product of olive oil production in the context of circular economy and functional feeds. Thus, the aim of this study was to investigate the effect of dietary inclusion of OC (DM 95.6; CP 8.6; EE 15.9; NDF 49.4; ADF 39.4; ADL 23.1; Ash 4.1; Starch 1.5% as fed) on metabolic response of cows highly specialized for milk, Holstein (HO), cows selected for meat and milk, Simmental (SI), and Sicilian local breed reared for milk, Modicana (MO). The experimental design of this study enrolled a total of 18 lactating cows (6 for each breed) from the same farm. Blood samples for the metabolic profile assessment were collected before (0 d; 150±67 d in milking) and after (30 d) the dietary treatment with 8% inclusion of OC in the concentrate. The response of plasma biomarkers was calculated using fold-change: 1+[(measure at 30 d–measure at 0 d)/measure at 0 d]. Data were subjected to MIXED model of SAS. MO had greater fold-change of cholesterol (1.53 vs 1.13; P=0.02) and creatine kinase (1.82 vs 0.82; P=0.02) compared with SI cows. A tendency for greater fold-change was detected for γ-glutamyl-transferase (1.70 vs 1.32), aspartate aminotransferase (1.21 vs 1.05), and Ca (1.12 vs 0.93) in MO than SI cows (P=0.10). Fold-change of lactate dehydrogenase was greater in HO (0.96) than SI (0.85) or MO (0.88) cows (P<0.05). A tendency (P=0.10) was detected for potassium levels, where HO (1.54) showed greater fold-change than SI (1.27) or MO (1.22) cows. These results unravel a breed-specific metabolic response between SI, HO, and MO cows, mainly at liver metabolism level, providing biological insights on the metabolic response among lactating cattle breeds when OC is used in the diet. It is noteworthy to report that no detrimental effect was obtained at metabolic status; thus, OC can be considered a valuable component in dairy cow diet, improving agriculture sustainability. Work supported by P.O. FESR SICILIA 2014/2020 – Project BIOTRAK.

Session 17 Poster 21

The effect of vitamin D3 supplementation on the liver transcriptome of domestic pig

A.S. Steg[1], A.W. Wierzbicka[1], M.O. Oczkowicz[1] and M.Ś. Świątkiewicz[2]
[1]National Research Institute of Animal Production, Department of Animal Molecular Biology, ul. Krakowska 1, 31-047 Kraków, Poland, [2]National Research Institute of Animal Production, Department of Animal Nutrition and Feed Science, ul. Krakowska 1, 31-047 Kraków, Poland; anna.steg@iz.edu.pl

There has been a significant increase in interest in supplementing the diet with vitamin D to improve health in recent years. One can find information about the influence of vitamin D on many diseases, including civilization diseases, obesity, cancer, etc. However, most published meta-analyses do not confirm such effect of vitamin D, except for the skeletal system and the immune system. The suggested supplementation dose of vitamin D is about 2,000 U, although there are indications of a therapeutic effect of much higher doses. Our study aims to check whether supplementation with high doses of vitamin D3 in the pigs' diet causes changes in the transcriptome of the liver and in which signalling pathways and biological processes are involved genes which expression is changed the most. In the experiment, 30 pigs were divided into three dietary groups, differing in the content of vitamin D3 in the feed (no supplementation, supplementation with 5,000 U/kg of feed, and supplementation with 10,000 U/kg of feed). Liver fragments were collected for gene expression analysis after three months of daily supplementation. RNA was isolated and after preparing 3' quant mRNA libraries (Lexogen), it was subjected to next-generation sequencing (NGS) on NextSeq5500 (Illumina). After filtering and mapping, differential expression analysis was performed with DeSeq2 software. Subsequently, data has been validated with the use of the qPCR method. The obtained results indicate that supplementation with high doses of vitamin D3 causes changes in the expression of 115 genes in the case of the dose of 5,000 U/kg of feed and 140 genes in the case of 10,000 U/kg of feed. The qPCR validated genes that were most altered in expression includes *HMGCS2* (P=0.0053), *LDHB* (P=0.0034), *SERPIN A3* (P=0.0135), *APOA4* (P=0.0280). The most enriched pathways are involved in fatty acids beta-oxidation (FDR=0.0015), fatty acid catabolic processes (FDR=0.0233), lipid modification (FDR=0.0062), carboxylic acid catabolic processes (FDR=0.0062) and oxidation-reduction processes (FDR=0.0257).

Lack of effectiveness of acidified diet in preventing hypocalcaemia in commercial dairy farms

T. Aubineau[1,2], A. Boudon[3] and R. Guatteo[1]
[1]INRAE, BIOEPAR, Route de Gachet, 44300 Nantes, France, [2]Innoval, Rue Eric Tabarly, 35538 Noyal-sur-Vilaine, France, [3]INRAE, PEGASE, La Prise, 35590 Saint-Gilles, France; thomas.aubineau@innoval.com

Hypocalcaemia prevention programs have been widely studied in experimental setting, but their effectiveness has not been assessed in field conditions. The objective of this study was to assess, in a context of typical dairy farms in western France, the preventive effects of prepartum diet acidification (AC) on the incidence of hypocalcaemia after calving (calcemia ≤2 mmol/l). A cohort of 547 Holstein cows from 32 French dairy farms was followed in a longitudinal study in 2020, including 315 cows (from 18 farms) supplied with anionic salts during late gestation, and 232 cows (from 14 farms) not being supplied. Blood samples were collected from 17 cows in each farm between 24 and 48h after calving and tested for total calcium. Effect of providing AC on incidence of hypocalcaemia was assessed using mixed linear model including cow-level (parity, 305-day milk yield, individual health events and treatments) and farm-level (forage composition of prepartum diet, duration of AC) factors, with farm as random effect. Effect of the level of prepartum diet DCAD (in mEq/kg DM) on incidence of hypocalcaemia was assessed using mixed linear model including cow-level (as above) and farm-level (delivered amounts of calcium, phosphorus and magnesium, duration of AC) factors, with farm as random effect. About 55% of cows were hypocalcaemia (n=299). The first model revealed a negative effect of increased parity on hypocalcaemia incidence (P≤0.01) but no effect of AC (P=0.36). The second model also revealed a greater risk of hypocalcaemia for increased parity (P≤0.01) and high phosphorus content of diet (≥39 g/cow per day, OR=1.7, P=0.07). We also observed a trend to a preventive effect of a high magnesium content of diet (≥3.3 mg/kg DM, OR=0.5, P=0.06). Controversial effects of diet DCAD were observed. Indeed, only mildly acidified diet (i.e. DCAD between 20 and 100 mEq/kg DM) tended to prevent hypocalcaemia (OR=0.7, P=0.09). Only few farms using AC provided a diet DCAD≤-50 mEq/kg DM (n=3). Our results highlight the need for reinforcement of advice to farmers in order to improve effectiveness of hypocalcaemia prevention programs.

Effect of a combination of three yeasts on body temperature of weaning piglets by thermal imaging

V. Perricone[1], S. Sandrini[1], V. Redaelli[2], F. Luzi[3], E.R. Parra Titos[4], G. Savoini[1] and A. Agazzi[1]
[1]Univeristy of Milan, Department of Veterinary Medicine and Animal Science (DIVAS), Via dell'Università 6, 26900 Lodi, Italy, [2]Freelancer, Frazione Castello 1, 23848 Oggiono, Italy, [3]Univeristy of Milan, Department of Biomedical, Surgical and Dental Sciences, Via della Commenda 10, 20100 Milano, Italy, [4]Vétoquinol Italia S.r.l., Via Piana 265, 47032 Bertinoro, Italy; vera.perricone@unimi.it

Yeasts supplementation showed to mitigate the weaning stress, improving animal health and performance, but little is known on their effects on piglets' welfare. This study evaluated the effect of a yeast mixture (YM, Levustim B0399, Vetoquinol; *Kluyveromyces marxianus fragilis B0399*, *Pichia guilliermondii*, and *Saccharomyces cerevisiae*) on body temperature of post-weaning piglets. Forty-eight male weaned piglets (27±1.7 d, 7.19±0.54 kg), were randomly allocated to two homogeneous groups and enrolled in a 28-days trial. Both groups received a basal diet with (T) or without (C) inclusion of 0.8% YM during weeks 1 and 2, and 0.6% during weeks 3 and 4. Individual skin temperatures were weekly recorded on the dorsal (DT), ocular (OT) and ear (ET) regions using an Avio thermoGear Nec G120EX microbolometer infrared camera (Nippon Avionics Co., Tokyo, Japan; 320×240 pixels). Data were acquired at 0.80 m from the animals and at a height of 1.60 m. The camera was automatically calibrated prior to any image and emissivity was set at 0.96. Thermal images were analysed using NEC InfRec Analyzer and Grayess IRT Analyzer software. For each day of sampling, the environmental temperature was measured at floor level and at piglet level. Data were analysed by means of ANOVA for repeated measures, including the effect of the treatment, time, and their interaction. Pearson correlations were tested among body and environmental temperatures. DT was negatively correlated with both environmental temperatures (P≤0.01). The dietary treatment significantly increased ET (P≤0.01) and DT (P=0.05), but not ET (P=0.75). Body temperatures were increased during the trial in both C and T groups (P≤0.01; P=0.03 and P≤0.01, respectively for DT, ET, and OC). In conclusion, yeast administration demonstrated to have significant effects on body temperature of post-weaning piglets.

Session 17 Poster 24

Effect of a new high fibre feed on blood biochemistry of outdoor finished male local pigs

J.M. Martins[1,2], R. Charneca[1,2], R. Varino[1,2], A. Albuquerque[1,2], A. Freitas[1,2], J. Neves[1,2], C. Marmelo[1], F. Costa[1], A. Ramos[1,3] and L. Martin[1,3]

[1]ECO-PIG Consortium, Z.I. Catraia, Ap. 50, 3441 S. Comba Dão, Portugal, [2]MED – Mediterranean Institute for Agriculture, Environment and Development, Un. Évora, Ap. 94, 7000 Évora, Portugal, [3]Inst. Politécnico Coimbra, S. Martinho do Bispo, 3045 Coimbra, Portugal; rmcc@uevora.pt

This work, within the framework of ECO-PIG Project, measured the effects of a high soluble dietary fibre feed on blood biochemistry parameters in outdoor raised Portuguese Alentejano (AL) male pigs (n=30) with access to *ad libitum* water and feed. Surgically castrated (group C) and intact pigs (groups I and IE) were fed commercial diets from 40 to 130 kg body weight. From 130 until 160 kg (slaughter), groups C and I were fed commercial feeds while group IE ate the isoproteic and isoenergetic experimental diet, including agro-industrial by-products. Blood samples were collected at 120 kg (before the start of the fattening period) and two days before slaughter. Serum levels of total protein, urea (U), glucose, triacylglycerols (TG) and cortisol (COR) were determined. At 120 kg, U levels were different among the groups (3.3±0.2 in IE, 4.0±0.2 in I, and 5.0±0.1 mmol/l in C pigs, P<0.001). At 160 kg, they were different between IE and C groups, again with lower values in intact pigs (3.1±0.2 in IE, 3.4±0.3 in I, and 4.0±0.2 mmol/l in C pigs, P=0.034). These overall lower U levels in intact pigs suggest a more efficient nitrogen use for lean tissue growth than in castrates. TG levels were lower in intact pigs at 160 kg (0.30±0.03 in IE, 0.37±0.04 in I, and 0.53±0.04 mmol/l in C pigs, P<0.001). Higher blood TG levels relate with fatter pigs, and C pigs produced fatter carcasses than intact ones (see 'The ECO-PIG project: Use of a new high fibre feed for outdoor finishing of intact male local pigs'). Finally, COR levels at 160 kg were lower in IE than in I and C pigs (79±13 in IE, 272±38 in I, and 204±22 nmol/l in C pigs, P<0.001). This suggests lower stress levels on the IE group and agrees with the number pigs with of skin injuries related to agonistic interactions observed in the last week of trial (4 in IE, 8 in I and 9 in C pigs). Further studies will test the effect of the experimental feed on pork boar taint and meat quality of intact AL pigs raised outdoors.

Session 17 Poster 25

The effects of external treatment with essential oils on milk quality: a lipidomics approach

F. Ceciliani[1], R. Nehme[2], P. Cremonesi[3], B. Castiglioni[3], F. Biscarini[3], M. Ciccola[1], S. Di Mauro[1], S. Andrés[4], R. Pereira[5], M. Audano[1], D. Caruso[1], D. Pereira[5], H. Falleh[6], R. Ksouri[6], C. Gini[1] and L. Abfennedbi-Najar[2]

[1]Università di Milano, Via dell'Univesrità 6, 26900 Lodi, Italy, [2]IDELE, 149, R. de Bercy, 75012 Paris, France, [3]CNR-IBBA, Via Enstein, 26900 Lodi, Italy, [4]CSIC, Finca marzanas, 24346 Leon, Spain, [5]University of Porto, R Jorge Viterbo Ferreir 228, 4050-313 Porto, Portugal, [6]CBBC, Tecnopole de Borj-Cédria, 901 Hamman-Lif, Tunisia; fabrizio.ceciliani@unimi.it

Mastitis is one of the major health issues in dairy farming. An increase in mastitis in a herd is generally followed by increased use of antibiotics, which in turn increases the risk for antibiotic residues in milk and the eventual increase in antibiotic resistance. The EU project addresses this issue by exploring alternative approaches such as essential oils (EO) to explore their antibacterial and immunomodulatory activity. The present *in vivo* study describes the results of external treatment of mammary gland affected by sub-clinical mastitis using natural essential oil of *Thymus capitatus*. The study was carried out on 12 animals. Mammary glands were treated at topic level with EO, whereas control was treated with the emulsifying solution sued as a vehicle. Milk samples were collected at Days T0, T8, T21 and T28, and untargeted lipidome analysis was carried out. 2 and 5 µl of sample for the positive and negative ion mode, respectively, were separated by liquid chromatography (LC) with a Kinetex EVO C18 – 2.1×100 mm, 1.7 µm (Phenomenex®) column at 45 °C connected to an ExionLC™ AD system (ABSciex) maintained at 15 °C. Separated lipid species were then ionized through an electrospray ionization (ESI) source and analysed in a TripleTOF 6600 (quadrupole time-of-flight, QTOF – ABSciex) mass spectrometer. A total of 2,450 lipids were identified. There were no differences between the samples collected at T0, T8, T28, whereas little differences were observed at T21, including DG 35:2, DG 36:8, DG 35:2, DG 33:3, DG 33.2, DGTS 22:0. The changes were limited but demonstrated that an external treatment over the surface of the mammary gland with EO may reflect, albeit in a limited way, on milk quality, at least for what concerns the lipid content.

Effect of dietary plant extracts in broiler chicken breast composition, lipid and protein oxidation

S. Vasilopoulos[1], S. Dokou[1], I. Giannenas[1], S. Savvidou[2], S. Christaki[3], A. Kyriakoudi[3], E. Antonopoulou[4], I. Mourtzinos[3] and T. Bartzanas[5]

[1]Aristotle University, Veterinary Medicine, University Campus, 54124 Thessaloniki, Greece, [2]Hellenic Agricultural Organisation-DEMETER, Animal Science, Paralimni, 58100 Giannitsa, Greece, [3]Aristotle University, Food Science and Technology, University campus, 54124 Thessaloniki, Greece, [4]Aristotle University, Zoology, University campus, 54124 Thessaloniki, Greece, [5]Agricultural University of Athens, Natural Resources Mgt and Agricultural Engineering, Iera Odos 75, 11855 Athens, Greece; igiannenas@vet.auth.gr

Chicken meat is highly nutritious with a steadily rising consumer interest. This study evaluated a dietary mixture of *Punica granatum* and *Allium cepa* aqueous and cyclodextrin extracts on broiler breast meat chemical composition and its oxidative stability as well as their interaction with Mitogen-activated protein kinases (MAPKs) signalling. MAPKs are major targets of redox dependent regulation of several cell functions in a stimulus-specific manner including nutrient factors. 120 one-day-old male Ross-308 chicks were randomly allocated to 3 treatments with 4 replicate pens (10 chicks per pen). Broiler chicks in the control group were fed typical commercial rations in mash form based on maize and soybean. The other two groups were further supplemented with the mixture of *Punica granatum* and *Allium cepa* aqueous and cyclodextrin extracts at 0.1% per kg of dry matter (DM) of feed respectively. On day 35 breast tissue samples were collected for chemical analysis, fatty acid profile, lipid stability and protein carbonyls. The phosphorylation of p38 and ERK-1/2 MAPKs were addressed using Western blot analysis. Diet supplementation increased $\sum$n-3 and $\sum$n-6 fatty acids. Both extracts favourably affected meat composition, TBARS and protein carbonyls compared to the CL group. p38 and ERK-1/2 MAPKs were significantly activated in response to the cyclodextrin extract, while the aqueous one had no influence on phosphorylated/total MAPKs ratio. Dietary inclusion of encapsulated pomegranate and onion can be regarded as a promising feed additive with functional properties. Acknowledgments: 'Phytobiotics' (MIS 5063328), 'Partnerships of industries with Research and Knowledge Dissemination Organizations' OP, RIS3Crete, region of Creta.

Effect of nutritional strategies at cow-calf on gene expression and marbling in Angus-Nelore cattle

R.A. Curi[1], G.L. Pereira[1], M.J.G. Ganga[1], W.A. Baldassini[1], L.A.L. Chardulo[1], M.R. Chiaratti[2], R.P. Nociti[3] and O.R. Machado Neto[1]

[1]São Paulo State University, Campus de Botucatu/SP, 18618-681, Brazil, [2]Federal University of São Carlos, Campus de São Carlos/SP, 13565-905, Brazil, [3]University of São Paulo, Campus de Pirassununga/SP, 13635-900, Brazil; rogerio.curi@unesp.br

Different strategies have been used to increase the intramuscular fat in beef produced in Brazil, mainly from Nellore (*Bos indicus*) and its crossings, to improve the final product quality. There are few researches evaluating the effect of the creep feeding (a cow-calf supplementation system) on global gene expression and intramuscular adipogenesis in beef cattle. Thus, this study aimed to analyse the differences in gene expression in a broad way in the weaning of crossbred cattle submitted to different treatments at cow-calf. Forty-eight male Angus-Nellore crossbred steers, not castrated, half-sibs, kept from 30 days of age until weaning under two different treatments (n=24/treatment): group 1 (G1) – no creep feeding; G2 – creep feeding, were used. The animals were weaned at 210 days, and then finished in feedlot for 180 days receiving a diet containing 12.6% roughage and 87.4% corn based concentrate. With the *Longissimus thoracis* (LT) muscle samples of the left side of the carcass, collected at slaughter, were determined the fat percentage/total lipids (TL), and the marbling index (MI) of all animals of the two groups. At weaning, aliquots of LT muscle were collected by biopsy and the differences in gene expression between treatments (G1 vs G2 – 12 individuals/group) were investigated by RNA-Seq. The animals of both groups showed significant differences ($P<0.05$) for TL (4.95±0.20 vs 5.80±0.23) and MI (321.50±13.65 vs 366.11±12.39). The weights at the beginning of the experiment, at weaning and at slaughter (390 days) did not differ. Nine hundred and forty-seven differentially expressed genes (DEG) were identified (FDR 5%; log2 foldChange 0.5) among the groups at weaning, 504 of which were up-regulated and 443 down-regulated in the G2 (high marbling). Functional enrichment analysis, performed with DEG, identified important KEGG pathways ($P<0.01$) narrowly related to adipogenesis such as: PPAR signalling and Adipocytokine signalling, with most up-regulated genes in G2.

Session 17 Poster 28

Milk yield from dairy ewes supplemented with pumpkin and chia seeds

L.E. Robles-Jimenez[1], E. Aranda Aguirre[1], A.J. Chay Canul[2], R.A. Garcia Herrera[2] and M. Gonzalez Ronquillo[1]
[1]Universidad Autónoma del Estado de Mexico, Instituto Literario 100, 50000, Mexico, [2]Universidad Juarez Autonoma de Tabasco, Carretera BHSA-Teapa Km 25, Villa Hermosa, Tabasco, 86290, Mexico; lizroblez@hotmail.com

The uses of chia seed (*Salvia hispanica* L.) and pumpkin (*Cucurbita moschata*) as a Mexican native product, could be a good ingredient in supplementation of diets for dairy ewes. Therefore, the objective of this study was to determine milk production and composition as well as nitrogen balance in dairy ewes supplemented with Chia and Pumpkin seeds as sources of fatty acids and protein. Fifteen primiparous Texel ewes (BW 78.7±0.5 kg) with 70±5 days in milk were fed three diets, distributed in a 3×3 Latin square design repeated 5 times, with 3 experimental periods of 20 days each. Dairy ewes were supplemented with 50 g/kg $BW^{0.75}$ of concentrate and corn silage *ad libitum*. The protein source was the only factor of diet variation. The control diet (SBM) included 160 g of SBM / kg DM, which was replaced by Pumpkin seed (PS) and Chia seed (CS) diets. Diets were formulated to be isoenergetic and isoproteic (ME, 11.7 MJ/kg DM, 14% CP). Milk yield (MY, kg/d) was recorded, and individual milk samples (100 ml) were taken on the last 5 days of each experimental period at 16:00h. Dry matter intake (2,488±25 g/d) was similar among diets (P>0.05). Milk samples were analysed for total solids (TS), lactose, protein, and fat. Fat-corrected milk (FCM 6.5%, kg/d) and fat-protein corrected milk (FPCM 6.5, 5.8%, kg/d) were determined. Nitrogen was determined in urine, faeces and milk samples. The inclusion of PS (0.356 kg/d) or SBM (0.390 kg/d) diets increased milk production (up to 43%), as well as FCM and FPCM (P<0.05) compared to CS diet. CS or PS seed diets did not affect milk fat concentration (5.76±0.3 g/100 g, P>0.05), but increased lactose (5.09), protein (5.41) and total solids (11.37) (P<0.05) in CS-fed dairy ewes. The supplementation of CS or PS seeds increased N intake (19%, P<0.05), in addition to presenting a lower negative balance (-0.462±0.16 g/d) than SBM (-2.8 g/d) diet. The PS could be used in dairy ewes diets as protein-lipid feed alternatives without affect milk yield compared with SBM diet.

Session 17 Poster 29

Does supplementation of pig diets with high doses of vitamin D3 alter transcriptome of adipose tissue?

A.S. Steg[1], A.W. Wierzbicka[1], M.O. Oczkowicz[1] and M.Ś. Świątkiewicz[2]
[1]National Research Institute of Animal Production, Department of Animal Molecular Biology, ul. Krakowska 1, 32-083 Balice k. Krakowa, Poland, [2]National Research Institute of Animal Production, Department of Animal Nutrition and Feed Science, ul. Krakowska 1, 32-083 Balice k. Krakowa, Poland; anna.steg@iz.edu.pl

While vitamin D functions related to the maintenance of skeletal function and the immune system are well documented, there are also indications for the role of vitamin D in obesity prevention. While studies on adipocytes indicate some changes in genes due to vitamin D supplementation, there are no studies on the effect of this compound on the entire transcriptome *in vivo* in pigs. The aim of our study is to check whether supplementation with high doses of vitamin D3 in the pigs' diet causes changes in the transcriptome of the adipocyte tissue. In the experiment, 30 pigs were divided into three dietary groups, differing in the content of vitamin D3 in the feed (no supplementation, supplementation with 5,000 U/kg of feed, and supplementation with 10,000 U/kg of feed). Following three months of supplementation, adipose tissue was collected. Isolated RNA after preparing 3' quant mRNA libraries (Lexogen) was subjected to next-generation sequencing (NGS) on NextSeq5500 (Illumina). After filtering and mapping, analysis was performed with DeSeq2 software. The analysis showed that in the case of adipose tissue, there were no significant changes (Differentially Expressed Genes-DEGs) in the transcriptome, regardless of the dose of vitamin D3 used (p adjusted>0.05). In order to verify in depth whether supplementation with high doses of vitamin D3 has any effect on a specific signalling pathway or functional groups of genes, the gene set enrichment analysis (GSEA) was performed with the use of STRING software analysis. From the whole dataset, genes with a baseMean value of less than 10 were removed, and by ranking the genes by the P-value, a list of genes with their corresponding log2FoldChange values was created. Analysis showed two KEGG pathway enrichments: oxidative phosphorylation (enrichment score=1,96, FDR=0.00052) and non-alcoholic fatty liver disease (enrichment score=1,53, FDR=0.00052). In conclusion, our results show that vitamin D supplementation, even with high doses, has little effect on adipose transcriptome in pigs.

Session 17 Poster 30

Transcriptomic and proteomic analysis of omega-3 fatty acids in the porcine enterocytes

T. Sundaram[1,2], C. Giromini[2], R. Rebucci[2], S. Pisanu[3], D. Pagnozzi[3], M.F. Addis[2,3], M. Bhide[1], J. Pistl[1] and A. Baldi[2]
[1]University of Veterinary Medicine and Pharmacy in Košice, Department of Microbiology and Immunology, Komenského 68/73, 04181 Košice, Italy, [2]University of Milan, Department of Veterinary and Animal Science (DIVAS), Via Trentacoste 2, 20134 Milano, Italy, [3]Porto Conte Ricerche Srl, Strada Provinciale 55, 07041 Alghero SS, Italy; tamil.sundaram@unimi.it

Dietary omega-3 fatty acids (n-3 PUFA) are reported to improve the intestinal barrier integrity and maintain the gut homeostasis. Small intestinal enterocytes are the primary site of ω-3 PUFA uptake, but till date n-3 PUFA-induced global changes in the gene/protein expressions of these cells remains unexplored. Earlier, we demonstrated the anti-inflammatory, antioxidative and proliferative properties of the n-3 PUFA as Eicosapentaenoic acid (EPA) and docosahexaenoic acid (DHA) using a non-transformed porcine small intestinal enterocyte model, IPEC-J2. In the present study, n-3 PUFA-induced changes in the gene and protein expression profiles of the IPEC-J2 were explored using bioinformatic analysis as transcriptomics and proteomics. To this, the IPEC-J2 cells were treated with EPA:DHA (1:2, 10 μM) for 24 h and subsequently samples were processed following the previously published methods for bioinformatic analysis. The cells without any treatment were included as the control. Transcriptomic analysis was performed using the Illumina Nextseq technology, while proteomic analysis was performed using the filter-aided sample preparation with liquid chromatography (LC)-mass spectrometry (MS)/MS techniques. Presently, 51 differentially expressed genes (DEGs) and 40 differentially expressed proteins (DEPs) were identified when compared against the control. Further, Gene Ontology analysis revealed the participation of these DEGs/DEPs in several biological process including the lipid metabolic process, and cytokine signalling. The outcome of this study could contribute to comprehensive understanding on the role of n-3 PUFA in intestinal barrier for better planning of n-3 PUFA-based nutritional strategies in mammalian diets.

Session 17 Poster 31

Effect of vitamin D3 supplementation on brain transcriptome in rats

M. Oczkowicz[1], A. Steg[1], A. Wierzbicka[1], I. Furgał[2], B. Szymczyk[2], A. Koseniuk[1], M. Świątkiewicz[2] and T. Szmatoła[3]
[1]National Reserach Institute of Animal Production, Department of Molecular Biology of Animals, ul Krakowska 1, 32-083 Balice, Poland, [2]National Reserach Institute of Animal Production, Department of Animal Nutrition and Feed Science, ul Krakowska 1, 32-083 Balice, Poland, [3]University of Agriculture in Kraków, Centre of Experimental and Innovative Medicine, Al. Mickiewicza 24/28, 30-059 Kraków, Poland; maria.oczkowicz@iz.edu.pl

Vitamin D3 supplementation has been widely recommended in humans and animals in recent years. Research shows that apart from the traditional role that vitamin D3 plays in the skeletal system, it can stimulate the immune, muscular, nervous and many other systems. Our study aimed to evaluate *in vivo* the effect of supplementation with two different doses of vitamin D3 of the diet of female rats on the expression of genes in the prefrontal cortex of the brain. Adult female Wistars rats (approximately one year of age) received a diet without supplementation (group I, n=6) and a diet supplemented with 1000 U / Kg of vitamin D3 (group II, n=6) and 5,000 U / Kg (group III, n=6). 6) for three months. After the experiment was finished, the animals were killed, and their brains were completely placed in RNAlater (Thermofisher Scientific). RNA isolation was performed from the same fragment of the prefrontal cortex from all individuals. NGS mRNA 3'UTR libraries were then generated using the kit (Lexogen). Next-generation sequencing on the Nextseq 5500 (Illumina) instrument yielded approximately 10 million raw reads per sample. Bioinformatic analysis using the Deseq2 program identified 1,115 differentially expressed genes (DEGs) between groups I and III (p adjusted <0.05, no fold change cut off), most of which (923) were downregulated in the group receiving 5,000 U / Kg vitamin D3 in the diet. Between groups II and III, 1,337 DEGs were identified, of which 1,045 were downregulated in group III. No genes with altered expression were found between groups I and II. These results suggest that while the standard dose of vitamin D3 (1000 U / Kg) has little effect on the brain transcriptome, a high dose of 5,000 U / Kg can induce large changes in the brain transcriptome.

Feeding the larvae: the first results of a large scale mealworm ring test

D. Deruytter[1], L. Gasco[2], M. Van Peer[3], C. Rumbos[4], S. Crepieux[5], L. Hénault-Ethier[6], T. Spranghers[7], W. Yakti[8], C. Lopez Viso[9], C.L. Coudron[1], I. Biasato[2], S. Berrens[3], C. Adammaki-Sotiraki[4], A. Resconi[2], S. Bellezza Oddon[2], N. Paris[6] and C. Athanassiou[4]

[1]Inagro, Ieperseweg 87, Rumbeke-Beitem, Belgium, [2]University of Turin, Largo P. Braccini 2, Grugliasco, Italy, [3]Thomas More, Kleinhoefstraat 4, Geel, Belgium, [4]University of Thessaly, Phytokou Str., 38446, Volos, Greece, [5]Invers, Champ de la Croix, Saint-Ignat, France, [6]INRS, Rue de la Couronne 490, Québec, Canada, [7]Vives, Wilgenstraat 32, Roeselare, Belgium, [8]Humboldt University of Berlin, Lentze allee 55/57, Berlin, Germany, [9]University of Nottingham, Sutton Bonington Campus, Sutton Bonington, United Kingdom; david.deruytter@inagro.be

Tenebrio molitor is one of the main insects used for food and feed. Unsurprisingly, this results in a plethora of studies assessing the optimal feed for this species. Due to the differences in experimental designs of the studies, it is not easy to compare the results and draw conclusions. To improve the comparability, an international consortium has created a standardized feed experiment protocol. Initially, a theoretical protocol was constructed based on the experiences of the partners ensuring that the protocol was scientifically sound, easy to execute and relevant for the industry. Some of the standardized elements were: larval age and density, control feed, amount of feed, etc. Summarised: the experiment starts with 10,000 four-week old larvae in a 60×40 cm crate with wheat bran and agar as feed (6 replicates). Growth progression is checked every week. The experiment is terminated when feed is depleted or when more than 10% pupae are present. In a second stage, the protocol was put to the test by the different partners to determine the practical feasibility, repeatability and reproducibility. This was done for two feeds: a standardized wheat bran (identical for all partners) and a locally-sourced wheat bran. Initial results indicate that the protocol is practical but there are noticeable differences in the data obtained among partners. The initial weight of the larvae plays herein an important role. Furthermore, for some partners a growth difference among mealworm growth performance on standardized wheat bran was observed. This indicates a further need for feed standardization.

Low-value agricultural by-products as ingredients for black soldier fly larvae

M.T. Klaassen[1], E. Spoler[1], L. Star[1], J.L.T. Heerkens[1] and T. Veldkamp[2]

[1]Aeres University of Applied Sciences, Department of Applied Research, De Drieslag 4, 8251 JZ Dronten, the Netherlands, [2]Wageningen University & Research, Wageningen Livestock Research, Postal code 338, 6700 AH Wageningen, the Netherlands; m.klaassen@aeres.nl

From the perspective of circular farming, the black soldier fly (BSF – *Hermetia illucens* L.) is able to convert low-value agricultural by-products into high-value food and feed ingredients. In this work we studied the effect of six different mixture diets composed of agricultural by-products consisting of: processed onion peels, mushroom growth medium, malt germ and beer brewers grains. Effects were studied in a single rearing trial that was carried out using three biological replicates per diet formulation. Growth, yield and mineral composition analyses were performed on BSF larvae and their diets. We observed that our diet formulations showed positive correlations between BSF larvae fresh weight yield and diet protein content (r=0.644), magnesium and calcium contents. Moreover, BSF larvae dry matter levels also showed significant ($P<0.05$) positive correlations with total contents of phosphorus, (r=0.692) potassium (r=0.636), magnesium (r=0.716) and calcium (r=0.771). The single BSF larvae weights of all six formulated diets were lower than the control diet. Total BSF larvae fresh weight yields were lower than the control diet. Overall our findings suggest that these low-value agricultural by-products may be applied as feasible ingredient options for rearing BSF for alternative protein applications in food and feed.

Session 18 — Theatre 3

Crate level climate control for BSFL rearing

J.W.M. Heesakkers
Bühler, Insect Technology, Bronland 10P, 6708 WH Wageningen, the Netherlands; janwillem.heesakkers@buhlergroup.com

Bühler Insect Technology provides solutions for rearing and processing of Black Soldier Fly Larvae (BSFL). Like other insect rearing solutions, also in our solution rearing of BSFL is done in crates which contain the feed for the larvae. Next to the nutritious composition of this feed, the climate conditions around the BSFL are an important factor to optimize growth. For example temperature and humidity need to be right to achieve the maximum growth rate and yield per crate. Several different biological and physical processes affect these climate conditions inside the rearing crate. In this presentation these processes and the effect on the climate conditions in a single crate will be elucidated. Also the way how to control these conditions inside the crate is further clarified. Finally our approach for scale up will be explained. Controlling the climate in a single crate can be different than controlling the climate in a large room completely filled with crates. By performing rearing trials in our Insect Technology Center we are able to collect data on a pilot scale and to define the right conditions for climate control at an industrial size.

Session 18 — Theatre 4

Novel heat-treatment sufficiently controls *Varroa* levels in honeybee colonies

C. Sandrock[1], J. Wohlfahrt[1], M. Hasselmann[2], W. Brunner[3] and P. Brunner[3]
[1]Research Institute of Organic Agriculture (FiBL), Livestock Sciences, Ackerstr. 113, 5070 Frick, Switzerland, [2]University of Hohenheim, Institute of Animal Science, Garbenstr. 17, 70599 Hohenheim, Germany, [3]Vatorex AG, Klosterstr. 34, 8406 Winterthur, Switzerland; christoph.sandrock@fibl.org

For decades, the honeybee, *Apis mellifera*, has suffered from severe colony losses due to the ectoparasitic mite *Varroa destructor*. Routine chemical treatments fail to mediate equilibrium and often comprise substantial trade-offs like side-effects on the host, parasite-resistance and residues in bee products. Research on alternative, minimally invasive yet resilient measures is scarce. Lower mite tolerance to high temperatures compared to its host has the potential to kill adult parasites or interrupt reproduction within bee brood cells, but readily implementable applications have so far been lacking. We investigated a high-tech approach based on controlled interval-heating of wires integrated into individual combs beeswax foundations. Three apiaries, each with an average of six control and treatment colonies were monitored for two years. Seasonal colony growth dynamics were documented (estimates of worker and brood populations plus pollen and honey stores) and complemented by regular tracking of *Varroa* infestations. Impacts on various brood stages were evaluated by comprehensive digital brood assessment. Moreover, we conducted dedicated transcriptomic and pathogen profiling for pupae and nurse bee samples. Our field study proves that this innovative heat-treatment permits maintaining mite levels below critical thresholds across seasons. Winter reference-curing (oxalic acid) applied to both groups revealed similar infestations of heat-treated and control colonies, the latter previously being subjected to summer-treatments (formic acid). However, increased mortality particularly of early non-target brood stages, corroborated by altered pupal gene expression, appears to trigger excessive brood compensation of heat-treated colonies, overall translating into relatively decreased worker populations during peak growth, and consequently lower honey harvests. Further refining computational sensitivity of the highly promising prototype may adequately enable this system to achieve widespread successful applicability.

The effect of rearing substrate on black soldier fly larval fatty acid profile: a meta-analysis

E.Y. Gleeson and E. Pieterse
Stellenbosch University, Animal Sciences, Mike de Vries Building, Merriman ave, Stellenbosch Central, 7600, Stellenbosch, South Africa; 19539037@sun.ac.za

The black soldier fly (*Hermetia illucens*) has been well researched as a protein source for both animal feed and human consumption, as it could augment more traditional protein sources such as soybean meal, which has a large environmental impact. The black soldier fly larvae are also high in lipids and have the potential to be utilized as a source of energy in the form of fatty acids. The use of black soldier fly larvae as a lipid source is however less well researched. Studies have shown that the larval fatty acid profile is affected by multiple factors including: nutrition, age and environmental conditions. Numerous studies have investigated the effect that the rearing substrate composition has on the larval fatty acid profile. However the extent to which individual fatty acids are affected in still unclear. Therefore a meta-analysis of published research was conducted to afford more clarity to the effect. Data were collected and consolidated using the systematic review protocol as described by the Cochrane Handbook for Systematic Reviews and Interventions. All quantitative data regarding the larval fatty acid profiles were collected from each publication. The software Review Manager version 5.4.1 was used to perform random-effects meta-analyses. Each meta-analysis consisted of a comparison of the fatty acid concentration differences reported by each publication. Due to heterogeneity the data were transformed using the Freeman-Tukey double arcsine transformation. The meta-analysis results were reported on the transformed scale. The results indicated that the concentrations of many of the fatty acids are significantly affected by the rearing substrate composition. The largest effect was found for lauric acid concentration. Subsequently the level of saturation, in other words the total amount of saturated-, monounsaturated- and polyunsaturated fatty acids, are also significantly affected by the rearing substrate composition. The secondary research data of this meta-analysis could contribute to more dynamic larval nutrition thereby tailoring the larval fatty acid profile to its intended purpose, whether that be animal feed, human consumption, biomedical or industrial such as biofuel production.

Effect of six isoenergetic, isonitrogenous, isolipidic, waste-based substrates on adult life traits

A. Resconi, S. Bellezza Oddon, I. Biasato, Z. Loiotine, C. Caimi and L. Gasco
University of Turin, Department of Agricultural, Forest and Food Sciences, Largo Paolo Braccini 2, 10095 Grugliasco, Italy; andrea.resconi@unito.it

Black soldier fly farms often use a closed-loop system, where the eggs for the new offspring are produced within the facility. The effects of the diet can have a strong impact on the adult flies and, therefore, the supply of eggs. This study aims to assess the effects of six isoenergetic, isonitrogenous, isolipidic waste-based diets (D1-D6), formulated using a combination of 21 ingredients, on flies. Gainesville diet (GA) was used as control. A total of 4,200 6-day-old larvae were weighed in groups of 5 larvae and divided in 42 boxes (6 replicates/diet, 100 larvae/replica) containing 1.8 g of diet per larva. Larvae were allowed to pupate, and when found, pupa weight (PW, g) and time of development from larva to pupa (L-P, days) were noted. Upon emergence, flies' weight (FLW, g), sex (S), pupation time (P-F, days) were recorded, and, at fly's death, the lifespan (FLS, days). Finally, emergence rate (ER, %) and the sex ratio (SR) were calculated. D2 and D5 yielded the heaviest pupae, but the L-P time was slower in D2 than D5 ($P<0.05$). Sex only affected FLS, where females lived longer than males ($P<0.05$). The FLW was higher in D5 and D2, and these treatments also had the longest P-F time ($P<0.05$). D4, D3 and D1 had intermediate results, while D6 had the smallest flies ($P<0.001$). FLW had an influence on the FLS, since D6 showed the shortest lifespan, while D1, D2, D4 and D5 had greater FLS results ($P<0.05$). D3 showed an intermediate result among D6 and the other treatments. The ER was highest in D1, D2, D3, followed by D4 and D5, which had an ER similar to D1 and D3. D6 showed a similar ER to D3, D4 and D5, ($P>0.05$). Sex ratio was always male-oriented, with D3 being the treatment with most males and D4 the one with least ($P<0.05$). In conclusion, even diets are comparable from a nutritional point of view, other diet-linked factors, such as granulometry, water holding capacity and pH, can play an important role on the development of larvae and, in turn, on adult performance. These factors should be taken into consideration when formulating a commercial substrate.

Growth and chemical composition of BSF larvae on biowastes containing similar crude protein content

T. Veldkamp[1], K. Van Rozen[2], E.F. Hoek-Van Den Hil[3], P. Van Wikselaar[1], A. Rezaei Far[1], I. Fodor[1], H.J.H. Elissen[2] and R.Y. Van Der Weide[2]
[1]Wageningen Livestock Research, De Elst 1, P.O. Box 338, 6700 AH Wageningen, the Netherlands, [2]Wageningen Research – ACRRES, Edelhertweg 1, 8219 PH Lelystad, the Netherlands, [3]Wageningen Food Safety Research, P.O. Box 230, 6700 AE Wageningen, the Netherlands; teun.veldkamp@wur.nl

A series of experiments were performed to study black soldier fly larvae (BSF; *Hermetia illucens*) growth performance and safety on different biowaste sources. In this experiment, five biowaste sources were used, solely or in combinations: fast food (FF), mushroom feet (stems) (MF), poultry meal (PM), slaughter waste (SW), and pig manure solids (PS). The following six diets were formulated based on 30% dry matter and 20% crude protein: FF, FF-MF-PM, FF-MF-SW, FF-PS, PS-MF-PM, and PS-MF-SW. Per kilogram (wet) diet, 1,850 BSF larvae (8 days of age, ~12.0 mg fresh weight) were incubated in 18 plastic containers (75×47×15 cm) (each diet in triplicate) in a climate chamber under dark conditions and were harvested at 15 days of age. Larval growth rate in mg/d was highest in FF (18.0^{a}), followed by FF-PS (14.9^{b}), FF-MF-SW (12.3^{c}), FF-MF-PM (10.0^{d}), PS-MF-PM (7.8^{e}), and PS-MF-SW (7.3^{e}). Different superscript letters indicate significant differences ($P<0.05$). Larval crude protein content (% in DM) was highest in FF (52.7^{a}), PS-MF-PM (50.7^{a}), and PS-MF-SW (50.5^{a}), followed by FF-MF-PM (43.7^{b}), FF-MF-SW (42.3^{bc}), and FF-PS (39.4^{c}), showing a variation despite the similar crude protein content in diets. Crude fat content in the diets varied from 31.2% in FF to 5.3% in PS-MF-PM. Larval crude fat content (% in DM) was highest in FF-PS (32.7^{a}), FF (30.3^{ab}), and FF-MF-SW (30.0^{ab}), followed by FF-MF-PM (27.1^{b}), PS-MF-SW (16.9^{c}), and PS-MF-PM (10.0^{d}). Black soldier fly larvae are well-known for their high lauric acid bioaccumulation/biosynthesis as demonstrated in this experiment: lauric acid content (% in DM) varied from 0.06 in FF to 0.24 in FF-MF-SW in the diets, but increased substantially in the larvae, ranging from 1.18 in PS-MF-PM to 9.81 in FF-PS (% in DM; 11 and 30% of crude fat content, respectively). Chemical composition of the diets heavily affected larval growth and chemical composition.

Lipid fraction of the black soldier fly larvae diet: consequences on adult parameters

S. Bellezza Oddon, I. Biasato, A. Resconi and L. Gasco
University of Turin, Department of Agriculture, Forest and Food Sciences, Largo Paolo Braccini, 2, 10095 Grugliasco, Italy; sara.bellezzaoddon@unito.it

The black soldier fly (BSF) nutritional requirements determination represents the basis for the production maximization and, consequently, can guarantee an effective upscale of the insect market. The present study aims to evaluate the effects on adult parameters of increasing lipid levels (L1, 1%, L1.5 1.5%; L2.5, 2.5%, L3.5, 3.5% and L4.5, 4.5% of EE on DM) in isoenergetic and isonitrogenous semi-purified diets. The Gainesville diet was used as environmental control and excluded from the statistical analysis, which was conducted by General Linear Mixed Model (SPSS V28.0, $P<0.05$). A total of 23 pupae per replicate of the larval trial (6 replicates/treatment) were placed individually in the emergence boxes. The pupa-fly (P-F) time, puparium weight (PW), fly live weight (FLW), fly weight reduction (WR), fly life span (FLS), emergence rate (ER) and sex ratio (SR) were recorded. The longest FLS was recorded in the L2.5 treatment and the shortest in the L1.5 group ($P<0.05$), while the L1, L3.5 and L4.5 did not differ from the other experimental diets. The L1, L1.5 and L4.5 showed a greater WR than L3.5 ($P<0.05$), while L2.5 had analogous WR to all the treatments. The L3.5 FLW was heavier than L1 and L1.5 ($P<0.01$). No differences were observed among the FLW of L1 and L1.5, while the L2.5 and L4.5 only varied from L1 ($P<0.01$). The P-F duration time was shorter in the L1, L1.5 and L2.5 diets when compared to the L4.5 group ($P<0.05$). Finally, the PW, ER and SR were not influenced by the dietary treatments. Contrary to what was observed in the larval stage, it is not possible to identify a lipid level that has optimal effects on the adult stage. Considering the differences herein found in the evaluated parameters, future research will have to be conducted to determine the effects also on production performance. Research was supported by the Fondazione Cariplo project CELLOW-FEEP: Circular economy: live larvae recycling organic waste as sustainable feed for rural poultry (ID 2019-1944).

Valorisation of local agricultural side-streams from Greece for *Tenebrio molitor* rearing

M. Vrontaki[1], C. Adamaki-Sotiraki[1], C.I. Rumbos[1], A. Anastasiadis[2] and C.G. Athanassiou[1]
[1]University of Thessaly, Department of Agriculture, Crop Production and Rural Environment, Laboratory of Entomology and Agricultural Zoology, Phytokou str., 38446 Volos, Greece, [2]Animal Feed Anastasiadi Single Member P.C., Akropotamia, 61100 Kilkis, Greece; crumbos@uth.gr

Agricultural side-streams represent a large pool of mainly untapped and underrated resources. An efficient way to exploit agricultural wastes and valorise them by converting them to high-value nutrient source is to use them as insect growth substrate. However, to enhance the sustainability and the environmental impact of their upcycling by insect bioconversion, locally available organic by-products should be used. Therefore, the objective of this study was to evaluate the suitability of nine agro-industrial by-products locally produced in Greece as feeding substrates for the larvae of the yellow mealworm, *Tenebrio molitor*. Specifically, we tested by-products of maize, sunflower, lucerne, oats, as well as rice hulls, rice bran, brewer's spent-grains, spent mushroom substrate and animal feed mill leftovers. In a first lab screening (in 7.5 cm diameter vials) 50 early-instar larvae were fed with each of the by-products tested and larval growth parameters were recorded until the emergence of the first pupa. Based on the results of the lab trial, the best performing by-products were further singly evaluated at pilot-scale in crates (60×40 cm) with approximately 10,000 larvae. In both trials wheat bran served as control. The results of both the lab and the pilot-scale bioassay showed the suitability of several by-products for the rearing of *T. molitor* larvae. Particularly, the oat and the maize by-product, as well as the brewer's spent-grains, efficiently supported the growth and development of *T. molitor* larvae, although in most cases a lower growth rate was recorded compared to control. In contrast, on the spent mushroom substrate larvae survived and grew poorly. These results aim to boost insect farming and the exploitation of agro-industrial side-streams as insect feeding substrates at a local level. This study is part of the project 'EntoFeed' (MIS 513644) that is co-funded by Greece and European Union under the 'RIS Innovation Call 2020'.

Practical guide to commercial BSF farming

B. Holtermans
Insect Engineers, Nijverheidsstraat 2a, 5961 PJ, Horst, the Netherlands; bob@insectengineers.com

Every day new entrepreneurs stand up wanting to start in the insect industry. They come with great enthusiasm looking how they can start this new business. Assuming they can obtain the right knowledge they also get a reality check. Because it's not a business for a quick win. Like every other business you first need to realize who is your customer? Which sounds logic but I see on a weekly basis that it's not common. Is your focus on food, feed or waste? Do you want to produce as many insect protein possible? Or is your focus to process the largest pile of organic waste? Considerations where every new entrepreneur should be pointed out to. In my abstract the focus is on BSF farming with no focus on food, focus is on feed and waste processing. The feed industry is a cost price driven market with demand of high volumes. Challenge for the BSF farmer is that you can't gradually grow your business. It starts with a pilot facility (f.e. 2 tons of wet waste per day) and there you make the step to f.e. 50 tons of waste per day. So from the moment you start, also with a pilot, no matter if you use trays or ZOEM racks. Figure out in advance your next steps. This again sounds logic. But when I started I've seen many examples of farmers who started with trays but after 2-3 years stopped because the plan on how it would look commercially was not taken care of and/or was not viable. That should be lessons learned. The commercial attractiveness of your BSF farm has of course many factors. Location, energy cost, available waste, cost price waste, sales market, etc. Some lessons learned I can share from Sanergy in Kenya. Who in practice today processes 160 tons of organic waste per day. Their focus is processing as much as possible organic waste each day. For the EU grower that focus is less attractive because the choice of which waste streams you're allowed to use is limited. At same time let that be a lesson for the decision makers in EU. A quick focus is putting as much as possible substrate per m^2 floor surface. Which in itself is not bad but everything has a tipping point. That is often a blind spot for the new to become farmer. Air circulation is very important in BSF farming and it also has that tipping point. Practical examples can be shared in Porto.

Bioconversion of olive pomace by black soldier fly larvae into a high-quality frass fertilizer

I.G. Lopes[1,2], T. Ribeiro[1,2], M.A. Machado[2], I. Vieira[2], R. Antunes[2], R. Nunes[2] and D. Murta[1,2]
[1]CiiEM – Centro de Investigação Interdisciplinar Egas Moniz, Campus Universitário, Quinta da Granja, Monte da Caparica, 2829-511 Almada, Portugal, [2]Ingredient Odyssey SA – EntoGreen, Research & Development, Rua de Santo António lote 7, 2005-544, Portugal; iva.lopes@entogreen.com

One of the most challenging organic wastes in Portugal is olive pomace (OLP), a highly phytotoxic sludge generated in olive oil production. Among multiple treatment technologies, the bioconversion process that uses larvae of the black soldier fly (*Hermetia illucens*, BSF) stands up for being a sustainable method that transforms waste into added-value products, namely a larval biomass and a fertilizer (frass). This study aimed at evaluating the bioconversion of OLP with BSF larvae and assessing the quality of the frass, regarding the concentrations of plant nutrients, microbial safety and final concentration of phytotoxic compounds. For this purpose, OLP was mixed with other vegetable by-products and used to feed larvae. Each formulated diet was placed inside 12 plastic boxes (40×60 cm boxes with 13.5 kg each) with 5-day-old larvae and monitored for 12 days. By the end of the process, larvae were separated from frass by sieving, which was chemically and microbiologically analysed according to standard methods. The concentration of phenols was also assessed in both OLP and frass. The chemical characterization of frass revealed that this product has an enormous potential to be used as an organic fertilizer, as it had mean concentrations (DM basis) of 3.1% N_T, 3.9% P, 4.5% K and 84% organic matter, and pH of 8.3. The initial concentration of phenols in the OLP (6.1 g/kg_{DM}) was reduced by more than 67% when bioconverted into frass (2.0 g/kg_{DM}), demonstrating that BSF larvae are able to reduce the phytotoxicity of this organic material. Finally, the process resulted in the absence of pathogenic microorganisms in the frass. It was possible to conclude that BSF larvae is able to bioconvert OLP and transform this phytotoxic waste stream into a nutrient-rich and safe biofertilizer that can be reintroduced in other production processes such as agriculture, promoting the circular economy. Acknowledgements: this study was funded by the project NETA – New Strategies in Wastewater Treatment – POCI-01-0247-FEDER-046959.

Is 16 hours of daylight better than 12 hours for the production of house crickets?

L. Holm and A. Jansson
Swedish University of Agricultural Sciences, Anatomy, Physiology & Biochemistry, Box 7011, 75007 Uppsala, Sweden; lena.holm@slu.se

Crickets are under photoperiodic control but ideal photoperiod for rearing has been less studied. A generally applied regime is 12 h light (L)/12h dark (D), but time to maturity and body weight in a population vary considerably. Variation in sensitivity to photoperiod may exist even within a species depending on which latitude they originate from, as seen in field crickets. Our house crickets (*Acheta domesticus*) originate from wild specimens caught in Sweden (55-65°N) with over 16h of light in the summer. In a pilot study crickets were reared in long days (L=16L/8D, 4 replicates, 25-32 ind/replicate) or short days (S=12L/12D, 3 replicates, 26-36 ind/replicate) at 30±1 °C and 55±5% RH with feed and water *ad libitum*. Days to maturation and number of mature crickets/day were recorded and sexes separated. Females, 7 days post last molt (4/replicate) were mated with males 3 days post last molt and given moist sand to lay eggs. Eggs were collected every third day for 3 times (A, B and C), washed from sand, counted and incubated for 10 days in new sand. Three days after hatch nymphs were photographed and counted. Crickets were euthanized and measured (length and weight) when 80% in a replicate reached maturity. Statistical analysis was performed using ANOVA (SAS, ver 9.4). Differences were considered significant when $P<0.05$. L crickets started to mature five days earlier than S crickets (33 vs 39 days) and reached 80% maturity in 66±14 days whereas it took S crickets 97±6 days to reach 80% mature. S crickets were longer (20.7±0.2 mm) than L crickets (19.9±0.2 mm) and heavier (S=0.45±0.01 g, L=0.40±0.01 g). S crickets laid more eggs in round A and B but not C, with better hatchability in round A compared to L crickets. The results of this study suggest that a species-specific photoperiod can be a tool to increase production by shortening time to maturation in house crickets from our region. Even if a 12L/12D cycle produced heavier crickets the number of batches/year could be increased from three to five with a 16L/8D schedule compared to 12L/12D. Breeding stock might however benefit from a 12L/12D cycle. Further studies are needed to evaluate the effects of photoperiod on both production and reproductive success in house crickets.

Optimizing insect rearing chain through the development of a digital twin for insect rearing

M. Tapio[1], S. Heiska[2], M. Karhapää[3] and J. Niemi[4]
[1]Natural Resources Institute Finland, Myllytie 1, 31600 Jokioinen, Finland, [2]Natural Resources Institute Finland, Yliopistokatu 6 B, 80100 Joensuu, Finland, [3]Natural Resources Institute Finland, Latokartanonkaari 9, 00790 Helsinki, Finland, [4]Natural Resources Institute Finland, Kampusranta 9, 60320 Seinäjoki, Finland; miika.tapio@luke.fi

The environmental footprint of food production and concerns of self-sufficiency has created interest in using novel production insect species as a protein source. To meet the industrial scale in insect production, it is however necessary to enhance production efficiency, reduce labour intensity and markedly decrease the production cost. CoRoSect project introduce a digitalized, integrated solution to insect production. The automated production flow requires monitoring of the farmed insect batches. Anomalies or situations that require action can be recognized e.g. through machine learning models or through modelling the biological processes or through their combination. The main challenge in machine learning based approach is the need for high volumes of data and generalizing the models, while the structural modelling may have difficulties in taking into account specific influential factors. Recently, combination of the two approaches have been suggested as a compromise. In the CoRoSect project, we explore the hybrid digital twin approach. Structural modelling simplifies the machine learning task and makes the transfer of the model easier. In this presentation we will provide an outline of the digital twin for insect rearing. The model uses sensor data from project developed intelligent crates utilizing printed sensors and aims to provide rearing predictions and form an explainable part of the anomaly detection tool. The model integrates with a wider model-driven decision support system and insect farm management system. The tool will be parametrized, tested, and validated with real cases. CoRoSect project has received funding from the European Union's Horizon 2020 research and innovation programme under grant agreement No 101016953.

Do vegetal compounds have a potential on yellow mealworm performances during rearing?

M.S. Fagnon[1], M.E. Gonzales[2,3], V. Ageorges[3], S. Kerros[1], L. Tournier[3], T. Chabrillat[1] and C. Carlu[1]
[1]PHYTOSYNTHESE, Innovation dpt, 57, avenue Jean Jaures, 63200 MOZAC, France, [2]VetAgro Sup, Campus agronomique, 89 Avenue de l'Europe, 63370 Lempdes, France, [3]INVERS, Research and Development, Champ de la Croix, 63720 Saint-Ignat, France; claire.carlu@phytosynthese.fr

This study aims to compare the growth performance of yellow mealworm larvae fed on wheat bran supplemented with eight (8) botanical ingredients. More exactly, 3 classes of botanicals were tested at various inclusion levels, i.e. essential oils (0.5, 5, 50 ppm), plant extracts (1, 10, 100 ppm) and crude dry powders (5, 50, 500 ppm). Each additive was added at 0.1% of the wheat bran control diet and tested in 3 replicates. The measurement of the growth performance of each rearing crate was performed on a weekly basis until the 5th week corresponding to the harvest of larvae. Data clearly showed a significant growth improvement of the crude plant technology on yellow mealworm larvae ($P<0.05$) whereas purified substances did not show significant results compared to the control diet ($P>0.05$). Based on these preliminary data, synergetic effect of combined crude dry powders will be evaluated on growth performance and potentially on larval nutritional values as well. In conclusion, this study highlighted the great potential of the use of plant secondary metabolites (whole plants) as a dietary supplement in mass rearing of yellow mealworm larvae.

Population dynamics during *Tenebrio molitor* pupation

C.L. Coudron and D. Deruytter
Inagro vzw, Insect Research Centre, Ieperseweg, 87, 8800, Belgium; carl.coudron@inagro.be

Mealworm production systems require a continuous replenishment of adults to compensate for fertility loss in older beetles and the loss of deceased adults. Part of the production must therefore be used to maintain the next generation of beetles. Due to biological and environmental variation, not all mealworms grow equally fast or pupate at the same age or body weight. As a result all developmental stages (mealworm, pupae and beetle) will eventually co-exist in a tray at the same time. To avoid this and to prevent mealworms from cannibalizing the freshly formed pupae, separation of pupae from mealworms is preferred. This however requires manual labour before all mealworms have undergone metamorphosis. Furthermore, each time the pupae are handled there is a risk of damaging them resulting in death or deformed beetles. The goal of this experiment was to get a better understanding of the inner-crate dynamics that lead to several weeks of pupation for one crate. To achieve this, mealworms of the same age from Inagro's standard reproduction line were divided based on the width of the larvae by sieving them on a bar sieve (sieve width 2, 2.5, 3 and 3.5 mm). This resulted in 5 different fractions: >3.5 mm (pure pupae), 3-3.5 mm (mealworms of 175 mg), 2.5-3 mm (140 mg), 2-2.5 mm (95 mg) and <2 mm (42 mg). From each fraction 3×100 mealworms were randomly picked and placed together in a container with *ad libitum* feed. Every few days the number of pupae were counted and the sex ratio of each mealworm fraction was determined. Initial results indicate that for each smaller fraction of mealworms approximately a 7 day delay was observed before 50% of pupae were observed. In other words, each crate contains a subpopulation of mealworms that is ready for pupation and other subpopulations that still need some further development.

***Hermetia illucens* and *Tenebrio molitor*: comparison between heat treatments for microbial reduction**

C. Lucchetti, A. Grisendi, P. Bonilauri and M. Scremin
IZSLER, Reggio Emilia, via Pitagora, 2, 42124, Italy; paolo.bonilauri@izsler.it

Insects have acquired increasing importance for both feed and food purposes. In fact, they are starting to be recognized as a possible alternative contribution to ensure world food security. Thus, the analysis of whether edible insects can be considered as safe both for feed and food has become impelling. The aim of the present study was to evaluate and compare the efficiency of several stabilization treatments in controlling the microbial load of several edible insects' products, and to analyse whether it is linked to the species of edible insects used. To do so, larvae of two very different orders of insects, respectively black soldier flies (*Hermetia illucens*) and yellow mealworm (*Tenebrio molitor*), were chosen. Both of them were reared in a controlled environment. Pre-pupae were harvested, rinsed, and frozen at -80 °C. Several stabilization treatments were performed, such as blanching, both alone and in combination with drying (static electric oven for 24 h). The microbial load of each sample was examined by performing total bacterial count (TBC), and enumeration of Enterobacteriaceae, *E. coli*, lactic bacteria, and spore-forming sulphite-reducing anaerobic bacteria. The two initial microbial profiles were quite similar. Their main difference was the presence in *H. illucens* of a conspicuous load of both *E. coli* and sulphite-reducing ana. bacteria, which were not found in *T. molitor*. Blanching was observed to be effective in reducing TBC of both insects of about 7 logarithms. Three different boiling times were tested: 30 sec, 1 min and 3 min. A complete microbial load clearance was observed only for *T. molitor*, when boiled for 3 min. Enterobacterial load dropped in both species starting from 1 min boiling time. When blanching was combined with drying, an additional reduction in TBC of *H. illucens* was observed, although it induced a slight increase in the level of Enterobacteriaceae. On the contrary, when applied to *T. molitor* it resulted in an increase in TBC, which reached 5.3 log vs the 2.3 logs obtained by blanching alone. Overall, all the stabilization treatments tested helped control microbial development. But while for *T. molitor* the most effective stabilization treatment was observed to be blanching, for *H. illucens* it was the combination of blanching and drying.

Session 18 Poster 17

Insect substrates: evaluation of efficiency indices

M. Ottoboni, M. Moradei, L. Ferrari, G. Ceravolo, R. Abbate, A. Luciano and L. Pinotti
University of Milan, Department Veterinary Medicine and Animal Science, DIVAS, Via dell'Università, 6, 26900 Lodi, Italy; matteo.ottoboni@unimi.it

Larvae of the black soldier fly (BSF), *Hermetia illucens* L. (Diptera: Stratiomydae), are able to convert organic material (i.e. waste and by products) into high-quality biomass, which can be processed into animal feed. Although, several studies have been conducted on evaluation of influence of growing substrate on the nutritional value of BSF larvae and prepupae, this field need more accurate investigation. The intent of this work was to collect, synthesize, discuss and review the available information on the substrates used for rearing BSF larvae in literature. Data collected were furtherly elaborated to estimate of the food conversion rate (FCR) and efficiency of conversion of ingested food (ECI) for BSF larvae. In light of this, a systematic review based on 16 publications published from 2008 to 2021 was carried out in order to evaluate the effects rearing substrate on deriving insect's biomass. Based on selected studies it can be suggested that BSF is able to convert organic material (i.e. waste and by products) into high-quality biomass. The role and effects of selected nutrients, such as ether extract/fats, carbohydrates and fibre in the substrate, seem to be key factors in defining the features of the biomass as well as the time needed to reach the harvesting state. With regard to FCR, a wide variability was observed ranging from 2 up to 20, while in the case of ECI value ranged between 3.4% and 54%. In the frame of the present study it has been observed that very low FCR or high ECI seems to be related to longer larvae developing time. The BSF larvae can be considered efficient 'farmed animal' when nutritionally appropriate substrates are used. Furtherly the high levels of protein produced by the BSF larvae make these insects a very important resource for the animal feed formulation.

Session 18 Poster 18

Bioconversion of olive pomace by black soldier fly larvae (*Hermetia illucens*)

R.F. Antunes[1], I.G. Lopes[1,2], T. Ribeiro[1,2], M.A. Machado[1], I. Vieira[1], R. Nunes[1] and D. Murta[1,2]
[1]Entogreen – Ingredient Odyssey S.A., Rua de Sto. António no. 7, 2005-544 Santarém, Portugal, [2]CiiEM – Centro de Investigação Interdisciplinar Egas Moniz, Caparica, Portugal, Caparica, Portugal, Portugal; iva.lopes@entogreen.com

The management of olive pomace (OLP) is a major challenge in the olive oil industry. Bioconversion using black soldier fly (*Hermetia illucens*, BSF) larvae has proven to be a sustainable solution to convert organic waste streams into protein and fertilizer (frass). However, many knowledge gaps exist regarding to this process. The aim of this study was to evaluate the efficiency of OLP bioconversion using BSF larvae and the impacts on development time, larvae yield and weight, and larvae nutritional composition. Diets containing OLP were tested at different inclusion levels (18.5, 46.5, 62 and 81%) and mixed with other locally available by-products as feed substrate for BSF larvae. Two other diets were used as controls, a mixture of vegetables and fruits (VF) and a Gainesville (GV) diet. Each treatment was distributed into 12 plastic boxes (40×60 cm box with 13.5 kg each) with 5-day-year-old larvae and was ended when the first prepupae appeared. Larvae were separated from frass by sieving and the outputs were measured separately (larvae and frass). Substrate reduction, bioconversion rate and feed conversion ratio (FCR) were also calculated for each diet. The development time differed between diets, with shorter times with GV and 18.5OLP and other OLP diets, and longer with the VF diet. The highest larvae yields were observed for the VF, 18.5OLP and GV diets with 1.77,1.51, and 1.50 kg/box, respectively. The larvae yield obtained from the other OLP diets ranged between 1.39-1.51 kg/box. The highest bioconversion efficiency was observed for the VF diet, which presented an 80.2% substrate reduction, 13.1% bioconversion rate and 7.6 FCR. OLP diets had a substrate reduction between 67.3-73.5%, with corresponding bioconversion and feed conversion ratios ranging from 10.3-11.2% and 8.9-9.7. Overall, the data suggest that OLP can be efficiently bioconverted by BSF larvae contributing to improve OLP management. Chemical composition of BSF larvae was still being analysed at the time of submission of this abstract.

Evaluation of plastic degradation capacity by black soldier fly (*Hermetia illucens*)

M.A. Machado[1], *I. Lopes*[1] *and D. Murta*[1,2]
[1]*Ingredient Odyssey-EntoGreen, ID, Rua de Santo António, Lote 7. Zona Industrial de Santarém, 2005-332 Santarém, Portugal,* [2]*CiiEM – Centro de Investigação Interdisciplinar Egas Moniz, ID, Campus Universitário, Quinta da Granja, 2829-511 Monte de Caparica, Almada, Portugal; iva.lopes@entogreen.com*

The project RECOVER focuses on biotechnological solutions using microorganisms, novel enzymes, insects and earthworms to remediate plastic pollution and to convert non-recyclable agri-food waste plastic (AWP) into biofertilizers and biodegradable plastics for agricultural and food packaging applications. In this sense, this study aimed at evaluating the degradation capacity of Black Soldier Fly larvae (BSF) *Hermetia illucens* L. (1758) in consuming LLDPE and LDPE plastics. The experiment was conducted using 5-day-old BSF larvae from EntoGreen's® colony, located in Santarém, Portugal, and the plastics tested was presented as powder (2 mm particle size), both as virgin and in post-industrial format. Four treatments with 4 replicates each were tested, a control treatment only with Gainesville (GV) and 3 others with plastic inclusion of 50, 90 and 100%, where the remaining part was constituted proportionally by GV. A total of 100 larvae were inoculated in each individual tray and keep at approximately 23 °C for 14 days. On days 1, 4, 7, 10 and 14 larvae samples (n=10) were collected to be weight. At the end of the experiments, the larvae were manually counted and weighted, as well as the quantity of material left in each tray. Larvae survival was evaluated by the Student's T-test, and a *one-way* ANOVA followed by a Tukey test were used to evaluate the remaining data (5% probability level). Considering all treatments, the plastic inclusion levels 90 and 100% did worse, with all larvae perishing before the end of the trial, while 50% treatment had a 94.2% survival rate. Larvae grew poorly among 50% plastic treatment (0.11 g/ larvae) with the control treatment generating the highest growth over time (0.190 g/larvae). We conclude that there was no evidence that BSF larvae are able to consume LLDPE and LDPE plastic powder, since all plastic treatments resulted in larvae with the lowest final weight and lower development rates. This work receives funding from the Horizon 2020 Research Innovation Programme granted by the Bio based Industries Joint Undertaking (GA 887648)

Mealworm rearing on substrates enriched with functional ingredients of aromatic and medicinal plants

M. Gourgouta[1], *S.S. Andreadis*[2], *E.I. Koutsogeorgiou*[2,3], *C.I. Rumbos*[1], *G. Skoulakis*[4], *I. Giannenas*[5], *K. Grigoriadou*[2], *E. Bonos*[6] *and C.G. Athanassiou*[1]
[1]*Laboratory of Entomology and Agricultural Zoology, University of Thessaly, Volos, Greece,* [2]*Institute of Plant Breeding and Genetic Resources, Hellenic Agricultural Organization – DEMETER, Thermi, Greece,* [3]*Laboratory of Applied Zoology and Parasitology, School of Agriculture, Aristotle University of Thessaloniki, Greece,* [4]*AgriscienceGEO, Thermi Industrial Area, Thessaloniki, Greece,* [5]*Laboratory of Nutrition, School of Veterinary Medicine, Aristotle University of Thessaloniki, Greece,* [6]*Laboratory of Animal Production, Nutrition and Biotechnology, University of Ioannina, Arta, Greece; crumbos@uth.gr*

The enrichment of insect diets with functional ingredients may greatly enhance insect development and growth. In this context, the objective of the present study was to study the development of larvae of the yellow mealworm, *Tenebrio molitor*, the superworm, *Zophobas morio*, and the lesser mealworm, *Alphitobius diaperinus*, on substrates enriched with functional ingredients of aromatic and medicinal plants of the Greek flora. Particularly, wheat bran enriched with 10, 20 or 30% of two mixtures (A and B) was evaluated as insect feeding substrate, whereas wheat bran alone served as control. Mixture A: Post-distillation residues of aromatic plants, e.g. oregano, thyme, sage and rosemary, linseed oil, rock-samphire, industrial cannabis and olive paste by-product. Mixture B: Mixture A amended with essential oils of the abovementioned aromatic plants. According to our results a 10% inclusion level of mixture A led to a 32.9% increase in the final individual *T. molitor* larval weight, whereas 10% inclusion of mixture B increased the final larval weight by 4.4%. A positive effect of the dietary inclusion of the mixtures was recorded in most cases also for *A. diaperinus* and *Z. morio.* Our results aim to promote insect production for food and feed by enriching insect feeds with functional ingredients with desired properties. This research has been co-financed by the European Regional Development Fund of the European Union and Greek national funds through the Operational Program Competitiveness, Entrepreneurship and Innovation, under the call RESEARCH – CREATE – INNOVATE (Project Code: T2EΔK-02356, Acronym: InsectFeedAroma).

Session 18 Poster 21

Counting of insect larvae using computer vision

F. Carvão[1], D. Monteiro[1,2], D. Alexandre[1,3] and V. Filipe[1,3]
[1]Universidade Trás os Montes e Alto Douro, QTA de Prados, 5000-801 Vila Real, Portugal, [2]Universidade Trás os Montes e Alto Douro, CECAV, QTA de Prados, 5000-801 Vila Real, Portugal, [3]INESC TEC, R. Dr. Roberto Frias, 4200-465 Porto, Portugal; f.carvao95@hotmail.com

Animal feed industry is increasingly using alternative animal protein sources increasing the demand of this product. In order to achieve higher productivity and efficiency in insect production systems, precision livestock farming can be used to solve some problems. One of the crucial issues related to insect production is the measuring (phenotypic selection) and counting of insect larvae, especially when they are transferred from breading to rearing stage. Since in insect farming this task can be performed by automated systems, there is a demand for both hardware and software to address the task. Some companies have introduced insect larvae counting technology on the market, based on different systems achieving high accuracy rates. However, the larvae counter technology faces some difficulties detecting and/or identifying dead, small or overlapping larvae. In this work, a computer vision system is proposed to identify and count the number of larvae in photos captured in a laboratory experimental setting. Random larvae (different ages and sizes) spread on retro illuminated conveyor were shoot with a digital camera (Nikon d3400) always using the same settings (distance – 54.5 cm, light – only retro illuminated light/ back light, focus – 24L) relative to larvae, on a dark room. The RGB images of the larvae, 4,000×6,000 pixels were processed using a combination of classical algorithms from the ImageProcessing toolbox from MatLab®. After reading the image file, a conversion to gray scale is made followed by a standard crop and global thresholding. This reduce the image to the area of interest improving software performance and reducing processing time and also improves contrast and allow better object identification. Then a sequence of erosion/dilatation steps are applied trying to separate overlapped larvae. Finally, the 'skeleton' for each object is determined and each one of those is labelled allowing to count the number of objects in the image among other properties like position, area, convexity. The algorithm was tested in 14 images having an average accuracy of 67% with a precision of 97% and a recall of 85%.

Session 18 Poster 22

Black soldier fly larvae as up-cyclers of organic foodwaste

I. Noyens, S. Goossens, L. Frooninckx, L. Broeckx, A. Wuyts and S. Van Miert
Thomas More University of Applied Sciences, Research – Radius, Kleinhoefstraat 4, 2440 Geel, Belgium; isabelle.noyens@thomasmore.be

The continuous growth of the global human population, associated with an increased demand for natural resources and the generation of global waste has a negative impact on our planet. In the last decades an evolution toward sustainable and environmental friendly systems has emerged. Whereas many people consider plants as sustainable and environment-friendly sources for food, feed and bio-materials, this perception is not always correct as plants require massive amounts of land for agriculture, negatively impacting deforestation and loss of biodiversity. Insects may fulfil an important role in a circular economy by up-cycling low-value organic side-streams into valuable biomass. Especially Black Soldier Fly larvae (BSFL) are intensively investigated for their role as a waste converter. BSFL are a non-pest species that can convert diverse organic waste streams into biomass that can be used for the production of feed, bio-materials for biodiesel production, technical applications such as surfactants and protein-based bio-plastics. BSFL can thus play an important role in organic waste reduction and renewable bio-material production, and can contribute to the EU Green Deal strategy of reducing food waste by 50% by 2030. However, LCA analyses indicate that the production of BSFL can only be sustainable and economically viable if they can be reared on low quality biomass such as manure and mixed organic wastes. Here we present our data on the use of BSFL for the conversion of food waste composed of post- and pre-consumer food waste from catering, food industry and supermarkets. Samples were taken every month for a whole year and were physico-chemically characterized. Proximate analysis, pH and sediment/free water were determined to understand the variability of swill composition throughout the year. Subsequently BSFL were reared on these samples with different compositions, and were characterized for growth, survival and food conversion capacity. The results of this experiment will be revealed in this presentation. This work is part of the Upwaste project (EU FACCE SURPLUS – Vlaio LA traject HBC.2019.0028).

Session 19 Theatre 1

Influence of different fixation times of sows in farrowing pens on their behaviour and piglet losses

B.-M. Baude[1], K. Krugmann[1], S. Diers[2], E. Tholen[3] and J. Krieter[1]
[1]Institute of Animal Breeding and Husbandry, Christian-Albrechts-University Kiel, Olshausenstr. 40, 24098 Kiel, Germany, [2]Chamber of Agriculture of Schleswig-Holstein, Gutshof 1, 24327 Blekendorf, Germany, [3]Institute of Animal Breeding Science, University of Bonn, Endenicher Allee 15, 53115 Bonn, Germany; bbaude@tierzucht.uni-kiel.de

Permanent fixation of sows in farrowing crates has long been criticized for animal welfare. Alternative farrowing systems have been researched but are often related to high piglet losses. This study investigated short-term fixation in farrowing pens with movable crates compared to permanent fixation in farrowing crates and aimed to examine whether crate opening time or previous experience of sows being fixed or not have an influence on piglet losses and crushed piglets after crate opening. Thus, movable crates were opened among different times in three groups (AM, PM, CON): Three days after farrowing at 6 am (AM: n=107) respectively at 6 pm (PM: n=108) or remained closed for control group (CON: n=88). All litters were standardized to 14 piglets within two days post partum. Results showed significant differences of piglet losses in AM- and PM-group compared to CON-group (AM: 17.1%; PM: 15.8% vs CON: 12.1%; P<0.05) while no significant differences were found for crushed piglets. However, in AM-group 22.9% of all crushed piglets caused within the first two days after crate opening were significant higher compared to the CON-group in this time (AM: 22.9% vs CON: 13.1%; P<0.05; AM: 22.9% vs PM: 19.8%; P>0.05). Piglet losses and crushed piglets did not differ significant for sows with or without previous experience in farrowing pens within groups or experience level (AM: 18.6% vs 14.6%; P>0.05; PM: 16.6% vs 14.6%; P>0.05; respectively AM: 7.94% vs 5.83%; P>0.05; PM: 7.69% vs 4.47%; P>0.05). However, results show that sows might adapt better to crate opening in the evening with no disturbance due to stockpersons. In the current study, video data is analysed to examine laying down and rolling over behaviour of sows in order to verify critical situations of crushing piglets and to prove the influence of different opening times. In addition, moving sensor data related to video data will be evaluated to investigate nest-building behaviour and predict farrowing date.

Session 19 Theatre 2

Does reducing time in gestation stalls in early pregnancy affect sow welfare and performance?

M.C. Galli[1], M.E. Lagoda[2,3], F. Gottardo[1], B. Contiero[1] and L.A. Boyle[3]
[1]University of Padova, Department of Animal Medicine, Production and Health, Viale dell'Università 16, 35020 Legnaro (PD), Italy, [2]Institute of Genetics and Animal Biotechnology of the Polish Academy of Sciences, Department of Animal Behaviour, ul. Postępu 36A, Jastrzębiec 05-552 Magdalenka, Poland, [3]Teagasc Moorepark, Animal & Grassland Research & Innovation Centre, Fermoy, P61 P302 Co Cork, Ireland; m.costanza.galli@gmail.com

Under EU legislation, sows can be kept in stalls for 28 days post-service but there is growing societal pressure to ban confinement systems. Sows fight to establish a dominance hierarchy at mixing. This poses welfare and productivity concerns particularly in early pregnancy when the pregnancy is less well established and sows are still recovering from the demands of lactation. The aim of this study was to compare the effects of day of mixing (2 (PEN2) or 28 (PEN28) days post-service) on sow welfare (measured at mixing, 1, 7 and 22 days), reproductive performance and litter size. Sows (n=48, parity 1-4) were housed in stalls from weaning until mixing into groups of 12 in pens (2.25 m^2/sow) with one full-length self-closing feeding stall each. Floors were fully-slatted and two blocks of wood on chains were provided. All occurrences of aggressive behaviour were counted directly for 2 hours. Skin lesions caused by fighting were counted at three locations on the left and right hand sides of the sows body. Sow locomotory ability was assessed before mixing and at the middle and end of pregnant using a visual analogue scale. Data were analysed using SAS. There were no treatment effects on aggression (P>0.001), skin lesions, or locomotory ability (P>0.05). Furthermore there were no treatment effects (P>0.05) on reproductive performance (PEN2 vs PEN28 – pregnancy rate: 88% vs 96%; farrowing rate: 88% vs 92%), or litter size (PEN2 vs PEN28 – total born: 15.3±0.63 vs 15.7±0.61; born alive: 14.2±0.80 vs 13.1±0.78; stillbirth: 8% vs 6%; mummified: 0% for both). Early mixing post-service did not have a negative impact on sow welfare or performance. It is likely that the feeding stalls protected sows from aggression. In conclusion, sows can be mixed into a housing system with feeding stalls two days post-service without any detrimental impact on welfare or herd productivity.

Welfare indicators for pigs within a national monitoring – results of survey and application on farm

K. Krugmann[1], J. Johns[2], D. Frieten[3] and J. Krieter[1]
[1]Institute of Animal Breeding and Husbandry, Christian-Albrechts-University Kiel, Olshausenstr. 40, 24098 Kiel, Germany, [2]Thünen Institute of Farm Economics, Bundesallee 63, 38116 Braunschweig, Germany, [3]Thünen Institute of Organic Farming, Trenthorst 32, 23847 Westerau, Germany; kkrugmann@tierzucht.uni-kiel.de

Well-founded reporting on the status quo of animal welfare in livestock husbandry is currently not available in Germany, e.g. in order to provide political actors reliable information for knowledge-based decisions. Therefore, the basis for regular indicator-based monitoring of the status quo and process of animal welfare is being developed in the project National Animal Welfare Monitoring. For this purpose, indicator concepts for assessing welfare status in pig husbandry are being evolved among other livestock species and topics. Thus, after intensive literature research which results are published as web application under www.ktbl.de, expert discussions and an online survey with associated professions, indicator concepts for fattening pigs, sows, suckling and rearing piglets were designed. To examine categorical relationships between answer options (yes/no/uncertain) of interviewed experts (n=64) whether the indicators are suitable or not chi-square test of equality ($P<0.1$) was performed. Thus, the indicators shivering and hernias revealed to be less appropriate indicators to assess the welfare status in rearing piglets as there was no significant majority of experts that approved their suitability (42.6/44.7/12.7% resp. 38.3/46.8/14.9%). In contrast, e.g. injuries at tails and ears were determined as suitable indicators since the significant majority of experts answered 'yes' to these indicators (91.5/2.1/6.4% resp. 93.6/2.1/4.3%). On the same principle, the interviewed experts evaluated rate of farrowing and social behaviour (41.2/49.0/9.8% resp. 39.2/49.0/11.8%) to be less appropriate for measuring welfare in sows within a national monitoring. For fattening pigs, the consulted experts rated e.g. hernias and human animal relationship tests as less useful indicators (40.6/43.8/15.6% resp. 39.1/45.3/15.6%). In consideration of these pre-studies, compiled indicator concepts are presently tested holistically on pig farms with regard to their reliability and practicability as part of national animal welfare monitoring in Germany.

Selection of applicable variables as indicators for tail biting using a partial least squares model

V. Drexl[1], I. Dittrich[1], T. Wilder[1], S. Diers[2] and J. Krieter[1]
[1]Christian-Albrechts-University, Institute of Animal Breeding and Husbandry, Olshausenstraße 40, 24098 Kiel, Germany, [2]Chamber of Agriculture of Schleswig-Holstein, Gutshof, 24327 Blekendorf, Germany; vdrexl@tierzucht.uni-kiel.de

Early detection of tail biting using applicable variables potentially helps solving one of the major welfare issues in rearing (REA) and fattening (FAT) of pigs. The aim of the study was to determine which variables potentially indicate tail biting. Sensor data such as water consumption, activity and climate data as well as daily animal checks were collected. Tail biting behaviour was investigated in six pens for 47 days in REA (12 batches) and 84 days in FAT (5 batches) with tail lesions being scored twice a week in both periods. The evaluations were carried out at pen level including 1,008 scoring days (SCD) in REA and 624 SCD in FAT. In a partial least squares (PLS) model, the prevalence of tail lesions (21.3% REA and 20.3% FAT) was used as dependent variable and the values of all variables of 3 to 4 days before a certain SCD were used as predictors. Therefore, all variables between two SCD were aggregated and assigned to the following SCD. 26 variables were tested for their influence on the prevalence of tail lesions in a PLS model for REA and FAT. In REA, twelve variables were selected and three factors were extracted that explained 57.1% of the variance. In FAT, ten variables were included in the PLS model resulting in four extracted factors and 73.5% explained variance. In both REA and FAT, selected variables are SCD, the number of hanging tails per pen, activity, water consumption, CO_2 concentration, the largest difference of minimum and maximum temperature-humidity index, the exhaust air rate and the number of lame animals and skin lesions. Additionally, the treatment index pre-weaning and in REA and the number of tear stains were included in REA, whereas in FAT, the maximum difference in temperature was selected. No explanation was provided by weight at weaning or end of REA, respiratory and gastrointestinal issues as well as indoor and outdoor temperature, humidity and NH_3 concentration. The results indicate the presence of a strong relationship between tail lesions and the selected variables, thus a neural network will be computed using these variables.

Session 19 Theatre 5

Behavioural differences of tail-biting pigs and their victims before a tail-biting incident

T. Wilder, V. Drexl and J. Krieter
Christian-Albrechts-University, Institute of Animal Breeding and Husbandry, Olshausenstr. 40, 24098 Kiel, Germany; twilder@tierzucht.uni-kiel.de

Tail-biting behaviour in pigs is one of the major animal welfare issues in modern pig husbandry. Because of its multifactorial genesis, it is difficult to prevent entirely and early identification is important to take measures in time. The aim of this study was to find differences in the behaviour of pigs which become a biter or a victim. The behaviour was analysed during the light hours of the previous 48h before a tail-biting incident using video footage. For each of the 24 tail-biting incidents, three animals were observed (biting pig, control pig with small tail lesion (TL), control pig with large TL). The individual behaviour of each day was summarised either as duration of behaviour per 60 min or as number of performed behaviour per 60 min which resulted in a total of 144 observations. Each behaviour was used as the dependent variable in a generalised linear model. In each model, the animal group (biter, small TL, large TL) and day (-2, -1) were used as fixed effects and the pen was used as a random effect. The interaction between group and day was tested and the age and weight at weaning were added as covariates. The results show significant differences ($P<0.05$) for active behaviour between biter and large TL (biter: 34.3 min, large TL: 26.9 min), for body posture changes between large TL and biter/small TL (biter: 14.9, small TL: 15.4, large TL: 22.8), for manipulating environment between small TL and large TL (small TL: 15.0 min, large TL: 10.4 min), for manipulating pen mates between biter and large TL (biter: 3.39 min, large TL: 2.03 min) and for receiving manipulative behaviour between biter and small/large TL (biter: 1.19 min, small TL: 2.26 min, large TL: 2.32 min). For the group small TL, there were significantly more body posture changes on day -1 (day -2: 12.7, day -1: 18.6). Thus, there are behavioural differences between the animal groups. However, these differences characterise mainly pigs with TL and less the biting pigs. Furthermore, the differences between the days were not as big as expected, as the changes may have occurred before day -2. Therefore, the behaviour should be further investigated, especially with a focus on changes over a longer time period.

Session 19 Theatre 6

***Ex vivo* model of porcine foot to assess biomechanical effects of flooring and claw trimming in sows**

E. Telezhenko and A.-C. Olsson
Swedish University of Agricultural Sciences, Department of Biosystem and Technology, Sundsvägen 14, 230 53 Alnarp, Sweden; evgenij.telezhenko@slu.se

The aim of the project was to develop an *ex vivo* model of porcine foot to investigate the biomechanical effects of claw trimming and flooring on claws in sows. The laboratory model used sows cadaver foot specimen mounted in a special-constructed rig with hydraulic actuator, where the vertical load as well as the load the deep and superficial flexor tendons were controlled with load cells and were simultaneously loaded to ensure emulation of foot loading during the middle of the stance phase. Pressure and load distribution at the floor-foot interface was determined using Tactile Force and Pressure Measurement system Tekscan and thin (0.10 mm), flexible pressure sensor with 15.5 sensing elements per cm^2 (5,101 CMP) which was protected by a 0.12 mm thick Teflon cloth and placed between the foot and flooring surface. For the studies, back feet from eight euthanized sows with overgrown and/or asymmetrical hooves were used. Between and within claw contact pressure and load distribution were measured when foot was vertically loaded with 480 and 1,060 N (that represented the maximal vertical load while standing and walking). The measurements were made on different floor materials (concrete and rubber mat), and before and after functional claw trimming. Untrimmed sow feet had highest load and contact pressure directed to rear part of the outer claw. Rubber mats resulted in lower average and peak contact pressure compared to concrete floor, but did not have any significant effect on load distribution within and between the claws. Functional claw trimming resulted in shorter claws with a steeper toe angle, with relatively greater load in the front part of the claws as a result. Claw trimming also resulted in less pressure on the posterior part of the claw, which can be beneficial in preventing claw lesions. The developed *ex vivo* model of porcine foot demonstrated to be a valuable tool for further biomechanical research in pigs. The project resulted in several recommendations for claw trimming for sows, which can contribute to better sustainability of pig production.

From biosecurity audit to tailor-made recommendations in pig farms: how to prioritize action points?

P. Levallois, M. Leblanc-Maridor, C. Belloc and C. Fourichon
INRAE, Oniris, BIOEPAR, 101, route de Gachet, 44300, France; pierre.levallois@oniris-nantes.fr

Different audits have been developed to assess biosecurity in pig farms. Recommendations formulated by veterinarians from these audits aim at improving disease prevention. A prioritization needs however to be done among unimplemented practices to enhance compliance on targeted action points. This leads to formulate biosecurity recommendations which are tailor-made (i.e. adapted to the farm context) and not systematic (i.e. selected and targeted among all the unimplemented practices). The process allowing to understand this prioritization is not explicit. The aim of this study was to analyse possible reasons allowing veterinarians to select recommendations and define a tailor-made biosecurity plan with specific action points. Biosecurity (internal and external) audits from the European project HealthyLivestock were performed in 20 farrow-to-finish pig farms in western France. Each audit was performed by one person in the presence of the farmer and the veterinarian. Prioritized recommendations were formulated among unimplemented practices by the farm veterinarian at the end of the visit and recorded. Information about pathogen statuses and current health disorders were also collected. Unimplemented practices were categorized according to their theoretical expected effect on prevention, financial cost and time cost. For each farm, unimplemented practices were classified according to these categorizations. Matching of recommendations with unimplemented practices was analysed, accounting for these categories and for the available health information. Recommendations can target unimplemented practices, with different expected effect (from high to low) on disease prevention according to the pathogen statuses and current health disorders of the farm on top of financial and time costs. This study underlines the importance of integrating the health context in audited farms to provide tailor-made biosecurity recommendations.

Influence of glyphosate residues in sow diets on piglet diarrhoea, kinky tails and stillbirth

J.F.M. Winters[1], L. Foldager[1,2], U. Krogh[1], N.P. Nørskov[1] and M.T. Sørensen[1]
[1]Aarhus University, Department of Animal Science, Blichers Allé 20, 8830, Denmark, [2]Aarhus University, Bioinformatics Research Centre, Universitetsbyen 81, 8000 Aarhus C, Denmark; jeanet.winters@anis.au.dk

Glyphosate is widely used for weed control in genetically modified (GM) glyphosate resistant crops and for pre-harvest non-GM crop desiccation, leaving glyphosate residues in livestock feed. Since glyphosate is known to inhibit bacteria to various degree *in vitro*, it can be hypothesized that feed glyphosate residues will affect the gut microbiota *in vivo*, potentially in an unfavourable direction resulting in diarrhoea. Furthermore, glyphosate is known to bind minerals, thus it can be hypothesized that feed glyphosate residues will decrease e.g. metalloenzyme availability and thereby potentially impair embryo/foetus development and lead to congenital malformations and stillbirth. Accordingly, we studied the association between glyphosate residue concentrations in sow diets and the frequency of diarrhoea in neonatal piglets and the frequency of congenital tail kinks and stillbirth. Rectal swaps were used to examine the faeces consistency of individual piglets between 0 and 5 days of age. A total of 5,526 piglets from 388 sows in four herds were examined and lactation diets were analysed for glyphosate content. The average feed glyphosate residue concentration was 0.31 mg/kg dry matter (DM); min 0.06 and max 0.87 mg/kg DM. An average of 19.4% of piglets from primiparous sows and 11.6% of piglets from multiparous sows was observed to have diarrhoea; we found no association between glyphosate residue concentrations in lactation diets and piglet diarrhoea. A total of 28,336 docked tails from seven herds were examined for congenital kinks and the number of kinky tails per herd per week was assessed for association with the glyphosate content in gestation diets during week 1 to 6 of gestation. The average feed glyphosate residue concentration was 0.45 mg/kg DM; min 0.08 and max 1.52 mg/kg DM. The ranges of percent kinky tails and stillborn piglets per week were 0 to 2.6% and 0.7 to 3.2%, respectively; we found no association between the glyphosate content in gestation diets and the frequency of kinky tails or stillborn piglets.

ANOXIA(TM) versus T61® euthanasia method in pigs reveals no differential gene expression

B. Chakkingal, R. Meyermans, W. Gorssen, S. Janssens and N. Buys
KU Leuven, Center for Animal Breeding and Genetics, Department of Biosystems, Kasteelpark Arenberg 30 – Bus 2472, 3001 Leuven, Belgium; bimal.chakkingalbhaskaran@kuleuven.be

Transcriptome studies in pigs have increased over recent years, aiming to unravel tissue specific processes that control significant traits. Additionally, many studies have been using pig as animal model for preclinical studies, due to their closeness to human physiology and anatomy. When animals are used as subjects in biomedical research, they often have to be euthanized. A criterion for selecting a particular euthanasia method is its compliance with animal welfare. Approved methods for euthanasia of pigs used in laboratory research include use of injectable or inhaled anaesthetics. Lethal injections of barbiturate or non-barbiturate derivatives are used more often than inhalant anaesthetics. A common method for euthanasia in pigs uses T61, an injectable, non-barbiturate anaesthetic. Nitrogen gas in foam, an inhalant anaesthetic used in the anoxia box (ANOXIA™), is an alternative method for euthanasia in pigs requiring no injection. Little is known on the effect of these euthanasia methods on subsequent RNA analyses, and whether ANOXIA™ would be a good alternative to T61. The current study compares gene expression patterns in brain and liver tissues collected from one week old male piglets (n=12), divided into two groups based on euthanasia by T61 or by ANOXIA™. RNA-seq (3'-mRNA) data were generated by following the Lexogen Quantseq library preparation protocol and subsequent sequencing on the Illumina HiSeq4000 platform. Differential gene expression analysis, using R-based Bioconductor package DESeq2 showed no differentially expressed genes in the evaluated tissues. However, small nuclear RNAs (snRNAs) namely U1, U2 and U4 that constitute the eukaryotic spliceosomal machinery were found to be downregulated (log2fold<-2 with an adjusted P≤0.02) in brain tissues from the T-61 group. Considering the significant role that snRNAs play in pre-mRNA processing, the above finding and its implication warrant further research. Given that no difference in gene expression was found between the two methods, ANOXIA™ can be used as an alternative euthanasia method in pigs.

Effect of zearalenone at concentrations below and above EU recommendation in piglets after weaning

V.C. Bulgaru[1,2], G.C. Pistol[2], D.E. Marin[2], I. Taranu[2] and A. Dinischiotu[1]
[1]Faculty of Biology, University of Bucharest, Splaiul Independentei 91-95, R-050095, Romania, [2]National Research-Development Institute for Animal Biology and Nutrition IBNA-Balotesti, Calea Bucuresti 1, 077015, Romania; cristina.bulgaru@ibna.ro

Zearalenone (ZEA) is a fusariotoxin well known for its negative effects on the reproductive system. ZEA has also negative effects on the digestive, immune and nervous system. Pig is known to be susceptible to ZEA especially at weaning when animals are confronted with many changes such as the switch from sow's milk to solid feed, changes in environmental conditions, undeveloped immune and digestive system that predispose piglets to pathogen infections and digestive disorders including diarrhoea. The quality of the feed is of great importance in this period for the development of the digestive and defence systems. Intake of mycotoxins contaminated feed in this period could amplify the decrease of animal resistance and increase the susceptibility to infection. According to the European Commission, the maximum allowed ZEA level in feed for piglets and young sows is 100 ppb and 250 ppb in feed for mature sows and pigs (CE/2006/576). To bring new scientific data leading to the establishment of a standardized limit for ZEA in young pigs, the aim of this study was to evaluate i*n vivo*, the effects of two concentrations of ZEA, one below (75 ppb) and one above (290 ppb) the CE thresholds for piglets immediately after weaning in the colon, one of the most vulnerable tissues to inflammation and oxidative stress after weaning. Our results showed that at a concentration below or above the EC thresholds did not affect, in the colon, the gene expression of proinflammatory cytokines *IL-1β, TNF-α, IL-8,* antioxidant enzymes *CAT, GPx*, and nuclear receptors *NF-kB, NRF2*. The only change induced by exposure to ZEA is the increase in the gene but not on the protein expression of antioxidant enzyme SOD and ZO1 junction protein. These results suggest that 100 ppb of ZEA can be considered a safe limit in feed for piglets after weaning.

Session 19 Theatre 11

Minor influence of feed glyphosate on gut microbial ecology of weaned piglets

S. Rani[1], M.T. Sørensen[1], S.J. Noel[1], N.P. Nørskov[1], U. Krogh[1], L. Foldager[1,2] and O. Højberg[1]
[1]Aarhus University, Dept. of Animal Science, 8830 Tjele, Denmark, [2]Aarhus University, BiRC, 8000 Aarhus, Denmark; martint.sorensen@anis.au.dk

Glyphosate possesses antimicrobial properties acting in microorganisms, like in plants, by inhibiting the shikimate pathway and thus synthesis of the aromatic amino acids (AAA: tryptophan, tyrosine and phenylalanine). We investigated potential effects of glyphosate on gut microbial ecology of piglets, weaned at 28 days of age and allocated one of four experimental diets with the following planned glyphosate content (mg/kg feed): 0 (CON), 20 as Glyphomax (GM20), 20 and 200 as IPA salt (IPA20 and IPA200). Piglets were sacrificed after 9 or 35 days in treatment, and digesta sampled from stomach, small intestine, caecum and mid colon to analyse glyphosate, organic acids (SCFA and lactate), pH and dry matter (DM) content and digesta microbiota composition (16S rRNA gene amplicons). Digesta glyphosate content reflected dietary levels; it increased from proximal to distal gut segments and it increased from day 9 to day 35. On day 35, average glyphosate levels in colon were 0.17, 16.2, 20.5 and 208 mg/kg digesta for the four treatments, respectively. Overall, we observed no significant treatment effects on digesta organic acid levels, pH, DM content and faecal score; watery faeces occurred, but across treatments. We observed only minor treatment effects on gut digesta microbiota on day 9. On day 35, we observed a significant decrease (IPA200 vs CON) in species richness in caecum. Further, at phylum level, we observed significant treatment effects on relative abundance of Firmicutes (CON: 57.7%; GM20: 66.5%; IPA20: 69.4%; IPA200: 66.1%) and Bacteroidetes (CON: 32.6%; GM20: 25.1%; IPA20: 23.5%; IPA200: 27.4%) in colon digesta as well as for a few bacterial genera. Some bacteria may utilise extracellular AAA, thereby overcoming glyphosate inhibition; in a second trial, we thus reduced feed tryptophan, however, obtained similar overall outcome. In conclusion, exposing weaned piglets to glyphosate for up to 5 weeks influenced gastrointestinal microbial ecology (e.g. microbiota diversity and composition) somewhat. Nevertheless, the nature and level of the observed perturbations were not considered as actual dysbiosis with potential adverse effects on animal health.

Session 19 Theatre 12

Reducing enterotoxigenic *E. coli* in young pig intestinal cell lines: the role of bacteriophages

A. Ferreira[1,2,3], C. Almeida[1], D. Silva[1], M.E. Rodrigues[2,3], M. Henriques[2,3], J. Azeredo[2,3] and A. Oliveira[2,3]
[1]ALS ControlVet, Zona Industrial de Tondela ZIMII, Lote 6, 3460-605 Tondela, Portugal, [2]LABBELS – Associate Laboratory, Braga, Guimarães, Portugal, AvePark – Parque de Ciência e Tecnologia, Zona Industrial da Gandra, Barco, 4805-017 Guimarães, Portugal, [3]CEB – Centre of Biological Engineering, University of Minho, LIBRO – Lab. de Investigação em Biofilmes Rosário Oliveira, Campus de Gualtar, Rua da Universidade, 4710-057 Braga, Portugal; anaoliveira@deb.uminho.pt

Enteric colibacillosis is a disease caused mainly by Enterotoxigenic *Escherichia coli* (ETEC) that affects young pigs. This pathogen is known to spread in pig farms and brings high producing costs. The rise of antibiotic selective pressure together with on-going limitation on their use demands news strategies to tackle this pathology. Bacteriophages, as obligatory bacterial parasites, are being widely explored for its potential in veterinary medicine. In this work, the efficacy of $CaCO_3$/alginate encapsulated T4-like Myoviridae phage vB_EcoM_FJ1 on reducing an ETEC O9:H9 (Sta, F5/F41) load was assessed on an intestinal porcine epithelial cell line (IPEC-1). IPEC-1 cells were grown in transwell inserts for 96 h prior infection. Encapsulated phage was administered 2 h after bacterial adhesion to epithelium, providing reductions of, approximately, 2.32 log cfu/cm^2 and 2.77 log cfu/cm^2 after 3 h and 6 h, respectively, comparing to the non-treated groups. Further studies on bacteriophage-insensitive mutants demonstrated that 70% of the new variants generated after challenge with phage vB_EcoM_FJ1 were more susceptible to pigs' serum complement system, with an average reduction of (1.1±0.1) Log cfu/ml comparing to the wild type. The adhesion of the new mutants to 24 h-old porcine intestinal epithelium also decreased by, approximately, 90% in comparison with the originating strain. Overall, encapsulated phage vB_EcoM_FJ1 is presented here as a potential tool to fight against ETEC infections in piglets.

Session 19 Theatre 13

Consequences of anaemia in sows on their reproductive performance and on the long-term pigs' growth

A. Samson[1], F. Guillard[2], E. Janvier[1], J. Martrenchar[2] and J. Pique[3]
[1]ADM Animal Nutrition, PD&A/R&D, Rue de l'Eglise – CS90019, 02402 Chateau-Thierry, France, [2]Wisium, Swine Technical Department, Talhouet, 56250 Saint-Nolff, France, [3]SETNA, Swine Technical Department, Rambla d'Egara, 08224 Terrassa, Spain; asamson8002@gmail.com

Anaemia is characterized by a low haemoglobin (Hb) concentration and is often the result of nutritional deficiencies, particularly iron. Anaemia is frequent in neonatal pigs and few evidences suggest that sows are also affected but its consequences are poorly described. For instance, there is no information about the consequence of the maternal anaemia on the growth of the offspring after weaning. A total of 40 gilts and sows and their progeny weaned at 21 days were followed in order to monitor their Hb with a non-invasive method. For 19 females, the litters were divided into two groups depending on the iron supplementation (none vs 200 mg of intramuscular iron). The growth performance of those piglets were recorded until slaughter. Performance of animals were analysed with ANOVA to assess the effect of the sows' status whereas proportion of anaemic animals were analysed with Chi^2 test. Anaemia (Hb<11 g/dl) has affected 37.5% of the sows at 65 days of gestation (G65) and 87.5% at weaning. New-born piglets from anaemic mothers at G65 tended to be more sensitive to anaemia (Hb<9 g/dl) than the others (56.6 vs 43.0%, P=0.09). Proportion of stillborn was significantly higher in the sows anaemic at G65 than others (13.3 vs 6.6%, P=0.03). Mortality of the suckling piglets was not affected by both the sows' status and the iron supplementation. Iron supplementation has significantly improved pigs' growth rate during the suckling and the post-weaning phases. The Hb of the piglets at weaning and at day 67 was not significantly affected by the sows' status. However, pigs born from anaemic sows at G65 grew significantly slower over the fattening period than the others (958 vs 910 g/d, P<0.05). Though the causal link was not established, this study highlights that anaemia in sows is associated with deleterious long-term effects on the growth of offspring.

Session 20 Theatre 1

Effect of nutrition in early life on the development and performance of beef heifers

I. Casasús, J.A. Rodríguez-Sánchez, J. Ferrer, A. Noya and A. Sanz
Centro de Investigación y Tecnología Agroalimentaria de Aragón (CITA) – IA2 (CITA-UniZar), Avda Montañana 930, 50059, Spain; icasasus@cita-aragon.es

Advancing the age at first calving of beef cows has been proposed as a strategy to increase lifetime productivity. Heifer nutrition in early life, known to affect growth and physiological development, can be compromised in extensive production systems, due to large seasonal variations in feed quality and availability. A series of experiments were carried out to determine how different planes of nutrition in the early pregnancy and in the pre- and the post-weaning phase influenced the metabolic and endocrine status, linear body measurements and the productive and reproductive performance of beef heifers with a target age at calving of 2 years. The maternal undernutrition of beef cows (65% vs 100% energy requirements) in the first 3 months of pregnancy resulted in a less mature hematopoietic system of their calves at birth and lower IGF-1 and gains during lactation. Heifers born to dams on the 65% diets had impaired metabolic status and lower follicle counts around puberty, but they did not differ in age or weight at puberty, fertility or the performance of their calves in their first lactation (age at calving: 26 months). Another study analysed the effects of feeding strategies targeting a high (1 kg/d) or low gain (0.7 kg/d) during the lactation (0-6 months) and the rearing phase (6-15 months). Heifers reached puberty at a similar weight but different age, which was associated to their growth rates and metabolic status (IGF-1, glucose, cholesterol). A consistently low gain (0.7 kg/d) from 0 to 15 months resulted in lighter weight, body size and pelvic area at calving (26 months), and higher calving assistance rates. Heifer feeding treatments did not affect their milk yield or the gains of their calves, but a genome wide expression profiling in the mammary gland showed an upregulation of genes related to the immune response in heifers on high nutrition planes during lactation, consistent with an increased somatic cell count. Apart from the economic aspects related to feeding costs in the different phases, the final decision on heifer management should also be driven by its potential impact on cow performance beyond this first lactation, which remains to be determined.

Session 20 Theatre 2

Integrative network analysis on the nutrigenomic control of reproductive development in cattle

K. Keogh and D.A. Kenny
Teagasc, Animal and Bioscience Research Department, Grange, Dunsany, Co. Meath, Ireland; kate.a.keogh@teagasc.ie

Advancements in next generation sequencing technologies have greatly increased our understanding and capability in deciphering the complex molecular control regulating important biological traits in cattle. More recently efforts have been focused towards evaluating 'omics' datasets, (genomic, transcriptomic, proteomic), in unison, through systems biology based analytical approaches. Such integrative analyses allow for a more comprehensive evaluation of the underlying biology governing traits of interest, determining discrete relationships between genes towards a particular trait as opposed to merely focusing on genes with the largest difference, evident through differential expression analysis alone. Moreover, through gene co-expression analyses, networks of highly correlated genes may be derived, with the additional utility of identifying 'hub genes'. Hub genes are of particular interest as such genes are key to the regulation of additional genes within a network and thus are key to controlling the expression of a particular trait. Furthermore, the importance of hub genes to the regulation of a particular trait may not be apparent through differential expression analyses alone, thus highlighting the functionality of gene co-expression analyses. Additionally, through the integration of various types of genomic data, for example, genetic and transcriptomic, the functional role of specific SNPs may be evaluated, based on the interaction between genes harbouring SNPs and the larger co-expression network. Recently, we have applied such integrative molecular analyses towards deciphering the effect of enhanced nutritional status on the regulation of reproductive development in prepubertal cattle. Transcriptomic and proteomic data from tissues of the hypothalamic-pituitary-testicular (HPT) signalling axis were integrated. In particular a putative role of the miR-2419 miRNA towards reproductive development within the HPT tissues of bull calves was apparent through enrichment of processes related to metabolism, GnRH signalling, and cholesterol biosynthesis across tissues of the HPT axis. Knowledge derived from such analyses may be used to identify more accurate genomic targets within the context of genomically assisted breeding programs.

Session 20 Theatre 3

Influence of trace minerals on metabolic processes related to reproduction in ruminants

T.E. Engle
Colorado State University, Fort Collins, CO, USA; terry.engle@colostate.edu

Trace minerals have long been identified as essential dietary components for domestic livestock species. Included in the category of essential trace minerals (or microminerals) are chromium, cobalt, copper, iodine, iron, manganese, molybdenum, selenium, and zinc. Numerous biochemical reactions require trace minerals for proper function. However, the interactions between trace minerals and metabolic processes are extremely complex. Trace minerals have been identified as essential for normal carbohydrate, lipid, protein, and vitamin metabolism, and have been shown to be involved in hormone production, immunity, and cellular homeostasis. In general, trace minerals function primarily as activators or components of enzyme systems within cells by assisting with enzyme structural integrity and substrate binding. Enzymes involved in protection of cells from oxidative stressors, electron transport, oxygen transport, bone metabolism, gene expression, and nutrient metabolism all have been shown to require certain trace elements for proper function. It has been well documented that deficiencies of various trace minerals can impact reproductive efficiency in domestic livestock species. Reproductive deficiency signs of trace minerals in ruminants include: decreased conception rate, reduced birth weight, infertility, increased duration of anoestrus, foetal deformities, altered reproductive hormone production, and foetal resorption. There are many factors that could affect an animal's response to trace mineral supplementation such as the duration and concentration of trace mineral supplementation, physiological status of an animal, the absence or presence of dietary antagonists, environmental factors, and the degree of stress. Therefore, the intent of this review is to discuss the impact of trace minerals on metabolic process related to reproduction in ruminants.

Impact of early life nutrition on the molecular control of sexual development in the bull calf

D.A. Kenny, S. Coen and K. Keogh
Teagasc Animal and Bioscience Research Department, Grange, Dunsany, County Meath, Ireland; david.kenny@teagasc.ie

Progress in genomic selection processes has accentuated interest in procuring saleable semen from young genetically elite bulls as early in life as possible.

Impacts of postweaning growth rate in beef heifers on reproductive development and productivity

K.M. Harvey
Mississippi State University, Prairie Research Unit, Prairie, MS, USA; kelsey.harvey@msstate.edu

In cattle there is a crucial period of mammary development from 3 to 9 months of age during which extensive growth of the duct network occurs and invades the stroma, providing the foundation for later alveolar development during gestation. Nutritional manipulation during this period can greatly impact future lactation potential of beef females. The most common period to manipulate heifer growth in a typical cow-calf operation is between weaning at 7 months of age and their first breeding season. This experiment evaluated the effects of postweaning body weight (BW) gain of replacement beef heifers on their reproductive development and productivity as primiparous cows. Seventy-two heifers were ranked on day −6 of experiment (17 d after weaning) by age and BW and assigned to receive 1 of 3 supplementation programs offered individually 6 d per week from days 0 to 182: (1) no supplementation to maintain limited BW gain (LGAIN); (2) supplementation to promote moderate BW gain (MGAIN); or (3) supplementation to promote elevated BW gain (HGAIN). On day 183, heifers were combined into a single group and received the same nutritional management until the end of the experimental period (day 718). Average daily gain from days 0 to 182 was greater ($P<0.01$) in HGAIN vs MGAIN and LGAIN (0.78, 0.60, and 0.37 kg/d, respectively; SEM=0.02), and greater ($P<0.01$) in MGAIN vs LGAIN heifers. Puberty attainment by the beginning of the breeding season was also greater in HGAIN vs MGAIN and LGAIN (87.5, 62.5 and 56.5%, respectively; SEM=7.1) but similar ($P=0.68$) between MGAIN vs LGAIN heifers. Ten heifers per treatment were assessed for milk production via weigh-suckle-weigh at 56.8±1.5 d postpartum, followed by milk sample collection 24 h later. No treatment differences were detected ($P\geq0.16$) for milk yield and composition. No treatment differences were detected ($P\geq0.44$) for offspring weaning BW. Collectively, results from this experiment indicate that HGAIN hastened the reproductive development of replacement heifers, without negatively affecting their milk productivity and offspring weaning weight as primiparous cows.

Nutritional programming of puberty in *Bos indicus*-influenced heifers

R. Cardoso, S. West, V. Garza and G. Williams
Texas A&M University, Department of Animal Science, Texas A&M University, USA; r.cardoso@tamu.edu

Approximately 5 million beef heifers enter the U.S. cow herd annually and their lifetime productivity is heavily dependent upon their ability to attain puberty and produce a calf by 24 mo of age. However, a significant proportion of heifers within existing U.S. production systems fail to achieve these goals, particularly in southern regions where *Bos indicus*-influenced cattle predominate. Therefore, approaches are needed that facilitate nutritional programming of the reproductive neuroendocrine axis, while minimizing feeding costs and optimizing the consistent attainment of puberty by 14-15 mo of age. Increased BW gain between 4 and 9 mo of age facilitates pubertal development by programming hypothalamic centres that regulate gonadotropin releasing-hormone (GnRH) secretion. Among the different metabolic hormones, leptin plays a critical role in conveying nutritional information to the brain and controlling puberty. Two hypothalamic neuronal populations that express the orexigenic peptide neuropeptide Y (NPY) and the anorexigenic peptide alpha melanocyte-stimulating hormone (αMSH) are key components of afferent pathways that convey inhibitory (NPY) and excitatory (αMSH) inputs to GnRH neurons. Our studies have demonstrated that short-term increases in dietary energy intake during juvenile development result in epigenetic, structural, and functional modifications in these hypothalamic pathways to promote high-frequency, episodic release of GnRH and luteinizing hormone. However, integrating the foundational knowledge of metabolic imprinting of the brain for early puberty with issues related to lifetime performance is complex. One approach has been to employ a novel stair-step nutritional regimen involving alternating periods of dietary energy-restriction and re-feeding during juvenile development. This approach is designed to support early onset of puberty by imprinting functional alterations in the hypothalamus during key periods of brain development while optimizing other aspects of growth and performance. Finally, our recent findings suggest that maternal nutrition during gestation can also induce neuroendocrine changes that are likely to persist and influence reproductive performance throughout adulthood in cattle.

Impact of sire nutrition on fertility

P.L.P. Fontes
Department of Animal and Dairy Science, University of Georgia, Athens, GA 30602, USA; pedrofontes@uga.edu

The first 30 days of gestation represent a pivotal period for pregnancy establishment and have considerable implications for fertility and beef production efficiency. Although beef females have high fertilization rates, pregnancy loss between fertilization and the first pregnancy diagnosis occurs in a considerably high proportion of cows and heifers. Most research efforts to address early pregnancy loss in cattle have focused on female-related factors that influence embryonic mortality. However, research from other species have shown that sperm plays an important role in early conceptus development. In addition to its genetic contributions, sperm-derived factors are involved in many post-fertilization events, such as syngamy, cleavage, and epigenetic regulation of the developing embryo. Large-scale field fertility studies have shown differences in pregnancy rates between sires with similar sperm parameters. However, few studies have investigated the paternal contributions to early embryonic development in cattle. In mice, embryos sired from diet-induced obese fathers had reduced cleavage and blastocyst development rates compared with embryos sired from control fathers. A variety of Extension programs and bull development stations have reported the general preference of cattle producers for bulls with high rates of average daily gain during their growth and development phase. In fact, beef cattle producers prioritize growth-related traits versus feed efficiency traits such as feed to gain ratio or residual feed intake. While bull over-conditioning is a common phenotype observed in the beef industry, the impact of paternal nutrition on embryo and conceptus development remains poorly understood in cattle. Our group has recently evaluated evaluate the effect of paternal high energy diets on blastocyst development during *in vitro* embryo production. Bulls exposed to highly anabolic diets had decreased cleavage rates, blastocyst rates as a percentage of oocytes, and blastocyst rates as a percentage of cleaved structures when compared with bulls fed a maintenance diet. These results corroborate with reports in other species; however, further research is required to better understand the contributions of paternal diets to conceptus development and pregnancy loss.

Influence of feed and seasons on semen qualities and their relationships with scrotal size in bucks

Y.E. Waba
Department of Animal Health and Production, Federal Polytechnic Bali Taraba State, Nigeria, Department of Animal Health and Production, Department of Animal Health and Production, Federal Polytechnic PMB OO5, Bali Taraba State, 672101, Bali, Nigeria; ezekielwaba@googlemail.com

The study evaluated the influence of feed on the reproductive parameters of West Africa Dwarf (WAD) and Sahel bucks at the Teaching and Research farm of the Federal Polytechnic Bali, Taraba State. A total of eight bucks of the two breeds that service does in a body of feeding experiments were used for this study. The eight bucks were moved to Vom for semen collection, using an electro-ejaculator and evaluation every two weeks. The study lasted for 12 calendar months. The buck were fed Gmelina and cassava peel meal, Gmelina and cowpea husk, Ficus and cassava peal meal and, Ficus and cowpea husk as experiments (T1-T4 respectively). The bucks by breed were allotted to the treatments randomly ie two bucks per treatment. The average scrotal circumference, semen volume, sperm concentration, PH, percentages motility, normal and live sperm cells for WAD goats were 17.70 cm, 424.11×10^6/ml, 6.65 0.50 ml, 84.88, 93.45 and 92.01% respectively. The corresponding values for Sahel goats were 18.12 cm, $405{:}73\times10^6$/ml, 6.92, 0.44 ml, 81.16 91.18 and 92.01. There were significant ($P<0.01$) breed, season and diet effects on most of the parameter tested except the effects of breed and season one percentage live sperm cells. There is however no season or diet that showed superiority for all semen/sperm qualities.

Influence of feed on the reproductive parameters of does WAD and Sahel goats in Bali, Taraba State

Y.E. Waba
Department of Animal Health and Production, Federal Polytechnic Bali Taraba State, Nigeria, Department of Animal Health and Production, Department of Animal Health and Production, Federal Polytechnic PMB OO5, Bali Taraba State, 672101 Bali, Nigeria; ezekielwaba@googlemail.com

The study evaluated the influence of feed on the reproductive parameter of West African Dwarf (WAD) and Sahel goats at the Teaching and Research farm of the Federal Polytechnic Bali, Taraba State. A total of 32 goats of the two breeds were used for the experiment. The study lasted for 26 months (October, 2017 to November, 2019). They were fed Gmelina and cassava peel meal, Gmelina and cowpea husk, Ficus and cassava peel meal and, Ficus and cowpea husk as experimental treatments (T1 – T4) respectively. Three does and a buck per breed were randomly allocated to each treatment. The average age at first oestrous, oestrous period, oestrous length, gestation period, age at first kidding and postpartum dam weight for WAD were 4.8 wks., 37.50 hrs, 20.25 days, 144.35 days, 15.92 months, and 7.67 kg respectively. The cool corresponding values for Sahel were 41.75 wks., 29.00 hrs, 20.17 days, 142.83 days, 15.83 month and 11.36 kg. Age at first oestrus, oestrous and gestation length and, age at first kidding did not differ significantly among breeds. However, oestrus period and postpartum dam weight varied significantly ($P<0.05$) with breed. Similarly age at first oestrus, oestrus period and length did not vary with diet, whereas gestation period, age at first kidding and postpartum weight varied ($P<0.05$) with diet. The cassava peel meal containing diets had longest gestation period and age at first kidding, while the cowpea husk counterpart had higher postpartum dam weight. For decrease in gestation length and age at first kidding and, increase in postpartum dam weight therefore the feeding of cowpea husk with browse is more recommended than the inclusion of cassava peal meal.

Dying for uniformity: is pre-weaning mortality the major cause of the decrease in variation?

K. Hooyberghs[1], S. Goethals[2], S. Millet[2], S. Janssens[1] and N. Buys[1]
[1]KU Leuven, Center for Animal Breeding and Genetics, Department of Biosystems, Kasteelpark Arenberg 30 – Box 2472, 3001 Leuven, Belgium, [2]ILVO, Scheldeweg 68, 9090 Melle, Belgium; katrijn.hooyberghs@kuleuven.be

Uniformity of fattening pigs may increase the efficiency and sustainability of meat production and reduce economic, environmental and social costs. This research aimed to evaluate a change in uniformity in the first 6 weeks after birth and to see if this is mostly due to pre-weaning mortality. Data was collected in one farm on 791 crossbred progeny born in 46 litters, sired by 4 Piétrain and 4 MaxiMus boars. From the 696 live-born piglets, 623 (89.5%) survived until the age of 6 weeks. To evaluate the uniformity, the coefficient of variation (CV% = (standard deviation/mean) ×100) of body weight was analysed at birth, at weaning (24-29 days), 1 week after weaning, and 2 weeks after weaning (after post-weaning dip). The average within litter CV% at birth of all live-born piglets (including piglets dying before 2 weeks after weaning) was 20.6%. In the first 6 weeks, the variation within litter slightly decreased (P=0.10) to 18.5%. However, the average CV% at birth of the surviving piglets only is 17.8%, which is significantly lower (P=0.028) than the CV% when including all live-born piglets. This indicates that the pre-weaning mortality causes a decrease in within litter weight variation. When considering only the surviving piglets, CV% did not change significantly from birth to 6 weeks (17.8% to 18.5%, (P=0.55)). Our findings suggest that pre-weaning mortality should be accounted for when investigating 'uniformity' of litters of piglets. However, additional data need to be collected to confirm our results and to study other effects on the CV% of body weight.

Are biters sick – studies on the health status of tail biters

I. Czycholl[1,2], D. Becker[3], C. Schwennen[4], C. Puff[5], M. Wendt[4], W. Baumgärtner[5] and J. Krieter[2]
[1] Pig Improvement Company (PIC), 100 Bluegrass Commons Blvd. Ste 2200, Hendersonville, TN, 37075, USA, [2]Kiel University, Institute of Animal Breeding and Husbandry, Olshausenstr. 40, 24098 Kiel, Germany, [3]Leibniz-Institute for Farm Animal Biology (FBN), Institute of Genome Biology, Wilhelm-Stahl-Allee 2, 18196 Dummerstorf, Germany, [4]University of Veterinary Medicine Hannover, Clinic for Swine, Small Ruminants and Forensic Medicine and Ambulatory Service, Bischofsholer Damm 15, 30173 Hannover, Germany, [5]University of Veterinary Medicine Hannover, Department of Pathology, Bünteweg 17, 30559 Hannover, Germany; iczycholl@tierzucht.uni-kiel.de

Tail biting is a multifactorial problem. Identifying the individual animals actually performing the behaviour and separating them from the group can be an effective management intervention strategy. As the health status is one of the factors commonly linked to tail biting, this study focuses on the health of identified biters. Within two years (May 2019-February 2021), 30 biters were identified on 3 farms in Germany with the help of a joint ethogram. With each identified biter, a control animal of the same age class, but of a pen not affected by tail biting, was chosen. Biters and controls were subjected to a clinical as well as a post-mortem examination. The persons carrying out the examinations were blind to the groups. Biters and controls were compared by a Chi^2-Test. In the clinical examination, 6 biters (2 controls, P=0.019) were noticeably agitated in the general behavioural evaluation, while 8 controls were noticeably calmer (2 biters, P=0.02). 13 biters had overlong bristles (4 controls, P=0.008). In the post-mortem examination, 5 biters (0 controls, P=0.018) were affected by inflammation of the pars proventricularis of the gaster (P=0.018). No further differences regarding the health status of the pigs were found. The results suggest that behavioural tests might be helpful in identifying biters. Moreover, stomach pain may be linked to becoming a biter, which fits to the observation that tail biting often occurs around two weeks after weaning, at which a significant and abrupt changeover in feeding occurs. Blood and liquor samples, as well as the brains of the pigs were also collected and will be analysed in the ongoing study.

Session 21 Theatre 3

Effect of various short-term events on behaviours of gestating sows

M. Durand, J. Abarnou, A. Julienne, C. Orsini and C. Gaillard
PEGASE, INRAE, Institut Agro, Le Clos, 35590 St-Gilles, France; maeva.durand@inrae.fr

Environmental conditions and technical disturbances may affect gestating sows' behaviour and performance, and their nutritional requirements. The aim of this study was to evaluate the effects of induced short-term events on gestating sows drinking, feeding, and social behaviours. Different events were induced on two groups of 20 gestating sows during a week for each event (test week), always preceded by a control week: a 'hot' and a 'cold' thermic stress (i.e. rooms on average at 32 and 15 °C for each stress), a pen enrichment event (providing ropes, bags), and a feeding competition (i.e. in each room one automated feeder closed over two). Automatons recorded individual feeding and drinking behaviours. Individual social behaviours classified as negative (head butts, bites) or positive were extracted manually by video analysis during the first 36h of the feeding competition, and during two periods of 5h for the other events (13.30-18.30h, 23.00-04.00h). The effects of the week (test vs control) on behaviours were analysed using a linear mixed-effect model, including the random effect of the sow. During the feeding competition and the enrichment weeks, the number of nutritive visits to the feeder was not affected. Compared to the control week, the sows did less non-nutritive visits (NNV) during the feeding competition, (4.7 vs 2.3 NNV/d/sow, $P<0.001$) and more during the enrichment week (4.6 vs 5.3 NNV/d/sow, $P<0.01$). During the 'hot' week, the sows visited more the water trough than during the control or 'cold' weeks (15.3 vs 13.4 or 12.9 visits/d/sow, $P<0.05$). Except during the enrichment, the number of negative behaviours (NB) increased during test weeks compared to control weeks (22.3 vs 6.9 NB/sow over 36h for the feeding competition $P<0.001$; and 1.9 vs 1.3 NB/sow over 10h for the cold stress, NS). During the enrichment period, the number of NB decreased compared to control week (7.3 vs 5.3 NB/sow over 10h, $P<0.05$). These preliminary results suggest that various shorts-term events during gestation modify differently the sows feeding, drinking, and social behaviours and that these behaviours could represent relevant indicators of welfare.

Session 21 Theatre 4

Contact voltages exposure lower than 0.5 V in feeders and drinkers affects the behaviour of piglets

T. Nicolazo[1], C. Clouard[2], G. Boulbria[1], E. Merlot[2], A. Lebret[1], C. Chevance[1], J. Jeusselin[1], V. Normand[1] and C. Teixeira-Costa[1]
[1]REZOOLUTION, Pig Consulting Services, Parc d'Activités de Gohélève, 56920 Noyal-Pontivy, France, [2]INRAE, PEGASE, Le Clos, 35590 Saint-Gilles, France; t.nicolazo@groupe-esa.net

Our study aimed to describe changes in the behaviour of weaned pigs exposed to stray currents of voltage <0.5 V on farm. The study was conducted on a nursery barn with two rooms of 12 pens of 38 28-day old weaned piglets on two batches. In each pen, stray voltages were measured for each drinker and feeder every two weeks (on days 9, 23, 37 and 50). 2 focal pigs per pen were randomly selected, and their behavioural activity was recorded during two 1-h sessions per day, two days a week for 7 weeks (from days 6 to 50) using 5-min scan sampling. Pens were allocated to 4 groups, with high (HV, >125 mV) or low (LV, <125 mV) voltage in drinkers and high (HV, >50 mV) and low (LV, <50 mV) voltage in feeders. Behaviours were averaged per pen and effects of voltage levels in drinkers, feeders, and their interactions were analysed using mixed linear models with pen and replicate as random effects. At 6 days post-weaning, pigs exposed to HV in drinkers spent more time manipulating pen mates ($P=0.02$) and less time sleeping ($P=0.02$) than pigs exposed to LV in drinkers. During the following weeks (from 2 to 7 weeks post-weaning), pigs exposed to HV in drinkers spent more time manipulating pen mates ($P=0.003$) than piglets exposed to LV. Piglets exposed to HV in feeders spent less time sleeping ($P=0.02$) and more time nosing pen mates while lying ($P=0.04$) and aggressing pen mates ($P=0.008$) than piglets exposed to LV. Significant interaction effects suggest that animals exposed to HV in drinkers spent more of their inactive time lying ($P<0.001$) and less standing ($P=0.002$), and spent more time exploring objects ($P=0.05$) and the floor while lying ($P=0.004$), but that these effects were highly influenced by voltage in feeders. Our results suggest that stray voltages in housing affect behaviour of pigs, with an increase in harmful social behaviours, and may be detrimental for pig welfare. Although further experiments should be considered to confirm our findings, this study is the first to highlight behavioural changes linked with stray voltage lower than 0.5 V in pigs.

Intra-uterine growth restriction affects growth performances but not the body composition at weaning

R. Ruggeri[1], G. Bee[1], P. Trevisi[2] and C. Ollagnier[1]
[1]Agroscope, Pig Research Unit, Animal Production Systems and Animal Health, Route de la Tioleyre 4, 1725 Posieux, Switzerland, [2]University of Bologna, Department of Agricultural and Food Sciences, Viale Giuseppe Fanin, 40-50, 40127 Bologna, Italy; roberta.ruggeri@agroscope.admin.ch

Intra-uterine growth restriction (IUGR) refers to the impaired growth and development of the mammalian embryo/foetus or its organs during pregnancy. IUGR may have permanent stunting effects on muscle growth and the development of the progeny. This study aims to investigate how IUGR affects piglets growth performances during lactation and body composition at weaning (age=25.2±1.2 d). Two days (±1) after birth, each piglet (n=268) was classified as either normal (score 1), mild IUGR (score 2) or IUGR (score 3) by two different observers. The classification was based on the morphological characteristics of the head. IUGRs can be recognized by their dolphin-like shape, bulging eyes, hair without direction of growth and wrinkles perpendicular to the mouth. Body weight was recorded individually once a week. Bone mineral content (BMC), lean mass (LM) and fat tissue mass (FM) of each piglet were determined right after weaning (age=29.6±0.7 d) using Dual-energy X-ray absorptiometry. IUGR score had a significant impact on the growth performances. At weaning, piglets with score 1 weighed 505 and 876 g more than those with scores 2 and 3, respectively ($P \leq 0.01$, P-value of the contrast). Additionally, the score affected the BMC, LM and FM, with score 1 showing 11% and 19% higher BMC, 9% and 17% higher TLM, 9% and 18% higher TFM than scores 2 and 3, respectively ($P \leq 0.01$). Nevertheless, there was no difference between scores in the proportions of BMC, LM, and FM on total mass. Likewise, the ratio between total fat mass and total lean mass did not differ between scores. In conclusion, IUGR condition affects growth performances from birth to weaning, but not the relative body composition at weaning.

The use of thermal imaging and salivary biomarkers during the periparturient period of sows

M. De Vos, S. Tanghe, Q. Duan, B. Loznik, M. Ubbink-Blanksma and R. D'Inca
Royal Agrifirm Group, Landgoedlaan 20, 7325 AW Apeldoorn, the Netherlands; m.d.devos@agrifirm.com

This study used infrared thermal imaging (IRT) and saliva sampling as non-invasive methods to gain information on the physiological status of sows in the period around farrowing. A total of 46 sows were followed during gestation and lactation. Sows were distributed in parity blocks, with parity block 1 containing sows with parity 1, 2 or 3 (n=19) and parity block 2 containing sows with parity above 3 (n=27). Infra-red heat measurements were made daily by using a portable FLIR heat camera between 5d antepartum and 5d postpartum. Standardized anatomical locations were identified in IRT digital images which were analysed using FLIR Tools for thermal analysis to determine the average temperature at each anatomical site of each sow at several time-points. Saliva was collected 2.1±0.9d antepartum and 2.5±1.2d postpartum in which haptoglobin was measured. For both the average ear and eye temperature, there was a significant effect of timepoint ($P<0.001$), with increasing temperatures starting 1d before farrowing. This is in line with previous studies that demonstrated a gradual increase of 1-1.5 °C in rectal temperature starting approximately 1d pre-farrowing. When calculating the delta between 1d before and 1d after farrowing, we could observe that sows in parity block 2 showed a higher increase in temperature then sows in parity block 1 in both ear (+2.1 vs 1.2 °C; P=0.09) and eye (+1.5 vs 0.9 °C; P=0.04) temperature. We hypothesized that temperature increase could be linked to a difference in total born piglets (parity block 1 = 15.3±0.84; parity block 2 = 18.0±0.75; P=0.01). Average eye temperature increase was positively correlated with total born piglets ($R^2=0.29$, P=0.05) whereas for average ear temperature no significant correlations were found. In addition, a significant increase in salivary haptoglobin levels was observed post-farrowing (+205%; $P<0.001$), which is in line with earlier observation in serum haptoglobin. The results of this study show that non-invasive biomarkers can be used to observe physiological changes during the periparturient period. Additionally, our observations suggest that parity number, and consequently litter size, has an influence on temperature increase.

Poster presentation

K. Nilsson[1] and S. Vigors[2]
[1]Swedish University of Agricultural Sciences, Ulls väg 26 Uppsala, 750 07 Uppsala, Sweden, [2]University College Dublin, Belfield, Dublin 4, Ireland; katja.nilsson@slu.se

In this time slot we allocate time to view the posters.

Effects of *Lactobacillus* spp. encapsulated probiotic on performance and faecal microbial in piglets

N.A. Lefter, M. Hăbeanu, M. Dumitru, A. Gheorghe and L. Idriceanu
National Research&Development Institute for Animal Biology and Nutrition, Balotesti, Calea Bucuresti no. 1, 077015, Romania; mihaela.dumitru22@yahoo.com

A total of eighty 30 day-olds weaning Topigs hybrid piglets with initial body weight (BW) of 8.52±0.13 kg were used in a 7-days trial to evaluate the effect of *Lactobacillus* spp. encapsulated probiotic supplementation on the growth performance and faecal microbial counts. Piglets were randomly distributed in one of four dietary treatments: (1) C (control diet), (2) LA (C + 1.0% *L. acidophilus*, 1×10^8 cfu/kg of diet), (3) LP (C + 0.1% *L. plantarum*, 3×10^8 cfu/kg of diet) and (4) LA: LP (C + 1.1% mix of *L. acidophilus* and *L. plantarum*). At the end of the study, the performances (BW; average daily feed intake, ADFI; average daily gain, ADG; gain: feed ratio, G: F ratio) and faecal microbial flora were evaluated. The piglets fed probiotic supplementation diets slightly increased ($P>0.05$), the BW (9.80±0.98 kg, LA: LP; 9.73±0.93 kg, LA), ADFI (0.258±0.23 g, LA: LP; 0.236±0.19 g, LP) and ADG (0.214±0.02 g, LA: LP) vs C diet (9.72±0.38 kg; 0.234±0.10 g; 0.199±0.09 g). The probiotic diet slightly decreased ($P>0.05$) the G: F ratio (0.829±0.04 g, LA: LP; 0.831±0.11 g, LP) vs C diet (0.849±0.11 g). Dietary addition of probiotics has improved ($P<0.05$) the faecal *Lactobacillus* counts. There was a significant decrease in the *E. coli* and *Clostridium* spp. counts among treatments as the effect of probiotic supplementation. In conclusion, the results indicate that dietary supplementation of encapsulated probiotic based on *Lactobacillus* spp. has a beneficial effect on growth performance and faecal microbial in 7-days post-weaning piglets. Acknowledgements: This research was supported by funds from the Ministry of Research, Innovation and Digitalisation, Romania [grant PN 19010104/2019] and [grant PFE 8/2021].

Session 22 Theatre 1

Carving a path across 'g-local' challenges: wildlife responses to human disturbance in the European 'anthroscape'

F. Cagnacci
Animal Ecology Unit, Research and Innovation Centre, Research and Innovation Centre, Fondazione Edmund Mach, Trento, Italy; francesca.cagnacci@fmach.it

Humans have occupied or affected with their activities all ecosystems across the planet. Climate change and the current plans to exponentially expand the road network, so increasing the continuum between areas occupied by humans and previously untouched habitats are dramatic threats that require deepening our understanding of how wildlife's use of habitat and resources is impacted by humans. In Europe, animals move in a highly anthropic context, where habitat fragmentation and limited ecological connectivity combines with other sources of anthropic pressure, such as human pervasive presence and disturbance on the landscape, management practices, and the aforementioned global issues i.e. climate change. Yet, the shift in land use has gone both ways in the last decades, with some productive areas being abandoned first and successively re-occupied by forest, showing an opposite trend to what observed globally. In this talk, I analyse these components on several species (mainly ungulates and large carnivores), using a niche-based interpretation of the human-wildlife relationships, to disentangle some of the challenges and adaptations of European wildlife. I will conclude commenting on possible mitigating actions to limit human impact on the European mammal community in particular- possibly representing a benchmark of solutions in other global contexts.

Session 22 Theatre 2

The coexistence of wildlife and livestock: sanitary aspects

C. Gortázar
Instituto de Investigación en Recursos Cinegéticos IREC (UCLM & CSIC), Ronda de Toledo 12, 13005 Ciudad Real, Spain; christian.gortazar@uclm.es

Animal disease, both emerging and endemic, is one of the main threats for the EU animal farming sector. This overview intends first to describe the problem making use of selected case studies on epidemiology and disease control at the wildlife-livestock interface in Europe and, second, to list and comment on a selection of innovative tools to improve the detection, monitoring, and control of infections shared with wildlife with a particular emphasis on farm biosafety. Indoor production systems are better prepared for disease emergence and disease management but are still at risk of direct and indirect (e.g. human-, or vector-mediated) infection spill over from wildlife. The ongoing African swine fever and avian influenza epidemics are relevant examples. In turn, due to the closer interaction with other livestock and with wildlife, open air livestock practices are at an even higher sanitary risk. For instance, the maintenance of endemic diseases such as animal tuberculosis is partly a consequence of increasingly complex host communities including both wildlife and livestock. Innovative tools for a better understanding of the epidemiology of infections shared with wildlife include whole genome sequencing (WGS) and environmental nucleic acid detection (ENAD), among others. Innovative disease control strategies at the wildlife-livestock interface include recent breakthroughs in vaccination and immunity stimulation and modern farm biosafety. Farm biosafety – i.e. the management and physical measures designed to reduce the risk of introduction, establishment and spread of infections – has emerged as a key tool, especially where no vaccine is available nor authorized, and when wildlife reservoirs increase the risks, as well as for new and emerging health risks such as antimicrobial resistance (AMR). Hence, there is a need to upgrade European farm biosafety and enable biosafety monitoring in a broad range of farming systems. Together, farm biosafety and innovation can help to bridge the health gap between indoor and outdoor farming systems, thereby benefitting both animal farming sustainability and biodiversity conservation in a context of resource limitation, societal changes, and increasing complexity.

Session 22 Theatre 3

Management and technologies for conflict mitigation between wildlife and breeding of livestock

K. Jerina

University of Ljubljana, Biotechnical Faculty, Department for Forestry and Renewable Forest Resource, Večna Pot 83, 1000 Ljubljana, Slovenia; klemen.jerina@bf.uni-lj.si

There is no real wilderness in Europe and even the widest habitat blocks are relatively small for the most mobile terrestrial wildlife species. Consequently, conservation of large species is possible only in coexistence and shared use of space with humans. Human-wildlife interactions are therefore inevitable and frequent, and they come in a multitude of forms. Although preserved wildlife populations and shared use of space with humans brings many benefits and opportunities (e.g. hunting, recreation, ecotourism, and various ecosystem services) they may also trigger significant problems and conflicts. Some of the major conflicts between domestic animals and wildlife include transmission of diseases and predation of domestic animals, damage to crops grown for domestic animals, damage to grasslands (rooting and grazing), and other types of competition for food and space. There are multiple means and methods available to prevent and mitigate conflict, ranging from the primitive to the cutting-edge, that differ by subject of protection, effectiveness, operators (breeders or wildlife managers), cost, feasibility, and social and legal acceptability. The lecture will focus on the following: (1) overview of principal interactions and drivers of conflict between animal husbandry in Europe and species of large carnivores/wild ungulates; (2) mitigation of conflict through the management of wildlife populations and their habitat; (3) species conation (of domesticated and wild animals); (4) sonic, visual, chemical and other deterrents, classical and electric fencing, nocturnal sheltering of animals, shepherd dogs and shepherding; (5) overview of effectiveness, legality and limitations in implementation. The main EU-driven mechanisms for the co-financing of mitigating measures and examples of best practice will be presented as well.

Session 22 Theatre 4

Wild or domesticated – cultural interpretation and contested boundaries

K. Skogen

Norwegian Institute for Nature Research – NINA, Sognsveien 68, 0855 Oslo, Norway; ketil.skogen@nina.no

What defines 'wildlife' and what distinguishes a wild animal from a domestic one? These boundaries are not only historically variable and contingent on cultural interpretations; they are also contested in contemporary societies. Yet, they may have very practical consequences for wildlife management, approach to animal diseases and indeed to human health challenges with zoonotic origin. The presentation will be based on these challenges and on researches about conflicts over wildlife and conservation in Norway, with a view overarching perspectives that are applicable globally.

Session 22 Theatre 5

Animal source foods and the nature-culture divide

F. Leroy
Industrial Microbiology and Food Biotechnology (IMDO), Faculty of Sciences and Bioengineering Sciences, Vrije Universiteit Brussel, Pleinlaan 2, 1050 Brussels, Belgium; frederic.leroy@vub.be

To quote George Orwell, 'it could be plausibly argued that changes of diet are more important than changes of dynasty or even of religion'. Orwell's argument seems to be particularly relevant for the position of animal source foods in the human diet, which is becoming increasingly a matter of anxiety, polarization, and confusion. This trend does not only parallel shifting human-animal interactions, but is also indicative of societal transformation and evolving human needs and wants. Using Foucauldian methods of *archaeology* and *genealogy*, changing attitudes towards the consumption of animal source foods can be uncovered as historical products of the *conditions of possibility* that prevail within a given period (*episteme*). Such an approach is paramount to understanding the origin of the problematic binaries that are now characterizing the post-industrial mindset (Good/Evil, Life/Death, Nature/Culture, etc.) Especially in the urban West, beliefs related to the role of animals in the food system are very different from the ones upheld by hunter-gatherer or pastoralist communities. In the former case, the categories of wildlife, pets, pests, and livestock are commonly used to rationalize and make sense of the various roles of animals, even if this comes with cognitive distortion and oftentimes moral crisis. Animal rights activism further complicates this situation and may potentially lead to instability of the current paradigm. Be that as it may, further deepening of the Nature-Culture divide will have important consequences for a number of topical debates ('Half-Earth' project, rewilding, land sparing vs sharing, veganism, etc.) In some of the more extreme scenarios, the Nature compartment (presented as innocent and strictly biological) and Culture compartment (assumed to be self-restrained and in control) would need to become hyper-separated after further acts of purification. Simultaneously, the Life-Death binary will continue to undermine a more balanced outlook on existence, with death already being perceived by many as a 'contaminant essence'. In contrast, a more holistic view on the matter would benefit humans, animals, and the planet.

Session 23 Theatre 1

Presentation techniques and open science

T. Wallgren[1], I. Adriaens[2] and M. Pszczola[3]
[1]Swedish University of Agricultural Sciences, Department of Animal environment and Health, Box 7068 Uppsala, 75007, Sweden, [2]Wageningen Livestock Research, Animal Breeding and Genomics, Droevendaalsesteeg 4, 6708 PB Wageningen, the Netherlands, [3]Poznan University of Life Sciences, Department of Genetics and Animal Breeding, Wolynska 33, 60-637 Poznan, Poland; torun.wallgren@slu.se

Welcome to this years Young EAAP session! This session is especially aimed at Young researchers, planned by the EAAP Young Club. The topics of this years session is Presentation and communication and Open Science. And as usual, there is plenty of room for networking with your peers! *Part I: Communication and presentation technique* Why, how and what, key questions to ask yourself when considering the transition from academic research to the start-up and business world. This is the premise to the talks of Ana Geraldo (PhD, MED and U.Évora) and Michael Odintsov Vaintrub (COO, regrowth), as they show us what makes a successful elevator pitch and what to expect when transitioning from research to industry. This interactive session will allow the participants to test their elevator pitches with each other and get feedback from their peers. *Part II: Open Science* How to make Open Science the new normal in animal science? The Open Science movement aims at ensuring accessibility, reproducibility and transparency of research. The adoption of Open Science practices in animal science, however, is still at an early stage. In this talk, Anna Olsson (researcher Laboratory Animal Science, i3S – Institute for Research and Innovation in Health, University of Porto, Portugal) and Rafael Muñoz-Tamayo (researcher at INRA, France) will discuss two Open Science topics. The first one is the preregistration approach that aims to ensure reproducibility of research. The second one is a novel model of publication called Peer Community In Animal Science. animal – open space is a new publishing initiative of the animal Consortium, a collaboration between the British Society of Animal Science (BSAS), the Institut National de la Recherche pour l'Agriculture, l'Alimentation et l'Environnement (INRAE) and the European Federation for Animal Science (EAAP). Animal – open space will be presented by Giuseppe Bee, (Research leader, Agroscope Posieux, Schwitzerland) and Isabelle Ortigues-Marty (Editor-in-Chief of animal)

Session 24 Theatre 1

Environmental impacts and services of ruminant livestock: overview and insights from Portugal

T. Domingos
Instituto Superior Técnico, Universidade de Lisboa, MARETEC/LARSyS, Av. Rovisco Pais, 1, 1049-001 Lisboa, Portugal; tdomingos@tecnico.ulisboa.pt

Ruminant livestock production is now taken as a main culprit in environmental pressures. This talk will strive towards a balanced assessment of its impacts and benefits, and will present ways forward in minimising the former and maximising the latter, taking Portugal as a main case study, covering: (1) The context of the climate change, biodiversity, nitrogen and phosphorus cycles and soil degradation challenges. (2) The increased competition for land, balancing its demand for food production, renewable energy (e.g. solar, wind, bioenergy), carbon sequestration and nature conservation. (3) The special role of ruminants in a sustainable food system, due to their capacity to process human inedible (namely fibre-rich) feed transforming it into high nutritional value milk and meat, but a corresponding trade-off with higher methane emissions from less digestible feed. (4) The land use requirements for ruminants, both directly in grazing and indirectly for the production of feed, especially for the more nutritionally demanding phases of their life cycle. (5) The re-integration of grazing livestock and food crops, leveraging the capacity of ruminants to accelerate and concentrate nutrient cycling. (6) The comparison between Haber-Bosch vs legume sources of nitrogen for the food system, and the relative role of forage vs food legumes. (7) The role of grazing ruminant livestock in Mediterranean landscapes, reducing fuel load and creating fire-resistant rural landscapes, and establishing a level of intermediate perturbation that may be maximally beneficial for biodiversity. (8) An intermediate intensification approach: sown biodiverse permanent pastures rich in legumes, with benefits for climate adaptation and mitigation, soil protection and the nitrogen cycle. (9) An integrated strategy, leveraging Fourth Industrial Revolution approaches: precision fertilisation on pastures and forage crops, precision rotational grazing (based on remote sensing of pastures and tracking of animals), precision feeding (based on grazing intake estimates), increased digestibility (namely with legumes), additives for methane reduction, decision support tools for LCA and GHG farm balances.

Session 24 Theatre 2

The role of farm animals in the agroecological transition: evidence from a global FAO tool

A. Mottet, D. Lucantoni, A. Bicksler and R. Sy
FAO, Vialle delle Terme di Caracalla, 00153 Rome, Italy; anne.mottet@fao.org

Livestock are found in all regions of the world and supply a wide range of products and services. They are kept by more than half of rural households and are essential to livelihoods, nutrition and food security. They can contribute to important ecosystem functions such as nutrient cycling, soil carbon sequestration and the conservation of agricultural landscapes. However, most of the sector's development has taken place in large-scale and intensive systems, with relatively little contribution from small-scale producers or pastoralists. Concerns are also growing over the impact of the livestock sector on climate and the environment, the role of livestock in sustainable and healthy diets, animal welfare and the impact of zoonotic diseases on public health. Many means of addressing these risks involve enhancing interactions between animals, plants, humans and the environment as part of an agroecological approach. Since 2019, FAO and partners have been building evidence in over 30 countries on the transition to agroecology and how it contributes to more sustainable food systems. The Tool for Agroecology Performance Evaluation (TAPE) collects information in rural households to propose a diagnostic on how far systems are in the agroecological transition and what their impacts are on different dimensions of sustainability. The 10 elements of agroecology proposed by FAO and approved by member nations are used to establish the diagnostic.10 core criteria of performance are then assessed to measure the impact of systems, namely: land tenure, productivity, income, added value, exposure to pesticides, dietary diversity, women's empowerment, youth employment, biodiversity, and soil health. This study is a cross-country analysis of TAPE's results looking specifically at the role of farm animals. Preliminary results show that the presence of livestock in farms and in territories can significantly improve the economic, social and environmental performances of farms, but under different conditions, which range from improved mobility and access to markets in pastoral systems in the Sahel to better recycling of residues and manure in mixed crop-livestock systems in South America or in Europe.

Session 24 Theatre 3

Exploring the 'indicator explosion' for the European livestock sector

B. Van Der Veeken[1], M. Carozzi[2], C. Barzola Iza[1,3] and F. Accatino[2]
[1]Radboud University, Science, Management & Innovation, Faculty of Science, Heyendaalseweg 135, 6525 AJ Nijmegen, the Netherlands, [2]INRAE, AgroParisTech, Université Paris-Saclay, UMR SADAPT, 16, rue Claude Bernard, 75005 Paris, France, [3]Escuela Superior Politecnica de Litoral, Facultad de Ciencias de la Vida, km 30.5 via Perimetral, 09000 Guayaquil, Ecuador; francesco.accatino@inrae.fr

Promoting the sustainability of the European livestock sector passes through its quantification. In the last decades, several studies explored sustainability using indicators of the environmental (ENV), economic (ECO) and social (SOC) pillars. However, this was referred to as 'indicator explosion' as the indicator development was unbalanced over the three pillars. We aimed at quantifying the 'indicator explosion' for European livestock systems through a literature review, to study the quantity and the diversity of the indicators calculated for each pillar. Starting from search results of 1,048 papers, we selected 44 studies that defined and measured sustainability indicators in Europe at the farm scale. The names of the indicators were harmonized across studies and classified into pillar themes. Indicators were counted as total (T) or unique (U), according to whether the same indicator, used in different studies, was counted multiple times or once. Out of T=736 indicators, most were used in ENV (T=282), followed by SOC (T=271) and ECO (T=183). ENV showed a low diversity of indicators (U/T=0.52), the most represented themes being GHG emissions, nutrient balance, and non-renewable resources, ECO had more diversity of indicators (U/T=0.64) but the great majority (65.5%) of them fell in the profitability theme and were mostly quantitative (94%). SOC had the higher diversity of indicators (U/T=0.72), most of them in the animal welfare theme, with a high percentage of qualitative indicators (50%). In addition, most of indicators (38.9%) were used for dairy cattle systems and the diversification over the three pillars of sustainability started to grow significantly in 2012, while before most efforts were focused only in ENV. This study highlights that more efforts should be used to expand the study of indicators of positive impacts of livestock systems and to make the social indicators more uniformly assessed.

Session 24 Theatre 4

Norwegian wood: a key to sustainable feed production and improved land use efficiency?

H. Møller[1], S. Samsonstuen[2] and H.F. Olsen[2]
[1]NORSUS, Stadion 4, 1671 Kråkerøy, Norway, [2]NMBU, P.O. Box 5003, 1432 Ås, Norway; hanne.fjerdingby@nmbu.no

Global pork production amounts to over 100 mill metric tons and is, together with poultry, the most commonly consumed meat in the world. To maintain a production volume to this extent, there is also a need for an enormous amount of feed ingredients, such as soy and grain. Studies have shown that the largest environmental impact in pig production is from the feed production. Also, the feed production constitutes a threat to biodiversity and to efficient use of land to maximize the production of human edible protein, due to feed-food competition. In Norway, the share of farmed land is only 3,5% due to large mountain areas and deep forests and it is highly relevant to utilize the resources properly and to be able to increase self-sufficiency and reduce the dependency of imported feed, which again is important to ensure food security. By using sugar from hydrolysed wood, yeast can be produced to serve as a protein source in the feed to pigs, as an alternative to the imported soybean meal. In the study, we compared two different domestic biorefinery processes; a demo plant with a complex biorefinery and a small-scale refinery process based on residues. The environmental impacts of the pig production system were assessed using life cycle assessment, with 1 kg of pork as functional unit. In addition, the biodiversity loss and land use ratio were estimated. The results showed that the current pig production with traditional feed was occupying areas suitable for food production and thus contributing negatively to feed-food competition. Replacing the soybean meal with the yeast from wood sugar, with both biorefinery methods, gave lower environmental impact, lower biodiversity loss and higher land use efficiency. The lower potential for food production contributes to the latter, even if the areas of wood were larger than the areas used for soy production. Also, there are possibilities to gain increased circularity in the production system by utilizing residuals and surplus CO_2 from the biorefinery process into for instance greenhouse production. Further studies on the sustainability impact of feeding yeast to pigs, with respect to health traits like gut function and post-weaning diarrhoea, is to be done during 2022.

Agro-industrial co-products to improve economic and environmental sustainability of ruminant livestock

F. Correddu, M.F. Caratzu and G. Pulina
Università di Sassari, Dipartimento di Agraria, Viale Italia, 39, 07100, Sassari, Italy; fcorreddu@uniss.it

Continuous increase of feed prices and the need to reduce the environmental impact (EI) of feed production force to pay attention on the importance of reduce feed and food wastes, to valorise residues, and to promote regenerative and circular economies. In this contest, the use of agro-industrial co-products (AIC) in animals' nutrition should receive more attention, considering that a large amount of edible matter is still lost every year. A systematic literature revision and analysis of official database were conducted to obtain data of some AIC production and their chemical composition, in order to estimate the potential contribution of AIC to reduce feed cost and environmental impact of ruminant livestock. The investigation focused on AIC originates from winery, olive oil, and beer industries in EU: grape pomace, olive cake, and spent grains, respectively. The amount of each AIC was calculated considering the mean total production of grape for wine, olive for oil, and beer in EU in the last years, and the mean percentage of each related AIC retrieved from literature. Contribution of AIC in reduction of feed cost and EI was estimated supposing the replacement of 1/5 of the daily amount of dietary protein from concentrates and an average loss of material of about 50% from the industry to the fed-trough, and considering typical ration for beef cattle, dairy sheep, and dairy goat. The carbon footprint reduction was calculated considering 0.83 and 1.00 kg CO_2eq/kg of concentrate for beef cattle (CP 15%), and sheep (CP 17%) and goat (CP 13%), respectively. The total amount of AIC and the relative protein content were about 18 million tons and 794,184 tons, respectively. The protein available from by-products could be included in the diet of about 13 million beef cattle, or 100 million dairy sheep or 76 million dairy goats. In addition, the use of co-products in the ration of ruminants, could reduce the carbon footprint related to the concentrates for about 20%. A reduction of the concentrate cost of 15% (±5%) was also estimated. In conclusion, the inclusion of AIC (not yet fully considered by feed industries) in ruminant nutrition is a promising strategy to reduce the price of feeds and environmental impact of their production.

Mapping indicators of farming systems' robustness to input constraints: a French case study

C. Pinsard and F. Accatino
INRAE, UMR SADAPT, 16 rue Claude Bernard, 75005 Paris, France; corentin.pinsard@inrae.fr

European farming systems (FS) are currently dependent on the transport of feed and the synthesis of mineral fertilisers. Perturbations, such as the global peak oil that could be reached by 2030, could result in supply shortages. It is necessary to investigate the robustness of European FSs, i.e. their capacity to maintain food production with their current crop-grassland-livestock composition. The aims were: (1) to assess, through indicators, the robustness of French FSs to a joint decline in the input imports; and (2) to explore the links between robustness and crop-grassland-livestock compositions. We simulated a progressive 30-years long decline of synthetic fertiliser and feed imports with a time dynamic FS nitrogen flow model. We defined and calculated two indicators of robustness: (1) the robustness window, i.e. the time period during which the FS food production does not fall below 95% of the current food production; (2) the robustness intensity, i.e. the share of the current total food production remained at year 30. We clustered French FSs according to the robustness indicators and put these clusters in relation to indicators of crop-grassland-livestock composition. Indicators were mapped over the whole France. French FSs had a median robustness window of 8 years and a median robustness intensity of 28%. The most specialised FSs (intensive monogastrics or field crops) have the shortest robustness windows (between 4 and 7 years), being the most dependent on inputs in relation to their needs, but on the contrary had the highest robustness intensity (40%), except for field crops (about 25%), due to feed that could be allocated to food when livestock numbers decreased. On the contrary, mixed FSs tended to have the longest robustness window (from 8 to 19 years) with the levels of robustness being dependent on the degree of integration between crops and livestock. This study presents the first map of robustness indicators of French FSs to potential consequences of global peak oil. Changes in practices or compositions that decrease import dependency, such as an increase in legume areas and a decrease in feed-food competition, should be implemented during the robustness window to increase the resilience of these FSs.

Session 24 — Theatre 7

Diversification of crop-livestock systems: nutrient cycling and food efficiency implications

T. Puech[1] and F. Stark[2]

[1]UR Aster, INRAE, 88500 Mirecourt, France, [2]UMR Selmet, University Montpellier, INRAE, Institut Agro- Montpellier, CIRAD, 34000 Montpellier, France; thomas.puech@inrae.fr

Mixed farming systems are of interest in the search for sustainability because of their species diversity and the potential for synergy from integrating crops and livestock, through nutrient cycling. However, their ability to maximize food production while maintaining their agroecological benefits has been little addressed in the literature and deserves to be explored further. The issue of nutrient cycling raises questions about resource allocation between food crops, feed crops and animal products. This study, based on a whole farm system experiment conducted for about fifteen years in north-eastern France, assesses the biotechnical processes and food production performance of two integrated mixed system configurations. These configurations differed both in their types of production (diversity in both livestock and crops), and in their overall strategies (dairy cattle striving for self-sufficiency vs diversified system maximizing food-crop output). Taking a Nitrogen metabolic approach, the study evaluates biotechnical processes (by ecological network analysis and nutrient balances) and food production efficiency. Our results show that: (1) the two systems metabolism is exclusively based on crop-livestock integration and renewable inputs (biological nitrogen fixation, atmospheric depositions); and (2) the configuration geared to maximizing food production is not the more productive but is the more efficient. In both cases, food efficiency at farm system scale is higher than the food efficiency of each production. This confirms the importance of combining systemic and analytical approaches to better understand and act on the development of performance at different levels. We also show the importance, for a self-sufficient system, of having stocks in reserve to cope with unfavourable years. Finally, our study confirms the value of integrated mixed farming systems in terms of agroecology, but highlights the need for a closer consideration of the role of livestock in those systems (ruminant vs monogastric) to produce food and the trade-offs between food production and nutrient cycling.

Session 24 — Theatre 8

Dual-purpose production of milk and beef – effect on climate change, biodiversity and land use ratio

S. Samsonstuen[1], H. Møller[2], B. Aamaas[3] and H.F. Olsen[1]

[1]NMBU, P.O. Box 5003, 1433, Norway, [2]NORSUS, Stadion 4, 1671 Fredrikstad, Norway, [3]CICERO, Gaustadalléen 21, 0349 Oslo, Norway; stine.samsonstuen@nmbu.no

The global demand for milk and beef has increased the last decades and is expected to have a further increase due to human population growth. Ruminants play an important role in the production of agricultural products. However, the production is facing challenges including GHG emissions, feed-food competition and biodiversity loss related to land use. The intensification of the agricultural sector with selection for milk yield have decreased the number of dairy cattle and increased the proportion of beef from beef breeds. In Norway, the consumption of dairy products regulates the domestic milk production, served by the dual-purpose breed Norwegian Red, also covering 62% of the beef market. The remaining demand for beef is covered from beef breeds (27%) and imported beef (11%). Assuming a constant demand for milk, changes in milk yield will have trade-offs for the specialized beef production. Using LCA, the sustainability of the domestic production of milk and beef in 2040 was assessed, with 2017 as a baseline, through three different scenarios: (1) a trend yield scenario (TY) based on the expected increase in milk yield; (2) a high yield scenario (HY) considering a further increase in milk yield; and (3) a low yield scenario (LY) where the milk yield was optimized to cover the demand for milk and beef exclusively from dual-purpose production. System boundaries were from cradle to farm gate, with focus on climate change, potential loss of biodiversity and land use ratio (LUR). Climate change was calculated using both a GWP100 and a GWP* approach. Considering climate change, the HY scenario showed the lowest impact for the system (CO_2 eq, GWP100). However, when taking the different behaviours of short- and long-term climate pollutants into account, the future scenarios re-rank in terms of warming equivalents (CO_2 we, GWP*), favouring LY. Also, when decreasing milk yield in the LY scenario, the proportion of concentrates in the dairy cow ration was decreased, thereby reducing the potential biodiversity loss associated with concentrate production and the land use efficiency (LUR) of the production system improves as a consequence of reduced feed-food competition.

Maximizing food production while minimizing inputs in farming systems: studying the trade-off

C. Pinsard and F. Accatino
INRAE, UMR SADAPT, 16 rue Claude Bernard, 75005 Paris, France; corentin.pinsard@inrae.fr

European farming systems (FS) are currently dependent on synthetic fertilisers and feed imports while facing environmental, economic and social challenges. Thus their resilience, i.e. their capacity to maintain food production facing these challenges, is questioned. Input autonomy is identified as increasing resilience but it involves trade-offs: e.g. a decline in inputs could lead to a decline in food production. The aims were to: (1) explore the trade-offs between food productions (crop- and animal-sourced) maximisation and input imports (synthetic fertilizer and feed) minimisation for three French FSs; and (2) to explore the set of crop-livestock composition changes that best alleviate this trade-off. We selected three FSs with different crop-livestock compositions: (1) field crops (Plateau Picard); (2) intensive dairy cattle and monogastrics (Bretagne Centrale); (3) extensive ruminants (Bocage Bourbonnais). We defined a FS nitrogen mass flow model and used multi-criteria optimisation techniques, formulating two scenarios: (1) maximisation of animal- and crop-sourced food productions, and minimisation of synthetic fertilizer and feed imports; (2) maximisation of food productions imposing zero input imports. Concerning (1), we found trade-offs between the objectives; the compositions that maximise total food production while minimising inputs cannot be too different from the current composition. The poultry number increased at the expense of cattle due to their better nitrogen use efficiency, except in the extensive ruminant FS as permanent grassland could only be valorised by ruminants. The feed-food competition for biomass use decreased a bit in all FSs. Concerning (2), compositions that maximise total food production showed high decreases of livestock numbers of all species. As a result, the feed-food competition for biomass use decreased significantly (below 10% in the three FSs in median). In both scenarios, these compositions had doubled in most cases the area of protein crops compared to current values. These results showed that livestock redesign is an essential lever to soften the trade-off between food production and input imports. This change will have to be accompanied by a diet change to avoid increasing animal-sourced food imports.

Financial and resource efficiency of Atlantic area dairy farms

L. Shewbridge Carter and M. March
SRUC, Dairy Research and Innovation Centre, Glencaple Road, SRUC Crichton Royal Farm, DG1 4AS Dumfries, United Kingdom; laura.shewbridge.carter@sruc.ac.uk

The efficiency of production systems can be measured using Data Envelopment Analysis (DEA), a programming technique stemming from economic research. Farms are assigned an efficiency score between zero and one, with 1 representing efficient farms, operating along a best practice frontier, and scores assigned to farms that do not operate on the frontier can be used to determine the level of improvement needed to operate efficiently. The dairy industry in the Atlantic Area (AA) represents 20% of EU-27 and UK milk production and, therefore, the economic performance and stability of dairy farms in these regions are incredibly important. This study gauged the financial performance of 100 innovative and/or highly profitable pilot farms in the AA region from the Dairy-4-Future project, which included farms from 12 regions across 5 AA countries, by assessing efficiency through the value of farm revenue and costs of production, applying DEA models to EU and UK farms. Two DEA models were run: Model 1 assessed the efficiency of inputs controlled by farmers (land, labour, and herd size) with outputs including milk yield and buttermilk and protein percentages; Model 2 assessed financial efficiency and included all revenue streams and variable costs. A Tobit model was applied to both DEA models to estimate the impact of variables related to herd management; age at first calving, calving interval, and replacement rate (RR). Model 1 identified that 9% of EU farms (mean: 0.48) and 25% of UK farms (mean: 0.59) were operating on the best practice frontier. Model 2 identified that 35% of EU farms (mean: 0.83) and 72% of UK farms (mean: 0.95) were operating on the best practice frontier. For the EU and UK, farms were on average more efficient financially than in terms of resource input efficiency. Herd management variables had no effect on Model 1 scores. Heard replacement rate had an effect on Model 2 UK farms ($P<0.01$), with an increased RR associated with a decrease in efficiency score, but not on EU farms ($P=0.9$). Overall, the efficiency of farmer-controlled inputs has more room to improve than financial efficiency measures on these pilot farms in the AA region, however, farm selection criteria for farms may have skewed these results.

Integrating ecosystem services in the LCA evaluation of meat of different species

F. Joly[1], P.K. Roche[2] and L. Boone[3]

[1]Université Clermont Auvergne, INRAE, VetAgro Sup, UMR Herbivores, 63122 St Genes Champanelle, France, [2]INRAE, UMR RECOVER, 3275 Route de Cézanne, 13182 Aix-en-Provence, France, [3]Ghent University, Department of Green Chemistry and Technology, Sint-Pietersnieuwstr 25, Ghent, Belgium; frederic.joly@inrae.fr

The environmental impact of meat production assessed from life cycle assessment (further LCA impact), is highly dependent on livestock species. To produce one kg of meat, energy consumption and greenhouse gas emissions increase from chicken to pork, and from pork to beef, partly because of differences in feed efficiency. Beef production has thus the highest impact according to LCA but it can be also beneficial to biodiversity and ecosystem services (ES), if it is grass based. To integrate these aspects, a method has been proposed to allocate the LCA impacts between the strictly productive activities, and the other types of services. The method proposes allocation factors based on the capacity of production systems to supply provisioning ES (PES) and regulating ES (RES). The factors represent the proportions of LCA impacts that can be allocated to these ES (sum of RES and PES factors =1). The method has been applied to compare organic and conventional crop productions and here we apply it to the production of 1 kg of chicken, pork and grass-based beef. We modified the method to account for the fact that feeding systems can involve several land uses. Without allocation, the differences in LCA impact between chicken and beef along the species gradient are two and a half higher for energy (~20 MJ/kg to ~50 MJ/kg), and six times higher for CO_2-eq (~5 to ~30 kg CO_2-eq/kg). When these differences are reallocated according to the PES factors, the energy gradient is modified with beef having the lowest impact (14 MJ/kg), and pork the highest (20 MJ/kg). The CO_2-eq gradient is not modified, but the difference of impact is reduced to two times (4 to 8 kg CO_2-eq/kg). These calculations indicate that depending on the species reared, feed efficiency and feeding system, the impact of one kilo of meat benefits more or less to services other than productive ones. It does not mean that the impact of producing 1 kg of beef should be considered lower than currently assessed, but that this impact can also contribute to the delivery of ES of interest.

Identification and characterization of agri-forestry system in Portugal through remote sensing

T.G. Morais, R.F.M. Teixeira and T. Domingos

MARETEC – Marine, Environment and Technology Centre, LARSyS, Instituto Superior Té, Avenida Rovisco Pais, 1, 1049-001, Portugal; tiago.g.morais@tecnico.ulisboa.pt

Sown biodiverse pastures (SBP) are a pasture system developed in Portugal to provide quality animal feed and offset concentrate consumption. This pasture system has been shown to increase ecosystems services provision, such as carbon sequestration. However, SBP are mostly located in agri-forestry montado regions. Nevertheless, the maintenance of this biodiverse and complex land cover system is threatened, among other causes, due to shrubs encroachment. In this work, we combined remote sensing data with machine learning algorithms to: (1) identify SBP areas; and (2) characterize Montado areas, we identified tree areas, shrub areas and covered soil and/or bare soil and other areas. For identification of SBP areas, we used Landsat-7 spectral data and climate data. Gradient boosted decision trees (XGBOOST) and artificial neural networks (ANN). The identification of SBP performance, on an independent validation dataset, was 90%, with 80% recall and F1-Score of 85%. The total estimated area of SBP in Alentejo region was 102,000 hectares in 2013, which is similar to the total known installed area (100,000 hectares). The estimated spatial distribution is in accordance with the known distribution at municipal level. For characterization of SBP landscapes, the deep convolution neural networks architecture used was U-net. For train models, 800 images with 10,000 m^2 each were used to train the model, which were divided between training and test set. In the best model, the overall classification performance (measured on the test set) was 94%, the recall 91% and the F1-Score 90%. Nevertheless, identification of shrubs had the lowest performance (accuracy of 85%), which are mainly confused with trees that have similar spectral signature. Obtained model make possible the identification of the status of SBP systems in Montado ecosystem, namely of shrub encroachment.

Session 24 Theatre 13

Complementarity between production basins to ensure optimal supply of organic lambs along the year

M. Benoit[1], V. Bellet[2], M. Miquel[3], S. Béziat[1] and C. Experton[4]
[1]INRAE, Theix, 63122, France, [2]IDELE, CS 45002, 86550 Mignaloux-Beauvoir, France, [3]IDELE, 9 allée Pierre de Fermat, 63170 Aubiere, France, [4]ITAB, 149 rue de Bercy, 75595 Paris, France; marc-p.benoit@inrae.fr

Even if it remains limited, the demand for organic sheep meat is a real one. In France, this production shows contrasting production calendars, with summer-autumn lamb production in the north and west areas, winter and spring production in the south. As we are considering here the long channels marketing, these regional specificities raise the question of the adequacy of the lamb supply to the demand of the sheep industry and leads to consider complementarities of supplies between regions throughout the year. On the basis of 11 typical farms in organic farming, contrasted in terms of organization of ewe reproduction and animal feeding, and distributed in these two large regions, we sought, by modelling, to determine the combination of these 11 farms which makes it possible to approach as closely as possible the demand throughout the year. By relaxing this constraint, we can introduce complementary objectives of maximizing the overall performance of the pool of farms, in economic terms (income), and from environmental (GHG and energy) and feed-food competition point of view. We show that it is not possible to reduce the deficit of the most deficient period (April) to less than 23% of the demand; this optimum resulted from the combination of 60% of farms in the North zone and 40% in the South. By 'opening' the calendar matching constraint to 41% of the allowed deficit between supply and demand in April, we show that the global income increases by 51%, GHG emissions decrease by 15% and energy consumption by 6%, and that the feed-food competition improves by 14%. There is thus a tension between the demand of the sector and consumers for a regular production of lambs all year round on the one hand, and the economic interest of the producers and that of the community in environmental and social terms on the other. This work allows us to quantify these antagonisms and opens a discussion on possible solutions while pointing out the possible methodological advances in terms of optimization of the combination of production systems.

Session 25 Theatre 1

Phenotyping pigs and dairy cows using PLF technologies: applications for animal welfare assessment

P. Llonch and X. Manteca
Universitat Autònoma de Barcelona, Department of Animal and Food Science, Campus UAB, 08193 Cerdanyola del Vallès, Spain; pol.llonch@uab.cat

Precision livestock farming (PLF) has emerged as a revolutionary tool to deepen our knowledge on farm animals and improve their welfare and husbandry. Sensor technologies are capable of monitoring a growing range of phenotypical variables such as animal behaviour (e.g. activity), physiological response (e.g. body temperature), health status (e.g. somatic cell count to detect mastitis) and productivity (e.g. daily milk production). PLF data potential is likely one of the more revolutionary advances that animal production has seen in decades. PLF data can be used for many purposes including evaluation of the genetic merit of individual animals, monitoring their performance and assessing their welfare. PLF technology can identify existing welfare issues but also to anticipate them (i.e. early warning signals), so that preventative measures can be implemented. Also, they provide continuous information, thus expanding the available information on the severity and extent of existing welfare problems. PLF is able to monitor the welfare status at an individual level, facilitating individualized strategies to improve animal welfare. Last but not least, data from PLF technologies is objective and quantifiable, which facilitates transparent and unbiased analysis. There are several issues that require further work before PLF technology can be widely used to assess animal welfare. Firstly, whereas some technologies have a high performance, others (e.g. body condition score) still appear to have lower performance. Furthermore, the majority of existing technological tools have been validated only on adult animals and there is a lack of studies on younger animals. Also, available PLF technologies focus on some domains of animal welfare (mostly feeding and health), whereas other domains, such as housing and appropriate behaviour, remain underrepresented, with less available technologies to assess them. Importantly, most -if not all- PLF technologies are potentially useful to identify negative welfare states but fail to provide information on positive welfare. Finally, existing data analysis tools focus on the current status of the animals and advanced mathematical should be developed to predict welfare problems at a longer term.

Session 25 Theatre 2

Bringing precision livestock farming data to consumers – evidence from a conjoint analysis

C. Krampe[1], J.K. Niemi[2], J. Serratosa[3] and P.T.M. Ingenbleek[1]
[1]Wageningen University and Research, Hollandseweg 1, 6706 KN Wageningen, the Netherlands, [2]Natural Resources Institute Finland (Luke), Kampusranta 9, 60320, Finland, [3]Universitat Autònoma de Barcelona, Campus UAB, 08193 Cerdanyola del Vallès, Barcelona, Spain; jarkko.niemi@luke.fi

Precision Livestock Farming (PLF) has demonstrated to improve management and optimisation processes on farms. While some have suggested that PLF can also have potential to promote sustainable consumer choices, a gap remain regarding how the connection with consumers can be made beyond traditional animal welfare labels. In the EU-funded Horizon 2020 projec 'ClearFarm' (862919), consumer preferences for the design attributes of a system that connects PLF-data to the choice context of consumers were investigated. Drawing on stakeholder interviews and consumer focus groups, potential attributes for the system were selected, like whether consumers could use the system to insert restrictions for the sets of products that they would be confronted with (like 'animal welfare and/or reduction of food waste'), the ability to interact directly with farmers or retailers, and whether the system would apply to online purchases, supermarket visits, or both. Attributes were tested in a conjoint study, a method to identify the importance that consumers attach to the different attributes, with 1,194 consumers from Finland, the Netherlands, Spain, and Italy. The results show that consumers attach the highest preference to use the system in the supermarket. They find it important that the system: offers additional benefits like price discounts or sustainability (e.g. CO_2-emission); informs them about the match between their own animal welfare standards and the actual welfare levels associated with the products; and is governed by a trustworthy organization. Other attributes like whether the system applies to pork or dairy, or whether it allows to send feedback to retailers or farmers, were deemed slightly less important. Additional analyses showed country- and segment-specific differences, suggesting to design flexible systems that can be geared to specific consumer groups. The results give guidance to design PLF-systems and to extend them towards consumers' exploitation. They also suggests that some persistent issues with 'traditional' animal welfare labels can be overcame.

Session 25 Theatre 3

Analysis of quality schemes: how far are we from data-driven and animal-based welfare assessment?

A. Stygar[1], C. Krampe[2], P. Llonch[3] and J. Niemi[4]
[1]Natural Resources Institute Finland (Luke), Latokartanonkaari 9, 00790 Helsinki, Finland, [2]Wageningen University and Research, Hollandseweg 1, 6706 KN Wageningen, the Netherlands, [3]Universitat Autònoma de Barcelona, Campus UAB, 08193 Cerdanyola del Vallès, Barcelona, Spain, [4]Natural Resources Institute Finland (Luke), Kampusranta 9, 60320 Seinäjoki, Finland; anna.stygar@luke.fi

Within the EU there is no harmonization of animal welfare quality schemes of meat and dairy products. Instead, there are several industry-driven initiatives and voluntary schemes, which aim to provide information on animal welfare for attentive consumer segments. The aim of this study was two-folded. First, we quantified how selected industry-wide quality schemes cover the welfare of pigs and dairy cattle on farms by comparing evaluation criteria selected by the schemes with animal-, resource- and management-based measures defined in the Welfare Quality protocol (WQ®). Second, we identified how data generated along the value chain (sensors, breeding, production, and health recordings) are used by quality schemes for animal welfare assessments. Twelve quality schemes applying 19 standards for certification (nine for dairy and ten for pig production), were selected for the analysis. Schemes originated from Finland, Sweden, Denmark, Ireland, the Netherlands, Germany, Austria, and Spain. Most of the analysed schemes were comprehensive in welfare assessment, covering four welfare domains including feeding, housing, health, and appropriate behaviour. When compared to WQ, only 5 evaluation protocols use predominantly animal-based measures, in front of 14 protocols using mainly environmental measures. The utilization of data generated along the supply chain by quality schemes remains also low as only one quality scheme allowed direct application of sensors for providing information on animal welfare. Though, several schemes used routinely collected data from farm recording systems, mostly on animal health. Our results suggest that quality schemes are failing to inform about the perception of the animal to the environment in a continuous manner. Future research should concentrate on enhancing standards for certification by broader utilization of animal-based data generated along the supply chain.

Session 25 Theatre 4

Individual feeding patterns as indicators of pig welfare

E.A.M. Bokkers[1], J.D. Bus[1], J. Engel[2], R. Wehrens[2] and I.J.M.M. Boumans[1]
[1]Wageningen University & Research, Animal Production Systems, P.O. Box 338, 6700 AH Wageningen, the Netherlands, [2]Wageningen University & Research, Biometris, P.O. Box 16, 6700 AA Wageningen, the Netherlands; eddie.bokkers@wur.nl

Electronic feeders (EF) are used in group-housed sows and fattening pigs. EF record animal identification, feed intake and time stamps of entering and leaving a feeder. Individual feeding patterns (e.g. feeding frequency, rate and time and meal duration) can be extracted from these data, though little is known about the variation across and within days of the basal diurnal patterns. We studied variation in diurnal feeding patterns between pigs and within a pig over time. Data was collected at a farm using 12 EF (Nedap) within one group of ~640 gestating sows, and at a farm using single-space EF (Hokofarm group) in pens with 11 fattening pigs. We included records of 239 sows (parity 1-9) with 100 days of feeding data of one gestation and of 110 fattening pigs during their growing period of 95 days. In sows, data was cleaned and aggregated to daily and full gestation period and in fattening pigs to the hourly level and scaled at pig level by dividing hourly by daily intake. Sows visited EF on average 3±2.8 times/day, but mostly ate their allowance at once in 18±3.7 min during the first visit with a feeding rate of 151±18.7 g/min. All feeding patterns showed differences in distributions between sows and effects were found for the hour of first visits, rank in feeding order, number of visits without feed allowance and feeding rate (all P<0.01; Kruskal-Wallis). In fattening pigs, self-organising maps (dissimilarity based on weighted cross-correlation, WCC) combined with hierarchical clustering showed that diurnal patterns could be summarised in 8 clusters, differing in the number, height, width and timing of intake peaks. WCC-coefficients at lag=1d were between 0.40-0.80 (median 0.61) for individual pigs, suggesting quite some pigs show similar patterns in adjacent days. As the lag increased, WCC-coefficients decreased (lag=28d, median 0.58, range 0.28-0.75), reflecting a slow change in diurnal patterns over time. We conclude that pigs show individual feeding patterns, which develop over time. This can help to identify normal and deviating patterns and improve detection of welfare problems via EF.

Session 25 Theatre 5

Validating play behaviour as a positive welfare indicator in pigs (*Sus scrofa*) around weaning

G. Amorim Franchi[1], M.L.V. Larsen[1,2], I.H. Kristoffersen[1], J.F.M. Winters[1], M.B. Jensen[1] and L.J. Pedersen[1]
[1]Aarhus University, Department of Animal Science, Blichers Allé 20, 8830 Tjele, Denmark, [2]KU Leuven, Department of Biosystems, Division Animal and Human Health engineering, M3-BIORES, Kasteelpark Arenberg 30, 3001 Heverlee, Belgium; amorimfranchi@anis.au.dk

Play behaviour in juveniles is reduced by threats to animal welfare such as hunger and perceived danger. Conversely, engagement in play behaviour has been linked to the presence of positive-valanced affective states. Hence, play behaviour has been proposed to be an indicator of positive animal welfare. However, the interpretation of play in animals remains challenging due to its variability between and within species as well as complex motivating factors. To clarify underlying factors, we investigated, as part of the H2020 project ClearFarm, the relationship between locomotor and social play behaviour and physiological (i.e. growth and saliva cortisol), clinical (i.e. faeces score and ear damage) and behavioural (i.e. visits to feeder and drinker, tail posture and tail motion) measures in conventional Yorkshire × Landrace pigs before and after weaning. Weaning is a stressful event in the life of pigs typically involving abrupt separation from the sow, dietary change, and regrouping in a novel environment. In total, 24 litters [pigs/litter: (mean ± SD) 13±2; age at weaning: 26±2 days] raised under conventional husbandry conditions were included in this study. Before weaning, increased performance of locomotor play was associated with increased growth rate, absence of diarrhoea (indicated by liquid faeces), and increased frequency of visits to the drinker. During the first 48 h post-weaning, increased locomotor play was observed in pigs with increased growth, absence of diarrhoea, absence of ear damage, and increased frequency of visits to the feeder. Social play positively related with growth pre- and post-weaning. No relations between play behaviour and saliva cortisol, tail posture and tail motion were observed. Our findings corroborate the idea that an animal in a more favourable condition (e.g. well-nourished and without sickness or injuries) are motivated to perform more play behaviour. The methods, results and discussion will be presented in detail.

Comparative study between scan sampling observations and automatic monitoring image systems in pigs

Q. Allueva-Molina[1], H.-L. Ko[1], Y. Gómez[1], P. Fuentes Pardo[2], X. Manteca[1] and P. Llonch[1]
[1]School of Veterinary Science, Universitat Autònoma de Barcelona, 08193 Cerdanyola del Vallès, Spain, [2]Department of I+D+i, CEFU, S.A., Paraje de la Costera, s/n, 30840 Alhama de Murcia, Spain; pol.llonch@uab.cat

Computer vision is progressively integrated as a technology to monitor animal behaviour on farm and can assist in applied research and decision making for farmers. However, the accuracy of computer vision needs to be assessed. This study compared a computer vision sensor developed by Copeeks (Copeeks SAS, France) with human-based observation on a commercial pig farm, including activity, posture and location within the pen, to determine the accuracy of the automatic monitoring image system. Two sensors (Peek_1 and Peek_2) were installed and each sensor recorded information from two pens, with 20 and 19 pigs each (n=39). One observer performed scan sampling for five consecutive days. The ethogram included behaviour (resting, walking, exploring, eating, drinking), posture (lying, sitting, standing), and location (resting area, feeder, drinker, enrichment tool). Each day was distributed into three intervals of two-hour slots (9-11, 13-15, and 16-18h), with 30 hours of observation in total. Each pig was observed six times per hour, with 7,018 observations in total (180 observations/pig). Data obtained through human observations and the sensors were contrasted using Pearson correlation tests. Considering human observation as the gold standard, the sensors showed a higher accuracy at recording posture in different areas of the pen (Peek_1: $r=0.80$, $P<0.001$; Peek_2: $r=0.75$, $P<0.001$). Conversely, correlation between automatic monitoring image system and human observation was low ($r<0.40$; $P<0.05$ for all comparisons), which could be due to differences on behavioural analysis by the two methods. For instance, human observation of pig walking considered the frequency (i.e. the number of times pigs were seen walking), whereas the image system may consider other features such as distance walked, but this could not be confirmed as the algorithm is a protected knowledge from the company. In conclusion, Copeeks sensors can provide accurate information regarding posture, whereas further investigation is needed to confirm whether computer vision on pig behaviour is accurate.

Classification of dairy cattle welfare status using sensor data – ClearFarm pilot studies

A. Stygar[1], L. Frondelius[1], G. Berteselli[2], Y. Gómez[3], E. Canali[2], J. Niemi[1], P. Llonch[3] and M. Pastell[1]
[1]Natural Resources Institute Finland (Luke), Latokartanonkaari 9, 00790 Helsinki, Finland, [2]Università degli Studi di Milano, Via dell'Università 6, 26900 Lodi, Italy, [3]Universitat Autònoma de Barcelona, Campus UAB, Campus UAB, 08193 Cerdanyola del Vallès, Barcelona, Spain; anna.stygar@luke.fi

Welfare assessment of dairy cows by in-person farm visits has been used for many years. However, this approach only provides a snapshot of welfare, is time-consuming and costly. Possible solutions to reduce the need for in-person assessment visit would be to exploit sensor data which are continuously collected on farms. The aim of this study was to develop an algorithm to classify dairy cow welfare based on sensor data. In total 300 cows from six commercial farms located in Finland, Italy and Spain were enrolled for pilot study data collection between February and June 2021. During this time, the welfare of dairy cows from piloted herds was closely monitored. At the beginning, the middle and the end of the pilot experienced observers scored cows using 12 animal-based indicators of good feeding, health, housing and appropriate behaviour based on Welfare Quality protocol. Indicators, such as body condition score, lameness, coughing, diarrhoea, nasal, ocular and vulvar discharge, udder, flank and lower legs cleanliness, hairless patches, lesions and avoidance distance were assigned with a severity score (min. 0, max. 1 or 2) and duration (min. 7, max. 45 days). Farm registrations on treatments concerning e.g. mastitis, ketosis were used to supplement the score for acute welfare issues. The final welfare index was obtained by summing up scores on daily basis for each cow. The lower the score, the better welfare. Parallel to manual registrations, data from two sensor types, namely milking robots (e.g. kg of milk produced, number of visits to the robot) and a neck mounted accelerometer (e.g. rumination, walking, lying time) were automatically collected. We intend to investigate whether it is possible to classify dairy cattle according to their welfare status using sensor data. For analyses we will employ machine learning algorithm based on decision trees. In the remaining time of the project, we hope to validate the developed algorithm internally and externally.

Session 25 Theatre 8

Behavioural traits from accelerometers to alert the potential occurrence of acidosis in dairy cows

Y. Gómez[1], G. Berteselli[2], E. Dalla Costa[2], E. Canali[2], D. Ruiz[3], X. Manteca[1] and P. Llonch[1]
[1]Universitat Autònoma de Barcelona, Department of Animal and Food Science, Campus UAB, 08193 Cerdanyola del Vallès, Spain, [2]Università degli Studi di Milano, Dipartimento di Medicina Veterinaria e Scienze Animali, Via dell'Università, 6, 26900 Lodi, Italy, [3]COVAP, C. Mayor, 56, 14400 Pozoblanco, Spain; yanethrocio.gomez@uab.cat

Identifying ruminal acidosis is possible by using rumen boluses. On the other hand, accelerometers can measure a wide variety of behaviours related to individual animal activity. Accelerometers are generally used to inform the farmer about reproductive traits and deviations of normal behaviour, but little is known about its potential to detect ruminal acidosis. Our aim is to evaluate the ability of three Behavioural indicators (feeding, rumination, and resting times), measured with an accelerometry collar, to alert the occurrence of acidosis. The study lasted three months in a farm located in Lombardy (Italy), and two farms located in Andalucía (Spain). 5 cases of acidosis were diagnosed though bolus in the Spanish farms (pH below 5.6 at least 50 min/day), and 3 more cases were confirmed in the veterinary records of the Italian farm. All cows presenting acidosis were included as experimental group (n=8). A control group of healthy cows (n=8) was balanced according to parity and lactation stage of the experimental group. A time window of six days before the diagnosis of acidosis was considered. One day before the presentation of clinical manifestations, animals suffering from acute or subacute ruminal acidosis showed a slight increase in resting time, and a slight decrease in feeding time, compared to control group (12±2 vs 11±1 h, and 2±1 vs 4±1 h, respectively), detectable by means of accelerometers. T-test was applied to compare the mean daily time spent on each behaviour between groups. One day before diagnosis, rumination time decreased significantly ($P \leq 0.05$). No difference was found on feeding nor resting time between groups the days before diagnosis. These results suggest that ruminating time could be a useful indicator to generate timely alerts of acidosis, and can help to prevent sickness development. Therefore, the construction of algorithms for early detection of acidosis, using behavioural indicators such as rumination, measured with accelerometry sensors, is promising.

Session 25 Theatre 9

Correlations between environmental data with activity level and respiratory health by PLF sensors

H.-L. Ko[1], P. Fuentes Pardo[2], X. Manteca[1] and P. Llonch[1]
[1]School of Veterinary Science, Universitat Autònoma de Barcelona, 08193 Cerdanyola del Vallès, Spain, [2]Department of I+D+i, CEFU, S.A., Paraje de la Costera, s/n, 30840 Alhama de Murcia, Spain; henglun.ko@uab.cat

Incorporating sensing technology into on-farm welfare assessment is a trend in modern livestock farming. Using sensors to collect quantitative information from the environment and the animals can assist the producers on decision making. The aim of the study was to investigate the correlation of the parameters collected by two commercial sensors on two pig farms in Spain (one after weaning and one during fattening). The two sensors used in the study were: Diagno'Peek (Copeeks SAS, France) and SoundTalks (SoundTalks NV, Belgium). The Diagno'Peek collected environmental data (temperature, humidity, NH_3 and CO_2; on average 710 data points/day) and animal-based data (activity level of the animals; on average 27 data points/day). The SoundTalks generated a Respiratory Health Status (ReHS) daily, in which the value ranges from 0 to 100 (i.e. the higher the value, the healthier the animals). Data were collected from two pens (one pen on each farm) from March to June continuously. The means of each environmental and animal-based parameter from Diagno'Peek were calculated to correlate with ReHS from SoundTalks. The correlations between the parameters were analysed by Pearson's correlation tests. Results showed that activity level was moderately correlated with environmental parameters. The activity level increased when NH_3 ($r=-0.42$, $P<0.001$), humidity ($r=-0.34$, $P<0.001$) and CO_2 ($r=-0.30$, $P<0.001$) decreased. On the other hand, the activity level increased when temperature ($r=0.36$, $P<0.001$) was higher. The correlations between ReHS and environmental data were weak (CO_2, $r=-0.27$; humidity, $r=-0.22$; NH_3, $r=-0.19$; temperature, $r=0.27$, all $P<0.01$). Lastly, there was no correlation between activity level and ReHS ($P>0.05$). Our study shows that the activity level and the respiratory health status of pigs are moderately and weakly associated with temperature and air quality of the environment, respectively. Quantitative information collected by sensors may be useful for on-farm welfare monitoring in real-time, but a larger sample size and a longer study duration are warranted.

Session 25 Theatre 10

Relationship between climatic and housing conditions and carcass diagnostics using sensor technology

G.B.C. Backus
Connecting Agri and Food, Oostwijk 5, 5406 XT Uden, the Netherlands; g.backus@connectingagriandfood.nl

This study presents a unified analysis of the relationship between climatic and housing conditions and pig carcass diagnostics. Real time sensor climate data was collected from 34 farms: CO_2, NH_3, humidity and temperature. Outside climate data were also collected. Carcass diagnostics were gathered from slaughterhouses: pleurisy, lung and liver problems, pericardium, and carcass remarks. We analysed data from 122 groups of growing pigs. For 23 farms also information on feeding (wet versus dry feed), building material (concrete, synthetic and plaster) and ventilation (door & air canal, trickle deck, slot ventilation) was collected. For each round, daily statistics (mean, min, max, median, std, differences between 2 consecutive hours) were created. Correlations between climate and housing conditions and carcass diagnostics were computed. Daily climate parameter patterns show most days a nearly constant indoor temperature, with a small maximum in early/middle afternoon. Maximum CO_2 values and humidity values were reached in the morning. Minimum CO_2 and humidity values were observed in the afternoon, before raising again at night. Positive correlations were calculated for: CO_2 and pleurisy; CO_2 and lung problems; temperature and pleurisy; temperature and liver problems. Correlations between carcass diagnostics and the percentage of observations outside a temperature range start to be significant (P=0.1) once the upper threshold 250 °C or higher. Prevalence of pleurisy and of lung problems were positively correlated at 3.000 and 3,500 ppm CO_2. Maximum correlation coefficients were calculated at 25 ppm for NH_3, around 3,000 ppm for CO_2, and 26 °C. Strong negative correlations were calculated for: average humidity and average difference of temperature for two consecutive hours; average humidity and average difference of CO_2 for two consecutive hours; average difference of humidity & average difference of temperature for two consecutive hours. Animals fed with wet feed had a higher prevalence of lung problems and pleurisy. Barns with trickle deck ventilation had higher NH_3 concentrations and more pig carcasses with drift tracks and abscesses. Barns with synthetic walls had higher NH_3 concentrations and higher prevalence of pleurisy and abscesses.

Session 25 Poster 11

Decrease in rumination as early warning indicator of clinical mastitis onset in high-producing dairy

G.V. Berteselli[1], E. Dalla Costa[1], Y. Gómez Herrera[2], M.G. Riva[1], S. Barbieri[1], R. Zanchetta[1], P. Llonch[2], X. Manteca[2] and E. Canali[1]
[1]Università degli Studi di Milano, Dipartimento di Medicina Veterinaria e Scienze Animali, via dell'Università, 26900 Lodi, Italy, [2]Universitat Autònoma de Barcelona, Campus UAB, 08193 Cerdany, Department of Animal and Food Science, Campus UAB, 08193 Cerdanyola del Vallès, Barcelona, Spain; greta.berteselli@unimi.it

Behaviour is recognized as an important indicator of health and welfare of dairy cattle. In dairy cows, changes in behavioural patterns can be automatically monitored by PLF technology and used to early detect diseases such as mastitis, the most crucial health issue of high-producing dairy cows. The aim of the present study was to identify whether behaviours, automatically monitored with PLF technology, change in dairy cows suffering from mastitis. Two groups of dairy cows balanced for stage of lactation and parity order, were recruited: mastitis group (n=19) and control group (n=19). Mastitis was diagnosed through somatic cell count in milk. For each cow, daily hours spent ruminating, eating and lying down were recorded using accelerometery collars. Behaviours were monitored for a total of 9 days: 5 days prior to diagnosis of mastitis, the day of diagnosis and starting of antibiotic treatment, and 3 days after; for the control group the same days were considered. Mann-Whitney test was used to assess differences between groups on each day. Three days before diagnosis, cows with mastitis ruminated significantly less than control cows ($P \leq 0.05$), in particular cows with mastitis ruminated for 7±1 hours, while control cows for 8±1 hours. After 3 days of antibiotic treatment, no differences between groups were found for time spent ruminating. Regarding the other recorded behaviours (eating and lying down), no differences were found between groups. Rumination may be a proxy behaviour to early detect mastitis. Our results show that cows with mastitis show a decrease in rumination 3 days before diagnosis, and that this behaviour recovers after antibiotic treatment. Further studies are needed to confirm the possible role of PLF in monitoring dairy cows' behaviours to identify automatically sick cows and to predict the onset of diseases that threaten animal welfare in the dairy industry.

Session 25 Poster 12

Correlation of physiology and behaviour traits from PLF sensors with salivary biomarkers in dairy cow

Y. Gómez[1], M. Contreras[2], J. Cerón[2], D. Ruiz[3], X. Manteca[1] and P. Llonch[1]
[1]Universitat Autònoma de Barcelona, Department of Animal and Food Science, Campus UAB, 08193 Cerdanyola del Vallès, Spain, [2]University of Murcia, Department of Animal Medicine and Surgery, Campus Mare Nostrum, Espinardo 30100, Spain, [3]COVAP, Valle de los Pedroches Livestock Cooperative, C. Mayor, 56, 14400 Pozoblanco, Spain; yanethrocio.gomez@uab.cat

Widely used PLF technologies allow to monitor physiological and behavioural parameters of individual animals. The aim of this study was to assess the predictive capacity of automatic measurements generated by PLF as potential indicators of cow welfare, taking as reference stress, metabolic and immunity biomarkers, as well as production and health records. During three months, 40 dairy cows from two commercial intensive dairy farms located in Andalucía (Spain), were fitted with ruminal boluses and accelerometery collars. Saliva samplings were taken from each animal on days 1, 45 and 90. Selected cows were clinically healthy, between 24 months to 8 years old, of various parities (from 1 to 5 calvings) and at different lactation stages. Correlation between acute stress salivary biomarkers (cortisol, total esterase activity (TEA), alpha-amylase, lipase), immunity (total adenosin-deaminase, adenosin-deaminase 1, adenosin-deaminase 2), and metabolic status (creatinkinase, total protein) with animal-based automatic measures was evaluated (ruminal pH and temperature, times standing, lying, walking, ruminating, and eating). When lying time decreased, cortisol increased. When walking and standing time decreased, lipase and TEA increased. Rumination time was positively correlated to rumen pH. Veterinary and production records of cows registering physiological and behavioural changes, evidenced the presence of lameness, mastitis, or ruminal acidosis. These findings point out that integration of physiological and behavioural indicators, monitored by means of PLF technologies, may have the capacity to inform about metabolic or inflammatory conditions that lead to pain and stress in dairy cows, which may be used to assess their welfare.

Session 26 Theatre 1

Climate care cattle farming initiative

P. Groot Koerkamp[1], D. Ruska[2], A. Cieslak[3], S. Koenig[4], V. Juskiene[5], N. Edouard[6], X. Verge[7], M. Barbari[8], P. Galama[1] and A. Kuipers[1]
[1]Wageningen University & Research, De Elst 1, 6708 WD Wageningen, the Netherlands, [2]Latvia University of Life Sciences and Technologies, 2 Liela Street, Jelgava 3001, Latvia, [3]Poznan University, Wolynska 33, 60-637 Poznan, Poland, [4]Justus Liebig Universität, Ludwigstraße 21 b 35390, 35390 Gießen, Germany, [5]Lithuanian University of Health Sciences, R. Zebenkos, 1282317 Baisogala, Lithuania, [6]PEGASE INRAE, Institut Agro, 35590 Saint Gilles, France, [7]Institute de l'Elevage, la Mouesonnais, 35650 le Rheu, France, [8]University of Florence, Via San Bonaventura 13, 50145 Firenze, Italy; abele.kuipers@wur.nl

The cattle sector is a major source of greenhouse gas (GHG = CH_4 + N_2O + CO_2) and ammonia (NH_3) emissions from rumination of cows, manure, and in a less extend, from grassland and corn cultivation including machinery. The www.CCCFarming.eu project focusses on the processes within the dairy farm itself, which is about two-thirds of the GHG emissions from this sector. Of GHG, the major component to deal with is methane in the barn and manure storage, of which about 80% is expected to come from the cow and 20% from the manure. The project aim is to develop climate smart cattle farming systems reducing GHG and ammonia emissions while maintaining the social-economic outlook of the farm business. The following activities are performed: Kitchen table interviews to learn about farm families' attitudes towards dealing with GHG and ammonia emissions on farm level in 8 European countries; Gathering field data about the nitrogen, phosphorus and carbon balances from 60 farms with three different NPC tools, and estimates of GHG and ammonia emissions by a so-called simplified measurement method (especially within the housing), and, in development, application of drones; Examine the socio-economic and political aspects of the possible interventions identified; Develop holistic farm systems which are targeted at meeting the socio-economic and environmental goals; Develop a set of good practices and techniques to be combined in farm systems, to be checked and evaluated on pilot and experimental farms. The progress of the project activities will be explained and illustrated by a series of presentations during this seminar.

Climate care research and applications in Portugal

H. Trindade[1] *and D. Fangueiro*[2]
[1]*Universidade de Trás-os-Montes e Alto Douro (UTAD), Department of Agronomy, CITAB, Quinta de Prados, 5000-801 Vila Real, Portugal,* [2]*Instituto Superior de Agronomia, Universidade de Lisboa, Tapada da Ajuda, 1349-017 Lisboa, Portugal; htrindad@utad.pt*

Mainland Portugal is mainly located in the NW coastal region and in the Tajo river plain and Alentejo regions where the most fertile soils are located and with access to irrigation systems. Under these conditions a specialized dairy system has evolved based on an intensive zero-grazing regime grounded on two forage crops per year for silage making, maize and a winter crop with high annual forage yields (ca. 30 t DM/ha). This dairy system with high stocking rates (5-7 LSU/ha) faces important environmental issues, which continue to act as a main driving force for research studies and adoption of measures to improve its sustainability. The first environmental studies began in the 1990s, covering the assessment of nitrogen fluxes and losses from soils throughout nitrate leaching, ammonia volatilization and emission of biogenic GHG (CH_4, N_2O and CO_2). In the last two decades, studies were carried out on NH_3, N_2O and CH_4 emissions from animal housing to obtain national emission factors and the evaluation of slurry management and treatment effects on nitrogen losses and on the slurry N and P use efficiency and agronomic value for forage crops. Mitigation management techniques investigated comprised: mechanical separation and composting of solid fraction; rate and time of slurry application and its balance with mineral N fertilizers, and the use of nitrification inhibitors and other slurry additives such as acidifier agents, beneficial microorganisms and biochar. The evolution of the environmental performance of the dairy farms of mainland Portugal has been followed by the evaluation of the farm gate balances at pilot farms and although positive results have already been achieved, there is still a lot of room for progress. Recent setting up of two eco-schemes within the framework of the new CAP (>2023), which establish aid linked to the 'Improvement of animal feed efficiency' and the 'Promotion of Organic Fertilization' can contribute directly and indirectly to improve the sustainability of Portuguese dairy farms. Other possible solutions are suggested to face the upcoming challenges.

Rethinking methane from animal agriculture

F. Mitloehner
University of California, Animal Sciences Department, Davis, California, USA; fmmitloehner@ucdavis.edu

As the global community actively works to keep temperatures from rising beyond 1.5 °C, predicting greenhouse gases (GHGs) by how they warm the planet – and not their CO_2-equivalence – provides information critical to developing short- and long-term climate solutions. Livestock, and in particular cattle, have been broadly branded as major emitters of methane (CH_4) and significant drivers of climate change. Livestock production has been growing to meet the global food demand, however, increasing demand for production does not necessarily result in the proportional increase of CH_4 production. We evaluate the actual effects of the CH_4 emission from U.S. dairy and beef production on temperature and initiate a rethinking of CH_4 associated with animal agriculture to clarify long-standing misunderstandings and uncover the potential role of animal agriculture in fighting climate change. Two climate metrics, the standard 100-year Global Warming Potential (GWP_{100}) and the recently proposed GWP Star (GWP*), were applied to the CH_4 emission from the U.S. cattle industry to assess and compare its climate contribution. Using GWP*, calculations show that CH_4 emissions from the U.S. cattle industry have not contributed to additional warming since 1986. The projected climate impacts show that the California dairy industry will approach climate neutrality in the next ten years if the current cow inventory holds constant, with the possibility to decrease warming if there are further reductions of methane emissions exceeding 1% annually. GWP* should be used in combination with GWP to provide informative strategical suggestions on fighting SLCPs-induced climate change. By continuously improving production efficiency and management practices, animal agriculture can be a short-term solution to fight climate warming that the global community can leverage while developing long-term solutions for fossil fuel carbon emissions.

Session 26 Theatre 4

Effective nutritional strategies to mitigate enteric methane in dairy cattle

A.N. Hristov[1], A. Melgar[1,2], D. Wasson[1] and C. Arndt[3]
[1]The Pennsylvania State University, Department of Animal Science, 109 AVBS Building, University Park, PA 16802, USA, [2]Agricultural Innovation Institute of Panama (IDIAP), Clayton, City of Knowledge 07144, Panama, [3]International Livestock Research Institute, P.O. Box 30709, Nairobi 00100, Kenya; anh13@psu.edu

Intensive research in the past decade has resulted in a better understanding of factors driving enteric methane (CH_4) emissions in ruminants. Meta-analyses of large databases, developed through the GLOBAL NETWORK project, have identified successful strategies for mitigation of CH_4 emissions. Methane inhibitors, alternative electron sinks, vegetable oils and oilseeds, and tanniferous forages are among the recommended strategies for mitigating CH_4 emissions from dairy and beef cattle and small ruminants. These strategies were effective in decreasing CH_4 emissions yield and intensity as well, but animal health concerns with some of them need to be addressed. Tannins and tanniferous forages may have a negative impact on nutrient digestibility and more research is needed to confirm their effects on overall animal performance in long-term experiments with high-producing animals. Research with the macroalga *Asparagopsis taxiformis* has revealed its significant CH_4 mitigation potential but has also posed questions about feasibility of its adoption in practice. More research is needed to determine additivity of enteric CH_4 mitigation strategies and their effect on manure composition and greenhouse gas emissions. A meta-analysis of studies with dairy cows fed the CH_4 inhibitor 3-nitrooxypropanol (3-NOP) at The Pennsylvania State University showed a consistent 28 to 32% decrease in daily CH_4 emission or emission yield and intensity. The inhibitor had no effect on animal productivity and body weight change but increased milk fat percent and yield (0.19%-units and 90 g/d, respectively). Long-term studies indicated a potential decrease in the efficacy of 3-NOP over time, which needs to be further investigated in full lactation or multiple-lactation experiments. It is concluded that widespread adoption of mitigation strategies with proven efficacy by the livestock industries will depend on cost, government policies and incentives, and willingness of consumers to pay a higher price for animal products with decreased carbon footprint.

Session 26 Theatre 5

Challenges for climate care dairy farming in the Netherlands

P.J. Galama, J. Zijlstra and A. Kuipers
Wageningen University and Research, Livestock Research, De Elst 1, 6700 AH Wageningen, the Netherlands; paul.galama@wur.nl

The successfully development of the Dutch Agri & Food sector in the last 70 years by maximizing the production at minimal costs has led to a situation where the Dutch dairy sector is now responsible for a high national contribution to environmental impacts. These are acidification of nature areas (mainly by NH_3), eutrophication of water bodies (losses of N and P), global warming (emissions of CH_4) and biodiversity loss. Improvements and reductions have been achieved, to various extents, by single-issue approaches (policy) and solutions in the biophysical production system. Further strong improvements on all environmental aspects are needed, without causing trade-offs and negative side effects. This is a huge challenge for typical highly populated regions as the Netherlands with intensive livestock production and very critical societies. Therefore the Dutch Dairy chain has set goals for 2030 to increase the sustainability on the topics of climate, welfare, grazing, biodiversity, environment, new business model, land based and safety in the yard. Also a coalition of several Dairy organization have set management goals about dilution of manure, grazing and protein in ration together with the Ministry of Agriculture to reduce the nitrogen losses, especially ammonia emission. These ambitions will be shown together with the needs and solutions of a group of dairy farmers who are involved in the Climate Care Cattle Farming project. They relate to best practices around environmental, technical and social & economic topics. The focus is on making the dairy sector more resilient by improving both the farm and the farmer by improving his skills to adapt and anticipate on threats and opportunities. The Ministry of Agriculture stimulates research & development and implementation on dairy farms in an integrated approach to reduce the emissions of nitrogen and green house gasses. We will present the topics around breeding, nutrition, grassland management, manure management, cattle housing & technology and groups of pilot farms who are working on implementation to reach the environmental goals.

First results of a screening method for GHG emission measurements in European dairy cattle barns

X. Vergé[1], P. Robin[2], V. Becciolini[3], A. Cieślak[4], N. Edouard[2], L. Fermer[5], P. Galama[6], P. Hargreaves[7], V. Juškienė[8], G. Kadžiene[8], H. Schilder[6], L. Leso[3], A.-S. Lissy[9], J. Priekulis[10], B. Rees[7], D. Ruska[10] and M. Szumacher-Strabel[4]
[1]IDELE, Monvoisin, 35652 Le Rheu, France, [2]INRAE, PEGASE & SAS, Institut Agro, 35590 Saint Gilles, France, [3]University of Florence, Via san Bonaventura 13, 50145 Firenze, Italy, [4]Poznan University, Wolynska 33, 60-637 Poznan, Poland, [5]Justus Liebig Universität, Ludwigstraße 2[1]b 35390, 35390 Gießen, Germany, [6]Wageningen Livestock Research, De Elst 1, 6708 WD Wageningen, the Netherlands, [7]SRUC, Barony Campus, Parkgate, DG1 3NE Dumfries, United Kingdom, [8]Lithuanian University of Health Sciences, R. Zebenkos, 1282317 Baisogala, Lithuania, [9]INRAE Transfer EnVisaGES, Bâtiment Bioclimatologie, Route de la Ferme, 78850 Thiverval-Grignon, France, [10]Latvia University of Life Sciences and technologies, Faculty of Agriculture, Institute of Animal Sciences, 2 Liela street, Jelgava, 3001, Latvia; xavier.verge@idele.fr

The effects of cattle farming systems and individual management practices on GHG emissions have been studied by applying a screening method on eight European countries totalizing 60 dairy farms and four seasonal measurements. This method is based on indoor and outdoor CO_2, CH_4 and N_2O concentration measurements and on a questionnaire developed to estimate the carbon mass balance at the building scale. The results of the questionnaire were also used to characterize the farm diversity. This international survey conducted with the same measurement method helped identifying the main factors influencing GHG emissions as well as the existing low-emitting systems: the calculated emissions based on concentration measurements showed that CH_4 emissions from manure could double those from the enteric fermentation of the animals highlighting the importance of manure management as a GHG mitigation approach; the detection level of the N_2O emissions was below 1 g N-N_2O/animal/day; in some farms, the emission results showed very few seasonal variations over the year while, for other ones, seasonal variations were well observed; the results showed well detectable GHG emission reductions related to changes in farm practices.

Genome wide associations for methane emissions in dairy cows

L. Fehmer, J. Herold, P. Engel, T. Yin and S. König
Justus-Liebig-University Giessen, Institute Animal Breeding and Genetics, Ludwigstr. 21b, 35390 Giessen, Germany; lena.fehmer@agar.uni-giessen.de

Among all livestock species, dairy cattle are the main producers of methane (CH_4). Several previous studies focused on the estimation of genetic parameters for methane traits. In the current study, we aimed on genomic analyses for different methane traits of cows kept in two different production systems, allowing studies on genotype by environment interactions. Production system A was cubicle housing comprising 1,370 cows from six herds. Production system B included five composted bedded pack barns with 590 cows. The cows were genotyped with the Illumina 50K SNP Bead Chip. After quality control, 38,933 SNP remained for the genomic analyses. Individual CH_4 emissions were recorded using a mobile laser methane detector (LMD), which measures methane in ppm×m with two values/second during the recording period of three minutes. Based on a protocol for CH_4 trait preparation including outlier analyses, differentiation and physiological aspects, methane traits reflecting respiration, eructation and total CH_4 emissions, were derived. For genomic analyses, a two-step strategy was applied. Step 1 implied a pre-correction of CH_4 phenotypes for the fixed effects of parity, days in milk and herd nested within feeding ration group. The generated residuals were used as dependent variables in step 2 in ongoing genome-wide associations considering the effect of the genomic relationship matrix. The genome-wide significance level according to Bonferroni (pBF=0.05/NSNP=44,046) was 1.14×10^{-6}. Furthermore, a less conservative normative significance threshold was used to identify potential candidate SNP defined as pCD=1×10^{-4}. We detected significantly associated SNP for the mean of the CH_4 emissions per day on BTA 4 and 13. The annotated potential candidate genes SCIN and ARL4A genes on BAT 4 regulate metabolic pathways. The significant SNP on BAT 13 is located in close distance to the genes PLXDC2 and NEBL, which have functions with regard to energy metabolism and muscle contraction. In enhanced genome-wide associations, we studied SNP × production system interactions, indicating specific genetic mechanisms of some chromosome segments either in system A (cubicle) or in system B (compost).

Nitrogen excretion and ammonia emissions in dairy cows fed low-N fresh grass and maize silage

M. Ferreira, R. Delagarde and N. Edouard
PEGASE, INRAE, Institut Agro, 16, Le Clos, 35590 Saint-Gilles, France; manon.ferreira@inrae.fr

Dairy cattle farms must limit their negative impacts and nitrogen (N) losses to the environment. In particular, they must decrease greenhouse gases and ammonia (NH_3) emissions. Cow's N excretion in faeces and urine and the related NH_3 emissions are affected by diet intake and composition, especially by N intake. In many farms, particularly in spring and summer, fresh grass and conserved forage are fed together. The influence of these mixed diets on N excretion at the cow level is poorly known. This study aimed at quantifying the effect of 4 maize silage proportions in a fresh grass diet (0:100, 17:83, 34:66, 51:49 of maize silage:fresh grass ratio, DM basis) on the N excretion in faeces and urine and on the total ammonia N (TAN) excretion in slurry, at the cow level. For this, 7 Holstein cows were offered the 4 dietary treatments according to a Latin square design, with individual measurements of intake, urine and faeces amounts, and their N concentration. The TAN concentration was analysed on slurry reconstituted from a urine and faeces mixture, in proportion to their production. Potential NH_3 emissions at barn level were estimated from the TAN excretion multiplied by the 0.24 emission factor for cattle housing (EMP/EEA national inventory guidelines). The unusual low grass crude protein (CP) concentration involved very low diet CP concentrations ranging between 107 and 86 g/kg DM. Dry matter intake decreased with increasing maize proportion, from 15.7 to 12.4 kg DM/d, along with the milk production, from 16.4 to 13.2 kg/d, and the N exported in milk, from 90 to 68 g/d. The N excretion decreased linearly from 172 to 119 g/d, with increasing maize proportion, with similar urinary N proportion in N excreted between diets (39%). The TAN excretion in slurry decreased with increasing maize proportion from 34.7 to 21.0 g/d, as the related NH_3 estimated emissions. The TAN excretion in reconstituted slurry corresponded to 19% of the N excreted in faeces and urine, which is widely outlying from the 60% commonly used in the EMEP/EEA methodology. This study suggests that the emission estimations from national inventory guidelines should better consider the variability in TAN excretion, which is the primary factor that influences NH_3 emissions, especially for poor N-diets.

Two experimental dairy systems built to decrease the carbon footprint

V. Brocard[1], S. Foray[1], E. Tranvoiz[2] and L. Morin[3]
[1]Idele, Monvoisin, 356252 Le Rheu Cdx, France, [2]Chambre d'Agriculture de Bretagne, 24 route de Cuzon, 29322 Quimper Cdx, France, [3]Station expérimentale, La Blanche-Maison, 50880 Pont-Hébert, France; valerie.brocard@idele.fr

Agriculture contributes to global greenhouse gases (GHG) emissions, particularly through methane and nitrous oxide emissions. In France, this amounts to 21% of the overall national GHG emissions (CITEPA, 2021), with around 10% attributable to ruminants. Thereby, dairy production is facing a challenge to reconcile food challenge, competitiveness, and environment. Many practices are identified in bovine production to reduce the GHG emissions and improve profitability. However, these individual practices are rarely integrated at whole system level. Monitoring experimental dairy farming systems aims at assessing to which level the implementation of best practices will lower the environmental impacts while maintaining current level of profitability in contrasted pedoclimatic conditions. CAP'2ER tool was used to determine the GHG gross emission, the carbon footprint of the milk, the nitrogen balance and associated indicators in 2 contrasted dairy systems – Trévarez in Brittany and La Blanche Maison in Normandy. The carbon footprint was measured using IPCC methodology, life cycle assessment, including carbon sequestration. The main levers used on these two farms to decrease the C footprint were: in Trevarez, suppression of production concentrate, replacement of soja by rapeseed for the protein concentrate, resort to early grass silage rich in proteins, decrease in the replacement rate and age at first calving, strong reduction of mineral fertiliser, increase in the share of grassland in rotations. In la Blanche-Maison, levers implemented were the following: reduction of age at first calving, increase in the share of grass in the cows' diet to decrease resort to protein concentrates, winter grazing for beef bulls, limited resort to mineral fertilisers. The main issues addressed by both research farms aim at improving the global sustainability of the farms and adapt to climate change, reduce the competition with humans for the use of food resources, better manage the nutrients resources and provide social and environmental services. This study was funded by Interreg Atlantic area through the Dairy-4-Future project.

Session 26 Poster 10

Monitoring indoor climate in a dairy barn

P. Massabie
IDELE French Livestock Institute, 42 rue Georges Morel – CS 60057, 49071 Beaucouzé Cedex, France; patrick.massabie@idele.fr

In dairy barn, during summer time, heat stress is generally evaluated through thermal humidity index (THI) calculated with outside temperature and humidity values. But depending on weather conditions, climate inside the barn could be quite different and heat stress could be underestimated. The objectives of this study were to determine links between weather conditions and inside climate of the barn, to analyse the effect of wind on CO_2 and ammonia levels and to determine exposition of cows to radiant heat from walls or roof by measuring black globe temperature. An 80 lactating cows experimental barn (60×17 m) with two row of cubicles was equipped with ten temperature probes, four black globe temperature, three humidity sensors and one ammonia and one carbon dioxide probes. All were 2.5 m above the floor. Weather parameters (temperature, humidity, wind speed (WS) and direction) were also recorded. Recordings were made on a complete year with an interval of 15 minutes. The barn was fully opened on the East side. THI was calculated and heat load index (HLI) was estimated with air velocity extrapolated with WS. Outside temperature and humidity were an average 12.7±6.9 °C and 77.5±18.1%. Inside temperature and humidity were an average 13.9±6.2 °C and 66.6±11.2%. When wind came from opened side of the barn, inside and outside temperatures were similar. At the opposite, when wind was coming from closed side, inside temperature was in average 1 °C above outside value. Black globe temperature was in average close to ambient temperature, but maximum values are 2.2 °C higher. During hottest day, sensor above cubicle near west side wall showed respectively an average value and a maximum 1.5 and 2.9 °C higher than temperature sensor. THI calculated from weather station and with inside temperature and humidity have respectively an average value of 54.8 and 56.9 and a maximum of 76.9 and 80.6. HLI calculated for temperature above 25 °C was in average 76.0 with a maximum value of 91.6. Both ammonia and carbon dioxide levels were minimum when wind was coming from the opened side and were respectively an average 0.7 ppm and 469 ppm Those results showed that when dairy barns are partially closed, wind effect is reduced and heat stress can occur before outside THI has reached critical level.

Session 26 Poster 11

NMVOC emissions from a naturally ventilated dairy housing – comparison of different diets

S. Schrade[1], M. Zaehner[1], K. Zeyer[2], S. Wyss[2], D. Steger[2], J. Mohn[2] and F. Dohme-Meier[3]
[1]Agroscope, Ruminant Research Group, Taenikon 1, 8356 Ettenhausen, Switzerland, [2]Empa, Laboratory for Air Pollution / Environmental Technology, Ueberlandstrasse 129, 8600 Dübendorf, Switzerland, [3]Agroscope, Ruminant Research Group, Tioleyre 4, 1725 Posieux, Switzerland; frigga.dohme-meier@agroscope.admin.ch

Agriculture, in particular cattle farming, not only causes emissions of ammonia and methane, it is also a considerable source of non-methane volatile organic compounds (NMVOC) emissions. Volatile organic compounds contribute, in combination with nitrogen oxide and sunlight, to the formation of ground-level ozone and further may act as greenhouse gases. According to previous studies, a large proportion of NMVOC emissions originate from feeding and predominantly from silage. However, the validity of the rare studies available is limited, as they were often carried out on a laboratory scale and mostly focused on individual NMVOC compounds, of which often only concentration values were determined. In order to improve the data basis on NMVOC emissions from dairy housing, Agroscope and Empa carried out investigations on different diets in the experimental dairy housing for emission measurements on a practical scale. The feeding treatments 'silage-free diet', 'mixed silage diet', 'grass silage diet' and 'sainfoin silage diet' were investigated at herd level. To determine emissions under natural ventilation, a dual tracer-ratio method with SF_6 and SF_5CF_3 was used. Concentrations of around 30 different NMVOC compounds were analysed by GC-FID and SF_6 and SF_5CF_3 by GC-ECD. Beside emission data, further parameters like animal-related data, feed and climate data were documented. Results from the summer measurements show that ethanol represented the highest proportion (by mass) of the NMVOC compounds in both mixed silage and grass silage diets, with at least 80%, and in the silage-free diet, with at least 40%. In the case of sainfoin silage diet, methanol emissions were slightly higher than ethanol emissions, at around 50% by mass. The NMVOC emissions of all feeding treatments with silage were significantly higher than those of the silage-free diet. The daily patterns of NMVOC emissions showed maxima especially in the silage-based diets during feed distribution and at the main feeding times.

Session 26 Poster 12

Simulation of enteric methane emission friendly management strategies in Danish dairy herds

V.M. Thorup, A.B. Kudahl, L. Chen and S. Østergaard
Aarhus University, Department of Animal Science, Blichers Alle 20, 8830, Denmark; vivim.thorup@anis.au.dk

Nearly 35% of the total greenhouse gas emission from Danish agriculture is from dairy cattle, and enteric methane emission makes up about 50% of the climate footprint of milk. Globally, milk and meat consumption are expected to increase. Production and management strategies for improved health, longevity and fewer young stock may be important to reduce the total climate impact from the production of milk and meat. We wanted to test a number of such management strategies specifically aiming to reduce enteric methane emission per kg milk and to test their impacts on emission, economy and animal welfare. We used herd simulation modelling to model the individual animal as well as the results per kg milk and meat from the herd. We applied the SimHerd model, with a module capable of estimating enteric methane emission from a dairy herd based on feed intake estimated for the individual animal. Our baseline was a conventional herd of 200 Danish Holstein cows producing 11,000 kg ECM/cow year. Our simulation strategies included: prolonged lactation, use of beef semen, increased insemination rate and improved health. The strategy with the largest methane reduction combined four changes: insemination rate increased from 38 to 85%, number of claw and leg disorders lowered from 30 to 13%, number of mastitis cases lowered from 31 to 14%; and beef semen used on all cows of parity 3 and above. Results from this methane-friendly combination were: enteric methane emission reduced from 17.0 to 15.7 g CH_4/kg ECM; milk yield increased from 10,969 to 11,217 kg ECM/cow year; milking years per cow increased from 2.6 to 5.1; number of calvings reduced from 214 to 204 per year; youngstock reduced from 191 to 93 animals; purebred bull calves reduced from 102 to 48 animals; cross-bred calves increased from 0 to 97 animals; gross margin per cow year increased from €1,785 to €1,939. In conclusion, the methane-friendliest management strategy reduced methane emission by 7.6% per cow year, whilst also achieving a better gross margin and a higher animal welfare through fewer disease events. Future research will advance methane estimation of the model further. We thank The Danish Agricultural Agency for supporting our study.

Session 26 Poster 13

***Asparagopsis taxiformis* infused vegetable oils: a solution to reduce ruminant methane emissions**

D.M. Soares[1,2], R. Torres[3], A.M. Campos[3], A.P. Portugal[4], M.T. Dentinho[2,4], L. Mata[3] and R.J. Bessa[2,4]
[1]Terraprima Serviços Ambientais, Avenida das Nações Unidas, no. 97, 2135-199 Samora Correia, Portugal, [2]CIISA, Faculdade de Medicina Veterinária, Universidade de Lisboa, Avenida da Universidade Técnica, 1300-477 Lisboa, Portugal, [3]CCMAR, Universidade do Algarve, Campus de Gambelas, 8005-139 Faro, Portugal, [4] INIAV, Estação Zootécnica Nacional, Quinta da Fonte Boa, 2005-048 Santarém, Portugal; diana_soares_07@hotmail.com

The red macroalgae *Asparagopsis taxiformis* contains several anti-methanogenic compounds, including bromoform ($CHBr_3$) which is a potent inhibitor of rumen methanogenesis. $CHBr_3$ is also a highly volatile, and biomass needs to be freeze-dried to retain its activity. Practical and cost effective alternatives to freeze-drying need to be developed to allow the *A. taxiformis* use in ruminant feeding. The aim of this work was evaluate the use of *A. taxiformis* infused oils to prevent rumen methanogenesis *in vitro*. *A. taxiformis* biomass was harvested at Algarve coast and kept at 4 °C for 20 min and then either frozen and freeze-dried (FD), immersed in sunflower oil (ISO) or immersed in linseed oil (ILO). The $CHBr_3$ concentration was determined after 15 days by gas chromatography coupled to mass spectrometry. The anti-methanogenic effects of *A. taxiformis* oil infusions (ISO and ILO) and FD biomass were compared using an *in vitro* gas production system (ANKOM Technology, USA) and a $CHBr_3$ dosage of 0.06 mg/g of dry matter. *In vitro* rumen fermentations last for 48h and were replicated in three consecutive weeks. Total gas production was registered and the methane concentration in the gas was measured by gas chromatography (GC). Relative to the control, all *A. taxiformis* presentations strongly reduced ($P<0.001$) CH_4 production, with 84% reduction for FD and more than 96% reduction for both ISO and ILO. Nevertheless, these differences observed among FD, ISO and ILO were not significant ($P>0.05$). *A. taxiformis* algae oil infusions show no loss of antimethanogenic activity freeze-dried biomass but are much more economical and practical, so its use as a post-harvest processing method is highly promising.

Polyphenols in apple pomace decrease ruminal methane production in dairy cows

A.C. Cieślak[1], J.A. Jóźwik[2], H.J. Horbańczuk[2], K.M. Kozłowska[1,2], G.M. Gogulski[3,4], S.M. Skorupka[1], W.B. Wyrwał[1], J.W. Jaworski[1], P.D. Petric[3] and S.Z.S.M. Szumacher-Strabel[1]
[1]Poznań University of Life Sciences, Department of Animal Nutrition, Wołyńska 33, 60-637 Poznań, Poland, [2]Polish Academy of Sciences, Institute of Genetics and Animal Biotechnology, Postępu 36A, 05-552 Magdalenka, Poland, [3]Centre of Biosciences, Institute of Animal Physiology, Soltesovej 4-6, 040-01 Kosice, Slovak Republic, [4]Poznań University of Life Sciences, Department of Preclinical Sciences and Infectious Diseases, Wołyńska 35, 60-637 Poznań, Poland; adam.cieslak@up.poznan.pl

Apple pomace (AP) is rich in bioactive compounds like polyphenols. The previous study showed that dietary components with polyphenols mitigate methane production by reducing the number of methanogens while improving the basic ruminal fermentation characteristics. Four multiparous cannulated Polish Holstein-Friesian dairy cows were assigned to two dietary treatments (CON vs APD – TMR diet with 6% apple pomace) with two cows in each treatment in a replicated 2×2 crossover design. The ruminal fluid was collected, and basic ruminal parameters and microbial characterization were tested. Additionally, methane emission from dairy cows was measured using respiratory chambers. The addition of the APD diet positively affected ruminal fermentation and the populations of bacteria. Whereas the population of methanogens was significantly reduced by 19%. A statistically significant increase in the pH value, ammonia concentration, and volatile fatty acids in the ruminal fluid was found. Changes in the content and profile of volatile fatty acids were also observed, expressed as an increase in propionic acid concentration with the reduction of acetic acid. There was also an increase in the sum of bacteria and a decrease in the population of methanogens, resulting in a reduction in the emission of methane to the environment by ca. 8%. Positive changes were also found in milk yield, which increased by 4.5% due to using 6% apple pomace in the TMR. The research results indicated that 6% of apple pomace in TMR could successfully be used as a component of the ration for dairy cows.

Session 26 Poster 15

Life cycle assessment of current and future pasture-based dairy production systems

J. Herron[1], D. O'Brien[2] and L. Shalloo[1]
[1]Teagasc, Paddy O'Keeffe Centre, Moorepark, Fermoy, Co.Cork, Ireland, P61 P302, Ireland, [2]Teagasc, Johnstown Castle, Co. Wexford, Ireland, Y35 TC97, Ireland; jonathan.herron@teagasc.ie

Dairy farmers must improve the efficiency of their systems to overcome current environmental challenges and ensure economic viability. An average dairy system must be determined to establish a benchmark from which the efficacy of proposed mitigation strategies can be assessed. Mitigation strategies can then be compiled to create targets for the sector to strive towards. The objective of this study was to determine the effect of updated life cycle assessment (LCA) methodology, and calculate the environmental performance of the current average dairy system (Current) and a target dairy system (Target). An existing dairy LCA model was updated with country specific emission factors and life cycle inventory data. The environmental impact categories assessed were global warming potential (GWP), non-renewable energy depletion (NRE), acidification potential (AP), and eutrophication potential (marine, MEP; freshwater, FEP). The Current dairy system was simulated twice, once with the previous version of the LCA model, and secondly with the updated LCA model. The addition of country specific emission factors and updated inventory data resulted in GWP and NRE reducing from 1.08 to 0.97 kg CO_2-eq/kg FPCM and 2.8 to 2.5 MJ/kg FPCM, respectively. The updates had negligible effect on AP, MEP, and FEP. The inclusion of assumptions around carbon sequestration in grassland, further reduced GWP by 16.4%. Moving towards the Target system reduced the environmental impact per kg FPCM across all impact categories investigated. When expressed per ha, transitioning towards the Target system reduced AP, FEP and NRE by 2.0%, 8.8% and 13.8%, respectively. In contrast, transitioning towards the Target system increased GWP per ha and to a lesser degree MEP per ha. The increase in GWP per ha was attributed to the increase in productivity per ha (9,950 vs 14,100 kg FPCM/ha). This study demonstrates that improving system efficiency will reduce the environmental impact per kg FPCM, however it is important to that improved system efficiency can be offset by associated increase in productivity. Further research is required to improve environmental performance beyond the Target system.

Water quality in dairy cattle farms: impact on animal production, reproduction and health

V. Resende
Universidade de Évora e MED, Departamento de Zootecnia, Pólo da Mitra, Ap. 94, Évora, Portugal., 7002-554, Portugal; vjgr33@gmail.com

Currently, agriculture is responsible for 70% of the world's water abstractions and animal production represents 29% of the water used in agriculture. Dairy production is estimated to be responsible for 4% of global water withdrawal. Therefore, it is increasingly relevant and necessary that the use of water for agricultural purposes be efficient. The objective is to verify: (1) the importance attributed by national producers to the monitoring of water and its quality on farms; and whether (2) the water quality affects the production, reproduction and health of animals. The 1st stage consists of the elaboration and application (online/face-to-face) of a questionnaire at national level (Continente), using the database of the APCRF and the EABL, in a total of 890 farms. The questionnaire was designed to collect: (1) monitoring, storage and water quality data; (2) productive and reproductive data from the farm; and (3) animal health data associated with water quality. Preliminary results of this step: (1) water monitoring, storage and quality data; - 5% surveyed farms; - 91% use their own water (hole); - 40% do not perform water quality analysis regularly; - 86% do not have water consumption monitoring. The 2nd stage consists of the case study. Group of animals (n=25) with access to treated water and a group with untreated water (n=25). All experimental groups are in the same conditions: farm, feeding, ambient temperature and humidity, number of drinkers, manger space, feeding time. Variation factor is the manganese concentration in drinking water. It was still possible to collect samples of blood, urine and milk for 4 months. In the surveyed farms, most farms use their own water, do not monitor water consumption and that only 40% of farms carry out (annual) water quality analyses. In the case of study, it appears that excess manganese affects reproduction (untreated group with an increase in the number of inseminations, increased incidence of embryonic mortality, increased calving interval) and animal health (higher incidence of metritis, kidney problems, diarrhoea), as well as its production (smaller quantity, longer drying time).

Quantitative genetics of infectious disease: much more heritable variation than we think

P. Bijma[1], A.D. Hulst[1,2] and M.C.M. De Jong[2]
[1]Wageningen University and Research, Animal Breeding and Genomics, P.O. Box 338, Droevendaalsesteeg 1, 6700 AH, the Netherlands, [2]Wageningen University and Research, Quantitative Veterinary Epidemiology, P.O. Box 338, Droevendaalsesteeg 1, 6700 AH Wageningen, the Netherlands; piter.bijma@wur.nl

Infectious disease status (0/1) typically shows low heritability. Current quantitative genetic theory, however, ignores that infections rely on transmission. Here we integrate quantitative genetics and epidemiology, focussing on R_0, on the difference between breeding value for individual disease status vs breeding value for response to selection, and on genetic variation and response to selection for the prevalence of an endemic infection. The theory rests on an additive linear model for the logarithm of the breeding value for R_0, which is the sum of the logarithms of breeding values for susceptibility, infectivity and recovery rate. Results show that infectious disease status responds very differently to selection than non-communicable binary traits, and that the genetic variance determining the potential response of prevalence to selection must be much larger than currently believed. Similar to the classical threshold model, heritability of individual disease status has a maximum at P=0.5. However, heritable variation for prevalence and response of prevalence to selection increase significantly when the prevalence decreases, ultimately leading to local extinction of the infection due to herd immunity. Specifically, without genetic variation in infectivity, the breeding value for prevalence is the product of the reciprocal of prevalence and the breeding value for individual disease status. For example, for a prevalence of P=0.2, the breeding value for prevalence is a factor 1/0.2=5 greater than the ordinary breeding value for individual disease status. This phenomenon occurs because individuals who are less susceptible also infect fewer others, simply because they are less often infected themselves. This effect increases strongly at lower prevalence, resulting in an accelerating response to selection as prevalence decreases. These results change our perspective on the prospects of genetic selection against infectious diseases.

Transcriptomic map of mastitis in dairy cows: clues on *S. agalactiae* and *Prototheca* infections

V. Bisutti[1], N. Mach[2], D. Giannuzzi[1], E. Capra[3], R. Negrini[4], P. Ajmone-Marsan[4], A. Cecchinato[1] and S. Pegolo[1]
[1]DAFNAE, University of Padova, viale dell'Università 16, 35020 Legnaro (PD), Italy, [2]IHAP, Université de Toulouse, INRAE, ENVT, 23, Chem. des Capelles, 31300 Toulouse, France, [3]IBBA, National Research Council, via Einstein, 26900 Lodi, Italy, [4]DIANA, Università Cattolica del Sacro Cuore, via E. Parmense 84, 29122 Piacenza, Italy; vittoria.bisutti@unipd.it

Emerging evidence indicates that *S. agalactiae* and *Prototheca* are two of the most frequent mastitis-causing pathogens in dairy cows, but the extent of their pathogenesis and underlying mechanisms in regulating the immune systems remain elusive. In this study, we performed RNA-Seq-based transcriptome profiling of the milk somatic cell and milk cytometry analysis of healthy cows and cows naturally infected by *S. agalactiae* or *Prototheca*. Bacteriological screening on 188 Holstein cows reared in one herd under similar conditions was made to select: (1) healthy individuals with no history of mastitis (n=9); (2) infected animals for *S. agalactiae* (n=11); and (3) infected animals for *Prototheca* (n=11). Milk production (P<0.01) and lactose content (P<0.05) were significantly lower in *S. agalactiae* and *Prototheca* infected cows compared to healthy ones. *S. agalactiae* induced an immune response higher in polymorphonuclear cells and macrophages, while *Prototheca* infection was mediated by lymphocytes proliferation. A total of 3,965 and 5,173 differentially expressed genes were identified when comparing *Prototheca* and *S. agalactiae* to healthy cows, respectively. Functional pathways analysis suggested that the response to *Prototheca* infection was dissimilar from that of *S. agalactiae*, spanning pathways associated with the immune system, such as PPARγ induction, monocytes and B cells proliferation and carbohydrate catabolism. On the other hand, local immunization against *S. agalactiae* mobilized molecular mechanisms that rely on the cellular catabolic process and organic acid and lipid metabolism. Complementary functional analyses are in progress to unravel expanded molecular pathways to fine-tune the immune response of the mammary gland, laying the basis for clinical interventions that could benefit cow performance and welfare. Acknowledgements. The study was part of the LATSAN project funded by MIPAAF.

Robustness indicators in a divergent selection experiment for birth weight variability in mice

L. El-Ouazizi El-Kahia[1], I. Cervantes[1], J.P. Gutiérrez[1] and N. Formoso-Rafferty[2]
[1]Dpto. Producción Animal, Facultad de Veterinaria, UCM, 28040, Spain, [2]Dpto. Producción Agraria, E.T.S.I.A.A.B, UPM, 28040, Spain; lailaelo@ucm.es

A divergent selection experiment for environmental variability of birth weight in mice was carried out successfully during 29 generations. Two divergent lines were created, the high variability line (H-line) and the low variability line (L-line), but also there were correlated responses in other important traits in animal production. The objective of this study was to compare the lines according to the performance of several traits at the first and the last generation of the selection experiment, and to evaluate the susceptibility to stress and diseases measuring two metabolites. Birth and weaning weight variability indicators, such as variance (WV), standard deviation (WSD) and coefficient of variation (WCV) were analysed, and the individual pup weight (W), litter size (LS) at birth and at weaning and fertility (F) were also studied. The stress and inflammatory response were studied measuring corticosterone (COR) and C-reactive protein (CRP) in gestated females. We used a linear model including the line and parturition number for WV, WSD, WCV, LS at birth and at weaning, and for birth and weaning W the model also included the sex and the LS at birth (linear and quadratic). The linear model for COR and CRP included line, generation, and microplate as effects. Fertility differences were assessed using a Chi-square test, comparing the number of females which performed one, two or none parturition from the 43 females per line that were mated each generation. The L-line showed a lower birth W variability in the last generation, while for the weaning W variability, there were differences since the beginning of the experiment in both lines. The pup birth and weaning W started to be different between lines from the first generation, being the L-line lower than the H-line. The LS was higher in the L-line as well as the F in the last generation. Regarding stress and inflammatory indicators, the COR was not different between lines, while the CRP was lower in the L-line. All these results trends suggested that animals from L-line were more robust and could have less susceptibility to diseases increasing the animal welfare.

Can resilience and reproduction be improved in paternal lines of rabbits selected for growth rate?

C. Peixoto-Gonçalves, E. Martínez-Paredes, L. Ródenas, E. Blas, M. Cambra-López and J.J. Pascual
Universitat Politècnica de València, Institute for Animal Science and Technology, Camino de Vera s/n, Valencia, 46022, Spain; capeigon@upvnet.upv.es

A total of 197 nulliparous rabbit females from three paternal lines (R, RF and RLP) were used to explore potential solutions to improve female resilience and reproductive performance. R was a line selected by average daily gain after weaning during 36 generations; RF was a line founded via a selection of elite animals of the R line (high growth rate with high reproductive traits); RLP was a line obtained by backcrossing RF with LP animals (maternal line characterised by its high robustness). Fertility, body weight, perirenal fat thickness (PFT), daily feed intake (DFI) and milk yield of females was controlled from 1st to 3rd parturition. RF females were significantly lighter than R and RLP females throughout the trial (on av. -5.0; $P<0.05$). Furthermore, RF animals had a higher fertility percentage than RLP females, at first cycle (+10.5 percentage points; $P<0.05$), which could indicate a more precocious reproductive development. However, RLP showed a higher fertility than RF females at second cycle (+21.5 percentage points; $P<0.01$), due to a high decrease of RF fertility (-32.6 percentage points), being RLP females more stable across the three reproductive cycles. In general, no significant differences were found for PFT among the three genetic lines. However, RLP females showed a higher PFT than R females at parturition (+3.0%; $P<0.05$). Daily feed intake of RLP females was higher than that of R and RF during gestation and at late lactation (on av. +9.7% and +8.7%, respectively; $P<0.05$). Throughout the experiment RLP females produced more milk than R and RF females (on av. +18.5%; $P<0.001$). Results suggest that foundation of a new paternal line using elite animals could generate lighter females with better early reproductive performance. However, the introduction of genetic resilience in this elite paternal line, through its backcrossing with a robust line, has increased the ability of females to obtain and effectively use resources. This strategy reconciles milk yield addressed to the current litter, fertility to ensure next offspring and female body condition as a driver of its reproduction and survival.

Genetic correlations of direct and indirect social dominance components vs production and fitness

B. Tuliozi[1], R. Mantovani[1], E. Mancin[1], I. Schoepf[2] and C. Sartori[1]
[1]University of Padova, DAFNAE, Viale dell'Università 16, 35020 Legnaro (PD), Italy, [2]Queen's University, Dept. of Biology, 99 University Ave, K7L 3N6, Kingston, ON, Canada; enrico.mancin@studenti.unipd.it

Traits measured under social interactions may be affected by indirect genetic effects (IGEs), that are the influence of conspecific(s)' phenotype(s) on the expression of phenotype of a focal individual. Accounting for IGEs may affect the genetic correlations among traits, but scarce evidence in literature currently exists, mainly because of the necessity of a large amount of phenotypic information and of individuals with known genetic relationships. The aim of the study was to assess both direct and indirect genetic correlation of the social dominance, i.e. a trait strongly affected by IGEs, with other productive traits. The study subject was the local Aosta Black Pied-Chestnut cattle breed. Target traits were both productive (milk yield, muscularity, udder) and functional (fertility, SCS), expected to be genetically linked to social dominance through functional constraints of various magnitude. The outcomes of more than 37,800 dyadic interactions involving 8,700 cows in 20 years were used to assess social dominance. Contests were ritualized bloodless interactions mimicking the same dynamics occurring at pasture. Data were joined with individual information for productive and functional traits recorded during routinely controls. Bi-trait animal models including IGEs for social dominance, as well as the direct and indirect permanent environmental effect of the individuals were run. Winning agonistic interactions was genetically correlated with more developed muscle mass, but also with marginally lower milk production, more somatic cells in milk, small udders, lower fertility. In reverse, being a loser showed positive genetic correlations with greater fertility and health. Results are a consequence of the expected genetic correlation of around -1 between direct and indirect components of social dominance. This is the first empirical evidence of the (roughly) equal in magnitude but opposite in verse genetic correlations of the direct and indirect components of social dominance with functional traits, suggesting the importance for animal welfare of accounting for IGEs when considering traits expressed in social contexts.

Session 27 Theatre 6

Repeatability as a criterion for the selection of welfare biomarkers in beef cattle production

C. Meneses, M.J. Carabaño, C. Gonzalez, A. Hernández-Pumar and C. Diaz
INIA-CSIC, Ctra de la Coruña Km 7.5, 28040 Madrid, Spain; cdiaz@inia.csic.es

Animal welfare, apart from being a necessary condition for livestock production in the European Union, is an increasing demand from consumers and citizens. One of the indicators of lack of welfare is the level of stress. Stress level is evaluated in many cases with physiological biomarkers that in order to be informative require to be repeatable at least under similar stressing circumstances. The aim of this study was to estimate the repeatability of a number of commonly used stress biomarkers in different samples sources. Blood (B), hair (H), faeces (F) and saliva (S) samples were taken monthly from 17 Avileña-Negra Ibérica male calves in a Testing Centre (Colmenar Viejo, Spain) in five occasions. Laboratory determinations of cortisol were done in all sample sources, corticosterone only in F and phosphokinase (CPK), glucose, lactate, lactate dehydrogenase (LDH) and globulin as well as albumin/globulin ratio, and total proteins (albumin and globulin) only in B. Repeatability of the same biomarker was calculated from a model including age of the animal as covariate and animal as a random effect. Estimation was performed under different scenarios provided that the first control implied the novelty, while the last one imposed a new situation with more people than previous controls. Four alternative scenarios were considered, use the five controls, ignore the first one, the last one, or both. Repeatability estimates for cortisol in S, corticosterone in F as well as CPK in B were very small (0-6%). Repeatability for cortisol in B, H and F were consistent across scenarios, ranging from 10 to 28%. For the first three combinations, albumin repeatability had a value of 100%, and the albumin/globulin ratio ranged between 83% and 99%, while in the last combination it showed a value of 77% for albumin, but for the albumin/globulin ratio remained high (86%). Repeatability of stress indicators obtained under the same experimental conditions showed a very high variability. This is particularly true for those indicators of psychological stress and muscle damage. More information is needed to elucidate if this is due to the biomarker in itself or to the protocol we followed. In any case, this is a call of attention to the use of biomarkers as welfare indicators.

Session 27 Theatre 7

Role of genetics in swine inflammation and necrosis syndrome – SINS

N.G. Leite[1], E.F. Knol[2], S. Nuphaus[2], R. Vogelzang[2] and D. Lourenco[1]
[1]University of Georgia, Department of Animal and Dairy Science, 425 River Rd, Athens, GA 30602, USA, [2]Topigs Norsvin Research Center, Schoenaker 6, 6641 SZ Beuningen, the Netherlands; nataliagaloro@uga.edu

Our objectives were to: (1) investigate the genetic basis of the swine inflammation and necrosis syndrome (SINS); and (2) estimate its genetic relationship with postweaning skin damage, growth, and carcass traits. A total of 5,960 piglets were scored for SINS at birth (SINS) and had their birth weight (BW) and weaning weight (WW) recorded. After weaning, piglets were evaluated for skin damage (SKD) at nine weeks, carcass backfat (BF), and loin depth (LOD). We used four multiple-trait animal models with combinations of SINS, SKD, and an alternative production trait (i.e. BW, WW, LOD, BF) to access trait heritability and genetic correlations. The maternal-effect was included in the models for BW, WW, and SINS. The direct heritabilities for SINS ranged from 0.15 to 0.24 and were significantly different from zero in all models. The direct genetic correlation between SINS and SKD ranged from 0.19 to 0.50, suggesting that piglets genetically more likely to present necrosis at birth are also more likely to suffer from skin damage after weaning. The direct genetic relationship between SINS and early growth traits (i.e. BW and WW) was negative (from -0.40 to -0.30), indicating that an increase in newborn necrosis is also associated, at the piglet genetic level, with a reduction in piglets' birth and weaning weights. However, when BW is selected at the dam genetic level, no significant genetic correlation might exist with SINS or SKD (from -0.07 to 0.07). For carcass traits (i.e. BF and LOD), the direct effect of SINS was lowly or not significantly correlated, with estimates ranging from -0.16 to 0.05. Our study suggests that around 20% of SINS phenotypic variation is explained by additive genetics and therefore reducing its incidence through genetic improvement is feasible. Selecting animals genetically less prone to developing SINS is also expected to positively affect piglet genetics for heavier weaning weight and reduce postweaning skin damage. Realizing that a large percentage of the animals is genetically compromised at conception or before weaning can help to improve piglet quality by reducing SINS susceptibility, which will minimize welfare issues.

Usefulness of the claw-position score for genetic evaluations to reduce lameness incidence

A. Köck[1], L. Lemmens[2], M. Suntinger[1], M. Gehringer[3], G. Berger[4], F.J. Auer[3] and C. Egger-Danner[1]
[1]ZuchtData EDV-Dienstleistungen GmbH, Dresdner Str. 89, 1200 Vienna, Austria, [2]University of Veterinary Medicine Vienna, Veterinärplatz 1, 1210 Vienna, Austria, [3]LKV-Austria, Dresdner Str. 89, 1200 Vienna, Austria, [4]Rinderzucht Austria, Dresdner Str. 89, 1200 Vienna, Austria; koeck@zuchtdata.at

Lameness is an important health and welfare issue that causes considerable economic losses. The objective of this study was to investigate whether the claw-position score can be used for genetic evaluations as an auxiliary trait for claw health. The claw-position score is evaluated by visual scoring of the position of both the hind-digits (angle formed by the line of interdigital space of each claw-pair) to the mid-line of the cow´s body (line along the vertebral column). The higher the heel height of the lateral claw, the higher is the score, and the higher is the risk for development of a clinical lameness. In total, 2,978 records from 1,051 Fleckvieh cows from 36 farms were available from September 2021 to January 2022. Data collection was carried out by the regional milk recording organizations. Claw-position was scored at each milk recording in the milking parlour using a 3 class scoring system. Physiologically the angle formed by the interdigital line and the body-midline ranges between 0° and <17° (Score 1) indicating a balanced heel height of both the medial and the lateral claw. A score 2 describes an angle of 17-24°, and score 3 an angle of >24°. Afterwards, lameness scoring was performed for each animal using the scoring system of Sprecher *et al.* (1997) with 1 = normal, 2 = slightly lame, 3 = moderately lame, 4 = lame, and 5 = severely lame. For genetic analyses, a bivariate linear animal model was fitted with fixed effects of herd, lactation, and lactation stage and random effects of animal and permanent environment. Heritabilities for the claw-position score and lameness were 0.05 and 0.09, respectively, and the genetic correlation between the two traits studied was 0.85. These results suggest that the claw-position score could be used for genetic evaluations to reduce lameness incidence in dairy cattle.

Breed-dependent genetic variants altered the immune capacity in dairy and beef cattle bulls

M. Saeed-Zidane, A. Yousif, I. Blaj and G. Thaller
Institute of Animal Breeding and Husbandry, Animal Genetics Group, Olshausenstraße 40, 24098 Kiel, Germany; mzidane@tierzucht.uni-kiel.de

Immunity is one of the fundamental and challenged traits needed for animal health and welfare sustainability. There are several molecular pathways involved in two major types of immune responses including antibody-mediated immune response and cell-mediated immune response. Unravel of molecular markers that concurrent with immune capacity is prerequisite to identify a sensing immune tool that could be applied in animal breeding strategies. In this regard, the current study shed light on genomic and transcriptomic analyses of genes involved in immune response of beef and dairy cattle. For that, eight bulls from beef (Charolais) and dairy (Holstein) breeds were subjected to DNA sequence analysis of genes involved in immune response pathways. Furthermore, transcription and protein expression levels were performed using RNA and protein isolated from different organs including spleen, intestine, lung, and liver. The genotype analysis revealed number of breed-dependent single nucleotide polymorphisms (SNPs) at different DNA sequence locations. For instance, NF-κB1 gene, stress/immune regulatory transcription factor, showed 219 and 114 SNPs that were found in Charolais and Holstein breeds, respectively. Furthermore, TNF-α gene of Charolais breed has 33 SNPs while, IL-1β gene indicated 29 SNPs in Holstein. Moreover, spleen compared with the other organs showed significantly higher mRNA and protein levels of immune genes in particularly genes involved in Toll-like receptor pathway. Although, the stress and infection marker genes did not show and transcription level differences between the two breeds however, bulls of Charolais breed revealed significantly higher mRNA and protein expression levels of immune genes (CD14, TLR4, NF-κB1, TNF-α, IL-1β, etc.) compared to the Holstein counterparts. In conclusion, private alleles could be associated with the activation of Toll-like receptor and NF-κB pathways which in turn induce higher immune capacity in different immune related organs. More investigations are ongoing to determine specific haplotype with linkage disequilibrium associated with higher immune response in breed dependent manner.

Breeding strategies for calf health and leukosis: challenges and opportunities

C.F. Baes[1,2], R. Bongers[2], C. Lynch[2], F.S. Schenkel[2], K. Houlahan[2], N. Van Staaveren[2], H. Oliveira[2,3] and F. Miglior[2,3]
[1]University of Bern, Bremgartenstrasse 109a, 3012, Switzerland, [2]University of Guelph, University of Guelph, 50 Stone Road East, N1E 2W1, Canada, [3]Lactanet, 660 Speedvale Ave W Unit 101, N1K 1E5, Guelph, ON, Canada; cbaes@uoguelph.ca

A number of key diseases threaten dairy farming profitability for which no genomic evaluations are currently available in Canada, including Leukosis and calf health. On Canadian dairy farms, respiratory infections in calves are treated with antibiotics. A test and cull strategy is implemented to manage Leukosis, however there is no treatment available. Heritabilities of disease resistance traits are generally low, which has historically impeded genetic progress through selection programs. Reliable breeding values for low-heritability traits, however, can be attained through single-step genomic evaluation. Here we discuss the development of genetic evaluations for Leukosis and calf health, including respiratory disease, scours, and calf survivability. Datasets for Leukosis and calf health were provided by Lactanet Canada and included 76,405 calf disease records from 1,890 farms (collected via herd management software), as well as 86,912 Leukosis records (collected via milk ELISA tests) from 988 herds between 2007 and 2021. Pedigree heritability for Leukosis was estimated at 0.09 (±0.01), whereas calf health trait heritabilities ranged from 0.01-0.04 (±0.01-0.04) using mixed linear methodology. While the available data for Leukosis could be used in a national breeding program, genotype collection needs to be expanded to ensure single-step evaluations can be conducted. Furthermore, protocols for disease description and for standardized recording with inputs from veterinarians, producers and geneticists are required to ensure accurate and consistent diagnosis.

Persistence of genomic evaluations with different reference populations in pigs and broilers

I. Misztal, M.K. Hollifield, J. Hidalgo, S. Tsuruta, M. Bermann and D. Lourenco
University of Georgia, Animal and Dairy Science, Athens GA 30605, USA; ignacy@uga.edu

Genomic information has a limited dimensionality (number of independent chromosome segments Me) related to the effective population size. Under the additive model, the persistence of genomic accuracies over generations should be high when the nongenomic information (pedigree and phenotypes) is equivalent to Me animals with high accuracy. The objective of this study was to evaluate the persistence of genomic accuracy with varying quantities of data and for traits with low and moderate heritability. The first dataset for pigs included 161k phenotypic records for a growth trait and 27k phenotypic records for a fitness trait related to prolificacy, with 404k animals in the pedigree, of which 55k were genotyped. The second dataset for broilers contained 820K phenotypes for a growth trait, 200K for two feed efficiency traits, and 42K for a carcass yield trait. The pedigree included 1.2M animals, of which 154k were genotyped. In both populations the dimensionality was around 5k. ssGBLUP models were used with sliding subsets of 1 to 5 generations of ancestral data. Estimated accuracies were calculated by the linear regression method. The validation population consisted of single generations succeeding the training population and continued forward for all generations available. When the number of genotyped animals were over Me in each generation, the decay of accuracy was approximately linear with future generations and was small for moderate heritability traits and higher for a low heritability trait. For traits with moderate heritability, improvement of accuracy with reference data beyond 2 generations was minimal. When the reference population is large enough to accurately estimate the effects of the independent chromosome segments, GEBV can be persistent, with minimal decay of accuracy over generations. In such a case, the impact of old data is minimal. The impact of old data is stronger for sparsely recorded low heritability traits.

Tail biting and its precursors have distinctive gene-expression profiles in protein-restricted pigs

L. Couteller[1,2], L. Roch[2,3], E.O. Ewaoluwagbemiga[2] and C. Kasper[2]
[1]École Supérieure d'Agricultures, 55 Rue Rabelais, 49007 Angers, France, [2]Agroscope, Tioleyre 4, 01725, Switzerland, [3]University of Bern, Veterinary Public Health Institute, Animal Welfare Division, Länggassstrasse 120, 3012 Bern, Switzerland; claudia.kasper@posteo.de

Tail biting is a common problem in pig production, but genetic studies are impeded by the difficulty of observing it. Even if observations or video recordings are possible, large-scale phenotyping is enormously complex and labour-intensive. Behavioural problems caused by various stressors usually manifest before escalating into damaging behaviour where serious injuries occur. Therefore, it useful to study the precursors of tail biting, which are behaviours termed 'abnormal' such as belly nosing, ear biting and 'tail-in-mouth' to gain insights into the molecular physiology of tail biting. It has been shown that poor sanitary conditions in combination with dietary protein reduction increase the prevalence of tail biting. Since restricting the protein content in the diet is a promising way to reduce nitrogen emissions in pig manure, it is important to investigate whether this measure can lead to a reduction in animal welfare. Even if essential amino acids are substituted, the overall reduction in dietary protein could lead to an increase of abnormal behaviours. Pigs differ in their ability to utilize dietary proteins, which has been shown to be heritable. Therefore, we expect that individuals are not equally affected by protein reduction, because more efficient pigs might better cope with the reduced availability of amino acids required for neurotransmitter synthesis. Here we show that gene expression profiles in the hypothalamus of 73 non-tail-docked pigs fed a protein-reduced diet differed between individuals showing abnormal and normal behaviours. We found that differences in the expression of genes involved in neurotransmission, notably in dopamine, serotonin and GABA, and G-protein coupled signalling, energy metabolism and appetite. Our results demonstrate that abnormal behaviours have distinct gene expression signatures in the brain, which could be partly caused by genomic variation. These insights can lead to a better understanding of the biological mechanisms involved and thus may ultimately inform genomic selection programmes.

Estimation of genetic parameters of days open in the Florida goat breed

C. Ziadi[1], E. Muñoz-Mejías[2], M. Sánchez[3], M.D. López[4] and A. Molina[1]
[1]University of Córdoba, Department of Genetics, Campus de Rabanales, 14014 Córdoba, Spain, [2]University of Las Palmas de Gran Canaria, Department of Animal Pathology and Animal Production, Campus Universitario Cardones de Arucas, 35413 Arucas, Spain, [3]University of Córdoba, Department of Animal Production, Campus de Rabanales, 14014 Córdoba, Spain, [4]ACRIFLOR, Departement of Animal Production, Campus de Rabanales, 14014 Córdoba, Spain; ziadichiraz4@gmail.com

The breeding program of Spanish dairy goats has focused on increasing milk production, together with morphology. Nevertheless, the known genetic antagonism between milk production and other traits such as fertility has been reported in various studies. Indeed, since the last years, both classical (e.g.: age at first kidding and the interval between kidding) and new (reproductive efficiency) criteria of female fertility are being evaluated in Spanish dairy breeds. Therefore, this study aimed to estimate genetic parameters of female fertility as expressed by days open (DO) calculated as the interval from last kidding to conception date. The database was provided by the National Association of Florida Goat Breeders. In total, 5,162 DO records from the first to the sixth parities from 4,758 Florida females were analysed and the number of animals in the pedigree was 9,454. Genetic parameters of DO were obtained by fitting an animal repeatability model using Bayesian inference as implemented in the GIBBS3F90 software. The model included the interaction of herd-year-season of parity, as well as parity number as non-genetic effects and the random additive genetic effect. The additive genetic variance was 301.77 and the heritability estimate was 0.16. This result shows that estimates of genetic variation and heritability observed for DO indicated that reasonable genetic improvement for this trait in Spanish dairy goats might be possible through selection.

Session 27 Poster 14

Temporal analysis of miRNAs expression revealing the molecular changes related to lameness recovery

W. Li[1], P. Zhe[1], K. Schwartzkopf-Genswein[2] and L. Guan[1]
[1]University of Alberta, AFNS, 116 St & 85 Ave, Edmonton, AB, T6G 2R3, Canada, [2]Agriculture and Agri-Food Canada, Lethbridge Research and Development Centre, 5403 1st Avenue South, Lethbridge, Alberta, T1J 4B1, Canada; lguan@ualberta.ca

Lameness is one of the major health issues in beef cattle, leading to significant economic losses to the industry and affecting animal welfare. Accumulating evidence indicates that circulating microRNAs (miRNAs) can be biomarkers for health diagnosis in cattle. Our previous findings have revealed that certain miRNAs in cattle blood were differentially expressed in cattle that had different lameness diagnoses. This study aimed to compare the blood miRNA profiles between cattle that had recovered from lameness (RE) and those that had not recovered (UNR) following treatment. Blood samples were collected from feedlot beef cattle diagnosed with different lameness phenotypes, including digital dermatitis (DD, n=38); toe tip necrosis (TTN, n=27) and foot rot & digital dermatitis combined (FRDD, n=41) at 3 time points: week 0 (W0); week 1(W1) and week 2 (W2) after the diagnosis and initial treatment. The recovery status of cattle was determined based on their gait score, which was separated into RE and UNR groups. Total RNA was extracted using a Preserved Blood RNA Purification Kit (Norgen) and for microRNA RNA-seq library construction. In general, significant overall miRNA profile differences were identified between RE and UNR cattle in TTNW1 and FRDDW2 groups, with differentially expressed (DE) miRNAs (Absolute log2 fold change<1, P<0.05) identified between RE and UNR cattle in DDW0 (n=2); TTNW1 (n=1); TTNW2 (n=3); FRDDW0 (n=3) and FRDDW1 (n=1). In addition, particular DE miRNAs, including bta-miR-1 and -206, were only expressed in blood samples collected from the UNR group, whose functions were related to bacterial infection and muscle cell self-repair. Furthermore, specific miRNA changing patterns were presented in only RE or UNR cattle within the same phenotype among 2 weeks, miRNAs in these patterns were related to infection and inflammatory responses during the treatment period. In conclusion, these findings provide an initial understanding of the role of specific miRNAs in lameness recovery, which lays the molecular basis for the underlying mechanism for recovering from lameness.

Session 27 Poster 15

Effect of weaning at the time of immunophenotyping on cell-mediated immune responses of beef calves

S.C. Beard, D.C. Hodgins, J. Schmied and B.A. Mallard
Ontario Veterinary College, University of Guelph, Pathobiology, 50 Stone Road E, N1G 2W1, Canada; sbeard@uoguelph.ca

Improvement of animal health using genetics is a recognized goal of the livestock industry, since the costs associated with infectious disease constitute a significant fraction of production expenses. In addition, animal well-being and consumer perceptions of the industry have become drivers for change. Despite pressure to improve disease resistance, most genetic programs for beef cattle focus on improving production, meat quality, efficiency, and reproduction traits. Genetic selection for enhanced immune responses in dairy cattle has been achieved using technologies such as Immunity+™, based on the University of Guelph's High Immune Response (HIR™) method. HIR™ may also offer the beef industry a new tool to improve sustainability by breeding cattle that are naturally healthier while maintaining productivity. Determining the optimal time to immunophenotype calves is crucial for selection and evaluation of traits to include in this breeding goal. In this study, time of weaning in relation to immunophenotyping was evaluated to determine the optimal time to phenotype beef calves for cell-mediated immune responses (CMIR). Delayed-type hypersensitivity (DTH) to a type-1 test antigen was used as an indicator of CMIR. Phenotyping was initiated either on the day of weaning (~5-7 months old, n=72) or 2 months later (n=41). Crossbred beef calves at the Ontario Beef Research Centre were injected intramuscularly with the test antigen, and received an intradermal injection two weeks later in the tail fold. DTH was assessed by the percent increase in double-skinfold thickness (DSFT) 24 hours later. Calves that were weaned on day 0 of the test protocol had significantly lower DTH than calves that were tested 2 months post-weaning (least squares means for percent change in DSFT of 33.23% [SE=10.32%] and 126.23% [SE=13.32%], respectively; P>0.0001). These results indicate that measuring DTH at the time of weaning stress may underestimate a calf's actual ability to mount a CMIR; immunophenotyping after weaning will be more effective in predicting their genetic ability to mount CMIR and resist disease.

Session 27 Poster 16

Effect of mild hypoxia on platelet activities in pregnant ewes with different haematocrit levels

P. Moneva, I. Yanchev and N. Metodiev
Institute of Animal Science, Kostinbrod, sp. Pochivka, 2232, Bulgaria; ijantcev@mail.bg

The object of the present study was to investigate changes in platelets activation due to mild hypoxia in pregnant Ile De France ewes with different haematocrit values. Thirty Ile De France ewes were selected from an experimental herd (n=110) according to their haematocrit level and were allocated into 3 groups as follows: low haematocrit (LHct, n=10) group (haematocrit range 19.7-27.9%), high haematocrit (HHct, n=10) group (haematocrit range 32.0-36.9%) and mean haematocrit (MHct, n=10) group (haematocrit range 28.3-29.8%). Immediately after the shearing, performed at the experimental unit of the Institute, (altitude of 500 m) the ewes were transported to a mountain pasture at altitude of 1440 m (Petrohan Pass, Balkan mountains). Blood samples were taken at the following time points: before transportation (baseline level), on day 7, 20 and 42 after the transport. Platelet count (PLT) and mean platelet volume (MPV) were measured via 5 diff VET haematology analyser. There was a significant increase of platelet count in LHct and MHct ewes at 20 d relative to 7 d (P<0.01). Mean platelet volume (MPV) declined in ewes of LHct group at 20d as compared to 7d (P<0.01). There was a difference of PLT count between LHct and HHct ewes after shearing (baseline level) which reached level of significance on 20d and 42d (P<0.05). We found an inverse relationship between mean platelet volume (MPV) and platelet count (PLT) as follows: r=-0.45798 (P<0.01); r=-0.59136 (P<0.001); r=-0.4403 (P<0.05) and r=–0.56996 (P<0.001) for baseline, 7d, 20d and 42d, respectively. Also, mean platelet volume to mean platelet count (MPV/PLT) ratio in LHct and MHct ewes, unlike HHct ewes increased at 7d as compared to baseline ratio. MPV/PLT ratio tended to be higher in the ewes of HHct group as compared to LHct ewes. MPV/PLT ratio in ewes of LHct group was significantly lower on 20d and 42d as compared to 7d (P<0.05). It was concluded that baseline haematocrit level is associated with platelet activity caused by stress-induced hypoxia. Acknowledgement: This work was financially supported by Bulgarian National Science Fund (Grant KP-06-H26/2, 04.12.2018).

Session 27 Poster 17

Native sheep breeds of Sicily genetic reservoirs against lentiviruses

S. Tumino[1], M. Tolone[2], P. Galluzzo[3], S. Migliore[3], S. Bordonaro[1], R. Puleio[3] and G.R. Loria[3]
[1]University of Catania, Via Valdisavoia, 95123 Catania, Italy, [2]University of Palermo, Viale delle Scienze, 90128 Palermo, Italy, [3]Istituto Zooprofilattico Sperimentale della Sicilia, Via Gino Marinuzzi, 90129 Palermo, Italy; serena.tumino@unict.it

Local breeds represent a precious reservoir of genetic diversity, crucial to adapting to environmental and climate changes and reacting to evolving diseases. In Sicily, four native dairy breeds, namely Valle del Belice, Comisana, Barbaresca and Pinzirita, have adapted to low-input farming systems and often difficult climatic conditions and semiarid environments, having an essential role in producing high-quality milk and dairy products. Maedi Visna (MV) is one of the most important chronic diseases affecting the sheep sector worldwide, causing production losses. Sheep susceptibility to MV infection is influenced by genetic variation within the ovine transmembrane 154 gene (*TMEM154*). Animals with either of *TMEM154* haplotypes (haplotypes 2 and 3) that encode glutamate at position 35 (E35) of the protein are at higher risk of MV infection than those homozygous with lysine at position 35 (K35) (haplotype 1). The present study reported for the first time the genetic investigation at *TMEM154 E35K* locus in the Sicilian autochthonous sheep breeds to provide preliminary data about the frequencies of the protective allele (K). A total of 572 animals (149 rams, 86 ewes) from 21 flocks were genotyped using the TaqMan allelic discrimination high-throughput method. In the total sample set, the putative protective allele (K) was observed with a frequency of 25%; within breed groups, the K allele was less frequent than the putative risk allele E, especially in Valle del Belice breed (14%), probably related to the selection strategies mainly addressed to get a highly productive dairy sheep. In contrast, Comisana Barbaresca and Pinzirita showed a more balanced distribution of the two alleles. Deviation from the Hardy-Weinberg equilibrium was only observed considering the total sample (P<0.001). These preliminary data could be useful to establish selection strategies aimed at controlling and eradicating MV infection in Sicilian sheep farming and could increase interest in breeding Sicilian native breeds, preserving the biodiversity on the island.

Enzootic bovine leukosis: surveillance measures and control programmes in European countries

M. Mincu[1], G. Van Schaik[2], C. Faverjon[3], E. Meletis[4], J.J. Hodnik[5], M. Guelbenzu-Gonzalo[6], M. Alishani[7], A. Gerilovych[8], Z. Acinger-Rogic[9], M. Rubin[9], I. Djadjovski[10], I. Cvetkovikj[10] and I.G.M.A. Santman-Berends[2]
[1]Research and Development Institute for Bovine, Balotesti, 077015, Romania, [2]Royal GD, Deventer, 7418, the Netherlands, [3]Ausvet Europe, Lyon, 69001, France, [4]University of Thessaly, Thessaly, 38221, Greece, [5]University of Ljubljana, Ljubljana, 1000, Slovenia, [6]Animal Health, Carrick-on-Shannon, 75, Ireland, [7]University of Prishtina, Prishtina, 10000, Kosovo, [8]Institute for Experimental and Clinical Veterinary Medicine, Kharkiv, 61023, Ukraine, [9]Veterinary and Food Safety Directorate, Zagreb, 020921, Croatia, [10]Ss Cyril and Methodius University, Skopje, 1000, Macedonia; madalinamincu8@gmail.com

The COST action 17110 SOUND control aims to harmonise the results of surveillance and control programmes (CPs) for cattle diseases to improve overall control of infectious cattle diseases. The aim of the study was to assess the existence and quality of data that could be useful for models that quantify the probability of freedom from Enzootic Bovine Leukosis (EBL) and for creating an overview of the EBL situation across several European countries. A data collection tool embedded in a Google form was designed and used to acquire standardized and high-quality data. The tool consists of 3 sections: A-general information and demographics of cattle populations, B-EBL CP and C-EBL testing strategies. Animal health experts from 8 European countries (ROU, GRE, SLO, CRO, UKR, KOS, IRL and MKD) completed the questionnaire. Regarding demographics, none of the respondents could provide all requested information, however, all respondents answered at least 50% of the questions. This could be attributed to the lack of access to specific data at a national or regional level in the participating countries. Regarding the existence of EBL CP, 7 out of 8 respondents specified that their country have a CP in place. Regarding the EBL individual country status, 5/8 countries have sporadic cases, 1/8 reported endemic status, while 2/8 reported a free status. Our study provides an overview of the current situation in several European countries regarding accessibility and quality of data on EBL. The data can be used in models that can assess the outputs of the different CPs for EBL.

Heritabilities of clinical mastitis, metritis and ovarian cysts in Polish HF cows

P. Topolski[1], W. Jagusiak[1,2] and B. Szymik[1]
[1]National Research Institute of Animal Production, Department of Cattle Genetics and Breeding, Krakowska 1, 32-083 Balice near Krakow, Poland, [2]University of Agriculture in Krakow, Department of Genetics, Animal Breeding and Ethology, Al. Mickiewicza 24/28, 30-059 Krakow, Poland; piotr.topolski@iz.edu.pl

Heritabilities of health traits such as mastitis, metritis and ovarian cysts in the population of Polish Holstein-Friesian (PHF) cows were estimated. The dataset was created on the basis of PLOWET database developed by the National Research Institute of Animal Production and included phenotypic data collected in 4 experimental dairy farms. Phenotypic evaluation of health traits of the cows was carried out between 2016 and 2021 and was based on veterinary registration. The dataset used in our study comprised 7,671 cows. REML algorithm implemented in the BLUPF90 package was used for the computations. The linear and threshold observation models included additive random effects of animal, fixed effects of herd-year-season of calving subclass (HYS) and lactation stage, fixed regressions on cow age at calving and fixed regressions on cow age at the time the disease. Cow pedigrees contained two generations of ancestors i.e. parents and grandparents. Heritabilities estimated based on both observation models were very similar and ranged from 0.01 (0.0063, 0.0072) for metritis and ovarian cysts to 0.03 (0.0067) for mastitis. The magnitude of the obtained coefficients of heritability are in the range of heritability coefficients published in the papers of other authors and indicate the possibility of including these traits in the national dairy cattle improvement program.

Integrating European cattle research infrastructures for innovation: outcomes from SmartCow project

R. Baumont[1], M. O'Donovan[2], R. Dewhurst[3], P. Lund[4], B. Kuhla[5], R. Carelli[6], C. Reynolds[7], C. Martin[1] and I. Veissier[1]
[1]INRAE, UMR Herbivores, Theix, 63122 Saint-Genès-Champanelle, France, [2]Teagasc, Moorepark, Fermoy, Cork, Ireland, [3]SRUC, West Mains Road, EH9 3JG Edinburgh, United Kingdom, [4]Aarhus University, AU Foulum, Blichers Alle 20, 8830 Tjele, Denmark, [5]FBN Leibniz, Wilhelm-Stahl-Allee 2, 18196 Dummerstorf, Germany, [6]EAAP, Via Giuseppe Tomassetti 3 A/1, 00161 Roma, Italy, [7]University of Reading, Earley Gate, RG6 6AR Reading, United Kingdom; rene.baumont@inrae.fr

The SmartCow project (H2020 N°730924; https://www.smartcow.eu/) integrated key cattle research infrastructures (RIs) for the first time at European scale to help their operators to develop synergies and complementary capabilities. This resulted in improved and harmonised research services, and supported 24 research projects led by researchers from academic and private sectors. Research activities in SmartCow aimed to increase phenotyping capabilities while implementing the 3R principles (refine, reduce and replace) in cattle physiology and behaviour studies. A ring test identified sources of measurement error in enteric CH_4 emissions from respiration chambers. Sources of errors in digestion and N balance studies were quantified through a meta-analysis and specific experiments to refine experimental procedures including number of animals and length of measurements. The development and validation of proxy indicators, among others, of feed efficiency, digestibility and methane emissions traits, was undertaken with the goal of minimising handling of experimental animals. As an example, NIR spectra from spot samples of faeces are able to predict total tract digestibility and CH_4 emissions in both dairy and beef cattle. Guidelines were provided to validate data from activity sensors and new algorithms were developed to predict health status of animals. Recommendations from the SmartCow consortium on ethics in experiments on live cattle and on best practices in cattle physiology and behaviour research are disseminated through the already published Book of Methods and through training courses. This will contribute to educate a new generation of researchers ready to exploit optimally essential tools for their research, and help the cattle sector to find innovative solutions to improve sustainability in cattle production.

Collaboration is imperial or how to address this critical issue when developing novel traits

N. Gengler
ULiège-GxABT, Passage des Déportés, 2, 5030 Gembloux, Belgium; nicolas.gengler@uliege.be

In the age of genomics, phenotype is king. This saying is fundamental when we try to consider novel traits in the context of animal breeding but also smart nutrition management in dairy and beef cattle. Projects like SmartCow are crucial to generate the critical mass for novel phenotypes. A rich source of novel phenotypes in dairy cattle is potentially milk mid-infrared (MIR) spectrometry that allows to define novel traits exploiting the variability of milk composition. Available reference data and MIR spectral data are joined to allow development of MIR based predictions. But data must be sufficient in quantity and in variability covering the expect ranges for future uses on existing MIR databases. International collaboration becomes here a critical challenge in terms of data acquisition and standardization to develop robust models. Joining data across many sources need innovative strategies called hereafter 'Open Consortium Building'. In this type of approach, first calibration building organizations will be defined. General consortium members retain full ownership and control of their data, only providing it to equation builders which can only use this data to improve equations under development. By helping improving equations, consortium members get access to calibrated equations, but also access to all future updates when additional data from new members is included. This 'Open' approach that avoids exchanges of data across all consortium members, allows everybody to join under the same conditions, even competing commercial structures that would normally not collaborate. The general context of collaboration extends to the needed standardisation of relevant data. Experiences with this approach will be presented, especially for methane emissions in dairy cattle. The final message is that in the age of challenging research, the needed collaboration and coordination are imperial.

Sources of variation in cattle N balance measurements from digestion trials at 15 research sites

Z.E. Barker[1], L.A. Crompton[1], H. Van Lingen[2], J. Dijkstra[2] and C.K. Reynolds[1]
[1]University of Reading, Earley Gate, RG6 6EU, United Kingdom, [2]Wageningen University, P.O. Box 338, 6700 AH, the Netherlands; z.e.barker@reading.ac.uk

To evaluate sources of variation in N balance measurements, a data set of 3,969 individual cow digestion measurements at 15 sites was collated. Data included animal status, collection methods, analytical methods, diet composition and intake, N excretion in faeces, urine, and milk, and N retention. Two smaller subsets of the data that included measurements of urine, faecal and milk (if lactating) N excretion, diet NDF, ADF, and ash content, bodyweight (BW), and diet forage proportion were analysed. Subset A included data from lactating and non-lactating dairy, lactating beef and growing beef cattle (1,704 observations) and subset B contained lactating dairy cattle only (1,277 observations). A series of mixed effect models were fitted for retained N (balance), urine N excretion, faecal N excretion and milk N excretion with random effects of experiment nested within site and with independent variables selected based on BIC and RMSPE. N intake was included in the models for all outcome variables and for both data subsets. For retained N and faecal N excretion the models from data subset A also included cow status, with lactating dairy cows having significantly lower retained N and higher faecal N excretion than non-lactating dairy cattle. For subset B, N intake, BW, and diet ash content were kept in models for retained N, and diet ash and ADF content kept in models for faecal N excretion. Models for urine N excretion for both data subsets included N intake and BW as explanatory variables. Milk N was best explained by N intake, BW, diet ash and NDF content, and parity, with primiparous cows having significantly lower milk N excretion. Despite significant variation associated with experiment (26-36%) and site (25-35%) no methodology variables were retained in the best models. This could be explained by a lack of variation across and within sites for some methods and insufficient detail available for some methodologies. The results confirm considerable variation in N balance measurements between research sites, which may relate to unidentified variation in how methods are used. Funding by the SmartCow project (H2020 N°730924) is gratefully acknowledged.

Implementation of the 3Rs in SmartCow

K.M.D. Rutherford
SRUC, Animal Behaviour and Welfare, West Mains Road, Edinburgh, EH9 3JG, United Kingdom; kenny.rutherford@sruc.ac.uk

There are many reasons for research studies using cattle, including animal health and welfare, performance, and environmental impact. However, whilst improving understanding in these areas is vital, concern about the use of animals in research is also an ongoing societal issue. Support for animal research is based on the idea that studies should be judged on the balance of harms imposed to animals, versus the benefits of the work. An important part of this is reducing the welfare harms imposed whilst still achieving scientific goals – a principle first outlined more than 60 years ago by Russell and Burch in their seminal work on the '3Rs' (Replacement, Reduction, Refinement) of humane animal research. At its inception the SmartCow project aimed to advance the conduct of cattle research including making improvements relating to the 3Rs. Over the course of the project various beneficial outcomes have been documented. These includes improvements to housing and restraint facilities that demonstrates that welfare can be improved – through reduced behavioural restriction, increasing cow comfort, and optimising sampling period duration – whilst continuing to ensure the collection of high-quality data. Attention has also been paid to approaches that allow for more refined procedures (by identifying appropriate proxies for key traits such as digestibility and nitrogen excretion) and the utilisation of reduced animal numbers (by identifying sources of measurement errors in studies involving feed digestion, nitrogen balance and emissions measurements). Progress has also been made on monitoring welfare in both study animals and in commercial settings. Research work using cattle is likely to remain as a critical source of information and solutions to key societal issues for the foreseeable future. As such, it remains important to develop scientific approaches that yield the necessary data whilst giving due respect to the 3Rs and to the experiences of the sentient individuals that are the subjects of the work. Whilst some problems remains to be solved, the work to date shows that livestock researchers can conduct high quality science that informs upon major global issues, whilst also improving animal welfare in their study animals. Funding by SmartCow project (H2020 N°730924) is acknowledged.

Beef cattle welfare in digestibility stalls

R. Bellagi, R. Baumont, P. Nozière and I. Veissier
INRAE, 63122, Saint-Genès-Champanelle, France; rahma.balegi@gmail.com

Animals are traditionally maintained in digestibility stalls for two weeks to measure feed digestibility or Nitrogen balance. Digestibility stalls prevent animals from moving and can thus impair their welfare. In the SmartCow project (H2020 N°730924; https://www.smartcow.eu/) we aimed at evaluating the impacts of being maintained in digestibility stalls on animal welfare. Sixteen Charolais bulls were moved in digestibility stalls for two periods of 15 days, including 6 adaptation days followed by 9 days with total urine and faeces collected. Bull posture (standing vs lying) and activity (inactive, ruminating, feeding, or active) were recorded by accelerometers from 7 days before to 7 days after the stay in stalls. The time budget was expressed in % (min/100 min). Hair concentration in cortisol was measured before the bulls were moved to stalls then either after 10 or 15 days in stalls. Analyses of variance were performed to assess the effect of being in stalls, the duration of the stay in stalls, and the repetition (SAS, mixed model for repeated data in case of behaviour). Bulls reacted to being moved to stalls with a significant increased proportion of time spent standing-inactive (+8%) and a significant decreased proportion of time spent feeding (-3%), standing-active (-1%) or lying-ruminating (-7%). During the first 6 days in stalls, time spent standing-inactive decreased across days (-0.8%/day) whereas time spent feeding, standing-active or lying-ruminating increased (0.001, 1.2, and 0.5%/day). Once urine and faeces collection started, opposite trends were observed (0.6, -0.07, -0.3, and -0.18%/day). Hair cortisol was increased after 10 days in stalls but did not increase further from 10 to 15 days in stalls (6.60 vs 9.12 vs 9.77 pg/mg). Bulls thus seem to react strongly to being moved to stall – with a release of cortisol and modifications of the time budget- and to progressively habituate to stalls. The collection of urine and faeces again trigger modifications of behaviour that tend to increase with time, therefore with no habituation. We conclude that animals would benefit of a reduction of the duration of the measurements in digestibility stalls.

A non-invasive sound technology to monitor rumen contractions and rumen gas fermentation

E. Vargas-Bello-Pérez, A.L. Alves Neves and A. Harrison
University of Copenhagen, Department of Veterinary and Animal Sciences, Faculty of Health and Medical Sciences, 1870, Frederiksberg, Denmark; evargasb@sund.ku.dk

This technology report demonstrates that sound recordings from the forestomach of ruminants is not only possible, confirming earlier work, but that it can be of such a quality that individual sounds can be identified and differentiated from each other. In this report, using a wireless device (CURO MkII) it has been shown that rumen sound recordings of a high quality are not only feasible, but they can also discern between ruminants of different production status (dry cow versus 100 d lactating cow). The sound recordings were made using a CURO unit (CURO-diagnostic ApS, Bagsværd, Denmark). The recording site was prepared with acoustic gel (CURO-diagnostics ApS) which was thoroughly rubbed into the overlying hair to ensure a good connection with the skin above the rumen, located on the left flank of the animal in between the last rib, paralumbar fossa, and lumbar vertebrae. Sensors (piezo ceramic – diameter of 50 mm) were likewise prepared with acoustic gel, before they were attached to the site of interest using flexible self-adhesive bandage (Animal Polster, Snögg Industry AS, Kristiansand, Norway). The sensors were then connected to the CURO unit and recordings were made in the form of a WAV file to an iPad via a wireless connection, so that the recordings could be monitored in real-time. Data collection used a sampling frequency of 2,000 Hz. The CURO units weighed 10.2 g and has a Bluetooth 4 connection to both tablet/smartphone supported software. The range on the Bluetooth connection was approx. 40 meters, but through a remote link it is also possible to place a CURO unit on a cow, record rumen sounds at a distance of many km´s and transmit them to a Client Cloud where they can be accessed from anywhere around the World. There is now a need for focused research into the correlation of rumen sounds with measurements of rumen CH_4 and CO_2 in animals at different stages of production, something of particular relevance in terms of global warming and the emissions of greenhouse gases. The use of this technology for the early prevention of digestive tract disorders is also of considerable importance when viewed in terms of ruminant health.

Effects of stage of lactation and parity on Δ15N as a biomarker for N use efficiency in dairy cows

R.J. Dewhurst, E. Hyles and A. Bagnall
SRUC, Dairy Research & Innovation Centre, Barony Campus, Parkgate, Dumfries, DG1 3NE, United Kingdom; richard.dewhurst@sruc.ac.uk

Fractionation of N isotopes between feed and milk (or plasma) has been proposed as a biomarker for N use efficiency (NUE) or feed efficiency of ruminants. To explore some of the variation in this relationship, we analysed milk samples from lactating Holstein cows of either first (n=56) or second parity (n=40). Cows were from the Langhill genetic experiment with a 2×2 factorial arrangement of two selection lines (for fat plus protein yield) and two total mixed rations with divergent energy content (12.6 or 12.0 MJME/kg DM). Days in milk ranged from 24 to 481 (mean=208; SD=119.2). Feed intakes and milk yield were recorded over one week, and milk samples taken from consecutive morning and evening samples (for compositional analysis by infra-red analysis) and evening samples (for N isotope analysis by isotope ratio mass spectrometer). DM intake (kg/day), milk yield (kg/day), milk protein % and N-use efficiency (g/100 g) averaged (SD): 20.7 (3.67), 26.4 (7.54), 3.48 (0.318) and 26.1 (5.30). Overall, and after correcting for a small (0.33‰) but highly significant diet effect, there was the expected negative relationship between N isotopic fractionation (y; Δ15N; milk 15N – diet 15N) and N-use efficiency (x; r^2=29.5; RSD=0.482; P<0.001). There was no significant relationship in cows during the first 100 days of lactation, which may be due to differences in N mobilisation during this period, as well as possible carryover effects from dry period feeding. Excluding cows in the first 100 days of lactation (leaving 70 cows) resulted in a stronger relationship (r^2=36.0; RSD=0.465; P<0.001). Changes in body N must be considered in addition to milk protein output when using Δ15N as a biomarker for NUE in lactating dairy cows. Funding from the SmartCow project (H2020 N°730924) is acknowledged.

Inclusion of seaweed in the dairy cattle diet and the effect on milk production and feed efficiency

E.E. Newton[1], M. Terré[2], K. Theodoridou[3], S. Huws[3], P. Ray[4], C.K. Reynolds[1], N. Prat[2], D. Sabrià[2] and S. Stergiadis[1]
[1]University of Reading, School of Agriculture, Policy and Development, Reading, RG6 6EU, United Kingdom, [2]Institute de Recerca i Tecnologia Agroalimentàries, Department of Ruminant Production, Caldes de Montbui, 08140, Spain, [3]Queen's University Belfast, Institute for Global Food Security, Belfast, BT9 5DL, United Kingdom, [4]The Nature Conservancy, North Fairfax Drive, Virginia 22203, USA; eric.newton@pgr.reading.ac.uk

Using seaweed as animal feed has recently received increased attention. However, cow diet affects product quality and animal health and assessing the implications of feeding is essential. This study examined the effect of partial replacement of corn meal with seaweed (*Ascophyllum nodosum*) on milk composition, efficiency, and haematological parameters. Lactating Holstein cows (n=48) were allocated into two experimental groups, balanced for parity, milk yield, contents of fat, protein, and somatic cell count (SCC), in a randomized block design. Diet treatments were: (1) control (CON); and (2) partial replacement of corn meal with *A. nodosum* (SWD; 330 g/day on dry matter (DM) basis). After a 2-week adaptation, cows consumed the experimental diets for 8 weeks. DM intake, feed, milk, and blood samples were collected daily, weekly, fortnightly and at the end of the trial respectively. Linear mixed models used diet, week, and their interaction as fixed factors, and cow (nested within diet) as random factor; with pre-treatment records as the covariate. Week was excluded from the model for haematological parameters (measured once). Milk yield, basic composition and efficiency parameters were similar between CON and SWD. When compared with CON, cows fed SWD had lower neutrophil concentrations ($2.65e^9$/l vs $2.06e^9$/l respectively, P=0.009) and a tendency for lower white blood cell count ($7.37e^9$/l vs $6.70e^9$/l, respectively, P=0.050); although their concentrations were within the normal range for healthy dairy cows in both diets. This study provides evidence that corn meal can be safely partially replaced with *A. nodosum* at 330 g/day, without negative implications to cows' productivity, milk quality and efficiency, while changes in measured haematological parameters did not indicate any potential health risks. Funding by SmartCow project (H2020 N°730924) is acknowledged.

Guidelines for using cow milk mid-infrared spectra to predict GreenFeed enteric methane emissions

M. Coppa[1], A. Vanlierde[2], M. Bouchon[3], J. Jurquet[4], M. Musati[1,5], F. Dehareng[2] and C. Martin[1]
[1]Université Clermont Auvergne, INRAE, VetAgro Sup, UMR 1213 Herbivores, Theix, 63122 Saint-Genès-Champanelle, France, [2]Walloon Agricultural Research Centre, Rue de Liroux 9, 5030 Gembloux, Belgium, [3]INRAE, UE1414 Herbipôle, Marcenat, 63122 Saint-Genès-Champanelle, France, [4]Institut de l'Elevage, 42 Rue Morel, 49071 Beaucouzé cedex, France, [5]Department Di3A, University of Catania, Via Valdisavoia 5, 95123 Catania, Italy; mauro.coppa@inrae.fr

Various methodological protocols were tested to identify the best approach to predict GreenFeed system (GF) measured enteric methane (CH_4) emissions by mid-infrared spectroscopy (MIR) on milk. Individual milk yield (MY), fat and protein corrected milk (FPCM), and dry matter intake (DMI) were recorded daily on 115 Holstein cows fed diets with different methanogenic potential. Milk samples were collected twice a week. Twenty CH_4 spot measurements with GF were taken as the basic measurement unit (BMU) of CH_4. Partial least squares regressions were validated on an independent datasets. Models based on single daily spectra (SD spectra) were calibrated using a CH_4 measurement duration of 1, 2, 3 or 4 BMU. Models built from the average of daily spectra (AD spectra) collected during the corresponding CH_4 measurement periods were also developed. Corrections of spectra by days in milk (DIM) and the inclusion of parity, MY, and FPCM as explanatory variables were tested. Long duration of CH_4 measurement by GF performed better than short duration: the R^2 of validation (R^2V) for CH_4 emissions in g/d were 0.60 vs 0.52 for 4 and 1 UBM, respectively. Coupling GF reference data with the corresponding milk MIR AD spectra gave better prediction than using SD spectra (R^2V=0.70 vs 0.60 for CH_4 as g/d on 4 BMU). Correcting the SD spectra by DIM improved R^2V compared to the equivalent DIM uncorrected models (R^2V=0.67 vs 0.60 for CH_4 as g/d on 4 BMU). Adding other phenotypic information as explanatory variables did not further improve the performance of models built on DIM-corrected SD spectra, whereas including MY (or FPCM) improved the performance of models built on the AD spectra (uncorrected by DIM) recorded during the CH_4 measurement period (R^2V=0.73 vs 0.70 for CH_4 as g/d on 4 BMU). Funding by SmartCow project (H2020 N°730924) is acknowledged.

Estimation of methane eructed by dairy and beef cattle using faecal near-infrared spectra

A. Vanlierde[1], F. Dehareng[1], A. Mertens[1], M. Mathot[1], A. Lefevre[1], I. Morel[2], G. Renand[3], Y. Rochette[4], F. Picard[4], C. Martin[4] and D. Andueza[4]
[1]CRA-W, Gembloux, 5030, Belgium, [2]Agroscope, Posieux, 1725, Switzerland, [3]INRAE, GABI, Jouy-en-Josas, 78350, France, [4]INRAE, UMR Herbivores, Saint-Genès-Champanelle, 63122, France; a.vanlierde@cra.wallonie.be

The relevance to consider faecal near-infrared spectra (NIRS) to predict digestible organic matter (DOM) has been established while methane (CH_4) eructation, is also related, among others, to the intake of DOM. Consequently, it seems also relevant to investigate the estimation of eructed CH_4 from faecal NIRS. A reference dataset collected in different countries (Belgium, France and Switzerland), including faecal NIRS and corresponding CH_4 values, has been constituted in the framework of the Smartcow project (H2020 N°730924). Automated Head-Chamber System (AHCS, GreenFeed) was used to collect CH_4 production references data. Only values including at least 20 visits of the AHCS during the period were considered as representative. A period of 3 weeks for CH_4 measurement has been fixed to maximize the dataset. In parallel spot samples of faeces were collected at the end of the period. Data from lactating cows (LC) [n=91, 361±86 g CH_4/day (mean ± SD), Holstein LC, diet based on grass silage, grass and/or hay] were considered separately from beef cattle (BC) data [n=346, 221±43 g CH_4/day, Charolais (heifer), Belgian Blue (reformed cow), Dual purpose Belgian Blue (suckling cow), diet based on grass, grass silage and/or hay]. From the data available, the best models presented for LC and BC respectively, a R^2 of calibration about 0.43 and 0.62, and a standard error of calibration about 65 and 26 g CH_4/day. The better results for BC than for LC were expected because this calibration set include data from more animals and conditions, consequently more variability at animal level but also regarding CH_4 and spectral information, which is needed to develop such models. These results suggest the feasibility of having a proxy based on faecal NIRS to estimate CH_4 eructated which is particularly interesting for non-lactating animals (milk-based proxies are inapplicable). Additional data have to be collected with the same sampling protocol but presenting other diets, breeds, physiological status etc. to confirm this trend and to improve the model robustness.

Identifying health and stress states in cows from their daily activities by machine learning

I. Veissier[1], Q. Ruin[1], R. Lardy[1], V. Antoine[2] and J. Koko[2]
[1]INRAE, UMR Herbivores, 63122 Saint-Genès-Champanelle, France, [2]Université Clermont Auvergne, Limos, BP 10125, 63173 Aubière, France; isabelle.veissier@inrae.fr

Precision livestock farming (PLF) systems help to detect when animals are under specific states such as illness, stress, oestrus. They often detect one or few disorders at a time or several but without distinguishing them. We aimed at predicting and classifying a wide range of health, stress and physiological states from cows' daily activity using Machine Learning on time series. We used five datasets (120,000 cow × days). Caretakers noted any event related to the state of cows – specific disease (lameness, mastitis, acidosis, accident, etc.), in oestrus, calving – or disturbances such as handling. We labelled a number of days before and after the events when the behaviour might be altered. Sensors (location systems or accelerometers) provided the duration of cows' activities per hour – eating, resting, in alleys –, from which we computed the activity level. The distribution of the activity level in all 24h time series was described by 32 statistical features from the time and the frequency domains. We applied the Random Forest algorithm to relate the statistical features of time series to the animal state (1000 trees). Features with a weight above 4% were: mean; STD; RMS; RMSSD; maximum; quantiles 10, 25, 50, 75, 90%; Kurtosis; Skewness; autocorrelations of order 1 to 5, Fourier harmonics 1 to 4. We classified correctly 99% of control series (i.e. with no event) and 42% series labelled with an event. Confusions occur between series labelled with an event and series control and disturbances or acidosis when these predominate in the dataset. Around a disease or a reproductive event, the probability to classify successfully at least one series was 91-100%. This detection occurred often 1-2 days before caretakers noticed the event. Machine Learning on time series allows detecting several welfare states of cows and could thus be implemented in PLF systems to facilitate herd management. SmartCow, Horizon 2020 Grant Agreement 730924.

Edge-computing to monitor cow behaviour in rangelands

J. Navarro García[1], I. Gómez Maqueda[2], D. Varona Renuncio[2], M. Gaborit[3], F. Launay[3], M.J. García García[4] and F. Maroto Molina[4]
[1]Rey Juan Carlos University, Data Science Laboratory, Tulipan St., 28933 Mostoles, Spain, [2]Digitanimal SL, Castilla Ave. 1, 28830 San Fernando de Henares, Spain, [3]INRAE, Domaine Expérimental du Pin, Borculo Exmes, 61310 Gouffern-en-Auge, France, [4]University of Cordoba, Animal Production, Madrid-Cadiz Rd. km 396, 14071 Cordoba, Spain; fmaroto@uco.es

The automatic monitoring of cattle behaviour is one of the main targets of precision livestock farming (PLF) tools, as deviations from normal behaviour patterns can be indicative of events of interest. Many PLF tools are commercially available to monitor cow behaviour, most of them based on accelerometery, but they are normally designed to be used indoors or in small paddocks, where connectivity constraints are limited. In rangelands, sensor data need to be transmitted wirelessly over long distances to allow the real-time monitoring of animals. Low power wide area (LPWA) networks have been deployed for the Internet of Things, allowing long-range data transmission, but the number and size of data packages that can be sent through them are small. High-resolution data gathered by accelerometers cannot be directly transmitted using LPWA networks. The objective of this study is to develop and finetune an edge-computing strategy aimed at estimating behaviour from sensor data directly in the PLF device and then sending to the cloud only the relevant information to monitor cow behaviour. Beef and dairy cows from INRA Le Pin (France) and La Blaqueria (Spain) were equipped with 3D accelerometers mounted on collars gathering high-resolution data. Cow behaviour was labelled *in situ* or videorecorded and then labelled using the software BORIS. A total of 78 observation hours were labelled for the following behaviours: walking, grazing, standing and lying. Two models were chosen to be implemented in the device board: decision tree and logistic regression. Harmonic F-score was 0.8398 for decision tree and 0.8828 for logistic regression. Regarding individual behaviours, F-score was maximum for walking (0.9757) and minimum for standing (0.7337). More complex models, e.g. neural networks, may have a better performance, but they cannot be implemented in existing boards due to energy and computation constraints. Authors thank the funding of SmartCow project, grant number H2020 N°730924.

Recording of eating time in rosé veal calves by use of an automated video system

M. Vestergaard[1,2], H.H. Bonde[1], M. Bjerring[2], P. Ahrendt[3], L.A.H. Nielsen[1] and A.M. Kjeldsen[1]
[1]SEGES, Livestock Innovation, 8200 Aarhus N, Denmark, [2]Aarhus University, Department of Animal Science, 8830 Tjele, Denmark, [3]AniMoni ApS, Åbogade 15, 8200 Aarhus N, Denmark; mogens.vestergaard@anis.au.dk

In some beef calf operations, feed intake might be compromised, especially in some calves in a pen, due to limited access to feeders and/or too little feed bunk area. This can reduce overall feed intake and thus limit calf growth. No practical tool is available for measuring feed intake in private settings. Automated video recording at the feed bunk area could be a tool to estimate feeding time in rosé veal calves at commercial farms. The objective of the present study was to develop and test a video-algorithm to predict eating time and test how type of diet and age of bull calf affected these measures. Four pens each with 22-25 Holstein male calves were equipped with video cameras, two cameras per pen. Accelerometers were ear-mounted on 10 calves per pen per block. Four blocks of 3-week recordings were made with new calves in the pens in each block. In each block, two pens housed 5.8 and two pens 8.1 months old calves at start of the block. Two types of total mixed ration, named DRY and WET, were offered to 2 pens per block. DRY consisted of a concentrate pellet and 15% alfalfa hay and WET was a corncob silage-based TMR. Net Energy (NE), protein and starch were similar in DRY and WET, but DM was 86 vs 53%. Using pen as the experimental unit, we had 16 observations. NE intake, average daily gain (ADG), Proportion of calves at the feed bunk, and Time spent eating were not affected by feed ration. Inactivity were 11% lower (P<0.01) and Rumination time 32% higher (P<0.01) for WET compared with DRY. NEI was 36% and ADG was 21% higher (P<0.01) and Proportion of calves at the feed bunk was 22% less (P<0.01) and Time spent eating was 20% shorter (P<0.01) for 8.1 compared with 5.8 months calves. The eating time estimated by the video-algorithm (100 to 130 min per day) is similar to eating times recorded by automated feed bins, and suggests the developed video-method gives reliable estimates.

Paving the way towards common standards and guidelines for experimental studies with cattle

S. Mesgaran[1], R. Baumont[2], L. Munksgaard[3], D. Humphries[4], E. Kennedy[5], J. Dijkstra[6], R. Dewhurst[7], H. Furguson[7], M. Terré[8] and B. Kuhla[1]
[1]FBN, Wilhelm-Stahl-Allee 1, 18196 Dummerstorf, Germany, [2]INRAE, UMR1213 Herbivores, 63122 Saint Genès Champanelle, France, [3]Aarhus University, P.O. Box 50, 8830 Tjele, Denmark, [4]University of Reading, P.O. Box 237, Reading RG6, United Kingdom, [5]Teagasc, Moorepark, Fermoy, Co. Cork, Ireland, [6]WUR, P.O. Box 338, 6700 AH Wageningen, the Netherlands, [7]SRUC, Barony Campus, Dumfries, DG1 3NE, United Kingdom, [8]IRTA, Ruminant Production, 08140 Caldes de Montbui, Spain; sadjad.danesh.mesgaran@kaesler.de

Many similar studies are performed in different countries, though they differ in the experimental techniques used and standards applied. This often makes the results difficult to compare. Thus, it is essential that common standards and guidelines for measurements and data recording are developed and applied. Researchers in the SmartCow project (H2020 N°730924) developed a common framework regarding routine and experimental measurements in cattle physiology and behaviour. Experimental guidelines are available as an Open Access living handbook on PUBLISSO entitled 'Methods in cattle physiology and behaviour research – Recommendations from the SmartCow consortium'. The book contains 19 chapters covering ethics in experiments on live cattle, intake and behaviour, body condition and anatomy, reproductive, stress and health assessment, rumen function, nutrient digestibility and balance studies, respiratory chamber facility and techniques to measure gas emissions. Each chapter lists specific animal traits referring to their identifiers in the Animal Trait Ontology of Livestock (ATOL) and the Environmental Ontology of Livestock (EOL). Currently, two more chapters regarding 'Guidelines to apply for ethical approval of animal experiments' and 'Validation of eating duration using an automatic feeding system' are in preparation and will be implemented to the book soon. Researchers interested in adding a new method to the book should contact the editors. Next, progressive implementation of the guidelines in the experimentation with cattle and referencing the guidelines in scientific publications is required. This will verify the quality of research and allow better comparison and re-use of research data.

Effect of days of collection and a faecal preservative on measured N balance of lactating cows

C. Reynolds, D. Humphries, A. Jones, P. Kirton, R. Szoka and L. Crompton
University of Reading, Earley Gate, RG6 6EU, United Kingdom; c.k.reynolds@reading.ac.uk

Our objective was to assess effects of diet crude protein (CP) level (14 vs 18% of dry matter [DM]), days of collection (4 to 8), and a faecal preservative on measurements of diet digestion, faecal and urine N excretion, and N balance of 4 midlactation dairy cows using a switch-back experiment with three 5-wk periods. Measurements of N intake and total faecal, urine, and milk N excretion were obtained over the last 8 days of each period. Daily samples of diet, orts, faeces, urine, and milk were analysed for N concentration immediately after each 24-h collection. Frozen bulk samples for the first 4 to 8 days of collection were also saved and analysed for N concentration within 2 weeks of sampling. Diet samples were chopped using dry ice to minimize N loss. During period 2, HCl-ethanol was added as a preservative to bulked faecal samples that were either frozen or refrigerated before analysis immediately after the last day of collection. Data were analysed for fixed effects of CP and period and random effects of cow. Urine N concentration was 5% lower for frozen vs fresh samples, but otherwise N concentration of bulk samples was similar to average concentrations measured on fresh samples. Frozen and refrigerated HCl-ethanol treated faeces N concentration was 6% lower than for fresh faeces analysed daily. DM and N digestion, urine, faeces and milk N excretion, and body N retention, and effects of diet CP, were similar for measurements obtained over 4 days compared to longer periods. For example, N (g/d) intake (422 vs 600±21.5), digestion (253 vs 410±17.5), urine (128 vs 232±11.0) and milk (135 vs 153±8.5) excretion, and balance (-9.0 vs 27.9±14.8) were increased by diet CP concentration ($P<0.001$) when measured using 4-d frozen bulk samples. Comparable values for 8-d frozen bulk samples were: intake (429 vs 607±22.2), digestion (261 vs 415±18.9), urine (125 vs 229±10.7) and milk (131 vs 160±9.0) excretion, and balance (3.8 vs 27.3±14.3) for 14 and 18% CP diets, respectively. In the present study there was little variation in diet intake and composition, which minimized variation in N balance measurements over time. Funding by the SmartCow project (H2020 N°730924) is gratefully acknowledged.

Accuracy and length of digestibility and N balance measurement in beef cattle

R. Bellagi[1], R. Baumont[1], L. Salis[1], S. Alcouffe[2], G. Cantalapiedra-Hijar[1] and P. Nozière[1]
[1]INRAE, UMR Herbivores, 63122 Saint-Genès-Champanelle, France, [2]INRAE, UE Herbipole, 63122 Saint-Genès-Champanelle, France; rahma.balegi@gmail.com

Digestibility and Nitrogen balance (NB) are classically measured in animals held in digestibility stalls for several consecutive days (d). These measurements are essential for feed evaluation and animal phenotyping but rise animal welfare issues. This study in the SmartCow project (H2020 N°730924; https://www.smartcow.eu/) quantified how the length of measurement period affects the measurement errors, the associated repeatability and the minimum detectable difference (MDD) for the following traits: voluntary dry matter intake (VDMI), dry matter digestibility (DMd), nitrogen intake (NI), urinary (UN) and faecal (FN) N excretions and finally NB. Sixteen Charolais bulls previously adapted to their respective experimental diets, were submitted to two 15d experimental periods (P1 and P2) with 5d of adaptation to the digestibility stalls and 10d of measurement. Eight bulls received a high crude protein diet (173 g/kg DM), and the other eight a low one (116 g/kg DM). From d1 to d10 of measurements, VDMI, faeces and urine excretion were measured daily. Pooled samples representative of d1-d3, d4-d5, d6-d7, d8, d9 and d1-d10 of offered feeds, refusals, urine and faeces were analysed for N content allowing to calculate N balance for several period lengths (from 3-d to 10-d). Repeatability of the measured traits during 3, 5, 7, 8, 9 and 10 consecutive days was assessed through their respective within-group (diet × batch) correlation coefficients between P1 and P2. The MDD for each dependent variable was calculated as the residual standard error from the model including the fixed effect of diet, period and its interaction and the random effect of animal multiplied by 2 times 1.96. For all variables, MDD increased between 10 and 3 d; UN and NB being the most impacted variables (respectively 4.40 and 7.2 g/d at 10 d; + 33.5% and +26.6% between 10 and 3 d), and FN the less (5.11 g/d at 10 d; +7.9% between 10 and 3 d). A decrease in repeatability was observed: -0.13 for UN, -0.07 for NB and -0.29 for DMd but only -0.03 for FN between 10 and 3 d. From these results it appears that reducing the duration of the measurement to less than 10 days necessarily leads to a loss of accuracy.

Session 28 Poster 17

Promising proxies to predict feed efficiency and its components in dairy and beef cattle

C. Martin[1], A. Vanlierde[2], F. Dehreng[2], R. Dewhurst[3], P. Lund[4], C. Reynolds[5], D. Andueza[1] and G. Cantalapiedra[1]
[1]INRAE, INRAE, UMR Herbivores, 63122, France, [2]CRAW, Rue de Liroux, 9, 5030 Gembloux, Belgium, [3]SRUC, West Mains Road, EH9 3JG Edinburgh, United Kingdom, [4]Aarhus Universitet, Nordre Ringgade 1, 8000 Aarhus, Denmark, [5]University of Reading, Earley Gate, RG6 6AR Reading, United Kingdom; cecile.martin@inrae.fr

Research activities in SmartCow ((H2020 N°730924; https://www.smartcow.eu/) aimed to increase phenotyping capabilities while implementing the 3R principles (refine, reduce and replace) in cattle nutrition and behaviour studies. Development and validation of non-invasive proxies of feed efficiency (FE) and its determinants were undertaken with the goal of minimizing handling of experimental cattle. A database of individual phenotypes and proxies from different easily accessible matrices (milk, faeces, blood, breath gas, urine) for beef and dairy cattle was built through a collaborative network among SmartCow collaborators. Models were tested for different proxies to predict phenotypes across diets and between-individuals. Meta-analysis demonstrated that the natural 15N abundance in animal proteins has a stronger predictive ability than plasma- or milk-urea to discriminate dietary treatments, as well as individual variation in FE of beef cattle and nitrogen (N) use efficiency of dairy cattle. Models based on faecal near-infrared spectra (NIRS) discriminated dietary treatments and extreme individuals in terms of organic matter digestibility. Milk mid-infrared spectra (MIRS) models predicted enteric methane (CH_4) emissions, and faecal NIRS showed potential for estimating CH_4 emissions of non-lactating animals. Potential of new proxies, like milk MIRS, breath volatolome and plasma metabolites, for rumen diagnostics and prediction of urinary N, have also been investigated using more limited datasets. First results highlighted the need to assess repeatability of proxies across time. Common and standardized protocols for reference measurements and aggregation of data from different sources will aid development of proxies. Open access guidelines for using the most promising proxies will be developed, which will strengthen cattle phenotyping capabilities and contribute to sustainability of the livestock sector.

Session 28 Poster 18

Prediction of daily dry matter intake using milk mid-infrared spectra data in Swedish dairy cattle

S. Salleh, R. Danielsson and C. Kronqvist
Swedish University of Agricultural Sciences, Animal Nutrition and Management, Ulls väg 26, 75007, Sweden; suraya. mohamad.salleh@slu.se

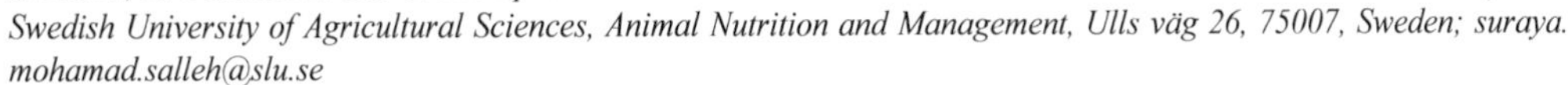

Generally, dry matter intake (DMI) is one of the most crucial parameters in feed efficiency evaluation in livestock production, although it is difficult to measure on an individual level in most farms. In order to keep track of the animal performances, in dairy production, milk mid-infrared reflectance spectroscopy (MIRS) data have been widely available in many farms. Milk MIRS data typically comprises measurement of the interaction of infrared radiation with milk liquid by absorption that able to determine the concentration of milk composition i.e. fat, protein, lactose, casein, fatty acids as well as somatic cell count. DMI is known to have a good relationship with milk yield (MY) and composition. Therefore, the milk parameters from the MIRS data could potentially be used to estimate daily DMI in dairy cows. The objective of the study was to develop prediction models in estimating DMI from the milk MIRS data in Swedish dairy cattle and evaluate its performances. Daily total feed intake (forage + concentrate) data were collected from the same herd from 2017 until 2021 and DMI was computed and averaged over 3 days. The average daily DMI was coupled with milk MIR spectroscopy data that were collected every two weeks for the prediction model development. Partial Least Squares (PLS) regression analysis was used to develop the prediction models for DMI from milk MIRS data. Data from 2017-2020 were used to develop a calibration model and the latest year (2021) data were used for validation. The coefficient of determination for validation datasets (R^2) were ranging between 0.17-0.47. A good prediction accuracy (R^2=0.47) was achieved by using only weekly averages of milk yield. Adding more variables (MIRS and lactation stage) did not improve the coefficient of determination of the prediction model however, the mean absolute error (MAE) was reduced. This study shows that the MIRS data can potentially be used to estimate daily DMI and other key additional information i.e. MY and lactation stage could improve the prediction accuracy.

Characterization of milk production in suckler beef cows

B. Sepchat, M. Barbet and P. D'Hour
INRAE, Herbipole, Theix, 63122 Saint-Genès-Champanelle, France; bernard.sepchat@inrae.fr

Maintaining and improving milk potential of suckler cows contribute to the reduction of feed costs. However, measuring milk production (MP) of suckler cows is difficult because it relies on mother-calf interaction and quantification of milk intake by calf. The reference method is to weigh calf before and after suckling twice a day. It is time consuming, labour intensive, subject to measurement error and rarely used except in a few experimental facilities. New phenotyping methods based on continuous measurement of calf weight open up possibilities for easier and more reliable measurements. We developed an automated calf weight measurement system to develop a prediction model for MP. A device based on electronic identification and automatic weighing was used when cows were in the barn, from calving to grazing (5 months). The principle is to let calf suckle freely its mother with a compulsory passage between its pen and that of its mother. Weights collected automatically are processed by a software to transform calves' weight over 24 hours into daily MP. During seven consecutive winters, we tested 101 cows (71 Charolais (CH) and 30 Salers (SA)). The average live weight of cows was 790±64 kg for CH and 750±83 kg for SA. Cows and calves were fed grass-based fodder, calves did not receive any concentrate. In order to validate the system, each winter was divided into 6 consecutive periods of 14 days, the MP was quantified alternatively according to the reference method (4 days, 2 suckling per day) or according to the 10-day free-suckling system with weighing at each pen change. The calves suckled 3.1±1.5 times per day during the first period compared to 4.5±1.7 during the last. The MP estimated by the device was 8.3±2.1 kg/d for SA vs 6.9±1.3 kg/d for CH, with a good correlation with the reference method (r^2 of 0.86 to 0.66 from period 1 to 6). ADG of calves for both breeds was 900 g per day. A better milk production leads to an ADG of 60 g/kg more milk drunk, corresponding to a 70 kg of body weight gain for a lactation of 2,300 kg vs 1,200 kg. This system will be extended to other suckler cattle experimental stations in order to extend the measurement of MP and to refine the indexation of maternal qualities that can be integrated into breeding strategies and selection schemes.

Effect of abomasal amino acid infusions on performances and metabolism in postpartum dairy cows

L. Bahloul[1], L.E. Hernández-Castellano[2], H. Lapierre[3], C.G. Schwab[4] and M. Larsen[2]
[1]CERN, Adisseo, Malicorne, France, [2]Aarhus University, Foulum, Denmark, [3]Agriculture and Agri-Food Canada, Sherbrooke, Canada, [4]Schwab Consulting LLC, Boscobel, WI, USA; lahlou.bahloul@adisseo.com

Objective was to investigate, through Smartcow grant N°730924, the effect of continuous abomasal infusion of total AA (TAA) or only essential AA (EAA) in early postpartum dairy cows on performances and metabolism. Nine multiparous Holstein cows were used in a randomized block design with repeated measurements at 5, 15, 29, and 50 days in milk (DIM). At calving day, TAA (n=4; casein profile) or only EAA (n=5; EAA portion of TAA) was initiated. The TAA was graduated with half of full dose at 1 day in milk (DIM), full dose (805 g/d) at 2 to 5 DIM, and followed by daily reductions until 0 g/d at 35 DIM. Cows received the same TMR (NE: 6.85 MJ/kg DM, MP: 102 g/kg DM). Feed intake and milk yield were recorded daily. Milk samples and six sets of tail and mammary venous plasma samples were obtained at sampling days. The DMI did not differ between treatments (P=0.55). Overall, with no treatment × DIM interaction (Trt×d), milk yield was greater with TAA compared with EAA (P<0.01; 47.9 vs 39.3 kg/d, SEM=1.4) as was milk protein yield (P=0.01; 1,635 vs 1,393 g/d, SEM=50). Milk fat content was lower with TAA compared with EAA (P=0.02; 41 vs 47 g/kg, SEM=1.5), but treatments did not affect milk fat yield (P=0.20). The arterial total EAA concentration was lower (Trt×d=0.01; SEM=0.07) with TAA compared with EAA at 5 (0.97 vs 1.24 mM) and 15 DIM (0.97 vs 1.13 mM). The arterial total non-EAA concentration was higher (Trt×d<0.01; SEM=0.04) with TAA compared with EAA at 5 DIM (1.35 vs 1.17 mM). Yet, plasma concentration differences across the udder of EAA and non-EAA did not differ between treatments (P>0.88), indicating that the intra-mammary utilization of both EAA and non-EAA was changed. Arterial urea concentration was greater (Trt×d=0.02; SEM=0.21) with TAA compared with EAA at 5 (2.87 vs 1.75 mM) and 15 DIM (3.06 vs 1.90 mM) indicating catabolism of some of the AA supplied with TAA in early but not later lactation. Results indicate that some of or all non-EAA are as important as EAA in the early postpartum period. Continued higher milk yield through 50 DIM with TAA after ceasing infusions indicate a carry-over effect.

Effect of grazing dairy system and breed on monthly variation of milk minerals and heat stability

A. Boudon[1], B. Horan[2], L. Delaby[1], J. Tobin[3], J. Flament[1], A. Jezequel[2], B. Graulet[4], M. O'Donovan[2], G. Maxin[4], M. Gelé[5], S. Lemosquet[1] and C. Hurtaud[1]

[1]PEGASE, INRAE, Institut Agro, Saint-Gilles, France, [2]Animal & Grassland Research and Innovation Center, Teagasc, Moorepark, Ireland, [3]Teagasc, Food Research Centre, Moorepark, Ireland, [4]Université Clermont Auvergne, INRAE, VetAgro Sup, UMR Herbivores, Saint-Genès-Champanelle, France, [5]Institut de l'Elevage, Idèle, Angers, France; anne.boudon@inrae.fr

In a context of increasing societal concerns around animal production, consumers asks for products coming from grazing dairy systems and. The segmentation of collection rounds to offer milk specifically from these systems results in higher seasonal variability in milk composition that needs to be better understood. Our objective was to characterize the monthly variation of milk and plasma mineral composition and heat stability, in a factorial arrangement of two breeds of cow and three grazing systems. The experiment involved 48 cows, half Holstein-Friesian (HF) and half F1 Jersey Holstein-Friesian crossbred (JFX) at Teagasc Moorepark. The grazing systems were based on either a monoculture of perennial ryegrass, or a perennial ryegrass and white clover sward, or a multispecies pasture. Milk, blood and pasture were sampled monthly between April and June 2021. Milk heat stability increased constantly between April and June and was unaffected by either the breed or the grazing system. Milk Ca, P, Zn and citrate contents were higher with JFX compared with HF, whereas that of Cl was lower. Milk soluble Ca and P contents were unaffected by the breed. Milk contents of a large part of elements (Ca, P, Mg, K, Fe, Cu, Zn, Mn), except Na and Cl, varied strongly according to the month and differently according to the considered element in interaction with the grazing system. No dominant parameters, among those prone to affect milk heat stability, could be clearly related to milk stability results. Plasma contents of all mineral elements, except Fe, strongly increased between April and June, in interaction with the grazing system for inorganic P, Mg and Zn. An important part of this variability could be related to herbage mineral content although plasma mineral contents were not clearly related to milk mineral contents. Funding by SmartCow project (H2020 N°730924) is acknowledged.

Inclusion of a high fat extruded pellet in starter feeds

L. Amado[1,2], L.N. Leal[2], W.J.J. Gerrits[1] and J. Martín-Tereso[1,2]

[1]Wageningen University, Animal Nutrition Group, P.O. Box 338, 6700 AH Wageningen, the Netherlands, [2]Trouw Nutrition, Research and Development, P.O. Box 299, 3800 AG, the Netherlands; liliana.amado.barrantes@trouwnutrition.com

During weaning, withdrawal of milk replacer is not completely compensated for by an increase in solid feed intake. Therefore, a greater fat inclusion in the starter might mitigate this dietary energy gap. However, fat in solid feeds may limit feed fermentability and rumen development. To increase fat in the starter, while preventing these limitations, two high-fat extruded starter pellets containing different hydrogenated fat sources, were mixed at 10% inclusion with standard starter pellet. This study contrasted fat source digestibility and their value to fill the energy gap of early weaned calves. Twenty-four Holstein weaned bull calves (80±4 d of age; 110±3 kg BW) were blocked by age and BW, and randomly assigned to 1 of 3 treatments: a standard control starter (CON; 3.1% fat) or two mixtures of CON with 10% inclusion of two different high-fat extruded pellets containing 85% hydrogenated fats: hydrogenated free palm fatty acids (PFA, 7.1% fat) or hydrogenated rapeseed triglycerides (RFT, 6.7% fat). For 18 days, calves had *ad libitum* access to starter treatments, straw, and water. Individual daily feed intakes and weekly BW were measured. Apparent total-tract digestibility was determined by 72-h total faecal collection. Overall, PFA calves consumed 690, 62 and 2.200 g/d more starter, straw, and water ($P<0.01$) than the CON and RFT group. Similarly, BW, and ADG of PFA calves was the greatest among all the treatments ($P<0.01$). The greatest total-tract apparent DM digestibility was observed in calves fed the CON diet (79.4%) when compared with RFT (75.1%) or PFA (76.0%; $P<0.01$). Fat digestibility was the highest for PFA calves (87.9%; $P=0.01$) when compared to RFT (56.8%) and CON (81.0%) calves. Dietary treatments influenced faecal pH, and it was observed that PFA calves had the lowest pH ($P=0.01$) when compared with CON and RFT treatments. Mixing a high fat extruded pellet in the starter can improve BW gain, solid feed and energy intake. However, these benefits are conditioned by fat source and its digestibility.

On farm evaluation of multi-parametric models to predict SARA on dairy cows

M. Coppa[1], C. Villot[2], C. Martin[3] and M. Silberberg[3]
[1]Université Clermont Auvergne, INRAE, VetAgro Sup, Independent researcher UMR 1213 Herbivores, Theix, 63122 Saint-Genès-Champanelle, France, [2]Lallemand, 19 rue des Briquetiers, 31702 Blagnac, France, [3]Université Clermont Auvergne, INRAE, VetAgro Sup, UMR 1213 Herbivores, Theix, 63122 Saint-Genès-Champanelle, France; villotclothilde@gmail.com

The aim of this research was to validate on farm and, if needed, re-calibrate the multi-parametric models based on non-invasive indicators to detect induced subacute ruminal acidosis (SARA) in dairy cows developed by Villot *et al.* (2019) under controlled trial. Fifteen farms were selected for their farming practices to cover a wide range risk of observing cows under SARA. In each farm, 4 Holstein early-lactating primiparous cows were selected based on the last on farm analysis of their milk composition (MY, fat/protein ratio, SCC). Each animal was equipped with a reticulo-ruminen pH sensor (eCow, Exeter, UK). The pH kinetics were analysed over a subsequent 7 days period and relative pH indicator (NpH) described by Villot *et al.* were used to classify cows as under SARA (22 cows) or not (37 cows) (1 cow missed). Meanwhile, milk, blood, faeces, and urine samples were collected for analysis of all indicators included in previous models. An external validation of Villot *et al.* models was performed on farm data. Then, the same indicators included in each model were used to build new models by linear discriminant analysis and leave-one out cross-validation using controlled trial and on farm datasets. The sensitivities (true positive rate) in external validation on farm were largely lower than those from cross validation by Villot *et al.* (range: 0.1-0.75 vs 0.79-0.96), and the specificities (true negative rate) showed a larger range with lower minimum values (range: 0.18-1.0 vs 0.62-0.97). The sensitivities of new models was lower than those of the models by Villot *et al.*, but higher than those of their external validation on farm data (range: 0.63-0.77). Blood cholesterol and milk n-6 fatty acids based model had the highest performance, whereas faeces sieving residual and urine pH based one had the lowest. Multi-parametric models based on non-invasive indicators to detect SARA in dairy cows seem promising.

Is optimization of dairy cow rations by linear programming possible with INRA 2018 system?

P. Chapoutot[1,2], G. Tran[1], V. Heuzé[1] and D. Sauvant[1,2]
[1]AFZ (Association Française de Zootechnie), 16, rue Claude Bernard, 75005 Paris, France, [2]Université Paris-Saclay, INRAE, AgroParisTech, UMR Modélisation Systémique Appliquée aux Ruminants, 16, rue Claude Bernard, 75005 Paris, France; patrick.chapoutot@agroparistech.fr

The new INRA 2018 system involves iterative calculations, because feed values (FV) and animal requirements (AR) are modified by ration characteristics. Moreover, animal responses to dietary strategies around their potential are not linear. The challenge of this study was to use linear programming (LP), based on linear combinations of variables with constant coefficients, to optimize dairy cow rations using INRA 2018 system. LP optimizes rations for animals considered at their potential by solving linear equations of variables (Q_j: quantity of feed j) so that, for each criterion i, supplies by all feeds [S_i: $\Sigma(V_{ij} \times Q_j)$] reach a $Rmin_i$ and a $Rmax_i$ value, with V_{ij}: constant value of feed j for criterion i; $Rmin_i$ and $Rmax_i$: constant requirement levels for criterion i, each variable j being restricted between $Qmin_j$ and $Qmax_j$ levels. The optimal solution is selected among all so that the cost of the ration [Z: $\Sigma(Cj \times Qj)$], with C_j: cost of feed j, is minimized. The prototype tested was implemented with Excel. It is organized in 3 interconnected modules, 'FV', 'AR' and 'Ration Optimization' (RO), and based on iterative processes. Calculation is initialized with INRA 2018 'Table' FV and ration-nondependent AR and begins with a first LP optimization. After each iteration step n of RO, FV and AR are recalculated with the new ration output data and reintroduced as new characteristics for the n+1 RO step. Several criteria of the new ration from n+1 step are compared to those obtained from previous n in a combined weighted deviation, and convergence process rapidly leads (in 5 to 15 runs) to an optimal solution. Opportunity prices and invariance ranges for feeds as well as dual costs and contributions for saturated constraints are calculated using LP dual analysis. Moreover, ration-cost margin is calculated based on non-linear performance responses of dairy cows around their potential. These interesting results encouraged the scaling up of the prototype as a real field tool ('Optim'Al v2') for technical and economic diagnosis of dairy cow rations.

Development rate variation in insects: what we currently understand

C.J. Picard
Indiana University Purdue University Indianapolis (IUPUI), Department of Biology, 723 W Michigan St., Indianapolis, IN 46202, USA; cpicard@iupui.edu

Development time is a classic quantitative trait that can be influenced by genetic and environmental factors and is associated with fitness. With insects, that development rate variation can be both a blessing and a curse for growing insects at scale for food or feed production. On the one hand, investigating the genetic basis of development rate of a range of strains of insects can lead to genetic markers that can assist in breeding strategies, yet on the other hand, reducing the amount of variation, especially if tied to selection for faster development times, can lead to fitness trade-offs. My talk will focus on what we have learned so far with development rate variation in one species of insect (Calliphoridae) and how we applied this knowledge to other species of interest for insect agriculture that includes the black soldier fly (Stratiomyidae) and the mealworm (Tenebrionidae).

Session 29 Theatre 2

Defining a founding population: evaluating the genetic diversity in a BSF commercial line

E.M. Espinoza
EnviroFlight LLC, Genetics R&D, 303 North Walnut Street, Yellow Springs, OH 45387, USA; eespinoza@enviroflight.net

Among the insect production industry, breeding programs have begun to improve insect rearing and quality. Specifically, most black soldier fly (BSF) breeding programs primarily use established commercial populations for artificial selection, as these domesticated populations are already thriving in captivity. The establishment of a founding population is one of the primary factors that defines the success of any long-term selection strategy. This study describes the comparison of different effective population sizes and their impact on the genetic diversity of a BSF commercial line. Three founding populations with similar genetic diversity and the same genetic background were used to establish three different scales of selection. The same selection approach was maintained for the three populations and genotyping at various generations was performed to assess genetic diversity. The three BSF lines yielded different levels of inbreeding and produced differentiable populations with varying genetic profiles, illustrating the importance of initial population characteristics to consider as breeding programs are launched into practice. Genetic diversity, inbreeding levels, and effective population size should be assessed prior to defining a founding population for BSF commercial selections.

Influence of strain genetics on larval performance and bioconversion efficiency of *Hermetia illucens*

L. Broeckx[1], L. Frooninckx[1], S. Berrens[1], A. Wuyts[1], C. Sandrock[2] and S. Van Miert[1]
[1]RADIUS, Kleinhoefstraat 4, 2440 Geel, Belgium, [2]FiBL, Ackerstrasse 113, 5070 Frick, Switzerland; laurens.broeckx@thomasmore.be

Due to increasing welfare and population, demands for more sustainable protein sources are rising in today's society. Insects are considered such an alternative as they have short life cycles, high feed conversion and can be grown on low-value feedstocks. Particularly the black soldier fly *Hermetia illucens* is able to convert low-value organic side streams into high-value biomass composed of proteins, lipids and chitin. Therefore, *H. illucens* larvae can be used for waste reduction paired with the production of high-value biomass, bringing more circularity in our food- and agricultural industry. Although the black soldier fly has been subject of extensive research and suggested as the crown jewel of an emerging insect-livestock sector, characterisations of its genetic resources, crucial for future breeding progress, have been neglected so far. Recent studies using wild and captive strains demonstrate that there is remarkable genetic variation across origins, including signatures of domestication. However, it still remains to be elucidated how genetic differentiation may translate into distinct phenotypic traits, such as economically interesting larval performance and bioconversion. In this study 10 captive *H. illucens* strains were obtained and reared using a standardized protocol. The strains were genotyped based on the 15 microsatellite markers developed by Kaya *et al.* (2021). Subsequently, larvae were reared on 3 different diets and larval performance and conversion efficiency was calculated. This allowed to investigate the influence and potential interactions of genotype and diet on these economically interesting traits.

The reproductive tract of the black soldier fly (*Hermetia illucens*) plays with sexual selection

C. Bressac and C. Labrousse
IRBI University of Tours, Parc de Grandmont, 37200 Tours, France; christophe.bressac@univ-tours.fr

The reproductive biology of living organisms is constrained by sexual selection. In insects, post mating sexual selection is considered as a strong evolutionary pressure, leading to highly differentiated female reproductive tracts that store sperm, and may select males for their sperm size and numbers. Black soldier fly (BSF) is a promising solution to produce highly valuable protein feedstuff from agri-food wastes, and then its reproduction is a crucial output. Both male and female tracts are long, and constituted different successive parts. In males, testes fill vas deferens with mature sperm, that are tangled as windrows, and sperm pass successively through a studded pipe and three phallic wands to be transferred to females during copulation. In females, sperm are deposited at the base of each three spermathecae in a smooth tube followed by a fishnet canal at the top of which they accumulate. Then, they pass through a ringed canal, an elbow and a rigid rod with glands, before reaching the spermathecal reservoir. In virgin males, sperm counts increase from approx. 1000 to 45,000 in 20 days, and testes gradually decrease, evidencing a continuous spermatogenesis. Females store approx. 680 spermatozoa in each spermathecal reservoir after one mating. Sperm cells are long, more than 3 mm. Such complex sperm storage organs of females may result from a strong post-mating sexual selection, and is a field of competition for long and numerous sperm. BSF is a seductive model for sexual selection investigations, and this evolutionary constrain may induce paternity biases in mass cultures.

A sweet smell: a search for the perfect attractant for female black soldier fly (*Hermetia illucens*)

D. Deruytter[1], C.L. Coudron[1], A. De Winne[2] and M. Van Der Borght[3]
[1]Inagro, Insect Research Centre, Ieperseweg 87, 8800 Rumbeke-Beitem, Belgium, [2]KULeuven, Centre for Aroma and Flavour Technology, Gebroeders De Smetstraat 1, 9000 Ghent, Belgium, [3]KULeuven, Department of Microbial and Molecular Systems, Kleinhoefstraaat 4, 2440 Geel, Belgium; david.deruytter@inagro.be

In nearly all black soldier fly (BSF, *Hermetia illucens*) reproduction facilities, the females must be lured to lay her eggs in one predefined area. Frequently used 'lures' or attractants are rotting animal (e.g. fish) or plant products, fermenting chickenfeed, wet BSF larvae frass, etc. All these products have one thing in common: they are biologically active. Therefore, they may pose a health hazard. For example, by spreading pathogenic bacteria among insects and humans. In addition, they are not standardized (one rotting fish is not the other) so their effectiveness may not be stable over time. The hypothesis of this study was that the female flies are attracted to one specific chemical compound (or a mixture of a few) and not to the complex mixture of all chemicals released by attractants. Furthermore, based on personal observations at Inagro's Insect Research Centre, it is clear they are also attracted to the smell of BSF eggs deposited earlier by other females. Headspace Solid Phase Microextraction (HS-SPME) as isolation technique was used to determine the chemical composition of the headspace volume above the attractants. Based on the resulting Gas Chromatographic - Mass Spectrometric (GC-MS) analysis, some volatile chemical components were selected for further analysis based on prevalence (and also price). The selected compounds will be evaluated as attractant at different concentrations and application methods, to determine the optimal application conditions. The outcome of this study may provide a safe and more consistent way to attract female flies to deposit her eggs.

Mating compatibility among different *Tenebrio molitor* strains

C. Adamaki-Sotiraki[1], D. Deruytter[2], C.I. Rumbos[1] and C.G. Athanassiou[1]
[1]Laboratory of Entomology and Agricultural Zoology, Department of Agriculture, Crop Production and Rural Environment, University of Thessaly, Phytokou Str., 38446 Volos, Greece, [2]Insect Research Centre, Inagro, Ieperseweg 87, Rumbeke-Beitem, Belgium; athanassiou@uth.gr

The fast-growing sector of insects as food and feed stimulates researchers and commercial insect-producers to explore uncharted territories, such as insect breeding, to improve fitness traits. The yellow mealworm, *Tenebrio molitor* L., is one of the most studied insect species for food and feed. Recent research has shown that strain decisively affects the growth and development of *T. molitor* larvae, as well as the performance of adults. However, little is known about the effect of cross-breeding among different strains on the performance and fitness of the offspring. Therefore, the aim of the present study was to comparatively evaluate the mating compatibility among different *T. molitor* strains. In a first series of lab bioassays, we evaluated 4 cross-mating combinations between 2 pure *T. molitor* strains (i.e. Greek and Italian) (2 intra- and 2 inter-strain mating). In a second series of bioassays, cross-mating combinations between 4 pure *T. molitor* strains (i.e. Inagro, Greek, Italian, USA) (4 intra- and 6 inter-strain mating) were evaluated at a pilot scale. The inter-strain mating was performed either by placing adult beetles into oviposition boxes at a 1:1 ratio (male: female) for the laboratory bioassay, or by mixing pupae of equal age from the different pure strains for the pilot-scale bioassay. Sexually mature beetles were left to mate and oviposit undisturbed. Laid eggs were collected at regular intervals, while larval hatching rate was also recorded. The 1st instar larvae of specific intervals were kept for further evaluation of their growth and performance. Based on the results of the lab trial, no significant differences were detected in the performance of intra- and inter-mated strains in terms of egg production, hatching rate and larval growth. In the pilot-scale bioassays significant differences were recorded with regard to the hatching rate, but not the egg production. The results of our study provide an empirical demonstration of the immediate fitness effects of cross-mating among different *T. molitor* strains.

Prospects of manipulating sex ratios of commercial insect populations

L.W. Beukeboom, L. Francuski, G. Petrucci, S. Ter Haar, A. Rensink and S. Visser
University of Groningen, Groningen Institute for Evolutionary Life Sciences, P.O. Box 11103, 9700 CC Groningen, the Netherlands; l.w.beukeboom@rug.nl

Proportions of females and males (sex ratios) in populations are typically equal. In commercial insect populations the two sexes often do not have equal value. In Sterile Insect Technique programmes only males are needed for release, whereas in biological control programmes only females parasitise pest species. In breeding programmes for feed and food, a shift towards females may improve production because only females lay eggs. This calls for manipulation of population sex ratios. Two general approaches for altering population sex ratios are phenotypically and genetically. Phenotypic separation of the sexes utilizes differences in male and female morphology, behaviour or life history traits. Genetic methods for sex separation exploit the sex determination system. Genetic sexing strains can be developed through genetic engineering techniques offering the possibility to produce one sex-only progeny. We present results of artificial selection experiments in the housefly on body size (sexual size dimorphism, SSD) and development time (sexual bimaturism, SBM) differences between males and females. We also present how the group sex ratio can affect mating behaviour and how this may have repercussions for mass rearing. We present the scope of genome-editing techniques to alter population sex ratios, using the housefly as an example. We discuss the prospects and limitations of sex ratio manipulation for production insect populations.

An extensive multi-approaches distribution map of *Apis mellifera* mitochondrial DNA lineages in Italy

V. Taurisano, V.J. Utzeri, A. Ribani and L. Fontanesi
University of Bologna, Department of Agricultural and Food Sciences (DISTAL), Viale Fanin, 46, 40127 Bologna, Italy; valeria.taurisano2@unibo.it

The conservation of the genetic integrity of honey bee subspecies is a matter of concern in Europe. Specific mitochondrial DNA (mtDNA) lineages characterize several *Apis mellifera* subspecies. In this study we produced a distribution map of the main *A. mellifera* mtDNA lineages in Italy by combining two approaches: (1) mtDNA analyses of individual honey bees; (2) mtDNA analyses from honey environmental DNA. This second approach was also applied to obtain a distribution map of mtDNA lineages over four years in the North of Italy. DNA was extracted from ~800 worker bees collected in 2020 and 730 pupae collected in 2021 from different colonies. DNA was also extracted from about 1,600 honey samples produced in the North of Italy over four years (2018-2021) and additional 400 honey samples produced in 2018 in the Central and South of Italy and in the two main Italian islands (Sardinia and Sicily). PCR amplified fragments were sequenced or separated using a size-based assay. Both approaches showed that the C lineages were the most frequent mitotypes all over Italy except in Sicily, where the A lineage was highly represented. Individual bee analyses showed that the C1 mitotype, which mainly characterizes the *A. m. ligustica* subspecies, was the most frequent mtDNA haplotype in the investigated cohorts. Nevertheless, mtDNA haplotypes associated with non-endemic subspecies (A and M lineages) were present in almost all Italian regions and constantly represented over years. The obtained information can be useful to design conservation strategies of *A. mellifera* genetic resources in Italy.

Deciphering the genetic architecture of honey production in a selected honeybee population

M.G. De Iorio[1], J. Ramirez-Diaz[2], S. Biffani[2], G. Pagnacco[2], A. Stella[2] and G. Minozzi[1]
[1]University of Milan, DIMEVET, Via dell'Università 6, 26900, Italy, [2]IBBA-CNR, Via Alfonso Corti 12, 20133, Italy; giulietta.minozzi@unimi.it

The *Ligustica* subspecies of *Apis mellifera* in Italy is the most well-known. However, little has been done in this species to improve productivity. Within the Beenomix 2.0 project, a selected population has been established with the overall aim to improve health and production traits. The population consisted in 108 hives measured each year and selected for 6 years for several traits including honey production. On this population we carried out a genome wide association study (GWAS) to identify candidate regions and genes associated with honey production using single nucleotide polymorphism (SNP) markers covering the whole genome using whole genome sequence data. The use of whole genome information is a powerful method to identify positional and functional genes. Genome sequences of the 100 *A. mellifera* colonies were obtained through Illumina NextSeq with a 150 bp paired-end module. Sequences were mapped to the HAv3.1 *A. mellifera* genome. Associations between genotypes and phenotypes were analysed the linear regression model adjusted for birth year and apiary as covariable using PLINK software. After data editing, the final dataset consisted of 100 records from 100 hives belonging to the same apiary and 4.233.599 SNPs. Significant associated SNPs were determined based on a 5% genome-wide Bonferroni-corrected threshold. Significant SNPs were identified on chromosome 10. Genes within the identified QTL region were determined using the NCBI Genome Data Viewer and the reference genome assembly Amel_HAv3. The Bifunctional heparan sulfatase N-deacetylase/N-sulfurotransferase gene (Chr10:LOC413242) was identified in the associated region as functional and positional candidate being involved in glycan metabolism. Further research is needed to confirm the results and to better understand the genetic basis of honey production. This research was funded by the BEENOMIX 2.0 project funded by the Lombardy Region (FEASR program), PSR (grant number 201801057971 – G44I19001910002).

High-density SNP development in black soldier fly via throughput DNA pool sequencing

A. Donkpegan[1], A. Guigue[2], F.X. Boulanger[3], S. Brard-Fudulea[1], P. Haffray[4], M. Sourdioux[1] and R. Rouger[1]
[1]SYSAAF, Centre INRAE, Val de Loire, UMR BOA, 37380, France, [2]INNOVAFEED, 85 rue de Maubeuge, 75010 Paris, France, [3]AGRONUTRIS, 35 Boulevard du libre Échange, 31650 Saint-Orens-de-Gameville, France, [4]SYSAAF, LPGP/INRAE, Campus de Beaulieu, Université de Rennes 1, 35042 Rennes, France; armel.donkpegan@inrae.fr

Use of genomic tools is now widespread in animal genetic breeding programs. In black soldier fly (BSF) breeding, which is a promising candidate for the development of the insect farming industry, these genomic tools are still non-existent, while they could accelerate and improve genetic gain of selection programs in increasing accuracy. The purpose of this study was to call SNPs from DNA pool sequencing to design high, medium and low-density SNPs chips. Pooled DNA library preparation was performed using several BSF populations from the breeding companies Innovafeed and Agronutris. Generated library pools were sequenced on Illumina platform in 2×150 bp and the sequences were assigned to the samples based on their index sequence. Over 300 million of Paired-End reads and 43X of sequencing depth was obtained on average per pool. The latest BSF genome with chromosome assembly level (≈1 Gb) was used as reference to call 52 million SNPs. Filters based on Quality, Minor Allele Frequence (MAF), genome coverage (depth) and non-informative sites were used to yield a final set of 9.7 million high quality SNPs that can be later used to develop the first genotyping chip for BSF. Beyond their use in genomic selection, these resources will be useful to address other questions in BSF: manage inbreeding with adapted panels for DNA chip, parentage and paternity assignment, improve selection or diffusion efficiency based on QTL detection after association genetic studies (GWAS).

Session 29 Poster 11

Haemolymph analysis as a potential method to assess the quality of insect nutrition

R. Gałęcki and T. Bakuła
University of Warmia and Mazury in Olsztyn, Department of Veterinary Prevention and Feed Hygiene, Faculty of Veterinary Medicine, Olsztyn Oczapowskiego 13, 10-718, Poland; bakta@uwm.edu.pl

In addition to its immunological role, the haemolymph performs many other functions: it carries metabolic products, is a water reservoir and distributes nutrients in the insects' organism. It can affect both the health condition of insects and the development and rate of weight gain. There are currently no rapid methods of assessing the nutritional value of insect feed that would allow farmers to achieve better economic success. The aim of this research was to assess the level of total protein in the haemolymph of the mealworm larvae kept on various feed materials. The research material were insects weighing 0.03-0.05 g. The insects were kept in a climatic chamber (temperature 28 °C, humidity 70%, without lighting) on 7 different forages. The seven feed formulas used in the research groups differed in the source and level of protein. Each container consisted of 100 g of insects and 300 g of feed. Every 4 days, 50 g of fresh feed was added. Three repetitions were performed for each group. The insects were kept for 30 days. 50 insects per group were used for the research. The haemolymph was collected intravenously using sterile needles from the cephalic region. Thereafter, total protein levels were assessed by the A280 method using NanoDrop™ One. Average levels of total protein for the seven groups were: (1) 63.22 mg/ml; (2) 56.53 mg/ml; (3) 63.69 mg/ml; (4) 57. 83 mg/ml; (5) 64.82 mg/ml; (6) 61.37 mg/ml; and (7) 73.48 mg/ml. The obtained results were statistically significant ($P<0.05$). The post-hoc test showed a statistically significant difference ($P<0.05$) between group 7 and other research groups. The presented research suggests the possibility of using haemolymph to determine the level of total protein as an evaluation of the nutritional value of feed for mealworms. Haemolymph studies using a spectrophotometer can be a simple and quick method to assess the quality of insect nutrition. Project financially co-supported by Minister of Science and Higher Education in the program entitled 'Regional initiative of Excellence' for the years 2019-2022, Project No. 010/RID/2018/19, amount of funding 12,000,000 PLN.

Session 29 Poster 12

Identification of selected haemocytes in the haemolymph of the mealworm (*Tenebrio molitor*)

R. Gałęcki and T. Bakuła
University of Warmia and Mazury in Olsztyn, Department of Veterinary Prevention and Feed Hygiene, Faculty of Veterinary Medicine, Oczapowskiego, 13, 10-719, Olsztyn, Poland; remigiusz.galecki@uwm.edu.pl

Haemocytic cells are an important component of the haemolymph, which plays an important immunological, catabolic and detoxification role. They are an important element of protecting the body against potential pathogens. There are currently numerous studies on the role of haemocytes in bees, but no such studies have been performed on edible insects. The aim of this research was to identify selected populations of haemocytes of mealworm larvae (*Tenebrio molitor*). The research material consisted of insects weighing 0.02-0.015 g. The mealworms were kept on the broiler starter feed. Each container consisted of 50 g of insects and 400 g of feed. The insects were kept without interference for 14 days in a climatic chamber (temperature 28 °C, humidity 70%, without lighting). Fifty insects were used in the research. Haemolymph was collected intravenously using sterile needles from the cephalic region. The insects were then euthanized at low temperature. The acquired haemolymph was assessed qualitatively. Then, from the drops of haemolymph, a smear was made, which was stained with the May-Grunwald-Giemsa method. A total of 50 smears were performed. Haemocytes were identified using a light microscope. Differences in cell morphometry and morphology, as well as differences in stainability, were demonstrated. A total of 8 cell types have been identified: proleukocytes, plasmocytes, neutrophiles, eosinophils, basophils, apidohemocytes, pycnoleukocytes, and hyalinocytes. The described studies indicate the possibility of assessing the health of insects, the influence of xenobiotics and pathogens on the immune and metabolic response of mealworms using haemocytes. The presented methodology can be easily implemented in laboratories dealing with edible insects and in the future, haemocytes might be one of the markers of the physiological assessment of insects. Project financially co-supported by Minister of Science and Higher Education in the range of the program entitled 'Regional initiative of Excellence' for the years 2019-2022, Project No. 010/RID/2018/19, amount of funding 12.000.000 PLN.

Welfare as preventative medicine in pig production

L.A. Boyle, M. Lagoda and K. O'Driscoll
Teagasc, Pig Development Department, Animal and Grassland Research and Innovation Centre, Moorepark, Fermoy, Co. Cork, Ireland; laura.boyle@teagasc.ie

The importance of the role animal welfare can play in preventing diseases is poorly appreciated but gaining momentum given concerns about antimicrobial resistance (AMR) which poses a major threat to human and animal health. The aim of this paper is to demonstrate how improving pig welfare through better handling, housing and husbandry techniques leads to better health and thereby reduced antimicrobial use. The association between animal welfare and animal health is primarily driven by the detrimental impact of stress on the immune system. Indeed there is an abundance of research and practical evidence illustrating how stress arising from deficient management and housing practices causes poor pig welfare and consequently poor pig health and the associated use of antimicrobials (AMU). In this paper we will use the examples of cross-fostering and 'All-in, All-out' or 'pig flow' to demonstrate how poorly practiced management strategies have detrimental implications for pig welfare and thereby pig health. We will also elucidate how 'tail-biting' the most obvious outcome of stress in grower/finisher pigs and the major welfare problem for these animals is causally related to pig health as well as to discuss the implications of managing tail lesions for AMU. Finally we will provide examples of how improving the housing and management of pigs and sows unsurprisingly improves pig welfare but is also associated with health and performance benefits. In conclusion, the magnitude of the potential contribution animal welfare can make to the global AMR threat is such that it should be considered an integral pillar of the sustainability of animal agriculture in its own right.

Session 30 Theatre 2

Optimal farrowing accommodation hygiene reduces pre-weaning antibiotic and anti-inflammatory usage

K.M. Halpin[1,2], P.G. Lawlor[2], J. Teixé-Roig[1,2,3], E.A. Arnaud[1,2], J.V. O'Doherty[4], T. Sweeney[4] and G.E. Gardiner[1]
[1]Dept. of Science Waterford Institute of Technology, Cork Road, Waterford, Ireland, [2]Teagasc Pig Development Dept., AGRIC, Fermoy, Co. Cork, Ireland, [3]Food Technology Dept., University of Lleida, Lleida, Spain, [4]School of Agriculture and Food Science, University College Dublin, Dublin, Ireland; keely.halpin@teagasc.ie

Pig herds with high internal biosecurity are likely to have reduced antibiotic usage. The aim of this study was to determine if optimising farrowing accommodation hygiene routines can reduce pre-weaning antibiotic and anti-inflammatory usage. Forty seven sows were blocked prior to farrowing on parity, previous litter size and body weight and assigned to one of 2 treatments: (1) sub-optimal hygiene: washing of pens with cold water, no use of detergent or disinfectant and minimal drying time (≤18 hrs); and (2) optimal hygiene: detergent application (Blast Off; Biolink Ltd, Hull, UK) allowing a contact time of 20 min, washing of pens with cold water, application of a disinfectant (Interkokask®; Interhygiene GmbH, Cuxhaven, Germany) 24 hrs later and pens allowed 6 days of drying time. Clinical cases of joint ill, lameness, diarrhoea and general malaise were monitored in suckling pigs during the study and when observed, injectable antibiotics and/or anti-inflammatories were administered to individual piglets. Data were analysed using the mixed models procedure in SAS (version 9.4; SAS Institute Inc.). The litter was the experimental unit. Mortality was 15.1 and 10.9 (SE=2.37%; P=0.22), the number of clinical cases was 1.6 and 0.4 (SE=0.30/litter; P<0.01) and the number of injections administered was 4.7 and 1.0 (SE=0.9/litter; P<0.001) for sub-optimal and optimal hygiene routines, respectively. Antibiotic usage was 2.34 and 0.53 (SE=0.435 ml/litter; P<0.001) and anti-inflammatory usage was 0.32 and 0.08 (SE=0.063 ml/litter; P<0.01) for sub-optimal and optimal hygiene routines, respectively. Adopting an optimal hygiene routine in farrowing accommodation reduces the number of pre-weaning clinical cases presented and the frequency and quantity of injectable medication administered and should therefore aid in reducing on-farm antibiotic usage and subsequent development of antimicrobial resistance in pigs.

Session 30 Theatre 3

Effects of early-life changes on health, welfare and performances of pigs in a commercial farm

S. Gavaud[1], K. Haurogné[1], A. Buchet[2], I. Garcia-Vinado[3], M. Allard[1], M. Leblanc-Maridor[4], J.M. Bach[1], C. Belloc[4], B. Lieubeau[1] and J. Hervé[1]

[1]Oniris, INRAE, IECM, Rte de Gachet, 44300 Nantes, France, [2]Cooperl, Rue de la gare, 22640 Plestan, France, [3]Hyovet, ZA Carrefour de Penthièvre, 22640 Plestan, France, [4]Oniris, INRAE, BIOEPAR, Rte de Gachet, 44300 Nantes, France; solenn.gavaud@oniris-nantes.fr

Weaning is one of the most critical periods during pigs' life. Indeed, piglets are separated from their mothers, moved to a novel environment and mixed with unfamiliar congeners. These events can lead to conflicts between piglets until the establishment of a new hierarchy. As a consequence, weaning is often characterized by growth slowdown and increased susceptibility to infectious diseases. In order to reduce the deleterious effects of a standard conventional weaning, we proposed an alternative management of juveniles, which associated birth in free-farrowing pens, absence of tail-docking, early socialization of piglets and maintenance of established social groups. We analysed the effects of this alternative practice on pigs' health, welfare and performances. We followed 75 pigs, in the standard group, and 80, in the alternative one, from birth to slaughter. Frequent visits allowed us to detect clinical signs as well as body and tail lesions. Blood and bristle samples were regularly collected to evaluate stress response, inflammation level and immune competence. As expected, just before weaning, we observed a higher number of lesions and a lower average daily gain (ADG) in piglets that were pre-socialized. After weaning, these piglets exhibited a greater number of circulating leucocytes, a higher ADG and less body injuries, which may result from a better preparation to the weaning transition. Unfortunately, cannibalism occurred mainly in the alternative group which obviously altered animal welfare. It was associated with higher concentrations of hair cortisol at day 36 and acute-phase proteins at day 36 and 66. If the alternative management of juveniles did not prevent cannibalism in undocked pigs, most of the parameters analysed during the post-weaning period and at slaughter were not significantly different between the two groups. Our results emphasize the importance of field studies to assess the transposability of alternative practices and their real relevance to pigs' health and welfare.

Session 30 Theatre 4

Weaning intact litters in farrowing pens for loose housed sows: diarrhoea and microbiota composition

J.F.M. Winters[1], A.A. Schönherz[1], N. Canibe[1], L. Foldager[1,2] and L.J. Pedersen[1]

[1]Aarhus University, Dept. of Animal Science, Blichers Allé 20, 8830 Tjele, Denmark, [2]Aarhus University, Bioinformatics Research Centre, Universitetsbyen 81, 8000 Aarhus C, Denmark; jeanet.winters@anis.au.dk

Post-weaning diarrhoea outbreaks are associated with the early and abrupt weaning strategy of 4-week-old pigs in conventional pig herds. Dietary, social and environmental changes increase the amount of stress on the immature pigs leading to gut malfunction and impaired immunity. Diarrhoea outbreaks have been associated with a reduced diversity of the intestinal microbiota. Social and environmental stress at weaning can be reduced trough weaning intact litters in the farrowing pens for loose housed sows and may potentially reduce the risk of post-weaning diarrhoea outbreaks and stress-induced alteration of the gut microbiome. Introduction of a genetic sow hybrid giving birth to heavier piglets may increase the resilience at weaning due to increased weaning weight. The study was a 2×2 factorial design: (1) two sow hybrid lines Topigs Norsvin TN70 (TN) or DanBred LY (DB), which have different litter characteristics on total born and birth weight; and (2) two weaning strategies, either keeping the intact litter housed in the farrowing pen (STAY) or moving two litters together into a conventional weaner pen (MOVE). Litters were on average 26 days of age (22-30 days) at weaning and were fed a weaner diet *ad libitum* without medical zinc oxide. In total 57 litters from four batches were included and distributed in the following number of pens: TN-STAY=17, TN-MOVE=8, DB-STAY=10 and DB-MOVE=7. Rectal swabs were collected on the day of weaning (d0), day 2 (d2) and day 7 (d7) post-weaning to determine individual faeces consistency (diarrhoeic or non-diarrhoeic). Differences in diarrhoea incidence were analysed using three separate generalised linear mixed effects models with faeces consistency as response variable and batch, litter and weaner pen as random effects. No differences in odds for diarrhoea between MOVE and STAY were detected on d0 and d2, but a tendency towards significance was found on d7, with MOVE showing higher diarrhoea incidence than STAY (odds ratio 2.3, 95% CI: 0.94 to 5.5, P=0.07). No difference between hybrids was found. Results on faeces microbiota composition will be presented.

Session 30 Theatre 5

Enrichment affects antibiotic consumption in pig farms in Great Britain

I. Kyriazakis[1], S.M. Matheson[2] and S.A. Edwards[2]
[1]Queen's University, Institute for Global Food Security, 19 Chlorine Gardens, Belfast, BT9 5DL, United Kingdom, [2]Newcastle University, King's Gate, Newcastle-upon-Tyne, NE1 7RU, United Kingdom; i.kyriazakis@qub.ac.uk

We investigated links between antibiotic use, production stages present on farm and pig farm characteristics using farm data collected through national recording systems in Great Britain. We focused on the provision of enrichment in the form of straw; providing straw within pig pens may reduce the need for antibiotics by enhancing both pig welfare and resilience to infection. Three different databases were used for this purpose: the electronic medicine book for pigs (eMB), Real Welfare scheme involving on farm assessment of pig welfare and Red Tractor Farm Assurance scheme which collects data on farm characteristics. When eMB and Real Welfare databases were upscaled to a year and combined, there were 1,370 matching unique farm IDs with 2,323 records. Analysis was performed through the use of GLMM with farm category, year and their interaction being fixed effects; farm characteristics were used as explanatory covariates. The amount of antibiotic used (mg/kg) reduced between 2017 and 2018 for Breeder–Finisher farms, but not for Nursery–Finisher or Finisher farms. Breeder–Finisher farms were more likely to use critically important antibiotics (CIA) compared with other production stages. Larger farms were more likely to use CIA, but farm size had no effect on mg/kg of antibiotic used. As the proportion of pens containing straw increased, the total use of antibiotics decreased for Breeder–Finisher, but not for Nursery–Finisher or Finisher farms. As the proportion of pens containing straw increased, the probability of using CIAs also decreased. Farms with a higher proportion of finisher pens with an outdoor space had a lower use of non-critical antibiotics and lower probability of use of CIA. Farms with a higher proportion of pens with automatically controlled natural ventilation (ACNV) had lower total use of antibiotics, although ACNV had no effect on the probability of using CIA. Consistent with our hypothesis, provision of straw was associated with reduced antibiotic use. There are several routes through which provision of straw may reduce antibiotic use in pig farms.

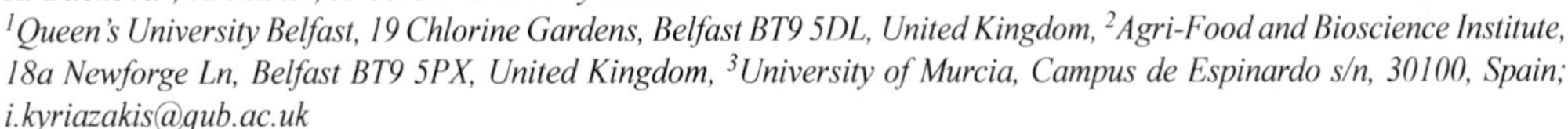

Session 30 Theatre 6

Effect of novel enrichment on pig resilience

K. Bučková[1], R. Muns[2], J. Cerón[3] and I. Kyriazakis[1]
[1]Queen's University Belfast, 19 Chlorine Gardens, Belfast BT9 5DL, United Kingdom, [2]Agri-Food and Bioscience Institute, 18a Newforge Ln, Belfast BT9 5PX, United Kingdom, [3]University of Murcia, Campus de Espinardo s/n, 30100, Spain; i.kyriazakis@qub.ac.uk

Most pigs in slatted systems are provided with enrichment meeting only minimum legal requirements. We aimed to explore the effects of novel enrichment for pigs in slatted systems, and to investigate the timing of enrichment provision on stress resilience and performance. We used 280 pigs allocated to a standard (S, meeting legal requirements) or enriched (E) treatment after weaning at 4w of age. S enrichment consisted of a plastic toy and wood. E pigs were in addition provided with fodder beet and jute bags. Each treatment was replicated on 14 groups balanced for BW and gender. At 10w of age, pigs were moved into a finisher accommodation and were either kept in the same enrichment treatment (EE, SS) or switched from enriched to standard (ES) and vice versa (SE). Each treatment was replicated on 5 groups. Pigs were weighted at the start and end of weaner and finisher stage, and feed intake was recorded. Pigs were assessed for body lesions twice a week until 21w of age. Ten males per treatment were sampled for saliva on days 1, 2, and 4 post-weaning, and again after housing switch. Saliva samples were analysed for adenosine deaminase, haptoglobin, amylase, and cortisol to measure stress response to weaning and housing switch. E weaners consumed less feed (P=0.04) and had better FCR than S weaners (P=0.03). E weaners had higher amylase concentrations than S weaners on day 1 post-weaning (P=0.01). During finisher stage EE and SE pigs consumed less feed than SS and ES finishers (P=0.04), but EE and ES pigs had better FCR (P=0.006) and higher BW (P=0.0001) at the end of finisher stage. Moreover, there was a significant effect of the interaction between enrichment treatment at weaner and finisher stage on body lesions (P=0.04). EE pigs had less body lesions than ES, SS and SE pigs during finisher stage. There were no other significant differences caused either by enrichment treatment at weaner, finisher stage or their interaction. We conclude that the beet and jute bags provided at weaner stage had positive effect on pig FCR at both stages and finisher BW; the enrichment at weaner as well as finisher stage reduced body lesions during finisher stage.

The PPILOW project: innovations improving welfare in low input and organic pig and poultry farms

A. Collin[1] and Ppilow Consortium[2]
[1]INRAE, Université de Tours, BOA, 37380, France, [2]https://www.ppilow.eu/wp-content/uploads/2022/03/co-authors.pdf, France; anne.collin@inrae.fr

The PPILOW project aims to co-construct innovations to improve Poultry and Pig Welfare in Low-input outdoor and Organic farming systems through a multi-actor approach. PPILOW implements a participatory approach for proposing and studying welfare-improvement levers. It will provide a combination of practical solutions that can be applied at a pan-European level with specific adjustments depending on citizen's expectations and the target market. The multi-actor approach consists in involving end-users including farmers, breeding companies, feed producers, consumer associations, retailers, advisers, processors, and scientists in National Practitioner Groups (NPG) in six participating countries. PPILOW partners facilitate the groups by connecting NPG at European level, transferring scientific information, interacting with partners engaged in animal experiments, and co-creating innovations rising from NPG-specific demands. They co-build with PPILOW partners welfare self-assessment tools (development of the PIGLOW app for pigs and refinement of the EBENE® app for poultry), and innovative breeding, feeding, and rearing strategies and techniques to improve the welfare of animals. They co-design protocols, test innovations on farm, and disseminate the results. In turn, they receive insights on methods and scientific results, and inputs from other NPG reinforcing the value of the expected outcomes. Approaches focus on avoiding physical damage and the elimination of layer male chicks, on reducing boar taint of intact male pigs, promoting positive behaviours, animal health, and robustness through field studies with pigs and poultry. Multicriteria analyses of the most effective levers of welfare improvement will be performed to evaluate their economic, social, and environmental impacts based on the 'One Welfare' concept; economic and business models will also be developed. To ensure the rapid uptake of the project results by end-users, the close involvement of PPILOW's NPG throughout the EU will ensure dissemination activities and the facilitation of change. The PPILOW project has received funding from the European Union's Horizon 2020 Research and Innovation Programme under grant agreement N°816172. www.ppilow.eu.

Transport at an older age: the key to a higher robustness of veal calves?

F. Marcato[1,2], H. Van Den Brand[1], M. Wolthuis-Fillerup[2], F. Hoorweg[2] and K. Van Reenen[2]
[1]Wageningen University & Research, Adaptation Physiology Group, Department of Animal Sciences, P.O. Box 338, 6700 AH Wageningen, the Netherlands, [2]Wageningen Livestock Research, P.O. Box 338, 6700 AH Wageningen, the Netherlands; francesca.marcato@wur.nl

Age at arrival at the veal farm may be an important determinant of the biological state of a calf and may affect subsequent health and performance. This study aimed to investigate effects of age of calves at transport from the dairy farm to the veal farm on measures of health and performance, including the use of antibiotics or other medicines. Over a 34-week period, calves (n=684) were transported at an age of either 14 or 28 days from 13 dairy farms to 8 veal farms. At each veal farm calves of both age groups and originating from all 13 dairy farms were present. Calves followed the same rearing procedure at the veal farm together with calves that were not part of the study. Health problems were scored at the dairy farm on a weekly basis until the day prior to transport, and at the veal farm in week 2, 6, 10, 18 and 24 post-transport. At both the dairy and the veal farms, mortality risk and the use of antibiotics and other medical treatments at individual calf level were recorded. Body weight (BW) was measured at arrival at the veal farm, and carcass weight was recorded at slaughter. Clinical health problems and individual medical treatments were expressed as binary variables and analysed with the GLIMMIX procedure in SAS 9.4. Body and carcass weights were analysed with a linear mixed model. Non parametric Wilcoxon signed rank test was performed to test for mortality risk differences. Health problems at the veal farm did not differ between the two transport age groups. Relative to calves transported at 14 days, calves transported at 28 days had a lower mortality rate (2.8% vs 5.9%) at the veal farm and a higher carcass weight (corrected for BW upon arrival Δ=12.8 kg). At the veal farm, calves transported at 28 days received less individual medical treatments other than antibiotics (Δ=-5.4% of calves) than calves transported at 14 days. Collectively, these results suggest that calves transported at 28 days are more robust and better able to cope with the conditions at the veal farm than calves transported at 14 days.

Session 30 Theatre 9

Antimicrobial susceptibility of *E. coli* and *Pasteurellaceae* in veal calves before slaughter

J. Becker[1], V. Perreten[2], A. Steiner[1], D. Stucki[1], G. Schüpbach-Regula[3], A. Collaud[2], A. Rossano[2], D. Wüthrich[1,2], A. Muff-Hausherr[1,2] and M. Meylan[1]

[1]Vetsuisse-Faculty, University of Bern, Clinic for Ruminants, Bremgartenstrasse 109a, 3012, Switzerland, [2]Vetsuisse-Faculty, University of Bern, Institute of Veterinary Bacteriology, Länggassstrasse 122, 3012, Switzerland, [3]Vetsuisse-Faculty, University of Bern, Veterinary Public Health Institute, Schwarzenburgstrasse 161, 3079 Liebefeld, Switzerland; jens.becker@vetsuisse.unibe.ch

Animal husbandry requires practical measures to limit antimicrobial resistance (AMR). Therefore, a novel management and housing concept for veal calf fattening was implemented on 19 intervention farms (IF) and evaluated regarding its effects on AMR in *Escherichia coli*, *Pasteurella multocida* and *Mannheimia haemolytica* in comparison with 19 conventional control farms (CF). Treatment intensity (-80%) and mortality (-50%) were significantly lower in IF than in CF, however, production parameters did not differ significantly between groups. Rectal and nasopharyngeal swabs were taken at the beginning and the end of the fattening period. Susceptibility testing by determination of the minimum inhibitory concentration was performed on 5,420 isolates. The presence of AMR was described as prevalence of resistant isolates (%), by calculating the Antimicrobial Resistance Index (ARI: number of resistance of one isolate to single drugs/total number of drugs tested), by the occurrence of pansusceptible isolates (susceptible to all tested drugs, ARI=0), and by calculating the prevalence of multidrug (≥3) resistant isolates (MDR). Before slaughter, odds for carrying pansusceptible *E. coli* were higher in IF than in CF (+65%, P=0.022), whereas ARI was lower (-16%, P=0.003), and MDR isolates were less prevalent (-65%, P<0.001). For *P. multocida*, odds for carrying pansusceptible isolates were higher in IF before slaughter compared to CF (+990%, P<0.009). No differences between IF and CF were seen regarding the prevalence of pansuceptible *M. haemolytica*. These findings indicate that improvement of calf management can lead to a limitation of AMR in Swiss veal fattening farms.

Session 30 Theatre 10

Effect of housing dairy calves, individually or in pairs, on performance, welfare, and behaviour

D. Espinoza-Valdes[1,2], A. Iritz[1,2], F. Salhab[1,2], S. Mabjeesh[1] and Y. Ben Meir[2]

[1]Hebrew University of Jerusalem, Animal Sciences, Rehovot, 7610001, Israel, [2]Agricultural Research Organization, Volcani Center, Ruminant Sciences, Derech HaMaccabim 68, Rishon LeTsiyon, 7505101, Israel; daniel.espinoza@mail.huji.ac.il

Housing practices for lactating dairy calves have recently become a topic of interest. The safest approach is to house suckling calves in individual pens, minimizing the risk of diseases infection and ensuring appropriate feeding of both milk and solids. This practice suppresses the expression of natural social behaviour. Therefore, it is highly recommended to improve animal welfare by socialize with at least one other calf. Although group housing of suckling calves became more popular, the need to adopt new practices requires changes in farms structure and routine inhibit the majority of Israeli dairy farmers from adopting the new approach. We hypothesis that a significant improvement in calves welfare can be made efficiently by adjusting two individual pens next to each other, enabling the calves to see and touch each other while each calf has access to food, water, and bedding. Therefore, this study aimed to compare among three housing treatments: individually housed (IH) – each calf housed in an individual pen distanced 1 m from the other; paired housing (PH) – two calves housed in a double space pen; and paired by individual pens adjusted (PIA) to each other. Calves (Israeli Holstein) were assigned alternatively to one of the treatments immediately after birth until we reached four blocks of 3 pairs each. Calves were fed 41 / days of fresh milk until age 55 days and reared gradually until age 60. Solids mixed ration and water supplied *ad libitum*. Calves were scaled and measured for height and length weekly. Feed was delivered daily, and refusals were collected and scaled every other day. Each pair's behaviour was recorded by time-lapse, taking 1 picture every 20 seconds from 0d to 60d, day and night, using GoPro Hero 8BLACK. Different behaviours were classified using a deep network. Average daily gain (ADG) and feed intake after rearing (days 56-63) of PH group tend (P<0.1) to be higher than that of the PIA and IH (ADG=0.78, 0.74, 0.71 kg/d; DMI=1,393, 1,265, and 1,130 g / day, respectively). There was no disease diagnosed in any of the calves during the study.

Session 30 Theatre 11

Effect of welfare standards and biosecurity practices on antimicrobial use in beef cattle

M. Santinello[1], A. Diana[2], V. Lorenzi[3], E. Magni[3], G.L. Alborali[3], L. Bertocchi[3], M. De Marchi[1] and M. Penasa[1]
[1]University of Padova, Department of Agronomy, Food, Natural resources, Animals and Environment, Legnaro (PD), 35020, Italy, [2]Purdue University, Department of Comparative Pathobiology, West Lafayette, IN, 47907, USA, [3]Istituto Zooprofilattico Sperimentale della Lombardia e dell'Emilia Romagna Bruno Ubertini, Italian National Reference Centre for Animal Welfare, Brescia, 25124, Italy; matteo.santinello@phd.unipd.it

Antibiotic use (AMU) in livestock species and the associated antibiotic resistance are global issues that affect human and animals' health, and strategies to reduce them in livestock species are necessary. There is a link between AMU reduction and animal welfare and biosecurity measures in most of livestock species, but there is a lack of knowledge in beef cattle. This study aimed to investigate the impact of welfare and biosecurity measures on AMU in beef cattle. Data on performance traits and AMU were collected from January 2016 to April 2019 from 27 specialised Italian beef farms. The BW at arrival and at the end of the fattening cycle, and the respective ADG were calculated per batch, defined as a group of animals of the same sex and breed that entered the same fattening farm. Four different treatment indexes were calculated using the Italian and EU defined daily doses: 2 treatment incidence (TI100) indexes and 2 highest priority critically important antibiotics (HPCIA) indexes. An on-farm assessment was carried out by assigning a score from 0 to 100% to 3 sections: biosecurity, emergency management and welfare. A generalized linear mixed model was used to investigate whether biosecurity and welfare measures affected the aforesaid indexes and performances. The model accounted for sex, season, total welfare, biosecurity, emergency management and the interaction between total welfare and sex as fixed effects, and farm and breed as random effects. Biosecurity had the lowest average score (24%), suggesting that more efforts in this area are needed. Statistically significant lower treatment indexes were observed with improved level of total welfare ($P<0.05$), whereas no differences were observed for the corresponding HPCIA indexes ($P>0.05$). Results are helpful for farm benchmarking and highlight the importance of improved animal welfare for an efficient antimicrobial stewardship.

Session 30 Theatre 12

Benefits for achieving positive dairy cattle welfare

S. Mattiello, S. Celozzi, F.M. Soli, L. Bava and M. Battini
University of Milan, Department of Agricultural and Environmental Sciences – Production, Landscape, Agroenergy, Via Celoria 2, 20133 Milan, Italy; silvana.mattiello@unimi.it

Aim of the present study was to identify potential benefits that can affect positive welfare. We evaluated animal welfare in 20 loose housing dairy cattle farms (6 on permanent litter, 14 in cubicles), based on positive indicators of welfare and animals' emotional state (feeding and resting synchronisation, rumination during resting, comfortable lying postures, no visible eye white, relaxed ear postures). Potential benefits in terms of housing, feeding and management were then related to these variables (Mann-Whitney U test). Qualitative Behaviour Assessment (QBA) was also carried out and analysed by Principal Component Analysis to explore the effect of factors that were not evenly distributed in our sample (access to pasture, presence of paddock or environmental enrichments, automatic milking systems, number of feed distributions). When hay was included in the diet, higher feeding synchronisation (93.7±1.6 vs 52.2±4.7%; $P<0.01$), percentage of cows with relaxed ear postures (35.8±5.4 vs 15.5±2.1%; $P<0.01$) and percentage of cows with no visible eye white (55.9±17.0 vs 36.6±4.1%; n.s.), were recorded. A higher level of feeding synchronisation was observed also when the feeding places/cow ratio was >1 (72.1±9.9 vs 53.8±5.8%), although differences were not significant ($P=0.14$). Permanent litter had a more positive effect than cubicles on comfort at resting, with a significantly higher percentage of ruminating cows (65.8±10.2 vs 34.2±3.7%; $P<0.01$), a higher percentage of cows with no visible eye white (55.6±9.9 vs 33.1±3.7%; $P<0.05$) and a higher percentage of cows in a more comfortable posture with stretched legs (14.3±5.1 vs 5.6±1.6%; $P=0.09$). No significant differences on comfort at resting were found depending on the space available either in permanent litter, or in cubicles. QBA could not highlight any trend in farm distribution depending on the considered factors, except for the only farm that gave access to pasture, which had the highest scores on PC1, indicating a positive emotional state. We suggest to consider permanent litter, a feeding places/cow ratio >1, hay in the diet and access to pasture as benefits for enhancing positive animal welfare.

Session 30 — Theatre 13

Stress in dairy calves suckled or not by their dam assessed from cortisol in hair

D. Pomiès[1], A. Nicolao[1], I. Veissier[1], K. Alvåsen[2] and B. Martin[1]
[1]INRAE, UMR Herbivores, 63122 St Genès Champanelle, France, [2]SLU, P.O. Box 7054, 750 07 Uppsala, Sweden; dominique.pomies@inrae.fr

Allowing suckling the dam is considered to improve the welfare of calves but separation from the dam at weaning can be highly stressful. We questioned if dairy calves reared with their dam experience more or less stress before and after weaning than calves artificially reared. Cortisol concentration in hair is assumed to reflect stress accumulated over days, weeks or even months. We compared hair cortisol in calves reared or not with their dam. Three groups of 9 calves were used at the INRAE Herbipôle experimental farm: 'Control' calves separated from dams immediately after birth and fed with an automatic milk feeder until weaning at 11 weeks of age, 'Dam' calves suckled by their dam between morning and evening milking until 11 weeks of age, 'Mixed' calves suckled by their dam for 4 weeks then reared as Control calves. Hair samples were taken from each calf's shoulder before weaning (calf age, 60.3±4.3 days) and 30 days after weaning. The hair was washed and dried before cortisol extraction and determination using Salimetrics ELISA kit at the Swedish University of Agricultural Sciences. Hair cortisol concentration was lower in Dam calves than in other calves both before weaning (19.9 pg/mg for Dam calves vs 26.9 pg/mg; P=0.02) and after weaning (11.2 pg/mg for Dam calves vs 15.5 pg/mg; P=0.08). Before weaning, the Dam calves grazed during the day whereas other calves remained in indoor pens. Cortisol increases with activity and should thus be higher in Dam calves if they were as stressed as the other calves. This reinforces the conclusion that calves with their dam are less stress than calves without their dam. During the week following weaning, Dam calves mooed more than other calves, suggesting intense stress, which did not reduce the difference in hair cortisol with other calves. Despite stress at weaning, rearing dairy calves with their dam seems thus to result in less stress than rearing them without the dam.

Session 30 — Poster 14

Browsing effect on natural control of the elimination of gastrointestinal strongyles eggs in goats

A.T. Belo[1], S.M. Almeida[2], S. Cavaco-Gonçalves[1], J.P. Barbas[1], J.M.B. Ribeiro[1], C.C. Belo[1], C. Costa[3], L.M.M. Ferreira[2] and L. Padre[3]
[1]INIAV IP, Pólo da Fonte Boa, 2005-048 Vale de Santarém, Portugal, [2]Univ. de Trás-os-Montes e Alto Douro, Depart. Zootecnia, 5000-801 Vila Real, Portugal, [3]Univ. de Évora, MED, Depart. Medicina Veterinária, 7002-554 Évora, Portugal; claudia.dc.costa@gmail.com

The aim of this study was to evaluate how browsing could affect the control of gastrointestinal strongyles elimination by naturally infected goats, and avoid the use of synthetic anthelmintics. Two groups of goats were allowed to browse Mediterranean shrubs for different periods of time, from early March to July (group A) or from late April to July (group B). The level of elimination was evaluated throughout the study and confronted with a group of goats maintained exclusively on irrigated pasture (Control). Group A started browsing for a 4 hours-period (first month) increasing to 10 hours for the remaining months. Group B was maintained (2 weeks) without access to grass or shrubs before starting to browse for 10 hours, with group A. Diet composition was estimated using the n-alkanes and long chain alcohols technique. All goats were monitored for blood parameters (packed cell volume, total protein, albumin, creatinine, glucose, total cholesterol and urea N, as well as aspartate and alanine aminotransferases) and compared to the Control goats. Both groups of browsing goats showed a steady and significant decrease in EPG (eggs/ g faeces) as time of browsing increased while elimination in Control goats was maintained during the same period. A more evident decrease was observed for group A as browsing increased from 4 to 10 hours per day. By July both browsing groups attained the same level of EPG. The diets' content in total phenolics and tannins increased from 31 to 84 mg GAE/kg DM for group A and from 10 to 68 mg GAE/kg DM, April to May respectively, and both reached 70 mg GAE/kg DM by July. The contribution of shrubs to their diets went from 27 to 79% of DM ingested for A and from 0 to 52% for B, reaching 75% in July for both groups. Blood parameters assured the goats' welfare regarding EPG levels and phenolics intake. Funding: ALENTEJO2020, project ALT20-03-0145-FEDER-000009 VegMedCabras: Mediterranean shrubs: natural anthelmintics in the diet selected by grazing goats.

Effect of layers' housing systems on physical characteristics and quality of commercial eggs

D. Amorim[1], *J. Costa*[1,2] *and J. Oliveira*[1,2]
[1]*ESAV, Polytechnic Institute of Viseu, Escola Superior Agrária de Viseu, Quinta da Alagoa, Estrada de Nelas, 3500-606, Portugal,* [2]*CERNAS Research Centre, Polytechnic Institute of Viseu, Escola Superior Agrária de Viseu, 3500-606, Portugal; joliveira@esav.ipv.pt*

The evolution of layer hens housing systems according to the EU legislation and the pressure from society for the hens′ welfare has led to an increase in cage-free systems. In a consumers′ point of view, there is a perception that free range eggs are healthier and have better quality than conventional ones. The aim of this study was to analyse the effect of layers′ housing systems, namely enriched cage (EC), barn (B), and free-range (FR) on internal and external physical egg quality parameters. A total of 180 eggs (M and L classes) were acquired from retail stores in different times considering three housing systems (enriched cage [EC], barn [B] and free-range [FR]). They were analysed for egg weight (EW), shell weight (SW) and shell thickness (ST), shell breaking strength (SBS) and shell deformation limit (SDL), albumen height (AH), vitelline membrane strength (VMS), yolk height (YH), yolk width (YW), yolk colour (YC), yolk ratio (YR) and Haugh unit (HU). A one-way ANOVA procedure was implemented to analyse the influence of housing systems in the egg's characteristics with Tukey post-hoc tests. The level of significance of $P \leq 0.05$ was considered. The EW don't differ from housing systems (61.95±4.29 g [53.7 to 72.7 g]). The ST is the only shell characteristic that is significantly different between housing systems (0.40±0.06 mm, 0.44±0.06 mm and 0.38±0.05 mm to EC, B and FR, respectively). Almost all the yolk characteristics are higher in the barn system, spite not always different across the three housing systems, with YW higher in EC system ($P \leq 0.05$). We can conclude that egg quality, as measured by physical characteristics, does not generally differ between housing systems. However, considering divergent results from the literature, we can state that other approaches, including layers age and specificities in feeding approach may strengthen the conclusions. In any case, the consumer's knowledge about the quality characteristics of eggs in different housing systems can help in the purchase decision, even if it means paying more for the product.

Assessment of active substances' cross-contamination in non-target feed in Portugal

A.C.G. Monteiro and A.S. Santos
FeedInov CoLab, Quinta da Fonte Boa, 2005-048 Vale de Santarém, Portugal; ana.monteiro@feedinov.com

The rules on medicated feed have a significant influence on animal production as well as on the production of products of animal origin. The publication of EU Regulation 2019/4 of the European Parliament and the Council established additional measures to strengthen the efficiency of the internal market and improve the rules for animal treatment with medicated feed, as well as the obligation to establish maximum levels of cross-contamination by active substances in non-target feed and the methods of analysis for their detection. This work aimed to analyse the levels of cross-contamination of eight active substances (chlorotetracycline, oxytetracycline, amoxicillin, doxycycline, lincomycin, tiamulin, tylosin and colistin) during the manufacturing process, to assess and define acceptable levels, safeguarding animal health and welfare, and ensuring the feasibility of its practical implementation. Samples were collected from the non-target feed produced immediately after the medicated feed (without cleaning of the production circuit), being the active substances analysed those used in the previously produced medicated feed. In total, 10 elementary samples (AE) were collected per feed. Of the 10 AE collected, 3 global samples (AG) were formed. The calculation of the level of cross contamination (NCC) was carried out considering the contribution of each of the AE to the three AG, which were analysed separately. Results obtained showed that NCC was lower when tylosin and colistin were used. In the case of colistin, the values obtained were always below the detection limit of the analysis method. The substances for which NCC levels were the highest or whose results were more variable between factories were oxytetracycline and tiamulin, respectively. We conclude that, considering legislative changes, industry must make the procedures implemented in factories more effective, namely establish circuit cleaning routines after the manufacture of medicated feed, to prioritise the use of coated substances, whenever possible, or having electrostatic properties that do not favour cross-contamination. The installation of lines dedicated to the production of medicated feed can also be an appropriate measure to prevent cross-contamination from occurring.

Session 31 Theatre 1

Effect of pork fat replacement by vegetable oleogels in quality of goat meat burgers

I. Ferreira, L. Vasconcelos, A. Leite, E. Pereira, S. Rodrigues and A. Teixeira
Centro de Investigação de Montanha (CIMO) Escola Superior Agrária, Instituto Politécnico de Bragança, Campus Sta Apolónia, 5300-253 Bragança, Portugal; ias.ferreira@hotmail.com

Healthy lifestyle has become a goal for most of the consumers nowadays. A well-known strategy used to improve nutritional characteristics of meat products is animal fat replacement, as pork back fat, which has a considerable amount of saturated fatty acids (SFA) that are considered a risk factor to cardiovascular diseases. Goat meat is an alternative to replace beef and pork, with a similar protein content and lower fat content, and also as an opportunity to commercialize goat meat from culled animals. The present study aimed to develop a goat meat burger, replacing the pork fat by olive oil or sunflower oleo gels. The formulations used had 87,9% of goat meat, 1.1% of NaCl, 7% of H_2O and 4% of pork fat (GPF) or olive oil (GO) or sunflower oil (GSF) Prosella® gels. The aw values differed statistically towards all formulations, the only burger that differentiated pH from the other two was GO, being lower than GPF and GSF. In colour parameters, GSF showed the lowest luminosity values and the highest heme pigments content. Ashes and protein content differed statistically towards all burgers, along with the fat, although the formulations could not affect those, as the ingredients were standardized. Concerning the lipidic quality indicators, GPF (40.84%) showed the higher content of SFA, while GO and GSF did not differentiate, moreover the MUFA content differed between all of formulations, with GO (57.54%) presenting the highest one, followed by GPF (49.19%) and GSF (40.30%). Atherogenicity and thrombogenicity indexes were also determined, GPF burgers had the highest value (0.52 and 1.26) while GO and GSF had the lowest values. Evaluating the results, is possible to say that the use of oleo gels improved the lipidic quality of the burgers and trough discriminant analysis the three formulations could be correctly identified by two FA (C18:2n-6 and C20:1n-9) with a high rate of accuracy.

Session 31 Theatre 2

A longitudinal study of the microbiological quality of goat's raw milk cheese during ripening

S. Coelho-Fernandes, A.S. Faria, G. Santos-Rodrigues, A. Fernandes, L. Barros, V. Cadavez and U. Gonzales-Barron
Instituto Politécnico de Bragança, Campus de Santa Apolonia, 5300-253 Bragança, Portugal; vcadavez@ipb.pt

Transmontano goat's cheese is a cheese with protected designation of origin, made of goat's raw milk produced in the *Trás-os-Montes* Northern region of Portugal. The objective of this study was to assess the between-batch variability in selected quality properties, and the evolution of hygiene/safety indicator microorganisms in cheese during ripening. Samples of goat's raw milk and cheeses were taken every 20 days along the 60-day ripening from four production batches surveyed from an artisanal producer. Longitudinal models showed that, between-batch variability in quality properties was high, with intra-class correlations initially varying between 30.4% (water activity) and 60.5% (*Staphylococcus aureus*), although such variability was largely reduced during ripening. While total viable counts [TVC] and lactic acid bacteria [LAB] counts increased (P<0.001) every sampling day (TVC: Day_0=9.78, Day_{20}=11.34, Day_{40}=11.93, Day_{60}=13.18 log cfu/g; LAB: Day_0=9.24, Day_{20}=10.87, Day_{40}=11.93, Day_{60}=13.06 log cfu/g), fastidious bacteria such as *S. aureus*, *Clostridium* spp. and *Listeria* spp. dropped during ripening to safe levels (2.28 [SE=0.284], 1.23 [SE=0.594] and <1.22 log cfu/g, respectively). In addition, the growth of TVC in cheese was affected by milk pH (P=0.001) and water activity (aw; P=0.009), TVC levels in milk (P=0.004), initial pH (P=0.001) and aw of cheese (P=0.021), and protein (P=0.037) and fat contents (P=0.036); whereas the development of LAB was mainly driven by milk pH (P=0.043) and aw (P=0.008), and the initial LAB counts in milk (P=0.004). The progressive drop of *Clostridium* spp. during ripening was regulated by all of the intrinsic properties mentioned above, whilst the inactivation of *Listeria* spp. until undetectable levels was highly associated to milk pH (P=0.006), initial pH of cheese (P=0.016) and lactic acid concentration (P=0.049). The absence of *L. monocytogenes* and *E. coli* O157:H7 reassured the microbiological safety of the Portuguese Transmontano cheese.

Consumer attitudes from different continents towards dairy products from sheep and goat

E. Vargas-Bello-Pérez[1], K. Tajonar[2], G. Foggi[3], M. Melle[3], S. Panagiotis[4], A. Mavrommatis[4], E. Tsiplakou[4], M.R. Habib[5], M. Gonzalez-Ronquillo[6] and P. Toro-Mujica[7]
[1]University of Copenhagen, Frederiksberg, 1870, Denmark, [2] Universidad Nacional Autónoma de México, Av. Universidad 3000, Ciudad de México, Mexico, [3]University of Pisa, Via del Borghetto, 80, 56124, Pisa, Italy, [4]Agricultural University of Athens, Iera Odos 75, Athens, Greece, [5]Bangladesh Agricultural University, Mymensingh, 2202, Bangladesh, [6]Universidad Autónoma del Estado de México, Instituto Literario 100 Ote., Toluca, Mexico, [7]Universidad de O'Higgins, San Fernando, 3070000, Chile; evargasb@sund.ku.dk

The objective of this study was to assess consumer knowledge, attitudes, and perceptions towards dairy products from sheep and goat. A web-based survey was conducted in America (Mexico and Chile), Europe (Italy, Spain, Greece, and Denmark) and Asia (Bangladesh). Adult participants answered the survey that was available online in five languages from March to June 2021. In total 1,879, surveys were completed online. Categorical and ordinal data was analysed as frequencies and percentages. To determine the relationship between the variables for purchasing and consumption behaviours of respondents that declared to consume dairy products, a multiple correspondence analysis (MCA) was carried out. Most surveys were from Mexico and Italy (30% and 33.7% respectively). The MCA defined 5 dimensions. Dimension 1 was associated with the geographic location of the respondent (country and continent), the type of milk (sheep or goat) and the consideration of well-being and health as a characteristic associated with the consumption of dairy products from sheep and goats. Dimension 2 was also associated with the respondent's country of origin and the frequency of consumption. Dimension 3 was associated with gender, education, and employment status. Dimension 4 was associated with the age of the respondents, the association of the 'Healthy' concept with sheep and goat dairy, and the consideration of the nutritional benefits of dairy as responsible for considering them healthy. Finally, Dimension 5, having a component in the respondent's country, incorporates the relationship of 'strong smell and taste' with sheep and goat dairy products. This study showed that consumer attitudes towards dairy products from sheep and goat vary between continents.

Dietary *Moringa oleifera* affects quality and health promoting traits of lamb meat

M. Cohen-Zinder, H. Omri, E. Shor-Shimoni, R. Agmon and A. Shabtay
Beef Cattle Unit and Model Farm for Sustainable Agriculture, Newe Ya'ar Research Center, Agricultural Research Organization, Volcani Center, 3009500 Ramat Yishay, Israel; mirico@volcani.agri.gov.il

It is well recognized, that different dietary conditions can alter the quality of ruminant food products. Special focus has recently been given to the effects of *Moringa oleifera*, an anti-oxidant rich feed, on ruminant growth and production. However, little attention has been paid to its effects on ruminant product quality. To this end, Moringa oleifera was grown in the model farm for sustainable agriculture (Newe Ya'ar, Israel), prior to its large-scale ensiling and supplementation as feed additive to weaned lambs (n=36). The lambs were randomly assigned to each of 3 dietary regimes: 10% of Moringa silage with soybean hulls and sugar-cane molasses (proportion of 72.5: 22.5: 5); 10% of ensiled Moringa with crushed corn grains and sugar-cane molasses; *Control diet*: 10% of wheat silage as roughage. All diets included 90% (DM) of concentrated pellets, and were balanced for crude protein, energy, DM and NDF. The potential of Moringa silage to improve lamb meat tenderness (by means of lower shear force; SF) was revealed ($P<0.01$). Sarcomere length (SL), yet another parameter positively correlated with meat tenderness, was significantly higher ($P<0.01$) in the muscles of the same lambs, in comparison to their counterparts. Surprisingly, the intra-muscular fat, which normally correlates positively with meat tenderness, exhibited lower content ($P<0.05$) in these meat samples. Moreover, vitamin E concentration was higher ($P<0.05$) in the meat of Moringa as compared with wheat silage-fed lambs. Malondialdehyde (MDA) content, a marker of lipid peroxidation, was lower ($P<0.01$) in the meat of Moringa silage-fed lambs, comparing to the control group. Both measures highlight the potential of Moringa silage to increase the accumulation of health promoting compounds in the meat, and to prolong its shelf life. The current study may promote future use of Moringa as a natural ruminant feed, for the production of lamb meat, with higher edible properties and improved health characteristics, for human consumption.

Detailed fatty acid profile of human milk and cow, buffalo, goat, sheep, dromedary, and donkey milks

N. Amalfitano[1], M.L. Haddi[2], H. Benabid[2], M. Pazzola[3], G.M. Vacca[3], A. Lante[1], F. Tagliapietra[1], S. Schiavon[1] and G. Bittante[1]

[1]University of Padova, DAFNAE, Viale dell'Universita 16, 35020 Legnaro, Italy, [2]Université des Frères Mentouri, LaMyBAM and INATAA, 325 Ain El Bey Way, 25017 Constantine, Algeria, [3]University of Sassari, DVM, Via Vienna 2, 07100 Sassari, Italy; nicolo.amalfitano@unipd.it

The present study aimed to explore and compare the detailed fatty acid profile of human milk and of milk from 6 farm species (cows, buffaloes, sheep, goats, dromedary camels, donkeys). This study is part of the GOOD-MILK project (D.M. 9367185 – 09/12/2020) which involved the collection of at least 10 samples/species of bulk milk from cows, buffaloes, sheep, and goats herds located in Italy, of individual camel milk from two different areas of Constantine Province (Algeria), and of individual human milk donated by the human milk bank (BLUD of Vicenza, Italy). The detailed fatty acid (FA) profile was obtained through a bi-dimensional gas chromatography technique and the data were expressed as g/100 g of total FAs. The FA profiles were analysed firstly with a hierarchical clustering analysis and then using a general linear model where the species was included as fixed effect. The results showed that the species clustered hierarchically according to their taxonomic distance. In detail, the milks from the monogastric species, human and donkey, were almost always different from the polygastric species. The former two species showed generally a lower proportion of SFA, except for the C8:0 and C12:0 higher in donkey milk, and a higher proportion of MUFA and PUFA compared to the other species; human milk was particularly high in ω-6 and long-chain FA. Within the polygastric species, the milk from the dromedary camel showed a higher total proportion of branched FA, of MUFA, and of conjugated linoleic acid (CLA) than the ruminants. Within the ruminants, there was no difference between the *Bovinae* and *Caprinae* sub-families in terms of SFA and MUFA, but within the two groups the goats and buffaloes were higher in SFA and lower in MUFA compared to ewes and cows respectively. The *Caprinae* sub-family had a higher percentage of PUFA than the *Bovinae.* This information can be useful to determine possible effects of these different milks on human health.

Professionals perception on lamb and kid meat quality characteristics in Greece

I. Vakondios and A. Foskolos

University of Thessaly, Animal Science, Campus Gaiopolis, 41500 Larisa, Greece; afoskolos@uth.gr

Greece is considered one of the main European countries of lamb and kid meat production. However, carcass evaluation systems, like SEUROP, are not the standard method to assess quality. In contrary, the criteria of selection are based on personal experience. This study aims to investigate the criteria that professionals, butchers and chefs, use to evaluate carcasses of lamb and kids. A questionnaire was developed and conducted by phone calls to professionals all around the country, to examine their criteria of carcass quality and evaluation. Among received information carcass quality was assessed with meat colour, carcass conformation, fat colour, fat cover, carcass weight and slaughter age. A total of 500 butchers and 500 chefs were contacted from which 72 butchers and 39 chefs replied to the survey. Regarding the meat colour there was no significant differences between butchers and chefs. Both groups preferred pink meat colour. On the other side, there was statistical difference between the meat types. The preferred colour for kid meat was between pink and red (2,31), but for the lamb meat was between light pink and pink (1,58). No significant differences were found at the carcass conformation preference between lamb and kid meat. There was a difference between the perception of professionals. Butchers showed preference on heavier carcasses than chefs (3,98 and 3,38 respectively). The fat colour parameter was for both meats type and profession the same (close to white). Fat cover was an aspect that was differentiated between meat types, but not between professions. The lamb meat was considered better if it was fatter (2,84) than kid meat (2,32). In the aspect of smell for both professions and meat types the smell was considered to be non existing to very light (1,20). Carcass weight and slaughter age showed the same preferences between butchers and professionals and between different types of carcasses. The preferred carcass weight was between 8 and 12 kg and the preferred slaughter age was between 3 and 5 months old. These characteristics can be further used to develop quality system for Greek lambs and kids.

Session 31 Theatre 7

YoGArt project: the use of algae as feed additives in ewes diet to promote functional dairy products

A. Mavrommatis, F. Zisis, P. Kyriakaki, F. Satolias, A. Skourtis and E. Tsiplakou
Agricultural University of Athens, Department of Animal Science, Iera odos 75, 11855, Greece; mavrommatis@aua.gr

An imperative challenge for nutritionists arises during the 21st century in order to produce highly nutritious and functional foods which promote human health. Polyunsaturated fatty acids (PUFA) and other bioactive compounds which are highly contained in microalgae biomass are capable to enrich milk and dairy products with beneficial molecules. For this purpose, 24 Lacaune dairy ewes, were separated into 4 groups (n=6) and were fed with different levels of microalgae *Schizochytrium* sp. [0 (CON, no microalgae), 20 (S20), 30 (S30) and 40 (S40) g/kg concentrates] for 60 days. The purpose was to examine the optimum inclusion level for ewes of such feed rich in ω3 PUFA, for enhancing milk and yogurt quality. The results showed that none of the 3 different *Schizochytrium* sp. levels impaired milk performance or its organoleptic features. Concerning the fatty acids (FA) profile, the proportions of CLA isomers, C20:5 (EPA), C22:5n-6 (DPA), C22:6n-3 (DHA), the total ω3 FA and the PUFAs of treated ewes' milk were significantly increased, while those of C18:0 and cis-9 C18:1 were decreased. Also, in the S40 group an oxidative response was induced, observed by the increased malondialdehyde (MDA) levels in milk and blood plasma. In conclusion, the dietary supplementation with 20 and 30 g *Schizochytrium* sp./ewe/day, promotes milks' fatty acid profile and seems to be a promising strategy for producing highly contained ω3 dairy products.

Session 31 Theatre 8

Impact of loin versus individual muscle weight on the eating quality prediction of several lamb cuts

S.M. Moyes, G.E. Gardner, D.W. Pethick and L. Pannier
Murdoch University, College of Science, Health, Engineering and Education, 90 South Street, Murdoch 6150 Western Australia, Australia; s.moyes@murdoch.edu.au

The lean meat yield percentage (LMY) of a lamb carcass heavily affects its financial value, hence Australian producers select for leaner, more muscular, faster growing animals. However, this selection causes reduced sensory scores, as shown by untrained consumers. Computer tomography is used to estimate the percentage of muscle and fat within a carcass, and this correlates strongly with loin and topside muscle weights when corrected for carcass weight, signifying these measures as good indicators of whole carcass muscling. If individual cut weights could be measured in an abattoir and used as indicators of muscling, this may enhance the eating quality prediction for that cut. In this study, the knuckle, loin, and topside cuts were collected from 3,119 lambs from the Meat and Livestock Australia Genetic Resource Flock 24 hours post-slaughter. Cuts were aged at 1 °C for 5 to 10 days before freezing and later used for grilled sensory assessment. Untrained consumers (n=14,280) assessed the samples for overall liking on a 100-point scale line (100=best). Ten individual consumer responses per cut were analysed using linear mixed effect models in R (version 3.6.1) with production factors, such as site and sire type, included as fixed effects, and hot carcass weight included as a covariate. Cut weights as indicators of muscling were tested in this base model in two ways. Firstly, consumer scores for each cut were predicted by loin weight included as a covariate. Secondly, consumer scores for the cuts were predicted by the individual weight of each cut included as a single covariate in the model. Preliminary findings showed increasing muscle weight exhibited a negative association with consumer overall liking scores for the knuckle and loin cuts, whereas the topside slightly increased in eating quality, with either loin or individual muscle weights included ($P<0.05$). These findings show muscle weight can be used to predict eating quality variability between cuts, emphasising the importance of selection for carcass traits, in combination with eating quality attributes, when establishing an eating quality prediction model to maintain eating quality.

Effect of *Camelina sativa* seeds in ewes milk quality and oxidative status in both organism and milk

C. Christodoulou[1], A. Mavrommatis[1], B. Kotsampasi[2] and E. Tsiplakou[1]
[1]Agricultural University of Athens, Department of Animal Science, Iera Odos 75, 11855, Greece, [2]Research Institute of Animal Science, Hellenic Agricultural Organization, Demeter, Paralimni, 58100, Greece; eltsiplakou@aua.gr

Camelina sativa seeds are rich in bioactive compounds such as polyunsaturated fatty acids (PUFA) and antioxidants, thus, their inclusion in ewes' diets, may be an effective way to develop high nutritional dairy products. For this reason, forty-eight dairy Chios ewes were divided into four homogenous groups and were fed individually. The concentrate of the control group (CON) had no Camelina seeds. In the concentrates of the treated groups, 6 (CS6), 11 (CS11), and 16% (CS16) of Camelina seeds were included, respectively. The highest supplementation level decreased the milk fat content. Supplementing Camelina seeds improved milk quality from a human health perspective by decreasing the content of saturated fatty acids and increasing the proportions of α-linolenic (C18:3 n-3), C18:2 cis-9, trans-11 (CLA), and the ω6/ω3 ratio. Furthermore, in the CS-fed ewes, milk oxidative stability was fortified, as suggested by the activities of superoxide dismutase (SOD), catalase (CAT), and glutathione peroxidase (GSH-Px), by the antioxidant capacity, and the oxidative stress biomarkers. In blood plasma, only the lowest supplementation level did not have a negative impact on the oxidative status. To summarize, the incorporation of 6% Camelina seeds in ewes concentrates improves milk quality and oxidative status. However, more research is required regarding the possible negative effects of the constant consumption of Camelina seeds by ewes.

Association of milk composition traits as indicators of feed efficiency in Chios dairy ewes

S. Vouraki, V. Papanikolopoulou, N. Siachos, V. Fotiadou and G. Arsenos
School of Veterinary Medicine, Aristotle University, University Campus, 54124 Thessaloniki, Greece; svouraki@vet.auth.gr

The objective was to investigate the association of milk composition traits as indicators of feed efficiency in Chios dairy ewes. A total of 38 Chios dairy ewes were randomly selected from a purebred flock. They were at the third month of first to fifth lactation. The ewes were housed in two pens (Group A, n=18; Group B, n=20) and fed a pelleted concentrate feed together with Lucerne Hay and wheat straw (1.5, 1.5 and 0.3 kg/animal/day, respectively). Pelleted feeds in Groups A and B had different physical composition but were equal regarding energy (1 UFL/kg dry matter) and protein content (158.6 g/kg dry matter). Data collection was performed every two weeks during a 60-day period. The body condition score of each ewe was assessed by palpation of the dorsal lumbar region. Individual ewe milk yield was recorded electronically; energy corrected milk yield was calculated. Individual milk samples were also collected to assess chemical composition (fat, protein, lactose, solids-non-fat). Feed refusals from each group were weighted to calculate total and average individual feed intakes. The chemical composition of feeds was also assessed. Feed efficiency was defined as the energy corrected milk yield to energy intake ratio. The association of each milk composition trait with feed efficiency was tested with mixed linear models accounting for the fixed effects of group, sampling number, milk yield after weaning and body condition score. Statistically significant associations ($P<0.05$) were reported with fat and lactose content; one unit increase of fat and lactose content was associated with an increase in feed efficiency by 8.7±2.02% and 21.7±8.33%, respectively. Results suggest that Chios dairy ewes with higher milk fat and lactose content utilize dietary energy more efficiently. Therefore, these milk composition traits could be used as indicators for selection towards increased feed efficiency. Further research with individual feed intake records could provide further support to these findings. This work was undertaken as part of the SMARTER project that has received funding from the European Union's H2020 research and innovation programme (772787) and Legusmes4Protein project (T1EΔK-04448).

Session 31 Theatre 11

New bone detection method for lamb carcass DXA scanning improves total composition predictions

S.L. Connaughton, A. Williams and G.E. Gardner
Murdoch University, 90 South Street, Murdoch 6150, Australia; s.connaughton@murdoch.edu.au

Dual Energy X-ray Absorptiometry (DXA) has been used as an on-line apparatus in multiple Australian lamb abattoirs as a precise predictor of carcass composition, however some small inaccuracies have been shown in high bone-content breeds. This is likely due to the existing method of bone detection within DXA images, which allocates all pixels with an R-value more than the mean of the whole carcass image as bone, resulting in the smaller, higher bone % lambs having more pixels classified as soft tissue rather than bone. A new method was developed to more accurately identify bone containing pixels, while still providing computational efficiency sufficient to keep up with abattoir chain-speed, thus providing results in real time. The updated method uses a function of the logarithmic square of pixel R-values and a proxy for pixel thickness, which in this case was used as the natural log of the low-energy image. A set value was then determined for the threshold between soft tissue and bone containing pixels, with all values above this value identified as bone containing pixels. This new image analysis method was compared to the existing method within a population of 200 phenotypically diverse lambs that were slaughtered at a commercial abattoir where a DXA was installed. These carcases were also scanned using computed tomography (CT) as a gold-standard reference method for composition. The precision of DXA predicting the CT determined % of each tissue type (fat, lean and bone) was analysed with general linear models, with the bone % predictions increasing in precision from R^2=0.23, RMSE=1.55% to R^2=0.72, RMSE=0.93%. Similarly, predictions of fat % had improved precision from R^2=0.84, RMSE=1.71% to R^2=0.89, RMSE=1.38%, and lean % from R^2=0.73, RMSE=1.89% to R^2=0.75, RMSE=1.77%. This improved precision for predicting carcase composition would decrease the inaccuracy seen at the extremes of predicting lamb bone %, giving confidence that lamb carcasses would be graded and sorted correctly based on their objective carcass composition measurement.

Session 31 Theatre 12

Intramammary administration of lipopolysaccharides at parturition on goat colostrum and milk quality

M. González-Cabrera[1], N. Castro[1], M. Salomone-Caballero[1], A. Torres[2], S. Álvarez[2] and L.E. Hernández-Castellano[1]
[1]Institute of Animal Health and Food Safety. Universidad de Las Palmas de Gran Canaria, Trasmontaña s/n, 35413 Arucas, Spain, [2]Department of Animal Production, Grassland and Forages. Canary Agronomic Research Institute (ICIA), Finca el Pico, 38260 La Laguna, Spain; lorenzo.hernandez@ulpgc.es

Twenty Majorera dairy goats were randomly allocated in one of the two experimental groups (TRT vs CON). The TRT group (n=10) received an intramammary administration (IA) of saline (2 ml) containing 50 µg of lipopolysaccharides (LPS) from *E. coli* O55:B5 in each quarter. The CON group (n=10) received an IA of saline (2 ml) without LPS. Rectal temperature (RT) was measured at d0 before the IA and then on d0.125, d1 and d7 relative to IA. Colostrum/milk yield as well as milk composition (fat, protein, lactose and total solids, and somatic cell count (SCC)) were measured on d0.125, d1 and d7. The data was analysed using the MIXED procedure from SAS (9.4) and the model included the IA as fixed effect (TRT vs CON), time (T) and the interaction between both (IA×T). The SCC data was log-transformed (log10) to comply with the model assumptions (i.e. variance homogeneity and normality of residuals). Statistical significance was set as $P \leq 0.05$, and tendencies were set as $0.05 < P \leq 0.10$. Animals from the TRT group increased RT after the IA, while the CON decreased RT ($P_{IA \times T}$=0.005). Colostrum/milk yield as well as colostrum/milk composition was not affected by IA. Somatic cell count was higher in the TRT group than in the CON group (3.3±0.09 and 2.9±0.09 cells×10^6/ml in the TRT and CON group, respectively; P_{IA}=0.001) and declined from d1 to d7 (3.3±0.10 and 2.9±0.10 cells×10^6/ml on d1 and 7, respectively; P_T=0.003). In conclusion, the intramammary administration of LPS at parturition caused increased SCC in colostrum without affecting either yield or composition.

Session 31 Poster 13

Effects of white button mushroom (*Agaricus bisposus*) on growth and meat quality of growing lambs

G. Kis[1], M. Popovic[2], L. Pajurin[2], D. Spoljaric[2], L. Kozacinski[2], B. Spoljaric[2] and K. Kljak[1]
[1]University of Zagreb Faculty of Agriculture, Department of Animal Nutrition, Svetosimunska 25, 10000, Croatia, [2]Faculty of Veterinary Medicine University of Zagreb, Heinzelova 55, 10000, Croatia; kis@agr.hr

The effect of supplementation of button mushroom (*Agaricus bisporus*) to animal feed as a promoter of the health and productivity of domestic animals is known. Adding high-quality mushrooms to sheep rations has shown to lower cholesterol but the effect on the lamb meat qualitative properties is unknown. Therefore, the aim of this paper is to determine the effect of supplementation of white button mushroom in the diet on growth performance and meat quality characteristics of growing Pramenka lambs. Lambs (n=18, 16.6±0.6 kg body weight; BW) were randomly assigned to one of three dietary treatments (6 lambs/treatment). All lambs received *ad libitum* forages with supplementation of compound feed for lambs contained: 0% of mushroom (MO) for the control group; 1.5% of dried mushrooms (MD) and 1.5% of fresh mushrooms on dry matter basis (MF), respectively. In addition to production, parameters samples of m. semimembranaceus were taken for analysis of lamb meat quality. Dietary treatments had effects (P≤0.05) on average daily gain (ADG) with highest gains for lambs ate dried mushrooms, then fresh and lowest values for ration without mushrooms. In addition to production parameters, the results of the qualitative traits of lamb meat show the lowest levels of fat in the group of lambs M0 or highest in MF, compound feed with the addition of fresh mushrooms. Regardless of the amount of fat, the most significant difference (P≤0.001) of cholesterol was in lambs fed with dry mushrooms, which was many times lower than in all other lambs (8.6 vs 28 mg/100 g). The thiobarbituric test (TBARS) as an indicator of the oxidative stability of the meat was used. TBARS values were expected to increase from day 0 to day 6 (0.1 vs 1.4) and differ significantly, while expected dry mushroom meal had no significant differences between other two lamb groups. Meanwhile, it can be concluded that supplementation white bottom mushrooms in growing lambs ration enhance growth performance and meat quality, but further investigations in explanation of the obtained results will be required.

Session 31 Poster 14

Effect of wellness and diet on ovine milk quality

M. Tognocchi, G. Conte, L. Turini, S. Tinagli, A. Silvi, L. Casarosa, R.E. Amarie, A. Serra and M. Mele
University of Pisa, DISAAA-a, via del Borghetto 80, 56124 Pisa, Italy; moni.tognocchi@gmail.com

Nowadays the quality of milk is not only defined by the hygienic-sanitary and nutritional point of view but also by a good state of health and appropriate living conditions of animals. The aim of this study was to evaluate the ability of linear discriminant analysis (LDA) applied on the milk composition (fat, protein, lactose, casein, fatty acid profile and urea), milk clotting properties (r, k20 and A30) and mammary health (somatic cell count and bacterial level) to discriminate the farming system adopted in 3 commercial dairy farms, considering the level of animal wellness and dietary linseed supplementation as described follow: Farm A (low level of wellness and linseed supplementation), Farm B (mild level of wellness and linseed supplementation), Farm C (high level of wellness and no linseed supplementation). Linseed was only supplement was during winter, while in spring the sheep are grazed on all three farms. The sampling was repeated on the same animals in three different periods: winter, spring and summer. Two canonical variables were extracted explaining more than 61% of variance. The first canonical variable clearly discriminated Farm C from the other two ones, while the second canonical variable discriminate milk samples on the basis of seasons. On the basis of these results it was possible to detect a discrimination of company B from A and C in winter, while in summer it is company C that distinguishes itself from the other two. In winter, the integration with flax seeds enhances the quality of the milk, especially by increasing the nutraceutical components. This effect, however, is more evident if the animals are raised in conditions of high welfare. This data would justify the fact that in winter company A does not discriminate against C even if it enriches the diet with flax. On the contrary, in spring and summer, when the sheep are grazed without supplementation of flax, the qualitative characteristics of the company emerge with a high level of well-being, in particular as regards the rheological properties of the milk. In conclusion, our results show that any strategy adopted in the company manages to express its effects on the quality of production, if it is complemented by a high level of animal welfare.

Diet inclusion of ensiled olive cake increases unsaturated lipids in goat milk

O. Tzamaloukas[1], M.C. Neofytou[1], C. Constantinou[1], S. Symeou[1], D. Sparaggis[2] and D. Miltiadou[1]
[1]Cyprus University of Technology, Department of Agricultural Sciences, Biotechnology and Food Science, P.O Box 50329, 3106 Limassol, Cyprus, [2]Agricultural Research Institute, P.O. Box 22016, Nicosia, Cyprus; ouranios.tzamaloukas@cut.ac.cy

This study aimed to evaluate the effect of dietary inclusion of ensiled olive cake (OC) on milk yield, composition, and fatty acid (FA) profile of goats. Seventy-two dairy Damascus goats in mid-lactation, were assigned randomly to three iso-nitrogenous and iso-energetic diets for 42 days. The diets contained 0, 10, and 20% of ensiled OC on diet DM for the OC0, OC10, and OC20 treatment, respectively, as a replacement of forages, whilst concentrate participation in diets remained at 60% (DM) in all treatments. During the wk 5 and 6 of the trial, dry matter intake, milk yield, milk composition, and FA profile were recorded and analysed using a complete randomized design with repeated measurements. No significant differences were observed between treatment groups concerning milk yield, 4% fat-corrected milk, fat or protein yield (kg/d). In contrast, milk fat percentage was gradually increased with increasing OC inclusion rates in the diets, while milk protein percentages were elevated in both OC groups but significantly only in the milk of OC20 group. The content of FA between C4:0 to C16:0 was reduced, while the concentration of long-chain (>16 carbons; LCFA) and mono-unsaturated FA (MUFA) concentration was enhanced in the goat milk of OC groups. Among individual MUFA, increments of C18:1 cis-9, C18:1 trans-10, and C18:1 trans-11 were demonstrated in both OC groups compared to OC0 group. No significant effect was reported on total poly-unsaturated FA levels, while the concentration of CLA cis-9, trans-11 was increased by 11 and 21% ($P<0.001$) with OC10 and OC20 diets, accordingly, compared to OC0 group. Overall, OC silage can be used up to 20% (DM) in goats' diets, as a forage replacement, since this could increase the milk protein and fat percentage and enrich its content with beneficial for human health lipids without adversely affecting milk production traits.

Nutritional composition of less valued primal cuts from lamb genotypes

F.P. Esteves[1], C. Oliveira[2], M. Pimpão[2], J. Santos-Silva[2,3], O. Moreira[2,3] and J.M. Almeida[2,3]
[1]UEvora, Universidade de Évora, 7006-554 Évora, Portugal, [2]INIAV, Quinta da Fonte Boa, 2005-048 Vale de Santarém, Portugal, [3]CIISA-FMV, Universidade de Lisboa, Av. Universidade Técnica, 1300-477 Lisboa, Portugal; joaoalmeida@iniav.pt

The production and consumption of low-weight lamb depends on various socio-economic and cultural factors. Since lambs' less valued primal cuts may be more difficult to prepare, one way to promote them on the market could be through the assessment of their nutritional composition. The aim of this study was to evaluate the proximate and fatty acid (FA) composition of three less valued retail chains primal cuts shoulder, breast, and neck from three lambs' genotypes. Two Portuguese pure breeds, Merino Branco (MB) and Saloia (S), and a commercial Ile France × Merino Branco (IF×M) crossbreed received the same farm management and were slaughter at four months of age. Lambs' genetics significantly affected all proximate parameters (humidity, fat, protein, and ash). Meat from IF×M had the highest protein content (20.14%; $P<0.001$). Moreover, IF×M had the lowest fat content (5.43%; $P<0.001$), when compared to Saloia (6.74%) and Merino (8.65%). Primal cuts also influenced all proximal contents ($P<0.001$), with only a 0.52% protein variation between the breast (18.96%) and the shoulder (19.48%). This cut was the one with the lowest fat content (4.88%) compared to neck (7.38%; more 50% fat) and breast (8.55%). Fatty acid composition (% total FA) of intramuscular fat was highly influenced by lamb genetic ($P<0.001$), with significant differences in 22 out of the 23 FA evaluated. Primal cut effect was only significant in seven FA and there was no significant interaction between main effects. The FA obtained in highest amounts for all breeds were C18:1c9, C16:0, C18:0 and C14:0. Saturated FA (SFA) concentrations were significantly different across crossbreds ($P<0.001$), with Saloia having the lowest levels of SFA (45.4%) when compared to Merino (50.3%), and IF×M (53.4%). Regarding the content of monounsaturated FA (MUFA), Saloia presented the highest concentration (47.9%), followed by Merino (42.9%) and IF×M (37.9%). Our findings show that crossbreed is an important factor in the nutritional composition of less valued lamb primal cuts. Funding: Project: PDR2020-101-031690 Child Lamb.

American, Chinese, and Australian consumers prefer sheepmeat from carcases not selected for leanness

R.A. O'Reilly[1,2], G.E. Gardner[1,2], D.W. Pethick[1,2] and L. Pannier[1,2]
[1]College of Science, Health, Engineering and Education, Murdoch University, WA, 6150, Australia, [2]Australian Cooperative Centre for Sheep Industry Innovation, NSW 2351, Australia; r.oreilly@murdoch.edu.au

Consumer preferences for leaner meat products has provided economic incentives for Australian producers to breed leaner, more muscular sheep. Increased lean meat yield can be achieved through selection for Australian Sheep Breeding Values of reduced post weaning fat depth (PFAT) and increased post weaning eye muscle depth (PEMD). However, Australian studies have shown increased carcass muscling and leanness has a negative impact on eating quality. It is unknown whether this same trend will be discerned by Chinese or American consumers – currently Australia's key sheepmeat export markets *Longissimus lumborum* (LL) and *semimembranosus* (SM) muscles were collected from 321 Australian lambs and yearlings. Lambs were the progeny of 12 Maternal, 20 Merino, and 43 Terminal sires, while yearlings were the progeny of 54 Merino sires. American, Australian and Chinese consumers (n=720 per country) scored grilled LL and SM steaks for tenderness, juiciness, flavour, and overall liking (scale 0-100). The association between sire traits of PFAT and PEMD, and sensory scores was assessed using linear mixed effects models (SAS). Sire estimates for PFAT ranged from -2.3 to 1.7 mm, and PEMD from -2.2 to 4.9 mm. On average, consumers across all countries associated a 6.9 unit increase in PEMD with a 4.7 score reduction in juiciness in the LL (P<0.01). Similarly, the association of flavour scores with PFAT was consistent across countries, with a 2.6 unit decrease in flavour across the reducing 4mm PFAT range in both cuts (P<0.05). In contrast, American and Chinese consumers did not perceive a negative association of tenderness with PFAT, despite Australian tenderness scores reducing by 9.2 units across the 4 mm PFAT range in the LL (P<0.01). Generally, these results highlight the consistency of consumer responses across nations to variation in selection for leaner and more muscular carcases. These results reinforce the need for sheepmeat producers targeting international markets to balance selection for lean meat yield with traits that improve eating quality.

Effects of lamb genotype and primal cuts on burgers' texture and sensory properties

N. Franco[1], M. Pimpão[2], C. Oliveira[2], M.R. Marques[2] and J.M. Almeida[2,3]
[1]UEvora, Universidade de Évora, 7006-554 Évora, Portugal, [2]INIAV, Quinta da Fonte Boa, 7006-554 Vale de Santarém, Portugal, [3]CIISA-FMV, Universidade de Lisboa, Av. da Universidade Técnica, 1300-477 Lisboa, Portugal; joaoalmeida@iniav.pt

Lambs' low slaughter weight, and thus, light carcasses are characteristic of Portuguese market. Such carcasses present lower muscle volume and high bone, and some primal cuts require laborious home preparation. Accordingly, lamb meat is appreciated due to its low fat and connective tissue content. The aim of this study was to evaluate the effects of lamb genotype and embedding ratio of meat from lesser economic value primal cuts on textural and sensory properties, and how it determines the overall acceptability of lamb burgers. Two groups of lambs, pure Merino Branco (MB) and Ile France × Merino Branco (IF×M) crossbreeding, received the same management at the farm and were slaughter at four months of age. Lean meat from shoulder, breast and neck primal cuts were separately minced (⌀ 3.5 mm), and a standard recipe was established for burgers consisting of meat (88%), water (10%), flavourings, and salt (0.75%). Three formulations were made with different proportions of meat: 40:30:30, 50:25:25 and 60:20:20, corresponding to shoulder, breast, and neck, respectively. Texture profile analysis was carried out on grilled burgers, and sensory evaluation was made by an expert panel on grilled and roasted burgers. Lambs crossbreed influenced (P<0.001) hardness, chewiness, and gumminess of burgers, with higher values from IF×M lambs. The proportion of meat from the three primal cuts significantly influenced all the texture profile parameters. Burgers with equal proportion of meat from shoulder as well as breast + neck (50:25:25) originated the lowest values (P<0.001) of all texture parameters. Juiciness and tenderness were not influenced by any factor. Meat from Merino lambs promotes burgers with higher flavour intensity (5.01 vs 4.36; P<0.001). Grilled vs roasted method, didn't affect any of the sensory properties. Interaction between genetics and meat from primal cuts revealed that five out of the six formulations didn't differ in flavour or global acceptability and that Merino burgers had the highest, thus the best, sensory scores. Funding: Project: PDR2020-101-031690 Child Lamb.

Effects of lamb genotype, carcass weight and primal cut on baby soup sensory properties

M. Pimpão[1], C. Oliveira[1], M.R. Marques[1] and J.M. Almeida[1,2]
[1]INIAV, Quinta da Fonte Boa, 2005-048 Vale de Santarém, Portugal, [2]CIISA-FMV, Universidade de Lisboa, Av. da Universidade Técnica, 1300-477 Lisboa, Portugal; joaoalmeida@iniav.pt

Lamb meat is often recommended as one of the first sources of animal protein to be included in babies' diets. This red meat has a high nutritional quality which is closely associated with the healthy diet of lambs based on sheep's milk. The aim of this study was to assess, through the evaluation of its sensory properties, the effects of lamb genotype, average carcass weight and less noble primal cuts embedding ratio on infant formula of carrot/lamb meat soups. Two groups of pure Merino Branco (MB) and two of crossbreed Ile-France × Merino Branco (IF×M), with 15 to 20 ram lambs each, grazed natural pastures together with their dams until weaning (three months of age) and were supplemented with commercial concentrate and hay. Two groups of different genotypes were slaughter at four months of age and the others two months later, to provide light (MB 12.5 kg; IF×M 13.5 kg) and heavy carcasses (MB 15.0 kg; IF×M 17.0 kg), respectively. Primal cuts were vacuum packed individually. Shoulder, breast, and neck were deep-frozen and sent to laboratory to be carefully separated into bone, fat and muscle tissues. The amount of lean meat in each soup was 10% (W/W) which varied between 40%, 50% and 60% from shoulder (S) and from breast + neck (BN) in equal proportion (40S/60BN; 50S/50BN; 60S/40BN). Twenty-four soups were prepared with a Thermomix Vorwerk and a standard formulation of vegetables, and subsequently sensory evaluation was made by an expert panel. The meat proportion from the various primal cuts (S or BN) did not influence any of the sensory characteristics evaluated. The weight of the carcasses significantly affected ($P<0.001$) the overall and meat odour, with the heavier carcasses providing soups with higher odour intensities. Genotype influenced ($P<0.001$) the acceptance of the meat flavour and was a determining factor in overall acceptability, with a preference for meat from crossbreed lambs. A positive interaction was found between crossbreed and weight with an increase in acceptability of less odour and less intense flavour soups, provided by meat from light carcass and from IF×M lambs. Funding: Project: PDR2020-101-031690 Child Lamb.

Pig-based bioconversion: keeping nutrients in the food chain

L. Pinotti[1], A. Luciano[1], M. Ottoboni[1], M. Manoni[1], S. Mazzoleni[1], G. Ceravolo[1], M. Tretola[1] and M. Rulli[2]
[1]University of Milan, Department of Veterinary Medicine and Animal Science, Via dell'Università 6, Lodi, 26900, Italy, [2]Politecnico di Milano, Department of Civil and Environmental Engineering, DICA, Milan, 20133, Italy; luciano.pinotti@unimi.it

With the diminishing availability of farming land, climatic changes and the threat of declining water resources, the challenge in agriculture is to how to meet the growing demand for food, feed, fiber, fuel, and industrial products using fewer resources. This will involve the reduced use and redistribution of resources by applying the principles of the circular economy, in other words, 'Do more with less!'. These aspects, together with the increase in the cost of traditional feed for food producing animals over the last few years, have motivated feed researchers and producers to search for sustainable solutions for feeding animals. Insects, algae, ex-food (also termed former food products) as well as duckweed are regarded as interesting alternative protein/energy sources for feed and are expected to be increasingly used around the globe as replacers for conventional nutrient sources. The use of non-human-edible by/co-products, such as former food products (FFPs) or ex-food, by the feed industry is evidence of the potential of this sector to make the best of the nutritional and economic added value of by-co-products by applying circular economy principles. As omnivorous animals, pigs are ideally the most suited species to convert several kinds of alternative ingredients into high-quality animal protein, thus keeping nutrients in the food chain. However, it is essential to thoroughly examine the compositional/safety/dietary features in order to provide new and fundamental insights aimed at efficiently reusing these ex-foods as value-added products for animal nutrition. Firstly this involves a nutritional and functional evaluation of these materials, followed by examining the impact of ex food on pigs' performance, wellbeing and ultimately product quality. In this scenario the present work will address these aspects in pig nutrition.

Total versus human-edible protein efficiency of Belgian pigs and poultry

C. De Cuyper[1], E. Delezie[1], S. Millet[1], J. Van Ginderachter[2], W. Wytynck[3], M. Wulfrancke[4], L. Van Heupen[5], K. D'hooghe[5], S. De Campeneere[1] and J. De Boever[1]
[1]ILVO, Scheldeweg 68, 9090 Melle, Belgium, [2]Vanden Avenne-Ooigem nv, Zwaantjesstraat 12, 8710 Ooigem, Belgium, [3]Boerenbond, Diestsevest 40, 3000 Leuven, Belgium, [4]Algemeen Boerensyndicaat vzw (ABS), Industrieweg 53, 8800 Beveren-Roeselare, Belgium, [5]Belgian Feed Association vzw (BFA), Gasthuisstraat 29, 1000 Brussel, Belgium; carolien.decuyper@ilvo.vlaanderen.be

Livestock converts plant protein into meat, milk or eggs. While animal protein has high biological value, this conversion is often considered inefficient. However, a large part of the feed consumed by livestock consists of by-products (e.g. DDGS, beet pulp, etc.) which are not edible by humans. Previous studies therefore tried to reassess the competition for plant protein between animal feed and human nutrition by distinguishing two types of efficiency: total and human-edible protein efficiency. While the total protein efficiency (TPE) is the ratio of total animal protein produced on total feed protein consumed, the human-edible protein efficiency (HPE) only considers the human edible protein fraction in animal products and feeds. In this study, we calculated the TPE and HPE for Belgian pigs and poultry. For this, 100 sows or breeders were selected as a starting point and the total input and output of these animals, throughout their entire lifespan, was mapped. Common commercial feed formulations were used and edible protein contents in feedstuffs were assumed as published by Laisse *et al.* (2018). The TPE for fattening pigs, broilers and laying hens amounted to 0.43, 0.54 and 0.34, whereas the HPE amounted to 0.87, 0.61 and 0.86, respectively. This HPE indicates that the conversion of plant protein into animal products is more efficient than generally perceived, particularly for the production of pork and eggs. In addition, the HPE seemed to be very sensitive to the choice of feed ingredients and the assigned edible protein content. For example, when wheat is assigned an edible protein content of 3.3 instead of 66%, considering the quality of Belgian wheat is for 90% feed grade and not suited for human consumption, the calculated HPE for fattening pigs, broilers and laying hens increases to 1.36, 0.96 and 1.30, respectively.

Session 32 Theatre 3

Can the inclusion of different olive oil cakes on diet affect the performances of Bísaro pigs?

D. Outor-Monteiro[1,2], J. Teixeira[2], F. Madeira[3], A. Teixeira[4], J. Ribeiro[1,2], C. Guedes[1,2], M. Gomes[1,2] and V. Pinheiro[1,2]
[1]Animal and Veterinary Research Centre (CECAV), Quinta dos Prados, 5001-900 Vila Real, Portugal, [2]Universidade Trás-os-Montes e Alto Douro, Departamento Zootecnia, Quinta dos Prados, 5001-900 Vila Real, Portugal, [3]MORE – Laboratório Colaborativo Montanhas de Investigação– Associação, Edifício do Brigantia Ecopark, Avª Cidade de Leon 506, 5300-358 Bragança, Portugal, [4]Centro de Investigação de Montanha (CIMO), Instituto Politécnico de Bragança, Campus de Santa Apolónia, 5300-253 Bragança, Portugal; jessicapaie@utad.pt

The olive sector produces considerable amounts of by-products, being some of them toxic and dangerous to the environment. The Bísaro pig is a Portuguese local breed raised in the same region of olive-oil production in the northeast of the country. The present work aimed to evaluate the effect of the different olive cakes in the diet of Bísaro pigs, and to study its effect on the animal´s growth performances (Live weigh, daily weight gain, feed intake and feed conversion ratio). The experiment was carried out on 40 animals (initial LW 103+-22 kg), housed in 20 pens, during 60 days on average. Five treatments with different olive cakes (basal diet (T1); and basal diet with 10% crude olive cake (T2); 10% olive cake two phases (T3), 10% dehydrated olive cake (T4); and 10% dehydrated olive cake + 1% olive oil (T5). No significant differences was found between the treatments for the final live weight (141 kg), daily weight gain (642 g/d), feed intake (3.38 kg/d) and feed conversion ratio (5.27). Results obtained are encouraging and indicates that the inclusion in this amounts of olive cake does not affect the performances of Bísaro breed pigs.

Session 32 Theatre 4

Transcriptome study in subcutaneous adipose tissue of Iberian pigs fed with olive-by-products

J. García-Casco[1,2], R. Peiró-Pastor[1], P. Palma-Granados[1,2], A. López-García[1], C. Óvilo[1], E. González[3] and M. Muñoz[1]
[1]INIA-CSIC, Animal Breeding Department, Crta de la Coruña, km 7.5, 28040 Madrid, Spain, [2]INIA-CSIC, Centro de I+D en Cerdo Ibérico, Crta. EX101 km 4.7, 06300 Zafra, Spain, [3]UEX, Instituto Universitario de Recursos Agrarios, Av. de la Investigación, 06006 Badajoz, Spain; garcia.juan@inia.es

The use of by-products from the agricultural industry for animal feed is an alternative that meets the requirements of the bioeconomy and sustainability. Previous studies have showed the usefulness of by-products from the olive industry in feeding during the growth period of Iberian pigs (from 40 to 100 kg), when the feed intake is restricted to obtain an adequate size and weight before the final period of free-range fattening based on acorn and pastures (*montanera*). These works have been completed with the study of the transcriptomic expression in backfat tissue samples obtained by biopsies at 100 kg. The whole transcriptome was sequenced by RNAseq of 18 pigs fed with three different diets (6 pigs × diet) with different energy and crude protein contents and different feed supply: (1) control diet (C); (2) compound feed based on dry olive pulp (DOP) supplied under restriction; and (3) a wet crude olive cake (WCOC) supplied *ad libitum* in silage form and supplemented with a small daily ration of regular feed. The results of the differential expression analyses revealed 776 differentially expressed genes (DEGs) between diets C and DOP and 1,064 between diets C and WCOC. Ingenuity Pathway Analysis (IPA) software revealed an enrichment of C vs DOP DEGs in pathways related to regulating growth and reprogramming metabolism (AMPk-Signalling), glycolysis (Glycolysis I) and metabolism of fatty acid in liver (PPARα/RXRα-Activation) among others. On the other hand, the main pathways enriched in the C vs WCOC comparison were related to energy processes in the mitochondria (Oxidative Phosphorylation), regulation of oxygen homeostasis (HIF1α Signalling) and cellular damage by reactive oxygen and nitrogen species in mitochondria (Mitochondrial Dysfunction) These results show that by-products based diets have an impact in pathways involved in growth, lipid metabolism and oxidation that could explain the phenotypic differences observed in previous studies.

Session 32 Theatre 5

Effect of three varieties of field beans on growth and carcass quality of grow-finisher pigs

F.M. Viard[1,2], J.V. O Doherty[1], G.E. Gardiner[3] and P.G. Lawlor[2]
[1]University College Dublin, School of Agriculture and Food Science, Belfield, Dublin, D04 C7X2, Ireland, [2]Teagasc, Pig Development Dept., Animal and Grassland Research and Innovation Centre, Moorepark, Fermoy, Cork, P61 R966, Ireland, [3]Waterford Institute of Technology, Dept. of Science, Cork Road, Waterford, X91 Y074, Ireland; florence.viard@teagasc.ie

There is growing interest in field (faba) beans as a replacement for soybean in pig diets, particularly from a carbon footprint perspective. This study aimed to determine if pig growth and carcass quality were affected by the variety of field bean fed and to compare these parameters with those of pigs fed a cereal/soybean-based diet. The varieties tested were a commonly grown variety (LYNX; Seedtech, Ferrybank, Waterford, Ireland), a low vicine and convicine variety (VICTUS; Seedtech), both grown in Ireland, and a low tannin bean (TAIFUN; Seedtech), grown in Germany. Two hundred and sixteen grow-finisher pigs (27.0 kg; ±1.03 s.e.), grouped in pens of six pigs each (n=36 pens), were blocked by weight and sex and liquid fed the treatment diets for 84 days until slaughter. The dietary treatments were: 1. CONTROL cereal/soybean diet; 2. 40% LYNX/cereal diet; 3. 40% VICTUS/cereal diet and; 4. 40% TAIFUN/cereal diet. The pen group was the experimental unit. Pigs were weighed at the beginning and the end of the experiment and feed disappearance was recorded. Carcass data were collected from individual pigs following slaughter. Data were analysed using the mixed model procedure in SAS. Final pig body weight prior to slaughter (118.1 kg; ±1.58 s.e.) was not affected by treatment. Carcass weight was 90.8^a, 88.1^b, 87.6^b, 88.0^b (s.e.=0.96 kg; P<0.001) for CONTROL, LYNX, VICTUS and TAIFUN diets, respectively. Carcass ADG was 871, 855, 836, 861 (s.e.=12.1 g/day; P=0.11) and carcass feed conversion efficiency (FCE) was 3.01^a, 3.15^b, 3.14^b, 3.01^a (s.e.=0.028 g/g; P<0.001) for CONTROL, LYNX, VICTUS and TAIFUN diets, respectively. Kill out yield was 75.3^a, 75.2^{ab}, 74.8^{bc}, 74.5^c (s.e.=0.19%; P<0.001) and lean meat yield was 59.0^a, 59.6^{bc}, 59.3^{ab}, 59.8^c (se=0.14%; P<0.001) for CONTROL, LYNX, VICTUS and TAIFUN diets, respectively. Feeding TAIFUN was advantageous compared to feeding LYNX and VICTUS because it improved carcass FCE and increased lean meat yield.

Session 32 Theatre 6

Poster presentation

G. Bee[1] and S. Millet[2]
[1]Agroscope Posieux, Rte de la Tioleyre 4, 1725 Posieux, Switzerland, [2]ILVO, Scheldeweg 68, 9090 Melle, Belgium; giuseppe.bee@agroscope.admin.ch

In this time slot, we take time to present the posters submitted in this session.

Session 32 Theatre 7

Exposure of free-range pigs to environmental contaminants via soil ingestion

C. Collas[1], J.L. Gourdine[2], D. Beramice[3], P.M. Badot[4], C. Feidt[1] and S. Jurjanz[1]
[1]URAFPA, Université de Lorraine-INRAE, Nancy, 54000, France, [2]URASSET, INRAE, Petit-Bourg, 97170, France, [3]UEPTEA, INRAE, Petit-Bourg, 97170, France, [4]LCE, Université Bourgogne Franche-Comté-CNRS, Besançon, 25000, France; claire.collas@univ-lorraine.fr

Ingested soil contributes to the exposure of free-range animals to environmental pollutants. Soil ingestion by pigs is few described whereas their burrowing behaviour suggests that it could be high. Although highly productive pigs are generally reared indoor, free-range farming is increasing in view of ethical considerations for animal welfare and organic farming criteria and is a common practice for subsistence agriculture systems such as in the Caribbean. The experiment lasted 8 weeks (2 weeks of habituation to the conditions, 6 weeks of measurements) with 40 growing pigs of Guadeloupean Creole (CR) or Large White (LW) breeds. Pigs were assigned to 5 treatments with 3 outdoors: low pasture LP (35 days of regrowth), high pasture HP (>60 days of regrowth), sweet potato SP (sweet potato field) and 2 indoors: indoor grass IG (same diet as HP pigs), indoor potato IP (same diet as SP pigs, except for potato leaves and other vegetation on the plot). The animals received a daily Cr_2O_3-enriched protein supplement to use Cr as a faecal output marker, and maize to balance nutritional supplies between the different treatments. 4 CR and 4 LW pigs were assigned to each treatment for the duration of the experiment. Animals were moved to a new paddock each Monday. Grass height and body weight were recorded. Ti (soil marker) and Cr contents of faeces, vegetation and soil samples were used to estimate individual daily soil ingestions. The average, 10th and 90th percentiles were 440, 200 and 726 g/100 kg BW respectively without significant differences between the 3 outdoor treatments or the 2 breeds but with a significant period × treatment interaction ($P<0.001$). In the French West Indies, animals ingesting soil may be exposed to chlordecone (CLD), a very persistent organochlorine insecticide. Simulations of CLD ingestion and tissue contamination were carried out and compared to the maximum residue limit. These results show that grazing management is an efficient way to limit soil ingestion by pigs and maintain the sustainability of pig systems in contaminated areas.

Effects of restricted feeding and sex on the feed to muscle gain ratio of growing pigs

G. Daumas, M. Monziols and N. Quiniou
IFIP-Institut du Porc, BP 35104, 35650 Le Rheu, France; gerard.daumas@ifip.asso.fr

The aim of the study was to determine tissular growth performance by using computed tomography and to test the effects of feeding level and sex. Two feeding levels, *ad libitum* vs restricted (85% of *ad libitum*), and two sexes, female vs castrated male, were compared on a sample of 134 pigs. Two batches, one in winter and the other in summer, were used with four pens of six gilts or castrates per treatment each. The pigs were crossbred between Pietrain boars and Large White × Landrace sows and were heterozygous for the halothane gene. Half-carcases were scanned with computed tomography to measure the muscle volume, calculate the lean meat content (LMC), and to assess the average daily muscle gain (ADGm) and the feed to muscle gain ratio (F:Gm). Least squares means of growth and feed efficiency criteria were assessed by a general linear model by pen, while least squares means of LMC were assessed by pig. The overall means of initial and final body weight were 26 and 120 kg respectively. The effect of feeding level was significant on the final body weight (BW), the daily feed intake (DFI), the average daily gain (ADG), ADGm, the feed to gain ratio (F:G), and LMC. Feed restriction decreased BW by 6.7 kg (116.8 vs 123.5; P<0.001), DFI by 0.40 kg/d (2.00 vs 2.40; P<0.001), ADG by 189 g/d (757 vs 946; P<0.001) and ADGm by 87 g/d (465 vs 552; P<0.001). It increased LMC by 2.3 percent points (61.7 vs 59.4; P<0.01) and F:G by 0.11 kg/kg (2.70 vs 2.59; P<0.01), but not F:Gm (4.39 vs 4.45). The effect of sex was significant on DFI, ADG, F:G, F:Gm, and LMC. Compared to females, castrates had a higher DFI (2.28 vs 2.12; P<0.01) and ADG (872 vs 831; P<0.05), but not ADGm (508 vs 509), a higher F:Gm (4.57 vs 4.27; P<0.01), but not F:G (2.67 vs 2.62), and a lower LMC (59.0 vs 62.2; P<0.001). This study showed that a severe feed restriction, meeting the amino acid requirements, improved carcase leanness without negative effect on the feed to muscle gain ratio. The use of computed tomography is to be recommended in feeding trials and more generally in animal production trials to determine tissue deposition, and associated feeding efficiency and final tissular body composition.

Fatty acid composition of immunocastrated vs female pigs in Germany

K. Götz and D. Mörlein
University of Göttingen, Department of Animal Sciences, Kellnerweg 6, 37077 Göttingen, Germany; katja.goetz@uni-goettingen.de

The ban on piglet castration without anaesthesia has been in effect in Germany since 2021. An alternative to this is the so called immunocastration of male pigs using the vaccine Improvac®. There is, however, widespread resentment with regard to the processing properties of meat and fat from immunocastrated males (IC). The fat tissue of fattening pigs should, for further processing, have a firm consistency, good oxidation stability and low water content in order to ensure optimal product quality and storage stability. The decisive factor for consistency and oxidation stability is essentially the fatty acid composition, i.e. the amount of saturated (SFA), mono-unsaturated (MUFA) and polyunsaturated (PUFA) fatty acids which is known also to be largely affected by diet. Therefore, this study assessed the variability of the fatty acid composition of in commercial fattening pigs as affected by sex and farm. We hypothesize, that farm effects outweigh differences (if any) between IC and female pigs. Neck fat samples were taken from a total of 840 pigs (414 gilts, 426 IC) all slaughtered at the same commercial slaughterhouse. The animals (average carcass weight 95.8 kg) came from a total of 40 farms and were sampled on 7 slaughter days over a period of 5 months. The neck fat samples were vacuum-packed and deep-frozen at -22 degrees. The fatty acid composition was analysed using gas chromatography (GC). Contents of MUFA, PUFA and SFA (given as % of detected fatty acid methyl esters) were subjected to statistical analysis. Therefore, a linear model considering fixed effects of sex, farm, sex × farm, and slaughter weight as covariate was applied. With regard to the proportion of SFA and MUFA, farm and sex have a significant influence. The proportion of MUFA was significantly lower (P<0.05) in IC (45.3%) than in female pigs (46.0%), while the IC showed 1.1% more SFA than the female (36.0%). The amount of PUFA was not effected by sex but varied remarkably between farms (11.9% till 23.5%). In conclusion, in terms of fat quality immunocastrated male pigs are considered on par with female pigs; the observed farm differences point to potential strategies to improving fat quality, e.g. by controlling feed.

Effect of hot or cold cutting conditions on pork belly shape and firmness

C. Zomeño, A. Brun, M. Albano, M. Gispert, B. Marcos and M. Font-I-Furnols
IRTA, Food Quality and Technology Program, Finca Camps i Armet s/n, 17121 Monells, Spain; cristina.zomeno@irta.cat

The belly is one of the most valued parts of the pig carcass. Firmness is an important quality trait in this cut because it influences the processing aptitude and consumer acceptability. It can be affected by intrinsic factors and extrinsic ones such as post-mortem (pm) conditions. The aim of this work was to evaluate the effect of the cutting conditions (hot vs cold) on belly morphological and mechanical traits. A total of 14 crossbred pigs were slaughtered at the IRTA abattoir using standard procedures. All carcasses were split and the bellies from the right half carcasses were immediately cut (hot cutting). Then, right bellies and left half carcasses were kept at -2 °C for the first 5.5 h pm and at 3 °C until the end of the trial. Bellies from the left half carcasses were cut at 24 h pm (cold cutting). Belly weight, length and width were recorded at 24 and 48 h pm. Belly firmness was determined by the flop test, fat separation and subjective pressure. Flop distance and angle were measured at 24 and 48 h pm using the bar-suspension method. Fat separation was determined in the centre of the cranial and ventral sides at 24 and 48 h pm by stretching the skin with tweezers, measuring the thickness before and after, and calculating the difference. Last, after removing the skin, two trained technicians applied pressure with a finger in the centre of the cranial and ventral sides and scored firmness using a 5-point scale from 1 (very firm) to 5 (very soft) at 48 h pm. Analysis of variance was performed with SAS Software (version 9.4) including the cutting condition as fixed effect. Belly weight did not change with the cutting condition (4,462±617 g), but the bellies cut in hot were shorter and wider than those cut in cold conditions at 24 h and 48 h. Flop distance and angle were higher in bellies cut in hot than in those cut in cold conditions at 24 h (20.8 vs 11.2 cm; 98.8 vs 34.5°; $P<0.001$) and 48 h (19.8 vs 12.2 cm; 91.1 vs 39.7°; $P<0.001$), indicating that the first were firmer. No significant differences were found for fat separation or the subjective firmness. In conclusion, cutting conditions modified belly shape and global firmness whereas those measures more related to fat firmness were not significantly affected.

National audit of pork pH in Australia

F.R. Dunshea[1,2], A. Lealiifano[3], M. Trezona[4], V. Gole[5] and R.J.E. Hewitt[6]
[1]The University of Leeds, Leeds, LS2 9JT, United Kingdom, [2]University of Melbourne, Parkville, 3030, Australia, [3]Rivalea, Corowa, NSW 2646, Australia, [4]Derby Industries, Wundowie, WA 6560, Australia, [5]Australian Pork Limited, Kingston, ACT 2604, Australia, [6]Sunpork, Eagle Farm, QLD 4009, Australia; fdunshea@unimelb.edu.au

Pork pH affects yield and quality and the ideal ultimate pH range is between 5.50 and 5.70. A total of 14,932 pH measures from 8 supply chains were analysed to determine off-farm factors that influence loin pH. The maximal model included supply chain, time post-slaughter, temperature, carcass weight, backfat, sex, chiller, month, transport time, lairage time and ambient temperature. The mean and median pH were 5.70 and 5.69, respectively with lower and upper quartile of 5.60 and 5.78. Thus, 50% of values were within 0.20 pH units with a relatively normal distribution. The final model accounted for 16.4% of the variation in carcass pH. The major off-farm factor contributing to variation in pork pH was supply chain followed by month, chiller, time after slaughter, carcass temperature and lairage time. Rapid chilling increased pH by 0.21 pH units and appears to be the most effective means of increasing carcass pH. Loin pH was lower in Autumn and Winter, suggesting that colder temperatures may result in low pH. These data indicate that low pH isn't a major problem for the Australian pork industry since only 14% of carcasses have a pH of 5.55 or below.

Studying intestinal metabolite profiles of piglets in different housing conditions

S.K. Kar[1], J. Engel[2], C.H. De Vos[3] and D. Schokker[1]
[1]Wageningen Livestock Research, De Elst 1, 6708 WD Wageningen, the Netherlands, [2]Biometris, Droevendaalsesteeg 1, 6708 PB Wageningen, the Netherlands, [3]Bioscience, Applied Metabolic Systems, Droevendaalsesteeg 1, 6708 PB Wageningen, the Netherlands; dirkjan.schokker@wur.nl

Pig housing and management conditions have an effect on disease susceptibility in early life. Novel strategies are being developed to increase animal health and welfare, and thereby reducing both therapeutic and prophylactic antibiotic use in swine production. In a previous study, housing and management conditions had a positively influence on pig welfare and their immune status, leading to a lower susceptibility to infections. However, the effects on the metabolome of pigs in the gut is poorly understood. In this study, we have compared the metabolome of two housing conditions, i.e. conventional housing (CH) and enriched housing (EH) in different intestinal segments, i.e. jejunum, ileum, colon, and faeces. We acquired samples of eight piglets per housing condition per intestinal segment. Subsequently, these samples were analysed by untargeted metabolomics using Liquid chromatography–mass spectrometry (LCMS). Subsequently, we have statistically tested whether metabolites were significantly associated to housing condition, sample origin, or a combined effect (interaction). These observed metabolite associations should be considered as exploratory, given the limited sample size in this pilot experiment. Effects of housing condition (CH vs EH) on the metabolome were tentatively shown, when focusing on interaction contrasts, i.e. a comparison of the differences in housing condition for different intestinal segments. Within each intestinal segment an effect of housing condition could only be shown for 1 to 4 putative metabolites. We observed large differences in the metabolome at the different intestinal segments. Moreover, a presence / absence analysis suggested that some metabolites are uniquely present at specific intestinal segments. When focusing on metabolites that were present at multiple locations clear significant differences in metabolite abundance were observed. In particular the jejunum and ileum metabolite profiles differed significantly from colon and faecal profiles.

Impact of calcium content and calcium to phosphorous ratio in diets for weaned pigs

P. Bikker[1], J. Fledderus[2] and M. Van Helvoort[3]
[1]Wageningen University & Research, Wageningen Livestock Research, P.O. Box 338, 6700 AH Wageningen, the Netherlands, [2]For Farmers N.V., Kwinkweerd 12, 7241 CW Lochem, the Netherlands, [3]De Heus Animal Nutrition, Rubensstraat 175, 6717 VE Ede, the Netherlands; paul.bikker@wur.nl

Recommendations for calcium (Ca) and phosphorous (P) content in pig diets are generally based on a factorial approach, including endogenous losses and requirements for retention in body tissues. In practice Ca content in weaning diets is often below recommendations to minimise potential adverse effects of Ca. This study was conducted to evaluate the impact of reducing dietary Ca content and Ca:P ratio compared to CVB (2017) recommendations. Two experiments with 0 or 2,000 FTU microbial phytase per kg (assumed equivalent to 1.5 g Ca and 1.4 g digestible P) were conducted to determine the influence of Ca content and Ca:P on performance, apparent total tract digestibility (ATTD), and urinary Ca and P excretion. Each experiment comprised 4 treatments in a 2×2 factorial arrangement with reduced Ca content and reduced Ca:P ratio relative to CVB as respective factors and 16 pens per treatment in a 4 to 5 week period after weaning. Spot samples of faeces and urine were collected during 3 days in week 4. Results were analysed using ANOVA with dietary phytase, Ca, Ca:P and their interactions as fixed effects and pen as random effect. Overall, ATTD of Ca and P were 18 and 33%-units higher in phytase-supplemented diets. In phytase-free diets, the reduction in Ca content enhanced feed intake ($P=0.027$) and daily gain ($P=0.007$), both with and without simultaneous reduction in Ca:P ratio. A Ca:P reduction in phytase-supplemented diets did not improve daily gain and impaired feed efficiency ($P=0.011$). ATTD of Ca was enhanced by reducing Ca content ($P<0.001$), whereas ATTD of P was primarily enhanced by the reduction in dietary Ca:P ratio ($P<0.001$). Reduction of both Ca content and Ca:P (practical diets versus CVB) enhanced the ATTD of Ca and P by 9 and 5%-units, respectively. Results indicate that dietary Ca content can be reduced below CVB recommendations without loss in performance. In phytase-free diets, this may improve growth performance. However, the reduction in Ca content and Ca:P enhanced urinary P content ($P<0.001$), indicating a lower P utilisation and bone mineralisation.

The effect of adding guar meal to the ration of pigs on the quality and chemical composition of pork

K. Karpiesiuk[1], W. Kozera[1], T. Daszkiewicz[1], K. Lipiński[1], J. Kaliniewicz[1], A. Okorski[1], A. Pszczółkowska[1] and G. Żak[2]
[1]University of Warmia and Mazury in Olsztyn, ul. Oczapowskiego 5, 10-719 Olsztyn, Poland, [2]National Research Institute of Animal Production, ul. Sarego 2, 31-047 Kraków, Poland; grzegorz.zak@iz.edu.pl

The present study investigated the effect of replacing SBM protein with guar meal protein in pig diets on carcass characteristics, meat quality and the fatty acid profile of the longissimus lumborum (LL) muscle. It was hypothesized that guar meal protein would not compromise meat quality. The experimental material comprised hybrid pigs produced by commercial crossbreeding [♀ (Polish Landrace × ♂ Polish Large White) × ♂ (♀ Pietrain × ♂ Duroc). The animals were divided into four groups (16 animals per group), based on their initial body weight, age and gender. Research efforts have focused on replacing expensive imported genetically modified soybean meal (SBM) with other feed components with similar nutritional characteristics, which could improve meat quality. Control group animals were fed diets containing SBM as the main protein source. In diets for experimental groups 2, 3 and 4 SBM protein was replaced with guar meal protein in 25%, 50% and 75%, respectively. An analysis of linear correlations revealed a strong negative correlation between the concentrations of monounsaturated fatty acids (MUFA) and saturated fatty acids (SFA) in the LL muscle in pigs fed diets containing 25% of guar meal, which is nutritionally desirable. The inclusion of 25% guar meal protein in pig diets had a beneficial effect on carcass weight, and pork carcasses in this dietary treatment had lower lean content and higher subcutaneous fat content. The meat of pigs fed diets containing 25% guar meal protein had the highest protein content. Pork quality was similar in all experimental groups and remained within the limits for normal meat, which indicates that guar meal had no negative effect on the qualitative attributes of pork. Dietary supplementation with guar meal protein affected the fatty acid profile of meat, but the present findings are inconclusive and require further investigation. The strongest negative correlation (R=-0.96) was found between the concentrations of MUFA and SFA in group 2 (25% guar meal protein) where the MUFA/SFA ratio was most beneficial to consumer health.

Prediction of corn metabolizable energy for pigs using near infrared reflectance spectroscopy

A.C. Figueiredo[1], N.T.E. Oliveira[2], R.V. Nunes[2], A.E. Murakami[1], S.M. Einsfeld[1], P.C. Oliveira[1], J.S. Martins[1] and P.C. Pozza[1]
[1]State University of Maringá, Animal Science, Av Colombo, 5790, 87020-900, Brazil, [2]Western Paraná State University, Rua Pernambuco, 1777, 85960-000, Brazil; pcpozza@uem.br

The objective of this study was to evaluate prediction equations (PEs) of metabolizable energy (ME) of corn for pigs using the chemical composition obtained from different NIRS. A metabolism assay was carried out with forty-four barrows, averaging 72.61 kg of initial weight, distributed in a randomized blocks design with 10 treatments and four replicates. The treatments consisted of 10 corn cultivars, replacing 25% the basal diet. The total collection method was used to determine the ME values of the 10 corn cultivars. At the same time, twenty PEs were used to estimate the ME of the same 10 corn cultivars, using the chemical composition of each cultivar obtained from four different NIRS(X). Each NIRS were calibrated with different curves, from one-another, to estimate the chemical composition of corn. Only organic matter, ash, crude protein, ether extract (EE) and crude fibre (CF) were considered, since the 20 PEs evaluated also presented these compounds as common regressors in the equations. The estimated ME (EME), using the PE, were evaluated by determining the standard deviation (SD), coefficient of variation (CV) and also by the approach based on the mean squared deviation (MSD), related with the observed values of ME (OME) obtained in the metabolism assay, for each corn cultivar. The OME ranged from 3,332 to 3,541 kcal/kg, as fed basis. The same PE (ME = 3,982.99 – 79.97 ash) showed the lowest values of SD and CV (EME × OME) for NIRS1, 2 and 4, but NIRS3 presented another PE (ME = 3,675.39 + 50.02EE – 25.62CF) with the lowest SD and CV (2.89 and 0.08%). Based on the MSD, different PEs performed better than the aforementioned ones. Based on the MSD it is concluded that the best PE for only one NIRS was ME = 16,482 – 35.7CF, as presented the lowest MSD, and the PEs that provided the better EME for all evaluated NIRS were ME = 16.81 – 0.031CF and ME = 3,675.39 + 50.02EE – 25.62CF.

Session 32 Poster 16

Transforming bakery residuals to high energy diets in pig production: a pilot study

E.N. Sossidou[1], M.Z. Kritsa[1], A. Dedousi[1], D. Kipourou[1] and G.F. Banias[2]
[1]Ellinikos Georgikos Organismos-DIMITRA, Veterinary Research Institute, ELGO Campus, 57001 Thessaloniki, Greece, [2]Center for Research and Technology Hellas (CERTH), Institute for Bio-economy and Agri-technology (iBO), Thessaloniki, Greece; sossidou@vri.gr

In a circular economy where food waste management is developed sustainably, bakery residuals may have great potentials to be recovered and reused. The aim of this pilot study was to investigate the effect of bakery meal inclusion in piglets' diets on animal health and welfare. Sixty (60) weaned piglets, 28 days old, with an average weight of 8.40 kg, were randomly divided into three (3) groups (treatments), 10 piglets/group, 2 replications/treatment. In treatment CON 0% no bakery meal was added to diet. In treatments CON 15% and CON 20% bakery meal was added to diet at a level of 15% and 20%, respectively. The experiment period was 2 months. Piglets were weighted at the end of each week at a group level and feed intake was calculated on a weekly basis. Health indicators have been also measured on a weekly basis. Preliminary results showed differences in feed intake (FI), weight gain (WG) and feed conversion rate (FCR) among treatments. More specifically, feed intake was lower in CON 20% (8.49 kg) than in CON 0% (8.64 kg) and CON 15% (8.74 kg). At the end of the trial, animals' weight was higher in CON 15% (45.8 kg) followed by CON 20% (44.5 kg) and CON 0% (43.72 kg), respectively. A significant increase of weight gain was also observed in the last week in dietary treatments supplemented with bakery meal in comparison to the control group. The FCR of the whole experiment period was lower in CON 0%, however the FCR in the last week was lower in CON 20%. No evidence of diarrhoea was observed while mortality was 5% for CON 0%. The pilot study gave first evidence that bakery meal can be included in pig formulas without any restraints. Acknowledgment: This research has been co-financed by the European Regional Development Fund of the European Union and Greek national funds through the Operational Program Competitiveness, Entrepreneurship and Innovation, under the call RESEARCH – CREATE – INNOVATE (project code: T2EDK-04537).

Session 32 Poster 17

Analysis of monitoring data on per- and polyfluoroalkyl substances (PFAS) in German wild boars

R.H. Mateus-Vargas, D. Maaz, A. Mader, R. Pieper, J. Steinhoff-Wagner and J. Kowalczyk
German Federal Institute for Risk Assessment (BfR), Max-Dohrn-Str. 8-10, 10589 Berlin, Germany; rafael.mateus-vargas@bfr.bund.de

Per- and polyfluoroalkyl substances (PFAS) are a group of anthropogenic compounds, which are ubiquitous distributed in the environment due to its use in a wide variety of industrial processes and consumer products. For monitoring purposes, wild boars (*Sus scrofa*) were reported to be suitable bio-indicators for PFAS. Within food of animal origin, wild boar livers may contain particularly high levels of PFAS. In the context of consumer health risk assessment, the knowledge about influencing factors is important for the risk analysis process. The aim of the study was to examine to what extent PFAS levels in wild boars are influenced by land-use systems like settlement, agricultural, forest and water areas. Data on concentration in wild boar livers of two German federal states (n=184), being provided by the national monitoring program, were used. Information on sample origin were broken down to district level (n=70). Districts were characterized using publicly available data of the respective State Statistics Offices. Overall, data concerning the sample (e.g. animal´s age or sex, or municipality of origin) differed between and within districts. Despite this fact, results showed significant differences in the average concentrations of PFAS and/or the predominant PFAS-profiles between the districts as well as the federal states. Statistical analyses revealed that the district of origin significantly influence the PFAS concentrations in wild boar livers. Furthermore, significant correlations were observed to particular parameters of land-use features of the districts (forest or settlement). In conclusion, land-use based evaluations of food monitoring data may be useful to characterize the PFAS levels/profiles in the food item wild boar livers. However, data used did not enable a certain confirmation of influences determined by land-use types, due to challenges involved in data preparation. Thus, we recommend the harmonization of sampling and analytical methods in the laboratories as well as gathering, refinement and complementation of the data sets reference laboratories to increase comparability of food monitoring data in future evaluations.

Session 33 Theatre 1

Development of genomic tools for horses and their potential impact on the equine sector

K.F. Stock
IT Solutions for Animal Production (vit), Heinrich-Schroeder-Weg 1, 27283 Verden, Germany; friederike.katharina.stock@vit.de

Genomic selection and other services relying on genomic data are of major interest for horse breeding. wider availability of genomic tools implying enormous potential for improved decision making in breeding and management collaborative approaches required to make best use of modern technology international collaboration and partnerships expected to meet the challenges and fulfil the expectations of breeders, owners, riders.

Session 33 Theatre 2

Genetic trends for performance and functionality in specialized breeding programs of riding horses

K.F. Stock[1], A. Hahn[2], I. Workel[2] and W. Schulze-Schleppinghoff[2]
[1]IT Solutions for Animal Production (vit), Heinrich-Schroeder-Weg 1, 27283 Verden, Germany, [2]Oldenburger Pferdezuchtverband e.V., Grafenhorststrasse 5, 49377 Vechta, Germany; friederike.katharina.stock@vit.de

Specialization of studbooks on either dressage or show-jumping is a way to increase breeding progress in sport horses. Distinct breeding goals and programs facilitate decision making and allow identifying developments which may relate to strong focus on particular aspects of performance. Knowledge about correlated selection responses can importantly contribute to responsible und sustainable breeding management if a wider range of performance related and functional conformation traits are considered. The aim of this study was to determine patterns of genetic trends in specialized breeding programs for dressage (D) and show jumping (J) referring to the active mare populations of the Oldenburg studbooks (>7,700 mares). Results from routine genetic evaluations 2021 for sport traits reflecting competition performance in D and J and for 46 linear conformation and performance traits were used. Genetic trends were determined considering mares with own performance and/or at least two progeny with phenotypic data for respective traits, resulting in sample sizes of up to 2,363 mares in the analyses. Across analysed subsets, 88-95% of the mares were born in 2000-2016. Reference was for all traits made to relative breeding values (RBV) standardized to a mean of 100 and a genetic standard deviation of 12. Substantial genetic improvement of 12-15 points for the target sport traits over the study were found within the respective specialized mare populations. These developments were paralleled with discipline-specific significant changes of RBV for most of the linear gait and jumping traits ($P<0.001$). Largest increases of >10 points were determined for freedom of shoulders and impulsion in trot (D) and for take-off power and jumping ability (J). Genetic trends for functional conformation traits indicated significant development towards larger frame, longer legs, shorter back and lower set tail in D and towards larger frame and lighter calibre in J. Analyses did not reveal distinct genetic trends for traits relating to correctness of limbs or functional aspects which would indicate the need for particular attention in the breeding program.

Specialized breeding in Swedish Warmblood horses

S. Bonow[1], S. Eriksson[1], E. Thorén Hellsten[2] and Å. Gelinder Viklund[1]
[1]Swedish University of Agricultural Science, Department of Animal Breeding and Genetics, P.O. Box 7023, 750 07, Uppsala, Sweden, Sweden, [2]Swedish Warmblood Association, P.O. Box 2, 247 93 Flyinge, Sweden, Sweden; sandra.bonow@slu.se

In many European warmblood studbooks, clear specialization towards either jumping or dressage horses is notable. Also in the Swedish Warmblood (SWB) there is an on-going specialization, raising the question of separate breeding programs and a discipline-specific Young Horse Test (YHT). This study investigated the specialization within the SWB breed and its potential consequences. In a population of 122,054 SWB horses born between 1980 and 2020, the individuals were categorized according to pedigree as: jumping (J), dressage (D), allround (AR), or Thoroughbred (Th). Data consisting of 8,713 J horses and 6,477 D horses assessed for eight traits in YHT 1999-2020 were used to estimate genetic parameters within and between J and D horses, and between different time periods. Future scenarios in which young horses are assessed as either a jumping or dressage horse at YHT were also analysed. More than 80% of horses born 1980-1985 were categorized as AR horses, while 92% of horses born in 2016-2020 belonged to either J or D category. The average relationship within J or D category has increased during the last decade, whereas the relationship between these categories decreased. Heritability estimates for gait traits were higher for D horses (0.42-0.56) than for J horses (0.25-0.38). For jumping traits, however, heritability estimates were higher for J horses (0.17-0.26) than for D horses (0.10-0.18). Genetic correlations between corresponding traits assessed in J and D horses varied between 0.48 to 0.81, with a tendency to be lower in the late study period. In the future scenarios, heritability and genetic variance both decreased for traits that were not assessed in all horses, indicating that estimation of breeding value and genetic progress for these traits could be affected by a specialized YHT. However, ranking of sires based on estimated breeding values (EBVs) and accuracy of EBVs were only slightly altered for discipline-specific traits. With continued specialization in SWB, a specialization of the YHT can be regarded as a possible option.

Factor analysis of young horse test traits in Swedish Warmbloods

A. Nazari-Ghadikolaei[1], W.F. Fikse[2], Å. Gelinder Viklund[1] and S. Eriksson[1]
[1]Swedish University of Agricultural Sciences, Dept. Animal Breeding and Genetics, Box 7023, 75007 Uppsala, Sweden, [2]Växa, Box 288, 75105 Uppsala, Sweden; susanne.eriksson@slu.se

Linear traits scored on a scale from one extreme to the other were introduced at young horse tests for Swedish Warmblood horses in 2013, in addition to the evaluating traits that are assessed in relation to the breeding goal. The linear traits describe various aspects of conformation, gaits, and jumping ability, and are assumed to be less subjective and more easily compared across populations than evaluating traits. The introduction of linear traits considerably increased the number of traits in the routine genetic evaluation, and several of the traits are assumed to be correlated. The aim of this study was to investigate the interrelationship between different evaluating and linear traits in Swedish Warmbloods using factor analysis. In total, 20,935 horses born 1996-2017 had information on evaluating traits, and 6,436 also had linear trait records. A factor analysis with varimax rotation was performed separately for evaluating and linear traits using the Psych package in R. Height at withers was included in both analyses. The factor analysis resulted in four evaluating and 13 linear trait factors. Missing trait values in linear trait records were imputed based on correlated traits. Thereafter factor scores were calculated using factor loadings. Genetic parameters for, and correlations between, the new factor traits were estimated using multi-trait animal models in the BLUPf90 package. For both evaluating and linear traits, separate factors were formed for jumping and gait traits, as well as for body size. Strong genetic correlations were estimated between such corresponding evaluating and linear trait factors. Heritability estimates were on a similar level as for traits in the genetic evaluation, ranging from 0.06 for the linear trait factor L.behavior to 0.59 for the evaluating trait factor E.size. In conclusion, factor analysis can be used to reduce the number of traits to be included in multi-trait genetic evaluation or in genomic analysis for warmblood horses. It may also contribute to a better understanding of the interrelationships among the assessed traits and be useful to decide on trait-groups to be used in separate multi-trait evaluations.

Adjusted fence height as a phenotype to assess jumping ability in Belgian Warmblood horses

L. Chapard, R. Meyermans, N. Buys and S. Janssens
KU Leuven, Center for Animal Breeding and Genetics, Department of Biosystems, Kasteelpark Arenberg 30 – Box 2472, 3001 Leuven, Belgium; lea.chapard@kuleuven.be

Belgian Warmblood (BWP) and Zangersheide horses (Z) are among the world's best horses for show jumping performance. The breeding goal of BWP and Z studbooks specifically mentions the objective of breeding successful horses that compete at the highest level. However, the current phenotype used for genetic evaluation is only based on rankings in show jumping competitions. Moreover, the rider effect is not included, although it is known to be an important effect. Consequently, research is needed to design a new precise phenotype that matches the studbooks' breeding goal. Here, we studied the 'adjusted fence height' (AFH) that combines the ranking and the fence height in competitions. To calculate AFH, first the ranking is transformed into a normally distributed variable using the Blom transformation. Second, the fence height (in cm) is added to the Blom score which is multiplied by an adjustment factor. This factor is used to scale the Blom score to the fence height. Genetic parameters for AFH were estimated based on almost 675,000 Belgian competition records and more than 80,000 informative horses in the pedigree using a restricted maximum likelihood method. Several univariate models were tested with fixed effects for sex and age at competition and different random effects: a permanent environment effect and a rider effect. The Akaike's Information Criterion (AIC) was used to assess the fit of the different models. Estimated heritabilities of AFH varied from 0.12 to 0.72. According to AIC values, the best fitting model had the lowest heritability and included both a permanent environment and a rider effect. This new phenotype showed a higher heritability than the trait currently used in the Belgian genetic evaluation (0.09). Hence, it may lead to a faster genetic progress in the population. It also better reflects the breeding goal by combining both ranking and level reached in competitions. Moreover, this study showed the importance of accounting for the rider effect as it improved greatly its fit. Therefore, developing a genetic evaluation for the trait AFH will help the two studbooks to safeguard their international standard.

Session 33 Theatre 6

Re-ranking of Coldblooded trotter stallions based on performance at different race lengths

P. Berglund[1], S. Andonov[1], T. Lundqvist[2], C. Olsson[2], E. Strandberg[1] and S. Eriksson[1]
[1]Swedish University of Agricultural Sciences, Animal Breeding and Genetics, P.O. Box 7023, 75007 Uppsala, Sweden, [2]Swedish Trotting Association, P.O. Box 20151, 16102 Bromma, Sweden; paulina.berglund@slu.se

In Coldblooded trotters, used for harness racing in Sweden and Norway, a few popular sires have dominated the breeding over the past decades. This has led to high levels of inbreeding and a fast inbreeding rate in the population. The aim of this study was to investigate if different race lengths would favour different stallions and thereby could promote the use of a larger number of breeding sires from different families. Performance data in the same format as used in the routine genetic evaluation were obtained for three to six-year-old horses born between 1978 and 2017. The analyses were based on racing performance as best annual time per km, adjusted for start type, and the corresponding length of the race. The races were divided into short-distance (1,640 m), middle-distance (2,140 m) and long-distance (2,640 m). In total, there were 27,515 individuals with 34,170 observations from short races, 49,019 observations from middle-distance races and 1,770 observations from long-distance races. Variance components were estimated with the f90 software using univariate and multi-trait mixed linear animal models with year of the race, age, country of registration and sex as fixed effects. Based on the univariate analyses, the estimated heritability for kilometre time was 0.72, 0.70 and 0.32 for short, middle-distance and long races, respectively. However, the latter heritability increased to 0.75 when a trivariate analysis was performed. The genetic correlation estimates between the three traits were strong (0.97-0.99). Despite this strong genetic correlation, the top-ranked stallions based on estimated breeding values (EBV) differed between the three race-lengths. Out of the stallions ranked as top 50 in short races, 44 were also found among the top 50 for middle-distance races, and 11 stallions had an EBV among the top 50 for long-distance races. These results indicate that the use of different race lengths in the genetic evaluation, and in the breeding goal, could favour the use of a larger number of stallions, but further studies using a more comprehensive dataset are needed.

Genetic analysis of temperament traits important for performance in Standardbred trotters

P. Berglund, M. Wilbe, G. Lindgren, M. Solé and S. Eriksson
Swedish University of Agricultural Sciences, Animal Breeding and Genetics, P.O. Box 7023, 75007 Uppsala, Sweden; paulina.berglund@slu.se

Trotting horses require not only physical ability but also mental qualities to perform. In this study, temperament traits important for performance in Swedish and Norwegian Standardbred trotters were assessed by a survey aimed at horse trainers from 2019 to 2021. The aim was to estimate genetic parameters for 13 temperament traits. In total, 376 horses from 121 trainers participated in the study, and the horses had started in at least one race. The trainers were asked to rate how often the horse expressed a specific temperament trait at competition on a linear 7-point scale (never to always). Factor analysis with orthogonal rotation was performed to find underlying temperament characteristics. Variance components for traits and extracted factors were estimated using univariate and bivariate mixed linear animal models. Fixed effects of sex and age group of the horse and level of trainers' license (amateur or professional) were included. The factor analysis revealed two main factors named anxiousness and tractability, which explained 38% and 35% of the common variance, respectively. Heritability estimates for the original assessed traits ranged from zero for will to win, concentration and stereotypies to 0.42 for learning and cooperation. For the factors anxiousness and tractability, heritability estimates of 0.13 and 0.22 were estimated, respectively. Genetic correlations from the bivariate analyses were estimated between the four traits with the highest heritability estimates: learning, cooperation, excitability, and nervousness. These correlations ranged from -0.62 between learning and nervousness to 0.82 between excitability and nervousness. The preliminary results indicate that there exists a genetic variation in some of the temperament traits analysed. These results need to be confirmed in a larger dataset and correlated to the traits used in the genetic evaluation of Standardbred trotters. In conclusion, the results from this study can serve as a starting-point for further research on how temperament influence performance and possibilities to select for desired temperament traits.

Genetic and environmental parameters for gaits linear profiling in the Lusitano horse

M. Mateus[1], A. Vicente[2,3,4,5] and N. Carolino[2,3,6,7]
[1]Universidade de Évora, Largo dos Colegiais 2, 7004-516 Évora, Portugal, [2]SPREGA, Estação Zootécnica Nacional – Fonte Boa, 2005-048 Vale de Santarém, Portugal, [3]CIISA, Faculdade de Medicina Veterinária, Av. Universidade Técnica, 1300-477 Lisboa, Portugal, [4]Escola Superior Agrária do Instituto Politécnico de Santarém, Quinta do Galinheiro. Apart.310, 2001-904 Santarém, Portugal, [5]APSL, Centro Empresarial De Évora, R. Circular Norte do Parque Industrial, 7005-841 Évora, Portugal, [6]Escola Universitária Vasco da Gama, Av. José R. Sousa Fernandes 197 Lordemão, 3020-210 Coimbra, Portugal, [7]INIAV, Fonte Boa, 2005-048 Vale de Santarém, Portugal; margaridamateus10@gmail.com

The present study is a preliminary approach regarding the morpho-functional linear profiling for the Lusitano horse and aims to estimate the genetic and environmental parameters of linear traits such as gaits, evaluated in order to consider the possibility of these traits to be included in the genetic evaluation of the breed, allowing a more objective selection of future breeding stock selection programme. Records of 3,200 animals scored from 2017 to 2021 and respective pedigrees, available in the Studbook of the Lusitano horse breed, were used, building up a relationship matrix with 12,109 individuals. The genetic parameters were obtained using BLUP – Animal Model by restricted maximum likelihood and univariate analysis. The model included the fixed effects of age, inbreeding, year of classification and gender and, as random effects, the direct genetic value and the residual error. The heritability estimates ranged between 0.057 (walk: correctness) and 0.453 (trot: stride length). For the gaits walk, trot and galop, the average heritability estimates were 0.18, 0.36 and 0.30, respectively. The estimates of correlations between breeding values were in the range of -0.766 and 0.857. It was found that the linear and quadratic effect of age at assessment was relevant, as well as the effect of gender and year of classification. On the other hand, the effect of inbreeding was not significant for most of the linear traits studied. The results obtained suggest that linear profiling can be used in a selection programme, leading to an improvement in morpho-functional traits in accordance with the objectives defined for the Lusitano horse breed.

Do we have breeding for performance and distinctiveness in traditional horse local breeds?

M. Ablondi[1], V. Asti[1], S. Capomaccio[2], C. Sartori[3], E. Mancin[3], A. Giontella[2], K. Cappelli[2], R. Mantovani[3], M. Silvestrelli[2] and A. Sabbioni[1]

[1]University of Parma, Department of Veterinary Science, Strada del Taglio, 10, 43045 Parma, Italy, [2]University of Perugia, Centro di Ricerca sul Cavallo Sportivo, Via San Costanzo, 4, 06126 Perugia, Italy, [3]University of Padova, Department of Agronomy, Food, Natural Resources, Animal and Environment, Viale dell'Università, 16, 35020, Italy; michela.ablondi@unipr.it

Today, horse local breeds traditionally used in agriculture face the challenge to modernize their breeding goals to meet current market demand while preserving their distinctive characteristics; otherwise, extinction is a solid possibility. Genetic diversity management is a powerful tool to prevent this risk. Nevertheless, by improving the breed's economic value through selective breeding might contribute as well to enhance its sustainability in the long term. The Italian equine gene pool is rich and diversified in more than 20 native breeds. Thus, this study aimed to evaluate the presence of signatures of selection in the five most representative Italian breeds, based on population size. Genotype data (GGP Equine 70k® SNP chip) of 915 horses from Bardigiano, Haflinger, Maremmano, Murgese and Italian Heavy Draught breed were analysed. Two methods were applied: Pairwise Fixation Index (Fst) and Runs of Homozygosity (ROH). The pairwise Fst was calculated in the R package SNPRelate and ROH island discovery (as ROH shared in over 70% of the horses within breed) was performed in the R package DetectRUNS. The pairwise Fst ranged between 0.06 to 0.11 showing for most comparisons Fst values above 0.50 for specific SNPs, which might underline regions potentially linked with the distinctive features of each breed. In all the breeds ROH longer than 16 Mb were found. A total of 32 ROH islands were detected: 19 were present in one breed only, whereas the others were shared among at least two breeds. Several mapped to known quantitative trait loci for morphological traits (e.g. body size and coat colour), disease susceptibility and gaits related traits. Further research is ongoing to unravel if those signs of selection play a role in keeping the traditional features of each breed while helping the conversion from agricultural uses to riding purposes.

Show attendance is heritable in Belgian Draft Horses: possible implications for breeding against CPL

R. Meyermans, W. Gorssen, E. Reynaert, L. Chapard, N. Buys and S. Janssens

KU Leuven, Center for Animal Breeding and Genetics, Department of Biosystems, Kasteelpark Arenberg 30 – Box 2472, 3001, Belgium; roel.meyermans@kuleuven.be

Chronic progressive lymphedema (CPL) is a systemic disease that plagues the Belgian Draft Horse. It affects the lower limbs and afflicts progressive swelling, nodulous formations and severe skin folds. Therefore, Belgian draft horses are screened for their CPL status by clinical examination of the four limbs. Currently, most of the CPL scores are collected at (local) shows. However, not all horses are competing in these contests which could present a bias in the data collection (censoring). Therefore, we investigate the heritability of show attendance. If show attendance is heritable, it could affect the breeding program against CPL in Belgian Draft Horses. All shows since 2015 (local and national level) were analysed for more than 4000 competing horses and a pedigree depth up to 11 generations (>8,000 animals). Genetic parameters for attending shows were estimated with a threshold model using thrgibbs1f90. The heritability estimates were high (±70%), indicating that breeders are highly selective on which animals they take to a contest. These results suggest a breeders' decision for show attendance is clearly based on pedigree records and/or highly heritable favourable traits. This censoring has major implications on the upcoming breeding program for breeding against CPL. Hence, we advise the studbook that a simple implementation of a breeding program is not straightforward with the current phenotyping strategy. Moreover, this research also shows that using data that are collected on contests may be biased and preselected, and therefore should be treated with caution.

Conformational hock defects in Pura Raza Española horses: risk factors, prevalence, genetic analysis

M. Ripollés, D.I. Perdomo-González and M. Valera
Universidad de Sevilla, Ctra. de Utrera, km.1, 41005, Sevilla, Spain; marriplob@alum.us.es

Diseases of the horse's limbs constitute an important group of defects, in particular including developmental and degenerative joint diseases. They are often underlaid by complex genetic effects, which may be breed-specific or common in certain types of horses. Disorders of the angle of hock have a significant impact on these specific performance characteristics and thus severely affect the use of horses in equestrian sports. The Purebred Spanish Horse (PRE) has its origins in the Baroque horses and has contributed to the formation of many other breeds both in the Americas and in Europe. The aim of this study was to establish the within-breed prevalence, possible associated risk factors and heritability of four different hock defects in PRE horses. Lateral view hock defect was divided into 'closed' and 'open' and rear view hock defect into 'convergent' and 'divergent'. Each defect has been classified in 3 classes: 1 having no defect, 2 slight defect and 3 serious defects. A total of 43,358 Pura Raza Española horses were evaluated. Prevalences of 22.1, 33.7, 74.1 and 21.0% were determined for closed hock, open hock, convergent and divergent, respectively. Genetic parameters were estimated using a Bayesian procedure with the BLUPF90 software. Fixed effects were included in the model according to significant results in analyses of variance (GLZ): age, gender, coat colour, geographical area, size breeder's stud farm, inbreeding, proportionality index and height at point of croup. The pedigree file information included a minimum of 4 generations (92,997 horses). Heritability estimates ranged from 0.20 (open hock) to 0.31(convergent hock). To study if there was a genetic relationship between the hock defects, bivariate mixed models were applied: closed hock vs convergent hock, closed hock vs divergent hock, open hock vs convergent hock, open hock vs divergent hock. The closest correlation was found between 'closed' and 'convergent' (0.44) and the weakest between 'closed' and 'divergent' (0.19). These results imply that selection against limb defects is possible and would allow reducing the genetic risk of the horses' offspring to suffer from them.

Analysis of genomic copy number variation in Pura Raza Español horses

N. Laseca[1], S. Demyda-Peyrás[2], M. Valera[3] and A. Molina[1]
[1]University of Cordoba, Department of Genetics, Edificio Gregor Mendel, Campus de Rabanales, 14014 Córdoba, Spain, [2]National University of La Plata, Faculty of Veterinary Sciences, La Plata, Argentina, [3]University of Sevilla, Department of Agronomie, ETSIA, Ctra. de Utrera, km. 1, 41013 Sevilla, Spain; ge2lagan@uco.es

Copy number variations (CNVs) are an essential source of genetic variation that could explain phenotypic variation in complex traits and diseases. In recent years, their study has increased in many populations of animal species. However, equine CNV studies are still limited. In this preliminary study, we developed a genome-wide CNV analysis with PennCNV software using high-density SNP genotype data from a large cohort of Pura Raza Español (PRE) horses (n=805). After quality control and the exclusion of animals with more than 56 CNVs, 19,902 segment CNVs (with at least 5 consecutive SNPs and a minimum size of 1 kb) and 3,361 CNV regions (CNVRs; overlapping CNVs by at least 1 bp) were identified in 654 horses. The length of the CNVs ranged from 1.024 kb to 4.55 Mb and coverage of the total genome by CNVRs was 7.35% with 62% of CNVRs containing genes involved in sensory perception of smell, immune response, locomotory behaviour, reproduction, and metabolic processes. Our results showed that duplications (2,853) were more abundant than deletions (334) and mixed CNVRs (174). Moreover, the distribution of CNVs in chromosomes was not uniform. ECA12 was the chromosome with the largest percentage of its genome covered (21.8%) while the highest number of CNVs were found in ECA20, ECA12, and ECA1. Comparative analysis between two PRE subpopulations (Carthusian strain horses and undifferentiated population of PRE horses) showed differences between them, with 18.9% of CNV segments differing. This could suggest that genetic variations due to CNVs contribute to the specific phenotypic and differential characteristics of the breed and the subpopulations. Overall, this preliminary study could contribute to the knowledge of CNVs in the equine species and facilitate the understanding of genetic and phenotypic variations in the equine genome. Future research is needed in order to confirm if the observed CNVs in breed are also linked to phenotypical differences or complex traits.

Quantitative analysis of imprinting effect of reproductive traits in Pura Raza Española mares

D.I. Perdomo-González[1], A. Molina[2], M. Valera[1] and L. Varona[3]
[1]Universidad de Sevilla, Carretera de Utrera km1, 41013 Sevilla, Spain, [2]Universidad de Córdoba, Ctra. Madrid-Cádiz km 396, 14071 Córdoba, Spain, [3]Universidad de Zaragoza, C. Miguel Servet 177, 50013 Zaragoza, Spain; davpergon1@alum.us.es

Parent-of-origin effect occurs when the expression of a gene varies with the sex of the parent. It is caused by the epigenetic effect of genomic imprinting, giving place to a partial or completely suppression of genes involved. With partial imprinting, genes from each parent are both differently expressed, but with complete imprinting, the gene inherited from one parent is silent. Imprinted genes have been described in mice, sheep, beef, dairy cattle, pig and even humans, highlighting the high importance when ignored in routine genetic evaluations, but few have been described for horses. The Pura Raza Española (PRE) horse is an autochthonous Spanish horse with a deep and large pedigree since its stud book creation in 1912 with a high pedigree completeness and more than 40 years of proven parental information which make it a perfect population for the analysis of quantitative effect of imprinted loci. The aim of this work was 1. to determine the existence of parent-of-origin effect for a reproductive efficiency trait in PRE mares and, 2. to determine the best model to estimate the variance component due to imprinting. A pedigree matrix of 39,074 individuals was generated from 22,790 mares with reproductive efficiency information. Then a parent-of-origin effects mixed model with paternal, maternal, and additive genetic effects was implemented. Results reveal that the best fitted model had coat colour, country, and stud size as fixed effects and established that both, maternal and paternal imprinting do affect to reproductive efficiency of PRE mares. Finally, maternal component of variance (19.42) was twice as high as for paternal component (10.94) but both were lower than the additive genetic effect (30.67) in the analysed trait. Genes are imprinted during gametogenesis so that even though a maternal (paternal) gene may be silenced or constrained in an individual animal, it will be expressed in its progeny if it is its sire (dam). For this reason, imprinting could be important in PRE breeding program as, if ignored, it could bias estimated breeding values and, estimates of genetic parameters.

Novel genomic regions of importance for quality of the gait tölt in Icelandic horses

H. Sigurdardottir[1,2], S. Eriksson[2], M. Rhodin[3], T. Kristjansson[1], E. Albertsdottir[4] and G. Lindgren[2,5]
[1]Agricultural University of Iceland, Faculty of Agricultural Sciences, Hvanneyri, 311 Borgarbyggð, Iceland, [2]Swedish University of Agricultural Sciences, Dept. Animal Breeding and Genetics, P.O. Box 7023, 75007 Uppsala, Sweden, [3]Swedish University of Agricultural Sciences, Dept. of Anatomy, Physiology and Biochemistry, P.O. Box 7011, 75007 Uppsala, Sweden, [4]The Icelandic Agricultural Advisory Centre, Hagatorgi 1, 107 Reykjavik, Iceland, [5]KU Leuven, Livestock Genetics, Department of Biosystems, Kasteelpark Arenberg 30, 3001 Leuven, Belgium; heidrun.sigurdardottir@slu.se

The Icelandic horse has captured interest in many countries largely because of its superior gaiting ability, especially the symmetrical four-beat ambling gait called tölt. Tölt is a very smooth and comfortable gait to ride and the ability to tölt at a wide range of speed makes the Icelandic horse an excellent riding horse for leisure and competition purposes. The quality of tölt is assessed based on several attributes, such as clarity of beat, suppleness of movements and speed ability, rendering tölt a complex trait. Furthermore, there is an extensive phenotypic variation within those factors that hypothetically can be genetically mapped. The aim of this study is to identify these genomic regions or genes for tölt by using a genome-wide association study. 173 Icelandic horses with phenotypic records were genotyped with the 670 K+ Axiom Equine Genotyping Array. No genome-wide significant SNPs, using Bonferroni correction for multiple testing, were identified for tölt. However, three suggestive genome-wide SNPs were identified on chromosome 8 (close to the GALNT1 gene) associated with the overall score for the trait tölt, and with clarity of tölt-beat. The results appear to be stable across different models. Furthermore, additional suggestive QTL's on chromosome 1, 4, 10 and 12 were identified. For further investigation, at least 190 additional horses are planned to be genotyped, and objective motion analysis for a subset of these horses will be performed. This study has revealed that there appear to be regions of interest, especially on chromosome 8, in relation to tölt quality in Icelandic horses. Further studies of these regions are needed to better understand the genetic control of gait quality.

Session 33 Poster 15

Genetic and population analysis of limb´s socks in Pura Raza Española horses

M. Ripollés, D.I. Perdomo-González, A. Encina and M. Valera
Universidad Sevilla, Ctra.Utrera, 41005 Sevilla, Spain; marriplob@alum.us.es

White markings on the horse's coat are characteristic of many equine breeds. The markings are present at birth and do not change throughout the horse's life, which helps identify the horse as a unique individual. The presence of stocking on the forelimbs and hindlimbs are considered by The Purebred Spanish Horse (PRE) stud book as serious defect in the coat colour when the breeches exceed part of the knees or hocks. In this research, the population of PRE horses were from Spanish stud farms, the coat colour was chestnut, bay or black and they were between 3-10 years old at the time of the morphological evaluation. The aims of this study were: (1) to establish the within-breed prevalence, possible associated factors, genetic parameters of the socks according to their extent and the affected limb, (2) to study the distances between the subpopulations, according to the number of limb' s socks. The socks of the four limbs of the horse have been studied: right forelimb (RF), left forelimb (LF), right hindlimb (RH), left hindlimb (LH). Each level of socks has been classified in 3 classes*:* 1 ha*ving* no socks, 2 socks and 3 stocking. A total of 14,314 Pura Raza Española horses were evaluated. Prevalences (classes 2 and 3) of 5.23%, 6.11%, 19.62% and 24.58% were determined for RF, LF, RH, LH respectively. Genetic parameters were estimated using a Bayesian procedure with BLUPF90 software. Fixed effects were included in the model according to significant results in Generalized Non-linear Model (GLZ*)*: coat colour, year of birth and number of limb´s socks. The pedigree file information included 49,496 horses. Heritability ranged from 0.10 (RH) to 0.38 (LF). To study whether there was a genetic relationship between the limb´s socks, multivariate mixed models were applied. The closest positive correlation was found between RF and LF (0.42) and the closest negative correlation between LF and LH (-0.55). Subsequently, 5 subpopulations were reconstructed according to the number of socks: (subpopulation 1: one socks, 2: two socks, 3: three socks, 4: four socks and 5: no socks). The subpopulations composed of animals with 4 socks, showed a greater Nei's distance respect for other subpopulations. However, further analysis including genomic information would be necessary to elucidate the mechanisms involved.

Session 33 Poster 16

Genomic diversity and structure of three autochthonous horse populations

S. Tumino[1], A. Criscione[1], S. Mastrangelo[2], E. D'alessandro[3], G. Chessari[1], R. Di Gerlando[2], A. Zumbo[3], D. Marletta[1] and S. Bordonaro[1]
[1]University of Catania, Via Valdisavoia, 95123 Catania, Italy, [2]University of Palermo, Viale delle Scienze, 90128 Palermo, Italy, [3]University of Messina, Piazza Pugliatti, 98168 Messina, Italy; serena.tumino@unict.it

Sicily preserves an invaluable equine heritage represented by three populations, namely Purosangue Orientale Siciliano, Sanfratellano and Siciliano. The genomic characterization is an essential tool to preserve the local breeds, limiting the reduction of genetic diversity and the excess of inbreeding that negatively affects fitness-related traits. This study evaluated the genetic diversity, population structure, and the pattern of autozygosity of Sanfratellano (n=17), Purosangue Orientale Siciliano (n=12), and Siciliano (n=17) horses using the Illumina Equine SNP70K BeadChip. The genetic relationships of Sicilian populations were analysed together with 17 other equine breeds, including Arab and Maremmano, that influenced their genomic variability. Multidimensional scaling analysis, model-based clustering, and Neighbor-Net showed close connections between the Purosangue Orientale Siciliano and the Arab as well as between Sanfratellano, Siciliano, and Maremmano and confirmed the historically existing relationships between these populations. The parameters of genetic diversity were highest in Siciliano and lowest in Purosangue Orientale Siciliano. The genomic inbreeding based on runs of homozygosity (ROH) reported high levels in Purosangue Orientale Siciliano and Sanfratellano already recognized as endangered populations. A total of 25 ROH islands harbouring 364 markers were identified in the data set. Among Sicilian horses, Purosangue Orientale Siciliano exhibited a SNP-within-ROH incidence exceeding 75%. In this population, the autozygosity rich regions on ECA9 and ECA18 overlapped with QTLs for temperament and altitude adaptation, and confirmed the influence of the Arab breed on the Purosangue Orientale Siciliano's traits. The genomic information plays a pivotal role in assisting the management of small populations with the primary target of planning the correct breeding strategy. Our results underline the importance of planning adequate conservation and exploitation programs to reduce inbreeding and the loss of genetic diversity.

Session 33 Poster 17

Runs of homozygosity in Estonian horse breeds

E. Sild, S. Värv, T. Kaart and H. Viinalass
Estonian University of Life Sciences, Institute of Veterinary Medicine and Animal Sciences, Fr. R. Kreutzwaldi 1, Tartu 51006, Estonia; erkki.sild@emu.ee

High-density genotype (670K SNP Axiom® Equine Genotyping Array) analysis was carried out to detect ROH patterns and inbreeding levels in five Estonian horse breeds. Samples from 160 horses from local (Estonian Native, Estonian Heavy Draught, Tori) as well from international breeds raised in Estonia (Trakehner, Hanoverian) were collected in 2020. Considering the Tori Studbook data and the analysis of structure, the purebred Tori Horse (universal type) was assessed apart from outcrossed/admixed (sport type) group resulting in the six study-populations. Overall 10,683 ROH-s longer than 1 Mb were detected of which 58% were shorter than 2 Mb while only 0.25% were longer than 32 Mb. Among the six populations fewer homozygous regions were detected in Estonian Native (38.6 ROHs per animal) and in Tori universal horse group (45.0) than in sport horses, Trakehner (114.1) and Hanoverian (93.5). However, the local breeds had longer average ROH lengths and the proportion of the ROHs at the longest size category (ROH>16 Mb). The genomic inbreeding in terms of F_{ROH} varied among the breeds ranging from 5.9% (Tori universal type) to 14.9% (Trakehner). The genome-wide location of ROHs was disperse except an island region found at *MC1R* location across the breeds. Additionally, ROHs at 110 Mb on chromosome 3 revealed putative signature of selection characteristic to Tori universal type horse. The number of unique recurrent ROHs (homozygous regions with the same first and final genomic position occurring only in one population) were highest in the small Estonian Heavy Draught Horse population and in Trakehner being the lowest in Estonian Native. Trakehner and the Tori admixed group as well as both Tori groups (universal and admixed) had the highest amount of shared recurrent ROHs reflecting the common ancestries. Estonian Research Council (PRG554) supported this work.

Session 33 Poster 18

Development of inbreeding and relation between inbreeding & performance in a German heavy warmblood

U. König Von Borstel[1], A.-M. Beckers[1] and P. Allhoff[2]
[1]Justus-Liebig-University Giessen, Animal Breeding and Genetics, Leihgesterner Weg 52, 35392 Giessen, Germany, [2]Zuchtverband für das Ostfriesische und Alt-Oldenburger Pferd e.V., Bahnbreede 25, 33824 Werther, Germany; uta.koenig@agrar.uni-giessen.de

The aim of the present study was to assess development of inbreeding coefficients as well as their relationship to sport and reproductive performance in the protected horse breed 'East-Frisian and Oldenburger Heavy Warmblood' (Ostfriesische und Alt-Oldenburger Pferd). Pedigree and performance data was made available from the breeding association and included 11,047 animals in the pedigree, 26% of which were founder animals, i.e. no parental information was available for these animals. Performance data was available for 160 (life-time earnings from competitions), 67 (1-year earnings from competition), and 60 (stallion performance test results) horses. Population parameters such as inbreeding coefficients were calculated using the program CFC, and the relationship between inbreeding coefficients and performance data were analysed using a mixed model. Minimum pedigree completeness for the years 1995-2015 was 97% in the first generation and 30% in the 6th generation. Average inbreeding coefficient for the entire population in the pedigree was F=0.006, maximum inbreeding was F=0.273. However, starting from 2005, a marked increase in average inbreeding values can be observed, rising to F=0.1 in 2011 for the entire population and F=0.04 for the paternal inbreeding coefficients. Effective population size remained with ~ne=200 (ne=214 in 2015) reasonably stable across the past 20 years, indicating that the breed is well-managed but nevertheless at risk. Inbreeding coefficients divided into classes had no significant effect on log-lifetime nor on log-yearly earnings (P>0.05). Similarly, Pearson correlations between stallion performance test grades and inbreeding coefficients were not significant (P>0.05), indicating that in the present population, no effect of inbreeding on performance is present. In conclusion, the East-Frisian and Oldenburger Heavy Warmblood is an endangered, however, at present reasonably stable breed that requires specialized breed protection management, but presently does not show marked inbreeding depression.

Session 33 Poster 19

Genetic parameters of morphofunctional traits in Purebred Menorca Horses based on their coat colour

D.I. Perdomo-González, M.D. Gómez, E. Bartolomé, R. García De Paredes and M. Valera
Universidad de Sevilla, Carretera de Utrera km 1, 41013, Spain; davpergon1@alum.us.es

The Purebred Menorca Horse (PRMe) is an endangered and versatile breed of the Menorca island, Spain. Since conformation traits are correlated with health and performance, influencing horses' economic value, breeders select horses based on a linear system. In addition, the quality of the black coat colour (QC) and the quantity of white marks (QW) are also very important. Therefore, the aim of this study was to evaluate the relationship of the morphofunctional traits with QC and QW. Linear type traits were scored in animals with 3 or more years-old, by 4 appraisals, using a structured score sheet of 7 classes. A total of 772 records (belonging to 333 animals,188 males and 145 females) for 35 morphological, 9 functional and 2 colour traits were analysed. A multi-trait REML animal model was used to perform the genetic analysis using BLUPF90. The effects of sex (2), appraisal-season (13), QC (3) and QW (5) were included as fixed effects, and the age as linear covariate. A pedigree with the last 5 generations was generated with a total of 757 animals (341 males and 416 females). Ranges evidenced the complete use of the scale by appraisals and the coefficients of variation ensure that analysed population have sufficient variability to obtain a selection response. Heritabilities for conformation traits ranged from 0.13 (loin length) to 0.78 (fore-hoof front view) and for functional traits from 0.45 (walk clarity) to 0.66 (trot suspension). High heritabilities were also obtained for coat colour, 0.64 for QC and 0.90 for QW, demonstrating that effective selection is feasible for these traits in the PRMe population. The 62% of the genetic correlations were positive and the 38% were negative, being 0.99 between QC-QW. Their values ranged between 0.02 (QW-height of the withers, QW-shoulder length) and 0.78 (QC-back length) for the correlations of coat traits and the other linear traits. The 68.89% of the genetic correlations between QC and the other analysed traits were higher than 0.25 and the 77.78% for the QW, confirming that selection of traits influenced each other and the values and signs have to be taken into account within the breeding program for morphofunctional selection of PRMe.

Session 33 Poster 20

Inbreeding depression on mare reproductive traits in the Arab horse in Spain

P. Álvarez-García, J.P. Gutiérrez and I. Cervantes
Departamento de Producción Animal, Facultad de Veterinaria, Universidad Complutense de Madrid, 28040, Spain; icervantes@vet.ucm.es

Reproductive traits are expected to be affected most by inbreeding depression. The Spanish Arab horse has a census of 17,920 and an effective population size of 55.5, but despite this healthy genetic variability status it is important to evaluate the possible effect of inbreeding depression in these traits. The inclusion of inbreeding depression in a genetic model is tricky because it does not have a linear affect. Individual increase in inbreeding (ΔF) has shown to be useful for this because it allows inbreeding coefficients to be standardized by the pedigree depth, thus becoming linear. Four reproductive traits were analysed using Studbook data from the Spanish Arab horse: age at first foaling (AFF), average interval between foals (AIF), reproductive live (RL) and reproductive efficiency as ratio between number of foals and maximum (RE). Data consist of 5,299 records for AFF, 3,620 for AIF, 3,350 for RL and 3,465 for RE, and pedigree data contained 27,452 animals. Mare's year of birth and country region were included as fixed effects in all traits, and parturition month also for AFF. The owner was included in all traits as random effect and the breeder also for AFF. The inbreeding coefficient (F) and the individual increase in inbreeding (ΔF) of the mare were tested as covariate, linear (ΔF, and F) and quadratic (F) to evaluate the inbreeding depression using VCE6.0 program. The heritability values were 0.07 (0.02) for AFF, 0.01 (0.02) for AIF, 0.05 (0.02) for RL and 0.12 (0.04) for RE. The breeder had a ratio of 0.08 (0.02) for AFF and the owner ratio ranged between 0.11 (0.02) and 0.20 (0.03). Genetic correlations were all significant except AIF/RL, being -0.92 (0.05) for AFF/RE, -0.73 (0.14) for AIF/RE, -0.56 (0.12) for AFF/RL, -0.52 (0.26) for AFF/AIF and 0.60 (0.10) for RF/RE. Regarding inbreeding depression, a ΔF of 0.01 equivalent to 6% of inbreeding given the average of equivalent complete generations in the data (6.06), increased AFF in 58.52 days, AIF in 10.77 days, and decreased RL in 387.37 days and RE in 0.03. The results showed that the genetic component of these traits was low mainly because the influence of the Stud management. The effect of inbreeding depression was moderate to low.

Session 34 Theatre 1

Big data in animal breeding

A. Granados, L. Schibler, L. Fontanesi, J. Solkner, E. Knol, M. Mc Grew and N. Van Hemert
FABRE TP, rue des Trèves 61, 1040 Bruxelles, Belgium; noraly.vanhemert@effab.info

With increasing societal and climate challenges, animal farming production must adapt to the demand for animal products in a sustainable and animal welfare-friendly manner. Advanced technologies such as sensors, cloud computing, machine learning, and artificial intelligence can significantly generate tonnes of data and change the scope and organisation of animal breeding and farming. Availability of real-time data, forecasting and tracking information can lead to automated and autonomous farm operations that can change the way we both produce and consume in the near future. Better efficiency and management and also more transparency are expected but also the privacy of breeders and farmers will be impacted. With funding bodies placing more and more emphasis on the open-access publication of both results and original data, discrepancies may arise between the needs of the funding source and the project partners. Moreover, new technologies are also beneficial for improving animal breeding traits and helping to explore new ones. With advanced technologies, we can increase our understanding of complex animal systems through deep phenotyping. Advanced technologies ultimately provide us with the tools to breed more efficient animals with increased animal welfare and sustainability. Yet, to do so, enormous amounts of data are available. Do you know what technology there is available to collect this data, and to whom is it accessible? Which data is really important for breeding? How and by whom is it collected? Who stores it, and more importantly, who owns this data and is it shared with any third parties? FABRE TP organises this challenge session to engage with research, academia, and the private sector in a discussion on the growing role of Big Data in livestock production and the challenges for breeding companies and research. Speakers of this session will be defined during the 2022 spring in coordination with the PLF and Genetics commissions.

Session 35 Theatre 1

Reimagining grazing systems: from thoughtscapes to ethical and sustainable pasture-based foodscapes

P. Gregorini
Lincoln University, Agricultural Science, Cnr Ellesmere and Spting Roads, 7647 Lincoln, New Zealand; pablo.gregorini@lincoln.ac.nz

Throughout different landscapes, livestock fulfil essential roles in ecology, agriculture, economies and cultures. Not only do they provide food and wealth, they also deliver ecosystem services. Grazing, as a descriptive adjective, locates livestock within a spatial and temporal pastoral context where they naturally graze or are grazed. In some cases, however, grazing driven by a single and myopic objective of maximising animal production and/or profit has transformed landscapes, diminished biodiversity, reduced water and air quality, accelerated loss of soil and plant biomass, and displaced indigenous flora, fauna and people. Such degenerative landscape transformations have jeopardised present and future ecosystem and societal services, breaking the natural integration of land, water, air, health, and social sphere, and even our own thoughtscapes. No wonder, some societies/consumer in pursuit of health are demanding foodscapes absent of animal products. There is a call for diversified-adaptive and integrative agro-ecological systems that simultaneously operate across multiple 'scapes'; i.e. thought- social- land, health- and foodscapes. There needs to be a paradigm shift in pastoral production systems and how grazing livestock are grazed within them. In this keynote speaker address, I will present and discuss how a change in paradigm, can be initially derived from a change of 'thoughtscapes', with them referring to our geography of the mind in where we locate ourselves not as observers but participants. Alternative thoughtscapes include paradigm shifts where graziers move away from the myopic view of contemporary pasture-based systems, where the animals are perceived as a 'source' of products, existing in isolation to the wider landscape and societal functions. Landscapes are the tables where humans and livestock gain their nourishment (i.e. foodscapes). Foodscapes and dietary perceptions (a component of our thoughtscapes) dictate dietary-choice actions and reactions. Ultimately, animal products reflect the history of our landscape, foodscapes and agricultural systems manifested though soil and plant chemistry, and thereby our health and that of the planet.

Biodiversity – the forgotten component of sustainable pasture based production systems

H. Sheridan
University College Dublin, School of Agriculture and Food Science, Belfield, Dublin 4, D04 V1W8, Ireland; helen.sheridan@ucd.ie

Globally it is recognised that the rate of species loss has reached critically high levels, leading to acceptance that the sixth mass extinction of species is currently ongoing. From a European perspective, much biodiversity has evolved with the agricultural systems that developed across this part of the world over the last 8,000 years. Therefore, the retention of this biodiversity is ultimately dependent on the continuation of particular agricultural activities, with land abandonment leading to decreased biodiversity. On the other hand, increased emphasis on maximising food production over recent decades has resulted in the excessive simplification and homogenisation of significant areas of agricultural land. This has largely been facilitated through reliance on monocultures coupled with high inputs of fertilisers and pesticides, which, until recently were readily available at relatively affordable rates. These factors, combined with the specialisation of agricultural systems, have placed very significant pressures on biodiversity and are major contributing reasons for its decline. There is a strong moral argument for the protection of biodiversity. But another more humancentric argument recognises the range of production, regulating, supporting and cultural ecosystems services, that it facilitates the delivery of. These services include biological processes involved in food production, nutrient and water cycling, carbon sequestration and storage, pollination, soil formation and retention, etc., and therefore, fundamentally underpin the sustainability of agricultural systems. It is now widely accepted that the continued existence of many species, including those that may still be relatively common, is heavily dependent on the maintenance of diverse agricultural practices at field, farm and landscape scale coupled with the retention of a matrix of semi-natural habitat within the farmed landscape. This presentation will focus on a re-imagining of livestock production systems, from those that have been intensified and simplified through the use of fossil fuel derived inputs, to ones that reinstate biodiversity and associated biological processes at their core.

Session 35 Theatre 3

Effect of alternative forages on ewe body condition and subsequent lamb weight during lactation

J.T. Higgins[1,2], G. Egan[2], F. Godwin[2], S. Lott[2], M.B. Lynch[3], M. McEvoy[1], F.M. McGovern[4] and T.M. Boland[2]
[1]Germinal Ireland Ltd, Horse and Jockey, Co Tipperary, E25 D286, Ireland, [2]School of Agriculture and Food Science, University College Dublin, D4, D04 V1W8, Ireland, [3]Teagasc Research Johnstown Castle, Johnstown Castle, Co Wexford, Y35 Y521, Ireland, [4]Animal and Grassland Research and Innovation Centre, Teagasc, Athenry, Co Galway, H65 R718, Ireland; jonathan.higgins@ucdconnect.ie

Environmental and financial pressures increasingly challenge the sheep meat industry as global sheepmeat demand rises. In Ireland, increasing efficiency within livestock production systems is critical. Ewe prolificacy and grazing animal performance are two key areas for improvement. A 2×3 factorial design experiment was established to examine the impact of sward composition or breed type on ewe performance. Two sward treatments (n=60) were investigated in a farmlet study; 5 ha of perennial ryegrass (Lolium perenne; CRT) only, compared to 5 ha of alternative forages (AF) composed of 3 ha five species mix, 1 ha of perennial ryegrass and red clover (Trifolium pratense), predominantly for silage conservation, and 1 ha of redstart (forage rape × kale hybrid), a forage crop for out-wintering ewes. The three (n=40) breed types used were, Belclare X, Lleyn X, and Mule (Bluefaced Leicester × Blackface Mountain) ewes. Pasture herbage mass was measured immediately pre and post grazing with surplus herbage conserved as silage. Data were checked for normality using PROC UNIVARIATE with model selection via linear regression set at $P<0.25$ and analysed using repeated measures and PROC GLIMMIX in SAS (Version 9.4) Ewes offered CRT tended to have lower body weight (BW; $P=0.09$) and body condition (BC; $P=0.07$) compared to ewes offered AF at the end of lactation (14 weeks postpartum). Ewes offered CRT displayed a loss in BW ($P<0.01$) and BC ($P<0.05$) over the 14 weeks of lactation whereas ewes offered AF remained unchanged for both variables ($P>0.05$). Ewes offered AF produced higher total lamb weaning weight ($P<0.001$) and supported higher lamb growth rates to weaning ($P<0.001$) with increased weaning efficiency (kg weaned lamb per kg ewe mature BW; $P<0.001$) compared to ewes offered CRT. Prolific breed type had no effect on any parameter ($P>0.05$). In conclusion, sward composition can affect ewe BW, BC, and lamb growth performance.

Can altering sowing rate increase the dry matter production of multispecies swards under grazing?

C. Hearn[1,2], M. Egan[2], M.B. Lynch[1,3] and M. O'Donovan[2]
[1]University College Dublin, School of Agriculture and Food Science, University College Dublin, Belfield, Dublin 4, D04 V[1]W8, Ireland, [2]Teagasc, Animal and Grassland Research and Innovation Centre, Moorepark, Fermoy, Co. Cork, P61 P302, Ireland, [3]Teagasc, Environmental Research Centre, Johnstown Castle, Co. Wexford, Y35 Y521, Ireland; ciaran.hearn@teagasc.ie

Irish dairy farmers are interested in the use of multispecies swards (MSS) for intensive grazing systems. Little information is available regarding the establishment and management of MSS which include the forage herb ribwort plantain (PL, *Plantago lanceolate* L.). A grazed plot experiment was established to investigate the dry matter (DM) production of MSS containing PL, perennial ryegrass (PRG, *Lolium perenne* L.) white clover (WC, *Trifolium repens* L.) and red clover (RC, *Trifolium pratense* L.) in the first year post sowing. Five sward types were established which included a PRG control and four MSS treatments; each of the four MSS were allocated to high (H) or low (L) sowing rate treatment. The sowing rates referred to PL, RC and WC only where the rate of PRG seed was reduced as other species were added to the sowing mixture. Plots were established in July 2020 and were grazed by lactating dairy cows on ten occasions in 2021. There was no association between sowing rate treatment and DM production for any of the swards. The sward mixture of PRG, RC, WC & PL produced the highest level of DM; 1,755±537 kg DM/ha more than the average of all other sward mixtures (P<0.05). Each of the remaining sward mixtures produced similar levels of DM. These results indicate that more complex MSS containing WC, RC and PL can contribute increased DM production in the first year post sowing. Sowing rate was associated with the level of RC in MS swards where there level of RC in swards was increased by an average of 65 kg DM/ha/year under the higher seeding rate treatment (P<0.05). The interaction of sowing rate and sward species mixture was associated with the level of WC in swards where the L seeding rate reduced WC DM by 543 kg DM/ha/year in the PRG, WC & PL treatment (P<0.0001) but no differences existed within other sward species treatments as WC seeding rates changed. Further work will be required to assess sward and component species DM production persistence over multiple grazing seasons.

Rotational grazing and multi-species swards increase pasture-based productivity – a meta-analysis

M.W. Jordon[1], K.J. Willis[1], P.C. Bürkner[2] and G. Petrokofsky[1]
[1]University of Oxford, Department of Zoology, Mansfield Road, Oxford, OX1 3SZ, United Kingdom, [2]University of Stuttgart, Cluster of Excellence SimTech, Pfaffenwaldring 5a, 70569 Stuttgart, Germany; matthew.jordon@zoo.ox.ac.uk

Regenerative Agriculture (RA) practices are increasingly promoted to improve forage production and livestock performance in temperate livestock systems. These practices include: (1) rotational grazing (RG) of livestock around multiple subunits of a pasture to achieve periods of rest; and (2) herbal leys (HL), where perennial forbs such as chicory, lucerne and trefoils are included as components in multi-species swards. While there are plausible mechanisms for adoption of these practices to improve agricultural productivity, quantitative syntheses of their impacts are required. Here, we present the results of a systematic review and meta-analysis of the effects RG and HL practices have on herbage dry matter (DM) production, animal daily liveweight gain (DLWG), and sheep wool growth in temperate oceanic regions. We identified 115 relevant studies across nine countries, extracting 586 observations for analysis. We used quantitative predictors in our Bayesian hierarchical models to investigate the role of rest period and stocking density in RG systems, and specific plant traits and sward diversity in HL. We found that herbage DM increased by 0.31 t/ha as the proportion of rest in an RG grazing system increased from 0 to 1. Stocking density significantly moderated the effect of rest period on sheep and cattle DLWG; at higher stocking densities, longer rest periods were required to maintain livestock growth rates. In HL studies, herbage DM yielded 1.63 t/ha more per metre of increased sward root depth, and a sward entirely comprised of legumes yielded 2.20 t/ha more than when no legumes were present. Sheep DLWG increased by 3.50 g/day per unit increase in leaf nitrogen concentration (mg/g), but we could not determine an effect of leaf condensed tannin content on animal performance. Although there remain differences between the RG and HL study treatments meta-analysed here and RA in practice, our results provide empirical support for some of the mechanisms attributed to increased pasture and livestock productivity following adoption of RA grazing practices.

Session 35 Theatre 6

The effect of multispecies swards on the sustainability performance of Irish grazing systems

M.C. Ayala[1,2], J.C.J. Groot[3], I.J.M. De Boer[2], J. Kennedy[1], C. Grace[1] and R. Ripoll-Bosch[2]
[1]Devenish Nutrition, Dowth, A92ER22, Ireland, [2]Wageningen University & Research, Animal Production Systems group, De Elst 1, 6708 WD Wageningen, the Netherlands, [3]Wageningen University & Research, Farming Systems Ecology group, Droevendaalsesteeg 1, 6708 PB Wageningen, the Netherlands; cecilia.ayala@wur.nl

The European agricultural sector faces major environmental and socio-economic challenges, as shown by the growing number of legislations urging the sector to reduce its negative environmental impacts while increasing productivity. Within the agricultural sector, ruminant systems are often identified as key drivers of the negative environmental impacts associated with agriculture (e.g. greenhouse gas emissions -GHGE- and land use/land-use change). Intensifying production by using monocultures and/or higher inputs (i.e. fertilizers) has usually been the main path to address agricultural socio-economic challenges. However, an approach based on ecological processes could tackle the sector's environmental challenges as well. Taking Ireland as a case study, and using a comprehensive bio-economic farm model, we analysed the farm-level productivity, profitability and environmental performance of a traditional beef and sheep co-grazing system (permanent pasture: PP) and compared it to two contrasting intensification scenarios: a monoculture of perennial ryegrass (PRG) and, as an ecological process-based approach, a multispecies sward (MS). Both scenarios resulted in a reduction of GHGE and land use, as well as an increase in productivity and profitability when compared to the original PP system. Nevertheless, and mostly due to a higher productivity of the MS sward, this one outperformed the PRG in all indicators analysed. Farm-level GHGE were reduced by 6.6% in the PRG scenario and by 31.9% in the MS scenario. The MS allowed for a 45% reduction of the original grazing area, whereas the PRG's reduction was of 16.2%. Operating profit and meat productivity per grazing hectare were increased by 79% and 83% respectively in the MS scenario and by 3.8% and 20% in the PRG scenario. These results, although of a limited scope, illustrate the prospective value of ecological-based approaches when trying to answer some of the environmental and socio-economic challenges of the agricultural sector.

Session 35 Theatre 7

Mitigation of greenhouse gas emissions from dairy beef production systems

M. Kearney[1,2], J. Breen[2], E. O'Riordan[1] and P. Crosson[1]
[1]Teagasc, Animal & Grassland Research and Innovation Centre, Grange, Dunsany, Co. Meath, C15PW93, Ireland, [2]University College Dublin, School of Agriculture and Food Science, Belfield, Dublin 4, Ireland, D04V1W8, Ireland; mark.kearney@teagasc.ie

Introduction Beef production has been identified as the most GHG emissions intensive livestock sector. Several studies have identified approaches to reduce emissions from suckler beef cow production systems but there is an absence of similar studies for dairy beef systems. The objective of this present study is to identify management practices which lead to lower greenhouse gas emissions from dairy calf to beef production systems. Material and Methods An existing dairy beef farm systems model was substantially augmented to permit quantification of GHG emissions, net energy and protein production and land use ratios. This new model was used to investigate the potential to mitigate GHG emissions from grass-based dairy beef systems. The baseline dairy beef system was based on the steer progeny of Holstein-Friesian sires bred to Holstein–Friesian dams slaughtered at 24 months of age. Mitigation strategies were based on alternative slaughter ages and feeding systems. Results Higher age at slaughter tended to increase total GHG emissions per animal; however, higher carcass weights from cattle slaughtered later in life offset this to some extent when emissions were expressed per kg beef carcass. Grass only systems resulted in lower 'embodied' GHG emissions in animal feed but led to older slaughter ages and higher total emissions. Judicious concentrate feed supplementation resulted in lower emissions per animal and per kg beef carcass, particularly where by-products from the food and brewing industries were fed. Conclusion Age at slaughter and feeding system are important factors influencing GHG emissions from dairy beef production systems.

Session 35 Theatre 8

Reducing N leaching with better N mineral management using forecasted weather and grass growth

E. Ruelle[1], L. Shalloo[1], M. O'Donovan[1], L. Delaby[2] and P. Dillon[1]
[1]Teagasc, Animal & Grassland Research & Innovation Centre, Moorepark, Fermoy Co. Cork, Ireland, [2]INRAE, Agrocampus Ouest, Physiologie, Environnement et Génétique pour l'Animal et les Systèmes d'Elevage, 35590 Saint-Gilles, France; elodie.ruelle@teagasc.ie

The ecological status of Ireland waterways is above the European average with 52.8% of national surface waters classified as good or high ecological status, however latest reports are showing a decline in water quality. A modelling exercise has been conducted to highlight the positive impact that could have precision fertiliser application on nitrogen (N) leaching compared to a blank approached based only on general recommendation. The simulation were ran from the year 2013-2020 using the MoSt grass growth model. Weather data were collected from the Moorepark met Eireann weather station, Co. Cork. The base fertiliser application timing was using the Teagasc recommendation guideline without taking into account weather condition or grass growth. The positive impact of precision fertiliser application has been shown on the 2018 year which was, in Ireland, a very challenging year with a very cold spring and a drought in the summer. The rules of the fertiliser application for the 2018 year have been based on the same Teagasc recommendation than for the base simulation but some adjustment have been done based on weather forecast and grass growth prediction: (1) if the predicted grass growth in spring for the week ahead was lower than 10 kg of DM/ha or if the rainfall in the days following the planed application was high; fertiliser N application for the week ahead was delayed; (2) 24 kg of N in March was not applied due to the very low soil temperatures/snow; no growth for almost 3 weeks; (3) during the main grass-growing season, as soon as the predicted grass growth for the week ahead went below 25 kg of DM/day no more N fertiliser was applied until grass growth rates recovered above this level. The results has shown that fertiliser application has been reduced by 79 kg/ha (171 kg/ha applied) but which only lead of a reduced grass production per hectare by 259 kg of DM and it did not reduce grass intake per cow. It did lead however to a reduced N leaching of 16% bringing it back to the average leaching of the 18 years simulated where 250 kg N/ha was applied.

Session 35 Theatre 9

Pasture based cow calf contact: effects of full-time, part-time and no contact on calf growth

A.M. Sinnott[1,2], E. Kennedy[1], S. McPherson[1,2] and E.A.M. Bokkers[2]
[1]Teagasc, Animal and Grassland Research and Innovation Centre, Moorepark, Fermoy, Cork, Ireland, P61C996, Ireland, [2]Wageningen University & Research, Animal Production Systems Group, Wageningen, Netherlands, 6708 PB Wageningen, the Netherlands; alison.sinnott@teagasc.ie

Preventing cow-calf contact in rearing systems has come under scrutiny recently. This study investigated growth patterns of calves with part-time contact housed indoors (PT-I), full-time contact outdoors on pasture (FT-O) or no contact housed indoors (NC-I). Cows (n=55) were balanced pre-calving by parity, previous milk production and expected calving date. Contact pairs bonded in individual calving pens for 48h. Part-time cows were milked once-a-day (8:00) and grazed outdoors post-milking until returning to their calves at 15:00-8:00 the following morning. Full-time calves moved outdoors at 2.7±3.29 days old, only separated from cows for milking twice daily. NC-I calves were offered 6-9.5 l/day milk replacer allowance via an automatic feeder. Calves had *ad libitum* access to water, concentrates and forage. Gradual weaning occurred at 8 weeks old. Weekly weighing occurred until 3 weeks post-weaning and fortnightly thereafter. Data is divided into 5 periods: early and late pre-wean (day 3-28 and 29-56), early and late post-wean (day 57-70 and 71-77) and carryover. Average birth date was Feb 12±16.6 days; average birth weight was 34.2±5.75 kg. No differences in weight were found from day 7-21 (day 21: 46.8±2.35 kg). From 28-77 days, NC-I calves weighed less than contact calves (CC)($P<0.05$), which were similar (day 77: 81.2 and 90.5±2.33 kg). From day 91-119, although no differences between NC-I and PT-I calves (day 119: 107.4±2.55 kg), calves in the FT-O system (day 119: 113.5±2.55 kg) weighed more than NC-I calves ($P<0.05$). The CC had higher ADG than the NC-I group early (0.75 and 0.50±0.08 kg/day; $P<0.05$) and late pre-weaning (0.96 and 0.73±0.07 kg/day, respectively). Early post-weaning, NC-I calves had greater ADG than PT-I calves (0.56 and 0.32±0.07 kg/day, $P<0.05$). Late post-weaning, the ADG of NC-I was higher than all CC (0.71 and 0.46±0.07 kg/day, $P<0.05$). Thus, CC had elevated growth pre-weaning, whereas NC-I calves had compensatory growth post-weaning. Low CC growth post-weaning indicates additional weaning stress due to de-bonding and dietary changes.

Farmer's knowledge and practices in mineral supplementation in Western France dairy farms

C. Manoli, M. Salaun and P. Gaignon
Groupe ESA, URSE Research Unit, Rue Rabelais, 49007 Angers, France; c.manoli@groupe-esa.com

Addition of minerals to a basic ration is a common practice in dairy farming for prevention of health problems or production losses. However, precise coverage of mineral requirements is a complex balance because of uncertainties in estimation of feed intake and because individual physiological requirements are very variable. Mineral supplementation is also a cost for farmers that should be minimized in an input reduction perspective. Objective of this study is to improve knowledge on farmers' mineral supplementation practices: how are minerals distributed, between coverage of individual requirements, simplification of practices to reduce work and cost limitation? What are the main advisors with whom they discuss mineral supplementation in dairy farming? Semi-structured surveys were conducted with 18 farmers in Western France, representing a diversity of farms (from more to less grazing; more or less robotization; more or less large herds). Semi-structured interviews were conducted in the Spring of 2020 during the French Covid lockdown and were therefore conducted by telephone or videoconference. A qualitative and thematic analysis of the interviews was then conducted. Three main types of results emerged. First, four categories of farmers' discourse can be distinguished according to the level of precision about mineral requirements and consequences of failures for animals. Secondly, mineral supplementation practices appear to be very diversified, in terms of: quantities and forms of distribution; and more or less individualized intakes (taking into account or not the specificities of dry cows). Third, mineral suppliers appear very diverse, and informational resources on minerals rely on a diversity of consultants, from technicians of performance control to sales representative, veterinarian, cooperative advisors, etc. These results confirm the special place of mineral supplementation in dairy farming: widespread but often with approximations in the precision of the intakes. The diversity of suppliers and advisors raises questions about the way farmers arbitrate between these different sources of advice: what place for empirical technical knowledge and commercial relationship in decisions concerning mineral purchases?

Removing concentrate supply in grazing goats: effects on milk production and grazing behaviour

R. Delagarde and C. Moreau
PEGASE, INRAE, Institut Agro, 16 Le Clos, 35590 Saint-Gilles, France; remy.delagarde@inrae.fr

Grazing high-quality pastures and reducing concentrate supply can increase feed and protein self-sufficiency in dairy goat production systems. How dairy goats are able to adapt to the total removing of concentrate supplementation at grazing is unknown. A grazing trial was carried out in spring 2021 to compare 3 concentrate supplementation levels: 0.72, 0.37 and 0.05 kg DM/day. The concentrate was fed individually twice per day at milkings times. A total of 36 lactating Alpine dairy goats (64 DIM and 4.0 kg/d of milk at the beginning of the experiment) were used in a 3×3 Latin square design replicated 12 times over three 3-week consecutive periods from April to June. Daily access time to pasture was of 11 h/d, including 8 h between milkings and 3 h after PM milking, with a daily herbage allowance of 2.6 kg DM/ d at 4 cm above ground level. The three treatments strip-grazed in separated herds, with front fences moved once daily in the morning. At night, goats were inside with no feed provided and *ad libitum* drinking water available. During the last 5 days of each period, individual milk production, milk composition and grazing behaviour were recorded daily. Behaviour was recorded thanks to the Lifecorder Plus device. Milk production averaged 2.65 kg/d and was reduced linearly with decreasing supplementation level (-67 g of milk /100 g DM of concentrate), as well as milk fat production, milk protein production and milk protein concentration (-27 g/d, -21 g/d and -0,1 g/kg of milk per 100 g DM of concentrate, respectively). Milk fat concentration was reduced on average by 0.5 g/kg of milk per 100 g DM reduction of concentrate supply. Grazing time averaged 453 min/d and increased linearly by 5 min/d per 100 g DM reduction of concentrate supply. As expected, the total suppression of concentrate supply to grazing dairy goats had negative impacts on milk, fat and protein productions, but with effects similar to those observed at greater supplementation levels (e.g. between 400 and 1000 g/d of concentrate). Unsupplemented goats showed good ability to increase their grazing time with fewer but longer meals. Feed-self sufficiency of goat feeding systems may be increased through grazing high-quality pastures with low concentrate supply.

Session 35 Theatre 12

Distinctive organoleptic characteristics and consumer acceptability of pastured poultry eggs

M. Odintsov Vaintrub and P. Di Giuseppe
Regrowth s.r.l.s, Contrada specola (snc), 64100, Italy; regrowth2020@gmail.com

Pastured Poultry (PP) is a chicken rearing system based on mobile coops positioned on open pasture and moved regularly in order to provide fresh forage. PP is becoming increasingly popular among small-scale farmers pursuing higher product value and frequently certified as 'Organic'. In the current work, a total of 24 eggs were collected from a PP setting and a traditional 'Bio' certified coop (Bio). Both were submitted to a blind sensory panel evaluation (ICS 67.240) within 24h of the collection. The eggs were boiled for 15 minutes, left to cool, and then sliced. All 48 evaluators were untrained in food sensory analysis, with ages ranging 25-58 and equal distribution between male/ female. The evaluation included a linear score (1-16) of the following parameters: (1) colour – a visually assisted score for yolk colour intensity; (2) aroma – overall smell; (3) taste – overall taste (3) Aroma+ Taste – combined impression; (4) aftertaste – lingering taste for 60 secs after the consumption; (4) unpleasant tastes – specific characteristics of the taste; (5) general acceptability of the product. Additional space was left for qualitative comments by the assessor. The results were analysed for descriptive statistics and t-test comparison using Microsoft Excel software while qualitative data was analysed following the principles of the inductive approach. A distinct difference was observed in all parameters, with PP eggs having a significantly narrower variance in most aspects ($1.25<V(x)<5.98$) in comparison with Bio eggs ($1.1<V(x)<7.17$). While overall acceptability was relatively similar (t=+2.99 for PP), several parameters had more noticeable differences. The PP eggs had a higher score in Aroma (t=6.24), Aftertaste (t=4.52), and Unpleasant taste (t=7.94) in comparison with Bio eggs. The trend was also supported by qualitative comments with a repeated pattern of words such as 'herbal taste', 'grassy aroma' and 'light' frequent in PP eggs description. These differences support the hypothesis that pasture has a significant impact on the perception of eggs by consumers. The generally higher score despite distinctive taste aspects also confirms the existence of consumer interest in poultry products with distinct pastured related characteristics.

Session 35 Poster 13

Effect of climate and soil physical variables on alfalfa production through big data analysis

J.Y. Kim[1], M.J. Kim[2], J.S. Choi[1], B.W. Kim[1] and K.Y. Sung[1]
[1]Kangwon National University, Department of Animal Industry Convergence, College of Animal Life Science, 24341, Korea, South, [2]Institute of Animal Life Science, Kangwon National University, 24341, Korea, South; kisung@kangwon.ac.kr

The purpose of this study was to evaluate the influence of climate and soil physical variables on alfalfa DMY as a contribution ratio. A yield prediction model was constructed to calculate the contribution ratio of climate and soil physics to the alfalfa dry matter yield (DMY). The process of constructing yield prediction model using Big data was alfalfa, climate and soil physical data collection, data processing, statistical analysis, and model construction. The number of data collected was 280, but the number of data used for the final analysis through data processing was 197. The climate variables were accumulated temperature (GDD, Base temperature 0, 5 and 10), accumulated sunshine duration (ADS, base temperature 0, 5 and 10), and days and accumulated temperature of summer depression (DSD and ASD), minimum temperature of January (MTJ and LTJ), Growing day (GD 0 and 5), and days and accumulated precipitation (NDP 0, and 5, AAP 0 and 5). Soil physical variables were clay content (CC), drainage class (DC), and slope (Sl). The partial-correlation analysis was used to intentionally select variables with overlapping associations among climate and soil physical variables. Factor analysis was used to generate new variables of Climate(C) and Soil Physical(SP). The contribution ratio of climate and soil physics to alfalfa DMY was confirmed by R^2 of multiple regression analysis. The contribution ratio of climate and soil physics was the sum of the contribution ratio of each variable. The climate variables selected through partial correlation analysis were GDD 0-25, 5-25, and 10-30, ADS 0, and ADS 10, DSD, ASD, LTJ, GD 0-25, NDP 0, and AAP 0. The climate and soil physical variables generated through factor analysis were C1, C2, and C3 and SP, respectively. The contribution ratio of C1, C2, C3, and SP to DMY through multiple regression analysis were 25.8, 21.4, 1.4, and 18.1%, respectively. The sum of the contribution ratio of C1, C2, C3, and SP was 66.7%. The contribution ratio of climate and soil physical to DMY using the Big data was 48.6 and 18.1%, respectively, which was higher in climate than in soil physics.

Session 35 Poster 14

Multifunctional Mediterranean AGF systems: constraints and opportunities for sustainable transition

N. Borràs[1], M.P. Romero[2], P. Gaspar[2], F.J. Mesías[2] and M. Escribano[2]
[1]Volterra Ecosystems S.L., Balmes 76, 1-2, 08007 Barcelona, Spain, [2]Department of Animal Production and Food Science, University of Extremadura, Avda. Adolfo Suárez, 06007 Badajoz, Spain; pgaspar@unex.es

To revitalise Mediterranean agroforesty (AGF) systems one of the management options is the implementation of regenerative agriculture techniques, since those have been recognised to have the capacity to produce high-quality food, improve profit margins, sustain biodiversity, restore degraded land and soils, and store carbon, with minimal dependence on external inputs. Dynamic and profitable operational practices for these agrosilvopastoral systems focus mainly on rotational livestock management, soil restoration with compost and biochar, keyline, and ectomycorrhizal fungal inoculations, and are being implemented in several Quercus-based silvopastoral farms in 3 different countries (dehesas in Spain, meriagos in Italy, and montados in Portugal) under LIFE Regenerate project (www.liferegenerate.eu). Over 4 years of implementation several barriers have been observed that hinder implementation of more regenerative practices. In order to identify and classify these difficulties and the opportunities that have also been observed, a focus group session was held with the participation of farmers that currently work with regenerative practices in the project. During the session, they were asked to carry out a joint quantitative SWOT analysis. The main difficulties identified were the lack of qualified human resources, poor support in the transition and in the initial investment (especially regarding infrastructure for rotational grazing), the shortage of public policies to promote and facilitate these practices or the ambiguity of the market insertion of products derived from regenerative agriculture. Despite this, farmers are content and determined to make the change because they see that it is economically viable in the long term. The need for action to ensure a stable, productive, and profitable food supply system is more than evident. However, to do so, the policy scene must create opportunities for market change that favour regenerative agriculture and stimulate its transition tailored to the real needs and concerns of farmers.

Session 35 Poster 15

Grazing behaviour and milk production in dairy cows as an effect of daytime and night-time grazing

H. Gonda, R. Danielsson and E. Ternman
Swedish University of Agricultural Sciences, Department of Animal Nutrition and Management, Ulls väg 26, Box 7024, 75007 Uppsala, Sweden; horacio.gonda@slu.se

In grazing cattle, previous studies have reported increases in body weight gain or in milk, and milk fat and protein yields when the daily grazing strip was allocated at afternoon rather than early in the morning. Those effects were related to changes in herbage chemical composition along the day, better ratio of rumen easily degradable carbohydrates to nitrogen as day progresses, as well as to cattle grazing behaviour as dusk grazing event appears as the most intense of all grazing events of the day. To test whether those changes would occur in high latitudes, with longer period of daylight, we conducted a grazing experiment at the Röbäcksdalen Research Station, Swedish University of Agricultural Sciences (63°48' N 20°14' E), Umeå, Sweden, during June 1-July 2, 2021 (average daylight 20.5 h). Forty-eight Swedish Red cows in mid-lactation, averaging 634±73 kg of body weight, and with an initial milk yield (MY) of 31.4±8.7 kg/cow/d, were included in the study and randomly assigned to two grazing treatments: Morning-Afternoon (MA) and Evening-Night (EN) grazing. Cows were milked at 06:00 h (AM) and at 16:00 h (PM), and the MA cows were outdoors between AM and PM milkings (approx. 8.5 h/d), and EN cows were outdoors between PM and AM milkings (approx. 13 h/d). Cows were provided a new grazing strip with access to water every day. Herbage allowance, grass-legume ley, was around 30 kg DM/cow/d. Even though the time cows spent outdoors differed between treatments, no difference was observed in grazing duration (302 and 296 min/d for MA and EN cows, respectively). EN cows spent more time eating indoors (P=0.008; 132 vs 150 min/d for MA and AE cows, respectively) and ruminating (P=0.001; 422 vs 477 min/d for MA and EN cows, respectively) compared to MA cows. Feed intake indoors and milk yield were similar between treatments (13.3±2.5 kg DM/cow/day for MA cows and 14.8±2.5 kg DM/cow/day for EN cows). MA cows yielded 27±7.2 kg and the EN cows 26±6.7 per day. Our results suggest that the possible benefits of evening-night grazing might be less pronounced in areas with long days and short nights than those situated in lower latitudes.

Session 35 Poster 16

The challenge of the new CAP: a Delphi study on extensive livestock farms in dehesa systems

P. Gaspar, A. Horrillo, F.J. Mesías and M. Escribano
Department of Animal Production and Food Science, University of Extremadura, Avda. Adolfo Suárez, 06007 Badajoz, Spain; pgaspar@unex.es

Extensive livestock farms located in *dehesa* agroforestry systems in the current context of climate change are vulnerable, but through proper management they can become self-sufficient and profitable based on the efficient use of resources and efficiently managing the public resources that they receive as subsidies. The main objective of the regional project MitigaDehex (IB20070) is to analyse extensive dehesa livestock farming systems in Extremadura (SW Spain) as models of sustainable production and agents against climate change. One of the specific objectives is to assess the possible impacts of the post-2023 Common Agricultural Policy (CAP) on these livestock production systems. These impacts may condition in the future, the adaptation of farms with important changes in their current management model. To achieve this objective, a preliminary study has been carried out with the participation of 34 experts with a variety of professional profiles applying a predictive technique: the Delphi method. In this technique, a series of statements related to the effect that the application of the CAP 2023 may have in the future in the area of the dehesa were presented to the participants in a questionnaire distributed online, organised into themes so that they could give their score according to their degree of agreement or disagreement with each one of them. The results show that the positive effects that the experts do not consider will be achieved are fundamentally of a social and economic nature, as the experts consider that there will be no increase in the number of livestock farms, nor will there be a greater number of farms benefiting from the CAP, nor will there be an increase in farmers' incomes. On the other hand, the effects that they do consider will occur are those related to the environment, considering that it will increase the carbon sink capacity of the soil through the promotion of better practices, that there will be an improvement in the management and conservation of soil quality, that there will be a boost to the of traditional agro-silvo-pastoral practices, and that it will improve the use of pastures through better grazing practices such as rotational grazing.

Session 35 Poster 17

DCAB of grass and NSBA in urine of suckler cows

H. Scholz[1] and G. Heckenberger[2]
[1]Anhalt University of Applied Sciences, Faculty LOEL, Strenzfelder Allee 28, 06406 Bernburg, Germany, [2]State Institute for Agriculture and Horticulture Saxony-Anhalt, Lindesntraße 18, 39606 Iden, Germany; heiko.scholz@hs-anhalt.de

Dietary anion-cation balance (DCAB) in grazing systems under German condition tends to decrease from May until September and often are measured DCAB lower than 100 meq per kg dry matter. Lower DCAB in grass feeding system can change the metabolic status of suckler cows and often are results in acidotic metabolism. The hypothesis was that metabolic imbalances could be identified by urine measurement in suckler cows. The farm study was conducted during the grazing seasons 2017 and 2018 and involved 7 suckler cow farms in Germany. Suckler cows grazing during the whole time of the investigation and had no access to other feeding components. Cows had free access to water and free access to minerals (loose). The dry matter of the grass was determined at 60 °C and were then analysed for energy and nutrient content and for the Dietary Cation-Anion Balance (DCAB). Urine was collected in 50 ml-glasses and analysed for net acid-base excretion (NSBA) and the concentration of creatinine and urea. Statistical analysis took place with ANOVA with fixed effects of farms (1-7), month and number of lactations using SPSS Version 25.0 for windows. An alpha of 0.05 was used for all statistical tests. During the grazing periods of years 2017 and 2018 was observed an average DCAB in the grass of 167 meq per kg DM. A very high mean variation could be determined from -42 meq/kg to +439 meq/kg. Reference values in relation to DCAB were described between 150 meq and 400 meq per kg DM. It was found the high chlorine content with reduced potassium level led to this reduction in DCAB at the end of the grazing period. Between the DCAB of grass and NSBA in urine of suckler cows was a correlation according to PEARSON of $r=0.478$ ($P\leq0.001$) or after SPEARMAN of $r=0.601$ ($P\leq0.001$) observed. The influence of several feeding components such as chlorine, sulfur, potassium and sodium and dry matter feed intake during the grazing period of suckler cows should be considered in further research. The results obtained show that up a decrease in the DCAB is related to a decrease in NSBA in urine of suckler cows. Monitoring of metabolic disturbances should include analysis of urine, blood, milk and ruminal fluid.

Nitrogen balance and use efficiency as indicator for monitoring the proper use of fertilizers

J.G. Oliveira[1], M.L. Santana Júnior[1], N.J. Costa Maia[2], A.H. Gameiro[3], P.C. Domingues Dos Santos[1] and F. Simili[1]
[1]Instituto de Zootecnia/APTA/SAA, Núcleo Regional de Pesquisa, de Ribeirão Preto, SP, 14030670, Brazil, [2]Faculdade de Ciências Agrárias e Veterinárias/ Universidade Estadual Paulista, Jaboticabal, SP, 14884-900, Brazil, [3]Universidade de São Paulo, Faculdade de Medicina Veterinária e Zootecnia, Departamento de Nutrição e Produção Animal, Pirassunga, SP, 13635-900, Brazil; flaviasimili@gmail.com

This study aimed to verify the nitrogen (NB) balance and use efficiency (NUE), in integrated crops-livestock systems, in comparison to conventional systems specialized in monoculture. This research evaluated six models of integrated systems and two conventional systems, in a randomized blocks design with three replicates: maize production (CROP), beef cattle under grazing (LS), and four ICLS (integrated crop-livestock system) for grain maize production and beef cattle. The NB indicator showed significantly negative results for CROP in relation to other treatments, possible due to a higher export of N (output), and in function of the high demands that maize crops have for grain production. The NB in LS was significantly higher in comparison to all other treatments. The indicator NUE showed higher efficiency of N use for CROP in comparison to other treatments, but in the LS it was significantly lower in comparison to the integrated systems, which did not differ among each Other. Integrated systems presented more balanced results in comparison to conventional systems, because while they were efficient and presented a NUE varying from 0.89 to 0.99, they managed to maintain the NB positive, with a little surplus of N and without having to appeal to the soil's emergency reserve. The use of NB as an estimate of N dynamics in integrated systems could be an efficient tool for the proper use of fertilizers.

Co-grazing of multispecies swards enhances growth performance of heifers and lambs

G. Beaucarne[1,2], C. Grace[2], J. Kennedy[2], H. Sheridan[1] and T.M. Boland[1]
[1]University College Dublin, School of Agriculture and Food Science, Belfield, Dublin 4, D04 V[1]W8, Ireland, [2]Devenish, Research and Innovation, Dowth Hall, Dowth, Co. Meath, A92 T[2]T7, Ireland; gaspard.beaucarne@devenish.com

The objective was to determine the effect of co-grazing perennial ryegrass, permanent pasture or multispecies swards on heifer and lamb performance. In 2020 and 2021, a grazing experiment was established with four sward treatments: a perennial ryegrass sward (PRG) receiving 170 kg N/ha/y, a permanent pasture sward (PP) receiving 135 kg N/ha/y, a six species sward (6SP) with 2 grasses, 2 legumes and 2 herbs receiving 70 kg N/ha/y and a 12 species sward (12SP) with 3 grasses, 4 legumes and 5 herbs receiving 70 kg N/ha/y. Each sward treatment (9 ha) was rotationally co-grazed from April to November by dairy cross heifers (n=20 per treatment per year, turned out at 397±1.13 days of age, mean ± SEM) and ewes plus lambs (n=22 ewes per treatment per year). Heifers were weighed monthly and drafted for slaughter when their estimated fat class on the EUROP grid scale reached 3-. Lambs were weighed fortnightly and drafted for slaughter at 42 kg (lambing to weaning), 44 kg (weaning to September) and 46 kg (after September 1st) to obtain a target carcass weight of 21 kg. All lambs and heifers were finished from forage only diet. Considering both years, average daily gain (ADG) was higher for the heifers grazing the 6SP sward (1.09 kg/day) compared to the 12SP (0.99 kg/day), the PRG (0.92 kg/day) and the PP swards (0.92 kg/day; P<0.001). ADG from lambs grazing the 6SP (0.39 kg/day) and 12SP (0.36 kg/day) were greater than the PP (0.30 kg/day) or PRG swards (0.29 kg/day; P<0.001). Kill out percentage was higher for lamb grazing the 12SP (48.7%) and the 6SP (48.4%) compared to lambs grazing the PRG (44.8%) and PP swards (44.7%; P<0.001). Lambs grazing the 6SP and 12SP swards had reduced number of grazing days from turnout to slaughter (82 and 93 days respectively) compared to the lambs grazing the PP (127) and PRG swards (132; P<0.01). Overall, co-grazing multispecies swards improved heifer and lamb performance.

Session 36 Theatre 1

Digital traceability and data acquisition in pastured poultry mobile system: preliminary results

M. Odintsov Vaintrub and P. Di Giuseppe
Regrowth s.r.l.s, Contrada specola (snc), 64100, Italy; regrowth2020@gmail.com

Technological progress is providing several useful tools for 'precision livestock farming' (PLF) in extensive livestock farming which fulfils specific zootechnical, health, ethical, and marketing needs. The present study aimed to evaluate PLF feasibility and value in a Pastured Poultry (PP) production system. This is a niche high-value production chain with a strong focus on animal welfare and direct marketing. Three PLF systems were positioned in different mobile coops (12 sqm/ coop) which were moved daily according to a grazing plan on a 3,000 sqm of pasture. An additional system was placed in a traditional 'Bio' certified coop. The field setting included the following components: (1) electronic identification (EID) of individual animals; (2) walk over weight (WOW) sensor station; (3) environmental sensors and VR unit; (4) data management and communication unit; (5) external server (cloud) and software (WebApp). Customer traceability was provided by a quick response code (QR) integrated on an experimental product label. A consumer panel assessment (ICS 67.240) with an integrated qualitative open questionnaire included a group of 48 untrained panellists and was conducted on two independent occasions. During the trial period (22 days) the average weight in the PP system was slightly increased (+0.35 kg), while the weight variance slightly decreased (V(x)=0.07, day 22). In comparison, Bio housed animals had no change in average group weight while having a slight increase in group variance (V(x)=0.11, day 22). Mobility remained similar for both groups, with 4.78 passages a day on average. Environmental data recorded a greater day/night difference in the PP system (0 °C<t<12 °C) than Bio housing (7 °C<t<12 °C), this may have contributed to the slight weight gain of pastured birds which increased their plumage mass. An inductive approach analysis highlighted the increased interest of consumers regarding the provided information. An increase in product acceptability score (21.4%) and attributed economic value (37.2%) were observed when farm data was added to the panel information base. In conclusion, product traceability of a single animal in PP production is a technically viable option and provides the farmer additional value as a marketing tool for high-value products.

Session 36 Theatre 2

Adaptation of cows to automatic milking in a pasture grazing system

B. O'Brien, J. Foley and P. Silva Bolona
Teagasc, Animal and Grassland Research and Innovation Centre, Moorepark, Fermoy, Co.Cork, Ireland; Bernadette.obrien@teagasc.ie

This study monitored the time during which a cow herd acclimatised to automatic milking (AM), i.e. reached a steady-state in terms of milking and performance parameters. Commencing AM with an inexperienced dairy herd presents significant challenges to both the operator and cows. A key objective is to minimize the time taken for the cow herd to become accustomed to the AM system, thus ensuring minimum stress to the cows and minimum reduction in production output. This study focused on the first eight weeks of transition from conventional to automatic milking for a dairy cow herd in a pasture grazing system. A herd of 59 cows were assembled and introduced to AM on 20 July, 2022. The cows were milked in 1 AM unit. Cows were grazed in a 3-way system and received 3 kg concentrate/cow per day. The herd average calving date was 9 Feb (162 days in milk at start-up); 26% of the herd were first lactation animals. These cows had not undergone any familiarization or training on the AM system prior to the study. Average number of milkings/ day were 117, 112, 94, 92, 92, 86, 89, and 86 over the eight weeks of the study. Average milking intervals (h) were 11.4, 12.7, 14.6, 16.0, 15.5, 16.4, 16.0 and 15.8, while milk yield (kg)/cow per day levels were at 20.6, 20.9, 18.1, 17.9, 17.2, 15.9, 16.3 and 16.1, respectively. These data indicate a relative steady state being reached at 21 days. The reduced number of milkings/day and increased interval during the first 21 days post-introduction of cows to AM may result from the gradual receding of human presence and training. However, more detailed 3-day average values showed that the coefficient of variation increased consistently as time progressed (from 0.27 in the first week to 0.51 in the last week of the analysis). Variation in milking frequency was also greater for cows in their parity 1 and 2 compared to 3 or over. Thus, the data is suggesting that as time progressed there was more reluctance of cows to approach the AM and that this effect was more marked in younger cows. This was strongly influenced by the advancing stage of lactation of this seasonally calved herd, and it is an important aspect to take into account when aiming to optimize AM operation in such a herd in the latter half of the lactation.

Session 36 Theatre 3

Can data from commercial PLF sensor give us information on dairy cow performances?

M. Bouchon[1], S. Scully[1], M. Coppa[2], E. Ollion[3], B. Martin[4] and P. Maitre[5]
[1]INRAE, UE Herbipôle, 63122 Saint-Genes-Champanelle, France, [2]Independant researcher at INRAE, UMRH, 63122 Saint-Genes-Champanelle, France, [3]Centre de Développement de l'Agroécologie, 5 Place Aristide Bouvet, 01500 Amberieu-En-Bugey, France, [4]INRAE, VAS, UCA, UMRH, 63122 Saint-Genes-Champanelle, France, [5]Montbéliarde Association, 4 rue des Epicéas, 25640 Roulans, France; matthieu.bouchon@inrae.fr

Animal behaviour as a proxy for animal performance, health or welfare is of growing interest. Precision Livestock Farming (PLF) sensors are increasingly more common on farms and allow for large behavioural dataset acquisition, that can be used to identify cattle that are better able to graze. To explore the potential of behavioural data acquired through PLF sensors to determine links between animal behaviour and individual performances a study was conducted on 20 dairy cows in mid-lactation, grazing on permanent grassland over 45 days in spring 2019, without supplementation. Cows were equipped with commercial sensors (Axel Medria®) that report behavioural activity data in 5 minutes intervals. Ingestion (ING), rumination (RUM) and resting (REST) times per day and their relative number of events (ev) were explored. All behavioural variables were expressed per unit of BW to limit individual size effect. Principal component analysis (PCA) and clustering were used to define 3 behavioural patterns (BP). The effect of the cluster on Fat Protein Corrected Milk (FPCM) and Body Condition Score (BCS) was tested through a general linear model. BP1 was characterised by more selective patterns (high ev_ING/BW and low RUM/BW), likely due to active seeking of more digestible feeds, diminishing rumination time. BP2 was characterised by non-selective patterns (low ING/BW and high RUM/BW) and BP3 by 'typical' behaviour for grazing cows. Cattle who expressed BP1 had a higher FPCM than BP2 and BP3 (17.6, 16.2, 17.2 kg respectively; $P<0.05$) whereas cattle performing primarily BP2 during the period had a lower BCS than BP1 and BP3 (2.04, 2.24, 2.24 respectively; $P<0.05$). In conclusion, cattle with more selective behaviour (BP1) seem to have a higher FPCM at pasture while able to maintain body reserves. This first approach confirms that accelerometer generated data could potentially be used as a proxy of cattles ability at pasture and opens the field to larger study.

Session 36 Theatre 4

Are GPS sensors suitable to ensure the traceability of dairy cows on pastures?

A. Lebreton, C. Allain, C. Charpentier, M. D'Introno, A. Fischer, W. Lonis, E. Nicolas and A. Philibert
Institut de l'Elevage, 149 Rue de Bercy, 75595 Paris, France; adrien.lebreton@idele.fr

Since 2007, the 'pasture milk' recommendation has grown further in Europe. It requires the cows to spend a minimal duration on pasture (>6 h). Our objective was to develop and compare two different algorithms that estimates the time dairy cows spend outside the barn (TOut).The first algorithm is based on an automatic detection of the barn and therefore of the pastures. With an analysis of cows' locations local density distribution, we identified a local density threshold to discriminate the cows' locations that are on pasture from the cows' locations that are in the barn. The second algorithm analyse cows locations regarding map data of the farm (barn and paddocks positions) and apply kinetics corrections of cows' locations. We defined TOut as the sum of locations labelled 'on pastures' multiplied by the time interval between location acquisitions (11 min). We tested the solution (GPS sensors + algorithm) on 2 experimental farms in 2019 (2 datasets) and 2020 (1 dataset). Farms were equipped with 8 and 9 GPS sensors. The TOut reference was recorded by an RFID antenna at the barn entry or manually by the farm staff. On experimental farms the estimations of the TOut were accurate with RMSEs from 17 min/d to 50 min/d. Both algorithms provide very similar results. However, the solution (GPS sensors + algorithms) needs to face on farm conditions to evaluate its reproducibility with a variety of farm systems and quality of GPS data.

Tracking cows with GPS: can we know when they are calving?

M.J. García García[1], F. Maroto Molina[1], D.C. Pérez Marín[1] and C.C. Pérez Marín[2]
[1]University of Cordoba, Animal Production, Campus de Rabanales, Ctra. Madrid-Cádiz km 396, Córdoba, 14071, Spain, [2]University of Cordoba, Animal Medicine and Surgery, Campus de Rabanales, Ctra. Madrid-Cádiz km 396, Córdoba, 14071, Spain; g42gagam@uco.es

Calving detection allows an early assistance in case of dystocia. Thus, different sensor-based solutions have been developed and are commercially available. Most of these technologies are devoted to intensive farming systems, where calving takes place in confined areas. In rangelands, animals are dispersed in large areas, which challenges data gathering and transmission. Nowadays, GPS collars are becoming increasingly popular among farmers, which poses an opportunity to gather behavioural data of grazing cattle. Nevertheless, the temporal resolution of GPS data is normally low due to battery limitations (most devices provide location fixes every 30 min to 2 h). The objective of this study was to assess the feasibility of low-temporal resolution GPS data to detect behavioural changes associated with calving. Forty-two cows were fitted with GPS collars (Digitanimal®, Spain) at least 10 days before and after the expected calving date. Collars provided fixes every 30 min when data coverage was available. Location data were analysed with R version 3.6.3, using the packages trajr and adehabitatHR. Trajectories were calculated for every animal and every day. Behavioural indicators were calculated for single animals (daily path length and home range) and in relation to other animals (distance to herd centroid). Significant differences were found between path length before and after calving, with a decreased path length after calving, but no significant differences existed between the day of calving and surrounding days. Significant differences among days were also observed for home range, with a peak in day -1, which corresponds to the exploring behaviour of cows searching for a suitable calving site. Distance to herd centroid also showed significant differences, with a maximum in calving day due to the isolation seeking behaviour of cows. Calving-related behavioural changes were identified on a daily scale, being the distance to the centroid of the herd the most adequate indicator to detect calving date. However, this indicator, contrary to path length and home range, requires the use of multiple collars.

Machine-learning techniques to help decision-making of replacement beef cows in extensive conditions

P. Guarnido-Lopez[1], L. Quevedo[2,3], C. Perez[3], L.O. Tedeschi[4] and M. Benaouda[5]
[1]INRA, UMRH, Saint-Genès-Champanelle, 63000, France, [2]Jandavet, C. Paterna, 4, 11170, Spain, [3]Universidad de veterinaria de Cordoba, Departamento reproduccion animal, Av. de Medina Azahara, 14005, Spain, [4]Texas A&M University, Department of Animal Science, College Station, TX 77845, USA, [5]Agrosup Dijon, Production animal, 26 bd Docteur Petitjean, 21079, France; pabloguarnido@hotmail.com

Replacement decisions have a significant effect on breeding farms' profitability. Some machine learning (ML) techniques have been used in dairy cattle to help decision-making of farmers in replacement decisions. However, these techniques have only been used in dairy cows because they require the cow value, estimated from parameters daily measured as DMI or milk yield. This work aimed to develop an algorithm to help with the decision-making of replacing cows in extensive beef conditions, where these technologies are still not implemented. For this, we used individual beef cow data (n=258) in extensive farms collected by veterinaries to develop an unsupervised machine learning model to cluster cows according to their reproductive performances. Best cows were chosen by; age (no more than ten years), first calving age (as early as possible), calving interval (as short as possible), number of calving (as more as possible), pre-weaning mortality (no more than 10%) and neospore diagnosis (yes=discard). We also conducted a density-based clustering analysis to determine the optimal number of clusters more decreasing intra-group variability. Every research was performed in R, and clusters were created through k-means method. Results showed the optimal number of clusters decreasing intra-group variability was four. We grouped animals according to 4 clusters, and the model resembled those with similar characteristics showing the variability of each character related to the average. We quickly identified the worst animals to replace, cluster 3 showed animals positives to neospore, and cluster 4 showed cows with higher pre-weaning mortality. Favourable animals to keep on the farm were those from cluster 1 and 2, especially those from the first one, which were younger. Therefore, unsupervised ML models could be a simple and easy tool to help-decision making of replacement cows in beef breeding farms.

Heifers, sheep, and grass adapt well to rotational grazing managed with virtual fencing

A. Fischer[1], D. Deleau[2], L. Depuille[1], G. Dufour[3], S. Fauviot[3], D. Gautier[1,4], T. Huneau[5], C. Lerond[2], P. Mangin[2], L.A. Merle[3], A. Philibert[1], A.S. Thudor[4], A. Jomier[2] and A. Lebreton[1]

[1]Institut de l'élevage, 149 rue de Bercy, 75595 Paris, France, [2]arvalis institut du végétal, Ferme expérimentale de Saint Hilaire en Woëvre, 55160 Saint Hilaire en Woëvre, France, [3]Chambre d'agriculture Pays de la Loire– Ferme expérimentale des Etablières, Route du Moulin-Papon, 85000 La Roche sur Yon, France, [4]CIIRPO, ferme expérimentale ovine du Mourier, 87800 Saint Priest Ligoure, France, [5]Chambre d'agriculture Pays de la Loire, Ferme expérimentale de Derval, 15 la touche, 44590 Derval, France; adrien.lebreton@idele.fr

Virtual fences (VF) have been developed to make fencing management less time-consuming for farmers. The current study aimed at testing the ability of dairy and beef heifers, and sheep to adapt to VF. Several tests were performed in 4 experimental farms of the Digifermes® network. In total, those tests were performed on 46 heifers and 19 sheep, in spring and autumn 2021. Each animal wore a neckcollar mounted with a GPS sensor. This sensor tracks each animal's position, plays a sound when the animal crosses the virtual boarder, and delivers up to three electric stimuli if the animal keeps moving off the paddock. Each test included an adaptation period of 2 to 4 days for the heifers and 2 for the sheep to teach them the VF functioning. For three farms, the experimentation included a control and a test group. Both groups were managed with a rotational grazing (3-5 d./paddock) during 1 to 4 months, but only the control group had physical fencing while the test group had VF. Both groups were monitored for average daily gain (ADG) or body condition score (BCS), grass growth and one farm monitored also behaviour. Overall, both groups had similar grass growth, ADG or BCS, and behaviour. All animals adapted quickly to the VF, as shown by the fast decrease of sound and electric alerts the first days of adaptation. However, we've observed a high variability between animals with some having very low sound or electric stimuli, and others maintaining a high number of sound and electric stimuli. Future research may consider to study the relationship between this diversity of behaviours, and hierarchy, leadership and exploration behaviour to improve VF management.

Deployment of models predicting compressed sward height on Wallonia: confrontation to ground truth

C. Nickmilder[1], A. Tedde[1], I. Dufrasne[2,3], F. Lessire[2,3], B. Tychon[4], Y. Curnel[5], N. Glesner[6], J. Bindelle[1] and H. Soyeurt[1]

[1]TERRA Research Centre, Uliège, Passage des Déportés 2, 5030 Gembloux, Belgium, [2]Centre des Technologies Agronomiques, Rue de la Charmille, 16, 4577, Belgium, [3]Uliège, Faculté de Médecine vétérinaire, Place du 20 Août 7, 4000 Liège, Belgium, [4]Uliège, Département des sciences et gestion de l'environnement, Avenue de Longwy 185, BE-6700 Arlon, Belgium, [5]Centre wallon de Recherches agronomiques (CRA-W), Rue de Liroux, 9, BE-5030 Gembloux, Belgium, [6]Fourrages Mieux ASBL, Horritine, 1, 6600 Michamps (Bastogne), Belgium; charles.nickmilder@doct.uliege.be

Currently, pasture management is of interest for economical or ecological reasons. The use of remote sensed data and the implementation of machine learning algorithms is growing. So, over the past two years, models predicting the available compressed sward height (CSH) in Walloon pastures using Sentinel-1, Sentinel-2, and meteorological data were published. Those models were developed to be integrated in a decision support system (DSS). A platform predicting CSH over Wallonia was therefore developed. The variability of the predicted CSH within parcels ranged from 0 to 287.7% once the non-finite values and the values out of the training range were discarded. Concerning the CSH values, the five developed models predicted CSH below 75 mm more than 75% of the time. These values were compared with an independent dataset including a total of 122 average measures of CSH were available and concerned 5 different parcels, grazed in 2019. These reference values ranged from 45 to 212.5 mm of CSH with a mean of 83,8±31.2 mm. The estimated root mean square error values estimated between predicted and reference values varied between 20 and 35 mm of CSH. The coefficient of determination ranged from 0.6 to 0.8 depending on the model and the parcel considered. The poorest performances were recorded on parcels that were split in sub-parcels managed differently during the year. So, there is a need for including flexibility in the parcel definition for the future DSS, the visual support and their corresponding analysis.

Improving the prediction of grass regrowth after a dry period with the MoSt GG model

L. Bonnard[1,2], E. Ruelle[2], M. O'Donovan[2], M. Murphy[1] and L. Delaby[3]
[1]Munster Technological University, Department of Process, Energy and Transport Engineering, Cork, T12 P928, Ireland, [2]Teagasc, Animal & Grassland Research and Innovation Centre, Moorepark, Fermoy, Co. Cork, P61 C996, Ireland, [3]INRAE, AgroCampus Ouest, Physiologie Environnement et Génétique pour l'Animal et les Systèmes d'Elevage, 35590 Saint Gilles, France; laeticia.bonnard@teagasc.ie

Knowledge of past and future grass growth is an important factor for good grazing management. It allows the farmer to estimate the evolution of the amount of available grass for their cattle and to adapt grassland management accordingly. In Ireland, the Moorepark St Gilles Grass Growth Model (MoSt GG) is used to predict weekly grass growth on 78 farms across the country. The growth is communicated to farmers and advisors directly, and also through media outlets (national TV, newsletter and journals). The MoSt GG model is a dynamic model operating at the paddock level that takes into account weather data, soil characteristics, grassland management, including nitrogen fertilization practices. One of the weakness of the model was the absence of regrowth or a delayed response of the regrowth after a drought or dry period. The model has been updated to address this issue. Before the update, the soil was considered homogeneous throughout its depth (up to 1 m). When a rain event occurred after a drought, the water was distributed evenly over the entire soil depth. The growth was limited because the rain was not sufficient to reconstitute the water reserve. In the updated version of the model, a new water pool corresponding to a small surface layer has been created. During a drought or post-drought period, the water deficit is calculated only on this new water pool and not on the total depth of the soil. This new pool is filling up more quickly which allows the growth to restart faster after a drought. Primary results show that for years such as 2006, 2013 or 2018, which were particularly dry for Ireland (Moorepark), the simulated annual yield from the updated model has increased on average by 0.9 t DM/ha/yr (+ 13.9%), compared to the previous version of the model. On free draining soil, the updated version of the model simulated over 30 days of additional growth and about 20 days of double growth compared to the previous version in 2013 (after the dry period).

Attitudes of European small ruminant farmers towards new digital technologies

A. McLaren[1], L. Depuille[2], N. Katzman[3], A. Bar Shamai[3], I. Halachmi[3], L. Grøva[4], T.W.J. Keady[5], B. McClearn[5], V. Giovanetti[6], P. Piirsalu[7], O. Nagy[8], J.-M. Gautier[2], F. Kenyon[9] and C. Morgan-Davies[1]
[1]SRUC, West Mains Road, Edinburgh, EH9 3JG, United Kingdom, [2]IDELE, Campus INRae, 31321 Castanet Tolosan, France, [3]ARO, The Volcani Centre, 7505101, Rishon LeTsiyon, Israel, [4]NIBIO, Gunners veg 6, 6630, Tingvoll, Norway, [5]Teagasc, Athenry, Co Galway, Ireland, [6]AGRIS, Viale Adua, 07100 Sassari, Italy, [7]EULS, Fr.R. Kreutzwaldi 1, Tartu 51006, Estonia, [8]UNIDEB, Egyetem Ter 1, Debrecen 4032, Hungary, [9]MRI, Bush Loan, Penicuik, EH26 0PZ, United Kingdom; ann.mclaren@sruc.ac.uk

Despite continued advances in the application of precision livestock farming (PLF) tools and digital technologies associated with some livestock species, small ruminant production industries often still lag behind. Before attempting to improve the profitability of these production systems (both meat and dairy), through the increased uptake of PLF tools and digital technologies, it is necessary to assess the current needs and barriers of implementation. An online survey was launched in April 2021 across all countries (and languages) participating in the Sm@RT: Small Ruminant Technologies project (UK, Ireland, France, Norway, Hungary, Italy, Estonia and Israel). The aim was to assess the opinions and needs of sheep and goat farmers and other stakeholders regarding the use and uptake of PLF tools and digital technologies on farm and beyond the farm gate. After data quality checks, a total of 669 surveys were included in the analyses, 68% of which were completed by farmers, shepherds or farm workers across the meat sheep, dairy sheep and dairy goat sectors. The most popular tools already used on farms related to flock management (weigh crate, 40%; flock/herd management software, 35%). Virtual fences ranked highest on the list of those deemed to be the most beneficial tool to their system. More details will be covered in this paper. Overall, the data has provided an invaluable insight into the needs and priorities of various stakeholders within the sheep and goat industries across the participating countries. It has also provided important information on the current level of technology in use as well as identifying barriers to uptake. Although production type brought some differences, there were notable commonalities across all the different participants.

Session 36 Theatre 11

Explore the potential & link opportunities by information technology (EXPLOIT) for the feed sector

D. Schokker and S.K. Kar
Wageningen University & Research, Droevendaalsesteeg 1, 6708 PB Wageningen, the Netherlands; soumya.kar@wur.nl

There is a wealth of information that is publicly available, but its potential is not yet fully exploited. To exploit this wealth of information, the management team of Wageningen Livestock Research has funded an initiative to develop a prototype that adds value to animal nutrition and agri-food. The prototype developed is called *EXPLOIT*, *Ex*plore the *P*otential & *L*ink *O*pportunities by *I*nformation *T*echnology. Our innovative pipeline is a blend of bio-IT and human intelligence. It focuses on integrating various publicly available resources and then leverages our expertise in animal and veterinary science to add value for our clients. Our software tool helps navigate to potential new feed applications. We envision that our software tool can help the animal nutrition industries to further diversify their product portfolios and open new routes to market. The results from *EXPLOIT* will enable research and development teams to make well-informed and rapid decisions. In doing so, *EXPLOIT* can help with both tactical and strategic decisions that will enable users to achieve maximum market penetration with their existing portfolio and/or open up new market niches by providing appropriate guidance for the development of new products with novel ingredients/additives.

Session 36 Poster 12

Development of a hybrid monitoring system for dairy cattle

J. Maxa[1], D. Nicklas[2], J. Robert[3], S. Steuer[2] and S. Thurner[1]
[1]Bavarian State Research Centre for Agriculture, Institute of Agricultural Engineering and Animal Husbandry, Vöttingerstr. 36, 85354 Freising, Germany, [2]University of Bamberg, Chair of Mobile Systems, An der Weberei 5, 96047 Bamberg, Germany, [3]University of Erlangen-Nürnberg, Institute of Information Technology (Communication Electronics), Am Wolfsmantel 33, 91058 Erlangen, Germany; jan.maxa@lfl.bayern.de

Animal tracking, especially in extensive grazing areas, has been practiced in the field of research for several decades using various techniques, mainly GPS. Due to the high power consumption for positioning and data transmission, GPS tracking systems are particularly suitable for extensive, wide-range pastures where low-frequency positioning is sufficient. However, such systems are less suitable for dairy cattle grazing on intensive pastures with regularly milking in the barn. There is currently a lack of animal monitoring systems that reliably provide data on the location of the animals with high frequency in the barn and on the pasture as well. Therefore, the main aim of this project is to develop and test a cost-effective and energy-efficient localization system for cattle in combined barn and pasture conditions. Furthermore, the workload on selected dairy farms will be investigated throughout the seasons 2022 to 2023. For monitoring dairy cows, Low Power Wide Area Network (LPWAN) technology is used on pasture and Bluetooth Low Energy (BLE) technology in the barn. Both technologies will be finally implemented as a hybrid localization system. The reference positions of the new monitoring system will be achieved from manually recorded videos and other barn facilities like the milking robot. The first tests of the localization system in the barn have been successfully started, the deployment on pasture begins in the 2022 grazing season. With the help of permanently available location information on all animals in the herd, it will be possible to implement targeted changes in the area of animal control and management, especially in dairy farming with pasture use. Furthermore, animal welfare could be improved by early detection of situations relevant to health or herd management, whereas the labour input needed to search the animals on pasture and in the barn should be reduced.

Displacement rate during grazing of dairy cows under different defoliation intensities

D.A. Mattiauda[1], G. Menegazzi[1], O. Fast[2], M. Oborsky[1] and P. Chilibroste[1]
[1]Universidad de la República Uruguay, Facultad de Agronomía, Animal production and pasture-EEMAC, Ruta 3, km 636. Paysandú, 60000, Uruguay, [2]Colonia El Ombú, Ruta 3 km 284, 65100, Uruguay; dma@fagro.edu.uy

An experiment was carried out to evaluate the effect of two defoliation intensities of a tall fescue based pasture on the animals displacement rate (DR) during grazing. Two treatments with four spatial replicates were compared in plots of 0.3 ha each; treatments were Lax (TL) and Control (TC), with a post grazing sward height of 12 and 6 cm, respectively. Twenty-four mid lactating Holstein cows were blocked according to parity, body weight, body condition score and calving date. Cows had access to the grazing plots from 8:00 to 16:00 with an initial herbage mass of 2,530±180 kg DM/ha. The criteria to start grazing was when the pasture reached the three-leaf physiologic stage. Between PM (17:00 h) and AM (05:00 h) milking's, cows remained separated in pens with individual feeders supplemented with 5.5 kg DM of concentrate (17% CP and 2.8 Mcal EM/kg DM), and extra 1.9 kg DM of concentrate were fed during AM milking. During the access to pasture, a GPS device (Polar® M400 Electro Oy, Finland) was fitted to one cow per plot on the 3rd (beginning) and last day (end) of occupation period (OP) that was 6 d for TL and 8 d for TC. Grazing time (GT) was determined by visual observation. The DR (m/min) was calculated by dividing total displacement while animals were grazing by the duration of the recorded GT in each grazing session (GS). The data was analysed with Proc Glimmix of SAS and means were declared different when Tukey test <0.05. The DR was higher for TC than TL (0.98 vs 0.55±0.021 m/min.), as was GT for TC than TL cows (374 vs 327 min/d.). The DR in the first GS was higher than in the second one (0.96 vs 0.56±0.012 m/min.), but it decreased throughout the OP and was higher at the beginning than at the end of OP (1.04 vs 0.52±0.015 m/min.). Even though we expected an interaction between day of the OP and treatment this was not found. The TL cows reduced their DR and probably dedicated more time to each bite while grazing. The TC cows increased the DR as a strategy to visit more feeding stations, searching for bites as they probably tried to compensate the low bite mass by increasing the bite rate.

Circadian rhythmicity of sheep body temperature changes during transhumance, measured by bio-loggers

J.A. Abecia[1], F. Canto[1], M.A. Saz[2], M. Ramo[3], J. Plaza[4], A. Bjarnason[5] and C. Palacios[4]
[1]IUCA, UNIZAR, Fac.Vet., 50013 Zaragoza, Spain, [2]IUCA, UNIZAR, Dept Geography, 50009 Zaragoza, Spain, [3]IA2, UNIZAR, Fac.Vet., 50013 Zaragoza, Spain, [4]USAL, Fac.Ciencias Agrar Amb, 37007 Salamanca, Spain, [5]Star Oddi, Skeidaras 12, 210 Gardabaer, Iceland; alf@unizar.es

A flock of 3,000 heads of Merino de los Montes Universales sheep practices a 23-d transhumance (Tr) in Nov. The flock walks from its home summer mountains in the north, to the winter pastures, in southern Spain (350 km). Seven days before the Tr, two ewes received a surgically implanted subcutaneous bio-logger (DST milli-HRT, Star Oddi, Iceland) (13×39.5 mm, 12 g) programmed to record body temperature (BT) every 15 min for 35 d. Bio-loggers were retrieved seven days after the end of the Tr, and the data were downloaded with the Mercury software v5.83. The flock was permanently kept outdoors. Sheep walked from 9 am to 17 pm. Mean BT (°C) recorded four days before Tr, during the whole Tr and four days after Tr, were compared by the T-test for related samples, as well as mean BT during walking and during rest, and during daytime (8 am-18 pm) and night-time (19 pm-7 am). Circadian rhythmicity of BT was tested, and mesor (average value around which the variable fluctuates), and acrophase (time of peak activity) were calculated. Mean (±SE) BT before Tr (38.64±0.02) was higher (P<0.001) that during Tr (37.71±0.04) and after Tr (37.71±0.04). No differences between day and night-time BT were observed the days before Tr (38.36±0.03 vs 38.35±0.02, resp.), although ewes had higher (P<0.001) BT during daytime compared with night-time during (38.06±0.05 vs 37.41±0.03) and after Tr (37.85±0.05 vs 37.65±0.03). BT during walking (38.19±0.03) was higher (P<0.001) than while resting (37.43±0.04) during Tr; this difference was maintained after Tr comparing the same periods of time (37.88±0.05 vs 37.66±0.03; P<0.001), but were similar before Tr (38.37±0.03 vs 38.36±0.02). BT did not present a circadian rhythmicity before Tr, but it did during (mesor 37.71 °C; acrophase 15:39 pm) and after Tr (mesor 37.75 °C; acrophase 17:13 pm). In conclusion, transhumance induces changes in BT when compared with the days previous to the migration, and these changes are maintained during the first days after arriving to the winter destination.

Session 36 — Poster 15

Assessing grazing behaviour of goats with GPS collars and sensors in a Mediterranean forest rangeland

Y. Chebli[1,2], S. El Otmani[1,2], J.L. Hornick[1], J. Bindelle[3], M. Chentouf[2] and J.F. Cabaraux[1]
[1]Faculty of Veterinary Medicine, University of Liège, Department of Veterinary Management of Animal Resources, Avenue de Cureghem 6, 4000 Liège, Belgium, [2]Regional Center of Agricultural Research of Tangier, National Institute of Agricultural Research, Avenue Ennasr, BP 415 Rabat Principale, 10090 Rabat, Morocco, [3]Gembloux Agro-Bio Tech, University of Liège, Precision Livestock and Nutrition Unit, TERRA Teaching and Research Center, Passage des Déportés 2, 5030 Gembloux, Belgium; youssef.chebli@inra.ma

Mediterranean forest rangelands are an important component of the extensive goat production systems. In northern Morocco, the diet requirements of goats are mainly provided by the mountainous forest rangelands. Data on individual behaviour, such as movement and activity patterns, are often important for the management of grazing animals. The recent development of precision livestock farming tools provides a real opportunity to understand the grazing behaviour of animals. In the Mediterranean forest rangeland of northern Morocco, eight local goats were selected to monitor their seasonal grazing behaviour using GPS collars and tri-axial accelerometers. Goats spent most of their daytime foraging budget grazing (including actual forage prehension, searching, and walking) during spring and autumn. The goats prolonged their lying time in summer at the expense of standing duration. The number of steps was numerically similar and significantly greater in both seasons of summer and autumn. According to the CART analysis, the time spent grazing (eating) was longer in spring and similar in summer and autumn. Resting while standing was similar between seasons. The time spent walking without grazing (eating) was greater during autumn and summer compared to the spring. Based on GPS collars results, the grazing area was estimated to be 8, 18, and 16 ha in spring, summer, and autumn, respectively. The combination of GPS collars and leg sensors proved very useful to explore the grazing area and behaviour of goats in southern Mediterranean forest rangelands. Monitoring the grazing behaviour of goats through new technologies constitutes a great potential to improve livestock and rangeland management.

Session 36 — Poster 16

A precision technology to optimize accuracy and efficiency of grass measurement on grazed pastures

D.J. Murphy[1,2], B. O'Brien[2] and M.D. Murphy[1]
[1]Munster Technological University, Department of Process, Energy and Transport Engineering, Bishopstown, Cork, Ireland, [2]Teagasc, Animal and Grassland Research and Innovation Centre, Moorepark, Fermoy, Co. Cork, Ireland; bernadette.obrien@teagasc.ie

Increasing grass utilisation can have major financial benefits, as fresh grass is the cheapest feed source on some production systems, e.g. Irish ruminant livestock farms. But maximising grass utilisation and optimal grassland management is highly dependent on frequent and accurate measurement of grass quantity and quality. Even though the rising plate meter (RPM) is the most established, objective and rapid tool for measuring grass, little research has been conducted on its accuracy in measurement of Irish swards in recent years. Measurements are typically carried out 30-50 times in a 'W' pattern or along transects of a paddock, which leaves possibility for operator error and bias. Traditionally it was considered that increasing the number of measurements taken within a paddock would, in turn, increase measurement precision. But labour requirement and availability are issues on many farms and there needs to be a balance between the benefit of increasing measurement frequency/accuracy and the cost of measurement time/effort. The reduction in measurement time/effort is also important to encourage more farmers to measure grass regularly. This paper presents a grass measurement optimization tool, designed to generate measurement protocols that optimize for time and accuracy of grass measurement for farmers. Rising plate meter (RPM) grass measurements and reference herbage cuts were performed on trial plots and grazed paddocks over three years. Survey error was estimated in terms of relative prediction error using Monte Carlo simulations that combined measurement and calibration error distributions for the RPM. Calibration error was the largest source of error (25.9%) compared to measurement error (8%). Actual error for the RPM decreased from 37% to 26% as measurement rates increased from 1 to 8 /ha and reductions in error were negligible (<1%) as measurements increased from 8 to 32 /ha. Optimal measurement value was achieved by performing 8 measures/ha and further increasing the measurement rate resulted in diminishing returns.

Nutrient circularity: the role of dairy systems and a solution for GHG and NH_3 mitigation

N. Edouard[1], K. Klumpp[2], X. Vergé[3] and J.-L. Peyraud[1]
[1]INRAE, Institut Agro, PEGASE, 35590 Saint-Gilles, France, [2]INRAE, VetAgro-Sup, UREP, 63000 Clermont-Ferrand, France, [3]Institut de l'élevage, 149 rue de Bercy, 75012 Paris, France; nadege.edouard@inrae.fr

To achieve climate neutrality in agriculture, whole system changes are needed. This involves the end of the linear resource-product-waste approach, essentially based on increasing production efficiency. Today, the circular approach prevails to optimise resource use efficiency throughout the agri-food system, integrating a recycling phase allowing waste to become a resource, while reducing environmental impacts. Livestock contribute to sustainable agri-food systems by valorising biomass and co-products not directly consumable by humans and transforming them into healthy food. Animal production also contribute to a more efficient agriculture via the diversification of cropping systems and the production of organic fertilizers, while promoting soil fertilization, generating co-products and energy, and providing service diversity. The (re)-integration of crop and livestock farming systems offers likewise the possibility to reduce pollutant emissions, be more efficient at using natural resources, limit the use of external inputs, maintain soil health benefits. However, in a circular system where plants, livestock and management for nutrients and water play central roles, significant challenges can be encountered. We propose to illustrate the linkage between resource, dairy ruminants, manure and soil management such as: very low N and C input systems especially those based on grass and pasture valorisation; the consequences of manure management practices from the animal to the soil on NH_3 and GHG emissions and the risk of reverse effects; the substitution of synthetic fertilizers by organic manure, entailing large amounts of manure storage, which may further increase GHG emissions and pollution. Particular attention will be directed to the circularity of diet changes related to sustainable livestock management, crop and soil management choices for animal diets, and at last subsequent manure change, their appropriate use and storage. These illustrations will help to evaluate the reintegration of livestock and cropping systems in a holistic systems-based approach, aiming to mitigate GHGs whilst reducing environmental footprints.

Environmental footprint of Holstein and crossbred cows using an individual LCA-derived method

M. Piazza, M. Berton, S. Schiavon, G. Bittante and L. Gallo
University of Padova, [1]Department of Agronomy, Food, Natural Resources, Animals and Environment, Viale dell'Università 16, 35020 Legnaro (PD), Italy; martina.piazza.1@studenti.unipd.it

This study aimed at assessing the effects of a 3-breed crossbreeding program on the environmental footprint of individual dairy cows (Life Cycle Assessment, LCA, method). Data concerned 564 cows, 279 purebred Holstein (HO) and 285 crossbreds (CR), originated from a 3-breed crossbreeding program (PROCROSS) based on the rotational use of Viking Red, Montebèliarde and HO sires and kept in two dairy herds of northern Italy (224 and 340 cows/herd, respectively). The reference unit of the LCA model was the lifespan of cows (from the birth to the farm-gate). Data were collected at different levels: individual animal-based data referred to the whole life (birth, calving, dry and cull dates, milk production); individual test-date collection of body measures and BCS, used to predict body weight and to estimate energy requirements; common farm-based data concerning herd management (diets composition, materials used). Data were used to compute or estimate dry matter intake, milk and milk components production, gross income (GI) and income over feed costs (IOFC) pertaining to the whole career of cows. Impact categories assessed were global warming (GWP), acidification (AP) and eutrophication (EP) potentials. Different functional units (cow; d of life; kg of milk; kg of fat + protein; € of gross income and IOFC) were used. Data were analysed using GLM including the fixed effects of farm, genetic line (CR vs HO) and their interaction. Compared to HO, CR completed more lactations (+12%), had earlier first calving (-2%), produced less milk per day of production (-3%), but more fat plus protein per d of life (+ 4%), and tended to provide a greater IOFC (+7%). Moreover, compared to HO herdmates, CR tended to have greater GWP per cow in the whole life, similar GWP per d of life and 4% lower GWP per kg of fat plus protein produced in the whole life (P=0.07). Also GWP per € GI tended to be lower (-3%) in CR compared to HO cows. Managing dairy cows within the crossbreeding scheme may therefore contribute to mitigate emissions of GHGs in dairy operations.

Session 37 — Theatre 3

Emissions and manure quality in future cow barn

P.J. Galama, H.J. Van Dooren and H. De Boer
Wageningen University and Research, Livestock Research, De Elst 1, 6700 AH Wageningen, the Netherlands; paul.galama@wur.nl

The design of a cow barn does not only have impact on animal welfare but also on manure quality, which is important in relation to use as fertiliser or soil improver on grassland and arable land and to decrease the ammonia emission. The most common housing system is a Freestall with cubicles and a slatted floor with slurry storage. New developments in housing systems are based on different floor types and Cowtoilet in Freestalls with cubicles to reduce ammonia emission and Freewalk systems to increase space per cow for more animal welfare, natural behaviour and to combine this with less ammonia emission. Preliminary results of ongoing research at Dairy Campus in Netherlands with case-control barn units of each 16 cows will be shown with three separation techniques: (1) The CowToilet collects about 35% of the total urine production of the cow and reduces the ammonia emission in the barn about 35%. The urine is stored outside the barn in a separate tank. (2) A permeable plate on a slatted floor separates the urine and faeces. The faeces is stored outside the barn and the urine is stored underneath the slatted floor. The urine is acidified with sulphuric acid to reduce ammonia emission in the barn and when applied on the field. The reduction of ammonia emission is ca. 50% if the permeability of the plate is improved by flushing the plate with the acidified urine. 3) A rubber floor with gutter and holes. Developments of Freewalk housing without cubicles will be shown, namely with a bedding of organic material like wood chips or straw and, an artificial floor which separates urine from faeces (primary separation) and a sand bedding in combination with bedding cleaner. The effect of these housing systems on manure quality of ' composted bedding material', slurry and urine and faeces fractions will be illustrated. Primary separation of manure with a specific floor type or Cowtoilet in a cubicle barn or with an artificial floor in a Freewalk barn can make a dairy farm more sustainable and may offer arable farmers more opportunities to select the right fertiliser or soil improver. The results of research in EU projects like FreeWalk and CCCfarming and other ongoing research at Experimental Station Dairy Campus will be shown.

Session 37 — Theatre 4

Comparing outcomes of three GHG emission calculation tools applied on dairy production systems

M. De Vries[1], M.H.A. De Haan[1], P.R. Hargreaves[2], X. Vergé[3], G. Kadžienė[4], A. Cieślak[5], P. Galama[1] and T.V. Vellinga[1]
[1]Wageningen Livestock Research (WLR), De Elst 1, 6708 WD Wageningen, the Netherlands, [2]Scotland's Rural College (SRUC), Crichton Royal Farm, Bankend Road, DG1 4TT, Dumfries, United Kingdom, [3]Latvia University of Life Sciences and technologies (LLU), Liela str. 2, 3001, Jelgava, Latvia, [4]Lithuanian University of Health Sciences (LSMU), R. Žebenkos str. 12, 82317 Baisogala, Lithuania, [5]Poznan University of Life Sciences (PULS), Wołynska 33, 60-637 Poznań, Poland; marion.devries@wur.nl

The use of greenhouse gas (GHG) emission calculation tools is of increasing importance to identify effective farm strategies and policy measures for mitigating GHGs from agriculture. The aim of this study in the context of the EU project CCCfarming was to compare outcomes of three calculation tools for estimation of cradle-to-farm gate GHG emissions from dairy production systems: Agrecalc, ANCA, and CAP'2ER-level 1. Data were collected from 3 dairy farms in Lithuania, the Netherlands, and Poland. Results showed that total GHG emissions were 16 to 52% and -8 to 50% higher in ANCA than Agrecalc and CAP'2ER, respectively, and -4 to 21% higher in CAP'2ER than Agrecalc. GHG emission intensity was 8 to 44% and -4 to 73% higher in ANCA than Agrecalc and CAP'2ER, respectively, and -20 to 12% higher in CAP'2ER than Agrecalc. Differences in tool outcomes were largest for emissions embedded in imported feeds and rumen enteric fermentation, and, to a lesser extent, manure storages and soil nitrogen inputs. These processes were also the largest sources of emissions, jointly contributing to at least 72% of total emissions. Main differences in outcomes could be explained by differences in embedded emissions of purchased feedstuffs, GWP characterization factors, calculated herd feed intake and composition, and the type and detail of input data. It was concluded that, despite compliance with international standards for GHG accounting, GHG calculation tools showed large differences in outcomes. Further harmonization is needed to reduce differences between outcomes of tools.

Fermented rapeseed cake in dairy cows' nutrition mitigates methane emission and production

A.C. Cieslak[1], G.M. Gao[1], H.H. Huang[1], G.M. Gogulski[2,3], R.D. Ruska[4], N.B. Nowak[1], S.Z.A. Szejner[1], P.J. Puchalska[1], S.J. Światłowski[1] and S.Z.S.M. Szumacher-Strabel[1]

[1]Poznań University of Life Sciences, Department of Animal Nutrition, Wołyńska 33, 60-637 Poznań, Poland, [2]Centre of Biosciences, Institute of Animal Physiology, Soltesovej 4-6, 040-01 Kosice, Slovak Republic, [3]Poznań University of Life Sciences, University Center for Veterinary Medicine, Szydłowska 43, 60-637, Poznań, Poland, [4]Latvia University of Life Sciences and Technologies, Institute of Animal Sciences, Liela str.2, Jelgava, 3001, Latvia; adam.cieslak@up.poznan.pl

The fight against high methane (CH_4) emissions from dairy cows are essential in reducing global gas production and achieving climate neutrality. A recent study showed that using the fermentation process improved the efficiency of feed utilization by animals in the gastrointestinal tract. Less methane may consequently be generated from fermented components than from unfermented. There is evidence that the digestibility of a fermented feed increases, also accompanied by the amount of energy available. Consequently, increased digestibility of diets often means less methane emission per production unit. A complex *in vitro* and *in vivo* study composed of 4 experiments aimed to evaluate the impact of fermented rapeseed cake (FRC) on ruminal fermentation, CH_4 emission/production, and milk production in lactating dairy cows. All data were analysed using SAS statistical software (Univ. Edition, version 9.4). Improved nutrient digestibility and promoted propionate formation during FRC Hohenheim Gas Test incubation caused a decrease in ruminal methane production. It was confirmed in batch culture and *in vivo* experiments where the amount of 2.65 kg/d/cows reduced the methane production by 26 and 10% (Hohenheim Gas Test or batch culture, respectively) and the methane concentration by 17 and 11% (experiments with cannulated cows or commercial dairy cows' condition, respectively). FRC in dairy cow diets can effectively mitigate rumen methane production and improve nutrient digestibility.

Association of microorganisms in rumen with methane emission intensity of cows in the Netherlands

D. Schokker[1], S. Roques[1], L. De Koning[1], M.N. Aldridge[1], A. Bossers[2], Y. De Haas[1], L.B. Sebek[1] and S.K. Kar[1]

[1]Wageningen Livestock Research, De Elst 1, 6708 WD Wageningen, the Netherlands, [2]Wageningen Bioveterinary Research, Houtribweg 39, Leylstad, the Netherlands; dirkjan.schokker@wur.nl

The increase in absolute methane emissions is one of the main factors contributing to the growing concentration of greenhouse gases. About one third of anthropogenic emissions come from livestock, mainly from enteric fermentation of ruminants, including cattle. The rumen microbiota aids in converting feedstuffs into useful nutrients, including the energy needed to maintain productivity. During this process, the rumen microbiota produces methane by anaerobic fermentation of feed. A plethora of scientific evidence supports this relationship between the rumen microbiota and enteric methane emissions. We have evaluated the relationship between rumen microbiota and methane emissions on 17 Dutch commercially dairy farms, where we had informative data on 176 cows. We have analysed both the rumen microbial diversity and composition. In addition, we investigated the environmental and host factors, i.e. ration fed, season and cow characteristics, that contribute to the change in rumen microbiota. The most variation in the rumen microbiota was explained by the dietary and seasonal factors, approximately 34% for richness and 10% for the Shannon index. In addition, cows with a high or low methane emission showed a divergent rumen microbiota composition. Lastly, a preliminary analysis for breeding purposes was performed on this cohort of cows to estimate microbial variances and these variances explained 7% of the methane emission intensity. Further investigation should focus on the combination of host characteristics and environmental conditions, including diet, to modulate the rumen microbiota related to enteric methane emissions.

Danish commercial farm reduces enteric methane by 35% with 3-NOP

N.I. Nielsen[1], M.Ø. Kristensen[1] and C. Ohlsson[2]
[1]SEGES Innovation, Livestock Innovation, Agro Food Park 15, 8200 Aarhus N, Denmark, [2]DSM Nutritional Products, Kirkebjerg Allé 88, 2605 Brøndby, Denmark; ncn@seges.dk

A newly established Green Deal in the Danish Parliament obliges Danish dairy farmers to reduce greenhouse gas (GHG) emissions by 0.2 and 1.0 mio tons of CO_2e in 2025 and 2030 respectively. The latter 2030 target corresponding approximately to 40% reduction of enteric methane being emitted from dairy cows in Denmark. To reach this reduction in GHG, without reducing the number of cows, methane mitigating feed additives are necessary. A commercial Danish Dairy herd with 320 VikingRed cows was given dispensation by Danish authorities to test Bovaer®-10 (containing minimum 10% of the active ingredient 3-NOP) during the winter of 2021-2022. The cows were loose housed in a cubicle stall with slatted floors in four pens with one AMS unit per pen. The Bovaer-10 was mixed with protein feed stuffs, silages, straw and sodium hydroxide treated wheat in a mixer wagon to reach a target dose of 58 mg 3-NOP/kg total DM intake and fed as a PMR. The total ration included (% of DM): maize silage (37.1%), grass silage (24.5%), wheat straw (2.8%) compound feed (17.6%), rape seed meal (6.9%), rape seed cake (6.0%), sodium hydroxide treated wheat (3.7%) minerals (1.3%). Cows were fed restrictively, i.e. there were no refusals and the feeding table was typically emptied 1-2 hours before feeding fresh PMR in the morning at ap-proximately 8:00 h. Enteric emissions (CH_4, CO_2, H_2) was measured with two GreenFeed units in one of the four pens during a control period, followed by a treatment period. Preliminary results show that 3-NOP reduced enteric methane by 35% (424±90 vs 275±78 (s.d.) g/cow/day), increased enteric hydrogen by 452% (1,24±0.74 vs 6.85±4.59 (s.d.) g/cow/day) and did not affect ECM yield (31.3 vs 31.6 kg/cow/day) nor total DMI (22.4 vs 22.3 kg/cow/day). There was significant diurnal variation for methane and hydrogen which was affected by 3-NOP and indicates that the effect of 3-NOP is dependent on feeding management.

Assessment of cattle manure acidification effects on ammonia and GHG emissions and crop yield

V. Juskiene, R. Juska, A. Siukscius and R. Juodka
Lithuanian University of Health Science, Animal Science Institute, R. Žebenkos 12, Baisogala, 82317, Lithuania; violeta.juskiene@lsmuni.lt

Manure acidification is recognized as one of best treatments available to reduce ammonia emission from manure. However, the potential to reduce the emissions of polluting gases, as well as the effects on crop yield have been poorly studied. Therefore, this study was conducted to evaluate the effect of acid treatment on NH_3 and other GHG emissions from manure and assess the effects on plant nutrient utilization. The effect of manure acidification on gas emissions was investigated on laboratory scale at LUHS Animal Science Institute. The study was performed using 20 litre manure tanks, gas emissions were measured by the passive chamber method. Field trials were conducted to evaluate the effects of acidified slurry on crop yield, using slurry acidification techniques in-field. Both acidified and non-acidified cattle slurry was applied to barley and spring wheat. Control fields were fertilized with mineral fertilizers. Slurry was acidified with H_2SO_4 to target pH 6.0. Our study showed that the emission rate for non-acidified cattle manure was as follows: NH_3: 11.6, CH_4: 1.33 and CO_2: 251.9 mg/(m^2/60s), respectively. Acidification of manure to pH7 has reduced NH_3 emissions by 39.6% and CH_4 emissions by 29.5%. Higher level of manure acidification (pH5) gave even better emission reduction results: by 75.6% for NH_3, and even by 91.8% for CH_4. Different results were found for CO_2 emission, which showed that the emission rate could be 22.2% higher with pH7, and 27.8% lower with pH5. Field trials have shown that acidification of slurry can have a positive effect on crop yield. For barley, acidified slurry resulted in 13.1% higher yield and 3.43% higher protein content in comparison with untreated slurry, and 14.1% higher yield and 9.15% higher protein content in comparison with mineral fertilization. Smaller differences were found in the spring wheat fields, where acidified slurry resulted in 3.7 and 4.9% higher yield compared to untreated slurry and mineral fertilization, respectively. In summary, it can be concluded that mild acidification of cattle manure and slurry can be a successful solution to help farmers reduce NH_3, CO_2 and CH_4 emissions and at the same time improve the fertilization value of manure.

Developing an in-flight network for gas and particulate emissions assessment in cattle dairy farms

V. Becciolini[1], L. Conti[1], G. Rossi[1], M. Merlini[1], D. Bedin Marin[1], G. Coletti[2], U. Rossi[2] and M. Barbari[1]
[1]University of Firenze, Department of Agriculture, Food, Environment and Forestry (DAGRI), via San Bonaventura 13, 50145 Firenze, Italy, [2]Project & Design s.r.l.s., via Livorno 8/28, 50142 Firenze, Italy; valentina.becciolini@unifi.it

Quantifying and reducing greenhouse gases (GHGs) emissions in livestock systems and, specifically, in the context of dairy farming, is a debated topic. International policies are supporting mitigation strategies to reduce the environmental impact of agricultural practices. Besides, the assessment of air quality in livestock buildings for ensuring human and animal safety and welfare is poorly addressed and no real-time monitoring systems are currently available. In this framework, automated and low-cost tools enabling a continuous monitoring of air concentrations of gases and particulate in cattle buildings, manure and feed stores would represent a significant advancement in the sector. As part of the *CCCFarming* project, we aimed to assess the feasibility of a UAV-based system (drone) for real-time measurements of air pollutants at farm level. Given that drones are increasingly used for air quality monitoring in several fields (e.g. atmospheric chemistry research, industrial emission monitoring), we present a first attempt to develop an integrated prototype system for gas and particulate monitoring using portable self-engineered measurement units at ground and on a small UAV. The goal is to detect emission hotspots and to provide real-time graphic alerts by means of a web-app. The system embeds low-cost commercial sensors for GHGs (CH_4, CO_2), NH_3 and particulate ($PM_{2.5}$, PM_{10}) into customized portable units located at ground and on a rotor-based drone. The sensors were calibrated in a specialised laboratory and the system was tested in a commercial dairy farm. Ground measurement units were located inside and close to the external boundaries of the cattle building, while simultaneous flights were carried out in the top atmospheric boundary layer up to 30 m a.g.l. Gas and particulate concentration measurements were timestamped and georeferenced with centimetre accuracy. The results confirmed the feasibility of the project at farm level, although further research is required to validate field measurements with reference instruments and techniques.

Air filtering as alternative approach to combat emissions from cattle facilities

A. Kuipers, P. Galama and R. Maasdam
Wageningen Livestock Research, De Elst 1, 6708 WD Wageningen, the Netherlands; abele.kuipers@wur.nl

Various strategies can be followed to reduce ammonia and methane emissions. One strategy is to adapt the animal to the environment and the other is to adapt the environment to the animal. Practices of the first category are feeding practices, e.g. adding methane blockers to the feed or lowering protein in feed, and genetics, e.g. selecting for low methane animals. We study the filtering of the air belonging to the second strategy. In pig husbandry air washers are common to filter ammonia from the air in barns. The air is sucked into a water basin where ammonia reacts with acid to a solid component. This technique has not been adopted in cattle housings, mostly because those facilities have an open structure and other source oriented approaches prevailed. For methane, the very low concentration in barns (measured on 60 farms spread over Europe from 5 to 80 ppm at 2 m height) and low solubility in water complicates the filtering of methane. In a dairy housing about 20% of methane comes from the manure and 80% from the mouth of cows by the natural physiological process of rumination. We studied the challenging possibility of simultaneously filtering methane and ammonia from the air in the housing including manure storage facilities. We examined three processes: (1) Air circulation: try to combine air containing methane and ammonia in one flow; air is sucked from manure storage plus above floor (close to mouth of cows). The effectivity of filtering air with varying degrees of openness of barn, and by increasing of methane concentration by recirculating the air. (2) Filter techniques: Study of the effectivity of absorbing material, bio-bed and land/soil filters for oxidation or conversion of methane and ammonia. (3) Re-use of filtered N and C material. Background information from literature and a variety of mostly experimental applications in practice will be demonstrated, which indicate that the filtering of methane from barns and storages forms the biggest challenge to solve.

Emissions, grassland and biodiversity and role of carbon footprint tool

P.R. Hargreaves[1], R.M. Rees[1] and J. Bell[2]
[1]Scottish Rural University College, Scotland, 12345, United Kingdom, [2]SAC Consulting, Edinburgh, Scotland, United Kingdom; paul.hargreaves@sruc.ac.uk

Agriculture is committed to large reductions in greenhouse gas emissions in order to achieve policy commitments needed to deliver net zero targets. Accounting for and mitigating these GHG emissions is crucial in reducing the industry's overall carbon footprint. This is especially true for the dairy industry with the associated GHG emissions from the livestock, as well as the forage crops. Methane (CH_4) emissions generally contribute the greatest proportion of GHG's with CH_4 contributing 60% of agricultural greenhouse gases in the UK (UK National Inventory Report 2021). Emissions of nitrous oxide (N_2O), an important greenhouse gas (third most persistent), contributes 36% of the UK emissions. It is therefore important to understand and control these GHG's. N_2O emissions are linked to applications of N to grasslands, and environmental conditions. Although reductions in milk production and efficiency in the UK dairy industry has resulted in GHG reductions, with total emissions falling by 16.1% (1.12 Mt CO_2-eq) between 1990 and 2020, further reductions are needed if a C neutral state is to be achieved. Tools such as AgreCalc allow farm scale assessments of the carbon footprint to identify where GHG reductions can be made. However, costs are associated with these mitigations and will only be taken up if either legislation changes or there is some other benefit to the farmer. The AgreCalc carbon footprinting tool has been used to run several different scenarios for two dairy farms to compare GHG emissions and outcomes for the herd and milk yield. These scenarios included a combination of mitigation factors including those related to the methane produced by the livestock (methane inhibitors in the diet, sale of excess stock) and forage production (enhanced grassland productivity and increased legumes to reduce fertiliser use). There is an opportunity with these GHG reductions to deliver increased efficiency. However, care must be taken that GHG mitigation doesn't have unintended consequences for biodiversity and ammonia emissions. Recent work has shown the main mitigation measures for GHG reduction in dairy farms will not be sufficient to achieve C neutrality and C storage must be enhanced.

Total and differential somatic cell count affect cheese-making traits of individual milk samples

E. Mariani[1], G. Stocco[1], C. Cipolat-Gotet[1], G. Niero[2], M. Penasa[2] and A. Summer[1]
[1]University of Parma, Department of Veterinary Science, Strada del Taglio 10, 43126 Parma (PR), Italy, [2]University of Padova, Department of Agronomy, Food, Natural resources, Animals and Environment, Viale dell'Università 16, 35020 Legnaro (PD), Italy; elena.mariani@unipr.it

The study aimed at investigating the effect of total (SCC) and differential somatic cell count (DSCC) on cheese yield and milk nutrients recovery using an individual cheese making model. A total of 800 milk samples (1.5 l/cow) were collected from Brown Swiss cows in 40 herds (20 cows/herd) and processed into cheese. Milk and whey were weighted and analysed for total solids, fat and protein content using MilkoScan FT3. Milk SCC and DSCC were obtained from Fossomatic 7 DC counter, and SCC were log-transformed to somatic cell score (SCS). A total of 7 cheese-making traits were computed: 3 measures of cheese yield (%CY) [fresh ($\%CY_{CURD}$), solids ($\%CY_{SOLIDS}$), and water ($\%CY_{WATER}$)] and 4 recovery traits (%REC) [milk fat ($\%REC_{FAT}$), protein ($\%REC_{PROTEIN}$), solids ($\%REC_{SOLIDS}$), and energy ($\%REC_{ENERGY}$) recovered in the curd]. The variability of cheese-making traits was assessed using a linear mixed model that included herd as random effect and parity, days in milk, SCS, DSCC, and their interaction as fixed effects. The interaction between SCS and DSCC affected all the cheese-making traits. Also, some main effects were important to explain the variability of the aforementioned features. In particular, milk samples with high SCS and low DSCC provided lower $\%CY_{CURD}$, $\%CY_{SOLIDS}$, and $\%CY_{WATER}$ compared to milk samples with higher DSCC. For %REC traits, different patterns were observed among DSCC classes and across SCS. Thus, the combined use of DSCC and SCS might be indicative not only of health status, but also of technological properties of milk, allowing for a better understanding of the relationships between somatic cell traits and the efficiency of cheese-making. Since the present study has been conducted at individual cow level and samples collection is still in progress, there is a potential for estimating genetic parameters of the investigated traits and evaluate the opportunity to improve them through genetic selection strategies.

Session 38 — Theatre 2

Milk mineral profile from cows fed seaweed (*Saccharina latissima*) and different protein feeds

S. Stergiadis[1], N. Qin[1], Á. Pétursdóttir[2], D. Humphries[1], N. Desnica[2], E.E. Newton[1], A. Halmemies-Beauchet-Filleau[3], A. Vanhatalo[3], L. Bell[1], I. Givens[1,4], H. Gunnlaugsdóttir[2] and D. Juniper[1]

[1]University of Reading, School of Agriculture, Policy and Development, Reading, RG6 6EU, United Kingdom, [2]Matis, Vínlandsleið 12, Reykjavík, 113, Iceland, [3]University of Helsinki, Department of Agricultural Sciences, Helsinki, 00014, Finland, [4]University of Reading, Institute for Food, Nutrition and Health, Reading, RG6 6EU, United Kingdom; s.stergiadis@reading.ac.uk

This study investigated the effect of seaweed supplementation in dairy diets on milk mineral concentrations, when feed proteins with contrasting glucosinolates content (which affect feed-to-milk iodine transfer) were used. 16 Holstein cows were allocated in a 4×4 Latin square design with four 4-week experimental periods (7d washout, 14d adaptation, 7d measurements). Diets (forage:concentration ≈75:25 in dry matter (DM)) were: (1) wheat distiller's grain (WDG)-based without seaweed (SWD); (2) WDG-based with SWD; (3) rapeseed meal (RSM)-based without SWD; (4) RSM-based with SWD. Dried SWD (*Saccharina latissima*) was offered at 38.6 g/cow/d. Fresh-cut grass or grass silage (as buffer) was fed *ad libitum* and concentrates were iso-nitrogenous (16% CP in DM). Dietary DM intake and chemical composition, and milk yield, basic composition and mineral profiles were recorded during the 7d measurement periods. Linear mixed models used seaweed supplementation (without, with), protein concentrate (WDG, RSM), and their interaction as fixed effects; period as the random effect; animal as within-subject effect; and a first-order autoregressive structure as covariate. SWD supplementation did not affect milk production and efficiency parameters but increased milk I contents (+487 ug/kg; +234%), and reduced milk Ca (-49 mg/kg) and Cu (-6.2 ug/kg), compared with milk from cows without dietary SWD. Milk from RSM-fed cows had less I (-143 ug/kg; -27%) and Mn (-3.6 ug/kg) than milk from WDG-fed cows. Feed-to-milk transfer efficiency was higher for Ca, Na, and Se and lower for Mg, P, I, Fe, Mn in RSM-cows, than in WDG-fed cows. Given that milk and dairy products are main dietary I source in several countries, feeding *S. latissima* can be used to enrich milk I contents for populations and demographics suffering from I deficiency.

Session 38 — Theatre 3

Combined effect of anti-methanogenic compounds (rapeseeds, nitrate and 3-NOP) on fatty acids in milk

G.M. Sirinayake Lokuge[1], L.B. Larsen[1], M. Maigaard[2], L. Wiking[1], P. Lund[2] and N.A. Poulsen[1]

[1]Aarhus University, Food Science, Agro Food Park 48, 8200 Aarhus N, Denmark, [2]AU Foulum, Aarhus University, Animal Science, Blichers Allé 20, 8830 Tjele, Denmark; gmadu@food.au.dk

Use of fat supplementation and feed additives, like nitrate (NIT) and Bovaer (3-NOP), have shown to be able to elicit significant reductions in enteric methane (CH_4) emission from dairy cows. However, the inhibition effect of 3-NOP may shift the fatty acid (FA) profile towards more SFA and less unsaturated fatty acids (USFA). Combining 3-NOP and whole cracked rapeseeds would be an effective way to increase USFA without compromising the reduction of methane emission. In this study, the combined effect of supplementing whole cracked rapeseeds, NIT, and 3-NOP on milk composition was determined, using 48 lactating Danish Holstein cows in a 8×8 incomplete Latin square design with 6×21-days experimental periods. The 8 diets were 2×2×2 factorial arranged, including 2 levels of fat by supplying whole cracked rapeseeds (low; 30 g and high; 63 g CF/kg DM) with or without NIT (10 g/kg DM) and 3-NOP (80 mg/kg DM). Milk samples were collected at the end of each period and FA profile was analysed by GC-FID. A mixed model, which included fixed and interaction effects of the treatments and parity, fixed effects of DIM, period, and a random effect of cow, was applied in data analysis. Rapeseed alone increased cis-9, trans-11 CLA, total MUFA and PUFA, and decreased SFA (P<0.01) in milk. 3-NOP alone decreased cis-9, trans-11 CLA, as well as total PUFA (0.72 vs 0.61 g/100 g of total FA, P<0.01 and 3.99 vs 3.86 g/100 g of total FA, P<0.01, respectively) in milk, whereas supplementation of 3-NOP with rapeseed increased both fatty acids (P<0.01). NIT alone and also in combination with rapeseed increased cis-9, trans-11 CLA in milk (P<0.01). However, NIT with 3-NOP did not alter SFA, MUFA and PUFA. The results indicate that supplementation of whole cracked rapeseeds, NIT and 3-NOP in combination with rapeseeds is an effective strategy to increase cis-9, trans-11 CLA and total PUFA in milk together with a lower CH_4 production. As higher PUFA and CLA concentrations in milk have been associated with health benefits, this change in milk composition may be perceived beneficial from the consumers.

Session 38 — Theatre 4

Mineral concentrations in milk from Finnish Ayrshire cows fed microalgae or rapeseed

E.E. Newton[1], M. Lamminen[2], P. Ray[3], A. MacKenzie[4], C.K. Reynolds[1], M.R.F. Lee[4], A. Halmemies-Beauchet-Filleau[2], A. Vanhatalo[2] and S. Stergiadis[1]

[1]University of Reading, School of Agriculture, Policy and Development, Reading RG6 6EU, United Kingdom, [2]University of Helsinki, Department of Agricultural Sciences, Helsinki 00014, Finland, [3]The Nature Conservancy, 4245 North Fairfax Drive, Virginia 22203, USA, [4]Harper Adams University, Department of Agriculture and the Environment, Newport TF10 8NB, United Kingdom; eric.newton@pgr.reading.ac.uk

This study investigated the effect of partial or total replacement of rapeseed with microalgae Spirulina platensis on milk, plasma, and faecal mineral concentrations. Lactating Finnish Ayrshire cows (n=8) were allocated into a replicated 4×4 Latin square design experiment (2w adaption, 7d sampling). Diets on a dry matter basis were: (1) standard (negative control, NEG); (2) pelleted rapeseed supplementation (RSS; 2,550 g/d); (3) a mixture of rapeseed and *Spirulina platensis* (RSAL; 1,280 g RSS/day + 570 g *S. platensis*/day); and (4) *S. platensis* (ALG; 1,130 g *S. platensis*/day). Linear mixed effects models used diet, square, period nested within square, and their interactions as fixed factors, and cow nested within square as a random factor. When compared with NEG and ALG, milk from cows fed RSAL and RSS had lower milk iodine (I) concentrations (-69.6 and -102.7 μg/kg milk, respectively). Transfer efficiency from feed to milk was also reduced by up to 11 g I/100 g I ingested (from CON to RSS). Plasma I concentration was not affected by diet. Faecal concentrations of Ca, Mg, Co, Cu, Fe, I, Mn, and Zn were increased when *S. platensis* was fed, and their apparent digestibility was decreased. Feeding *S. platensis* did not affect milk mineral concentrations or cow mineral status. However, dietary rapeseed inclusion reduced milk I concentrations significantly. As milk is a major source of I in human diets in several countries a reduction of I supply in the food chain via milk products should be avoided by increasing I supplementation in cow diets.

Session 38 — Theatre 5

Sub-clinical mastitis affects milk fatty acid profile in Holstein cattle

S. Pegolo[1], A. Toscano[1], V. Bisutti[1], M. Gianesella[2], D. Giannuzzi[1], S. Schiavon[1], L. Gallo[1], F. Tagliapietra[1] and A. Cecchinato[1]

[1]University of Padova, DAFNAE, Viale dell'Università 16, 35020 Legnaro, Italy, [2]University of Padova, MAPS, Viale dell'Università 16, 35020 Legnaro, Italy; sara.pegolo@unipd.it

Mastitis is a highly prevalent disease, which negatively affects animal health and welfare, production performances, and profitability. Herein, we evaluated the associations between subclinical mastitis infection from *Streptococcus agalactiae*, *Staphylococcus aureus*, *Streptococcus uberis* and *Prototheca* spp. and milk fatty acids (FA) profile in 450 Holstein cows from three dairy herds. After an initial bacteriological screening (T0), 85 positive cows were identified and followed at the quarter level two weeks (T1) and six weeks (T2) after T0. In total, 600 single-quarter records were available at T1 and T2. Target traits were 70 individual FA, 12 FA groups and six desaturation indices analysed by gas-chromatography. Data were analysed with a hierarchical linear mixed model including: the fixed effects of days in milk (DIM), parity, herd, somatic cell count (SCC), bacteriological status (BACT, positive and negative), and SCC × BACT interaction; the random effect of the individual cow/replicate nested within herd, DIM and parity. The most significant effects were detected at T2. In particular, mastitis decreased the proportions of several individual short chain FA (SCFA), especially those from the de novo synthesis in the mammary gland. The proportions of SCFA and Trans 18:1 groups were also decreased (P<0.01 and P<0.05, respectively) while the proportion of medium chain FA (MCFA) was increased (P<0.05). The 10:1 desaturation index was decreased (P<0.01). Somatic cell count affected several individual SCFA and MCFA. In particular, milk samples with SCC≥400,000 cells/ml had lower proportions of 10:0, 11:0 and 12:0 respect to those having SCC<200,000 cells/ml and lower proportions of 13:0 respect to those having SCC<50,000. The effect of SCC × BACT interaction was significant for the 14:1 desaturation index which was the lowest for positive samples with SCC ≥50,000 and <200,000 cells/ml. This study showed that changes in milk FA composition might be indicators of subclinical mastitis in dairy cattle. Acknowledgements. The research was part of the LATSAN project funded by MIPAAF (Italy).

Impact of milking interval on milk spontaneous lipolysis in dairy cows

C. Hurtaud[1], L. Bernard[2], D. Taillebosq[1] and C. Cebo[3]

[1]PEGASE, INRAE, Institut Agro, 35590 Saint-Gilles, France, [2]UMR Herbivores, Université Clermont Auvergne, INRAE, Vetagrosup, 63122 Saint-Genès-Champanelle, France, [3]GABI, Université Paris-Saclay, INRAE, AgroParisTech, 78350 Jouy-en-Josas, France; catherine.hurtaud@inrae.fr

Milk lipolysis is defined as the hydrolysis of triglycerides, the major component of milk fat, resulting in the release of short-chain fatty acids responsible for rancid flavour and partial glycerides that impairs milk functional properties such as foaming and creaming abilities. Milk lipolysis is a complex phenomenon that depends on both animal parameters and breeding factors. Regarding milking rhythms, milk spontaneous lipolysis (SL) is higher in milk from evening milking. This may be related to the length of milking intervals or to nychtemera. The objective of the experiment was thus to study the impact of milking intervals on milk SL in a 2 milkings per day system: 10-14 (6.30 a.m. and 4.30 p.m.); 14-10 (6:30 a.m. and 8:30 p.m.); 12-12 (6:30 a.m. and 6:30 p.m.). To achieve this goal, 21 multiparous dairy cows in mid-lactation were used in a 3×3 Latin square design over 3 periods. The experiment lasted 5 weeks (3 experimental periods of 1 week alternating with a classic 10-14 milking week). The time interval between milkings influenced milk SL. Indeed, we observed more SL in the evening milk for 10-14 (+0.20 mEq/100 g fat) and in the morning milk for 14-10 (+0.22 mEq/100 g fat). No significant difference was seen for 12-12 even if SL was numerically higher in evening milk (+0.10 mEg/100 g fat). Higher SL was associated with lower milk production and higher fat content. Milk protein, total phosphorus and citrate contents increased with the duration of mammary gland storage of milk (from 10 to 14 hours). There was no effect of milking intervals on milk fat globule diameter. The Na/K ratio indicating an opening of tight junctions in the mammary gland only increased in evening milking with 12-12 and 14-10 intervals. In conclusion, the effect of milking intervals on SL seems to be more important than the effect of nychtemera, which was partly confirmed for 12-12. The underlying mechanisms will be further explored in the framework of the LIPOMEC project funded by French National Research Agency (ANR-19-CE21-0010), in which this study was conducted.

Indirect prediction of grass-based diet from milk MIR traits to assess the feeding typology of farms

H. Soyeurt[1], C. Gerards[1], C. Nickmilder[1], S. Franceschini[1], F. Dehareng[2], D. Veselko[3], C. Bertozzi[4], J. Bindelle[1], N. Gengler[1], A. Marvuglia[5], A. Bayram[5,6] and A. Tedde[1,7]

[1]TERRA, GxABT – ULiège, Passage des déportés 2, 5030 Gembloux, Belgium, [2]CRA-W, Chaussée de Namur 24, 5030 Gembloux, Belgium, [3]CdL, Route de Herve 104, 4651 Battice, Belgium, [4]Awé, Rue des Champs Elysées 4, 5590 Ciney, Belgium, [5]LIST, Rue du Brill 41, 4422 Belvaux, Luxembourg, [6]University of Luxembourg, Avenue de l'Université, 4365 Esch-sur-Alzette, Luxembourg, [7]FNRS, Rue d'Egmont, 5, 1000 Bruxelles, Belgium; hsoyeurt@uliege.be

Grassland provides local and low-cost feed. Some specifications asked to farmers is to put their cows on pasture for a minimum period or feed them mainly with a grass-based diet. Due to the fingerprint of this feed in milk, this work aims to predict indirectly the grass level in cows' diet using the changes in milk composition. We used over 3 million records collected from 2011 to 2021 on 2,449 farms and included 48 features reflecting the milk composition. As no detailed feed composition were available, the innovation consisted of predicting the grass-based diet using a trait defined from the month of analysis of bulk milk. Records collected from 5th to 8th months were assumed coming from a grass-based diet ('GRASS') as nearly all herds were on pasture. Those from the 1st, 2nd, 11th and 12th months, and the remaining ones were assigned as 'NOGRASS' and 'OTHER'. Using records from 30% of the studied farms (n=593,096), an analysis discriminating GRASS and NOGRASS gave a sensitivity of 86.91%±0.18. From the remaining GRASS and NOGRASS records (n=778,942), the accuracy reached 89.35%. To highlight the feeding typology, a dataset was created including the averaged probability of belonging to the GRASS modality for each month for a specific farm and year. A hierarchical clustering using Ward distance was applied on the records without missing values (n=25,832) and identified 6 clusters. One of them was associated to farms given a diet rich in long-chain fatty acids throughout the year. The remaining 5 clusters show a grazing period but different degrees of grass quality and quantity. In conclusion, using the MIR analysis of bulk milk, it is feasible to detect the presence of grass in the cow diet and estimate potentially the number of grazing days during a year.

Differential somatic cell count affects gelation, curd-firming and syneresis of bovine milk

G. Stocco[1], A. Summer[1], G. Niero[2], M. Ablondi[1], M. Penasa[2] and C. Cipolat-Gotet[1]
[1]University of Parma, Department of Veterinary Science, via del Taglio, 10, 43126, Parma, Italy, [2]University of Padova, Department of Agronomy, Food, Natural resources, Animals and Environment, Viale dell'Università, 16, 35020, Legnaro, Italy; giorgia.stocco@unipr.it

The aim of this study was to investigate the effect of differential somatic cell count (DSCC, %) on traditional milk coagulation properties (MCP) and modelled curd firmness over time (CF_t) parameters. The MCP were: rennet coagulation time (RCT, min), curd-firming time (k_{20}, min), and curd firmness (CF, mm) at 30, 45 and 60 min (a_{30}, a_{45}, and a_{60}, respectively), and the CF_t parameters were: estimated rennet coagulation time (RCT_{eq}), curd-firming (k_{CF}), and syneresis (k_{SR}) instant rate constants, maximum CF (CF_{max}), time at which the CF_{max} value is achieved, and potential CF. A total of 800 individual milk samples were collected during evening milking from cows of 40 herds located in the North of Italy. The DSCC was determined using Fossomatic 7 DC. Values of DSCC were multiplied by total somatic cell count and then log-transformed to achieve a normal distribution (DSCS). For each cow, MCP were measured in duplicate using 2 mechanical lactodynamographs (MA.PE. System, Firenze) and 240 CF values were recorded (60 min analysis; 1 CF value every 15 s). To fit the observed CF values, a 4-parameters model was applied to estimate the CF_t parameters. The effect of DSCS on MCP and CFt parameters was studied using a linear mixed model which included days in milk, parity and DSCS as fixed effects, and herd, cow, and the measuring unit of the lactodynamograph as random effects. The DSCS significantly affected the coagulation pattern of bovine milk. In particular, the RCT and RCT_{eq} were about 3 min longer, and k_{20} was about 1 min delayed in high DSCS milk samples compared to low DSCS samples. Curd firmness decreased by about 5, 4, and 3 mm for a_{30}, a_{45}, and a_{60}, respectively, and k_{CF} decreased by 1%/min, at increasing DSCS. Finally, the k_{SR} was slightly greater moving from the lowest to the highest values of DSCS. These results might help to further understand the relationships between indirect indicators of udder health and milk processing.

The Effect of feeding system on the fatty acid profile of three muscles from Arouquesa cattle breed

L. Sacarrão-Birrento[1], S.P. Alves[2], A.M. De Almeida[1], L.M. Ferreira[3,4], M.J. Gomes[4,5], J.C. Almeida[4,5] and C.A. Venâncio[4,5]
[1]ISA, LEAF, Universidade de Lisboa, Tapada da Ajuda, Lisboa, Portugal, [2]FMV, CIISA and AL4AnimalS, Universidade de Lisboa, Lisboa, Portugal, [3]CITAB, UTAD, VR, Portugal, [4]ECAV, Universidade de Trás-os-Montes e Alto Douro (UTAD), Vila Real (VR), Portugal, [5]CECAV, UTAD, VR, Portugal; laurasvbirrento@isa.ulisboa.pt

Beef production in Portugal mountain regions has an important socio-economic role improving production sustainability and valuing the genetic heritage by using autochthonous breeds such as the Arouquesa that is known for its high quality PDO beef. The aim of this work was to compare the meat fatty acid (FA) profiles of Longissimus thoracis (LT), Semitendinosus (ST) and Infraspinatus (INF) from 64 Arouquesa yearlings of five different groups: a traditional (T) group (n=11) fed with hay and ground maize, weaned and slaughtered at 8 months; a T+S1 group (n=13) where a starter concentrate was added (S1), a S1+S2 group (n=10) where the animals were fed with S1 until weaning at 5 months and then reared with a growth concentrate (S2) until slaughtering at 8 months, a T+S3 group (n=15) that, after the traditional weaning, animals were reared with a finishing concentrate (S3) until slaughtering at 12 months; and the S3 group (n=15) that was produced like S1+S2 group but finished with S3 until slaughtering at 12 months. FA methyl esters were prepared from freeze-dried muscles and quantified through gas chromatography with flame ionization detection. There were no differences (P>0.05) in the total FA content of ST and INF muscles among groups, but there were in LT muscle where T+S1 had the highest content of 66,4 mg/g muscle dry matter. The total of omega-3 FA content was significantly higher in T in the three muscles. However, the total of omega-6 FA was higher in T+S3 and S3 in all muscles. The FA CLA-c9t11 and C18:1t11 were higher in groups T and T+S1, whereas C18:1t10 was higher in groups S1+S2, T+S3 and S3 for all muscles. These results showed that supplementation affects negatively the FA acid profile, as supplemented groups had lower proportion of omega-3 FA and C18:1t11 that can desaturate in CLA isomers with anticancerogenic properties and had higher proportion of trans FA. PDR2020-101-031094, FCT PhD grant 2021.07638.BD.

Comparison of growth performances of Arouquesa steers of different production systems

L. Sacarrão-Birrento[1], M.J. Gomes[2,3], D. Moreira[3], A.M. De Almeida[1], S.R. Silva[2,3], J. Carlos Almeida[2,3] and C.A. Venâncio[2,3]
[1]Instituto Superior de Agronomia, LEAF, Universidade de Lisboa, Tapada da Ajuda, Lisboa, Portugal, [2]Veterinary and Animal Research Centre (CECAV), UTAD, VR, Portugal, [3]ECAV, Universidade de Trás-os-Montes e Alto Douro (UTAD), Vila Real (VR), Portugal; laurasvbirrento@isa.ulisboa.pt

The Arouquesa is a Portuguese autochthonous cattle breed known for the PDO beef label that has an important role in product valorisation. This breed has relatively small numbers, albeit different production systems exist in a small geographical area. They are poorly characterized. The aim of this work was to compare the productive parameters – final live weight (LW) and average daily gain (ADG) – between animals from different feeding systems. Two trials were carried out differing in slaughter and weaning age. Trial 1 (n=64), comprised a traditional (T) group (n=11) fed with hay and ground maize, weaned and slaughtered at 8 months; a T+S1 group (n=23) where a starter concentrate was added (S1), and finally an S1+S2 group (n=30) where the animals were fed with S1 until weaning at 5 months and then reared with a growth concentrate (S2) until slaughtering at 8 months. In Trial 2, 2 groups were used: the T+S3 group (n=10) in which, after the traditional weaning, animals were reared with a finishing concentrate (S3) until slaughtering at 12 months; and the S3 group (n=15) that was produced like S1+S2 group but finished with S3 until slaughtering at 12 months. The LW and ADG were significantly different ($P<0.05$) between groups. In Trial 1, the T+S1 group had the highest LW (267.5±41.6 kg), and the S1+S2 group had the lowest (232.4±24.1 kg). In Trial 2, the T+S3 group had a higher LW (343.3±34.0 kg). About the ADG, the T+S1 and S1+S2 groups had the highest values (1,006.3±118.0 g and 1,004.2±178.0 g) compared to the T group (883.0±114.0 g). The T+S3 group had the lowest ADG value (985.0±129.0 g) compared to S3 group (1,145.0±169.0 g). In conclusion, the inclusion of concentrate feed in traditional feeding systems and prolonging the rearing improves the productive parameters that can be an asset for the Arouquesa PDO beef production. PDR2020-101-031094, FCT PhD grant 2021.07638.BD.

Live yeast supplementation improves the performance and carbon footprint of beef cattle production

S.A. Salami[1], C.A. Moran[2] and J. Taylor-Pickard[3]
[1]Alltech (UK) Ltd., Ryhall Road, Stamford, United Kingdom, [2]Alltech SARL, Rue Charles Amand, Vire, France, [3]Alltech Biotechnology Centre, Summerhill Road, Dunboyne, Ireland; saheed.salami@alltech.com

Supplementation of live yeast (LY, *Saccharomyces cerevisiae*) in ruminant diets has the potential to improve rumen function and animal performance. However, there is considerable variation in the response of ruminants to dietary LY, particularly in beef cattle. Thus, a meta-analysis was conducted to examine the effect of a LY product (Yea-Sacc® [YS], Alltech Inc., USA) on rumen fermentation and the performance of beef cattle. Data were extracted from 17 and 36 experiments for the rumen fermentation and performance database, respectively, and a random-effects model was used to compare the basal diets without or with dietary YS. Additionally, the meta-analysis results of cattle performance were used to conduct a scenario simulation in a life cycle assessment (LCA) model that evaluates the impact of feeding YS on the carbon footprint (CFP) of beef production in a semi-intensive system. There was no effect of YS on overall rumen pH; however, YS increased rumen pH by +0.12 units when cattle were fed diets containing >50% concentrate, which could be partly attributed to lower ruminal lactate (-9.1%). YS improved rumen fermentation by increasing the NDF digestion (+9.2%), microbial cell yield (+3.5%) and the concentration of total volatile fatty acids (VFA, +5.7%), mainly due to increased acetate (+6.0%) and valerate (+5.5%). No effect of YS on ruminal NH_3-N, molar proportions of individual VFA, acetate/propionate ratio and predicted methane yield. YS increased overall dry matter intake (+1.5%), but this effect was exhibited predominantly in growing cattle (+1.9%). Consequently, YS enhanced the growth and slaughter performance of cattle by increasing the overall final body weight (BW, +1.2%), average daily gain (+5.1%), feed efficiency (+3.5%), hot carcass weight (+1.1%) and dressing percentage (+0.6%). The LCA simulation revealed that feeding YS reduced the CFP of beef production when expressed as emission intensities of BW gain (-4.3%; -0.33 kg CO_2e/kg BW gain) and carcass gain (-4.8%; -0.61 kg CO_2e/kg carcass gain). Overall, these results demonstrate that YS is a viable natural feeding technology that improves beef production efficiency and sustainability.

Endoplasmic reticulum response to pre-slaughter stress: discovering new biomarkers of beef quality

L. González-Blanco[1,2], Y. Diñeiro[1,2], M.J. García[1,2], V. Sierra[1,2] and M. Oliván[1,2]
[1]Instituto de Investigación Sanitaria del Principado de Asturias (ISPA), Av. del Hospital Universitario, 33011, Spain, [2]Servicio Regional de Investigación y Desarrollo Agroalimentario (SERIDA), Sistemas de Producción Animal, Ctra. AS-267, PK19, 33300, Spain; lgblanco@serida.org

One of the most well-known factors causing negative effects on final meat quality is pre-slaughter stress (PSS), that in cattle is related to the appearance of defective meat (DFD), with abnormal *post-mortem* muscle pH decline (pH24≥6) resulting in beef with poor processing characteristics which produces food waste and significant economic losses for the beef industry. PSS increases significantly the oxidative stress altering the normal function of cellular organelles, leading to mitochondrial and endoplasmic reticulum (ER) dysfunction. Despite the important role of the ER in the cellular response to oxidative stress, its role in the process of conversion of muscle into meat has not been studied to date. Therefore, the aim of this work was to study the differences in biomarkers of oxidative stress (antioxidant defence, the Heat Shock proteins (HSPs) expression and the Unfolded Protein Response (UPR) of the ER, between normal (Control) and DFD beef from 'Asturiana de los Valles' at 24h *post-mortem*, in order to understand the changes that PSS causes in the process of conversion of muscle into meat. Our results indicate that DFD meat showed lower antioxidant capacity (P<0.001), higher expression of large HSPs (P<0.05) and higher UPR activation (P<0.05), which together seem to indicate that PSS increases the cellular response to stress during the early *post-mortem* by delaying apoptosis and leading to an abnormal process of conversion of muscle into meat that results in defective meat. Biomarkers of these cellular processes are putative biomarkers of meat quality.

Diet supplemented with olive cake as a model of circular economy: metabolic response in beef cattle

A. Bionda[1,2], V. Lopreiato[2], P. Crepaldi[1], V. Chiofalo[2], E. Fazio[2] and L. Liotta[2]
[1]University of Milan, Department of Agricultural and Environmental Sciences, Production, Landscape and Energy, Via Celoria 2, 20133 Milan, Italy, [2]University of Messina, Department of Veterinary Sciences, Viale Palatucci, 13, 98168 Messina, Italy; ariannabionda95@gmail.com

Integrating by-products in livestock diet is a great opportunity for implementing circular economy while reducing feed costs. Olive cake (OC) is an agro-industrial waste, but the high content of valuable metabolites makes it a promising feed supplement. This study investigated the effect of OC integration in beef cattle diet on some blood parameters. 48 Limousine young growing fattening –24 bulls (body weight 350±15 kg) and 24 heifers (280±10 kg)– were randomly allocated to a diet: concentrate at 0% (CTRL), 10% (LOW), or 15% (HIGH) of OC inclusion. Blood samples were collected before administrating the supplemented diet (0 d) and at the end of the stocker growing phase (56 d) and of the fattening (147 d). A linear regression model was fitted for each blood parameter with the 0 d as covariate and diet, time, sex, diet × time, and diet × sex as factors. g-glutamyl-transferase (GPT) was decreased by both the OC integrations (CTRL: 28±9, LOW: 31±8, HIGH: 29±5 UI/l; P=0.003), with the highest level in CTRL group at 56 d (P=0.03). LOW group showed the lowest creatine kinase (CTRL: 304±100, LOW: 243±140, HIGH: 280±170 UI/l; P=0.01), whereas HIGH group had higher bilirubin than CTRL (0.21±0.05 vs 0.18±0.06 mg/dl; P=0.002). Overall, heifers had higher glucose, cholesterol, and triglycerides, but lower insulin levels than bulls. Aspartate aminotransferase (GOT), GPT, urea (higher in heifers than bulls of LOW group), and LDH (higher in bulls than heifers of HIGH group) were affected by diet × sex interaction (P<0.01). Overall, from 56 to 147 d, GOT, bilirubin, urea, cholesterol, and insulin increased, whereas GPT and glucose decreased (P<0.05). These data support the OC inclusion up to 15% of the concentrate with no detrimental effect on beef cattle metabolic status. However, the different response between heifers and bulls should be further investigated. In conclusion, OC can be considered as component in beef diet giving an opportunity to improve agriculture sustainability. Work supported by P.O. FESR SICILIA 2014/2020 – Project BIOTRAK.

Can haylage-based diets be a sustainable alternative to concentrate-based diets for finishing veals?

J. Santos-Silva[1,2], A.E. Francisco[1,2], A.P. Vaz Portugal[2], M.T. Dentinho[1,2], A.J. Rodrigues[3], N.R. Rodrigues[4], T. Domingos[4] and R.J.B. Bessa[1]
[1]CIISA, Av Universidade Tecnica, 1300-477 Lisboa, Portugal, [2]INIAV, Polo de Santarém, 2005-048 Vale de Santarém, Portugal, [3]ELIPEC, Av. de Badajoz, 3, 7350-903 Elvas, Portugal, [4]Terra Prima Serviços Ambientais, Avenida das Nações Unidas 97, 2135-199 Samora Correia, Portugal; jose.santossilva@iniav.pt

High quality forage-based diets for finishing cattle will reduce competition for human edible food resources, Portuguese dependence on cereal imports and protein rich commodities and potentially the carbon footprint of the operation. Sixteen crossbred Limousine × Alentejana 7 months weaned veals, were individually housed and randomly assigned to two diets during 64 days of finishing. The control diet (C) was based on a commercial concentrate and the Total Mixed Ration diet (TMR) contained 54% DM (dry matter) haylage, 36% DM concentrate and 10% DM sunflower seed. The haylage for TMR used was from a biodiverse mixture of annual species. The diets were formulated to have 16% DM of crude protein and were offered *ad libitum*. Intake was controlled daily and live weight fortnightly. Methane production was individually measured, using a GreenFeed unit (C-Lock, Rapid City, SD, USA). A Life Cycle Analysis was conducted to estimate the carbon footprint of the feeding systems. DM intake was 7.6% higher with TMR, corresponding to higher fat and fibre and lower starch intakes. Daily weight gain was not affected by the diets, averaging 1,598±g/day. Dry matter conversion ratio was 43% higher for TMR but feeding cost was not affected by the diet when normalized by live weight gain (2.37 €/kg). Concentrate consumption was 46% lower with TMR. Although methane production was 75% higher, the total carbon footprint of the TMR diet was slightly lower when compared to the Control. So, haylage-based diets did not compromise growth performance or meat quality of crossbred veals and had small impact on the carbon footprint of veals finishing phase when compared to conventional concentrate-base diets. Work funded by PDR2020 program through the FEADER under the project LegForBov (PDR2020-101-031179).

Improving sustainability of dairy cattle farms through the use of Limousin and Charolais beef sires

D. Gavojdian, M. Mincu and I. Nicolae
Research and Development Institute for Bovine, Cattle Production Systems, sos. Bucuresti-Ploiesti, km 21, Balotesti, Ilfov, 077015, Romania; gavojdian_dinu@animalsci-tm.ro

The elimination of milk quotas at EU level resulted in restructuring of the dairy sector, intense cross-border milk commerce, increased volatility in production and prices, including a reorientation of some farmers towards beef production. Moreover, male veal-calves from Holstein-Friesian dairy breeds are regarded as uneconomical for fattening, given their slow growth rates and lower carcass attributes. The aim of the current research was to evaluate the use of Limousin and Charolais French beef sire breeds for crossbreeding with the Romanian Black and White Spotted HF dairy breed (n=20), with the ultimate purpose as to test their F_1 progeny performances for meat production and reproduction efficiency. At the age of 14 months the F_1 Limousin × Romanian Black and White Spotted heifers had on average 376.6±23.73 kg, while the F_1 Charolais × Romanian Black and White Spotted heifers had on average 459.67±21.63 kg, with differences between the two genotypes being significant ($P \leq 0.01$). Age at first conception was on average of 16.7±1.05 months for the Limousin sired heifers and of 20.3±2.63 months for the Charolais crossbreeds, differences between genotypes being significant ($P \leq 0.05$). Based on current preliminary results, when crossbreeding dairy breeds, the Limousin sires are recommended for farms which desire to transition to beef production, throughout the use of maternal composites, given the better performance of the Limousin crosses under semi-intensive systems when reproduction efficiency of heifers is concerned. Furthermore, in order to develop a beef composite breed, the F_1 Limousin and Charolais sired heifers were crossed Piemontese breed, to produce a three-way cross, in order to obtain animals with high genetic potential for growth rates and prime quality meat. The use of French beef sires for crossbreeding in Romanian dairy enterprises could represent a feasible alternative for cattle farmers to diversify their production, to produce calves with higher growth rates and improve the overall farm returns, while taking advantage of the heterosis effects and breeds complementarity.

Session 38 Poster 16

Genome-wide association studies for milk production quality traits in Romanian Spotted breed

D.E. Ilie, A.E. Mizeranschi, C.V. Mihali and R.I. Neamt
Research and Development Station for Bovine, Calea Bodrogului 32, Arad, 310059, Romania; danailie@yahoo.com

The Romanian Spotted (RS) is an autochthonous cattle breed that has been selected for milk traits since 1981. In this study, we conducted a genome-wide association study (GWAS) to identify genomic regions associated with milk production quality traits in the first 3 lactations of the RS breed. Cattle were genotyped with Axiom Bovine v3 SNP-chip and phenotypic data consisted of 24,988 records from 540 RS cattle. After imputation and quality control, a data set of 40,305 SNPs was used for GWAS, which was performed in two stages. First, estimated breeding values were computed using the Blupf90 software, implementing a test-day model with herd-year-season as contemporary groups and 2nd order Legendre polynomials representing both the fixed and random lactation curve elements (genetic + permanent environment effects). Breeding values were then deregressed and used as phenotypes in the second stage, implemented using the R package rrBLUP. The Bonferroni threshold was set to $-\log 10(p) \geq 5.906$, corresponding to a 0.05 significance level, and suggestive SNPs were considered based on a false discovery rate (FDR) threshold of 0.25. Phenotypic data included fat percent (FP) and protein percent (PP). In GWAS analysis, 22 and 6 SNPs associated with FP and PP in all lactations were identified. Significant SNPs, including AX-117081655 (rs109234250/*DGAT1*), AX-115114815 (rs109968515), AX-115116727 (*MAF1*), AX-106763216, AX-106749350, AX-117086956 (nearby *ZNF623*), AX-185112146 (rs135739471) and AX-106758724 (rs109350371, nearby *PLEC*) were found for FP in first lactation. One SNP, AX-117081655/rs109234250 overlapping *DGAT1* (K232A), was significantly associated with both FP and PP in all lactations. This SNP was previously reported to be associated with milk traits. Based on the FDR, 13 suggestive SNPs located nearby genes *ZNF34, ZNF7, GRINA, TLR4, C14H8orf33, RFX7, LY6E, RPL8, SPATC1, TRIM32, ZNF16* and *LY6H* were identified as associated with FP and 4 suggestive SNPs as associated with PP. This study is the first GWAS on milk traits in RS breed and the identification of promising associations will allow to better understand milk quality traits in this breed. This research was funded by the project ADER 8.1.6/2019.

Session 38 Poster 17

Survey of the trans fatty acid composition of Portuguese beef

S.G. Muchecha[1], T. Fernandes[1], R.J.B. Bessa[1,2] and S.P. Alves[1,2]
[1]CIISA-Centro de Investigação Interdisciplinar em Sanidade Animal, Faculdade de Medicina Veterinária, Universidade de Lisboa, Lisboa, 1300-477, Portugal, [2]Associate Laboratory for Animal and Veterinary Sciences (AL4AnimalS), Lisboa, 1300-477, Portugal; susanaalves@fmv.ulisboa.pt

Ruminant-derived foods are becoming the main source of trans-FA (TFA) in food due to the acknowledgment of the role of industrial TFA from partially hydrogenated vegetable oil (PHVO) in the promotion of cardiovascular diseases and the consequent regulations to restrict them in foods. In ruminant foods, the major TFA can be the vaccenic acid (18:1*t*11), which has the potential to promote human health. But, in certain circumstances, the 18:1*t*11 is replaced by 18:1*t*10 that may have negative effects on health similar to PHVO. Thus, a survey of the FA composition of the commercially available Portuguese beef was performed using samples (n=62) from certified (Mertolenga-DOP, Carnalentejana-DOP, Carne dos Açores-IGP) and non-certified beef purchased at supermarkets in Lisbon region. Lipids were extracted and FA methyl ester were analysed by gas chromatography with flame ionization detection. Statistical analysis was carried out using the SAS software with a linear model considering the meat origin as the single fixed factor. Intramuscular fat content did not vary (P=0.855) among origins, averaging 1.8 g/100 g meat. Regarding the TFA, it was observed a high variability among meats, although the proportion of total TFA content did not differ (P=0.791) averaging 2.7 g/100 g of total FA. However, there were significant differences (P<0.05) in the proportions of 18:1*t*11 and 18:1*t*10 among origins. The highest proportion of 18:1*t*11 was observed in Açores-IGP (1.26 g/100 g FA), and the highest proportion of 18:1*t*10 was observed in Carnalentejana-DOP (1.28 g/100 g FA) but it did not differ from Mertolenga-DOP and non-certified beef. Açores-IGP beef also showed the greatest proportion of 18:2*c*9*t*11 (P<0.001). In brief, the lipid value of Portuguese beef appears to be variable and affected by its origin, moreover some meats presented 18:1*t*10 values above the limits of 2 g/100 g of total fat recommended by the EU Regulation (no. 1925/2006) for TFA in food. Financial support was provided by PTDC/CAL-ZOO/29654/2017, UIDB/ 00276/2020 (CIISA) and LA/P/0059/2020 (AL4Animals) projects.

Effect of dietary protein quality on performance and carcass characteristics of feedlot cattle

L.J. Erasmus, M. Liebenberg, R.J. Coertze, C.J.L. Du Toit and E.C. Webb
University of Pretoria, Dept Animal Science, Lynnwood rd, 0001 Pretoria, South Africa; lourens.erasmus@up.ac.za

The objective was to evaluate the effect of rumen undegradable protein (RUP) and rumen protected lysine (RPLys) and methionine (RPMet) supplementation on growth, blood parameters and carcass characteristics of feedlot cattle over a 134 d feeding period. Bonsmara type steers (120) were blocked by BW and randomly allocated to one of four treatments in a complete randomised block design. Each treatment consisted of 5 replicates with 6 animals per replicate. The treatments were: (1) basal diet supplemented with urea (CON); (2) basal diet supplemented with RUP (RUP); (3) basal diet supplemented with RUP and RPLys (RUP+L); and (4) Basal diet supplemented with RUP and RPLys and RPMet (RUP+L+M). Data was statistically analysed using the PROC MIXED model. The average daily (ADG) gain and feed conversion ratio (FCR) did not differ between treatments and ranged from 1.84 to 1.90 kg/d and 4.84 to 5.04 kg feed / kg gain respectively (P>0.05). Treatments RUP+L and RUP+L+M resulted in lower serum blood urea nitrogen concentrations on three sampling days when compared to the CON treatment which is indicative of better N utilisation (P<0.05).Serum blood AA were reduced for Ala, Glu, Pro and Tyr on sampling day 99 and reduced for Ileu, Lys, Leu, Val, Pro and Tyr when treatment RUP+L+M is compared to CON suggesting an increased uptake and utilisation of these AA for protein and muscle deposition (P<0.05).The 13th rib subcutaneous fat thickness was lower for treatment RUP+L+M compared to the other treatments (P<0.05) and accordingly the carcass channel fat mass from animals supplemented with both RPLys and RPMet was lower than the CON group (P<0.05).This suggest that Lys and Met supplementation as well as the ratio of Lys : Met may be important factors to consider if consumers demand leaner carcasses. In conclusion, a diet formulated according to NRC standards with sufficient RDP, energy and effective NDF that promotes optimal rumen fermentation can achieve above average growth performance and do not need to be supplemented with additional RUP or RPAA to meet MP requirements. Supplementation of RPLys and Met has shown potential for production of leaner carcasses.

Relationship between the content of selected bioactive compounds and antioxidant activity of milk

J. Król, A. Brodziak and M. Stobiecka
University of Life Sciences in Lublin, Department of Quality Assessment and Processing of Animal Products, Akademicka 13, 20-950 Lublin, Poland; jolanta.krol@up.lublin.pl

The aim of the study was to assess the relationship between the content of substances with antioxidant properties in milk and its antioxidant activity. The study was carried out on a farm specializing in dairy cattle, where Holstein-Friesian cows were raised. The cows were fed in a TMR system. Concentrate feed was supplied in amounts depending on the cows' demand for nutrients at a given level of yield and in a given stage of lactation. The material for analysis consisted of milk collected from 90 cows (30 for each analysed lactation number – 1st, 2nd and 3rd) during three periods: (1) up to 100 days of lactation; (2) 101-200 days of lactation; and (3) 201-305 days of lactation. The following were determined in all milk samples: crude protein, fat, lactose, dry matter (Infrared Milk Analyzer; Bentley Instruments, USA), casein and somatic cell count (SCC) (Somacount 150; Bentley Instruments, USA). In addition, the content of selected whey proteins (α-lactalbumin – α-LA, β-lactoglobulin – β-LG, lactoferrin and bovine serum albumin – BSA) and fat-soluble vitamins (A, D_3 and E) was determined by RP-HPLC (Varian, USA). Total antioxidant status (TAS) was determined by using ABTS assay. With each successive lactation, there was a significant decrease (P≤0.01) in the content of antioxidant substances, i.e. vitamins A and E and albumins (β-LG and α-LA). At the same time, there was a decrease in the TAS level in the milk. The stage of lactation, on the other hand, had no significant effect on antioxidant status. The significant (P≤0.01) positive values for the correlation coefficients between the TAS level and the content of vitamins A (r=0.687) and E (r=0.664) and of β-LG (r=0.515) indicate that the content of these compounds largely determines the antioxidant potential of milk. Negative correlations were noted between the TAS level and daily milk yield (r=–0.347 at P≤0.05), which suggests that high productivity negatively affects the antioxidant value of milk. This paper was funded through the 'Regional Initiative of Excellence' in 2019-2022, project number 029/RID/2018/19, funding amount: PLN 11,927,330.00.

Session 38 Poster 20

Changes in fatty acid profile from milk to cheese

S. Segato[1], G. Marchesini[1], S. Tenti[1], M. Mirisola[1], L. Serva[1], I. Andrighetto[1], S. Balzan[2] and E. Novelli[2]
[1]Padova University, Dept of Animal Medicine, Production and Health, Viale dell'Università, 16, 35020 Legnaro (PD), Italy, [2]Padova University, Dept of Comparative Biomedicine and Food Science, Viale dell'Università, 16, 35020 Legnaro (PD), Italy; severino.segato@unipd.it

In cheese, the fatty acid (FA) profile is related to the parameters of cheese-making process, even if it mainly depends on the milk production system. Furthermore, during ripening of cheese, a series of reactions occurs to lipids. Triglycerides undergo hydrolysis by the action of endogenous and/or exogenous lipases resulting firstly in the liberation of FA and secondary in the production of many volatile flavour compounds. The study aimed at evaluating the changes of FA from milk to cheese produced in an Alpine dairy system. Samples of raw milk were collected before starting the cheese-making process of the Asiago cheeses (n=36) that were made and annual ripened in a mountain dairy plant. In milk and cheese samples, the analyse of FA methyl esters (FAME) were performed by two dimensional GC and data were submitted to one-way ANOVA. After 12 months of ripening period, a significant decrease of SFA was observed, that lead to the increasing of MUFA. The main differences were found in C6:0, C8:0, C10:0, C18:0, C18:1 n9, C18:1 n7 and CLA. No significant difference was detected in the ratio n-6 to n-3. Milk contains many lipolytic agents which have an optimum of activity at the sn-1 and sn-3 position of triglycerides where, preferentially, the short- and medium chain saturated FAs are located. Thus, during the ripening process the FAs progressively accumulate, especially in cheese made with milk that has been mild thermized. The significant increase of CLA in cheese may be due to its de novo synthesis associated with the activity of propionibacteria and lactic acid bacterium strains throughout the ageing process. Summarizing, the ripening process seemed to modify the FA profile affecting the nutritional value of the dairy products fat component.

Session 38 Poster 21

Total versus human-edible protein efficiency of Belgian cattle

J. Van Mullem[1], C. De Cuyper[1], J. De Boever[1], S. De Campeneere[1], K. D'Hooghe[2], D. Schoonhoven[3], M. Wulfrancke[4], S. Millet[1], E. Delezie[1] and L. Vandaele[1]
[1]ILVO, Animal production, scheldeweg,68, 9090, Belgium, [2]Belgian Feed Association vzw, Gasthuisstraat 29, 1000 Brussel, Belgium, [3]Boerenbond, Diestsevest 40, 3000 Leuven, Belgium, [4]Algemeen Boerensyndicaat vzw, Industrieweg 53, 8800 Beveren-Roeselare, Belgium; carolien.decuyper@ilvo.vlaanderen.be

The efficiency of livestock to convert proteins from feed into human-edible food is usually estimated to be low. However, many feed ingredients that are used as livestock feed are by-products, which are not consumed by humans. This latter is an important element in the sustainability debate and should not be ignored when protein efficiency is assessed. Previous studies tried to re-evaluate the competition for plant proteins between livestock and humans by comparing total and human-edible protein efficiency. Total protein efficiency (TPE) is calculated as the ratio of total animal protein produced to the total feed protein consumed by livestock. Human-edible protein efficiency (HPE) only considers the fraction of the proteins which are used in human nutrition. In this study, we calculated the TPE and HPE for Belgian beef and dairy cattle. First, all inputs and outputs connected to 100 dairy or beef cattle were identified over their entire life cycle. In collaboration with stakeholders, typical Belgian rations where formulated for all animal categories. The edible protein contents of feed ingredients were based on Laisse *et al.* (2018). For dairy cattle, TPE and HPE where 0.24 and 1.22, 0.22 and 1.71, and 0.21 and 3.13, for an intensive dairy farm type with a maize-based diet, an intensive dairy farm type with a grass-based diet and an extensive dairy farm type, respectively. The TPE and HPE for beef cattle farms amounted to and 0.20 and 0.93 for an intensive beef farm and 0.18 and 1.34 for an extensive beef farm. Thus, all cattle farm types are net producers of human-edible protein with exception of an intensive beef farm. However, it is important to mention that the choice of the feed ingredients and their human-edible fraction has a large impact on the HPE.

Evaluation of carcass electrical stimulation on meat quality traits of F1 Angus-Nellore cattle

C.P. Oliveira[1], B. Torin[1], M.A.S. Coutinho[2], G.A. Rovadoscki[2], L.A.L. Chardulo[1], O.R. Machado Neto[1], A.E. Pedro[1], R.C. Rodrigues[1], P.Z. Rattes[1] and W.A. Baldassini[1]
[1]São Paulo State University (FMVZ UNESP), Animal Breeding and Nutrition, Avenida Universitária, no, 3780 Altos do Paraíso, Botucatu, SP, 18.610-034, Brazil, [2]Brazil Beef Quality, Avenida Comendador Pedro Morganti, 4965 Monte Alegre, Piracicaba, SP, 13405-240, Brazil; otavio.machado@unesp.br

Carcass electrical stimulation (ES) of medium voltage (120 V, 60 Hz) was evaluated in carcass of F1 Angus-Nellore bulls and heifers (30 per group) feedlot finished. At slaughterhouse, the carcasses were properly identified, washed and, before cooling, the left half carcass of each animal was submitted to ES for 10 seconds. The right half carcass of the same animal did not receive ES (control). Before deboning, the carcasses of all animals were evaluated for carcass weight, ossification, ribeye area, fatness, marbling score and final pH (48 h). Subsequently, between the 12th and 13th ribs (*Longissimus thoracis* – LT) meat samples were collected for physical-chemical analyses at four aging times (7, 14, 21 and 28 days). Other LT samples were used in sensory quality tests with the consumer, aged for 14 and 21 days. The ES did not influence (P>0.05) meat quality traits such as purge loss, cooking loss or meat colour (L, a*, b*, Chroma and Hue). On the other hand, shear force (SF) tended (P<0.10) to change in response to ES, with lower SF values (more tender meat) being observed more frequently in the ES treatment when compared to the control. Additionally, it was found that consumers tended (P<0.10) to attribute higher tenderness scores to meats from ES treatment. However, the same trend was not observed for flavour in meats aged for 21 days. The juiciness and overall satisfaction variables were not influenced by treatments (P>0.05) in the two aging times evaluated (14 and 21 days). Significant (P<0.01) and positive correlations were found between sensory tenderness, general satisfaction and flavour (r>0.70). In general, correlations between objective and sensorial variables of meat quality traits were of low magnitude or not significant (P>0.05). In this study, ES in half carcass did not result in consistent change in the meat quality traits of crossbred cattle (bulls or heifers) feedlot finished.

Biodiverse haylages in beef fattening diets – Effect on lipid oxidation of cooked meat

L. Fialho[1,2], L. Cachucho[1,2], A. Francisco[1,3], J. Santos-Silva[1,3] and E. Jerónimo[2,4]
[1]Centro Investigação Interdisciplinar em Sanidade Animal, CIISA, Lisboa, Portugal, [2]Centro de Biotecnologia Agrícola e Agro-Alimentar do Alentejo, IPBeja, Beja, Portugal, [3]Instituto Nacional de Investigação Agrária e Veterinária, Quinta da Fonte Boa, Santarém, Portugal, [4]Mediterranean Institute for Agriculture, Environment and Development, MED, Beja, Portugal; leticia.fialho@cebal.pt

In Portugal, bulls and heifers are fattened with concentrate-based diets mainly composed of cereals and oilseed derivatives, raw materials in which country is strongly dependent on the external market. Currently is possible produce high-quality haylages, even when the climate conditions in Spring are unfavourable. Under the project 'LegForBov – Alternative feeds in beef production', haylages were used to replace part of conventional raw materials and, thus, reduce external dependence and the environmental impact of beef production systems. In 3 experiments, haylage produced from a biodiverse mixture (Speedmix – Fertiprado, Portugal) was included at different levels (50, 54, 60 and, 67.5% DM) in bull and heifer diets (64 animals). The diets with 54 and 67.5% DM haylage contained sunflower seed (10% DM). The effect of haylage on meat lipid oxidation was compared with a conventional concentrate-based diet. Lipid oxidation (TBARS) was evaluated in muscle samples stored under vacuum at 2 °C from 3 to 14 days after slaughter and maintained at 4 °C during 0 and 3 days after cooking. On cooking day, the meat lipid oxidation was lower in haylage diets (P<0.001; 0.12 mg malonaldehyde (MDA)/kg muscle) than in concentrate-based diet (0.18 mg MDA/kg muscle). In cooked meat preserved for 3 days, lipid oxidation was not affected by the diet, averaging 3.09 mg MDA/kg muscle. The higher contents of antioxidant compounds, such as phenolic compounds in haylage diets than in concentrate diet (8.93 vs 1.84 mg gallic acid equivalents/g DM, respectively), may have contributed to the meat lipid protection. However, in more severe oxidative conditions (cooking and storage) the antioxidant effect of haylage was not enough to limit the lipid oxidation of meat. Funding: Project LegForBov (PDR2020-101-031179; PDR2020-101-031184) – FEADER; Projects UIDB/05183/2020 (MED) and UIDB/00276/2020 (CIISA), and PhD grants awarded to LC (2020.05712.BD) and LF (2020.04456.BD) – FCT.

Session 38 Poster 24

Effect of diet intensification on growing parameters and carcass quality of Holstein-Friesian steers

T. Moreno[1], S. Crecente[1], R. Alonso[2] and C. García-Fontán[2]
[1]Centro Investigaciones Agrarias de Mabegondo (CIAM)-AGACAL., Apdo 10., 15080 A Coruña, Spain, [2]Centro Tecnológico Carne., Avda. Galicia no. 4., 32900 Ourense, Spain; teresa.moreno.lopez@xunta.es

Demand of premium quality meat from major beef is experiencing a notable increase in Spain, particularly 2017 emerged PGI Galician Cow&Steer. Meat production from fattened steers and cull cows has a great potential, especially Holstein-Friesian (HF) breed in Galicia, by its high census. 17 HF male calves were selected from CIAM dairy herd and castrated in the first month. Experimental trial started when calves were 6-7 months-old and finished with 52 months-old. From 6-7 months up to 2 years-old steers were fed on grass in spring and autumn, and on grass-silage during summer and winter, receiving a concentrate supplement of 1-2.5 kg/head/day. After 2 years, steers were divided into two groups balanced by weight and age to assess the effect of diet intensification: -Extensive-system(ES):fed with grass and grass-silage.8 months before slaughtering with grass silage and 5 kg high-energy concentrate. -Intensive-system(IS):fed with grass and grass-silage and 1-2.5 kg/head/day concentrate supplement. 14 months before slaughtering with grass silage and 5-12 kg high-energy concentrate. Animals were monthly weighed and average daily gain(ADG) was calculated. No significant differences were found between groups at slaughter-weight (ES:1,165 vs IS:1,209 kg; n.s), neither ADG (ES:0.647 vs IS:0.708 kg;n.s), although there was a large difference in concentrate intake along fattening period (IS:4,640 vs ES:1,569 kg). No significant differences were found in carcass characteristics: carcass weight (ES:603.9 vs IS:631 kg; n.s.),dressing percentage(ES:51.85 vs IS:52.17%; n.s),conformation-score (ES:R-vsIS:O+; n.s) and fatness-score (ES:4 vs IS:4;n.s).The loin, most valued cut, entails over 15% carcass weight. Following EUROP classification, average carcass classification was R-4.Considering the minimum carcass classification for PGI Galician Steer certificate is R-4, then 40% of experiment animals did not reach PGI requirements. Consequently, HF steers showed low conformation-score for this label. It might be interesting to cross HF breed with beef strains to produce steers which could reach conformation requirements.

Session 38 Poster 25

Genetic polymorphism of milk proteins in Holstein cattle population from Serbia

M. Šaran, M. Despotov, S. Trivunović, M. Ivković, D. Janković and L.J. Štrbac
Faculty of Agriculture, Department of Animal Science, Trg Dositeja Obradovića 8, 21000 Novi Sad, Serbia; momcilo.saran@stocarstvo.edu.rs

Previous research has shown that genetic polymorphism of milk proteins significantly affects the quality of milk, milk yield, and composition. Cows with the BB genotype of κ-casein and β-lactoglobulin produce milk that is more favourable for cheese production, compared to cows of the AA genotype. Consumption of milk from cows that possess the A1 allele of the β-casein gene can have negative health effects on human health. For this reason, the aim of this study is to examine the polymorphisms of genes encoding the synthesis of κ-casein, β-casein, and β-lactoglobulin in the population of the Holstein cattle breed from Serbia. Polymorphisms were determined on a sample of 32 cows using the PCR-RFLP method. Depending on the locus, the following number of animals was successfully genotyped: 31 (κ-casein), 27 (β-casein), and 29 (β-lactoglobulin). In the examined population sample, it was found that the genotype frequencies of κ-casein were as follows: AB (0.51), AA (0.39), and BB (0.10). The frequency of alleles A and B was 0.64 and 0.36, respectively. After analysis of the β-casein locus, it is determined that only the A2A2 genotype was present in the examined cows. Analysis of the β-lactoglobulin locus showed that the heterozygous AB genotype was the most common with a frequency of 0.41. It was followed by the AA and BB genotypes with a frequency of 0.35 and 0.24, respectively. The frequency of alleles A and B was 0.55 and 0.45, respectively, which indicates high variability in the study population. Using an exact test, it was determined that the population was in Hardy-Weinberg equilibrium at the loci for κ-casein and β-lactoglobulin. Equilibrium was not calculated for β-casein because only allele A2 was present. Although a relatively small sample was used, this research is a good basis for further, more extensive research of this type and selection in the Holstein cattle population from Serbia. Acknowledgements: This research is part of the PROMIS project: 'A Bioinformatics Approach to Dairy Cattle Breeding Using Genomic Selection', No. 6066512, funded by the Science Fund of the Republic of Serbia.

Session 38 — Poster 26

Impact of Chlorine-free cleaning of milking equipment on bulk tank milk quality

D. Gleeson and L. Twomey
Teagasc, Moorepark, Fermoy, Co. Cork, Ireland; david.gleeson@teagasc.ie

The use of chlorine for cleaning milking equipment has the potential to cause residues such as trichloromethane (TCM) and chlorate in raw milk and dairy products. Little is known about the impact that removing chlorine-based (CB) detergents (sodium hydroxide combined with sodium hypochlorite) and using alternative chlorine-free (CF) detergents (sodium hydroxide) may have on the microbiological quality and residue levels of bulk tank milk. The influence of CF milking equipment cleaning protocols were compared to traditional CB protocols by evaluating the microbiological quality and chlorine related residue levels in bulk tank milk on three occasions during the milk production season. Commercial dairy farms (n=51) using different cleaning protocols were identified by four milk processors. Twenty-eight farms used CF cleaning products, thirteen farms used CB cleaning products and ten farms used CF products for bulk milk tank cleaning only (BTCF). Bulk tank milk was tested for a range of bacteria including total bacteria, psychrotrophic, thermoduric, thermophilic, presumptive *Bacillus cereus* and enterococci counts. TCM residues in milk samples were quantified using static head-space gas chromatography (HS-GC) and chlorate analysis of milk was performed by high-performance liquid chromatography coupled with tandem mass spectrometry (LC/MS-MS). Total bacteria counts were lower with CF protocols (3,168 cfu/ml) as compared to CB protocols (12,454 cfu/ml) ($P<0.001$). Similarly, psychrotrophic and thermophilic counts were lower for CF as compared to CB protocols ($P<0.001$). Whilst differences were observed for these parameters, they were not considered biologically important. Chlorine based cleaning protocols had higher mean TCM levels (0.0013 mg/kg) compared to CF (0.0004 mg/kg) and BTCF (0.0005 mg/kg) protocols ($P<0.001$). The percentage of milk samples with chlorate detected (≥0.001 mg/kg) were 23, 6 and 11% for CB, CF and BTCF respectively. Farms using CF wash protocols had milk of a higher microbiological quality than farms using CB cleaning; this may be due to these farms having received specific guidance from milk quality advisors on the new protocols. Chlorine-free protocols demonstrated a positive impact on milk quality when implemented on the commercial farms in this study.

Session 38 — Poster 27

Using 'chlorine free' detergents for factory CIP: is dairy product bacterial quality compromised?

L. Twomey[1,2] and D. Gleeson[2]
[1]Munster Technological University, Bishopstown, Cork, Ireland, [2]Teagasc, Moorepark, Fermoy, Co. Cork, Ireland; lorna.twomey@teagasc.ie

The Irish dairy industry has recently moved away from chlorine based detergents for cleaning in place (CIP), but concern has been raised about the impact that removing chlorine from CIP may have on the bacterial quality of dairy products. The aim of this study was to investigate the bacterial quality of salted sweet cream butter and rennet casein powder; along their manufacturing chains when produced in a facility using 'chlorine free' detergents for CIP. All samples were taken during a single manufacturing run; at peak operating capacity. Samples of cream (n=2), buttermilk (n=2) and butter (n=3) were taken aseptically during the churning process. Skim milk samples (n=2) were taken prior to rennet casein formation. Samples of rennet casein powder (n=3) were aseptically taken across the ring drying process. Total bacteria (TBC) and thermoduric counts were conducted on all samples using Petrifilm Aerobic Count plates in accordance with standard methods. Skim milk and rennet casein powder samples were also tested for mesophilic, thermophilic and thermo-resistant thermophilic spores using plate count agar. Samples for mesophilic and thermophilic spores were heated at 80 °C for 10 minutes and incubated at 37 and 55 °C respectively. Samples for thermo-resistant thermophilic spores were heated at 100 °C for 30 minutes and incubated at 55 °C. All spore plates were incubated for 48 hours. All cream, buttermilk & butter samples displayed TBC and thermoduric counts of ≤10 cfu/g. The skim milk had a TBC of 3,005 cfu/g and a thermoduric count of 30 cfu/g. The rennet casein powder had a TBC of 200 cfu/g and a thermoduric count of 250 cfu/g. Only mesophilic (57 cfu/g) and thermophilic spores (12 cfu/g) were found in skim milk. Mesophilic (1,120 cfu/g), thermophilic (1,311 cfu/g) and thermo-resistant thermophilic spores (250 cfu/g) were all enumerated in rennet casein powder. As all bacterial counts were less than guided limits it can be concluded that using 'chlorine free' detergents for CIP did not compromise the bacterial quality of dairy products in this study.

Herbal mixture with antimicrobial activity in cow diet: effect on quality of milk and dairy products

A. Brodziak, J. Król, R. Klebaniuk and M. Stobiecka
University of Life Sciences in Lublin, Akademicka 13, 20-950 Lublin, Poland; aneta.brodziak@up.lublin.pl

The ban on use of antibiotic growth promoters in EU countries and the growing problem of drug-resistant bacteria have led to the search for natural substances and their use in veterinary treatment and prevention. Therefore a study was conducted to determine the effect of herb mixture addition to the diet of cows on nutritional value, health-promoting properties, and microbiological quality of raw milk and yoghurt prepared from it. The study included 30 Holstein-Friesian cows (15 each in the control group and experimental group) in their 3rd lactation. Cows were fed in a TMR system. The experimental factor was the addition of standardized mixture of dried herbs (oregano 25%, thyme 25%, cinnamon 15%, and coneflower 35%) in the amount of 3% DM of feed/day/cow. Content of bioactive compounds with antimicrobial activity was determined in an experimental concentrated mixture with the addition of herbs, obtaining: carvacrol – 1,1, carvone – 2,3, cymene – 7,0, linalool – 8,3 and thymol – 5,8% in total of ingredients. Milk from cows receiving the herb mixture was shown to have higher casein content (2.63 vs 2.88%, $P \leq 0.05$). In addition, raw milk from the experimental group had significantly ($P \leq 0.05$) higher content of antimicrobial whey proteins (β-lactoglobulin – 10% more, lactoferrin – 32% and lysozyme – 20%). In yoghurts, the differences were at a similar level. Somatic cell count in milk was below 200,000/ml in both groups and total bacteria count – below 15,000/ml. Milk from the experimental group contained significantly less or no bacteria of the genus: *Clostridium* sp., coli, faecal coliforms, *Salmonella* and *Campylobacter* bacilli, and fungi and moulds. Significantly ($P \leq 0.05$) higher total antioxidant status (TAS) of milk (4.02 mg Trolox/100 ml) and yoghurts (6.01) compared to controls (3.08 and 4.14, respectively) was obtained. The addition of herbs to the diet of cows had a positive effect on the quality of raw milk and yoghurts, increasing the level of antimicrobial compounds and antioxidant activity. Project financed under the program of the Minister of Science and Higher Education under the name 'Regional Initiative of Excellence' in 2019-2022 project number 029/RID/2018/19 funding amount 11 927 330.00 PLN.

Optimizing amino acids dairy cows' diets improves performances and reduces environmental impacts

D. Militello[1,2], S. Lemosquet[3], Y. Mathieu[4], L. Bahloul[1], D. Andrieu[5], M. Rolland[6], S. Rouverand[7], G. Trou[8], N. Draijer[9], L. Thiaucourt[10] and A. Hocini[1]
[1]Adisseo France S.A.S., Antony, France, [2]Turin University, Turin, Italy, [3]PEGASE, INRAE, Institut Agro, Saint-Gilles, France, [4]Seenovia, Nantes, France, [5]CCPA, JANZÉ, France, [6]Vision Lait, Muizon, France, [7]Valorial, Rennes, France, [8]Chambres d'agriculture de Bretagne, Rennes, France, [9]Blonk Consultants, Gouda, the Netherlands, [10]Animal Nutrition Consultant, Molsheim, France; davide.militello@outlook.it

The cradle-to-farmgate environmental performance of dairy production was evaluated using Life Cycle Assessment (LCA). Three treatments using INRA 2007 were compared: unbalanced AA diets (Con; CP: 16.1%, NEL: 1.62 Mcal/kg DM, MP: 99 g/kg DM of MP), balanced AA diets substituted by 15% SBM with cereal (Trt1; CP: 15.6%, NEL: 1.62 Mcal/kg DM, MP: 96±4 g kg/DM, Met: 2.4% MP, Lys: 7.2% MP), fed in EU Project Dy+Milk, with theoretical optimized Met balanced diet (Trt2; CP: 15.8%, NEL: 1.66 Mcal/kg DM, MP: 97 g/kg DM, Met: 2.3% MP, Lys: 6.9% MP). A total of 444 lactating dairy cows (188±101 DIM) from five farms were used in an ABA reversal design with 3 successive periods (A – one month, B – two months). DMI was measured per period. Milk yield and composition were analysed five times. LCA were performed in the dairy module of the Animal Production Systems tool from Blonk Sustainability Tools following the methodological approach of PEFCR. Climate change impact, in kg CO_2 equivalent per kg fat and protein corrected milk (FPCM), was evaluated considering inputs, outputs, and direct farm emissions. Associated GHG emissions (CH_4, N_2O and CO_2) were characterized with the JRC Environmental Footprint impact assessment method. Data were analysed using Proc Mix of SAS. DMI was impacted only in Trt2, +1.0 kg DMI predicted. Milk yield and composition increased ($P<0.01$) by 0.6 and 0.5 kg milk/d/cow, 0.5 and 1.1 g milk true protein/kg and 39 and 43 g milk true protein/d/cow with Trt1 and 2, respectively. Gross MP efficiency increased ($P<0.05$) by 4 and 6.5%, nitrogen use efficiency increased by 9 and 7% with Trt1 and 2, respectively. Consequently, CO_2 eq emission decreased by 10% and 14%, respectively. Optimizing and balancing Met and Lys improved cow performances and reduced environmental impacts.

Session 38 Poster 30

Herbal bolus to boost Holstein-Friesian cow's health status and reproduction traits

C.L. Manuelian[1], M. Simoni[2], F. Callegari[1], C. De Lorenzi[1], M. De Marchi[1] and F. Righi[2]
[1]University of Padova, DAFNAE, Viale dell'Università 16, 35020 Legnaro, Italy, [2]University of Parma, Department of Veterinary Science, Via del Taglio 10, 43126 Parma, Italy; carmenloreto.manuelianfuste@unipd.it

Herbs are rich in bioactive compounds with anti-inflammatory and antioxidant properties that can improve animals' health status. These types of diet supplementation are of particular interest in organic production and for the reduction of antimicrobial use in dairy cattle. This study aimed to evaluate the effect of administering 3 herbal boluses in early lactation (<30 DIM) on milk composition, metabolic profile and fertility traits in Holstein-Friesian dairy cows. A total of 105 cows with high (>250.000 cells/ml) or low (<250.000 cells/ml) somatic cell count (SCC) in the previous lactation were enrolled in the study across the year. Cows received 3 boluses in three different periods (7-10 DIM, 15-20 DIM, 25-30 DIM). The boluses for the Treatment Group (T; n=53) consisted of 20 mg of a mixture of barley meal, calcium carbonate and 4 herbs: *Salix alba* (bark), *Echinacea purpurea* (herbs-root), *Silybum marianum* (dry-plant extract) and *Cynarae folium*. Whereas in the Control Group (C; n=52), the herbs were replaced by barley meal. Groups were balanced for the level of SCC. Milk (yield, energy-corrected-milk, fat, protein, casein, lactose, pH and somatic cell score) and plasmatic (α-tocopherol, total protein, globulin, AST-GOT and creatine kinase) samples were collected at 30 DIM and fertility traits (days at first AI, days open and calving to conception interval) were recorded. Data were analysed with the PROC MIXED of SAS ver. 9.4 with treatment, level of SCC and their interaction as fixed effects, and animal nested within treatment as random effect. Results showed a similar (P>0.05) production level between groups (High SCC: C, 45.81±2.01 kg/d; T, 45.62±1.74 kg/d; Low SCC: C, 42.48±1.65 kg/d; T, 42.54±1.58 kg/d). Moreover, groups (T vs C or Low vs High SCC level) did not differ in any of the traits evaluated. In conclusion, the herbal administration post-calving and before the 30 DIM did not improve or impair milk production and composition, nor the metabolism or fertility traits in healthy cows. The study was funded by EU H2020 No 774340 – Organic-PLUS.

Session 38 Poster 31

Analysis of mastitogenic agents in the buffalo species in Campania from 2014 to 2019

G. Di Vuolo, G. Cappelli, E. De Carlo, M. Orrico and D. Vecchio
Istituto Zooprofilattico Sperimentali del Mezzogiorno, UOSD Animal Science and Welfare, National Reference Centre on Water Buffalo Farming and Production, Via Salute 2, 80055 Portici (Napoli), Italy; gabriele.divuolo@izsmportici.it

Mastitis is one of the biggest problems on dairy farms. Reg (EU) No.2016/429 on communicable animal diseases enshrines and regulates the prevention and control of animal diseases that are transmissible to animals or humans Despite the important negative effect of mastitis, in buffaloes it continues to be underestimated especially by farm managers as it is less symptomatic than bovine. The aim of the present work was to analyse the results of bacteriological investigations carried out on milk samples conferred to the Istituto Zooprofilattico Sperimentale del Mezzogiorno (IZSM) for microbiological investigations between 2014 and 2019. The mastitis, compromises animal welfare and cause significant direct (milk waste, decreased productivity) and indirect (increased labour, diagnostic costs, drugs, early culling); economic losses, in dairy breeds, mastitis is the first reason for the use of antimicrobials, whose abuse must be averted to avoid the emergence of dangerous phenomena of antimicrobial resistance (AMR). The fight against AMR and animal welfare assurance, are the core of the new 'Farm to Fork Strategy'. In the period between January 1, 2014 and December 31, 2019, 2,650 milk samples were conferred to IZSM. The milk came from 105 farms in the Campania region (of which 37% Caserta; 40% Salerno and 23% between Naples, Benevento and Avellino provinces). Mastitogenic agents were isolated from 88.57% of the conferring farms. In detail in 67.74% at least one contagious pathogen was isolated, in 26.88% of farms environmental pathogens were isolated and in 22.58% opportunistic microorganisms. Analysing the contagious group in detail, 93.06% were positive for Staphylococcus aureus. For environmental pathogens, the most frequent was E.coli with 60.33%, for opportunists the most frequent were SCN (81%). Mastitis in buffalo species continues to be underestimated; It is important to train breeders and technicians on the subject in order to implement strategies more suited to the physiology of this species.

Session 38 Poster 32

Milk to cheese – phenotypic profiles of antibiotic resistance in coagulase-positive staphylococci

J. Barros[1], A.M. Pereira[2], S.P.A. Câmara[2], C. Pinto[1], H.J.D. Rosa[2], C. Vouzela[2], A.E.S. Borba[2] and M.L.E. Dapkevicius[2]
[1]Faculdade de Ciências Agrárias e do Ambiente, Universidade dos Açores, Rua Capitão João d'Ávila, 9700-082 Angra do Heroísmo, Portugal, [2]Instituto de Investigação em Tecnologias Agrárias e do Ambiente, Universidade dos Açores, Rua Capitão João d'Ávila, 9700-082 Angra do Heroísmo, Portugal; ana.mb.pereira@uac.pt

Coagulase-positive staphylococci are important mastitis pathogens in dairy cows, which can harbour and disseminate genetic determinants of antibiotic resistance in the milk chain. Prevalence and resistance to 23 antibiotics, from the carboxylic acids, aminosides, cephalosporins, streptogramins, phenicols, fluoroquinolones, fusidanes, macrolides, oxazolidinones, penicillins, aminocoumarins, rifamycins, sulphamides, and tetracyclines families, has been investigated in the first, aseptically collected, jets of milk from the four udder quarters of 60 lactating dairy cows belonging to the same farm and in raw-milk cheese made from bulk tank milk obtained from this farm. Californian Mastitis Tests were performed for each udder quarter and somatic cell counts in bulk tank milk were obtained from the farm's records. All cows were healthy and devoid of relevant signs of mammary gland inflammation. Within the scope of the One Health approach, monitoring the antibiotic resistance status in the production of animal-associated microbiota is important to design strategies aimed at minimizing the spread of (multi-) resistance to antimicrobial drugs of relevance in human and veterinary medicine. Funding: AM Pereira received funding from FCT (UIDP/00153/2020). SPA Camara received a POST-DOC grant from Regional Fund for Science and Technology (M.3.2.DOCPROF/F/044/202). This work was funded by FCT/MCTES (UIDB/00153/2020; UID/CTV/00153/2019) and by the Regional Government of the Azores (M1.1.a/008/Funcionamento/2019 (IITAA); M1.1.a/008/Funcionamento/2018/RTF/010.

Session 38 Poster 33

Quantification of CMP in A1 and A2 milk contaminated with *B. cereus* s.s. and *P. fluorescens*

A. Vasconcellos, A. Santos, D. Fonseca, A. Vaz, A. Saran Netto and A. Vidal
FZEA – USP, Medicine Veterinary, Duque de Caxias, 225 Jardim Elite, Pirassununga, SP, 13635-900, Brazil; anavidal@usp.br

Little is known about the action of microbial proteases in milk from cows with contrasting genotypes for β-casein and, given that consumer demand has been growing for innovative and quality products, and the benefits that A2 milk can bring to health and well-being are still under discussion in the literature. It is also noteworthy that the quantification of CMP (caseinomacropeptide) in relation to the action of microbial proteases can also bring relevant discussions to the current dairy production chain. The aim of the study was to standardize the experimental contamination of milk from cows with genotypes A1A1, A1A2 and A2A2 for β-casein with *Bacillus cereus* s.s. and *Pseudomonas fluorescens* and later to quantify the CMP at different incubation times (0h, 1h30, 3h, 4h30, 6h and 12h), in order to verify if there is an influence of the three genotypes on the microbial degradation of κ-CN. Over time, there was an increase in CMP concentration regardless of genotype or contamination with *Bacillus cereus* s.s. or *P. fluorescens*. Comparing the rates of sialic acid, for A1A1 we observed lower rates (mg/ml) in the initial times, contaminated with *B. cereus* (5.18) or not (4.75) or contaminated with *P. fluorescens* (5.44) or not (4.51). As for the A1A2, we found averages for samples contaminated with *B. cereus* (7.94) or not (7.65) or contaminated with P. fluorescens (7.89) or not (7.67), while for the A2A2, we observed averages for *B. cereus* (6.03) or not (5.78) or contaminated with *P. fluorescens* (6.46) or not (2.44). Comparing genotypes, it was possible to notice lower degradation rates (mg/ml) of κ-CN for milk in the end time from genotype A1A1 with *B. cereus* (7.73) or not (5.86) and with *P. fluorescens* (6.85) or not (4.85), followed by A2A2 with *B. cereus* (28.03) or not (8.8) and with *P. fluorescens* (10.29) or not (9.70) and finally, A1A2 with B. cereus (29.79) or not (5.78) and with *P. fluorescens* (7.19) or not (6.98). Thus, it is important to maintain the quality of a value-added product such as milk from cows with the A2A2 genotype for β-CN, which is more sensitive to κ-CN degradation.

A study on Italian consumers' perception towards local dairy products

V.M. Merlino[1], M. Renna[2], J. Nery[2], A. Muresu[1], A. Ricci[2], A. Maggiolino[3], G. Celano[4], B. De Ruggieri[4] and M. Tarantola[2]
[1]University of Turin, L.go Braccini 2, 10095 Grugliasco, Italy, [2]University of Turin, L.go Braccini 2, 10095 Grugliasco, Italy, [3]University of Bari Aldo Moro, Strada prov.le per Casamassima, Km. 3, 70010 Valenzano, Italy, [4]University of Bari Aldo Moro, Via Amendola 165/a, 70126 Bari, Italy; manuela.renna@unito.it

Many definitions have been proposed for 'local food', with connotations ranging from political boundaries, sustainability dimensions, quality, and emotional criteria. This study aimed to understand whether the perception of Italian consumers towards sustainability aspects of local milk and cheese (related to environmental, social, and ethical dimensions) is related to the socio-demographic characteristics of individuals and their preferences towards products quality attributes. A total of 543 consumers from Apulia (South Italy, a region characterized by ancient dairy tradition) filled in a structured questionnaire developed to collect information about consumers' socio-demographic traits, purchasing and consumption habits, preferences towards dairy products and opinions towards sustainability and market availability of local dairy products. Data were subjected to a Principal Component Analysis, from which four Principal Components were defined, named as Responsive to quality attributes, Local is better, Local is sustainable, and Availability request. We found that the attitudes of consumers are driven by quality attributes of dairy products rather than by sustainability aspects or perceived higher quality of local dairy products. The results of non-parametric tests showed that gender, age, place of residence and educational level of consumers significantly affected the above-mentioned Principal Components. On the contrary, the financial situation and family size did not influence the definition of the perception of sustainability. Such results clearly highlight the need of targeting production and developing *ad hoc* communication strategies for different consumer profiles. Finally, dissatisfaction emerged about the availability and visibility of local dairy products at the market level, which suggests the need for better promotion (e.g. using specific labels) of local milk and cheese. Funded by EIT FOOD, 2020: SUDAPS (project number: 20181).

Feed efficiency, body composition and fatty acid profile in Nellore cattle

S.F.M. Bonilha, G.K. Bassi, C.D.A. Batalha, R.H. Branco, R.C. Canesin and M.E.Z. Mercadante
Instituto de Zootecnia, Rodovia Carlos Tonani, km 94, 14.174-000 Sertãozinho/SP, Brazil; sarah.bonilha@sp.gov.br

The association between feed efficiency, body composition and fatty acid (FA) profile in beef cattle is relevant due to the need to reduce costs and environmental impacts of production systems, taking into account human health when consuming beef. This study aimed to evaluate the effect of classifying Nellore males by residual feed intake (RFI), residual weight gain (RWG) or residual intake and gain (RIG) on performance traits, Longissimus fat content and FA profile. Eighty-seven Nellore males were classified as efficient (E) or not efficient (NE) according to RFI, RWG and RIG after 98 days of confinement. Longissimus fat content and FA profile were determined after animals were slaughtered. Statistical model included the fixed effects of efficiency class and age at slaughter as covariate, and the random effect of year. Least square means were compared using *t* test at 5% of probability. There was no significant effect of RFI, RWG or RGI classes, E or NE respectively, on dry matter intake during the confinement (RFI: 6.48 vs 6.92 kg/d-P=0.132; RWG: 6.64 vs 6.75 kg/d-P=0.691; RIG: 6.31 vs 7.15 kg/d-P=0.004). For average daily weight gain, a significant effect of efficiency classification, E or NE respectively, based on RWG and RIG were detected (RFI: 0.980 vs 1.04 kg/d-P=0.453; RWG: 1.16 vs 0.860 kg/d-P<0.001; RIG: 1.11 vs 0.890 kg/d-P=0.002). Longissimus fat content was similar between efficiency classes (E or NE, respectively) determined based on RFI (14.1 vs 13.4%; P=0.488), RWG (14.5 vs 12.9%; P=0.107), or RIG (13.8 vs 13.5%; P=0.792). Animals classified as E in terms of RFI, when compared to the ones classified as NE, had lower concentration of vaccenic FA (1.61 vs 1.80 g/100 g; P=0.010) and docosatetraenoic FA (0.568 vs 0.718 g/100 g; P=0.044). When the classification was performed based on RWG or RIG, no significant differences were detected between E or NE animals for vaccenic FA (RWG 1.69 vs 1.72 g/100 g-P=0.645; RIG 1.69 vs 1.73 g/100 g-P=0.603) or docosatetraenoic FA (RWG 0.639 vs 0.610 g/100 g-P=0.715; RIG 0.623 vs 0.626 g/100 g-P=0.967). Nellore animals classified as efficient based on RFI produce health Longissimus in terms of fat content and FA profile.

In silico analyses reveals a deletion in SCD5 promoter gene: a possible role in lipid synthesis

L.L. Verardo[1], J.I. Gomes[1], M.A. Machado[2], J.C.C. Panetto[2], I. Carolino[3], N. Carolino[3] and M.V.G.B. Silva[2]
[1]Universidade Federal dos Vales do Jequitinhonha e Mucuri, Animal Science, Campus JK, 39100-000, Diamantina-MG, Brazil, [2]Brazilian Agricultural Research Corporation (EMBRAPA), Dairy, Rua Eugênio do Nascimento, 610, 36038-330, Juiz de Fora-MG, Brazil, [3]Instituto Nacional de Investigação Agrária e Veterinária, Fonte Boa, 2005-048, Vale de Santarém, Portugal; lucas.verardo@ufvjm.edu.br

The stearoyl-coenzyme A desaturase-5 protein gene (SCD5) has been cited to be expressed in mammary gland of cattle and related with lipid metabolism in goats. Milk lipids are important for derived milk products, being a relevant indicator of milk solids. We aimed to *in silico* evaluate its promoter region in cattle to identify possible functional variants affecting the biding of transcription factors (TF). First, we retrieved all known insertions and deletions (InDels) in the promoter sequence (3,000 bp upstream and 300 bp downstream from the transcription start site) of SCD5 gene using the Ensembl Biomart tool based on the ARS-UCD1.2 assembly of the bovine genome. FASTA files with the wild type sequence of promoter region and sequences with each identified InDel were used as input in TFM-explorer program to search for locally overrepresented TF binding sites (TFBS) using weight matrices from the JASPAR and TRANSFAC vertebrate database. A total of 15 InDels were observed, and thus used to retrieve the FASTA sequences of the promoter regions of SCD5 gene containing those variants. The wild type of promoter region presented 4 TF (AP-1, MZF1, GATA-2 and MYF). From the promoter sequences with the identified InDels, one with a deletion of two base pairs (from 97637549 to 97637550) promoted the biding of SP1. It is a specific protein 1 transcription factor and has been cited to play a role in maintaining milk-fat droplet synthesis in goat mammary epithelial cells. In cattle, it has been cited to modulate lipid synthesis of the mammary gland. In this study, we observed a deletion of two base pairs in the promoter sequence of SCD5 gene, more precisely, in a TFBS for SP1. Since both SCD5 and SP1 have been cited to be involved in lipid metabolism, we suggest that this deletion may have a role in mammary gland of cattle, and thus should be used for further *in vitro* and *in vivo* analyses.

A new method of assessing protein-energy balance of feed based on milk urea concentration

K. Rzewuska and K. Bączkiewicz
Polish Federation of Cattle Breeders and Dairy Farmers, Centre for Genetics, Żurawia 22, 00-515 Warsaw, Poland; k.rzewuska@cgen.pl

Information about urea content in milk is used to assess the energy-protein balance in the ration and as an indicator of the energy status of a cow. Urea is a commonly used indicator, but not very precise. Increasing the usefulness of urea as a parameter indicating the efficiency of feed protein utilization is possible by taking into account the influence of non-nutritional factors when analysing the urea content in milk. The aim of the project is to provide breeders with a tool useful in managing the herd, thanks to which they will reduce production costs and nitrogen pollution of the environment. In this study, 36,873,079 test-day records collected in years 2014 to 2021 from 2,215,168 cows from 23,175 herds were analysed. The average milk urea (MU) concentration was 218 mg/l. MU was not converted to milk urea nitrogen (MUN) because this is the form of phenotypic information used by Polish dairy farmers for herd management. The influence of such effects as lactation stage and daily milk production as well as individual genetic potential were estimated. In the next stage of the project, nine nutrition advisers assessed the nutritional balance in dairy herds during each advisory visit from December 2020 to November 2021. In the case of 855 visits, they assessed that the herd had balanced nutrition. These data were used to calculate reference value of corrected MU by taking into account MU and the adjustments for day of lactation, milk production and breeding value of each cow. Deviation from the reference value can be calculated for test day records from the registered Holstein herds. It is largely dependent on nutritional factors and therefore a more accurate indicator of the energy and protein balance than MU.

A study on quality of maize silages from Dairy-4-Future project Portuguese farms

M. Gomes[1], N. Ferreira[1], H. Trindade[2] and L. Ferreira[2]
[1]CECAV-UTAD, Quinta de Prados, 5000-801, Portugal, [2]CITAB-UTAD, Quinta de Prados, 5000-801, Portugal; mjmg@utad.pt

In the frame of Interreg Atlantic Area Program project Dairy-4-Future, a study was conducted to evaluate the chemical composition and fermentation profile of the maize silages used by 18 pilot dairy farms of the North region of Portugal involved in this project. Maize silage is the major diet component in this double-cropping forage system (zero-grazing). Data obtained by wet chemistry in UTAD laboratory and NIR analysis performed by a conventional laboratory were also compared using data of only 9 maize silages. In average maize silages had DM 35.0±3.0% (mean ± sd) pH 3.46±0.10, and in DM, CP 7.0±0.6, NDF 40.7±6.4, ADF 24.3±2.1, ADL 2.24±0.37, Ash 3.44±0.44, acetic acid 1.63±0.68, propionic acid 2.26±2.19, butyric acid 0.91±1.58 and lactic acid 5.91±1.10%. The high starch content, varying from 24.8 to 42.9% (15 samples with values higher than 30%) combined with fibre fractions (i.e. NDF and ADF) content within the desirable range, indicate expectable high energy value of these silages. However, it should be pointed out that, although the majority of standard chemical parameters indicate a general good quality of these maize silages, other indicators not routinely determined by conventional laboratories, should be considered. In fact, in general these laboratories do not carry out N-NH_3 assessment for maize silages. We observed that all maize silages had a content of N-NH_3 higher than 10% (14.3±1.74%), which indicate high proteolysis during the ensiling process, that could impair desirable levels of feed intake. Regarding the organic acids content, 5 samples had noticeable contents on butyric acid (1.6-4.5% DM). Concerning NIR and wet chemistry analyses comparison, despite limited sample size, significant coefficient of correlation where obtained for all parameters, except for ash and ADL content, indicating the main parameters were accurately estimated by NIR. In conclusion, maize silages used in these pilot farms, could be considered having in general good quality, but farmers should consider complementary analyses that enable proper adjustments in ensilage process.

***Thymus vulgaris* essential oil in dairy cows with subclinical mastitis**

T.M. Mitsunaga, L. Castelani, W.V.B. Soares, L. El Faro and L.C. Roma Junior
Nova Odessa, SP, 13380-011, Brazil; luiz.roma@sp.gov.br

Subclinical mastitis is commonly related with the increase on somatic cell count (SCC) and decrease in milk production. In general, the lack of effective therapy for its control drives the searches for alternative treatments without residues levels in dairy milk, which increases bacterial resistance. *Thymus vulgaris* essential oil (TEO) has shown in vitro antimicrobial activity and there are reports stating its anti-inflammatory activity. The aim of this study was to evaluate TEO effect over SCC of dairy cows, as well as microbiological culture of each mammary quarter. Twenty-six Holstein dairy cows were used, with SCC values above 500×10^3 cells/ml, with no symptoms of clinical mastitis and without any veterinarian intervention in the previous 14 days. The animals were randomly distributed in two groups: one group (T1) using of 12 ml of TEO per animal/day and second group (T2) without TEO using 12 ml of soybean oil (placebo), through oral gavage administration for 7 days. From both groups it was monitored SCC while offering treatment, and for the next seven days, resulting in 14 days of trial. In addition, it was collected fresh milk samples in sterile bottles from each mammary quarter individually, in the first and last day of the trail, for microbiological exam. Statistical analysis was performed comparing the animals (before and after treatment) for each group. The results showed that in the T2 group there was a significant increase (P=0.002) in average SCC, from 1.285×10^3 to 3.702×10^3 cells/ml and an increasing 8% of number of samples with bacterial growth from day 1 to day 14. However, there was a significant decrease (P=0.001) in average SCC of cows treated with TEO, from 4.675×10^3 to 2.097×10^3 cells/ml and reducing 10% of number of samples with bacterial growth. Such results showed that the use of TEO via oral gavage had an antimicrobial effect, by means of the reduction on bacterial growth. Moreover, the results showed potential anti-inflammatory effect, due to SCC reduction. In conclusion, TEO can contribute to subclinical mastitis control, without using any conventional medications or any milk disposal. Financial support: FAPESP 2014012124.

Session 38 Poster 40

Blood parameters and immune responses of dairy cows to supplementation with thyme essential oil

L.C. Roma Junior,[1], M.S.V. Salles[1], F.A. Salles[1], E.S. Castro Filho[2], E.H. Van Cleef[3] and J.M.B. Ezequiel[2]
[1]Instituto de Zootecnia, CPDBL, Rua Heitor Penteado 56 Nova Odessa, SP, 13380-011, Brazil, [2]UNESP, Via de Acesso Prof Paulo Donato Castelane Castellane S/N, Vila Industrial, 14884-900, Brazil, [3]IFTM, Av. Rio Paranaíba, 1229, Centro, Iturama, MG, 38280-000, Brazil; luiz.roma@sp.gov.br

The use of essential oils (EO) in animal production has increasingly stimulated research for alternatives to its use as antimicrobial medication or as ruminal modulator. However, there are only few research that indicates all the aspects in which EO can contribute in animal origin products' quality and production. This study was carried out to evaluate the effects of thyme (*Thymus vulgaris*) essential oil (TEO) in the blood parameters and immune response of lactating dairy cows. Twenty Jersey cows (days in milk = 57.4±12.6, and milk yield = 21.4±3.4 kg/d) were fed for 28 days a corn silage-based total mixed ration. The experiment was conducted in a completely randomized design with two treatments: control and with the addition of 8 g/animal/d of thyme (TEO). Animals were milked three times daily (06:00, 15:00, and 22:00h). After third daily milking, the TEO capsules were delivered by oral gavage. On the last day of the trial, blood samples were collected from each cow's mammary vein. These samples were taken immediately to the laboratory for analysis of complete blood count, aspartate aminotransferase, alanine aminotransferase (ALT) and quantitative evaluation of cytokines such as: interleukins (IL2, IL4, IL6, IL8, IL10), tumor necrosis factor alpha, gamma-interferon, colony stimulating factor and immunoglobulins (IgG, IgA, IgM). As results, it was possible to highlight the significant effect of treatment on the variables AST (P=0.07), IL2 (P=0.04), IL4 (P=0.07) and IgG (P=0.09). The values (means±SD) of ALT, IL2, IL4 and IgG for CON and TEO groups are respectively 19.5±3.8 and 17.2±4.2 U/l, 40.5±12.2 and 94.5±17.4 pg/ml, 129.2±72.0 and 295.0±93.6 pg/ml, and 77.6±3.3 and 80.6±3.2 pg/ml. In conclusion, the TEO supplementation of lactating dairy cows for 28 days presented positive effects over some blood and immune system parameters. Financial support: FAPESP 2014/01212-4.

Session 39 Theatre 1

Amaranth leaf meal inclusion on blood profiles and gut organ characteristics of Ross 308 chickens

T.G. Manyelo[1,2], N.A. Sebola[1], J.W. Ng'ambi[2] and M. Mabelebele[1]
[1]University of South Africa, Department of Agriculture and Animal Health, Florida, 1710, South Africa, [2]University of Limpopo, Private Bag X1106, Sovenga, 0727, South Africa; grace.manyelo@ul.ac.za

The aim of this study was to determine the effect of varying inclusion levels of amaranth leaf meal on the performance, blood profiles and gut organ characteristics of Ross 308 broiler chickens. A total of 200, day-old, Ross 308 broiler chicks were randomly allocated to five dietary treatments in a complete randomised design, with each group having four replicates with ten chicks. Amaranth leaf meal (ALM) inclusion levels used in this study were 0, 5, 10, 15 and 20%, body weight and feed intake were measured on a weekly basis to calculate the feed conversion ratio. Gut organ weights, lengths, organ pH and blood profiles were measured, and the general linear model of statistical analysis software was used to analyse collected data. ALM had no effect (P>0.05) on feed intake, body weight or the feed conversion ratio of Ross 308 broiler chickens aged one to 42 days, respectively. Furthermore, ALM inclusion levels had no effect (P>0.05) on dry matter, (DM), crude protein (CP), or gross energy (GE) digestibility of Ross 308 broiler chickens. Ross 308 broiler chickens which were fed with 5 and 15% ALM inclusion levels, had higher (P<0.05) overall blood profiles than those fed with diets containing 0, 10 and 20%. ALM inclusion of 5% had higher (P<0.05) essential and non-essential amino acid digestibility in Ross 308 broiler chickens. ALM inclusion levels had no effect (P>0.05) on gut organ lengths or weights of Ross 308 broiler chickens aged 21 and 42 days. In conclusion, 5, 15 and 15% ALM inclusion levels can be included in broiler chickens without having any adverse effect on the chickens' performance. This indicates that ALM can be added to broiler diets at any inclusion rate without negatively affecting their performance.

Productive performances of normally feathered and naked neck slow growth broilers fed with Spirulina

E.A. Fernandes, C.F. Martins, A.A. Chaves, L. Louro, A. Raymundo, M. Lordelo and A.M. Almeida
ISA, LEAF – Linking Landscape, Environment, Agriculture and Food, Tapada da Ajuda, 1349-017 Lisboa, Portugal; eafernandes@isa.ulisboa.pt

The increasing demand for human protein food sources has resulted in the need for alternative feedstuffs, such as microalgae. Spirulina has an interesting nutritional profile especially due to high protein, being a sustainable alternative to soybean meal. The objective of this study was to evaluate the effect of 15% Spirulina dietary inclusion on growth performance, carcass yield and organ measurements in two slow- growing broiler strains: FF (normally feathered) and Na (naked neck). Twenty, one-day old, male chicks per strain were used in 84 days experiment. Each genotype received either a Control diet (C) or a diet with 15% Spirulina (SP). Each broiler was individually housed with *ad libitum* access to water and feed. Broilers were weighed weekly to determine animal performance parameters. At the end of the experiment animals were slaughtered and organs collected, weighed, and measured. The effects of different dietary treatments were appraised by one-factor analysis of variance – ANOVA followed by Tukey – multiple comparison tests. At the end of the trial, average daily intake and feed conversion ratio were not significantly affected ($P>0.05$), contrasting between 144-154 g/d and 4.4-4.6 respectively. The final weight of the animals was 3,360, 3,182, 2,894 and 2,622 g for CFF, CNa, SPFF and SPNa groups respectively. These parameters were negatively influenced by Spirulina diet incorporation ($P<0.001$) in both strains. CFF animals had the highest carcass yield (78.9%) while CNa animals had the lowest (76.8%). Relative crop weight ranged between 2.33 and 4.02, being higher in SPFF animals ($P<0.05$), when compared to controls. The incorporation of Spirulina in the diet increased relative length (cm/kg) of the gastrointestinal tract, from 20.73 to 25.82 and 5.57 to 7.01 ($P<0.05$) in ileum and cecum, respectively. The incorporation of SP negatively influenced the growth performance of broilers and was more pronounced in Na animals. Future information provided by digestibility coefficients and meat quality traits will complement these results, allowing a better knowledge of the nutritional values of Spirulina for broilers.

Session 39 Theatre 3

Effects of grape seed oil, walnut oil and the two combined on hens' performances and egg quality

G. Cornescu, T. Panaite, A. Untea, M. Saracila, R. Turcu and A. Vlaicu
INCDBNA Balotesti, Animal Nutrition and Physiology, Calea Bucuresti nr 1, 077015, Romania; gabriela_cornescu@yahoo.com

The present study evaluates the effects of dietary grape seed oil, walnut oil, and the two oils combined on laying hens' performances, yolk antioxidant capacity, oxidative stability, and polyphenols content. The experiment was conducted on 128 Lohmann Brown laying hens (55 weeks of age) assigned to 4 treatments with 8 replicates of 4 birds each. A corn-soybean meal basal diet with 2.760 kcal/kg metabolizable energy (ME) and 16.8% crude protein (CP) was formulated to supply the nutritional requirements. The dietary sunflower oil (0.5%; C) was replaced by grapeseed oil (0.5%; E1), walnut oil (0.5%; E2), and the two oils combined (0.5% grape seed oil + 0.5% walnut oil; E3). Average daily feed intake (g/hen/day) was statistically significant ($P<0.05$) on C and E1 compared to E2. The egg production and weight were significantly higher ($P<0.05$) on the E3 group compared to C, E1, E2. A significantly higher ($P<0.05$) feed conversion ratio was registered on E2 compared to E3. The yolk antioxidant capacity analysis (mmol/kg Equiv. vitamin C) was statistically significant ($P<0.05$) on E2 at T_0 days of storage time compared to T_{28} days of storage time. The same significant difference ($P<0.05$) could be noticed when determining the yolk antioxidant capacity (mmol/kg Equiv. vitamin E) at initial T_0 days storage time, on E2 compared to its registered values at T_{28} days of storage time. The thiobarbituric acid reactive substances (TBARS) values at T_{28} days for groups E1, E2, E3 (on average 0.055 ppm) were similar to those registered on T_0 days initial (on average 0.048 ppm), whereas for the C group TBARS values were statistically significant ($P<0.05$) at T_{28} days (0.076 ppm) compared to those registered at T_0 days (0.037 ppm). The yolk total phenolic compounds (mg gallic acid equivalent [GAE]) values registered at T_0 days initial on E1 and E2 were statistically significant ($P<0.05$) compared to their registered values at T_{28} days of storage time. Therefore, we concluded that the inclusion of 0.5% grapeseed oil, 0.5% walnut oil, and their combination has a beneficial effect on productive performances, and also could enhance the eggs antioxidant status during the storage period.

Nutritive value of arborescent alfalfa and immature rye in free-ranged organic growing rabbit

J.P. Goby[1], C. Bannelier[2], O. Faillat[2] and T. Gidenne[2]
[1]University of Perpignan, IUT applied agronomy, IUT, 66962 Perpignan, France, [2]INRAE Occitanie Toulouse, GenPhySE, BP 52627, 31326 Castanet-Tolosan, France; thierry.gidenne@inrae.fr

Free-ranged organic rabbit meets the consumer's demand of outdoor rabbit farming and improved animal welfare, but lacks in technical references and nutritive value of forages available at farm. We thus studied the nutritive value of two feeds: arborescent alfalfa (*Medicago arborea*) 'AA' and immature rye 'IR' (herbaceous stage), in comparison to a pelleted feed 'P' for 5 groups of 5 growing rabbits (74 days old, mean live weight=1,802 g) housed individually in movable cages on pasture and following organic regulation for rabbit farming. Each cage had a shelter of 0.4 m^2 and a pasturing surface of 1.2 m^2. The groups were: group P fed a pelleted feed only, AA fed only with arborescent alfalfa (400 g fresh per day), AA+P was fed with alfalfa (200 g fresh/d+ 60 g/d of pellets), IR fed only with immature rye (500 g fresh per day), IR+P fed 400 g fresh/d rye + 60 g/d of pellets. After one week of adaptation to movable cage and feeds (74-81 d old), the digestibility was measured for four days, with feed intake and faecal total collection measurement (a board was installed on the cage floor to collect the faeces). During digestibility period, the dry matter (DM) intake of AA and IR fed alone averaged respectively 88.6 and 56.3 g/d per rabbit, corresponding resp. to a daily fresh intake of 239 and 378 g/d. Pellet intake of the P group averaged 121 g DM /d, while in IR+P group the IR intake averaged 34.0 g DM/d and in AA+P group the AA intake averaged 47.1 g DM/d. The AA contained 134 g/kg of crude protein for 291 g ADF/kg, while IR contained 289 g/kg of crude protein for 215 g ADF/kg. Digestible protein and energy content were 37 g PD/kg fresh and 4.68 MJ DE/kg for AA (fed alone), and 31 g PD/kg and 1.34 MJ DE/kg fresh for IR (fed alone). When the AA and IR were fed freely and complemented by 60 g/d of pellets (groups AA+P, IR+P), the DM digestibility was reduced proportionally to the pellet intake, and averaged 55.4 and 56.8%, similar to that recalculated from IR and AA groups. Thus, mixing forages (AA or IR) with pelleted feed did not alter the nutritive value of the forages.

Silage juice in diets to weaner pigs and pregnant sows

M. Presto Åkerfeldt[1], J. Friman[1], F. Dahlström[2], A. Larsen[2] and A. Wallenbeck[2]
[1]SLU, Animal Nutrition and Management, Box 7024, 75007, Sweden, [2]SLU, Animal Environment and Health, Box 234, 53223 Skara, Sweden; magdalena.akerfeldt@slu.se

Liquid fractions from green bio-refinery contains nutrients with high availability for pigs and have potential as a local feed resource in liquid feeding to pigs. The objective of this study was to evaluate the applicability of silage juice in liquid diets to weaner pigs and dry sows and its' effects on production and health. In total, 96 weaner pigs (LY×H) and 24 sows (LY) from four and three batches in an organic pig production system, respectively, were included. During the weaning period (six wk. age-delivery to fattening unit) and pregnancy (six wk. after insemination-one wk. prior to farrowing) weaners and sows were allocated to either a control diet (C) or an experimental diet with silage juice (SJ). The C-diet consisted of a commercial feed for growing pigs and dry sows, respectively, mixed with water prior to feeding. The SJ-diet consisted of a lower ration of the same commercial feed as in the C-diet but mixed with SJ instead of water, theoretically replacing 10% (weaners) or 15% (dry sows) of the crude protein content. In weaners, growth, cleanliness of the pens and pigs and clinical health indicators was registered, while sow weight during pregnancy and litter characteristics at farrowing were registered in the sows. Weaners fed the SJ and C-diets had similar growth (16.0 and 15.9 kg, respectively) and feed conversion ratio based on the commercial feed. Pigs fed the SJ-diet were dirtier on their back and head than pigs fed the C-diet ($P<0.001$), but cleanliness in the rectum area and in the pen did not differ. Very preliminary results show that sows fed the SJ-diet had a higher body weight during pregnancy than sows fed the C-diet although their growth development was similar. Also, higher number of total born (19.3 vs 15.8), live born (16.8 vs 15.0)/litter and the litter weight at birth (27.4 vs 23.7 kg) was found in the SJ compared to the C sows, resulting in a higher share of dead born piglets in SJ sows. This study concludes that silage juice can contribute with nutrients in liquid diets but possible effects on reproduction calls for an awareness. More studies on nutrient digestibility and potential biological effects on sow reproduction is needed.

The use of Chlorella v. in weaned piglet diet: growth performances and digestive tract development

A.A.M. Chaves[1], C.F. Martins[1,2], D.M. Ribeiro[1], D.F.P. Carvalho[1], R.J.B. Bessa[2], A.M. Almeida[1] and J.P.B. Freire[1]
[1]Instituto Superior de Agronomia, University of Lisbon, LEAF Linking Landscape, Environment, Agriculture and Food, Tapada da Ajuda, 1349-017 Lisboa, Portugal, [2]Faculdade de Medicina Veterinária, Universidade de Lisboa, CIISA – Centro de Investigação Interdisciplinar em Sanidade Animal, Av. da Universidade Técnica, 1300-477 Lisboa, Portugal; andreiachaves@isa.ulisboa.pt

Chlorella vulgaris (CH) is an eukaryotic microalga, rich in protein and n-3 polyunsaturated fatty acids, being a promising alternative feedstuff for monogastric diets. The aim of this study was to evaluate the effect of dietary CH on growth performances, digestibility and development of the digestive tract of recently weaned piglets. Animals were randomly assigned to four experimental groups (n=6): Control (no CH) and three groups fed with 5, 10 or 15% dietary incorporation of CH, as a replacement of the basal diet. Piglets were individually housed in metabolic cages, fed equal amounts of the experimental diets and had free access to water. After an adaptation period of 4 days, the trial lasted two weeks. Piglets were weighed every week and feed intake controlled daily. At the end of the trial, piglets were slaughtered. Stomach, small and large intestine, liver, gallbladder and pancreas were collected, weighed and measured (only small and large intestines). Polynomial contrasts analysis was used to test the linear, quadratic or cubic effect of CH level in the diet. The average daily gain was 497, 501, 526 and 531 g/day and the daily intake was 650, 651, 640 and 648 g for diets with 0, 5, 10 e 15% of CH, respectively (P>0.05). Feed conversion ratio was 1.32 for the control diet, 1.30 for 5% CH diet and 1.22 for 10 or 15% CH diets (linear, P<0.05). Moreover, CH level had a quadratic effect increasing the empty stomach weight (P<0.05). Piglets fed with 10% CH incorporation level had higher weights of the empty small and large intestine. No significant effect for diet was found on liver, gallbladder and pancreas weights. Further information, to be provided by digestibility coefficients analysis, will complement the results described above, allowing further knowledge of the nutritional values of CH for the weaned piglet.

Dietary *Ulva lactuca* in weaned piglet diets: effects on performance and gastrointestinal contents

D.M. Ribeiro[1], D.F.P. Carvalho[1], C.F. Martins[1,2], A.M. Almeida[1], J.A.M. Prates[2] and J.P.B. Freire[1]
[1]Instituto Superior de Agronomia, Universidade de Lisboa, LEAF – Linking Landscape, Environment, Agriculture and Food, Tapada da Ajuda, 1349-017, Lisboa, Portugal, [2]Faculdade de Medicina Veterinária, Universidade de Lisboa, CIISA – Centro de Investigação Interdisciplinar em Sanidade Animal, Avenida da Universidade Técnica, 1300-477, Lisboa, Portugal; davidribeiro@isa.ulisboa.pt

The objective of our work is to evaluate the effect of the dietary inclusion of *Ulva lactuca*, with and without CAZyme supplementation, on the performance and gastrointestinal health of weaned piglets. A total of 44 piglets (Large White × Duroc, initial live weight of 8.63±0.87 kg) were randomly assigned to one of each experimental diets: control (maize, wheat and soybean meal – based diet), UL (7% *U. lactuca* replacing the basal diet), ULR (UL + 0.005% Rovabio® Excel AP) and ULU (UL + 0.01% ulvan lyase). They were individually housed in metabolic cages with free access to water and fed on a pair-feeding basis. The trial consisted of 4 days of adaptation and two weeks where piglets were weighed weekly, feed intake and faecal scores were measured daily. At the end of trial all piglets were slaughtered, and gastrointestinal contents sampled for pH, viscosity and volatile fatty acid analysis. Data was analysed with ANOVA for the effect of diet using the SAS software. Diets had no effect on average daily gain (P>0.05) and average daily feed intake (P>0.05). Control, UL, ULR and ULU piglets grew 374.6, 358.6, 362.1 and 371.4 g/day and ingested 563.2, 549.0, 590.2 and 569.8 g feed/day, causing feed conversion ratios (P>0.05) of 1.55, 1.54, 1.67 and 1.75, respectively. Faecal consistency was not significantly different between groups (P>0.05). Viscosity of intestinal contents was also unaffected (P>0.05). Caecum and colon contents pH were significantly lower in ULU (5.57 and 5.82) compared to control (5.77 and 6.11) and UL (5.8, only in caecum). In the colon, the ratio of isovaleric acid (iso-C5):total VFA was significantly lower (P<0.05) in ULU compared to control. Overall, these results indicate a dietary effect over large intestinal fermentation, particularly higher fermentation rates of carbohydrates, and leucine, in ULU piglets.

Sweet vs Salty Former Food Products in Piglets slightly affect subcutaneous adipose tissue quality

M. Ottoboni[1], A. Luciano[1], M. Tretola[1,2], S. Mazzoleni[1], N. Rovere[1], F. Fumagalli[1], R. Abbate[1], L. Ferrari[1] and L. Pinotti[1]

[1]University of Milan, Department Veterinary Medicine and Animal Science, DIVAS, Via dell'Università, 6, 26900, Lodi, Italy, [2]Agroscope, Institute for Livestock Sciences, Rte de la Tioleyre, 4, 1725, Posieux, Switzerland; matteo.ottoboni@unimi.it

Former food products (FFPs) have a great potential to replace conventional feed ingredients. This study aimed to investigate the possibility to partially replace standard ingredients with two different types of FFPs: bakery (FFPs-B) or confectionary (FFPs-C) FFPs and their effects on growth performances, and fatty acid profile of subcutaneous fat in post-weaning piglets. Thirty-six post-weaning piglets were randomly assigned to three experimental diets (n=12 per diet) for 42 days: a standard diet (CTR), a diet where 30% of standard ingredients were replaced by confectionary FFPs (FFPs-C) and a diet where 30% of standard ingredients were replaced by bakery FFPs (FFPs-B). Individual body weight was measured weekly. Feed intake (FI) was determined daily. Average daily gain (ADG), average daily feed intake (ADFI) and feed conversion ratio (FCR) were calculated. Subcutaneous abdominal fat samples were collected from 12 selected piglets immediately after slaughtering (day 42). No significant differences (P>0.05) between groups were found in growth performances. No significant differences (P>0.05) between groups were found for saturated fatty acids, however lower percentage of monounsaturated fatty acids were observed in CTR compared to FFPs-C and FFPs-B group. While in the case of polyunsaturated fatty acids FFPs-C group was lower compared to CTR group. This study confirmed the possibility to formulate homogeneous diets integrated with 30% of both categories of FFPs. Furtherly the inclusion of different FFPs types, namely deriving from bakery or confectionary food product may affect the fatty acid profile of subcutaneous fat in post-weaning piglets.

Session 39 Theatre 9

Dietary tall oil fatty acids with resin acids improve the performance of seabass and white shrimp

H. Kettunen[1], J. Vuorenmaa[1] and O. Jintasataporn[2]

[1]Hankkija Oy, Peltokuumolantie 4, 05801 Hyvinkää, Finland, [2]Kasetsart University, 50 Ngamwongwan Rd, Bangkok 10900, Thailand; hannele.kettunen@hankkija.fi

Supplementing feeds with tall oil fatty acid (TOFA) has improved the performance of broiler chickens, piglets and sows, by its positive effects on intestinal integrity and microbiota. The potential of TOFA aquaculture has remained unexplored. Here we investigated the effects of TOFA with 9% resin acids (Progres®, Hankkija Oy, Finland) on the growth performance and survival rate of Asian seabass (*Lates calcarifer*) and white shrimp (*Litopenaeus vannamei*) in Thailand. Juvenile sea bass (~50 g) were allocated into 20 freshwater cages of 2 m^3, 15 fish/cage. Commercial-type seabass feed was amended with 0, 0.35, 0.7, or 1.0% of TOFA for treatments T1-T4. Feed was applied to seabass 3 times/day at 3-5% of body weight for 16 weeks. The shrimp trial was carried out in 30 aquariums with 120 l of 15 ppt saline water. Juvenile white shrimp (~1-1.5 g) were stocked at 200 shrimp/m^3, 25 individual/aquarium. The water in each aquarium was changed 20% every 3 days. Commercial-type shrimp feed was amended with TOFA at 0, 0.5, and 1.0% for dietary treatments T1-T3. Feed was applied to shrimp 3 times/day at 3-5% of body weight for 8 weeks. One hour after feeding, the uneaten feed was siphoned out to dry by hot air oven and record the weight for calculated the feed consumption. For both species, the water was aerated to maintain DO >5 mg/l; and the parameters measured included weight gain, feed intake, feed conversion ratio for every 2 wk, and 4-wk survival rate. Data was analysed with ANOVA, using P<0.05 as a limit for statistical significance. For both species, TOFA improved weight gain, FCR, and survival rate, and reduced feed intake. For example, seabass FCR (14-16 wk) was 2.36, 2.15, 2.09, and 1.98, and survival rate (12-16 wk) was 91.7, 93.2, 94.7, and 96.0% for T1-T4, respectively. For white shrimp, FCR (6-8 wk) was 1.46, 1.28, and 1.20, and survival rate (4-8 wk) was 78.7, 81.3 and 85.3, for T1-T3, respectively. The results indicate that dietary TOFA amendment improves the growth performance, feed efficiency and survival rate of Asian seabass and white shrimp. Further research is needed to reveal the mechanism-of-action of TOFA on aquaculture species.

Session 39 Theatre 10

Amaranth grain for animal feeding: a high-quality wholesome crop

M. Oteri, D. Scordia, R. Armone, F. Gresta and B. Chiofalo
University of Messina, Department of Veterinary Sciences, Via Palatucci, 98168, Italy; marianna.oteri@unime.it

The present research aims to explore the agronomic traits, proximate composition and oil and fatty acid content of five amaranth species, namely *Amaranthus cruentus*, *A. hypochondriacus*, *A. hybridus*, *A. caudatus* and *A. tricolor*, grown in a semiarid Mediterranean area, with the purpose of increasing the knowledge of these plants, as a source of nutrients and essential fatty acids, for animal feeding. *A. cruentus* resulted the most productive species (382 g/m^2) even though not differentiated from *A. hypochondriacus* and *A. hybridus* (273 and 256 g/m^2, respectively). The highest thousand seed weight was recorded in *A. hypochondriacus* and *A. cruentus* (0.84 and 0.71 g, respectively). The crude protein showed the significant highest value in *A. tricolor* (18.5%) while the significant highest starch content was ascertained in *A. cruentus* (60%) and *A. caudatus* (59.9%). *A. hybridus* showed the highest oil content (7.06%) and the highest crude fibre content (17.2%). Among the fatty acids of nutritional interest, *A. hypochondriacus* and *A. cruentus* emerged as the best healthy ingredient for animal feeding. *A. hypochondriacus* seeds showed a high content of polyunsaturated fatty acid of the n6 (linoleic acid) and n3 (linolenic acid) series, while, *A. cruentus* showed a high content of monounsaturated fatty acid of the n9 series (oleic acid). As regards the nutritional indices, results showed significant differences for atherogenic and thrombogenic indices, and hypocholaesterolemic and hypercholaesterolemic ratio among the species, giving the best values for *A. hypochondriacus*, while *A. hybridus* and *A. caudatus* showed the lowest peroxidation index and therefore the highest stability. According to present findings, grain amaranth can be recommended as a promising non-conventional source for animal healthy diets provided that further studies will ascertain the best species for Mediterranean environmental conditions and animal species, and sustainability of the cultivation phase.

Session 39 Theatre 11

Cactus cladodes inclusion in lactating goat diet and their effects on milk production and quality

S. El Otmani[1], Y. Chebli[1], M. Chentouf[1], J.L. Hornick[2] and J.F. Cabaraux[2]
[1]INRA Morocco, Regional Center of Agricultural Research of Tangier, 78, Boulevard Sidi Mohamed BenAbdellah, Tangier, 90010, Morocco, [2]Faculty of Veterinary Medecine, University of Liège, Department of Veterinary Management of Animal Resources, Avenue de Cureghem 6, B43, Liège, 4000, Belgium; samira.elotmani@inra.ma

In the Southern Mediterranean region, goat farming is widely practiced, due to this species adaptation ability. Generally, goat diet is based on forest pasture, characterized by their seasonal forage availability. The pastures over-use conducts to their degradation. Cactus is a multi-purpose shrub, widely available in the harsh environment. Many studies recommended their cladodes use as feed resources to diversify ruminant diets and reduce the forest dependence. This study aims to evaluate the cactus cladodes (CC) supplementation effects on goat milk production and quality. Twenty-two lactating goats from the local population of northern Morocco 'Beni Arous' were divided into two groups. The control group (Co) received a conventional diet (barley and faba bean), while test group was supplemented by 30%CC. Fortnightly, daily milk production was recorded and samples were collected to determine the composition (fat, protein, lactose, fat-free solids and mineral matter). Milk fat was extracted and fatty acids were esterified and identified using Gas Chromatograph. According to the results, daily milk production was similar for all animal groups with an average of 370 g/day. The produced milk contained on average 2.5, 3.7, 4.5, 9.2 and 0.8% fat, protein, lactose, solid non-fat and mineral matter, respectively. The CC incorporation was not affected yield per lactation of milk, fat, protein, lactose, solid non-fat and mineral matter with an average of 44.4, 1.2, 1.6, 2, 4.1, and 0.35 kg/lactation, respectively. The incorporation of 30% CC had an effect on some fatty acids. Their introduction in goat diet decreased C4:0, 9t-C18:1, 6t-C18:2and C20:0, and increased C15:0, C18:1n-9, C21:0 ($P<0.05$). In conclusion, the introduction of CC in lactating goat diet had no effect on milk production and chemical composition, however, they slightly affected fatty acid profile. The CC could be introduced as alternative feed resource into goats' diet in order to diversify their feed and to reduce rangeland degradation.

Methane emission and energy partitioning in Hanwoo Steers supplemented with seeds of *Pharbitis nil*

R. Bharanidharan[1], K. Thirugnanasambantham[2], R. Ibidhi[2], T.H. Kim[3], W.H. Hong[3], G.H. Bang[2], P. Xaysana[3] and K.H. Kim[2,3]
[1]College of Agriculture and Life Sciences, Seoul National University, Gwanak-Gu, Seoul 08826, Korea, South, [2]Institutes of Green Bio Science and Technology, Seoul National University, Pyeongchang, Gangwon-do 25354, Korea, South, [3]Graduate School of International Agricultural Technology, Seoul National University, Pyeongchang, Gangwon-do 25354, Korea, South; bharanidharan7@snu.ac.kr

Two *in vivo* experiments were conducted to evaluate the potential of *Pharbitis nil* seeds (PA) as an anti-methanogenic feed additive to ruminants. In experiment 1, six Hanwoo steers (411±57 kg) were fed either total mixed ration (TMR; Control) or TMR supplemented with PA at 5% dry matter intake (DMI; PA5) for two consecutive periods of 35 days in pairwise comparison. Faecal and urine output were measured in apparent digestibility trial. The methane (CH_4) and heat energy (HE) were measured using respiratory chambers equipped with gas analysers. Although no differences (P>0.05) in nutrients or gross energy intake between the periods, an increase (P<0.05) in apparent digestibility of DM (8%) and neutral detergent fibre (18%) was observed in PA5. A pronounced decrease (P<0.05) in CH_4 yield (16%) and urinary energy (30%) and nitrogen (N) excretion (25%) was noted in PA5 which led to an increased metabolisable energy intake by 16%. However, only numerical increase (P>0.05) in retained energy and average daily gain was noted due to increase in HE loss. In experiment 2, five rumen cannulated Holstein steers (744±35 kg) were assigned to PA5 diet for 40 days to study the rumen fermentation characteristics. Supplementing PA decreased (P<0.05) rumen ammonia concentration and pH, associated with an increased (P=0.091) rumen short-chain fatty acid concentrations at 3 h post feeding. A 29% increase (P<0.05) in propionate proportion clearly reflected a shift in the ruminal H_2 sink in PA5. A 40% reduction (P=0.067) in relative abundance of *Entodinium caudatum* supported by the *in silico* binding of secondary metabolites from PA to the *cyclic GMP (cGMP)-dependent protein kinase* (cGK) of *E. caudatum* was also noted. Overall, seeds of *P. nil* could be a potential alternative for ionophores in reducing CH_4 and N emission from livestock production.

Silages of agro-industrial by-products in lamb diets – effect on performance and methane emissions

C. Costa[1], K. Paulos[1], J. Costa[1], L. Cachucho[2,3], A. Portugal[1], A.T. Belo[1], E. Jerónimo[2,4], J. Santos-Silva[1,3] and M.T. Dentinho[1,3]
[1]INIAV, Quinta Fonte Boa, Santarém, Portugal, [2]CEBAL, Centro de Biotecnologia Agrícola e Agro-Alimentar do Alentejo, Beja, Portugal, [3]CIISA, Avenida Universidade Técnica, Lisboa, Portugal, [4]Mediterranean Institute for Agriculture, Environment and Development, Évora, Évora, Portugal; claudia.dc.costa@gmail.com

The use of agro-industrial by-products in animal feed is an opportunity to value these products and ensure low-cost feedstuffs. We aimed to produce silages based on by-products of carrot, sweet potato, potato, and tomato and integrate them in lamb diets replacing 50% of a conventional feed. The effect on animal performances and methane production was evaluated. Three silages were prepared with 35% tomato pomace + 20% wheat bran + 15% grass hay associated with 30% of potato (Psil) or 30% sweet potato (SPsil) or 30% of carrot (Csil). Thirty-two lambs were divided into four groups fed with: 85 concentrate/15% hay (control); 50% of concentrate and 50% of Psil or SPsil or Csil. After 6 weeks animals were slaughtered, and rumen content collected. The rumen fluid of each animal was incubated for 48 hours with 1 g DM of the lambs' diets in fermentation bottles with gas detectors (Ankom Tech.). The pressure and temperature in the bottles were recorded every 20 minutes during the incubation period and converted into ml of gas produced. Methane was quantified in gas by gas chromatography. No differences among diets were detected in dry matter (DM) and crude protein (CP) intake, averaging 1,046 g and 170 g respectively. Animals fed silage diets ingested more NDF (184 g/d) than those fed the control diet (312 g/d). The ADG was lower with silage diets than in control (316 vs 348 g/d, respectively). Among diets, no differences were found in DM conversion ratio (2.84). Silage diets did not affect the gas and methane production, averaging 204 ml and 46 ml/g DM, respectively. By-product silage can be a good strategy to replace part of the concentrate feed in lamb diets without increase methane production.
This work is funded by project SubProMais (PDR2020-101-030988; PDR2020-101-030993), FCT – Foundation for Science and Technology projects UIDB/05183/2020 MED and UIDP/CVT/00276/2020 (CIISA), and the PhD grant awarded to LC (2020.05712.BD).

Foetal programming in ewes fed with chrome propionate on the body weight of the progeny

D. Lázara De Almeida[1], F. Mallaco Moreira[1], V. Nassif Monteiro[2], A. Rua Rodrigues[2], T. Bianconi Coimbra[2], I. Tomaz Nascimento[2], F. Sesti Trindade[2] and S. Bonagurio Gallo[2]
[1]University of Sao Paulo, Nutrition and Animal Production, Av. Duque de Caxias Norte, 225, Pirassununga, São Paulo, 13635900, Brazil, [2]University of Sao Paulo, Department of Animal Science, Av. Duque de Caxias Norte, 225, Pirassununga, São Paulo, 13635900, Brazil; saritabgallo@usp.br

There is a high energy demand at the end of pregnancy and the beginning of lactation, with mobilization of lipid reserves, metabolic stress, and, possibly, the formation of ketone bodies. Chromium (Cr) is used in nutrition to enhance the use of glucose. However, the recommended dose of chromium is not established for ruminants. The objective was to evaluate the effects of maternal diet with chromium propionate supplementation on the progeny's body weight (BW) and average daily weight gain (ADG). We used 49 ewes, Dorper × Santa Ines, 60±3 kg of body weight, 3±2 years, pregnant with males. The experimental design was entirely causal and studied doses of 0, 0.5, 1.0, and 1.5 mg Cr/animal/day (KemTRACE Chromium®). The experiment took place from 100 days of gestation until the 80th day of lactation. Lambs were weighed at birth, 30 and 80 days of age. For statistical analysis, SAS (9.4), Proc Mixed, Tukey's test, and $P<0.05$, a fixed effect for treatment and type of delivery (single or multiple pregnancies) were used. There was an effect of maternal diet (Cr) on all BW of the lambs ($P<0.05$), and the dose of 1.5 mg of Cr resulted in heavier animals, with values of 5.18, 15.49, and 33, 96 kg for PN, 30 and 80 days, respectively. And the treatment without Cr the lighter lambs, weighing 4, 11.85, and 28.03, for BW, 30 and 80 days, respectively. The treatments did not change the ADG. Lambs from single births were heavier than lambs from multiple births. However, the type of pregnancy did not affect the weight gain of lambs ($P>0.05$). There was no interaction ($P>0.05$) between treatments and type of pregnancy. It is concluded that the supplementation of 1.5 mg of chromium propionate/animal/day, during the phases of pregnancy and lactation, resulted in heavier progenies from birth to 80 days of life. Thanks to FAPESP for the financial support (2020/06433-0) and CAPES.

Effects of incorporation of grape stalks in rabbit diets on carcass and gastrointestinal parameters

V. Costa-Silva, V. Pinheiro, M. Rodrigues, L. Ferreira and G. Marques
UTAD, Zootecnia, UTAD, Departamento de Zootecnia, Quinta de Prados, 5000-801, Portugal; valeriasilva@utad.pt

Recently, there is a growing interest in reducing the impact of agro-industrial wastes by using them as possible raw materials for new feed products and other applications, thus enhancing its contribution to the general policy framework inherent to the implementation of a circular economy. Grape stalks are one of those by-products, in average, 4 kg of grape stalks are generated per each hl of produced wine. This study aimed to evaluate the effects of the inclusion of grape stalks on diets for growing rabbits, as an alternative source of fibre. Three diets were provided: a control diet (C) without grape stalk incorporation, two diets with the incorporation of different levels of grape stalks (5% and 10%, 5GS and 10GS, respectively) Fifty hybrid (New Zealand × Californian) male rabbits weaned at 35 days of age with an average body weight of 1,091±56.3 g, were randomly assigned to the three experimental diets groups (10 rabbits/diet). Animals were kept in a closed air-conditioned building maintained between 18 and 23 °C and received 12 h of light daily. The grow trial was conducted between 35 and 66 days of age and no animals died during the experimental period. At 66 days of age (2,574±120 g), the animals were slaughtered by cervical dislocation. The slaughtering and carcass dissection procedures followed the World Rabbit Science Association recommendations. Different carcass parameters were analysed such as slaughter weight, hot and chilled carcass parts weight, internal organs weight (thymus, trachea, oesophagus, lungs, heart, kidneys, stomach, colon, small intestine, caecum) and carcass fat weight (perirenal, scapular, pelvic and dissectible). In addition the length of colon, small intestine and caecum were also measured. No differences were shown on carcass and gastrointestinal parameters between the experimental diets. The results show the potential of using grape stalks as a source of fibre and an alternative raw material in rabbits' diets without compromising animal performances.

Session 39 Poster 16

Performance of ewes supplemented with chromium propionate at the end of pregnancy and lactation

F. Mallaco Moreira[1], D. Lázara De Almeida[1], A. De Carvalho[2], T. Bianconi Coimbra[2], V. Nassif Monteiro[2], I. Tomaz Nascimento[2], F. Sesti Trindade[2], J. Biagi Veronez[2] and S. Bonagurio Gallo[2]
[1]University of Sao Paulo, Nutrition and Animal Production, Av. Duque de Caxias Norte, 225, Pirassununga, São Paulo, 13635900, Brazil, [2]University of Sao Paulo, Department of Animal Science, Av. Duque de Caxias Norte, 225, Pirassununga, São Paulo, 13635900, Brazil; saritabgallo@usp.br

Chromium propionate (Cr) supplementation in ruminant nutrition has shown positive responses related to lipid and carbohydrate metabolism, immune system, and stress resistance mechanisms. At the late of gestation and lactation, sheep go through situations of high energy demand, suppression of the immune system, and increased stress. However, the chromium recommendation for ruminants has not been established. The aim was to evaluate the effect of doses of Cr for ewes during late gestation and lactation (G - L). Sixty-nine Dorper × Santa Ines ewes distributed in a completely randomized design were used. The treatments were doses of 0, 0.5, 1 and 1.5 mg of Cr/animal/day (KemTRACE Chromium). The experimental period corresponds to the last 50 days of gestation and 60 days of lactation. Body weight (BW, kg) and body condition score (BCS, scale from 1 to 5) were performed in G - L. In the statistical analysis, treatments (chromium doses) and type of gestation (single or multiple), and the interaction between these factors were considered as fixed effects. The SAS program (9.4), PROC MIXED, and Tukey's test at 5% probability were used. There was no effect of Cr doses on the BW and BCS of ewes in the period of G – L (P>0.05). The same occurred for the mean BW between types of gestation (P>0.05). There was a difference (P<0.05) for the BCS between the type of gestation, with single ewes obtaining scores of 3.5 and 3.25 during gestation and lactation respectively, while ewes' multiple gestations showed values of 3, both phases. There was no interaction between Cr and type of gestation (P>0.05). It can be concluded that the doses of Cr studied did not change the BW and BCS of the ewes, at the end of gestation and lactation. Thanks to FAPESP for the financial support (2020/06433-0) and to CAPES for the scholarship.

Session 39 Poster 17

Efficacy of yeast culture and garlic extract supplementation on broilers performance

S. Biswas, Q.Q. Zhang, M.D.M. Hossain, S. Cao and I.H. Kim
Dankook University, Department of Animal Resource and Science, 119, Dandae-ro, Cheonan, 31116, Korea, South; sarbani.dream@gmail.com

Garlic (Allium sativum) has been considered an effective plant with a lot of beneficial bioactive substances. This experiment was conducted to find the supplemental effect of yeast (Saccharomyces cerevisiae) and garlic (Allium Sativum) mixture in broiler diets on growth performance, nutrient digestibility, faecal microorganisms, blood profiles, and meat quality. A total of 640 mixed Ross 308 broilers (one day old) with average body weight (BW) of (41±0.5 g) were used in a 5-week trial. Birds were blocked based on body weight and randomly assigned to 1 of 4 dietary treatments (10 replicates, 16 birds/ replicate). Treatments consisted of basal diet (CON) supplemented with 0, 0.1, 0.2 and 0.3% yeast-garlic mixture (YGM). The broiler body weight gain (BWG) linearly increased (P<0.05) and feed intake (FI) showed a tendency of the increment (P=0.095) during the overall period by dietary inclusion of YGM supplementation. However, feed conversion ratio (FCR), mortality, and nutrient utilization were not influenced. YGM supplementation exhibited the lowest Salmonella (P<0.05) counts in treatment groups, but Lactobacillus and E. coli count remained unaffected. Excreta CO_2 emission was linearly reduced (P<0.05) in YGM supplemented groups compared to the CON group; nevertheless, other noxious gas emission was not affected. Furthermore, YGM supplementation elicited a tendency of improved lymphocyte (P=0.065), and linearly increased IgG (P<0.05). A linear reduction in the weight of the bursa of Fabricius (P<0.05), a trend in increment in the weight of liver (quadratic, P=0.069) and gizzard (cubic, P=0.069) was observed by YGM inclusion. The tendency of decrease in lightness (L*) (quadratic, P=0.069) was also observed in the breast muscle colour, however, other meat quality indices were unaffected by YGM inclusion. Based on the positive effects, this research has provided the basis and new perception for future research on the usage of the YGM as a substitute for antibiotics in broiler diets.

Session 39 Poster 18

Efficacy of Achyranthes japonica extract as phytogenic feed additive in broiler diet

M.D.M. Hossain, Q.Q. Zhang, S. Biswas, S. Cao and I.H. Kim
Dankook University, Department of Animal Resource and Science, 119, Dandae-ro, Cheonan, 31116, Korea, South; mortuza40275@bau.edu.bd

Achyranthes japonica Nakai (A. japonica) is a medicinal herb found widely distributed throughout Korea. The biological activities of A. japonica are well-documented and include anti-fungal, anti-inflammatory, and immunity enhancement. The purpose of this study was to see how the addition of Achyranthes japonica extract (AJE) as a natural feed additive affects growth performance, nutrient utilization, caecal microbiota, excreta noxious gas emission, meat quality, and relative organ weight in broilers fed a corn-soybean meal diet. In total, three hundred sixty-one-day-old male Ross 308 broilers were used in a 35-d feeding trial. All of the broilers were assigned randomly to one of the four treatments. Each treatment comprises five replication pens, each with 18 birds per cage. Dietary treatments were composed of corn-wheat- soybean meal-based diets along with the addition of 0, 0.02, 0.04, and 0.06% of AJE. Bodyweight gain was linearly ($P<0.05$) improved by the supplementation of AJE during days 8 to 21, 22 to 35, and the overall experiment. In addition, feed intake also increased linearly during days 22 to 25 and the overall experiment with the increased AJE doses in the broiler diet. At the end of the experiment, the digestibility of dry matter was linearly improved with an increasing amount of AJE. At the same time, nitrogen and energy utilization tended to increase in response to increasing AJE supplementation. In summary, the inclusion of AJE in the corn-soybean meal diet started to improve growth performance by increasing nutrient utilization through AJE supplementation, which confirmed the applicability of AJE as a phytogenic feed additive in broilers.

Session 39 Poster 19

Effect of genetically modified organism supplementation on egg production and egg quality in layers

M.D.M. Hossain, S. Cao, S. Biswas, Q.Q. Zhang and I.H. Kim
Dankook University, Department of Animal Resource and Science, 119, Dandae-ro, Cheonan, 31116, Korea, South; mortuza40275@bau.edu.bd

Concerns have been expressed regarding the safety of using biotechnology-derived feeds in diets of livestock animals and in regard to human consumption of products from species-fed transgenic crops. As a consequence, we intend to examine the effect of feeding genetically modified organisms (GMO) compared with non-GMO diets on the performance of layers. One hundred and ninety - two Hy - line brown laying hens were randomly assigned to one of two dietary treatments in an experiment that was conducted for four weeks. There were 8 replicates for each treatment with 12 adjacent cages (1 hen / per cage, 38.1 cm-width, 50 lengths, 40 height) representing a replicate. We observed that no significant difference has been found in egg production in layers fed GMO feed in the diet compared with the non-GMO treatment in the trial period. Whereas in terms of egg quality, a significant reduction was observed in yolk colour and eggshell thickness of layers fed non - GMO diet compared with layers fed GMO diet at 1st week. Additionally, there was a significant increase in eggshell thickness of layers fed GMO diet compared with layers fed non-GMO diet at 2nd, 3rd, and 4th week. However, no significant differences were found in other profiles such as egg quality including egg weight, albumen height, Haugh units, yolk colour, shell colour, strength, and eggshell thickness from 1st to 4th week since layers fed GMO. In conclusion, layers fed GMO diet had increased eggshell thickness and yolk colour. However, no significant difference was found in egg production. Layers fed GMO diets may enhance the egg quality, and further research needs to be done.

Session 39 Poster 20

Comparative effects of zinc oxide and zinc aspartic acid chelate on performance of weaning pigs

S. Cao, M.D.D. Hossain, S. Biswas, Q.Q. Zhang and I.H. Kim
Dankook University, Department of Animal Resource and Science, 119, Dandae-ro, Cheonan, 31116, Korea, South; lecaoshanchuan@126.com

Zinc oxide (ZnO) at pharmacological doses is extensively employed in the pig industry as an effective tool to manage post-weaning diarrhoea (PWD), a condition that causes huge economic losses because of its impact on the most pivotal phase of a piglet's production cycle This study was conducted to compare the effects of pharmacological zinc oxide (ZnO) and zinc aspartic acid chelate (Zn-Asp) on growth performance, nutrient digestibility, faecal bacteria counts, and faecal score in weaning pigs. Based on the average initial body weight (7.01±0.65 kg), a total of 60 21-day-old weaning pigs [(Yorkshire × Landrace) × Duroc] were randomly assigned to 3 groups with 5 replicate pens and 4 pigs (mixed sex) per pen. The experimental period was 42 days (phase 1, days 1-21; phase 2, days 22-42). Treatments were as follows: (1) CON, basal diet; (2) TRT1, basal diet + 3,000 ppm ZnO; and (3) TRT2, basal diet + 750 ppm Zn-Asp. Pigs in TRT1 and TRT2 groups had higher final body weight, average daily gain during days 22-42 and 1-42, average daily feed intake during days 22-42, feed efficiency during days 22-42 and 1-42, and faecal lactic acid bacteria counts, lower faecal score and faecal coliform bacteria counts than those in CON group. In addition, the apparent dry matter digestibility in the TRT2 group was higher than in the CON group. However, no significant differences among measured parameters were observed between TRT1 and TRT2 groups. Therefore, the pharmacological ZnO and Zn-Asp feeding strategies could regulate faecal bacteria counts, and subsequent improvement in nutrient digestibility and reduction in faecal score, thus improving growth performance. Thus Zn-Asp can replace pharmacological ZnO to achieve similar performance. In addition, the lower dose of Zn in chelated form may contribute to the reduction of Zn to the environment.

Session 39 Poster 21

Lipolysis of structured triacylglycerols

C.C.E. Manuel, R.J.B. Bessa and S.P. Alves
Faculdade de Medicina Veterinária, Centro de Investigação Interdisciplinar em Sanidade Animal, Avenida de Universidade Técnica, 1300-477 Lisboa, Portugal; icaernesto@gmail.com

Supplementing ruminant diets with vegetable oils rich in polyunsaturated fatty acids (PUFA) is a strategy used to improve the nutritive value of ruminant edible fats. Vegetable oils are composed of three fatty acids (FA) attached to glycerol forming the triacylglycerols (TAG). It is known that TAG are extensively hydrolysed (lipolysis) in the rumen liberating the PUFA, which becomes available for biohydrogenation (BH). During the BH, the PUFA are isomerized and hydrogenated by microbial enzymes, forming saturated and trans FA. The information about the lipolysis kinetics of structured TAG in the rumen is scarce. Thus, the main objective of this work was to evaluate the lipolysis of five structured triacylglycerols (TAG-16:0/16:0/16:0, TAG-18:0/18:0/18:0, TAG-18:1/18:1/18:1, TAG-18:2/18:2/18:2 e TAG-18:3/18:3/18:3). The specific aim was to evaluate the effect of the FA length and level of unsaturation in the lipolysis. Thus, *in vitro* batch incubations were performed using rumen inoculum and 60 mg of TMR with 3 mg of each TAG during 0, 0.5, 2, 4 and 6 hours under CO_2. Lipids were extracted, fractioned into TAG, free-FA, and phospholipids, and FA methyl esters were prepared from each lipid fraction and analysed by gas-chromatography. We observed that the disappearance of all TAG increased ($P<0.05$) with increasing incubation time from 0.5 to 6h, with the exception of the TAG-18:3 ($P=0.726$). There were no great differences in the FA composition of free-FA fraction during the 6h of incubation with TAG-16:0 and TAG-18:0, whereas with the TAG-18:1, TAG-18:2 and TAG-18:3, the free-FA fraction showed a FA profile consistent with the ruminal BH pathways of the 18:1c9, 18:2n-6 and 18:3n-3, respectively. The results of dry matter disappearance, the concentration of dimethyl acetals, and the branched-chain FA, suggest that there was no inhibition on the growth or activity of the microbial population. Concluding, the lipolysis of TAG-18:3 seems to be very fast compared with the other TAG. Financial support was provided by PTDC/CAL-ZOO/29654/2017 and UIDB/ 00276/2020 research projects.

Fatty acid profile of milk from ewes fed a diet including broccoli or cauliflower

J. Mateo[1], C. Saro[1,2], I. Mateos[1,2], I. Caro[3], F.J. Giráldez[2] and M.J. Ranilla[1,2]
[1]Universidad de León, Campus Vegazana, s/n, 24071 León, Spain, [2]Instituto de Ganadería de Montaña (CSIC-Universidad de León), Finca Marzanas, s/n, 24346 Grulleros, Spain, [3]Universidad de Valladolid, Avda. Ramón y Cajal, 7, 47005 Valladolid, Spain; cristina.saro@unileon.es

Recent research has suggested that brassicas could be included in the diet of ruminants without negative effects on animal performance but its effects on milk composition remain unknown. The aim of this study was to assess the effects of the inclusion of broccoli or cauliflower in the diet of dairy sheep on the fatty acid (FA) profile of milk. Thirty Assaf ewes divided in three groups and in the middle phase of lactation (2.07±0.20 kg) were fed *ad libitum* on a mixture of 1:1 forage:concentrate. Control (CON) ewes received no supplement, but broccoli (BRO) and cauliflower (CAU) groups received a supplement of 1.5 kg of fresh chopped vegetable, respectively, for 6 weeks (3 for adaptation and 3 for sampling. Milk yield was recorded daily and on days 22, 28, 35 and 42 milk samples were taken for chemical analysis. Samples for fatty acids (FA) determination were frozen and lyophilized. FA concentration was expressed as relative proportions (% of the sum of area of the all FA methyl esters identified). Data were analysed as repeated measures using the lmr4 package of R. Diet, sampling day and the interaction were included as fixed effects in the model. Feeding with brassica vegetables at the levels used in this study significantly ($P<0.05$) increased the proportion of saturated fatty acids and decreased that of monounsaturated fatty acids as compared to control. The ratio n6:n3 was also affected ($P<0.05$) by the diet, being lower for BRO (3.18) than for CON (3.57) treatment; CAU milk showed values in-between values (3.33). These differences probably were related in part to differences in dietary intake of fatty acids, as brassica supplementation decreased the content of oleic acid and increased those of palmitic acid and linolenic acid. The inclusion of brassica vegetables in the diet of dairy ewes modified the fatty acid profile of milk. Grant AGL2016-75322-C2-2-R funded by MCIN/AEI/ 10.13039/501100011033 and by the European Union 'ERDF A way of making Europe'.

Introducing tomato pulp in a dairy ewes diet does not affect rumen microbial growth or structure

I. Mateos[1,2], C. Saro[1,2], T. De Evan[3], A. Martin[1,2], R. Campos[2,4], M.D. Carro[3] and M.J. Ranilla[1,2]
[1]Instituto de Ganadería de Montaña (CSIC-Universidad de León), Finca Marzanas, s/n, 24346 Grulleros, Spain, [2]Universidad de León, Campus Vegazana, s/n, 24071 León, Spain, [3]Universidad Politécnica de Madrid, Ciudad Universitaria, 28040 Madrid, Spain, [4]Universidad Nacional de Colombia, Carrera 32, 76531 Palmira, Colombia; mjrang@unileon.es

The industrial use of tomato generates high quantities of by-product, such us the tomato pulp (TP), a feedstuff for ruminant nutrition with a medium protein and energetic contents. The aim of this study was to evaluate the effect of partially replacing alfalfa hay, soybean and beet pulp by tomato pulp in a dairy sheep diet on rumen microbial population and microbial protein synthesis (MPS) in Rusitec fermenters. Two diets were incubated in 4 Rusitec fermenters in a cross-over design with two 14-day incubation periods. 30 g of diet (1:1 forage:concentrate) were incubated in each fermenter; 2 received a control diet (CON) and the other 2 received the diet containing TP (17.3% TP). ^{15}N was used as a microbial marker. On days 8 and 9 of incubation, samples of liquid and solid content were obtained from the fermenters for DNA extraction and MPS determination. DNA was used to quantify the abundance of bacteria and protozoa and the relative abundance of fungi, archaea and 3 fibrolytic bacteria by qPCR. Solid and liquid digesta were processed, microbial pellets obtained and both were analysed for ^{15}N content. In the liquid phase, protozoa were more abundant ($P<0.05$) in TP fermenters than in control ones. In solid phase none of the microbial groups was affected by the diet, but the relative abundance of *Fibrobacter succinogenes*, *Ruminococcus albus* and *R. flavefaciens* was higher ($P<0.05$) in CON fermenters than in TP ones. Neither MPS nor its efficiency were affected ($P>0.05$) by the incubated diet. However, the ^{15}N enrichment in liquid associated bacteria and the ^{15}N enrichment of NH_3-N were higher ($P<0.05$) in TP fermenters than in the CON. Replacing common feedstuff by tomato pulp a in dairy sheep diet incubated in the Rusitec system slightly influenced rumen microbial structure and did not affect microbial protein systhesis. Grant AGL2016-75322-C2-2-R funded by MCIN/AEI/10.13039/501100011033 and by the EU 'ERDF A way of making Europe'

Difference and regression methods give similar energy values for dried brewers' spent grains in pigs

L. Piquer[1], A. Cerisuelo[1], S. Calvet[2], C. Cano[1], D. Belloumi[1] and P. García-Rebollar[3]
[1]Instituto Valenciano de Investigaciones Agrarias, CITA, Polígono de la esperanza 100, 12400 Segorbe, Spain, [2]Universitat Politècnica de València, ICTA, Camí de Vera s/n, 46022 Valencia, Spain, [3]Universidad Politécnica de Madrid, Dpto. Producción Agraria, ETSIAAB, Av. Puerta de Hierro, 2-4, 28040 Madrid, Spain; piquer.lai@gmail.com

The use of local fibrous by-products dehydrated by environmentally friendly drying procedures can contribute to increase sustainability of pig feed. The digestible energy (DE) content of brewers' spent grains (BSG) dried with two different low temperature drying processes, based on biomass or solar energy sources for air heating, were determined through the difference and the regression methods. Thirty growing pigs of 59.9±3.66 kg body weight were distributed in five experimental diets, allowing for two replicates in two periods and six replicates per diet. The experimental diets included a basal and four additional diets in which the energetic part of the basal diet was replaced by 150 or 300 g/kg of each BSG. Experimental periods consisted of 14 days of adaptation to diets and 4 days of total collection of faeces per animal. Feed intake and faeces production were quantified per animal and analysed for dry matter (DM) and gross energy content. The coefficient of total tract apparent digestibility of energy was then determined by diet. The final amount of DE in both BSG was calculated by difference (using 0 and 300 g/kg treatments; MIXED procedure in SAS) or regression (using all treatments; REG procedure in SAS) and compared. The DE values obtained using the difference procedure were considered not different from those obtained using linear regression if the values were within the 95% confidence interval (CLB statement in SAS) for the DE estimated using linear regression. The DE calculated using the difference method were 12.6 kJ/kg DM for biomass and 11.4 kJ/kg DM for solar dried BSG. The values obtained for DE using the regression method were 13.1 kJ/kg DM for biomass and 11.1 kJ/kg DM for solar dried BSG. The difference and regression procedures do not give different DE values for each type of BSG.

Agro-industrial by-products in silages for lactating ewes

E. Jerónimo[1,2], L. Cachucho[2,3], D. Soldado[2,3], O. Guerreiro[1,2], H. Alves[4], S. Gomes[5], N. Alvarenga[5], K. Paulos[5], C. Costa[5], J. Costa[5], J. Santos-Silva[5] and T. Dentinho[5]
[1]Mediterranean Institute for Agriculture, Environment and Development, MED, Beja, Portugal, [2]Centro de Biotecnologia Agrícola e Agro-Alimentar do Alentejo, IPBeja, Beja, Portugal, [3]Centro Investigação Interdisciplinar em Sanidade Animal, CIISA, Lisboa, Portugal, [4]Sociedade Agropecuária Carlos e Helder Alves, Funcheira, Ourique, Portugal, [5]Instituto Nacional de Investigação Agrária e Veterinária, INIAV, Santarém, Portugal; liliana.cachucho@cebal.pt

The use of by-products in animal feed can be a strategy to reduce external dependence on conventional raw materials and associated feeding costs. The objective of this work was to evaluate the effect of using silages containing by-products in lactating ewes diets on milk quality and lamb growth compared to a concentrate-feed based diet. Fifty-four Merino crossbred ewes at the end of gestation and during lactation were fed either concentrate or two silages supplemented with concentrate. Silages were composed by alfalfa hay, brewers grains and sweet potato or almond hull. All groups received hay *ad libitum*. Eighteen ewes from each diet and their respective lambs were randomly distributed among 3 pens. Lambs were weighed at birth and weekly throughout 8 weeks of trial. Milk samples were collected for analysis. The intake was monitored daily, and the feeding cost was determined. The use of silages with by-products increased (P=0.017) the milk fat content compared to concentrate (7.81 vs 7.12%). Protein, lactose, dry residue and fat-free dry residue content, pH and somatic cell count of milk were not affected by the dietary treatments. Lambs average daily gain was not affected by the diet fed to ewes (P=0.231, 0.271 g/day). Replacement of concentrate by silages reduced up to 37% the feeding costs of ewes during lactation. Results show that the use agro-industrial by-products silages are a promising approach for its use in lactating ewes' diets, with reduced feed costs without compromising the lamb growth. This work is funded by project SubProMais (PDR2020-101-030988, PDR2020-101-030993) through FEADER and by the projects UIDB/05183/2020 (MED) and UIDB/00276/2020 (CIISA), and the PhD grant awarded to LC (2020.05712.BD) funded by national funds through FCT – Foundation for Science and Technology.

Use of almond hull in lamb diets – effect on growth performance

L. Cachucho[1,2], M. Varregoso[3,4], C. Costa[4], K. Paulos[4], S.P. Alves[1,5], J. Santos-Silva[1,4], M.T.P. Dentinho[1,4] and E. Jerónimo[2,6]
[1]Centro de Investigação Interdisciplinar em Saúde Animal, CIISA, Lisboa, Portugal, [2]Centro de Biotecnologia Agrícola e Agro-Alimentar do Alentejo, IPBeja, Beja, Portugal, [3]Instituto Superior de Agronomia, Universidade de Lisboa, ISA, Lisboa, Portugal, [4]Instituto Nacional de Investigação Agrária e Veterinária, INIAV, Santarém, Portugal, [5]Associate Laboratory for Animal and Veterinary Sciences, AL4AnimalS, Lisboa, Portugal, [6]Mediterranean Institute for Agriculture, Environment and Development, MED, Lisboa, Portugal; liliana.cachucho@cebal.pt

Almond production is growing in the Mediterranean region, particularly in southern Portugal, with increased availability of its by-products, such as almond hull (AH). Almond hull, which consists of the green outer covering of the almond, represents 52% of the total fruit fresh weight and contains high sugar content (18-30% dry matter (DM)). The main objective of this work was to evaluate the effect of partial replacement of cereals in the diet by increasing levels of AH on the growth performance of lambs. Twenty-four ram lambs were individually housed and randomly assigned to the 3 diets (8 lambs per diet), with feed offered *ad libitum*. All diets included 40% dehydrated lucerne, 6% soybean oil, soybean meal 44, sunflower meal 28, and cereals (maize, barley, and wheat), which were stepwise replaced by AH, reaching 0% (control), 9% (AH9) and 18% (AH18) of AH in diets. All diets contain 14% crude protein. The trial started after an adaptation period of 7 days, and Average daily gain (ADG) and feed intake were evaluated over 6 weeks of the experiment. Preliminary results showed that partial replacement of cereal in the diets by the AH up to levels of 18% DM did not affect the growth of lambs (P=0.283; ADG=383 g/day) and the dry matter intake (P=0.391; 1,745 g/day). Diets with 18% of AH increased the feed conversion ratio (P=0.013; 4.96) compared to the diets with 0 and 9% of AH (4.37). The use of AH in animal nutrition could be a promising strategy to replace conventional raw materials, with no relevant effect on growth performance. Funding: Project SubProMais (PDR2020-101-030988, PDR2020-101-030993)-FEADER; Projects UIDB/05183/2020 (MED) and UIDB/00276/2020 (CIISA), and PhD grant of LC (2020.05712.BD)-FCT.

Colour stability and a-tocopherol content of meat from lambs fed *Cistus ladanifer* plant and extract

D. Soldado[1,2], L. Fialho[1,2], O. Guerreiro[2,3], L. Cachucho[1,2], A. Francisco[1,4], J. Santos-Silva[1,4], R.J.B. Bessa[1,5] and E. Jerónimo[2,3]
[1]Centro Investigação Interdisciplinar em Sanidade Animal, CIISA, Lisboa, Portugal, [2]Centro de Biotecnologia Agrícola e Agro-Alimentar do Alentejo, IPBeja, Beja, Portugal, [3]Mediterranean Institute for Agriculture, Environment and Development, MED, Beja, Portugal, [4]Instituto Nacional de Investigação Agrária e Veterinária, Quinta da Fonte Boa, Santarém, Portugal, [5]Associate Laboratory for Animal and Veterinary Sciences, AL4AnimalS, Lisboa, Portugal; david.soldado@cebal.pt

Cistus ladanifer L. (CL) is an abundant shrub in Mediterranean countries, containing high levels of condensed tannins (CT) and other antioxidant compounds, such as a-tocopherol. Incorporation of CL aerial part or its CT extract in the lamb diets can improve the oxidative stability of meat. This study aimed to evaluate the effect of the inclusion of aerial part and CT extract from CL in lamb diets on meat a-tocopherol content and colour stability. Six diets were formulated considering three levels of CL CT (0, 1.25 and 2.5% CT) and two ways of CT supply (CL aerial part vs CL CT extract). Basal diet was composed of dehydrated Lucerne supplemented with soybean oil (60 g/kg). Thirty-six lambs were housed individually (six lambs/diet). The trial lasted for 35 days. Content of a-tocopherol was analysed in Longissimus thoracis (LT) muscle samples, immediately collected after slaughter. Meat colour was analysed, using CIELAB system, in LT muscle collected seventy-two hours after slaughter and stored at 2 °C for 0, 3 and 7 days. Muscle a-tocopherol content was not affected by dietary treatments (P>0.05). Meat colour coordinates L* and b* were affected by dietary treatments, with higher L* values in diets containing 2.5% CT than in diets with the lower CT levels (P=0.018). *C. ladanifer* CT extract resulted in meats with lower values of b* than CL aerial part (P=0.028). During storage time, colour parameters L*, b*, C* and H* increased, while a* decreased (P<0.001). In this work, the inclusion of CL aerial part and CT extract in lamb diets did not affect the a-tocopherol content or improve the colour stability of meat. Funding: Project CistusRumen (ALT20-03-0145-FEDER-000023) – FEDER; Projects UIDB/05183/2020 (MED) and UIDP/00276/2020 (CIISA), and PhD grant awarded to DS (SFRH/BD/145814/2019) – FCT.

Effect of dietary supplementation with sodium butyrate and aromatic plants on chicken meat quality

A. De-Cara[1,2], B. Saldaña[2], J.I. Ferrero[2], G. Cano[3] and A.I. Rey[1]
[1]Universidad Complutense de Madrid, Animal Production, Avda. Puerta de Hierro s/n, Madrid, 28040, Spain, [2]Nuevas Tecnologías de Gestión Alimentaria S.L, Nutrition, C/ Marconi, 9, Coslada, Madrid, 28823, Spain, [3]Imasde Agroalimentaria S.L, C/ Nápoles, 3, Pozuelo de Alarcón, Madrid, 28224, Spain; bsaldana@nutega.com

Due to antibiotics restriction and social concern about the use of synthetic antioxidants, other alternatives are being investigated not only to improve bird's health and performances but also meat quality. The aim of this assay was to study the effect of protected sodium butyrate (PSB), aromatic plants mixture (AP-mix), and their interaction on meat quality (colour and lipid stability over time) in broilers from 1 to 40 days of age. A total of 1,320 one-day-old male Ross 308 broilers were distributed in 60 pens with 4 dietary experimental treatments 2×2 factorial structured with 2 levels of PSB (0 vs 0.2%) and 2 levels of AP-mix (0 vs 0.2%). There were 15 replication per pen. Fresh breast of one broiler per pen were taken for measurements. Colour parameters were evaluated by using a chromameter following the CIELAB system. The oxidation degree of the breast was determined by thiobarbituric acid reactive substances (TBARs) method by spectrophotometric measurement. The statistical analysis of data was carried out using SAS program v9.2. TBARs of meat increased with time, being PSB group the one which presented less oxidation in all the measurements days (0, 3, and 6 days of evaluation) ($P<0.001$), followed by the group that combined PSB+AP-mix and AP-mix. PSB supplementation also tended to increase a* value of meat, whereas AP-mix administration resulted in higher yellowness (b*) and intensity of the breast colour ($P<0.05$). Other authors using higher supplementation PSB doses in hostile situations found positive effects on meat stability as well as, extracts present in plant material has also been reported to have antioxidant activity to control meat lipid oxidation. The PSB dose of the present study without any other supplementation would be enough to control lipid stability of chicken meat.

Optimisation of spent coffee grounds nutritional value: thermal conditions and enzymatic hydrolysis

M. Medjabdi[1], I. Goiri[1], R. Atxaerandio[1], J. Ibarruri[2], B. Iñarra[2], D. San Martin[2] and A. Garcia-Rodriguez[1]
[1]NEIKER – BRTA, Animal Production, Campus Agroalimentario de Arkaute s/n, 01192 Arkaute, Álava, Spain, [2]AZTI, Food Research, Basque Research and Technology Alliance (BRTA), Parque Tecnológico de Bizkaia, Astondo Bidea, Edificio 609, 48160 Derio, Bizkaia, Spain; aserg@neiker.eus

The valorisation of agri-food by-products is one of the solutions proposed to face the high costs of animal feed today. Coffee grounds are one of the food wastes with potential for use in animal feed, providing beneficial effects on milk yields at limited inclusion levels. However, it has been observed that their introduction in ruminant diets at levels higher than 10% can lead to a decrease in digestibility, thus limiting the productive performance of the animals. Therefore, the objective of the current study was to subject coffee grounds to a enzymatic hydrolysis process combined with thermal pretreatment with the aim of improving the digestibility of this raw material. To this end, the effect of an enzymatic process with 4 commercial enzymes focused on fibre digestion and their interaction with a thermal pretreatment by autoclaving at 120 °C for 15 minutes on the chemical composition of coffee grounds as well as on the digestibility was tested. The samples were incubated in triplicate and in four different *in vitro* incubation series during 24 hours. A not significant interaction between pretreatment and enzymatic hydrolysis was found except for the cellulose content ($P=0.021$). The thermal pretreatment showed a tendency to increase ($P=0.055$) the crude fat content of the residue without affecting any of the other studied variables. The results indicate that enzymatic hydrolysis reduced ash ($P<0.001$), phosphorus ($P=0.006$) and potassium ($P=0.012$) content of coffee grounds. The enzymatic hydrolysis did not affect the crude protein content but it increased the acid detergent fibre crude protein content ($P<0.001$). As a consequence, the enzymatic hydrolysis process resulted in a reduction in organic matter digestibility ($P<0.001$). In conclusion the pretreatment conditions and the enzymatic hydrolysis tested/selected were not adequate to improve the nutritional value of coffee grounds.

Impact of *Nannochloropsis gaditana* diet on *Sparus aurata* metabolism by using an omics approach

D.E.A. Tedesco[1], K. Parati[2], T. Bongiorno[2], R. Pavlovic[3] and S. Panseri[3]
[1]Department of Environmental Science and Policy, University of Milan, via Celoria 2, 20133 Milano, Italy, [2]Experimental Institute Lazzaro Spallanzani, Loc. La Quercia, 26027 Rivolta d'Adda (CR), Italy, [3]Department of Veterinary Medicine and Animal Sciences, University of Milan, Via dell'Università 6, 26900 Lodi, Italy; doriana.tedesco@unimi.it

Microalgae are recognised today as a promising functional feed and their inclusion in aquaculture feeding represents a sustainable environment-friendly alternative. One of the more promising microalgae in terms of nutritional content and easily integrable in biorefineries is *Nannochloropsis gaditana.* The main objective was to perform a metabolomic untargeted profiling on *Sparus aurata* fillets to distinct following feeding trails performed: diet without microalgae (Control) and four different diets with *N. gaditana* either with 5% raw microalgae or with 5% hydrolysed microalgae, from inorganic nutrient or biorefinery. The fish fillets were analysed by liquid chromatography coupled with high-resolution mass spectrometry (LC-HRMS, Q-Exactive – CD). Applying omics software, 320 molecules were identified. The majority of recognized compounds belong to the class of amino acids and their metabolites, and products of lipid metabolism. The presence of various lipids, products of glycolysis and pyruvate metabolism, principal mono/ oligosaccharides, nucleotides and nucleosides, and vitamins were confirmed as well. The content of some amino acids and their metabolites revealed significant differences between experimental diets. Specifically, in the muscle of fish fed with hydrolysed microalgae, there was an increase in essential amino acids: valine, methionine and arginine. These compounds can be used as potential markers of higher nutritional quality of diets with hydrolysed microalgae. The muscle of fish raised with the experimental diets also contains the highest amounts of glutathione and carnosine. The incorporation of the microalgae in the fish diet had a significant and progressive impact on the fish metabolome profile. The metabolites modulated by microalgae meal were involved in protein, lipids and energy metabolism. The findings provide new insights into how the dietary food metabolome affects fish metabolism.

Food industry leftovers slightly affect gut microbiota and blood metabolites in growing pigs

M. Tretola, A. Luciano, M. Manoni, M. Ottoboni, R. Abbate, G. Ceravolo and L. Pinotti
University of Milan, Department of Veterinary Medicine and Animal Sciences, Via dell'Università, 6, 26900 Lodi, Italy; marco.tretola@unimi.it

Worldwide, the amount of wasted food is around 1.3 billion tons per year. At the same time, the sustenance demand is expected to increase significantly. The recovery of food loss as animal feed addresses both waste reduction and zero-hungry challenges. Food industry leftovers, also called former foodstuff products (FFPs) can be differentiated into sugary confectionary FFPs (FFP-C) and salty FFPs from bakery production (FFP-B). The present study intends to test the impact of FFP-C and FFP-B in growing pig's diet on the large intestinal microbial community composition and biodiversity, together with their metabolic status. Thirty-six post-weaning female piglets (Large White × Landrace, body weight 8.52±1.73 kg) were randomly assigned to a standard diet (CTR), or diets in which traditional ingredients were partially replaced by the 30% inclusion (w/w) FFP-C or FFP-B for 42 days. Growth performance were measured. The faecal samples were collected after 42 days for 16S rRNA gene sequencing (NGS). Blood serum samples were collected at day 0 and 42 and analysed by UHPLC/MS-MS in ionization mode to quantify serum metabolites. All data were analysed in R (v 4.1.2). Data about serum metabolites were analysed through the software MetaboAnalyst (version 5.0). The three diets did not evidence any effect ($P>0.05$) on growth performance, gut microbial composition, alpha or beta diversity. Few bacteria differed in their abundance. Despite several metabolites were influenced by the age, only two were significantly affected by the interaction diet × age. The FFP-C strongly increased ($P<0.001$) the serum concentration of theobromine and caffeine compared to the CTR and FFP-B. No significant correlations between blood metabolites and bacterial taxa were found. Performances, faecal microbiota, and the metabolic status of the pigs were slightly affected by the partial replacement of standard ingredients with FFPs-C or FFP-B. Those sustainable products can be safely used in post-weaning and growing pig diets. The effect of the FFPs-diets for a longer feeding trial on pig physiology and product quality needs to be further investigated.

Session 39 Poster 32

Preliminary remarks on tobacco cv. *Solaris* seed cake in growing cattle diet

A. Fatica[1], F. Fantuz[2], C. Sacchetto[3], A. Ferro[4] and E. Salimei[1]

[1]Università del Molise, via de Sanctis, 1, 86100 Campobasso, Italy, [2]Università di Camerino, via Gentile III da Varano, 62032 Camerino, Italy, [3]Associazione Professionale Trasformatori Tabacchi Italiani, via Monte delle Gioie, 1C, 00199 Roma, Italy, [4]Deltafina s.r.l., via Appia, 81050 Francolise, Italy; a.fatica@studenti.unimol.it

Fourteen growing Friesian and Crossbreed heifers, average weight 214.9 (±50.9, SD) kg and average age 313.4 (±152.7) days, were selected and divided in two homogeneous groups, *Solaris* (SOL) and Control (CTR) group. Isoenergetic and isonitrogenous diets were formulated based on the nutritional needs of animals. Each group received daily 21 kg of legume and grass haylage, and groups were *ad libitum* fed mature grass hay and concentrates mixture made by 40% of corn, 40% of barley, 20% of commercial feed for CTR group, and by 40% of corn, 40% of barley, 10% of commercial feed and 10% of *Solaris* (PCT/IB/2007/053412) seed cake for SOL group. During 34 days of trial, feedstuff consumption was monitored. At 0- and 34-days animals were scored for body condition (BCS), locomotion capacity (LS) and faecal consistency (FS). At 34 day of the trial, *Solaris* seed cake resulted palatable as the daily group DM intake from concentrate was 85.8 (±1.46) kg and 82.8 (±1.58) kg, respectively for SOL and CTR group. At 0- and 34-days BCS was 3.34 (±0.21) and 3.50 (±0.35) in SOL group, and 3.39 (±0.31) and 3.60 (±0.30) in CTR group. All animals showed good locomotion score (average LS 1 in both groups), and faecal score (average FS 3.5 in both groups) with manure visibly darker in SOL group compared to yellow-olive colour in CTR group. Based on the preliminary observations, the introduction of this alternative vegetal protein source as feedstuff deserves to be more in depth studied as further opportunity for an innovated tobacco cultivation, in those marginal areas traditionally suited to tobacco and devoted to animal breeding. The recovered know-how of tobacco cultivation could represent a resilient strategy against the smoking tobacco sector crisis, recently experienced by the small farming system of inner Mediterranean areas. Research funded by AGRIFOOD, PON MISE.

Session 39 Poster 33

Can dietary algae blend modulate the fatty acids composition of lamb meat?

C.S.C. Mota[1], A.S. José[2], V.A.P. Cadavez[2], R. Domínguez[3], J.M. Lorenzo[3], U. Gonzales-Barron[2], A.R.J. Cabrita[1], H. Abreu[4], J. Silva[5], A.J.M. Fonseca[1] and M.R.G. Maia[1]

[1]REQUIMTE, LAQV, ICBAS, Universidade do Porto, R. Jorge Viterbo Ferreira 228, 4050-313 Porto, Portugal, [2]CIMO, Instituto Politécnico de Bragança, Campus de Santa Apolónia, 5300-253 Bragança, Portugal, [3]Centro Tecnológico de la Carne de Galicia, R. Galicia 4, 32900 Ourense, Spain, [4]ALGAplus, PCI, Via do Conhecimento, 3830-352 Ílhavo, Portugal, [5]ALLMICROALGAE – Natural Products, SA, R. 25 de Abril, 2445-413 Pataias, Portugal; catiam04@gmail.com

Lambs raised in extensive systems have healthier fatty acid (FA) profile of their meat, enriched in *n*-3 polyunsaturated fatty acids (PUFA) with anti-inflammatory effects, whereas those raised in intensive systems have higher saturated fat, associated with increased incidence of metabolic diseases. Dietary supplementation of oilseeds and algae have been proposed to promote the unsaturation and *n*-3 PUFA content of meat. Thus, this study aimed to evaluate the supplementation of a commercial algae blend composed of macro- and microalgae (Algaessence™) on meat FA composition of lambs reared in intensive production system. To achieve it, three groups of 10 Bordaleira-de-Entre-Douro-e-Minho lambs were distributed to one of three systems: (1) traditional extensive production (pasture by day, indoors by night); (2) intensive production fed a concentrate diet, and (3) intensive production fed a concentrate diet with 5% algae blend. All animals had *ad libitum* access to water and meadow hay indoors. The growth trial lasted for 60 days. Lambs were slaughtered and FA composition of *Longissimus lumborum* evaluated. Hot carcass weight was the lowest in pasture fed lambs yet intramuscular fat content was similar among groups. Extensive reared meat had the most unsaturated profile, *n*-3 PUFA being 3-fold higher than intensive meat. The algae blend lowered the *n*-6/*n*-3 ratio of intensive reared meat but the proportion of health promoting 20:5*n*-3 (EPA), 22:5*n*-3 (DPA) and 22:6*n*-3 (DHA) was similar. Principal component analysis revealed a strong dissociation between extensive and intensive systems; algae inclusion having a subtle effect. Financial support of FCT, ALGAplus and Allmicroalgae to CSCM (PD/BDE/150585/2020) and of FCT to MRGM (DL 57/2016; SFRH/BPD/70186/2010) is acknowledge.

Fatty acid profile of oil and cake of four hemp varieties cultivated in Italy along three years

S. Arango[1], M. Montanari[2], N. Guzzo[1] and L. Bailoni[1]

[1]University of Padova, Dipartimento di Biomedicina Comparata e Alimentazione, Viale dell'Università, 16, 35020 Legnaro PD, Italy, [2]Centro di cerealicoltura e colture industriali CREA-CI, Consiglio per la ricerca in agricoltura e l'analisi dell'economia agraria, Viale Giovanni Amendola 82, 45100 Rovigo RO, Italy; sheylajohannashumyko.arangoquispe@studenti.unipd.it

Hemp (*Cannabis sativa* L.) is an annual plant, globally distributed and cultivated in the past as a source of fiber. Recently, the interest in hemp cultivation has significantly increased, considering its positive environmental impact and several application fields. This study aims to assess the fatty acid profile of oil and cake of four varieties (CS, Carmaleonte, Codimono, and Futura 75) of hemp cultivated in Italy along three consecutive years (2019, 2020 and 2021). The outdoor cultivation took place at CREA-CI Rovigo (northeast Italy). Samples of each variety, year, and products (oil and cake) were subjected to analysis of fatty acids profile, after accelerated solvent extraction (ASE), by gas-chromatographic way. The fat content of hemp cake was highest in CS variety and increased along the three years (from 16.6 to 16.3 to 23.7% on DM in 2019, 2020, and 2021 resp.) but it remained stable in the other varieties (on average 11.8±1.2% on DM). The total saturated fatty acids (SFA), the monounsaturated FA (MUFA) and the polyunsaturated FA (PUFA) of oil and cake were resp. 12.0, 16.0 and 72.0% of total FA. The overall mean of n-6 to n-3 fatty acids was 3.9. The CS variety showed the highest value of PUFA (74.0% of total FA). Among PUFA, the linoleic acid (C18:2 n6) and alpha-linolenic acid (C18:3 n3) are the most representative (58.2 and 17.5% of total FA resp.). On opposite, Carmaleonte showed the lowest value of PUFA (67.8% of total FA). As expected, the FA profile of hemp oil and cake was very similar. Along the 3 years of cultivation, also the values for SFA, MUFA and PUFA were similar as SFA and PUFA decreased (-2.2 and 1.4% resp.) and MUFA increased (3.6%). In conclusion, the FA profile of hemp oil and cake is very interesting when these products are used as feed of farm animals, considering the high level of PUFA and the good ratio between n-6 and n-3 FAs. Some differences on FA profile were observed among varieties whereas the effect of the year was meaningless.

Incorporation of *Malvaceae* into weaned piglet diets increases faecal dry matter and immunoglobulin A

O.P.R. Ashton[1,2], M.D. Scott[1] and B.D. Green[2]

[1]Devenish Nutrition, Lagan House, 19 Clarendon Road, BT1 3BG Belfast, United Kingdom, [2]Institute for Global Food Security, School of Biological Sciences, Queen's University Belfast, 19 Chorine Gardens, BT9 5DL Belfast, United Kingdom; oashton01@qub.ac.uk

Piglets are susceptible to infectious disease at weaning which is typically treated with antibiotics and/or pharmaceutical zinc oxide (ZnO). Species of the *Malvaceae* family are commonly used in human ethnobotanical systems to treat infections. In this investigation powdered root from a *Malvaceae* spp. and/or pharmaceutical ZnO were incorporated into weaned piglet diets. The aim was to assess for the first time the tolerability and potential effectiveness of plant material as a replacement for ZnO. Ninety-six piglets at 28-days of age, weighing 7.63±0.107 kg were weaned into 24 pens on a commercial farm. Piglets were offered diets following a 2×2 factorial design with 2.5 kg/tonne ZnO and 10 kg/tonne *Malvaceae* spp. for 11 days. ZnO treatments were maintained for a further 11 days, without *Malvaceae* spp. Growth rates and feed intakes were calculated at 11 and 22 days. Faecal dry matter was determined. Faecal immunoglobulin A (IgA) content was quantified by ELISA in 10 pigs per treatment at day 11. Formally identified faecal IgA outliers were excluded. Differences were assessed by 2-way ANOVAs. Growth rates and feed conversion ratios (FCR) did not differ between any groups averaging 431.74±6.755 g/day and 1.12±0.030, respectively. Feed intake averaged 389.55±11.178 g/day across groups but ZnO reduced feed intake by 9% (days 0-11; $P=0.041$). *Malvaceae* spp. increased faecal dry matter by 20% ($P=0.037$). ZnO and *Malvaceae* spp. increased faecal IgA content by 239% ($P<0.001$) and 75% ($P=0.016$), respectively. This study demonstrates that *Malvaceae* spp. inclusion is tolerable in piglets, does not negatively affect performance, but increases faecal dry matter and increases faecal IgA content. ZnO did not increase dry matter and had larger effect on IgA. Further studies over longer feeding periods are necessary to assess whether the *Malvaceae* spp. could be used as a routine replacement for pharmaceutical ZnO. Faecal microbiome analysis is ongoing and may provide further insights.

Session 39 Poster 36

Rumen microbial community and meat quality of goat kids under cactus cladodes inclusion

S. El Otmani[1], Y. Chebli[1], B. Taminiau[2], M. Chentouf[1], J.L. Hornick[3] and J.F. Cabaraux[3]
[1]INRA-Morocco, Regional Center of Agricultural Reserach-Tangier, 78, Boulevard Sidi Mohamed BenAbdellah, Tangier, 90010, Morocco, [2]University of Liège, Department of Food Science, Food Microbiology, Avenue de Cureghem 6, B42, 4000, Belgium, [3]University of Liège, Department of Veterinary Management of Animal Resources, Avenue de Cureghem 6, B43, 4000, Belgium; samira.elotmani@inra.ma

In harsh environment, goats' herd presents an essential source of proteins by providing large quantities of meat. This livestock is the most dominant, thanks to their adaptation capacity, and their poor nutritional quality valuation. Their diet is composed essentially of forest rangelands. The pressure exercised on these pastures leads to a serious degradation. The cactus is a multi-purpose shrub which the use of their cladodes (CC) are recommended in the ruminants' diet. However, there are no researches were conducted on meat quality and ruminal microbiota of goat kids. This work aims to evaluate the CC effects on ruminal bacteria diversity and meat quality of goat kids. Twenty-two goat kids of 3 months were divided into two groups. The control group received a conventional concentrate, while test group was supplemented by 30%CC. After 3 months, goats' kids were slaughtered. The rumen liquor was collected to determine pH and to extract DNA and identify the microbial community. The pH of meat was evaluated on longissimus Dorsi at (0) and after 24h, and colour. Meat samples were collected to determine humidity, ash, water retention capacity, proteins, and fat. Bacterial community composition was affected at the genera. Control liquor contained more *Aeriscardovia*, a lactic acid producer using the high starch in this diet, which reflected on low ruminal pH (5.7 vs 6.22). *Defluviitaleaceae* was significantly higher with 30% CC that is correlated negatively to obesity that explains the low fat in meat. For meat quality, there was no difference in pH0 and pH24, and meat colour, moisture and water retention capacity. CC inclusion was positively affected meat quality by increasing proteins and reducing fat contents. The introduction of CC in goat kids' diet did not strongly change the bacterial composition of rumen liquor, however they improved meat quality. The CC could be introduced as an alternative feed resource into goat kids' diet.

Session 39 Poster 37

Avian intestinal epithelium metabolome influenced by genetic strain and seaweed supplementation

S. Borzouie, B.M. Rathgeber, C. Stupart and L.A. MacLaren
Dalhousie University, Animal Science & Aquaculture, Truro, Nova Scotia, B2N 5E3, Canada; leslie.maclaren@dal.ca

We investigated the effects of seaweed supplementation on the intestinal epithelial cell metabolome in two commercial chicken strains and examined the cellular and molecular pathways potentially regulated by these factors to complement our previous study of blood metabolome response. In a short-term trial, 100 Lohmann Brown and LSL White laying hens at 55 weeks of age were supplemented with 0 (Control) or 3% red seaweed *Chondrus crispus* (CC) for 21 days prior to sampling. In a long-term trial, 240 hens were assigned to 0%, 3% CC or 0.5% brown seaweed *Ascopyllum nodosum* (AN) from weeks 31 to 72. Jejunum samples were collected for epithelial cell isolation and cell metabolites were identified and quantified using ^{1}H nuclear magnetic resonance spectroscopy (NMR), followed by multivariate data analyses to discriminate between metabolites. Metabolite set enrichment and pathway analyses were performed to predict the metabolic pathways affected by treatments. Seaweed supplementation impacted 44 epithelial cell metabolites in the short-term trial and 35 metabolites in the long-term trial ($P<0.05$), of which 29 overlapped. Five of these also overlapped in the previous blood metabolome response. The largest change was observed in birds supplemented with 0.5% AN, which increased levels of 27 epithelial cell metabolites. In both trials, seaweed supplements affected glycine, serine and threonine metabolism, alanine, aspartate and glutamate metabolism, as well as aminoacyl-tRNA biosynthesis, a metabolic pathway that was also altered in blood. There were significant differences ($P<0.05$) in epithelial cell metabolites among the two strains, including 45 metabolites identified in the short trial and 49 metabolites identified in the long-term study, of which five had also been affected in the blood metabolome. The majority of impacted metabolites were amino acids, potentially affecting 15 metabolic pathways. Glycine, serine and threonine metabolism was significantly altered in both intestinal epithelium and plasma samples. This study suggests that NMR will be useful for identifying the mechanisms by which genetic or environmental changes can affect bird health.

The effect of extrusion on the RUP fraction of canola oilcake meal and lupins

T.S. Brand[1,2], O. Dreyer[1] and L. Jordaan[1]
[1]Stellenbosch University, Animal Sciences, Merriman street, 7600 Stellenbosch, South Africa, [2]Department of Agriculture: Western Cape, Animal Science, Muldersvlei Street, 7607 Elsenburg, South Africa; tersb@elsenburg.com

Highly degradable protein sources such as canola oilcake meal and lupins could possibly be extruded to decrease the rumen degradability and increase the rumen undegradable protein (RUP) fraction thereof. This study was conducted to determine the effect of a certain extrusion method with molasses on the *in situ* degradability of canola oilcake meal and crushed sweet lupins. Six Dohne Merino wethers (±80 kg live weight) fitted with rumen cannula were used during this trial. The sheep had *ad libitum* access to clean water and a basal diet of wheat straw and lucerne hay (50:50). Locally produced canola oilcake meal and crushed sweet lupins with addition of 6% molasses were extruded at 116 °C with a Millbank extruder. The following feeds were tested in this trial: canola oilcake meal not extruded (CM), canola oilcake meal extruded (CME), crushed sweet lupins not extruded (CL); crushed sweet lupins extruded (CLE). Samples were incubated in the rumen of the sheep in polyester bags at intervals of 0, 2, 4, 8, 16, 24; 48 hours, with an all-out approach. The dry matter (DM) and crude protein (CP) degradabilities of the samples were determined by the *in situ* technique described by Ørskov & McDonald. Extrusion significantly lowered the CP soluble fraction of CM by 62.2%, while not differing from CL (46.0%) and CL not differing from CLE (38.2%). Extrusion also increased the CP potential degradable fraction by 43.5%, with the highest rate of degradation found for CL (0.130%/h). At each outflow rate, namely 0.02, 0.04, 0.05, 0.06; 0.08/h, extrusion significantly lowered the effective degradability for both protein sources. While the largest effect being seen at 0.08/h where the effective degradation was lowered by 25.6%. Overall, the combined RUP fraction was increased by 85.4%. Therefore, extrusion with molasses modified ruminal degradation parameters of both canola oilcake meal and crushed sweet lupins, while also decreasing effective rumen degradation at faster outflow rates. The study also indicated that benefits of extrusion could be obtained at a relatively low temperature of 116 °C, with the addition of 6% molasses.

Effect of lactic acid bacteria on high dry matter forage preservation

I. Nikodinoska[1], C. Moran[2] and H. Gonda[3]
[1]Alltech's European Bioscience Centre, Regulatory Affairs, Sarney, Summerhill Road, Dunboyne, Ireland, [2]Alltech SARL, Regulatory Affairs, Rue Charles Amand, 14500 Vire, France, [3]Swedish University of Agricultural Sciences, Department of Animal Nutrition and Management, Faculty of Veterinary Medicine and Animal Science, Ulls väg 26, 756 51 Uppsala, Sweden; cmoran@alltech.com

The efficacy of 4 silage additives on high dry matter (DM) forage preservation was examined. A grass-clover dominated ley at Lövsta Farm (Sweden) was harvested in mid-July (2nd cut) and wilted to achieve high DM content forage (40%) and 2 different levels of water-soluble carbohydrates (WSC) (low = 2.1-3.4% and high = 3.6-5.3% in fresh matter). Low WSC content was achieved by combining the effects of shade and cutting time. The crop was chopped to 3 cm and inoculated at 1×10^6 cfu/g with *Lactiplantibacillus plantarum* (IMI 507027 or IMI 507028), *Pediococcus pentosaceus* IMI 507024 or *Lacticaseibacillus rhamnosus* IMI 507023, or water (Control). Five replicates per treatment were prepared in 1.75 l micro-silos (5 treatments × 5 replicates = 25 silos) at a density of 106 kg DM/m^3. The NH_3-N, % Total Nitrogen (TN) was determined via Flow Injection Analysis and Kjeldahl method, Lactic Acid (LA) and Ethanol (EtOH) via high-performance liquid chromatography, DM concentration after drying at 60 and 103 °C and DM loss was determined from the weight difference at start and end of ensiling. Data were collected after 90 days of ensiling at 20 °C and analysed according to Wilcoxon Signed-Rank Test (Real Statistics Resource Pack for Excel 365) to compare the difference between the measured variables for the control and treatments ($P<0.05$). All treatments significantly improved the silage quality according to low pH value and related high LA production. The proteolytic marker, NH_3-N, %TN, was significantly lower in inoculated silages compared to control. All inoculants except IMI 507024 showed significantly lower EtOH production compared to control and all inoculants except the *L. plantarum* strains improved the DM loss, in both crops. These findings suggest that the homofermentative pathway was dominant and that all treatments efficiently improved the quality of two types of high dry matter forage.

Session 39 Poster 40

Effect of lactic acid bacteria on low dry matter forage preservation

I. Nikodinoska[1], C. Moran[2] and H. Gonda[3]

[1]Alltech's European Bioscience Centre, Regulatory Affairs, Summerhill Road, Dunboyne, Ireland, [2]Alltech SARL, Regulatory Affairs, Rue Charles Amand, 14500 Vire, France, [3]Swedish University of Agricultural Sciences, Animal Nutrition and Management, Ulls väg 26, 756 51 Uppsala, Sweden; cmoran@alltech.com

The efficacy of 4 silage additives on low dry matter (DM) forage preservation was examined. A grass-clover dominated ley at Lövsta Farm (Sweden) was harvested in mid-July (2nd cut) and wilted to achieve low DM content forage (24-27%) and 2 different levels of water-soluble carbohydrates (WSC) (low = LWSC = 2.1-3.4% and high = HWSC = 3.6-5.3% in fresh matter). Low WSC content was achieved by combining the effects of shade and cutting time. The crop was chopped to 3 cm and inoculated with *Lactiplantibacillus plantarum* (IMI 507027 or IMI 507028), *Pediococcus pentosaceus* IMI 507024 or *Lacticaseibacillus rhamnosus* IMI 507023 at the rate of 1×10^6 cfu/g, or with water (Control). Five replicates per treatment were prepared in 1.75 l micro-silos (5 treatments × 5 replicates = 25 silos) at a density of 106 kg DM/m^3. The NH_3-N, % Total Nitrogen (TN) was determined via Flow Injection Analysis and Kjeldahl method, Lactic Acid (LA) and Ethanol (EtOH) via high-performance liquid chromatography, DM concentration after drying at 60 and 103 °C and DM loss was determined from the weight difference at start and end of ensiling. Data were collected after 90 days of ensiling at 20 °C and analysed according to Wilcoxon Signed-Rank Test (Real Statistics Resource Pack for Excel 365) to compare the difference between the measured variables for the control and treatments ($P<0.05$). All treatments improved ($P<0.05$) the silage quality of both forage types according to a low value for the pH and the proteolytic marker, NH_3-N, %TN. All inoculants reduced ($P<0.05$) the DM loss in both forage types, except IMI 507028 which reduced the DM loss in HWSC but not in LWSC forage. All inoculated silages showed higher LA concentration ($P<0.05$) than the control, except the *L. plantarum* strains which increased ($P<0.05$) the LA in HWSC but not in LWSC forage. The EtOH production was lower ($P<0.05$) than the control in both treated forage types. These findings suggest that all treatments efficiently improved the quality of low dry matter forage.

Session 39 Poster 41

Bacteria profile of water and gills of *Mytilus galloprovincialis* from a North Italian lagoon

G. Zardinoni[1], A. Trocino[2] and P. Stevanato[1]

[1]University of Padova, Department of Agronomy, Food, Natural resources, Animals and Environment, Viale dell'Universita' 16, 35020 Legnaro, Italy, [2]University of Padova, Department of Comparative Biomedicine and Food Science (BCA), Viale dell'Universita' 16, 35020 Legnaro, Italy; giulia.zardinoni@phd.unipd.it

This study aimed to assess the microbiota in water and gills of *Mytilus galloprovincialis* collected at Scardovari lagoon (Po Delta, Italy). The seasonal and spatial dynamics microbial communities were investigated with a DNA metabarcoding approach, which simultaneously amplifies seven hypervariable regions of the bacterial 16S rRNA gene. Mussels and water were sampled in two areas of the lagoon (North-West and South-East) three times between April and June 2021. The most abundant bacterial taxa (*Methylobacterium*, *Burkholderia*, and *Sphingomonas*) in gill tissue found by DNA metabarcoding were confirmed using Sanger analysis and qPCR. Our results highlight a characteristic bacterial profile in mussels compared to water and the Principal Component Analysis (PCA) revealed the presence of separate community structures within gills and water samples. In gills tissues, the families of *Sphingomonadaceae*, *Beijerinckiaceae*, and *Burkholderiaceae* were abundant, while in seawater samples *Rhodobacteraceae*, *Flavobacteriaceae*, and *Microbacteriaceae* families were the most represented. The gills of mussels sampled in the northern and southern areas presented a high similarity of microbial communities. The seasonal dynamic revealed structural variation of bacterial profile and alfa-diversity composition both in gills and water ($P<0.05$) and the unique presence of some pathogenic bacteria orders was highlighted in May and June when temperature raised. These findings enhance our understanding of the gill-associated microbiota with a space and time microbiota analysis approach. The relationship of gill bacteria communities with the surrounding environment was pointed out, enabling the development of a transferable method to other economically important seafood.

Session 39 Poster 42

Marine sources in dog feeding: nutritive value and palatability of squid meal and shrimp hydrolysate

J. Guilherme-Fernandes[1], T. Aires[2], A.J.M. Fonseca[1], M.R.G. Maia[1], S.A.C. Lima[3] and A.R.J. Cabrita[1]
[1]LAQV, REQUIMTE, ICBAS, Instituto de Ciências Biomédicas Abel Salazar, Universidade do Porto, R. Jorge Viterbo Ferreira 228, 4050-313 Porto, Portugal, [2]SORGAL, Sociedade de Óleos e Rações S.A., Lugar da Pardala, 3880-728 S. João Ovar, Portugal, [3]LAQV, REQUIMTE, Faculdade de Farmácia, Universidade do Porto, R. Jorge Viterbo Ferreira 228, 4050-313 Porto, Portugal; jmgmfernandes@gmail.com

The growing population of companion animals is raising awareness of the quality and sustainability of pet food. Among macronutrients, protein is the most expensive in economic and ecological terms. The study of alternative and functional protein feed sources will contribute for the pet food sustainability, providing optimal nutritional value and promoting animal health. The present study evaluated the palatability and nutritive value of alternative protein sources of marine origin: squid meal and shrimp hydrolysate. Both sources presented high levels of crude protein (81.0% dry matter basis, DM, in squid meal; 65.8% DM in shrimp hydrolysate), with arginine, lysine, and leucine being the essential amino acids presented at higher concentrations. Based on the minimum recommended levels for adult dogs, amino acid scores were above 100, except for methionine+cystine in both sources, and threonine in shrimp hydrolysate. Shrimp hydrolysate presented the highest antioxidant activity. The palatability tests on adult beagles showed preference for a commercial diet over 15% dietary inclusion of either source, with no differences for first approach and taste. Dietary inclusion of 5, 10, and 15% of either marine sources had no effect on *in vivo* diet digestibility and metabolizable energy content, having these feeds high values of DM (80%) and protein (87%) digestibility. The results suggest high potential of these alternative protein sources in dog feeding. Further research is needed to unveil their functional value. Financial support of Fundação para a Ciência e a Tecnologia (FCT) and Soja de Portugal to JGF (PD/BDE/150527/2019), and of FCT to SCL (CEECIND/01620/2017), MRGM (DL 57/2016–Norma transitória), LAQV (UIDB/50006/2020), and NovInDog – Novos Ingredientes Proteicos Funcionais para a Alimentação de Cães: uma Abordagem Sustentável (POCI-01-0247-FEDER-047003) are acknowledged.

Session 40 Theatre 1

INSECT DOCTORS: a European training program focusing on pathogens that threaten mass reared insects

M.M. Van Oers
Wageningen University and Research, Laboratory of Virology, Droevendaalsesteeg 1, 6708 PB Wageningen, the Netherlands; monique.vanoers@wur.nl

Mass production of insects is a core activity to address food security and the global demand for animal proteins, one of the top societal and environmental priorities worldwide. Mass-reared insects are also crucial for other applications, including pollination, fishery, biocontrol, human and animal disease prevention, and waste management. Successful application of mass-reared insects heavily relies on culturing healthy insect colonies. However, insect pathogens easily emerge in insect mass rearing facilities, often leading to colony collapses. To be sustainable and cost-effective, insect production urgently needs to become more resilient to a variety of insect pathogens. The major drawback is though, that pathogens that threaten mass reared insects have hardly been studied. This means that the knowledge needed to prevent infectious diseases in insect rearing facilities is almost non-existent and specific diagnostic tools are not available, leaving the industry at serious risk for re-occurring outbreaks with large economic costs. To overcome this knowledge gap and to increase the number of researchers with the relevant scientific background to support the insect mass rearing industry in managing insect pathogens, the INSECT DOCTORS programme was initiated. This European Joint Doctorate (EJD) programme started in 2019 and aims to deliver researchers that are well-educated in the breadth of disciplines within insect pathology. Nine European universities, five research institutes, and four insect rearing companies contribute to the training and research programme of INSECT DOCTORS. The fifteen research projects within the programme serve to: (1) increase our understanding on pathogens that (may) affect the production of diverse mass reared insects; (2) develop tools and techniques for the diagnosis and management of infectious diseases in large scale insect rearing facilities; and (3) make insect cultures more resilient to infectious disease. In this special session, the introduction to the programme will be followed by presentations, in which seven of the PhD candidates active in the programme will present the aim of their projects and highlight the research data obtained thus far.

Nanopore-based diagnostics and surveillance of pathogens in mass-reared insects

F.S. Lim[1,2], J. González Cabrera[2], R.G. Kleespies[1], J.A. Jehle[1] and J.T. Wennmann[1]
[1]Julius Kühn-Institut, Heinrichstraße 243, 64287, Darmstadt, Germany, [2]Universitat de València, Department of Genetics and Institut Universitari de Biotecnología i Biomedicina (BIOTECMED), Dr Moliner 50, 46100, Burjassot, Spain; fang-shiang.lim@julius-kuehn.de

Rapid and reliable detection of pathogens is crucial to complement the growing industry of mass-reared insects, in order to safeguard the insect colonies from outbreak of diseases causing significant economic loss. Current diagnostic methods are mainly based on conventional PCR and microscopic examination, which require prior knowledge of disease symptoms and are limited to detecting known pathogens. Here we evaluate the feasibility of portable nanopore based sequencing technology for the detection of insect pathogens in selected diseased insects, such as *Acheta domesticus*. Metagenomic sequencing approaches allow the reveal of organisms of all domains in an insect sample, which may reveal unsuspected and untargeted pathogens. Taking into account the portability of the device, Oxford Nanopore Technologies MinION device was coupled to a Nvidia Jetson AGX Xavier Developer Kit to allow accelerated real-time base calling and sequence analysis. Taxonomic assignment of sequencing reads from diseased *A. domesticus* revealed the presence of viral sequences of *A. domesticus* densovirus (AdDNV), a well-known member of the *Parvoviridae* virus family that causes widespread mortality on cricket populations. *De novo* assembly of these sequences yielded a single virus contig for each of the sequenced individuals. The completeness and quality of the assembled viral genomes was checked with CheckV software, with the coding sequences all found to be present. Also, AdDNV sequences were detected near the beginning of each sequencing run, demonstrating the capability of this methodology to obtain results rapidly. These findings demonstrated the potential of nanopore-based metagenomic sequencing as a powerful addition to the diagnostic toolkit for routine pathogen surveillance and diagnosis in the insect rearing industry.

Probiotics impact on *Tenebrio molitor* performance, microbial composition and pathogen infection

C. Savio[1,2], P. Herren[3,4,5], A. Rejasse[2], A. Bruun Jensen[5], J.J.A. Van Loon[1] and C. Nielsen-Leroux[2]
[1]Wageningen University, Laboratory of Entomology, Droevendaalsesteeg 1, 6708 PB, Wageningen, the Netherlands, [2]INRAE, Micalis, 4 avenue Jean Jaures, 78350, Jouy en Josas, France, [3]University of Leeds, School of Biology, Woodhouse, Leeds LS2 9JT, United Kingdom, [4]UK Centre for Ecology & Hydrology, Maclean Building, Benson Lane, OX10 8BB, Wallingford Oxfordshire, United Kingdom, [5]University of Copenhagen, Department of Plant and Environmental Science, Thorvaldsensvej 40, 1871, Frederiksberg C, Denmark; carlotta.savio@inrae.fr

The yellow mealworm *Tenebrio molitor* is an insect model for infection and immunity studies and is mass-produced as feed and food. The industrial rearing of *T. molitor* on agricultural by-products may expose larvae and adults to microbes applied as biocontrol agents, e.g. spores of *Bacillus* bacteria and fungal conidia that could impact the performance of *T. molitor*. Therefore, the project deals with experiments analysing different outcomes of single and co-infections of *Bacillus thuringiensis*, and the fungal pathogen *Metarhizium brunneum* on the larval stages of *T. molitor*. Furthermore, as for other animals, the possible benefits of addition of active and tantalized probiotic bacteria to the feed is investigated. The pathogenicity of *B. thuringiensis* serovar *tenebrionis* (*Btt*) and *Metarhizium brunneum* KVL 12-30 has first been tested by single infection of *T. molitor*. Then targeted co-infections were performed to determine additive, synergistic or antagonistic interactions between these pathogens. Alongside infections, growth rate and survival rate are recorded and insect microbiota composition was analysed by 16S rRNA sequencing to measure how probiotics and pathogens modify the microbial composition. The hypotheses are: (1) *M. brunneum* and *Btt* have different mechanisms of infection, therefore dose and timing of pathogen exposure will influence the outcome; (2) the presence of probiotics may help the insect to cope with the infection by improving immunity, by presenting a shorter period for pathogen clearance and by expressing higher performance. Preliminary results show evidence of positive effects of adding a vital probiotics strain to the feed on *T. molitor* performance and survival highlighting the role of probiotic metabolites and microbiota relationship for maintaining host health.

Impact of heat stress on immunity, fitness & susceptibility to a fungal pathogen in *Tenebrio molitor*

P. Herren[1,2,3], A.M. Dunn[2], N.V. Meyling[1] and H. Hesketh[3]
[1]University of Copenhagen, Department of Plant and Environmental Sciences, Thorvaldsensvej 40, 1871 Frederiksberg, Denmark, [2]University of Leeds, Faculty of Biological Sciences, LS2 9JT Leeds, United Kingdom, [3]UK Centre for Ecology & Hydrology, Maclean Building, Benson Lane, Crowmarsh Gifford, OX10 8BB Wallingford, United Kingdom; pasher@ceh.ac.uk

The mass-rearing of insects for food and feed is a growing sector because insects are a promising sustainable source of proteins. The yellow mealworm, *Tenebrio molitor*, is one of the most important species produced for these purposes. Pathogens spreading in insect populations reared at high densities can cause devastating losses in commercial rearing facilities. In addition to biotic stressors, mass-reared insects are exposed to abiotic stressors, including temperature, which is one of the most important environmental factors determining insect growth and metabolic rate. Mass-reared insects can be exposed to heat stress in the course of the production process during transport or due to metabolic heat production. Moreover, exposing insects to temperature treatments has been suggested to increase their immunity and make them less susceptible to pathogens. However, this has not been tested in *T. molitor* and it remains unclear for how long beneficial effects of temperature stress on immunity persist. Moreover, long-term effects of temperature stress on fitness and development must be assessed to ensure that it has no negative impacts. We exposed *T. molitor* larvae to either short (2 h) or long (14 h) periods of heat stress. Thereafter, we measured: (1) the susceptibility of the larvae when exposed to a fungal pathogen (*Metarhizium brunneum*) immediately after, or five days after the heat stress; (2) immune responses of larvae exposed to *M. brunneum* and heat stress; and (3) fitness and development of heat stressed larvae. Our results show that a short heat stress has a beneficial effect on immunity and pathogen susceptibility of larvae immediately after the heat stress. However, these beneficial effects wane if larvae are exposed to the pathogen five days after the heat stress. Moreover, we found that heat stress can affect growth negatively. We discuss benefits and risks of heat stress to maintain insects that have an increased resistance to pathogens.

Effect of co-stressing on the viral abundance of the *Acheta domesticus* densovirus in house crickets

J. Takacs[1,2,3], V. Ros[2], J.J.A. Van Loon[1] and A. Bruun Jensen[3]
[1]Wageningen University and Research, Laboratory of Entomology, Droevendaalsesteeg 1, 6708 PB Wageningen, the Netherlands, [2]Wageningen University and Research, Laboratory of Virology, Droevendaalsesteeg 1, 6708 PB Wageningen, the Netherlands, [3]University of Copenhagen, Department of Plant and Environmental Sciences, Thorvaldsensvej 40, 1871 Copenhagen, Denmark; tjoci88@gmail.com

The house cricket *Acheta domesticus* is a promising candidate for food and feed purposes, but pathogens can cause epizootics in its rearing. Crickets are usually reared in high densities to maximize cost effectiveness, but this creates a highly favorable environment for disease outbreaks. The *A. domesticus densovirus* (AdDV) is one of the major pathogens that can cause serious losses in production of house crickets. The virus is present in most *A. domesticus* stocks in a covert state, but can become overt. Co-infection with other bacterial and fungal entomopathogens and abiotic stresses are suggested to be the triggers for transition to the overt state of AdDV. Rearing density is hypothesized to be an important factor for AdDV outbreaks, but the correlation between rearing density and AdDV titers has yet to be confirmed. To explore this we designed an experimental setup with three different rearing densities, combined with exposure to two different fungal pathogens. We recorded the effect of co-stressing on the mortality and weight gain of individuals and measured AdDV viral abundance on group level using qPCR. High rearing density reduced survival rate, but did not affect average individual weight. Results on the effect of the two stressors on AdDV load will be presented. Producers need to consider chances of higher mortality with higher densities, but the losses due to mortality might be outweighed by the higher weight output from the same surface area. An optimal rearing density leading to high productivity while keeping the risk of collapse of the production batch minimal needs to be established.

Expanding the medfly virome: viral diversity, prevalence, and sRNA profiling in medflies

L. Hernández-Pelegrín[1], A. Llopis-Giménez[1], C. Maria Crava[1], F. Ortego[2], P. Hernández-Crespo[2], V.I.D. Ros[3] and S. Herrero[1]

[1]University of Valencia, Biotechnological Control of Pest, Dr Moliner 50, 46100 Burjassot (Valencia), Spain, [2]Centro de Investigaciones Biológicas Margarita Salas, Department of Microbial Biotechnology and Plants, Ramiro de Maeztu, 9, 28040 Madrid, Spain, [3]Wageningen University, Laboratory of Virology, Droevendaalsesteeg, 6708 PB Wageningen, the Netherlands; luis.hernandez.pelegrin@gmail.com

The Mediterranean fruit fly (medfly) *Ceratitis capitata* is an agricultural pest of a wide range of fruits. Its control by the Sterile Insect Technique (SIT) relays in mass-rearing of sterile males, which are released in the field. Lately, the advent of high-throughput sequencing has boosted the discovery of RNA viruses infecting insects, including medfly, but their implications in mass-rearing warrant further investigation. To start to shed light to this question we characterized the RNA virome of the medfly and the response that viral presence elicits in the host. By means of transcriptome mining, we expanded the medfly RNA virome to 13 viruses, including two novel positive ssRNA viruses and the first two dsRNA viruses reported from the medfly. Our analysis across multiple laboratory-reared and field-collected medfly samples showed the presence of a core RNA virome comprised of two positive ssRNA viruses, and a higher viral diversity in field-collected flies compared to laboratory-reared flies. Based on small RNA sequencing, we detected small interfering RNAs mapping to all the viruses present in each sample, except for Ceratitis capitata nora virus. Although the identified RNA viruses do not cause obvious symptoms to medflies, the outcome of their interaction may still influence the medfly´s fitness and ecology, becoming either a risk for mass-rearing or an opportunity for SIT applications.

Identification and tissue tropism of a newly identified iflavirus and negevirus in tsetse flies

H.-I. Huditz[1,2], I.K. Meki[2], A. Strunov[3], R.A.A. Van Der Vlugt[1], H.M. Kariithi[2,4], M. Rezapanah[5], W.J. Miller[3], J.M. Vlak[1], M.M. Van Oers[1] and A.M.M. Abd-Alla[2]

[1]Wageningen University and Research, Laboratory of Virology, RADIX, building nr. 107 Droevendaalsesteeg 1, 6708 PB Wageningen, the Netherlands, [2]Joint FAO/IAEA Programme of Nuclear Techniques in Food and Agriculture, IAEA, Insect Pest Control Laboratory, International Atomic Energy Agency, Vienna International Centre, P.O. Box 100, 1400, Vienna, Austria, [3]Center for Anatomy and Cell Biology, Medical University of Vienna, Department Cell & Developmental Biology, Lab Genome Dynamics, Schwarzspanierstraße 17, 1090, Vienna, Austria, [4]U.S National Poultry Research Center, Agricultural Research Service, USDA-ARS, Southeast Poultry Research, 934 College Station Rd, Athens, GA 30605, USA, [5]Iranian Research Institute of Plant Protection (IRIPP), Agricultural Research Education and Extension Organization (AREEO), Rashid-Aldin-Fazlollah Blvd, District 1, Tehran 19853, Iran; hannah-isadora.huditz@wur.nl

Tsetse flies cause major health and economic problems, as they transmit trypanosomes causing sleeping sickness in humans (Human African Trypanosomosis) and nagana in animals (African Animal Trypanosomosis). One option to control the spread of these flies and their associated diseases is the Sterile Insect Technique (SIT). For successful application of SIT, it is important to establish and maintain healthy and competitive insect colonies. However, mass production of tsetse can be affected by covert virus infections. Recently, we demonstrated the presence of two new viruses, an iflavirus and a negevirus, in *Glossina morsitans morsitans*. These viruses both contain positive-sense, single-stranded RNA and are named GmmIV and GmmNegeV. We analysed the tissue tropism of these viruses by RNA-FISH to understand their mode of transmission. Our results demonstrate that both viruses can be found not only in the host's brain and fat body, but also in salivary and milk glands, and in their reproductive organs. These findings suggest potential horizontal viral transmission during feeding and vertical viral transmission from parent to offspring. Although the impact of GmmIV and GmmNegeV in tsetse rearing facilities is still under study, none of the currently infected flies show any signs of disease.

InsectFeed – insects as sustainable feed for a circular economy: value chain development

M. Dicke
Wageningen University, Entomology, Droevendaalsesteeg 1, 6700 AA Wageningen, the Netherlands; marcel.dicke@wur.nl

Current production of feed for livestock competes with food production for humans (e.g. cereals and soymeal) or relies on resources that threaten biodiversity (e.g. overfishing for fishmeal). For a novel circular and sustainable approach to feed production, insects provide excellent opportunities because various species can be reared on organic residual streams. Producing insects for feed provides animal proteins through a sustainable production process, with low-value input, high-value output and low environmental impact. The large-scale production of insects as 'mini-livestock' can yield animal proteins of high nutritional quality via conversion of organic waste streams. Moreover, insect production has lower land and water requirements and much lower greenhouse gas output compared to traditional sources of animal protein. Insect production can, therefore, become an important component of a circular economy, by closing nutrient and energy cycles, fostering food security while minimising climate change and biodiversity loss. In the development of this new sector, various fundamental and applied aspects require investigation, including welfare of insects and of livestock fed with insects, disease risk of insects, economic aspects and biology of the insects that are reared as 'mini-livestock'. The NWO-funded programme InsectFeed investigates: (1) insect production, especially focussing on insect health, insect welfare and the intrinsic value of insects; (2) health and welfare of livestock that is fed with insects; and (3) economic aspects of the value chain development, i.e. the development of an integrated set of activities to deliver a valuable product. To achieve our aim, we take an interdisciplinary research approach (ethics, economics, biology), and involve a wide range of stakeholders. We focus on fly larvae (black soldier fly and housefly) for poultry production.

Preference of black soldier fly larvae for feed substrate previously colonised by conspecific larvae

Y. Kortsmit, J.J.A. Van Loon and M. Dicke
Wageningen University and Research, Laboratory of Entomology, P.O. Box 16, 6700 AA Wageningen, the Netherlands; yvonne.kortsmit@wur.nl

The black soldier fly *Hermetia illucens* L. (Diptera: Stratiomyidae) is gaining wide use as alternative protein ingredient in livestock feed. With the growth of insect production, ethical debates about insect welfare are increasing. Insect welfare is currently rarely addressed within the mass-rearing industry due to the focus on parameters such as developmental time and yield. To address welfare, knowledge about species-specific behaviour is essential. A specific behaviour frequently observed within rearing crates is larval aggregation. This behaviour, also known as clustering, is common in larvae of various dipteran species. However, the sensory mechanisms guiding aggregation behaviour are unknown in black soldier fly larvae. In the context of the InsectFeed research programme, behavioural research was conducted to gain understanding of the initiation of aggregation behaviour within different larval stages. Cues attracting larvae in substrate colonised by conspecific larvae and the effect of age on attraction were examined via behavioural observations in dual-choice tests in the dark. Larval choices were identified based on first direct substrate contact, cumulative number of substrate contacts and entry of the substrate. The results of this study suggest the ability of black soldier fly larvae to discriminate between similar substrates with or without cues released by conspecifics.

Session 40 — Theatre 10

Two behaviours related to welfare in housefly (*Musca domestica*) production

M. Van Der Bruggen, E.I. Michail, J. Falcao Salles, B. Wertheim and L.W. Beukeboom
University of Groningen, Nijenborgh 7, 9747 AG Groningen, the Netherlands; m.van.der.bruggen@rug.nl

Fly larvae have large potential as a sustainable feed source for poultry and fish. This requires large-scale production of flies in mass-rearing facilities, which raises concerns about animal welfare. To make a welfare protocol for both adult and larval houseflies in mass-rearing systems, more knowledge is needed on the behaviour of this species. Two behaviours of interest are aggregation and the response to noxious thermal stimuli. Aggregation is an important larval behaviour in many fly species. Potential benefits of this behaviour are a decreased risk of predation, increased food intake by social digestion, and faster development due to increased temperatures. In mass-rearing systems aggregations might form, or they might be broken up due to periodical mixing of the substrate. We designed an experimental set-up to measure aggregation behaviour and to determine which factors influence aggregation in housefly larvae. This included whether housefly larvae are attracted to substrates previously visited by conspecifics, and the influence of substrate disturbance on aggregation formation and fitness. Specific behaviours to noxious stimuli are known from other fly species. These specific behaviours may be used to monitor animal welfare. We studied the behavioural response of larval and adult houseflies to a noxious thermal stimulus, to determine which specific behaviours occur. We discuss how our results can be used in a welfare protocol for mass-rearing practices.

Session 40 — Theatre 11

Innate immune system of house flies and black soldier flies

M. Vogel[1], P.N. Shah[2], A. Voulgari-Kokota[1], S. Maistrou[2], Y. Aartsma[2], L.W. Beukeboom[1], J. Falcao Salles[1], J.J.A. Van Loon[2], M. Dicke[2] and B. Wertheim[1]
[1]Groningen University, Groningen Institute for Evolutionary Life Sciences (GELIFES), Nijenborgh 7, 9747 AG Groningen, the Netherlands, [2]Wageningen University, Laboratory of Entomology, Droevendaalsesteeg 1, 6708 PB Wageningen, the Netherlands; m.vogel@rug.nl

Housefly (*Musca domestica*) and Black Soldier Fly (*Hermetia illucens*) are receiving considerable attention as sustainable protein source for (poultry) feed, due to their ability to be reared on residual streams from the food industry. Keeping insects healthy is important for setting up a robust mass rearing system. This requires a good understanding of their immune system. In contrast to mammals, insects only possess an innate immune system, consisting of a complex web of immune pathways. Although the immune pathways themselves are well conserved between insect species, the activators and effectors of these pathways are very diverse. The innate immune system of various dipteran species, including *Drosophila melanogaster* and several mosquito species have been well characterized. We will present a synthesis of the literature on species-specific innate immune system of the housefly and black soldier fly, and make a comparison with other dipteran species for which more knowledge is available. This comparative study reveals that housefly and black soldier fly have undergone expansions of particular classes of immune genes. We will outline what challenges should be considered when comparing and extrapolating research from one insect species to another. We will also highlight some inconsistencies and contradictions in the findings for housefly and black soldier fly. Finally, we argue that species-specific research on the immune system of these two production insects is important, to promote their health in mass rearing facilities.

The effect of entomopathogenic *Bacillus thuringiensis* on housefly development

A. Voulgari Kokota, R. Slijfer, L.W. Beukeboom, J.F. Salles and B. Wertheim
University of Groningen, Groningen Institute of Evolutionary Life Sciences, Nijenborgh 7, 9747 AG Groningen, the Netherlands; a.voulgari.kokota@rug.nl

The mass rearing of insects to satisfy the protein requirements of livestock is a fast-growing sector and the ability of housefly larvae to grow on many different substrates makes *Musca domestica* a good candidate species for this practice. Although large scale culture of houseflies is highly promising for producing high quality feed for livestock, industrialization of housefly rearing has numerous challenges. One of those challenges is to safeguard insect health against potentially pathogenic microbial agents. To test the impact of such a microbe on the rearing of housefly larvae, we used a sublethal entomopathogenic *Bacillus thuringiensis* strain and monitored the effects of its introduction into the rearing substrate on housefly health. We observed acute effects on larval biomass and pupation rates, depending on the concentration of the introduced pathogen, as well as changes in larval behaviour. We quantified the expression of genes coding for antimicrobial peptides (AMPs) and other stress-response genes and observed changes in expression, albeit with high fluctuations, among infected larvae. We discuss how gene expression monitoring could serve as a valuable diagnostic tool to monitor insect health in mass rearing systems.

Effect of feeding black soldier fly products on performance and health of slow-growing broilers

A. Dorper[1], H. Berman[1], G. Gort[1], M. Dicke[1] and T. Veldkamp[2]
[1]Wageningen University & Research, Droevendaalsesteeg 1, 6700 AA Wageningen, the Netherlands, [2]Wageningen University & Research, De Elst 1, 6708 WD Wageningen, the Netherlands; marcel.dicke@wur.nl

Insects have gained increasing popularity as replacement for fishmeal, soybean meal or oil in poultry diets. However, rather than being a novelty, insects celebrate a comeback in poultry diets. Already the ancestors of our modern chickens, living in the forests, consumed them to meet their dietary needs. Unlike past times, today's industry offers a bandwidth of products such as live insects, insect meal and oil. In this context, black soldier fly larvae (BSFL) products have been suggested as promising feed ingredients. Moreover, components of BSFL such as, lauric acid, antimicrobial peptides, and chitin are assumed to improve poultry health. However, their structure and abundance might be altered by processing steps and positive effects might be related to dietary inclusion levels. The present trial investigated the effect of BSFL products and inclusion levels in boiler diets on performance and health. At arrival, 1,728 1-day-old slow growing broilers (Hubbard JA757) were randomly assigned to 9 treatment groups with 8 replicates. A commercial broiler diet was used as the control group. In the second group, 5% of the dry matter feed intake (DMFI) was replaced by live BSFL. In the third group, 5% DMFI was replaced by a mix of 2/3 BSFL meal and 1/3 BSFL oil, to mimic the nutritional composition of live BSFL. The fourth and fifth group received the same quantity of BSFL meal or oil as group three, respectively. The same pattern of treatments was repeated in group six to nine with a replacement of 10% DMFI by live BSFL as a baseline. All dietary treatments were iso-caloric and formulated on digestible essential amino acids. Until d 49 the broiler performance and health was quantified by several parameters. The results support the adoption of utilizing insects as feed from the forest to the feed trough. We will present the data of this trial evaluating the effects of BSFL products on broiler performance and health.

Economic modelling of Insect-fed broiler value chains: processed quantities and feasibility analysis

M. Leipertz, H.W. Saatkamp and H. Hogeveen
Wageningen University and Research, Business Economics, Hollandseweg 1, 6706 KN Wageningen, the Netherlands; mark.leipertz@wur.nl

Insect-fed broiler value chains offer opportunities regarding sustainability, but to be economically feasible, the value chains need to be balanced with equal profits for all actors. In this research, the feasibility of the value chains is analysed by a novel value chain model consisting of three main components: A Technical module simulates the transition of quantities and qualities between the stages of the value chain, Economic modules analyse the profitability for each value chain stage, and In between stage cost modules simulate the storage and transportation between different stages as well as the cooperation/integration within the chain. The model input parameters were collected through expertise and experiments within the INSECTFEED consortium. Our provisional results show that one medium sized insect rearing company (50,000 t substrate handling capacity) is capable to provide enough black soldier fly (BSF) larvae to produce 24.91 kt broiler life weight at a 10% alive larvae inclusion level in broiler feed, within a one star better life (BLS) certified broiler system. The total Dutch BLS broiler production (estimated to be 785 kt in 2025) could be served by 32 insect rearing companies. These insect rearing companies would require a total amount of 1,615 kt raw substrate, which resembles only 0.1% of the Dutch food waste in 2019. Thus, procuring could be fairly possible, but legislation and quality demands are of course hurdles. Although including insects in broiler feed improves the conversion rate of broilers, current high costs of alive larvae (2,017€/ t DM) lead to an increase of 29% in broiler life weight production costs compared to a farming system without the use of insects (production costs being 1.34€/kg compared to 1.04€/kg). Feeding defatted BSF larvae meal even further increases the production costs of broiler life weight to 1.48€/kg. Respecting all changes of production parameters, the break even costs of alive larvae are 419€/t DM, assuming a soybean meal price of 406€/t DM. Thus, further innovations and smart business models are required for lower costs and economic feasibility of the value chain.

Effects of diet on the expression of immune genes in the house fly (*Musca domestica*)

M. Vogel[1], F. Boatta[2], A. Voulgari-Kokota[1], R. Slijfer[1], L.W. Beukeboom[1], J. Falcao Salles[1], J. Ellers[2] and B. Wertheim[1]
[1]University of Groningen, Groningen Institute for Evolutionary Life Sciences (GELIFES), Nijenborgh 7, 9747 AG Groningen, the Netherlands, [2]Vrije Universiteit Amsterdam, Amsterdam Institute for Life and Environment, De Boelelaan 1085, 1081 HV Amsterdam, the Netherlands; m.vogel@rug.nl

Insect meal is considered a sustainable alternative to soy or fishmeal as feed for agricultural stock, especially because some insects can be reared on waste from the food industry. However, the nutritional composition of the industrial food waste can vary substantially. Therefore, it is important to understand the effects various food wastes as insect rearing substrate can have on the health of production insects. The innate immunity of insects is known to be influenced by their diet, with both high protein content and high sugar content linked to upregulation of immune genes in *Drosophila melanogaster*. Expression of antimicrobial peptides is significantly affected by diet in black soldier flies (*Hermetia illucens*), even in the absence of an infection. The housefly is of particular interest for rearing on waste streams, as this fly is well-adapted to living on organic waste. Species-specific literature on the effects of diet on the immunity of houseflies is however still lacking. We studied the effects on the immune response of housefly larvae when reared on three different diets (sugar rich, fat rich and control) for up to 12 generations, through qPCR on a panel of genes coding for antimicrobial peptides. We will present the (preliminary) results of these experiments, to give insight into which diets are preferable when optimizing health of houseflies in mass rearing facilities and the adaptation of the immune system of housefly larvae to their diet.

Evaluation of the physical activity of a group of gestating sows using an artificial neural network

M. Durand[1], M. Simon[2], J. Foisil[2], J.Y. Dourmad[1], C. Largouët[3] and C. Gaillard[1]
[1]PEGASE, INRAE, Institut Agro, Le Clos, 35590 St-Gilles, France, [2]DILEPIX, 3 avenue G. Tillion, 35136 St-Jacques-de-la-Lande, France, [3]Institut Agro Rennes-Angers, Univ Rennes, Inria, IRISA, 35000, Rennes, France; maeva.durand@inrae.fr

Physical activity influences the energy requirements of group-housed gestating sows, and changes in their activity or behaviour patterns may be signs of welfare or health disorders. Ear tag accelerometers usually used for assessing the activity are fragile, costly, and invasive sensors. Instead, cameras can record videos of the group of sows but require countless hours to manually analyse the different activities of the sows. In the present work, the performances of a deep-learning algorithm developed to automatically detect the different activities of the gestating sows on images are evaluated. Two groups of 18 sows, housed in two pens, were included in the experiment and followed during two consecutive gestations. Two cameras recorded each pen continuously. Six activities (lying ventrally and laterally, sitting, standing, eating and drinking) were manually annotated by animal behaviour experts, on the 1,331 images extracted from the videos. This annotated set of images was used to train the algorithm, an object detection model that uses convolutional neural networks to detect and classify objects in an image. Another set of 403 images was used to validate the performance of the algorithm which proved to be reliable. The classification accuracy of sows lying ventrally and eating were respectively 82% and 87%. The lowest accuracies were generated by sitting (47%) and drinking (53%) activities, probably partly due to the lack of images including these activities in the training dataset. Indeed, sitting represented 3% of the activities labelled on the training dataset, and drinking only 1%. On a daily basis, the sows spent 75% of their time lying laterally (53%) or ventrally (22%). They were more active (i.e. standing) between 00:30 and 09:00, due to the start of the new feeding day at 00:00. Variation in the physical activity appeared between the two sows' groups and the different gestation weeks. To improve this algorithm, new data will be collected and a tracking module will be integrated to be able to detect the walking activity and also to work at the individual level.

Automatic classification of agonistic behaviour in groups of pigs – different modelling approaches

F. Hakansson and D. Børge Jensen
Copenhagen University, Department of Large Animal Sciences, Grønnegårdsvej 2, 1870 Frederiksberg C, Denmark; fh@sund.ku.dk

The majority of commercial pigs in the EU are raised under intensive conditions that are likely to increase the development of agonistic behaviour. Because of its association with stress, pain and an increased risk of infection of inflicted wounds, agonistic behaviour is a major welfare and economic challenge in pig production. Timely detection of the behaviour within groups of pigs could aid farmers in their decision making on prevention and intervention strategies. In the Code Re-Farm project (Horizon 2020, Proposal no. 101000216) we are developing deep learning-based machine vision models to detect and classify animal behaviour based on video images. We expect that these methods will be broadly applicable to video data on various different types of behaviours of various different species of animals, including detecting agonistic behaviour in groups of pigs based on video recordings from above the pen. The long-term goal for automatic detection of agonistic pig behaviour is to be able to provide this information to the farmer to aid on-farm decision-making. To detect and classify these behaviours, we will use a two-step approach. The first step will use a pre-trained convolutional neural network (CNN) to transform individual video images into latent space representations. The second step will use a secondary model to classify the behaviour seen in the video images. As the input data are visually complex with added temporal features, we intend to implement and compare three different modelling approaches for handling the temporal features in the second step. These three methods will be: (1) a fully connected artificial neural network (FC-ANN); (2) a long-short term memory (LSTM) neural network; and (3) a novel approach for combining feature vectors into a time-series representation. We will systematically optimize each of the secondary models using a subsample of the data. The optimized models will subsequently be applied on data from 25 different pens, thus ensuring a fair cross-validation based comparison between our novel method and the two well-established methods.

Early detection of diarrhoea in weaned piglets from individual feed, water and weighing data

J. Thomas, Y. Rousselière, M. Marcon and A. Hémonic
IFIP-Institut du Porc, La Motte au Vicomte, 35650 Le Rheu, France; johan.thomas@ifip.asso.fr

This study, part of HealthyLivestock project, aims to analyse individual water and feed consumption related to weight of weaned piglets and their link to diarrhoea. Data were collected from 15 batches of 102 piglets using specific automata (connected feeders and drinkers, automatic weighing scales, RFID ear tags) and observations of piglets' health status made by specialized technicians. Analyses were carried out every week on the 138 healthy animals compared by weight category (light, medium, heavy). The average feed consumption had no significant difference between categories whatever the week and was close to 4% of the live weight. For the average water consumption according to weight, it was close to 10% and there was no significant difference between groups. However, at the end the water consumption of one heavy pig gave significant difference with the light group. These overall stable averages in healthy pigs, were promising for distinguishing deviant behaviours of sick pigs. However, they shaded the high intra-individual variabilities, around 40% at the beginning of post-weaning for both feed and water consumption and almost 16% for feed and 25% for water at the end. Then, the comparison between healthy and diarrheic piglets showed no statistical difference for average water consumption on the day of the first clinical signs and even 1 and 2 days before. In contrast, the average feed consumption had a very significant difference ($P \leq 0.001$) for days 5-7 after the weaning and a significant difference for day 8 ($P \leq 0.05$). These differences were also significant for data collected 24-48h before first clinical signs. This means either those diarrheic piglets decreased their feed consumption the first days after weaning or that it is because they eat less when they become diarrheic. As feed consumption seemed to be an interesting indicator to detect early diarrheic weaned animals, we continued the study by using machine learning methods. But they all failed in detecting individually diarrheic animals from water and feed consumption related to weight, probably because of considerable individual variability of feed consumption, even in healthy pigs. To improve these results, new data are collected, like location of piglets in the pen by image analysis.

Effect of an automatic enrichment device on the behaviour of weaner pigs

P. Heseker[1,2], J. Probst[2], S. Ammer[1], I. Traulsen[1] and N. Kemper[2]
[1]Georg-August-University Göttingen, Department of Animal Sciences, Livestock Systems, Albrecht-Thaer-Weg 3, 37075 Göttingen, Germany, [2]University of Veterinary Medicine Hannover, Foundation, Institute for Animal Hygiene, Animal Welfare and Farm Animal Behaviour, Bischofsholer Damm 15, 30173 Hannover, Germany; philipp.heseker@agr.uni-goettingen.de

Providing pigs with organic enrichment material has become a proven method to satisfy natural explorative behaviour, to prevent tail biting and to increase animal welfare in general. To enable the daily supply of adequate amounts of organic material in regular interval, an automatic enrichment device could represent a promising measure. The aim of this study was to investigate how the behaviour of weaner pigs changed before and after the provision of material by an automatic enrichment device. The first batch included one compartment with weaner pigs (n=120) with undocked tails housed in five pens (2.6×5.5 m). They were offered alfalfa pellets on a plastic mat (0.6×1.2 m) four times daily (40 g/day/pig) by an automatic enrichment device. Behaviour analysis was performed by evaluating video recordings from 5 minutes before until 15 minutes after the device released, one day per week over six weeks. The results showed that the attractiveness of alfalfa pellets increased with rising age. In week 1, on average 28.3% of the animals used the organic material one minute after it was released, versus 34.1% in week 6. However, the exploration duration was longer in week 1, with 9.7% of all animals still actively exploring the alfalfa after 15 min, versus 5.7% in week 6. These results showed the potential of an automatic enrichment device of enabling pigs to perform their natural explorative behaviour in a conventional housing system. The benefit for reducing tail biting outbreaks will be analysed in the next steps. The study is part of the project DigiSchwein, which is supported by funds of the Federal Ministry of Food and Agriculture (BMEL) based on a decision of the Parliament of the Federal Republic of Germany. The Federal Office for Agriculture and Food (BLE) provides coordinating support for digitalisation in agriculture as funding organisation, grant number 28DE109E18.

Digital drug registration and its value for herd management of pigs

H. Görge[1], I. Dittrich[1], N. Kemper[2] and J. Krieter[1]
[1]Kiel University, Institute of Animal Breeding and Husbandry, Olshausenstr. 40, 24098 Kiel, Germany, [2]University of Veterinary Medicine Hannover, Foundation, Institute of Animal Hygiene, Animal Welfare and Farm Animal Behavior, Bischofsholer Damm 15, 30173 Hannover, Germany; hgoerge@tierzucht.uni-kiel.de

Digitalising treatments provides the possibility to review the status of disease prevalences or rather therapeutic prevalences on pig farms and simplifies the legally required documentation. Such data enable the farmer to initiate measures and optimise health management. The aim of the study was to implement a digital tool on a pig farm, collecting treatment data at animal level and analysing the status quo of sickness occurrences on farm. The tool combines a self-filling syringe, a radio frequency identification-reader and an application, which is used in the stables whilst medicating individual animals. The data is stored on a server and transferrable to farm managing systems. Data was collected between August 2020 and December 2021 on a combined farm where the replacement is done with self-produced gilts. During this period 147 active sows were counted each day with 2.3 litters per year, 39 piglets born alive and an average of 30.6 weaned piglets per sow per year. All sows were analysed towards the most frequent disease patterns: mastitis-metritis-agalactia (MMA) and lameness. Overall, 181 sows were included for MMA and grouped into seven farrowing groups. The mean prevalence of MMA was 19% (±15%), with the largest prevalence at 55% in the farrowing groups. From 187 sows, 59 lameness treatments were registered, which leads to a prevalence of 12% (±10%). From 5,613 suckling piglets the most frequent disease patterns that required treatment were gastro-intestinal diseases (2%) and lameness (1%), while for the weaned piglets (n=5,328) respiratory diseases (7%) and lameness (3%) were most frequently diagnosed. For the fatteners (n=4,786) lameness (4%) and tail biting lesions (2%) were observed most often. Overall, the tool is useful for digitalising treatments and provides good data quality, which is supportive for the farm's health management at animal and herd level. Future steps are to connect individual treatments with slaughterhouse data and production data of the individual animal using an interface to the sow planner.

Heart rate of sows obtained by image photoplethysmography under farm conditions

W. Kuiken[1], P.P.J. Van Der Tol[1], A.J. Scaillierez[2], I.J.M.M. Boumans[2], H. Broers[3] and E.A.M. Bokkers[2]
[1]Wageningen University & Research, Farm Technology Group, Droevendaalsesteeg 1, 6708 PB Wageningen, the Netherlands, [2]Wageningen University & Research, Animal Production Systems Group, De Elst 1, 6708 WD Wageningen, the Netherlands, [3]Signify, High Tech Campus 48, 5656 AE Eindhoven, the Netherlands; rik.vandertol@wur.nl

The heart rate (HR) and respiratory rate (RR) of a sow provides valuable information of the physiological status, subsequently stress might affect cardiac activity. Image PhotoPlethysmoGraphy (IPPG) is available that non-invasively measures the HR of sows, hence not interfering with their behaviour. A pilot study was conducted to test two light conditions (fluorescent and fluorescent-halogen combination) and two methods of IPPG (green and green-red difference) aiming at the abdomen of resting sows to determine HR and RR. The experiment consisted of video recordings of three individually housed resting pigs. A standard RGB-camera recording at 50 fps was positioned 1.7 m above the centre of the pen. Raw video data was stored on an external drive. Simultaneously, reference HR was measured using belt-type heart rate monitor around the chest (BioHarness 3, Zephyr Technology). Signals were not synchronically recorded, hence reference HR and skin colour changes were aligned using cross-correlation. Recorded lag times are the sum of Pulse Transit Time (<400 ms) and the difference in timing between recording methods. The HR was subtracted from 1-min videos of the abdominal skin using a region of interest optimized for epidermal thickness and view consistency. Data was filtered (FIR Butterworth bandpass; 1.25-2.25 Hz) for motion artefacts and environmental interference. The baseline wander was removed using a 0.05 Hz Notch filter. HR and RR were assessed by the root-mean-square-error (RMSE) per minute. Recorded lag times (range: 0.06-1.3s) showed Pearson correlations >0.99. The RMSE of HR of both IPPG methods were close to similar for fluorescent (4.25 beat/min) and halogen light (4.21 beat/min). In addition the RR was determined accurately (RMSE 0.6 breaths/min). In conclusion, IPPG is a feasible method to obtain HR and RR in practise given the two light conditions. The next steps are to estimate HR variability based on IPPG and to perform HR measures in group housed pigs.

Session 41 Theatre 7

Smart textiles biotechnology for heart rate variability monitoring as welfare indicator in sheep

L. Turini[1], F. Bonelli[2], A. Lanatà[3], V. Vitale[2], I. Nocera[2], M. Sgorbini[2] and M. Mele[1]
[1]Univerity of Pisa, Department of Agriculture, Food and Environment, Via del Borghetto 80, 56124 Pisa, Italy, [2]Univerity of Pisa, Department of Veterinary Sciences, Via Livornese, 56122 Pisa, Italy, [3]University of Florence, Department of Information Engineering, Piazza San Marco 4, 50121 Firenze, Italy; luca.turini@phd.unipi.it

Animal-based measures (ABMs) are essential for assessing animal welfare but collecting ABMs in semi-extensive sheep farming systems is challenging. Heart rate variability (HRV), a sensitive indicator of the functional regulatory characteristics of the autonomic nervous system, is a non-invasive method for determining stress levels in animals. HRV has been used to measure animal welfare in dairy cows, but sheep research is limited. Furthermore, measuring HRV in small ruminants at pasture is crucial due to the unavailability of a solution that can be used in the field. The aim of this study was to examine if a smart textiles biotechnology could be compared to a Standard base-apex ECG for measuring HRV in small ruminants. Eight healthy adult crossbreed sheep were recruited. Standard base-apex ECG and Smartex ECG were simultaneously acquired for 5 minutes in the standing, unsedated, unclipped sheep. The ECG tracings were taken while the animals stood still. After parameter extraction, many tests in time, frequency, and nonlinear approaches were used to compare Smartex to traditional base-apex ECG systems. The Bland-Altman test was used to assess the level of agreement between the two instruments, and a linear regression analysis was used to assess the association between them. The smart textiles biotechnology is easy to wear and clean with an elastic belt. It may be worn without the need of adhesive or shaving the sheep's fleece, which reduced animal handling and stress. The Bland Altman test revealed some interesting results in terms of agreement between the two systems and the regression analysis of HRV parameters showed that some parameters had an R^2 coefficient more than 0.75, indicating the most stable relationship between the two systems. Smart textiles biotechnology can be employed for HRV measurement in sheep species, and it might be used as a possible ABM for animal welfare assessment both in the barn and on the pasture.

Session 41 Theatre 8

Early and automated detection of BRD disease in young bulls using activity and rumen temperature

M. Guiadeur[1], C. Allain[1], S. Assie[2], D. Concordet[3], L.A. Merle[4], B. Mounaix[1], M. Chassan[3] and J.M. Gautier[1]
[1]IDELE, 149 Rue de Bercy, 75012 PARIS, France, [2]ONIRIS, 101 Rte de Gachet, 44300 Nantes, France, [3]ENVT, 23 Chemin Des Capelles, 31300 Toulouse, France, [4]Ferme Experimentale des Etablieres, Barrage Du Moulin Papon, 85000 La Roche Sur Yon, France; jean-marc.gautier@idele.fr

Bovine respiratory disease (BRD) is a major health issue in fattening young bulls (TIMSIT, 2011). Previous studies showed that visual detection of BRD performed by farmer are not accurate whereas accurate treatment is the key for its efficiency. The BeefSense project aims to develop an innovative automated early detection system of BRD, involving a multi-sensor approach associated with the development of detection algorithms. One hundred and four 243-day-old Charolais cattle (332 kg in average) were equipped with pedometers, collars and intraruminal thermometers (boluses) and followed during the first month of fattening in the experimental farm of Etablières (Vendée, France) in 2019 and 2020. Throughout the study period, a veterinarian performed daily visual examination of clinical signs of BRD and weekly rectal temperature measurement. A clinical score was defined in order to differentiate between affected and healthy animals. Approximately 160 signals describing the behaviour and intraruminal temperature of individual animals were derived from the sensor signals. A mixed-effects logistic regression model was developed to predict the occurrence of clinical signs from the combination of signals. The model minimising $(1\text{-}Se)^2+(1\text{-}Sp)^2$ was selected from several models using a simulated annealing algorithm (Se and Sp are the sensitivity and specificity of detection respectively). The prediction performance of this model was corrected for overfitting using a cross-validation process (Leave One Out). The model using only the signals given by the collar and pedometer led to a detection of BRD 24h before the onset of clinical signs with Se=72% and Sp=71%. This performance increased to Se=74% and Sp=74% when the ruminal temperature signal was added. This study shows that early detection of BRD can be reliably achieved using sensors fitted on animals. Further studies should evaluate the interest of using this algorithm as a decision tool for farmer in order to choose the best treatment strategies.

Session 41 — Theatre 9

Automatic behaviour assessment of young bulls in pen using machine vision technology

C. Mindus[1], J. Manceau[2], V. Gauthier[2], C. Dugué[3], L.A. Merle[4], X. Boivin[5], and A. Cheype[1]
[1]Institut de l'élevage, 149 rue de Bercy, 75595 Paris, France, [2]NeoTec-Vision, 7 allée de la Planche Fagline, 35740 Pacé, France, [3]France Limousin Sélection, Pôle de Lanaud, 87220 Boisseuil, France, [4]Ferme des Etablières, route du Moulin-Papon, 85000 La Roche-sur-Yon, France, [5]Université Clermont Auvergne, INRAE, VetAgro Sup, UMR 1213 Herbivores, 63122 Saint-Genès Champanelle, France; claire.mindus@idele.fr

Changes in animal's behaviour may be good indicators of health and welfare variations. However, human observation is time-consuming and labour-intensive. Development of video technology and image processing represents a complementary tool to human observation. This non-invasive method may offer the opportunity for a better prevention by detecting behavioural and welfare issues continuously and automatically and therefore at an early stage. We are developing a tool using video recordings and 'deep learning' algorithms to analyse routinely the behaviour of young bulls reared in a fattening station and a qualifying station. Bulls originating from two different breeds (Limousine, 6 bulls/pen; Charolais, 13±1 bulls/pen) were housed accordingly to the standard management conditions of their respective stations (Pôle de Lanaud, Ferme des Etablières). Two cameras (2 D colour) were installed above each pen with different angular views. Video recordings were carried out every third day during daylight hours for the whole fattening and qualifying period. To construct an image and video database, video sequences are currently being annotated by a trained observer using an established ethogram and submitted to an algorithm of deep learning. Both posture (standing, lying) and behaviours (eating, moving, etc.) are considered. Preliminary training of the algorithm with 419 standing bulls and 373 lying bulls' pictures are promising with 88% sensitivity and 79% precision. According to current validation standards, the objective is to collect 1000 labelled pictures for postures and 250 sequences for behaviours (75% for training, 25% for validation) in several light conditions. This project will contribute to the current need for on-farm, operational behavioural welfare indicators that can be easily used to assess not only the individual welfare but also the welfare of the whole group.

Session 41 — Theatre 10

Automatic video-based classification piling behaviour in poultry

D.B. Jensen[1], M. Toscano[2], J. Winter[2], A. Stratmann[2], E. Van Der Heide[1], M. Grønvig[1] and F. Hakansson[1]
[1]University of Copenhagen, Department of Large Animal Sciences, Faculty of Health and Medical Sciences, Grønnegårdsvej 2, 1870 Frederiksberg C, Denmark, [2]University of Bern, Center for Proper Housing: Poultry and Rabbits (ZTHZ), Division of Animal Welfare, VPH Institute, Burgerweg 22, 3052 Zollikofen, Switzerland; daj@sund.ku.dk

Piling is an unusual behaviour in flocks of laying hens, where individuals suddenly aggregate in dense clusters. The behaviour occurs especially in loose group-housing systems harbouring a large number of individuals and can result in birds suffocating at the bottom of the pile. Piling is a major concern for egg producers because of its welfare implications and production consequences. In a collaboration between the Code Re-farm (Horizon 2020, Proposal no. 101000216) and a Swiss-based effort jointly funded by GalloSuisse and the Swiss Federal Food Safety and Veterinary Office [Grant number: 2.17.05], we are developing deep learning-based machine vision models to detect piling behaviour in group-housed poultry based on video data recorded at 12 commercial Swiss loose-housed flocks. The long-term goal with these models is to automatically detect the onset of piling behaviour. Such models may be used to induce a signal initiating a pile disruption before it reaches a hazardous size. To develop a piling detection model, we will use a two-step approach. First, a pre-trained convolutional neural network (CNN) will transform video images into latent space representations. The second step will use a secondary model to classify the behaviour. We will implement and compare three methods, namely: (1) a fully connected artificial neural network (FC-ANN); (2) a long-short term memory (LSTM) network; and (3) a novel approach for combining feature vectors into a time-series representation. We will systematically optimize each of the secondary models using data from one flock. The optimized models will then be applied to data from 11 different flocks ensuring fair cross-validation-based comparison of our novel method and the two well-established methods.

Using ultra-wideband backpacks to track broiler chickens in commercial housing

M. Baxter and N.E. O'Connell
Queen's University Belfast, 19 Chlorine Gardens, Belfast, BT9 5DL, United Kingdom; m.baxter@qub.ac.uk

Due to the size of commercial flocks, broiler chicken welfare is almost exclusively considered at the flock level. Very little is understood about how individual broilers use the space available and whether health and welfare parameters are influenced by individual characteristics. Here, to the best of our knowledge, we report the first use of ultra-wideband (UWB) technology to track individuals in a flock of indoor-reared commercial broilers. During initial testing, we established that broilers tolerated wearing UWB tags in backpacks well and typically resume normal behaviours within 24 hours. We also developed an algorithm to trim and refine the large amounts of data produced and improve our movement estimates. The aims of this project were to: (1) determine how far broilers travel around the space available; and (2) examine the association between movement and elements of their physiology, including body weight and gait score. For this purpose, a commercial broiler house, stocked with 28,000 birds per production cycle, was fitted with an UWB system. In one production cycle, 24 broilers (final n=17; seven tags removed) were tagged on either Day 21 or 24 until Day 38. The system proved to be an effective method of tracking individuals, with all tagged birds easily located for daily checks. Movement data revealed significant individual variation between tagged birds, both in terms of space use and the effect of various aspects of their physiology. When split into 100 equal zones, tagged broilers occupied between 97 and 21% of the house, with individuals recorded in more of the house not necessarily travelling the furthest distance. Start and final weights, gait score and age all had a mixed effect on the movement of tagged birds. We did see some evidence of location preference, with 82% of broilers spending most of their time in the same main area (front, middle or back) of the house that they were initially tagged in, although how close they stayed to this initial area was inconsistent. In conclusion, the individuals tagged demonstrated a wide range in their levels of space use and activity, with further investigation needed to explore the source of this variation.

Automatic detection of health risks in weaner pigs – practical approaches and limitations

J. Probst[1], M.-A. Lieboldt[2], G.-F. Thimm[3], P. Hölscher[3], P. Heseker[1] and N. Kemper[1]
[1]University of Veterinary Medicine Hannover, Foundation, Institute for Animal Hygiene, Animal Welfare and Farm Animal Behaviour, Bischofsholer Damm 15, 30173 Hannover, Germany, [2]Chamber of Agriculture Lower Saxony, Hermann-Ehlers-Straße 15, 26160 Bad Zwischenahn, Germany, [3]Thünen Institute of Agricultural Technology, Bundesallee 47, 38116 Braunschweig, Germany; Jeanette.Probst@tiho-hannover.de

Early diagnosis by sensor-based computer systems can improve animal health in weaner pigs, increase farm productivity and minimize the use of antibiotics. In this study, a combination of sensors was installed in six pens of one nursery compartment, housing 257 weaner piglets for 38 days per batch. Following sensors were used: climate sensors (for air temperature, humidity, NH_3, CO_2, light intensity), cameras and microphones (for activity, noises), drinking stations with RFID-antenna, waterflowmeter, optical weighing system and infrared camera (for water consumption, weight, body temperature). Sensor-based data was assessed continuously and compared to a daily health check on group level, and weekly clinical examinations and climate checks. The comparisons showed that the climate sensors generated reliable results. For video and sound analyses, the corresponding algorithm had to be adapted to detect deviations in movement behaviour and vocalisation. In the RFID stations, the sensors generated implausible values if the animals left too quickly or two animals tried to enter. Raw data of the infrared cameras was not directly useable, because a threshold value for 'suspected fever' had to be defined first. On farm, measuring errors occurred due to dirt, connectivity problems and wrong object presentation. These problems have to be solved, sensor data synchronicity has to be ensured and error variance should be known before combinations of sensors can support farmers in improving animal health by early detection of deviations. The study is part of the project DigiSchwein, which is supported by funds of the Federal Ministry of Food and Agriculture (BMEL) based on a decision of the Parliament of the Federal Republic of Germany. The Federal Office for Agriculture and Food (BLE) provides coordinating support for digitalisation in agriculture as funding organization, grant number 28DE109E18.

Relationship between packed cell volume and gastrointestinal nematodes in sheep through NIRS

A.C.S. Chagas[1], I.B. Santos[2], A.U.C. Ferreira[1], M.D. Rabelo[1], L.A. Anholeto[1] and S.N. Esteves[1]
[1]Embrapa Southeastern Livestock (CPPSE), Rod. Washington Luiz, Km 234, 13560-970, São Carlos, SP, Brazil, [2]Faculty of Agriculture and Veterinary Sciences, São Paulo State University (UNESP), Rod. Prof. Paulo Donato Castellane, 14884-900, Jaboticabal, SP, Brazil; carolina.chagas@embrapa.br

Haemonchus contortus is the most prevalent and important gastrointestinal nematode (GIN) in small ruminants. Since it reduces the packed cell volume (PCV), causing anaemia, early diagnosis can be used for targeted selective treatment (TST) of sheep, reducing antiparasitic use and anthelmintic resistance. This study aimed to predict PCV values through near-infrared reflectance spectroscopy (NIRS) and to develop a classification and diagnosis model of GIN infection using PCV values, eggs per gram of faeces (EPG) count and mean daily weight gain (DWG). A total of 1728 spectra were collected from blood samples of 216 lambs with a portable NIRS. In parallel, other parameters indicative of infection were measured: PCV by haematocrit, FAMACHA grade, EPG and DWG. To evaluate the relationship between NIRS spectra and the evaluated parameters, principal component analysis (PCA) was used for an exploratory analysis, regression by the partial least squares method (PLS) for the prediction of PCV values via NIRS, and linear discriminant analysis (PCA-LDA) as a classification model for diagnosis. The absorption peaks in the NIRS region associated with the excitation of overtones of NH functional groups of proteins had strong impact on the principal components (PCs), indicating that blood proteins, especially haemoglobin, can be estimated by the NIRS technique. The model for predicting PCV by PLS presented a standard error of prediction of 2.526%, root-mean-square error of 2.48% and coefficient of determination of 0.837, indicating good correlation between the PCV values predicted by the model and the PCV obtained by haematocrit. The PCA-LDA model presented 93.33% sensitivity and 82.18% accuracy, both higher than those of the FAMACHA method. The multivariate models associated with the NIRS technique reported here can be used in the future as a quick and versatile tool for GIN infection diagnosis and to apply TST in lambs.

Precision livestock farming to study relationship between feed intake and milk production in sheep

A. Ledda, S. Carta, L. Falchi, A.S. Atzori, A. Cesarani, F. Correddu, G. Battacone and N.P.P. Macciotta
University of Sassari, Dipartimento di Agraria, viale Italia 39, 07100 Sassari, Italy; anledda@uniss.it

Precision livestock farming (PLF) plays an important role in the technological innovation of farms. PLF involves innovation tools regarding feeding, management, and milking, aimed at improving the productivity and the economic sustainability of the livestock industry. PLF technologies are mostly used in dairy cow farms because their intensive management is particularly suitable for the implementation of innovative tools. However, dairy sheep farmers could also benefit from the introduction of PLF. The aim of this study was to present an application of PLF in dairy sheep farms gathering individual records on dry matter intake (DMI), milk yield (MY) and feed efficiency (FE, expressed as MY to DMI ratio) in dairy sheep. For this purpose, 24 Sarda ewes were kept for 30 days in a barn with individual automatic feeders and milking measurement devices. Animals were fed a total mix ration. Individual DMI was recorded daily, whereas milk yield was recorded two time a week (8 samplings) Animals were daily allocated into two classes according to the mean of MY, low (LMY) and high (HMY). DMI and FE of the day of the milk test, the day before, or the two days before were analysed by a mixed linear model that included date as covariable, the fixed effect of MY class, and the random effect of the animal. The correlation between MY, DMI, and FE were also calculated. The DMI of one day before the milk test was statistically different between the MY classes, showing that MY is more influenced by the intake of the day before than others. As expected, animals belonging to the HMY had higher average DMI than animals of LMY (2.21 vs 1.91 kg/d per head, respectively). In the whole period, the FE was significantly larger (i.e. desirable) in the HMY than LMY class (1.11 ± 0.49 vs 0.73 ± 0.19, mean $\pm$ SD). This result could be associated to the higher genetic level of high producing animals, which showed a greater ability to convert feed input. A positive correlation was found between MY and DMI ($P>0.05$, $r=0.17$). In conclusion, the PLF allowed phenotyping dairy sheep flocks relatively to nutritional inputs and feed efficiency.

Session 41 Poster 15

Introducing the aWISH project: animal welfare indicators at the slaughterhouse

J. Maselyne, T. Coppens, A. Watteyn, A. De Visscher, E. Kowalski, M. Aluwé and F. Tuyttens
ILVO (Flanders Research Institute for Agriculture, Fisheries and Food), Burg. Van Gansberghelaan 92, 9820 Merelbeke, Belgium; jarissa.maselyne@ilvo.vlaanderen.be

The objective of aWISH (animal Welfare Indicators at the SlaughterHouse) is to develop and offer a cost-efficient solution to evaluate and improve the welfare of meat-producing animals at a large scale, across Europe. This approach will be developed and evaluated in close collaboration with all actors involved, from primary producers up to policy makers and citizens. At the heart of the aWISH solution is the automated assessment at the slaughterhouse of complementary animal-based indicators for monitoring welfare on-farm, during (un)loading, transport and slaughter. Besides that, existing or routinely collected data (slaughterhouse data, antibiotics usage, farm data, etc.) and needed technologies on-farm or on-transport to complement the measurements at slaughter will be exploited. Piloting and development activities will be done in 6 broiler chicken and fattening pig production chains across Europe (FR, PL, ES, NL, AT, RS), using a lean multi-actor approach, in order to test and validate the project results. Novel sensor technologies and artificial intelligence algorithms will be developed, and a feedback tool and interface will allow each actor in the chain to get direct feedback of each slaughter batch, visualize trends and benchmark animal welfare outcomes. An Animal Welfare Indicator Catalogue will disseminate all validated indicators and standardized data collection methods. From the pilot data, animal welfare initiatives taken at operator, chain, regional or national level will be assessed alongside their environmental and socio-economic impact at operator and sector level. Next to that, 9 Best Practice Guides will be developed to improve key welfare issues in pigs and broilers, and to help external actors deploy the aWISH technologies and feedback tool. How the feedback loop guides and motivates each party to take actions to improve animal welfare will be tested in a longitudinal study, and the needs, perceptions and barriers of all actors from farm to fork incl. the consumer will be researched to maximize impact of the aWISH results. aWISH is a 4-year Horizon Europe project (nr. 101060818) starting autumn 2022 with 24 partners across the EU.

Session 41 Poster 16

Stakeholders' perceptions of precision livestock farming to improve small ruminant welfare

E.N. Sossidou[1], E.G. Garcia[2], M.A. Karatzia[1], L.T. Cziszter[3], A. Elhadi[4], L. Riaguas[4], G. Caja[4], A. Barnes[5], J.M. Gautier[6], T. Keady[7], I. Halachmi[8], G. Molle[9], L. Grova[10] and C. Morgan-Davies[5]
[1]Ellinikos Georgikos Organismos-DIMITRA, Veterinary Research Institute, ELGO Campus, 57001 Thessaloniki, Greece, [2]INRAE SELMET, Centre des Recherches Occitanie, Campus INRA-Monpellier, 34060 Monpellier, France, [3]USAMVB, Regele Mihai I al Romaniei, Timisoara, Timisoara, Romania, [4]Universitat Autonoma de Barcelona, Faculty Veterinaria, Campus Universitari de la UAB, 08193 Barcelona, Spain, [5]Scotland's Rural College, West Mains Road, Edinburgh EH9 3JG, United Kingdom, [6]Institut de l'Elevage, Castanet-Tolosan, France, [7]Teagasc, Athenry, Co. Galway, Galway, Ireland, [8]The Agricultural Research Organization, Rishon LeZion, Rishon LeZion, Israel, [9]AGRIS Sardegna, Bonassai, Sassari, Italy, [10]Norksk Institute for Biookonomi, NIBIO, Norway; sossidou@vri.gr

Within the TechCare Project, a list of Precision Livestock Farming (PLF) tools, with potential for monitoring animal welfare in small ruminant production, was formulated in TechCare Countries. The opinions of stakeholders were taken into account after consultations, following a multi-actor approach. The OPERA method was used to reach consensus between stakeholders. Overall, stakeholders' opinions appear rather uniform. Regarding meat sheep, the highest appreciated PLF tool was the automated weighing and low frequency identification (LF) system, followed by localisation (GPS) and 3-axial accelerometers, as well as new ultra-high-frequency (UHF) eartags and readers for several uses (e.g. water intake). The most suitable PLF tools selected for dairy sheep were environmental-air quality sensors (e.g. weather stations), followed by automated milk recording, LF and automatic weighing. As for dairy goats, in addition to those selected for dairy sheep, the new UHF eartags and readers were prioritized. Regardless of the production system, the most important traits of selected PLF tools were low cost and ease-of-use. Most stakeholders also expressed concerns on their ability to collect and handle the data generated from PLF tools for monitoring individual animals. TechCare (www.techcare-project.eu) receives funding from the European Union's H2020 research and innovation programme grant no. 862050.

Session 41 — Poster 17

The endogenous and viral microRNAs in the nasal secretions of buffaloes during BuHV-1 infection

C. Lecchi[1], A. Martucciello[2], S. Petrini[3], G. Cappelli[2], C. Grassi[2], A. Balestrieri[2], G. Galiero[2], G. Salvi[1], F. Panzeri[1], C. Gini[1] and F. Ceciliani[1]

[1]University of Milan, Lodi, 26900, Italy, [2]Istituto Zooprofilattico Sperimentale del Mezzogiorno, Salerno, 84132, Italy, [3]Istituto Zooprofilattico Sperimentale dell'Umbria e delle Marche, Perugia, 06126, Italy; cristina.lecchi@unimi.it

The bubaline alphaherpesvirus 1 (BuHV-1) is a ubiquitous pathogen of buffaloes responsible for economic loss worldwide. MicroRNAs (miRNAs) are regulators of gene expression produced by BuHV-1 and Bovine (Bo)HV-1 and by the host. This study aimed at: (1) unravelling the ability of BuHV-1 to produce miRNAs (hv1-miR-B6, hv1-miR-B8, hv1-miR-B9) by sequencing and RT-qPCR; (2) quantifying the host immune-related miRNAs (miR-210-3p, miR-490-3p, miR-17-5p, miR-148a-3p, miR-338-3p, miR-370-3p) by RT-qPCR; (3) pointing out candidate markers of infection by ROC curves; (4) exploiting the biological functions by pathway enrichment analyses. Since the diagnosis of BuHV-1 respiratory tract infection involves isolation of the virus from nasal secretions, these matrices were selected as a potential source of miRNAs. Five buffaloes devoid of BuHV-1/BoHV-1-neutralizing antibodies were vaccinated via the intramuscular route with the inactivated glycoprotein E deleted marker vaccine. Five additional buffaloes served as controls. Sixty days after the first immunization, all animals were challenged with a wild-type (wt) BuHV-1 via the intranasal route. Nasal swabs were obtained from each buffalo at different post-challenge days (PCDs) and were used for virus isolation. After the challenge, the animals of two groups shed wt BuHV-1 up to 7 PCD. Results demonstrated that: (1) the sequences of hv1-miR-B6-5p and hv1-miR-B9 are conserved between BoHV-1 and BuHV-1; (2) miRNAs produced by host and BuHV-1 can be efficiently quantified in the nasal secretion up to 60 and15 PCDs, respectively; (3) the levels of host and BuHV-1 miRNAs are different between vaccinated and control buffaloes; (4) miR-370-3p discriminated vaccinated and control animals with excellent diagnostic accuracy; (5) host immune-related miRNAs may modulate genes involved in the cell adhesion pathway of the neuronal system. Overall, the present study demonstrated that miRNAs can be detected in the nasal secretion of buffaloes and that their expressions are modulated by BuHV-1.

Session 42 — Theatre 1

Diet-induced unresolved metabolic inflammation in fattening Holstein bulls?

A. Kenez[1], S.C. Baessler[2], E. Jorge-Smeding[1] and K. Huber[2]

[1]City University of Hongkong, Department of Infectious Diseases, 31 To Yuen Street, Hong Kong, China, P.R., [2]University of Hohenheim, Institute of Animal Science (46^{0}d), 35, Fruwirthstrasse, 70599, Germany; korinna.huber@uni-hohenheim.de

High dietary energy and protein intake is needed to gain the maximum growth performance in fattening bulls. However, nutritional surplus is well known to increase obesity, promote insulin insensitivity and generate a low-grade inflammation, detrimental for metabolic health. Thirty Holstein bulls were randomly divided into 2 groups, low energy and protein (LEP) and high energy and protein (HEP) intake, provided from 13th to 20th month of life. Life weight, carcass composition and laminitis score were determined. Plasma insulin and glucose concentrations were measured by ELISA and colorimetric assays, respectively. In the liver, muscle and adipose tissue, expression and extent of phosphorylation of insulin signalling proteins were detected by Western Blotting. Sphingolipid metabolome was quantified by a targeted LC-MS based metabolomics approach. Data were analysed by unpaired student`s t-test and by partial least square discriminant analysis, respectively. HEP bulls were obese, and expressed hyperinsulinemia with euglycemia. Clinically, all HEP bulls expressed signs of chronic laminitis. In the liver, protein kinase B (PKB) phosphorylation was decreased and this was associated with higher ceramide (Cer) 16:0 concentrations. Cer 16:0 is well known to diminish insulin signalling by dephosphorylating PKB. In adipose tissue, insulin receptor expression was lower in HEP bulls, and this was associated with higher hexosylceramide concentrations. Hexosylceramides are able to disturb membrane-receptor interactions, thereby reducing the abundance of functional insulin receptors. These major findings indicate that metabolic inflammation induced by dietary surplus induces ceramide accumulation and consequently, disturbs insulin signalling. As insulin insensitivity is known to exacerbate metabolic inflammation, a self-reinforcing cycle was established leading to further increase in ceramide levels and insulin resistance in fattening bulls. Chronic laminitis is a visible signal of this detrimental metabolic situation.

Effect of concentrate feeding on adipose prostanoids and oxylipins in dairy calves

R. Khiaosa-Ard, S. Sharma and Q. Zebeli
University of Veterinary Medicine Vienna, Veterinärplatz 1, 1210 Vienna, Austria; ratchaneewan.khiaosa-ard@vetmeduni.ac.at

Dietary polyunsaturated fatty acids (PUFA) affect the PUFA of storage lipids and membrane lipids. The latter source can be oxygenated on demand into numerous bioactive lipid mediators that influence many biological processes of tissue and organs. We hypothesized that concentrate feeding high in n-6 PUFA increases adipose n6 PUFA-derived oxylipin and prostanoid contents. From the first to 15 weeks of life, 20 Holstein-Friesian calves were adapted to one of the four diets: 100% medium-quality hay (MQH), 30% medium-quality hay plus 70% concentrate (MQH+C), 100% high-quality hay (HQH), and 30% high-quality hay plus 70% concentrate (HQH+C). Intakes of dry matter and PUFA were examined. Two locations of kidney fat: proximal and distal to the kidneys were taken for analyses. After weaning (week 12-15), MQH+C led to the average n6 to n3 PUFA intake of 5.2, followed by HQH+C (2.6), MQH (0.45), and HQH (0.23) ($P<0.01$). MQH decreased dry matter and PUFA intake compared to the other groups ($P<0.05$). For both adipose locations, concentrate-fed groups had higher n6 to n3 ratios in both neutral and phospholipids mainly due to greater enrichment of 18:2 n6 compared to the hay-only groups ($P<0.05$) but did not affect n3 PUFA. Adipose lipidomics targeted 29 oxylipins and 6 prostanoids. Concentrate feeding increased the contents of total oxylipins (pmol/mg tissue) by 4-5 times the contents found with the hay-only groups ($P<0.05$). The distal location showed higher contents of several 18:2 n6 metabolites compared to the proximal location. Concentrate-fed groups increased their contents in both locations ($P<0.05$) and three of which (9-hydroxyloctadecadienoic acid (HODE), 13-HODE, and 12, 13-epoxyoctadecenoic acid) were location-dependent showing a drastic increase only in the distal location. MQH had the lowest content of most of the oxylipins from 20:4 n6 and 20:5 n3. Prostaglandins PGD2 and PGE2 were higher with MQH+C in the distal location and PGE3 tended to be higher with HQH. This study underlines the significance of the feeding regimen of young calves on the PUFA intake and the release of various derivatives of 18-carbon PUFA, which can be heterogeneous within the same depot. Their involvement in adipose metabolism and inflammation awaits future research.

Reducing endocannabinoid system activation affects insulin sensitivity in peripartum dairy cows

G. Kra[1,2], J.R. Daddam[2], U. Moallem[2], H. Kamer[2], R. Kocvarova[3], A. Nemirovski[3], G.A. Contreras[4], J. Tam[3] and M. Zachut[2]
[1]the Robert H. Smith Faculty of Agriculture, Food and Environment, t, Department of Animal Science, Rehovot, 761000, Israel, [2]ARO, Volcani Center, Department of Ruminant Sciences, Institute of Animal Sciences, Rishon Lezion, 7528809, Israel, [3]School of Pharmacy, Faculty of Medicine, The Hebrew University of Jerusalem, Obesity and Metabolism Laboratory, The Institute for Drug Research, Jerusalem, 9112001, Israel, [4]College of Veterinary Medicine, Michigan State University, Department of Large Animal Clinical Sciences, College of Veterinary Medicine, MI 48824, USA; mayak@volcani.agri.gov.il

Dietary omega-3 fatty acids may reduce endocannabinoid system (ECS) activity, thus we examined the effects of omega-3 on ECS and insulin sensitivity. Thirty-five late pregnant cows were individually fed: (1) CTL – prepartum and postpartum (PP) common diets; (2) FLX – prepartum diet containing 700 g/d/cow of extruded flaxseed supplement containing C18:3n-3 (Valomega 160, Valorex, France), and PP at 6.4% of diet (DM basis). At 5-8 DIM, a glucose tolerance test was conducted on 15 cows, and blood was collected every 5 min. At 20 min after glucose injection, adipose tissue (AT) was biopsied. Glucose, insulin and endocannabinoids (eCBs) levels were examined in blood. mRNA and protein abundance were examined by RT-PCR and immunoblots. Milk production during the first 21 DIM was similar, while DMI ($P=0.001$) and calculated energy balance ($P=0.02$) were lower in FLX than in CTL. Plasma 2-arachidonoylglycerol, palmitoylethanolamide and oleoylethanolamide were lower in FLX ($P=0.03$), and arachidonic acid tended to be lower in FLX than in CTL ($P=0.10$). Insulin concentrations were lower in FLX than CTL ($P=0.05$). In AT, mRNA abundance of *N*-acylphosphatidylethanolamine-phospholipase-D (*NAPEPLD*) was lower ($P=0.01$), and toll-like-receptor-4 (*TLR4*) tended to be lower in FLX vs CTL ($P=0.10$). Phosphorylation of protein-kinase-B tended to lower in FLX than in CTL ($P=0.06$). Protein tumour-necrosis-factor-alpha ($P=0.03$) and interleukin-10 ($P=0.006$) were lower in FLX compared to CTL, and monoglyceride-lipase tended to be lower in FLX than in CTL ($P=0.06$). Reducing ECS activation by omega-3 affects AT inflammation and systemic insulin sensitivity in early postpartum dairy cows.

Session 42 Theatre 4

Effect of calfhood nutrition on insulin kinetics and insulin signalling pathway genes in muscle

A.K. Kelly[1], C. Byrne[2], K. Keogh[2], M. McGee[2] and D.A. Kenny[2]
[1]School of Agriculture and Food Science, University College Dublin, Belfield, Dublin 4, Ireland, [2]Teagasc Animal and Grassland Research and Innovation Centre, Teagasc, Grange, Dunsany, Co Meath, Ireland; alan.kelly@ucd.ie

The objectives of this study were to examine systemic insulin and glucose responsiveness to a glucose tolerance test (GTT) and transcript abundance of genes of the insulin signalling pathway in skeletal muscle in calves offered an enhanced plane of nutrition from 3 to 21 weeks of life. Angus × Holstein-Friesian heifer calves (19±5 days of age, BW: 51.2±7.8 kg, mean ± SD) were offered either a high (HP, n=15) or moderate plane of nutrition (MP, n=15) from 3 to 21 weeks of age. Target growth rates were 1.2 kg/d and 0.5 kg/d, for HP and MP groups, respectively. At 19 wk of age a GTT was performed and at 21 weeks of age all calves were euthanized and skeletal muscle tissue was harvested and RNA-Seq analysis was performed. No difference was detected between treatments for area under curve (AUC) at 60 min for blood glucose concentrations. However, clearance rate of blood glucose was faster (P<0.05) for calves from the HP versus those on the MP diet. Calves on HP diet had a greater AUC at 60 min for blood concentrations of insulin compared MP diet. The change in systemic insulin from max to basal levels were greater for HP than MP dietary treatments, and a faster clearance rate of blood insulin (P<0.05) was observed in HP compared to MP diet. There was no evidence of insulin sensitivity differences (P>0.10) between the dietary treatments, nor an association between insulin sensitivity index with early life calf growth. Genes of the insulin signalling pathway involved in fatty acid synthesis were up-regulated, while genes involved in glucose metabolism were down-regulated in HP compared to MP calves, suggesting a greater requirement of nutrients in the skeletal muscle of MP treatment calves. Taken together, these results indicate that a high plane of nutrition up to 21 week of age does not negatively impact insulin responsiveness in calves.

Session 42 Theatre 5

Effects of dietary P and Ca supply on phosphate homeostasis and insulin signalling in laying hens

A. Abdi, F. Gonzalez-Uarquin, V. Sommerfeld, M. Rodehutscord and K. Huber
Institute of Animal Science, Fruwirthstraße, 35, 70593, Germany; a.abdi@uni-hohenheim.de

Objective Metabolic performance of laying hens is based on an adequate energy metabolism. The latter needs phosphate (Pi) to generate sufficient ATP and depends on effective cellular insulin actions. Thus, a close interrelationship between phosphate homeostasis and insulin signalling pathway can be assumed. It was hypothesized that low dietary phosphorus (P) and calcium (Ca) supply may affect insulin signalling, and this might be due to a lack of intracellular Pi. A sufficient cellular Pi availability is maintained by Pi importers and exporters. Furthermore, the impact of dietary mineral supply might depend on the genetic background of the hen. Methods 40 brown and 40 white Lohmann laying hens were randomly divided into four groups fed with either adequate or reduced concentrations of dietary Ca and P (P+Ca+, P+Ca-, P-Ca+, P-Ca-). Protein expression and extent of phosphorylation of hepatic insulin signalling proteins such as insulin receptor ß (IRß), phosphatidylinositol 3-kinase (PI3K), serine/threonine protein kinase (PKB), AMP-activated protein kinase (AMPK) and expression of Xenotropic and polytropic Retrovirus Receptor 1 (XPR1) were semiquantified by Western blot. The data analysis was performed by using the software JMP Pro version 15. Significance was set at P<0.05. Results As a major result, low P treatment demonstrated a clear impact on insulin signalling proteins. As first step of the cascade, the insulin receptor expression showed a strong P supply×hen strain interaction with lower amounts of receptor in the P- treatment of brown, but not white hens. Furthermore, low dietary Ca significantly upregulated the expression of XPR1 (P<0.05). Conclusion In laying hens, dietary Ca and P supplements can have a significant effect on intracellular Pi homeostasis and insulin sensitivity in the liver of laying hens indicating an interrelationship between pathways. The genetic background is modulating the response to a reduction of dietary P and Ca. The underlying mechanisms are unknown so far.

Session 42 — Theatre 6

Targeted metabolomics reveal differences between Holstein and Simmental cows adapting to lactation

L.F. Ruda[1], C. Straub[2], A. Scholz[2] and K. Huber[1]
[1]Uni Hohenheim, Fruwirthstr. 35, 70599, Germany, [2]LMU Munich, Sonnenstrasse 16, 85764 Oberschleissheim, Germany; lena.ruda@uni-hohenheim.de

Metabolic robustness defined as the ability to cope with metabolic stress without developing disease, might be a strategy of dairy production to integrate sustainability and economics. The dual-purpose breed German Simmental (SI) is generally considered to be more robust than the single purpose dairy breed German Holstein (HF). The present study compared plasma metabolite profiles of these breeds in a targeted metabolomics approach to identify differences in biochemical pathways associated with metabolic health around parturition. SI (n=6) and HF (n=6) cows – similar in milk yield and kept under the same management and feeding conditions – were examined at day -42 ante partum (ap) and day 21 post partum (pp). (German Animal Welfare Act; permit number: 55.2-1-54-2532.0-66-2016). Classical clinical chemical parameters were analysed in plasma (glucose, BHBA: Cobas c 311, Roche Diagnostics, Germany; insulin: ELISA Mercodia, Sweden) and proxies of insulin sensitivity were calculated. A targeted metabolomics approach (AbsoluteIDQ p180, Biocrates Life Science AG, Austria) was applied. Data were evaluated by univariate data analysis, and metabolomics data were also subjected to multivariate data analysis using the MetaboAnalyst 3.5 software. $P<0.05$ was considered significant. At day 21 pp, concentration of BHBA in HF was higher and RQUICKI-BHB was lower than in SI, whereas concentrations of insulin and glucose did not differ. This was paralleled by a greater body weight loss in HF until day 60 pp. Concerning branched chain amino acids (BCAA), α-amino adipic acid (α-AAA) and creatinine, time and breed showed an interaction with higher concentrations at day 21pp in HF. High levels of BCAA and creatinine might point to a higher protein turnover in HF than in SI adapting to lactation. Further BCAA and α-AAA are associated with insulin resistance in other species and therefore possibly indicate more effective insulin action in SI compared to HF. This is supported by lower RQUICKI-BHB in HF at day 21pp. Higher α AAA in HF cows at this day additionally suggest differences in liver mitochondrial functionality between breeds.

Session 42 — Theatre 7

Effects of systemic inflammation on liver telomere length and mitochondrial DNA copy numbers

K.D. Seibt[1], M.H. Ghaffari[1], J. Frahm[2] and H. Sauerwein[1]
[1]University of Bonn, Institute of Animal Science, Katzenburgweg 7, 53115 Bonn, Germany, [2]Friedrich-Loeffler-Institute (FLI), Institute of Animal Nutrition, Bundesallee 37, 38116 Braunschweig, Germany; morteza1@uni-bonn.de

Telomeres protect chromosomal integrity. They are shortening with each cell division; telomere length (TL) may reflect cellular aging processes. The mitochondrial DNA copy number (mtDNAcn) reflects the abundance of mitochondria in a cell. Both parameters are affected by various factors including energy status. In dairy cows, the changes in energy balance (EB), the concomitant lipolysis, and systemic inflammation during early lactation were demonstrated to be related to TL and mtDNAcn. Carnitine, besides its role in facilitating the mitochondrial uptake of fatty acids for β-oxidation, may reduce inflammation. The aim of this study was to investigate the effect of a standardized inflammatory challenge by administering bacterial lipopolysaccharide (LPS) in pluriparous Holstein cows during a positive EB on TL and mtDNAcn in liver cells. Half of the cows studied received L-carnitine (CAR, 25 g/d; n=26) from day (d) -42 relative to calving onwards, the cows that did not receive the supplement served as a control (CON; n=24). Liver biopsies were performed on d -42, +100, and +112 relative to calving. All cows were challenged by an injection of LPS (Escherichia coli O111: B4, 0.5 µg/kg BW) on d +111. Further samples were collected 24 h later. The hepatic TL and mtDNAcn were determined by multiplex qPCR, i.e. either TL or mt12s rRNA were amplified together with β-globin as reference gene. Strict validation criteria were applied. Statistical analyses (linear mixed model) using SPSS were performed with time, treatment, and their interactions as fixed effects. The TL values in liver on d -42, +100, and +112 were not different and not affected by CAR, but by parity. The mtDNAcn values obtained on d +100 after calving and +24 h after LPS were higher than on d -42 before calving ($P<0.003$) irrespective of CAR or parity. The LPS challenge had no effect on TL or mtDNAcn in liver cells. The greater mtDNAcn in liver at mid lactation as compared to prepartum values might be associated with the change in energy status.

Session 42 Theatre 8

Metabolic response to LPS challenge in early postpartum cows maintained at eucalcemia

T.L. Chandler, T.A. Westhoff, T.R. Overton and S. Mann
Cornell University, Ithaca, NY, USA; tlc236@cornell.edu

Cows experience changes in energy metabolism and hypocalcaemia following immune activation during the early postpartum period. Calcium therapy alters adaptations in Ca metabolism, but the effect of Ca therapy on energy metabolism during immune activation is unknown. Our objective was to describe the metabolic response following an intravenous (IV) lipopolysaccharide (LPS) challenge in postpartum cows with or without IV Ca to maintain eucalcemia. Cows (n=14, 8±1 DIM) were enrolled in a matched-pair randomized controlled design and received an IV LPS challenge (40 to 45 ng/kg BW over 1 h) either with IV Ca (IVCa) in a eucalcemic clamp for 12 h, or 0.9% NaCl (CTRL). All cows were fasted during the 12 h infusion period. Blood was collected at 0, 2, 4, 6, 12, and 24 h relative to the start of LPS challenge to measure glucose, non-esterified fatty acids (NEFA), β-hydroxybutyrate (BHB), and urea nitrogen (BUN) concentrations. Repeated measures ANOVA with baseline covariates (0 h) were analysed in PROC MIXED (SAS v. 9.4) with fixed effects of treatment (trt), time, trt×time, LPS dose, DIM, parity, and BW, and random effect of pair. Glucose concentration decreased (P=0.004) 10% from baseline (63.9±2.6 mg/dl), but did not differ (P=0.37) between 2, 4, 6, 12, and 24 h. Concentration of NEFA changed over time (P=0.009), increasing at 12 h before decreasing and reached a nadir at 24 h. Concentration of BHB increased (P=0.004) at 24 h and was 12% of baseline values (0.81±mM). Concentration of BUN increased (P<0.001) during challenge, reaching a maximum at 12 h. Glucose, NEFA, BHB, and BUN concentration during the 24 h following LPS challenge did not differ by trt (P>0.15). Following LPS challenge, cows experienced decreased glucose concentrations and increased BUN concentrations within 2 h, but increases in NEFA and BHB concentrations were delayed to 12 and 24 h, respectively. Eucalcemia was not associated with differences in the metabolic response to LPS in the current experiment. Given our experimental design, metabolic changes following LPS infusion could be due to either immune activation, fasting, or both.

Session 42 Theatre 9

Urolithin A as new strategy to rescue the quality and developmental potential of aged oocytes

E. Fonseca[1], C.C. Marques[1], M.C. Baptista[1], J. Pimenta[1,2], A.C. Gonçalves[3] and R.M.L.N. Pereira[1,2]
[1]Instituto Nacional de Investigação Agrária e Veterinária, Santarém, Unidade de Biotecnologias e Recursos Genéticos, Quinta da Fonte Boa, 2005-048 Vale de Santarem, Portugal, [2]CIISA, Reproduction and Development, Avenida da Universidade Técnica, 1300-477 Lisboa, Portugal, [3]Universidade de Coimbra, iCBR, Azinhaga Santa Comba, Celas, 3000-548 Coimbra, Portugal; rosa.linoneto@iniav.pt

Female gamete aging impairs the reproductive capacity through several mechanisms not fully understood. Oxidative stress and mitochondrial dysfunction have been identified as major contributing factors to the age-related decline of oocyte quality and poor developmental potential. Urolithin A (UA), a natural metabolite with pro-apoptotic and antioxidant effects, has been shown to prevent the accumulation of dysfunctional mitochondria with age in different cells, by inducing mitophagy. In this study, we tested for the first time, the effect of UA in the quality and developmental potential of cumulus-oocyte-complexes (COCs), during the aging process. Maturation progression rates, mitochondrial membrane potential (MMP) ratios, and blastocyst formation rate in physiologically mature and *in vitro* aged oocytes obtained from bovine prepubertal and adult females, supplemented or not with UA, were compared. Our study confirmed the harmful effect of oocyte aging on the nuclear maturation progression, MMP and developmental competence. Results showed that UA treatment during *in vitro* maturation effectively enhanced (P≤0.05) the percentage of matured oocytes and significantly promoted subsequent developmental capacity of aged oocytes. A positive effect (P≤0.05) of UA on physiological maturation in adult females, MMP and embryonic development was also identified. These results strongly suggest that UA supplementation is an effective way to prevent oocyte aging and improve the subsequent embryonic development, which provides a potential new therapeutic approach to prevent or delay gamete aging, and improve fertility outcomes. Funded by ALT20-03-0246-FEDER000021, UIDB/CVT/00276/2020 and PDR2020-101-03112.

Assessment of exosomal microRNA profiles in bovine colostrum with different IgG concentrations

T.M. Ma[1,2], W.L. Li[1], Y.C. Chen[1], E.R.C. Cobo[3], C.W. Windeyer[3], L.G. Gamsjäger[4], Q.D. Diao[2], Y.T. Tu[2] and L.L.G. Guan[1]

[1]Department of Agricultural, Food and Nutritional Science, University of Alberta, 4-16 Agriculture Forestry Center, T6G 2P5, Edmonton, Alberta, Canada, [2]Institute of Feed Research, Chinese Academy of Agricultural Sciences, No.12 Zhongguancun south st., 100081, Beijing, China, P.R., [3]Department of Production Animal Health, Faculty of Veterinary Medicine, University of Calgary, 2500 University Drive NW, T2N 1N4, Calgary, Alberta, Canada, [4]Department of Ruminant Medicine, Vetsuisse Faculty of Veterinary Medicine, University of Zurich, Winterthurerstrasse 260, 8057 Zürich, Switzerland; lguan@ualberta.ca

Exosome microRNAs (miRNAs) are one of non-IgG biomolecules in bovine colostrum. However, whether their profiles differ between high and low IgG colostrum is unknown. The present study investigated the profiles of exosomal miRNAs in bovine colostrum using RNA-sequencing and compared them between colostrum with high (average of 256.5 mg/ml, n=4) and low (average of 62.8 mg/ml, n=4) concentrations of IgG. Different combination of exosome extraction methods and bioinformatic pipelines (miRDeep2 and sRNAbench) for miRNA analysis were evaluated. The differential expression analysis of miRNA between high- and low-IgG samples was conducted using DESeq2. Significant differences were declared at Benjamini-Hochberg adjusted $P<0.05$. For functional annotation analysis, a pathway was significantly enriched when it passed the count threshold of three genes per annotation term and P values with Benjamini-Hochberg correction set to <0.05. For miRNA-seq data analysis, 389 miRNAs were identified in bovine colostrum exosome using both bioinformatic pipelines. These core exosomal miRNAs were predicted to target 2,655 genes, which regulate 78 KEGG level-3 pathways including PI3K-Akt and MAPK signalling pathway, axon guidance, and focal adhesion. The expression profiles of exosomal miRNAs were similar between high- and low-IgG colostrum samples, despite that the abundance of miR-27a-3p was higher in colostrum with high concentrations of IgG. In conclusion, a core miRNAome in bovine colostrum may play a role in regulating health and developmental stages in neonatal calves, independent of IgG concentration.

Session 42 Poster 11

Reducing endocannabinoid system activation affects phosphoproteome of adipose tissue in dairy cows

G. Kra[1,2], J.R. Daddam[2], U. Moallem[2], H. Kamer[2], R. Kočvarová[3], A. Nemirovski[3], G.A. Contreras[4], J. Tam[3] and M. Zachut[2]

[1]the Robert H. Smith Faculty of Agriculture, Food and Environment, the Hebrew University of Jerusalem, Department of Animal Science, Rehovot, 76001, Israel, [2]ARO Volcani Center, Department of Ruminant Sciences, Institute of Animal Sciences, Rishon Lezion, 7528809, Israel, [3]School of Pharmacy, Faculty of Medicine, The Hebrew University of Jerusalem, Obesity and Metabolism Laboratory, The Institute for Drug Research, Jerusalem, 9112001, Israel, [4]Michigan State University, Department of Large Animal Clinical Sciences, College of Veterinary Medicine, MI 48824, USA; gititk@volcani.agri.gov.il

We examined the effects of reducing ECS activity by omega-3 fatty acids on the phosphoproteome and proteome in adipose tissue of peripartum cows. Thirty-five late pregnant cows were individually fed: (1) CTL – prepartum and postpartum (PP) common diets; (2) FLX – prepartum a diet containing 700 g/d/cow of extruded flaxseed supplement (Valomega 160, Valorex, France) providing C18:3n-3, and PP at 6.4% of diet (DM basis). At 5-8 DIM, a glucose tolerance test was conducted on 15 cows. At 20 minutes after glucose injection, AT was biopsied to obtain insulin-stimulated tissue. Proteomics and phosphoproteomics analyses of AT were performed in 5 AT from each treatment using GC-MS/MS and nanoUPLC-MS/MS. A total of 2,309 proteins and 3,502 phosphopeptides were identified, in which 144 proteins and 169 phosphopeptides were differential between CTL and FLX, respectively [$P\leq0.05$ and fold change (FC)±1.5]. In phosphoproteomics, the enriched pathways in FLX vs CTL AT were: Protein Kinase A Signalling, RHOA Signalling, Glycolysis I, p38 MAPK Signalling, LXR/RXR Activation, ERK/MAPK Signalling, Calcium Signalling, Insulin Receptor Signalling and AMPK Signalling. In proteomics, the enriched pathways were: Oxidative Phosphorylation, Acute Phase Response Signalling, LXR/RXR Activation, Sirtuin Signalling, FXR/RXR Activation and Protein Ubiquitination Pathway related to insulin signalling. We present a complete map of phosphoproteome and proteome of AT in dairy cows with reduced ECS activation, and identified many unknown phosphorylation sites suggestive of increased insulin sensitivity in AT.

MitoCow: L-carnitine supplements alter the response of mid-lactation cows to systemic inflammation

M.H. Ghaffari[1], H. Sadri[2], S. Häußler[1], J. Frahm[3] and H. Sauerwein[1]
[1]University of Bonn, Katzenburgweg, 53115, Germany, [2]University of Tabriz, Tabriz, 516616471, Iran, [3]Friedrich-Loeffler-Institute (FLI), Bundesallee 37, 38116 Braunschweig, Germany; morteza1@uni-bonn.de

The acute phase reaction (APR) and inflammation are energetically demanding processes. L-carnitine (CAR) is part of the fatty acid uptake shuttle in mitochondria and can counteract inflammation-associated oxidative stress. Our objectives were to compare: (1) the response of hepatic mRNA abundance of key genes associated with inflammation and oxidative stress [ceruloplasmin (CP), superoxide dismutase 1 (SOD1), Nuclear Factor Kappa B Subunit 1 (NFKB1), and Signal transducer and activator of transcription 3 (STAT3)]; and (2) the concentration of lactoferrin (Lf), Haptoglobin (Hp), fibrinogen, reactive oxygen metabolites (d-ROM), and of arylesterase activity (AEA) in blood as well as Lf and HP in milk to an inflammatory stimulus in mid-lactation cows either supplemented or not with CAR. Pluriparous Holstein cows were assigned to a control group (CON, n=26) or a CAR-supplemented group (CAR; 25 g L-carnitine/cow/d; d 42 ante partum to d 126 post partum (pp), n=27). On d 111 pp, each cow was injected i.v. with lipopolysaccharide (LPS *Escherichia coli* O111: B4, Sigma-Aldrich, 0.5 µg/kg). Plasma and milk samples were frequently collected before and after the challenge. The mRNA abundance was examined in liver biopsies of d -11 and +1 relative to LPS. Circulating concentrations of plasma fibrinogen (increased after LPS), serum d-ROM (increased after LPS), and AEA (decreased after LPS), and Hp in serum and milk changed over time independent of CAR supplementation ($P<0.01$). The Lf concentrations increased ($P<0.01$) in both groups after LPS, with the CAR group showing higher levels in serum (1-4 h after LPS) and milk (24 and 72 h after LPS) than the CON group. These findings suggest that CAR supplementation increased lactoferrin, likely to prevent tissue damage via pro-inflammatory cytokines. There was no significant difference in the mRNA abundance of NFKB1 between treatments. In contrast to the CON group, the CAR group had higher mRNA abundance of CP (after LPS), SOD1 (after LPS), and STAT3 (before and after LPS). Overall dietary CAR altered some variables of the response of mid-lactation cows to systemic inflammation.

Session 42 Poster 13

MitoCowLiver gene expression of mitochondrial dynamics related to inflammation and dietary carnitine

M.H. Ghaffari[1], H. Sadri[2], S. Häußler[1], J. Frahm[3] and H. Sauerwein[1]
[1]University of Bonn, Katzenburgweg, 7, 53115, Germany, [2]University of Tabriz, Faculty of Veterinary Medicine, Tabriz, 516616471, Iran, [3]Friedrich-Loeffler-Institute (FLI), Bundesallee 37, 38116 Braunschweig, Germany; morteza1@uni-bonn.de

Mitochondria (Mt) are dynamic organelles that fuse and fission to maintain mitochondrial function during metabolic or environmental stress. Mt proteins are mainly synthesized in the cytosol and imported into the protein import system in the outer and the inner Mt membrane. L-carnitine (CAR) is part of the fatty acid uptake shuttle in Mt and can counteract inflammation-associated oxidative stress. Our objective was to compare the response of hepatic mRNA abundance of nine key genes associated with Mt dynamics [mitofusins 1 and 2, and optic atrophy protein-1, and fission 1] and the Mt protein import system, including the TOM (TOMM70, TOMM20) and TIM (TIMM17B, TIMM22, TIMM23) complexes, to an inflammatory stimulus in cows that either supplemented or not with L-CAR. Pluriparous Holstein cows were assigned to a control group (CON, n=26) or a CAR -supplemented group (CAR; 25 g L-carnitine/cow/d; d 42 ante partum to d 126 post partum (pp), n=27). On d 111 pp, each cow was injected i.v. with lipopolysaccharide (LPS *Escherichia coli* O111: B4, 0.5 µg/kg). The mRNA abundance was examined in liver biopsies of d -11 and +1 relative to LPS. Statistical analysis was performed using mixed models (SAS). Of the genes involved in Mt dynamics, only fission 1 was affected by time and increased in both groups after LPS. Interactions were observed for optic atrophy protein-1 (increased in CAR, decreased in CON) and for mitofusin-1 (decreased in CAR, increased in CON). For protein import genes, group but no time differences were found for TIMM17B (increased in CAR), whereas for TIMM23 (increased with LPS) only time was significant. Of the genes targeted here, treatment of CAR affected the LPS response mainly in terms of Mt dynamics and protein import, which may be relevant to Mt function.

Steroid hormones and acute-phase proteins in lactating dairy cows fed with olive by-products

A. Bionda[1,2], E. Fazio[2], V. Chiofalo[2], D. La Fauci[2], P. Crepaldi[1] and L. Liotta[2]
[1]University of Milan, Department of Agricultural and Environmental Sciences, Production, Landscape and Energy, Via Celoria 2, 20133 Milan, Italy, [2]Università degli Studi di Messina, Department of Veterinary Sciences, Viale Palatucci 13, 98168 Messina, Italy; ariannabionda95@gmail.com

Steroid hormones influence inflammatory response and thus acute phase proteins (APPs) throughout lactation. APPs are often inflammation markers, but they were also found in healthy bovines, positively related to milk production. This study compares blood concentrations of 17β-estradiol (E_2), progesterone (P_4), cortisol, serum amyloid A (SAA), and C-reactive protein (CRP) in 24 healthy dairy cows fed a diet integrated with olive cake (8% of concentrate) throughout lactation. Kruskal-Wallis and Steel-Dwass tests were used to assess variations among the different 60-day-long phases of lactation. Spearman's correlation coefficient (ρ) between the parameters was calculated. The lactation phase affected P_4 concentration ($P<0.02$), with the lowest values at the beginning of the lactation, and cortisol ($P<0.003$), with higher concentrations at ≥60-120 d than at ≥300 d. E_2, SAA, and CRP did not significantly differ among the lactation phases. SAA and CRP ($P<0.0001$, $\rho=0.83$), SAA and cortisol ($P<0.05$, $\rho=0.29$), and CRP and cortisol ($P<0.02$, $\rho=0.29$) were positively correlated, whereas cortisol and P_4 were negatively correlated ($P<0.01$, $\rho=-0.32$). The low P_4 concentration during the first weeks of lactation is consistent with the negative energetic and protein balance, which makes P_4 elimination faster than its synthesis, and with its anti-lactogenic effect. The stress occurring during this period also increases circulating cortisol, which enhances the lactogenic action of prolactin. The opposite effects of P_4 and cortisol on milk production account for their negative correlation. In this study, APP concentration was very variable, as already reported, showed the highest values around the lactation peak, and was positively correlated with cortisol; this supports that APP may be lactation-associated proteins with a role in physiological mechanisms. The concentrations found during this study are in line with those reported in the Literature, suggesting that the olive cake had no adverse effects on steroid hormone and APP profile in dairy cows.

Serum acylcarnitine profiles in dairy cows from late gestation through early lactation

H. Sadri[1], M.H. Ghaffari[2], J.B. Daniel[3] and H. Sauerwein[2]
[1]Faculty of Veterinary Medicine, University of Tabriz, Tabriz, Iran, [2]Institute of Animal Science, Physiology Unit, University of Bonn, Bonn, Germany, [3]Trouw Nutrition, R&D, Amersfoort, the Netherlands; sadri@uni-bonn.de

Carnitine and its acyl esters (acylcarnitines; ACC) are crucial for mitochondrial β-oxidation of fatty acids (FA) through enabling the transfer of long-chain FA from the cytoplasm to the mitochondrial matrix across the mitochondrial membranes. We aimed to characterize the changes in carnitine and ACC concentrations in blood serum of dairy cows to address potential changes in the intramitochondrial acyl-CoA patterns over a wide transition period, including both dry-off and calving. Blood was collected from 12 Holstein dairy cows at weeks (wk) -7 (before dry-off), -5 (after dry-off), -1, 1, 5, 10, and 15 relative to calving. Serum FA were measured using the Randox Kit and ACC profiles were quantified through targeted metabolomics using the BIOCRATES MxP Quant 500 Kit. The FA serum concentrations were highest at the first wk of lactation, pointing to increased lipolysis as a response to a substantial demand of energy to accomplish milk synthesis. Free carnitine and 39 ACC were detected. Time-dependent changes ($P<0.05$) were observed in the serum concentrations of carnitine, 11 short- (C2-C5), 10 medium- (C6-C12), and 7 long-chain (C14-C18) ACC. The serum concentrations of carnitine increased 2-fold after drying-off (wk -5), declined to nearly before dry-off values by wk 1, and then remained unchanged, likely reflecting the carnitine excretion pattern in milk and its uptake by peripheral tissues during the preceding and current lactation. Acetylcarnitine, deriving from acetyl-CoA via the action of carnitine acetyltransferase for transport out of the mitochondria, increased from wk -7 to -5, decreased at wk -1, and remained unchanged thereafter. This is probably due to an excess generation of acetyl-CoA in the mitochondrial matrix relative to the flux into the TCA cycle around dry-off. Amino acid-derived ACC, i.e. C3, C4, and C5, that were elevated from wk -7 to -5 and decreased thereafter, seem the most relevant ACC during dry-off. Most lipid-derived ACC such as C12:1, C14:2, C18, C18:1, and C18:2 increased around parturition, which is consistent with the lactation-induced rise in circulating FA.

Session 43 Theatre 1

Challenges in poultry nutrition, health, and welfare

D. Korver
University of Alberta, Edmonton, AB, Canada; dkorver@ualberta.ca

The poultry industry has benefited greatly from advances in genetics, nutrition, housing and management. Geneticists have made welfare and health traits important components of selection programs, and in general, modern, high-producing poultry are healthier than 30 years ago. However, increased productivity means that the birds are closer to their physiological limits, and nutrition, environment and management have become increasingly important. The move away from antibiotic growth promotors has resulted in challenges in maintaining gut health and consequently, bird performance. Various alternatives, with different mechanisms of action, have provided the industry with viable choices. However, the ideal combination of alternatives may be different for different locations, seasons or even ages within a flock. As genetic selection increased broiler production traits, it became necessary to restrict parent stock nutrient intake in order to prevent excessive muscle and fat deposition, to reduce metabolic disease, and maintain ovarian control. With continued selection for broiler production traits, the degree of restriction implemented has become a welfare issue. Additionally, recent research suggests that highly efficient broiler lines may have limited fat deposition and therefore energy reserves to support sexual maturation and egg production, especially if typical broiler breeder body weight targets are maintained. A re-examination of broiler breeder feeding programs is necessary to maintain productivity and welfare. Modern laying hens are capable of laying cycles in excess of 100 weeks of age. This has reduced the use of stress-inducing forced molting programs, and reduces the total number of hens needed to meet the demand for egg production. The long egg production cycles can put pressure on the ability of the hen to deposit adequate eggshell material on the egg; reduced shell quality, rather than impaired production is often the reason a flock is depopulated. The important role of the skeletal system in eggshell deposition demands that skeletal development during rearing be carefully managed to avoid shell and skeletal problems at the end of the production cycle. As the production potential of modern poultry continue to increase through genetic selection, even greater care must be paid in order to maintain bird health and welfare.

Session 43 Theatre 2

Can monitoring of keel bone and skin damage at the slaughterhouse tell about hen welfare on-farm?

L. Jung[1], B. Kulig[2], H. Louton[3] and U. Knierim[1]
[1]University of Kassel, Farm Animal Behaviour and Husbandry, Nordbahnhofstr. 1a, 37213 Witzenhausen, Germany, [2]University of Kassel, Agricultural Engineering, Nordbahnhofstr. 1a, 37213 Witzenhausen, Germany, [3]University of Rostock, Animal Health and Animal Welfare, Justus-von-Liebig-Weg 6b, 18059 Rostock, Germany; lisa.jung@uni-kassel.de

Measuring welfare indicators at the bottle neck slaughterhouse can be more labour efficient and easier to automatize compared to measurements on-farm, but it needs to be investigated to which degree it reflects the on-farm welfare status. We therefore assessed skin lesions at the cloaca and back and keel bone damage (KBD) in 20 commercial non-cage laying hen flocks 'on-farm', at the 'arrival' at the slaughterhouse and at the 'slaughter-line' (except cloaca skin lesions) and compared results by paired t-tests and correlation analysis. On-farm, all flocks were affected by KBD with a mean prevalence of 55.9%, and nearly half of the flocks by skin lesions with mean prevalences of 9.5% for cloaca and 15.1% for back lesions. While detected KBD prevalences did not significantly differ at arrival (52.0%, P=0.10), they were significantly lower than on-farm at the slaughter line (41.7%, P<0.0001). This was probably due to the different recording methods (palpation versus visual inspection). Nevertheless, a substantial linear correlation (r=0.794) between on-farm and slaughter-line assessments was observed. Despite non-significant differences between the different skin lesions assessments, prevalences were numerically higher at arrival (cloaca: 12.4%, back: 21.7%) and at the slaughter line (back: 22.1%), probably due to additional impacts during catching, transport, waiting time and the slaughter procedure. Again, substantial correlations between on-farm and arrival (cloaca lesions, r=0.712) or slaughter-line assessments (back lesions, r=0.617) were found. For further evaluations of the concordance of results also sample size and the associated prevalence estimation error need to be considered. However, these first results suggest that on-farm welfare regarding KBD and skin lesions can be monitored at the slaughterhouse as long as the under- or overestimation of prevalences is considered.

Efficacy of dietary selenium-loaded chitosan nanoparticles in rabbits and broiler chickens

G. Papadomichelakis[1], S. Fortatos[1], E. Giamouri[1], A.C. Pappas[1] and S.N. Yannopoulos[2]
[1]Agricultural University of Athens, Department of Animal Science, 75 Iera Odos Street, 11855 Athens, Greece, [2]Foundation for Research and Technology-Hellas, Institute of Chemical Engineering Sciences, Stadiou Street, Platani, 26504 Patras, Greece; gpapad@aua.gr

The present study aimed to investigate the potential of selenium (Se)-loaded chitosan (CS) nanoparticles (CS-SeNPs) as dietary Se source in comparison with the commonly-used organic (Se-enriched yeast; SY) and inorganic (sodium selenite; SS) sources. First, the CS-SeNPs were synthesized using a chemical reducing method and their physicochemical properties were characterized by dynamic light scattering (DLS), X-ray diffraction (XRD) and X-ray photoelectron spectroscopy (XPS). Subsequently, one feeding trial in rabbits (FT1) and one in broiler chickens (FT2) were carried out. The same batches of SS, SY and CS-SeNPs were used in both trials. Ninety-six growing rabbits (in FT1) and 200 broiler chickens (in FT2) were allocated into 4 dietary treatments; one control (C) with no added Se to the diet, and 3 treatments with 0.4 mg added Se/kg diet either from SS+SY (1:1 ratio; T1), SY+CS-SeNPs (1:1 ratio; T2) or CS-SeNPs alone (T3). Meat Se content and oxidative stability were determined by hydride (vapor) generation atomic absorption spectroscopy and iron-induced lipid oxidation, respectively. The results showed that spherical monodispersed CS-SeNPs of 80.5±20 nm average diameter were obtained. The Se nanoparticles were exclusively composed of amorphous elemental Se and were totally encapsulated in CS. No effects on the growth performance indices were observed between the dietary treatments in both livestock species. Meat Se content and oxidative stability was similar in T1, T2 and T3 rabbits and broiler chickens, but significantly higher ($P<0.05$) when compared to the control ones. In conclusion, the CS-SeNPs delivered dietary Se efficiently to rabbits and broiler chickens, and directly enhanced meat oxidative stability. This efficacy was similar to that observed for the inorganic and organic Se sources in both rabbits and broiler chickens. This Research Project (EDBM103-MIS 5048474) was co-financed by Greece and ESF through the Operational Program 'Human Resources Development, Education and Lifelong Learning 2014-2020'.

Animal welfare in broilers free-range and organic systems: economic implications at farm level

P. Thobe[1], M. Almadani[1], M. Coletta[2], J. Hercule[3], A. Collin[4] and J. Niemi[5]
[1]TI, Thuenen-Institute of Farm Economics and Institute of Organic Farming, Bundesallee 63, 38116 Braunschweig, Germany, [2]AIAB, FVG, Via dei tigli 2, 43[8]Q+[3]Q Fagagna (UD), Italy, [3]ITAVI, 149 rue de Bercy, 75595 Paris cedex 12, France, [4]INRAE, Université de Tours, BOA, 37380 Nouzilly, France, [5]LUKE, Kampusranta 9, 60320 Seinäjoki, Finland; mohamad.almadani@thuenen.de

As broiler production is highly cost-price driven, costs incurred by adopting welfare practices, are of great importance. The study aims to identify and evaluate a set of welfare practices regarding broilers' health and behaviour, with particular focus on the economic viability at farm level in organic and low-input outdoor systems. By adopting a participatory multi-actor approach involving National Practitioner Groups and including a farm-level economic evaluation, the study screened the best practices from the welfare and economic point of view to be further investigated by the PPILOW project for their sustainability according to the One Welfare concept. Measures of improving outdoor run quality were identified the most promising. Enriching range with vegetation stimulates the range use and thus enhances expression of natural behaviour. The use of guard animals and improving fences help to limit contacts with predators and wildlife that may transmit pathogens. Despite high labour input, mobile housing systems are effective to prevent parasitism and improve range quality. Alternative drugs to reduce the use of antimicrobials and controlling the bacteriological content of the drinking water are highly recommended health and biosecurity measures. Incubation light and on-farm hatching were observed as promising measures in terms of early life management practices. Costs incurred in the implementation of such practices varies considerably between production systems and countries. Increased production costs, however, can be compensated by additional revenues and gains obtained from product branding. PPILOW project has received funding from the European Union's Horizon 2020 Research and Innovation Programme under grant agreement no. 816172.

Riboflavin supply of different turkey lines fed diets with an organic-compliant riboflavin product

B. Thesing[1], P. Weindl[1], S. Göppel[1], S. Born[2], P. Hofmann[2], C. Lambertz[3] and G. Bellof[1]
[1]Weihenstephan-Triesdorf University of Applied Sciences, Am Staudengarten 1, 85354 Freising, Germany, [2]Poultry Competence Centre of the Bavarian Institute for Agriculture, Mainbernheimer Str. 101, 97318 Kitzingen, Germany, [3]Research Institute of Organic Agriculture (FiBL), Walburger Str. 2, 37213 Witzenhausen, Germany; benedikt.thesing@hswt.de

A sufficient riboflavin supply of turkeys is necessary for optimal health and performance. For organic production, the use of GMO-derived vitamins is prohibited. Therefore, an alternative riboflavin source based on bakery yeast has been certified and proven as suitable. The present study aimed to determine the effect of different riboflavin concentrations in organic diets on growth performance of an intensive (B.U.T. 6) and semi-intensive (Auburn) turkey line. 768 one-day old female turkey poults (384 Auburn and 384 B.U.T. 6) were raised at two locations for two 28 day-phases (P1 and P2). Organic diets were offered with four riboflavin levels ranging from 5.4 mg/kg to 9.2 mg/kg (P1) and 4.6 mg/kg to 7.2 mg/kg (P2). All diets were similar in AME_N and amino acid content. The B.U.T. 6 animals had higher feed consumption, body weight and feed conversion rate than the Auburn animals ($P<0.05$). For the Auburn line, the varying riboflavin concentrations did not affect growth performance ($P>0.05$). The semi-intensive line was adequately provided with riboflavin even at the lowest level in both phases. The B.U.T. 6 animals fed the diets with lower riboflavin level showed higher feed consumption in P2 and in consequence, higher body weight. The increased feed consumption may be related to the lower riboflavin supply of the diet. The turkey poults consumed more feed to reach the riboflavin storage capacity of the body (represented by the liver), liver riboflavin content did not differ between feeding groups ($P<0.05$). Hence, the B.U.T. 6 hens receiving the diets with lower riboflavin concentration still had sufficient riboflavin supply to achieve maximum growth. In conclusion, the semi-intensive line Auburn requires less dietary riboflavin than intensive line B.U.T. 6 when using an organic-compliant riboflavin product. Further, B.U.T. 6 hens compensated lower dietary riboflavin concentration by higher feed consumption.

In vitro* anthelmintic evaluation of petroleum ether plant extracts against *Ascaridia galli

I. Poulopoulou[1], M.J. Horgan[2], B. Siewert[2], L. Palmieri[3], E. Martinidou[3], S. Martens[3], P. Fusani[4], V. Temml[2], H. Stuppner[2] and M. Gauly[1]
[1]Free University of Bolzano, Faculty of Science and Technology, Piazza Università 5, 39100, Italy, [2]Institute of Pharmacy/Pharmacognosy, Center for Chemistry and Biomedicine, University of Innsbruck, Innsbruck, 6020, Austria, [3]Edmund Mach Foundation, Food Quality and Nutrition Department, Via E. Mach 1, 38010, Italy, [4]Consiglio per la ricerca in agricoltura e l'analisi dell'economia agraria, Centro di ricerca Foreste e Legno, Nicolini 6 loc. Villazzano, 38123, Italy; ioanna.poulopoulou@unibz.it

This study aims to find efficient alternatives among ethno-veterinary herbs. *Ascaridia galli* eggs isolated from the worm uterus were exposed *in vitro* to 9 petroleum ether extracts (PE) in dimethyl sulfoxide (1%) from the cultivated plant species *Achillea millefolium* (AM), *Artemisia absinthium* (AA), *Artemisia vulgaris* (AV), *Cicerbita alpina* (CA), *Cichorium intybus* (CI), *Inula helenium* (IH), *Origanum vulgare* (OV), *Tanacetum vulgare* (TV), *Tanacetum parthenium* (TP), positive (flubendazole) and negative controls. The ability of different PE concentrations (0.5, 0.325, 0.2 mg/ml) to affect the embryonation rate of *A. galli* was assessed in duplicate/PE/concentration. Eggs' embryonic development (ED) was evaluated (560 eggs/replicate) from the day of egg isolation until day 28 resulting in the examination of 40,320 eggs. Analysis performed using generalized linear mixed model, stating a negative binomial distribution, having plant species, concentration, and week as fixed effect. IH and CA showed significant lower ($P<0.05$) ED the first experimental week, estimated at approximately 26%. The highest PE concentration of the species IH and CA showed a significant lower ($P<0.05$) ED, estimated at 40 (±3.04) and 41% (±2.72), respectively. Lower tested concentrations had similar patterns with IH and CA having the best performance, showing a dose dependent effect. PE extracts obtained from the reported plant species have promising results in inhibiting ED, contributing to the identification of alternative anthelmintic treatments against *A. galli*. This study is part of the 'HERBAL' project funded by the GECT 'Euregio Tirolo-Alto Adige-Trentino, 3rd call.

Effect of key husbandry factors on chicken meat and carcass quality

J. Marchewka[1], P. Sztandarzki[1], M. Solka[1], H. Louton[2], E. Rauch[3], K. Rath[4], L. Vogt[4], W. Vogt-Kaute[4], D. Ruijter[5] and I.C. De Jong[5]
[1]Institute of Genetics and Animal Biotechnology, Animal Welfare, Jastrzębiec, ul. Postępu 36A, 05-552 Magdalenka, Poland, [2]University of Rostock, Universitätsplatz 1, 18055 Rostock, Germany, [3]LMU Munich, Geschwister-Scholl-Platz 1, 80539 München, Germany, [4]Naturland, Association for Organic Agriculture e.V., Kleinhaderner Weg 1, 82166 Graefelfing, Germany, [5]Wageningen Livestock Research, Animal Health and Welfare, De Elst 1, 6708 WD Wageningen, the Netherlands; j.marchewka@ighz.pl

Various production systems for meat chickens exist in practice, ranging from conventional broilers in intensive systems to organic broilers or dual purpose chicken production. These production systems vary in the degree of extensiveness, such as the applied stocking density, genetic strain (fast- versus slower- or slow-growing breeds), presence of environmental enrichment and the type of diet. In addition to differences between production systems with respect to sustainability, meat quality may also be affected by the production system. Meat quality refers to meat quality aspects, safety, authenticity, but also the extrinsic value of the product. As an example, slower-growing broiler strains are often housed in more extensive systems. Although these slower-growing strains may have a lower breast meat yield compared to fast-growing chickens, meat quality concerns typical for intensive production systems have reported to be smaller. Diet composition is another important factor for meat quality, affecting for example the meat fatty acid profile, but also meat yield and other quality aspects. In the present review the effect of four key husbandry factors on chicken carcass characteristics and meat quality has been studied to get more insight in the effect of these factors on meat quality aspects. The key husbandry factors were: stocking density, genetic strain, environmental enrichment and diet, and literature analysis included any possible interactions between these. This knowledge will help to improve intrinsic meat quality in meat chicken husbandry systems by applying specific husbandry aspects. Scientific literature from 2012 and further has been screened. Most scientific papers were found for diet, followed by breed, enrichment and stocking density. Results will be presented at the conference.

Effects of hatching system and enrichment on broiler chickens' behaviour

J. Malchow[1], E.T. Krause[1], R. Molenaar[2], M.F. Giersberg[3] and L. Schrader[1]
[1]Friedrich-Loeffler-Institut, Institute of Animal Welfare and Animal Husbandry, Dörnbergstr. 25/27, 29223 Celle, Germany, [2]Wageningen University and Research, Adaption Physiology Group, P.O. Box 338, 6700 AH Wageningen, the Netherlands, [3]University Utrecht, Faculty of Veterinary Medicine, Department Population Health Science, Yalelaan 1, 3584 CL Utrecht, the Netherlands; julia.malchow@fli.de

After hatching in commercial hatcheries, chicks usually have no access to water and feed for 24 to 48 hours. This can lead to decrease post-hatch development and impair animal welfare. One alternative to avoid this problem is on-farm hatching. Enrichments of the housing environment can be offered to further promote animal welfare. To investigate the influence of hatching system and additional enrichment on behaviour, 320 mixed-sex broilers were kept in a 2×2 design (hatchery-hatch vs on-farm hatching × barren environment vs enriched). Environmental enrichment was provided by elevated structures (plastic grids). Individual chickens were tested twice, in an arousal test (test arena in the pen, duration: three minutes) on day 5 and in an open field test (test arena outside the pen, duration: five minutes) on day 25 of life. Latency to first move, number of field changes, and amount of defecation were recorded. Additionally, enrichment usage in the pens was analysed throughout the cycle. Statistical analyses were done with linear mixed-effect models using nlme package in R and we tested for correlations of test parameters between day 5 and 25. The usage of elevated structures was higher in on-farm hatched than in hatchery-hatched chickens (P=0.0312). On day 5, chickens hatched on-farm showed a longer latency to first move compared to the hatchery-hatched birds (P=0.0052). At day 5, chickens from barren environment tended to show more field changes than animals kept with enrichment (P=0.08) in the arousal test. On day 25, chickens from the enriched environment showed more field changes (P=0.016). Only a weak correlation was found between day 5 and day 25 in the latency to first move (r_p=0.134, P=0.017). Both treatments affected different behavioural traits, parameters in various extent: on-farm hatched chickens showed more fear in early life. Furthermore, chickens kept in a barren environment were more active.

Personality traits and exploratory behaviour of free-range slow growing broilers

C. Bonnefous[1], L. Calandreau[2], E. Le Bihan-Duval[1], V.H.B. Ferreira[2,3], A. Barbin[1], A. Collin[1], M. Reverchon[4], K. Germain[5], L. Ravon[5], N. Kruger[1], S. Mignon-Grasteau[1] and V. Guesdon[3]
[1]INRAE, Université de Tours, BOA, 37380 Nouzilly, France, [2]INRAE, CNRS, IFCE, Université de Tours, PRC, 37380 Nouzilly, France, [3]Junia Hauts de France, ISA Lille, 48 Boulevard Vauban, BP 41290, 59014 cedex, Lille, France, [4]SYSAAF, 37380, Nouzilly, France, [5]INRAE, UE EASM, Le Magneraud, CS 40052, 17700, Surgères, France; claire.bonnefous@inrae.fr

Free-range systems provide an outdoor range for broilers to give them the possibility to express more behaviours, like exploration. Nevertheless, high variability of outdoor range use between individuals of the same flock is often reported. Range use shows individual-consistency over time (early and late range access), thus suggesting that exploratory behaviour is a personality trait. Besides, foraging behaviour, locomotion, and social motivation were also linked to range use. However, these results focus on one broiler strain, while multiple strains are used in free-range farms; therefore, genetic influences and their relationship to range use remain understudied. In this study, we investigated range use in relation with behavioural traits in four slow-growing broiler strains. On that purpose, we recorded the behaviour and range use of chickens, both in the poultry house and during individual test situations, before and after range access. Our results show a strong correlation between range use measured at 37 to 46 days-old and at 56 to 67 days-old for all four strains, confirming that range use is a consistent behavioural trait. However, we found no robust links between range use and chick social motivation, foraging, and locomotion (measured before range access). Moreover, social motivation seemed to be inconsistent over time as no strong correlation (r_s<0.5 when P<0.01) was detected between the social motivation tests performed before (at 22 to 25 days-old) and after range access (at 50 to 53 days-old). Our results suggest that in very young animals, when personality traits are probably not yet established, early behavioural patterns may be poor indicators of later range use. PPILOW project has received funding from the European Union's Horizon 2020 research and innovation programme under grant agreement No 816172.

Diagnostic performance of a copro-antigen ELISA to assess nematode infections in chickens

O.J. Oladosu[1], M. Hennies[2], M. Stehr[1], C.C. Metges[1], M. Gauly[3] and G. Daş[1]
[1]Research Institute for Farm Animal Biology, Institute of Nutritional Physiology 'Oskar Kellner', Wilhelm-Stahl-Allee 2, 18196 Dummerstorf, Germany, [2]TECOdevelopment GmbH, Marie-Curie-Str. 1, 53359 Rheinbach, Germany, [3]Free University of Bozen, Faculty of Science and Technology, Universitätsplatz 5, 39100 Bolzano, Italy; oladosu@fbn-dummerstorf.de

A non-invasive method of diagnosing nematode infection is beneficial for the control of nematodes and poultry welfare. In two separate experiments (E1, E2), the performance of a copro-antigen ELISA to assess nematode infection was investigated and its accuracy with faecal egg counts (FEC) and a plasma antibody ELISA compared. E1 comprised of 179 birds (24 weeks old) and E2 had 635 chicks (1 week old). Birds were either experimentally infected with two nematode species (*Ascaridia galli* and *Heterakis gallinarum*) or kept as uninfected control. Faecal and blood samples, FEC, and worm burden data were collected from all individual birds that were necropsied at different time points in each week post-infection (wpi) to assess the development of worm antigens, antibody, worm burdens and FEC. Antigen and antibody concentrations were higher in infected birds than in uninfected ones (P<0.001) and the differences were wpi dependent. In E1 with hens, antigen concentration was significantly higher in infected birds from wpi 6-18 (P<0.001) and tended to be higher (P=0.09) in wpi 4, but not in in wpi 2. Whereas data from E2 showed a higher antigen concentration in the infected birds in wpi 2 and in wpi 6-9, no significant differences were found for wpi 3, 4 and 5. Copro-antigen ELISA had a higher overall accuracy (AUC=0.93 and 0.73 in E1 and E2, respectively) when discriminating between infected and uninfected birds than classical FEC (AUC=0.91 and 0.57, respectively) and antibody ELISA (AUC=0.83 and 0.67, respectively) as determined by the ROC analysis. While the performance of antibody ELISA and FEC appears to be heavily dependent on the changes in worm burden and maturation, performance of copro-antigen ELISA is less dependent on these changes, likely due to capturing antigens of immature and mature worms of both sexes. This project has received funding from the EU-Horizon 2020 programme under the Marie Sklodowska-Curie grant agreement No 955374.

Feather pecking and cannibalism of non-beaktrimmed turkeys in organic husbandry

D. Haug[1], B. Thesing[1], S. Göppel[1], S. Born[2], R. Schreiter[3], C. Lambertz[4], G. Bellof[1] and E. Schmidt[1]
[1] University of Applied Sciences Weihenstephan-Triesdorf, Am Staudengarten 1, 85354 Freising, Germany, [2]Bavarian LfL, Mainbernheimer Str. 101, 97318 Kitzingen, Germany, [3]HTW Dresden, Bergweg 23, 01326 Dresden-Pillnitz, Germany, [4]FiBL Deutschland e.V., Walburger Strasse 2, 37213 Witzenhausen, Germany; desiree.haug@hswt.de

The occurrence of feather pecking and cannibalism continues to challenge turkey production in terms of animal welfare. The trigger for this behaviour is not finally clarified, but is due to multifactorial causes. The aim of the study was to investigate the effects between genotype, husbandry and feeding on behavioural abnormalities of turkeys with intact beaks. For the assessment scoring of the integument condition were used. 1,344 males of a slow- (Auburn) and fast-growing (B.U.T. 6) genotype were raised under organic conditions. At two facilities (each with 24 pens) the animals were kept in three different systems (indoor housing with / without environmental enrichment / free-range system) in groups of 20 and 36 animals. Four different feeding diet formulations varied in amino acid content from a high level (according to Aviagen recommendations) to a lower level (F1-F4; 3 replications/subgroup). The study was divided into five 4-week feeding phases (P1-P5). At the end of each phase, scoring was performed to assess animal welfare. In different body areas the plumage, injuries, soiling and footpad changes were assessed using scores from 0 (no damage) to 3 or 4 (severe damage). Preliminary results show that the genotype and the husbandry conditions had an effect on injuries ($P<0.001$). B.U.T. 6 showed more injuries than Auburn (20th week: 84.4%, resp. 63.8%). At the 20th week less injuries were found in the indoor system with enrichment (53.8%) compared to the free-range system (82.5%). There is a clear trend of increasing severeness of injuries with increasing age, especially at the head and neck (20th week: 74.1%). Few abnormalities were found at the wings and back. In summary, agonistic behaviour increased with age. Based on these results, feather pecking and cannibalism depended on genotype, age and husbandry. The amino acid concentration showed no effect. For husbandry, a strong impact of external environmental factors can be suspected.

Trends in mountain sheep farming in Europe over the past 40 years – challenges and opportunities

C. Morgan-Davies[1], T. Zanon[2], M.K. Schneider[3], C.M. Pauler[3], Ø. Holand[4], A. Bernués[5], P. Dovc[6], M. Gauly[2] and G. Cozzi[7]
[1]Scotland's Rural College (SRUC), West Mains Road, EH9 3JG, United Kingdom, [2]Uni Bolzano, Universitätsplatz 5, 39100 Bolzano, Italy, [3]Agroscope, Reckenholzstr. 191, 8046 Zürich, Switzerland, [4]Norwegian University of Life Science (NMBU), Universitetstunet 3, 1433 Ås, Norway, [5]Agrifood Res & Tech Ctr Aragon CITA, Avda. Montañana 930, 50059, Zaragoza, Spain, [6]Uni Ljubljana, Groblje 3, 1230 Domžale, Slovenia, [7]Uni Padova, Viale dell'Università 2, 35020 Legnaro, Italy; claire.morgan-davies@sruc.ac.uk

The role and capacity of sheep farming in mountain areas have evolved greatly over the past decades, often shaped by farming policies, research orientation, site conditions and societal demand. This paper presents changes and opportunities faced by sheep farming in mountain areas in Europe from the 1980s to present. Therefore, it gives: (1) a review of sheep numbers trend over that period; (2) a thematic review of European research funding; and (3) a scientific literature review over that period, screening for *sheep in mountain* and hill**, giving a total of 978 publications. Results from these three strands are compiled and compared with either national or European farming policies over that same period. Three main geographical areas are used as case studies, based on their different climate, namely the North West of Europe (Norway, Scotland), Central Europe and alpine areas (Switzerland, Northern Italy, Slovenia) and South Western Europe (Pyrenees, Northern and Central Spain). Products and services provided by sheep farming in these mountain areas will be discussed. The paper will conclude on challenges and opportunities for the future of sheep farming in mountain areas, as shaped by policy orientations and megatrends.

Session 44 Theatre 2

Mountain Sheep in Polish Carpathians – centuries of tradition and present day

A. Kawęcka, M. Puchała, M. Pasternak and J. Sikora
National Research Institute of Animal Production, Department of Sheep and Goat Breeding, ul. Sarego 2, 31-047 Kraków, Poland; michal.puchala@iz.edu.pl

Shepherding and sheep breeding in the Polish mountains have a long tradition. Sheep in the Polish Carpathians were derived from the Wallachian Zackel which were brought there by the Wallachians shepherds. Pastures were used jointly by several villages by the thirteenth century and since the fourteenth century sheep were milked and lump cheese was made. The Wallachians blended with the indigenous people and spread their own customs, beliefs, pastoral culture and terminology; these customs are present to this day in the highlander names and rituals. Mountain sheep (currently three breeds – Podhale Zackel, Polish Mountain sheep, Coloured Mountain Sheep) constitute 15% of all ewes entered in the herd books and are covered by the genetic resources protection program. Mountain sheep are used comprehensively, in a way that has not changed for centuries, in line with the centuries-old pastoral tradition. Traditional mountain sheep's milk cheeses (Oscypek, Bryndza, Redykołka) have been entered on the EU list as the Protected Designation of Origin. The status of EU protection is also given to Jagnięcina Podhalańska from mountain lambs, and wool and skins are used to produce regional and traditional highlander costumes used during shepherding ceremonies. The traditional system of keeping sheep on mountain pastures is a collective, large-scale grazing in the period from May to October, and for the winter they return from the mountains to their owners' sheepfolds. Extensive grazing of mountain sheep serves the preservation of naturally valuable mountain areas, and its special form is cultural grazing understood as limited, collective sheep grazing and comprehensive pastoral management, carried out in all mountain national parks. The use of mountain sheep is accompanied by the development of the market of sheep products, traditional crafts and tourism, which is conducive to the revival of local entrepreneurship. Mountain shepherd is therefore an example of multifunctional agriculture, optimal for mountain and foothill areas, with low capital intensity, offering numerous market and public goods, while taking care of biodiversity and the environment.

Session 44 Theatre 3

Attacks of wolves (*Canis lupus*) on the herds of native breeds of sheep in the Polish Carpathians

M. Pasternak, M. Puchała, A. Kawęcka and J. Sikora
National Research Institute of Animal Production, Department of Sheep and Goat Breeding, ul. Sarego 2, 31-047 Kraków, Poland; michal.puchala@iz.edu.pl

In Poland, the gray wolf is a strictly protected species (Directive 92/43/EEC), and its population in 2020 reached 3,530. Wolves prey on deer, wild boars and smaller wild mammals, but sometimes they attack farm animals, mainly sheep. This is due to the specificity of sheep keeping: long-term grazing, mainly in mountain areas. Since 1999, a genetic resources conservation programs of sheep has been operating in Poland, which currently covers 17 breeds. In case of loss of an animal due to a wolf attack, the breeder sends to the National Research Institute of Animal Production – the Coordinator of genetic resources programs, a replacement document informing about the incident. The aim of this study was to analyse the scale of wolf attacks on native sheep herds kept in mountain areas in Poland in 2020. The material for the research was official „Declarations on the replacement of animals' sent by breeders from Lesser Poland and Subcarpathia for sheep covered by the protection program: Podhale Zackel, Coloured Mountain Sheep (CMS), Polish Pogórze Sheep and Black-headed Sheep. In 2020, sheep replacements caused by wolf attacks accounted for 10.41%. The remaining replacements resulted from the selection (76.44%) and deaths (13.15%). Attacks in mountain areas accounted for 94.82% (165/174) of all reported attacks on native sheep herds in Poland. Attacks in Lesser Poland accounted for 86.66%, which concerned the Podhale Zackel 114 (79.73%), the CMS 17 (11.88%) and the Black-headed sheep 12 (8.39%). Attacks in Subcarpathia accounted for 13.34%, which concerned Polish Pogórze Sheep 18 (81.82%) and Black-headed Sheep 4 (18.18%). Most of the attacks (84.33%) took place during the most intensive grazing period, from May to October. The average age of attacked sheep was 5-7 years. Wolves most often attacked ewes (81.59%), then lambs (11.64%), and finally rams (6.77%). The loss of sheep as a result of an attack by wolves is a threat that breeders must consider during grazing them in mountain areas. The injured breeders receive financial compensation, because the Polish State Treasury is responsible for damages caused by wolves in the livestock population.

Welfare assessment of dairy goats extensively reared in mountain ranges

S. Mattiello[1], M. Renna[2], L. Battaglini[3] and M. Battini[1]
[1]University of Milan, Department of Agricultural and Environmental Sciences – Production, Landscape, Agroenergy, Via Celoria, 2, 20133 Milan, Italy, [2]University of Turin, Department of Veterinary Sciences, Largo Paolo Braccini, 2, 10095 Grugliasco (TO), Italy, [3]University of Turin, Department of Agricultural, Forest and Food Sciences, Largo Paolo Braccini, 2, 10095 Grugliasco (TO), Italy; silvana.mattiello@unimi.it

The evaluation of goat welfare in extensive farming systems can be particularly challenging due to adverse conditions during the assessment and the relative lack of validated measures compared to those developed for intensive systems. Following a previous study conducted in semi-extensive condition, we recorded 16 welfare indicators (seven group indicators at pasture and nine individual indicators during milking) in six herds of dairy goats (Alpine and Valdostana breed) in extensive conditions during alpine summer grazing. Inter-observer reliability among three assessors was also checked. The welfare of goats resulted acceptable with some exceptions. The prevalence of some indicators differed from the one registered in semi-extensive conditions. Unexpectedly most of the goats showed a normal body condition (90.8±2.45%). This could be explained by the fact that goats were in mid-late lactation stage and that no abrupt change of feed occurred when they were moved to the summer pasture, as they had daily access outdoor in spring. However, faecal soiling was higher than in semi-extensive conditions (only 85.6±5.72% of animals with no signs of diarrhoea), possibly because of the presence of fresh grass and high parasite load due to a very rainy season. Parasite infestation may also explain the low prevalence of goats with normal hair coat conditions (45.8±2.43%). As in semi-extensive conditions, synchrony at resting was low (14.4±8.85%). The use of shelter in alpine summer range (35.2±8.75%) was lower than in semi-extensive conditions (95.1±4.86%), probably because of the insufficient availability of protected areas. The assessors reported many constraints in observing the animals at pasture (e.g. presence of woods, fog, need of binoculars for long distance observation): this explains why the reliability varied from poor for indicators collected at pasture to good for some indicators collected individually during milking.

Prediction of sheep daily gain in Swiss sheep populations

A. Burren[1], C. Aeschlimann[2] and H. Joerg[1]
[1]Bern University of Applied Sciences, School of Agricultural, Forest and Food Sciences, Länggasse 85, 3052 Zollikofen, Switzerland, [2]Swiss Sheep Breeding Association, Industriestrasse 9, 3362 Niederönz, Switzerland; hannes.joerg@bfh.ch

In 2021 fix breed effects of average daily gain up to 45 days of age were estimate for the Swiss sheep breeds Brown Headed Meat (BFS), Charolais (CHS), Dorper (DOP), Nolana (NOS), Ile de France (OIF), Rouge de l'Ouest (RDO), Black Brown Mountain (SBS), Shropshire (SHR), Valais Black Nose (SN), Suffolk (SU), Texel (TEX) and White Alpine (WAS). For this purpose, litters from 130,224 ewes with 558,307 records were used (Records by breed: BFS=89,950; CHS=6,497; DOP=8,034; NOS=2,696; OIF=5,367; RDO=1,710; SBS=95,802; SHR=5,279; SN=108,287; SU=10,368; TEX=14,394; WAS=209,923). The data were collected in the years 2010-2021. The model for predict daily gain across all breeds included the fixed effects litter size (1 female, 1 male, 2 lambs, >2 lambs) sex of lamb, litter number (1. litter, 2. litter, 3.-5. litter, ≥6. litter), age of dam (<779 days, 780-1,139 days, ≥1,140 days) and random effects of herd, year, ewe and residual effect. The coefficient of determination of the model was (R^2=0.571). The estimated fixed main breed effects ± standard errors were 387.4±2.0 (Intercept), -4.7±2.5, -58.0±2.9, -62.3±7.2, -10.7±2.0, -53.5±8.6, -28.3±1.2, -75.3±5.9, -19.6±1.9, -7.3±1.8, -24.1±1.6 and -4.4±1.2 g/day for BFS, CHS, DOP, NOS, OIF, RDO, SBS, SHR, SN, SU, TEX and WAS, respectively. Estimated fixed effects and standard errors across litter number were 9.4±0.3, 10.2±0.5 and -1.6±0.5 for 2. litter, 3.-5. litter and ≥6. litter, respectively. The effects of litter size were 11.6±0.3, -53.9±0.3 and -85.7±0.4 for 1 male, 2 lambs and >2 lambs, respectively. The effects of age of dam were very similar with 9.9±0.4 and 9.8±0.5 for 780-1,139 days and ≥1,140 days, respectively. The results are consistent with other studies and show that the most important effects on daily gain were litter size and breed.

Association of farming descriptors with sheep bulk-milk quality from semi-extensive flocks

S. Caddeo[1], D. Brugnone[2], N. Amalfitano[3], G. Bittante[3], G.M. Vacca[4] and M. Pazzola[4]
[1]Agenzia LAORE Sardegna, Servizio politiche regionali per il benessere animale, via Caprera 8, 09123 Cagliari, Italy, [2]Regione Autonoma della Sardegna, Assessorato della difesa dell'ambiente, via Roma 80, 09123 Cagliari, Italy, [3]University of Padova, Department of Agronomy, Food, Natural resources, Animals and Environment, viale dell'Università 16, 35020 Legnaro (PD), Italy, [4]University of Sassari, Department of Veterinary Medicine, via Vienna 2, 07100 Sassari, Italy; pazzola@uniss.it

Semi-extensive dairy sheep farming is based on natural cycles of animals, outdoor pasture and the use of traditional techniques. These are often considered as beneficial by the citizens and consumers. However, the same characteristics can have a negative impact on animal health and productions. The present study is included in the activities of the project GOODMILK, funded by the Italian Ministry of Agriculture. The purpose was to investigate the influences of a large panel of farming factors on bulk milk produced in sheep farms located in a central-west area of Sardinia (Italy), as a case study of the typical semi-extensive sheep farming. A total of 2,141 bulk milk samples were collected from 96 farms. Milk composition and farms descriptors were achieved. Farms with 200-399 ewes, showed a higher concentration of milk urea and conjugated linoleic acid (CLA) evidencing the best feeding management for medium-size flocks. The load of grazing animals per hectare influenced two indicators of udder health, pH and SCS, which were significantly higher in farms with overcrowded pastures, i.e. 5-10 sheep/ha, and still more evident with 11-66 sheep/ha. The mineral fertilization of pasturelands was significantly associated with a decrease of lactose concentration, another indirect indicator of udder health, and the increase of urea. Conversely, the management score, time at pasture in hours per day, distance between the farm and pasture, crossing roads, presence of grazing competitors, pasture shading, and prevalent pasture species were not significantly associated with bulk milk quality. The obtained information are useful to improve the management of dairy sheep farms and stimulates further investigations to be conducted on milk coagulation and cheese-making traits.

Relationships between the rumen ciliate protozoa, carcass characteristics and lamb meat quality

A.E. Francisco[1,2], J. Santos-Silva[1,2], A.P. Portugal[2], J.M. Almeida[1,2] and R.J.B. Bessa[1]
[1]CIISA, FMV-ULisboa, FMV-ULisboa, Avenida da Universidade Técnica, 1300-477 Lisboa, Portugal, [2]INIAV, Pólo de Inovação da Fonte Boa, Quinta da Fonte Boa, Vale de Santarém, 2005-048 Santarém, Portugal; alexandra.francisco@iniav.pt

Rumen protozoa (PTZ) comprise a very active microbial community within rumen ecosystem presenting symbiotic links with rumen methanogenic archaea. As PTZ are not essential for host´s survival, their elimination from rumen has been proposed to reduce the rumen methane production. However, the relationships between PTZ taxa and many aspects of rumen metabolism, including methanogenesis, which in turn, may influence meat and milk quality, are still scarcely understood. In the present study we analysed a dataset derived from five feeding experiments carried out by our research team involving 120 growing lambs, fed with complete diets containing forage:concentrate ratios ranging from 20:80 to 50:50, in order to identify possible links between rumen PTZ and carcasses and meat quality variables. Rumen fluid samples were collected at slaughter, and PTZ counting and identification at genus level were obtained with optical microscopy. A regression analysis, corrected for variation associated to the diets and considering the experiment as a random effect was conducted. Carcass dressing increased linearly (P=0.029) with the total of PTZ counts, the PTZ Entodiniomorphida order counts (P=0.033) and the *Entodinium* genus (P=0.041) counts (log10 cells/ml rumen fluid). The yellow coordinate (b*) of the subcutaneous fat colour decreased linearly with total PTZ (P=0.027) and *Entodinium* (P=0.046) counts. Both meat protein and the a* coordinate of meat colour increased linearly (P=0.008 and P=0.023, respectively) with the abundance of *Epidinium* genus, while intramuscular fat decreased (P=0.011). The results indicate that, in lambs from the intensive meat production system, the rumen PTZ community can be related with some carcasses and meat quality variables. Financial support was provided by the Fundação para a Ciência e a Tecnologia (FCT) (PTDC/CVT/103934/2008 I&D project and A. E. Francisco PhD grant – FCT SFRH/BD/68773/2010) and by the European Fund for Regional Development (ValRuMeat project – ALT20-03-0145-FEDER-000040).

Effects of ewes' GH2-Z genotypes on milk parameters and fresh cheese traits

M.R. Marques[1,2], J.M. Almeida[1,2], J.M.B. Ribeiro[2], A.P.L. Martins[3], C.C. Belo[2] and A.T. Belo[2]
[1]CIISA-FMV, Universidade de Lisboa, Avenida da Universidade Técnica, 1300-477 Lisboa, Portugal, [2]INIAV, UEISPSA, Quinta da Fonte Boa, 2005-048 Vale de Santarém, Portugal, [3]INIAV, UTI, Avenida da República, Quinta do Marquês, 2780-157 Oeiras, Portugal; rosario.marques@iniav.pt

Growth hormone (GH) is involved in milk production and composition regulation. It affects milk protein content by regulating the expression of casein genes, which might affect milk properties for cheese-making (MPC) and cheese traits. Aiming to evaluate the impact of GH2-Z genotypes on MPC and cheese traits in Serra da Estrela ewes, fresh cheeses have been made from milk batches of each of the GH2-Z genotypes [AA (R9R/S63S), AB (R9C/S63S), and AE (R9R/S63G)] at days 35, 45 and 75 of lactation. Milk batches' composition [fat, protein, lactose, total solids (TS), and fat-free TS (FFTS) contents], and MPC parameters acidity (pH), clotting time (R; min), curd firmness (AR), curd firmness after 2×R (A2R), curd firmness after 20 (A20) and 40 min (A40), and firming rate (0K20, min) were evaluated at the same lactation days. After 24 and 96 h of cheesemaking, weight loss has been calculated, as well CIE lab colour (L*a*b*), chroma (C), and hue angle (H), and texture profile analysis parameters (fracturability, hardness, adhesiveness, cohesiveness, springiness, resilience, gumminess, and chewiness) were calculated. Data were analysed using SAS Proc Mixed. Genotypes had no significant effects nor in milk batches, composition or in its MPC parameters ($P>0.05$), however, they significantly affected colour (L*, b*, and C*; $P<0.01$), and all texture parameters ($P<0.001$), but resilience. Lactation days significantly ($P<0.05$) affect milk fat, protein, lactose and TS contents, pH, curd firmness A20 and (A40, and firming rate (0K20, min), all cheese colour ($P<0.001$), and texture traits, which are also influenced by the hour at which texture measurements were made ($P<0.0001$). In conclusion, although ewes' GH2-Z genotypes had no effect on milk traits, they significantly influenced fresh Serra da Estrela cheeses' colour and texture parameters. Funding: Project PTDC/CVT/112054/2009 funded by FCT.

The role of the lactose in the milk coagulation properties of Sarda dairy sheep

S. Carta, F. Correddu, A. Cesarani and N.P.P. Macciotta
Università degli Studi di Sassari, Dipartimento di Agraria, Viale Italia 39, 07100, Italy; scarta2@uniss.it

Milk coagulation properties (MCP), defined by rennet coagulation time (RCT), curd-firming time (k20), and curd firmness (a30), are variables of importance for the dairy sheep industry. It is well known that some milk components (i.e. proteins, caseins, and fat) affect MCP. However, also lactose is important for the milk coagulation proves even if it does not play a direct role in the coagulation process and about 90% of it ends up in the whey. The aims of this study were: (1) to investigate the effects of days in milk (DIM), parity, and lambing month (LM) on milk components, MCP, and individual laboratory cheese yield (ILCY); (2) to highlight the effect of lactose, protein, and fat classes on MCP and ILCY. For this purpose, 2,358 milk samples from 509 ewes were analysed with Fourier transform midinfrared for milk components, and with the Formagraph instrument for MCP. ILCY was determined by a micromanufacturing protocol. Data were analysed with a mixed linear model that included DIM, parity, LM, and milk components classes as fixed effects and animal and herd-test-day as random effects. Milk component (fat, protein and lactose) classes were created according to the quartiles of the distribution of each parameter, i.e. the first quartile represents the lowest values. Almost all parameters were influenced by DIM, whereas only milk yield, lactose, somatic cell count, and NaCl were influenced by parity. RCT was influenced by DIM, k20 by parity and LM, and a30 by DIM and parity. ILCY was affected only by DIM class. An increase of the ILCY (desirable) was found with high content of lactose, protein, and fat. Lactose classes influenced ILCY and all MCP parameters ($P<0.001$). In particular, the higher the lactose content, the lower (i.e. the better) the RCT; k20 was also improved by the high concentration of lactose. Finally, a30 was lower in the lower classes of lactose and fat. In conclusion, lactose content represents one of the main factors influencing the milk coagulation properties and, therefore, it may be used to predict the cheesemaking aptitude of sheep milk.

Session 44 Poster 10

Relationship between bacterial communities and volatile aroma compounds of a ewe's raw milk cheese

G. Santamarina-García, G. Amores, I. Hernández and M. Virto
Lactiker Research Group, University of the Basque Country (UPV/EHU), Paseo de la Universidad 7, 01006 Vitoria-Gasteiz, Spain; gorka.santamarina@ehu.eus

Cheese microbiota contributes to various biochemical processes that lead volatile compounds formation and flavour development during cheese ripening. Nonetheless, the role of these microorganisms in volatile aroma compounds production is little understood. This work discusses the relationship between the dynamics and odour impact of volatile compounds and bacterial succession, analysed by high-throughput sequencing (HTS), during ripening of a ewe's raw milk cheese (Idiazabal cheese, PDO). By means of SPME-GC-MS, 81 volatile compounds were identified, among which organic acids predominated, followed by esters, ketones and alcohols. Through Odour Impact Ratio values, esters and acids were reported as the predominant odour-active families and individually, ethyl hexanoate, ethyl 3-methyl butanoate, ethyl butanoate, butanoic acid or 3-methyl butanal were notable odorants, providing fruity, rancid, cheesy or malt odour notes. Using an O2PLS approach with Spearman's correlations, 12 bacterial genera previously identified by HTS were reported as key bacteria for the volatile and aromatic composition of Idiazabal cheese. Specifically, *Psychrobacter, Enterococcus, Brevibacterium, Streptococcus, Leuconostoc, Chromohalobacter, Chryseobacterium, Carnobacterium, Lactococcus, Obesumbacterium, Stenotrophomonas* and *Flavobacterium*. Lactic acid bacteria (LAB) were highly related to acids, esters and alcohols formation, whereas environmental and/or non-desirable bacteria were related to ketones, hydrocarbons and sulphur compounds. These results provide novel knowledge to help understand the aroma formation in a ewe's raw milk cheese.

Session 44 Poster 11

Goat milk fatty acid profile as affected by the inclusion of cocoa bean shell in the goat diet

M. Renna[1], C. Lussiana[2], L. Colonna[2], V.M. Malfatto[2], A. Mimosi[2] and P. Cornale[2]
[1]University of Turin, L.go Braccini 2, 10095 Grugliasco, Italy, [2]University of Turin, L.go Braccini 2, 10095 Grugliasco, Italy; manuela.renna@unito.it

Agro-industrial by-products can be included in diets destined to ruminants with expected positive outcomes in terms of environmental, economic, and ethical sustainability of animal-derived food production. This study was designed to assess the effects of the dietary inclusion of cocoa bean shell, a by-product of the cocoa (*Theobroma cacao* L.) industry, on milk fatty acid (FA) profile of dairy goats. Twenty-two Camosciata delle Alpi goats were divided into two balanced groups and fed mixed hay *ad libitum*. One group (control, CTRL) also received 1.2 kg/head × day of a commercial concentrate, while the other group (cocoa bean shell, CBS) received 1.0 kg/head × day of the same concentrate and 0.2 kg/head × day of pelleted cocoa bean shell [per kg dry matter: 173 g crude protein; 61 g ether extract; 495 g neutral detergent fibre; 177 g lignin; 8.76 g tannins; net energy for lactation: 4.27 MJ]. The two diets were formulated to be isonitrogenous and isoenergetic. After 10 days of diet adaptation, individual milk samples were collected four times every 10 days and analysed for their milk FA profile. Data were statistically analysed with a mixed model for repeated measures over time and significance was set at $P<0.05$. The inclusion of CBS in the goat diet induced only minor changes to the milk FA profile. These changes, mainly driven by lower PUFA and higher fibre intakes from the CBS diet when compared to the CTRL diet, included increased concentrations of both *iso*- and *anteiso*-branched-chain FA, total monounsaturated FA, stearic and oleic acids, and a decreased $\sum n6/\sum n3$ FA ratio. Other groups of FA, such as *de novo* saturated FA, total polyunsaturated FA and total conjugated linoleic acids were not affected by the dietary treatment. The amount of tannins (almost equally represented by hydrolysable and condensed forms) in CBS was too low to exert a significant effect on the concentration of the majority of ruminal biohydrogenation intermediates (e.g. *trans*-octadecenoic and *trans*-octadecadienoic acids). This suggests no impairments of ruminal biohydrogenation pathways or steps when including CBS in the goat diet.

Session 44 — Poster 12

The impact of bacterial shifts on several quality and safety parameters of a ewe's raw milk cheese

G. Santamarina-García, G. Amores, I. Hernández and M. Virto
Lactiker Research Group, Faculty of Pharmacy, University of the Basque Country (UPV/EHU), Paseo de la Universidad 7, 01006 Vitoria-Gasteiz, Spain; gorka.santamarina@ehu.eus

The microbiota inhabiting cheese is of great importance since it contributes to the production of several compounds related to cheese quality and safety. Nonetheless, so far no studies have been developed in cheese to elucidate the relationship between the bacterial succession characterized by high-throughput sequencing (HTS) and the evolution during ripening of several quality and safety parameters, such as gross composition, free fatty acids (FFA) and biogenic amines (BA). In this study, Idiazabal PDO cheese was analysed, which is a semi-hard or hard cheese produced with ewe's raw milk. By means of HTS, it was observed that lactic acid bacteria (*Lactococcus, Lactobacillus, Leuconostoc, Enterococcus, Streptococcus* and *Carnobacterium*) predominate during ripening, whereas the relative abundance of non-desirable and/or environmental bacteria (such as *Pseudomonas, Staphylococcus* or *Chromohalobacter*) is reduced. In terms of quality and safety parameters, 8 gross parameters were monitored (pH, dry matter, protein, fat, Ca, Mg, P and NaCl) and a total of 21 FFA and 8 BA were detected. Through an O2PLS approach with Spearman's correlations, the non-starter LAB *Lactobacillus*, *Enterococcus* and *Streptococcus* were reported as positively related to the evolution of gross composition and FFA release, while only *Lactobacillus* was positively related to BA production. On the other hand, environmental or non-desirable bacteria showed negative correlations, which could indicate the negative impact of gross composition on their growth, the antimicrobial effect of FFA or their ability to degrade BA. Even so, *Obesumbacterium* and *Chromohalobacter* were positively related to FFA release and BA production, respectively. This study provides novel information to help understand the functional relationships between bacterial succession and the evolution of several cheese quality and safety parameters.

Session 44 — Poster 13

May dietary fatty acids with antilipogenic effect influence milk fat content in sheep?

A. Della Badia[1], G. Hervás[1], R. Gervais[2], P. Frutos[1] and P.G. Toral[1]
[1]Instituto de Ganadería de Montaña (CSIC-Universidad de León), Finca Marzanas, 24346, Grulleros, León, Spain, [2]Université Laval, Département des Sciences Animales, 2425 Rue de l'Agriculture, Québec G1V 0A6, Canada; a.dellabadia@csic.es

The milk fat concentration in dairy ewes seems more stable than in cows. Sheep are not prone to milk fat depression (MFD) induced by plant oil supplements, but marine dietary lipids decrease milk fat content in this species. Because this latter response would not be fully explained by changes in biohydrogenation metabolites with a presumed or confirmed antilipogenic effect (e.g. t10-18:1 or t10c12-CLA), other fatty acids (FA) may play a role. Thus, this study aimed at examining relationships between milk FA and milk fat content in dairy ewes. We compiled a database comprising 23 trials conducted by our team and including 103 lot observations, 55 dietary conditions (using different lipid supplements, such as sunflower or fish oils) and 78 individual milk FA. Pearson correlation coefficients were generated using the proc CORR of SAS, and linear and quadratic relationships between milk fat and FA contents were examined using the proc MIXED. Surprisingly, FA showing the strongest correlation with milk fat concentration were c11-18:1, c9-16:1 and 22:6n-3 ($r=-0.81$, -0.79 and -0.76, respectively; $P<0.001$), which are partly or mostly provided by marine oils. As observed previously, t10-18:1 only showed a moderate correlation with milk fat content ($r=-0.48$; $P<0.001$) and no association between this trait and t10c12-CLA was detected ($r=-0.06$; $P=0.59$). Prediction models indicated that inverse relationships with milk fat concentration were linear for c11-18:1 and c9-16:1 ($P<0.001$) and quadratic for 22:6n-3 ($P<0.001$). Overall, results suggest that milk fat content in dairy sheep may be influenced by certain dietary FA, consistent with their reported antilipogenic activity in laboratory models and meat sheep. Further research would then be necessary to confirm if these FA that are highly correlated with milk fat content also have inhibitory effects on mammary lipogenesis in ruminants in general and in sheep in particular, supporting a contribution to diet-induced MFD. Acknowledgements: project PID2020-113441RB-I00, MCIN/AEI; grant PRE2018-086174, MCIU/AEI/FSE, EU.

Session 44 Poster 14

The growth, feed intake and back fat deposition of South African Boer goats

T.S. Brand[1,2], J.P. Van Der Westhuizen[1] and J.H.C. Van Zyl[1]
[1]University of Stellenbosch, Animal Sciences, Merriman Street, 7607 Stellenbosch, South Africa, [2]Department of Agriculture: Western Cape, Directorate Animal Sciences, Private Bag x1, 7607 Elsenburg, South Africa; tersb@elsenburg.com

Increases in the prices for mutton, lamb meat and chevon in South Africa have resulted in farmers finishing lambs or kids in feedlot systems in order to optimize growth rates and achieve a desirable slaughter weight at an earlier age with an optimal degree of fatness. Growth curves as well as other mathematical models may be use to describe the production traits of animals. Analysis of such growth models can be helpful in establishing precision feeding strategies, as well as optimal slaughter age. For this study, the growth of 20 Boer goat does and 18 Boer goat castrates were monitored in order to assess growth, intake and back fat traits. Male goat kids were castrated by use of Burdizzo at 7 days of age. The goat kids were weaned at about 16 weeks of age (average weight 25.6 kg) housed in individual pens, where they were reared on a feedlot diet (12.0 ME MJ/kg and 14.28% protein) supplied *ad libitum*. Feed intake, back fat thickness and growth of the goat kids were monitored weekly and daily dry matter intake was calculated for each animal. Production was monitored from weaning up to 252 days of age. Back fat was first visible and measured via ultrasound at 138 days of age (33.31±0.659 kg). Traits measured included cumulative feed intake, average daily intake (ADI), average daily gain (ADG), feed conversion ratio (FCR) and fat accumulation. No significant difference was found between the cumulative intake (~121.04±2.408 kg) and ADI (~1.297±0.026 kg/day) of castrates and does respectively. No difference was found in ADG of castrates and does (~183±6.54 g/day), the FCR for castrates and does similarly (~7.16±0.162) did not differ. An average back fat thickness of 0.21±0.01 cm was measured at 138 days of age (33.31±0.66 kg) and an average back fat thickness of 0.290±0.002 cm was measured at 252 days of age (49.25±1.01 kg). The study mathematically described the growth, feed intake as well as the fat accumulation of South African Boer goats. These results will be use in an optimization model to predict the ideal slaughter age of Boar goats in a feedlot.

Session 45 Theatre 1

Breeding strategies to reduce methane emission from dairy cattle in 10 years

O. González-Recio[1], J. López-Paredes[2], A. Saborío-Montero[1], A. López-García[1], M. Gutiérrez-Rivas[1], I. Goiri[3], R. Atxaerandio[3], E. Ugarte[3], N. Charfeddine[2], J.A. Jiménez-Montero[2] and A. García-Rodríguez[3]
[1]INIA-CSIC, Animal Breeding, Ctra La Coruña km 7.5, 28040, Spain, [2]CONAFE, Valdemoro, Valdemoro 28340, Spain, [3]Neiker- BRTA, Animal production, Campus Agroalimentario de Arkaute s/n, Arkaute 01192, Spain; gonzalez.oscar@inia.es

The recent global methane pledge aims to limit methane emissions by 30% in 10 years compared with 2020 levels. Decreasing enteric CH_4 from ruminants without altering animal production is desirable both as a strategy to decarbonise the livestock economy and to improve feed conversion efficiency. Selective breeding can contribute to reduce methane emissions from ruminants. We used data for dry matter intake (DMI) obtained from 551 dairy cows in 5 farms, as well as 4,624 methane emission measurements from 1,501 cows in other 14 farms. In addition, rumen content was extracted from 437 cows with methane phenotypes. Data from traits in the routinely milk recording scheme were available for all cows. The heritability estimates were 0.17 (±0.05) for methane production and 0.18 (±0.04) for methane concentration. Genetic correlations between methane and milk yield traits ranged between -0.05 (±0.11) and 0.50 (±0.11). Genetic correlations between methane and body capacity trait ranged from 0.14 to 0.31. The estimated heritability for DMI was moderate (0.16±0.03), with a genetic correlation with milk yield of 0.41 (±0.11), and with capacity index of 0.20±0.09. Genetic correlations between methane traits and DMI were positive and ranged from 0.20 (±0.48) to 0.27±0.43. The heritability of the core microbiota composition was >0.30 at all taxonomic levels and showed strong genetic correlation with methane traits (0.42±0.19 and 0.83±0.19), and dry matter intake (0.32±0.36). The selection responses were assessed under different possible scenarios. All evaluated scenarios resulted in reduced global emissions in 10 years. However, a partial reduction of 20% would only be possible by applying ad-hoc weights for methane in the selection index to achieve such a desired genetic gain. This would come at an expense of a slower genetic progress in milk yield and its components, and thus the economic response. These results should be considered when applying policies to reduce methane emissions from livestock.

Breeding for carbon neutrality, the role of the rumen

S.J. Rowe[1], T. Bilton[1], M. Hess[1], S. Hickey[1], R. Jordan[1], C. Smith[1], H. Henry[1], L. McNaughton[2], P. Smith[3], P. Paraza[4], G. Noronhe[5], H. Flay[6], J. Budel[1] and J.C. McEwan[1]

[1]AgResearch, Puddle Alley, Mosgiel, 9092, New Zealand, [2]Livestock Improvement Corporation, Private Bag 3016, Hamilton 3240, New Zealand, [3] Teagasc, Animal and Grassland Research and Innovation Centre, Dunsany, Ireland, [4]Instituto Nacional de Investigación Agropecuaria (INIA), Estación Experimental Las Brujas, Canelone, Uruguay, [5]Universidade Federal do Pará (UFPa), Graduate Program in Animal Science, Castanhal, Brazil, [6]DairyNZ Limited, Private Bag 3221, Hamilton, New Zealand; suzanne.rowe@agresearch.co.nz

Reducing methane emissions from ruminants via breeding requires indirect or proxy measures, as direct measures are often impractical. Rumen microbial community (RMC) profiles may be a suitable proxy. We explored a low-cost, high-throughput method for profiling the rumen microbiome in sheep and cattle to determine how predictive RMC are for rumen related traits. We describe a dataset of 3640 sheep across 8 NZ flocks, and a diverse dataset of 1015 cattle from an Enteric Fermentation Flagship project, funded by the Global Research Alliance. We used restriction enzyme reduced representation sequencing to generate RMC. In sheep, we predicted methane emissions from RMC profiles and obtained higher prediction accuracies than for host genomic prediction. Methane traits measured directly using portable accumulation chambers (PAC) and predicted from RMC profiles were highly genetically correlated (0.77). The cattle samples were from animals across a diverse range of environments and production systems, and methane emission and feed intake were measured on 62% and 90% of the animals respectively. We explored the variation in the RMC across all samples and found large differences not only between countries but also between different systems within a country. We obtained prediction accuracies around 40%-60% for methane and 20%-30% for residual feed intake when predicting across cohorts. This work suggests that there is potential for using RMC to predict important livestock traits in diverse production systems used around the world and that RMC profiles have the potential to be used as a proxy for methane emissions in ruminants. Finally, there is evidence that direct and indirect methane measures could be combined within a breeding scheme.

Building the capacity to reduce enteric methane emissions in Canadian dairy cattle

H.R. Oliveira[1,2], F. Malchiodi[2,3], S. Shadpour[2], G. Kistemaker[1], D. Hailemariam[4], G. Plastow[4], F.S. Schenkel[2], C.F. Baes[2,5] and F. Miglior[1,2]

[1]Lactanet Canada, 660 Speedvale Avenue West, N1K 1E5 Guelph, Canada, [2]University of Guelph, Centre for Genetic Improvement of Livestock, Department of Animal Biosciences, N1G 2W1 Guelph, Canada, [3]Semex, Ontario, N1H 6J2 Guelph, Canada, [4]University of Alberta, Alberta, T6G 2H1 Edmonton, Canada, [5]University of Bern, Institute of Genetics, Vetsuisse Faculty, 3012 Bern, Switzerland; fmiglior@uoguelph.ca

Together with other major international dairy organizations, Dairy Farmers of Canada has pledged to reach net-zero greenhouse gas (GHG) emissions from farm-level dairy production by the year 2050, with a milestone to be reached by 2030. Consequently, Canada is continuously building the capacity to measure and/or predict enteric methane (CH_4) emissions to be used for both herd monitoring and genetic tools. Data collection of CH_4 emissions started in 2016 in two research herds using the GreenFeed System. Genetic analysis of this data from 330 cows has resulted in heritability estimates of 0.16 (±0.10), 0.27 (±0.12) and 0.21 (±0.14) for daily methane production, methane yield, and methane intensity, respectively. Additionally, CH_4 emissions were predicted using milk mid-infrared (MIR) spectra data at various DIM ($r \geq 0.70$). There is an opportunity for Canada to exploit this option as 12M milk MIR records from 1.6M cows (130,000 genotyped) have been accumulated so far since 2018. Furthermore, within the Resilient Dairy Genome Project (RDGP), emissions data has currently been collected from 661 Canadian cows, and over 6,000 CH_4 phenotyped cows are expected to be available by the end of the project across multiple international partners. Future plans include the recording of methane emissions in robotic farms using the emission sniffers, which will enable a higher throughput of CH_4 measurements across the whole lactation and potentially multiple breeds.

Session 45 Theatre 4

Breeding for feed efficient and low methane-emitting dairy cows is feasible

C.I.V. Manzanilla-Pech[1], R.B. Stephensen[1], G.F. Difford[2], P. Løvendahl[1] and J. Lassen[1,3]
[1]Aarhus University, Center for Quantitative Genetics and Genomics, Blichers Allé 20, Postboks 50, 8830, Denmark, [2]Norwegian University of Life, Department of Animal and Aquacultural Sciences, Faculty of Biosciences, Norway, P.O. Box 5003, 1433 Aas, Norway, [3]Viking Genetics, Ebeltoftvej 16, 8960 Randers, Denmark; coralia.manzanilla@qgg.au.dk

Methane (CH_4) is the second largest GHG, with a short half-life but more heat-trapping power than CO_2. In Europe, the dairy cattle sector contribute with 37.5% of the total GHG emissions from livestock. Thus, the EU Commission has the aim of reducing GHG emissions to at least 55% by 2030 and reach neutrality by 2050. Using genetics to select for low methane emitting cows is a sustainable and permanent approach. In the last decade, several countries have included a feed efficiency index (including residual feed intake; RFI) in their breeding goal. Several studies have showed that RFI is favourably correlated with CH_4 emissions. Consequently selecting for RFI could help reducing CH_4 emissions. Could RFI alone or in combination with methane records help to reduce methane emissions in dairy cattle without compromising yield? Hence, the aim of this study was to: (1) evaluate the expected correlated response of CH_4 and milk production, when selecting for feed efficiency with or without including methane; (2) quantify the economic impact of reducing CH_4 emissions in the Danish Holstein population; (3) compare our results with a previous study with a collation of 4 countries and 3 different CH_4 measuring methods. Measurements of CH_4 on 650 Holstein cows recorded between 2013 and 2020 at the Danish Cattle Research Center were available. Records on dry matter intake (DMI), body weight (BW), and energy corrected milk (ECM) were also available, and they were used to calculate RFI. Methane emissions and RFI were strongly correlated. Selecting for feed efficiency had a positive impact on reducing CH_4 emissions; though, adding a negative economic value for methane would accelerate the reduction of emissions, although with a small impact in the genetic gain for milk production. These results confirmed earlier results with an international dataset with CH_4 measurements from different methods.

Session 45 Theatre 5

A novel microbiome-driven breeding strategy for feed intake and mitigation of methane emissions

R. Roehe[1], M. Martínez-Álvaro[1], J. Mattock[2], Z. Weng[3], R.J. Dewhurst[1], M.A. Cleveland[3] and M. Watson[2]
[1]Scotland's Rural College, Easter Bush Campus, EH259RG, United Kingdom, [2]The Roslin Institute, Easter Bush Campus, EH259RG Edinburgh, United Kingdom, [3]Genus plc, DeForest, WI 53532, USA; rainer.roehe@sruc.ac.uk

Individually recorded feed intake is one of the most important traits in animal breeding to improve feed efficiency since feed costs comprise 60% to 80% of variable costs of beef production. Environmentally, beef production is associated with a substantial carbon footprint, mainly due to methane (CH_4) emissions. Measuring these traits is costly, therefore, proxy traits such as rumen microbial biomarkers are essential for cost-effective breeding strategies. Data from 359 beef cattle obtained in factorial design experiments including breeds (Aberdeen Angus, Limousin, Charolais, and Luing) and basal diets (forage or concentrate), and with records of host genomics (68,871 SNPs), ruminal whole metagenomics (abundances of 3,631 microbial genes; MG), dry matter intake (DMI) and methane yield (CH_4Y, g CH_4/DMI) measured in respiration chambers. Bivariate Bayesian host genomic analyses of DMI or CH_4Y with MG abundances were performed fitting trial-breed-diet as a fixed effect and host genomics as a random effect. Of the 653 heritable MG, 75 showed relevant genomic correlations (r_g) to DMI (Prob>0.95), of which 65 were negatively (ranging from -0.39 to -0.81) and only 10 positively (ranging from 0.41 to 0.63) correlated, suggesting that most of these host genomically controlled MG were suppressing DMI. The 10 MG positively genomically correlated to DMI are mainly involved in amino acid and carbohydrate metabolism. One of those 10 MG (*mfnA*) is reported in *Methanocaldococcus jannaschii* to be involved in β-alanine metabolism required for coenzyme A biosynthesis, which would also explain its positive genomic correlation (0.69) with CH_4Y (Prob=0.96). Sixteen of the 75 MG showed r_g in opposite directions with DMI than with CH_4Y, suggesting that genetic improvements in DMI and CH_4Y can be achieved simultaneously. This microbiome-driven breeding provides a new opportunity to select MG, which e.g. increase DMI and reduce CH_4Y in a cost-effective breeding strategy considering functional biological processes.

Lactation modelling and effects of crossbreeding on milk production and enteric-methane-emissions

G. Martínez-Marín[1], H. Toledo-Alvarado[2], N. Amalfitano[1], L. Gallo[1] and G. Bittante[1]
[1]University of Padova, DAFNAE, Viale dell'Università, 16, 35020, Legnaro (PD), Italy, [2]National Autonomous University of Mexico, FMVZ, Department of Genetics and Biostatistics, Av. Universidad 3000, 04510, CDMX, Mexico; gustavojavier.martinezmarin@phd.unipd.it

The production traits (PT) through lactation depend on several factors, which can be estimated within a population to know their impact, establish strategies to increase their performance and reduce the environmental impact. Therefore, the objective of this study was to evaluate the effects of crossbreeding, herd, and parity on PT such as milk (MY), fat (FAT), and protein (PRO) and enteric methane emissions (EME using FTIR spectra) using a non-linear function for lactation curve modelling. The data includes 1,059 test day records from cows of the Holstein breed (HOL) and from a 3-breed rotational crossbreeding scheme involving HOL, Montbeliarde and Viking Red bulls (CRO), 2 herds (BON, CAP) managed for different cheese productions (Grana Padano and Parmigiano, respectively), and 3 parity categories (1, 2 and ≥3). The EME traits studied were daily methane production (dCH_4, g/d), methane yield per unit of dry matter intake (CH_4/DMI, g/kg), and methane intensity per kg of corrected milk (CH_4/CM, g/kg) predicted from milk spectra. A NLIN model was fitted including the Wilmink-derived modelling function and the fixed effects of breed, herd and parity. All traits except FAT were affected by herd and parity (MY and PRO increased while the other traits decreased with increasing parity number). The CRO presented lower MY and higher FAT and PRO than HOL, whereas EME were not different. All PT and EME lactation models showed significant differences (P<0.0001), with zenith models (positive peak of lactation) for MY and CH_4/DMI, nadir models (negative peak) for FAT and PRO, and upward models (continuously growing) for CH_4/CM and dCH_4. In conclusion: lactation curve modelling is very different for different traits, and herd has a stronger effect than breed and parity. Crossbreeding do not affect EME traits of cows during lactation, but could affect EME during herd life through the increase of fertility and longevity.

Links between gut microbiome functions and feed efficiency in two divergently selected pig lines

A. Cazals[1], O. Zemb[2], A. Aliakbari[2], Y. Billon[3], H. Gilbert[2] and J. Estelle[1]
[1]Université Paris-Saclay, INRAE, AgroParisTech, GABI, Domaine de Vilvert, 78352 Jouy-en-Josas, France, [2]Université de Toulouse, INRAE, ENVT, GenPhySE, 24 Chem. de Borde Rouge, 31320 Auzeville-Tolosane, France, [3]INRAE, GenESI, INRAE Le Magneraud, 17700 Surgeres, France; anais.cazals@inrae.fr

The gut microbiota plays a major role in the digestive, absorptive and metabolic processes in pigs. The study of the combined impact of host genetics and the gut microbiota composition and its functions could contribute to new strategies to improve feed efficiency traits in livestock. Based on 16S sequencing data, previous results on 588 faecal samples from two Large White pig lines divergently selected for residual feed intake (RFI) showed that some features of the gut microbiota composition have relevant genetic correlations with feed efficiency traits. In this study, we predicted KEGG Orthologs (KO) functions using PICRUSt2 software on these datasets, in order to gain insights on the microbiota functions potentially relevant for feed efficiency. A non-metric multidimensional scaling on KO abundances showed a large difference between the two lines, and differential abundance analysis revealed over 3,000 differentially abundant (DA) KO functions, confirming the major microbiota differences already found in 16S analyses. Using RFI values, differential analysis revealed over 5 DA KO within the high RFI line associated with feed efficiency. In contrast, no KO was significant in low RFI line, suggesting that different mechanisms could explain the variability of RFI within each line. Finally, KO enrichment analysis highlighted functions and pathways consistently associated with feed efficiency. Currently, we are validating these results with whole meta-genome sequencing performed on a subset of samples. Overall, this study will provide new insights to better understand how gut microbiome functions contribute to the variability of feed efficiency in pigs.

Genomic predictions for residual feed intake in Italian Simmental

A. Cesarani[1], L. Degano[2], A. Romanzin[3], D. Vicario[2], M. Spanghero[3] and N.P.P. Macciotta[1]
[1]University of Sassari, Dipartimento di Agraria, viale Italia 39, 07100, Sassari, Italy, [2]Associazione Nazionale Allevatori Pezzata Rossa Italiana (ANAPRI), Via Ippolito Nievo 19, 33100 Udine, Italy, [3]University of Udine, Dipartimento di Scienze Agroalimentari, Ambientali e Animali, via Palladio 8, 33100 Udine, Italy; acesarani@uniss.it

The residual feed intake (RFI), defined as the difference between the actual and predicted intake based on animal performances, is the most widely used feed efficiency index, especially in beef cattle. Some breeding programs started to include it as selection goal, due to its genetic variability and heritability. Aim of this work was to study the feasibility of including RFI in the Italian Simmental (IS) breeding scheme. A total of 468 genotyped steers born from 2017 to 2020 were performance tested in the IS genetic station from 2018 to 2021. Genotypes at 42,141 loci were available for other 11,879 IS animals. The genetic background of RFI was analysed though: (1) variance components and heritability estimation; (2) genetic predictions with the EBV and GEBV estimation; (3) indirect predictions, with the DGV estimation. For all these analyses, a mixed animal model with contemporary group, twin status, and parity of the mother as fixed effects was carried out. The random additive effect was modelled using the numerator relationship matrix based on pedigree (BLUP) or using the relationship matrix H which combines pedigree and genomic matrices (ssGBLUP). Genetic and indirect predictions were validated using the youngest 18 steers. Two scenarios were analysed: WHOLE, where the candidates had phenotypes; REDUCED, where phenotypes of the candidates were removed. Heritability estimates were 0.27±0.19 and 0.28±0.16 for BLUP and ssGBLUP, respectively. Correlations between breeding values estimated using or removing the phenotypes of the young candidates were 0.41 (BLUP) and 0.69 (ssGBLUP). Correlations between adjusted phenotypes (WHOLE) and truncated breeding values (REDUCED) were 0.06 for BLUP and 0.45 for ssGBLUP. Correlation between GEBV and DGV was 0.70. The use of the extra genotypes never improved these correlations. Results of the present study suggested the possibility to genetically improve RFI in the IS population. The use of genotypes in the ssGBLUP would lead to more accurate predictions.

Session 45 Theatre 9

Identification of genomic variants for feed efficiency in pigs using whole genome sequencing

A. Parveen[1], T. Sweeney[2] and S. Vigors[1]
[1]University College Dublin, School of Agriculture & Food Science, Dublin 4, Ireland, [2]University College Dublin, School of Veterinary Medicine, Dublin 4, Ireland; staffordvigors1@ucd.ie

Feed efficiency is a critical trait in the pig production industry for its economic and environmental significance. Residual feed intake (RFI) is a metric for feed efficiency and responds to selection, but to this point reliable genetic markers of the trait have yet to be identified. An alternative approach is to develop predictive genetic markers such as single nucleotide polymorphisms (SNPs) for selecting highly feed-efficient animals to be more rapidly incorporated into breeding programs. The objective of this study, therefore, was to use whole-genome sequencing to identify SNPs associated with feed efficiency in a pig population divergent for RFI. Twelve DNA samples from each low feed efficient (HRFI) and high feed efficient (LRFI) line, were subjected to next-generation sequencing and were mapped onto Sscrofa11.1. An average of 247 million reads were generated with ~97.3% of reads successfully assembled onto the reference genome. Potential Loci were identified based on the single nucleotide polymorphism frequency difference between the two pig lines. The data identified 54 regions on a total of 15 chromosomes where the size ranged from 0.001 Mbp to 0.15 Mbp. Identified SNPs were in quantitative trait loci (QTLs) associated with genes such as HLCS, CLYBL, BNIP3L, GALNTL6, ELOVL6, PRKAG2 and LGMN. The genes from the QTLs were related to pathways such as metabolism, insulin regulation, nervous system, immune system, signal transduction and other cellular processes. Out of the 54 loci, HLCS on chromosome 13, was reported as a potential marker for RFI in a previous genome-wide association study. This work established that whole-genome sequencing is an efficient and direct method for identifying candidate genes for complex traits like feed efficiency. Ongoing work involves the validation of these SNPs through further genotyping in a larger catalogue of DNAs from divergent RFI pig populations.

Session 45 Theatre 10

Genetic correlations between feed efficiency, production and fertility in Nordic Red Dairy cattle

T. Mehtiö[1], E. Negussie[1], E.A. Mäntysaari[1], G.P. Aamand[2] and M.H. Lidauer[1]
[1]Natural Resources Institute Finland (Luke), Production Systems, Myllytie1, 31600, Finland, [2]Nordic Cattle Genetic Evaluation, Agro Food Park 15, 8200 Aarhus, Denmark; terhi.mehtio@luke.fi

Studies on genetic correlations between feed efficiency (FE), production and fertility are scarce. The correlations between traits are of interest for designing a selection index and to assess the economic and environmental impact of including FE into dairy cattle breeding programs. The objective was to estimate genetic correlations between FE, production, and female fertility traits. The data were collected from 731 primiparous Nordic Red dairy cows at four research farms in Finland during 1998-2020 and included 20,533 repeated records. The FE traits were regression on expected feed intake (ReFI) and residual feed intake (RFI). For ReFI dry matter intake (DMI) is regressed on the expected DMI that is calculated based on energy requirements and realized production. This approach describes FE as multiplicative effect by fitting random regression coefficients for the animal effects. For the RFI model DMI is regressed on energy sinks and animal effects are modelled as additive effects. The production traits in this study were milk yield (MY), protein yield (PY), fat yield (FY) and metabolic body weight (mBW). The female fertility trait was interval from calving to first insemination (ICF) and the ICF observations used were pre-corrected for fixed effects and were obtained from the official genetic evaluation for female fertility. These ICF observations were available for 2,560 cows in the pedigree. There were in total 5,044 animals in the pedigree for variance component estimation. Heritability estimates for ReFI, RFI, MY, PY, FY, mBW and ICF were 0.31, 0.17, 0.29, 0.19, 0.25, 0.75, and 0.02, respectively. Genetic correlations between ReFI and mBW, MY, PY, FY and ICF were 0.13±0.17, 0.05±0.24, 0.06±0.29, 0.27±0.27 and 0.03±0.38, respectively. Genetic correlations between RFI and MY, PY, FY and ICF were 0.51±0.25, 0.65±0.29, 0.29±0.26 and 0.02±0.45, respectively. The estimated genetic correlations indicated that ReFI is uncorrelated with MY and PY whereas RFI has an unfavourable correlation with MY and PY. Both ReFI and RFI were negatively correlated with FY and uncorrelated with fertility.

Session 45 Theatre 11

A transgenerational study on the effect of Great Grand Dam birth month on production traits in Itali

N.P.P. Macciotta[1], L. Degano[2], D. Vicario[2], C. Dimauro[1] and A. Cesarani[1]
[1]Università degli Studi di Sassari, Agraria, Viale Italia 39, 07100 Sassari, Italy, [2]ANAPRI, Via Ippolito Nievo n. 19, 33100 Udine, Italy; macciott@uniss.it

Heat stress (HS) exerts direct negative effects on welfare, productive and reproductive performances of dairy cattle. The occurrence of HS during the various stages of pregnancy of a dairy cow could affect also future generations, due to existence of either direct effect of climate conditions on the foetus but also of epigenetic modifications that could be further transmitted. Impact of HS on the dairy cattle industry has become a concern for the dairy cattle industry also for temperate regions due to the global warming and to metabolic heat produced by high producing animals. In this work, the effect of granddam month of birth on EBV for production and functional traits were investigated in Italian Simmental cattle. Data were EBV for milk, fat and protein yields of 128,437 Italian Simmental cattle provided by the Italian Simmental Association. For each cow, data from the dam (D), granddam (GD), and great granddam (GGD) were available. Data were analysed with a linear model. that included the fixed effects of cow birth month and GGD calving month, and the fixed covariables of GD birth date and GD EBV for the considered trait. All the effects included in the models significantly affected EBV of the three traits. The GDD calving month that showed the largest positive effects were May and June both for milk and protein yields, whereas it was October for fat. The lowest LSmeans for GDD calving month were observed for January for all the three traits. Results of the present work partially agree with previous reports on Israeli Holsteins literature, where positive effects of GGD that had their late pregnancy during winter and spring on their Great Grand Daugthers were reported. Less clear is the effect on fat yield. As far as negative effects are concerned, results of the present work seems to indicate a more relevant effect of temperatures during the month of conception rather that in had the last third of pregnancy, as previously observed in US Holsteins.

Genetic parameters for methane emission and its relationship with milk yield and composition

H. Ghiasi[1], D. Piwczyński[2], B. Sitkowska[2] and M. Kolenda[2]
[1]Payame Noor University, Department of Animal Science, P.O. Box 19395-3697, Tehran, Iran, [2]Bydgoszcz University of Science and Technology, Department of Animal Biotechnology and Genetics, 28 Mazowiecka St., 85-084 Bydgoszcz, Poland; darekp@pbs.edu.pl

Methane is a major source of greenhouse gases. Ruminants are one of the main methane emitters from anthropogenic sources. Measuring methane emitted by a large number of animals is expensive and requires specialized equipment, therefore direct selection on methane emission trait for reducing methane production by dairy cow in large-scale is difficult. A total of 38,342 raw milk yields (MY) and milk composition traits (dry matter (DM), protein (MP), fat (MF) and lactose yields (ML)) of 17,468 Polish Holstein-Friesian cows in parity from 1 to 6 were used in the study. Three methane production (g/lactation per cow) equations (MPE) developed by Niu *et al.* (2018) were used to estimate indirect methane production based on milk yield and composition: MPE1 = 299 + 2.73 × MY; MPE2 = 259 + 3.86 × ECM (energy corrected milk); MPE3 = 150 + 4.31 × ECM + 28.3 × MP. The AIC of the models indicate that the MPE1 model was superior to MPE2 and MPE3 for estimating genetic parameters of methane production. The heritability estimated for methane production based on MPE1 was 0.29, which was greater than MPE2 and MPE3 (0.24). The repeatability of methane production for all models was similar (0.15 for MPE1 and 0.16 for MPE2 and MPE3). For all traits and models, except MF and MPE1, high positive genetic (0.76 to 0.82) and phenotypic (0.85 to 0.98) correlations were estimated between methane production, milk yield and milk composition. The estimated genetic and phenotypic correlations with MPE2 were similar to obtained using MPE3. This material has been supported by the Polish National Agency for Academic Exchange under Grant No. PPI/APM/2019/1/00003.

Session 45 Poster 13

Skeletal muscle miRNA profiling in Charolais steers divergent for feed efficiency potential

K. Keogh[1], M. McGee[2] and D.A. Kenny[1]
[1]Teagasc, Animal and Bioscience Research Department, Grange, Dunsany, Co. Meath, Ireland, [2]Teagasc, Livestock Systems Research Department, Grange, Dunsany, Co. Meath, Ireland; kate.a.keogh@teagasc.ie

Efficient utilisation of feed resources in beef production systems is a major determinant of overall financial and environmental sustainability. Thus, identifying genes implicated in feed efficiency may allow for the identification and subsequent breeding of feed efficient cattle, benefiting sustainability. Moreover, it is crucial that genes contributing to feed efficiency are robust across varying management settings including for example dietary source. RNAseq analysis was employed to profile the skeletal muscle tissue miRNAome of Charolais steers divergent for residual feed intake (RFI) over consecutive contrasting dietary phases. During phase 1, steers were offered zero-grazed grass, followed by a high-concentrate diet for dietary phase 2. Muscle biopsies were collected at the end of each dietary phase from cattle most divergent for RFI. RNAseq was subsequently undertaken on muscle biopsies followed by bioinformatic analysis to determine differentially expressed (DE) miRNA between steers divergent for RFI across each dietary phase. Over each dietary phase, growth rates were not different ($P>0.05$) between RFI groups, however High-RFI (feed inefficient) steers consumed more feed ($P<0.05$). In total, 8 miRNA were identified as DE (P-value<0.05) between steers divergent in RFI, 6 were DE following the zero-grazed grass diet and 2 DE following the high-concentrate dietary phase. Of particular interest were miR-2419-5p and miR-2415-3p, both of which were up-regulated in the Low-RFI (feed efficient) steers compared to their High-RFI contemporaries across each dietary phase. The predicted target mRNA genes of miR-2419-5p and miR-2415-3p revealed an involvement in processes related to growth and metabolism. These results provide insight into the skeletal muscle miRNAome of beef cattle and their potential molecular regulatory mechanisms relating to feed efficiency. This research was funded by the Irish Department of Agriculture, Food and the Marine (RSF13/S/519) and by the Research Leaders 2025 programme, co-funded by Teagasc and the European Union's Horizon 2020 research and innovation programme under the Marie Skłodowska-Curie grant agreement number 754380.

In silico analyses reveals a deletion in TRIM40 promoter gene: a possible role in feed efficiency

L.L. Verardo[1], M.A. Machado[2], J.C.C. Panetto[2], I. Carolino[3], N. Carolino[3] and M.V.G.B. Silva[2]
[1]Universidade Federal dos Vales do Jequitinhonha e Mucuri, Animal Science, Campus JK, 39100-000, Diamantina-MG, Brazil, [2]Brazilian Agricultural Research Corporation (EMBRAPA), Dairy, Rua Eugênio do Nascimento, 610, 36038-330, Juiz de Fora-MG, Brazil, [3]Instituto Nacional de Investigação Agrária e Veterinária, Fonte Boa, 2005-048, Vale de Santarém, Portugal; lucas.verardo@ufvjm.edu.br

Recently, the tripartite motif containing 40 protein (TRIM40) has been suggested as candidate gene for residual feed intake in cattle and pigs. In this study we aimed to *in silico* evaluate its promoter region in cattle to identify possible functional variants affecting the biding of transcription factors (TF). We first retrieved all known insertions and deletions (InDels) in the promoter sequence (3,000 bp upstream and 300 bp downstream from the transcription start site) of TRIM40 gene using the Ensembl Biomart tool based on the ARS-UCD1.2 assembly. FASTA files with the wild type sequence of promoter region and sequences with InDels of two or more base pairs were used as input in TFM-explorer program to search for locally overrepresented TF binding sites (TFBS) using weight matrices from the JASPAR and TRANSFAC vertebrate database. A total of 52 InDels was observed, from which 34 presented two or more base pairs, which were used to retrieve the FASTA sequences of the promoter regions of TRIM40 gene. The wild type of promoter region presented biding sites for 5 TF (HSF, PBF, XFD-1, STE11 and DOF3). From the promoter sequences with the identified InDels, one with a deletion of three base pairs (from 28841320 to 28841322) affected the biding of HSF. It is a heat shock transcription factor and has been cited to be related with feed intake in pigs. HSF is cited to be necessary for thermo tolerance through the expression of classical heat stress genes in cattle, and thus having a role in feed intake. In this study, we observed a deletion of three base pairs in the promoter sequence of TRIM40 gene, more precisely, in a TFBS of HSF affecting its binding. Since both TRIM40 and HSF has been cited to be involved in feed efficiency, we suggest that this deletion may have a role in cattle feed intake, and thus should be better studied in further *in vitro* and *in vivo* analyses.

Associating gut microbiome with residual energy intake in dairy cows using extreme gradient boosting

M. Tapio, P. Mäntysaari and I. Tapio
Luke, Myllytie 1, 31600 Jokioinen, Finland; miika.tapio@luke.fi

The gut microbiota in ruminants is fundamental for feed digestion, but the role of microbiota in defining animal feed efficiency phenotype is still not fully understood. The objectives here were to: (1) assess rumen microbial community composition in dairy cows; (2) generate machine learning model for feed efficiency prediction by Extreme gradient boosting (xgboost) method; (3) identify bacterial taxa that have strong impact on feed efficiency. Residual energy intake (REI) was calculated for 100d period after calving and was based on a multiple linear regression model with energy corrected milk, metabolic body weight, and piecewise regressions of BW change on the metabolizable energy intake. Rumen bacterial community composition was determined by 16S rRNA gene V4 amplicon sequencing and processed using Qiime 2. In xgboost modelling, bacterial composition was included as dimension reduced components regularized to positive values only. The number of predictors reduced by two thirds while the omitted variation was only one third of the total variation. Simpson diversity statistics were included as ecological predictors. Data from 87 primiparous Nordic Red cows was randomly split in proportions 8:1:1 to training, validation and test data, respectively. Training and validation data were used to fit and tune xgboost regression models. Final model was used to predict REI in the test data. Overall rumen community differences had no simple association with REI estimates. The model explained 1/3 of the variation in REI in the test data. In total, 19 predictors, including the diversity values, had non-zero marginal contributions to the predictions. The four most important features were dimension reduction components demonstrating approximately monotonous impacts on REI predictions. Each of the top component had non-zero contribution from 10-20% of the total OTUs, that were affiliated with *Acetitomaculum*, *Bacteroidales* RF16, *Christensenellaceae* R-7, *Lachnospiraceae* NK3A20, *Methanobrevibacter*, *Paraprevotella*, *Prevotella*, *Rikenellaceae* RC9, *Succiniclasticum* and *Succinivibrionaceae* UCG-002 genera. The results can be used to infer microbial metabolic networks involved. However, the component ranks are sensitive to hyperparameter settings, and caution is necessary.

Session 45 Poster 16

Genome wide association study of compensatory growth in Irish and Canadian cattle

T. Carthy[1], A.K. Kelly[2], K. Keogh[1], Y. Mullins[1], D.A. Kenny[1], C. Li[3] and S.M. Waters[1]
[1]Animal and Grassland and Research Innovation Centre, Teagasc, Grange, Co. Meath, Ireland, [2]College of Health and Agricultural Science, University College Dublin, Belfield, Dublin 4, Ireland, [3]Department of Agricultural, Food and Nutritional Science, University of Alberta, Edmonton, AB, Canada; sinead.waters@teagsc.ie

Compensatory growth is a naturally occurring physiological process whereby an animal has the potential to display accelerated growth and enhanced efficiency upon re-alimentation following a prior period of restricted feed intake. The objective of this study was to identify regions of the genome associated with residual compensatory growth in cattle. Two populations of cattle were used where compensatory growth response profiles were available, incorporating two breeds; 720 Irish Holstein-Friesian bulls and 973 Canadian Angus bulls. Imputed whole genome sequence data was available on all animals. Genome wide association analyses were performed for each of the populations separately using linear mixed models in GCTA software. A meta-analysis between the two populations was also conducted to identify any overlapping regions across the two populations. SNPs in close proximity (<500 kb) were combined into a single QTL and genes within a 250 kb span of the most significant SNP in the QTL were determined. For the Irish animals, 18 QTLs where identified as associated with compensatory growth ($P \leq 1 \times 10^{-5}$); these regions contained 25 genes. While, for the Canadian population, 7 QTLs where associated with 13 genes located in these regions. The meta-analysis between the two populations identified 26 SNPs ($P \leq 1 \times 10^{-5}$) associated with compensatory growth located in 9 QTLs. The most significant QTL identified during the meta-analysis was on BTA7 from 106,204,994 bp to 106,257,531 bp; this region contains one gene; EFNA5. Following validation, SNPs identified in this study may be included in genomic selection programmes as DNA based biomarkers for the selection of cattle with an improved compensatory growth potential.

Session 45 Poster 17

Selection for growth and the impact on production efficiency indicators in Nellore cattle

J.N.S.G. Cyrillo, S.F.M. Bonilha, R.C. Canesin, R.H.B. Arnandes, L.F. Benfica and M.E.Z. Mercadante
Instituto de Zootecnia, Centro Avançado de Pesquisa de Bovinos de Corte, Rodovia Carlos Tonanni, 14 175-000, Brazil; jgcyrillo@sp.gov.br

The aim of the study was to evaluate the impact of selection for yearling body weight on feed efficiency traits and biological production indexes in Nelore herds. A total of 526 weaned calves belonging to selection line (NeS), n=340 and mean of 345.81±2,398 kg and control line, n=186 and mean of 269.80±3.163 kg were evaluated. The animals remained in collective pens equipped with automated troughs (GrowSafe®) for recording individual daily intake and were classified as low residual feed intake (RFI <0) and high residual feed intake (RFI>0).Significant differences (P<0.001) were found between NeS and NeC lines, in average daily gain (ADG): 1.100±0.011 vs 0.802±0.015; dry matter intake (DMI): 7.528±0.057 vs 6.712±0.073 and feed conversion (FC): 7.261±0.104 vs 7.668±0.138, respectively. No significant differences (P>0.01) were found between NeS and NeC for residual feed intake (RFI) (0.014±0.045 vs -0.132±0.059), DMI as a percentage of body weight (BW) (% DMI / BW), (2.216±0.017 vs 2.203±0.023) and biological efficiency, as kilograms of feed intake per arroba (kg DMI / @) (1 @ = 15 kg; meat production unit) (0.330±0.003 vs 0.332±0.002) indicating that DMI consumption is proportional to the animal weight. When evaluating the biological efficiency of animals classified as RFI>0 and RFI <0 within the selection lines, it was observed that animals of the NeS line, classified as more efficient (RFI <0) had consumption 15% lower than the less efficient animals (RFI>0) of the NeS lines and NeC and 0.8% lower than the animals in the NeC line classified as more efficient (RFI<0). Finally, it was concluded that the selection for yearling body weight affected positively the parameters of growth performance, DMI and FC, however, did not affect the indexes RFI, % DMI / PV and the biological efficiency. Animals from NeS classified as RFI <0 shown higher biological efficiency compared to animals of the same classification from the NeC line. Grant #2017/50339-5, São Paulo Research Foundation (FAPESP)

ATP1A1 and NPY gene expression associated with heat stress and residual feed intake in beef cattle

B.V. Pires[1], N.B. Stafuzza[2] and C.C.P. Paz[1,2]
[1]University of São Paulo (USP), Department of Genetics, Av. Bandeirantes, 3900, 14049-900, Brazil, [2]Animal Science Institute, Rod. Carlos Tonani, km 94, 14174-000, Brazil; claudiacristinaparopaz@gmail.com

Heat tolerance is an important trait in beef cattle and can influence feed efficiency. The ATP1A1 and NPY genes are associated with residual feed intake (RFI) and heat tolerance in cattle. This study compared the expression of ATP1A1 and NPY gene expression in Nelore (*Bos indicus*) and Caracu (*Bos taurus*) in the sun and shade environment. The study was carried in Sertãozinho, SP, Brazil during October 2017 (28.5 °C and 53.5% relative humidity), and evaluated 35 Caracu and 30 Nelore steers (12-15 months of age). Three different environments were evaluated: in the morning, all animals remained in a pen with some trees in the sun, in the afternoon the steers were separated and evaluated in shade and sun. Feed intake was measured during 88 days for each animal using automated feeding stations. The RFI for each animal was calculated based on a linear regression model of feed intake on mean metabolic live weight and average daily gain. All steers were classified into the most efficient (RFI<0) and least efficient (RFI>0). The blood samples were collected in the morning (7:30 am) and the afternoon (3:00 pm) from all animals and the relative expression of ATP1A1 and NPY genes were identified by qPCR. The relative change in mRNA expression levels was calculated using the delta Ct method using the GAPDH housekeeping gene. The data were analysed by SAS program. A total of 33 steers showed RFI>0, and 32 steers showed RFI<0. In sun treatment, Nelore steers obtained higher NPY expression than Caracu steers (3.15 and 0.55, respectively). The NPY expression in three treatments was not different (P>0.05) in Nelore breed, while in Caracu breed the shade treatment resulted in the highest expression (2.75). The treatments and breeds did not influenced the ATP1A1 expression (P>0.05). Animals' most efficient showed lower ATP1A1 expression than animals least efficient (2.20 and 2.77, respectively). ATP1A1 and NPY expression in Caracu is more likely to have alterations in comparison to Nelore breed in the hot climate. Financial support: São Paulo Research Foundation (FAPESP) grant 2016/19222-1. C.C.P. Paz was the recipient of a productivity research fellowship from CNPq.

Increased reproductive performance and lamb productivity as mitigation options in sheep production

B.A. Åby, S. Samsonstuen and L. Aass
Norwegian University of Life Sciences, Box 5003, 1432 Ås, Norway; bente.aby@nmbu.no

Sheep currently account for approx. 4% of global GHG emissions from livestock and human population growth are likely to increase demand for sheep products. To limit GHG emissions from sheep production, it is essential to reduce emission intensities, i.e. GHG emissions from producing one unit of product, e.g. sheep and lamb carcass. In this study, the mitigation options were increased ewe reproductive performance (i.e. increased no. of lambs born per ewe, reduced lamb mortality), and lamb productivity (i.e. higher lamb growth rates and carcass weights), investigated by using the whole-farm model HolosNorSheep. HolosNorSheep is based on IPCC methodology and estimate GHG emissions from dual-purpose meat and wool production, using a cradle to farm gate approach. The model considers direct CH_4 from enteric fermentation and manure management, direct and indirect N_2O from manure management and soils, and CO_2 emissions from production and use of farm inputs. Soil carbon balance is estimated using the ICBM-model. The number of ewes were kept constant, i.e. total carcass production was allowed to increase above the baseline, when ewe reproductive performance and lamb productivity were improved. Given a constant domestic production target for sheep meat, the required number of sheep herds needed to meet this target will thus be reduced. Compared to ewe average reproductive performance in the Sheep recording system in 2020 (1.79 weaned lambs per adult ewe) increasing performance to the level of best third performing herds for number of weaned lambs (2.00 weaned lambs per adult ewe) reduced emission intensity per kg of sheep and lamb carcass by 4.43%. Compared to the average levels (282 g/day and 19.7 kg), increased growth rates from birth to weaning and increasing carcass weights by 10% (310 g/day and 21.67 kg) reduced emission intensities by 4.70%. A combination of the mitigation options gave a total reduction of 9.20%. Given the domestic Norwegian production of sheep and lamb carcass of 24,561 tons in 2020, this equals a reduction in total GHGs of 35,689 tons CO_2-eq. per year. Thus, increased reproductive performance, lamb growth rate and carcass weights are efficient mitigation options in dual purpose sheep meat and wool production systems.

Genetic effect on the feed to muscle gain ratio of growing purebred pigs

G. Daumas[1], C. Hassenfratz[1], M. Monziols[1] and C. Larzul[2]
[1]IFIP-Institut du Porc, BP 35104, 35650 Le Rheu, France, [2]GenPhySE, Université de Toulouse, INRAE, ENVT, 24 chemin de Borde-Rouge, 31320 Castanet-Tolosan, France; gerard.daumas@ifip.asso.fr

The aim of the study was to compare tissular growth performance between 6 purebred populations, females from three paternal lines, and uncastrated males from 3 maternal lines. The paternal lines and their sample sizes were two Pietrain lines, either homozygous for the N allele (P_NN; n=40) or the n allele (P_nn; n=46) of halothane gene, and a Duroc (Du; n=33) line. The maternal lines and their sample sizes were Large White (LW; n=55), and two Landrace lines (LR; n=30, and LR_M6; n=29). Pigs were raised over 4 batches, distributed in 9 pens of 14 pigs. Feeding was *ad libitum* in a biphase sequence and consumption (DFI, g/d) was measured individually. Half-carcases were scanned with computed tomography to measure the muscle volume, calculate the Lean Meat Content (LMC, %), and to assess the average daily muscle gain (ADGm, g/d) and the feed to muscle gain ratio (F:Gm, g/g). Least squares means within paternal or maternal lines were compared with a Tukey test. Means with a different superscript letter differ significantly with a P-value less than 5%. The overall means of initial and final body weight were 33 and 123 kg. Regarding paternal lines, LMC was the highest in P_nn (64.2[a]), the lowest in Du (55.6[b]) and intermediate in P_NN (61.1[c]). ADGm was higher in P(757[a]) than in Du (655[b]), while ADG was higher in P_NN (1,050[a]) than in P_nn and Du (1,004[b]). DFI was the highest in Du (2,723[a]), the lowest in P_nn (2,438[b]) and intermediate in P_NN (2,586[c]). P_nn got the best F:Gm (3.22[a]), Du the worst (4.18[b]) and P_NN an intermediate position (3.45[c]), while P got the best F:G (2.45[a]) and Du the worst (2.73[b]). Regarding maternal lines, LMC was the highest in LR_M6 (61.6[a]), the lowest in LR (57.3[b]) and intermediate in LW (59.2[c]). ADGm was higher in LW (757[a]) than in LR (638[b]), as well as ADG (1,050[a] vs 987[b]). DFI was higher in LW (2,673[a]) than in LR_M6 (2,433[b]). LW and LR_M6 got a better F:Gm (3.74[a]) than LR (4.20[b]), while there was no significant difference of F:G (2.53). This study showed that F:Gm discriminated breeds better than F:G. Computed tomography seems a suited tool to determine tissue deposition and associated feeding efficiency in test station for breeding purposes.

Session 46 Theatre 1

Feeding practices and resilience in smallholder systems in Sub-Saharan Africa

A.J. Duncan[1,2], M. Bezabih[1], A. Mekasha[3] and S. Oosting[4]
[1]International Livestock Research Institute, Addis Ababa, 5689, Ethiopia, [2]University of Edinburgh, Easter Bush Campus, EH25 9RG Midlothian, United Kingdom, [3]Ethiopian Institute of Agricultural Research, Addis Ababa, 2003, Ethiopia, [4]Wageningen University, Wageningen, the Netherlands; a.duncan@cgiar.org

Improving the nutrition of livestock could transform the lives of smallholder farmers and consumers in Sub-Saharan Africa. However, improving feed supply and quality has proved challenging so far. Farms in Sub-Saharan Africa are diverse sitting on a continuum from subsistence mixed production right through to specialized commercial livestock units. Feeding practices vary along this continuum. In smallholder systems and especially at the subsistence end of the continuum, livestock have multiple functions: manure, traction, security, storage of capital as well as production of meat and milk. This multi-functionality of livestock production has implications for feeding practices. Farmers may feed as much for manure production as for milk production. Draught animals require energy at certain times of year. Farmers may prefer to feed multiple unimproved animals than one improved animal to spread risk. Depending on context, livestock keepers adopt various feeding practices based on indigenous knowledge, that seem counter to scientific wisdom. In many cases these practices rely on smart use of local resources that minimise the need for expensive inputs. Examples of such practices include tolerating weed growth in food crops, high planting density of cereals to allow use of thinnings, and feeding for manure production. Some apparently sensible 'imported' or external technologies can be problematic depending on the type of farmer in question. Examples include forage conservation technologies developed for temperate environments, chemical treatment of straws, sowing of short-strawed cereals and use of feed blocks. Such technologies often fail because they do not take enough account of economic realities, labour issues and the need for technical backup. To realise the transformational potential of improved feeding, development practitioners need to tailor feed interventions to farmer type, take account of, and build on, indigenous practices involving local feed resources and carefully consider the economics of imported technologies.

The role of camels as a lever enhancing the pastoral households resilience around N'Djamena (Chad)

M.A. Mahamat Ahmat[1], G. Duteurtre[2], M.O. Koussou[1] and C.H. Moulin[3]
[1]IRED, BP 433, N'Djamena, Chad, [2]CIRAD, UMR SELMET, TA C-112/A, 34398 Montpellier cedex 5, France, [3]L'Institut Agro Montpellier, UMR SELMET, Place Viala, 34060 Montpellier, France; charles-henri.moulin@supagro.fr

The climatic conditions, the political instability and the 70s and 80s war upset the trajectories of pastoral households. Some pastoral households left Batha in central Chad to settle around N'Djamena. During this migration, they profoundly transformed their farming systems. Our goal is to show how the camel, a local animal resource, has been a lever of transformation of local production systems, with their resilience, robustness and adaptability. We rely on a survey conducted in 2018 among 173 households, randomly selected in the 27 camel pastoralists camps around N'Djamena, i.e. 10% of the households. These households have specialized in camel breeding at the expense of cattle breeding. Camels now represent 80% of the TLU. Previously, these households moved between an attachment site in the north of Batha during the rainy season, and a host site in the south during the dry season. For this long-distance mobility (~400 km), the whole household accompanied the herd. Today, they organised the mobility between three sites. In the dry season, the herds are split in two. The household keeps the lactating females around N'Djamena for milk sale as a young adult drives the rest of the animals to the south (up to 500 km). In the rainy season, camels cannot stay around N'Djamena due to sanitary constraints but also to land occupation by crops. The household and the whole herd go north (around 300 km). The well-known robustness of camels enables these mobilities over very long distances. The sale of milk was once considered a social taboo. Today, it is a major strategy for diversifying the pastoralists' livelihoods. Milk remains a commodity for self-consumption, but it is also a major source of income. Forty-two percent of households generate a gross milk margin per worker greater than the minimum wage. The camel capacities to produce milk in the dry season (3 to 4 litres milked per female and per day) enabled this conversion to milk sale during the presence near the urban market of N'Djamena, on the contrary of cows for which the rainy season remains the most favourable to produce milk.

Determine breed proportion and suitable percentage of dairyness in Ethiopian smallholder dairy farm

S. Meseret[1], R. Mrode[2,3], J.M.K. Ojango[3], E. Chinyere[3], G. Gebreyohanes[1], A. Hassen[1], A. Tera[4], B. Jufar[4] and A.M. Okeyo[3]
[1]International Livestock Research Institute, Box 5689, Addis Ababa, Ethiopia, [2]Scotland's Rural College, Easter Bush, EH25 9RG, Edinburgh, United Kingdom, [3]International Livestock Research Institute, Box 30709-00100, Nairobi, Kenya, [4]Livestock Development Institute, Box 22692, Addis Ababa, Ethiopia; s.meseret@cgiar.org

Ethiopian dairy sector is determined by smallholder farming system and strongly conditioned by the environment. The study aimed to identify optimum level of dairyness for specific environment. Ethiopian dairy animal was genotyped using medium density SNP array and merged with reference cattle population of Holstein and Jersey genotyped data. Breed proportion was estimated using maximum likelihood method in unsupervised model and reference population ancestries summed up to determine the exotic proportion. Study location was grouped into dry-wet and moist-wet-cold highland agroecological zones based on herd elevation data. The final dataset consisted of 1,935 genotyped animals from 894 herds and 12,254 test day milk records. Effects of breed proportion (grouped into five) versus agroecological zone on milk yield were analysed using mixed model equation at $P<0.005$. The population structure revelled optimal number of $k=3$. About 61% of study animal shared >75% of exotic breed proportion and Holstein breed contributed considerable proportion in Ethiopian population. Overall, exotic breed proportion had significant effects on test day milk yield. Though, animals grouped to exotic breed proportion 75-87.5% and >87.5% had the same milk yield in two agroecological zone. Likewise, 50-75% and 75-87.5% breed group were not significantly different for milk yield in dry-wet agroecology. Milk yield response to an increase in exotic proportion from 25-50% to 50-75% group was not significant under moist-wet-cold highland agroecology. Result suggested that exotic breed group 75-87.5% and 50-75% were optimum in moist-wet-cold and dry-wet agroecological zones, respectively. Thus, matching right level of exotic inheritance with specific agroecology is determinant to build sustainable dairy crossbreeding program. The findings from this study should be further supported with smallholder dairy farming management practices and larger dataset.

Analysis of key typologies for integrated dairy-fodder crop systems in Europe: A NUTS2 approach

X. Díaz De Otálora[1,2], F. Dragoni[2], A. Del Prado[1], F. Estellés[3], V. Anestis[4] and B. Amon[2,5]
[1]Basque Center for Climate Change, B/Sarriena s/n, 48940 Leioa, Spain, [2]Leibniz-Institute for Agricultural Engineering and Bioeconomy, Max-Eyth-Allee 100, 14469 Potsdam, Germany, [3]Universitat Politècnica de València, Camino de Vera s/n, 46022 Valencia, Spain, [4]Agricultural University of Athens, Iera Odos 75, 11855 Athens, Greece, [5]University of Zielona Góra, Licealna 9, 65417 Zielona Góra, Poland; bamon@atb-potsdam.de

Dairy production systems (DPS) face significant economic, social and environmental challenges. Given the great diversity of DPS across Europe, region-specific and holistic approaches to improve their sustainability are needed. Integrated livestock and crop production systems emerge as more resilient and sustainable farming approaches than highly specialized livestock production. In this context, identifying these production systems is presented as a first step in transitioning towards future optimized food production systems. However, the definition of the currently existing DPS typologies is often insufficient for applying sustainability measures due to the lack of consideration of the farm's structural, socioeconomic, and environmental aspects. This work aims to identify, describe and compare key DPS typologies and their interrelation with fodder crop production at the European NUTS2 scale. A multivariate statistical approach was followed to assess the diversity of DPS according to their land-use practices, farm structure, socioeconomic characteristics, and the intensity of greenhouse gas and ammonia emissions. In addition, the share of the area occupied by each of the main fodder crops in the region was analysed. The results show how the diversity of farm structure characteristics, socioeconomic attributes, and emission intensities condition the integrated dairy-fodder crop typologies in Europe. Significant differences were observed in the level of intensification and the use of fodder crops across the regions analysed. This joint assessment could facilitate decision-making by analysing the different components of DPS. Identifying interactions between the components of these integrated systems could contribute to designing and implementing measures targeted to mitigate climate change and promote sustainability.

Session 46 Theatre 5

Farm to fork: the story of local Israeli 'dairy beef'

M. Cohen-Zinder, E. Shor-Shimoni, R. Agmon and A. Shabtay
Beef Cattle Unit and Model Farm for Sustainable Agriculture, Newe Ya'ar Research Center, Agricultural Research Organization, Volcani Center, 3009500 Ramat Yishay, Israel; mirico@volcani.agri.gov.il

Global animal production systems are often criticized for lack of sustainability, and in times of crisis, for insufficient resilience to ensure food security. The 'farm-to-fork' approach is an emerging strategy, aiming, amongst others, at accelerating the food systems towards a positive environmental impact, nutritious, healthy, safe and sufficient foods, and fairer economic returns, in particular for primary producers. Within the global beef production system, countries that do not produce sufficient fresh meat, rely to a great extent, on imported supply of live animals, to fulfil their needs. In Israel, ~60% of the sources for fresh beef comes from import of live animals. The rest stems from free ranging beef herds (8%) and dairy farms (33%). Thus, in order to encourage sustainable beef production in Israel, the local beef share should be raised at the expense of imported animals. However, for that to be achieved, the superior performance of local over exotic breeds should be justified. We compared meat quality characteristics between local (Israeli Holstein; n=205) and imported (Australian *Bos indicus* × *Bos taurus* crosses; n=169) animals. Generally, while the imported calves presented higher production parameters, as carcass weight ($P<0.0001$) and dressing percentage ($P<0.0001$), the local animals were characterized by improved meat quality phenotypes. These included lower Shear Force (SF) values ($P<0.0001$), longer sarcomeres ($P<0.0001$), preferential colour and pH ($P<0.001$), and superior cooking ($P=0.002$) and thawing loss ($P<0.0001$) measures. With respect to intra muscular fat (IMF) characteristics, while its content and PUFA profile were higher ($P<0.01$ and $P<0.0001$, respectively), SFA tended to be lower ($P=0.13$), bringing about a higher PUFA: SFA ratio in the meat of the Israeli Holstein calves. Our findings, presented herein, indicate that the local Holstein is superior over the imported calves, in terms of meat quality characteristics. In a broader sense, these results may provide sound arguments for stakeholders and policy makers, to facilitate sustainable local beef production in Israel.

Reducing concentrates supply level in Alpine dairy farms: a consequential-based LCA model

M. Berton[1], S. Bovolenta[2], M. Corazzin[2], L. Gallo[1], M. Ramanzin[1] and E. Sturaro[1]
[1]DAFNAE, University of Padova, Viale dell'Università 16, 35020 Legnaro, Padova, Italy, [2]DI4A, University of Udine, via delle Scienze 206, 33100 Udine, Italy; marco.berton.1@unipd.it

This study aimed to assess the consequences of reducing the concentrates supply level (CSL) in lactating cows' diets on the carbon footprint of the north-eastern Alps dairy systems by adopting a consequential-based Life Cycle Assessment. Data originated from 40 dairy farms (31±21 livestock units - LU, 1.4±0.8 LU/ha, 21.0±5.5 kg fat- and protein-corrected milk – FPCM, and 4.5±2.8 kg DM concentrates per lactating cow/d). The impact category assessed was global warming potential (kg CO_2-eq), without (GWP) and with (GWP_LUC) land-use change emissions. Three scenarios were tested: 100% (t_0), 75% ($t_1$75) and 50% ($t_1$50) of the initial CSL, with a hay:concentrate substitution rate of 0.45:1 and milk yield (MY) modelled with regression analysis based on feed intake and ingredient composition. System expansion was used to handle the farm bodyweight coproduction and the consequences due to the change in milk provision level. Due to the cheesemaking destination of the alpine milk, a reduced availability of milk for mountain cheeses production has been considered. Functional units were 1 kg FPCM (farm level) and 1 kg of protein (CP; expanded system). Impact values at the farm and expanded system levels were analysed with GLM model that included CSL (3 classes: t_0, $t_1$75, $t_1$50) as fixed effect, tested on the variance of the farm. From t_0 to $t_1$75 and to $t_1$50, MY decreased by 10% and 20%, respectively. At t_0, the production of 1 kg FPCM caused the emission of 1.06 kg (GWP) and 1.28 kg (GWP_LUC) CO_2-eq. At the farm level, impact values increased by 2-5% at $t_1$75 and 5-12% at $t_1$50, with no significant differences with respect to t_0. At t_0, the production of 1 kg CP generated 32.1 kg (GWP) and 38.8 kg (GWP_LUC) CO_2-eq. In the expanded system, the three scenarios of CSL did not affect the GWP and GWP_LUC values (+0-5%). In conclusion, the reduction in the CSL fed to lactating cows in Alpine dairy farms could favour the decoupling of the milk production from human-edible resources without worsening the related GWP but can also enhance the use of local resources and advance the resilience and sustainability of the local farming systems.

Resilience of animal genetic resources in the face of climate change and extreme events

R. Baumung[1], P. Boettcher[1], C.A. Reising[2], C. Okore[3] and G. Leroy[1]
[1]Food and Agriculture Organization, viale delle Terme di Caracalla, 00153 Roma, Italy, [2]Instituto Nacional de Tecnología Agropecuaria, Estación Experimental Agropecuaria Bariloche, EEA, Modesta Victoria 4450, 8400 San Carlos de Bariloche, Argentina, [3]State Department of Livestock, P.O. Box 34100-00100 Nairobi, Kenya; roswitha.baumung@fao.org

Climate change impacts livestock production systems in different ways. Direct consequences include impacts that more-frequent extreme weather events have on livestock (e.g. stress, increased morbidity and mortality). Indirect impacts include decreased availability of inputs and changes in the natural range of diseases and pests. The resilience of animal genetic resources at all levels (species, breed, individual) is therefore expected to be a powerful leverage to support the necessary adaptation of the sector. Differential mortality and shifts in species distribution has already been observed in several African countries over the last decades, a plausible consequence of variable resilience across species. At breed level, changes are less well documented, however some case studies illustrate better resilience (mortality, production, etc.) of locally adapted breeds when facing extreme events. Improving knowledge on adaptation traits and genotype-by-environment interaction could lead to the development of breeding strategies for better resilience, although the application is currently limited. At farm level, livestock diversification in terms of species and breeds can be considered as effective strategies to help farmers cope with extreme events.

Session 46 Theatre 8

Whole-genome sequencing reveals genes associated with heat tolerance in South African beef cattle

K.S. Nxumalo[1,2], M.L. Makgahlela[1,2], J.P. Grobler[1], J. Kantanen[3], C. Ginja[4], D.R. Kugonza[5], N. Ghanem[6], R. Gonzalez-Prendes[7], A.A. Zwane[2] and R.P.M.A. Crooijmans[7]

[1]University of the Free State, Department of Genetics, Bloemfontein, 9300, South Africa, [2]Agricultural Research Council-Animal Production Pretoria, Irene, Old Olifantsfontein road, Animal Genetics Building, 0062, South Africa, [3]Agricultural Research Centre (MTT) Jokioinen, Animal Production Research, Jokioinen, 31600, Finland, [4]Universidade do Porto, CIBIO-InBIO, Centro de Investigação em Biodiversidade e Recursos Genéticos, Universidade do Porto, 4050-290, Portugal, [5]Makerere University, Agricultural Production, School of Agricultural Sciences, College of Agricultural and Environmental Sciences, Kampala, Uganda, 759125, Uganda, [6]Cairo University, Animal Production Department, Giza, 12511, Egypt, [7]Wageningen University and Research Wageningen, Animal Breeding and Genomics Group, the Netherlands; nxumalok@arc.agric.za

South African (SA) indigenous beef cattle are well known for their adaptation to high altitudes and harsh environmental conditions. Knowledge of selection signatures can provide better understandings on the mechanisms of selection and reveal genes related to biological functions. The aim is to identify genes associated with heat tolerance in SA Tuli, Afrikaner, Bonsmara, and Nguni cattle, using within-population iHS and between population Rsb signatures of selection analysis. Strong signals of candidate selection signatures were detected in Nguni and Afrikaner pair underlie by genes previously associated with thermogenesis including HSPB9, DNAJC7 and UCP1. Enrichment analysis further revealed the key biological processes in GO:00048583 (response to stimulus) and GO: 0050896 (heat response) that were highly enriched within thermogenesis gene regions in Nguni and Afrikaner as well as Bonsmara and Tuli populations pairs, respectively. The current findings provide a basis of understanding of genomic variations associated with thermogenesis in South African indigenous beef cattle.

Session 46 Theatre 9

The role of animal genetic resources in the resilience of livestock farming systems: ERFP experience

E. Sturaro, D. Bojkovsky, E. Charvolin, C. Danchin, S.J. Hiemstra, C. Ligda, M. Castellanos Moncho, N. Svartedal and F. Tejerina

ERFP, 149 rue de Bercy, 75595 Paris Cedex 12, France; enrico.sturaro@unipd.it

The European Regional Focal Point for Animal Genetic Resources (ERFP) is the regional platform to support *in situ* and *ex situ* conservation and sustainable use of animal genetic resources (AnGR) and to facilitate the implementation of the FAO Global Plan of Action for AnGR. In the recent years, ERFP has been active in several initiatives and projects on the added value of AnGR, and on the relationships between AnGR, farming systems and agroecosystems. This contribution presents the outcomes of case studies related to the links between AnGR and their agro-ecosystems, with particular emphasis on the role of local breeds in the resilience of farming systems. The case studies cover different species and different countries (France, Greece, Italy, Norway, Slovenia, Spain). This approach is in line with the Strategic Priority 5 of the Global Plan of Action for Animal Genetic Resources, which aims to promote agroecosystem approaches to AnGR management. The outcomes are in line with the priorities in the Animal Genetic Resources Strategy For Europe, developed in the frame of the EU Horizon 2020 GenResBridge projects. The aim of this strategy is to minimize the loss of livestock genetic diversity, to support breeding, diversification and innovation, and strengthen the resilience of the livestock sector. The conservation of AnGR diversity is essential for ensuring sustainable livestock development and use, rural livelihoods, environmental stewardship and food security.

Session 46 Theatre 10

Performance and breeding goals for Dutch local cattle breeds in relation to agroecology

G. Bonekamp[1], M.A. Schoon[1,2], S.J. Hiemstra[2] and J.J. Windig[1,2]

[1]Wageningen University & Research, Animal Breeding and Genomics, P.O. Box 338, 6700 AH Wageningen, the Netherlands, [2]Centre for Genetic Resources, P.O. Box 338, 6700 AH Wageningen, the Netherlands; gerbrich.bonekamp@wur.nl

Agroecology and a transition towards circular and nature-inclusive livestock systems is gaining attention both in science and policy. Simultaneously, there is a renewed interest in local cattle breeds due to their potential suitability for agroecological systems. Local Dutch cattle breeds, now largely replaced by international transboundary high input/high output breeds, have originally been bred under conditions requiring less inputs. Focus group discussions involving experts, farmers and stakeholders, indicated robustness and an efficient conversion of low and variable quality feed into high quality product as important traits to incorporate into the breeding goal of local cattle breeds. This underlined the (potential) added value of the genetic diversity in local Dutch cattle breeds for agroecological production systems. To further substantiate this claim, we performed a national inventory of (dairy) herds with local Dutch cattle breeds using data on production, health and fertility, across a range of environments. Results were contrasted with performance of Holstein Friesian herds in the same regions. While some farms only have cows of local Dutch breeds, many herds consist of combinations of local and other breeds. Milk production varied considerably but was on average of a higher level in Holstein compared to the local Dutch dual-purpose breeds. The variation in somatic cell count and calving interval was higher within the different breeds than between. Clearly, farm management and environment largely determine performance within and across breeds. In the next phase, breed by environment interactions will be determined to investigate under which circumstances local Dutch breeds perform best and how local genetic resources can contribute to agroecological farming systems.

Session 46 Theatre 11

Tenderness, fatty acids and marbling qualities in six endangered cattle breeds native to Norway

N. Sambugaro[1,2], B. Egelandsdal[2] and N. Svartedal[3]

[1]Università degli studi di Padova, Via 8 Febbraio 2, 35122 Padova (PD), Italy, [2]Norwegian University of Life Sciences, P.O. Box 5003, 1432 Ås, Norway, [3]Norwegian Institute for Bioeconomy Research (NIBIO), Norwegian Genetic Resource Centre, P.O. Box 115, 1431 Ås, Norway; nicola.sambugaro@studenti.unipd.it

The six local and endangered cattle breeds in Norway were close to extinct thirty years ago, whereas over the last ten years the populations have had a steady growth in sizes. The growth is in meat production even if the breeds traditionally were dairy breeds. From 2012 to 2021 the total population of the six breeds increased from 2,042 to 4,489 breeding cows. Cows from these breeds that were kept in dairy production decreased by 92 in this period, from 1,384 to 1,292, whereas cows kept in suckler/meat production increased from 658 to 3,197 during the same decade. Figures from Statistics Norway show that the total amount of cows in Norway decreased by 20% from 2011 to 2021. The increase in the populations of the endangered local breeds in a decade where the cattle population at national level decreased by 20%, indicates clearly that the local endangered dairy breeds are important resources in the resilience of livestock farming systems. The above underline that identifying marketing advantages related to the meat quality is of high relevance for the sustainable utilization of these breeds. This is obviously a challenge in an area where red meat has been linked to many lifestyle diseases first and foremost colon cancer, but also diabetes and cardiovascular diseases. A project has been dedicated to compare the meat quality in Norway's dominating breed Norwegian Red (NRF) with the meat quality in the endangered local breeds. The project aimed at 15 loins per breed and to characterize tenderness of these breeds using the Warner Bratzler Shear cell. Tenderness is regarded as the most sought-after positive quality characteristics for beef meat and the hypothesis is that the modern breeding is away from tenderness and possibly also juiciness. Juiciness is related to the fat content and in some markets a high intramuscular fat content (>3%) is regarded as a positive quality characteristic. On the other hand, cattle fat is also questioned regarding healthiness despite no final causal agreement. The results will be discussed within this quality frame.

Session 46 Theatre 12

Fatty acid profile of meat and milk of local sheep breeds in Italian Eastern Alps

E. Benedetti Del Rio, M. Teston, M. Ramanzin and E. Sturaro
University of Padova, DAFNAE, Viale dell'Università 16, 35020 Legnaro PD, Italy; elena.benedettidelrio@phd.unipd.it

This study is part of the 'Sheep Up' project (RDP, Veneto Region), which aims to sustain the conservation of four local sheep breeds ('Alpagota', 'Brogna', 'Foza' and 'Lamon') creating and sharing knowledge and tools between farmers for: (1) optimizing breeding schemes for the conservation of genetic variability; and (2) promoting the added value in terms of non-market services and products quality connected to their typical extensive farming management. For this purpose, since animal-derived products are an important issue for human health, the project aimed also to assess whether the peculiar management of these breeds confers an added nutritional value to their products. Specifically, we compared the fatty acid (FA) profiles of meat and milk derived from the above mentioned four local sheep breeds and investigated the effect of their feeding management, based on the seasonal availability of forage: hay in the winter and pasture in the summer. Forty-six meat samples of *longissimus lumborum* were collected from the four breeds in the different feeding conditions. Twenty bulk milk samples from 2 herds of Alpagota and Brogna sheep were collected every week for 10 weeks. All samples were determined using gas chromatography-flame ionization detector (GC-FID). Meat FA profiles showed an optimal omega-6/omega-3 ratio for the four breeds, remaining below 3 in both the pasture and hay feeding managements. The differences among breeds were in the total content of saturated and unsaturated fatty acids. In hay management Brogna averaged 50.8% of unsaturated fatty acid content, while the other breeds showed a range from 47.0 to 48.5%. The milk FA profile presented an omega-6/omega-3 ratio between 2.1 and 3.7, with greater values for Alpagota, 3.1±0.22, with respect to Brogna breed, 2.4±0.16. The results of this study showed that products from these breeds have a good fatty acid profile, independently from the type of feeding management. This information can help the promotion of local typical products and complement the other actions of the Sheep-up project to increase the awareness of local stakeholders and consumers about the importance of animal genetic resources and their link with traditional rural activities.

Session 46 Poster 13

Dry season effect on selected for yearling weight heifer's: resilience based on productive aspects

V.T. Rezende[1], J.N.G.S. Cyrillo[2], A.H. Gameiro[1], M.E.Z. Mercadante[2], S.F.M. Bonilha[2] and R.C. Canesin[2]
[1]Universidade de São Paulo, Faculdade de Medicina Veterinária e Zootecnia, Avenida Duque de Caxias Norte, 255, 13635-900, Pirassununga, SP, Brazil, [2]Instituto de Zootecnia, Rod. Carlos Tonani, 94, 14175-000, Sertãozinho, SP, Brazil; vanessatrezende@usp.br

Brazilian beef cattle production is based on pastures. Thus, seasons have impact on food supply and it can influence livestock performance. This study aimed to assess the influence of yearling weight selection on the resilience of Nellore heifers in the growing phase. Data from Nellore heifers born between 2008 and 2017 belonging to selection experiment, divided into Selection line (NeS), n=494, and Control line (NeC), n=215, selected for higher and mean yearling weight (YW), respectively, were analysed. Weightings were performed from birth to 586±26 days of age. Local climatic data were used, and pasture production was estimated to establish the challenge period. Difference from the YW, based on the weaning weight and ADG until 550 days of age (W550), with mean of 287.71±42.56 kg, and the same trait estimated excluding the ADG of dry season period (EW550), with mean of 309.03±46.63 kg, were enforced to indicate the resilience of animals (DFW550). Animals were classified as susceptible (S) or resilient (R) based on the first and the third quartile of DFW550, respectively. The model included fixed effects of selection line (NeS; NeC), birth year (2008 to 2017), and age at W550 as covariate. The average annual rainfall and estimated pasture production were 1,301.42±269.01 mm and 8,685.80±1,704.11 kg DM/ha/year, respectively. In the dry season (May to August), were 30.51±10.99 mm, and 101.30±40.29 kg DM/ha/year, in the same order. There was no significant difference (P=0.222) in the resilience of lines. Means of DFW550 were -22.31±13.97 and -19.07±12.97 for NeS and NeC. The results of DFW550 for susceptible animals were -4.23±0.70 and -5.90±0.93, and for resilient animals -38.86±0.64 and -38.35±1.15 for NeS and NeC, respectively. The dry season equally affected the resilience of NeS and NeC heifers, there is no impact of selection by yearling weight on the ability to maintain the performance standard, under adverse nutritional conditions, in the growth phase.

Session 46 Poster 14

Sardo-Bruna breed valorisation throughout enhancement of meat yield

M.R. Mellino, M. Lunesu, R. Rubattu, A. Marzano, G. Battacone, S.P.G. Rassu, G. Pulina and N. Nudda
University of Sassari, Dipartimento di Agraria, Viale Italia 39, 07100 Sassari, Italy; mrmellino@uniss.it

The valorisation of autochthonous cattle is advantageous to preserve local genetic resource in Sardinia. Sardo-Bruna (SB) is one of the minor Italian cattle breeds of limited diffusion recognized and protected by the Italian ministry of agriculture. However, to attain the sustainability of this breed is to improve the meat yield and quality. Aim of this study was to evaluate and compare breed effects on production traits, carcass and meat features and methane emission efficiency of Sardo-Bruna (SB), Limousine (L) and crossbreed Sardo-Bruna×Limousine (SB×L) maintained with the same management and feeding condition for six months. Thirty beef were used 10 SB, 10 L and 10 hybrids balanced for sex. Average daily gain (ADG), carcass yield (CY), proximate composition, and colour traits were measured. The methane emission was calculated and expressed as CH_4 emission per carcass weight. Data were analysed using general linear model with R. ADG was similar among to SB, L and SB×L (1.14, 0.98 and 1.02 kg/d, respectively, P>0.10). The CY of SB was lower than L and SB×L (58.63 vs 62.36 and 63.55%, respectively, P<0.001). However, composition of minced meat obtained to m. rectus Abdominis did not differs among breeds respect to dry matter, protein, and fat (P>0.10). The meat colour parameters measured using CIEL Lab coordinates were similar among breed (P>0.10). The cooking losses was similar (P>0.10) with losses between from 32.6% for SB×L to 34.1% for SB. The relationship between methane emission and carcass weight highlighted more efficiency of L and SB×L respect to SB (149.23, 146.03 vs 163.42 g/kg respectively, with P<0.001). The results of this study showed that the Limousine can be used to enhance carcass yield of autochthonous breed without effects on meat quality. Furthermore, crossbreed would improve the efficiency of methane emissions per carcass weight. However, preliminary results highlighted that purebreed SB fed balanced diet can give interesting results of carcass quality and average daily gain too. These results could be useful inputs to evaluate economic outcomes from local and crossbreed beef and to help beef producers in crossbreeding program. Research funded by PSR Sardegna 2014-2020-misura 16.2 – Probovis project.

Session 46 Poster 15

Weighted single-step genome-wide association study for somatic cell score in Valle del Belice sheep

M. Tolone[1], H. Mohammadi[2], A.H.K. Farahani[2], M.H. Moradi[2], S. Mastrangelo[1], R. Di Gerlando[1], M.T. Sardina[1], M.L. Scatassa[3] and B. Portolano[1]
[1]University of Palermo, Scienze Agrarie, Alimentari e Forestali, Viale delle Scienze, 90128 Palermo, Italy, [2]Arak University, Animal Sciences, Faculty of Agriculture and Natural Resources, Shahid Beheshti St., 38156879 Arak, Iran, [3]Istituto Zooprofilattico Sperimentale della Sicilia, Via Marinuzzi 3, 90129 Palermo, Italy; marco.tolone@unipa.it

Somatic cell count (SCC) or log transformed SCC (somatic cell score, SCS) have relatively higher heritability compared to mastitis and are used as the first trait to improve mastitis resistance. The objective of this study was to uncover genomic regions explaining a substantial proportion of the genetic variance in somatic cell score in a Valle del Belice dairy sheep. Weighted single-step genome-wide association study (WssGWAS) was conducted for somatic cell score (SCS). In addition, our aim was also to identify candidate genes within genomic regions that explained the highest proportions of genetic variance. Overall, the full pedigree consists of 5,534 animals of which 1,813 ewes have milk data (15,008 records) and 481 ewes were genotyped with 50K single nucleotide polymorphism (SNP) array. The effects of markers and the genomic estimated breeding values (GEBV) of the animals were obtained by five iterations of WssGBLUP. We considered the top 10 genomic regions in terms of their explained genomic variants as candidate window regions for SCS trait. The results showed that top ranked genomic windows (1 Mb windows) explained 5.24% of the genetic variances for SCS. Among the candidate genes found, some known associations were confirmed while several novel candidate genes were also revealed, IL26, IFNG, PEX26, NEGR1, LAP3, and MED28 for SCS. These findings increase our understanding of the genetic architecture of the examined trait and provide guidance for subsequent genetic improvement through genome selection.

A study of monogenic traits in the alpine autochthonous Aosta Breeds

F. Bernini[1], C. Punturiero[1], R. Milanesi[1], M. Vevey[2], V. Blanchet[2], A. Bagnato[1] and M.G. Strillacci[1]
[1]Università degli Studi di Milano, Department of Veterinary Medicine and Animal Sciences, Via dell'Università 6, 26900 Lodi, Italy, [2]ANABoRaVa, Fraz. Favret 5, 11020 Gressan (AO), Italy; francesca.bernini@unimi.it

The Aosta breed represents a key element in the economy of the valley for its production (milk-meat-cheese), its cultural value (Bataille des Reine) and the maintenance of the mountain territory thanks to the summer pasture practice. The reproductive management of the three Aosta breeds, Valdostana Red Pied (VRP), Valdostana Black Pied / Chestnut (VBPC) and Chestnut/Herèn (CH) has a key role in maintaining the biodiversity of the breeds and their productive and adaptive characteristics. The aim of this study was to estimate the genotypic frequencies for disease, fertility, and production mendelian traits in the three Aosta breeds. The study has been conducted on 2,930 Valdostana females genotyped with the GeneSeek Genomic Profiler (GGP) Bovine 100K by Neogen. The Principal Component Analysis highlighted that breed VRP clusters separately from breeds VBPC and CH that are, instead, similar. Genetic frequencies permitted to define that this population it's free from the main genetic disorders identified in other more cosmopolitan bovine populations (e.g. BLAD, Brachyspina, Citrullinemia, Holstein fertility haplotypes, Weaver syndrome, etc.) The genetic frequencies of loci related to production traits vary according to breed: breed VRP has a stronger milking attitude having as most frequent genotypes the BB and AB of the k-casein and β-lattoglobulin: these genotypes stand for a better dairy attitude and yield fundamental to produce the Fontina DOP cheese. Given the growing consumer interest in A2 β-casein, some studies identify it as more digestible, it is worth underlining that the most frequent genotypes are A2A2 and A1A2 with the respective frequencies of 47.74% and 43.14 for the VRP, 40.61% and 46.26 for the VBPC, and 39.60% and 50.50% for the CH breed. For meat production the genomic variation provides the possibility to select for better meat quality because the negative alleles in Calpain (CAPN_316, CAPN_4751) and Calpastatin (CAST_2870, CAST_2959 and UoGCAST1) genes aren't fixed in the population thus giving space to work on meat tenderness. Funded by PSRN DUALBREEDING – Fase 2.

Local breeds as a support for the protected area resilience

D. Bojkovski, T. Flisar and M. Simčič
University of Ljubljana, Biotechnical Faculty, Jamnikarjeva 101, 1000, Slovenia; danijela.bojkovski@bf.uni-lj.si

In the landscape of the south-eastern Alps is located the largest National Park in Slovenia, the Triglav National Park (TNP), which is classified in the IUCN category II, a protected landscape area, an alpine biosphere reserve and a Natura 2000 site. The alpine climate, with cold and long winters and short summers, provides a distinctive and a unique ecosystem with traditional alpine pastures and meadows, dairy cottages, forests and limited cultivated areas in valleys. A characteristic Alpine landscape prevails with typical soluble rock types such as limestone and alpine aquifers as an important sources of drinking water. The TNP is geographically and climatically divided into two distinct parts, Upper Carniola with a harsher climate and Bovec Region, with influences of sub-Mediterranean climate. Several agricultural plant varieties and local breeds have been preserved in the area, which are used for various purposes and thus can deliver ecosystem services in alpine area without irreversibly damaging the environment. Traditionally, alpine pastures are used for grazing in the summer and farmers have grazing rights, which means that the number of grazing animals per farm is limited. There are two opposing problems, one is overgrowth due to abandonment pastures and the other is very intensive use and overgrazing of active alpine pastures. In order to prevent abandonment of mountain pastures, keeping of local breeds such as Cika cattle, Drežnica goat and Bovec sheep should be further supported and promoted. Local breeds are adapted to the specific environments in the LFA areas, such as the slope pastures, the extreme climatic conditions and production systems within the TNP. They shape the appearance of alpine ecosystems, their hardiness, grazing behaviour and forage preference, as well as their body weights and frame size, which differ among breeds play important role. Well managed grazing of local breeds can increase land cover, plant productivity and species diversity, which positively affects the water infiltration and filtration, reduces soil erosion, and increases the grassland's ability to sequester carbon. It also helps control weeds and invasive species, and in high mountain areas, grazing is used to reduce the risk of avalanches.

Life Cycle Assessment of broiler chicken production using different genotypes and low-input diets

M. Berton[1], A. Huerta[2], A. Trocino[2], F. Bordignon[1], E. Sturaro[1], G. Xiccato[1] and M. Birolo[1]
[1]University of Padova, DAFNAE, viale dell'Università 16, 35020, Italy, [2]University of Padova, BCA, viale dell'Università 16, 35020, Italy; angela.trocino@unipd.it

This study evaluated the effect of diet (standard, ST: 3,050 kcal/kg metabolizable energy; 18.5% crude protein; low input, LI: 2,921 kcal/kg ME; 17.5% CP), genotype (a fast-growth genotype; two local breeds – Bionda Piemontese, BP and Robusta Maculata, RM; and their crosses with Sasso strain, SA – BP×SA and RM×SA), and sex on the environmental footprint of broiler production computed by Life Cycle Assessment (LCA). A total of 441 chickens (half males, half females) were housed in 40 pens (2 pens/genotype/sex/diet), and fed from 20 d of age until slaughtering (47 d for Ross, 105 d for all other genotypes) the ST diet or LI diet. The LCA used the pen as the reference unit and considered impacts from animal and manure management and dietary ingredients production to assess the following impact categories: global warming (without and with land-use change; i.e. GWP and GWP_LUC), acidification (AP) and eutrophication (EP) potentials, cumulative energy demand (CED) and land use (LU). The functional unit was 1 kg body weight gained (BWG). Impact values were submitted to ANOVA with diet, genotype, and sex as fixed effects using PROC GLM of SAS. Both diet (P<0.01) and genotype (P<0.001) affected almost all the impact categories (P<0.01). In details, LI diet produced greater GWP, EP and CED (from 5% to 43%) and lower LU (-8%) and GWP_LUC (-55%) than ST diet. Ross showed the lowest impact values, RM×SA and RM intermediate, and BP and BP×SA the greatest ones. Compared to females, male showed a lower impact, from -8% (LU) to -11% (AP), while GWP_LUC was not affected. When the local breeds were compared to Ross, higher environmental impact values were found: differences were greater with ST diet (+82% to +145% in local breeds vs Ross depending on the category) than with LI diet (+26% to +77%) (significant interaction diet × genotype). In conclusion, the higher the broiler performance the lower the environmental impact, but the use of LI diet reduced the difference in environmental footprint between fast-growing chickens and slow-growing local breeds.

Geographical distribution of Slovenian local breeds over time

T. Flisar, Š. Malovrh and D. Bojkovski
University of Ljubljana, Biotechnical faculty, Department of Animal Science, Groblje 3, 1230, Slovenia; tina.flisar@bf.uni-lj.si

Slovenia is geographically a small country, but it is a point of convergence for a number of different landscapes, each of which has its own characteristics and unique features. Consequently, the conditions for agriculture vary across country. The majority of utilised agricultural area is classified as less favoured area for agriculture, mainly located in mountainous regions. Agricultural land is important for food production, environmental protection, conservation of cultural landscape and population in rural areas, and for fulfilling ecological functions. Local breeds play a key role in efficient use of natural capital and utilisation of environmental resources in regions where conditions for agriculture are not optimal. Slovenian agriculture has been facing economic, environmental, and societal challenges, resulting in downward trend in the number of agricultural livestock holdings. To assess the resilience of farming system with local breeds, we analysed changes in population size and geographic distribution between 2010 and 2021. In the past, local breeds were reared closed to the area of their origin, while traditional and foreign breeds were concentrated in the central and east region of Slovenia. We have shown examples of increase and spread of local breeds populations such as Cika cattle and Krškopolje pig, confirming their adaptability to different ecosystems. The successful expansion of farms, keeping local breeds in the last decade is connected with the incentive measures for *in vivo* conservation of local breeds, raising awareness, promotion activities and the effective breeding programs.

Session 47 Theatre 1

Automated detection of dairy cows' lying behaviour using spatial data

I. Adriaens[1], W. Ouweltjes[1], M. Pastell[2], E. Ellen[1] and C. Kamphuis[1]
[1]Wageningen University and Research, Animal Breeding and Genomics, Droevendaalseweg, 6708 PB Wageningen, the Netherlands, [2]Natural Resources Institute Finland (Luke), PLF group, Production Systems, Latokartanonkaari 9, 00790 Helsinki, Finland; ines.adriaens@wur.nl

Health and welfare of animals is often translated in behaviour, which is traditionally observed visually by farmers at specific time points. The time-consuming and subjective nature of this limits the information that can be derived, raising the need for automation. Automated monitoring of animal behaviour with sensors allows to analyse health and welfare parameters over time, objectively and with little manual labour, as long as the recorded time series are appropriately interpreted. We developed a method to monitor lying behaviour of dairy cows in 2 freestall barns using an ultra-wide band positioning system. From 332 cow-days, time series of the distance to the centre of the barn, calculated from the x,y-coordinates, and the height of the tags (z-coordinate) were analysed. Data of each cow-day were first segmented with a statistical changepoint analysis, after which each segment was classified as either 'lying' or 'non-lying' with a bootstrapped-aggregated decision tree ensemble. We assessed the accuracy of both the segmentation (i.e. correct identification of the moment of getting up and lying down) and the classification (lying vs non-lying). Lying bouts as indicated by accelerometer data were taken as the ground truth. The classifier was trained on 33% of the data, with one data split based on time and one on cow identity. 66% of the segments were used for model evaluation. With the changepoint segmentation, 85.5% of the lying down and getting up were detected within 5 minutes of the ground truth. Classification performance was high, with in both test sets, over 91% of the segments correctly classified as lying or non-lying behaviour, and consistent over animals. Mainly data quality issues prevented higher performance (e.g. when large gaps were present in the data). The small training set and robustness of the algorithm regardless of split demonstrate the high potential of the developed methodology. Future research can focus on analysing behavioural parameters over time, e.g. time budget allocation, and its link with health and welfare.

Session 47 Theatre 2

Social network analysis of dairy cow contacts inside a free-stall barn

H. Marina[1], I. Hansson[1], W.F. Fikse[2], P.P. Nielsen[3] and L. Rönnegård[1]
[1]Swedish University of Agricultural Sciences, Department of Animal Breeding and Genetics, Box 7023, 750 07 Uppsala, Sweden, [2]Växa, Ulls väg 26, 756 51 Uppsala, Sweden, [3]RISE Research Institute of Sweden, Department of Agriculture and Food, RISE Ideon, 223 70 Lund, Sweden; hmarg@unileon.es

The social interaction between cows is important, both for production and animal welfare. In this study, the movements of nearly 200 dairy cows inside a free-stall barn were followed for two weeks. The cows were divided into two groups inside the barn and the following characteristics were recorded: parity, lactation stage, breed, oestrus and pregnancy status. The individual position of each cow was registered once a second with an ultra-wideband system, and the amount of time pairs of cows were in close proximity (<2.5 m) to each other during a day were used to perform a social network analysis. The aim of the study was to identify characteristics for pairs of cows that affected the probability of being in contact with each other. Different areas (cubicles, feeding area, and walking alleys) were analysed separately. Preliminary results show that cows of the same parity have a significantly higher tendency to be in contact than cows from different parities. Our study highlights non-random indoor movement of cows and is expected to be of importance in prediction of disease transmission and milk production.

Machine learning methods to predict blood metabolites in dairy cows using milk infrared spectra

D. Giannuzzi[1], E. Trevisi[2], L.F.M. Mota[1], S. Pegolo[1], F. Tagliapietra[1], S. Schiavon[1], L. Gallo[1], P. Ajmone Marsan[2,3] and A. Cecchinato[1]

[1]University of Padua, Department of Agronomy, Food, Natural Resources, Animals and Environment (DAFNAE), viale dell'Università 16, 35020 Legnaro (PD), Italy, [2]Università Cattolica del Sacro Cuore, Faculty of Agricultural, Food and Environmental Sciences, Department of Animal Science, Food and Nutrition (DIANA), via E. Parmense 84, 29122 Piacenza, Italy, [3]Nutrigenomics and Proteomics Research Center, via E. Parmense 84, 29122 Piacenza, Italy; diana.giannuzzi@unipd.it

Preventive management decisions are crucial to handle metabolic disorders in dairy cattle. Diverse serum metabolites are known to be valuable indicators of cows' healthy status. Hence, the capability to predict these metabolites would allow the early detection of metabolic impairments and to enact preventive interventions at farm level. In this study, we used various machine learning (ML) methods to implement prediction equations for a panel of 29 blood metabolites related to energy metabolism, liver function/hepatic damage, oxidative stress, inflammation/innate immunity, and minerals using milk Fourier-transform mid-infrared (FTIR) data. The dataset comprised 1,204 Holstein cows reared in 5 herds located in Northern Italy. The blood metabolic profile was assessed using the standard analytical method. The best predictive model was developed using an automatic ML algorithm that tested diverse methods, including penalized regression, random forest, gradient boosting machine, artificial neural network and stacking ensemble. We evaluated the performance of the model using two cross-validation (CV) scenarios: five-fold random and herd/date-out. Results showed that for all the blood metabolites, random CV acquired higher prediction accuracy than herd/date-out and results were consistent over traits. Prediction accuracies (R^2) ranged between 0.44 and 0.48 of haptoglobin to 0.85 and 0.89 of globulins, for random and herd/date-out CV, respectively. In conclusion, our study indicates that FTIR milk spectral data, routinely used for determining milk components, could also be used for the prediction of some blood metabolites for the assessment of the metabolic status of dairy cows.

Relevance of sensoring data from automated precision supplementation for breeding and management

H.L. Foged

Organe Institute, Skødstrupbakken 64, 8541, Denmark; henning@organe.dk

Automated Precision Supplementation (APS) is an IoT system for allocation of extra supplements, such as mineral feed supplements or feed additives to individual animals, including dairy cows, in periods of the production cycle, where this is relevant. For dairy cows, APS offers the possibility to move excess amounts of minerals and vitamins from the TMR ration, compared to needs in mid and late lactation, to the critical period of the lactation, from about 3 weeks before and until 3 months after calving. In this way, 25-50% of the mineral feed supplements given via TMR rations can be moved to the critical period of the lactation to the benefit of cow health and productivity, animal welfare and dairy production load on environment and climate. Use of APS has in trials shown a 23% reduction in disease events that require veterinarian involvement, substantial better milk quality in the form of lower somatic cell counts, and improved fertility, measured as the conception rate. The APS system is also providing the opportunity for allocation of advanced feed additives in periods of the lactation, where this is relevant. Advanced feed additives are developed with effects against, among other, acetonaemia and insufficient amino acid supply among fresh cows. APS function in the way that the animals by own choice visit the APS feeders and either is given a feeding of supplements, or no feeding in case they already have reached a maximal supplementation or if they are not entitled to get extra supplements at all. APS is thus sensing cow activity towards the feeders, and analyses of the collected data reveals large variations between individual cows' activity levels, as well as its variation over time, which could be correlated to cows' fertility and health status. By adding sensors for gas concentrations in cow breath, the APS system gives e.g. early warnings of acetonaemia by a non-invasive method for strengthened cow management. Enteric methane production shows a large heritability of around 0.4 and use of APS fitted with methane sensors has potentially an important role in breeding for a reduced climate footprint of dairy production.

The compensatory effect of quarters for milk yield perturbation in dairy cows

W. Xu, Y. Song, I. Adriaens and B. Aernouts
KU Leuven, Department of Biosystems, Division of Animal and Human Health Engineering, Kleinhoefstraat 4, 2440 Geel, Belgium; wei.xu@kuleuven.be

The milking compensatory effect (MCE) is the ability of quarters to produce more milk than expected, partially or fully compensating the milk loss of disordered or dry-off quarters in lactating dairy cows. The MCE is hypothesized to relieve the milk yield perturbation (MYP) at the cow level, when a cow suffers from the MYP, or inflammation, or (sub-)clinical diseases at the quarter level. However, the description and quantification of the MCE are limited. To study the MCE in depth, the data of 2,763 lactations of 1,321 cows from 3 farms equipped with automatic milking systems were involved. The MYP were defined as periods of at least 5 days of negative residuals and at least 1 day that the total daily milk yield at the cow or quarter level was below 85% of the estimated unperturbed lactation curve obtained by iteratively fitting the Wood model. Quantifying the MCE was based on the difference between the actual milk yield during the perturbation and the estimated unperturbed lactation curve. The MYP was hypothesized with the existed MCE, thus, perturbation period data were removed in Wood model fitting. To eliminate the effect of milk yield fluctuation and fitting bias by the Wood model, strict conditions were applied to define the MCE in each perturbation: (1) the MYP presented at the quarter level without the overlapped MYP at the cow level; (2) the MYP started later than 15 DIM and ended early than 290 DIM; (3) the total compensated milk yield based on 1 to 3 quarters was more than 5 kg. In total, 858 perturbations (5 to 67 days length) were detected with the quantified compensated milk yield (5.0 to 66.8 kg) from 1 (25.8%), or 2 (38.7%), or 3 (36.5%) quarters. The total compensated milk yield accounted for a maximum of 21.90% total milk yield at the cow level during perturbation. Our study confirms the existence of the milking compensatory effect during milk yield perturbation at the quarter level while no perturbation presents at the cow level. Further study is expected to link the MCE with more productive data and health indicators to investigate the resilience to challenges at the quarter level of dairy cows.

Analysing lying behaviour using real-time location system data in a dairy cow group

N. Melzer[1] and J. Langbein[2]
[1]Research Institute for Farm Animal Biology (FBN), Institute of Genetics and Biometry, Wilhelm-Stahl-Allee 2, 18196 Dummerstorf, Germany, [2]Research Institute for Farm Animal Biology (FBN), Institute of Behavioural Physiology, Wilhelm-Stahl-Allee 2, 18196 Dummerstorf, Germany; melzer@fbn-dummerstorf.de

Real-time location systems (RTLS) are becoming increasingly popular for analysing the behaviour of livestock. Currently, RTLS are primarily evaluated on how well animals are assigned to corresponding zones such as resting area. However, it is not only important to know when cows are in the resting area, but also whether they are lying or standing in their lying boxes. The duration of lying behaviour is an important indicator of animal welfare. To investigate how accurately lying behaviour of cattle can be analysed using RTLS measurements, we compared this data with annotated video data (3 half-days) of a group of 14 lactating Holstein cows. Two half-days (t1, t2) of annotated video data were used to develop an R-pipeline involving various processing steps to determine lying times from RTLS data for individual cows. The third half-day (t3) was used for testing. The observed total lying times were in average on 391 min (±88 min) in t1, 322 min (±88 min) in t2, and 364 min (±93 min) in t3. We calculated sensitivity and precision to test the quality of the assignment of the lying behaviour. There was a high level of agreement between video and RTLS data across all data sets (sensitivity and precision were in average ≥0.95). Moreover, we calculated the Spearman rank correlation coefficients (Rs) between total lying times based on video and RTLS data. In t1 and t2, the Rs≥0.96 and the mean absolute error was 15 min in t1 and 13 min in t2. Finally, we obtained Rs=0.86 and a mean absolute error of 15 min in t3. Better results could be achieved, if the R-pipeline is more closely adapted to individual behavioural differences. In summary, our results suggest that RTLS data might be useful to reveal lying behaviour of cattle with high precision and low effort.

Classification of mastitis in cows using deep learning approach with model regularisation

K. Kotlarz[1], M. Mielczarek[1,2] and J. Szyda[1,2]
[1]Wroclaw University of Environmental and Life Sciences, Kozuchowska 7, 51-631 Wroclaw, Poland, [2]National Research Institute of Animal Production, Krakowska 1, 32-083 Balice, Poland; krzysztof.kotlarz@upwr.edu.pl

Bovine mastitis is a disease that is one of the most common disorders in dairy cows causing animal well-fare problems and economic losses. Our study aimed to build a deep-learning classifier into mastitis-resistant and mastitis-prone individuals using single nucleotide polymorphisms (SNPs) from the whole genome sequence. The full dataset comprised SNPs identified based on whole-genome sequences of 32 Polish Holstein-Friesian cows that were 16 paternal half-sibs discordant in their mastitis status (a mastitis-resistant and a mastitis-prone). Genomes were sequenced using the Illumina HiSeq2000 platform. The whole-genome sequence pipeline for SNP identification consisted of: (1) quality control; (2) raw data filtering; (3) alignment to the reference genome; (4) SNP calling as well as SNP filtering. The training data set was composed of 12 mastitis-resistant and 13 mastitis-prone cows, while the test data set consisted of 4 mastitis-resistant and 3 mastitis-prone cows. After filtering, 16,618,983 SNPs were considered in classification. First, a Logistic LASSO Regression (L1) regularization was used for SNP pre-selection so that only SNPs with nonzero estimates were used in training the deep-learning-based classifier. The regularization lambda parameter was chosen using the Optuna software with the Parzan algorithm to maximize the training accuracy. Second, using only SNPs selected in the previous step, the Optuna was used for the optimization of the number of layers and neurons in deep neural network models implemented in Keras. Each model in the SNP pre-selection and hyperparameters tuning steps was 4-cross-validated. Three penalty values, resulting in the best accuracy of the classification of the training data set, were selected for the final deep-learning model: 0.563, 0.928, and 1.532. The accuracy of classification of the test data set was 0.428 for the first penalty, 0.574 for the third penalty, and 0.714 for the second penalty. This varying classification accuracy demonstrated the importance of a proper feature (SNP) selection in the high-dimensional data with a low number of records available for training.

Automated activity estimates of finishing pigs during weighing are repeatable and heritable

W. Gorssen[1], C. Winters[2], R. Meyermans[1], S. Janssens[1], R. D'Hooge[2] and N. Buys[1]
[1]KU Leuven, Centre for Animal Breeding and Genetics, Department of Biosystems, Kasteelpark Arenberg 30 – bus 2472, 3001 Leuven, Belgium, [2]KU Leuven, Laboratory for Biological Psychology, Tiensestraat 102 – Box 3714, 3000 Leuven, Belgium; wim.gorssen@kuleuven.be

Pig breeding is changing fast due to the rise of new technologies and societal developments. Precision livestock farming technologies, such as computer vision systems, allow the automated phenotyping for novel traits, like pig behaviour. Results from several behavioural genetics studies have shown that behavioural traits in pigs are heritable, and hence can be used in breeding programs. To implement this on large scale, an easy, robust and labour efficient phenotyping remains the main issue for implementation in breeding programs. In this study, we expanded pig weighings with automated measurements of pig activity levels using a tailor-made body pose estimation model via DeepLabCut software. Pigs' activity levels were quantified as *mean speed*, *straightness index* and *sinuosity index* using the travelled trajectory during weighing with the *trajr* package in R. Moreover, weight and activity data were coupled with pedigree information to estimate genetic parameters. The dataset consisted of 1,556 finishing pigs with up to eight measurements per pig and 7,428 records in total. Recordings on the same pigs took place with a two week interval. Successive estimates for activity traits had Pearson correlations ranging from r=0.16-0.39. Correlations were low but consistently positive over longer periods of time (r=0.01-0.27), indicating that activity traits had a low to moderate repeatability over time. Moreover, activity traits were estimated to be moderately heritable (22-35%), whereas estimated genetic correlations between activity and production traits and ear/tail biting were low. We demonstrated an automated method to phenotype activity traits in pigs during weighing. These traits were estimated to be repeatable and heritable, with no adverse genetic correlations. Hence, this method possibly offers a way to automatically phenotype activity traits on an individual level for pig breeding.

Analysis of social behaviours in large groups simulation and genetic evaluation

Z. Wang, H.P. Doekes and P. Bijma
Wageningen University and Research, Droevendaalsesteeg 1, 6708 PB Wageningen, the Netherlands; piter.bijma@wur.nl

Harmful social behaviours, such as injurious feather pecking in poultry and tail biting in swine, reduce animal welfare and efficiency. While these traits are heritable, breeding is still limited due to a lack of individual phenotyping methods for large groups and proper genetic models. In the near future, large-scale longitudinal data on social interactions will become available thanks to developments in computer vision and artificial intelligence. We will soon need genetic models to analyse such data. Here we investigate prospects for genetic improvement of social traits recorded in large groups. We present models to simulate and analyse large-scale longitudinal data, and present accuracies of EBVs Latent traits were defined representing tendency of individuals to be engaged in behavioural interactions, distinguishing performer from recipient. Binary interaction records were simulated by agent-based modelling, and analysed using generalized linear mixed models. A total of ~200,000 interaction events were simulated for 2,000 animals with a half sib family structure. Accuracies of EBV were high, despite the low observed-scale heritability (h^2) of the binomial trait (0.61, 0.70 and 0.76 for h^2=0.05, 0.1 and 0.2 respectively). However, 2000 individuals each with only ~100 interactions already yielded promising accuracies (0.47, 0.60 and 0.71 for h^2=0.05, 0.1 and 0.2 respectively). This number of ~100 interactions corresponds to a few weeks or months of recording, depending on the frequency of social interactions in the population. In conclusion, we show that selection for social traits recorded in large groups is promising. Our model is applicable to large-scale longitudinal data on animals kept in large groups, which is expected to become available in the near future.

CT scanning whole yellowtail kingfish accurately estimates fillet fat %

D. Milotic[1], A. Lymbery[1], G. Gardner[1], G. Partridge[2], F. Anderson[1] and R. Lymbery[3]
[1]Murdoch University, Harry Butler and Food Futures Institute, 90 South St, Murdoch WA, 6150, Australia, [2]Harvest Road Oceans, The Swan, 171-173 Mounts Bay Road, Perth, Western Australia 6000, Australia, [3]University of Western Australia, School of Biological Sciences, 35 Stirling Hwy, Crawley WA, 6009, Australia; dino.milotic@murdoch.edu.au

Yellowtail Kingfish (YTK) is emerging as an excellent candidate species for the breeding programs in aquaculture. The high fat % within YTK fillets is a highly desirable trait as it allows it to target premium grade sashimi markets. The long-term goal is to breed high fillet fat % fish and drive the deposition of fat within the fillet genetically, which is cheaper and more effective than relying purely on feed to drive fat deposition into the fillet. Traditionally, fillet fat is quantified destructively via chemical methods. However, if fillet and fish quality could be accurately assessed non-invasively, it would allow aquaculture farms to retain breeding candidates while assessing fish quality. In the current study, the CT scanner has been used to scan 20 fish to quantify composition (fat, protein and bone), which has then been verified chemically. It was found that CT scanning whole fish allows accurate estimations of fillet fat %, suggesting quality of live fish may be determined non-invasively.

Session 47 Theatre 11

Novel phenotypes in the honeybee, *Apis mellifera*, based on real-time hive telemetry

G.E.L. Petersen[1], M. Post[1], D. Gupta[1], R. Mathews[1,2], A.C. Walters[1], P.F. Fennessy[1] and P.K. Dearden[2]
[1]AbacusBio Ltd, 442 Moray Place, Dunedin 9016, New Zealand, [2]University of Otago, Department of Biochemistry, 710 Cumberland Street, Dunedin 9016, New Zealand; gpetersen@abacusbio.co.nz

As a species that has co-existed with humans for millennia and frequently been translocated into new environments by human settlers needing easily accessible high-calory additions to their diet, the Western Honeybee, *Apis mellifera*, has been extensively modified throughout its history. Today's honeybees are often managed intensively and play a crucial role in the production of pollination-dependent crops like fruit, nuts and vegetable, but honeybee breeding efforts in intensive beekeeping operations still tend to be based on *ad hoc* selection decisions. In an effort to reduce the risk to beekeeping-adjacent production systems which depend on beehives as a source of mechanical pollination and to counter the global pollinator decline, substantial investment has gone into the development of sophisticated (and sometimes over-engineered) remote beehive monitoring and -management systems, often without a clear use case or value proposition in mind. Working with a minimum viable product for hive telemetry, which only collects data on total hive weight and bees entering and leaving the hive entrance, we have investigated the feasibility of hive telemetry data as the basis of defining new phenotypes in honeybees. By analysing the patterns of activity throughout the day and the changes in weight, we can define different foraging types as well as determine the efficiency of individual colonies at the task on hand, while correcting for beekeeper interventions such as feeding or harvesting. These individual differences in phenotype can be used both for management purposes, for example by matching honeybee colonies with early / later foraging patterns to forages blooming at particular times of the day. In addition, these phenotypes can be used as the basis for the selection of bees more suited for their production system, e.g. the pollination of specific crops, and thus allow the breeding of more sustainable bee populations.

Session 47 Theatre 12

An international trading language for intramuscular fat% in sheep meat

G.E. Gardner, C.L. Alston-Knox and S.M. Stewart
Murdoch University, Murdoch, 6150, Australia; g.gardner@murdoch.edu.au

A new Meat Standards Australia grading system (MSA) is being implemented to predict the eating quality of sheep meat using predictors such as intramuscular fat% (IMF%). Technologies have been invented to measure this trait, yet initially their deployment was limited as IMF% was not recognised in the AUS-MEAT international trading language. Thus, abattoirs could neither trade upon this value, nor use devices accredited to predict this trait. This abstract details how this limitation was rectified in 2021 through the introduction of IMF% into the AUS-MEAT language. Four key elements were required to establish IMF% as a trait within the AUS-MEAT language and enable predictive technologies to seek accreditation. Firstly, a gold standard method was proposed to define the trait. This was based upon laboratory near-infrared spectrophotometry which was calibrated against Soxhlet fat extraction using a chloroform solvent. Secondly, population sampling was specified whereby at least 20 observations must be sampled at each IMF% increment across the proposed accreditation range. Thirdly, error tolerances for prediction accuracy were established, requiring 67% of all predictions from a technology to be within ±1 IMF% of the laboratory value, and 95% of all predictions to be within ±2 IMF% of the laboratory value. In addition, each device must demonstrate 'within' device repeatability 3 times, and 'between' device repeatability across 3 devices, in both cases meeting the above accuracy standards. The analysis of accuracy is undertaken using a Markov Chain Monte Carlo stochastic simulation to characterise the distribution of residuals for the technology seeking accreditation (Predicted value) compared to the IMF% reference value (Observed value), assessed within each quarter of the IMF% accreditation range. Lastly, the commercial application of the technology, its calibration method, and corresponding reference values must be described, enabling AUS-MEAT to establish routine auditing protocols. In conclusion, the Australian sheep industry now uses IMF% as a trading language. This has enabled accreditation of technologies that predict this trait, an auditing structure to ensure its robustness, and provides integrity for the commercial roll-out of the cuts-based MSA grading system.

Session 47 Theatre 13

Digital breeding and assisted management in organic rabbit farming: the first results

Y. Huang[1], M. Gigou[1], J.P. Goby[2], D. Savietto[1] and T. Gidenne[1]

[1]UMR 1388 GenPhySE, Université de Toulouse, INRAE, INPT, ENVT, 24, chemin de Borde Rouge, 31326 Castanet-Tolosan, France, [2]Université de Perpignan, IUT, 77, Chemin de la Passio Vella, 66962 Perpignan, France; thierry.gidenne@inrae.fr

The development of organic production is growing significantly, but the organic rabbit farming remains a niche market in France. This lack of technical references is an obstacle to the development of the 'alternative' rabbit sector. A smartphone application GAELA combining decision support (breeding management) and performance recording (single, direct and secure entry on a public server) for rabbit farming using individual monitoring of breeders was created. Performance of reproduction were compiled for 6 farms over 3 years of production (2018-2021). Preliminary results indicated that the livestock size averaged 30 does and varied largely among the farms. With 3.9 matings, 2.6 parturitions per female/year were obtained (fertility rate averaged 66.8%). Total number of kits born by parturition averaged 7.8 and total number of kits born alive averaged 7.1. At weaning, a low survival rate was recorded (69.3%). If it could be increased to 85%, the sales revenue could increase 3,564 euros for a farm having 50 females producing 3 litters per year (based on an average of 1.6 kg carcass/rabbit at 15€/kg). The result confirmed the existence of a progress margin in the management of the maternity unit, by improving the survival rate before weaning (housing management, prophylaxy, etc.), while reducing the parturition interval, and without impairing the survival rate after weaning. This first study validated the usefulness of GAELA and highlighted the potentialities of organic rabbit farming. GAELA will be updated in 2022, with more features for daily management of rabbits, and data management. Moreover, a new web service 'GAELA-Web' will be available in late 2022, to provide performances analysis for breeders. In 2023, GAELA intends to provide animal prophylaxis and genealogy tracking. Thanks to GAELA, a national reference system for all 'non-conventional' rabbit farming is now in progress.

Session 47 Poster 14

Digitalizing livestock management to improve production

J.L.M. Caçador

Digidelta Software, Rua Lino António, lote 44 R/C, Cruz D'Areia, 2410-055, Portugal; joao.cacador@wezoot.com

A market research from 2019, reported that precision livestock market is worth 2.7 billion euros, and it is predicted that it will reach 4,2 in 2024. And according to Costa (2015), which visited a lot of farms in Portugal, the worst in terms of production are the ruminant's ones, like cattle and lamb. The reasons to explain this problem are the lack of investment in technical-scientific level and reluctance to make any changes in the way the farms work, to change to precision livestock farming. In the other hand, a study made by the European Parliament, verified that there is a 30% decrease in agricultural employment due to a mechanization of the farms, which has made possible an increase of the production in all members, including Portugal. FAO (2014) says that to increase production, records made in farms with new technologies, are essential to ensure profitability, helping farmers to choose the best inputs (land, animal, etc.), and the best business strategy, always following food safety and traceability requirements. All this signs point to a need of investment in the Portuguese ruminant livestock production, in specialized labour workforce and helpful technology, in order to boost the production without more animals or resources. Wezoot is a livestock management system that allow producers to maintain all the records of the farm in one software, web-based, allowing to work offline in the field. It gives automated alerts to perform tasks in animals and has connections with Bluetooth and RFID (radio frequency identification) technologies like tag readers, auto drafters and weighing scales, allowing farms to immerse in the world of livestock 4.0, where everything can be made faster, smarter, and more efficiently. To show the results of using a software, Lince (2022), said that with the help of Wezoot, he gained more money with half of the animals, by making decisions based in data. And Leão (2022) said that without the data, it is impossible to know where a producer is heading. All this enables the producer to spend more time monitoring the animals and not only work with them, helping preventing sanitary and productivity problems. By gathering and organizing the information, it also helps to make difficult decisions regarding the farm.

Session 47 Poster 15

Combining real-time data and InraPorc® simulations to perform precision feeding in pigs

L. Brossard[1], C. Largouët[2] and L. Bonneau De Beaufort[2]
[1]PEGASE, INRAE, Institut Agro, 35590 Saint Gilles, France, [2]Institut Agro, Univ Rennes1, CNRS, INRIA, IRISA, 35000 Rennes, France; ludovic.brossard@inrae.fr

Applying precision feeding in growing pigs requires methods and models for real-time analysis of performance and prediction of nutrient requirements. This study compares two methods for these calculations and evaluates their potential to reduce nutrient input in pig feeding. A dataset of individual daily bodyweight (BW) and feed intake (FI) kinetics of 285 growing pigs, reared from 81 to 156 days of age in an experimental station in *ad libitum* feeding conditions, was used. The first approach (PF1) was developed during the Feed-a-Gene project, and uses Holt-Winters and MARS methods to daily forecast individual FI and BW, respectively. Standardised digestible lysine (dLys) requirements were computed daily according to the factorial method based on performance forecasting. The second method (PF2) relies on a set of 2,200 virtual pigs which *ad libitum* performances where simulated using InraPorc. Based on the comparisons of past FI and BW kinetics between real and virtual pigs, a set of up to 10 virtual nearest neighbours was determined daily for each real pig. The expected individual performances and dLys requirements were then obtained by averaging InraPorc data of these 10 virtual pigs. The precision feeding for each pig was then simulated. For each approach, the proportions of two premix diets (A and B, 9.7 MJ NE/kg, crude protein content of 16.9 and 9.3% and dLys content of 1.0 and 0.4 g/MJ NE, respectively) were computed daily to reach calculated requirements. Nitrogen (N) and dLys intakes and N excretion were computed individually using daily real performance and A:B ratio, considering equal performances independently of feeding method. A classical two-phase feeding (2-P) was also simulated (A:B ratio=83:17 before 65 kg mean BW, 50:50 afterwards). Compared to 2-P, N and Lys intakes and N excretion were respectively reduced by 6.6%, 9.6% and 11.9% with PF1 and 9.1%, 13.1% and 16.2% with PF2. Performance predictions were similar for PF1 and PF2 compared to real performance, but PF2 allows a better day-by-day stability, allowing a smoother decrease in N and dLys supply along growing period. The potential of the new method has now to be tested *in vivo*.

Session 47 Poster 16

New insights to predict seminal quality in young Limousine bulls

A. Noya[1], A. Montori[2], O. Escobedo†[3], J. Cancer[4], A. Echegaray[5], L. Gil[6], S. Villellas[6] and A. Sanz[1]
[1]CITA de Aragón – IA2, Montañana 930, 50059 Zaragoza, Spain, [2]AARLIM, Baja 16, 22192 Tierz, Spain, [3]In Memoriam, [4]CTA de Aragón, Movera 580, 50194 Zaragoza, Spain, [5]Humeco, De la Mecánica 11, 22006 Huesca, Spain, [6]Universidad de Zaragoza, Miguel Servet 177, 50013 Zaragoza, Spain; anoya@cita-aragon.es

The attainment of sexual maturity in bulls is reflected in testicular and seminal quality traits. Obtaining and analysing seminal samples requires specific techniques that could make it difficult in field conditions. This study aimed to relate testicular and seminal parameters with several traits associated to the performance of young bulls as an indirect indicator of sexual maturity. Nine Limousine bulls (10.4±1.3 months old) were studied. Testicular echogenicity values (Ecotext, Humeco, Huesca, Spain) and intramuscular fat content (MeatQ-Text, Humeco) were obtained by ultrasonography. Sperm samples were obtained by electroejaculation and were processed in fresh and frozen-thawed for assessment by the integrated semen analysis system (ISAS, Proiser, Valencia, Spain). Grade of nervousness was evaluated counting the number of movements of the animal in the scale during 10 sec. Correlation coefficients were calculated by SAS Statistical package. All correlation coefficients presented were significant ($P<0.05$). Preliminary results indicated that: (1) scrotal circumference was related with intramuscular fat content of longissimus thoracis ($r=0.79$); (2) sperm concentration was related with the subcutaneous fat thickness at P8 point ($r=0.76$) and negatively with the animal nervousness ($r=-0.76$); iii) sperm progressive motility was related with the grey-pixel intensity of testicular parenchyma ($r=0.72$) and negatively with the diameter of the hypoechogenic areas of seminiferous tubes ($r=-0.71$); iv) rapid-classified spermatozoa were related with the feed conversion ratio ($r=0.79$) and negatively with the diameter of the hypoechogenic areas ($r=-0.76$); v) sperm progressive motility of frozen-thawed straws was related with the age of bulls ($r=0.87$). Although performance parameters and non-invasive new techniques cannot be used solely as indicators of sexual maturity for the time being, they can help to determine the period when a bull can be used as a sire. Acknowledgements to PDR RecríaINNOVA.

Session 47 Poster 17

Bodyweight loss estimation of Holstein dairy cows using milk-analysis prediction of bodyweight

A. Tedde[1,2], V. Wolf[3], M. Calmels[3], J. Leblois[4], C. Grelet[5], P.N. Ho[6], J.E. Pryce[6], G. Plastow[7], D. Hailemariam[7], Z. Wang[7], N. Gengler[2], E. Froidmont[5], I. Dufrasne[8], C. Bertozzi[4], F. Dehareng[5], M.A. Crowe[9], A. Bayram[10] and H. Soyeurt[2]

[1]National Funds for Scientific Research, 5 Rue d'Egmont, 1000 Brussels, Belgium, [2]Research and Teaching Centre (TERRA), Gembloux Agro-Bio Tech, University of Liège, Agrobiochem, Passage des Déportés, 5030 Gembloux, Belgium, [3]Seenovia, 141 Bd des Loges, 53940 Saint-Berthevin, France, [4] Walloon Breeding Association, 32 chemin du Tersoit, 5590 Ciney, Belgium, [5]Walloon Agricultural Research Center, 9 Rue de Liroux, 5030 Gembloux, Belgium, [6]Agriculture Victoria Research, Centre for AgriBioscience, AgriBio, 5 Ring Road, Bundoora, VIC 3083, Australia, Australia, [7]University of Alberta, Department of Agricultural, Food and Nutritional Science, 116 St & 85 Ave, Edmonton, AB T6G 2P, Canada, [8]University of Liège, Faculty of Veterinary Medicine, 2 Quartier Vallée, 4000 Liège, Belgium, [9]University College Dublin, UCD School of Veterinary Medicine, Stillorgan Road, D04 V1W8 Dublin, Ireland, [10]Luxembourg Institute of Science and Technology, 5 Av. des Hauts-Fourneaux, 4362 Esch-sur-Alzette, Luxembourg; anthony.tedde@doct.uliege.be

We proposed assessing the bodyweight (BW) change (BWC) through time using predicted BW of Holstein dairy cows from partial least square equations based on milk production, stage and period of lactation, and milk mid-infrared spectral analysis. We calibrated the BW equation using 12,240 samples split into 27 exploitations from 7 countries, whose performances were measured using a country-independent set comprising 1,300 records. Our model showed calibration and validation performances of 50 kg. We computed smoothed trendlines (BW ~ days in milk (dim)) on the observed and predicted validation's BW to estimate the BWC. Because of the negative energy balance dairy cows must face every early lactation, we compared predicted with observed BW loss before 100 dim. The model gave 68% sensitivity and 75% specificity, meaning that among the cows losing weight, 68% were correctly classified as such. In conclusion, this study showed the potential contribution of milk analysis in association with other easily measurable traits of dairy cows to provide a somewhat accurate indicator of BWC in early lactation.

Session 47 Poster 18

Differences and similarities in the phenotypic evaluation of European Red dairy cattle

Š. Marašinskienė, R. Šveistienė and V. Juškienė

Lithuanian University of Health Sciences, Institute of Animal Science, R. Žebenkos str. 12, Baisogala, Radviliškis distr., Lithuania, 82317, Lithuania; sarune.marasinskiene@lsmuni.lt

This study was implemented as part of the project 'Biodiversity Within and Between European Red Dairy Breeds – Conservation through Utilization' (ReDiverse). European Red Dairy Breeds (ERDB) represent a unique source of genetic diversity and are partly organized in trans-national breeding programs. The objective of the study was to analyse the phenotype recording schemes for the ERDB and to provide an overview on the existing phenotype recording procedures. The questionnaire and information from Genetic Evaluation Forms, as available on the Interbull website, were used. The results showed that international evaluation for all 7 traits (production, conformation, udder health, longevity, calving, female fertility and workability) of Red cattle population were performed in Germany, Netherlands, Denmark, Sweden, Finland and Norway. Lithuania, Latvia and Estonia participate only in the international evaluation for production and udder health while in Poland 5 out of 7 traits are evaluated internationally although the evaluation is strictly limited to the Holstein breed. Estonia evaluate conformation trait only for Holstein. Only Dutch evaluation system is focused on analysing data from herd book. The most important traits, according to the goals of breeding organizations, were production, conformation and udder health traits. Other traits, such as longevity and female fertility, shows different level of importance in different countries. Calving and Workability traits are subjectively scored by the farmers. All weights of traits in a breeding objective should have a uniform basis of units of expression that is difficult to achieve for traits that are evaluated subjectively. If breeding organizations want to expand their breeding programs towards the inclusion of new traits, their decision will depend on the correct characteristic of the trait. Furthermore, the outcome from breeding programs is visible only after many years. Therefore, investments in breeding programs are often related to trait measurement and genetic evaluation.

Session 47 Poster 19

Use of artificial neural networks for prediction of milk yield of dairy cows

M. Matvieiev[1], Y.U. Romasevych[2] and A. Getya[3]
[1]National University of Life and Environmental Sciences of Ukraine, General Rodimtsev str., 19, 03041, Ukraine, [2]National University of Life and Environmental Sciences of Ukraine, General Rodimtsev str., 19, 03041, Ukraine, [3]National University of Life and Environmental Sciences of Ukraine, General Rodimtsev str., 19, 03041, Ukraine; mykhaiylo_17@i.ua

Systematic registration of productivity of animals is essentially important to improve genetic value of animals. It is recommended to register milk productivity on farms on a monthly basis (monthly recording milking – MRM). However, for various reasons, milk productivity is often not recorded every month, which causes missing values in the database and makes the calculation of productivity not possible (milk yield – MY). In this case, there are several ways to predict productivity of cows. In our work, we decided to use approaches based on artificial neural networks (ANN). The study was performed on Holstein cows. In cows, milk productivity (MY, fat, protein, lactose content, and somatic cell score) was taken into account. Four variants of data sets were selected: (1) 1st, 2nd, 5th, 8th, and 10th MRM; (2) the first three, the 9th and the 10th MRM; (3) the first five MRM; (4) 2nd, 5th and 10th MRM. Training of the ANN (data on 95 cows in the training sample) and testing of the quality of the prediction of MY by lactation made by trained ANN (a sample of data of 49 cows) were conducted. Training of ANN was performed using individual data of ten MRMs during lactation and taking into account the value of MY for lactation, calculated by a traditional Test Interval Method (TIM). For each of the 4 variants of data sets, the training and testing procedure of ANN was conducted separately. After calculation was established that coefficients of variations of differences between predicted (ANN) and calculated (TIM) MY for different variants data sets varied from 5.16 to 7.91%, and Spearman's rank correlation coefficients between predicted and calculated MY for different variants were 86.6-91.8% (p <0.001). In this study, the most accurate prediction of MY between four variants of data sets using ANN was done applying data on the 1st, 2nd, 5th, 8th, and 10th MRM (r_s=91.8%). So, the usage of ANN to predict the level of MY during lactation can reduce cost of estimation of the milk productivity of cows.

Session 47 Poster 20

Digital technology adopted in customized heifer growth management

R. Kasarda[1], P. Chudej[2] and N. Moravčíková[1]
[1]Institute of Nutrition and Genomics/Slovak University of Agriculture in Nitra, Tr. A. Hlinku 2, 949 76 Nitra, Slovak Republic, [2]Vitafort Zrt., Szabadság u. 3, 2370 Dabas, Hungary; radovan.kasarda@uniag.sk

Various scientifically validated approaches form the basis of precision livestock farming. Evaluation of growth intensity in dairy cattle represents an important part of economically efficient precision management on farms. Growth intensity, either genetically determined, is often influenced by production environment, nutrition and feeding. However, objective measurement and scaling of heifers, especially in dairy farms, is not performed on a routine basis because it is often associated with missing equipment, laborious manipulation with animals, use of mobile cattle scales and barriers. PLF tools provide the solution to the majority of these constraints. Agroninja Beefie3D™ digital camera scale mobile application has been used to evaluate the individual growth intensity of 99 heifers during the farm year 2021, and data has been associated with customized heifer growth chart protocol. Calculations were based on the mature size of animals in the herd and used the target growth system described in the 2001 Nutrient Requirements of Dairy Cattle. Animals were grouped according to their age from weaning to 6 months, 6-9., 9-12. and over 12 months (inseminated). Individual differences in growth intensity were observed according to the custom weight (35.95-66.12 kg) and custom height (3.58-5.58 cm) growth curves. Relative growth intensity of body weight/stature was expressed based on mature size and goal for the average age at first calving. Higher growth intensity compared to the custom growth curve is associated with increased feed cost and increased insemination index. The use of digital scale application in farm practice is influenced by the measurement technique and skills of the evaluator. Still, in the case of regular usage, deviation could be minimized. However, measurements could deviate between evaluators. Understanding these relationships could help assess the return on investment associated with the increase in either stature or body weight gain. Using a digital scale app for measuring and weighting cattle based on image analysis increases welfare by eliminating animal manipulation, labour, and stress. This work was supported by APVV-17-0060 and APVV-20-0161.

Session 47 Poster 21

Use of digital image analysis in evaluation of claw traits in dairy cattle

R. Kasarda[1], P. Polák[2], J. Tomka[3] and N. Moravčíková[1]
[1]Institute of Nutrition and Genomics/Slovak University of Agriculture in Nitra, Tr. A. Hlinku 2, 949 76 Nitra, Slovak Republic, [2]Slovak Association of Pig Breeders, Záhradnícka 4148/21, 811 07 Bratislava 1, Slovak Republic, [3]National Agricultural and Food Centre, Hlohovecká 2, 951 41 Lužianky, Slovak Republic; radovan.kasarda@uniag.sk

Claw disorders are associated with milk performance depression and represent the third most important culling reason in the dairy industry with significant influence on the farm economy. The work aimed to evaluate the influence of claw formation on the production and reproduction performance of Holstein dairy cows. Claw formation was observed during functional claw treatment, and digital images were collected on individual cows. In addition, the presence of claw diseases was observed as Digital Dermatitis, Interdigital Dermatitis, Heel Erosion and Sole Ulcer, including the stage of the disease. Using NIS software, image analysis has been made to obtain the following measures: claw angle, claw height, claw width, claw diagonal, toe length, heel depth, total claw area and functional claw area. Milk recording data has been analysed associated with claw traits and claw disorders. 164 high producing cows (9,365 kg milk) born between 2015 and 2019 were evaluated at the university farm. Observed was average claw angle 48.94±5.13°, toe length 7.91±0.81 cm, heel depth 3.84±0.71 cm, claw height 6.57±0.72 cm, claw diagonal 12.44±0.99 cm, claw width 5.35±0.53 cm, total claw area 43.22±7.62 cm^2, functional claw area 25.63±8.15 cm^2. A more significant positive correlation was observed between milk yield and functional claw area than the total claw area. Milk yield is in positive correlation with heel depth. Higher milk yield is also correlated with a shorter and lower claw in general. Claw diseases lead to differences in milk production and affect reproduction parameters. Significant differences were observed in claw formation between healthy and sick animals. Exact data on claw formation and association with disease development and milk yield could be used to select more robust cows, better health and less culling and therefore improve farm profit. This work was supported by APVV-17-0060 and APVV-20-0161.

Session 47 Poster 22

Romanian sheep farmers' welfare priorities and their knowledge on precision livestock farming

L.T. Cziszter[1], C.M. Dwyer[2], S.O. Voia[1], E.N. Sossidou[3], S.E. Erina[1] and C. Morgan-Davies[2]
[1]USAMVBT, Calea Aradului 119, 300645 Timisoara, Romania, [2]SRUC, West Mains Road, EH9 3JG Edinburgh, United Kingdom, [3]ELGO-DIMITRA, VRI, Campus Elgo-Dimitra, 57001 Thessaloniki, Greece; ludoviccziszter@usab-tm.ro

The aim of this study was to prioritize the welfare issues of the dairy sheep industry in Romania, and to determine farmers' perceptions of precision livestock farming (PLF) approaches that could be applied to sheep welfare. Data were collected during the 1st National Workshop of the TechCare project (https://techcare-project.eu/) with stakeholders from Romania (online, April 2021), using the OPERA method. A list of 45 dairy sheep welfare issues identified within the project, were translated into Romanian, and sent via email to 47 stakeholders 10 days before the meeting. Thirty stakeholders participated at the meeting, divided into three groups, and each participant had the opportunity to discuss all welfare issues. Each group produced a list of 5 priority welfare issues, that was further discussed after reuniting the groups. At the end of the meeting, three dairy sheep welfare issues were found to be important for all participants: gastrointestinal parasites, lameness and mastitis. Two groups considered inadequate or contaminated water to be very important, while in other groups respiratory infection, heat stress and ectoparasites were prioritised. For PLF perception, a questionnaire was used, together with terms definitions and explanations that were sent to all stakeholders. We present only the opinion of the Romanian sheep farmers (14 valid responses). A large proportion of farmers were aware of existing indicators of animal welfare (78%), and had some knowledge of using PLF technologies (64%). However, 42% of farmers had never used PLF for welfare assessment, while the same percentage reported using it occasionally. Most Romanian farmers agreed that PLF would improve consumer knowledge of animal welfare (93%), and PLF should be promoted along the value chain (86%). A majority of farmers agreed that PLF enables effective income increase (71%) and improve production efficiency (78%), and 71% did not believe that PLF is too costly to implement. Overall, the study suggests that farmers in Romania are positive towards the use of PLF to manage sheep welfare issues.

FOLLOW PIG software: a tool for breeders for the conservation of indigenous swine genetic resources

S.G. Giovannini and F.M.S. Sarti
University of Perugia, Department of Agricultural, Food and Environmental sciences, via Borgo XX giugno 74, 06121, Italy; samira.giovannini@studenti.unipg.it

Nowadays, we look with interest at the role of animal genetic resources as a font required for achieving the meat production more sustainable for the environment and more ethical in terms of animal welfare. Breeding pigs in semi extensive breeding systems required many efforts by breeders, especially when they are involved in conservation programs or genetic reconstruction projects, they also have to follow coupling programmes and phenotypic detection activities, in addition to productive economic objectives. In these cases, the main issue for breeders lies precisely in the impossibility of preventively identifying the detection of managerial errors. This situation often affects the productivity of the farm causing economic losses and sometimes suffering to the animals. A software has been developed in order to identify all the most delicate breeding phases, both from the management and animal welfare point of view, aiming to develop strategies that can facilitate the activity of breeders while increasing the control on the animals. FOLLOW PIG software intents to bring the farmer closer to the needs and the physiology of animals: each breeder, through its own graphical interface within the software, can actively interact by entering information (dates of heats, number of births, scheduled matings) and receiving qualitative statistical feedback. The software will be programmed with timetables related to all stages of breeding (including genetic selection) organized in blocks so there is no possibility of counterfeiting. Moreover, all pigs are univocally identified with a RFID microchip and so enrolled inside of the software. Anytime the breeder can access to the animals dedicated section where can find all the information: genealogical, morphological, physiological and every event in which the animal was involved. The ambition is to make easy and possible for everyone to be the actor in safeguard of local genetic resources and at the same time the territory, which, especially in marginal areas, is subject to abandonment mainly by young people.

Aligning CT Scanner data used in determining carcass composition via the XTE-CT phantom

A.J. Williams, K.L. Mata and G.E. Gardner
Murdoch University, 90 South Street Murdoch, 6150 Perth, Australia; andrew.williams@murdoch.edu.au

In 2022 Carcass composition in lamb was included as a new trait into the international AUS-MEAT trading language. The actual trait reports the carcass lean, fat and bone percentage determined via a computed tomography (CT) scanner. Industry deployable technologies have been trained to predict this trait, but the gold-standard measure itself requires supporting evidence to show its capacity to produce reference values and correct its predictions using a set of known standards. The commercially available XTE-CT test-piece contains a selection of materials covering the density ranges of the different tissue types within a carcass. These can be used to calibrate the composition determined by a CT scanner providing confidence that data acquired is uniform both within and between different scanners. There are 2 stages in the composition calculation where adjustments stemming from the XTE-CT test-piece are imposed. Firstly, the Hounsfield unit thresholds that differentiate fat, lean, and bone are adjusted relevant to a reference value for one of the plastics (ABS) within the test piece. Secondly, the known density of the selection of reference materials within the test-piece are regressed against their scanned Hounsfield unit values creating a linear equation to estimate tissue mass. To demonstrate this method, fifty lamb carcasses were scanned over several days using 2 separate CT scanners routinely calibrated against air and water. We then compared the carcase fat%, lean%, and bone% from the 2 scanners both before and after adjustment using the XTE-CT test piece. Although differences were small, we found that a better alignment between the 2 scanners was obtained after performing the calibration.

Session 47 Poster 25

Carcase sectioning and computed tomography slice width have little effect on estimated lean and fat%

K. Mata, S. Connaughton, G.E. Gardner, F. Anderson and A. Williams
Murdoch University, Murdoch, 6150, Australia; katherine.mata@murdoch.edu.au

In 2022, new traits describing the lean%, fat%, and bone% of a sheep carcase were introduced into the AUS-MEAT international trading language. This enables technologies that can predict these traits to seek accreditation within the AUS-MEAT language. These traits are based upon estimates from computed tomography (CT) scanning which is used as a 'gold-standard' measurement for carcase composition. Given the commercial importance of this gold-standard, it is imperative to define measurement protocols that ensure the reliability of data acquired. On this basis we explored the impact of two variables that may influence the CT composition estimates of sheep carcases, the CT image slice width defined as part of the scanning protocol, and the effect of carcase sectioning prior to scanning. Two groups of 30 lambs with a large phenotypic range were CT scanned 36 hours post-mortem at a scan voltage of 120 kV and 5 mm slice width. The first group was first scanned whole, and then cut into 3 sections, separated between the 4th and 5th ribs, and at the lumbosacral junction. These sections were then re-scanned using the same CT settings. The second group was repeat scanned firstly at 1 mm and then at 5 mm slice widths, with all other factors kept the same. Compared to whole carcase scans, carcase sectioning had a small effect on CT estimated carcase composition (0.85% higher for fat, 0.75% lower for lean muscle, 0.15% lower for bone) Similarly, carcase composition estimates from images acquired at 1 mm or 5 mm slice widths showed marginal differences. However, volume estimation, and therefore CT prediction of cold carcase weight, was marginally better at 5 mm slice widths when compared to 1 mm. Furthermore, carcase sectioning has the added advantage of providing standardised composition estimates within carcase regions. Therefore, to ensure the robustness of the CT scanning protocols within the AUS-MEAT language, we have specified a 5 mm slice width and the scanning of sectioned carcases.

Session 48 Theatre 1

Estimation of THI critical threshold affecting milk production traits in Italian Water Buffaloes

A. Maggiolino[1], N. Bartolomeo[2], A. Tondo[3], A. Salzano[4], G. Neglia[4], V. Landi[1] and P. De Palo[1]
[1]University of Bari, Department of Veterinary Medicine, S.P. Per Casamassima km 3, 70010, Italy, [2]University of Bari, Department of Biomedical Science and Human Oncology, Piazza Giulio Cesare 11, 70124, Italy, [3]Italian Breeders Association, Via Ventiquattro Maggio, 44/45, Roma, 00187, Italy, [4]University of Naples, Department of Veterinary Medicine and Animal Production, Via Delpino 1, Napoli, 80137, Italy; aristide.maggiolino@uniba.it

Many studies have been conducted in order to evaluate how heat stress affects dairy cows' production, reproduction, welfare and behaviour, but little is known about its effects on buffaloes. The aim of the work was to detect the THI thresholds, from the day of test-day sampling until 5 days before, for multiple milk production traits in Italian dairy Buffaloes. A 10-yr data set (2009-2018) of test-day records was used, composed by 442,354 test-day records of 42,651 buffaloes. All data were matched with the maximum and the minimum daily THI calculated by temperature and humidity hourly recorded. First, a mixed linear model was fitted to obtain least squares estimates of THI effect on production traits. In a second set of analyses, the solutions for the THI class per parity class effect in the first model were used as the dependent variables to estimate change points in the relationship between production parameters and heat/cold load, applying a 2-phase regression analysis. Buffaloes showed no heat stress THI breakpoint for protein yield and cheese yield ($P>0.05$). Differently, protein yield showed THI thresholds between 18 and 19 during the three days before test-day for first, second and third parity buffaloes with losses ranging from 16.89 to 26.73 g ($P<0.0001$). Fourth parity buffaloes showed THI thresholds of 19 with losses ranging from 18.32 to 62.96 g ($P<0.001$). The lack of fitting of the 2-phase regression model on protein yield in buffaloes highlights the different metabolic asset of this species compared to cows in warm conditions. Nevertheless, the most interesting result is buffaloes' susceptibility to low values of THI. This can be explained by some anatomical features of buffalo, such as a hairless and thicker skin and by the tropical origin of the species.

Physiological, productive and metabolic parameters during heat stress in Brown Swiss and Holstein

V. Landi, A. Maggiolino, F. Giannico, G. Calzaretti and P. De Palo
University of Bari Aldo Moro, Veterinary Medicine, via marina vecchia, 75, 70019, Italy; vincenzo.landi@uniba.it

Holstein Frisian (HF) and Brown Swiss (BS) are the two most represented in Italy for milk production. Generally, the BS is highly appreciated because it is considered more resilient to more unfavorable environmental conditions. In animals subjected to intense metabolic efforts, the management of heat stress has become one of the most pressing issues for the farmer. Global climate change has not only raised the average level of temperatures but, has expanded the number of hours and days in which animals are above the neutrality threshold, in latitude. Active cooling strategies are continuously developing but there is a need to identify news indicators of stress on the animal to reduce their economic impact or to be able to act strategically on the genetic selection of more resilient animals. In line with the practices of precision zootechnics, the objective of this work was the evaluation of the influence of thermal stress on physiological, productive and metabolic parameters. 77 multiparous cows from the same farm equipped with artificial ventilation (2nd or 3rd parity and within the 200 day in milk) were selected. Two 4-day replicates, in the absence of ventilation, were carried out with 3 days of rest, monitoring the THI (temperature humidity index) through digital data loggers at 5-minute intervals. Daily, at 3 intervals (4AM, 15PM, 20PM) the following parameter were collected: respiration rate, rectal, vaginal, iliac, eyes and snout temperature were monitored, and during the two daily milking the daily milk/kg day, protein, fat, acetone, β-hydroxybutyrate, citrate, and urea content. The maximum THI during the three moments was respectively 74.50, 84.07 and 82.62 during the test, and 64.95, 74.85 and 71.40 with ventilation. The results show a strong inflexibility of the breed and a significant ($P<0.01$) of the time of day on the physiological parameters and limited to the kg of milk and BHB. Some parameters such as percentage of protein and fat were significantly influenced by the day of the test. Physiological parameters showed a significant effect of consecutive day duration of heat stress, greater susceptibility of the HF breed and greater nocturnal resilience for the BS.

Effect of acclimatisation in physiological parameters of high-yielding dairy cows

F. Silva[1,2], L. Cachucho[2,3,4], C. Matos[2], A. Geraldo[2], E. Lamy[2], F. Capela E Silva[2], C. Conceição[2] and A. Pereira[2]
[1]Veterinary and Animal Research Centre, UTAD, Quinta de Prados, 5000-801 Vila Real, Portugal, [2]Mediterranean Institute for Agriculture, Environment and Development, U. Évora, Polo da Mitra, 7006-554, ÉVORA, Portugal, [3]Centre for Interdisciplinary Research in Animal Health, U. Lisbon, Avenida da Universidade Técnica, 1300-477 Lisboa, Portugal, [4]Alentejo Biotechnology Center for Agriculture and Agro-food – IP Beja, Rua Pedro Soares, 7801-908 Beja, Portugal; fsilva@uevora.pt

Heat stress alter the physiological status and the energetic balance in high producing animals. Acclimatisation is a thermoregulatory adaptation to heat stress with detrimental effects on productivity. We hypothesised that high-yielding dairy cows (HP; ≥9,000 kg – 305 days in milk; n=7) suffer a more significant influence of elevated environmental temperatures than low-yielding cows (LP; <9,000 kg – 305 DIM; n=6). Physiological and milk composition data was collected in summer (5 days with mean environmental temperatures of 23.5 °C – heat stress) and winter (5 days with mean environmental temperature of 6.6 °C – thermoneutrality). Respiratory rate (HP: 63.95±12.35; LP: 64.34±13.67 movements/minute), sweat rate (HP: 77.70±48,90; LP: 75.86±45.02 g/m^2/h) rectal temperature (HP: 38,87±0,72; LP: 38,76±0,63 °C) were significantly higher in summer than in winter across both groups, indicating a response to mild heat stress. Plasma triiodothyronine levels were lower in HP than LP in summer, indicating a higher degree of acclimatisation in HP cows. Haematocrit and hemoglobulin were significantly higher in summer but not different between groups. Regarding milk production, HP produced more milk than LP, but the difference between groups was shorter in summer than winter (17.90% and 22,30%, respectively). There were no differences in milk parameters within groups, except urea in the summer period (293.62 mg/kg and 253.69 mg/kg for HP and LP, respectively). Milk fat and protein were significantly lower in summer than winter. These results showed that elevated environmental temperatures alter the physiological status in both groups. Cows with different milk yield had similar first responses to heat stress. However, during the acclimatisation process, HP decreased metabolism rate while alterations in nitrogen pathways were observed.

Session 48 Theatre 4

Livestock emissions and the COP26 targets – main uptakes

M. Lee
Harper Adams University, Newport, Shropshire, TF10 8NB, United Kingdom; mrflee@harper-adams.ac.uk

Highlights from the parallel seminar about 'Livestock emissions and the COP26 targets' will be presented. Emphasis will be on GHG emissions. Main policies will be shortly explained. Also circular approaches will be outlined. What can we expect in years to come, and which routes to follow using a systems approach?

Session 48 Theatre 5

Emission research in dairy cattle barns

M. Klopčič, M. Bric, E. Selak and N. Valcl
University of Ljubljana, Biotechnical Faculty, Dept. of Animal Science, Groblje 3, 1230 Domžale, Slovenia; marija.klopcic@bf.uni-lj.si

In Slovenia a national project plan is being executed studying the emissions from cattle barns. As part of the EIP-AGRI project 'Circulation of nutrients, organic matter, processes and information in agriculture', we perform measurements of ammonia (NH_3) and other greenhouse gases (CH_4, CO_2, N_2O) concentrations on 10 dairy farms in Slovenia with different housing systems (free barn with cubicles, compost bedded pack barn, barn with artificial floor) for dairy cows and with different farming system (conventional / organic). Measurements of greenhouse gas concentrations on selected dairy farms has been performed once a month with the use of an FTIR gas analyser for ambient air measurements, with the possibility of measuring 25 gases at different, predetermined locations inside and outside the barn. Simultaneously with these measurements, we also perform measurements of the micro-climate parameters of the barn (temperature, humidity, air flow) at these same locations. The ventilation debit is estimated on basis of the situation in practice. The set-up of this applied field experiment and some preliminary results will be presented. Moreover, opinions of farmers are gathered concerning the attitudes towards emission reduction practices on farm level. Existing regulations or regulations in development will be explained. Also, impressions gathered about the attitude of farmer groups towards lowering emissions will be discussed.

Greenhouse gases and ammonia emissions from naturally ventilated dairy buildings of NW Portugal

A.R.F. Rodrigues[1], M.E. Silva[2], V.F. Silva[3], A. Gomes[4], L. Ferreira[5], M.R.G. Maia[1], A.R.J. Cabrita[1], H. Trindade[6], A.J.M. Fonseca[1] and J.L. Pereira[6,7]

[1]REQUIMTE, LAQV, ICBAS, School of Medicine and Biomedical Sciences, University of Porto, R. Jorge Viterbo Ferreira 228, 4050-313 Porto, Portugal, [2]LIADD-INESC TEC, Faculty of Economics, University of Porto, R. Dr. Roberto Frias, 4200-464 Porto, Portugal, [3]CRACS-INESC TEC, Faculty of Sciences, University of Porto, R. Campo Alegre, 4169-007 Porto, Portugal, [4]Cooperativa Agrícola de Vila do Conde CRL, R. Lapa 293, 4480-757 Vila do Conde, Portugal, [5]AGROS UCRL, R. Cidade Póvoa Varzim 55, 4490-295 Argivai, Portugal, [6]CITAB, University of Trás-os-Montes and Alto Douro, Quinta de Prados, 5000-801 Vila Real, Portugal, [7]Agrarian School of Viseu, Polytechnic Institute of Viseu, Quinta da Alagoa, 3500-606 Viseu, Portugal; anaferodrigues@gmail.com

Variability of animal housing, feed diets, and climate parameters among countries lead to different greenhouse gases (GHG) namely, methane (CH_4) and nitrous oxide (N_2O), and ammonia (NH_3) emissions. Monitoring those emissions allows the improvement of emission factors (EF) used in national inventories and to adopt specific abatement techniques. To date, few studies have determined NH_3 emissions from dairy buildings in Portugal and none assessed GHG. This work monitored GHG and NH_3 emissions in naturally ventilated dairy buildings of three farms with distinct feeding systems for at least 7 days in each season, for two years. In each building, air samples from 5 indoor locations were drawn by a multipoint sampler (INNOVA 1409) to a photoacoustic infrared multigas monitor (INNOVA 1412). Indoor temperature, relative humidity and production parameters were also recorded. Concentrations (mg/m^3) varied among buildings during 2017 from 1.63 to 30.9 for CH_4, from 0.73 to 1.03 for N_2O and from 0.34 to 1.80 for NH_3. During 2018 from 2.52 to 37.9 for CH_4, from 0.50 to 0.84 for N_2O and from 0.25 to 2.49 for NH_3. This work allows the accomplishment of EF for Portuguese dairy cows buildings needed to pursue the carbon abatement goals. Funding of FCT, AGROS and CAVC to ARFR (PDE/BDE/114434/2016), FCT to VFS (SFRH/BD/139630/2018), MRGM (DL57-Norma Transitória), REQUIMTE (UIDB/50006/2020), CITAB (UIDB/04033/2020) and LIAAD-INESCTEC (LA/P/0063/2020) is acknowledge.

Climate-neutral policy in the dairy sector expectations and current situation an example of Latvia

D. Ruska[1], K. Naglis-Liepa[2], D. Kreismane[3], J. Priekulis[4], A. Lenerts[2], A. Dorbe[3], S. Rancane[3], L. Degola[1] and D. Jonkus[1]

[1]Latvia University of Life Sciences and technologies, Institute of Animal Sciences, Faculty of Agriculture, 2 Liela street, 3001 Jelgava, Latvia, [2]Latvia University of Life Sciences and technologies, Faculty of Economics and Social Development, 18 Svetes Street, 3001 Jelgava, Latvia, [3]Latvia University of Life Sciences and technologies, Institute of Soil and Plant sciences, Faculty of Agriculture, 2 Liela street, 3001 Jelgava, Latvia, [4]Latvia University of Life Sciences and technologies, Faculty of Engineering, 5 J. Cakstes Blvd., 3001 Jelgava, Latvia; diana.ruska@llu.lv

The European Union (EU) has taken a leading role in mitigating climate change and preserving the environment, to which farmers must also make a significant contribution. The new Common Agricultural Policy 2023-2027 provides strong political support to achieve these goals. Latvia has prepared its capabilities and vision for the implementation of this policy. At the same time, various market measures are being discussed alongside this policy. They could be described in three directions. The first relates to carbon farming, which involves the monetization of carbon sequestration and avoidance. The second is related to possible changes in the demand side of the food market resulting in additional climate or environmental fields for milk purchasers that are appropriately labelled as carbon footprint or similar. The third direction is market developments related to various regulatory measures, such as the CBAM (Carbon Border Adjustment Mechanism), which will have an impact on fertilizer prices. At the same time, there is growing concern among farmers about how these different initiatives will affect farmers themselves. The study analysed the opportunities and losses of eight Latvian dairy farms related to the implementation of these initiatives, in the framework of the EU project CCCfarming. Using multiple interviews with farm managers and analysis of farm management data, it has been found that farms are interested in moving towards climate- and environment-friendly agriculture, while there are concerns about the economic development and efficiency of farms, as well as knowledge management problems.

Session 48 Theatre 8

Understanding the greenhouse gas and ammonia mitigation strategies in French dairy farms

C. Evrat Georgel and X. Verge
Institut de l'Elevage, 149 rue de Bercy, 75012 Paris, France; xavier.verge@idele.fr

Mitigating greenhouse gas (GHG) emissions (CH_4, N_2O and CO_2) and ammonia (NH_3) is now a major concern for the cattle sector, and especially for dairy farming since it represents two thirds of the GHG emissions of this sector. The Climate Care Cattle Farming project (CCCFarming) has therefore been built to study the gas emissions on dairy farms in 8 European countries, and to understand how farmers are facing this environmental issue. The aim is to identify the practices implemented on farms to reduce the environmental impact of dairy farms, and to link them with both their real efficiency and the motivations/barriers to their application on the farm. In this context, surveys and gas emission measurements were carried out on 60 European farms. This article analyses the data from the surveys conducted in the 8 French farms, each context having its own national specificities. Semi-guided interviews were conducted individually by the same interviewer with the 8 farm managers. The French panel stands out for its sensitivity to environmental issues, its curiosity to seek information in this domain and its openness to innovative practices that can improve the environmental footprint of systems. These observations can be explained in particular by the profile of the farmers, who come from experimental farms or are true entrepreneurs, as well as by the diversified consulting panorama existing in France and the impetus generated by environmental policies and regulations.

Session 48 Theatre 9

View of farmers on GHG and ammonia emissions by survey in eight countries

V. Eory[1], P. Hargreaves[1] and V. Becciolini[2]
[1]Scottish Rural University College, Scotland, 12345, United Kingdom, [2]University Firenze, via San Bonaventura 13, 50145 Firenze, Italy; vera.eory@sruc.ac.uk

In the framework of the European Green Deal, substantial efforts are addressed towards the objectives of sustainable agriculture and reduction of GHG emissions. Within this context, making resources available to researchers and farmers for the study and adoption of mitigation strategies is crucial. However, a likewise relevant matter is to invest in understanding farmers' perception of the problem and raising awareness to ensure their adequate engagement when practical actions are requested and implemented at enterprise level. As part of the CCCfarming project, interviews were conducted in over 40 study farms from 8 European countries. The aim was to unveil farm management strategies to reduce GHG and NH_3 emissions as well as farmers' knowledge, perception and goals about the topic. Overall, the survey revealed a group of above average size farms run by middle aged farmers with a good background education and close to research, environmentally involved and willing to test new technologies. The majority of farmers were engaged in changes entailing a reduction of N-losses (and ammonia) in the recent past. In addition, changes were mainly related to improvement of manure management in the whole chain and to energy use. Furthermore, most farmers are considering adopting further measures in the future concerning the use of renewable energies, changes in feeding, and improvements in manure management and livestock housings. The practices related to reducing GHG emissions (especially methane) were less clearly articulated and often overlapped with the practices related to ammonia reduction. It is questionable if methane reduction practices are actually known in the field in present time. The primary reasons behind the adoption of mentioned practices were the improvement of farm economy and efficiency, besides environmental outcomes. Differences between countries will be shown.

Session 48 — Theatre 10

Climate care dairy farming – highlights and discussion

A. Kuipers[1], N. Edouard[2], D. Ruska[3], R. Keatinge[4], A. De Vries[5] and P. Galama[1]

[1]Wageningen University and Research, Wageningen Livestock Research, De Elst 1, 6700 AH Wageningen, the Netherlands, [2]PEGASE, INRAE, Institut Agro, Saint Gilles, 35590 Saint Gilles, France, [3]Latvia University of Life Sciences and Technologies, Liela Street, Jelgava 3001, Latvia, [4]AHDB, Stoneleigh Park, Kenilworth, Warwickshire CV8 2TL, United Kingdom, [5]University of Florida, P.O. Box 110910, Gainesville, FL 32611, USA; abele.kuipers@wur.nl

The cattle sector is a major source of greenhouse gas and ammonia emissions from rumination of cows, manure, and in less extend, from grassland and corn cultivation including machinery. The www.CCCFarming.eu project focusses on the strategies and practices reducing emissions on farm level. Strategies and practices to deal with emissions may not always deliver the same effects for both methane and ammonia, and my vary among regions. Practices may be more or less simple to apply and may have a different cost. The presentation will summarize the highlights, barriers, opportunities and knowledge gaps which emerged during this climate care dairy seminar. The time period will be utilized to discuss the highlights and gaps with involvement of the audience.

Session 49 — Theatre 1

Organic livestock production in the USA: facts and figures

L. Bignardi Da Costa

Ohio State University, 1920 Coffey Rd, 43210, USA; da-costa.2@osu.edu

Organic production – 'a system aimed at producing food with minimal harm to ecosystems, animals or humans- in United States (US), as worldwide, continues to grow partially in response to consumer demand. Consumers which are concerned that food produced meets safety, environmental and social requirements. Organic producers understand this demand and can benefit society providing access to high quality organic food by implementing production systems that support water, soil health, and community health in their farms. Also, by using integrated pest management, weed control and techniques such as cropping rotations. Likewise, organic production and marketing seems as a feasible economic option with potential to improve quality of life that are attractive to beginning farmers and diverse/underrepresented groups in agriculture. However, the challenges of an increasing organic industry in US are complex, as there are many gaps about animal welfare, agroecological dynamics, economics and optimal management strategies associated with successful organic farming. As an example, transitioning to organic system where dairy farmers (also mentioned in organic beef cattle) have higher production costs, mainly from feeding, during this conversion period. Another issue relates to the US organic standards on antibiotic use which requires sick dairy cows to be treated, but those animals lose their organic status, so must be removed from the herd and their milk never be retailed as certified organic. Specific to the dairy industry, organic milk it is a large contributor to the overall economy in many states as California, Wisconsin, New York, Idaho and Pennsylvania. Consumer demand for organically produced agricultural products in the US increased by 31% from 2016 to 2019. Total fat reduced organic milk generated the largest number of sales in the US in 2020, with approximately 1.6 billion US dollars. In second and third place were whole organic milk and reduced fat organic milk (2%). In 2021, the total sales volume of organic milk in the US amounted to about 2.56 billion pounds. Even though organic farming in US has a potential growing market it still requires additional funding for research, maintenance of the organic standards integrity, and providing organic farmers with management tools for production with quality.

Productivity gains of organic ruminant farms: farm size and feed self-sufficiency matter

P. Veysset, E. Kouakou and J.J. Minviel
INRAE, UMR Herbivores, 63122 Saint-Genès-Champanelle, France; patrick.veysset@inrae.fr

This study focuses on the technical and economic analysis of 58 ruminant farms in organic agriculture at the Massif Central scale, in constant sample from 2014 to 2018. Over this five-year period, these farms expanded without increasing their labour productivity or animal density per hectare of forage area. While animal productivity has been maintained, repeated droughts have led to a decrease in feed self-sufficiency, and thus an increase in feed purchases. Overall, the selling prices of the products remained stable, but the increase in the cost of purchased feed as well as the increase in mechanization costs had a negative impact on the economic results, with the farm income per farmer falling by 40%. Over the period, the volume of inputs used has increased more rapidly than agricultural production, resulting in a decline in the overall factor productivity surplus (PS) at a rate of -2.6%/year. As the prices of products and inputs are relatively stable, this decrease in PS is financed at 41% by an increase in public aid (drought aid, agri-environmental climate measures) and at 49% by a decrease in profitability for the farmer. A binary choice estimation model (semi-nonparametric approach), i.e. which variables determine the positive or negative sign of the PS, shows that farm size is a negative determinant of the PS, as is system specialization, while feed self-sufficiency is a positive determinant. More statistically robust references on price indices of OF products and inputs, as well as long-term follow-ups of OF farms, are needed to validate these original results, which are based on a small sample size and a short period of time.

Organic woodchip versus conventional straw bedding on dairy cows production, behaviour and cleanline

M. Simoni, R. Pitino, G. Mantovani, G. Carone, L. Mochen, T. Danese and F. Righi
Parma University, Department of Veterinary Science, via del taglio Parma, 43122, Italy; marica.simoni@unipr.it

Conventional straw bedding (S) is one of the contentious inputs that needs to be phased out in organic livestock. The aim of the study was to compare organic poplar woodchip (OW) as alternative to conventional wheat S on dairy cows' production, behaviour and cleanliness. The trial was conducted in July 2021 in a tie stall farm. A total of 38 lactating Holstein cows (204±119 DIM, 26.9±6.5 kg MY) blocked by productivity were allocated to two bedding groups, S and OW, for 10 days. Each group was composed of 3 subgroups of 6 or 7 animals. Both bedding materials were provided in the amount of 7 kg/cow/day. At the beginning and during the last 5 days of trial, MY was measured twice daily from 3 cows per subgroup and sampled to determine composition, total bacterial count (TBC), coliform bacterial count (CBC) and spore-forming bacteria (SFB). During the sampling days, faecal (FS), udder (US) and cleanliness score (CS) were evaluated individually twice daily; meanwhile, faecal and bedding samples were collected before milking from each subgroup to measure humidity, N, aNDFom and ash. The individual behaviour of each cow was registered in the last 2 sampling days every 15 min for 48 h recording the frequency of standing, laying, eating, drinking, ruminating, sleeping or others. In the same days, DMI was calculated by subgroups as difference between hay and concentrate supplied and refused. The parameters MY (27.5 vs 26.4 kg), CBC (1,760 vs 2,449 UFC/ml), SFB (216 vs 208 MPN/1), US (1.7 vs 1.9) and CS (2.18 vs 2.18) were similar between groups (OW and S respectively), while TBC was higher in the S group (133 vs 84 UFC/ml, P=0.039). Despite faecal DM did not differ among groups, bedding moisture and N were the lowest in the OW. Laying behaviour was not affected by the bedding materials, while S group had a highest sleeping frequency compared to OW, whereas DMI was not different between groups. The S and OW appears comparable if productive performances are considered. However, the lower humidity of OW could have positively affected TBC. This project has received funding from the European Union's Horizon 2020 research and innovation programme under grant agreement No 774340.

Composition of organic and conventional Italian cheeses

C.L. Manuelian, M. Pozza, M. Franzoi and M. De Marchi
University of Padova, DAFNAE, Viale dell'Università 16, 35020 Legnaro, Italy; carmenloreto.manuelianfuste@unipd.it

This study aimed to compare the gross composition, minerals content and fatty acid profile of organic (ORG) and conventional (CON) Italian cheeses. Four Italian cheese varieties were evaluated: Asiago PDO 'fresco' (ORG, 9; CON, 9), Caciotta (ORG, 8; CON, 8), Latteria (ORG, 9; CON, 10) and Mozzarella TSG (ORG, 14; CON, 14). The ORG and CON samples (100 g) of each cheese variety belonged to the same cheese factory, production day, cheesemaking process and ripening time, and were collected from September 2020 to August 2021. Gross composition (moisture, fat, protein, salt, ashes, lactic acid; %), minerals (Ca, Na, P, S, K, Mg, Zn, Fe, Cu; mg/kg) and fatty acids (SFA, UFA, MUFA, PUFA, n-3, n-6, C4:0, C6:0, C8:0, C10:0, C12:0, C14:0, C15:0, C16:0, C18:0; g/100 g of cheese) were determined using infrared spectroscopy applying pre-existing prediction models. The ORG and CON samples were compared within each cheese variety using a non-parametric Mann-Whitney U test. Results showed differences ($P<0.05$) for all four cheese varieties. Asiago PDO 'fresco' and Caciotta were the cheese varieties that showed less differences, and Latteria and Mozzarella TSG were the ones that showed more differences between ORG and CON. Latteria showed lower lactic acid, Mg, UFA, PUFA, n-3 and n-6 content, and greater Fe, K, C10:0 and C12:0 content in ORG than in CON. Asiago PDO 'fresco' showed lower protein, Na and MUFA content, and greater ashes content in ORG than CON. Caciotta showed lower n-3 content, and greater K, C14:0 and C16:0 content in ORG than CON. Mozzarella TSG showed lower fat, SFA, UFA, MUFA, C4:0, C6:0, C10:0, C12:0, C14:0, C15:0 and C16:0 content, and greater ashes and Mg content in ORG than CON. In conclusion, differences on ORG and CON cheese chemical composition were observed in all four cheese varieties. However, being ORG or CON could be confounded with other factors at farm level such as breed and feeding. Funded by EU H2020 No 774340 – Organic-PLUS.

Effect of feeding system on SPME-GC-MS profile of mountain Asiago cheese

S. Segato[1], A. Caligiani[2], G. Marchesini[1], I. Lanza[1], G. Galaverna[2], L. Serva[1], B. Contiero[1] and G. Cozzi[1]
[1]Padova University, Dept of Animal Medicine, Production and Health, Viale dell'Università, 16, 35020 Legnaro (PD), Italy, [2]University of Parma, Dept of Food and Drug, Parco Area delle Scienze, 27/A, 43124 Parma, Italy; severino.segato@unipd.it

Flavour is an organoleptic property that contributes to define the quality of cheese and it originates during the ripening process by the action of non-volatile and volatile chemical compounds. Free fatty acids highly contribute to the development of cheese aroma and their concentration and composition are influenced by cows feeding system. The aim of this study was to evaluate the effect of the upland and lowland feeding system on the volatile profile of PDO Asiago cheese. Summer bulk milk samples came from dairy cows fed as (feeding system): upland pasture (alpine pasture, AP); upland barn TMR based on dried forage (barn upland, BU) and lowland barn TMR based on maize silage (barn lowland, BL). A total of 50 Asiago Allevo (6-mo ripening) samples were manufactured in a mountain (n=20 for AP; n=16 for BU) and a lowland (n=14 for BL) dairy plant in compliance with the PDO Asiago disciplinary. Volatile compounds (VCs) were analysed by a solid phase micro-extraction (SPME) coupled with gas-chromatography/mass spectrometry (GC/MS). The VCs were identified by their mass spectra retention indices and by comparison with commercial reference standards. Since VCs (relative concentration respect to the internal standard) data were not normal distributed, the effect of the feeding system was evaluated using the non-parametric Kruskal-Wallis test. Multiple post-hoc pairwise comparisons were performed adopting Bonferroni correction. Among the 45 VCs detected only 10 seemed to be affect by the feeding system probably because of the high variability among samples. The lowland feeding system based on maize silage (BL) thesis increased the 2-nonanone (ketone). The adoption of hays or pasture to feed dairy cows in the upland farming highlighted highest level of octanal (saturated fatty aldehyde) or 2-butanol (secondary alcohol), respectively. The outcomes of the study suggest that SPME-GC-MS could be reliable to authenticate mountain cheeses even though the VCs profile is characterized by a large variability also within thesis.

Session 49 Theatre 6

Beef carcase DEXA can predict lean trim weights

H.B. Calnan, B. Madlener and G.E. Gardner
Murdoch University, SHEE, 90 South St, 6150, Australia; Honor.Calnan@murdoch.edu.au

The quantity of retail product procured from beef carcases is an important determinant of their value. The ability to predict retail cut and trim weights from entire beef sides would improve the valuation of carcases in the beef supply chain and improve the ability to sort carcases for optimal fabrication. This experiment assessed the ability of a commercial dual energy x-ray absorptiometer (DEXA) system to predict the composition of beef sides scanned at abattoir line speed and thereby determine the lean trim weights procured from each beef carcase side. As DEXA can measure whole beef carcase composition with higher precision than the current Australian industry standard of P8 fat depth, we hypothesise that beef DEXA will predict side trim weights with higher precision and accuracy than P8 fat measures. Beef carcases (n=250) representing a wide range in carcase weight and fatness (measured by P8 fat depth) were selected for DEXA scanning at line speed before being boned out into a comprehensive set of retail meat cuts and trim. The fat content of the trim was visually assessed by the boner for designation into a trim category (65, 85 or 90% chemical lean or CL) before the sums of trim were weighed. The weight of each trim category was predicted using general linear models with hot side weight (kg) and P8 fatness or DEXA variables as covariates. Given the anticipated difficulty of a person visually differentiating 85 and 90% CL trim, these categories were also summed and predicted using the same models. In line with our hypothesis, DEXA variables predicted beef trim weights with better precision than P8 fatness, though the differences were small. Prediction models demonstrated variable precision, dependent predominately on trim category. DEXA variables predicted the weight of fatter trim (65% CL) with an R-square of 0.78 and a root mean square error of prediction (RMSEP) of 106 g, and predicted the combined leaner trim (85+90% CL) weight with an R-square of 0.92 and an RMSEP of 60 g. The DEXA system differentiated a greater range in trim weight than P8 fatness and therefore has potential as a novel technology to predict beef trim weights in a commercial setting.

Session 49 Theatre 7

How to assess beef and dairy products qualities according to farming systems: a review

B. Martin[1], I. Legrand[2], M. Coppa[1] and J.F. Hocquette[1]
[1]INRAE, UMR Herbivores, 63122 Saint Genes Champanelle, France, [2]Idele, Bd des Arcades, 87000 Limoges, France; bruno.martin@inrae.fr

Consumer and citizen expectations regarding animal products are related to both extrinsic quality such as animal welfare, environmental footprint and farmer income, as well as intrinsic quality encompassing product safety, nutritional value and sensory features. Regarding the latter, the very significant progress achieved during the last decades brought a deep knowledge of the individual livestock practices determining beef and dairy products intrinsic quality. Results obtained in controlled trials have mostly been validated on farm, with similar differences according to farming practises, but tighter than under controlled trials. Nevertheless, the question of how livestock practices or systems can influence simultaneously the various traits of the intrinsic quality of cattle products is less documented. For example, many studies revealed that grazing systems may improve the lipid profile of beef and dairy products, reduce the risk of pesticide residues, but the time spent outdoors coupled with a longer rearing period increases the risk of bioaccumulation of environmental contaminants in products and increases the variability of sensory features. In this review, a specific emphasis is put on the concurrently study of quality traits related to safety, nutritional value and sensory features of beef and dairy products in view of developing multi-criteria assessment for the characterisation of the global intrinsic quality of products. We will review the different databases, devices, methodologies and tools (including modelling approaches using existing databases) recently developed to assess and predict beef and dairy products global intrinsic qualities according to livestock practices and systems. This review underlines that further developments of tools and methodologies taking into consideration the synergies and trade-offs between quality traits depending on livestock systems are required. Such tools are crucial for agri-food actors to provide reliable and robust information meeting consumer expectations in relation to the multiple aspects of intrinsic quality of livestock products from the various European livestock systems.

Session 49 Theatre 8

Sensory and instrumental characterization of dry-aged loins of Bruna d'Andorra breed

N. Panella-Riera[1], C. Zomeño[1], G. Martínez[2], M. Gispert[1], M. Gil[1] and M. Font-I-Furnols[1]
[1]IRTA, Food Quality and Technology Program, Finca Camps i Armet, 17121 Monells, Spain, [2]Ramaders Andorra S.A., Camí de la Grau, AD500 Andorra la Vella, Andorra; cristina.zomeno@irta.cat

The application of dry ageing protocols provides a good strategy to improve sensory quality of beef, especially in terms of toughness during the first weeks. The aim of this study was to characterize sensory and instrumental attributes of beef from Bruna d'Andorra breed after different dry ageing periods. A total of 10 commercial pistol cuts from Bruna d'Andorra bulls were aged at 2 °C in a commercial cooling room for 3, 5, 10 and 21 days. After each period, a portion of 10 cm was cut and frozen immediately at -20 °C, while the rest of the pistol went on with the ageing process until the following sampling period. When all samples were collected, they were sent to IRTA's facilities and 3 slices (2.5 cm width) were cut, individually vacuum packaged and kept frozen at -20 °C until further analyses. Warner-Bratzler test (shear force) and sensory assessment by a trained panel were carried out. Analysis of variance was performed with SAS Software including the ageing time as fixed effect, and for sensory data the effects of session and panellist within session were also considered. The shear force values were similar in samples aged for 3, 5 and 10 days and significantly decreased in those aged 21 days ($P=0.003$). The ageing process also decreased the variability among samples after 21 days, which may contribute to commercialize more homogeneous meat. The sensory profile and textural traits were also affected by the ageing period. The odour intensity ($P=0.021$) and liver flavour ($P=0.004$) increased after 21 days, and the ageing flavour after 10 days ($P<0.001$). Juiciness, chewiness and stringiness were significantly lower in samples aged for 10 and 21 days than in the initial ones (3 days) ($P<0.001$). Toughness assessed by the panel was lower in samples aged for 21 days ($P<0.001$) than in the other, in line with the instrumental toughness, and showing the positive consequences of ageing beef on sensory traits. The evolution of chewiness and stringiness may also contribute to the sensory improvement of the meat during the ageing process. In conclusion, although juiciness could be lower, a dry ageing period of 21 days produced a sensory enhanced meat.

Session 49 Theatre 9

Meat quality parameters in Pajuna breed young bulls and steers – preliminary results

M. Cantarero-Aparicio[1], F. Peña[1], C. Avilés[2], J. García-Gudiño[3], J. Perea[1] and E. Angon[1]
[1]Universidad de Córdoba, Animal Production, Campus de Rabanales, 14071 Córdoba, Spain, [2]Universidad de Córdoba, Food Technology, Campus de Rabanales, 14071 Córdoba, Spain, [3]CICYTEX, Guadajira, 06187, Spain; eangon@uco.es

The use of local breeds raised under traditional low-input systems has gained great importance in recent years due to environmental and rural development issues; and it also offers opportunities to compete in differentiated markets that are better connected to new consumer preferences. This is the case of the Spanish Pajuna breed that is traditionally used in semi-arid mountainous areas of the Sierra Nevada Natural Park. Farmers and marketers lack the objective information necessary to promote the nutritional and sensory quality of the meat obtained from these breeds in their traditional systems, in order to achieve product differentiation. In this work we analysed the meat from ten entire males and ten steers from Pajuna breed slaughtered at 25 months of age. The animals were raised on a traditional low input system and fattened in a feedlot with concentrate and cereal straw *ad libitum*. Samples for analysis were obtained by removing the loin joint between the 6th and 7th rib from the left side of the carcass at 24 h post-mortem. Preliminary results showed that bulls had a higher percentage of muscle (61.9 vs 55.2%), while the steers had a higher percentage of total fat (23.3 vs 16.4%). The pH values and water holding capacity (drip and cooking losses) showed no significant differences. Colour only showed significant differences for the red index -higher in Bulls- (12.8 vs 15.1). Finally, maximum Warner-Bratzler shear force was better in steers (3.67 kg/cm^2), while bulls' meat was harder (5.19 kg/cm^2).

Investigation of beef eating quality through MSA grading system

M. Santinello, N. Rampado, M. De Marchi and M. Penasa
University of Padova, Department of Agronomy, Food, Natural resources, Animals and Environment, Viale dell'Università 16, Legnaro (PD), 35020, Italy; matteo.santinello@phd.unipd.it

Beef quality is assessed through different methodologies and one of the most recent is the Meat Standards Australia (MSA) system. This study aimed to investigate the post-mortem traits and measures of Italian Charolais cattle through the MSA approach. Data consisted of 798 young bulls and 945 heifers imported from France, fattened in 30 Italian commercial farms and slaughtered in an abattoir of Northern Italy. Live weight at slaughter and carcass weight were collected at the abattoir. The MSA evaluation and pH measurement on M. longissimus dorsi were performed on the 5th rib 24 h post-mortem following official MSA checklist by a trained technician. The MSA system evaluate the marbling on a scale from 100 (no intramuscular fat) to 1,190 (extreme intramuscular fat), ossification on a scale from 100 to 590 (low and high spinous process calcification, respectively), height of the hump (cm), fat and meat colour, and carcass damages. A generalised linear model was used to investigate sources of variation of MSA traits and ultimate pH. Sexes were analysed separately. Season of slaughter, category of slaughter age, category of carcass weight, and the interaction between category of slaughter age and category of carcass weight were included as fixed effects in the model, and slaughter batch and the residual as random effects. The marbling score of both young bulls and heifers was significantly affected by season of slaughter and carcass weight (higher marbling in winter than summer and spring; $P<0.05$). Animals with heavier carcasses had higher marbling score compared with animals with lighter carcasses (353 vs 336 for young bulls, and 407 vs 387 for heifers). Marbling score of bulls was not significantly affected by slaughter age, whereas it was affected for heifers ($P<0.05$). The ossification and hump height were higher for young bulls and heifers with heavier carcasses ($P<0.05$). The ultimate pH was not significantly affected by carcass weight, slaughter age, and season of slaughter. Results of this study suggest that carcass weight of the animals and season of slaughter had an important effect on marbling. This work was funded by Regione Veneto (Project 'SustaIn4Food', POR FESR, azione 1.1.4).

Across countries NIRS technology to predict MSA marbling score

A. Goi[1], M. Kombolo[2], M. Santinello[1], N. Rampado[1], J.-F. Hocquette[2], M. Penasa[1] and M. De Marchi[1]
[1]University of Padova, Department of Agronomy, Food, Natural resources, Animals and Environment, Viale dell'Università 16, 35020 Legnaro, Italy, [2]INRAE, Route de Theix, 63122 Saint Genès Champanelle, France; arianna.goi@unipd.it

This study aimed to investigate the ability of an *on-line* handheld near-infrared (NIR) spectrometer to predict the Meat Standards Australia (MSA) marbling score for Italian and French beef cattle. The MSA marbling score indicates the amount of marbling, but also size, fineness and distribution of intramuscular fat, and is determined by officially accredited graders on a scale of 100 (no intramuscular fat) to 1,190 (extreme amount of intramuscular fat) in increments of 10. A total of 277 Limousine females were analysed in a French abattoir, and 400 Charolais (213 females and 187 males) in an Italian one. The assessment was performed 24 h *post-mortem*, from 20 minutes up to 3 h after the cutting, by an accredited grader in each country. The reference MSA grading was on average 327.69±109.27 in France and 339.40±93.45 in Italy, with a coefficient of variation of 33.4% and 27.5%, respectively. The same 2 graders subsequently collected the NIR spectra following the same protocol, taking 5 measurements at different points of *Longissimus dorsi* at the 5th rib of the left half of the carcass. Reflectance NIR spectra (740 to 1,070 nm) were recorded by a portable NIR device and prediction models were developed using modified partial least squares regression and validated by both cross and external validation. Several approaches were tested by creating different calibration and validation sets within and across countries. Overall, the best prediction model was obtained by merging data from the 2 countries with coefficient of determination in external validation of 0.44 and standard error of prediction of 78.04. Overall, results were satisfactory and despite the predictions did not allow precise evaluation of this trait, the correlation between spectra and reference data might be reasonable for genetic purposes or for the discrimination of MSA classes directly in slaughterhouse and especially in countries with no official carcass grader.

Session 49 Theatre 12

Organ lesions in beef cattle – Reliability between authorised inspectors and an independent observer

A. Pitz, M. Krieger, M. Trilling and A. Sundrum
University of Kassel, Department of Animal Nutrition and Animal Health, Nordbahnhofstrasse 1a, 37213 Witzenhausen, Germany; sundrum@uni-kassel.de

The results of the official meat inspection, carried out by authorised inspectors at the slaughterhouse, can provide farmers with important feedback on the health of their animals and support farm health management. However, the basis for its usability is the validity of the collected data. Between September 2019 and March 2020, a total of 906 beef cattle were examined as part of the official meat inspection by authorised inspectors and, in addition, by a trained, independent observer (gold standard assessor) in a slaughterhouse in Northern Germany. Examination by authorised inspectors (veterinarians or official assistants) was based on an internal slaughterhouse protocol, whereas the independent observer used a more detailed catalogue. Inter-observer agreement was assessed using Cohens-Kappa (CK). As protocols were not identical, CK was calculated for different combinations of findings. Yules Y (YY) was used to estimate how much the CK value was reduced by the prevalence of a feature. The McNemar test was used to test for marginal homogeneity, with the P-value indicating the significance of the results. The independent observer (IO) documented 980 findings in 665 animals. Most findings were lung lesions (62.1%), followed by liver (29.7%) and heart lesions (0.1%). Within the same sample, authorised meat inspectors (AI) documented 186 findings in 151 animals with lung, liver and heart findings making up 7.3%, 12.9 and 0.2%, respectively. Interrater-agreement for lung findings was highest for the combination 'pneumonia' (IO) and 'dirty lung' (AI) but still very low (CK=0.029, YY=0.080, P=0.000). Interrater-agreement concerning the liver was highest if 'hepatitis' (IO) was combined with 'altered liver' (AI), resulting in slight agreement (CK=0.189, YY=0.359, P=0.010). In terms of heart lesions, there was no agreement at all between raters (CK=-0.001, YY=-1.000, P=1.000). Interrater-agreement in this study must be regarded as insufficient, especially if results are to be used within farm health management. Measures should therefore be taken to increase the validity of slaughterhouse findings, e.g. improved evaluation schemes, personalised training programs, etc.

Session 49 Poster 13

The INTAQT project: tools to assess and authenticate poultry, beef and dairy products

B. Martin[1], C. Laithier[2], F. Leiber[3], R. Eppenstein[3], E. Sturaro[4], F. Klevenhusen[5], S. De Smet[6], M. Petracci[7], C.L. De Marchi[4], C.L. Manuelian[4], J.F. Hocquette[1], D. Lopes[8], F. Faria Anjos[8], M. Bondoux[9] and C. Berri[10]
[1]INRAE, UMRH, 63122 Saint Genès Champanelle, France, [2]IDELE, 23, rue Jean Baldassini, 69364 Lyon, France, [3]FiBL, Ackerstrasse 113, 5070 Frick, Switzerland, [4]University of Padova, DAFNAE, Viale dell'Università 16, 35020 Legnaro, Italy, [5]BfR, Max-Dohrn-Straße 8-10, 10589 Berlin, Germany, [6]Ghent University, Coupure Links 653, 9000 Ghent, Belgium, [7]University of Bologna, Piazza Goidanich 60, 47521 Cesena, Italy, [8]CONSULAI, Rua da Junqueira 61G, 1300-307 Lisboa, Portugal, [9]INRAE Transfert, 5 ch de Beaulieu, 63000 Clermont-Ferrand, France, [10]INRAE, UMR BOA, 37380 Nouzilly, France; bruno.martin@inrae.fr

Actors of the agri-food chain lack reliable and robust information to meet consumer expectations in relation to the multiple aspects of intrinsic quality of livestock products from the various European livestock systems. The INTAQT project aims to assess the relationships between animal production systems and products quality in order to improve husbandry practices complying with high quality animal products and sustainability. This is the 'One Quality' concept. The project focuses on chicken meat, beef, and dairy products and applies a multi-actor participatory approach which involves all actors of the agri-food chain. The challenges are to: (1) develop comprehensive models to quantify the impact of livestock systems on product safety, nutritional value and sensory attributes; (2) propose, together with the agri-food chain actors, fast, easy and cost-effective analytical tools to predict the intrinsic quality of livestock products and authenticate the associated livestock systems; (3) propose together with the same actors multi-criteria scoring tools for the intrinsic quality of products; and (4) promote farming practices which can allow the production of safe, healthy and tasty animal products while ensuring a decent income to farmers and respecting animal welfare and the environment. The INTAQT project (EU H2020 No 101000250 – https://h2020-intaqt.eu/) started on June 2021 for 5 years.

Regulation of myogenesis and lipid deposition in bovine muscle satellite cells by miR-100 and -101

B.A. Mir, E. Albrecht and S. Maak
Research Institute for Farm Animal Biology, Institute of Muscle Biology and Growth, Wilhelm-Stahl-Allee 2, 18196 Dummerstorf, Germany; elke.albrecht@fbn-dummerstorf.de

Our previous investigations identified microRNA-100/-101 (miR-100/-101) and their putative mRNA targets insulin-like growth factor receptor-1 (IGF1R) and prospero-related homeobox 1 (PROX1), respectively, as differentially expressed in bovine musculus longissimus dorsi with varying intramuscular fat content. While the IGF1R signalling and PROX1 are implicated in myogenesis and lipid metabolism in muscle, the underlying regulatory mechanisms are poorly understood. The present study aimed to investigate regulation of above mentioned target genes by predicted miRNAs during bovine primary muscle cell proliferation and differentiation. Luciferase reporter assay confirmed miR-100/-101 to target IGF1R and PROX1 seed sequences, respectively. Furthermore, expression of miR-100/-101 and, IGF1R and PROX1 was reciprocal during bovine primary muscle cell differentiation, suggesting a cross-talk between microRNAs and target genes. Overexpression of miR-100 reduced ($P<0.05$) IGF1R and MYOG mRNA as well as IGF1R and MYH7 protein abundances in differentiating bovine muscle cells. Inhibition of miR-100, however, had no clear effect on mRNA and protein abundances of IGF1R and myogenic genes, but tended to increase ($P<0.1$) MYH8 protein. Overexpression or inhibition of miR-101 had no clear effects on generally low expressed PROX1 and on myogenic gene expression. Proliferation of satellite cells was determined by Bromodeoxyuridine incorporation and was not influenced by microRNA treatment. However, oleic acid induced lipid deposition was promoted by both miR-100/-101 overexpression and reduced expression of lipid oxidation-related genes. In conclusion, the results demonstrate modulatory roles of miR-100/-101 in bovine primary muscle cell development, lipid deposition and metabolism.

Pasture-raised Maremmana beef: effect of cooking with superheated steam oven on lipid oxidation

S. Tinagli, M. Tognocchi, G. Conte, L. Casarosa and A. Serra
University of Pisa, Agriculture, Food and Environment, Via del Borghetto 80, 56124, Italy; s.tinagli@studenti.unipi.it

The Maremmana is a cattle breed typically raised on pasture in southern Tuscany. Meat from Maremmana breed shows excellent organoleptic and nutritional characteristics. Pasture gives to the meat a specific flavour and contain more alfa-linolenic acid than cereals. As a result, meat from pasture-raised animals shows a higher content of n-3 fatty acids and lower n-6/n-3 ratio than meat from animals raised in the barn. Maremmana beef has enhanced nutritional value due to the well known positive effects of n-3 fatty acids on human health, however, high content of PUFA n-3, due to high number of bis-allylic carbons, makes meat more prone to oxidation. On the other hand, the lipid oxidation is the primary cause that negatively affects the quality of meat, seconds only to microbial contamination. One of the most critical points before consumption of meat is cooking. Some traditional cooking methods, for example grilling and frying induce potential risks for human health, that's why it is important to be careful to use cooking methods that are able to enhance meat nutritional value without negative effects. One of the most promising cooking methods is cooking with superheated steam oven (SHS). In fact, in SHS meat is cooked in an oxygen free environment, thus, making meat less susceptible to lipid oxidation. The aim of this work is to verify the effects of cooking with SHS oven on lipid oxidation of pasture-raised Maremmana beef. The study was conducted on 18 samples of beef cuts taken at the level of the first thoracic vertebrae. The samples have been divided into 3 groups: uncooked (U), cooked with SHS (SHS), cooked with a traditional steam oven (TSO). Composition of fatty acids, cholesterol, cholesterol oxidation products (COPs), thiobarbituric acid reactive substances (TBARs), volatile organic compounds (VOCs) have been determined. The cooking method SHS proved to be efficient at limiting the oxidative process in fatty acids; this is confirmed by results of TBARs: U 0.22 ppm, SHS 0.56 ppm, TSO 1.05 ppm with $P<0.01$. The cooking method didn't significantly affect cholesterol oxidation products (COPs) and volatile aldehydes.

Session 49 Poster 16

Pelvic suspension of carcass improve meat quality traits of zebu cattle

L.A.L. Chardulo, W.A. Baldassini, M.A.S. Coutinho, G.A. Rovadoscki, C.P. Oliveira, J.A. Napolitano, O.R. Machado Neto, R.A. Curi, G.L. Pereira and M.A. Tagiariolli
São Paulo State University, Animal Breed and Nutrition, DMNA, FMVZ, Unesp – Botucatu, SP, 18618-000, Brazil; luis.artur@unesp.br

The traditional hanging method (HM) of bovine carcasses in meat industry is suspension by the Achilles tendon (AT). However, HM by pelvic bone (PH) may improve beef quality. The effects of the AT versus PH methods on the sensory and objective meat quality traits were evaluated. Twenty bovine carcasses from Brangus heifers (n=10) and Nellore bulls (n=10) feedlot finished were used. After slaughter, the carcasses were longitudinally divided into two parts. The half carcasses of each group were randomly suspended by the AT (n=20) or by the PH method (n=20) and cooled (1 °C) for 48h. At deboning, Longissimus thoracis samples were collected for sensory analysis with untrained consumers (tenderness, flavour, juiciness and overall satisfaction), shear force (SF), colour (L*, a*, b*, chroma and hue), pH, cooking losses (CL) and purge losses, after two aging times (5 and 15 days). There was a positive effect ($P<0.01$) of the PH on the sensory tenderness of meat from Nellore and Brangus submitted to 5 days of aging. When the meat was aged for 15 days, there was no effect on sensory tenderness between the PH and AT methods ($P>0.05$) in both groups. Additionally, there was an interaction between the HM and meat aging of Nellore bulls ($P<0.05$) for sensory variables, while the same effects were not observed in Brangus heifers ($P>0.05$). In Nellore group, there was a tendency ($P=0.06$) in carcasses submitted to the PH method to produce more tender meat in relation to the AT method (SF=4.55±0.71 vs 5.14±0.82 kg, respectively). Furthermore, lower CL ($P<0.05$) were observed in the Nellore group submitted to the PH compared to the AT (27.7 vs 30.9%, respectively). The HM did not influence the other objective variables in both experimental groups ($P>0.05$). The results suggest that the PH method can reduce aging time from 15 to 5 days and improve meat quality traits of Nellore bulls, mainly tenderness.

Session 49 Poster 17

Saponins levels in Fenugreek seeds and their effect on palatability

M. Le Bot, A. Maniere, H. Bui and A. Benarbia
Nor-Feed SAS, R&D, 3 rue Amedeo Avogadro, 49070 Beaucouzé, France; hoa.bui@norfeed.net

Palatability enhancers are commonly used in ruminants to ensure proper consumption of mineral feed and to increase the milking frequency to the robot (Albright, 1993). Fenugreek (*Trigonella foenum-graecum*) is an annual plant that belongs to the family of the Fabaceae mainly cultivated in India. Due to the secondary metabolites, as steroidal saponins providing therapeutic properties and appetite stimulation, fenugreek seeds are commonly used in human and animal diets to stimulate appetite and weight gain. Depending on the part seed used, i.e. cotyledon or whole seed, the content of saponins can vary significantly which may lead to variable effects on palatability. In this study, we measured the saponins level in fenugreek seeds and compare it to a commercial product based on fenugreek cotyledons (Norponin cotyl®). The effect on dairy cow feed intake of both products was assessed. Briefly, three lick blocks were proposed simultaneously at the same place to fifty lactating cows (Holstein) during 18 days. The lick blocks differed by the addition of sepiolite (CTL), 280 mg/g of fenugreek whole seed (SEED) and 140 mg/g of fenugreek cotyledon (COTYL). Lick blocks consumption was evaluated by weighing once a day, after a 3-day long period of adaptation. Results showed that levels of total saponins ranged between 5.0 and 6.0% in whole seeds while in cotyledon it reached 12.3%. Although the results were not statistically significant a numerical increase of feed intake for lick blocks containing fenugreek was observed with +5.9% for SEED and +10.5% for COTYL. In addition, results showed that lick block intake was higher for COTYL than SEED despite a supplementation twofold higher for this latter. These results could be explained by the high saponins content of the cotyledons. Consequently, cotyledons may be interesting because of their high saponins content to ensure proper consumption of mineral feed or to increase the frequency of the visits to the milking robot. Further studies are however necessary to confirm these results.

Session 49 Poster 18

Proteomic investigation of longissimus muscle from beef cattle fed corn wet distillers grains

W.A. Baldassini, J.D. Leonel, A.J.R.A. Vieira, R. Scapol, R.C. Rodrigues, M.A. Tagiariolli, J.A. Torrecilhas, R.A. Curi, O.R. Machado Neto and L.A.L. Chardulo
São Paulo State University (FMVZ UNESP), Animal Breeding and Nutrition, Avenida Universitária, no. 3780 Altos do Paraíso, Botucatu, SP, 18.610-034, Brazil; rogerio.curi@unesp.br

Wet distillers grains (WDG) can be used in feedlot diets replacing corn and soybean meal. However, only few studies investigate possible proteome changes of longissimus muscle (and meat quality) on animals fed with corn WDG. This study evaluates the protein expression of *Longissimus thoracis* (LT) muscle in crossbred cattle fed control versus 450 g/kg de-oiled WDG (4% EE). Fifty non-castrated F1 Angus-Nellore bulls, aged 20-24 months, were used. After 120 days on feed, the animals were slaughtered and LT muscle samples were collected for molecular biology assays. Proteins were investigated combining two-dimensional gel electrophoresis (2D-PAGE) and electrospray ionization mass spectrometry (ESI-MS/MS). Images of 2D-PAGE gels (n=3/ animal) were scanned and analysed to obtain parameters such as number of spots, percentage of correlation (*matching*), isoelectric point and molecular mass. Protein spots (control vs WDG) were compared on their distribution, volume, and relative intensity. Correlation between the gels of each treatment was 63% and 65%, respectively. The mean number of protein spots found in the replicates of the control and WDG gels were 167±25 and 162.2±15.5, respectively. Subsequently, the ESI-MS/MS and bioinformatics procedures reveals 30 and 32 proteins in control versus WDG treatments, respectively. The biological processes, molecular function and cellular components of proteins differed ($P<0.05$) between experimental groups. Protein-protein interactions among treatments were analysed, which indicates in control animals, proteins related to energy metabolism (DNAJ, ANXA, HSPs, SOD, TPI, MDH, GPD, ENO, ALB, LDH and GAPDH), while proteins related to muscle contraction (MB, MYL1, MYL3, MYL6, AK1, and TNNT) were found in WDG treatment. Overall, this proteomic approach helps for understanding the mechanisms involved in product quality and nutrigenomics of crossbred cattle fed WDG.

Session 49 Poster 19

Antibiotic resistant bacteria in organic poultry meat: is this possible?

J. Matos
FeedInov Colab, Health and Biotechnology, Quinta da Fonte Boa, 2005-048, Portugal; jorge.matos@feedinov.com

Antimicrobial Resistance (AR) is now one of the main concerns for the WHO. The AR phenomenon is aggravated by the abuse of antimicrobial drugs both in human and veterinary health. General public perceive organic products as healthier and safer, however, organic production is not specifically designed to reduce the presence of bacteria showing antimicrobial resistant traits thus also presenting a potential microbiological safety risk for consumers. Precise data is still scarce about the microbiological status in organic meat products, so, the goal of the present study was to investigate the presence of resistance genes in different bacterial species appear in conventional and organic poultry meat. Bacterial isolation was performed on forty-five frozen whole chicken carcasses (30 conventional and 15 organic). Bacterial species identification was performed in all 157 isolates by MALDI-TOF MS followed by biochemical confirmation. Antimicrobial Susceptibility Testing (AST) was carried out according to recommendations of the Clinical and Laboratory Standards Institute. All isolates were tested for 26 antimicrobial drugs. Identification of the antimicrobial resistance encoding genes was obtained by Polymerase Chain Reaction (PCR). From the 15 poultry meat samples, 157 isolates were identified within 9 genus and it was clear that antimicrobial resistance was higher in conventional farming than organic, although 28% of all samples from organic producers showed resistant to at least one antibiotic drug. Furthermore, only this type of production had isolated susceptible to all 26 tested antibiotics. The most common isolated species was *E. coli*, and it was in the organic production meat that the highest number of this species was isolated. Within all the *E. coli* isolates, in conventional production 95% of these isolates showed a multi-drug resistance (MDR) profile while in organic production only 37% of them showed this profile. Organic farming relies on using better biosecurity practices, less medication, and the best animal well-being practices, nevertheless, organic poultry meat is still possible to present resistant bacteria to a multitude of antibiotic drugs, so it is advised to use good care while manipulating and cooking all chicken meat, including organic meat products.

Session 49 Poster 20

Effects of Cu, Zn and Mn hydroxide minerals supplementation in cattle diets: a quantitative review

M. Ibraheem[1], B. Bradford[1], D. Brito De Araujo[2] and K. Griswold[3]
[1]Michigan State University, Department of Animal Science, 1290 Anthony Hall, 48824, East Lasing, Michigan, USA, [2]Nutreco Nederland BV, Stationsstraat 7, 3811 MH Amersfoort, Utrecht, the Netherlands, [3]Micronutrients USA LCC, 1550 Research Way, Indianapolis, IN 46231, USA; davi.araujo@trouwnutrition.com

The objective of this study was to answer the question 'Does the complete substitution of sulfate trace mineral (TM) source by hydroxide sources of Cu, Zn and Mn improve DMI and NDF digestibility in cattle diets'? A quantitative summary was completed independently for each one of the two variables to demonstrate the impact of the complete replacement of the sulphate sources of Cu, Zn and Mn by hydroxide TM. Seven studies providing 11 comparisons were used in this quantitative summary with the objective to determine whether TM source of Cu, Zn and Mn affected NDF and DM digestibility. To be included, studies had to have some measure of digestibility, comparing sulphate and hydroxide trace mineral sources. Factors included to account for variability and possible effects on results included method of digestibility analysis, study design, beef vs dairy cattle use, trace mineral levels and %NDF in diet. For standardization, the 24-hour time point was used for studies where *in situ* methods were employed for digestibility analysis. The quantitative summary was carried out independently for each variable of interest. Only digestibility of DM and NDF were available from all studies included. Overall, NDF digestibility was significantly increased by hydroxide minerals. For this variable, digestibility assessment method was retained as the only significant factor in the final model (P=0.029). Total collection studies showed a mean increase in NDF digestibility of 2.81% (P=0.003), whereas studies using uNDF240 as a marker for total-tract digestibility tended to point to a 1.70% increase in NDF digestibility (P=0.06). Overall, there was no clear evidence of altered DM digestibility (P=0.13). However, model assessment revealed heterogeneity across studies. Among the explanatory factors evaluated, only sector (beef vs dairy) was retained as a significant factor in the final model (P=0.026). These results revealed a significant positive impact of hydroxide minerals in cattle diets.

Session 50 Theatre 1

Tannins and essential oil mixtures reduce rumen methane and ammonia formation *in vitro* (RUSITEC)

G. Foggi[1], S.L. Amelchanka[2], M. Terranova[2], S. Ineichen[3], G. Conte[1], M. Kreuzer[4] and M. Mele[1]
[1]University of Pisa, Dipartimento di Scienze Agrarie, Alimentari e Agro-ambientali, Via Borghetto, 80, 56124 Pisa, Italy, [2]ETH Zurich, AgroVet-Strickhof, Eschikon 27, 8315 Lindau, Switzerland, [3]Bern University of Applied Sciences, School of Agricultural, Forest and Food Sciences HAFL, Länggasse 85, 3052 Zollikofen, Switzerland, [4]ETH Zurich, Institute of Agricultural Sciences, Universitätstrasse 2, 8092 Zurich, Switzerland; giulia.foggi@phd.unipi.it

Supplementation of tannin extracts and essential oil compounds (EOC) have been widely reported as a promising feeding strategy to reduce the environmental impact of ruminants. A previous Hohenheim Gas Test screening demonstrated an enhanced mitigating potential on methane and ammonia formation when these compounds were supplemented as a mixture, in contrast to the result when supplemented alone. From that study, the two most efficient mixtures in terms of mitigation, A and B, were selected and their effects were quantified in the semi-continuous rumen simulation technique (Rusitec) in comparison to a negative control (NC) and a positive control (PC). The NC fermenters were supplied daily with a basal diet common to all treatments. Treatment PC additionally contained the commercial EOC-based additive Agolin Ruminant®. A total of 120 samples were analysed, considering six replicates of each of the four treatments in three independent runs, in which the measurements were taken for 5 consecutive days after 5 days of adaptation. The methane production declined by 12% with A and PC, whereas ammonia formation decreased by 35% with B and PC compared to NC. Moreover, PC was the only treatment in which the dry matter degradability decreased (-9%), probably due to an almost complete elimination of the rumen protozoa (-98%). Compared to NC, the total VFA production significantly decreased with PC and B treatment as acetate, propionate and valerate productions was reduced. Treatments B and PC also resulted in a slightly higher pH (7.1) in comparison to A and NC (7.0). All the parameters studied did not vary over days 5-10 of incubation, as the promising mitigating effects did not decline in this period. It now has to be clarified *in vivo* whether mixture A and B are really efficient against either methane or ammonia formation or both.

Session 50 — Theatre 2

Production performance of growing lambs fed with Backswimmer (*Noctonecta* spp.)

L.E. Robles-Jimenez[1], O.A. Castelan Ortega[1], M.F. Vazquez Carrillo[1], E. Vargas-Bello-Pérez[2] and M. Gonzalez-Ronquillo[1]
[1]Universidad Autónoma del Estado de Mexico, Instituto Literario 100, 50000, Mexico, [2]University of Copenhagen, Grønnegårdsvej 3, 1870, Denmark; mrg@uaemex.mx

One of the main protein sources in intensive ruminant systems is soybean meal, however, it has large environmental impact due to the use of land and water for large-scale cultivation. The use of water insects such as Backswimmer (*Noctonecta* spp.) as a source of protein in ruminant feed could be a feeding alternative. The objective of this study was to evaluate the effect of soybean meal (SBM), fish meal (FM) or Blackswimmer (BS) meal supplementation on production performance of growing lambs. Twenty-four Texel lambs (LW 34.6±3 kg) were randomized distributed (8 lambs / treatment) to one of three isoprotein and isoenergetic diets (15% CP; 11 MJ ME / kg DM). The protein source was the only factor of diet variation. The control diet (SBM) included 150 g of SBM / kg DM, which was replaced by FM or BS meal in the other diets. The study lasted for 28 d, in which 21 d were used for diet adaptation and the last 7 d for sample collection. Live weight (kg), intake (g/d), digestibility (g/kg) and N balance (g/d) were recorded. Data were analysed for a completely randomized design; significant effects were declared at P<0.05. Dry matter (DM) and organic matter (OM) intake (1,387±60 and 1,309±50) were similar among treatments (P>0.05), DM and OM digestibility(760 and 763 g/kg respectively) was higher for SBM than FM and BS diets (731±2 and 739±4 g/kg respectively). N intake was similar (43±4 g/d) among treatments (P>0.05), N retention was higher (P<0.05) for SBM diet (14.05 gN/d) than FM and BS diets (10.64±1.2 gN/d). The inclusion of BS in growing lamb diets could be an alternative feeding strategy without negative effects on animal performance.

Session 50 — Theatre 3

Milk yield, milk composition and N balance from dairy ewes supplemented with backswimmer

L.E. Robles-Jimenez[1], E. Cardoso Gutierrez[1], E. Aranda Aguirre[1], O.A. Castelan Ortega[1], E. Vargas-Bello-Pérez[2] and M. Gonzalez-Ronquillo[1]
[1]Universidad Autónoma del Estado de Mexico, Instituto Literario 100, 50000, Mexico, [2]University of Copenhagen, Grønnegårdsvej 3, 1870, Denmark; mrg@uaemex.mx

Notonectidae is a family of aquatic insects in the order Hemiptera, commonly called backswimmers which could be considered as an alternative protein source for ruminant feeding. Therefore, the objective of this study was to determine milk production and composition as well as nitrogen balance from dairy ewes supplemented with Backswimmer (BS). Fifteen primiparous Texel ewes (BW 86.8±1.3 kg) with 180±5 days in milk were fed three diets, distributed in a 3×3 Latin square design repeated 5 times, with 3 experimental periods of 20 days each. The protein source was the only factor of diet variation. Diets contained corn silage/concentrate (50/50), the control diet (soybean meal -SBM) included 120 g of SBM / kg DM, which was replaced by BS 60 g/kg DM and BS 120 g/kg DM. Milk yield (MY, kg/d) was recorded, and individual milk samples (100 ml) were taken on the last 5 days of each experimental period. Dry matter intake (2,247±25 g/d) and DM digestibility (671 g/kg) was similar among diets (P>0.05). Milk samples were analysed for total solids (TS), lactose, protein, and fat. Fat-corrected milk (FCM 6.5%, kg/d) and fat-protein corrected milk (FPCM 6.5, 5.8%, kg/d) were determined. Nitrogen was determined in urine, faeces, and milk samples. The inclusion of BS60 (0.441 kg/d) or BS120 (0.408 kg/d) diets increased milk production (up to 60%) compared with SBM diet (0.257 kg/d), as well as FCM and FPCM (P<0.05). BS60 or BS120 diets did not affect milk fat concentration (6.80±0.2 g/100 g, P>0.05), but decrease lactose (9%), and protein (8%) (P<0.05) in BS60 and BS120 diets compared with SBM. The supplementation of BS60 decreased N intake (15%, P<0.05) compared. with SBM diet, in addition to presenting a higher N excretion in faeces and milk than SBM and BS120 diet. The inclusion of BS60 could be used in dairy ewe's diets as protein-lipid feed alternatives with positive effects on milk yield compared to SBM.

Session 50 Theatre 4

Assessment of nutritional strategies to mitigate enteric methane emissions and nitrogen excretion

M.A. Eugène[1], M. Benaouda[1], A. Bannink[2] and T.E.A.M. Ceders Project Partners[2]
[1]INRAE, PHASE, Université Clermont, VetAgroSup, UMR1213 Herbivores, 63122 Saint-Genès-Champanelle, France, [2]Wageningen Livestock Research, Wageningen University & Research, Wageningen, the Netherlands; maguy.eugene@inrae.fr

The supplementation of ruminant diets with lipids or starch has been proposed to mitigate enteric methane (CH_4) emissions. However, consequences of these nutritional strategies on N excretions should be evaluated jointly for integral assessment on emissions. In the present work, we assessed the impact of dietary lipid supplementation (LS) or increased dietary starch concentration (IS) on emission or excretion (g/d) and yield (g/kg dry matter intake (DMI)) of CH_4, and faecal N (FN) and urinary N (UN). Data from 133 published *in vivo* studies (346 treatment means) with cattle (dairy and beef) and small ruminants were used. The effect of LS or IS was evaluated with univariate and multivariate models to examine how variation in ether extract (EE) or starch (STA) concentration in the diets influenced CH_4 emissions, and FN and UN excretion, with or without using further explanatory parameters obtained from diet composition and animal characteristics. The univariate models indicated that an increase of 10 g EE/kg DMI reduced CH_4 emission and yield by 13.6 g/d and 0.80 g/kg DMI, respectively, without affecting N excretion. With IS studies, an increase of STA concentrations of 10 g/kg DMI reduced CH_4 emission and yield by 1.10 g/d and 0.22 g/kg DMI, respectively, without changing N excretion. When LS and IS studies were modelled together, CH_4 emissions were best predicted by DMI, neutral detergent fibre and EE concentrations, whereas N excretion was best predicted by DMI and dietary crude protein concentration. Multivariate models indicated that DMI is a key predictor of CH_4 emissions and N excretion, whereas CH_4 and N yield were best predicted by both STA concentration and the percentage of concentrate in the diet. Although a combination of low CH_4 yield- low TN (sum FN and UN) yield was observed, in most situations there was a trade-off between CH_4 emission and TN excretion with LS as well as with IS strategy.

Session 50 Theatre 5

Feeding by-products from the olive industry modifies pig slurry characteristics and ammonia emission

L. Piquer[1], S. Calvet[2], P. García-Rebollar[3], C. Cano[1], D. Belloumi[1] and A. Cerisuelo[1]
[1]Instituto Valenciano de Investigaciones Agrarias, CITA, Pol. la Esperanza 100, 12400 Segorbe, Spain, [2]Universitat Politècnica de València, ICTA, Camí de Vera s/n, 46022 Valencia, Spain, [3]Universidad Politécnica de Madrid, Dep. Producción Agraria, ETSIAAB, Av. Puerta de Hierro, 2-4, 28040 Madrid, Spain; piquer.lai@gmail.com

The use of agro-industrial by-products such as those from the olive oil industry in pig feeding contributes to the circular economy in the livestock sector. Two different by-products, partially defatted (PDOC) and cyclone (COC) olive cake, from a common drying batch were included in fattening pig diets to evaluate the consequences on slurry composition and ammonia (NH_3) emission. Three diets were fed to 24 fattening pigs: a basal (corn, wheat and soybean meal) and two experimental diets in which 200 g/kg of the energetic part of the basal diet was replaced with PDOC or COC, respectively. The total amount of faeces and urine produced was collected per animal for three days and mixed to constitute artificial slurry. Slurry was analysed for dry matter (DM), total ammonia nitrogen (TAN), total Kjeldhal nitrogen (TKN) and pH. Additionally, slurry was subjected to an assay to measure *in vitro* potential NH_3 emission. Results were analysed employing PROC GLM of SAS®. Slurry excretion (kg/d) did not differ among diets, but the slurries of animals fed olive by-product diets showed a higher DM and a lower pH and TAN concentration compared to those fed the basal diet ($P<0.05$). No differences were observed in TKN. Additionally, the pH was significantly lower ($P<0.05$) in the slurry coming from animals fed COC compared with the slurry from animals fed PDOC. Regarding NH_3 emission, the amount of N-NH_3 (g/kg initial TKN) emitted was significantly higher in the slurry of animals fed the basal diet compared to animals fed olive by-products (+39%, $P<0.05$) and significantly higher in the slurry of animals fed PDOC compared to animals fed COC (+24%, $P<0.05$). A similar reduction occurred when the potential NH_3 emission was expressed in mg NH_3/animal and day. The reduction in NH_3 emission can be explained by the differences found in TAN concentration (lower in olive compared to basal slurries) and pH (lower in COC compared to PDOC slurries).

Biorefineries and livestock: an uncommon match made in heaven

M. Taghouti[1,2]
[1]*CITAB – Centre for the Research and Technology of Agro-Environmental and Biological Sciences, University of Trás-os-Montes and Alto Douro, Quinta de Prados, Edificio Reitoria Room D2.30, 5000-801 Vila Real, Portugal,* [2]*FeedInov CoLab, Integrated Production Systems, FeedInov, Estação Zootécnica Nacional, Qta da Fonte Boa, 2005-048, Portugal; myriam.taghouti@feedinov.com*

The International Energy Agency defined biorefining as 'the sustainable synergetic processing of biomass into a spectrum of marketable food and feed ingredients, products (chemicals, materials) and energy (fuels, power, heat)'. Thus, with growing concerns about the environmental impact of food and energy productions, biorefinery processes offer an opportunity to enhance synergy level and implement complementary approaches between these two important sectors. To be used as feed ingredients, nutritional and anti-nutritional compounds from biorefining side-streams should be assessed. Protein content and amino-acids profile are among the most important aspects to consider, especially for monogastric feeding while fibre-rich products are more useful for ruminants. The most studied example of biorefinery by-products in animal feed is the Distiller's Dried Grains with Solubles (DDGS), which are well characterized and are often used in compound feed formulation. Also, yeasts and other microorganisms give interesting solutions for amino-acids supply such as single-cell protein (SCP) and bioactive compounds namely prebiotics and probiotics. With its great capacity to valorise several biorefining coproducts, livestock provides high-quality protein for human consumption and generates wastes that can be further upcycled as feedstock in biorefineries for energy production and bio-based industries. Consequently, biorefining helps redeem animal husbandry as a strong part of the food and energy production chains; and as the crucial link to establish a sustainable circular bio-economy model.

Novel probiotics as oral supplements for sustainable dairy cow performance

A.S. Chaudhry
Newcastle University, Natural and Environmental Sciences, Newcastle University Agriculture Building, NE1 7RU, United Kingdom; abdul.chaudhry@ncl.ac.uk

Various probiotics are endorsed to modify rumen function to optimise nutrient use efficiency and improve dairy cow performance. Forty lactating cows were divided into 2 equal groups of 20 cows each for cubicle housing with self-locking yokes. The cows had *ad libitum* access to a total mixed ration (170 g CP & 13 MJ ME /kg DM) of grass silage (55%) and concentrate (45%). Each cow also received 1.25 kg of a pelleted concentrate (160 g CP & 12.5 MJ ME /kg DM) at milking. Also, each Treatment cow was drenched with 10 ml probiotic whereas each Control cow received 10 ml water once daily over 42 days. The cows were weighed (LW), condition scored (BC), and monitored for weekly milk yield (L /day) and composition (fat, MF; protein, MP and somatic cell count, SCC). The data were statistically analysed and means declared significant if $P<0.05$. Treatment cows gained on average 5 kg live-weight but the Control cows lost 9 kg during this study ($P>0.05$). Although milk yield tended to reduce less in the Treatment than the Control cows over 6 weeks, the week effect and the treatment × week interaction were not significant ($P>0.05$). While the Treatment cows had 2.1 litre less milk daily than the Control cows before the trial, they produced 0.2 litres more milk than the Control cows in the last week of this study. The MF, MP and SCC contents did not differ between the cow groups during trial weeks ($P>0.05$). However, MP in week 2 was greater for the Treatment than the Control cows ($P<0.05$). While MF tended to vary with the cow group and the week, the mean MP was always greater whereas the SSC were always lower in the Treatment than the Control group. Treatment cows tended to maintain their live-weight, condition and milk yield when compared to the Control cows. While mean MF was variable, the MP content tended to increase and the SCC tended to decrease in the Treatment cows than the Control cows during 6 weeks of this study. Further studies should examine the safety and long-term effect of probiotics on rumen function, health and efficiency of cows in different situations. It is vital to explore efficient ways to deliver most suitable liquid probiotics for sustainable dairy and other ruminant animal production systems.

Effects of guanidinoacetic acid on rumen microbial fermentation in continuous culture system

R. Temmar[1], M.E. Rodríguez-Prado[1], A. Kihal[1], V.K. Inhuber[2] and S. Calsamiglia[1]
[1]Universidad Autónoma de Barcelona (UAB), Animal Nutrition and Welfare Service (SNiBA), Edifici V, Travessera dels Turons, Cerdanyola del Vallès, 08193 Barcelona, Spain, [2]Alzchem Trostberg GmbH, Dr.-Albert-Frank-Straße 32, 83308 Trostberg, Germany; temmarrokia@gmail.com

The objective of the study was to evaluate the effect of guanidinoacetic acid (GAA) on rumen microbial fermentation and nutrient digestion. The study was a randomized block design with 8 dual flow continuous culture fermenters and 2 periods. Treatments (n=4) arranged in a 2×2 factorial, with factors being the type of fermentation environment: dairy (pH between 5.8 and 6.8; diet 50:50 forage:concentrate, 17.1% CP and 30.0% NDF) or beef (pH between 5.5 and 6.5; diet 10:90 forage:concentrate, 16.3% CP and 17.6% NDF); and GAA: 0 vs 2 g/d. Temperature (38.5 °C), liquid (0.10/h) and solid (0.05/h) dilution rates were kept constant. Diets (90 g/d DM) were fed 3 portions/d. Effluent samples were collected from a composite of the 3 sampling days, and bacteria were isolated on the last day of each period from fermenters. Fermenter samples were taken 3 h after the morning feeding for microbiome analysis. Fermentation data were analysed with the PROC MIXED of SAS and the microbiome diversity and composition with R-Studio. Significance was set at $P<0.05$. No differences were observed on true OM (59.4±7.94%) degradability. Degradability of NDF (49.9 vs 31.2±3.79%), the proportions (mol/100 mol) of acetate (51 vs 44±1.7) and butyrate (19 vs 11±1.5), the acetate to propionate ratio (2.02 vs 1.12±0.091), NH_3-N concentration (8.18 vs 2.92±0.662 mg/100 ml), the flow (g/d) of total (2.83 vs 2.67±0.023) and ammonia (0.27 vs 0.09±0.021) N, the efficiency of microbial protein synthesis (32 vs 22±2.3 g N/kg OM truly digested) and the relative abundance and diversity of microbial population were higher in dairy than in beef. Total VFA (113 vs 100±1.9 mM) and the propionate proportion (41 vs 26±1.8 mol/100 mol) were higher in beef than in dairy. The GAA increased NH_3-N concentration (7.31 vs 3.78±0.662 mg/100 ml) and the flow (g/d) of total (2.97 vs 2.71±0.023) and ammonia (0.24 vs 0.12±0.021) N. The degradation of GAA was higher in dairy (69.8%) than in beef (6.3%). The use of GAA in dairy cow, but not in beef, may require protection.

Silages of agro-industrial by-products in lamb diets – effect on carcass and meat quality

K. Paulos[1], C. Costa[1], J. Costa[1], L. Cachucho[2,3], P.V. Portugal[1], J. Santos-Silva[1,3], J.M. Almeida[1,3], E. Jerónimo[2,4] and M.T.P. Dentinho[1,3]
[1]INIAV, Fonte Boa, Santarém, Portugal, [2]CEBAL, Centro de Biotecnologia Agrícola e Agro-Alimentar do Alentejo, Beja, Portugal, [3]CIISA, Avenida Universidade Técnica, Lisboa, Portugal, [4]MED, Mediterranean Institute for Agriculture, Environment and Development, Évora, Portugal; katia.paulos@iniav.pt

In Portugal large amounts of agro-industrial by-products are generated which can ensure low-cost feedstuffs. We aimed to evaluate the effect of silages produced with by-products of potato, sweet potato, tomato pomace and carrot on lamb growth performance, carcass composition and meat quality. Three silages were performed using 35% tomato pomace + 20% wheat bran + 15% of grass hay + 30% potato (Psil) or 30% sweet potato (SPsil) or 30% carrot (Csil). Good quality silages were obtained, with pH 4.16 (P), 3.97 (SP), and 4.10 (C). Thirty-two lambs were housed individually in parks and divided into four groups (8 animals/group) each group with the following diet: Control – 85% commercial concentrate and 15% oat/vetch hay; P- 50% concentrate and 50% Psil in DM; SP – 50% concentrate and 50% SPsil in DM; C – 50% concentrate and 50% Csil in DM. After 6 weeks of trial, the animals were slaughtered and the carcass and the meat were analysed for chemical, physical and sensorial characteristics. Colour and lipid stability over 7 days of storage at 4 °C was also evaluated. Carcass traits and shoulder tissues composition were similar among diets. With silage diets the L* of dorsal subcutaneous fat was higher than control, averaging 78.2 and 75.3 respectively. The Longissimus lumborum colour was affected by diets with a* and C* higher with C diet. No differences among diets were found on shear force, cooking losses, oxidative stability after 7 days of storage and sensorial characteristics. We can conclude that the partial replacement of concentrate-feed by by-product silages in lamb diets is a good option not compromising the quality of the lambs' carcasses and meat. This work is funded by project SubProMais (PDR2020-101-030988; PDR2020-101-030993) and by National Funds through Foundation for Science and Technology projects UIDB/05183/2020 (MED) and UIDP/CVT/00276/2020 (CIISA), and the PhD grant awarded to LC (2020.05712.BD).

Estimation of carbon footprint and sources of emissions of an extensive alpaca production system

G. Gómez Oquendo[1,2], K. Salazar-Cubillas[3] and C.A. Gomez[4]

[1]Universidad Nacional Mayor de San Marcos, Department of Animal Production, Av Circunvalación 28, San Borja, 15021, Peru, [2]Universidad Científica del Sur, Faculty of Veterinary Medicine and Zootechnics, Panamericana Sur Km 19, Villa – Lima, 15067, Peru, [3]Christian-Albrechts-Universität zu Kiel, Institute of Animal Nutrition and Physiology, Kiel, 24118, Germany, [4]Universidad Nacional Agraria la Molina, Department of Animal Nutrition, Av. La Universidad s/n, 15024, Peru; gjanetgomez@gmail.com

Following international standards procedures, a cradle-to-gate life cycle assessment of an extensive alpaca production system was conducted to determine its carbon footprint (CF)[1]. The study was conducted during the dry season in a typical Peruvian alpaca production system for meat and fibre production comprising 1,492 alpacas (young, tuis, and adults) with an average fertility rate of 80%, a birth rate of 54%, and an offtake rate of 20%. Alpacas grazed native grassland for 8-10 h daily with supplementation of oat hay. Emissions from outside the system (fuel, electricity, and fertilizers) and within the system (methane (CH_4) from enteric fermentation, and nitrogen dioxide (NO_2) and CH_4 from manure management) were quantified. Following that, CF was calculated based on mass, economics, and biophysical allocations and expressed in kg of carbon dioxide equivalents (CO_2-e). The functional unit for economic and mass allocations was 1 kg of live weight (LW) while for the biophysical allocation was 1 kg of LW and 1 kg of fibre. The CF uncertainty was calculated using a Monte Carlo simulation with a prediction interval of 2.5 to 97.5% of the uncertainty distribution. The largest source of greenhouse gas emissions was CH_4 emissions from enteric fermentation (67%), followed by direct and indirect NO_2 (29%), emissions outside the system (3%), and CH_4 from manure management (1%). For economic and mass allocations, CF was estimated at 24.0 and 29.5 kg of CO_2-e of LW, respectively, while for the biophysical allocation was 22.6 and 53 kg CO_2-e per kg LW and fibre, respectively. These results provide information that can be used to develop strategies for reducing greenhouse gas emissions from alpaca production systems. [1]ISO 14040: 2006. Environmental management – life cycle assessment – principles and framework.

Influence of feeding technique of silage on behaviour in growing pigs

J. Friman[1], E. Verbeek[2] and M. Åkerfeldt[1]

[1]Swedish University of Agricultural Sciences, Dept. of Animal Nutrition and Management, Box 7024, 75007, Sweden, [2]Swedish University of Agricultural Sciences, Dept, of Animal Environment and Health, Box 7068, 75007, Sweden; johanna.friman@slu.se

Long straw silage is provided as enrichment to organically raised pigs, but seldom included as an ingredient in their feed rations. Grass and legumes show favourable protein and amino acid composition and are of interest as sustainable locally produced feed ingredients. Research have emphasized that feeding silage as a pellet or fresh with a short straw length in total mixed rations (TMR) increase feed intake and nutrient utilization compared to long straw silage. However, feeding pelleted or short chopped silage might not fulfil the pigs need for foraging and exploration. The aim of this study was to evaluate how pre-treatment (chopped or dried) and feeding technique (pelleted or TMR) of silage affected pig behaviour. In total 126 growing pigs were randomly divided into three dietary treatments, either a commercial control diet without silage inclusion (Pellet-C) or a cereal-based pellet containing dried and milled silage (Pellet-S) or a TMR containing fresh short chopped silage, mixed with commercial pelleted feed (TMR-Ch). It was hypothesized that feeding the TMR diet would increase activity levels and feed-related behaviours. Pigs were observed on three occasions at two-week intervals with both instantaneous and continuous sampling. Preliminary results show that diet affected activity level and feed related behaviours ($P<0.05$). Pigs fed the TMR-Ch diet spent significantly more time eating and rooting (8.3% and 6.8%, respectively) compared to pigs fed the Pellet-C diet (5.6% and 3.8%, respectively)($P<0.001$). Activity level were higher for both pigs fed TMR-Ch diet (19.8%) and the Pellet-S diet (14.4%) compared to the Pellet-C diet (13.8%) ($P<0.001$). These preliminary results indicate that short chopped silage fed as TMR significantly increased foraging and exploratory behaviours. Pigs that received pelleted silage were generally more active compared to the control diet, suggesting that pelleted silage may have a small benefit over a standard diet. Further analysis within the study will evaluate the effect of feeding silage on the gut microbiota composition and its influence on pig's behaviour.

Bovine total tract apparent digestibility of protein fractions and nutrients as affected by diet

M. Simoni[1], T. Danese[1], G. Esposito[1], I. Vakondios[2], L. Karatosidi[2], A. Plomaritou[2], M. De Marchi[3] and F. Righi[1]
[1]Università di Parma, Dipartimento di Scienze Medico-Veterinarie, Via del taglio 10, 43126 Parma, Italy, [2]University of Thessaly, Department of Animal Science, Campus Gaiopolis, 41222 Larissa, Greece, [3]Università di Padova, Animal Science, Department of Agronomy, Food, Natural resources, Animals and Environment, Viale dell'Università 16, 35020 Padova, Italy; tommaso.danese@unipr.it

Forage source and characteristics deeply affect diets formulation and consequently nutrients digestibility. We studied the effects of two different forages on protein fractions and nutrients total-tract apparent digestibility (ttaD) of dairy cows' diets. Eleven herds (60-200 lactating cows) located in Northern Italy and fed total mixed ration (TMR) were involved in the trial. Five were fed hay-based diets (HB; 40.34±4.94% aNDFom, 14.69±1.54% CP, 19.25±5.20% starch –DM basis) whereas six were fed silage-based diets (SB; 35.11±5.24% aNDFom, 15.37±2.24% CP, 25.79±4.93% starch –DM basis). Diet and faeces were sampled 8 times per farm during a whole year. Representative samples of TMRs and faeces were collected and chemically analysed to determine the following parameters: DM, CP, aNDFom, ADFom, lignin, cellulose, hemicellulose, N-NDF, N-ADF, ash and starch. The undigested NDF (uNDF) was determined by 240 h *in vitro* fermentation, and used as marker to estimate the ttaD of the above-mentioned nutrients. Data on dietary and faecal composition, protein fractions and nutrients ttaD were analysed using a generalized linear mixed model with the diet as fixed factor, and month of sampling and farm as random variables. Diet typologies were similar only for CP, N-ADF and uNDFom. Fecal composition differed among diet tipologies exception made for N-NDF, N-ADF, ash, starch and uNDFom. The SB diets showed higher ttaD of DM (64.2 vs 60.2%, P=0.004) and of CP (61.5 vs 57.7%, P=0.004) but lower ttaD of CP-NDF (34.8 vs 45.5%, P≤0.001) compared to HB diet. In conclusion, being generally more digestible in terms of DM and CP, SB diets seems to be more suitable to support high milk yielding cows. This project has received funding from the European Union's Horizon 2020 research and innovation programme, under the grant agreement No 777974.

Could *Arthrospira platensis* supplementation promote health status of post-weaning calves?

A. Marzano[1], R. Cresci[1], R. Abis[2], C. Ledda[3] and A.S. Atzori[1]
[1]University of Sassari, Sassari, Italy, Dipartimento di Agraria, viale Italia, 39, 07100 Sassari, Italy, [2]Ruminant nutritionist, Arborea (OR), 09092, Italy, [3]Livegreen S.r.L., Arborea (OR), 09092, Italy; amarzano@uniss.it

Microalgae supplementation is cultivated as potential source of bioactive compounds able to promote health, immunity, and performances of livestock. *Arthrospira platensis* (commonly called Spirulina) belongs to the oxygenic photosynthetic bacteria as microscopic, filamentous cyanobacteria, rich in phycocyanin and phenolic compounds, amino acids and PUFA, with potential uses as a natural antioxidant and an immunostimulant with less side effects than synthetic antioxidants. This work aimed to evaluate effects of the Spirulina supplementation on performances and metabolic profile in post-weaning calves. Eighteen calves at age of 76.9+4.2 days and weight of 97.80±8.6 were split in two groups called Control (C; n=9) and Spirulina (SP; n=9). Each group was housed in 3 pens (n=3) and fed a dry total mixed ration, whereas calf SP and C groups were individually supplemented with 10 g Soybean meal or 10 g/d of dried Spirulina for 40 days, respectively. Intake was measured weekly and blood samples were collected every two weeks. Spirulina did not significantly affect weight gain (1.22 kg/d per head) or dry matter intake vs control diet. Otherwise, selection of physically effective fibre was significantly higher in SP vs C. (P<0.05). Spirulina improved erythrocytes turnover through the increase of haemoglobin: calves fed SP showed higher mean cell haemoglobin concentration over time (P<0.01, average values 34.43 vs 33.94 g/dl±0.10), higher MCV (Mean Cell Volume) and mean cell haemoglobin (P<0.001). Spirulina reduced Bilirubin (0.07 vs 0.17 mg/dl±0.01; P<0.01) and affected lipid metabolism by reducing Cholesterol (61.85 vs 69.71 mg/dl±2.14; P<0.01), decreased creatine phosphokinase (P<0.05), β-Globulin (P<0.05) and increased Albumin (P<0.01). In summary, Spirulina positively affected blood parameters of calves suggesting that could be involved in modulating metabolic status facing pathological conditions. Further investigations need to study dose effects, supplementation conditions and possible interactions with metabolism and immune function.

Porcine intestinal mucosa hydrolysate Palbio 50RD® improves weaner pig production and profitability

A. Middelkoop[1], S. Segarra[2] and F. Molist[1]
[1]Schothorst Feed Research, R&D, Meerkoetenweg 26, 8218 NA Lelystad, the Netherlands, [2]Bioiberica S.A.U., R&D, Av. Països Catalans 34, planta 2a, 08950 Esplugues de Llobregat, Spain; amiddelkoop@schothorst.nl

Porcine intestinal mucosa hydrolysate (PIMH) products are high-quality, sustainable, and inexpensive protein sources obtained from porcine intestinal mucosa as by-product of the heparin manufacturing process. PIMH can be used as alternative to other protein sources, such as soybean ingredients, spray-dried plasma and fish meal. The present study aimed to determine the effect of incorporating PIMH as replacement for other protein sources on weaner pig performance and profitability. The trial consisted of two treatment groups, with n=16 pens/treatment with 6 piglets/pen. Pigs in the negative control (NC) group were fed a weaner I diet (crude protein of 19.4%) containing 3.5% skimmed milk powder and a weaner II diet (crude protein of 18.4%) containing 2.5% soy protein concentrate. In the PIMH group, these protein sources were replaced by 5% PIMH (Palbio 50RD, Bioiberica S.A.U.) in weaner I and 2.5% PIMH in weaner II, resulting in iso-energetic and iso-protein pelleted diets, with similar lactose content. Pig performance was analysed between day 0-13 and day 13-34 post-weaning and for the total 5-week weaner period using a general linear model. Income over feed costs (IOFC) was calculated as revenues minus feed costs. Piglets weighed 9.13±0.063 kg at weaning, when piglets were 31.61±0.132 days old. Inclusion of PIMH stimulated average daily feed intake between day 0-13 post-weaning by 32 g/piglet compared to the NC group (349 vs 317 g/piglet, P=0.004). In addition, piglets fed PIMH had an improved feed conversion ratio between day 13-34 post-weaning versus piglets in the NC group (-3%, P=0.002), with no effect on body weight and average daily gain in the first 5 weeks post-weaning. Together, this allowed a €0.49 per pig advantage in IOFC at day 34 post-weaning when PIMH was used as protein source substitute. In conclusion, replacing commonly used protein sources by PIMH increases weaner pig production and profitability. Palbio 50RD is therefore suggested as high-quality protein source to be used in piglets post-weaning, especially as replacement for other, more expensive alternatives, such as skimmed milk powder and soy protein concentrate.

***P. undulatum*, *H. gardnerianum* and *C. japonica* essential oils effects on the reduction of ruminal biogas**

H.P.B. Nunes and A.E.S. Borba
University of the Azores, Pico da urze, 41, Rua Capitão João d'Ávila – São Pedro, Rua Capitão João d'Ávila – São Pedro, 9700, Portugal; alfredo.es.borba@uac.pt

During digestion, ruminants inevitably produce greenhouse biogases such as methane and carbon dioxide. Currently, the application of essential oils (EO) in cattle feed fits into the strategies used to make ruminal fermentation more efficient and consequently mitigate the production of biogases, without resorting to the use of synthetic compounds. The aim of this study was to determine the effect of different EO concentrations of *Pittosporum undulatum* (PU), *Hedychium gardnerianum* (HG) and *Cryptomeria japonica* (CJ) on *in vitro* biogas production and *in vitro* gas production kinetics of pasture. EO of three plants were used for additives: PU, HG and CJ. Four different volumes of EO were added: 15 µl (T1), 30 µl (T2), 60 µl (T3) and 120 µl (T4) for each 0.2 g of dry matter of pasture substrate for *in vitro* biogas production. Each treatment was performed in triplicate. The biogas produced by treatments incubation was recorded at 4, 6, 8, 12, 16, 24, 36, 48, 72 and 96h after incubation. The kinetic interpretation was performed by the logistic model p = a + b(1-exp-ct) proposed by Ørskov and McDonald (1979). The results showed that there was a decrease in the biogas production curve during the 96h of incubation in all treatments applied when compared to the control treatment. The EO of PU and HG, in treatments T3 and T4, were the ones that showed a bigger reduction in the production of biogas, this decrease was significant (P<0.05) in the T4 treatment, regardless of the EO used. Regarding the gas production kinetics, it was observed that the T1 treatment allowed an increase in the gas production potential (a+b) when compared to the control. A significant reduction (P<0.05) in the gas production potential was verified when 120 µl of PU EO was used. No changes in pH were recorded during the study. Considering the conditions under which the study was carried out, the inclusion of PU and HG EO in ruminant feed on pastures have potential to mitigate biogas emissions. Moreover, their careful combination may be a useful tool to effectively manipulate rumen fermentation.

Nutritional valorisation of *Musa* spp. with urea treatment as a fibre source for ruminants

S.M.P. Teixeira, C.F.M. Vouzela, J.S. Madruga and A.E.S. Borba
University of the Azores, FCAA, Institute of Agricultural and Environmental Research and Technology, INV²MAC (MAC2/4.⁶a/229), Rua Capitão João D'Ávila, 9700-042, Portugal; alfredo.es.borba@uac.pt

The ability of ruminants to efficiently use fibre makes it possible to explore agricultural by-products and wastes from tropical crops, like banana (*Musa* spp.). In this work, we propose to increase *Musa* spp. nutritional value, through a treatment with urea, with the goal of making it a sustainable and environmentally friendly animal feed alternative. *Musa* spp. was collected, divided by leaves and pseudo-stems, and dried at 65 °C in an oven with controlled air circulation. Then they were dried and sprinkled with a 5 DM% urea solution and placed in a leak-proof container for 4 weeks. The chemical composition, *in vitro* digestibility, and *in vitro* gas production were determined in leaves and pseudo-stems. The obtained results indicate that the urea treatment leads to a significant increase (P<0.05) of DM in pseudo-stems (from 6.38 to 9.42 DM%) and protein content also increased, in leaves and pseudo-stems (from 14.17 to 20.4 DM% and 7.52 to 11.5 DM% respectively) and produce a non-significant decrease (P>0.05) in NDF in leaves (from 77.9 to 67.51 DM%). Regarding ADL, ADF, Ash and DM digestibility, the values increased non-significantly (P>0.05) in leaves and pseudo-stems. Despite not showing many significant differences, the study shows that this by-product, when treated with urea 5 DM%, can be used for animal feed, since the NDF values decrease with the treatment in leaves and the digestibility values increase, being the pseudo-stem with the highest % of digestibility. Cumulative gas production was recorded at 4, 8, 12, 24, 48, 72, and 96 h of incubation. The results showed that the gas production in leaves was lower (8.53 ml and 9.11 ml/200 mg DM) when compared to pseudo-stem (26.52 ml and 32.73 ml/200 mg DM), in control and with urea 5% respectively, being the results higher with urea, which goes in accordance with the digestibility results. However, it is necessary to carry out tests to improve its nutritional value, with another concentrations of urea or applying treatments with NaOH.

Session 50 Poster 17

Meta-analysis of the relationship between dietary condensed tannin and methane emission by cattle

A.S. Berca[1,2], R.A. Reis[1] and L.O. Tedeschi[2]
[1]Sao Paulo State University, Animal Science, Jaboticabal, 14884-900, Brazil, [2]Texas A&M University, Animal Science, College Station, TX 77840, USA; andressa.berca@tamu.edu

The use of condensed tannins (CT) has shown benefits in ruminant nutrition, especially regarding a better dietary protein utilization and a decrease in methane (CH_4) emission by manipulating ruminal fermentation. A meta-analysis was conducted to evaluate the effects of diet CT inclusion and other chemical components on CH_4 emissions by cattle. A database was developed from 40 published studies that measured cattle CH_4 emission using *in vitro* (iVt) or *in vivo* (iVo) methods, and reported the CT inclusion level and the diet characteristics, including crude protein (CP), organic matter (OM), and neutral detergent fibre (NDF). Meta-analysis was conducted by multiple linear regression, using the nlme package of R 4.1.1, in which studies were assumed random variables, affecting the intercept and slopes. The CH_4 emission, expressed in l/kg of dry matter intake, decreased with increasing levels of dietary CT when iVt and iVo were analysed together (CH_4=26.89–2.55×CT; P<0.001; AIC=455) or for iVo alone (CH_4=29.55–3.65×CT; P<0.001; AIC=328), but not for iVt method (P=0.1324). When analysing all chemical components of the diet and for iVt and iVo together, there were effects of NDF (P=0.0104) and CT (P=0.0043) concentrations (CH_4=40.55–4.10×CT+0.38×CP+0.29×NDF–0.33×OM; AIC=332). For iVo only, there was an interaction between CT and CP (CH_4=35.64–6.87×CT–0.49×CP+0.33×CT×CP; AIC=334). For iVt only, there was an isolated effect of CT (P=0.303) and CP (P=0.0061), but no interaction (P=0.2691): CH_4=61.16 – 0.81×CT – 3.25×CP; AIC=111). The decrease in CH_4 promoted by CT is indirectly attributable to a reduction in ruminal fibre fermentation, which decreases H2 and acetate formation, and is directly related to the inhibition of methanogens growth. Additionally, CT can bind to proteins, forming insoluble complexes in the rumen, which reduces the ruminal protein degradability and the microbial growth, then decreasing the CH_4 production. Across many experiments, increasing the level of dietary CT leads to a decrease in ruminal CH_4 emissions, measured iVt and iVo. In addition to CT, CP had an impact on iVt studies but not on iVo studies.

Session 50 Poster 18

Effects of *Thymus capitatus* essential oil and its compounds on *in vitro* ruminal fermentation

M.J. Ranilla[1,2], S. López[1,2], I. Mateos[1,2], C. Saro[1,2], A. Martín[1], F.J. Giráldez[1] and S. Andrés[1]
[1]Instituto de Ganadería de Montaña (CSIC-Universidad de León), Finca Marzanas, s/n, 24346 Grulleros, Spain, [2]Universidad de León, Departamento de Producción Animal, Campus Vegazana, s/n, 24071 León, Spain; mjrang@unileon.es

Over the last few decades, efforts have been made in the research of alternative options to the prophylactic use of antibiotics in animal farming. The MILKQUA project is focused on exploring the use of essential oils as antibiotics replacement in dairy cows. The effects of *Thymus capitatus* essential oil and its main compounds on the *in vitro* ruminal fermentation of two different diets (high forage (F) and high concentrate (C)) were investigated using batch cultures of mixed ruminal microorganisms. Six different treatments were tested: no additive (control), natural (NEO) and synthetic (SEO) essential oil of *T. capitatus* at 75 mg carvacrol/l, and three bioactive compounds added to the vials according to the proportions present in NEO at such rate (carvacrol 70.62%; p-cymene 7.06%; γ-terpinene 7.58%). After 24 h of incubation, the main fermentation parameters were determined. When C diet was incubated with the additives, only subtle effects on fermentation parameters were detected, but NEO seemed to slightly inhibit ruminal fermentation, as total VFA production was lower ($P<0.05$) compared to the control. SEO and carvacrol also decreased ($P<0.05$) total VFA production with F diet, NEO and carvacrol decreased ($P<0.05$) propionate molar proportion, and the three of them increased ($P<0.05$) Ac:Pr ratio, suggesting that carvacrol was the main responsible for the effects on VFA production and proportions. Methane production was unaffected ($P>0.05$) by the additives. The higher Ac:Pr ratios observed when using NEO, SEO and carvacrol with F diet, probably related to a higher fibre degradation, seem to corroborate the lack of potential of this EO to reduce methane production with the diets and doses assayed. According to the results obtained it seems that, under the conditions of the study, NEO offers no nutritionally or environmentally benefits when tested *in vitro* as an additive.

Session 50 Poster 19

Impacts evaluation of dietary crude protein reduction for finishing pigs using life cycle assessment

P.C. Oliveira[1], A.N.T.R. Monteiro[2], L.A.C. Esteves[1], J.S. Martins[1], A.C. Figueiredo[1], S.M. Einsfeld[1], F.A. Cancian[1] and P.C. Pozza[1]
[1]State University of Maringá, Animal Science, Av Colombo, 5790, 87020900, Brazil, [2]Animine, Sillingy, 74960, France; pcpozza@uem.br

The objective of this study was to evaluate the impacts of reducing crude protein (CP) and supplementing industrial amino acids in finishing pig diets using the life cycle assessment (LCA). Two experiments were carried out. Experiment I (Metabolism assay): Twenty crossbred barrows averaging 79.68 kg of initial weight were distributed in a randomized blocks design, consisting of four treatments (15.1%, 13.8; 12.5 and 11.2% CP), five replicates and one animal per experimental unit. Pigs were allotted in metabolic cages for urine and faeces collection. Nitrogen (N) and phosphorus (P) were analysed in diets, faeces and urine to calculate the phosphorus and nitrogen balance. Experiment II (Growth performance): Fourty barrows averaging 70 kg of initial weight were distributed in a randomized blocks design, consisting of the same aforementioned treatments, 10 replicates and one animal per experimental unit. The animals were weighed at the beginning and end of the experiment, as well as the diets provided to the animals, to determine the daily weight gain, daily feed intake and feed conversion. The LCA calculations were evaluated according to the data obtained in experiments I and II and the environmental profile of each system was constructed. The categories of global warming potential (GWP), acidification potential (PA), eutrophication potential (EP), terrestrial ecotoxicity (TE), cumulative energy demand (CED) and land occupation (LO) were determined. The dietary CP reduction did not affect ($P>0.05$) the performance but provided a decreasing in N intake ($P=0.00003$), N in faeces ($P=0.0382$), N in urine ($P=0.0025$) and N excreted ($P=0.0025$). A reduction in P intake ($P=0.0001$) and P in faeces ($P=0.0022$) were also observed. Increases in CED ($P=0.00002$) and TE ($P=0.0414$) were observed as CP decreased in diets, but a reduction in LO ($P=0.0292$) was also observed. In conclusion, the CP reduction associated with industrial amino acid supplementation in finishing pig diets (70-100 kg) reduced N and P intake, N excretion and also increased CED, TE and reduced LO, without affecting growth performance.

Session 50 Poster 20

Are all the rumen ciliate protozoa equally related with the enteric methane emissions of ruminants?

A.E. Francisco[1,2], J. Santos-Silva[1,2], A.V. Portugal[2], K. Paulos[2], M.T. Dentinho[1,2] and R.J.B. Bessa[1]
[1]CIISA, FMV-ULisboa, Avenida da Universidade Técnica, 1300-477 Lisboa, Portugal, [2]INIAV, Pólo de Inovação da Fonte Boa, Quinta da Fonte Boa, Vale de Santarém, 2005-048 Santarém, Portugal; alexandra.francisco@iniav.pt

Due to their high fermentative activity, ciliate protozoa (PTZ) are the main rumen producers of hydrogen, that is then used by the methanogenic archaea to produce methane (CH_4). The CH_4 has an impact as greenhouse gas 25 times higher than CO_2, and as ruminant digestive emissions contribute relevantly to the total anthropogenic CH_4 emissions, strategies for their mitigation are required. The elimination of PTZ from rumen has been proposed to reduce the CH_4 emissions from ruminants. However, there is still a strong lack of knowledge on the links between PTZ taxa and rumen methanogenesis. This study evaluated the relationships between the PTZ genera present in the rumen fluid of 35 young crossbred beef bulls, used in two experiments and fed 6 different diets and the correspondent quantity of gas and CH_4 produced. Rumen fluid was collected at slaughter and PTZ enumerated and identified at genus level by optical microscopy. The total gas and CH_4 production were evaluated *in vitro* in glass bottles with gas detectors (Ankom system), incubating for 48 h the individual rumen fluid samples with the diet provided to the donor during the feeding experiment. PTZ data were subjected to a regression analysis, removing the variation due to the diet and considering the experiment as a random effect. Total abundance of PTZ (log10cells/ml rumen fluid) was not related with total gas or CH_4. However, total gas production increased linearly with *Entodinium* (P=0.017) and decreased with *Epidinium* (P<0.001) counts. The CH_4 production decreased linearly with *Epidinium* counts (P<0.001) and the CH_4 proportion (% of total gas) decreased linearly with *Isotricha* counts (P<0.001). Ruminal total gas and CH_4 production seems to depend on the PTZ genera present in the rumen fluid, being *Epidinium* and *Isotricha* related with lower CH_4 emissions. Work funded by the PDR2020 program through FEADER (LegForBov project- PDR2020-101-031179) and by the Fundação para a Ciência e a Tecnologia (FCT) (PtzR´Methane project – EXPL/CAL-ZOO/0144/2021).

Session 50 Poster 21

Zinc-methionine alters the oxidative and inflammatory status of dairy cows under heat stress

M. Danesh Mesgaran[1], H. Kargar[1], R. Janssen[2], S. Danesh Mesgaran[2] and A. Ghesmati[1]
[1] Ferdowsi University of Mashhad, Department of Animal Science, Azadi Square, Mashhad, Iran, [2]Kaesler Nutrition GmbH, Zeppelinerstraße 3, 27472 Cuxhaven, Germany; sadjad.danesh.mesgaran@kaesler.de

Modern dairy cows undergo thermal stress starting at an average temperature-humidity index (THI) of 68, whereas at THI>72 markedly hinders animal's productivity. Heat stress exacerbates oxidative stress and reactive oxygen species production in dairy cows. Alteration of oxidative metabolites in cows would influence the cytokine production during heat stress. Present work aimed to observe impact of a commercially available rumen-protected Zinc-Methionine complex (RPZM; Loprotin, Kaesler Nutrition GmbH, Cuxhaven, Germany) supplementation in early high producing Holstein cows during environmental heat stress. Sixty-two multiparous lactating Holstein cows [balanced by days in milk (mean ± SD)=28±7 d; lactation number = 2.9±0.6] were randomly assigned to one of two dietary treatments [total mixed ration with RPZM (LP group) or without the RPZM inclusion (CON group)]. RPZM was included as 0.131% diet DM for a total period of 6 weeks. Blood sampling was conducted bi-weekly via puncture of the coccygeal vessels and obtained serum was subjected to analyses for non-esterified fatty acids (NEFA), calcium, albumin, haptoglobin, total antioxidant status (TAS), malondialdehyde (MDA) and interleukin-1 beta (IL1-B). Data was statistically analysed using the Proc Mixed procedure of SAS for a completely randomized design with repeated measures. Circulating NEFA concentrations were not significantly different between the experimental groups (P>0.05). Blood calcium concentration in the LP group was clearly higher in comparison with the control (P<0.01). RPZM supplemented cows tended to have higher albumin concentration than control (P=0.08). Blood serum haptoglobin was evidently lower in cows in the LP group (P=0.001). Dairy cattle in the LP group had significant (P=0.017) higher blood TAS along with lower MDA concentration (P=0.08). Circulating IL1-B concentration in animals supplemented with RPZM was markedly lower in comparison with the CON (P=0.001). Current work underlines improved anti-oxidative capacity and lower systematic inflammation in dairy cattle fed the RPZM under environmental heat stress.

Session 50 Poster 22

Evaluation of the presence of mycotoxins in corn silage from dairy cow farms

J.L. Cerqueira[1,2], C.P.C. Nogueira[2], A. Gomes[3] and J.P. Araújo[2,4]
[1]Centro de Ciência Animal e Veterinária (CECAV), Quinta de Prados, Apartado 1013, 5001-801 Vila Real, Portugal, [2]Escola Superior Agrária do Instituto Politécnico de Viana do Castelo, Agronomic and Veterinary Science, Rua D. Mendo Afonso, 147, Refóios do Lima, 4990-706 Ponte de Lima, Portugal, [3]Cooperativa Agrícola de Vila do Conde (CAVC), Rua da Lapa, no. 293, 4480-757 Vila do Conde, Portugal, [4]Centro de Investigação de Montanha (CIMO), Rua D. Mendo Afonso, 147, Refóios do Lima, 4990-706 Ponte de Lima, Portugal; cerqueira@esa.ipvc.pt

Corn silage, one of the most valuable forages for dairy cows worldwide, requires good crop and harvest management as well as careful ensiling practices to maintain nutritional value as much as possible. Forages can be contaminated with several mycotoxins in the field pre-harvest, during storage or after ensiling. Exposure to dietary mycotoxins adversely affects the performance and health of dairy cows. Studies indicate that dairy cows are often exposed to mycotoxins and other fungal secondary metabolites, via silage ingestion. The objective of this study was to evaluate the presence and type of mycotoxins in corn silage. Samples of 500 grams of corn silage (71) were collected in trench silos from 39 dairy cattle farms in Vila do Conde council on northwest of Portugal. Subsequently, laboratory analyses were carried out to detect the presence of mycotoxins. Statistical analysis was performed with o Microsoft Office Excel. The presence of mycotoxins in corn silage was very low, with 9.3% of samples presenting mycotoxin levels above the recommended reference limit. The incidence of positive samples by fungus genus resulted in 96% of *Fusarium*, 2% of *Penicillium* and 2% of *Aspergillus*. *Fusarium* are categorized as field fungi, therefore their high incidence is associated with agricultural practices and with the climatic conditions. The main corn silage mycotoxins detected (52.5%) belong to fumonisin group. The second most common is trichothecenes (30.9%), followed by zearalenone (12.9%), ochratoxin (2.3%) and aflatoxin (1.4%). Only three farms with mycotoxin levels above 5% of the recommended reference values were detected. As conclusion, the detection of the presence of mycotoxins in corn silage was very low, indicating that most producers comply with good agricultural practices of harvest transporting and storage the silage.

Session 50 Poster 23

Effects of calf early weaning on the productive and reproductive performance of Nellore heifers

R.S. Goulart, M.S.P. Carlis, T.K. Nishimura, G. Abitante, A.G. Silva, M.H.A. Santana, A.S. Netto, S.L. Silva, G. Pugliesi and P.R. Leme
University of São Paulo, Faculty of Animal Science and Food Engineering, Department of Animal Science, Duque de Caxias ave. 225, 13635900, Brazil; prleme@usp.br

The beef breed that has the largest cattle population in Brazil is Nellore, a *Bos indicus* breed that is late maturing resulting in heifers' first pregnancy later than 2-years old. Efforts have been made to reduce the age of conception, mainly through genetic improvement and nutrition. This experiment is part of a project to evaluate the effect of time of weaning (150 days [Early group] vs 240 days [Conventional group]) on foetal programming and calve and cows' performance. A total of 95 calves from Nellore primiparous (n=32) and multiparous (n=63) cows, born from August to November 2020, were used (47-48/group respectively). After weaning, the calves were supplemented on pasture up to 13 months of age and then allocated into feedlot pen (Intergado® equipment) receiving a total mixed ration (70:30 corn silage:concentrate) for 113 days. The body weight of calves at 10 months and the initial and final body weight at feedlot for the Early and Conventional groups were, respectively, 240 and 264 kg ($P<0.01$), 281 and 300 kg ($P<0.01$) and 434 and 446 kg ($P=0.14$). The weight gain during the feedlot's period, feed intake and feed efficiency were, respectively, 1.38 and 1.28 kg/day ($P=0.25$), and 7.38 and 7.43 kg DM/d ($P=0.70$), and 187 and 172 g DWG/kg DM ($P=0.17$), in the same order. The rate of puberty at 16-17 months for heifers in the Early and Conventional groups was, respectively, 6.4% and 14.6% ($P=0.32$). Heifers were submitted to a hormonal protocol for timed-artificial insemination at 74 days after beginning of feedlot and resulted for Early and Conventional groups, respectively, in a conception rate of 74.5% and 70.1% ($P=0.81$). In conclusion, the anticipation of weaning at 150 days in Nellore cow results in reduced body weight of weaned heifers up to 14 months of age, but does not impact on its final productive and reproductive performance. These results emphasize the importance of good feeding conditions to anticipate pregnancy, especially when the weaning was anticipated to improve reproductive performance of the dams.

Partial replacement of soybean meal with processed soybean meal in Karagouniko dairy sheep

A. Foskolos, T. Michou, A. Plomaritou, S. Athanasiadis, K. Gatsas, K. Droumtsekas and S. Changli
University of Thessaly, Animal Science, Campus Gaiopolis, 41500 Larisa, Greece; afoskolos@uth.gr

Milk nitrogen (N) use efficiency (MNE) of dairy sheep is low compared with that of dairy cattle. Besides the lower milk yield of traditional, local, dairy sheep breeds under extensive or semi-intensive systems, the use of alternative feeds that may overcome rumen N limitations might be important to increase MNE. This study investigated the effects of soybean meal replacement with processed soybean meal on milk yield and MNE. A total of forty eight (48) lactating dairy sheep we involved in a randomized block design with two treatments: a basal diet containing soybean meal as the main concentrate protein source (SBM) and the same basal diet where 2/3 of soybean meal were replaced with processed soybean meal (SPS). Following lamb weaning (45-50 days post lambing), lactating sheep were grouped in 8 blocks based on their milk yield, body weight and body condition score with the goal to form uniform groups. Then, each group was allocated to one of the treatments and remained in the treatment for 8 weeks. Allowing 2 weeks of adaptation to the diet, measurements of milk yield, milk composition and dry matter intake were performed every 3 weeks (week 5 and 8). Data was analysed with JMP using a fixed effects model, with fixed effects those of treatment and week of lactation. Dairy sheep received processed soybean meal had higher milk yield compared with those received only soybean meal (0.911 vs 0.765 kg/d for SPS and SBM, respectively; P<0.0001). Even though no effect was observed in milk fat, protein and lactose yield, SPS sheep had lower fat (7.7 and 8.0% for SPS and SBM, respectively; P<0.01) and protein (6.2 and 6.4% for SPS and SBM, respectively; P<0.01) milk composition. Dry matter intake and MNE were not affected by treatment. In conclusion, the inclusion of processed soy bean meal in lactating dairy sheep diet increased milk yield by 19% but reduced fat and protein milk composition by 3.7 and 3.1%, respectively.

Effects of diet density in protein and energy on feed intake, yields, digestibility, and efficiency

U. Moallem, L. Lifshitz, H. Kamer and Y. Portnick
Volcani Center, Department of Ruminant Science, 68 HaMaccabim Road, Rishon LeZion 7505101, Israel; uzim@volcani.agri.gov.il

The objectives of this study were to examine the production and efficiency of high producing dairy cows fed diets with different densities of net energy for lactation and (NEL) and crude protein. 42 cows were randomly stratified based on milk production, DIM, parity, and BW. All treatment groups were fed lactating-cow total mixed rations, designated as: (1) low-density diet (LDD) – cows were fed a diet containing 1.74 Mcal NEL/kg DM and 16.2% crude protein (DM basis); (2) medium density diet (MDD) – cows were fed a diet contained 1.78 Mcal NEL/kg DM and 16.5% crude protein; and (3) high-density diet (HDD) – cows fed a diet contained 1.81 Mcal NEL/kg DM and 17.1% crude protein. Diets were formulated to contain similar forage (34.7%), NDF (30.5%) and forage NDF (17.7%) content. Rumen and faecal samples were taken for VFA and digestibility measurements, respectively. Production data were analysed with PROC MIXED, and rumen and digestibility data with GLM models of SAS. No differences were observed among groups in milk yield, but fat and protein percentages were lower in the HDD cows. Energy corrected milk (ECM) was lowest in the HDD (P<0.008) and 4% fat corrected milk (FCM) tended to be lower in the HDD than MDD diet. Dry matter intake was lower in the HDD cows, and the milk to DMI ratio was higher in the HDD diet (P<0.0004). The marginal efficiency for milk production was higher in the HDD diet (P<0.02). Rumen pH was higher in the LDD, and acetate was lower in the HDD than in both diets. Total VFA was lower in the HDD than in both diets. Digestibility of dry matter, organic matter, and crude fat were higher in the HDD than in the LDD cows. In conclusion, milk yields were not different among groups but milk solids were lower in the HDD cows. Efficiency (milk/DMI) and marginal efficiency for milk production were higher and digestibility was higher in the HDD than in the LDD. The high-density diet improved the efficiency and digestibility but reduced the milk solids content.

Session 50 Poster 26

Barn dried hay quality by feeding on dairy farms

J. Mačuhová, D. Schmid and S. Thurner
Institute for Agricultural Engineering and Animal Husbandry, Vöttinger Str. 36, 85354 Freising, Germany; juliana.macuhova@lfl.bayern.de

Less information is available on the quality of barn dried hay or on changes in its quality when hay leftovers of one feeding group are fed to another by silage free feeding on dairy farms. The aim of this study was to evaluate the quality of barn dried grass hay fed from storage to one feeding group and the quality of hay leftovers of this group when fed to other feeding groups. Therefore, hay samples were taken for feed analysis during 5, 7, 8, and 10 feedings on four dairy farms. The feeding was performed once daily and twice daily on three farms and on one farm, resp. On all farms, the hay taken from the storage was fed *ad libitum* to lactating cows, i.e. feeding group 1 (sampling time 1; n=28) and hay leftovers of this feeding group (sampling time 2; n=28) were fed to other feeding groups (dry cows or also older calves and heifers). In addition to testing the effect of sampling time, the effect of farm and interaction of both factors on hay quality were tested. No significant differences (P>0.05) in hay quality between sampling time 1 and sampling time 2 were observed for crude fibre (CF), crude protein (CP), and usable crude protein (uCP). The net energy for lactation (NEL) and dry matter (DM) were significantly lower (P<0.05) and crude ash significantly higher (P<0.05) at the sampling time 1 than at the sampling time 2. However, the values of these parameters (5.85±0.34 MJ/kg DM and 5.62±0.40 MJ/kg DM NEL, 89.1±0.9% and 85.7±2.5% DM, and 84.09±10.60 g/kg DM and 94.93±20.85 g/kg DM crude ash at sampling time 1 and sampling time 2, resp.) and missing differences in firstly mentioned parameters (CF, CP, and uCP) indicate that hay leftovers have maintained good feed quality. A significant effect (P<0.05) of the farm was observed on NEL and DM. Considering the simple effect of farm at sampling time, the significant effect of farm was observed on NEL and DM at sampling time 1 and sampling time 2, resp. This means that hay differed in NEL before feeding (one farm differed to two farms) and in DM in hay leftovers (one farm differed to other three farms). In conclusion, the barn dried hay fed to lactating cows on tested dairy farms had a very good quality. Also, hay leftovers from lactating cows fed to other feeding groups were still good quality.

Session 50 Poster 27

Chemical composition of cowpea (*Vigna unguiculata*) – forage for small ruminants

F.I. Pitacas[1], M. Cristóvão[2], C. Reis[1,3], C. Martins[2], M. Resende[2], C. Espírito Santo[2,4] and A.M. Rodrigues[1,2,3]
[1]School of Agriculture, Polytechnic Institute of Castelo Branco, 6001-909, Castelo Branco, Portugal, [2]CATAA, Castelo Branco Agro-Food Technological Center, 6000-459, Castelo Branco, Portugal, [3]CERNAS-IPCB (project UIDB/00681/2020 funding by FCT), 6001-909, Castelo Branco, Portugal, [4]CFE-UC, Centro de Ecologia Funcional, Universidade de Coimbra, 3000-456, Coimbra, Portugal; amrodrig@ipcb.pt

The cowpea (*Vigna unguiculata*) (L.) Walp) is a legume tolerant to water stress. Is one of the crops most proficient of thriving in irregular rainfall and arid conditions in the rainfed regime. This plant is able to capture atmospheric nitrogen and transform into biological nitrogen. The cowpea is common in Portugal in the regions of Beira Baixa and Alentejo. Seeds are often for human consumption, and whole plant material used as straw or fresh pasture for ruminants, particularly important on those regions (dry during summer), for sheep and goat feed at the end of gestation period (autumn). This work aimed to evaluate the chemical composition profile of three different cowpea varieties: cara verde [CG], cara preta [CB] and bago-de-arroz [BA]. Whole plant samples were collected in Beira Baixa Region during October 2021. The results showed significant differences between cowpea cultivar (P<0.05) for dry matter (DM), ash, crude protein (CP), ether extract (EE), non fibre carbohydrates (NFC), Total digestible nutrients (TDN), metabolizable energy (ME), metabolizability (qm), neutral detergent fibre (NDF), acid detergent fibre (ADF) and cellulose. The best source for protein was CB (17.8%CP) and CG (16.2%CP) cowpea variety with higher CP (P<0.05) than BA cultivar (14.8%CP). However, CG cowpea cultivar has higher figures of TDN, ME and qm and lower values of NDF, ADF and ADL. This allows us to consider CG the best forage/pasture for small ruminant supplementation in the autumn. Acknowledgments: CERNAS – FCT project UIDB/00681/2020 and CULTIVAR – Network for sustainable development and innovation in the agri-food sector CENTRO-01-0145-FEDER-000020.

Session 50 Poster 28

Nutritional strategies using probiotics to improve immune response and health of beef cattle

W.Z. Yang

Agriculture and Agri-Food Canada, Lethbridge Research and Development Centre, 5403 – 1 Ave. S, P.O. Box 3000, T1J 4B1, Canada; wenzhu.yang@agr.gc.ca

Most research using live yeasts (LY) has focused on rumen fermentation such as stabilizing rumen pH, scavenging oxygen in feed particles, promoting fibre digestion. Several mechanisms whereby the LY may improve immune response and animal health of ruminants, but few have been directly examined in experiments with cattle. This abstract summarize the results that were recently obtained in our lab using finishing beef cattle to determine the effect of adding original or rumen protected LY (RPLY) on immune response and animal health in comparing to control (no LY) and antibiotics (ANT; monensin + tylosin). The RPLY was encapsulated using barley-based protein that is resistant to rumen degradation but digested in intestine. Cattle were fed a high-grain diet containing 10% silage and 90% concentrate that is typically used in western Canadian feedlots. Supplementation of LY did not improve rumen pH status (averaged 5.92), but it reduced (P<0.05) proportion of severely abscessed liver from 26.7 to 6.7%. Steers had lesser (P<0.05) faecal *E. coli* counts ($\times 10^7$) for RPLY (3.27) than control (8.47) and ANT (10.1), suggesting an anti-pathogenic activity of LY in the lower gut of cattle. Plasma haptoglobin (HP; mg/ml) was greater (P<0.01) with LY (0.37) than control (0.20) and ANT (0.16). However, blood serum amyloid A (SAA; µg/ml) was lower (P<0.01) with RPLY (4.1) than control (6.7). There was also greater (P<0.01) blood lipopolysaccharide binding protein (LBP; µg/ml) with RPLY (33.8) than control (17.1) and ANT (25.4). These results suggest that feeding LY to feedlot cattle may exert potential beneficial health and food safety effects that reduce liver abscess and possibly pathogen excretion. The supplementing RPLY, thus released in the intestine altered acute protein response and may have immune-stimulation activity. Thus, yeast can be used as an alternative to in-feed antibiotics in natural beef cattle production systems.

Session 50 Poster 29

Effect of breed type on ewe colostrum production and lamb birth weight during late gestation

J.T. Higgins[1,2], G. Egan[2], F. Godwin[2], S. Lott[2], M.B. Lynch[3], M. McEvoy[1], F.M. McGovern[4] and T.M. Boland[2]

[1]Germinal Ireland Ltd, Horse and Jockey, Co Tipperary, E25 D286, Ireland, [2]School of Agriculture and Food Science, University College Dublin, D4, D04 V[1]W8, Ireland, [3]Teagasc Research Johnstown Castle, Johnstown Castle, Co Wexford, Y35 Y521, Ireland, [4]Animal and Grassland Research and Innovation Centre, Teagasc, Athenry, Co Galway, H65 R718, Ireland; jonathan.higgins@ucdconnect.ie

Increased litter size (LS) is a key factor influencing the profitability of sheep meat production systems. Changing the breed type is the simplest method to increase LS. However, increased LS drives an increase in nutrient requirement to support both the ewe's maintenance demand and the growing foetus. A 3×2 factorial design experiment examined the impact of breed type and diet on ewe performance during the final eight weeks of gestation. Ewes (n=48) were offered 100% of predicted metabolisable energy (ME) requirements. Three (n=16) breed types, Belclare X (BX), Lleyn X (LX), and Mule (M) ewes were offered one (n=24) of two silage types, perennial ryegrass or Italian ryegrass and red clover. Concentrates were offered when silage intake was insufficient to meet 100% of predicted ME requirements on an individual animal basis. Colostrum yield and composition were measured at 1, 10, and 18 h post-partum. Lambs were fed colostrum via stomach tube at a rate of 20-50 ml/kg birth weight depending on the colostrum yield of the dam. Data were checked for normality using PROC UNIVARIATE with model selection via linear regression was set at P<0.25 and analysed using repeated measures and PROC GLIMMIX in SAS (Version 9.4). Belclare X ewes had higher dry matter and ME intakes (P<0.05) compared to M ewes. No differences in body weight, body condition, or combined litter weight (CLW; P>0.05) were observed with M ewes tending to be more efficient (predicted ewe energy requirements for actual CLW / actual ewe energy intake) compared to BX ewes (P=0.07). Colostrum yield of BX ewes over the first 18 h was higher compared to M ewes (P<0.05) with lambs born to BX ewes receiving higher volumes of colostrum compared to lambs from both LX and M ewes (P<0.05). In conclusion, breed type affects energy intake and subsequent ewe performance during the late pregnancy/parturition period.

Session 50 Poster 30

Comparative study on using dietary bilberry leaves in broiler reared under thermoneutral conditions

M. Saracila, T.D. Panaite, I. Varzaru and A.E. Untea
National Research-Development Institute for Animal Biology and Nutrition (IBNA), Feed and Food Quality Department, Calea Bucuresti, No.1, Balotesti, Ilfov, 077015, Romania; mihaela.saracila@yahoo.com

The aim of the present study was to evaluate the influence of dietary bilberry leaves (BL) on intestinal microflora of broiler reared under thermoneutral conditions (TN) vs high heat stress (HS). The first experiment was conducted on 60 Cobb 500 broiler chicks assigned in two groups (C-TN, BL-TN) with 30 chickens/group and kept in thermoneutral conditions. In the second experiment, other two groups (C-HS and BL-HS) with 30 chickens/group were kept in heat stress (32 °C). The structure of control groups diet (C-TN and C-HS) was the same. Compared with the control diets (C-TN; C-HS), the experimental diets included the addition of 1% of bilberry leaves (BL-TN; BL-HS). The performance parameters were recorded during the experimental trials (0-42 days). At 42 days, 6 broilers/ group were slaughtered and samples of intestinal content were collected for bacteriological assessment (Enterobacteriaceae, *E. coli*, staphylococci, Lactobacilli, *Salmonella* spp.). Dietary bilberry leaves (1%) increased bodyweight and average daily feed intake in BL-TN group compared to C-HS. The group supplemented with BL and reared in HS had a significantly lower ADFI and average daily weight gain compared to C-TN. Thus, in TN, the number of staphylococci colonies decreased significantly in the group that included BL in the diet (BL-TN) compared to C (C-TN). Lactobacilli increased significantly in BL-TN compared to C-TN. In conclusion, dietary bilberry leaves can be used as feed additive in broiler diet, with positive effect on performance and intestinal microflora in thermoneutral conditions.

Session 50 Poster 31

Cellular agriculture for cultivated meat production

C. Giromini[1], R. Rebucci[1], D. Lanzoni[1], F. Bracco[2,3], B.M. Colosimo[3], D. Moscatelli[2], A. Baldi[1] and F. Cheli[1]
[1]Università degli studi di Milano, Department Veterinary and Animal Science, via dell'università 6, 29600 Lodi, Milano, Italy, [2]Politecnico di Milano, Department of Chemistry, Materials and Chemical Engineering 'Giulio Natta', Piazza Leonardo da Vinci 32, 20133 Milano, Italy, [3]Politecnico di Milano, Department of Mechanical Engineering, via La Masa 1, 20156 Milano, Italy; carlotta.giromini@unimi.it

For the following decades, growing demand for food is expected due to environmental changes and an increase in the world population. In this perspective, the food industry will need to be transformed. Biotechnological techniques as lab-grown meat and fish can help in this direction, allowing for the expected meat demand fitting while reducing the bioresources utilization, animal exploitation, and the ecological footprint of the food production sector. Moreover, healthier and customized *in vitro* meat analogue can be produced with improved chemical and nutritional properties. Within this context, 3D bioprinting can be exploited to achieve repeatable and reproducible products of competitive quality on the market, and at the same time enhancing customization and personalized nutrition as well as scale-up possibilities for mass production. In this work, C2C12 muscle cells have been used to bio-fabricates meat-like constructs using bioprinting approaches. Further, sustainable ingredients (e.g. hemp and whey proteins) were tested as alternative to FBS for cell cultivation and used in the formulation of edible inks. The present study discusses open issues for cell growth media, biomaterial selection, optimal bioprinting condition influencing cell health and differentiation as well as the cultivated meat nutritional properties.

Session 51 Theatre 1

Workshop on multi-stakeholder views on using insects as feed

M. Dicke[1] and T. Veldkamp[2]

[1]Wageningen University & Research, Droevendaalsesteeg 1, 6700 AA Wageningen, the Netherlands, [2]Wageningen Livestock Research, De Elst 1, 6700 AH Wageningen, the Netherlands; marcel.dicke@wur.nl

Insects such as fly larvae are a valuable source of protein for feed for aquaculture, poultry and pigs. The larvae can be reared on residual streams of food production and thus contribute to a circular economy. Moreover, the production of insects has a small ecological footprint in terms of land use and greenhouse gas emission. The production of insects for feed is increasing rapidly and their use for aquaculture, poultry and pig feed has been approved by the European Commission. The production of insects for feed is part of a value chain that includes stakeholders such as: (1) insect producers; (2) providers of substrate for insect production; (3) feed producers; (4) poultry producers; (5) slaughterhouses; (6) retail; and (7) consumers. These stakeholders are directly involved in the value chain. Moreover, the value chain operates within a legal and institutional context that is determined by legal actors, i.e. governments. In addition, a range of Non-Governmental Organisations (NGOs), including e.g. farmer, consumer and animal welfare organisations, influence the legal and institutional context. Food production is currently subject to societal debate, in the context of e.g. food security, biodiversity, environment, climate change and animal welfare. In the EU this is, for example, relevant in the Farm to Fork Strategy of the European Commission. In this workshop, we aim to approach the use of insects as feed from a multi-stakeholder perspective to discuss the views from a diversity of stakeholders with the aim to assess a SWOT analysis that may guide future developments related to the novel approach of producing insects for feed for a circular agriculture. The workshop will consist of three introductions, representing the global view of the value chain, the producer point of view and the ethical point of view, followed by a discussion on the basis of provocative statements. The conclusions of the workshop will be used to define future needs for a sustainable feed production within a multi-stakeholder context.

Session 51 Theatre 2

Societal issues raised on the use of insects as feed

M. Dicke

Wageningen University & Research, Droevendaalsesteeg 1, 6700 AA Wageningen, the Netherlands; marcel.dicke@wur.nl

The production of insects as feed provides a novel source of protein for animal feed for a circular agriculture. Insects can replace fishmeal or soymeal, two protein sources that are under pressure because their use is not sustainable. In contrast, many insect species can be reared on organic left-over streams. Insects are now approved by the EC for use in feed for aquaculture, pigs and poultry. The technical processes in the value chain that focus on transforming resources into goods for people's benefit are managed by production-involved actors. A constraint is that the decision space in which they operate is determined by actors who are not involved in production. The legal and institutional context is the basis for production. These minimal requirements are determined by legal actors (i.e. governments). However, a range of NGOs influence the legal, beyond legal and institutional context of production. NGOs, particularly consumer and animal welfare oriented NGOs, voice public concerns regarding societal and political issues. Growing knowledge and understanding of animals has led to clear changes in the public perception and values of animals, causing extensive criticism of intensive and industrial husbandry systems. Similarly, moral concerns have recently extended to invertebrates. The use of insects as 'mini-livestock' for feed production raises questions similar to those that have previously been asked with regard to conventional livestock. For example, do insects undergo subjective states of pain and suffering, and if so, should insects be attributed moral status? Do insects experience welfare and if so, to what extent is this welfare harmed under mass-rearing conditions? Should we aim for rearing conditions more akin to their natural environment and behaviour in the first place? For NGOs to determine their standpoint, and whether insect products merit labelling, e.g. for sustainability or animal welfare, more knowledge about insect welfare, monitoring of welfare, animal-oriented design of housing systems and criteria are required. The latter is necessary for NGOs to carefully analyse and evaluate this novel agricultural practice. Leaving these questions unanswered poses a risk for insect production because it could lead to societal resistance and backlash.

Session 51 Theatre 3

Propositions and Introduction open dialogue

M. Dicke
Wageningen University & Research, Laboratory of Entomology, Droevendaalsesteeg 1, 6708 PB Wageningen, the Netherlands; marcel.dicke@wur.nl

Propositions are presented as input for the open dialogue and the open dialogue is introduced to the participants.

Session 51 Theatre 4

Open dialogue

M. Dicke[1] and T. Veldkamp[2]
[1]Wageningen University & Research, Laboratory of Entomology, Droevendaalsesteeg 1, 6708 PB, the Netherlands, [2]Wageningen Livestock Research, De Elst 1, 6700 AH Wageningen, the Netherlands; marcel.dicke@wur.nl

An open dialogue on multi-stakeholder views on using insects as feed.

Session 52 Theatre 1

Challenges of integration and validation of farm and sensor data for dairy herd management

K. Schodl[1,2], B. Fuerst-Waltl[2], H. Schwarzenbacher[1], F. Steininger[1], M. Suntinger[1], F. Papst[3], O. Saukh[3], L. Lemmens[4], D4dairy-Consortium[1] and C. Egger-Danner[1]

[1]ZuchtData EDV-Dienstleistungen GmbH, Vienna, Austria, [2]Univ.Nat.Res.Life Sci. (BOKU), Vienna, Austria, [3]TU Graz/Complexity Science Hub, Graz, Austria, [4]Univ.Vet.Med., Vienna, Austria; schodl@zuchtdata.at

Dairy farms are increasingly using precision technologies such as sensor systems for herd health monitoring. Various companies offer different sensors, e.g. accelerometers in collars or boli measuring cow activity and rumination, for different purposes, e.g. fertility or health management. The D4Dairy project aimed, amongst others, to investigate the potential of sensor and other farm and cow-specific data for disease prediction and genetic improvement of metabolic, udder and claw health. Results should lay a foundation for herd management tools and genetic health indices, which are expected to work across farms and sensor systems. First, validation of sensor measurements is necessary. Some companies validated their sensor technologies in scientific studies by comparing sensor measurements to behavioural observations. Another aspect comprises validation of patterns in sensor variables for a desired outcome, e.g. heat or health alarms. To our knowledge no studies investigated how outputs of different sensor systems correspond to one another, which will be a prerequisite for the development of herd management tools across farms, breeds and sensor systems. Furthermore, for the implementation of monitoring of any kind, reliability of measurements is a crucial aspect. Erroneous measurements due to hardware or software malfunctioning have to be identified correctly and outliers have to be distinguished from true deviations. The latter is even more difficult for sensor data without possibilities for plausibility check. Dimensionless sensor outputs (e.g. activity indices) lack established reference values and may differ even between animals equipped with the same sensor type whereas plausibility of rumen temperature or milk yield can also be assessed based on empirical knowledge. These issues had and partly still have to be overcome in the D4Dairy project. In our contribution we present our approaches to sensor data validation, the problems we encountered and how we dealt with them including general recommendations for future studies.

Session 52 Theatre 2

The promises of 3D imaging for phenotyping the morphology and innovative traits in ruminant sectors

A. Lebreton[1], A. Fischer[1], L. Depuille[1], L. Delattre[2], M. Bruyas[3], C. Lecomte[3], J.-M. Gautier[1], O. Leudet[1] and C. Allain[1]

[1]IDELE, 149 Rue de Bercy, 75012 Paris, France, [2]3D Ouest, 5 Rue de Broglie, 22300 Lannion, France, [3]FCEL, 149 Rue de Bercy, 75012 Paris, France; adrien.lebreton@idele.fr

Monitoring of body weight variation, body condition and/or morphological changes allow optimal management of animal health, production and reproduction performances. However, due to implementation difficulties (handling, time consumption, investments), this type of monitoring is not very common within commercial farms. The development of three-dimensional imaging technologies is an interesting solution to meet these needs. Since 2015, the French Livestock Institute and its partners have carried out numerous projects on the possibilities offered by 3D imaging in the dairy, beef cattle, sheep and goat sectors. Initially used to estimate the body condition score of dairy cows, it was then tested to scan the animals in their entirety and to estimate new morphological parameters. This presentation first reviews the performances of the tools developed in recent years to estimate the morphological and growth traits of dairy cattle, the udder conformation of goats or the body condition score and weight of sheep. Then the possibilities offered by 3D imaging to assess new measures such as body volume and surface and by extension new parameters involved in feed efficiency are presented. Finally, the challenges generated by the automation of image processing will be discussed and illustrated with a project to automate the morphological scoring of beef cattle.

An automated pipeline for real-time data integration, health monitoring and validation in dairy cows

M.J. Gote[1], B. Aernouts[1] and I. Adriaens[1,2]
[1]KU Leuven, Biosystems department, Kleinhoefstraat, 4, 2440 Geel, Belgium, [2]WUR, Animal Breeding and Genomics, P.O. Box 338, 6700 AH Wageningen, the Netherlands; martin.gote@kuleuven.be

Precision livestock farming technologies offer a crucial solution to cope with increasing demands on high yielding, sustainable production and animal health and welfare. In the past decades an increasing variety of sensors and technologies have been developed that generate real-time high-frequency data. The value of these technologies highly depends on the (near) real-time processing of the data, and the reliability of its subsequent alerts which is oftentimes insufficient. Hence, the potential impact and on-farm adoption remains below expectations. Additional complexity is added with the large range of farm management preferences and subjective evaluations making the acquisition of a 'ground truth' for validation difficult. An automated data pipeline, analysis and feedback tool was developed, allowing for real-time monitoring, feedback integration and validation. It consists of 4 components: (1) automated, semi-real time extraction of data from farm management systems, preprocessing and integration into a SQL database; (2) monitoring of the health status in dairy cows based on milk yield perturbations and the subsequent generation of alerts; (3) a web-application to display alerts and additional information to the farmer and to collect direct feedback on the alerts; (4) integration of the feedback and optimization of the automated pipeline. After initial implementation and optimization on an exemplary farm, the pipeline was implemented on 10 additional farms and maintained for 3 months. Collected feedback on the detected milk yield perturbations was analysed and combined with a trust factor used to optimized the perturbation algorithm and the data pipeline overall. The implementation of such a pipeline promises high value for future technology development and validation.

Effect of management on milk losses caused by mastitis on modern dairy farms

L. D'Anvers[1], M. Gote[1], Y. Song[1], K. Geerinckx[2], I. Van Den Brulle[3], B. De Ketelaere[1], B. Aernouts[1] and I. Adriaens[1,4]
[1]KU Leuven, Biosystems department, Kasteelpark Arenberg 30, 3001 Heverlee, Belgium, [2]Hooibeekhoeve, Province of Antwerp, Hooibeeksedijk 1, 2440 Geel, Belgium, [3]University of Ghent, M-team, Salisburylaan 133, 9820 Merelbeke, Belgium, [4]Wageningen University Research, Animal Breeding and Genetics, Droevendaalsesteeg 1, 6708 PB Wageningen, the Netherlands; lore.danvers@kuleuven.be

Mastitis-associated milk losses in dairy cows have a detrimental impact on farm profitability. Understanding how management practices affect the milk losses allows us to raise awareness among farmers and better substantiate management decision support. To this end, this study aimed at describing the relation between mastitis-caused milk losses at farm level and farm management. A selection of 43 commercial farms with an automated milking system (AMS) in Belgium and the Netherlands were visited for a farm audit and management survey. Back-up files of the farm management software were taken regularly to collect uninterrupted longitudinal sensor data of at least one year around the survey date. To estimate the theoretical unperturbed lactation curve and, subsequently, the quarter level milk losses, an iterative procedure using the Wood model was implemented. Perturbations in the milk losses were selected based on the duration and amplitude of the milk losses and from those, presumable mastitis cases were identified by detecting a 'deviating quarter' based on quarter level electrical conductivity and milk loss measurements. Finally, the farm level milk loss characteristics were calculated over a one year period. To reduce the number of dimensions in the survey dataset, a principal component (PC) analysis was performed. Next, a linear regression model was fitted on the PCs for each of the milk loss characteristics. The optimal number of PCs to be retained in the model was determined in a cross-validation step. AMS settings and maintenance, preventive mastitis management, mastitis treatment, farm productivity, test-day records, bacteriological testing strategy and farm biosecurity have the impact on calculated farm level milk loss characteristics (high PC loadings and high regression coefficients).

Improving the value of PLF tools for animal health and welfare metrics

D. Foy[1], J. Reynolds[1,2] and T. Smith[1]
[1]AgriGates, Philadelphia, 19146, USA, [2]Western University, CVM, Pomona, 91766, USA; d.foy@agrigates.io

As livestock agriculture progresses into a digital agricultural revolution and the use of Big Data, there is a growing need to develop PLF tools to quantify and assess the welfare and health of animals on dairy farms. This assures animal welfare standards for international sales of milk products can be globally consistent. Exploring and building on Bulk Tank Somatic Cell Count (BTSCC) and lameness prevalence as examples of such metrics to develop improved data that can be captured on-farm and yield more consistent, improved welfare and health decision support. True welfare issues associated with mastitis are the incidences and duration where BTSCC reflects the prevalence of mastitis in a herd. Because prevalence = incidence × duration (SCC = new case rate × duration) the BTSCC can artificially be maintained within standards (e.g.: <400,000 cells/ml) by culling cows, not milking affected quarters into the bulk tank, or by amputating affected teats. Capturing and reporting the new case rate (incidence) and the duration provides a better welfare metric than the prevalence of mastitis. Using mastitis case rate captured by inline PLF tools that more accurately reflect the individual cow SCC through conductivity, or other inflammatory markers. The current method for assessing lameness relies on prevalence and a subjective assessment. Incidence rates for lameness provide better and more accurate associations with welfare. New cases of lameness can be captured by sensing movement, stride length, posture, and lying time using accelerometers. Using PLF tools for determining when an individual cow first gets mastitis and lameness exist but needs refinement regarding sensitivity and specificity, where an industry data standard needs to also be adopted. Combining and integrating multiple sensor data from individual and herd level will improve the sensitivity and specificity of both mastitis and lameness case rates. By taking a critical look at SCC, lameness, and other Big Data outputs that are widely produced on dairy farms and demonstrate that these could be added to a future first generation virtual welfare audit and increase the value of a once a year in-person audit, that enhances consumer confidence in welfare and product quality.

Analysis of the longitudinal activity data to forecast perturbations in dairy cows' milk yield

M. Taghipoor[1], S. Bord[2], L. Sansonnet[2], Q. Bulcke[3] and J. Kwon[2]
[1]Université Paris-Saclay, INRAE, AgroParisTech, UMR Modélisation Systémique Appliquée aux Ruminants, 16, rue Claude Bernard, 75005, France, [2]Université Paris-Saclay, INRAE, AgroParisTech, UMR MIA-Paris, 16, rue Claude Bernard, 75005, France, [3]AgroParisTech, Ferme expérimentale AgroParisTech, route de la ferme, 78850, France; Masoomeh.taghipoor@inrae.fr

In the current context of agroecological transition, farm animals are facing perturbations due to changing environmental and farm conditions, which affect their performance and health. Forecasting the perturbation of the production performance is a prerequisite to propose adequate farm management strategies for precision livestock farming systems (PLF). Variations of animal activity is one of the indicators of the degradation of the environment, that may affect the performance of the production and health. PLF technologies enable the access to the longitudinal, time-series data of animal activity and performance. This makes possible to study the potential of the activity data to forecast the extent of perturbations in milk yield (MY) of dairy cows. Mathematical models are powerful tools to explore such heterogeneous and multi-level data. The objective of this work is to develop a modelling procedure, that combines a dynamic model of perturbed MY with a random forest algorithm for the early detection of perturbations of MY based on activity data. Activity data and daily MY of 230 dairy cows from experimental facility Thierval- Grignon during three years (2019-2021) were used in this work. Dairy cows were equipped with pedometers that recorded their activity (lying, changes in position and number of steps) every hour (min/hours). Sensors were also placed close to feeders recording the number of visits by each cow to the feeder and the duration of the visit. This is an ongoing work and the first results showed the potential of this hybrid procedure to forecast MY perturbations of dairy cows. The detected periods could be compared to morbidity reports, to check the capacity of the model to identify activity profiles correlated with health problems. Detected deviations with no associated morbidity could be either due to the health problem not being detected by the farmer, or to problems with no impact on animal health but only on animal performance.

Biometrical identification and real-time face recognition of dairy-cows

N. Bergman[1,2], Y. Yitzhaky[1] and I. Halachmi[2]
[1]Ben Gurion University of the Negev, Department of Electro-Optics Engineering, School of Electrical and Computer Engineering, David Ben Gurion Blvd 1, Be'er Sheva 84105, Israel, [2]Agricultural Research Organization (A.R.O.), Volcani Institute, Precision Livestock Farming (PLF) Lab, Agricultural Engineering Institute, 68 Hamaccabim Road, P.O. Box 15159, Rishon Lezion 7505101, Israel; halachmi@volcani.agri.gov.il

Biometrics methods, currently identify humans, have a potential to identify dairy-cows and cattle. Deploying animal biometrics recognition system faces two main challenges: (1) identification accuracy; and (2) system robustness, given that animal movements cannot be easily controlled. Our proposed system performs an individual cow held-in-a-group biometrical identification using recent state-of-the-art face-recognition techniques. In this research we developed, experimented, and validated this system on 77 individual Holstein cows from Israel ARO institute dairy-farm. In order to conduct this research, we created a dataset of ~7K cow-faces images, which were acquired by a unique system we built, and were annotated according to a semi-automatic methodology we developed. COW77 is a unique collection of cow-face images that we created and are applicable to future studies. Both system and data-annotating method are approved by typical mid-size dairy-farm and therefore can be used at a larger scale farms in future. We have developed a system using a face-detection tracking algorithm combined with an image classification algorithm, which can track and recognize multiple cows simultaneously while watching a video in real-time. With twenty-five hundred face images in our test set, we have achieved more than 97% accuracy in the face-classification task, while our face-detection algorithm has achieved 96.6% mean-average-precision in tracking and detection of faces.

Data visualization of individual-specific drinking behaviours in dairy cattle

N. Sadrzadeh[1], T. Munzner[2], B. Foris[1], M.A.G. Von Keyserlingk[1] and D.M. Weary[1]
[1]University of British Columbia, Animal Welfare Program, 2357 Main Mall, Vancouver, BC, V6T 1Z4, Canada, [2]University of British Columbia, Department of Computer Science, 201-2366 Main Mall, Vancouver, BC, V6T 1Z4, Canada; negar.sadr@ubc.ca

Access to water is essential for the health and welfare of dairy cattle. Water access on dairy farms has received little attention and what has been done is mainly at the herd level. There is insufficient available information on whether drinker systems provide all individuals with appropriate water access. Cows differ in how they use key resources, especially when required to compete for access. Thus, individual differences may affect water access of different cows within a group. We followed 53 lactating Holstein cows, each for at least 4 months, while living in a dynamic group of 48 cows with access to 5 electronic water bins. We measured between-individual variation and within-individual consistency in time spent drinking (min/d), the number of drinker visits (no./d), and water intake (l/d). We found that drinking time and number of drinker visits were repeatable, with most of the variation explained by differences between individuals (Intraclass correlation coefficient; ICC=0.68 and 0.55). Daily water intake showed lower consistency and varied within cows (ICC=0.206). To better understand these between-individual differences, we developed a visualization interface using Tableau that illustrates daily changes in all three measures, and facilitates testing the relationships with other factors such as individual feed intake, milk production, dominance, or air temperature. We conclude that data visualization tools such as the one developed for this study can help provide a clearer understanding of how individual cows use resources.

Monitoring individual behaviours and the social hierarchy of dairy cows using electronic drinkers

E. Nizzi[1], C. Gérard[1], B. Foris[2], C. Hurtaud[1], J. Lassalas[1] and B. Anne[1]
[1]INRAE, UMR1348 PEGASE, 16 Le clos, 35590 Saint-Gilles, France, [2]Animal Welfare Program, Faculty of Land and Food Systems, University of British Columbia, Vancouver, BC, V6T 1Z4, Canada; ellyn.nizzi@inrae.fr

Freedom from thirst is a key criterion for dairy cattle welfare but little is known about how competition at the drinker influences the water access of different individuals in a group. Recently, automated methods had been validated to record the individual drinking behaviour of cows and competition at the drinker. However, it is not known if competition at the drinker provides suitable data for identifying the dominance hierarchy in cow groups. We monitored agonistic interactions and drinking behaviour in a group of 22 lactating Holstein dairy cows for 9 consecutive days by video recording and with 6 electronic drinkers. 691 agonistic interactions were identified by video observation during this period, including 299 events at the drinkers. We used the Normalized David's Score to calculate the dominance score of each cow based on all observed agonistic interactions (*Complete DS*) and then based on those only at the drinker (*Drinker DS*). The social ranks of individuals were highly correlated between hierarchies (Pearson's r=0.82), suggesting that the social hierarchy can be identified using agonistic interactions at drinkers. We further categorized agonistic interactions at drinkers that can potentially be detected based on automated drinker data (i.e. replacements; aggressive contact from the actor cow that resulted in the recipient leaving the drinker and the actor occupying the place within 60s). Replacements (n=235) were used to determine each cow's social rank (*Replacement DS*) and found that it was highly correlated with the *Drinker DS* (Pearson's r=0.75). Our results indicate that drinkers can be used to identify the dominance position of cows, allowing for the monitoring of the drinking behaviour of cows with a low social rank.

Manoeuvrability test to assess slipping risk in cows on concrete floors with different patterns

E. Telezhenko[1], I. Svensson[2], M. Magnusson[1], E. Karlsson[1] and A. Lindqvist[1]
[1]Swedish University of Agricultural Sciences, Department of Biosystem and Technology, Sundsvägen 14, 230 53 Alnarp, Sweden, [2]Lund University, Department of Biomedical Engineering, Box 118, 221 00 Lund, Sweden; evgenij.telezhenko@slu.se

Concrete is the most common floor surface in passageways in loose housing systems for cattle. To avoid damage and premature culling it is important with non-slip passageways. Patterns in concrete floors during new production or grooves that are cut in the concrete are considered to reduce the risk of slipping. The purpose of the study was to develop an efficient model for assessing slipping safety of different patterns on the concrete floors. A manoeuvrability test was developed to use on passageways at commercial dairy free-stall barns. During the test a cow walked in the passageway in a path with two straight sections and two 180-degree turns. The kinematics and locomotion behaviour were registered with help of inertial measurement unit (IMU) sensors (on every foot and on sacrum) and high-speed motion cameras. The IMU sensors provided data about proportion of two- and three-feet support during the straight movement and the curve, as well as protraction and retraction foot angles, while video recording provided frequency of big and small slips. The dairy cows' manoeuvrability was assessed on 9 different floors (3 with patterns made in green concrete, 3 with cut grooves in one direction and 3 with cut grooves as a grid). Overall results suggest that floors with cut grooves as well as floors with patterns formed as a grid resulted in better slip resistance than floors with patterns made in green concrete or floors with grooves in only one direction. The age of the grooves and the degree of manure soiling of the floor significantly affected floor slipperiness, where newer grooves gave less slipping risk than older grooves and high amount of manure gave a higher risk of slipping than clean floor. The possibility to use IMU sensors for estimating slipping risk will be discussed. The produced knowledge could be helpful for cattle farmers who are considering improving slipping resistance in their barns.

Session 52 Theatre 11

Analysing the behaviour of dairy cows to improve their health and welfare

K. Chopra[1], H.R. Hodges[2], Z.E. Barker[2], J.A. Vazquez Diosdado[1], J.R. Amory[2], T.C. Cameron[3], D.P. Croft[4], N.J. Bell[5] and E.A. Codling[1]

[1]University of Essex, Department of Mathematical Sciences, University of Essex, Colchester, Essex, CO4 3SQ, United Kingdom, [2]Writtle University College, Chelmsford, Essex, CM1 3RR, United Kingdom, [3]University of Essex, School of Life Sciences, University of Essex, Colchester, Essex, CO4 3SQ, United Kingdom, [4]University of Exeter, College of Life and Environmental Sciences, University of Exeter, Exeter, Devon, EX4 4PY, United Kingdom, [5]Royal Veterinary College, Hatfield, Hertfordshire, AL9 7TA, United Kingdom; km19088@essex.ac.uk

Understanding the behaviour of housed dairy cows has the potential to detect changes indicative of illness and stress, to increase their welfare and optimise farm management. This research uses data collected on two commercial dairy herds in 2014 and 2015, via a wireless local positioning system. Firstly, we investigate the structure and consistency of proximity interaction networks, determining herd-level networks from sustained proximity interactions (pairs of cows continuously within three meters for 60 s or longer). We assessed for social differentiation, temporal stability and the influence of individual attributes including lameness. Secondly, we analyse the potential influence of heat stress, using the Temperature-Humidity Index (THI), on the clustering behaviour of the herd. As a part of this, we analyse relationships between THI and range size, inter-cow distance and nearest neighbour distance. We found the proximity networks to be highly connected and temporally unstable, with significant preferential assortment, and no social assortment by individual attributes. Furthermore, it appears that the herds are maladaptively increasing clustering behaviour in response to increasing THI, above given thresholds. Our research demonstrates the potential benefits of automated tracking technology to monitor commercially relevant groups of livestock. Specifically, the proximity interactions of individuals and their clustering responses to heat stress, the latter of which is particularly relevant due to climate change.

Session 52 Theatre 12

Learning behaviour and welfare of dairy cows using a virtual fencing system in rotational grazing

P. Fuchs[1,2], J. Stachowicz[3], M. Schneider[2], R. Bruckmaier[4] and C. Umstätter[5]

[1]University of Bern, Graduate School for Cellular and Biomedical Sciences, Mittelstr. 43, 3012 Bern, Switzerland, [2]Agroscope, Research Division Animal Production Systems and Animal Health, Rte de la Tioleyre 4, 1725 Posieux, Switzerland, [3]Agroscope, Research Division Sustainability Assessment and Agricultural Management, Tänikon 1, 8356 Ettenhausen, Switzerland, [4]University of Bern, Vetsuisse Faculty, Veterinary Physiology, Bremgartenstr. 109a, 3001 Bern, Switzerland, [5]Johann Heinrich von Thünen Institute, Institute of Agricultural Technology, Bundesallee 47, 38116 Braunschweig, Germany; patricia.fuchs@agroscope.admin.ch

GPS-based virtual fencing systems enclose grazing animals in a specific area without physical barriers and may thus support wildlife, facilitate targeted grazing and optimize labour efficiency in grassland management. The present study investigated whether dairy cows are capable of learning a virtual fencing system in rotational grazing management and how their well-being is affected by its use. Twenty dairy cows were divided into 4 groups (2× treatment, 2× control) of 5 individuals. Groups were balanced according to age. Each group grazed four separate rectangular plots during 21, 14, 14 and 7 days. All plots were conventionally fenced and of equal size. Within the plots of the treatment groups, a straight virtual boundary was set on one side. For 59 days, GPS collars recorded the cow's location per min and any interaction with the virtual boundary. Milk samples were collected individually at the beginning, middle and end of each period to determine milk cortisol concentrations. The total number of stimuli per cow decreased over time, with the number of audio tones (AT) far exceeding the number of electric pulses (EP) in each period. The mean ratio of AT to EP was 12.4 and 15.3 of treatment group one and two, respectively. The total number of electric pulses decreased from 64 (SD±1.02) in the first, to 0 in the last period. An ANOVA revealed that the cortisol concentrations did not significantly differ between control and treatment groups ($F(3)=0.583$, $P=0.626$). Provided that animals are given sufficient time to learn, all cows were successfully kept in their assigned area using a virtual fencing system.

Infrared thermographic detection in dairy calves in immunological challenge by *Anaplasma marginale*

M.S.V. Salles[1], G. Cunha[1], F.R. Narciso[1], F.J.F. Figueiroa[1], J.E. Freitas[2], L.E.F. Zadra[1] and J.A.G. Silveira[3]
[1]Animal Science Institute, Ribeirão Preto, 14030-670, Brazil, [2]Federal University of Bahia, Salvador, 40170-110, Brazil, [3]University of Minas Gerais, Belo Horizonte, 31270-901, Brazil; marcia.saladini@gmail.com

The good rearing practices of dairy calves is an important phase for the production system because it has an impact on the future milk productive potential. Infrared thermography can be an important non-invasive technology in the early detection of infectious diseases. This study aims to evaluate the applicability of infrared thermography (IRT) in capturing differences in temperature due to the presence of infectious process caused by *Anaplasma marginale*. Male Holstein calves (n=10) in the suckling phase were inoculated with *A. marginale* at 40 days of age. Infrared thermographic photos, infrared temperatures and physiological parameters were evaluated before the onset of the disease and on the day of disease manifestation (rectal temperature above 40 °C) and statistically compared. Calves with clinical signs of *A. marginale* had higher heart rate (40.03 with and 31.52 without), higher rectal temperature (40.57 with, 39.38 without), and higher left flank infrared temperature (38.33 with and 35.70 without). Animals with *A. marginale* had higher maximal right eye IRT (38.77 with and 37.61 without) and higher mean right eye (37.63 with and 36.47 without) and left flank temperature (38.03 with and 33.34 without) and tendency to higher body IRT on the right side (36.10 with and 33.78 without), behind the right ear (38.33 with and 36.62 without), left paw (35.99 with and 34.08 without), left side of the body (35.83 with and 33.13 without) and behind the left ear (38.20 with and 36.44 without). Infrared thermography proved to be efficient in detecting the physiological state of temperature increase caused by *A. marginale,* presenting the left flank with greater temperature differences, and should be further studied to be applied in dairy production systems. Financial support: FAPESP 2017/04165-5.

Cortisol determination in dairy cows hairs by near, mid infrared and Raman spectroscopy

O. Christophe[1], C. Grelet[1], V. Baeten[1], J. Wavreile[2], J. Leblois[3] and F. Dehareng[1]
[1]Walloon Agricultural Research Centre (CRA-W), Henseval, Chaussée de Namur, 24, 5030 Gembloux, Belgium, [2]Walloon Agricultural Research Centre (CRA-W), Vissac, Rue du Liroux, 8, 5030 Gembloux, Belgium, [3]Elevéo asbl, AWE groupe, rue des Champs Elysées, 4, 5590 Ciney, Belgium; o.christophe@cra.wallonie.be

Chronic stress in dairy cows is likely to affect emotional state, immunity, fertility and milk production of cows. Measuring and assessing chronic stress in herds would be beneficial to objectify the welfare status and to investigate strategies when problems are detected or suspected. Robust, easy-to-use and cost-effective methods to measure biomarkers are needed. The cortisol concentration in hair has been highlighted as a good biomarker for chronic stress. Its analysis through enzyme-linked immunosorbent assay (ELISA) is robust but time consuming and expensive. Alternatively, the development of spectroscopic methods has become more and more interesting due to their rapidity, non-destructive aspect and cost-effectiveness. So, the objective of this study was to assess the potential of three spectroscopic techniques: Near infrared (NIR), Mid-Infrared (MIR) and Raman to develop a method to predict the cortisol content in hairs of dairy cows. A total of 134 hair powder samples from 30 cows were analysed on bench-top spectrometers to acquire near infrared (NIR), mid-infrared (MIR) and Raman spectra. The hair cortisol concentration of theses samples was determined by ELISA analysis. Models linking spectra and ELISA data were performed using partial least square regressions (PLS) and assessed in a 10-folds cross-validation. Hair cortisol was ranging from 8.3 to 91.5 pg/mg of hair, with an average of 27.1 pg/mg of hair. The PLS model provided R^2cv of 0.63, 0.62 and 0.52 and RMSEcv of 9.5, 9.2 and 11.1 for NIR, MIR and Raman data respectively. The NIR and MIR give better performance than Raman spectroscopy. However, the spectroscopic methods are unable to predict accurately the hair cortisol concentration. This is a preliminary study with a relatively low number of samples and the accuracy of models could be improved through the addition of samples in the dataset.

Session 52 — Poster 15

Dairy cow's heart rate and heart rate variability during visits of a cow toilet for urine collection

M. Wijn[1], A. Pelzer[2] and U. König Von Borstel[1]
[1]Justus-Liebig-University Giessen, Animal Breeding and Genetics, Leihgesterner Weg 52, 35392 Giessen, Germany, [2]Agricultural Research and Education Center Haus Düsse, Haus Duesse 2, 59505 Bad Sassendorf, Germany; uta.koenig@agrar.uni-giessen.de

Training farm animals to use a latrine has significant potential to reduce environmental impact of livestock farming as well as to improve barn hygiene and thus animal welfare any health. However, in practice, toilets for cows are not yet widely used, and little is known on the effects on the cows. The aim of the present study was to assess stress in dairy cows during visits of an automated cow toilet compared other daily situations. The cow toilet investigated in the present study is composed of a compartment which cows can enter one at a time after individual recognition, an automatic feed dispenser for luring the cows by providing small amounts of concentrates and a pneumatic arm with a stimulation fitting and a bowl for urine collection, from where the urine is pumped to the storage container. When a cow is recognized by the system and she has a right to enter the toilet, the gate will give her access and a small concentrates are dispended. While the cow eats, the system uses tactile stimulation in the perivulvar area to stimulate urination. After collection of the urine, the cow is released. Heart rate data were collected from 6 out of 17 cows that had access to the cow toilet, and that were all familiar with the barn and the cow toilet for 5-6 months. Focus cows were selected based on frequency of toilet visits to cover a wide range of 0-10 visits/day on a random reference day. Focus cows were equipped with a Polar heart rate monitor for a period of 12 hours. Video analysis was used to classify cow's behaviour into categories including 'toilet visit', 'lying', 'feeding', 'standing', 'urinating outside the toilet' and 'milking'. While lactation number (1,3,4) had a significant influence on mean heart rate, there were neither for mean heart rate nor for heart rate variability parameters (PNS index, SNS index) any significant differences between visits of the cow toilet and other activities such as milking, standing or feeding (F-test, all P>0.05). These preliminary results suggest that after habituation cows do not experience stress due to this type of cow toilet.

Session 53 — Theatre 1

Responses of Merino, Dorper and Meatmaster ewes to an increased heat load

S. Steyn[1], S.W.P. Cloete[1,2] and T.S. Brand[1,2]
[1]Stellenbosch University, Private Bag X1, 7602 Matieland, South Africa, [2]Western Cape Department of Agriculture, Directorate Animal Sciences: Elsenburg, Private Bag X1, 7607 Elsenburg, South Africa; schalkc2@sun.ac.za

Twenty ewes each of the most numerous South African wool (Merino) and meat breeds (Dorper), as well as a recently established composite (Meatmaster) were recorded for their responses to heat during the South African summer. Observations extended across morning and afternoon sessions on three consecutive days with forecast maximum temperatures exceeding 30˚C, i.e. six sessions in total. Maximum and minimum temperatures obtained from a Tinytag Temperature and Relative Humidity Data Logger for the three days data were respectively 33.4, 34.4 and 35.8 as well as 13.0, 15.5 and 18.2 °C. Traits recorded were respiration rate per minute, rectal temperature and eye temperature. Means for all traits recorded during the hotter afternoons exceeded those recorded in the cooler mornings. The magnitude of this difference varied between days, as suggested by a significant day × session (morning or afternoon) interaction in line with the maximum and minimum temperatures provided above. Overall, Merino ewes had a higher rectal temperature and respiration rate than their Dorper and Meatmaster contemporaries, while eye temperature was unaffected by breed. Breed interacted with recording session for rectal temperature, increasing by respectively 1.2, 2.3 and 1.8% from the morning to the afternoon in Merinos, Dorpers and Meatmasters. As no distinct breed differences were found during the afternoon, this interaction appeared to be driven by breed differences in the morning rectal temperatures, amounting to respectively 38.5, 38.1 and 38.2˚C. Overall, the respiration rate of 98.4 breaths per minute (bpm) in Merinos exceeded those in Dorpers (79.3 bpm) and Meatmasters (82.4 bpm), which did not differ. Rectal temperature was moderately repeatable at 0.24±0.06, but repeatability estimates for eye temperature and respiration rate were low and not significant at respectively 0.02±0.04 and 0.07±0.04. Both behavioural (respiration rate) and physiological (rectal and eye temperatures) indicators suggested that heat stress increased from cooler mornings to hotter afternoons. There was an indication that Merinos were affected to a greater extent than the other breeds.

Session 53 Theatre 2

Response of pre-weaned calves to an enhanced milk replacer under heat stress conditions

A.A.K. Salama[1], S. Serhan[1], S. Gonzalez-Luna[1], S.A. Guamán[1], A. Elhadi[1], L. Ducrocq[2], M. Biesse[2], J. Joubert[2] and G. Caja[1]

[1]Universitat Autonoma de Barcelona, Research Group in Ruminants (G²R), Edifici V, Campus de la UAB, 08193, Spain, [2]Bonilait protéines, 5 Route de Saint-Georges, 86361 Chasseneuil-du-Poitou, France; ahmed.salama@uab.cat

Heat stress (HS) experienced before weaning may negatively affect performance and health. Angus×Holstein heifer calves (n=32) were used from d 12 to 56 (weaning) to evaluate the effects of milk replacer (MR) fortified with a combination of antioxidants (vit E, selenium, flavonoids, terpenes) under HS. From d 14 to 27 all calves were kept in thermal-neutral (TN) conditions (THI=62), with 20 animals receiving a control MR (CON) and 12 calves fed the fortified MR (FRT). From d 28 to 56, 12 CON and the 12 FRT calves were exposed to HS [THI=82 (day) and 77 (night)], whereas the remaining 8 CON animals continued in TN. This resulted in 3 treatments (ambient-diet): TN-CON, HS-CON, and HS-FRT. Both MR were produced by the same supplier (Bonilait-Protéines, Chasseneuil-du-Poitou, France) and were fed twice daily (125 g/l). Calves had free access to starter, straw, and water. Intake (DMI), rectal temperature (RT), respiratory rate (RR), and faecal score (FS; scale 0 for normal to 3 for diarrhoea) were daily recorded. Body weight was measured weekly and blood samples were collected biweekly. Data were analysed by PROC MIXED of SAS for repeated measurements. From d 14 to 27 under TN no differences were detected between CON and FRT. From d 28 to 56, HS increased ($P<0.01$) RT (+0.52 °C) and RR (+51 breaths/min) regardless MR type. Both DMI ($P<0.10$) and average daily gain ($P<0.01$), respectively, were lower in HS-CON (1,862 and 927 g/d) and HS-FRT (1,712 and 929 g/d) than in TN-CON (2,064 and 1,170 g/d). Gain:feed ratio in TN-CON (0.57) tended to be greater ($P<0.10$) than in HS-CON (0.50) but was similar to HS-FRT (0.56). Additionally, prevalence of FS≥2 was greater ($P<0.05$) in HS-CON (1.42) than HS-FRT (0.67) and TN-CON (0.38). By the end of the HS period, HS-CON tended ($P<0.10$) to have greater levels of blood haptoglobin (0.125 mg/ml) than TN-CON (0.111 mg/ml). Haptoglobin levels in HS-FRT (0.117 mg/ml) were similar to TN-CON. In conclusion, feeding heifer calves with MR fortified with antioxidants improved feed efficiency and faecal score, and reduced inflammation under HS conditions.

Session 53 Theatre 3

Heat wave effect on milk yield and concentrate intake in dairy farm with low and high s/w ratio

R. Cresci and A.S. Atzori

University of Sassari, Department of Agriculture, Viale Italia 39, 07100, Italy; rcresci@uniss.it

Heat waves (HW) negatively impact on dairy cattle performances causing increased mortality, reproductive problems, reduced milk production, and resulting in increased seasonality of calving and milk deliveries. The aim of this work was to study the effect of a given heat wave on farms with different degree of seasonality measured as summer to winter S/W ratio. Data from 11 intensively managed dairy farms in Italy (121±55 cows), all equipped with AMS were gathered focusing on HW identified between 8 and 23 August 2021. Temperature humidity index (THI), daily individual milk yield (MY) and concentrate intake (CI) in kg/d per cow were recorded. After August 8, THI_{max} and THI_{min} increased by 7 points in the first 3 days (August 7 to 10) then decreased in the following 15 days (Figure 1). It caused a reduction cattle performances which resulted significantly lower than the baseline from August 12 to 17 and from 11 to 18 for MY and CI, respectively ($P<0.05$; unreported data). the minimum of MY and CI was recorded 3 days after the peak of THI_{max}, and 6 days from the beginning of HW. When data were analysed considering the S/W ratio of each farm, Seasonal farms (S/W<0.97) were associated to lower performances. During the whole HW, MY was reduced by 1,03 and 1.63 kg/d (equal to 97.3% and 95.4% of baseline) for Not Seasonal and Seasonal farms, respectively ($P<0.001$). It suggests that animal response to HW deserve more investigation to better understand long term effects of heat stress.

Session 53 — Theatre 4

Organic acids and botanicals supplementation improves protein metabolism in heat-stressed dairy cows

A.B.P. Fontoura[1], V. Sáinz De La Maza-Escolà[1,2], A. Javaid[1], E. Grilli[2,3] and J.W. McFadden[1]
[1]Cornell University, Department of Animal Science, 507 Tower Road, 14853 Ithaca, USA, [2]Università di Bologna, Dipartimento di Scienze Mediche Veterinarie, via Tolara di Sopra, 50, 40064 Ozzano dell'Emilia, Italy, [3]Vetagro S.p.A, via Porro 2, 42124 Reggio Emilia, Italy; victor.sainzdelamaz2@unibo.it

To evaluate the effects of dietary organic acids and pure botanicals (OA/PB) on protein metabolism in heat-stressed dairy cattle, we enrolled 48 Holstein cows (208±4.65 days in milk [mean ± SD], 3.0±0.42 lactations, 122±4.92 d pregnant) in a completely randomized design trial. Following a 7-d acclimation in thermoneutral conditions (temperature-humidity index [THI] 68±0.32), cows were assigned to 1 of 4 groups (n=12/group): thermoneutral conditions (TN-Con), HS conditions (HS-Con; diurnal THI 74 to 82), TN conditions pair-fed to match HS-Con (TN-PF), or HS fed OA/PB (HS-OAPB; 75 mg/kg of body weight; 25% citric acid, 16.7% sorbic acid, 1.7% thymol, 1.0% vanillin, and 55.6% triglyceride; AviPlus R,Vetagro S.p.A, Italy) for 14 d. Cows were fed a corn-silage based total mixed ration top-dressed with OA/PB. All cows received top-dress equivalent for triglyceride used for microencapsulation. Blood was collected for plasma insulin quantification using a radioimmunoassay. Cows were milked twice daily. Data were analysed using a mixed model including fixed effects of treatment, time, and their interaction. Planned contrasts included HS-Con vs TN-Con, HS-Con vs TN-PF, and HS-Con vs HS-OAPB. Higher plasma insulin concentrations were observed in HS-Con, relative to TN-PF (P=0.03). Notably, HS-OAPB displayed similar plasma insulin concentrations compared to HS-Con (P=0.38; 0.97 and 0.86 ng/ml, respectively). HS-OAPB had greater milk protein yields and less milk urea-N (MUN) concentrations, and greater energy-corrected milk (ECM) yields, relative to HS-Con (P<0.05). The tendency for higher dry matter intakes (P=0.15; +1.4 kg/d) and increased water intakes (P<0.01; + 25 l/d) was observed for HS-OAPB, relative to HS-Con, may explain in part the increases in ECM. Such observations, combined with the elevated plasma insulin concentrations strongly suggest that OA/PB supplementation may be a means to partially restore milk production and improve N use efficiency in heat-stressed dairy cows.

Session 53 — Theatre 5

Understanding the genetic architecture of the thermotolerance – production complex in beef cattle

R.G. Mateescu[1], F.M. Rezende[1], K.M. Sarlo Davila[2], A.N. Nunez[1], A. Hernandez[1] and P.A. Oltenacu[1]
[1]University of Florida, Animal Sciences, 2250 S Shealy Dr, 32611, USA, [2]ORISE/NADC, 1920 Dayton Ave, Ames, IA 50010, USA; raluca@ufl.edu

Heat stress is a principal factor limiting production of animal protein in subtropical and tropical regions, and its impact is expected to increase dramatically. Development of effective strategies to improve the ability to cope with heat stress is imperative to enhance productivity of the livestock industry and secure global food supplies. However, selection focused on production and ignoring adaptability results in beef animals with higher metabolic heat production and increased sensitivity to heat stress. The goal of this research is to describe novel traits which can be used to characterize genetic pathways for thermotolerance which are independent or positively associated with production performance. Variance components, heritabilities, additive genetic correlations, and phenotypic correlations were estimated for skin histology characteristics, hair characteristics, body temperature under high THI conditions, and ultrasound carcass traits on 330 heifers from the University of Florida multibreed herd. A high heritability of 0.69 was estimated for the sweat gland area. The heritability for body temperature under high THI conditions was estimated to be 0.13 which is similar the heritability estimated reported for rectal temperature in a Brahman × Angus crossbred population (0.19) and dairy cattle (0.17). Sweat gland area had a negative genetic correlation with sweat gland depth (-0.49), short and long hair length (-0.45 and -0.28, respectively), and body temperature under high THI conditions (-0.65). These negative correlations suggest a similarity in the genetic control underlying these traits which would allow for selection of animals with large sweat glands, short hair (both topcoat and under coat), and able to maintain a lower body temperature under high THI conditions. More importantly, although weak, the genetic correlations between sweat gland area and the two production traits (backfat and intramuscular fat) were favourable (0.22 and 0.20, respectively).

Single-step genomic predictions for heat tolerance of US Holstein and Jersey bulls

T.M. McWhorter[1], M. Sargolzaei[2], C.G. Sattler[2], M.D. Utt[2], S. Tsuruta[1], I. Misztal[1] and D. Lourenco[1]
[1]University of Georgia, Department of Animal and Dairy Science, 425 River Road, Athens, GA 30602, USA, [2]Select Sires, Inc., 11740 US-42, Plain City, OH 43064, USA; taylor.mcwhorter@uga.edu

Data from 923,311 Holstein (HO) and 153,714 Jersey (JE) cows included 12.8 and 2.1 million test-day records, respectively. Milk, fat, and protein yield (kg) records were collected from 2015 to 2021 in 27 different states and included the first five lactations. Cows were required to have at least 5 test-day records per lactation. The pedigree contained 1.2 million (16,629) and 240,813 (4,516) animals (bulls) in the HO and JE evaluations, respectively. Genotypes were available for 76,481 HO and 46,046 JE animals. General (u) and heat tolerance additive genetic values (u_{HT}) for each animal were calculated using single-step genomic best linear unbiased prediction. A multi-trait random regression repeatability model on a function of a temperature-humidity index (THI) was implemented where f(THI)=max(0, THI_{TD}-$THI_{threshold}$), THI_{TD} was a 5-day average THI for each test-day, and $THI_{threshold}$ was 69 for HO and 72 for JE. The Algorithm for Proven and Young with 15K HO and 11K JE core animals was used to obtain a sparse inverse of the genomic relationship matrix. From the onset of heat stress to the maximum THI observed, heritability for production increased for HO (JE) from 0.20 to 0.35 (0.19 to 0.23) for milk, 0.15 to 0.16 (0.10 to 0.12) for fat, and 0.16 to 0.31 (0.12 to 0.13) for protein. Correlations between u and u_{HT} ranged from -0.53 to -0.38. Bulls were ranked by u_{HT}; the 100 HO (JE) bulls with the highest heat tolerance average u was -3.47 (-3.27) for milk, -0.13 (-0.14) for fat, and -0.06 (-0.11) for protein whereas those with the lowest heat tolerance average u was 2.28 (1.52) for milk, 0.11 (0.09) for fat, and 0.06 (0.05) for protein. Heat-tolerant bulls have below-average genetic merit for production traits, consequently, bulls with high-producing genetic ability are expected to have lower heat tolerance. The negative correlations and average genetic merit for production of ranked bulls indicate an antagonistic relationship between u and u_{HT}. However, the increasing heritability over THI shows potential to improve genetic merit for heat tolerance using single-step genomic predictions.

Session 53 Theatre 7

Host transcriptome and microbiome data integration in Chinese Holstein cattle under heat stress

B. Czech[1], J. Szyda[1,2] and Y. Wang[3]
[1]Biostatistics Group, Department of Genetics, Wroclaw University of Environmental and Life Sciences, Kozuchowska 7, 51-631 Wroclaw, Poland, [2]National Research Institute of Animal Production, Krakowska 1, 32-083 Balice, Poland, [3]College of Animal Science and Technology, China Agricultural University, No 2, Yuanmingyuan West Rd, 100193, Beijing, China, P.R.; bartosz.czech@upwr.edu.pl

Climate changes affect animal physiology. In particular, rising ambient temperatures reduce animal vitality due to heat stress. This phenomenon can be observed on various levels – genome, transcriptome, and microbiome. In a previous study, we identified microbiota highly associated with changes in cattle physiology under heat stress conditions – rectal temperature, drooling score, and respiratory score. In the present study, we select genes differentially expressed between individuals representing different additive genetic effects towards heat stress response. Moreover, we performed a correlation network analysis to identify interactions between transcriptome and microbiome for 71 Chinese Holstein cows sequenced for mRNA from blood samples and for 16S rRNA genes from faecal samples. We performed bioinformatics analysis comprising: (1) clustering and classification of 16S rRNA sequence reads; (2) mapping cows' transcripts to the reference genome and their expression quantification; and (3) statistical analysis of both data types – including differential gene expression analysis and gene set enrichment analysis. A weighted co-expression network analysis was carried out to assess changes in the association between gene expression and microbiota abundance as well as to find hub genes/microbiota responsible for the regulation of gene expression under heat stress. We found 1,851 differentially expressed genes that were shared by three heat stress phenotypes. Those genes were predominantly associated with cytokine-cytokine receptor interaction pathway. The interaction analysis revealed three modules of genes and microbiota associated with rectal temperature – two hubs of those modules were bacterial species, demonstrating the importance of microbiome in gene expression regulation under heat stress. Genes and microbiota from the significant modules can be used as biomarkers of heat stress in cattle.

Session 53 Theatre 8

Liver transcriptomics of sensitive and heat tolerant dairy sheep phenotypes

S. González-Luna[1,2], B. Chaalia[2], X. Such[2], G. Caja[2], M. Ramon[3], M.J. Carabaño[4] and A.A.K. Salama[2]
[1]FESC, Universidad Nacional Autónoma de México, Ctra. Cuautitlán-Teoloyucan km 2.5, 54714 Cuautitlán Izcalli, Mexico, [2]Group of Research in Ruminants (G²R), Universitat Autònoma de Barcelona, Edifici V, Campus UAB, 08193 Bellaterra, Spain, [3]IRIAF, CERSYRA, Avenida del Vino 10, 13300 Valdepeñas, Spain, [4]INIA, Ctra. de A Coruña km 7.5, 28040 Madrid, Spain; sandra.gonzalezl@autonoma.cat

The objective was to evaluate liver transcriptomics response to heat stress (HS) in dairy ewes differing in heat tolerance phenotype. Manchega dairy ewes in late lactation (153±8 DIM) previously classified as tolerant (T; n=5) or sensitive (S; n=5) to HS were individually penned in controlled climatic chamber. The design was a crossover of 2 periods (3 wk each) with 2 treatments: (1) thermo-neutral (TN; 15 to 20 °C throughout the day); and (2) HS (37 °C d; 30 °C night). Humidity (50%) and dark-light (12-12 h) were constant. Liver biopsies were obtained at the end of each period, snap frozen in liquid N, and stored at -80 °C until RNA sequencing (2×75 bp). Raw reads were trimmed (TrimGalore), aligned (HISAT2) to the ovine reference genome and assembled and quantified (StringTie). The differential gene expression (DGE) analysis was performed with DESeq2 (R environment). The effects of treatment (TN-S, TN-T, HS-S, and HS-T), period (1 and 2), and their interaction were included in the model with thresholds of $\log_2$ fold change>1.5 and adjusted P-value<0.05. Functional enrichment analyses of the differential gene expression (DGE) were carried out using the bioinformatic DAVID database. Under HS conditions, T ewes differentially expressed 1,318 genes (893 downregulated and 425 upregulated) compared to the S ewes. The downregulated pathways included N-glycan biosynthesis, protein processing in endoplasmic reticulum, protein export, RNA transport, Hippo signalling, nicotinate and nicotinamide metabolism. Whereas Ca and MAPK signalling were the upregulated pathways. In conclusion, the liver transcriptome profile differed between T and S ewes under HS conditions, where T ewes seemed to attenuate cellular mechanisms related to protein synthesis and transport as an adaptive response to HS.

Session 53 Theatre 9

Heat stress effects on the metabolome profile of blood and semen plasma of Iberian pigs

M. Muñoz[1], P. Palma-Granados[1,2], F. García[1], G. Gómez[3], G. Matos[3], C. Óvilo[1] and J.M. García-Casco[1,2]
[1]INIA-CSIC, Animal Breeding, Carretera de la Coruña, km 7,5, 28040 Madrid, Spain, [2]Centro de Investigación en cerdo Ibérico INIA-Zafra, Carretera EX101 km 4,7., 06300, Zafra, Badajoz, Spain, [3]Sánchez Romero Carvajal, Carretera de San Juan del Puerto, s/n, 21290, Jabugo, Huelva, Spain; mariamm@inia.es

One of the consequences of temperature rise is heat stress (HS), which modifies the physiology of animals, altering reproduction among other animal functions. Although Iberian pigs are well adapted to high temperatures, negative effects of HS on fertility and prolificacy have been observed in this rustic breed. The aim of the present study was to analyse the metabolome profile in blood plasma samples of 24 Iberian sows and in blood and semen plasma samples of seven sires at two time points: June (HS; average temperature = 23.84 °C) and November (without heat stress; average temperature = 9.84 °C). Blood samples were collected from sows in oestrus and blood and semen samples from sires just before inseminating females in two consecutive cycles. Plasma samples of both tissues were obtained by centrifugation. Afterwards, metabolites were extracted and subjected to mass spectrometry analyses (LC-MS). Molecular features were obtained using MassHunter Qualitative Analysis Software and MassHunter Mass Profiler Professional and Metabolanalyst software were used to carry out metabolomics analysis. After baseline correction and peak picking and alignment, more than 50,000 molecules were detected in both sows and sire blood plasma samples and 110,746 in semen plasma. After quality control assessment and filtering, 1,972 (sows' blood), 2,296 (sires' blood) and 2,381 (sires' semen) features remained for statistical analyses. A total of 117, 169 and 83 molecular features were differentially abundant between HS and NHS periods for sows' blood and sires' blood and semen, respectively. Overall, as a consequence of HS, blood plasma from males and females has a reduced content of amino acids and, specific modulations in fatty acids, glycerophospholipids and sphingolipids in the three fluids were observed. Therefore, some of the metabolites detected as differentially abundant could be used as biomarkers of heat stress in Iberian sows and sires, however, further experiments are required.

In utero heat stress and post-natal feeding behaviour, growth and carcass performance in pigs

D. Renaudeau[1], A.M. Serviento[1], B. Blanchet[2], M. Monziols[3], C. Hassenfratz[3] and G. Daumas[3]
[1]INRAE UMR PEGASE, 16 le clos, 35590 Saint Gilles, France, [2]INRAE UE3P, 16 le clos, 35590 Saint Gilles, France, [3]IFIP – Institut du Porc, BP 35104, 35651 Le Rheu, France; david.renaudeau@inrae.fr

Heat stress (HS) experienced *in utero* could have long-term effects on pig performance and especially affects lean deposition rate during the growing finishing phase. The aim of the study was to compare the effects of two climatic conditions during gestation on feeding behaviour, growth performance and carcass composition. A total of 24 pregnant sows (12 gilts and 12 multiparous sows) were exposed to thermoneutral conditions (GTN; cyclic 18 to 24 °C) or HS conditions (GHS; cyclic 28 to 34 °C) from day 9 to 109 of gestation. Four male offspring (two entire males, EM and two castrated males, C) per sow were selected at 75 d of age and raised in 8 pens of 12 animals based on *in utero* temperature treatment and sex type. During the growing finishing period, feed was provided *ad libitum* and pigs were kept in thermoneutral conditions. Feed intake and feeding behaviour traits were recorded on a daily basis using single-place SKIOLD Genstar electronic feeders. Pigs were slaughtered at 156 d of age and carcass composition was measured by X-ray tomography. The interaction between sex and *in utero* environment was not significant (P>0.10) for any of the measured traits. The average daily gain (ADG) did not differ between EM and C pigs (1,082 vs 1,085 g/d; P>0.05). However, EM consumed less feed and had a better feed conversion ratio and leaner carcass than C (P<0.01 for each). As the daily number of meals was similar in EM and C, the lower feed intake in EM was related to a reduced meal size (359 vs 379 g/meal; P<0.01). Lower meal size (348 vs 390 g/meal; P<0.01), partially compensated by a higher number of meals (8.2 vs 7.6 meals/d; P<0.05), resulted in reduced feed consumption (2,597 vs 2,690 g/d; P<0.05) in GHS pigs compared to GTN pigs. GHS pigs tended to have a lower ADG (1,069 vs 1,098 g/d; P=0.08) than GTN pigs. *In utero* environment had no effect on carcass composition. In summary, heat stress during prenatal period appear to reduce performance and to modify feeding behaviour during the growing finishing phase but did not impact lean or adipose tissue accretion.

Water sprinkling during lairage reduces impact of heat stress in slaughter pigs

A. Van Den Broeke[1], L. De Prekel[2], B. Ampe[1], D. Maes[2] and M. Aluwé[1]
[1]ILVO, Scheldeweg 68, 9090 Melle, Belgium, [2]University Ghent, Internal medicine, reproduction and population medicine, Salisburrylaan 133, 9820 Merelbeke, Belgium; alice.vandenbroeke@ilvo.vlaanderen.be

In the last decade, global warming has been associated with an increase in the frequency and severity of hot periods during summer. Therefore, pigs transported to the slaughterhouse during summer experience increasingly more heat stress, even in European regions with a mild climate. This has implications for animal welfare as well as meat quality. As economic margins in the pig industry are small, measures to counter heat stress effects should be cheap and easy to implement. Installing a sprinkler system is one of the cheap measures a slaughterhouse can deploy to alleviate heat stress. Some slaughterhouses already have sprinklers in the lairage, but these are not systematically used during hot summer days. The present study assessed the effects of sprinkling slaughter pigs with water in the lairage area of a slaughterhouse. Fifteen trailer loads were equally divided into a control and a sprinkling group. The latter group was sprinkled with water 15 minutes after unloading the trailer during 5 minutes. Twenty percent of the pigs (control n=233; sprinkling n=228) were individually marked and observed immediately after unloading and 30 minutes later. Apart from recording panting and open mouth breathing, a continuous pig heat stress scale from 0 to 150 was used to evaluate the level of heat stress. Immediately after unloading, heat stress indicators were comparable for both groups. The percentage of pigs with an open mouth breathing decreased significantly from 6.4% to 0.6% (P<0.001) after sprinkling while the control group remained at the same level (4.3% versus 3.6%; P=0.568). The percentage of panting pigs decreased significantly from 5.8% to 1.2% (P<0.001), while the control group remained at the same level (4.4% versus 4.6%; P=0.879). In the sprinkling group, the pig heat stress score decreased significantly from 39 at unloading to 29 at the second observation (P<0.001), while there was a much smaller decrease from 39 to 36 in the control group (P=0.010). These results show that sprinkling pigs helped to reduce the heat stress of slaughter pigs. The research was done within the VLAIO project HBC 2019.2877.

Session 53 — Theatre 12

A feed additive package mitigate heat-induced productivity losses in lactating sows

E. Jiménez-Moreno[1], D. Carrion[1], D. Escribano[2], C. Farias-Kovac[1] and Y. Lechevestrier[3]
[1]Cargill Animal Nutrition and Health, Poligono Industrial Riols s/n, 50170 Mequinenza, Zaragoza, Spain, [2]University of Murcia, Animal Production, Campus de Espinardo s/n, 30100 Espinardo, Murcia, Spain, [3]Cargill Animal Nutrition and Health, Provimi France, PA Ferchaud, 35320 Crevin, France; carloseduardo_farias@cargill.com

Heat stress is responsible for having detrimental consequences to animal's health and performances. The aim of this study was to evaluate the effect of the inclusion of a supplement based on heat stress-mitigating feed additives package on lactating sow feed intake and litter performance under commercial heat stress conditions in Spain. A total of 183 Landrance × Large White lactating sows with 3.47 average parity and 262±43.5 kg body weight at the entrance of the farrowing room were used and allotted to one of two treatments (Control and Treatment group). Treatment group received the control diet with the supplement for 30 days (6 days pre-farrowing plus 24 days of lactation). Supplement contains specific components (natural antioxidants, plant extracts and essential oils, vitamins and potassium source) to improve feed intake, oxidative status, preserve gut integrity, and help to support a lower respiratory alkalosis. Diets were offered *ad libitum* from day 10 of lactation until weaning. Sows fed with treatment diet consumed 9.1% more feed (6.06 vs 6.61 kg/day from farrowing to weaning; P<0.01). Total born alive was of 16.51 and 16.97 piglets/litter and stillborn, of 12.7 and 11.8% for Control- and Treatment group, respectively (P>0.1). Plasma thyroxin level was reduced (1.46 vs 1.35 µg/dl) while glutathione peroxidase increased (3,637 vs 3,952 U/l) in Treatment group (P<0.001). Treatment group showed a tendency to have a greater litter weight gain in gilts (1.86 vs 2.22 kg/day; P=0.076). Piglet mortality during lactation was of 9.05 and 6.85% (P>0.1), and the number of weaned piglets per litter tended to be higher for Treatment- than Control group (12.48 vs 12.92; P=0.056). In conclusion, the use of the novel heat stress-mitigating feed additives package into the diet for lactating sows increased feed intake and improved litter performance probably due to a better energy efficiency for milk production and oxidative status.

Session 53 — Theatre 13

The effect of a summer feed with selenium, vitamin E, vitamin C and betaine in fattening pigs

L. De Prekel[1], D. Maes[1], A. Van Den Broeke[2], B. Ampe[2] and M. Aluwé[2]
[1]University Ghent, Unit of Porcine Health Management, Salisburylaan 133, 9820 Merelbeke, Belgium, [2]ILVO, Scheldeweg 68, 9090 Melle, Belgium; lotte.deprekel@ugent.be

Heat stress can cause different adverse effects in pigs, leading to impaired animal welfare, health and production. Osmolytes and antioxidants such as Selenium (Se), vit E and C, and betaine may protect tissues and alleviate damage during heat stress. The present study investigated whether a feed enriched with osmolytes and antioxidants can reduce the negative effects of heat stress in fattening pigs. A total of 60 fattening pigs (crossbred Piétrain × hybrid sow) were randomly allocated into two groups: a control (CG) and test group (TG). Each group was divided into two pens of 15 pigs each, housed in the same compartment. Both groups received different feeds (*ad libitum*) during the fattening period from June 30 to October 12, 2021. The control feed contained 0.4 mg/kg inorganic Se, 100 ppm vit E, and the dEB was 195; the test summer feed contained 0.3 mg/kg inorganic Se, 0.1 mg/kg selenomethionine, 200 ppm vit E, 200 ppm vit C, 0.2% betaine and the dEB was 240. Breathing frequency, rectal and skin temperature, passive behaviour and average daily gain (ADG) were assessed weekly in both groups. These parameters were also measured daily on heat days (HD): when the outside temperature was ≥25 °C for three days (Sept 6-9), or artificial heating was induced (Sept 28-30). Individual water and feed intake were measured daily using an RFID system. In the end, carcass- and meat quality were also compared. The TG had a lower skin temperature than the CG (35.1 °C vs 35.2 °C) during HD due to the interaction of the summer feed (P<0.05). Before the natural HD, ADG from the TG was lower than that of the CG (895 g/day vs 1,018 g/day). During and after the natural HD, the TG grew better than the CG (1,076 vs 957 g/day and 1,020 vs 899 g/day, respectively). These differences were due to the interaction of the summer feed with the HD (P<0.05). The same effect was seen before and during the artificial HD, except in the post-HD when the ADG in the TG was lower than in the CG (P<0.05). No other significant differences were found for the other parameters. The summer feed had some minor effects during HD but did not have sufficient results on normal days.

Session 53 Poster 14

Milk fatty acids of horned and disbudded dairy cows at thermoneutral and mild heat stress conditions

A.-M. Reiche[1], B. Bapst[2], M. Terranova[3] and F. Dohme-Meier[1]
[1]Agroscope, Tioleyre 4, 1725 Posieux, Switzerland, [2]Qualitas AG, Chamerstrasse 56, 6300 Zug, Switzerland, [3]AgroVet-Strickhof, Eschikon 27, 8315 Lindau, Switzerland; anna-maria.reiche@agroscope.admin.ch

The role of horns in thermoregulation has not yet been studied *in vivo*, albeit being often discussed. At low ambient temperatures, milk fatty acids (MFA) of dairy cows differed by horn status, and it was presumed that heat loss via the horns would cause cold stress and alter the cows' metabolism and thus MFA. Moderate heat stress (HS) may also alter the MFA profile of cows. Therefore, this work studied the potential thermoregulatory function of horns via the MFA profile of horned (H+) and disbudded (H-) dairy cows at thermoneutral and mild HS conditions. In a feeding crossover study, 9 H+ and 10 H- mid-lactating Brown Swiss dairy cows with similar genetic indices were housed twice during 5d in respiration chambers at varying temperature-humidity indices (THI). The THI was 52 (considered as thermoneutrality; TN) for the first 48 h and 74 (considered as mild HS) for the last 48 h of the 5d-periods. Feed intake, milk yield and composition were measured daily. MFA were determined in samples of the last milking of each 48-h THI condition. Data were analysed using linear mixed models, with the fixed effects diet, THI condition, horn status, their interactions, and the random effects crossover, replicate and animal. Effects related to diet are not presented. THI condition and horn status did not alter feed intake, milk yield and composition; except the lactose concentration was lower under HS than TN ($P<0.05$). Independently of THI, milk of H+ cows had greater contents of the MFA C8:0, C10:0, C17:0 iso and C18:0 and lower contents of C16:1 cis9 n-7 than milk of H- cows (all $P<0.05$). No interactions were present ($P>0.10$). In accordance with earlier studies, mild HS lowered milk lactose contents, but did not alter the MFA profile, suggesting that it is not an indicator for mild HS. No thermoregulatory function of horns as expressed by changes in the FA profile was observed, as effects of horn status did not depend on THI. The direction of the slight differences between H+ and H- cows are mainly in contrast with the earlier study and their biological relevance is questionable.

Session 53 Poster 15

Transcriptomic comparison under different temperature humidity index in growing pig

J.-E. Park[1], H. Kim[1], Y. Kim[2] and M. Song[2]
[1]National Institute of Animal Sience, Division of Animal Genomics and Bioinformatics, Wanju, 55365, Korea, South, [2]Chungnam National University, Division of Animal and Dairy Science, Dajeon, 34134, Korea, South; jepark0105@korea.kr

Change of air temperatures and humidity by climate change triggers decline of products, and it causes stress in organism. Especially, exposed animals under heat stress shows problem in immune system and productivity, which results in negative effect in livestock farm. In this study, we performed transcriptome analysis using RNA-seq for understanding mechanisms of heat stress responses in pig blood samples. The animal experiments were performed for 14-days, and a total of 6 growing (10- and 12-weeks) pig was grouped according to THI (Temperature–Humidity Index) into Normal (23-24 °C and humidity 35%, THI 68-69) of 3 animals and Severe (32-33 °C and humidity 80%, THI 86-88) of 3 animals respectively. A total of 6 blood sample was isolated by manufacturer's instructions, and generated FASTQ files through RNA-seq were used to analysis. Raw reads were trimmed with FASTQC, and raw reads were mapped to reference genome (Sus_scrofa 11.1.105) using Hisat2. Then, generated sam files were annotated using FeatureCounts in subread package. Finally, edgeR in R package was used to obtain DEGs (differentially expressed genes) for comparison into Normal and Severe groups. As stress index, cortisol hormone level was significantly high ($P<0.05$) in severe condition. And the number of DEGs with false discovery rate 0.05 and fold change 1.0 were 254 up-regulated genes with *MAPK4*, *MMP8* and etc. and 241 down-regulated genes with *BPGM*, *SLAs* and etc., respectively. In function annotation by gene ontology terms, inflammatory response and toll-like receptor signalling were enriched in up-regulated genes, and cellular oxidant detoxification and oxygen transporter activity were enriched in down-regulated genes. KEGG pathway showed PI3K-Akt signalling pathway and Inflammatory bowel disease pathway in up-regulation, Antigen processing and presentation and Intestinal immune network for IgA production pathway in down-regulation. In conclusion, high-THI in growing pigs as stressor, could decline cellular oxidant detoxification and cause inflammatory response, which will lead intestinal damage and ionic imbalance in body.

Session 53 Poster 16

A meta-analysis on the impact of heat stress on ruminants' welfare and production

J.N. Morgado[1,2,3], E. Lamonaca[1], F. Santeramo[1], M. Caroprese[1], M. Albenzio[1] and M.G. Ciliberti[1]
[1]University of Foggia, Department of Agriculture, Food, Natural Resources, and Engineering (DAFNE), Via Napoli, 25, 71122 Foggia, Italy, [2]University of Lisbon, Lisbon School of Economics and Management (ISEG), Rua Miguel Lupi, 20, 249-078 Lisboa, Portugal, [3]University of Lisbon (FMUL), Nutrition Laboratory, Environmental Health Institute, Faculty of Medicine, Avenida Professor Egas Moniz, 1649-028 Lisboa, Portugal; maria.ciliberti@unifg.it

The livestock sector is highly affected by climate change, particularly by global warming phenomenon, with the increase to heat stress exposition. The decrease of livestock welfare and production due to the effects of heat stress events are causing severe economic losses. To help mitigate these effects different adaptation and mitigation measures may be applied in different climatic heat scenarios. Nevertheless, little is known as to which extent heat stress may negatively affect livestock performances and how effective different mitigation measures may be. Through a multidisciplinary approach and using a systematic literature review and meta- regression analysis, the overall aim of our study was to investigate the effects on impact in animals' performances of different heat stress scenarios with or without mitigation measures. To the best of our knowledge, this is the first time in which the assessment of climate impact was based on Temperature-Humidity-Index variations; this index was used to quantify ruminants' performance and assess the influence of different indicators studied in the analysed papers. A total of 231 articles were analysed, according to the criteria of comparing control with heat stress conditions, and the occurrence or not of mitigation measures. Our results showed that the impact of applying mitigation measures halved the negative impacts on the livestock's performance induced by heat stress. However, when animals were in severe heat stress conditions, the mitigation measures failed in sustaining animals' performances. These findings highlight the need for further investigation on heat stress adaptation and or mitigation measures in ruminants.

Session 53 Poster 17

Effect of a standardized dry grape extract supplementation in late gestating sows in a hot climate

P. Engler[1], A. Frio[2] and M.E.A. Benarbia[1]
[1]Nor-Feed, 3 rue Amedeo Avogadro, 49070 Beaucouzé, France, [2]1st Ten Consulting Asia Pacific, Bay, Laguna, Philippines; paul.engler@norfeed.net

Heat stress has detrimental effect in livestock. Excessive thermal stress around parturition can have detrimental on reproduction success and litter growth in swine. Antioxidants are commonly used in various species to limit oxidative stress caused by excess environmental heat. Grape polyphenols have been extensively studied for the last decades for their powerful antioxidant properties and have shown valuable benefits in mitigating oxidative stress in various species. The aim of the present trial was to evaluate the effect of a low dose of a standardized dry grape extract (SDGR) in late gestation and lactation in sows raised in a hot climate. 72 sows were randomly divided in 2 groups (36/group) of equal parity ranks, a control group (CTL) and a supplemented group receiving an added 20 ppm of a SDGE (Nor-Grape 80, Nor-Feed, France) for the end of gestation (D84-farrow) and the whole lactation (21D) on top of the CTL diet. Litter size, farrowing time and piglet expulsion time, litter lactation performances and sow feed intake were recorded. Temperature Humidity Index (THI) ranged from 75.0 to 98.2 during the trial, showing moderate to severe heat stress conditions (alert THI=72.0). Numerical but not statistical reduction of farrowing time was observed. Piglets from NG sows tended to be bigger (+0.05 kg, $P<0.10$) and sows from the NG group tended to have a higher feed intake than CTL sows ($P<0.10$). The inclusion of a small dose of SDGE in sows diet thus could prove beneficial in sows raised in hot climate conditions, with improvement of piglet birthweight and sows feed intake during lactation. More research is needed to better understand the full extent of benefits from such a solution.

MOET efficiency in a Spanish Japanese black herd and effects of environmental and metabolic markers

A. Fernández-Novo[1], J.M. Vázquez-Mosquera[2], M. Bonet-Bo[3], N. Pérez-Villalobos[1], J.L. Pesántez-Pacheco[4], A. Heras-Molina[2], M.L. Pérez-Solana[5], E. De Mercado[5], J.C. Gardón[6], A. Villagrá[7], F. Sebastián[8], S.S. Pérez-Garnelo[5], D. Martínez[3] and S. Astiz[5]
[1]UEM, C. Tajo, s/n, 28670 Madrid, Spain, [2]UCM, Avda Puerta de Hierro s/n, 28040 Madrid, Spain, [3]Embriovet, P8-IA Pol Ind de Piadela, 15300 A Coruña, Spain, [4]UC, Av. 12 de Abril, 010107 Cuenca, Ecuador, [5]INIA-CSIC, Av.del Padre Huidobro 7, 28040 Madrid, Spain, [6]UCV, C de Quevedo 2, 46001 València, Spain, [7]IVIA, CV-315, Km 10.7, 46113 Valencia, Spain, [8]Cowvet, Titaguas, 46178 Valencia, Spain; anaherasm@ucm.es

Environmental conditions and metabolic parameters of donors may influence MOET efficiency. The Japanese Black breed is characterized by a high intramuscular fat content, and different metabolic issues. From 137 *in vivo*, non-surgical embryo flushings performed in full blood, Japanese Black heifers (aged 21.8±6.79 mo. and BCS 3.5±0.57) kept in Spain, 102 in cool seasons (74.5%, 102/137), and 35 under heat stress (25.5%, 35/137), 1,015 embryos were obtained (7.4±5.18 embryos/collection). From these embryos, 724 were viable embryos/flushing (71.3%; 5.3±4.34), 165 degenerated embryos (1.3±1.56/flushing), and 126 non-fertilized oocytes. The assessed metabolic parameters were BHB; NEFA; TC; HDL; LDL; GLU; LAC; TG; UR; FRU; and LEP, determined by bioanalyzer and ELISA. No effect of the donor genetic line, bull, year or heat stress was detected. The donor concentrations of metabolites or leptin at the moment of superovulation, neither affected MOET efficiency. However, BHB, TC and HDL decreased with the years of study, while glucose and TG increased ($P<0.01$). Regarding flushing order, TC, HDL and leptin increased, while glucose decreased with flushing order ($P<0.01$). This could reflect an effect of age of the heifer, which increased with flushing order. We can conclude that an adequate efficiency of MOET programs can be achieved with foreign breeds such as the Japanese Black, when kept and managed under very different conditions as those described in Japan. The special metabolic characteristic of this breed, mainly related to a high intramuscular fat and different metabolic parameters can be kept under physiological levels, and did not affect the MOET performance of the heifers.

SCALA-MEDI: sheep and chicken sustainability in the Mediterranean area

P. Ajmone-Marsan, B. Benjelloun, M. Benlarbi, A. Carta, G. Chillemi, S.S. Gaouar, H. El-Hentati, A. Jannoune, C. Jerrari, H. Kemiri, V. Loywyck, N. H'hamdi, N. Tabet-Aoul, A. Stella, M. Tixier-Boichard, R. Valentini and G. Zitouni
SCALA-MEDI Consortium, via E. Parmense, 84, 29122 Piacenza, Italy; paolo.ajmone@unicatt.it

Sheep and chicken are the most important livestock species in the Southern Mediterranean countries. They are the basis of the local diets and are reared in marginal areas of the Northern Mediterranean. In North Africa local and regional populations and breeds are poorly defined while possessing unique adaptation to harsh environment. Accelerated selection for thermal tolerance and resilience to new endemic diseases is becoming urgent to counteract the detrimental effect of climate change on livestock welfare, as it is the valuation and conservation of local breeds as reservoir of unique gene variants. Genomics plays a key role in this respect, together with phenotype recording, and the collection of epidemiological and environmental data. The SCALA-MEDI project will characterise the genetic and phenotypic diversity of Mediterranean local breeds of sheep and chicken and study their ability to adapt to harsh environments and management systems. The project will leverage data produced in EU projects and generate new data, including traditional production traits and using new technologies for remote phenotyping of adaptation related traits, genotyping, and to explore the genome methylation status of animals reared in different environmental conditions. Data and samples will be collected on local breeds from Tunisia, Algeria and Morocco (16 sheep breeds and village chicken populations from different bio-climatic environments), taking advantage of local expertise in addition to that available in Italy and France. Local resources will be characterized for farming system, diversity, distinctiveness and adaptive traits. Genomic data will be analysed to identify loci controlling adaptation traits and product authenticity and will be used to create decision-making tools to improve conservation and selection programmes and management strategies for Mediterranean livestock production system to face future climate change scenarios.

Session 53 Poster 20

Mapping Italian caprine genetic resources: local distribution and environmental risk

M. Cortellari[1], A. Bionda[1], A. Negro[2], S. Grande[2] and P. Crepaldi[1]
[1]University of Milan, Dipartimento di Scienze Agrarie e Ambientali – Produzione, Territorio, Agroenergia, via Celoria 2, Milan, 20133, Italy, [2]Associazione Nazionale della Pastorizia, Via XXIV maggio 44, Rome, 00187, Italy; matteo.cortellari@unimi.it

Italy is characterized by a wide panorama of indigenous goat breeds, a fundamental part of its genetic biodiversity heritage. Unfortunately, some of these populations are at short- or long-term risk of extinction, according to the number of individuals and the effective population size (Ne). However, in the light of the strong connection between extensively reared autochthonous populations and their territory, it is of paramount importance to consider the distribution of these breeds and their level of environmental risk, given the recent increase of the extreme weather events, ascribable to the climate change and the land abandonment in marginal areas. We analysed the pedigrees of 34 Italian autochthonous goat populations and classified them on the basis of Wright's Ne. We geolocated the farms in which these animals are bred and calculated the number of heads of each breed by province, based on the data provided by the Italian Sheep and Goat Breeders Association. Lastly, we compared this information with the risk maps of hydrogeological and seismic events in Italy. 12% of the breeds are at short-term risk of extinction ($Ne \leq 50$), 47% at long-term risk ($50 < Ne \leq 500$) and the remaining 41% are not at risk ($Ne > 500$). Almost one third of the populations are bred only within one or two adjoining provinces, thus increasing the potential risk of irreparable damages following catastrophic climatic events. Most of the animals belonging to breeds at short-term risk of extinction are located in provinces at medium-to-high risk of landslides or earthquakes. However, populations that are not at risk of extinction but bred exclusively in limited areas at high environmental risk could be severely affected by natural disasters as well. The safeguard of autochthonous goat populations and the preservation of the territory are closely related. In fact, livestock breeding prevents land abandonment, especially in marginal areas; on the other hand, our results show that caprine biodiversity management should also consider the risk of extreme weather events like floods, landslides, and earthquakes.

Session 53 Poster 21

Milk microbiota variability in healthy Holstein and Brown Swiss cows exposed to hyperthermia

C. Lecchi[1], P. De Palo[2], S. Di Mauro[1], M. Ciccola[1], A. Bragaglio[2], P. Cremonesi[3], B. Castiglioni[3], F. Biscarini[3] and F. Ceciliani[1]
[1]University of Milan, Lodi, 26900, Italy, [2]University of Bari, Valenzano (Ba), 70010, Italy, [3]National Research Council, Lodi, 26900, Italy; cristina.lecchi@unimi.it

The milk of lactating cows presents a complex ecosystem of interconnected microbial communities which can be influenced by biotic and abiotic stressors. Hyperthermia exerts a negative impact on dairy cows' health, milk production, reproductive performance, and immune defences. The present study aimed at profiling the microbiota of milk collected from healthy multiparous Holstein (H) and Brown Swiss (BS) cows exposed to different thermal conditions – hyperthermia (HS) versus thermal comfort condition (TC). A total of 44 samples were included: 22 were collected under TC (THI=68) -11 from H and 11 from BS- and 22 after a natural four-day heatwave (HS, THI_{max}=84, THI_{min}=69) – 11 from H and 11 from BS. The study was carried out in the same farm. Both groups were at the same stage of lactation (DIM 102.42±8.65 vs 105.62±8.46 in H and BS, respectively) and in TC showed the same milk production level (milk yield, kg: 34.78±2.67 vs 33.11±2.11; energy corrected milk yield, kg: 37.43±3.40 vs 37.76±3.40; Fat Protein Corrected milk: 34.12±2.78 vs 34.53±2.73 in H and BS, respectively). After the heatwave, both groups had a drop in milk production although lower in BS than in H ($P<0.002$). Particularly H reduced milk yield of 3.73±0.74 kg, ECM yield of 6.37±1.62, FPCM of 5.89±0.97), while BS lost 1.93±0.46 kg of milk yield, 3.09±1.00 kg of ECM and 2.86±0.81 kg of FPCM). Bacterial DNA was purified and the 16S rRNA genes were individually amplified and sequenced. Significant differences were found in milk samples from HS and TC. Alpha diversity analysis showed that the microbiota of BS was richer under HS than under TC (776.72 vs 583.82 observed OTUs; 9.167 vs 8.80 Shannon index; $P<0.01$), while no significant differences were found in H. Beta diversity analysis showed clear clustering by thermal condition (TC vs HS), both overall and within breeds ($P=0.001$), while it was not possible to discriminate breeds with milk microbiota data. In conclusion, we found that HS can modulate the composition of the milk microbiota in dairy cow.

Session 53 Poster 22

Organic acids and botanicals supplementation on lactation performance in heat-stressed dairy cows

V. Sáinz De La Maza-Escolà[1], R. Paratte[2], A. Piva[1,2] and E. Grilli[1,2]
[1]Università di Bologna, Dipartimento di Scienze Mediche Veterinarie, via Tolara di Sopra, 50, 40064 Ozzano dell'Emilia, Italy, [2]Vetagro S.p.A., via Porro, 2, 42124 Reggio Emilia, Italy; victor.sainzdelamaz2@unibo.it

A field trial was conducted on a commercial Parmigiano-Reggiano farm to evaluate the effects of dietary supplementation of a microencapsulated organic acids and botanicals product (AviPlus R®, Vetagro S.p.A, Italy) on lactation performance during summer season (July to September 2020). Twenty multiparous non-pregnant Holstein cows (43±13.6 days in milk [mean ± SD], 2.7±0.86 lactation, and 43.3±3.87 kg of milk yield) were enrolled in a study with a completely randomized design. Cows were assigned to one of two groups (n=10/group) fed a basal diet supplemented with 25 g/d of AviPlus R (TRT; 75 mg/kg of body weight; 25% citric acid, 16.7% sorbic acid, 1.7% thymol, 1.0% vanillin, and 55.6% triglyceride) or basal diet supplemented equivalent triglyceride used for microencapsulation (CTR). Environmental temperature-humidity index (THI) was recorded hourly and individual milk yield was recorded daily. Rectal temperature measurement and milk components analysis were conducted every 2 weeks. Cows were milked twice daily. Data were analysed under a mixed model with the random effect of cow and the fixed effects of parity, date and days in milk, treatment and their interaction. Both groups were under heat stress conditions (THI: diurnal change 72 to 82) for the whole study period with no differences between them (P=0.46). Rectal temperature was lower in TRT group compared to CTR (38.18 vs 38.57 °C, respectively; P<0.05). Milk yield was greater for TRT compared to CTR (41.86 vs 39.51 kg/d, respectively; P<0.01). Milk protein, casein and lactose yields were also greater for TRT relative to CTR (1.31 vs 1.22, 1.04 vs 0.95 and 2.12 vs 1.91 kg/d, respectively; P<0.01). Although no differences were detected for milk fat yield (P=0.29), yields of energy-corrected milk tended to be higher for TRT group compared to CTR (39.80 vs 38.32 kg/d; respectively, P=0.12). These results suggest that dietary supplementation of microencapsulated organic acids and botanicals is a means to sustain lactation performance in dairy cattle experiencing heat stress.

Session 53 Poster 23

Does Zn hydroxychloride supplementation improve intestinal integrity in dairy cows?

E.A. Horst[1], P.J. Gorden[2], L.H. Baumgard[1], M.V. Sanz-Fernandez[3] and D. Brito De Araujo[4]
[1]Iowa State University, Department of Animal Science, 1221 Kildee Hall, 50011, Ames, Iowa, USA, [2]Iowa State University, Department of Veterinary Diagnostic and Production Animal Medicine, 2203 Lloyd Veterinary Medical Center, 50011, Ames, Iowa, USA, [3]Trouw Nutrition, Global R&D Department, Boxmeerseweg 30, 5845 ET Boxmeer, Noord-Brabant, the Netherlands, [4]Nutreco Nederland BV, Global Selko Feed Additives, Stationsstraat 7, 3811 MH Amersfoort, Utrecht, the Netherlands; davi.araujo@trouwnutrition.com

The objective of this study was to answer the question 'Does the substitution of Zn sulphate by Zn hydroxychloride affect positively gut permeability, metabolism, and inflammation during feed restriction? Holstein cows (n=24; 159±8d in milk; parity 3±0.2) were enrolled in a 2×2 factorial design and randomly assigned to 1 of 4 treatments: (1) *ad libitum* fed (AL) and CON (ALCON; 75 mg/kg Zn from Zn sulphate; n=6); (2) *ad libitum* fed and HYD (ALHYD; 75 mg/kg Zn from Zn hydroxychloride; n=6); (3) 40% of *ad libitum* feed intake and CON (FRCON; n=6); or (4) 40% of *ad libitum* feed intake and HYD (FRHYD; n=6). Trial consisted of 2 experimental periods (P) during which cows continued to receive their respective dietary treatments. P1 (5 d) served as the baseline for P2 (5d), during which cows were fed AL or FR of P1 feed intake. *In vivo* total-tract permeability was evaluated on d4 of P1 and on d2 and 5 of P2, using Cr-EDTA as marker. All cows were euthanized at the end of P2 to assess intestinal architecture. FR cows lost body weight (46 kg), entered into calculated negative energy balance (-13.86 Mcal/d), and had decreased MY. Circulating glu, ins, and gluc decreased, and NEFA and BHBA increased in FR relative to AL cows. Relative to AL cows, FR increased LPS-binding protein, SAA, and Hp concentrations (2-, 4-, and 17-fold); and peak SAA and Hp concentrations were observed on d5. Circulating SAA and Hp from FRHYD tended to be decreased (47 and 61%) on d5 relative to FRCON. Relative to FRCON, ileum villus height tended to increase in FRHYD cows. Feed restriction tended to decrease jejunum and ileum mucosal surface area, but the decrease in the ileum was ameliorated by dietary HYD. In summary, FR induced gut hyperpermeability to Cr-EDTA, and feeding HYD appeared to benefit some key metrics of barrier integrity.

Session 53 Poster 24

Plasma and milk metabolome in cows fed low-protein and balanced amino acids under heat stress

W. Xu[1], E. Jorge-Smeding[2], Y.H. Leung[2], A. Ruiz-Gonzalez[3,4], B. Aernouts[1], D.E. Rico[4] and A. Kenez[2]
[1]KU Leuven, Department of Biosystems, Geel, 2440, Belgium, [2]City University of Hong Kong, Department of Infectious Diseases and Public Health, Kowloon, Hong Kong, China, P.R., [3]CRSAD, Deschambault, QC G0A 1S0, Canada, [4]Université Laval, Québec, QC G1V 0A6, Canada; wei.xu@kuleuven.be

Heat stress (HS) is a challenge to the metabolic health and production performance of dairy cows. Supplementation of Lys, Met, and His may be needed to meet lactation requirements while cows ingest less metabolizable protein (MP) under heat stress. We have previously shown that balancing Lys, Met, and His supply reduces hyperthermia in dairy cows. Herein, we aimed to investigate the effects of such strategy on the plasma and milk metabolome. Twelve lactating Holstein cows (42.2±10.6 kg/d milk yield; 83±28 DIM) were randomly allocated to the following treatments in a Latin square design with 14-day treatment periods: (1) Heat stress (HS; Max. THI 82; 17% crude protein (CP); 1,715 MP, 107 Lys, 34 Met, and 37 His (g/d)); (2) pair feeding in thermo-neutrality (Con; Max. THI 64; CP, MP and AAs supply equal to HS group); or (3) HS with a diet lower in CP and MP (HS+AA; THI 82; 17% CP, 1,730 g/d MP, balanced for AAs: 178 Lys, 64 Met, and 43 His (g/d)). Heat stress was induced using a cycling pattern (THI 72-84) and resulted in decreased DMI (34%) and milk yield (40%) from d0 to d7. Blood and milk were sampled on the last day of each period and were analysed by the AbsoluteIDQ p400 metabolomics kit of Biocrates (Innsbruck, Austria). The effect of HS and AAs supply were evaluated by partial least square discriminate analysis (PLSDA). In total, 322 and 212 metabolites were quantified in plasma and milk, respectively. Both HS and the balanced AAs diet affected the plasma metabolome yielding valid PLSDA models with high predictive ability for both Con vs HS (predicted variation Q2=0.84) and HS vs HS+AA (Q2=0.76). In the milk metabolome, a valid model was found for HS vs HS+AA (Q2=0.79), but not for Con vs HS (Q2=-0.10). Significant alterations in milk and plasma metabolomes were observed in response to both HS and AAs supply in diets balanced for equal metabolizable protein supply. These changes may provide insight into the adaptions of cows to both HS and the diet.

Session 54 Theatre 1

Early life nutritional interventions in broilers as strategies to optimize performance and health

N. Everaert
KU Leuven, Department of Biosystems, Kasteelpark Arenberg 30, 3001 Heverlee, Belgium; nadia.everaert@kuleuven.be

Chickens are precocial, i.e. well-developed when they hatch, immediately being able to walk, and therefore can be reared without any contact with a laying hen. The yolk sac is a crucial nutrient source, also providing immunoglobulins for passive immunity, and continues to being used after hatch. From hatching onwards, gut colonization starts and a unique ecological niche, known as the microbiota, develops thanks to the constant temperature, availability of nutrients and the absence of oxygen. In return, these microorganisms facilitate the digestion of feed components which may otherwise remain unavailable for the host. Bacteria colonizing the gut immediately after hatch are crucial for optimal performance and health. Differences in bacterial species abundance for broilers with high versus low growth and feed efficiency exist. More and more, we consider the gut microbiota as an additional organ, impacting gut and general health, and affecting the gut-brain axis. Targeting the gut microbial colonization should therefore be done in early life, to give them a good start and to optimize performance and health. Several interventions in the perinatal period of the chickens have been developed, like *in ovo* supplementation of nutrients or health-related components like pre- and probiotics. Hatching-on-farm, allowing the chicks to have access to water and feed, immediately after emergence from the egg, is another strategy affecting gut maturation and performance of the broiler chick. This then opens opportunities to optimize the composition of a pre-starter diet. Moving one step backwards, strategies can target the diet of the breeders, like the supplementation of essential fatty acids, or altering the macronutrient levels in the diet, resulting in a maternal programming of the broilers, due to an altered availability of nutrients in the egg. During this talk, I will provide an overview of recent research outcome in this field.

Applied microbiology in poultry industry: from nutrition to the gut microbiota

N.M. Carvalho[1], D.L. Oliveira[1,2], C.M. Costa[1], M.E. Pintado[1] and A.R. Madureira[1]
[1]Universidade Católica Portuguesa, Escola Superior de Biotecnologia, CBQF – Centro de Biotecnologia e Química Fina – Laboratório Associado, Rua Diogo Botelho 1327, 4169-005 Porto, Portugal, [2]Amyris Bio Products Portugal, Unipessoal Lda, Rua Diogo Botelho, 1327, 4169-005 Porto, Portugal; ncarvalho@ucp.pt

Poultry products (i.e. meat and eggs) are one of the major protein sources for the human diet. The animal's diet is one of the key elements that the poultry industry has been focused on, to improve the animal's performance, maintaining their healthy growth and, ultimately, high quality end products. The incorporation of functional ingredients in feed formulations, aiming to provide extra benefits and/or prevent diseases, has been considered efficient in maintaining the animal's productivity and simultaneously ensure its well-being. Nutrient's bioavailability varies throughout digestion and absorption within the poultry's gastrointestinal tract (GIT). A reliable *in vitro* model, as the one developed and used in this study, capable of mimicking all digestion, absorptive and caecal fermentation processes, is a useful tool to study the potential benefits of feed supplemented with functional and/or bioactive ingredients. The developed *in vitro* gastrointestinal model simulates the chemical, enzymatic, and mechanical conditions prevailing in the chicken´s GIT, from beak to cecum. Fresh broiler's caecal samples were used as inoculum for batch caecal fermentation and the impact of different feed formulations, on bacteria modulation, organic acids, and total ammonia nitrogen production, were assessed. Overall, this approach enables to evaluate, as close to reality as possible, the potential of target additives, providing a trustworthy tool for the development of functional feeds.

Effect of sex on the performance and welfare indicators of broilers treated with vitamin D3 sources

A.C. Ogbonna[1,2], M. Mabelebele[3], L. Asher[2] and A.S. Chaudhry[2]
[1]Department of Animal Production and Livestock Management, Michael Okpara University of Agriculture, Umudike, Nigeria, [2]Agriculture Building, School of Natural and Environmental Sciences, Newcastle University, NE1 7RU, United Kingdom, [3]Department of Agriculture and Animal Health, University of South Africa, South Africa; a.ogbonna2@ncl.ac.uk

The genetic selection of broilers focused mainly on improving economic traits for faster growth rates and feed efficiency to facilitate intensified production. This has led to significant welfare issues resulting in poor bird performance. Effects of sex on performance and welfare indicators of broilers has seldom been reported for its impact on male broilers only. Therefore, this study investigated the effect of sex on growth performance, carcass, bone morphometrics and welfare indicators of broilers treated with vitD_3 sources. One day old Ross 308 broiler chicks (n=300; 150 males and 150 females) were weighed and allocated equally to the treatments to evaluate body weight gain (BWG), feed intake (FI), feed conversion ratio (FCR), meat-parts yield, bone morphometrics, gait (GS) and feather score (FS) of broilers. Each treatment had 15 replicated pens containing 10 broilers per pen. Published methods were used to observe GS and FS, while meat parts and bone of broilers were cut and weighed. The lighting schedule of 23L:1D (1-7 days) and 18L:6D (8-42 days) was used in all the treatment groups. The result presented here focused only on the sex effect. Data set were analysed using a one-way analysis of variance of SAS 9.4 software. Female broilers had a lower BWG ($P<0.05$), lower FI ($P<0.05$) but similar FCR ($P>0.05$); lower ($P<0.05$) meat parts yield, inferior ($P<0.05$) tibia and femur morphometrics, better feathering ($P=0.0001$) and better ($P=0.0012$) walking ability than the males. This study provides evidence that differences existed between males and females on growth and welfare. This appeared to be a function of differences in body weight of chickens and their innate distinct characteristics and that a rapid growth rate is negatively correlated with welfare. Further studies are needed to test the association between sex and vitD_3 for improving the behaviour, welfare and growth efficiency of broilers.

Searching for the optimal synbiotic to modulate chicken development

N. Akhavan[1], K. Stadnicka[1], G. Gardiner[2], P. Lawlor[3], K.J. Guinan[4], A. Walsh[4], J.T. O'Sullivan[4] and K. Hrynkiewicz[5]
[1]Bydgoszcz University of Science and Technology, Faculty of Animal Breeding and Biology, Mazowiecka 28, Bydgoszcz, Poland, [2]Waterford Institute of Technology, Department of Science, Cork Rd, Waterford, Ireland, [3]Teagasc Animal and Grassland Research and Innovation Centre, Pig Development Department, Moorepark, Fermoy, Co. Cork, Ireland, [4]Bioatlantis Ltd., Clash Industrial Estate, Clash East, Tralee, Co. Kerry, Ireland, [5]Nicolaus Copernicus University, Department of Microbiology, Gagarina 11, Toruń, Poland; nilakh000@pbs.edu.pl

There are two possible time-points to (immuno) modulate the chicken gut and microbiome in ovo: on day 12 of egg incubation by injecting a bioactive compound into the egg air chamber or day 18-19.5, by injection directly into the amnion. Such stimulation has a lifelong impact on chicken immunity and helps in combating the post-hatch stress. We have aimed to select the most potent prebiotics and probiotics that will guarantee an immunomodulatory effect and changes in microbiome post-hatch. Non-commercial probiotic bacteria (B1-B12) and prebiotic compounds (P1-P12) were tested to optimize candidate synbiotic combinations for in ovo application. Five *Bacillus* strains, three *Lactobacillus*, *Bifidobacterium lactis*, *Carnobacterium divergens*, *Propionibacterium thoenii* and *Clostridium butyricum* were grown separately *in vitro*, each in media with addition of the prebiotic at 2%, for 24h. The growth rate of probiotics was measured every 2h (optical density at 600 nm) using a microtiter plate reader (SpectraMax® ID3 Multi-Mode Microplate Reader by Molecular Devices®). *Lactobacillus* and *Bacillus* species revealed growth improvements with selected prebiotic compounds and may be candidates for subsequent in ovo trials. There was a strong correlation of synbiotic components with their source of origin. The post-hatch results will show how the synbiotic compositions optimized in this study, will affect in ovo development and early colonization of the embryonic intestines. Metabolomic analysis will provide a means of further understanding the function of the selected synbiotic combinations. Research was funded by the National Science Centre UMO-2019/35/B/NZ9/03186- OVOBIOM and the travel grant from EcoSET project (NAWA, Poland)

Effects of feeding strategies on faecal microbiome and endotoxin excretion in broiler chickens

F. Marcato, D. Schokker, V. Perricone, S.K. Kar, J.M.J. Rebel and I.C. De Jong
Wageningen Livestock Research, P.O. Box 338, 6700 AH Wageningen, the Netherlands; francesca.marcato@wur.nl

The gastro-intestinal tract of broilers is colonized with complex microbial communities, which play important roles in health, productivity, and diseases susceptibility of the animal. A change in the balance in the intestinal microbiome between Gram-positive and Gram-negative bacteria can result in a difference in endotoxin release. Endotoxins are cell-wall components of Gram-negative bacteria and high concentrations in faeces of broiler chickens can be a threat to both animal and human health. Two studies were carried out to investigate ways to modulate the microbiome with the ultimate goal to reduce endotoxin concentration in faeces of broilers. In addition, effects on performance, immunity, behaviour and welfare were also determined. The first study aimed at affecting the microbiota of 1,344 male broiler chickens through the use of 6 different feeding strategies in a completely randomized block design. The dietary treatments consisted of a basal diet (CON), the basal diet supplemented either with sodium butyrate (BUT), inulin (INU), medium-chain fatty acids (MCFA), Original XPC LS (XPC), and a diet with higher fibre and lower protein (HF-LP) content compared to CON. At d14, 21, and 35, pooled cloacal content was collected from 5 chickens per pen for microbiota (16S) and endotoxin concentration determination by Limulus Amebocyte Lysate assay. At d35 BUT and HF-LP tended to have higher Gram-/Gram+ ratio and higher relative abundance of Gram-negative compared to CON. Significant higher values were observed at d35 for endotoxin concentration (P=0.02) in BUT compared to CON, whereas HF-LP was only numerically higher. This study showed the modulatory effect of BUT, and HF-LP based diets on microbiota of broiler chickens, which in turn can affect the endotoxin excretion. The aim of the second study was to investigate the impact of three different factors (genetic strain, addition of probiotics to the diet and early feeding) on microbiota and endotoxin excretions of 1,248 male broiler chickens. Analyses of pooled cloacal swabs and endotoxins were conducted in the same way to the first study. Results of this second study will be presented at the conference.

Session 54 Theatre 6

Summary by the chairs of the posters in sessions 43 and 54

G. Das and K. Stadnicka
EAAP, Session, 46, Portugal; gdas@fbn-dummerstorf.de

A presentation of the posters abstracts submitted to session 46a-b will be made by chairs and authors.

Session 54 Theatre 7

Use of winter gardens with and without an automatic enrichment device by laying hens

A. Riedel[1,2], L. Rieke[1], N. Kemper[1] and B. Spindler[1]
[1]University of Veterinary Medicine Hannover, Foundation, Institute for Animal Hygiene, Animal Welfare and Farm Animal Behaviour, Bischofsholer Damm 15, 30173 Hannover, Germany, [2]University of Veterinary Medicine Hannover, Foundation, WING (Science and Innovation for Sustainable Poultry Production), Bischofsholer Damm 15, 30173 Hannover, Germany; anna.katharina.riedel@tiho-hannover.de

Laying hen housing systems offering a free-range area are perceived as animal-friendly, but the free-range is often poorly frequented. Offering an attractive winter garden (WG) with suitable enrichment could increase the using frequency. In this study, an automatic enrichment device was tested under practical conditions on an organic farm. It was installed in three different winter gardens (WGs) dosing grain on rough coated pecking plates (PPs), while another WG stayed without enrichment device. The number of laying hens (Lohmann Brown Lite) per m^2 and near the enrichment device were determined via photo records in four phases in the hens' production period. The usage behaviour differed in the four WGs and with the animals' age. Over the whole husbandry period (60 weeks), on average 1.48 hens/m^2 were detected in the WG without enrichment device, and a mean of 2.27 hens/m^2 in the enriched WGs. At the end of the husbandry period, less animals (1.43 hens/m^2) used the WGs than in the phases before (2.05-2.15 hens/m^2; $P<0.05$). Considering the time of day, the WGs were well frequented throughout the whole day. In the evening, the number of hens per m^2 continuously lowered. Regarding the number of hens around one pecking plate (PP), significantly more hens (3.16 hens/PP) were counted during the beginning of lay than during the middle and end of lay ($P<0.05$), independent of the WG. During the laying peak, a significantly higher ($P<0.05$) utilization of the enrichment device (5.11 hens/PP) was observed compared to all other phases. Maximum 14 hens used one PP simultaneously. The results indicate that the enrichment device is suitable to serve as enrichment in the WG and is well accepted by the hens. This work is financially supported by the Federal Ministry of Food and Agriculture based on a decision of the Parliament of the Federal Republic of Germany, granted by the Federal Office for Agriculture and Food (Grant number: 2817MDT200/201).

Effects of dietary supplement with *Urtica urens* on performances and serum cholesterol of broilers

V. Pinheiro[1,2,3], J. Teixeira[3], P. Nunes[3], D. Outor-Monteiro[1,2,3] and J. Mourão[1,2,3]
[1]Associate Laboratory of Animal and Veterinary Science (AL4AnimalS), Quinta dos Prados, 5001-900 Vila Real, Portugal, [2]Animal and Veterinary Research Centre (CECAV), Quinta dos Prados, 5001-900, Portugal, [3]Universidade Trás-os-Montes e Alto Douro, Departamento Zootecnia, Quinta dos Prados, 5001-900 Vila Real, Portugal; joseteixeira@utad.pt

The use of some plants with medicinal effects in animal feed has become more and more a common practice, sometimes allowing the reduction of antibiotics in animal production, due to their effects. In this work, we aimed to evaluate the effects of *Urtica urens* supplementation in broiler feed on performances, serum cholesterol and development of the digestive tract. For this purpose, 90 day-old chicks (commercial hybrids) were reared in 30 small pens (3 animals per each) and randomly divided into 3 treatments; the control diet (CT) and the control diet with dehydrated and ground *U. urens* at concentrations of 1% (UU1) or 2% (UU2). Individual live weight, feed intake, feed conversion ratio, daily feed intake and average daily gain per group were monitored between days 1 and 36 of age. On day 36 of the trial, 10 chickens of each treatment were slaughtered and was determined slaughter yield, pH, breast colour and serum cholesterol. Also, some segments of digestive tract were weighed and measured. There were differences (P=0.014) in the final live weight between the UU1 (2,522 g) vs CT (2,374 g) and UU2 (2,327 g). Also, the average daily gain was better in UU1 than CT and UU2 (68.9 vs 64.8 and 63.5 g/d, respectively). For the feed intake, conversion ratio, pH and colour there were no differences. Except in the colon length, which was lower in the UU2 treatment vs CT (P=0.045), the *U. urens* did not show differences in the others parameters of the chicken viscera. The HDL-cholesterol show differences (P=0.046), between CT (96.1 g/l) vs UU1 (88.8g/l) and UU2 (88.9 g/l). In summary, *U. urens* undoubtedly deserves more attention and study, as the positive differences recorded may have a very important economic role.

The effect of *Urtica urens* in the diet of broilers with different starting live weights

P. Nunes[1], D. Outor-Monteiro[1,2,3], J. Teixeira[1], J. Mourão[1,2,3] and V. Pinheiro[1,2,3]
[1]Universidade Trás-os-Montes e Alto Douro, Departamento Zootecnia, Quinta dos Prados, 5001-900 Vila Real, Portugal, [2]Associate Laboratory of Animal and Veterinary Science (AL4AnimalS), Quinta dos Prados, 5001-900 Vila Real, Portugal, [3]Animal and Veterinary Research Centre (CECAV), Quinta dos Prados, 5001-900, Portugal; divanildo@utad.pt

The use of *Urtica urens* in poultry diets as possible alternative to antibiotics concerning their natural properties like growth promotion and anti-oxidative effects. This plant are known as phytobiotics or botanicals considering the positive effects on animal performance and health. Poultry producers are trying to decrease the use of antibiotics and find potential alternatives in poultry diets. Thus, this research is to evaluate the potential use of *Urtica urens* in poultry feeds. For this purpose, 90 day-old chicks (commercial hybrids) were reared in 27 small pens with 3 animals and randomly divided into 3 treatments: the control diet (CT) and the control diet with dehydrated and ground *Urtica urens* at concentrations of 1% (UU1) or 2% (UU2). The groups were also divided in 3 different weights groups: light, medium and heavy. The data were analysed in two different periods, first phase (day1 until day 15) and second phase (day 16 until day 36). Individual live weight (PV), feed intake, conversion ratio, daily feed intake and average daily gain per group were monitored between days 1 and 36 of age. At day 36, the light weight group of chicks (39.01 g PV) in all evaluation parameters showed no significant differences. On the other hand, in the medium weight group (41.53 g PV) there is a positive trend in the treatment (UU1) versus (CT) at the level on final weights and average daily gains. Also, differences were observed between UU1 and CT in daily feed intake (P=0.002) and conversion ratio (P=0.006). In the heavy weight group (44.65 g PV) significant differences were found between (UU1) vs (CT) in final live weight (P=0.011), daily food consumption (P<0.0001) and feed conversion ratio (P=0.014). These results suggests positive effects of *Urtica urens* in 1% inclusion levels, on medium and heavy day-old chicks weight.

Session 54 Theatre 10

Identification of transcription factor co-operations underlying feather pecking in laying hens

S. Hosseini[1], M. Gültas[2], C. Falker-Gieske[1], B. Brenig[1], J. Bennewitz[3], J. Tetens[1] and A.R. Sharifi[1]
[1]University of Goettingen, Department of Animal Sciences, Burckhardtweg 2, 37077 Goettingen, Germany, [2]South Westphalia University of Applied Sciences, Faculty of Agriculture, Lübecker Ring 2, 59494 Soest, Germany, [3]University of Hohenheim, Institute of Animal Science, Garbenstr. 17, 70599 Stuttgart, Germany; shahrbanou.hosseini@uni-goettingen.de

Feather pecking (FP) is a behavioural disorder in laying hens that causes a serious problem for animal welfare and performance and leads to economic losses in poultry production. The causes for the propensity to perform this abnormal behaviour is influenced by a variety of factors, including environmental conditions, genetic makeup, and genotype-environment interaction. Although several genomic regions and candidate genes have been identified as being associated with FP, the precise genetic mechanisms regulating this behaviour are still not well understood. Hence, this study aimed to identify transcription factor (TF) co-operations underlying FP in laying hens under light and dark conditions in lines selected for high (H) and low (L) FP behaviour. RNA-Seq data of 48 brain samples of HFP and LFP lines of White Leghorn strains were used in this study. Differentially expressed genes (DEG) in HFP versus LFP lines under the two different lighting conditions were then used to investigate the underlying transcriptional regulatory mechanism of FP by identifying the cooperative TF pairs. We applied bioinformatics approaches using the promoter regions of DEG. The cooperation networks for significant TF pairs in each lighting condition were created using Cytoscape platform. The results showed 18 significant specific cooperative TF pairs under light condition and 15 significant specific cooperative TF pairs under dark condition, in which D-Box binding PAR BZIP transcription factor (DBP), a product of clock-controlled genes, was a highly connected TF in both networks. We also identified unique TF pairs e.g. DBP-PAX8 and DLX5-JUND under light condition and e.g. DBP-MYOD1 and DBP-TCF4 under dark condition. Our results provide new insights into the transcriptional regulatory mechanism of genes involved in FP, which may play an important role in the biological process regulating behaviour responses to environmental conditions.

Session 54 Theatre 11

Effects of dietary tall oil fatty acids on broiler performance and intestinal immunology

H. Kettunen[1], Z. Hayat[2], S. Hasan[1], J. Vuorenmaa[1] and M.Z.U. Khan[3]
[1]Hankkija Oy, Peltokuumolantie 4, 05801 Hyvinkää, Finland, [2]University of Sargodha, University Road Sargodha, Punjab 40100, Pakistan, [3]Agri-Food Research & Sustainable Solutions ARASS, Bahria Town, Lahore, Punjab, Pakistan; hannele.kettunen@hankkija.fi

Tall oil fatty acid (TOFA) contains fatty acids and resin acids of coniferous trees. When added to broiler chicken diets, resin acids reduce the inflammation-associated breakdown of collagen in gut mucosa and improve the quality of microbiota. The present 35-day study investigated the effects of TOFA with 9% resin acids on broiler chickens. Day-old male Ross 308 chicks were randomly allocated into 3 dietary treatments: T1) Positive control with antibiotic growth promoters (AGPs), T2) No AGPs, TOFA at 0.5 kg/ton, T3) No AGPs, TOFA at 1.0 kg/ton; 7 pens of 1.10 m^2/tr., and 17 birds/pen. Negative control without amendments was not included due to background challenge in this trial facility in Pakistan. Bird weight and feed intake were recorded weekly, and feed conversion ratio (FCR) was calculated. On day 21, 2 birds/pen were sampled for duodenal and ileal tissue for measuring villus length and crypt depth by routine histology, and the density of CD3+ T-cells by immunohistochemistry. Litter quality of each pen, and foot pad dermatitis (FPD) scoring of every bird were scored on day 34. Statistical analysis was conducted by ANOVA and Duncan's test. Bird weight gain, feed consumption, FCR, and mortality were similar in all treatments for days 1-35. Litter quality was improved by TOFA, with a significant ($P<0.05$) difference between T1 and T3, enforcing earlier similar findings from broiler studies with TOFA and resin acids. All treatments had similar FPD scores. Ileal crypts were deeper by TOFA ($P<0.05$) but other significant effects on histomorphology were not observed. Density of CD3+ T-cells in duodenum and ileum was significantly reduced by TOFA ($P<0.01$), indicating less inflammatory activity in gut wall in TOFA-fed birds than control birds. The results suggest an equal performance of TOFA-fed and AGP-fed broilers. Dryer litter and reduced number of CD3+ T-cells in small-intestinal mucosa suggest improved intestinal condition of TOFA-fed birds. For reducing the usage of AGPs in broiler diets, TOFA may be a promising option.

Session 54 — Poster 12

Assessment of performance and plumage condition of laying hens in a mobile house

L. Rieke[1], L. Raederscheidt[2], F. Kaufmann[2] and N. Kemper[1]
[1]Institute for Animal Hygiene, Animal Welfare and Farm Animal Behaviour, University of Veterinary Medicine Hannover, Foundation, Bischofsholer Damm 15, 30173 Hannover, Germany, [2]Faculty of Agricultural Sciences and Landscape Architecture, University of Applied Sciences, Emsweg 3, 49090 Osnabrück, Germany; lorena.rieke@tiho-hannover.de

In Germany, during the past few years, a steady rise of the number and the offer of mobile houses for laying hens is remarkable. Mobile houses are seen as an animal friendly and environmental sound alternative to static housing systems. Due to small flock sizes and increased use of the runout, mobile houses may address common problems of intensive farming regarding animal welfare, such as feather pecking. However, to date, scientific results about the benefits and challenges of keeping laying hens in mobile housing systems are rare. Therefore, in this study, we investigated the performance of laying hens (Lohmann Brown Light) in a commercial mobile house for 320 hens during two consecutive laying periods under practical conditions. Furthermore, we continuously assessed the plumage and integument condition by scoring a representative sample of 50 hens biweekly because feather loss and skin injuries are valid animal-based indicators for behavioural disorders. Data was also assessed under the obligation to keep the flocks indoors to prevent an outbreak of Avian Influenza. The impact of the absence of daily access to the runout on performance and plumage condition was also comprised in this study. First results of the study indicate that keeping conventional layer hybrids in a mobile house is promising. The performance of the Lohmann Brown Light hens in the mobile house corresponded to or exceeded the guidelines and performance goals of the breeder (Lohmann Tierzucht). However, data analysis of the plumage scoring suggested a rise in plumage loss with increasing age and during the time when the hens had to be kept indoors. The study is part of the project Hyg-MobiLe, which is supported by funds of the German Government's Special Purpose Fund held at Landwirtschaftliche Rentenbank.

Session 54 — Poster 13

The effect of untrimmed beaks on plumage condition and mortality in turkey hens

M. Kramer[1], K. Skiba[1], P. Niewind[2], C. Adler[3], N. Kemper[1] and B. Spindler[1]
[1]Institute for Animal Hygiene, Animal Welfare and Farm Animal Behaviour, University of Veterinary Medicine Hannover, Bischofsholer Damm 15, 30173 Hannover, Germany, [2]Haus Duesse, Agricultural Chamber of North Rhine-Westphalia, Haus Duesse 2, 59505 Bad Sassendorf, Germany, [3]Poultry Management and Welfare Lab, Department of Animal & Poultry Science, University of Saskatchewan, 51 Campus Drive, SK S7N 5A8 Saskatoon, Canada; marie.kramer@tiho-hannover.de

Beak trimming is a common non-curative intervention to minimize severe pecking injuries in poultry. This study aimed to show the effect of untrimmed beaks on plumage condition and mortality rate of turkey hens. Therefore, B.U.T. 6 hens with untrimmed (UT: 572 hens) and trimmed beaks (T: 572 hens) were housed in four compartments each, on one farm. In the 8th, 11th, and 14th week of life, 10 birds per compartment were scored to detect injuries, featherlessness and soiling. Losses due to mortality and separation were recorded daily. Separation management may have influenced the observed plumage condition in the herd. The highest prevalence of injured birds (UT: 27.5%; T: 10.0%) was recorded in the 11th week of life. The observed injuries (UT+T, n=15) occurred to the snood (60%), the head (27%) and the tail (13%). The highest prevalence of featherlessness (UT: 10.0%; T: 20.0%) was recorded in the 14th week of life, with a group difference at the wing (UT: 2.5%; T: 15%). The highest prevalence of soiled birds (UT: 95.0%; T: 97.5%) was recorded in the 11th week of life, with a group difference at the wing (UT: 35.0%; T: 62.5%). The predominantly minor soiling (UT+T, n=181) occurred mainly to the breast (38%), the tail (37%) and the wings (22%). The mean cumulative mortality rate in UT- compartments was 7.7% (4.9-9.8%), while in T- compartments it was 2.8% (2.1-4.1%). Concluding, keeping UT hens was accompanied by a higher mortality rate. The data suggested that a trimmed beak is more likely to lead to featherlessness in pecking events, while an untrimmed beak leads to injury. This work (Model- and Demonstration Project for animal welfare: #Pute@Praxis) is financially supported by the Federal Ministry of Food and Agriculture based on a decision of the Parliament of the Federal Republic of Germany, granted by the Federal Office for Agriculture and Food.

Session 54 Poster 14

Beaks trimmed or intact – how do turkey hens react to novel stimuli in a farm-related setting?

K. Skiba[1,2], M. Kramer[2], P. Niewind[3], J. Stracke[4], W. Büscher[1], N. Kemper[2] and B. Spindler[2]
[1]Institute of Agricultural Engineering, University of Bonn, Nussallee 5, 53115 Bonn, Germany, [2]Institute for Animal Hygiene, Animal Welfare and Farm Animal Behaviour, University of Veterinary Medicine Hannover, Bischofsholer Damm 15, 30173 Hannover, Germany, [3]Haus Duesse, Agricultural Chamber of North-Rhine-Westphalia, Haus Duesse 2, 59505 Bad Sassendorf, Germany, [4]Institute of Animal Science, Ethology, University of Bonn, Endenicher Allee 15, 53115 Bonn, Germany; karolin.skiba@tiho-hannover.de

In this study, the behaviour of turkey hens was analysed using avoidance distance (ADT) and novel object tests (NOT), which, according to the Welfare Quality Assessment Protocol, are proposed to assess fear behaviour in laying hens. The aim was to compare the behaviour of turkeys, which were either beak-trimmed (T) or non-beak-trimmed (N). 572 beak-trimmed and 572 non-beak-trimmed B.U.T.6 turkey hens were housed under identical conditions and divided evenly into eight groups. Both behavioural tests were carried out consecutively four times (2nd to 15th week). On average, there was no distinguished difference between both groups regarding the latency until the first animal was within range of the ADT-observer (T=0.1s; N=0.0s), as well as the number of animals within reach after 10 seconds (T=34; N=33). The highest average number of animals around the NOT (after 10-120s) was 55.8 at 20s for the N-, and 52.4 at 10s for the T-group. The latency until the first animal approached the object was 0.5s on average for both groups. However, the average latency until the first animal pecked at the object was 1.4s later in T (3.0s) than in N (1.6s). While the results of the ADT were comparable across groups, results of the NOT suggest a difference between T and N in the latency to peck at novel objects. Further studies are needed to confirm this assumption, as the behaviour in the applied tests might serve as an indicator for animal welfare. This work (Model- and Demonstration Project for Animal Welfare: #Pute@Praxis) is financially supported by the Federal Ministry of Food and Agriculture based on a decision of the Parliament of the Federal Republic of Germany, granted by the Federal Office for Agriculture and Food.

Session 54 Poster 15

MicroRNA expression in immune-related tissues after in ovo stimulation with bioactive substances

A. Dunislawska, E. Pietrzak, R. Wishna Kadawarage and M. Siwek
Bydgoszcz University of Science and Technology, Department of Animal Biotechnology and Genetics, Mazowiecka 28, 85-084 Bydgoszcz, Poland; aleksandra.dunislawska@pbs.edu.pl

MicroRNA is a fraction of small RNA molecules with a fundamental impact on gene expression. Mature miRNA binds to the 3'-UTRs end of the regulated mRNA molecule of the target gene, destabilizing it and preventing translation. In this way miRNA affects the target genes silencing. Single miRNA molecule is capable of regulating hundreds of target genes. Our previous research proved that modulation of the intestinal environment via bioactive substances on the day 12 of egg incubation has a significant impact on the regulation of epigenetic mechanisms in adult chicken. Based on this knowledge, we hypothesized the effect of host-microbial interaction on miRNA modulation. Therefore, the main aim of this study was to analyse miRNA activity in the spleen and caecal tonsils of two distinct chicken strains stimulated *in ovo* with bioactives. On the day 12. of incubation, eggs of Ross 308 broilers and Green-legged partridgelike were injected with probiotic- *Lactococcus lactis*, prebiotic-galactooligosacharide and synbiotic- combination of both. RNA was isolated from tissues collected post mortem on the day 42 of rearing. Analysis of miRNA expression was performed by LNA method for the selected panel of miRNA (miR-1612, miR-204-5p, miR-1674, miR-1652, miR-1598, miR-1996) and calculated using ddCt formula. *In ovo* administration of the probiotic significantly increased the expression of the analysed miRNAs in chicken broilers in the caecal tonsils and in the spleen. Analysis of Green-legged partridgelike showed a significant increase in miRNA expression in the caecal tonsils after the administration of all bioactive substances. Interestingly, in the spleen, the probiotic and the prebiotic caused a significant decrease of the miRNA expression. The increase in expression was noticeable after administration of the synbiotic in Green-legged partridgelike. It can be assumed that the probiotic component may be substantially responsible for the change in miRNA profile. Research was financed by grant UMO-2017/25/N/NZ9/01822 funded by National Science Centre (Poland). This mobility has been supported by the Polish National Agency for Academic Exchange under Grant No. PPI/APM/2019/1/00003.

Session 54 Poster 16

Effect of group size on dust bathing and reactivity of laying hens in a cage-free system

C. Ciarelli[1], G. Pillan[2], G. Xiccato[1], V. Ferrante[3], F. Bordignon[1], M. Birolo[1], F. Pirrone[2] and A. Trocino[2]
[1]University of Padova, Department of Agronomy Food Natural Resources Animal and Environment, Legnaro, 35020, Italy, [2]University of Padova, Department of Comparative Biomedicine and Food Science, Legnaro, 35020, Italy, [3]University of Milan, Department of Environmental Science and Policy, Milano, 20133, Italy; angela.trocino@unipd.it

Cage-free systems, such as aviaries, allow free movements and species-specific behaviours of laying hens, but group size in these systems can challenge the stability of hens' relationships and their behavioural patterns. Thus, this preliminary study compared the expression of dust bathing (an essential comfort behaviour) at different hours of the day and the response to a novel object (on farm measure of fear) in 1,800 brown laying hens in an aviary when in small groups (225 hens/pen) (at 43 and 45 weeks of age) and soon after grouping in a unique flock (at 46 and 48 weeks). Data were submitted to ANOVA by a mixed model with group size, observation hour (for dust bathing) as main effects, week as a random effect, and pen as a repeated measure. No effect of group size was observed on the total number of animals that touched the novel object during the test. On the other hand, the number of hens on the floor increased from the small groups to the unique flock (17.3 vs 28.1; P<0.001). This could explain the reduction of hens dust bathing (3.21% vs 1.64% of hens in small groups vs flock; P<0.001) likely because of a reduction of the individual space available for this behaviour. On average of observations on the small groups and the unique flock, the number of hens on the floor changed during the day (P<0.001) from a minimum (3.5) in the early morning (05:00) to a maximum (29.6) at 10:00, when most hens had already laid. Moreover, few hens played dust bath in the first day hours (0 at 5:00 to 0.30% hens on the floor at 7:00); dust bathing started to increase around 9:00 (1.41% hens) to reach the highest values at 11:00 and 13:00 (8.09% and 7.48%, respectively). Finally, the rate of hens dust bathing sharply decreased after 15:00. In conclusion, despite hens' reactivity did not change soon after grouping birds in a unique flock, the reduction of dust bathing in the flock compared to small-groups deserves further investigations along the production cycle.

Session 54 Poster 17

Effects of in ovo synbiotic and choline administration on transcriptomes of chicken tissues

E. Grochowska[1], Z. Cai[2], K. Stadnicka[1] and M. Bednarczyk[1]
[1]Bydgoszcz University of Science and Technology, Department of Animal Biotechnology and Genetics, Mazowiecka 28, 85-084 Bydgoszcz, Poland, [2]Aarhus University, Center for Quantitative Genetics and Genomics, Blichers Allé 20, Postboks 50, 8830 Tjele, Denmark; grochowska@pbs.edu.pl

Epigenetic modifications are essential mechanisms that fine-tune the gene expression in response to extracellular signals and environmental changes. Synbiotics and choline are bioactive substances, which are considered potential epigenetic factors. The research aimed to investigate the transcriptomes of different chicken tissues to identify those that responded to the applied epigenetic factors. The experiment was conducted on Green-legged Partridgelike chickens. Synbiotic PoultryStar® (Biomin) and choline were administrated *in ovo* on the 12th day of egg incubation. Three groups were established: 1/control (C); 2/synbiotic (S), 3/combined synbiotic and choline (SCH). Three tissues, intestinal tonsils, liver, and brain, were sampled from 7-day-old chickens (n=3 for each group). RNA-seq libraries were sequenced on an Illumina NovaSeq 6000 sequencing platform. Following quality control and trimming, the paired-end reads were mapped to the chicken reference genome (*Gallus gallus* bGalGal1.mat.broiler.GRCg7b) using a STAR v2.7.10a software. Differential expression analysis was conducted using DESeq2 v.1.34.0, an R/Bioconductor package. Effect of synbiotic and choline on tissue transcriptome was found only in the liver. The highest number of DEGs (n=95), including 36 up- and 59 downregulated genes, were determined between SCH and S groups. A total of 18 DEGs, including 8 significantly up- and 10 downregulated, were detected between the S and C groups. Moreover, 9 up- and 30 downregulated DEGs were identified between SCH and C groups. These findings indicate that early response for synbiotic and choline *in ovo* injection on the 12th day of egg incubation can be expected in the liver of 7-day-old chickens, but not in intestinal tonsils and brain. This experiment will be further continued for adult hens as well as in the next three successive bird generations. This research was supported by National Science Centre, Poland (grant no. 2020/37/B/NZ9/00497) and by the Polish National Agency for Academic Exchange (grant no. PPI/APM/2019/1/00003).

The behaviour of turkey hens during an aggressive pecking attack: a pilot study

N. Volkmann[1,2], M. Canci[1], N. Kemper[1] and B. Spindler[1]
[1]University of Veterinary Medicine Hannover, Foundation, Institute for Animal Hygiene, Animal Welfare and Animal Behaviour, Bischofholer Damm 15, 30173 Hannover, Germany, [2]University of Veterinary Medicine Hannover, Foundation, Science and Innovation for Sustainable Poultry Production (WING), Heinestraße 1, 49377 Vechta, Germany; nina.volkmann@tiho-hannover.de

Pecking is a natural investigating behaviour in poultry. This behaviour can escalate to a stage where the birds peck their conspecifics in an aggressive way and the victims can be injured seriously or even perish. Thus, such aggressive pecking is a serious animal welfare problem in turkey husbandry. The aim of this pilot study was to determine how victim birds were engaged and what kind of reactions existed. One single compartment (5.5×6 m) with turkey hens (B.U.T. 6; n=132) with untrimmed beaks was filmed two days a week (9:00-16:00 h), showing a picture excerpt of about 3×4 m. The videos were evaluated from the 10th to the 15th week of one fattening period. Data was used to evaluate the recognizable pecking attacks regarding the engaged body region as well as the reaction of the victims. In total, 130 victims were observed, which were attacked by 91 animals. In 86.2% of the offensives, the head of the victim was attacked, in 12.3% of the cases the body, and in 1.5% both regions. The most violent attack observed was that one victim was pecked by up to eight conspecifics over 9.36 minutes, resulting in a large bloody wound. This offensive finally ended with the victim got up and fled. Also for the other attacks, the most common reaction by victims was to flee the situation (67.7%). On the other hand, one third of the animals (32.3%) counterattacked or defended themselves. In this pilot study, aggressive pecking attacks occurred frequently in the observed turkey hens with numerous animals involved. Since the main reaction of the victims was to flee, the housing conditions should offer both opportunities to retreat and enough space to escape. However, further research should examine the behaviour of victims and perpetrators in larger scale studies to validate these first results.

Effect of water-soluble acidifier products on broiler crop fermentation and bacterial populations

P.T. Ward[1], S. Vartiainen[2], J. Apajalahti[2], R.A. Timmons[3], C.A. Moran[4] and I. Nikodinoska[1]
[1]Alltech Biotechnology Centre, Meath, A86 X006, Ireland, [2]Alimetrics Ltd, Espoo, 02920, Finland, [3]Alltech Inc., Kentucky, 40356, USA, [4]Alltech, Vire, 14500, France; cmoran@alltech.com

Chickens consume drinking water at ~1.6 to 2.0 times that of their feed intake. This makes water-soluble products attractive delivery routes. The objective of this study was to determine how the water-soluble acidifier product, Acid-Stat®, affects the fermentation and microbial populations of the chicken crop. In the *ex vivo* crop fermentation model, the diet was wheat-soya based feed. Microbial inoculum for the fermentation was fresh crop digesta collected from 26 to 28 day old broilers in a commercial farm. The test product was added to tap water at the recommended dose. Feed was mixed in the water to make a slurry in the fermentation vials. Fermentation was initiated when crop digesta (5%) was mixed with the feed slurry. Inoculation of the vials was conducted in random order to avoid any bias resulting from the time or freshness of the inoculum and the vials were incubated (6 hours at 38 °C). The results indicated that the test product stimulated *ex vivo* crop fermentation. All doses of Acid-Stat increased the gas production in a dose dependent manner but total microbial numbers were not affected. This indicated that the increased gas production was due to a shift in the fermentation pathway and/or changes in proportions of bacterial species. To study these alterations, the bacterial numbers from test treatments were compared with samples from the negative control. Selected DNA samples from the crop simulation were analysed with quantitative real-time PCR. The following bacterial assays were performed; *Lactobacillus* spp., *crispatus*, *salivarius*, *reuteri*; *Enterobacteriaceae* family and *Enterococcus* spp. Upon analysis, Acid-Stat did not affect total bacterial numbers in the *ex vivo* simulation, but significant changes in microbial communities were quantified. The dominating crop bacteria belonged to genera of *Lactobacillus* and *Enterococcus*. Both potentially pathogenic bacterial groups, genus *Enterococcus* and *Enterobacteriaceae*, were significantly lowered with the test product and no statistically significant differences were detected in the total lactobacilli numbers between treatments.

Session 54 Poster 20

Metabolic footprints of probiotic function in the chicken intestine

S. Zuo[1], E. Pietrzak[1], W. Studziński[1], G. Gardiner[2], P. Lawlor[3], P. Kosobucki[1] and K. Stadnicka[1]
[1]Bydgoszcz University of Science and Technology, Mazowiecka, 85-084 Bydgoszcz, Poland, [2]Waterford Institute of Technology, Department of Science, Main Campus Cork, Waterford, Ireland, [3]Teagasc, Animal and Grassland Research and Innovation Centre, Moorepark, Fermoy, Co. Cork, Ireland; sanzuo000@pbs.edu.pl

This study aimed to elucidate the mechanism of action of probiotics and prebiotics in chicken intestinal cells, by identifying the metabolic profile of probiotic function in the intestine cells *in vitro*. The candidate probiotics (n=12) having potential to modulate chicken microbiome, were selected based on previous findings, and cultivated *in vitro* with the addition of prebiotics at 2%. Two types of intestinal cell lines were used: the new chicken cell line Chick8E11 and the caco-2 cell line, as a reference. The cell lines were activated and upon reaching 80% confluence, they were co-cultured with the probiotics at a ratio of 30:1 (bacteria:cell) in 96-well plates, for 48h. The condition of cell culture was monitored with MTT, and the cells were screened to determine expression of basic epithelial markers including E-cadherin, claudin-1, and occludin. In order to determine the metabolic footprints, cell supernatants were analysed by gas chromatography-mass spectrometry (GC-MS). Targeted analysis of production levels and changes in the profile of short-chain fatty acids (SCFAs) was conducted. The obtained raw data were matched with the NIST17.L database to identify changes in the metabolite profile of intestinal cells co-cultured with probiotics. A significant change in the production of acetic acid in the probiotic-treated cells was found. SCFAs play an important role in maintaining the normal function of the intestine. The results obtained *in vitro* will be used to understand the relationship between the metabolic profile of probiotic function and the physiological response in chickens, after application of the candidate probiotic/prebiotic or synbiotic in ovo. The *in vitro* analysis of probiotic function may also be used to anticipate their effect in a host and to develop metabolomic strategies to specifically modulate chicken gut health. Research funded by the National Science Centre grant UMO-2019/35/B/NZ9/03186- OVOBIOM and the Polish National Agency for Academic Exchange (EcoSET NAWA).

Session 54 Poster 21

Use of probiotic to control *Campylobacter* in broilers

R.J.B. Bessa[1], E. Batista[1], F. Moreira[2], N. Santos-Ferreira[3], P. Teixeira[3] and M.J. Fraqueza[1]
[1]Faculdade de Medicina Veterinária, Universidade de Lisboa, CIISA, AL4Animals, Av. Universidade Técnica, 1300-477, Portugal, [2]CECA, Universidade do Porto, R. J. V. Ferreira, 228., 4050-313 Porto, Portugal, [3]Escola Superior de Biotecnologia, Universidade Católica Portuguesa, CBQF, R. Diogo Botelho, 1327, 4169-005 Porto, Portugal; rjbbessa@fmv.ulisboa.pt

Contamination of poultry carcass with *Campylobacter* is the main cause of Human campylobacteriosis. *Campylobacter* species are often found in the normal digestive microbiota of poultry. It has been hypothesized that dietary probiotic supplementation might prevent the colonization of the digestive tract by *Campylobacter*. We tested the effects of the dietary incorporation 0.5 g/kg of a commercial probiotic (Gallipro®, CHR Hansen with a *Bacillus subtilis* strain) to broilers on production traits and on microbiome patterns and Campylobacter counts. Twelve hundred ROSS 308 chicks were sexed and randomly distributed (25 males and 25 females per pen) to 24 pens with 1.2 m^2. Chicks were fed sequentially four types of diets (pre-starter, starter, growth and finishing) either without (Control) or with probiotic. Live weight and intake were registered and at the slaughter caecal content of 10 animals per pen were collected to form a composite sample for 16S RNA microbiome analysis and Campylobacter counts according to the ISO 10272-2:2017 method. The Probiotic treatment led to a slight but significant ($P<0.03$) reduction of live weight gain (-1.5 g/d) and a trend ($P=0.09$) to decrease intake (-11.5 g/d) with no effect on feed efficiency and mortality. The caecal microbiota was quite similar between treatments that did not differ on diversity (Shannon index) and on the abundance of the main species, families and phyla present. Campylobacter spp. were present in the microbiome of both treatments comprising about 0.003% of the OTUs. *Campylobacter* plate counting also did not differ between treatments and averaged 6.7 log cfu/g. Concluding, the dosage of 0.5 g/kg of the probiotic used was ineffective in preventing broiler´s digestive tract colonization by *Campylobacter*. Financial support was provided by PDR2020-1.0.1-FEADER-PDR2020-101-031254 and UIDB/00276/2021 projects.

Session 54 Poster 22

Light intensity and colour of LED lamps on performance and carcass characteristics of broilers

S. Silva[1,2], A. Tavares[2], E. Andrade[2,3], V. Pinheiro[1,2] and J. Mourão[1,2]

[1]CECAV-Associate Laboratory of Animal and Veterinary Science (AL4AnimalS), Quinta dos Prados, 5000-801, Portugal, [2]University of Trás-os-Montes e Alto Douro, Quinta dos Prados, 5000-801, Portugal, [3]Federal University of Bahia, Department of Animal Science, Salvador 40170110, Bahia, Brazil; ssilva@utad.pt

This study was carried out to investigate the effects of the intensity and colour of LED lamps on the performance and carcass characteristics of broilers. A total of 300 day-old broilers (Ross 308) were evaluated (150 females and 150 males) for 40 days. The broilers were randomly distributed in 15 independent pavilions, with 2 m^2 each, equipped with a temperature, ventilation, photoperiod, and light intensity control system; it also had a feeder and drinkers. Three different treatments, light-emitting diode (LED) lamps with different colours and intensities, were defined: white, 20 lux (W20); green, 20 lux (G20) and white, 100 lux (W100). The feeding program (*ad libitum*), luminosity, ventilation and temperature were carried out according to the management guide Ross 308. Individual live weight and feed intake per pavilion were measured weekly (days 7, 14, 21, 28, 35 and 40) to determine the average daily gain (ADG), average daily intake (ADI) and feed conversion (FC). On the 40th day, two animals per pavilion (10 animals per treatment of both sexes) were slaughtered. The carcasses were weighed and refrigerated at 4 °C for 8 hours. After cooling, the carcass, wing and leg were weighed. Subsequently, pH (thing and breast) and colour (CIE System, breast) were measured. Slaughter weight, carcass yield, and leg and wing yield were also determined. An ANOVA was performed to analyse the effects of the colour and intensity of LED light on performance and carcass traits. Means were compared by Tukey's test ($P<0.05$). The results showed that the W100 treatment presented lower results ($P<0.021$) for body weight at 7, 28, 35 and 40 days. Broilers submitted to the W20 treatment had higher ($P<0.002$) slaughter and chilled carcass weights when compared to the W100 treatment. In general, no effect was observed for ADG, ADI, FC, and carcass traits. In conclusion, controlling the light intensity using dimmable LED lamps can be an economical and productive solution for broilers.

Session 54 Poster 23

Effects of probiotic dietary supplementation on broiler performance in normal and low protein diets

V. Dotas[1], S. Savvidou[2], I.A. Giantsis[3], A. Athanasiou[4], M. Müller[5] and G.K. Symeon[2]

[1]Aristotle University of Thessaloniki, School of Agriculture, Thessaloniki, GR54124, Greece, [2]ELGO-DEMETER, Research Institute of Animal Science, Giannitsa, 58100, Greece, [3]University of Western Macedonia, Faculty of Agricultural Sciences, Florina, 53100, Greece, [4]Ravago Chemicals Hellas SA, Neratziotissis 115, 15124 Maroysi, Greece, [5]Evonik Operations GmbH, Nutrition & Care, Rodenbacher Ch 4, 63457 Hanau, Germany; gsymewn@yahoo.gr

Probiotics are live microorganisms that when administered in adequate amounts confer a health benefit on the host. The current study evaluated the dietary inclusion of *B. amyloliquefaciens* on broiler performance, gut health, and expression of genes encoding digestive enzymes in normal and low-protein diets. 384 male, day-old broilers were randomly allocated to 4 treatments, consisting of 8 replicates of 12 birds each. Experimental treatments were: C, fed a standard commercial diet throughout the experiment; CE, where the ratios were supplemented with the probiotic; LP, where the broilers received grower and finisher ratios with 2% reduced protein content; LPE, where the ratios offered to the previous group were supplemented with the probiotic. Live weight and feed intake were measured pen-wise at the end of each feeding period (days 10, 24, and 42). 24 hours after slaughter (42 days) the weights of cold carcass, carcass yield, breast, thigh, drumstick, wing, and abdominal fat were recorded. Average daily gain and feed conversion ratio were calculated respectively. Data were statistically processed by means of variance and Tukey's test. At 42 days of age, the probiotic supplemented groups (CE & LPE) were heavier than their non-supplemented counterparts while the LPE had comparable weight to the C and CE groups, despite the lower dietary protein. The reduction of protein levels increased feed intake, which was not affected by the probiotic supplementation. Both treatments had little effect on carcass parameters. Digestive genes' mRNA levels exhibited a decrease in probiotic treatments, implying a reduction in metabolic activity in line with the microorganisms' activity increase. Overall, the probiotic dietary supplementation resulted in better utilization of feed nutrients and thus improved broilers' performance.

Tight junction protein expression in response to seaweed supplementation in two laying hen strains

L.A. MacLaren, J. Wang, S. Borzouie and B.M. Rathgeber
Dalhousie University, Animal Science & Aquaculture, Truro, Nova Scotia, B2N 5E3, Canada; leslie.maclaren@dal.ca

Diet supplements such as seaweed may improve the integrity of intestinal epithelium by increasing the expression of selected tight junction proteins, including occludin and ZO-1. We used immunohistochemistry and western blotting to test this hypothesis by examining jejunum of two commercial strains of laying hens (Lohman LSL-lite (White) and Lohman Brown-lite (Brown)) supplemented or not with 3% *Chondrus crispus* or 0.5% *Ascophyllum nodosum* seaweeds from 31 to 42 weeks of age. Cryosections were scored for intestinal thickness and examined for occludin and ZO-1 distribution in the villus and crypt epithelia. Intensity of staining was scored from 1 (no staining) to 5 (most intense). Occludin was observed along the lateral surfaces of epithelial cells in all animals, with weaker reactivity observed closer to the lumen. Cryptal epithelium appeared to stain more intensely than villus epithelium in most sections, but the Mann-Whitney comparison did not confirm this observation ($P>0.05$). Reactivity for ZO-1 was observed in all animals regardless of treatment or strain. The pattern was slightly different than for occludin, with lateral epithelial cell borders staining most intensely close to the apical cell surface. Reactivity for ZO-1 was less evident in the deeper crypts than villi. There were no seaweed treatment effects on intestinal thickness or staining intensity detected for either occludin or ZO-1 by immunohistochemistry. The Mood's median test indicated that the villus epithelium of White hens may express more occludin (median intensity 3.5 vs 2.5 in Brown hens, $P=0.06$) but less ZO-1 in the deep cryptal epithelium (median intensity 1.5 vs 2.5 in Brown hens, $P=0.06$). Western blotting of jejunal protein extracts also showed higher levels of occludin in White than Brown hens ($P<0.05$). A decrease in ZO-1 expression in jejunal protein extracts was associated with *Chondrus crispus* supplementation in comparison to controls ($P<0.05$), but not with *Ascophyllum nodosum* supplementation. In conclusion, genetic strain and dietary seaweed supplements may affect tight junction protein expression levels in jejunum but effects are modest and do not impact the anatomical distribution as seen in cryosections.

Effect of botanical compounds fed to breeders on laying and chicks' performances

C. Carlu[1], T. Chabrillat[1], C. Alleno[2], C. Manceaux[2], R. Bouvet[2] and S. Kerros[1]
[1]Phytosynthese, 57 avenue Jean Jaurès, 63200 Mozac, France, [2]Zootest, 5 rue Gabriel Calloet-Kerbrat, 22440 Ploufragan, France; claire.carlu@phytosynthese.fr

The study aims to evaluate the effect of Phyto Ax'Cell (PAC), a plant based feed supplement, on breeder performances, hatching and chick quality. PAC is a natural mixture composed of green propolis, salicylates derivatives, curcuminoids and polyphenols. This product was distributed on broiler breeders during 10 weeks from weeks 30 to 40. Experiment was carried out on 2 treatments, each treatment was divided into 15 repeats of 24 females and 56 males/treatment. Experimental group received PAC at 0.1% in feed and control group (C) did not received any supplementation. Artificial insemination was implemented. Reproductive parameters were recorded on males and females. At 35 and 40 week-old, 2,400 eggs/treatment were sampled in order to compare hatching performance; uncleared eggs and hatchability rate. Uncleared eggs are eggs with embryo development. Day old chick performance (DOC) was measured on 50 chicks × 10 pens per group during the first 7 days of age coming from hens at 2 dates (35 and 40 weeks of parents' age). Relative growth (RG) was calculated according to pens weight between day 0 and 7. Next, immunoglobulins (IGG) were determined by blood sample at day 1 and bursa of Fabricius weight was measured on 12 chicks at day 7. Laying rate was similar with respectively 81.1 and 81.2% for C and PAC. Uncleared eggs rate was significantly improved with PAC (89.1 vs 87.5%; $P=0.013$). Hatching rate was significatively increased with PAC (84.0 vs 82.5%; $P=0.049$). For DOC, RG showed a better trend in group fed PAC (157.8 vs 153.2%; $P=0.040$). IGG was also increased with 3.21 g/l for chicks from PAC parents and 2.81 g/l for the control group. Moreover, bursa of Fabricius weight was numerically increased in comparison to the control group (0.225 vs 0.218 g; $P=0.39$). From all above, PAC showed a significant improvement of fertility and hatching. In addition, a positive impact on RG of chicks was observed in PAC group. The antioxidant properties of PAC polyphenols associated with immune modulation effect of green propolis may explained these results. Other studies could be performed to confirm PAC benefices on breeder all long the production and on chicks' performances in stress conditions.

Session 54 Poster 26

Productive characterization of the Portuguese autochthonous 'Branca' chicken breed

M. Meira[1], I.M. Afonso[1,2], J.C. Lopes[1,2], V. Ribeiro[3], R. Dantas[3], J.V. Leite[3] and N.V. Brito[1,2,3,4]
[1]ESA-IPVC, Rua D. Mendo Afonso, 147, 4990-706 Refóios do Lima, Portugal, [2]CISAS-IPVC, Rua Escola Industrial e Comercial de Nun'Álvares, 34, 4900-347 Viana do Castelo, Portugal, [3]AMIBA, R. Domingos Marques, 40, 4730-260 Vila Verde, Portugal, [4]TOXRUN, University Institute of Health Sciences (IUCS), Gandra, 4585-116 Gandra, Portugal; marciomeira@ipvc.pt

The conservation of autochthonous breeds is a topic of great importance for the maintenance of local animal genetic resources, contributing to sustainable production systems, and the preservation of a unique genetic heritage. Portuguese chicken breeds are almost extinct, with Branca presenting the most worrying situation. Bred on small-scale farms, double purpose, the characterization of products is fundamental to the survival of this population. The hens laying performance was conducted under field conditions, for 4 years, comprising 501 hens and 13 flocks. Meat quality parameters: pH, colour, protein, and lipid contents were estimated in 20 animals of each sex, in the 2 most economically valued pieces, breast and drumstick. The Branca breed showed a laying performance of 82±25.6 eggs/year. The laying activity differed (P≤0.001) between months, with a noticeable laying peak between March and June (40.5%). Significant (P≤0.001) egg production reductions were observed in June (21.2%), the consequence of natural hatching, and between November and January (21.3%), due to environmental factors (temperature and photoperiod). Related to meat, no sex significant effects were observed related to pH, reviling drumstick significant higher value. Hens showed significantly (P≤0.001) lighter breast, and no significant effect of sex was observed concerning redness (a*) and yellowness (b*). Colour is influenced by muscle type, showing drumstick low L* value and high red index (a*) in both sexes. The breast meat had higher protein and lower lipid contents when compared to the drumstick (P≤0.001) in both sexes. Sex significantly (P≤0.001) affected the lipid content in both muscles (males<hens). The preservation and characterization of local breeds, as Branca breed, in a low-input environment, is highly relevant, offering unique features and valuable quality traits, in the demand for alternative products with non-conventional quality.

Session 54 Poster 27

Influence of mixing and storage on the quality of components in the laying hen feed mix

M. Grubor[1], S. Zjalic[2], A. Matin[1] and T. Kricka[1]
[1]University of Zagreb, Faculty of Agronomy, Svetosimunska cesta 25, 10000 Zagreb, Croatia, [2]University of Zadar, Mihovila Pavlinovica 1, 23000 Zadar, Croatia; szjalic@unizd.hr

The production of feed mixture must meet the requirements set by the recipe. The basis of this is, from a technological point of view, the correct ratio, i.e. particle size uniformity, accurate metering device and mixing device. Often such devices do not follow the requirements of the recipe, so the aim of this paper is to monitor the movement of the variable coefficient throughout the process and based on the results to determine the layering that occurs in the production of feed mixtures. The research was conducted in the feed factory on the feed mixture for laying hens, and in the process samples were taken in the bunker above the mixer, the bunker below the mixer, on the chain conveyor at the silo cell entrance and at the silo cell exit. The values of particle size, moisture, protein, ash and fibre in the mixture and the mixing of the mixture with salt (NaCl) at 75 t of the mixture were examined. From the uniformity model, it was obtained that the particle size of the middle fraction was 70%, which can determine that the mill is working properly, but the ratio of coarse and fine fraction is not completely accurate. In the case of laying hens feed mixture, the coefficient of variation decreased, i.e. it was 4.37% for bunker under the mixer, 4.37% at the entrance to the silo cell, and 3.22% at the exit from the silo cell. Differences in chemical composition, i.e. the quality of the mixture between the individual sampling points on the way from the mixer to the exit from the silo cell is not statistically significant, which proves that the individual points on this transport path, in the observed factory, are properly adjusted. Mixture decomposition did not affect the chemical composition of the mixture. Differences in the weights of individual components of the mixture are due to insufficient adjustment of the dispenser and scales, which can be eliminated by installing newer dispenser systems. Based on all the above, it is possible to conclude that in the observed factory the technological processes were set up correctly and thus achieved quality assurance of the feed mixture during the entire production process from mixing to storage.

Session 54 — Poster 28

Growing performance of the Portuguese autochthon 'Branca' chicken breed

C.M. Maia[1], J.L. Cerqueira[1,2,3], J.P. Araújo[1,2,4], F.M. Fonseca[5], F.J. Mata[2] and M.L. Soares[1,2]
[1]ESA, Instituto Politécnico de Viana do Castelo, Refóios, 4990-706 Ponte de Lima, Portugal, [2]C. Invest. e Des. Sistemas Agro-alimentares e Sustentabilidade (CISAS), Viana Castelo, 4900-347 Viana Castelo, Portugal, [3]Centro de Ciência Animal e Veterinária, Q. Prados, 5001-801 Vila Real, Portugal, [4]Centro de Investigação de Montanha, Refóios, 4990-706 Ponte de Lima, Portugal, [5]De Heus, Trofa, 4785-682 Trofa, Portugal; cerqueira@esa.ipvc.pt

Many local poultry breeds are on the brink of disappearing. One of those is 'Branca', one of the four Portuguese autochthonous chicken breeds. This study aimed to contribute to the characterisation of its growing patterns, namely live weight (LW) and average daily gains (ADG) over a 195-day period. Gompertz models were fitted to the average LW, and 95% confidence intervals were also calculated, and model adjusted. Data were also analysed with ANOVA type models. n=40 identified birds induced with a starter, were randomly allocated at day 13, into 2 different flocks (FP_1, FP_2). These were subjected to the same husbandry and stocking density (0.45 m^2/beak) but were fed different feeding programmes (FP) (maize – FP_1 11♂ e 9♀ and a standardly designed ration – FP_2 12♂, 8♀). Birds were always fed *ad libitum* and were weighted at day 13, and weekly thereafter. Growth diversion occurred from day 34 (P<0.05) for gender and from day 41 for FP (P<0.01), FP_1 –724.0±81.8 g♂ and 604.2±85.0 g♀ vs FP_2 – 699.8±66.2 g♂ and 594.6±75.0 g♀. ADGs at this stage of growth are respectively FP_1 – 22.0±3.2♂ and 17.8±3.2♀. vs FP_2 – 20.9±2.7♂ and 17.3±2.6♀ g/day. The inflection points of the Gompertz curves, take place at 48, 56, 60 and 49 days respectively for FP_2♂, FP_2♀, FP_1♂, and FP_1♀. From these points onwards, the birds grow with decreasing accretes. At day 195 the differences enlarged (P<0.001) FP_1 – 2,311.0±409.8♂ and 1,973.3±428.8♀ vs FP_2 – 3,105.7±210.6♂ and 2,096.0±505.9 g♀. From day 41 to 195 ADG are also different (P<0.001), FP_1 – 10.4±2.3♂ and 8.9±2.4♀ vs FP_2 – 15.6±1.3♂ and 9.8±3.0♀ g/day. Males on ration grow at faster paces and mature earlier, while females fed maize have slower growth and mature later. Males on maize and females on ration have intermediate growing patterns while comparing with the previous groups, and similar while comparing with each other.

Session 54 — Poster 29

Physical, chemical and sensory analysis of brown-shelled vs white shelled eggs

P. Sousa, C. Martins, M. Mourato, L. Louro, S. Ferreira-Dias and M. Lordelo
Instituto Superior de Agronomia, LEAF Linking Landscape, Environment, Agriculture and Food, Instituto Superior de Agronomia, 1349-017 Lisboa, Portugal; pedroborgessousa@hotmail.com

Brown eggs have been perceived by the consumer to be more natural or healthier than white eggs. The current study intends to clarify the scientific community and consumers about the quality of brown and white eggs. This is important in markets where brown eggs are favoured, because white eggs and white laying hens, are more suitable for industrial egg production than brown laying hens. A total of 90 brown and white eggs were collected from a commercial farm. Eggs were laid on the same day and flocks were of the same age and both housed in furnished cages. Different feed formulations were given to the white and brown breeds since nutritional requirements were also different. 10 brown and 10 white eggs were individually weighed and candled to determine the percentage of eggs with pre-cracks after being laid. The egg shape index, air cell height, and shell, yolk and albumen percentage were recorded and the yolk colour was evaluated. The height of the thick albumen was determined for Haugh unit´s calculation. Albumen and yolk pH were analysed. Protein, fat, and mineral content of the albumen and yolk were examined, as well as the mineral content of the eggshells. A sensory evaluation was performed with brown and white egg samples by a panel of 70 consumers in blind triangular tests. Results indicated that, for the same flock age, brown eggs were heavier than white eggs (P<0.05). No differences were found in the egg shape index, yolk colour and Haugh units between the white and brown eggs. The pH of the yolk and albumen were also not different between the two genotypes. Moreover, the percentage of albumen was higher in the brown eggs and, surprisingly, the percentage of shell was higher in the white eggs (P<0.05). On the other hand, the percentage of fat in the yolk was higher in the brown eggs but the percentage of protein in the albumen and in the yolk were not different between white and brown eggs. Consumers preferred white eggs (P<0.05). Overall, the main differences were a greater percentage of shell in white eggs and a higher percentage of fat in the yolk of brown eggs, which are differences that can be attributable to the breed and/or the feed.

Session 54 Poster 30

Effect of *Urtica urens* in broiler feed on the development of the digestive tract, serum cholesterol

J. Teixeira[1], V. Pinheiro[1,2,3], P. Nunes[1], D. Outor-Monteiro[1,2,3] and J. Mourão[1,2,3]
[1]Universidade Trás-os-Montes e Alto Douro, Departamento Zootecnia, Quinta dos Prados, 5001-900 Vila Real, Portugal, [2]Associate Laboratory of Animal and Veterinary Science (AL4AnimalS), Quinta dos Prados, 5001-900 Vila Real, Portugal, [3]Animal and Veterinary Research Centre (CECAV), Quinta dos Prados, 5001-900, Portugal; joseteixeira@utad.pt

From a medicinal point of view, *Urtica urens* can be an excellent additive in animal feed as it has antibacterial, antioxidant and anti-inflammatory properties. Some studies suggest a positive impact of *U. urens* on the immune system, development of the digestive tract, regulation the appetite and voluntary intake, stimulation of basal metabolism, increasing the quality of meat and eggs and is a good source of vitamins. In this work, we aimed to evaluate the effect of adding *U. urens* in broiler feed and its effect on the development of the digestive tract and serum cholesterol content. For this purpose, 90 day-old chicks (commercial hybrids), were reared in 30 small pens and randomly divided into 3 treatments; control (CT), control diet with dehydrated and ground *U. urens* at concentrations of 1% (UU1) and 2% (UU2). On day 36 of the trial, 10 chickens of each treatment were slaughtered and the different segments of the digestive tract were weighed and measured. At the time of slaughter, blood was collected and total HDL and LDL cholesterol were analysed. Except in the colon length, which was lower in the UU2 treatment vs CT (P=0.045), the *U. urens* did not show differences in the remaining parameters of the chicken viscera. However, in HDL-cholesterol there were differences (P=0.046), between CT (96.1 g/l) vs UU1 (88.8 g/l) and UU2 (88.9 g/l). This result, although it needs a more robust scientific support, suggests that this cholesterol reduction due to the use of *U. urens*, may also be applied in the future to other species.

Session 54 Poster 31

Supplementing layers with Lactobacillus rhamnosus influences feather damage and the immune system

C. Mindus[1], N. Van Staaveren[1], D. Fuchs[2], S. Geisler[2], J. Gostner[2], J. Kjaer[3], W. Kunze[4], A.K. Shoveller[1], P. Forsythe[5] and A. Harlander[1]
[1]University of Guelph, Guelph, N1G2W1, Canada, [2]Medical University, Innsbruck, 6020, Austria, [3]Institute of Animal Welfare, Celle, 29223, Germany, [4]McMaster University, Hamilton, L8S4L8, Canada, [5]University of Alberta, Edmonton, T6G2R7, Canada; aharland@uoguelph.ca

Human disorders like depression and anxiety are associated with a dysregulated microbiome and immune system. Microbial and neuroimmune-mediated factors in behavioural problems is thus receiving greater attention in animal research. Severe feather pecking (SFP) has been associated with changes in the microbiome which has revealed an exciting field of research investigating targeted microbiome treatments, like probiotics, on pecking behaviour. We studied the effect of an oral Lactobacillus rhamnosus supplement on stress-induced feather damage (FD), fearfulness, and T cell populations in laying hens. Eighty-six hens (33 weeks of age/woa) received L. rhamnosus in water (Lacto, 6 pens) or a placebo (Placebo, 6 pens). Three pens from each group received a 3-week chronic social stress regimen to induce SFP. Tonic immobility was used to assess fearfulness at 36 woa, and FD was scored from 0 (no or slight wear, nearly intact feathering) to 2 (featherless area ≥3 cm) at 37 woa. Proportions of T helper cells, cytotoxic T cells and regulatory T cells in the spleen and caecal tonsils were measured at 38 woa. Generalized linear mixed models were used to assess treatment (Lacto, stressors) effects on FD, fearfulness, and T cell populations. Social stress increased FD (P<0.05); 74% of stressed birds had FD (score>2) compared to only 50% of non-stressed birds. Birds in the Lacto treatment tended to have less FD under stress as only 51% of Lacto birds exhibited FD (score>2) relative to 74% of Placebo birds (P=0.074). No effect of Lacto on fearfulness was observed. Lacto increased the regulatory T cell proportion in the caecal tonsils (P<0.001) and spleen (P<0.001) compared to Placebo. The T helper and cytotoxic T cell proportions were unaffected by Lacto or stress (P>0.05). Results suggest that plumage damage from SFP may be influenced by the immune system. We show the beneficial effect of L. rhamnosus on the avian immune response which could be used to deter plumage damage in laying hens.

Session 55 Theatre 1

Appropriate drone flight altitude for horse behavioural observation

T. Saitoh and M. Kobayashi
Obihiro University of Agriculture and Veterinary Medicine, Field Center of Animal Science and Agriculture, Inada-cho, Obihiro, Hokkaido, 080-8555, Japan; tsaitoh@obihiro.ac.jp

Recently, drone technology advanced, and its safety and operability markedly improved, leading to its increased application in animal research. Drones are used extensively in several fields, including remote sensing, determining ecosystems' complexity, and disaster countermeasures. Drone application has many benefits; drones are inexpensive and relatively easy to operate compared with aerial photography, which requires airplanes and helicopters. Therefore, their application is increasing in the field of animal research. This study demonstrated drone application in livestock management, using its technology to observe horse behaviour and verify the appropriate horse–drone distance for aerial behavioural observations. In this experiments, recordings were conducted from September to October 2017 on 11 horses using the Phantom 4 Pro drone. Eleven horses kept at the Obihiro University of Agriculture and Veterinary Medicine (Obihiro, Hokkaido) were used for experiment. These horses were recorded with only one other pair in a pasture. Four flight altitudes were tested (60, 50, 40, and 30 m) to investigate the reactions of the horses to the drones and observe their behaviour; the recording time at each altitude was 5 min. Direct continuous observations were made by one observer familiar with the focal horses. None of the horses displayed avoidance behaviour at any flight altitude, and the observer was able to distinguish between any two horses. Recorded behaviours were foraging, moving, standing, recumbency, avoidance, and others. Foraging was the most common behaviour observed both directly and in the drone videos. The correlation coefficients of all behavioural data from direct and drone video observations at all altitudes were significant ($P<0.01$). These results indicate that horse behaviour can be discerned with equal accuracy by both direct and recorded drone video observations. In conclusion, drones can be useful for recording and analysing horse behaviour. However, further study is needed to clarify behaviours that are not targeted in this study, and experiments that seek a more direct way to the practical application of drones to horse management are required.

Session 55 Theatre 2

Biothermo ID microchips for temperature e-monitoring in horses

C.B.L. Scicluna
EquInstitut, Haras du Plessis, 60300 Chamant, France; clinvetplessis@wanadoo.fr

Body temperature is crucial in inflammation, infection and other stress conditions in equine medicine, but rectal measurement is challenging and time consuming. Microchiping horses is mandatory in Europe for identification and Allflex introduces temperature monitoring with microchips, readable with ID number and recordable in a cloud through a simple application. This preliminary study checked interest, sensitivity and reliability of body temperature monitoring of horses through ID microchips positioned IM in the horses'neck. Rectal (RT) and Chip Temperature (MCT) were compared in 4 hospitalized horses, 5 to 18 times/day for each. The first 210 collected records were analysed all together but also individually for means, SD and t test. Mean values and SD were 37.64 °C (±0.19) for RT and 37.77 °C (±0.24) for MCT, and total mean difference – 0.14 °C (±0.22). There was no significant difference between RT and MCT values ($P<0.001$). Individual temperatures varied similarly daily, which could be interesting for follow up. Identified variations seemed to be correlated with special conditions (exercise, manipulations, other medical reasons). Collecting data from chips was easier, safer and faster than taking RT. The application allowed an easy, distant and complete connected follow up and analysis. Data are still collected and analysed, but Biothermo microchips monitoring already appears to be a nice interesting reliable tool to be used in many medical, and equine surveillance or emergency situations. Biothermo and its application will certainly contribute to the veterinary e-health development in horses and enrich medical knowledge of equine diseases.

Digitalization and animal welfare in horse husbandry

M. Pfeiffer and L.T. Speidel
HfWU Nürtingen-Geislingen, Hechingerstraße 12, 72622 Nürtingen, Germany; linda.speidel@hfwu.de

Horse husbandry in Baden-Württemberg has become a mainstay of operations for many farms, and an oversupply of horse boarding facilities has been observed over the past 20 years. The development of the horse population has also led to the fact that in the densely populated areas of Baden-Württemberg, about every fifth farm keeps horses. The organization of a horse stable is very complex and extensive. In addition, keeping horses requires a lot of energy, time and labour. Digitally controlled mechanization of the labour-intensive operating processes of a horse farm could provide considerable relief for the responsible operators. At the same time, automated and digital systems also benefit horses, as procedural solutions can improve the quality of horse husbandry. With the help of digitalization, horse husbandry can also be made more animal-friendly, which plays a role not only against the background of the introduction of the farm manager's self-monitoring obligation based on animal welfare indicators. Discussions about the animal welfare of different husbandry systems and the well-being of the animals have steadily increased in recent years. There is a general call for greater consideration of animal welfare and the animal as an individual. In the research on digitalization in horse farms from an animal welfare perspective, the focus is on ensuring the welfare and health of horses with the help of digitalization and on the response of horses to the use of technical systems based on animal welfare indicators. The research involves the recording of the daily time budget as well as the chewing and walking activity of horses in different housing systems at different times of the year on the practice farms. In addition, the stress level of horses during manual and automated feeding is investigated by heart rate measurements and direct observation of alarm signals according to Zeitler-Feicht (2011).

Donkey comfort at work

D. Fouvez[1], C. Bonnin[1], C. Briant[2] and S. Biau[3]
[1]INET, 17 Crs Xavier Arnozan, 33000 Bordeaux, France, [2]PRC, INRAE, CNRS, IFCE, Université de Tours, 37380 Nouzilly, France, [3]IFCE, Av. de l'ENE-BP 207, 49411 Saumur, France; domitille.fouvez@gmail.com

In France, donkeys are increasingly present in farms as workers for market gardening. As far as we know, few studies have focused on the comfort/discomfort of the donkey at work. In this context, it's relevant to focus on the donkeys' equipment, with a final objective of designing an innovative collar for donkeys at work. First step of this project was to collect missing data about donkeys at work, i.e. to objectively describe mechanical effort (pulling force), energy expenditure (heart rate HR) and behaviour of donkeys at work. The main hypothesis is that the more the pulling force required increases, the more we expect the HR to increase and discomfort indicators to appear. Also, the effort required by donkeys at work seems mainly moderate (low HR and pulling force) with mostly comfort indicators. Eight donkeys from market gardening farms were tested under their regular working conditions (12±5,9 years old, 302±74 kg, mean ± SD). Donkeys were equipped with a HR monitor (Polar®) to measure HR. Pulling force was measured with a dynamometer (Sensel®/CAIPS). Behaviour was analysed from video recordings by scan sampling method based on published ethograms for donkeys and horses at work. The results showed that the mean heart rate at work remained low (83.5±15.7 bpm, mean ± SD) even when the pulling force was high. Two types of works require an important pulling force: preparing the soil and hoeing+ridging. For hoeing alone, the mean pulling force required represents 11±3.2% of the donkey's weight whereas for preparation of the soil and hoeing+ridging the pulling force required represents 21±4.6% of its weight. In the literature, an effort requiring more than 15% of donkey's weight is important for a full day of work. In all the measures, only few behavioural discomfort indicators were observed. The most observed behavioural comfort indicators were: beyond vertical forehead, opened eyes, relaxed nostrils, base of the ears above the withers, closed mouth and tail lifted from the hindquarters.

Session 55 — Theatre 5

Combining novel motion technology and genotype data in an Italian horse native breed

V. Asti[1], A. Sabbioni[1], S. Menčik[2], A. Piplica[2] and M. Ablondi[1]

[1]University of Parma, Department of Veterinary Science, Strada del Taglio 10, 43126, Parma, Italy, [2]University of Zagreb, Department of Animal Breeding and Livestock Production, Heinzelova 55, 10000 Zagreb, Croatia; vittoria.asti@unipr.it

With the advent of genomics, nowadays there is a possibility to further understand the genetic background of economically relevant traits in horses. However, in the case of highly polygenic and complex traits, such as movement traits, this achievement might be more challenging. Application of Genome-Wide Association Studies (GWAS) was successful, especially for highly heritable traits, such as conformation and disease-related traits. Another important aspect to consider is the availability of data collected in a large cohort of animals; indeed the GWAS approach has been mainly applied in studies with great population size. Nevertheless, combining novel motion technology, which allows more objective measurements, with genotype data might help to further unravel the genetic background of movement traits also in breeds with unavailability of large number of genotyped horses. Thus, in this study, we aimed to perform a GWAS analysis in the Bardigiano horse breed, an Italian breed traditionally used in agriculture. We used the Equisense motion® sensor, which consists of a gyroscope and an accelerometer to objectively record several gaits-related traits linked with the horse's health, performance, and physical condition. A total of 185 horses were genotyped with the GGP Equine 70k ® SNP chip. The quality control of the genotype data was performed in Plink v 1.9 and the GWAS analysis in R Software. To assess the association with genotype information, a linear mixed model was used, corrected for sex, the sampling date, and for biometrical measurements. We mostly found significant results on the trot gait, for example, for the symmetry we found 9 significantly associated SNPs. In the proximity of those SNPs (windows size±250 kb) 22 protein-coding genes, 3 miRNA, and 7 lncRNA were identified. Some of those genes have known function in collagen structure and the control of the synapsis response. Further analysis of those genes might permit to enhance the use of genomics in the Bardigiano horse breed which is facing the conversion from agricultural to riding purposes.

Session 55 — Theatre 6

Growth characteristics of warmblood horses as predicted by non-linear functions

M.J. Fradinho[1,2], D. Assunção[3], A.L. Costa[3,4], C. Maerten[3], V. Gonçalves[3], A. Teixeira[3], M. Bliebernicht[3,4] and A. Vicente[1,2,5]

[1]Associate Laboratory for Animal and Veterinary Sciences (AL4AnimalS), UTAD, 5000-801 Vila Real, Portugal, [2]CIISA, Centro de Investigação Interdisciplinar em Sanidade Animal, Faculdade de Medicina Veterinária, Universidade de Lisboa, 1300-499 Lisboa, Portugal, [3]Pôle Reproduction–Haras de la Gesse, Boulogne-sur-Gesse, 31150 Boulogne-sur-Gesse, France, [4]Embriovet, Lda., Muge, 2125-348 Muge, Portugal, [5]ESAS, Santarém, 2001-904 Santarém, Portugal; mjoaofradinho@fmv.ulisboa.pt

The knowledge of adequate growth rates for each breed and purpose, is of high concern for breeders and users, in order to support the best management practices. The present study aimed to characterize growth patterns of warmblood horses (Hanoverian and Oldenburg) born and raised in a reference stud-farm in the south of France, using non-linear functions. A total of 1,102 records for body weight (BW) and withers height (WH) were obtained from 58 foals (32 colts and 26 fillies). Data were regularly collected between birth and 58 months of age, when horses were already on regular work. Several sigmoid growth functions (Brody, Logistic, Gompertz, von Bertalanffy and Richards) were adjusted using the NLIN procedures of SAS. However, the Richards equation $y = A(1 - b.exp(-kt))^M$ was chosen for further analysis because it was the best fit model for both variables. Growth rates (ADG, kg/d or cm/d) were obtained from the first derivative of the equations and the effect of sex was also evaluated. The mean mature BW was 635.3±16.5 kg and the average mature size was 172.0±1.3 cm. According to the models, the proportions (%) of mature BW at 6, 12, 24, 36 and 48 months of age were, respectively, 38, 54, 72, 82 and 89%. The proportions (%) of mature WH for the same ages were 81, 87, 94, 97 and 98%. In the present study, sexual dimorphism was not observed for BW. However, significant differences were found between males and females ($P<0.05$) in what concerns A and k parameters in the WH functions. Growth rates were in the range of those found for a moderate growth in other sport breeds. The present study shows that non-linear functions can be used to accurately describe growth and development in the warmblood horse. Acknowledgement: Projects UIDB/00276/2020; LA/P/0059/2020 – AL4AnimalS.

Self-organized equestrian practitioners: key characteristics and needs

C. Eslan[1,2,3], C. Vial[2,3], S. Costa[3] and H. Pham[3]
[1]FFE, Parc Fédéral Équestre, 41600 Lamotte-Beuvron, France, [2]IFCE, pôle développement et recherche, 61310 Exmes, France, [3]MoISA, Univ' Montpellier, INRAE, Institut agro, IRD, Montpellier, France; celine.vial@inrae.fr

French horse riding is organized around professional centres such as riding schools or livery stables. As other sports, outdoor practices are booming throughout the current pandemic. Thus, the French equine industry is interested in people who organize themselves to care for their horse(s) outside of professional centres. In this context, this research aims to understand the characteristics and behaviours of self-organized equestrian users. A quantitative survey was conducted online in France. The 660 respondents are aged 15 years and over with a mean of 34 years. The main motivation leading them to self-organize is to offer better living conditions to their horse. They are quite experienced riders and generally take care of or practice with their horses alone or with their family. The average duration of self-organization is 8 years and 85% are satisfied with their current organization. Only 5% of the respondents have never used a professional equestrian centre (usually to learn riding). Using a hierarchical classification with R, we identified 6 types of self-organized users. The 'hackers' (1st type representing 39%) are moderately experienced riders who prefer hacking. They have spent less time in professional structures, own fewer horses, and have been self-organized for longer than the sample average. The 'classical' profile (2nd type) has a conventional type of practice (competitions and Olympic disciplines), a higher equestrian and academic level, and they are members of a professional structure. 16% are considered 'multidisciplinary' (3rd type) as they seek many different types of experiences. The 'former professionals' (4th type) are for half competitors, ¾ of them offer a home retirement to their horses and they often breed. The 'non-practicing' and the 'observers' (5th and 6th types) represent 7% of the sample. These results offer an opportunity for equestrian managers to target their activities to the needs of these clients, making better use of their facilities and coaching offer. The identification of like-minded equestrian centres would allow the self-organized to access support from professionals and co-create a viable business model for all stakeholders.

Studies on the training and marketing of Haflinger horses in South Tyrol

S. Hornauer[1], L. Stammler[1], T. Zanon[2], M. Gauly[2] and D. Winter[1]
[1]Nürtingen-Geislingen University of Applied Sciences, Faculty of Agricultural Economics, Economics and Management, Neckarsteige 6-10, 72622 Nürtigen, Germany, [2]Free University of Bolzano, Faculty of Science and Technology, Piazza Universitá 5, 39100 Bolzano, Italy; thomas.zanon@unibz.it

Even though South Tyrol is the region of origin of the Haflinger horse breed and still plays a significant role in the breeding and spreading of the breed, there is a lack of local structures that can be used for the training of horses, riders, and breeders and as a central contact point for marketing. In order to be able to identify ways of improving the training and marketing situation, the current situation with regard to training and marketing was first recorded within the framework of a comprehensive online survey including 107 Haflinger horse breeders in order to create a fact-based foundation for innovative changes. The results of the online survey clearly reveal the breeders' desire and need to establish a training and marketing structure that could allow for a more specialized training of the horses which in combination with an appropriate marketing strategy could improve the revenue during horse sales. In addition, the following study could demonstrate the need for an improved marketing strategy, by which respective horses should be better described and positioned for attracting more attention and a higher willingness to pay during auctions and horse sales. Therefore, the foundation of such a central training and marketing centre might offer the potential to sustainably improve and secure the competitiveness of South Tyrolean Haflinger breeding.

Which farms raise small equine herds in France?

G. Bigot and J. Veslot
INRAE, UMR Territoires, Centre Clermont-Auvergne-Rhône-Alpes, 9 avenue Blaise Pascal, CS 20085, 63178, France; genevieve.bigot@inrae.fr

Despite an increase of socio-economic studies on the horse industry in recent decades, it remains difficult to know the importance of each type of actors (breeder, rider and other user, private or professional) and in particular, keepers of small equine herds. In a first approach, we analysed the latest French FSS (Farm Structural Survey) to characterize farms holding equines. A first typology elaborated on the 54,700 French farms with equine, highlighted 8 specific groups and a large one gathering 41,000 farms that kept an average of 3 equine heads. We broke down this set into 10 groups based on: the presence of a dominant type of breed (blood horses, draft horses or donkeys), the presence of broodmares and the size of the equine herd (more or less than 5 heads). As concerns economic size, half of the farms had a standard gross product (SGP) lower than €25,000, placing them in the category of small French farms. These farms had an average agricultural area of 15 ha, with more than 85% of grassland and raised mainly equines. They mobilized more than 1/2 AWU (Annual Work Unit), essentially family workers. Farms of medium to large economic sizes (more than €25,000 of SGP) had an average agricultural area of 85 to 110 ha, varying according to the type of breed and the presence of broodmares. These farms were larger than the national average, mobilizing an average of 2 AWU, essentially family workers. Draft horses usually represented less than a quarter of the livestock on farms, while blood horses could represent from 20 to 40% of the total livestock. Regardless of their economic size, farms with blood horses were mostly located on plains, and near urban centres or small towns. Conversely, structures with draft horses were mainly located in mountainous areas and near small isolated towns. In conclusion, this analysis highlighted 2 types of farmers keeping small equine herds: (1) specialized farmers with equines on small agricultural areas, and (2) farmers with large farms who held equines in addition to other agricultural production. The type of equines kept varied more with location than with farm size.

The horse meat market in France: towards a typology of consumers and non-consumers

C. Vial[1,2], M. Sebbane[2,3] and A. Lamy[2,3]
[1]Ifce, pôle développement innovation et recherche, Le Pin au Haras, 61310 Exmes, France, [2]MoISA, Univ Montpellier, CIHEAM-IAMM, CIRAD, INRAE, Institut Agro, IRD, Montpellier, France, [3]Centre de recherche de l'Institut Paul Bocuse, 1 Chem. de Calabert, 69131 Ecully, France; celine.vial@inrae.fr

Hippophagy, historically fragile in France, is currently declining significantly, as the consumption of horsemeat has been divided by 8 over the past 40 years. In 2019, the average consumption of horse meat in France was 0.1 kg per capita per year. Only 4% of French households bought horse meat at least once a year, which represents around 6 million people. Nevertheless, it seems to exist a substantial potential to develop this market. First, this meat has many nutritional, organoleptic and environmental qualities. Second, our initial work showed that approximately 15% of the French population that not currently consume horsemeat would represent potential consumers. For these people the reasons given for non-consumption are mainly related to the availability and visibility of the offer. In this context, our research aims to understand the French horse meat market and to study the determinants of its consumption. Specifically, the goal of this communication is to shed light on different profiles of consumers and non-consumers of horse meat. Results are based on a quantitative survey among 493 meat consumers in France, analysed through clusters analysis. They underline for all respondents a lack of knowledge about the product and access difficulties. Data allow us to distinguish 4 classes of individuals: the 'reluctant', the 'distant', the 'amateur', and the 'potential'. The two first types show a low level of acceptability towards horse meat, the 'reluctant' because of a strong moral opposition, the 'distant' because of a globally distant relationship with meat in general. The 'amateur' and 'potential' individuals show characteristics that are favourable to horse meat consumption, which could be improved through a better availability and visibility of the offer for the 'amateur' group, and thanks to the development of the offer in restaurant to encourage the 'potential' group to discover the product. Targeted managerial strategies can be proposed and discussed in light of these results, differentiating non-priority or priority customer targets.

Equid milk: an overview of the French sector

D. Fouvez[1], C. Bonnin[1], C. Burdin[2] and J. Auclair-Ronzaud[3]
[1]INET, 17 Crs Xavier Arnozan, 33000 Bordeaux, France, [2]AgroSup Dijon, 26 db Petitjean, 21000 Dijon, France, [3]IFCE, Plateau technique de Chamberet, 1 Impasse des Haras, 19370 Chamberet, France; domitille.fouvez@gmail.com

The equid milk sector is under development in France. To gain a better understanding of this sector, the INET communicated an online survey to the French equid milk producers. Over the 33 mare and the 82 asses milk producers 20 (60% respondents) and 37 (45% respondents) answered the survey, respectively. The survey covered various subjects from farming and milking strategies to market opportunities. Shortly, for 90% of the respondents, equid milk is not their only activity. Others activities might by directly linked to equids (tourism, breeding) or be another animal production. It might also be unrelated to agriculture like lodging. Donkey owners develop tourism more than horse owners do while horse owners more often have another agriculture related activity. A majority of equid milk producers have between 1 and 6 females in lactation per year. The choice of the breed is different for donkey and horse owners. The former choose mainly breeds producing more milk or local breed. The later choose the breed regarding their liking of a specific breed or regarding the valuing of the foal. Both donkey and horse owners, however, seek animals with an easy temper (75% of the respondents for both populations). Number of milking per day varies from 1 to 4 in both populations. It has to be noted, however, that all the asses milk producers practice hand milking while only 26% of the mare milk producers use this milking method, a large portion of the population using mechanical milking equipment instead (74%). Regarding market opportunities, a majority of producers sell milk based cosmetics (97% of asses owners and 95% of mares owners). Among them, some producers are selling human food products (24% of asses owners and 37% of mares owners). They mostly sell their products directly on the farm, on internet, on small markets and/or specialized shops. Overall, 89% of the asses milk producers and 83% of the mare milk producers think that their activity is cost-effective or that their costs met their expenses. In conclusion, both productions have close practices. Plus, even though asses milk is more broadly known by consumers, mare milk is a product that is gaining visibility.

Mare milk composition and antioxidant capacity: effect of farm and lactation stage

A. Blanco-Doval, L. Moran, L.J.R. Barron and N. Aldai
University of the Basque Country (UPV/EHU), Lactiker Research Group, Department of Pharmacy and Food Sciences, Paseo de la Universidad, 7, 01006 Vitoria-Gasteiz, Spain; ana.blancod@ehu.eus

Horse extensive management has been proven beneficial for rural areas and the environment, but its husbandry in western Europe is mainly limited to meat production and recreational purposes, while milk is still an unexploited product. Mare milk has a unique chemical composition that makes it very interesting for human consumption. Moreover, it seems to have several functional properties, but knowledge on this sense remains scarce, particularly regarding its antioxidant capacity. Basque Mountain horse is an autochthonous equid breed from northern Spain that is currently being bred under extensive management systems, but its milk has never been commercialized, neither has it been studied from a scientific perspective. Therefore, in this research, the composition and antioxidant capacity of Basque Mountain mare milk has been studied, as well as the influence of factors such as farm and lactation stage. We found that this type of milk is very poor in fat, and that, overall, concentration of fat, protein and dry matter slightly decrease during lactation, while lactose concentration increases, as stated in other studies. On another hand, there is a statistically significant influence of the farm (which is related with management and feeding of the mares) on lactose and dry matter content. The antioxidant activity also varies significantly along the lactation period, decreasing from the beginning to week 16, and increasing again after. In this case, the farm did not have any influence in the results. Total antioxidant capacity was similar to other studies on mare milk, but higher than results reported in some ruminant milks, suggesting that mare milk can be a particularly interesting dietary antioxidant source. This research provides relevant information about the composition and bioactivity of milk from an equid breed never studied before.

Rennet coagulation aptitude of donkey milk as affected by ultrafiltration

G. Natrella[1], A. Maggiolino[2], P. De Palo[2] and M. Faccia[1]
[1]University of Bari Aldo Moro, Department of Soil, Plant and Food Sciences, via Amendola 165/a, 70125, Italy, [2]University of Bari Aldo Moro, Department of Veterinary Medicine, S.P. per Casamassima, km 3, Valenzano (BA), 70010, Italy; giuseppe.natrella@uniba.it

Making cheese by enzymatic coagulation of donkey milk is considered not feasible due to poor response to rennet. However, a protocol for preparing fresh donkey cheese has been recently reported, and the main chemical and sensory characteristics of the cheese obtained have been described. Unfortunately, the event of coagulation was not deeply investigated, and the biochemical reasons of successful cheesemaking remained unclear. In the present paper, a study was undertaken in order to shield light on this topic. A series of coagulation trials were carried out on donkey milk retentate obtained by ultrafiltration (UF) by using the protocol described in the literature with some modifications. The milk was divided into three parts: one was used as control, one was concentrated by UF (50% reduction of the volume), and the third part was subjected to thermal concentration at low temperature (20% reduction of the volume). Successively, the three milk samples were divided into two aliquots: one was adjusted to pH 6.1 by starter fermentation, the other one had pH unchanged (around 7.2). Finally, all 6 aliquots were heated to 40 °C and added with microbial rennet; coagulation was measured by using a viscometer. All milks without pH adjustment did not coagulate, demonstrating that milk concentration was not sufficient for renneting. Differently, all milks at pH 6.1 evidenced clotting: the firmer coagulum was obtained in the case of milk concentrated by ultrafiltration, the weakest in that concentrated by heat. In conclusion, the investigation suggested that the poor response of donkey milk to rennet is not only due to the low casein concentration, but is connected to the high pH value, which is responsible of high electric charge of the casein micelle. At pH 7.2, the solvation shell should be too thick to allow sufficient inter-micellar interaction, whereas at pH 6.1 the decrease of the net charge allows the micelles to bound each other. The combination of concentration by ultrafiltration and pH lowering led donkey milk to behave almost like cow milk.

Session 55 Poster 14

Digitalization in equine management

L.T. Speidel
HfWU Nürtingen, Hechingerstraße 12, 72622 Nürtingen, Germany; linda.speidel@hfwu.de

Every day, horse-keeping operations must process extensive information and face a variety of managerial challenges. These include, for example, the time-consuming work processes, the limited availability of skilled labour, and the management of customers in boarding horse farms and riding schools. The economic success of horse farms depends in particular on the experience and commitment of the farm manager and the resulting management of the farm. Furthermore, the economic efficiency of a farm can be improved by technical or digital systems. Technification of feeding, manure removal, farm management, as well as health and safety monitoring in the stable, on pasture, and the entire farm can help to relieve the farm manager and employees significantly in terms of time and also physically. There are already various individual solutions for this on the market. However, these are only sporadically used on horse farms and an interface compatibility of the systems is not yet given. In the context of this research work it is examined how the individual systems can be established on the practice enterprises, whether they can be connected over interfaces and which positive or negative effects their installation has both on the necessary working time in the enterprise and its environment. For this purpose, in addition to surveys of the practical farms and the partner companies, various investigations are carried out with regard to the possible effects on emissions, working time savings and cost optimization. These are realized by evaluations of sensors, time recording experiments as well as with the help of surveys and expert interviews on topics relevant for this – such as customer management, horse feeding, manure removal, herd safety and the importance of sustainability on the horse farms. In addition, it is being determined which information from the individual systems is to be exchanged in order to make it usable for other digital-technical applications. It is being investigated how the compatibility of the systems can be ensured and which data are necessary to establish interfaces in the horizontal and vertical operational process and to enable communication.

Changes in lactate levels and blood cell composition due to exercise in Hokkaido native horses

T.J. Acosta, Y. Okamoto and T. Saitoh
Obihiro University of Agriculture and Veterinary Medicine, Inada Cho Nishi 2 Sen 11, Obihiro, 080-8555, Japan; tjacosta@obihiro.ac.jp

Yabusame (Japanese ancient style horseback archery) has been originated about 1,000 years ago as a sport designed to improve the skills of horseback archery. The history of Yabusame as modern sport is short, and physiological studies on horses used have not been sufficiently conducted. Therefore, this study investigated the changes in blood cell components and blood levels of lactate in Hokkaido native horses used for Yabusame, and whether the amount of exercise for horses is appropriate. Experiments were conducted at two horseback riding facilities used for sport Yabusame. Because the experimental procedures were partially different at each facility, they were designated as Experiments 1 and 2. Both Experiment 1 and 2 used four Hokkaido native breed. In Experiment 1, the first blood samples were taken before the exercise, the second samples were taken after the preparatory exercise, and the third samples were taken after the horses had run five times at a canter on a 170 m track to simulate Yabusame competition. In Experiment 2, blood samples were taken before the exercise, the second samples were taken after moving to the track for about 10 minutes by walk, the third samples were taken after the preparatory exercise, and the fourth samples were taken after running at a canter five times on a 150 m track to simulate a competition. Lactate Pro was used immediately after arrival at the laboratory to measure lactate levels in blood. Blood cell components were measured in refrigerated blood using a blood cell counter. In Experiment 1, lactate concentration, red blood cell count, and haemoglobin were significantly ($P<0.05$) higher after exercise than before. In Experiment 2, haemoglobin increased significantly ($P<0.05$) after exercise compared to after preparatory exercise. However, even after the increased exercise, each measurement remained within the normal range, especially lactate concentration, which did not exceeded 4 mmol/l, which is an indication of fatigue in horses. The results of this study suggests that the degree of exercise imposed in both experiments were within the physiologic range and did not impose an excessive burden on the horses.

Portuguese Sorraia horse dynamics study with innovative technology: stance and swing of the walk

R.A.S. Faria[1], J.A.I.I.V. Silva[2], L.Y. Rodrigues[2], G. Alberto[2], J. Lopes[2], C. Zucatelle[2] and A.P.A. Vicente[3,4]
[1]HT Equine, Escusa, 7330-313 Marvão, Portugal, [2]UNESP, FMVZ, Lageado, 18618-681 Botucatu, Brazil, [3]ESA do IP Santarém, S. Pedro, 2001-904 Santarém, Portugal, [4]CIISA – FMV, UL, Lisboa, 1300-477 Lisboa, Portugal; fariasky@gmail.com

Sorraia horse is an endangered native primitive breed from Portugal, highly inbred with a small census of around 300 animals worldwide. The aim of the study was to present the difference between stance (St) and swing (Sw), when the limbs are moving on the ground or in the air, respectively, during stride duration. Results were obtained for the total, forelimbs and hindlimbs, left and right sides, and each limb separately. Using innovative technology (EquiMoves), six Sorraia stallions were studied for the variables St and Sw (complete stride duration) at medium Walk. The averages of the variables for a complete stride duration (1.16±0.10 s) were 0.70±0.08 s (St) and 0.46±0.02 (Sw), for the forelimbs, 0.72±0.09 s (St) and 0.44±0.03 (Sw), for hindlimbs 0.69±0.08 s (St) and 0.47±0.02 (Sw), for the left side of 0.71±0.09 s (St) and 0.45±0.03 (Sw), and for the right side 0.70±0.07 s (St) and 0.46±0.02 (Sw). The distribution by four limbs was, for left front 0.71±0.09 s (St) and 0.45±0.03 (Sw), right front 0.72±0.08 s (St) and 0.44±0.02 (Sw), left hind 0.69±0.08 s (St) and 0.47±0.02 (Sw) and right hind 0.68±0.06 s (St) and 0.48±0.01 (Sw). Differences between variables St and Sw were significant ($p\text{-}v<0.05$), indicating that, on average, stride duration of Sorraias stallions is distributed by 60.6% with the limb on the ground (St) and 39.4% with the limb in the air (Sw). On the other hand, the differences within the variables were not significant ($p\text{-}v>0.05$) for the forelimbs, hindlimbs, both sides and isolated limbs. The analysis of the medium Walk for variables St and Sw from Sorraia stallions, indicated that, for a complete stride duration, limbs spend more time on the ground and present identical distribution among all limbs. Sorraia horses are animals with a balanced and symmetrical distribution of the Walk. New studies, including trot and canter, will allow to better understand the distribution of the variables St and Sw for Sorraia horse dynamics and possible differences for the Walk.

Session 55 Poster 17

Portuguese Sorraia horse dynamics study with innovative technology: three natural gaits and strides

R.A.S. Faria[1], N. Marques[2], P. Ferreira[2], C.V.R. Sousa[2], R. Correia[2], J.A.I.I.V. Silva[3] and A.P.A. Vicente[2,4]
[1]Hi-Tech Equine, Escusa, 7330-313 Marvão, Portugal, [2]Escola Superior Agrária do IP Santarém, S. Pedro, 2001-904 Santarém, Portugal, [3]UNESP, FMVZ, Lageado, 18618-681 Botucatu, Brazil, [4]CIISA – FMV, Universidade de Lisboa, 1300-477 Lisboa, Portugal; fariasky@gmail.com

Portuguese native Sorraia horse is considered a primitive breed and ancestor of horses in Iberian Peninsula, where they inhabited since the Pleistocene. A closed mating system allowed to maintain the breed standard since its discovery, one century ago. The aim of this study was to characterize the dynamics of 6 Sorraia Stallions, considered very endangered (±300 living animals), using innovative technology equipment (EquiMoves). Three natural gaits, walk (W), trot (T) and canter (C), were evaluated using the variables stride duration (sD) in seconds (s), length (sL) in meters (m) and speed (sS) in kilometres per hours (km/h), considering the stride as a complete movement of the four limbs. Differences in LS mean between W, T and G within each variable were significant ($P<0.05$). The values of the variable sD were 1.16±0.10 s (W); 0.71±0.06 s (T) and 0.58±0.02 (C), with a decrease in stride duration from W to T by -41.8% and half the time from W to C (-100%) and from T to C was -18.3%. Values with opposite trends from W to T to C were observed in the variables sL and sS. Variable sL provided means of 1.44±0.06 m (W); 1.85±0.14 m (T) and 2.22±0.77 m (C), indicating increases in sL of 28.5% (W to T), 53.7% (W to C) and 19.9% between T and C. The values of the sS variable were equal to 4.39±0.18 km/h (W), 9.79±0.77 km/h (T) and 14.15±0.33 km/h (C), with increases from W to T of 223.0%, W to C of 322.3% and from T to C of 144.5%. 52 (W), 85 (T) and 103 (C) strides were observed per minute, and 74.4 (W), 156.3 (T) and 229.7 (C) meters were covered. To run 100 m (s) Sorraia horses need an average of 69 (W), 54 (T) and 45 (C) strides and take 80.0 (W), 38.3 (T) and 26.1 (C) seconds. Sorraia horses are considered ponies (1.46 m mean height at withers) and their height influences the variables evaluated and should not be compared with horses of different structures and sizes. Evaluation of variations within Walk, Trot and Canter traits should be observed in future studies.

Session 55 Poster 18

Equine oral Gram-negative multidrug resistant microbiota

J. Pimenta, A. Pinho, M.J. Saavedra and M. Cotovio
University of Trás-os-Montes e Alto Douro, Veterinary Sciences Department, Quinta de Prados-Folhadela, 5000-103, Vila Real, Portugal; josepimenta@utad.pt

Horses are carriers of antimicrobial resistant bacteria, creating a potential public health concern. The identification of equine pathogens and their multidrug resistance (MDR) pattern help to understand the inherent potential public health problem. This study aimed to identify Gram-negative (GN) bacteria from oral cavity of healthy horses and study their antimicrobial resistance. During several routine oral examinations in Portugal, 18 healthy horses without antimicrobial therapy ≥6 months were selected. From a subgingival swab, GN bacteria were identified using the Vitek® Compact 2 (bioMérieux). Antimicrobial susceptibility to 27 antimicrobials from 8 classes was evaluated using the disc diffusion method. 43 GN isolates were identified, including *Escherichia coli* (n=14), *Pseudomonas fluorescens* (n=7), *Enterobacter cloacae complex* (n=4), *Klebsiella pneumoniae spp.* (n=3), *Serratia rubidaea, Pantoea agglomerans, Pseudomonas putida* (n=2), *Aeromonas salmonicida, Enterobacter aerogenes, Pantoea spp, Pasteurella pneumotropica, Pseudomonas paucimobilis, Kluyvera intermedia, Serratia plymuthica, Shigella sonnei, Sphingomonas paucimobilis* (n=1). 38 isolates showed MDR. The antimicrobials resistance was higher to macrolides (81%) and β-lactams (55%) and lower to sulfonamides (26%) and tetracyclins (26%). Meropenem and ertapenem had 67% of resistance. *Pseudomonas putida* presented 100% resistance to macrolids, quinolones, sulfonamides, tetracyclines, phenicols and phosphomycin. *K. pneumoniae* presented 98% resistance to β-lactams and *E. coli* presented 55% resistance to amynoglicosids. The high percentage of resistance to β-lactams observed can complicate antimicrobial therapy in equine medicine. Sulfonamids and tetracyclins appear better choices to antimicrobial therapy on the equine oral cavity. Results enhance the importance of veterinarian's good individual protection practices during dentistry procedures. Resistance to ertapenem and meropenem is a potential public health concern. MDR in ≥50% of the isolates enhances the importance of equine microbiota's continuous study in the One Health concept.

Session 55 Poster 19

Effect of different cryopreservation protocols on donkey's sperm viability/motility after thawing

M.H. Silva, L. Valadão, M. Anet, M.A.D. Saleh and F. Moreira Da Silva
University of the Azores, Faculty of Agrarian Sciences and Environment – IITAA, R. Joao Capitao D'Avila, Angra do Heroísmo, 9700-042, Portugal; joaquim.fm.silva@uac.pt

The aim of this study was to establish a protocol for freezing donkey semen, allowing its trade for artificial insemination. For such purpose, three different cryoprotectants and four different cooling times before freezing were evaluated on the viability/motility of donkey spermatozoa. Semen of six males of the 'Burro Anão da Graciosa', an Azorean Donkey breed, was collected three times each, using an artificial vagina. After collecting, semen was diluted in an extender composed by glucose and milk powder in water and the sperm was evaluated for volume, appearance, consistency, pH, motility (1-5) and percentage of dead/alive spermatozoa. Then a second extender, constituted by distilled water, glucose monohydrate, lactose monohydrate, sodium citrate dihydrate, HEPES, gentamycin, powdered milk, and egg yolk, in which the cryoprotectant ethylene glycol (30 mM); Glycerol (22 mM) and L-glutamine (80 mM) has been added, and 60 straws were filled (20 per cryoprotectant). Each group of 20 straws was divided into 4 subgroups of five straws each, and refrigerated at 5 °C for 1, 2, 3 and 4 hours. At the end of each time, straws were placed at 3 cm above liquid nitrogen for 12 minutes and then immersed in liquid nitrogen. Thawing was carried out at 37 °C for 30 seconds and tests afore described were again performed. After collecting, on average pH was 7.4 (±0.2), motility 4.7 (±0.2) and viability 85.3% (±3.5). After thawing, the best results ($P<0.05$) were obtained for the 2 hours of refrigeration protocol in which the sperm viability and motility were 69.2% (±3.5) and 3.7 (±0.8), respectively. For the straws refrigerated for 1, 3 and 4 hours, viability was 58.2% (±3.2); 48% (±6.4) and 36% (±6.8), respectively, while for motility results was 3.4 (±1.2); 2.9 (±1.2) and 2.1 (±1.8). No statistical differences were observed among cryoprotectants used besides the results with L-glutamine originated better results at the 2 hours refrigeration. Further research, however, still needs to be performed to ensure the effectiveness of these two aspects (cooling time and cryoprotectant), including, in addition to the characteristics evaluated, other aspects of sperm fertilization capacity.

Session 55 Poster 20

Health parameters of horses affected by headshaking syndrome

L.M. Stange[1], J. Krieter[1] and I. Czycholl[1,2]
[1]Institute of Animal Breeding and Husbandry, University Kiel, Olshausenstraße 40, 24098 Kiel, Germany, [2]Pig Improvement Company (PIC), 100 Bluegrass Commons Blvd. Ste 2200, Hendersonville, TN 37075, USA; lstange@tierzucht.uni-kiel.de

The equine headshaking syndrome (EHS) is a disorder in horses consisting of spontaneous movements of the horses head. The syndrome can be clearly identified from clinical symptoms, but reliable diagnosis remains an exception and aetiopathogenesis is often unclear. This study aims at gaining better understanding of causal development of EHS. In this context, we examined 23 headshakers and 19 control horses and analysed the role of health parameters e.g. eye condition, coughing while riding, back pain, lameness and stereotypes. Furthermore, blood samples from 19 headshakers and 14 control horses were tested with regard to Borna-Virus antibodies. For statistical comparison of the two groups, Chi^2-test and power analysis were carried out. For the study, 28 geldings and 14 mares were available, of which 26 horses were Warmbloods. 14 ponies and 2 draft horses participated in the study. The average age of the headshakers was 13 years. The control group was slightly older (15 years). Examination of the eyes revealed abnormal results in 10 horses (7 headshakers, 3 control horses). Throughout data collection, 6/19 control horses exhibited coughing during riding and 11/23 headshakers. Palpation of the horse showed back pain in 27 of all participating horses. Of these, 15 horses were affected by EHS. We observed stereotypes in 6 horses, one of these belonging to control group. However, no significant differences could be found between headshakers and control group regarding health parameters. By a power analysis, it could be determined that a larger sample of 142 horses would be necessary in order to reach significance. However, this number of horses could not be reached in this study. The evaluation of blood samples yielded seven positive test results from 19 headshakers tested. Borna and EHS were tested as significant ($P=0.01$). In contrast, no Borna-Virus antibodies could be detected in 14 control horses. The results suggest that more research needs to be carried out on rare diseases such as EHS to ensure large sample sizes in order to gain statistical reliable results. Overall, a connection between Borna-Virus and EHS is likely.

Volatile profile of dry-fermented foal sausages formulated with healthy oil emulsion hydrogels

A. Cittadini[1], M.V. Sarriés[1], P.E. Munekata[2], M. Pateiro[2], R. Domínguez[2] and J.M. Lorenzo[2,3]
[1]IS-FOOD. Universidad Pública de Navarra, Campus de Arrosadia, 31006 Pamplona, Spain, [2]Centro Tecnológico de la Carne de Galicia, Avd. Galicia 4, 32900 San Cibrao das Viñas, Ourense, Spain, [3]Universidade de Vigo, Área de Tecnología de los Alimentos, Facultad de Ciencias de Ourense, 32004 Ourense, Spain; aurora.cittadini@unavarra.es

The purpose of this work was to study the influence of partial replacement of animal fat by healthy oils on volatile profile of dry-fermented foal sausages. Three distinct batches were elaborated: control (CON) – 100% of pork back fat as fat source; treatments 1 and 2 (T1 and T2) – 50% of animal fat was replaced using oil mixture emulsions, tiger nut (T1) or sesame (T2) oils (35.05 g/100 g emulsion) mixed with algal oil (2.25 g/100 g emulsion). The sausages were fermented for 1 day at 20 °C and 80% relative humidity (RH) and then dried-cured for 55 days at 8-12 °C and 65-80% of RH. The extraction of the volatile compounds was carried out using solid-phase microextraction (SPME), while separation, identification and quantification was performed using a gas chromatograph coupled to a mass selective detector. The data were examined using a one-way ANOVA with the SPSS statistical software. A total of 96 compounds were identified and grouped into eleven chemical families. The replacement of animal fat by the healthy emulsion hydrogels increased ($P<0.001$) the total volatile compounds (VOC) and most of individual VOC ($P<0.05$). Terpenes and terpenoids represented the main family in all treatments, as typical in this type of product, probably due to the spices and the vegetable oils employed. However, it was the T1 group to report the highest values ($P<0.05$) in the major part of VOC families detected. In addition, fat degradation compounds showed the greatest amounts in T1 samples. On the other hand, T2 seemed to reduce the generation of lipid-derived VOC and minimized off-flavours. This trend could be related to the presence of higher quantity of natural antioxidant compounds in the sesame oil. Hence, it is evident that the type of fat source had a significant effect on the VOC profile of this product. Moreover, the T2 formulation showed encouraging outcomes and could improve the aromatic perception of the dry-fermented foal sausages.

Study of the milk composition of mares located in Goiás southwest, Brazil

A.M. Geraldo[1], R.M. Ribeiro[2], D.S.F. Ribeiro[2], H. Wajnsztejn[3], A.R. Amorim[2] and P.D.S.R.G. Wajnsztejn[3]
[1]Universidade de Évora, MED – Instituto Mediterrâneo para a Agricultura, Ambiente e Desenvolvimento, Polo de Mitra, Apartado 94, 7006-554 Évora, Portugal, [2]Centro Universitário de Mineiros, Department of Veterinary Medicine, Rua 22 esq. c/ Av. 21 – St. Aeroporto, 75833-130, Mineiros, GO, Brazil, [3]Universidade de Sorocaba, Department of Veterinary Medicine, Rod. Raposo Tavares, km 92.5, Vila Artura, 18023-000, Sorocaba, SP, Brazil; ana.de.mira.geraldo@gmail.com

The mares' milk is the only source of foals' nutrition in the first weeks of life, being extremely important for their satisfactory physical and social development. The production and composition of mares' milk can be influenced by several factors, such as age, calving order, mares' live weight, diets, environmental conditions and stage of lactation, and it is necessary to know the particularities and reference values of its composition to diagnose and correct nutritional deficiencies and potential signs of pathologies. The objective of this experiment was to evaluate mares' 90 days of lactation milk composition. Four, mixed breed, with 3 months of lactation, multiparous healthy mares, with an average of 8.8 (±4.81) years and 376 (±55.91)kg weight, located in studs in southwest Goiás, Brazil, were evaluated. Milk composition such as somatic cell count (SCC), fat analysis (FAT), protein (PT), total solids content (TS), lactose (LACT) and defatted dry extract (DDE) were determined in samples before the first feeding. The mean values (± standard deviations) were: CCS 20 (±5.35), FAT 1.16 (±0.65), PT 1.81 (±0.22), ST 10.59 (±1.28), LACT 6.43 (0.28) and DDE 9.43 (±0.64). Data obtained in this experiment are within the reference values when compared to other works. It is important to note that the feeding, handling and physiological response of each mare can influence milk composition's variables. However, the parameters evaluated in this experiment are within the values already observed in the literature, even when conditions such as diet composition, management, body condition score and climate differs from previous studies, which raises the need for further studies for a possible variable individualization effects on equine milk's composition.

Use of extruded linseed to improve the nutritional quality of Sarda Donkeys milk

F. Correddu, S. Carta, A. Mazza, A. Nudda and S.P.G. Rassu
Università degli studi di Sassari, Dipartimento di Agraria, Viale Italia 39, 07100, Italy; fcorreddu@uniss.it

The donkey milk represents an alternative to cow milk for children with allergy to bovine proteins, and, thus, its consumption is increasing worldwide. This trend could be a promising strategy for the valorisation and conservation of some local donkey breeds, which milk is often of poor quality because they are reared in marginal areas. This quality, especially in terms of fat content and fatty acid (FA) profile, can be improved by a dietary lipid supplementation. The aim of this study was to evaluate the use in donkeys' diets of extruded linseed on milk yield, composition, and FA profile. Eight Sarda donkeys were assigned to two groups homogeneous for body weight (110±15 kg), milk production (354±124 g/d) and days in milking (123±41 d): (1) control diet (CON); (2) CON supplemented with 100 g/day per head of extruded linseed (EL). The trial lasted 7 wks: adaptation (2 wks) and samples collection (5 wks). Individual milk samples were collected weekly and analysed for chemical composition and FA profile. Data were analysed with a linear mixed model with diet, sampling, and their interaction as fixed effects and animal as random effect. Milk yield (339.2 vs 263.2 g/d per head for CON and EL, respectively) and milk composition were not influenced by the diet. Fat content was low in both groups (0.36 vs 0.43% for CON and EL, respectively) and it did not change among the sampling time. Short- and medium-chain FA, especially C16:0, were lower in EL. The diet, sampling and their interaction significantly affected the concentration of alpha-linolenic acid. At the end of the trial, the average value of this FA was double in EL than CON (15.76 vs 7.83 g/100 g of FAME). The lipid supplementation decreased saturated and increased polyunsaturated FA. The inclusion of linseed in the diet of donkeys improved the nutritional indexes of milk: n6/n3 ratio and thrombogenic index decreased, whereas the hypocholesterolemic to hypercholesterolemic ratio increased. In conclusion, linseed in donkey diet might be a good strategy to maintain a high milk nutritional quality, as demonstrated by the increase of FA of the omega3 family and by the improvement of fat nutritional indices.

Resilience biomarkers in grazing Holstein cows

A.L. Astessiano[1], A.I. Trujillo[1], A. Kenez[2], A. Mendoza[3], M. Carriquiry[1] and E. Jorge-Smeding[1,2]
[1]FAGRO, UDELAR, Montevideo, 12900, Uruguay, [2]City University of Hong Kong, Hong Kong, NA, China, P.R., [3]INIA, Colonia, 39173, Uruguay; ejorgesmeding@gmail.com

Biomarkers are required to understand the mechanisms underlying resilience (Res) traits, possibly differing between grazing and confined systems. Sixteen multiparous grazing Holstein cows were ranked according to the resilience index (LnVar) proposed by Poppe *et al.* (2020) and the 8 most extreme animals were selected leading to two groups: high (High-Res, n=4) and low resilience (Low-Res, n=4) significantly differing in their LnVar (0.84 vs 2.17±0.27, mean ± SEM, P<0.01). Plasma metabolomic data at 21 and 180 days in milk (DIM) was analysed through ANOVA including the group (High-Res, Low-Res), the DIM and its interaction as fixed effects, and the cow as the random effect using MetaboAnalyst platform (https://www.metaboanalyst.ca). The P-values were adjusted by false discovery rate (FDR). No metabolite was affected by the group or the interaction group × DIM after FDR correction. However, 3 metabolites (homocysteine, uracil and glutaric acid) had a raw-P<0.05 for the interaction as all of them were greater for Low- vs High-Res cows only at 21 DIM. Additionally, 6 metabolites (sucrose, fructose, 1-kestose, malonic acid, methionine sulfoxide, malic acid) had a raw-P<0.05 for the group effect as all of them were lower for High- than Low-Res. Despite the low number of animals used for comparing Res groups, our results suggest that High- and Low-Res cows differed in carbohydrate metabolism and that Res biomarkers at early lactation should be linked with energy metabolism. Future works are warranted to confirm these results for better understanding the physiological mechanisms underlying resilience in grazing dairy cows.

Session 56 Theatre 2

Calibrating simulations of dominance variation in animal breeding: case study in layer chickens

I. Pocrnic[1], C. Gaynor[1], J. Bancic[1], A. Wolc[2,3], D. Lubritz[3] and G. Gorjanc[1]
[1]The University of Edinburgh, The Roslin Institute, Easter Bush Campus, EH25 9RG Edinburgh, United Kingdom, [2]Iowa State University, Department of Animal Science, Kildee Hall, Ames, IA 50011, USA, [3]Hy-Line International, P.O. Box 310, Dallas Center, IA 50063, USA; ivan.pocrnic@roslin.ed.ac.uk

In this contribution, we present the building blocks of a framework for stochastic simulations of additive and dominance genetic variation that reflect variation found in real-world datasets. Stochastic simulations are a cost-effective method for testing any novelty within a breeding programme *in silico* before actual experimental validation and eventual deployment in the real-world. Accordingly, to be informative as a hypothesis-generation and decision-making tool, the simulations should reflect real-world variation as close as possible. While additive genetic variation is a staple to many breeding methods and underlying simulations, dominance genetic variation is typically neglected, oversimplified, or simulated in an ad hoc manner. However, dominance variation is a vital genetic component of many breeding programmes, especially in the terms of inbreeding depression and heterosis. Here we showcase a framework for calibrating stochastic simulation of additive and dominance variation to reflect variation in a real-world dataset. To this end, we used SNP marker data and egg production phenotypes from a commercial layer chicken population to estimate additive and dominance genetic variances and inbreeding depression. We fitted both the full genomic directional dominance model and reduced dominance model (without the marker heterozygosity as a covariate) via Bayesian ridge regression. Furthermore, we built a framework of formulae and algorithms that use the real-world genetic parameters estimates to fine-tune a simulation, in particular the mean and variance of dominance degrees (relative magnitude of biological dominance effects compared to biological additive effects) and the number of quantitative loci. We evaluated a full grid across the parameter space to find the most credible inputs so that the resulting simulation reflected variation found in a real-world dataset. This work will enable fine-tunning of future simulation of animal breeding programmes influenced by additive and dominance variation.

Session 56 Theatre 3

Simulation study on the optimization of pasture-based German Merino sheep breeding programs

R. Martin[1], T. Pook[2], M. Schmid[1] and J. Bennewitz[1]
[1]Institute of Animal Science, University of Hohenheim, Garbenstrasse 17, 70599 Stuttgart, Germany, [2]University of Goettingen, Center for Integrated Breeding Research, Department of Animal Sciences, Albrecht-Thaer-Weg 3, 37075 Goettingen, Germany; rebecca.martin@uni-hohenheim.de

The Merino sheep is the predominant sheep breed in Germany where it is mainly used for grazing to preserve extensive grasslands. In the basic breeding program progeny testing for meat production traits is done with concentrated feed on station in which eight male lambs per ram are tested. This, however, does not represent the actual production environment on pasture thus limiting possible genetic gain. Regarding the maternal side, adequate fertility and nursing are required to produce vital lambs. A new trait has recently been added to the breeding program which records the 42-day-weights of the lambs born and acts as an indicator trait for the ewes' nursing ability. Since the willingness for recording is low among breeders, currently solely around 25% of all ewes are phenotyped for nursing ability. In this study, a lifelike simulation of the basic German Merino sheep breeding program was conducted using the R-package MoBPS. The simulation included all relevant selection steps, a realistic age structure and genetic parameters along with a pedigree-based BLUP breeding value estimation. An alternative scenario was simulated over ten generations in which, in addition to station testing, 20 male lambs per ram were phenotyped for meat production traits (average daily gain, meat and fat) on pasture and the proportion of phenotyped ewes for nursing ability was set to 100%. In the alternative scenario, an increase in genetic gain in average daily gain, meat and fat as well as in nursing ability was observed. This simulation study highlights the need to further increase the proportion of phenotyped animals. The additional implementation of testing progeny under field conditions demonstrates the potential to increase genetic gain by adjusting the breeding program to its actual production environment.

Effect of β-casein variants on cheese yield and quality measured in commercial cheese-making trials

E. Boschi, S. Faggion, P. Carnier and V. Bonfatti
University of Padova, Comparative Biomedicine and Food Science, viale dell'Università 16, 35020 Legnaro, Italy; valentina.bonfatti@unipd.it

Dairy companies in different countries have started to capitalize on the marketing of A2 milk, i.e. milk produced by the sole homozygous *BCN* A2 animals. As a consequence, there is an emerging interest in breeding only homozygous A2 animals, but a change in β-casein composition might have unfavourable consequences on cheese yield and quality. The aim of this study was to investigate the effect of the relative content of β-casein genetic variants in milk on cheese yield and quality. Based on individual β-CN genotype, protein composition, and milk composition, milk of small groups of cows (n=2-4) was collected to create milk pools to be compared in a set of experimental cheese-making trials. Trials were performed in an on-farm small-scale commercial dairy. For each cheese-making day, two experimental milks diverging for β-CN composition were processed for the manufacture of Caciotta, a soft cheese, with spherical-cylindrical wheels weighting between 0.6 and 0.8 kg, white-pale yellow paste with small and rare round-eyes. A total of 66 cheese milks (~25 Kg/each) were processed. Each experimental milk resulted in 2 wheels that were subsequently seasoned for 2 weeks and analysed for composition and texture. Cheese yield and quality were evaluated by a mixed model including the random effect of the day of cheese-making, and the linear effect of the content of fat, total protein, Ca, temperature of milk at the beginning of processing, content of κ-CN and β-CN, and content of β-CN A1 and β-CN A2. An increase in the relative content of β-CN A2 was associated with a significantly lower cheese yield at the end of processing (-0.94% for each g of β-CN A2, $P<0.05$), but no effects of the A2 variant were detected after 15 d of ageing. No significant effects of β-CN composition were detected on cheese composition and texture. Our results indicate that the use of homozygous A2 cows is not expected to exert unfavourable effects on the efficiency of the cheese-making process and product quality. *Acknowledgements*. Rural development 2014-2020 Submeasure 16.1 – Marche region n. 29228 (I-MilkA2).

Relationship between body condition score and backfat thickness in dairy cattle

C. Schmidtmann, C. Harms, I. Blaj, A. Seidel and G. Thaller
Christian-Albrechts-University Kiel, Institute of Animal Breeding and Husbandry, Hermann-Rodewald-Straße 6, 24118 Kiel, Germany; cschmidtmann@tierzucht.uni-kiel.de

Assessing the body condition score (BCS) of cows is widely used as management tool on dairy farms to estimate changes in body reserves during lactation. Backfat thickness (BFT) is mainly recorded on research farms, but could provide a better proxy for the nutritional status of animals than BCS. Knowledge on the association between these two traits in dairy cattle is still limited. Therefore, the aim of the study was to analyse the relationship of BCS and BFT. Data of 1,001 Holstein Friesian cows for body condition score (BCS; 17,265 monthly records; visual assessment on a 5-point scale) and backfat thickness (BFT; 63,050 weekly records; ultrasound measurements) was recorded between 09/2005 and 06/2021 on a research dairy farm. In addition, 896 animals had genotypic information (45,613 SNPs). Lactation curves of BCS and BFT were fitted using Legendre polynomials of 3rd degree. SNP-based heritabilities and genetic correlations between traits were computed for primiparous cows within 3 defined lactation stages (1 to 90 DIM; 91 to 180 DIM; >180 DIM) using genomic REML in GCTA. Results showed similar shape of curves for the two traits but nadir was later for BFT than BCS in all lactations. After calving, BCS decreased with nadir of lactation curve at 50 DIM in 1st lactation, at 56 DIM in 2nd lactation and at 80 DIM in 3rd lactation. For lactation curves of BFT, minimum values were reached later at 115 DIM, 102 DIM and 111 DIM in 1st, 2nd and 3rd lactation, respectively. Interestingly, BCS and BFT showed opposite orders of lactation curves for the different lactation numbers, indicating that primiparous cows had the highest BCS but the lowest BFT. Heritability was between 0.286 (±0.06) and 0.435 (±0.07) for BCS and between 0.417 (±0.07) and 0.486 (±0.07) for BFT. Genetic correlation between traits was high and ranged from 0.699 (±0.09) to 0.836 (±0.05) for cows in 1st lactation depending on lactation stage. Although strong genetic correlations were found between BCS and BFT, the results of this study suggest that different body fat reserves are assessed by the two traits. Thus, BFT can provide additional information on the nutritional status of dairy cattle.

Session 56 Theatre 6

Holstein effective population size reducing

J.B.C.H.M. Van Kaam[1], B. Lukić[2], M. Marusi[1] and M. Cassandro[1,3]
[1]National Association of Holstein, Brown and Jersey Breeders (ANAFIBJ), Via Bergamo 292, 26100 Cremona (CR), Italy, [2]University of J.J. Strossmayer of Osijek, The Faculty of Agrobiotechnical sciences Osijek, Vladimira Preloga 1, 31000 Osijek, Croatia, [3]University of Padova, Department of Agronomy, Food, Natural Resources, Animals and Environment, Viale dell'Università 16, 35020 Legnaro (PD), Italy; jtkaam@anafi.it

Since genomic selection became the new standard in Holstein breeding, genetic progress has vastly improved. Current progress is between 2 and 3 times the progress obtained before with the progeny testing system. The increased progress however comes at the cost of an increase in inbreeding. Generation intervals have been shortened dramatically especially for bulls. The Anafibj genomic database was used for analysing SNP heterozygosity across all birthyears. In 1990 average SNP heterozygosity was at 0.3518. In the pre-genomics period 1990-2010, with progeny testing, the average annual decline was -0.0003 arriving at 0.3451. During the genomics period 2010-2021, the average SNP heterozygosity arrived at 0.3237, which means an average annual decline of -0.0019. A sixfold increase in the decline of heterozygosity. If this linear trend would continue than in 170 years there would not be any variation left. The decline of heterozygosity can be converted in an inbreeding coefficient F which leads to 7.7% from 1990 till 2021. The inbreeding coefficient in turn can be transformed in an estimate of the effective population size. Effective population size (N_e) was estimated in 1990 at 120, whereas in 2021 it is around 70. Effective population size is, however, a measure which can lead to weird negative values, if heterozygosity increases. Overall effective population size can be estimated in many ways, leading to different results, therefore not simply comparable. Modest changes in heterozygosity can lead to large changes in effective population size. More attention and studies are needed to monitor and possibly improve the N_e for livestock populations under genomic selection.

Session 56 Theatre 7

Estimating additive genetic variance over time in chicken breeding programmes

B. Sosa-Madrid[1,2], G. Maniatis[3], S. Avendano[3] and A. Kranis[2,3]
[1]Institute for Animal Science and Technology, Universitat Politècnica de València, Valencia 46071, Spain, [2]The Roslin Insititute, The University of Edinburgh, Easter Bush, EH25 9RG, United Kingdom, [3]Aviagen Ltd, Newbridge, EH28 8SZ, United Kingdom; bosamu05@gmail.com

In populations under directional selection, monitoring the genetic variance of traits is a key priority in order to ensure the substantiality of breeding programmes. Studies monitoring changes in genetic variation through time have typically used long-term data from small experimental populations selected for a handful of traits. Using farm animal data, especially in species such as chickens where the generation interval is small and the breeding populations are large, offers the opportunity to analyse big datasets that enable retrospective insights into the dynamics of genetic variance under an holistic and balanced breeding objective. In this study, we focus on body weight (BWT) and egg production (EP) as they are both key traits in the breeding goal and also capture the genetic antagonism between growth and reproduction in broilers. A long-term dataset spanning over 23 years and consisting of over 2 million records was split into 22 overlapping windows (each one including 3 years' of data). A bivariate Bayesian model was applied to estimate genetic variance and covariances in each period for both traits and account for the uncertainty of the predictors over time. Results show that while the additive variance showed some variation between time windows, it remained steady over the whole period in both traits. Similarly, while heritabilities (h^2) for both traits showed some differences between windows (BWT h^2 ranged between 0.32 and 0.41; EP h^2 between 0.21 and 0.34), the analysis suggests that they have also been stable over time. Overall, our 'divide-and-conquer' approach for splitting the data was computationally tractable and enabled the temporal analysis of key traits used in broiler breeding. This study provides empirical evidence of the maintenance of genetic variation in a poultry breeding population under selection.

Session 56 Theatre 8

Purebred breeding goals optimized for crossbreeding improves crossbred performance

J.R. Thomasen[1], H.M. Nielsen[2], M. Kargo[2] and M. Slagboom[2]

[1]VikingGenetics, Ebeltoftvej 16, 8960 Randers SØ, Denmark, [2]Aarhus University, Centre for Quantitative Genetics and Genomics, Blichers Alle 20, 8830 Tjele, Denmark; jotho@vikinggenetics.com

In contrast to other species, production animals in dairy cattle are mainly purebreds. However, crossbred cows are becoming more widespread due to better profitability from heterosis of economically important production and health traits. Currently, breeding goals (BGs) for purebreds used in crossbreeding systems are optimized for maximizing genetic gain in purebred cows. This might not be optimal when the aim is to increase heterosis and breed complementarity in order to maximize profit in crossbred cows. We hypothesize that differentiated BGs in purebreds that complement each other can improve the performance of crossbred production animals. To test this hypothesis, we set up a pure- and crossbreeding design using the simulation tool ADAM with two breeds, Danish Jersey (DJ) and Nordic Holstein (NH). Real marker genotypes from these two breeds were used as reference to resemble the real population structure of the two breeds. A progeny testing scheme was simulated for 20 years to build large genomic reference populations in each breed followed by a genomic testing scheme of 5 years. Finally, one generation of crossbreeding between the two purebreds was generated. The BG in the purebreds included two traits, a production trait and a health trait, with a genetic correlation between the 2 traits of -0.46 for DJ and -0.35 for NH. Three BGs with different economic values of the traits were set up to test the hypothesis: (1) a reference scenario resembling the weights in the present Nordic total merit index for DJ and NH; (2) the same BG in NH as in 1 but in DJ the BG had a higher weight on health; and (3) BGs only weighing production in NH and health in DJ. The three scenarios was evaluated in the crossbred population according to genetic level and dominance. Preliminary results show a higher degree of dominance in the crossbred animals compared to the purebred animals, but a lower genetic level, especially in scenario 3. The outcomes of this study can be used to evaluate the value of setting up BGs in purebreds specified for crossbreeding.

Session 56 Theatre 9

Trade-off between animal resilience and energy use in grazing cows of two Holstein genetic strains

E. Jorge-Smeding[1,2], A.I. Trujillo[2], A. Kenez[1], D. Talmón[2], A. Mendoza[3], G. Cantalapiedra-Hijar[4], M. Carriquiry[2] and A.L. Astessiano[2]

[1]City University of Hong Kong, Hong Kong, NA, China, P.R., [2]FAGRO, UDELAR, Montevideo, 12900, Uruguay, [3]Programa Nacional de Lechería, INIA, Colonia, 39173, Uruguay, [4]INRAE, VetAgroSup, Saint-Genès-Champanelle, 63122, France; ejorgesmeding@gmail.com

A trade-off between animal resilience (Res) and feed efficiency has been evoked in dairy cows, limiting the genetic progress if both traits are not considered together. New Zealand Holstein (NZH) has a greater energy and feed efficiency than the North American Holstein strain (NAH) under grazing conditions, and a greater adaptive capacity also for NZH than has been suggested, but no quantitative comparison is reported yet. The Res of multiparous 39 cows was estimated as the log-transformed variance (LnVar) of the daily deviation from the modelled lactation curve between 10 and 200 DIM. Based on hierarchical cluster analysis of LnVar and regardless of the strain, the cows were grouped into 2 groups of Res: low (Low-Res, n=25, 14 NAH + 11 NZH) and high (High-Res, n=14, 6 NAH + 8 NZH). The LnVar, and productive (persistence, milk peak yield, 200 DIM accumulated milk yield) and energy use traits [residual heat production, (RHP), energy efficiency (EF = milk retained energy/metabolizable energy intake), measured at 180 DIM] were analysed by t-test comparing Low- and High-Res groups. The LnVar was similar (P=0.28) between Holstein strains. High-Res had a lower LnVar (1.09 vs 1.88±0.09, mean ± SEM; P<0.01), greater milk persistency (+23%; P=0.03) but similar milk peak (32.1 vs 33.3±1.78 kg/d, P=0.49), and 200 DIM-accumulated milk yield (5,443 vs 5,557±288 kg, P=0.67) than Low-Res cows. Both RHP and EF did not differ between Low- and High-Res. While LnVar and RHP were not correlated, LnVar and EF tended to be negatively correlated (r=-0.26, P=0.09) suggesting that greater EF is associated to lower animal Res. Our results indicate that High-Res individuals can be found in both Holstein strains under grazing conditions, and that increased Res would be associated with decreased EF. Further studies are warranted to confirm these preliminary results.

Session 56 Poster 10

Relationships between conformation traits and health traits in Czech Holstein cows

J. Vařeka, L. Zavadilová, E. Kašná, M. Štípková and L. Vostrý
Institute of Animal Science, Gnetics and Breeding of Farm Animals, Přátelství 815, 104 00, Praha, Czech Republic; vareka.jan@vuzv.cz

The relationships among exterior, the incidence of clinical mastitis (CM) and somatic cell score (SCS) were analysed in Czech Holstein cattle. The cases of CM from 27,709 first lactations of 17,622 cows were recorded on 15 farms from 1993 to 2020. The CM was considered an all-or-none trait with 0 (no CM case) and 1 (at least 1 CM case). The incidence of CM was monitored in seven lactation periods, 50 days long, and for the whole lactation. SCS was monitored in 10 periods after 30 days and on average for the whole lactation. Effect levels for SCS were estimated by the linear animal model. The logistic regression model was used for CM analysis. A fixed herd-year-period effect, a fixed age effect at the first calving and a fixed effect of the linear type trait the random effect animal were included. The incidence of CM and the average SCS was high at the beginning of lactation; around day 100-150 slightly decreased and then increased until the end of lactation. The incidence of CM and SCS significantly decreased with decreasing depth of udder. A slightly upward trend incidence of CM was observed for udders 10-15 cm above the hook between days 301-350 of lactation. Shallow udders were associated with the lowest incidence of CM and SCS. Cows with udders below the hook had the highest incidence of CM and SCS. Narrow and wide udders were associated with the highest risk of infection. The lowest incidence of CM and the lowest SCS was observed for intermediate udders (10 cm). The more the value deviated from udder width 10 cm, the greater the risk of CM and increased SCS. The logistic regression model predicted udder width 11 cm as the lowest risk of CM disease for the first part of lactation, between days 51-100 and for the whole lactation. Cows with shallow udders were protected from bacterial infection, but they tended to low yield. Suitable selection criterium could be relatively less deep udders but with appropriate udder width around 10 cm. The work was supported by the project QK22020280 and the project MZE-RO0718 of the Ministry of Agriculture of the Czech Republic.

Session 56 Poster 11

Dominant expression of beta-casein variant A1 in milk of Holstein-Friesian cows

S. Kaminski[1], A. Cieslinska[2], K. Olenski[1], T. Zabolewicz[1] and A. Babuchowski[3]
[1]University of Warmia and Mazury Olsztyn, Poland, Departament of Genetics, Michal Oczapowski str 5, 10719, Poland, [2]University of Warmia and Mazury Olsztyn, Poland, Departament of Biochemistry, Michal Oczapowski str 2, 10719, Poland, [3]Dairy Industry Innovation Institute Ltd., Kormoranow 1, 11700 Mragowo, Poland; stachel@uwm.edu.pl

Beta-casomorphin-7 (BCM7) is an opioid-like peptide released during gastrointestinal digestion exclusively from cow's beta-casein (CSN2) variant A1. Consumption of milk with variant A1 is associated with impaired digestive wellbeing and some neurological disfunctions in children. The aims of study were, 1/ to find how frequent is allele A1 within Holstein-Friesian cow population; 2/ to measure how much A1 and A2 protein variants occur in milk from cows of different CSN2 genotypes; 3/ how much BMC7 is released from milk of cows having or not undesirable variant A1. Beta-casein variants A1/A2 and BCM7 were measured using ELISA tests. CSN2 genotypes were identified by Illumina Bovine EuroG_MDv2 BeadChip. Within 1,511 Holstein-Friesian cows kept in 23 herds, the frequency of A1A1, A1A2 and A2A2 genotypes accounted to 12%, 43% and 45%, respectively. Hydrolysed A1A1 fresh milk was a source of ca. 9-10-fold higher quantity of BCM-7 than milk produced by A2A2 cows and unexpectedly, also by A1A2 cows. It suggests that allele A1 is dominantly expressed over A2. When fresh milk was pasteurized or treated by ultra-high temperature (UHT), disproportion between the quantity of BCM7 from A1A1 and A2A2 decreased but still the differences between them were highly significant. Assuming that A1 variant is correlated with possible deterioration of human health, 55% of milk (based of the frequency of A1A1 + A1A2 genotypes) – could be consider as undesirable in nutrition of people suffering from gut problems and autism. More research is necessary to explain the nature of the dominance of allele A1 over A2 as well as whether this phenomenon is specific to dairy cattle breed. Acknowledgments: Project financially supported by Minister Education and Science in the range of the program entitled 'Regional Initiative of Excellence' for the years 2019-2022, Project No. 010/RID/2018/19. We are thankful to all breeders who helped us in taking the samples from cows.

Relationships between lameness, conformation traits and claw disorders in Czech Holstein cows

L. Zavadilová[1], E. Kašná[1], J. Vařeka[1], J. Kučera[2], S. Šlosárková[3] and P. Fleischer[3]
[1]Institute of Animal Science, Genetics and Breeding of Farm Animals, Přátelství 815, Prague 104 00, Czech Republic, [2]Czech Moravian Breeders' Corporation, Inc., Benešovská 123, Hradištko pod Medníkem 252 09, Czech Republic, [3]Veterinary Research Institute, Hudcova 296/70, Brno 621 00, Czech Republic; zavadilova.ludmila@vuzv.cz

The relationship among lameness (L), conformation traits (C) and foot and claw disorders (D) was analysed in Czech Holstein cattle. The 51,074 lactations of 25,581 cows were monitored on 29 farms from 2017 to 2021 for L and infectious claw disorders (ICD) such as dermatitis digitalis (DD), interdigital phlegmon (IP), dermatitis interdigitalis and heel horn erosion. The D was considered an all-or-none trait with 0 (no case) and 1 (at least one case). The official web application Diary of Diseases and Medication was a source for L, D records. Linear type traits (LT) all scored on a 9-point scale included: rear legs rear view, rear leg set (side view), foot angle, legs&feet, locomotion and BCS (body condition score). Genetic parameters and standard errors were estimated using the AIREML method and bivariate animal models with fixed effects of herd-year-season of calving (L, D), parity (L, D); herd-year-season of classification (LT), classifier (LT), linear and quadratic regressions on day of classification (LT); random effects of the permanent environment of a cow (L, D) and animal (L, D, LT). The lactation incidence of L was 8%; DD 8%; IP 5% and ICD14%. The strongest genetic correlation occurred between L and locomotion -0.23. Those between L and other traits were lower: legs&feet -0.11; foot angle 0.12; BCS 0.09; rear legs rear view 0.08, rear legs side view 0.11. On dataset sustained from 12 herds, the genetic correlations were between L and DD 0.21; IP 0.76; ICD 0.52; between locomotion and DD -0.20; IP -0.73; ICD -0.15. We conclude that locomotion is usable in the genetic improvement of lameness in Czech Holstein cows. We affirmed that lameness is genetically linked to DD and IP. With genetic selection against lameness, we can also expect a reduction in the incidence of DD and IP; all on the condition of breeders' monitoring of L and D. The study was supported by the Ministry of Agriculture of the Czech Republic, institutional support MZE-RO0718 and project QK1910320.

Culling reasons in Czech dairy cattle

E. Kašná, L. Zavadilová, J. Vařeka and Z. Krupová
Institute of Animal Science, Přátelství 815, 104 00 Prague, Czech Republic; kasna.eva@vuzv.cz

Resilience as the ability of cow to maintain the performance despite various disturbances is reflected by her longevity. A high proportion of involuntary culling indicates an inefficient use of animal resources, which oppose sustainable dairy production. In the Czech Republic, two major breeds are used for milk production: Czech Fleckvieh (C, dual purpose) and Holstein (H) cattle. Their average milk production was 7,767 kg of milk per lactation in C and 10,254 kg in H in 2020. Our aim was to analyse the causes of culling in both breeds in cows born between 2000-2015. The data were retrieved from the central performance recording database (Plemdat, s.r.o.). They included 826,444 records of C and 1,202,472 H cows. The culling codes included termination of milk recording (H3%, C6%), low production (H8%, C14%), other causes (poor exterior, temperament, milkability; H5%, C4%), high age (H1%, C1%), udder diseases (C9%, H9%), fertility disorders (H21%, C20%), difficult calving (H11%, C9%) and other health reasons (foot&legs disorders, metabolic disorders, injuries, infections; C37%, H43%). The culling patterns were similar in both breeds, with the exception of low production (higher proportion in C) and other health reasons (higher proportion in H). The most frequent specified causes (difficult calving DC, fertility disorders F, and udder diseases U) were analysed with single trait animal model, which included fixed effects of herd, year-season of the last calving, number of the last parity (1st-6th and later), age at first calving (21-35 months) and random effects of animal and residuum. The analysis showed, that later parities were associated with higher proportion of culling due to DC and U, but lower proportion of culling due to F. The higher age at first calving was associated with lower proportion of culling due to DC, F and U. These traits had a genetic component, expressed in heritability of 0.01 (DC, U) and 0.02 (F). The significant Spearman rank correlations were found between sires based on breeding values for retained placenta and DC or F, for metritis and DC, and for clinical mastitis and U, indicating genetic correlation between traits. The study was supported by the Ministry of Agriculture of the Czech Republic, project QK22020280 and institutional support MZE-RO0718.

Session 56 Poster 14

Breed-specific heterosis and recombination effects for weaning weight in composite beef cattle

L.T. Gama[1], F.S. Baldi[2], R. Espigolan[3], J.P. Eler[3], E.C.M. Oliveira[3], R. Nuñez-Dominguez[4] and J.B. Ferraz[3]
[1]CIISA – Faculty of Veterinary Medicine. University of Lisbon, Alameda Universidade Tecnica, 1300-477 Lisbon, Portugal, [2]FCAV/UNESP, UNESP, 14884-900, Jaboticabal, SP, Brazil, [3]College of Animal Science of University of São Paulo, USP, 13635-900, Pirassununga, Brazil, [4]Universidad Autónoma de Chapingo, Estado de México CP 56230, Mexico, Chapingo, CP 56230, Mexico; ltgama@fmv.ulisboa.pt

Montana was developed in Brazil as a composite beef cattle breed intended to perform well in tropical climates. It was established by combining biological groups representing *Bos indicus* cattle (N), adapted *Bos taurus* breeds (A), British breeds (B) and continental breeds (C). Currently, the contribution of the N, A, B and C breed-groups to the Montana gene pool is about 52, 13, 25 and 10%, respectively, with a retention of maximum heterozygosity of nearly 64%. Records produced by 386,961 animals were analysed to investigate the impact of heterosis and recombination effects on weaning weight in Montana cattle. Assuming a scenario of maximum heterozygosity, a linear model including a pooled effect of heterosis and recombination on weaning weight resulted in estimates of +16.4 and +11.4 kg for direct and maternal heterosis, respectively. The pooled direct and maternal recombination effect on weaning weight was -14.8 and -6.0 kg, respectively. When breed-specific heterosis and recombination effects were included in the statistical model, the maximum direct heterosis was observed for the N-B, N-C and A-B breed combinations, while for maternal heterosis the maximum was observed for the N-B combination. On the other hand, recombination direct effects were more pronounced for the N and A breed groups, but the maternal recombination effect was only relevant for the C breed-group. Our results indicate that heterosis and recombination loss for both direct and maternal effects on weaning weight are important, and should be considered when assessing different mating systems. Still, a higher level of heterosis and recombination loss is achieved in crosses involving the N group, as a consequence of its higher genetic distance relative to the other breed-groups and revealing the possible existence of positive epistatic combinations retained in purebred N animals.

Session 56 Poster 15

Genetic parameters for wool traits in Portuguese White Merino sheep

R. Cordeiro Da Silva[1], T. Perloiro[1], A. Carrasco[1] and N. Carolino[2,3,4]
[1]ANCORME – Associação Nacional de Criadores de Ovinos da Raça Merina, Travessa João Rosa, 1A, 7005-665 Évora, Portugal, [2]Centro de Investigação Interdisciplinar em Sanidade Animal (CIISA), Faculdade de Medicina Veterinária-Universidade de Lisboa, 1300-477 Lisboa, Portugal, [3]Escola Universitária Vasco da Gama, Lordemão 197, Lordemão 197, 3020-210 Coimbra, Portugal, [4]Instituto Nacional de Investigação Agrária e Veterinária, Fonte Boa, 2005-048 Vale de Santarém, Portugal; rsilva@ancorme.com

The White Merino is a native Portuguese sheep breed, mostly raised in the south and interior center regions of Portugal (Alentejo and Beira Baixa), and whose main purpose is the production of meat and wool. The National Association of Merino Breeders (ANCORME) is responsible for managing the breed's flockbook and its breeding programme. The aim of this study was to estimate the genetic parameters for wool traits in the Portuguese White Merino sheep breed. The analysed data consisted of 12,051 records of wool fibre thickness, 12,054 records of wool fibre length and 3,155 records of fleece weight, obtained respectively in 8,479, 8,481 and 2,161 animals, and collected by ANCORME between the years of 2015 and 2020. These animals with wool records and their respective pedigrees, available in the White Merino flockbook, were used to make up a relationship matrix with 15,077 individuals. The genetic parameters were estimated using the BLUP-Animal Model, by REML and univariate analysis, including fixed effects of contemporary group, defined as flock and year of measurement, month of wool measurement, gender and age of measurement (linear and quadratic covariate). As random effects, the animal's breeding value, permanent environmental effect and residual effect were considered. The heritability estimates for wool fibre thickness, length and fleece weight were moderate, respectively 0.34±0.02, 0.25±0.02 and 0.37±0.04. The permanent environmental effect estimates were, respectively, 0.06±0.02, 0.11±0.02 and 0.22±0.04. The genetic parameter estimates obtained in the current study agree with heritability estimates for Merino wool traits published in other countries. Results show the possibility of improving wool traits in the Portuguese White Merino breed.

Genetic resources of Greek sheep breeds: insights into population structure and breeding strategies

S. Michailidou[1], E. Pavlou[1], M. Kyritsi[1] and A. Argiriou[1,2]
[1]CERTH, INAB, 6th km. Charilaou-Thermi Rd, 57001, Thessaloniki, Greece, [2]University of the Aegean, Food Science and Nutrition, Metropolite Ioakeim 2, 81400 Lemnos, Greece; kyritsimaria@certh.gr

Greece is a country where sheep breeds are reared under different production systems. In most cases, the selection of the breed to be reared is based on the ability to exploit the maximum of the breeding system and the target of sheep farm (feta PDO or cheese production). Hence, evaluation of sheep genetic resources is crucial, to explore genes associated to productivity, breed traceability and resilience. In the present study we assessed the population structure, ancestry and inbreeding levels of Greek breeds reared under semi-extensive systems. In addition, through GWAS we discovered a panel of 41 SNPs to be used for traceability purposes. A total of 216 animals were analysed using OvineSNP50K v2 bead array. For comparison reasons, we included breeds that we had previously described, thus forming a total of 311 animals. Samples per breed were selected from different farms to capture most of each breed's genetic diversity. After removing SNPs for call rate (<0.98), MAF (<1%), HWE (≤1.0E-6), X-linked SNPs or those lacking chromosomal coordinates, 49,634 SNPs remained for the genomic evaluation. Results revealed that Greek autochthonous breeds maintain high levels of genetic heterogeneity according to the indices used (Ho, He, F_{IS}), with Serres and Pelagonias breeds being the most heterogeneous. Analysis of genetic differentiation between breeds (Fst, PCA) revealed that Serres and Lesvou breeds are closely related whereas Pelagonias and Assaf breeds (the latter was used as reference population) presented the greatest genetic differentiation among the studied breeds. Our analysis also confirmed the genetic proximity of Kalarritiko and Boutsko breeds, a hypothesis that was to be confirmed. Inbreeding levels showed that some Greek populations maintain high levels of inbreeding that need to be controlled whereas others preserve rich gene pools that can be used to enrich species genetic diversity. With this study, we expand our knowledge on the genetic resources of Greek sheep breeds to be used in livestock breeding schemes for improving productive traits as well as for breed and cheese traceability purposes.

Litter size variability in a divergent selected mice population for birth weight variability

L. El-Ouazizi El-Kahia[1], N. Formoso-Rafferty[2], I. Cervantes[1] and J.P. Gutiérrez[1]
[1]Dpto. Producción Animal, Facultad de Veterinaria, UCM, 28040, Spain, [2]Dpto. Producción Agraria, E.T.S.I.A.A.B, UPM, 28040, Spain; lailaelo@ucm.es

Litter size (LS) is an important trait for breeding programs in prolific species such as pigs and rabbits. Moreover, selection for uniformity around an optimum is possible. Mice could be a suitable animal model for these prolific species. A divergent selection experiment was done to modify birth weight variability in mice. The homogeneous line (L-line) showed more advantages in terms of survival, reproductive longevity, feed efficiency and fertility than the heterogeneous line (H-line), and it performed better under some environmental stressors such as feed restriction or climate changes. The objective of this study was to evaluate the genetic component of LS and its variability analysing also phenotypic and genetic trends. It was used a total of 3,687 LS records from 2,239 females after 29 generations of selection and a pedigree of 32,885 records. The analysis was carried out using a heterogeneous model that included the generation and the parturition number as fixed effects in both, the trait and its variability. The additive genetic variance obtained for LS was 2.16 (±0.27) and for LS variability was 0.23 (±0.05). There was a negative genetic correlation between the LS mean and its variability (-0.84±0.08), which indicated that the higher litter size, the smaller variability. The heritability of LS was 0.25 (±0.04), the heritability across level of effects showed an irregular trajectory through 29 generations changing in a range between 0.18 (±0.04) and 0.37 (±0.08). Regarding the parturition number, the LS heritability was higher in the first parturition (0.29±0.04) than in the second (0.21±0.03). The genetic trends showed an increase in LS mean in L-line reaching an average breeding value of 1.16 pups in the last generation while the H-line got a value of -0.50 pups, obtaining a higher LS in L-line (9.02 vs 7.59 pups) in the last generation. The correlated response also affected the LS variability that showed a reduction of 9.3% in the L-line residual variance and an increase of 10.1% in the H-line. In conclusion, after the selection performed for birth weight variability, the L-line showed more uniformity and higher LS than H-line.

Estimation of the genetic (co)variance components of reproductive efficiency in Retinta beef cattle

R.M. Morales[1], J.M. Jiménez[1,2], A. Molina[1], S. Demyda-Peyrás[3] and A. Menéndez-Buxadera[1]
[1]Cordoba University, Genetics, Rabanales Campus, N-IV, km 396, 14014, Spain, [2]CEAG The Council of Cádiz, Jerez de la Frontera, 11400, Spain, [3]Veterinary School, Animal Production, National University of La Plata, 1900, Argentina; v22mocir@uco.es

Fertility is one of the most important traits in any livestock production system. But it is one of the most complicated to analyse from a genetic point of view since the obtention of large and reliable phenotypic datasets is difficult and this trait is subject to a great environmental influence. Both situations are particularly important in extensive beef production, in which reliable fertility data of cows is extremely scarce. For that reason, Reproductive efficiency (Re), estimated as the deviation in the calving number of a cow at a certain point of its productive life in comparison with the calving number of an 'ideal cow', is being suggested as an interesting trait to improve in that sense. In this study, we analysed the variation of Re across the life of 12,554 cows belonging to the Retinta Spanish cattle breed using random regression models. Re was estimated per cow at each calving. The regression coefficients modelled by a Legendre polynome representing the (co)variance variations for Re by caving number (Np (10 classes)). Results showed that h^2 increased across the parity, from 0.24 at np 1 to 0.53 at np 10. The genetic correlation between the different calving ranged between 0.599 and 0.996 in function of the moment of the animal's life and the degree of adjacency between both calving. In addition, the evaluation of the individual curves of the breeding values throughout the calving trajectory, shows animals with very different patterns, existing animals with high levels of reproductive efficiency in the first years of reproductive life, which quickly lowered their productivity, compared to others with a inverse pattern, and robust animals that maintained a similar level of efficiency throughout their lives. This finding suggests that evaluating the fertility of the cow by indirect methods at early ages should be made with caution to avoid the selection of individuals showing a negative trend in reproductive efficiency across life, and that early selection based on the Re could bias the genetic gain of the population.

Software for the automated correction of pedigree data

N. Burren and A. Burren
Bern University of Applied Sciences, School of Agricultural, Forest and Food Sciences, Länggasse 85, 3052 Zollikofen, Switzerland; alexander.burren@bfh.ch

Mistakes in pedigree may occur due to animals having the wrong pedigree or administrative mistakes. Applying parentage control animals having the wrong pedigree can be identified. Administrative mistakes comprise for example, assignation of the wrong sex to animals, cycling pedigrees or animals recorded with birth dates earlier than their parents. Manual corrections are difficult, time-consuming or even impossible, depending on the error. Several software tools (CFC, Endog, PEDIG, POPREP) have problems loading data with administrative pedigree mistakes. Manual pedigree corrections are difficult, time-consuming or even impossible, depending on the error. We have therefore searched for an effective solution for the correction of common errors in pedigrees. For this purpose, we have developed the tool PediCorr 2.0 using visual studio, which can be used to correct all common errors in a pedigree. The tool processes pedigree files that include a minimum of the three columns, progeny, sire and dam. IDs can be read in as character, numeric or integer variables. There is also the option of processing additional columns, such as date of birth and sex. The following errors are corrected automatically after the data have been imported: multiple entries in the column on progeny, missing entry on progeny for parent animals, animal entered as the father and mother, wrong date of birth, wrong sex, cycling pedigree. For animals that appear simultaneously as the father and mother, the correction is made based on frequency. For example, if an animal appears once as a father and once as a mother, then both entries are deleted. Conversely, if it appears more frequently as a father, then we assume it is a male, and the entries are only deleted from the dam column. The checks on dates of birth are carried out in a comparison with the dates of birth for the parents. This allows us to establish which data are to be deleted when the dates of birth are wrong (progeny, sire, dam, progeny + sire or progeny + dam). The sex can only be checked for animals that are either entered into the column sire or dam. We hope that with this tool, PediCorr 2.0, we have contributed towards simplifying the correction of errors in pedigrees.

Genetic analysis of productive life in the Florida goat breed using a Cox proportional hazards model

C. Ziadi[1], J.P. Sánchez[2], R. Morales[1], E. Muñoz-Mejías[3], M. Sánchez[4], M.D. López[5] and A. Molina[1]
[1]University of Córdoba, Department of Genetics, Campus de Rabanales, 14014 Córdoba, Spain, [2]IRTA, Department of Animal Breeding and Genetics, Caldes de Montbui, 08140 Barcelona, Spain, [3]University of Las Palmas de Gran Canaria, Department of Animal Pathology and Animal Production, Campus Universitario Cardones de Arucas, 35413 Arucas, Spain, [4]University of Córdoba, Departement of Animal Production, Campus de Rabanales, 14014 Córdoba, Spain, [5]ACRIFLOR, Department of Animal Production, Campus de Rabanales, 14014 Córdoba, Spain; ziadichiraz4@gmail.com

Productive life is an economically important trait directly related to the profitability of dairy farms, and it has hardly been studied how increasing milk production can affect longevity. In this study, we genetically analysed longevity considering the length of productive life (LPL) defined as the age at first kidding until the last collected record, culling/death or data censorship. The database was provided by the National Association of Florida Goat Breeders. Data consisted of 27,102 LPL records of Florida females collected between 2006 and 2020 and the pedigree included 56,952 animals. A preliminary analysis was performed using a Weibull hazard survival model to determine the significance of the effects included in the model using a likelihood ratio test. The effects evaluated were the age at first kidding, herd-year-season of doe birth as time-independent effects, and age at kidding, and herd-year-season of kidding as time-dependent effects. Estimation of genetic parameters was inferred from a full Bayesian analysis using Gibbs sampling. Data were analysed with a Cox proportional hazards model with the COXF90 program of the BLUPF90 family programs. The marginal posterior mean estimates of additive genetic variance and heritability of LPL were 0.28 and 0.17, respectively. The magnitude of heritability obtained for LPL suggests that a substantial response to selection may be expected for longevity in the Florida dairy breed.

Evaluation of stature in Polish dairy cattle using a single-step method

M. Pszczola[1,2], K. Bączkiewicz[1], W. Jagusiak[3], M. Skarwecka[4], T. Suchocki[4,5], M. Szalanski[1], J. Szyda[4,5], A. Zarnecki[4], K. Zukowski[4] and S. Mucha[1]
[1]Centre for Genetics, Polish Federation of Cattle Breeders and Dairy Farmers, Dabrowskiego 79A, 60-529 Poznan, Poland, [2]Poznan University of Life Sciences, Department of Genetics and Animal Breeding, Wołyńska 33, 60-637 Poznań, Poland, [3]University of Agriculture in Krakow, Department of Genetics, Animal Breeding and Ethology, al. Mickiewicza 24/2, 30-059 Kraków, Poland, Poland, [4]National Research Institute of Animal Production, Department of Cattle Breeding, Krakowska 1, 32-083 Balice, Poland, [5]Wrocław University of Environmental and Life Sciences, Department of Animal Genetics, Kożuchowska 7, 51-631, Wrocław, Poland; marcin.pszczola@puls.edu.pl

Currently, genomic breeding value estimation in Poland is performed as a multi-step evaluation. Many countries are currently investigating or implementing a single-step evaluation, since it allows incorporating all sources of information on individuals' additive genetic merit. Stature is often used as a test trait. The goal of this study was to compare the breeding value estimates from conventional and single-step models. We used data collected from the routine evaluation on 926,937 first lactation Polish-Holstein Friesian cows born between 1992 and 2019 and genotypes from 107,536 animals. The variance components were obtained from national evaluation. The model included squared regression on month of calving, herd-year-season-classifier class and lactation stage class as fixed effects as well as random additive animal and residual terms. Genotypes, after quality checks, were combined with the pedigree information into the H matrix. Validation was based on a reduced dataset created by removing the last four years of phenotypic observations. The accuracy was assessed as correlation of de-regressed bulls' breeding values from the conventional model on a full data set with breeding values from other scenarios. Single step method based breeding values estimated with the full data set yielded similar results to the conventional evaluation. For the reduced data, the single-step method was clearly superior to the pedigree based evaluation. The results show that implementing single-step would increase the accuracy of young bulls' evaluation for stature.

Population structure of Istrian and Croatian Coloured goat

I. Drzaic[1], V. Orehovački[2], L. Vostrý[3], I. Curik[1], N. Mikulec[4] and V. Cubric Curik[1]
[1]University of Zagreb Faculty of Agriculture, Department of Animal Science, Svetošimunska c. 25, 10000 Zagreb, Croatia, [2]Ministry of Agriculture, Department of the Livestock Gene Bank, Poljana Križevačka 185, 48260 Križevci, Croatia, [3]Czech University of Life Sciences Faculty of Agrobiology, Food and Natural Resources, Kamycka 129, 16500 Prague, Czech Republic, [4]University of Zagreb Faculty of Agriculture, Department of Dairy Science, Svetošimunska c. 25, 10000 Zagreb, Croatia; ikovac@agr.hr

Native Croatian goats are local breeds kept in an extensive system with low production. Local breeds are part of the genetic and socio-cultural heritage of the local community. For this reason, it is very important to preserve the genetic diversity of the remaining breeds, most of which are trapped in unselected autochthonous breeds. This study aimed to analyse the genetic diversity and conservation status of two indigenous Croatian goat breeds, the Istrian goat and the Croatian Coloured goat, using genome-wide SNP data. Population structure was analysed using 11 Istrian goats, 33 Croatian Coloured goats, and 506 additional animals belonging to 19 European goat breeds and 7 Iranian Bezoar. We used a graphical approach to illustrate the relationship between the analysed breeds in terms of their population structure. In addition, we determined the migration patterns of all goat breeds presented in this study. Our analyses revealed a fine population structure of Croatian goat breeds and their dependence on geographical origin and barriers. The mean observed heterozygosity for the Istrian goat and Croatian coloured goat was 0.39 and 0.40, respectively. The mean fixation index (F_{ST}) between the Istrian and Croatian coloured goats was 0.046, indicating an intermediate level of genetic differentiation between these two populations. Principal component analysis (PCA) showed distinct clustering of the two breeds, while some Istrian goats showed a clear clustering with the Saanen breed. The Croatian coloured goat was closest to the Carpathian goat. The results obtained with Treemix software confirmed these results and showed gene flow from Croatian coloured goat to Istrian goat. Our results will contribute to the breeding and conservation strategies for the native Croatian sheep breeds.

Session 56 Poster 23

Inbreeding depression on reproductive traits in Mertolenga beef cattle

A. Vitorino[1,2], J. Pais[3], N. Carolino[1,2] and Consórcio Bovmais[4]
[1]CIISA, Av. Universidade Técnica, 1300-477 Lisboa, Portugal, [2]INIAV, I.P., Fonte Boa, 2005-048 Vale de Santarém, Portugal, [3]ACBM, R. Diana de Liz Horta do Bispo, Apartado 466, 7006-806 Évora, Portugal, [4]Consórcio BovMais, Portugal, Portugal, Portugal; andreia.vitorino93@gmail.com

The aim of this study was to evaluate the inbreeding effects for calving interval (CI) and productive longevity (PL) in Mertolenga beef cattle. The data set of 234,190 pedigreed individuals over sixty years, available in the Mertolenga Herdbook was analysed. The inbreeding coefficients (Fi) were derived from an additive relationship matrix, registering an average of 0.059±0.088. This study focused on 260,821 records of CI from 47,640 cows and with an average of 436 days and 35,179 records of PL with an average of 113.1 months, both collected between 1970 and 2021. The inbreeding depression was expressed as a partial linear regression coefficient estimated via a single trait BLUP- Animal Model. For the IC analysis it was included the fixed effects of herd × year of calving, season, gender and sex of the calf, age of calving as a linear and quadratic covariable, and individual inbreeding coefficient; as random effects, the additive genetic, permanent environmental and residual error. For the PL analysis it was included the fixed effects of herd × year of first calving, age of first calving (linear and quadratic covariable) and individual inbreeding coefficient; as random effects, the additive genetic and residual error. The estimated linear regression coefficients of the CI and PL on the inbreeding of the cows were, respectively, +0.92±0.04 days/1%Fi and -0.46±0.02 months/1%Fi. These study found an important decline on the reproductive abilities of Mertolenga cattle due to inbreeding effect and confirmed that the inbreeding depression can reduce the mean phenotypic value of important reproductive traits in beef cattle.
Funded by PDR2020-101-3112 and 2020.09881.BD (Bolsa FCT)

Session 56 Poster 24

Importance of valid and representative phenotypes in novel traits using single-step GBLUP

H. Schwarzenbacher[1], J. Himmelbauer[1] and B. Fuerst-Waltl[2]
[1]ZuchtData GmbH, Dresdner Straße 89/B1/18, 1200 Vienna, Austria, [2]University of Natural Resources and Life Sciences, Gregor-Mendel Str. 33, 1180 Vienna, Austria; schwarzenbacher@zuchtdata.at

Currently many countries work on the implementation of single step GBLUP for their cattle populations. The strengths of the approach are particularly important for novel traits, since phenotypes are only available for young birth cohorts, information on SNP effects must be derived directly from phenotyped and genotyped animals. Often, novel phenotypes are difficult and expensive to measure, which might limit recording to dedicated farms or sub populations. In theory, these farms should be representative for the whole population. However, in practice novel phenotypes often come from well-managed farms with animals of superior genetic merit as selection is usually based on genomic information in these farms. With the limited amount of available phenotypes from a few birth years, complete and accurate recording is of paramount importance. Often these phenotypes are collected directly by the farmer. This might lead to incomplete, inaccurate or even erroneous information. In this study, we use simulations to analyse the impact of suboptimal phenotyping schemes on the reliability and bias of subsequent single-step genomic evaluations. Evaluated scenarios include: (1) the impact of superior or inferior management of farms in the phenotyping regime; (2) the impact of non-random missing data; and (3) the impact of manipulated data on the quality of evaluations.

Session 56 Poster 25

Significant contribution of parent-of-origin effects to genetic variation of milk production traits

R.E. Jahnel, I. Blunk and N. Reinsch
Research Institute for Farm Animal Biology (FBN), Institute of Genetics and Biometry, Wilhelm-Stahl Allee 2, 18196 Dummerstorf, Germany; jahnel@fbn-dummerstorf.de

There is ongoing research on the importance of parent-of-origin effects (POEs), including gametic imprinting, on the genetic variability of quantitative traits. Genes are genomically imprinted when their expression is limited to one of the two inherited gametes depending on their parental origin. So far, POEs were found mainly in growth and body composition traits. Milk production traits, in contrast, have been let under investigated. Furthermore, none of the few previous studies on milk production traits employed a random regression approach for test day records. This study aimed at estimating an imprinting variance for each day in milk, by analysing the traits milk yield, percentages and yields of fat and protein, milk urea content and somatic cell score from a dataset of 256,089 milk samples of 31,345 German Holsteins in Mecklenburg West-Pomerania. Statistical Analysis using POE random regression models (RRM) was performed with Echidna MMS. The results revealed significant imprinting variances in all traits with the exception of protein percentage. We attributed shares between 3.7% and 15.5% of the total additive genetic variance due to POEs. Largest shares of POE variance were found in the expression of traits in the most economically important lactation stage during peak lactation. Genetic correlations between paternally and maternally transmitted gametes ranged from 0.87 to 0.97 for the six traits with significant POEs. In conclusion, POEs proved to be significant in traits other than growth and body composition. Our POE-RRM may also be useful for investigations of other traits, where repeated observations over the age trajectory are available as e.g. for growth, feed intake or egg production traits.

Session 57 Theatre 1

Is sustainable transition compatible with value chains industrialization in developing countries?

G. Duteurtre
CIRAD, UMR Selmet, Campus de Baillarguet, 34398 Montpellier Cedex 5, France; duteurtre@cirad.fr

In many developing countries, livestock value-chains are undergoing rapid transformations. Those changes represent both challenges and opportunities for a transition towards more sustainable food systems. We discuss those challenges and opportunities with reference to various situations taken in developing countries. Following the liberalization of national and regional economies, international trade and long distances transport of raw materials and livestock products have increased, which has had a strong impact on carbon emission. In parallel, major agri-food corporation have invested in livestock genetics, feed industry, milk and meat processing units, or food distribution in developing countries, leading to a growing capital concentration in the agri-food sector, and to the emergence of mega-farms with high environmental and social costs. In the same time, the growing importance of quality standards for sustainable and equitable foods has supported value chains initiatives leading to organic products marketing, food safety certification, geographical indications, or other sustainable labels and trademarks. Public policies focused on local development dynamics have also supported social business-models such as micro-credit schemes, cooperatives, producers' organizations, or small-scale processing units with interesting returns in terms of shared added value. We conclude by underlying the importance of local territorial governance to promote sustainable trade-offs between positive and negative impacts. Local organizations such as innovation plate-forms, municipal bodies, decentralized public services, etc. can support local entrepreneurs, producers' organization and services providers to build sustainable value chains initiatives. The challenge of those local organizations is to promote simultaneously economic, social and environmental impacts of value-chains transformation. We illustrate the concept of territorial governance by some example taken in Africa, Southeast Asia and Amazonia.

Session 57 Theatre 2

Consumer's expectations about quality of livestock products: focus groups in 4 European countries

E. Sturaro, F. Bedoin, V. Bühl, A. Cartoni Mancinelli, C. Couzy, R. Eppenstein, S. McLaughlin and F. Pagliacci
INTAQT Consortium, Viale dell'Università 16, 35020 Legnaro, Italy; enrico.sturaro@unipd.it

Food chain actors lack objective, robust and reliable information to meet consumers' expectations in relation to the global quality of livestock products from the various European husbandry systems. This contribution presents some preliminary results of INTAQT project (EU Horizon 2020), which aims to perform an in-depth multi-criteria assessment of the relationships between animal husbandry and qualities of livestock products. In specific, this study aims to collect insights on the expectations of consumers about intrinsic and extrinsic quality of chicken meat, beef and dairy products. To achieve this goal, four focus groups with consumers were organized in Italy, France, Germany and the UK. The main objectives of all focus groups were to assess the type of information consumers seek for when purchasing chicken, beef and dairy products, and the extent to which a multicriteria scoring tool would be useful to them. A common methodological approach was developed, based on 3 main questions: (1) what is important when purchasing chicken meat, beef or dairy products; (2) which are the missing pieces of information on products' quality; (3) what is the interest in including some criteria to develop a multicriteria scoring tool. The 4 focus groups were organized as virtual meetings (due to the COVID-19 limitations) with an average of 10 participants per focus group in each country. The selected participants were living in urban, peri-urban and rural areas, and they were used to consume livestock products. Results evidenced that consumers are very interested in both origin and production systems of the livestock products and ask for the indication of extrinsic attributes (e.g. sustainability and animal welfare). In general, the consumers showed interest on the development of a standardized multicriteria scoring tool, but they also raised critical points on its reliability. The following criteria are of particular interest for the development of such tool: traceability, nutritional values, husbandry systems and animal welfare. The results of this study are part of the multi-actor approach of INTAQT project, and they will be used to support and orientate the next research activities.

Session 57 Theatre 3

Antibiotic-free pig supply schemes in France: a lever for valorisation and progress

C. Roguet and H. Hemonic
IFIP – Institut du Porc, Economy, La Motte au Vicomte, 35651 Le Rheu, France; christine.roguet@ifip.asso.fr

The European project ROADMAP promotes transitions for prudent and responsible antimicrobial use in livestock farming. In this project, we analysed the antibiotic (AB)-free schemes in pig industry in France. In addition to documentary research, our analysis was based on interviews in 2020 with 18 breeders, half of them in an AB-free scheme, and 12 veterinarians or quality managers. In the early 2010s, the French pig sector started to develop AB free private schemes to meet the demand of some retailers and to communicate on the efforts made by the breeders. In the absence of a collective and standardized response, these specifications have proliferated as a way for companies to stand out from their competitors and improve their image. Unlike GMO-free, the AB-free claim is not subject to any legal definition, leading to very diverse specifications and labelling. The specifications are evolving, in a process of progress. The claim 'AB-free from 42 days' is being replaced or supplemented by 'since birth'. It is always included in a more global communication including animal welfare, environment protection, etc. In 2020, the AB-free lines represent about 15% of French pig production. This market is 'mature'. The first motivation of breeders to enter a scheme is economic: they perceive a bonus, which varies according to the constraints and their degree of supply from their cooperative. They also appreciate the security of prices and outlets. According to them, the AB-free scheme makes it possible to promote their existing good practices of AB low use, without necessarily being the trigger for change. Breeders are also motivated to restore 'the image of breeding' and to 'produce a more rewarding product'. The commitment of the breeder in an AB-free line is materialized by a contract of 3 to 5 years which fixes the volume of pigs and the added value. Depending on the scheme, pigs treated with an antibiotic benefit or not from the bonus. This can lead the farmer not to treat sick pigs properly, a problem raised in the interviews. Thanks to changes in regulations, vet prescriptions and breeders practices, antibiotics use in the French pork sector fell by 55% between 2011 and 2020. The creation of the AB-free private specifications has made it possible to promote this strong reduction.

Session 57 Theatre 4

Inclusion of citizens in the design of livestock farming: case of mid-mountain dairy system

P. Coeugnet[1], G. Vourc'h[1], J. Labatut[2] and J. Duval[1]
[1]INRAE, Site de Theix, 63122, Saint Genès Champanelle, France, [2]INRAE, 5 Boulevard Descartes, 77454, Marne-la-Vallée, France; philippine.coeugnet@orange.fr

For several years, livestock farming has been the subject of increasing concerns for citizens. One of the ways to reinforce dialogue between society and agriculture would be to include citizens in the design of farms. If participatory research has increased in recent years in agriculture it does not seem able to support a new collective in generating innovative solutions in a context where the issues are multiple. Various innovative design methods, initially developed for the industrial sector, make it possible to explore new path of innovation based on interactions between stakeholders but these methods have paid no attention to how to include citizens. In this context, our study aims to adapt one of them, the DKCP method, in order to include citizens in the design of sustainable dairy farming systems with dairy stakeholders and researchers in a mid-mountain region in France. The DKCP method is divided into a phase D of diagnosis of the stakeholder system, a phase K of exchange of knowledge, a phase C of creativity and a phase P of project development. During phase D, we carried out 30 surveys and 2 focus groups with a diversity of stakeholders in the territory and researchers in order to understand the context. Then, 2 one-day workshops, alternating the 3 phases K, C and P were organized. About thirty participants were invited to these workshops: researchers working on the themes related to the topics discussed, dairy stakeholders and citizens. The first workshop aimed to rethink the relationships between livestock farming and society considering solutions allowing citizens to participate in the evolution of livestock farming system. During the second workshop, the participants worked on the design of farms respectful of overall health and welfare through the subject of the dairy calves. They explored possible solutions at different scales. Finally, a working meeting was held to select the projects and develop an action plan for their implementation. The analysis of this design process produced knowledge on how to include citizens in the design of future livestock farming and on the benefits of this inclusion.

Session 57 Theatre 5

Developing local protein resources in monogastric feeds to promote the agroecological transition in

C. Escande, M. Grillot and V. Thénard
INRAE, AGIR, Univ. Toulouse, Castanet Tolosan, France; vincent.thenard@inrae.fr

In France, the intensification and specialisation of territories and farms have led to a dependence of livestock farms on the purchase of feed, particularly proteins, produced outside their territory, or even outside the country. This is particularly the case for monogastric animals: pigs, chickens and ducks. The main protein food used for animal feed is oil cake, a co-product of the processing of soya, rape or sunflower seeds. In the hillsides of the Lauragais Tarnais, crop and livestock were formerly integrated. Now, as specialization towards cereal crop occurred, livestock farming is residual. Livestock are located on the rare grasslands (herbivores) or associated with field crop farms (monogastric animals). However, despite a growing number of cereal-growing farms and strong crop diversification, crop and livestock are disconnected. Livestock farms and particularly those with monogastrics are strongly dependent from imported proteins. Strengthening protein autonomy for monogastric farms at farm or region level is a strong regional issue. It is supported by the local agricultural actors, for economic and ecological reasons but also as a challenge for the production of local quality products. It is an opportunity to initiate the agroecological transition of the region. Reconnecting crop and livestock productions implies changes for the livestock farms (e.g. rations), for crop farms (e.g. new crops, other rotations) and agri-chains (e.g. toasting soya). Identify the different levers and limits of a reconnection between crops and livestock as well as its local stakeholders could allow to strengthen protein autonomy. There is also a need to provide knowledge on the potentialities of the Lauragais territory in order to inform and inspire the different actors(/stakeholders ?). All that could be done through interviews with experts on livestock feeding, local experts on monogastric agri-chains and local farmers. The participatory approach will therefore be particularly important in this work to ensure the implementation of a dynamic that favours the territory's protein autonomy. This presentation explores how strengthening protein autonomy on mixed crop-livestock farms in the Lauragais Tarnais contributes to the agroecological transition.

Session 57 Theatre 6

The mix of species within farms contributes to secure and regularise the supply of the sectors

S. Cournut[1], S. Mugnier[2], C. Husson[2] and G. Bigot[1]
[1]Université Clermont Auvergne, AgroParisTech, INRAE, VetAgro Sup, Territoires, Clermont-Ferrand, VetAgro Sup, 89, avenue de L'Europe – BP 35, 63370 Lempdes, France, [2]Université Clermont Auvergne, AgroParisTech, INRAE, VetAgro Sup, Territoires, Clermont-Ferrand, Institut Agro Dijon, 26 Boulevard du Docteur Petitjean, 21079 Dijon cedex, France; genevieve.bigot@inrae.fr

In the current context of increasing uncertainties and incentives for agro-ecological transition, the mixing of animal species within farms is one of the solutions put forward. Studies have shown that the combination of sheep and suckler cattle on a farm makes it possible to make better use of resources, secure income and diversify work. But few studies have considered the sector scale. The purpose of this communication is to show that the mix of species can also be of interest to the sectors. On-farm (37) and online (105) surveys of mixed sheep and cattle farms and specialised suckler sheep farms in the northern Massif Central were carried out. We showed that to adapt their farm to climatic, economic and workforce-related hazards, farmers used mechanisms related to the combination of the two species: modifying the ewe/cow ratio, breeding periods, worker versatility, grazing management and allocation of resources between species. The first lesson of this study is that, by adapting to different types of hazards, mixed farming systems contribute to the security of supply of lambs in the sector. We have also highlighted three profiles of lamb sales on farms: the first corresponding to sales spread over 5 to 7 months centred on the summer, the second with sales spread over 8 to 12 months and the last with sales over short periods (2 to 4 months) outside the summer. The first profile seems to be more specific to mixed farms. These sales profiles are complementary at the yearly scale, in particular because the first profile, found essentially on mixed farms, allows for the supply of lambs in the summer at a time when fewer farms in the other profiles are supplying them. This complementarity of sales profiles can be observed among all the operators. Thus, the second lesson of this study is the interest of mixed farming systems for the regularity of the supply of lambs to the sector.

Perceived resilience in Veneto beef farmers

G. Marchesini[1], I. Andrighetto[1] and F. Accatino[2]
[1]Università degli Studi di Padova, Viale Dell'Uiversità, 16, 35020 Legnaro (PD), Italy, [22]UMR SADAPT, INRAE, AgroParisTech, Université Paris Saclay, 16 rue Claude Bernard, 75005 Paris, France; giorgio.marchesini@unipd.it

Veneto's beef production is strongly characterized by the presence of specialized fattening units relying on imported animals. The purpose of this study was to analyse how farm resilience is perceived by Veneto breeders in a period of climate, social and economic challenges. We submitted a questionnaire to about sixty farmers, obtaining 25 replies. The questionnaire was composed of a first part asking the structural and managerial characteristics of the farm, and a second part related to the importance attributed by the breeder to the functions of his own farm (e.g. food production or valorisation of natural resources), to the risk management strategies applied until now, the main challenges that the farm will have to face in the near future, and the perception that the farmer has about the resilience of his farm in terms of robustness, adaptability and transformability. Farms were classified according to their size, in terms of average heads raised per year, into small (≤500), medium (501-1,800) and large (>1,800). Data on function and resilience capacities were analysed through ANOVA, whereas the frequency of management strategies used and the challenges perceived by farmers were analysed by using Fisher exact test. The main function for all farm sizes was to 'provide a sufficient agricultural income' (28.6% of importance assigned). Large farms gave more importance to providing good working conditions to employees ($P<0.05$), whereas for small farms it was more important to care about rural areas attractiveness ($P<0.05$). Risk management strategies applied in the last 5 years deeply differed among farm different sizes ($P<0.01$), whereas no difference in resilience, that was on average 3,76 in a 1 to 7 scale, was perceived by farmers. Main challenges also significantly differed between small, medium and large farms ($P<0.001$), but on the whole economic challenges were considered the most important (45.3%), followed by social (20.3%), institutional (18.7%) and environmental (10.9%) ones. The low resilience perceived can be explained by the high specialization of farms. More care on the adoption of suitable risk management strategies in different farms must be warranted.

Valuation of local breeds: methodological questions following interviews with Pyrenees goat farmers

A. Lauvie[1], F. Thuault[2], M.O. Nozières-Petit[1], B. Dupuis[3] and N. Couix[4]
[1]INRAE UMR SELMET, 2 place Viala, 34060 Montpellier Cedex 01, France, [2]Pyrenees goat breed association, 32 avenue du Gl de Gaulle, 09000 Foix, France, [3]ISTOM and Montpellier Institut Agro student, 2 place Viala, 34060 Montpellier Cedex 01, France, [4]INRAE UMR AGIR, Chemin de Borde rouge, 31326 Castanet Tolosan Cedex, France; anne.lauvie@inrae.fr

Animal domestic biodiversity, and particularly local breeds, are considered as an interesting potential lever for agroecological transition. It is important to understand the role played by local breeds in livestock farming systems, and the perception of this role by concerned actors. To do so, it is a key point to take into account individual and collective values associated to the animals of a given breed, and valuations of those local breeds and local breeds animals. Moreover, It is necessary to consider values without restricting the approach to economic values, as those values can be very diverse (for instance social, esthetical, etc.). As a consequence, taking into account those diverse values and valuations raise methodological questions. How approaches using semi structured interviews, quite common in livestock farming system researches, can tackle this question of values and valuation? Does this question involve specific ways to analyse qualitative data? Based on a reflexive analysis of Pyrenees goat farmers' semi structured interviews, we show: (1) the interest of gathering qualitative data about personal trajectories of the farmers to identify valuation moments during those trajectories; and (2) the limits of interviews to identify valuation processes. We finally put in perspective this analysis with bibliography from diverse scientific fields. We make hypothesis about the interest of narratives production and practices observation, to go further in understanding values and valuations associated with the use of local breeds. We conclude that to tackle such questions, enlarged interdisciplinary approaches are needed.

Session 57 Theatre 9

Implementing GHG models as advisory tools for milk, beef, pig, sheep, poultry, and crop production

S. Samsonstuen[1], H. Bonesmo[2], B.A. Åby[1], E.G. Enger[3], E. Kjesbu[4], M. Bergfjord[5], R. Okstad[5], S. Skøien[6] and T. Barman[7]

[1]Norwegian University of Life Sciences, P.O. Box 5003, 1432 Ås, Norway, [2]NIBIO, P.O. Box 115, 1431 Ås, Norway, [3]Norsvin SA, Storhamargata 44, 2317 Hamar, Norway, [4]Landbrukets Dataflyt SA, Hollendergata 5, 0190 Oslo, Norway, [5]Systor Trondheim AS, Bromstadvegen 2, 7045 Trondheim, Norway, [6]NLR, Osloveien 1, 1433 Ås, Norway, [7]Landbrukets Klimaselskap SA, P.O. Box 9354, 0190 Oslo, Norway; stine.samsonstuen@nmbu.no

Through the joint project Climate Smart Agriculture, the agricultural sector in Norway have successfully implemented the whole-farm models HolosNor models as farm advisory tools for milk, beef, pig, sheep, poultry, and crop production. The HolosNor modes are empirical models based on the methodology of the Intergovernmental Panel on Climate Change with modifications to Norwegian conditions. The models estimate direct emissions of methane (CH_4), nitrous oxide (N_2O), and carbon dioxide (CO_2) from on-farm livestock production and includes indirect emissions of N_2O and CO_2 associated with inputs used on the farm in addition to including soil carbon balance through the ICBM model. The digital GHG Calculator automatically collects data from sources the farmer already uses for farm management, such as herd recording systems, manure planning systems, farm accounts, concentrate invoice, dairy, slaughterhouse, in addition to site-specific soil and weather data. Based on the collected data, both total emissions from the production and emission intensities for the different products are estimated. The emission intensities are shown by source relative to a reference group consisting of farms with the same type of production and production volume. Using the GHG Calculator, the farmers have the unique opportunity to have tailor-made mitigation plans to reduce emissions from the farm trough certified climate advisors. Participation and results from the GHG Calculator will be presented in addition to experiences from implementation of a GHG model as a farm advisory tool for commercial farms.

Session 57 Theatre 10

Improving demonstration activities to foster change towards sustainable organic livestock farming

D. Neumeister and C. Evrat Georgel

Institut de l'Elevage, 149 rue de Bercy, 75595, France; delphine.neumeister@idele.fr

The NEFERTITI project aims to stimulate the uptake of innovation, improve peer to peer learning and connect farming networks across Europe to facilitate the transition to more sustainable livestock farming. It is organised around 10 interactive thematic networks, which are divided into regional groups (called hubs), composed of farmers, advisors, trainers, researchers and policy makers. They organise open days on regular farms or experimental farms and farmer exchange days. This article focuses on the learnings of the French hub on 'robust organic dairy farming systems'. After 3 demonstration campaigns and more than 15 events organised, the main success identified lies in the maintenance of a real cohesion within the hub, thanks to a dynamic operation, regular and constructive exchanges and phases of debate and capitalisation on technical subjects. The events were varied, some being in small committee to maximize exchanges between peers, others being intended for a large public in farm to maximize the awareness and the contribution of knowledge by on-farm visualization. All the topics studied (single milking, nursing cows, etc.) were systematically chosen according to the expectations of the hub farmers, in order to meet a need in the field. Among the main lessons learned, the Covid-19 pandemic strongly challenged the face-to-face events and forced the hub to adapt, notably by developing virtual demos. Despite technical uncertainties (handling of online platforms, connection problems, etc.), breeders as well as facilitators and operators in the sector have gradually become accustomed to online tools in order to continue exchanging. The Nefertiti project has provided substantial support for this transition, with the publication of methodological guides and the training of project members. The main areas of progress lie in the evaluation of the demonstration days: it is not easy to know what the participants have retained from the exchanges and, above all, what they will actually put into practice on their farms. A better consideration of the impact of the demonstration activities would allow for even better support of organic dairy farming systems towards greater resilience.

Session 57 Poster 11

The sustainability of buffalo livestock

A. Fierro[1], A. Forte[1], G. Di Vuolo[2], G. Cappelli[2], E. De Carlo[2], L.F. Zuin[3], M. Giampietro[4] and D. Vecchio[2]
[1]University of Naples Federico II, Laboratorio di Urbanistica e Pianificazione Territoriale (LUPT), Naples, 80134, Italy, [2]Istituto Zooprofilattico Sperimentale del Mezzogiorno, Unit Animal Science and Welfare- CReNBuf, Portici, 80055, Italy, [3]Universidade de São Paulo (FZEA-USP), Departamento de Engenharia de Biossistemas da Faculdade de Zootecnia e Engenharia de Alimentos, São Paulo, 13635-900, Brazil, [4]Universitat Autònoma de Barcelona, Institut de Ciència i Tecnologia Ambientals, Barcelona, 08193, Spain; domenico.vecchio@izsmportici.it

The sustainability of animal husbandry represents a topic, often with explicit accusatory arguments, on which a substantial part of the narrative is focusing. The following contribution intends to present a relational model of representation of animal husbandry which partially rehabilitates the sector but which in any case highlights the many metabolic criticalities of the sector. The approach used is that of MuSIASEM (MultiScale Integrated Accounting of Societal and Ecosystem Metabolism), a relational multicriteria approach that allows to evaluate different metabolic features of the system. Through a structural and managerial characterization of the system (including animal welfare strategies) and of the related matter, energy and monetary flows, it is possible to represent the system in a more comprehensive way. The accounting method allows to evaluate the metabolic performance of the system under observation by means of 4 level: (1) feasibility, shows the constraints imposed by the ecosphere (both resources availability and the ability to receive wastes); (2) viability, characterizes techno-economic strategies adopted by the system; (3) desirability, is about normative and institutional behaviour of the several social actors involved; (4) externalization, describes the opening of the system and therefore its dependence on external resources, with the related environmental and social externalized impacts. Specific analytical examples will be presented for the buffalo livestock sector analysed for the Campania Region, Southern Italy. Funding This work was supported by RC IZS ME 8/18 RC Financed by the Italian Ministry of Health.

Session 57 Poster 12

The keys to Reine Mathilde's success, a development program for the organic dairy sector

D. Neumeister and C. Evrat Georgel
Institut de l'Elevage, 149 rue de Bercy, 75595 Paris Cedex 12, France; delphine.neumeister@idele.fr

Faced with the sustainable growth of global demand for organic food products, some operators try to meet the demand while respecting the technical and ideological foundations of organic farming. This is what the *Reine Mathilde* program is all about: this multi-actor program was created 12 years ago in Normandy and designed to develop organic dairy farming. Nationally recognized as a truly innovative and replicable initiative to develop a sustainable supply chain, the power of this model is based on several original features. First of all, the multi-partner dimension is one of the characteristics of the program, since it brings together all the organic dairy actors in the region: livestock farmers (both conventional and organic), development actors, vets, technical institutes, public partners (local authorities, national agencies, education, etc.) and private economic actors (dairy & agri-food supply companies). It coordinates and federates them to develop their skills and communicate jointly on technical subjects such as food autonomy, forage quality, etc. This type of collaboration, which is still insufficiently developed in agricultural consulting, is one of the ways to better meet the needs of farmers in line with the expectations of the dairy sector. The second originality of the program lies in the diversity of the targets reached through a wide range of actions, both in terms of content and form. Technical innovations are varied, both on fundamental organic topics and on promising issues for the future (soil health, biodiversity, carbon storage, etc.). References acquired are disseminated through demonstration activities in real conditions with original cultivation devices (e.g. checkerboard test platform which a wide variety of associations tested simultaneously on a small surface) and to benefit from the producers' feedback. Finally, *Reine Mathilde* is notable for its approach of opening up to all producers, whatever their dairy company, whereas the initiative is mainly led by two of them and financed partially by private funds. This program is part of a collective and open source sector approach, with a common objective to optimize the conversion process and ensure the economic performance of organic dairy farms.

Session 57 Poster 13

Perception of Brazilian agrarian sciences students on sustainability indicators for dairy farming

M.F. Silva[1], F.I. Bankuti[1] and T.T.S. Siqueira[2]

[1]Universidade Estadual de Maringá, Department of Animal Science, Av. Colombo, 5790, Jd. University, 87020-900, Maringá PR, Brazil, [2]Cirad – Selmet, Environment and Societies, TA C-112/A – Campus International de Baillarguet, 34398 Montpellier Cedex 5, France; tiago.teixeira_da_silva_siqueira@cirad.fr

The perception of future professionals dealing with farms about the importance sustainability indicators is fundamental for the adoption and implementation of better agricultural practices. The objective of the research was to assess the perception of Agrarian Sciences* students from Brazilian higher education institutions about the importance of sustainability indicators for dairy farming. We used online surveys to find out students' perception. The total number of survey respondents was 351 students. Among these, 70% were students from public institutions and 30% from private institutions. Students from public educational institutions rated the technical and social indicators as among the most important. And for students from private institutions, the technical and social indicators were pointed out as those of lesser importance. Students who had previous contact with rural production evaluated technical indicators with lower scores (less important), and economic and social indicators with higher scores (more important). Students who performed internships or work with dairy cattle considered economic and social indicators as the most important, compared to environmental and technical indicators. The students of the Agronomic Engineering, Biosystems Engineering and Veterinary Medicine courses evaluated the technical indicators as more important than the other courses. Finally, we conclude that the plurality of students' perceptions on sustainability indicators for dairy farming and the difficulty to building up a shared vision in their ponderation is mainly related to concrete experiences on farms, to the differences on the syllabus of agrarian science courses and to the public or private nature of the educational systems. *Agribusiness, Agronomy, Agronomic Engineering, Agricultural Engineering, Biosystems Engineering, Veterinary Medicine, and Animal Sciences.

Session 57 Poster 14

How to better assess the carbon balance of pastoral and agropastoral systems in the Sahel region?

H.N. Rakotovao[1], P. Salgado[2], M.H. Assouma[3] and A. Mottet[4]

[1]CIRAD – UMR SELMET, Univ Montpellier, INRAE, Institut Agro-Montpellier SupAgro, ES, Campus International de Baillarguet, 34398 Montpellier, France, [2]CIRAD, UMR SELMET, INRAE-SupAgro, Dakar Étoile, BP 6189, Senegal, [3]CIRAD, UMR SELMET, INRAE-SupAgro, Avenue du gouverneur Louveau, Bobo Dioulasso, Burkina Faso, [4]FAO, Viale delle Terme di Caracalla, 00153 Rome, Italy; narindra.rakotovao@cirad.fr

In the Sahel region, assessing the carbon balance of pastoral and agropastoral ecosystems requires considering the main characteristics of these systems, including animal productivity, resources seasonality, and herd mobility in time and space. Integrating GHG emissions and carbon storage resulting from land management and livestock activities is also important as it is an integral part of these ecosystems. Sahelian countries face great challenges to assess their climate commitments in the livestock sector, mainly related to mitigation measures. These include the absence of a clear carbon balance methodology taking into account these specificities and a scarcity of local reference data (baseline data and specific emission factors). In the present work, we have proposed: (1) to improve the carbon balance mechanism of pastoral and agropastoral systems at national and territorial levels considering these specificities, and according to international methodologies and recommendations in terms of GHG emissions estimates and estimation of carbon storage; (2) to make available an improved and updated model, GLEAM-*i*, which will take into account these Sahelian specificities. The work consisted in updating the GLEAM-*i* model by integrating GHG emissions and carbon storage of land management related to the livestock sector in the Sahel, and by enhancing the specific reference data (baseline) and emission factors produced by the CaSSECS project (Carbon Sequestration and greenhouse gas emissions in (agro) Sylvopastoral Ecosystems in the Sahelian CILSS States).

Session 57 Poster 15

Is it possible to improve farms' economic results with crop-livestock system integration?

F.F. Simili[1], G.G. Mendonça[2], A.H. Gameiro[2], J.G. De Oliveira[1], P.C. Domingues Dos Santos[1] and D.F.L. Santos[3]
[1]Instituto de Zootecnia/APTA/SAA, Núcleo Regional de Pesquisa, de Ribeirão Preto, SP, 14030670, Brazil, [2]Universidade de São Paulo, Faculdade de Medicina Veterinária e Zootecnia, Departamento de Nutrição e Produção Animal, Pirassununga, SP, 13635-900, Brazil, [3]Faculdade de Ciências Agrárias e Veterinárias/ Universidade Estadual Paulista, Economia, administração e educação, Jaboticabal, SP, 14884-900, Brazil; flaviasimili@gmail.com

The objective of this study was to evaluate the potential of improve economic value of integrated crop-livestock systems in comparison to conventional systems specialized in monoculture. Empirical studies have demonstrated the environmental benefits of integrated crop-livestock systems, however the potential for creating economic value these systems is controversial, especially in emerging countries, where the necessity to expand the food supply needs be associated with better land use. This research evaluated six models of integrated systems and two conventional systems (corn grain production and pasture beef cattle production) in the south-eastern region of Brazil for two years. The models were conducted in an experiment to replicate the main management possibilities in the integrated systems. The economic impact analysis combined the risk optimization and discounted cash flow techniques based on Monte Carlo simulation, considering the price and productivity uncertainties of each system. Results indicated that, for the indicators of added value and return on investment, integrated crop-livestock systems had an economic advantage when compared to conventional systems. It was also found that integrated crop-livestock systems needed a smaller operational area for the economic break-even point to be reached. Our results may contribute to a greater adoption of integrated crop-livestock systems, contributing to the expansion of food supply in a more sustainable way and with better use of the soil.

Session 57 Poster 16

Importance in the development of short supply chains of local breeding and processing potential

P. Radomski and P. Moskała
National Research Institute of Animal Production, Team for cooperation with practice, knowledge transfer and innovation, Sarego 2, 31-047 Kraków, Poland; piotr.moskala@iz.edu.pl

Short food distribution chains are a very important element of rural development, as they are not only economic, but also social and environmental. The beneficiaries of short chains are therefore producers, consumers, local communities and the environment. It turns out that it is extremely difficult to find products from native breeds produced locally and bringing income to local producers on the market. The reasons for this have been identified as follows: (1) Breeders, processors and gastronomy often act on an individual basis, without even knowing about each other. (2) Processors / gastronomes do not see and experience a clear link between the income and the processing of the raw material from the native breeds mentioned. (3) Competition and strong marketing effectively prevent more expensive products, although of a much higher quality, from breaking through. The consumer is often unwilling to pay the higher price if he is not convinced that it is a product with unique qualities, having the features of a branded product. (4) Lack of credibility – neither in processing nor in catering, the consumer de facto does not receive a credible, convincing message about where the raw material and additives from which the product or dish is made come from. As a result, the consumer does not enjoy systematic access to the native, local product in fresh, semi-finished or processed form. On the other hand, the local breeder / farmer does not gain a proper systematic interest in the market in the high-quality product he produces, which is strongly demotivating for him. When taking actions to activate rural environments, its participants should be indicated how important it is in the development of the region to shorten the distribution channels of raw materials / products, especially which is of particular importance in the distribution of products from native breeds. The most optimal distribution channels should be within 150 km of the grower / processor. Valuable initiatives supporting local farmers and entrepreneurs are the supply of their products by administration and institutions at organization of local and regional events, hosting guests (especially foreign ones), etc.

Session 58 Theatre 1

Is precision livestock farming a way to improve farmers' working conditions?

N. Hostiou[1] and J. Fagon[2]
[1]UMR Territoires INRAE, Route de Theix, 63122 Saint Genès Champanelle, France, [2]Idele, BP 42118, 31321 Castanet-Tolosan, France; nathalie.hostiou@inrae.fr

Is precision livestock farming (PLF) able to allow more sustainable livestock farms, especially on the social dimension? PLF is often presented as a lever to reduce the working time of farmers who have more and more animals to raise with less workers. These technologies are often presented as assets, but the consequences on farmers' working conditions remain little known. In some cases, time savings are observed because PLF simplifies the monitoring of animals (heat, health problems, etc.). But time savings are very variable, as they depend on the initial state of the equipment and how farmers use the time saved. Time saving can be reinvested in animal production, new farming activities or farm management tasks, but also in private activities. The time saved can be reduced when it is accompanied by an increase of the number of animals raised on the farm. Furthermore, farmers appreciate the additional flexibility in organizing their work. PLF also leads to a reduction in the physical arduousness of the work by relieving the famers of constraining tasks. PLF can improve the mental workload due to the anticipation of events (insemination, health problems) that are sometimes not very visible to the human eye (temperature change, heart rate, etc.). But managing alerts to monitor animals can also generate stress and increase farmers' feeling of being permanently connected to the farm. The relationship between farmers and their animals is also modified. Most farmers feel that they know their animals better, although some express a deteriorated relationship with less time spent in contact with the animals. We have shown that farmers' working conditions with the adoption of precision livestock farming are not all and always in a more positive or more negative direction. Working conditions are impacted differently depending on the animal species raised, the tools adopted, and farmers' perception and identities. It is therefore critical to consider the different dimensions of farmers' work to facilitate their adoption of these new technologies.

Session 58 Theatre 2

Digital sustainability in livestock farming: a digitization footprint

M. Sozzi, A. Pezzuolo and F. Marinello
Univerisity of Padova, TeSAF, Viale dell'università 16, 35020 Legnaro, Italy; marco.sozzi@unipd.it

The growing availability of animal as well as environmental sensing technologies, IoT, tracking systems, standard communication protocols, etc. along with apparently decreasing costs of the same technologies are enabling widespread collection, transmission and implementation of digitalized information in livestock farming. In general, such developments are regarded as beneficial by society as well as the scientific community, having positive impacts on overall livestock management efficiency. On the other hand, the digital issues are typically not included in sustainability considerations: both data transmission and data processing might introduce challenges, especially in the case of farms or environments not ready for the digital transformation. The aim of the present work is to discuss the issue of digital sustainability, introducing the idea of a digitalization footprint, which parameterizes the amount of digital information to be collected, transferred and processed. As for carbon or water footprints, the DF is intended to quantify the specific or general use of digital or processing information in terms of volumes/time/efforts/costs required for data storage, transfer and processing. In order to support the approach, considerations based on historical evolution in automation and sensing in livestock farming and the present situation regarding telecommunications technology coverage will be presented and discussed.

Economic value of the data-driven environment of precision livestock farming

M. Van Der Voort and H. Hogeveen
Wageningen University & Research, Social Sciences, Business Economics Group, Hollandseweg 1, 6706 KN Wageningen, the Netherlands; mariska.vandervoort@wur.nl

Digitalization in livestock production is shaped by the use of Precision Livestock Farming (PLF) technologies, consisting of the use of technologies to continuously, real-time and automatically monitor animal production, health and welfare. The traditional labour-intensive forms of production, where farmers often use past experience and intuition to make decisions, are changing to technology-intensive, data-supported forms of precision production, where field-specific data are used to help farmers make appropriate decisions on the production process. As such PLF works toward a more efficient use of farm resources, improve or maintain animal health and welfare, support farm economic decision making, and mitigate environmental emissions. Studies on various PLF technologies have shown their potential for earlier and more precise animal disease detection. However, many technologies do not yet ensure full decision support and lack proven economic value for the farmer. Only under certain management practises these technologies are expected to be economically viable, reduce costs and improve the financial performance of the farm. A challenge in evaluating the economic value of PLF technologies is the scarce availability of economic, production and animal health data before and after the implementation of the technology. Besides, it is expected that sensors will also have potential value for other actors up- and downstream the livestock value chain. Quantification of this value is even more difficult than at the farm level. The scope of PLF transcends beyond the farm and includes the entire value chain where data is increasingly shared between feed companies, breeders, farmers, and food processors. With PLF technologies, livestock production has become a data-driven sector, which recast conventional process-driven agriculture for smarter data-driven farming. In this presentation we will focus on the question to which extend and for whom this data-driven environment of PLF results in added value.

Session 58 Theatre 4

Panel discussion – PLF for social, environmental and ethical impacts

J.M. Gautier[1] and C. Morgan-Davies[2]
[1]Institut de l'Elevage, BP42118 Castanet Tolosan Cedex, 31321, France, [2]Scotland's Rural College, Kirkton, Crianlarich, Scotland, FK20 8RU, United Kingdom; jean-marc.gautier@idele.fr

Precision Livestock farming (PLF) is frequently seen as a solution to reduce livestock footprint, strengthen social networks among farmers, and advisors and free-up time on farms. But what is the reality? How new technologies affect farmers' relation with their animals, with their working networks and how they view their role as a livestock keeper? What about the digital footprint of PLF, is-it in line with the global objective of decreasing livestock footprint? And is the data sharing made in an ethical way in agriculture? Globally, how is PLF (including new technologies) impacting on the social, environmental and ethical worlds. This panel discussion aims to tackle all these challenging topics by sharing experiences and identified key factors to consider for digital tool assessment. You are interested by those topics? This panel discussion is made for you, so don't hesitate to participate to it.

Session 59 Theatre 1

Perception of cultured meat of Italian, Portuguese and Spanish speaking consumers

J. Liu[1], J.M. Almeida[2], N. Rampado[3], B. Panea[4], É. Hocquette[5], M.P. Ellies-Oury[1,6], S. Chriki[5] and J.F. Hocquette[1]
[1]INRAE, UMR1213, Saint-Genès-Champanelle, 63122 Theix, France, [2]INIAV, Quinta da Fonte Boa, 2005-768 Vale de Santarém, Portugal, [3]University of Padova, Legnaro (PD), 35020, Italy, [4]CITA-Universidad de Zaragoza, C. de Pedro Cerbuna, 50059, Zaragoza, Spain, [5]Isara – Agro School for Life, 23 rue Jean Baldassini, 69364 Lyon cedex 07, France, [6]Bordeaux Sciences Agro, 1 cours du Général de Gaulle, 33175 Gradignan, France; liujingjing1003@126.com

This study was aimed at investigating how consumers originating from southern Europe perceive cultured meat (CM) and if demographic characteristics (origin, gender, age, education, activity area and meat consumption) are related to their willingness to try (WTT), to eat regularly (WTE) and to pay for (WTP) cultured meat. The 2,071 respondents were Italian (46.7%), Portuguese (31.0%) and Spanish (22.3%) speaking people, and 48.8% of them perceived CM as 'promising and/or acceptable' whereas 28.5% considered it 'absurd and/or disgusting' and 22.7% 'fun and/or intriguing'. In total, 65.5% and 24.7% would be respectively willing and not willing to try CM, 43.3% had no willingness to eat regularly and, 94.3% would not pay more for CM compared to conventional meat. In general, origin, gender, age, activity area and meat consumption had significant effects on WTT and WTE. Young people (18-30 yrs) had the highest WTT, WTE and WTP compared to mid-aged (31-50 yrs) and older people (51-yrs), whereas the effect of gender is more variable. Spanish-speaking consumers had the highest WTT and WTE. Scientists (within or outside the meat sector) had the highest WTT, non-scientific people within the meat sector had the lowest WTT. People outside the meat sector had higher WTE and people working within the meat sector had lower WTE. People with the lowest income had higher WTE. People with vegan and vegetarian diets would pay more for CM but generally no more than for conventional meat. People who heard about CM had higher WTT but lower WTE. The perceptions that CM may be more eco-friendly, ethical and healthy than conventional meat tend to be drivers for the current respondents to try and eat CM, whereas emotional resistance and the negative impacts on livestock farming systems caused by CM production might be the barriers for the current respondents to accept CM.

Session 59 Theatre 2

Consumer perception of cultured meat in certain African countries

M. Kombolo[1], S. Chriki[2], M.P. Ellies-Oury[3] and J.-F. Hocquette[1]
[1]INRAE, UMR1213, Route de Theix, 63122 Saint-Gènes, France, [2]Isara, 23 rue Jean Baldassini, 69364 Lyon, France, [3]Bordeaux Sciences Agro, CS40201, 33175 Gradignan, France; moise.kombolo-ngah@inrae.fr

African population is projected to grow by 1.76% to reach about 2.5 billion in 2050. This is creating an unprecedented boom in the demand for animal products over the coming years that would have to be properly managed. Industry players worldwide have continually shown interest in cultured meat in the recent years claiming it would be a more sustainable way to provide animal proteins. Cultured meat is therefore taking a global stage and this study aimed at investigating how African meat consumers perceive it depending on their country. Between 5,400 and 11,013 respondents from 12 different countries participated in this survey (Cameroon, Congo, DRC-Democratic Republic of Congo, Ghana, Ivory Coast, Kenya, Morocco, Nigeria, Senegal, South-Africa, Tanzania and Tunisia). The survey was divided into 2 sets of questions. The first set concerning questions on meat consumption and opinions regarding rural life, ethical and environmental problems. The second set concerned questions regarding the willingness to try, to eat and to pay. The first set had 5,485 respondents and the second set had 5,528 respondents. More than 60% out of the total respondents (11,013) had already heard of cultured meat. Out of the first set of respondents, 31.2% totally agreed that cultured meat will have a negative impact on the rural life and 32.9% were not ready to accept this novel product as a viable alternative to conventional meat in the future but were still ready to eat other meat alternatives. Furthermore, only 8.9% out of the second set of respondents were definitely willing to try this product, mostly males middle-aged (31-50 years old). From all results, we identified a significant interaction between the willingness to try cultured meat and several factors such as: country of origin, sex, income, age and education levels. For example, the richest and most educated countries tend to be readier to try cultured meat. A similar pattern was observed for willingness to pay except that sex has no significant effect and age only a low effect.

Nutrient composition of milk, dairy and plant-based alternatives and implications for consumer diets

S. Stergiadis[1], A. Tarrado Ribes[1], R. Reynolds[2], K.E. Kliem[1] and M.E. Clegg[2,3]
[1]University of Reading, School of Agriculture, Policy and Development, Reading, RG6 6EU, United Kingdom, [2]University of Reading, Department of Food and Nutritional Sciences, Reading, RG6 6DZ, United Kingdom, [3]University of Reading, Institute for Food, Nutrition and Health, Reading, RG6 6EU, United Kingdom; s.stergiadis@reading.ac.uk

Milk and dairy products are the main source of macronutrients (protein, fat), minerals (I, Ca, P, Zn), and vitamins (B_2, A, D, B_9, B_{12}) in human diets. An increasing number of consumers replace milk/dairy with plant-based dairy alternatives (PBDA) either by necessity (allergy, intolerance) or choice (due to concerns around saturated fat, environmental footprint and animal welfare). However, replacing nutrient-dense foods in the diet may reduce essential nutrients' intakes. This study aimed to: (1) compare the label nutrient composition PBDA and equivalent dairy products; and (2) model the impact on consumer nutrient intakes by substituting milk/dairy with PBDA. Data for all available PBDA (n=303) and corresponding milk (n=51), yogurt (n=78) and cheese (n=38) products were collected from six UK supermarkets. PBDA were categorised as per primary ingredient and label nutritional information was recorded. Data were analysed by linear models using primary ingredient as fixed factor. Pairwise comparisons were done by Fisher's Least Significant Difference test ($P<0.05$). When compared to most PBDA ($P<0.05$; per 100 g): (1) liquid milk contained more energy (9-20 kcal), protein (0.4-3.2 g), sugars (2.5-3.3 g), I (21 ug), B12 (35-41 ug) and vitamin E (2 mg), and less fibre (0.2-0.6 g); (2) yogurt contained more protein (1.4-4.5 g) and Ca (43 mg), and less fibre (0.2-0.9 g), fat (2.9-3.4 g) and salt (40-80 ug); and (3) cheese contained more energy (29-72 kcal), protein (10-16 g), fat (3-5 g) and Ca (299 mg), and less carbohydrates (3.6 g) and fibre (2.2-2.9 g). The nutritional modelling showed that PBDA are not a nutritional replacement of cows' products and substitution of milk/dairy with PBDA may increase the risk of nutrient deficiencies, especially in vulnerable groups (protein in children, I in adolescent women, vitamin B_{12} in vegetarians); and PBDA fortification can be recommended for improving their nutritional value.

Keep the herd, rear the future: prospects and challenges of hybrid, plant-based and cultured meat

M. Banovic
The MAPP Centre, Department of Management, Aarhus University, Fuglesangs Allé 4, 8210 Aarhus, Denmark; maba@mgmt.au.dk

The sustainability of the livestock sector is constantly tested by its direct and indirect influence on environment and public health, while at the same time being pillar of the global food system and a contributor to food security and agricultural development. However, despite the recommendations for a transition towards more plant-based diets, the latest flexitarian, vegetarian, and vegan trends, and the stagnation in meat consumption patterns, Europe still records high levels of meat per capita consumption, almost two times the world average. Meat still seems to be a preferred option for many consumers, and although consumers report their intention to stop eating or cut down on meat, they still prioritize on tradition, taste, nutritional value, food safety, convenience, and price, over sustainability concerns. Considering that reduction of meat consumption is necessary to fulfil European sustainability and health goals, and that the livestock sector needs to adapt its tactics to these new signals, more flexible and simpler strategies are needed for both demand and supply side. The strategies as producing 'less but better' and 'less but varied' and allowing behavioural change through information frames and nudges, while using hybrid, plant-based, and cultured meat that are at the same time accepted and adopted by the consumers could help the transition, where hybrid meat could be the opening bridge towards more radical transitions of completely eating plant-based and/or cultured meat. With recent advances in cultured meat technology and possibility of its commercialization, coupled with already existent hybrid and plant-based products on the market, and possible interventions in the environment that could steer consumers towards a desirable behaviour, this could lead to a reduction in both meat production and consumption, indorsing the healthier and more sustainable dietary habits, while at the same time preserving the livestock industry and the environment. This is particularly significant in the light of the massive stress the livestock sector is already facing in providing more choices and value-added solutions that are healthier and/ or more environmentally friendly, while at the same time winning the consumer support.

Session 59 Theatre 5

High isoflavones exposure through soy-based food in France and their reduction by new food-processes

S. Bensaada[1,2], G. Peruzzi[1], F. Chabrier[3], P. Ginisty[4], C. Ferrand[5], M. Vallat[6] and C. Bennetau-Pelissero[2]
[1]Biopress, 2 Rue Edouard Branly, 47400 Tonneins, France, [2]Université de Bordeaux, ARNA U1212 Inserm, 146 rue Léo Saignat, Carreire Faculté de Pharmacie, 33076, France, [3]Agrotec, Agropole, 47310 Estillac, France, [4]IFTS, 3 Rue Marcel Pagnol, 47510 Foulayronnes, France, [5]Université de Bordeaux, BFP UMR 1332, 71 Av. Edouard Bourlaux bâtiment A4, 33140 Villenave-d'Ornon, France, [6]Université de Bordeaux, I²M UMR 5295, Avenue d'Aquitaine, 33170 Gradignan, France; souad.bensaada@u-bordeaux.fr

The flexitarian trend is increasing the consumption of soy-based products in western countries. In addition to its very interesting protein profile, compared to other legumes, soy contains one of the highest concentrations of isoflavones (IFs): natural compounds that have shown estrogenic, anti-estrogenic, anti-androgenic and thyroid effects. An intake over 40 mg/day induces an impairment of menstrual cycles in women, proliferating effects on women's breast cancer and a reduction of men's sperm quality. Exposure to IFs is estimated from eating habits questionnaires and dosages on 140 foods available in France by ELISA method. The average IFs exposure of French soy consumers is estimated at 6,615±1,220 µg/day (mean ± SEM). The purpose of this study is thus to achieve non-worrying IFs levels in soy-matter according to the reprotoxic LOAEL (Lowest Observed Adverse Effect Level) for genistein which was published in 2008 by the US-NTP (National Toxicology Program). To reduce IFs, some food-processing steps mimicking those found in traditional Asian recipes were added to the production process of soy-proteins. at both laboratory and pilot scales. The parameters optimization resulted in 48% IFS removal which cumulated with previous results (obtained on final products) gives an adequate intake for the whole population. This study showed that the current exposure to IFs is alarming and that it is possible to significantly reduce them without impacting main soy-protein properties: size, water, lipid or protein content. This, provides a safer raw material for the food industry using soy proteins. Pre-industrial scale-up should confirm the accuracy of the selected parameters.

Session 59 Theatre 6

Are vegetable analogues of meats real competitors – elements of comparisons

P. Cayot
Institut Agro Dijon, PAM, 26 bd Dr Petitjean, 21079 Dijon, France; philippe.cayot@agrosupdijon.fr

The nutritional quality of animal products is often overlooked in the consumer's imagination. Sustainability, for example, must take into account the sustainability for humans, starting with the satisfaction of their nutritional needs as naturally as possible. To begin with, the protein intake, in quantity per 100 g of edible food (usually cooked), largely exceeds that of the vegetable equivalents, the meat analogues based on vegetable proteins. It is clear that animal products are sources of protein of high nutritional quality (Digestible Ileon Amino Acid Score, DIAAS>80) in contrast to the vegetarian equivalents (DIAAS<80). Animal proteins are more easily digestible and the contributions in essential amino acids are more important. The combination of legume and cereal proteins, complementary in theory in their essential amino acid content, is not satisfactory for a quality intake for humans (DIAAS <100 very generally). In addition to the quality of the proteins, the quality of a food should also be evaluated on its contribution in trace elements. Animal products are the only natural sources of vitamin B12. The quantities of other B vitamins (B1, B2, B3, B5, B6) are much higher in meat products. The mineral content is also higher in cooked meat or cooked meat products compared to a ready-to-eat vegetarian analogue: 2 to 9 times more zinc, and up to 3 times more for iron. The quantity is not the only criterion to take into account. Iron and magnesium are not very bioavailable in plant products. The efficiency of iron absorption when consuming meat products depend to the form of the ion (ferric or ferrous iron, or heme iron – haemoglobin or myoglobin) and to the absence of complexes such as phenols and phytate present in plants. Food does not only meet nutritional needs but also hedonistic ones. The sensory quality of analogues is generally inferior to their meat reference, but increasingly sophisticated strategies tend to reduce this gap in sensory appreciation. Nowadays, researchers are developing strategies to compensate these weaknesses in plant analogues of meat products, bringing advances that will eventually spread to analogue producers.

Session 59 Theatre 7

Milk alternatives from an environmental and nutritional point of view

B. Silva[1,2], V. Heinz[2] and S. Smetana[2]

[1]Universidade Católica Portuguesa, CBQF-Laboratório Associado, Escola Superior de Biotecnologia, Rua Diogo Botelho 1327, 4169-005 Porto, Portugal, [21]German Institute of Food Technologies (DIL e.V.), Professor-von-Klitzing-Str. 7, 49610 Quakenbrück, Germany; b.silva@dil-ev.de

Milk consumption in humans lasts longer than in other mammals' species. Today consumers' awareness of the environmental burden that some products carry keeps growing. Thus, they look for alternatives that are more environmentally friendly and nutritionally similar. This work explores data available in the literature, comparing the nutritional profile of several milk alternatives and milk from different mammals, as well as their environmental impact. For this, the Google Scholar search engine was used, and the search was structured into two phases using two sets of keywords. The first was aimed at LCA and the second was for the nutritional properties of the beverages. The research was limited to studies published in scientific journals from the last 10 years and available in English. The initial search yielded more than 231 articles. Further analysis of the articles and data available narrowed down the articles used in this review to 66. The values for the analysed macronutrients (proteins, fat, fibres and carbohydrates) and most micronutrients – Fe, Mg, P, K, Na, vit. B1, vit. B3, vit. B6, vit. B6, vit. B9, vit. B12, vit. E and vit. D – is higher (in g/100 g of product) in plant-based products. Mammals' milk had higher values of calcium, zinc, vit. C and vit. A. On the environmental footprint of these products, dairy has a higher impact on the categories of Global Warming Potential, Ozone Depletion Potential, Marine Eutrophication and Freshwater Eutrophication; while the substitutes have a higher energy and water consumption. Overall plant-based beverages appear to be nutritionally richer than animal milk: their profile shows a possible fortification in some nutrients, which is a normal practice during processing. Overall, the environmental impact of plant-based milk is lower than milk, with exceptions in some categories. However, a lot of data is missing, making it impossible to make a full comparison across environmental categories. This study has many limitations since data for the different products is limited, for both nutritional profile and environmental impact.

Session 59 Theatre 8

Alternative proteins: environmental impact of meat substitution from plants to cultured meat

S. Smetana

German Institute of Food Technologies (DIL e.V.), Prof.-von-Klitzing-str. 7, 49610, Germany; s.smetana@dil-ev.de

Modern food system is highlighted with high environmental impact, in many cases associated with increased rates of animal production and overconsumption. The adoption of alternative to meat proteins (insects, plants, mycoprotein, microalgae, cultured meat, etc.) might influence the environmental impact and human health in a positive way but could also trigger indirect impacts associated with higher consumption rates. A review analysis of own results and those available in the literature published in the last 10 years provides a condensed summary on potential environmental impacts, resource consumption rates and rebound effects associated with integration of alternative proteins in complex global food system in the form of meat substitutes. Meat substitutes can be differentiated according to processing functionality into a few groups with different level of processing, with different impacts. Impacts of both animal and plant-based ingredients can vary widely, and there is a range in which results of impact assessment. Beef is typically taken as a product with high environmental impacts, higher than most meat substitute ingredients. Still, for some protein sources like microalgae, the analysis shows that global warming potential and the non-renewable energy demand can be much higher than beef. Cell-based cultures and insects also tend to increase the environmental impacts when added as meat substitute ingredients. The water footprint was not indicative with results being different in a few orders (due to use of different assessment methodologies). Plant-based meat substitutes at the same time are significantly lower in carbon footprint than beef and hypothetical cultured meat; however cultured meat has a potential to have a lower impact than beef and farmed crustaceans. Meat analogues can be a great strategy for food system impact reduction, if assured they substitute meat on the market instead of adding another product and impact to the existing diet. Less processed ingredients from plants and insects should be used for substitution.

Simplified LCA in sheep farming: comparison of carbon footprint estimates from different tools

A.S. Atzori[1], O. Del Hierro[2], C. Dragomir[3], M. Decandia[4], M. Acciaro[4], C. Buckley[5], M. Habeanu[3], J. Herron[5], R. Ruiz[2], T.W.J. Keady[5] and S. Throude[6]

[1]Dipartimento di Agraria, University of Sassari, Sassari, Italy, Viale Italia 39, 07100, Italy, [2]Neiker – Basque Inst. for Agricultural Research and Development, Spain, Parque Tecnológico de Bizkaia, Parcela 812. Berreaga 1., 48160 Bizkaia, Spain, [3]Institutul National de Cercetare-Dezvoltare pentru Biologie si Nutritie Animala, IBNA Balotesti, Şoseaua Bucureşti-Ploieşti 1, Baloteşti, 077015, Romania, [4]AGRIS Sardegna, Loc. Bonassai, 07100, Italy, [5]Teagasc, Mellows Campus, Athenry, Co. Galway, H65 R718, Ireland, [6]Institut de l'Elevage, 23 Rue Jean Baldassini, 69007 Lyon, France; asatzori@uniss.it

LIFE Green Sheep (LIFE19 CCM/FR/001245) has been targeting a common Carbon Footprint (CF) assessment methodology at European level. This study aimed to compare 3 European tools to estimate the CF of dairy and meat sheep farming: CAP'2ER (C2E; 'Institute de l'Elevage, France); ArdiCarbon (AC; Neiker, Spain); CarbonSheep (CS; Univ. of Sassari, Italy). It was performed by collecting data from 12 dairy farms (Fr, Es, Ro, It) and from 9 meat farms (Fr, Es, Ir) on which 37 and 27 model runs were performed, respectively. Tools were fed yearly farm inputs and outputs from their life cycle inventory. Emissions from animals, manure, crops, purchased feeds, fuel and electricity were allocated 100% on milk or meat, and expressed per kg of CO_2eq./kg of fat and protein corrected milk (FPCM) or carcass weight. Differences between C2E vs AC, C2E vs CS and AC vs CS as root mean square error of prediction (RMSPE) were evaluated. Dairy sheep farms had 499±123 ewes, 187±118 ha, and produced 87.2±87.8 tons/yr of milk whereas meat farm had 756±682 ewes, 103±57 ha, with 17.2±19.1 tons/yr of meat. The mean CF for farm where: 4.85, 3.57 and 3.75, CO_2eq./kg of FPCM and 33.3, 30.9, and 31.2 CO_2eq./kg of meat for C2E, AC and CS. All tools estimated high incidence of enteric and manure emissions. RMSPE for dairy farms was 1.7, 1.7 and 1.2 kg of CO_2/kg of FPCM and 6.55, 4.62 and 3.52 kg of CO_2/kg of meat for C2E-AC, C2E-CS and AC-CS, respectively. The tools, developed for specific countries, gave different estimates when applied at European level suggesting a future alignment of inputs and impact assessments before broad applications.

Cocoa by-products as source of antioxidants and nutrients to dairy sheep

S. Carta[1], E. Tsiplakou[2], G. Pulina[1] and A. Nudda[1]

[1]University of Sassari, Dipartimento di Agraria, Animal Science Unit, Viale Italia 39, 07100, Italy, [2]University of Athens, Department of Animal Science, Aquaculture, Agricultural, IeraOdos 75, 11855, Greece; scarta2@uniss.it

Cocoa by-products are a valuable source of nutrient and antioxidant substances for animal diet. This study aim to evaluate the effects of cocoa husk (CH) on milk production traits and blood and milk oxidation status. Twenty-four Sarda ewes, divided in three groups were supplemented with 100 g/d of soybean hulls (CON group), and two groups in which soy was replaced with 50 g/d (CH50 group) or 100 g/d per head of cocoa husk (CH100 group). Milk samples were analysed for chemical composition and fatty acid profile. Blood and milk samples were analysed for antioxidant parameters. Data were analysed using a mixed model with diet, sampling and diet × sampling as fixed effects and pen as random effect. In addition, the average number of ewes that could receive an integration of 100 g/d of CH and the reduction of carbon footprint related to the concentrates were estimated. Chemical analyses of CH showed a good content of NDF (45.99% of DM), NFC (23.08% of DM) and CP (16.77% of DM). The lipid content of CH was 5.6% and the most abundant fatty acid found was C18:1 cis-9, followed by C18:0 and C16:0. CH had a great amount of polyphenols and high theobromine content (6,850 mg/kg of DM). The inclusion of CH on sheep diet did not affect milk yield and milk composition but a significant effect diet × sampling was found for milk fat, protein and SCC, suggesting a positive influence of CH on quality milk over time. As regard FA, a decrease of C16:0 and an increase of C18:0 was found in CH50, while an increase of C18:1 cis-9 was found in both groups fed CH. Blood and milk antioxidant analyses showed a positive effect of CH against oxidative damage. The global amount of CH could be recycled to integrate the diet of 16 million of sheep. The carbon footprint of concentrates could be reduced for about 20%. In conclusion, the results suggested that the introduction of 100 g/d of CH in sheep diet could be implemented without negative effects on animal performance and milk composition. Acknowledgements: Research funded by the University of Sassari with resources of 'Fondazione di Sardegna annualità 2017'.

Inclusion of sainfoin in the fattening concentrate: meat quality of light lambs

C. Baila, M. Joy, M. Blanco, I. Casasús, G. Ripoll, J.R. Bertolín and S. Lobón
Centro de Investigación y Tecnología Agroalimentaria de Aragón (CITA)- IA2 (CITA-Unizar), Department of Animal Science, Avda. Montañana, 930, 50059 Zaragoza, Spain; cbaila@cita-aragon.es

The inclusion of forages in animal diets has been encouraged to increase the sustainability of farms, however, it can modify meat quality. The objective of this study was to evaluate the effect of the inclusion of sainfoin (*Onobrychis viciifolia*) at different rates in the concentrate of fattening lambs on meat quality. After weaning, twenty-six Rasa Aragonesa male lambs (BW: 14.0±0.49 kg; age: 30±0.66 d) were assigned to 3 treatments. They were fed isoproteic (170 g/kg DM) and isoenergetic (18.4 MJ/kg DM) concentrates for 40 days (slaughter BW: 25.0±0.78 kg). The concentrates were: commercial with 0% of sainfoin (Control), with 20% of sainfoin (20SF) and with 40% of sainfoin (40SF). After 24h, the *longissimus thoracis et lumborum* muscles were removed to analyse the chemical composition and the evolution of colour, haeme pigments and lipid oxidation at 0, 2, 5, 7, 9, and 12 days of storage. The inclusion of sainfoin in fattening concentrate did not affect the meat chemical composition ($P>0.05$), except for the intramuscular fat content, which was higher in the Control than in the 40SF group ($P<0.05$), whereas the 20SF lambs had intermediate contents. Regarding the colour traits, redness and metmyoglobin (MMb) were affected by the interaction between treatment and day of storage ($P<0.05$), although values were similar among treatments within each day, only a trend was observed in the MMb to be higher on days 9 and 12 in the Control than the 20SF group ($P<0.10$). The rest of the colour traits were similarly affected by the day of storage among treatments. Lipid oxidation was affected by the interaction between the inclusion of sainfoin and the day of storage ($P<0.001$), being lower in 20SF lambs than Control group at day 9 (0.172 vs 0.394 mg malondialdehyde/kg FM; $P<0.05$), whereas at day 12, both sainfoin treatments were lower than Control lambs (0.215, 0.373, and 0.592 mg malondialdehyde/kg FM, 20SF, 40SF and Control, respectively; $P<0.05$). In conclusion, the inclusion of sainfoin in the fattening concentrate improved lipid oxidation of meat, which could extend the shelf life of the product.

Genetic parameters and trends in components of reproduction of Merino ewes

C.L. Nel[1,2], K. Dzama[2], A.J. Scholtz[1] and S.W.P. Cloete[1,2]
[1]Western Cape Department of Agriculture, Directorate Animal Sciences: Elsenburg, Private Bag X1, 7607 Elsenburg, South Africa, [2]Stellenbosch University, Animal Sciences, Private Bag X1, 7602 Matieland, South Africa; schalkc2@sun.ac.za

Selection for composite traits such as number of lambs weaned or total weight weaned risks unpredictable outcomes in genetic responses. The consideration of the underlying components of composite traits, such as fertility, fecundity, or rearing ability for number of lambs weaned, could deliver gains in reproductive performance tailored to a specific situation. The use of component traits in South African Merinos requires further evaluation, though. This study therefore investigated genetic parameters and trends of component and composite reproduction traits in the Elsenburg Merino flock that has been divergently selected in favour of (H-Line) or against (L-Line) performance in number of lambs weaned per ewe mated (NLW/EJ). Genetic parameters and trends were derived for component traits of NLW/EJ, namely number of ewes lambed per ewe joined (EL/EJ), number of lambs born per ewe lambed (NLB/EL), as well as ewe rearing ability per ewe lambed (ERA/EL). Additionally, the composite traits number of lambs born per ewe joined (NLB/EJ) and NLW/EJ were also analysed. All traits were heritable, but h^2 was generally low, at 0.04 for EL/EJ, 0.12 for NLB/EL, 0.09 for NLB/EJ, 0.03 for ERA/EL and 0.05 for NLW/EJ. All traits were affected by animal permanent environment with variance ratios ranging from 0.05 for ERA/EL to 0.12 for EL/EJ. All traits except for NLB/EL were affected by service sire, but this effect was also small with the highest estimate of 0.04 for EL/EJ. Genetic correlations of component traits with NLW/EJ were generally favourable, but ERA/EL and NLB/EL were unfavourably related. Genetic trends for all traits were divergent between the H- and L-Lines. Expressed relative to overall trait means, genetic improvements in the H Line amounted to 0.92% for NLW/EJ, 0.30% for EL/EJ, 0.45% for NLB/EL, and 0.15% for ERA/EL. Corresponding downward trends in the L Line were respectively -0.57%, -0.29, -0.10% and 0.23%. Selection for NLW/EJ thus resulted in worthwhile direct gains as well as in favourable correlated responses in component traits in the flock studied.

A protocol for meta-analysis on the impact of current pig farming systems on animal welfare

U. Gonzales-Barron and V. Cadavez
Instituto Politécnico de Bragança, Centro de Invesigação de Montanha, Campus de Santa Apolonia, 5300-253 Bragança, Portugal; vcadavez@ipb.pt

Pig farming is highly diverse, with rearing methods across Europe which go from extensive organic farming to conventional intensive production. However, despite the economic competitiveness of the sector, it is known that breeding and fattening pigs may give rise to issues linked to animal welfare. A protocol for systematic review and meta-analysis is proposed that addresses the welfare impact of current pig farming systems that effectively summarises the reality, challenges and hindrances of the various pig farms in the EU. An eligible primary study should be an observational field study containing information on: (1) type of pig farming system: conventional (intensive units), adapted conventional (deep litter, bedded indoor), organic (outdoor, indoor) or traditional (sylvopastoral Mediterranean); (2) management features, such as pig growing stage, bedding availability, space allowance, group size, floor type, type of feed, etc.; and (3) prevalence of welfare indicators, grouped by farming system, such as behaviour, health, housing and feeding. First, a categorisation of welfare outcomes should be developed according to animal-based indicators, grouped by the principles of animal welfare (behaviour, feeding, housing, injuries and disease) and disaggregated according to growing stage (weaners, growers and finishers, sows and piglets). Study characteristics to be extracted for meta-analysis would be of the following kind: general features, pigs sampled, farming system and sampled farms, housing, feeding and manure, welfare evaluation, and bias potential (i.e. selection, detection and reporting). Hierarchical meta-analysis models are proposed to be adjusted on data subsets of welfare outcome × farming system to assess the effects and heterogeneity caused by growing stage of pigs, purpose of farm, type of management, use of bedding, etc. To verify external validity of pooled welfare outcomes, animal welfare experts shall assess if the pig farms in the different studies appropriately represent a farming system in the EU, and to what extent the type of study, farm characteristics, types of housing/management/feed can acceptably differ among studies, and still render the meta-analysed effect size meaningful.

A meta-analysis on the tenderising effect of ageing beef muscle

V. Cadavez and U. Gonzales-Barron
Instituto Politécnico de Bragança, Campus de Santa Apolonia, 5300-253 Bragança, Portugal; vcadavez@ipb.pt

While tenderness is the most important quality trait of meat, it is a variable property affected by multiple factors; such as animal breed, sex and age, pre-slaughter stress, muscle type, meat's ultimate pH, ageing, etc. The objective of this work was to understand, through meta-analysis, how intrinsic and extrinsic factors modulated the effect of ageing time on beef meat tenderness, as measured by the Warner-Bratzler (WB) shear-force (SF). After methodological quality assessment of studies retrieved from Scopus and PubMed, forty studies were kept for meta-analysis, from which the following information was extracted: animal gender, age class, breed, muscle type, carcass weight, meat ultimate pH, pre-treatment of meat sample before SF measurement (chilled vs frozen), time of ageing, SF mean and standard error, WB crosshead speed and WB shear area (1.13-1.77 cm^2 for cylinder and 1 cm^2 for parallelepiped subsamples). A meta-analysis model was adjusted to a data set of 462 SF values as a mixed-effects linear regression testing for the effects of intrinsic/extrinsic factors and ageing time on SF. Ageing strongly favours tenderisation of meat ($P<0.0001$); however, ageing time has an accelerating tenderisation effect during ~2-3 weeks, after which the increase in tenderness slows down. Beef tenderness is also affected by cattle breed, having *Bos indicus* a tendency to produce tougher meat ($P<0.0001$) than *Bos taurus*. The type of muscle exerts an effect on tenderness ($P<0.0001$): Psoas major and Longissimus lumborum produce the most tender beef meat, and are significantly different from L. thoracis, which has an intermediate tenderness. Semimembranosus and Semitendinosus muscles produce the toughest meat. Freezing of meat, and posterior defrosting for analysis, has also a negative impact on tenderness ($P=0.0060$). Finally, SF measurements are affected by meat subsample shape ($P=0.0098$), crosshead speed ($P=0.0006$) and their interaction ($P=0.0002$). On average, the minimum ageing time for meat tenderisation at very low refrigeration temperatures is 2 weeks. Furthermore, this meta-analysis demonstrated the importance of standardising the measurement of WB SF, since higher crosshead speeds and the lowest shear areas of 1×1 cm^2 from parallelepiped-shaped meat subsamples produce lower SF values.

Session 60 Poster 7

Rumen fatty acids linked to phenotypes in Romane lambs selected for feed efficiency

F. Touitou[1], S. Alves[2,3], F. Tortereau[1], R. Bessa[2,3] and A. Meynadier[1]
[1]INRAE, INPT-ENVT, INPT-ENSAT, GenPhySE, 23 Chemin de Borde Rouge, 31320 Castanet-Tolosan, France, [2]Associate Laboratory for Animal and Veterinary Sciences (AL4AnimalS), UTAD, 5000-801 Vila Real, Portugal, [3]CIISA, FMV, University of Lisbon, Av. Universidade Técnica, 1300-477 Lisboa, Portugal; florian.touitou@envt.fr

Improving feed efficiency (FE) is a central tool to make animal production more sustainable. However, the mechanisms underlying the individual variation in FE remain unknown. Links between the ruminal microbiota and FE have been proposed. Among the ruminal processes, microbial metabolism of the dietary lipids is responsible for the formation of bioactive intermediates that could be involved in FE variability. To test this hypothesis, 277 Romane lambs divergently selected for Residual Feed Intake (RFI) were fed a 100% concentrate diet during 6 weeks when 4-month old (CONC phase) and 167 of them, the most extreme in terms of RFI genetic values were then fed a mixed diet during 6 weeks when 9 to 11-month old (MIX phase). During both phases, phenotypes (Body Weight (BW), Average Daily Feed Intake (ADFI) and Body Composition (BC) parameters) were recorded and RFI retrieved from a multiple linear regression of ADFI on metabolic BW, average growth (ADG) and BC. At the end of each period ruminal fluid was sampled and 70 long-chain fatty acids (FA) were analysed by gas chromatography. Fatty acid profiles, expressed as percentages, were CLR-transformed after zero-replacement by geometric Bayesian method and finally corrected for environmental effects. A PLS-DA approach was used to discriminate the RFI lines or phenotypic groups based on RFI on their rumen FA profiles. No difference was found in either case whatever the diet. Similarly, multiple regression models performed on phenotypes to search for links with either specific FA or groups of FA did not highlight any association despite the variability of the FA proportions and phenotypes. A PLS analysis revealed that the strongest link was between C18:1t10 and ADG during the CONC phase and between C18:0 and ADFI during the MIX phase but Pearson correlation coefficients were low (r=0.30 and r=0.36, respectively). Contrary to milk production, meat production does not vary according to ruminal FA profile. Nevertheless, variations in the quality of this meat can be expected.

Session 60 Poster 8

Methane measurements in PAC do not result in elevated milk somatic cell counts in Norwegian goats

J.H. Jakobsen[1], K. Lunn[2], R.A. Inglingstad[2] and T. Blichfeldt[1]
[1]The Norwegian Association of Sheep and Goat Breeders, Box 104, 1431 Ås, Norway, [2]TINE SA, Department of Research and Development in Dairy Production, 1430 Ås, Norway; jj@nsg.no

Goats are ruminants that emits methane. The actual emission has however never been measured on goats in Norway. Portable Accumulation Chambers (PAC) has shown to be adequate to capture individual emissions of enteric methane of sheep. The aim of the current study was to test if PAC equipment can be used for measuring methane emission on Norwegian dairy goats, and further if the placing of animals in the individual chambers for one hour resulted in an elevated milk somatic cell count due to stress. The experiment took place in two commercial herds, with equal experimental design in both herds. The measurements took place in July, 4-6 months after kidding, and goats were on pasture. Ten goats in first and ten goats in second parity were randomly selected. Five goats in first parity and five goats in second parity were allocated to the control group, and likewise for the methane group. Goats were weighed and sixty-minute methane emissions were captured in PAC using an Eagle 2 instrument for animals from the methane group on the second of the three days. Methane concentration was converted to gram per hour. Average body weights of goats measured on the two farms were 40.5 kg and 51.3 kg, respectively, and average methane emissions were 0.95 gram/hour and 1.24 gram/hour. Level of emissions were similar to sheep of the same body weight. Milk samples were taken from all twenty animals in each herd at three consecutive evening milkings, and number of milk somatic cells (SCC) were determined from the milk samples. Log(SCC) was compared between herds, and between the methane group and the control group using the GLM procedure in SAS. There was a significant difference between herds in log(SCC), while there were no significant difference in log(SCC) between the control group and the methane group. In conclusion, this study indicates that it is possible to measure methane emission on Norwegian Dairy goats using PAC equipment. Also, based on this study, methane measurements do not result in a stress response leading to elevated milk somatic cell counts.

Session 60 — Poster 9

Inclusion of sainfoin in lamb concentrate on blood metabolites and oxidation status

C. Baila, M. Joy, I. Casasús, M. Blanco, J.R. Bertolín and S. Lobón
Centro de Investigación y Tecnología Agroalimentaria de Aragón (CITA)-IA2, Department of Animal Science, Avda. Montañana, 930, 50059 Zaragoza, Spain; cbaila@cita-aragon.es

Light lambs are usually fed with cereal-based concentrates during fattening. However, interest in the use of forages in animal diets has grown to increase the self-sufficiency and sustainability of farms. The aim of this study was to evaluate the effect of increasing amounts of sainfoin (*Onobrychis viciifolia*), a legume forage, in the fattening concentrate of light lambs on their blood metabolites and oxidation status. Twenty-six Rasa Aragonesa weaned male lambs (14.0±0.49 kg; 30±0.66d) were divided in 3 groups depending on the inclusion of dry sainfoin in the concentrate: 0% of sainfoin (Control), 20% of sainfoin (20SF), and 40% of sainfoin (40SF). Blood samples were obtained once every two weeks (week 0, 2, 4, and 6) until slaughter (25.0±0.78 kg; 70.6±0.66 d). Plasma glucose, urea and non-esterified fatty acid concentrations were affected by the interaction between the diet and the week ($P<0.05$), although concentrations were similar among treatments within each week. The 20SF and 40SF lambs decreased their urea, glucose and NEFAS blood concentration from week 0 to 2 ($P<0.05$), whereas control lambs only shown a reduction of urea concentration during these weeks ($P<0.001$). Regarding the blood antioxidant capacity, measured with ABTS (ethylbensothiazoline-6-sulfonic acid) method and polyphenol concentration, both were only affected by the week ($P<0.05$). The blood antioxidant capacity increased from week 0 to 2 and from week 4 to 6 ($P<0.05$) when following the ABTS method, while polyphenol concentration raised only from week 0 to 2 ($P<0.001$). The concentration of malondialdehyde (MDA) was affected by the week ($P<0.001$), tending to raise between week 2 and 4 ($P<0.1$) and increasing from week 4 to 6 ($P<0.001$), and was also affected by the diet with lower MDA values (lower blood oxidative stress) of 40SF group than Control group ($P<0.05$) and a trend towards lower MDA values in 20SF compared to the Control group ($P<0.10$). To conclude, the inclusion of sainfoin in the fattening concentrate for lambs did not affect the metabolic state of the animals; however, the lower oxidative stress in blood of sainfoin-fed lambs may be associated with an improvement in animal welfare.

Session 60 — Poster 10

The effect of chromium propionate on growth performance and meat quality of lambs

C. Garrine[1,2,3], T. Fernandes[3], S.P. Alves[1,3], N.R. Ferreira[2], J. Santos-Silva[4] and R.J.B. Bessa[1,3]
[1]Associate Laboratory for Animal and Veterinary Sciences (AL4AnimalS), Lisboa, 1300-477, Portugal, [2]Universidade Eduardo Mondlane, Maputo, 257, Mozambique, [3]CIISA, Centro de Investigação Interdisciplinar em Sanidade Animal, Faculdade de Medicina Veterinária, Lisboa, 1300-477, Portugal, [4]Instituto Nacional de Investigação Agrária e Veterinária, Santarém, 2005-048, Portugal; carmengarrine@fmv.ulisboa.pt

Chromium (Cr) is required as an essential nutrient for ruminants. There are some reports of Cr effect on animal metabolism, as an increment in cellular sensibility to insulin; consequently, there is an influence on lipids, protein, and carbohydrate metabolism. The magnitude of the Cr effect on animal metabolism is dependent on chemical form. However, no study has reported the effect of organic form on the ruminal metabolism of lipids. Additionally, there are only a few reports on the effects of organic Cr supplements on lamb growth, carcass characteristics, and blood biochemistry. Thus, in this study we evaluated the effect of chromium propionate (CrC_3) on growth, carcass characteristics, blood parameters and rumen pH in an experiment with 32 weaned male lambs distributed in a randomized blocks design to four treatments with eight repetitions each. The treatments consisted of a factorial scheme 2×2, with two basal diets (concentrate and forage) and CrC_3 supplementation (0 or 800 ppb) from Cr – KemTRACE® Chromium. Supplementation with CrC_3 had no effects on the performance of lambs, but as expected, lambs fed concentrate diets presented higher average daily gain and slaughter body weight than those fed forage diets. Regarding blood parameters, CrC3 supplementation decreased the serum aspartate aminotransferase concentration but insulin was not affected. The supplementation with CrC_3 did not affect the carcass traits but increased the meat pH measured at 24h post-mortem, and that increase was greatest with the concentrated-based diet. However, CrC_3 supplementation did not affect meat colour, and meat dry matter, crude protein, or intramuscular fat. Financial support was provided by UIDB/ 00276/2020 project and Kemin is great acknowledge for providing the chromium propionate.

Session 60 Poster 11

Effects of lamb genotypes and carcass weight on primal cuts tissues distribution

C. Oliveira[1], M. Pimpão[1], J. Santos-Silva[1,2], O. Moreira[1,2] and J.M. Almeida[1,2]
[1]INIAV, Quinta da Fonte Boa, 2005-048 Vale de Santarém, Portugal, [2]CIISA-FMV, Universidade de Lisboa, Av. da Universidade Técnica, 1300-477 Lisboa, Portugal; joaoalmeida@iniav.pt

The availability of fresh lamb on the Portuguese market is highly seasonal with little supply outside Easter and Christmas seasons. Outside these periods some demand is maintained in retail chains but mainly for primary primal cuts. Knowing the proportion of the various tissues of the less valued primal cuts is one way to enhance them for the meat industry. The aim of this study was to evaluate the effect of lamb genetics and carcass weight in the proportion of tissues components. Two groups of pure Merino Branco (MB) and two groups of Ile-France × Merino Branco (IF×M) crossbreed lambs, with 15 to 20 lambs each. All lambs grazed natural pastures together with their dams until weaning (three months of age) and were supplemented with commercial concentrate and hay. Low weight groups of different genotypes were slaughtered on average at four months of age and the others two months later, to provide light (LC) (MB 12.5 kg; IF×M 13.5 kg) and heavy carcasses (HC) (MB 15.0 kg; IF×M 17.0 kg), respectively. Shoulder, breast, and neck primal cuts were individually vacuum packed, deep-frozen and sent to laboratory to be carefully dissected into muscle, bone, intermuscular, and subcutaneous fat tissues. Genotype influenced intermuscular fat content ($P<0.001$), with 11.3% in IF×M and 9.1% in Merino cuts. The muscle tissue content was neither affected by genetics nor by carcass weight, but its interaction was highly significant ($P<0.001$). The fat tissues suffered the biggest influence of the increase on carcasses weight, presenting the heaviest always higher values ($P<0.001$). Subcutaneous fat increased from 4.21% (LC) to 6.69% (WC), intermuscular fat from 8.85% (LC) to 11.52% (WC), and total fat content from 13.06% (LC) to 18.21% (HC). Muscle content was higher in shoulder (63.3%), followed by neck (54.9%) and breast (50.9%). The highest fat content was found in the breast (22.0%), however no differences were found for total and intermuscular fat between shoulder and neck. Finally, the greatest differences in tissues proportion were found among the three primal cuts. Funding: Project: PDR2020-101-031690 Child Lamb.

Session 61 Theatre 1

Safety of black soldier fly (*Hermetia illucens*) larvae reared on different biowaste substrates

E.F. Hoek- Van Den Hil[1], K. Van Rozen[2], Y. Hoffmans[1], P.G. Wikselaar[3] and T. Veldkamp[3]
[1]Wageningen University & Research, Wageningen Food Safety Research (WFSR), P.O. Box 230, 6700 AE Wageningen, the Netherlands, [2]Wageningen University & Research, Wageningen Plant Research (WPR), Edelhertweg 1, 8219 PH Lelystad, the Netherlands, [3]Wageningen University & Research, Wageningen Livestock Research, De Elst 1, 6700 AH Wageningen, the Netherlands; elise.hoek@wur.nl

Black soldier fly larvae (BSF, *Hermetia illucens*) can convert low quality organic waste into protein-rich ingredients for food and feed. The underutilized biowaste streams could be a source of valuable nutrients, however, they could also contain chemical residues. For example, antibiotics and antiparasitic drugs are regularly found in pig or chicken manure and could be present in slaughterhouse waste. Pesticides residues can be found in organic waste of for instance vegetal origin. Under current European law, it is not permitted to use biowaste substrates that contain animal products or manure for insect rearing, therefore it is needed to complete the necessary food and feed safety data to enforce possible legalization. This study aimed to investigate the effects of the presence of chemical contaminants in biowaste substrates, including BSF growth and survival, as well as the presence of residues in the larvae. Eight-days old larvae were reared on five biowaste sources, solely or in combinations: fast food (FF), mushroom feet (stems) (MF), poultry meal (PM), slaughter waste (SW), and pig manure solids (PS). These six diets were formulated based on 30% dry mater and 20% crude protein: FF, FF-MF-PM, FF-MF-SW, FF-PS, PS-MF-PM, and PS-MF-SW. Results of insect growth and analysis of the presence of chemical contaminants such as heavy metals, and veterinary drugs, in the substrates and larvae will be shown. Also the presence of bovine, pig and chicken DNA in the insects was analysed. When rearing BSF larvae on biowaste substrates, the possible presence of some chemical contaminants such as heavy metals and veterinary drugs should be controlled to ensure optimal insect growth and safety of the insect products.

Session 61 Theatre 2

Effects of insecticide residues in feed on reared insects

N.P. Meijer[1], H.J. Van Der Fels-Klerx[1] and J.J.A. Van Loon[2]
[1]Wageningen Food Safety Research, Agrochains, Akkermaalsbos 2, 6708 WB Wageningen, the Netherlands, [2]Wageningen University, Laboratory of Entomology, Droevendaalsesteeg 1, 6708 PB Wageningen, the Netherlands; nathan.meijer@wur.nl

Crops used for feed production may contain residues of insecticides after application in the field or storage. Legal limits have been set in the EU for feed material/pesticide combinations, in accordance with Good Agricultural Practice (GAP) and to protect sensitive humans. However, insecticides may affect insects reared for food and feed – even at low concentrations. We conducted experiments to determine the effects on survival and yield, and bioaccumulation of insecticide residues on black soldier fly larvae (BSFL, *Hermetia illucens*) and lesser mealworm (LMW, *Alphitobius diaperinus*). For both species, we observed significant effects on survival and yield of certain insecticides when tested at concentrations equal to the respective legal limit in the EU, in particular for the substances spinosad, imidacloprid, and cypermethrin. For instance, mean survival as low as 38% was observed for spinosad for the maximum residue limit in the EU (2.0 m mg/kg). Consecutive experiments focused on the effects of treatments containing multiple insecticides, to determine potential cumulative or synergistic effects, and to assess long-term effects of chronic exposure of insects to insecticide residues. We will present the results of these experiments and provide recommendations for the insect rearing industry and policymakers.

Session 61 Theatre 3

Insect disease monitoring: a proposal for a structural approach

E. Van Engelen, D. Van Der Merwe, W. Dekkers, P. Vellema and S. Luttikholt
Royal GD, Arnsbergstraat 7, 7400 AA Deventer, the Netherlands; e.v.engelen@gdanimalhealth.com

Insect farming is becoming increasingly important as deliverance of proteins in animal feed and human food. Diseases can have dramatic impact on the future of insect farming. This creates the urge for a monitoring structure to limit the dangers of diseases or toxic components that might harm the condition of insects, the animals that eat them or humans. Royal GD has more than 100 years of experience in fighting animal diseases and for farm animals has the governmental given task to monitor animal diseases for which a monitoring and surveillance system was implemented 20 years ago. We propose to use such a system to be used in insect farming. The goals of this system are: (1) the detection of outbreaks of diseases that are not endemic; (2) the detection of previously unknown disease conditions; (3) keeping track of trends and developments in animal health. The system has both pro active and reactive components. The tools are: (1) a helpdesk that advices farmers and consultants with regards to health problems and generates knowledge about the field; (2) laboratory and pathologic investigation results; (3) acting within a knowledge network; (4) data analysis and dissemination; (5) in dept search for diseases in cases of suspicion. This generates data that is analysed and reported regularly. This system has proven to work efficient for farm animals for many years and had picked up several diseases and disease status changes. For insects, this system might be useful to apply. However, in that case, it needs to be adapted to the specific situation of insect farming at this moment, such as the knowledge of diseases, impact of diseases, and size and structure of the sector.

Economic value of frass fertilizer from black soldier fly farming

D. Beesigamukama[1,2], S. Subramanian[2] and C.M. Tanga[2]
[1]Busitema University, Department of Crop Production and Management, P.O. Box 236 Tororo, 256, Uganda, [2]International Centre of Insect Physiology and Ecology, P.O. Box 30772, Nairobi, 00100, Kenya; dbeesigamukama@icipe.org

Although black soldier fly (BSF) farming is becoming a rapidly growing agribusiness, studies on BSF farming's economic aspects are limited. This study analysed the economic implications of farming BSF for animal feeds and frass fertilizer (FF) production to provide evidence of the economic benefits using experimental data. The BSF larvae were reared using brewery spent grain amended with sawdust, biochar, and gypsum, to determine the best waste combination for cost-effective production. The agronomic performance of FF on the maize crop was assessed using field experiments. Our results demonstrated that the rearing substrate accounts for 81-90% of the total BSF production cost. The utilization of frass fertilizer as an additional value-added product would increase farmer's net income by 5-15 folds compared to BSF farming alone. Feedstock amended with 20% biochar increased net income by 10-64% for BSF larvae and FF production compared to other feedstocks. Production of 1 tonne of dried BSF larvae (US$ 900) would generate 10-34 tonnes of frass fertilizer worth US$ 3,000-10,200. Maize grown on plots treated with FF yielded 29-44% higher net income than plots amended with commercial organic fertilizer. Furthermore, smallholder insect farmers' direct use of FF for maize production would generate 30-232% higher net income than farmers purchasing similar FF. Our results demonstrate the role of insect farming in circular economy and justify the opportunities for future investments that would lead to enhanced sustainability of agriculture and food systems, especially for smallholder farmers in low- and middle-income countries.

Life cycle assessment of *Hermetia illucens* meal inclusion in diets for trout reared in aquaponics

F. Bordignon[1], A. Trocino[2], E. Sturaro[1], G. Xiccato[1], L. Gasco[3], M. Birolo[1] and M. Berton[1]
[1]University of Padova, DAFNAE, Viale dell'Università 16, 35020 Legnaro (PD), Italy, [2]University of Padova, BCA, Viale dell'Università 16, 35020 Legnaro (PD), Italy, [3]University of Torino, DISAFA, Largo Paolo Braccini 2, 10095 Grugliasco (TO), Italy; francesco.bordignon@unipd.it

This study analysed the effect on global warming potential (life cycle assessment method, LCA) of the inclusion of a partially defatted *Hermetia illucens* larvae meal (HI) in diets for rainbow trout (*Oncorhynchus mykiss*) reared in a low-tech aquaponic system. Three inclusion levels were tested: 0%, 6% and 12% (HI0, HI6, HI12, respectively), for a fishmeal replacement of 0%, 25% and 50%, respectively. A total of 173 rainbow trout (initial live weight: 156±39.8 g) were distributed among nine experimental aquaponic units and fed the experimental diets for 76 days. At the end of the trial, no differences among fish fed the different diets were found in terms of mortality (3% on average), final weight (303 g) and feed conversion ratio (1.53). The LCA reference unit was the aquaponic unit and the system boundaries were set to include the impact due to aquafeed production; 1 kg increase of rainbow trout was used as functional unit (FU). Impact categories were global warming potential, without (GWP) and with (GWP_LUC) emissions due to land-use change. Emission factors related to HI production (EFHI) were derived from literature (9 studies; 20 values). Considering the control diet (HI0), the mean GWP and GWP_LUC generated to produce 1 FU were 1.32±0.08 and 3.36±0.21 kg CO_2-eq, respectively. Mean GWP per 1 FU (calculated using the different EFHIs) increased by 32% and 48% with HI6 and H12 diets compared to HI0 diet; mean GWP_LUC increased by 10% and 13%, respectively. The GWP values obtained in HI12 compared to HI0 diet changed according to EFIH from -1% to 31% (GWP) and from -10% to +55% (GWP_LUC). In HI12 diet, only 5 of 20 (GWP) and 8 of 20 (GWP_LUC) EFHI determined GWP equal or lower than that associated with HI0 diet. In conclusion, based on literature EFHI, a reduction in GWP in rainbow trout fed HI still need to be confirmed and likely requires the identification of technical hotspots in HI production.

Session 61 — Theatre 6

CO_2 and ammonia emitted by mealworms, crickets and black soldier fly larvae

C.L. Coudron and D. Deruytter
Inagro vzw, Insect Research Centre, Ieperseweg 87, 8800, Belgium; carl.coudron@inagro.be

Insect production systems are often proposed as more sustainable compared to conventional livestock farming. However, sustainability can be a vague umbrella term. This study covers one part, being emissions and more specifically emissions to the air. Some emissions are greenhouse gasses, such as CO_2, methane and nitrous oxide, and have a global impact on our climate. Other emissions, such as ammonia, can have a more local or regional impact as the nitrogen deposition can cause eutrophication of the soil and surface waters. The impact of insect production on this issue is poorly studied, leading to unsubstantiated claims and caution among local policy makers. This is one of the reasons why insect producers often struggle to obtain the necessary permits in the start-up phase. In addition, all insects are often treated the same and there is little nuance among policy makers. In order to partially overcome this knowledge gap, research was performed to determine CO_2 and ammonia emitted by *Tenebrio molitor*, *Acheta domesticus* and *Hermetia illucens* throughout their production processes. Two strategies were followed. One strategy was to rear insects in a climate controlled cabinet with a controlled air inlet and outlet during the entirety of their growth. Air flow at the inlet was measured with an anemometer and ammonia and CO_2 concentration were monitored in the incoming and outgoing air. A second strategy involved an airtight cabinet in which a known amount of live insects are held for a short period of time (a few hours at most). During their time in the cabinet, the accumulation of emission gasses is monitored. By repeating this at several different life stages, emissions for the entirety of their growth can be modelled.

Session 61 — Theatre 7

Sustainable perspectives of insect production

S. Smetana[1], A. Bhatia[1,2], N. Mouhrim[3], A. Tonda[3], A. Mathys[4], A. Green[4], D. Peguero[4] and V. Heinz[1]
[1]German Institute of Food Technologies (DIL e.V.), Prof.-von-Klitzing-Str. 7, 49610, Germany, [2]University of Osnabrück, Neuer Graben 29, 49074 Osnabrück, Germany, [3]UMR GMPA, AgroParisTech, INRA, rue de l'université 147, 75338 Paris, France, [4]ETH Zurich, Schmelzbergstrasse 9, 8092 Zurich, Switzerland; s.smetana@dil-ev.de

Mass production of insects is calling for environmentally optimized and economically efficient insect value chains. When insects are used as feed and food, multiple functions come into picture along with sustainability that includes regulations, nutritional factors, economic feasibility, social acceptance and reduction of greenhouse gases emissions. Thus, it is challenging to consider all sustainability indicators (SI) and to select transformative factors that can improve one function without unintended consequences on other objectives. Multi-objective optimized (MOO) decision system can identify trade-off between different objectives for management. That is why one of the aims of SUSINCHAIN project is the development of a framework for the application of MOO algorithms to determine the sustainability potential of insect production chains. The analysis of SI indicated that impacts can vary widely depending on the scope and boundaries of assessment, methodologies, assessed insect species, scale of production and other aspects. In order to generate results of the analysis of multiple potential chains for the MOO a modular assessment system was applied. It consists of three stages: (1) determination of system boundaries of insect production chains including relevant comparable studies via graphical mapping; (2) the modularization of insect production chain according to the insect production scheme and impact assessment methodologies; (3) the consideration of a functional unit, production scale and geographical location which may affect the results and alter the final outcomes and conclusions. Such approach allows for the running of multiple assessment options resulting in the data pool required for MOO. The developed framework demonstrated its effectiveness in identifying hotspots in the insect production system that require further improvement as general strategies for the improvement, but also applicable in every specific insect production chain. Specifics of the application of the system are presented.

Small-scale produced insects in hens' diet as tools for decrease of environmental impact of eggs

D. Ristic, K. Wiesotzki and S. Smetana
German Institute of Food Technologies (DIL e.V.), Prof.-von-Klitzing-Str. 7, 49610 Quakenbrück, Germany; d.ristic@dil-ev.de

This life cycle assessment study was based on pilot data of *Hermetia illucens* production (Essen, Oldenburg, Germany). The goal of this study was the assessment of environmental impacts of eggs produced by chicken grown on a diet with inclusion of insects. The results identified that inclusion of insects produced at the small scale significantly increased the environmental impact of eggs production. Contribution analysis of the most important categories indicated the highest contribution on the environmental impact of fresh *H. illucens* was energy use (55-70%). Feed production for insect breeding was the second most impacting factor for the environmental impact being responsible for 12-50% of impacts (except land use and water use categories). Therefore, egg production was also modelled for a barn with hens fed on 10% fresh *H. illucens* larvae, produced by employment of full capacity of insect production. Despite the highest share of larvae in the feed, this resulted in the lowest environmental impact of eggs coming from insect-fed hens. Eggs from both the experimental and benchmark chicken fed on commercial feed still had lower environmental impact. Comparison of environmental impact of chicken production (without insects in the diet) to other studies indicated that it is in the range of impacts present in literature. The use of solar panels as source of energy was very important for keeping the impacts of experimental laying hens' rearing in the range of large scale, highly efficient industrial productions. On the other hand, environmental impact of *H. illucens* production was very high, higher than indicated in the literature, mostly due to high electricity use. Based on the results of the study, it is recommended for the insect production to increase scale of production, reduce energy consumption, rely on alternative sources of electricity or heat, and rely on side streams for feed, which can reduce the environmental impact by 60-80%. Small scale insect rearing may only bring environmental benefits if it mostly relies on locally available and unused resources for energy and feed. Additionally, for laying hens' rearing, it is recommended to rely on renewable sources of energy supply.

Generic multi-objective optimised decision support tool for SUstainable INsect CHAINs production

A. Bhatia[1,2], N. Mouhrim[3], A. Green[4], D.A. Peguero[4], A. Mathys[4], A. Tonda[3] and S. Smetana[2]
[1]University of Osnabrück, Neuer Graben 29, 49074 Osnabrück, Germany, [2]German Institute of Food Technologies (DIL e.V.), Prof-von-Klitzing-Str. 7, 49610 Quakenbrück, Germany, [3]UMR 518 MIA-Paris, INRAE, Université Paris-Saclay, AgroParisTech 16, rue Claude Bernard, 75231 Paris CEDEX 05, France, [4]Laboratory of Sustainable Food Processing, ETH Zurich, Schmelzbergstrasse 9, 8092 Zürich, Switzerland; a.bhaita@dil-ev.de

Determination of insect production sustainability is a complex and challenging task. Insect production life cycle sustainability assessment (LCSA) can estimate social, economic, and environmental impacts. To use insects as feed and food, several aspects of sustainability must be considered, including food and feed production regulations, nutrition factors, social acceptance, and greenhouse gas regulation. The multi-objective integrated decision support system (DSS) based on: (1) modular LCA and scenario-based life cycle inventory (LCI) databases that analyse the environmental, social, and economic performance of multiple insect species (*Acheta domesticus*, *Musca domestica*, *Hermetia illucens*, *Tenebrio molitor*); (2) analytic hierarchy process (AHP) to evaluate and rank different insect production scenarios based on decision-makers; and (3) non-dominated sorting genetic algorithm (NSGA-II) to obtain pareto-optimal solutions covering all sustainability indicators to compare insect species production scenarios to test the sensitivity of results with different feeds, processing and utilities, type of end product, packaging, and scaling options. The beta version of the DSS allowed to define that the most eco-efficient scenario in Europe was for house cricket (*A. domesticus*) reared on plant residues subject to blanching and microwave drying and packaged in polyethylene foil (LDPE). *T. molitor* had the lowest carbon emissions despite using more non-renewable energy than *H. illucens*. A web-based multi-objective integrated decision tool can be used to identify hotspots in the insect production system. Environmental accounting methods that include holistic environmental, economic, social, and technical information are required for the concept of insect production sustainability to be implemented on a large scale.

Session 61 Theatre 10

Carbon footprint of eggs produced by laying hens fed with soybean or *Hermetia illucens* meal

A. Vitali[1], G. Grossi[1], M. Meneguz[2], F. Grosso[2], E. Sezzi[3], N. Ferrarini[4], E. Batistini[5] and N. Lacetera[1]
[1]Università della Tuscia, Via s. Camillo de Lellis, 01100 Viterbo, Italy, [2]Bef Biosystems S.r.l., Via Tancredi Canonico 18c, 10156 Torino, Italy, [3]Istituto Zooprofilattico Sperimentale del Lazio e della Toscana, Str. Bagni 4, 01100 Viterbo, Italy, [4]Azienda Sanitaria Locale, Via Enrico Fermi 15, 01100 Viterbo, Italy, [5]SE.CO.M. S.r.l., Via dell'Artigianato 3, 06089 Torgiano, Italy; g.grossi@unitus.it

The impact of dietary replace of soybean meal with insect meal on carbon footprint (CF) of layer hens' eggs was assessed. *Hermetia illucens* (Hi) meal was produced by BEF Biosystems in a prototypal bugs-farm based on the breeding of adults and fattening of larvae. Bugs-farm data of feeding operations, energy consumption, waste management and transports were collected. After one week, larvae were recovered, sacrificed, dried and ground to produce a meal tested in a feeding trial that lasted 56 days involving 72 laying hens divided into 18 groups (4 hens each) and housed into arches with an artificial light regime of 16L:8D. Groups were randomly assigned (9 replicates) to one of the two experimental diets isoenergetic, iso-proteic and balanced for amino acids: soybean-based meal as control (C) and *H. illucens* based meal (Hi) where soybean was totally replaced. Data on feed consumption and eggs production were recorded daily. Supply chain data of eggs produced by C and Hi groups were modelled with Simapro and GHG emissions were expressed in kg of carbon dioxide equivalent (kgCO_2e) and associated with one kg of eggs. Results of CF were 3.09 and 3.63 kg CO_2e/kg eggs for group C and Hi, respectively. Soybean meal accounted for 39.7% of total CF, with Land Use Change (LUC) resulting as main hot spots of its supply chain (86.8%). The production of Hi meal accounted for 49% of total CF, the main hot spots were energy consumption and feeding of larvae, in this latter case although were used and valorised by-products of agro-industry, some of these had still high CO_2e at plant gate. The progress of insect farming on an industrial scale may increase energy efficiency and stimulate the development of a circular economy to support by-products supply. In this context, the use of insect meal may become a sustainable alternative protein source in livestock feeding.

Session 61 Poster 11

Impacts of cypermethrin residues on mealworm rearing performances

F. Dupriez, A. Samama and T. Lefebvre
Ynsect, 1 rue Pierre Fontaine, 91000 Evry, France; florent.dupriez@ynsect.com

Due to the development of world population, production of more proteins will be needed either for the livestock or for human food. However, in the context of climatic crisis, sustainable solutions are required to feed the planet. Cereals consumption by livestock already accounts for a third of global world production and this is going to increase. Similarly, the aquaculture production uses soya flours and wild fishes which respectively provokes deforestation and overfishing. A new industry proposing alternative by using insect proteins in animal feed. Insect rearing have positive environmental impacts by using less land surfaces, produce less carbonic gazes and using less water. Ÿnsect a French company is rearing their mealworms on cereal-based by-products. Nowadays, pesticides like pyrethroids are widely used in agriculture to protect the production, especially storage grains. These chemical compounds have been designed to have lethal effects on insects. In the R&D laboratory, a study has been conducted to highlight effects of cypermethrin in an industrial context of *Tenebrio molitor* rearing. The study mainly focused on adults, to assess the impact on reproduction, and on early stages of larvae to determine the dose-effect of this residues on the growth. Experiments have been conducted under the legal dose limit call 'residual maximum limit (RML)' which is 2 mg for cypermethrin per kg of feed. According to the results obtained, cypermethrin had some physiological effects on young larvae growth even under the legal dose limit, by limiting the biomass product until 27% compared to the control. Regarding adults, the insecticide caused a decrease of 50% in the number of eggs laid per female for one week. These results show that even if the residual maximum limit is respected by the supplier, insect rearing fed by this conventional raw material won't be able to develop in good conditions. These conclusions could involve changes of main supplier's specifications in terms of raw materials. It could move the supply of insect industry to an agriculture with new methods to avoid the use of these types of residues. Finally, insect industry will lead to the development of a new market of insect's food cereal by-products by encouraging producers in this way.

Session 62 Theatre 1

Pain expression as an indicator of orthopaedic disease in dairy cattle

M. Söderlind, K. Ask, A. Granlund, A. Leclercq, E. Hernlund, P. Haubro Andersen and M. Rhodin
Swedish University of Agricultural Sciences, Department of Anatomy, physiology and biochemistry, Box 7011, 750 07 Uppsala, Sweden; maja.soderlind@slu.se

Orthopaedic disease causing pain and lameness is common in dairy cows and negatively affect animal welfare and economy. In other species, certain facial expressions associated with pain, 'pain face' (PF) is well recognized. Although less studied, a cow pain face has been suggested. This study aimed to explore if a multimodal pain scale, which assessed position of ears, head and back, response to approach, PF and attention to surroundings (total score 0-10), could be used as an indicator of orthopaedic disease and if knowledge of the lameness status biased the observer. PF was scored if an angular eye contour and/or increased facial muscle tension appeared during the observation period. An initial assessment (IA) of 28 lame loose-housed dairy cows was done, including video recording of the cows' body and face for pain evaluation by blinded observation (BO), pain evaluation by direct assessment (DA), lameness evaluation by objective (inertial sensors) and subjective methods, and determination of primary lesion by claw examination. When lameness was improved 1-5 months later, a follow up assessment (FA) was done similarly to the IA. Cows were included in further analysis (n=23) if they showed reduced asymmetry and a visually regressed primary lesion at FA. The data was tested for normal distribution (Shapiro Wilks test $P<0.05$) and nonparametric, descriptive and agreement statistics were applied. Pain scores from BO and DA were decreased at FA (median improvement BO: 3; DA: 2, $P<0.001$, Wilcoxon signed rank test), with an agreement of 0.604 at IA and 0.964 at FA (weighted kappa). A PF was present in a higher number of cows at IA compared to FA for both BO and DA ($P<0.05$, Wilcoxon signed rank test), with an agreement of 0.481 at IA and 1 at FA (weighted kappa). The results indicate that a multimodal pain scale, which include presence or not of PF, can be used to assess healing of orthopaedic disease. The BO and DA showed moderate agreement at IA, and almost perfect agreement at FA, suggesting that observers' awareness of lameness status only influence the pain evaluation to a minor degree. Yet, more studies on pain expression in cattle are needed.

Session 62 Theatre 2

Novelty and human approach tests in the Mertolenga breed of cattle

A. Vitorino[1,2,3], G. Stilwell[2], J. Pais[1] and N. Carolino[2,3,4]
[1]ACBM, R. Diana Liz Horta Bispo, Aptd.466, 7006-806 Évora, Portugal, [2]CIISA, Av. Univ. Técnica, 1300-477 Lisboa, Portugal, [3]INIAV, I.P., Fonte Boa, 2005-048 V.Santarém, Portugal, [4]EUVG, Av. José R. S. Fernandes 197, 3020-210 Coimbra, Portugal; andreia.vitorino93@gmail.com

Cattle, as prey, are more prone to fear, which can be assessed by tests such as the novelty test and the human approach test. The importance of human-animal relationship and its influence on welfare is clear. The aim of this work is to evaluate the behaviour of Mertolenga breed steers and the correlations between fear tests. This study was conducted at the Mertolenga Performance Testing Center from May until November of 2021 with 29 Mertolenga young bulls. In the novelty test, the following data was collected: animal identification, farm of origin, exit from the chute, latency to touch a novelty object (ball) and the number of contacts with it. Since the human approach test depended on the cooperation of the animals to perform it, it was not possible to collect data from all the animals on each of the dates. In this performance test, the test length and the final distance to the feed trough were recorded. All data was collected in Excel tables and analysed with different statistical routines with the SAS program. Data from 29 animals from 16 farms were analysed over two test days. The analysis file obtained from the novelty test after data validation consisted of 58 data from Mertolenga's animals, with an average age of 328.57±45.66 days, an average touch latency of 148.90±95.07 seconds, an average of 1.97±2.60 contacts with the ball and an average chute exit of 1.6±0.75 (with 1 corresponding to a walk and 2 trot out of the chute). The farm of origin did not influence ($P>0.05$) the latency of touch nor the number of ball contacts, but the chute score was different ($P=0.04$). The age of the animals influenced ($P<0.01$) the latency to contact and the number of contacts. In the human approach test, 42 records were obtained, giving an average of 18.93±9.88 seconds of test duration, with an average of 1.74±0.77 meters distance from the feed trough. It was intended to relate the two tests in the study, but this was not possible due to missing data from most of the animals in the human approach test. However, this study will continue beyond this work in order to draw more concrete conclusions.

Session 62 Theatre 3

Can stress induced by social challenge be contagious?

R.D. Guevara[1,2], S. López-Vergé[3], J.J. Pastor[3], X. Manteca[2], G. Tedo[3] and P. Llonch[2]
[1]AWEC Advisors S.L., Research Park UAB, Campus UAB, 08193, Spain, [2]Universitat Autònoma de Barcelona, Department of Animal and Food Science, Barcelona, Bellaterra (UAB), 08193, Spain, [3]Lucta S.A., Innovation Division, Research Park UAB, Campus UAB, 08193, Spain; raul.guevara@awec.es

Changes in the social hierarchy during weaning and other re-grouping events in pig production may induce aggressive behaviour, harming pig welfare. The aim of this study was to assess the performance of an experimental model to induce a stress response through a social challenge. Thirty-six transition pens (4 pigs/pen) from one room were randomly allocated either to a control group (piglets maintained in their social group, Ctrl) or a social stress group (piglets subjected to a social challenge, SS). The social challenge consisted of daily moves of 3 out of 4 pen mates to different pens for 3 consecutive days, while the other one stayed in the original pen. Before the 1st mixing day (day 25 post-weaning, p.w.) and at the end of the 3rd mixing day (day 28 p.w.), saliva and blood samples were collected from 2 random piglets per pen, and faecal samples from all the piglets. Plasma and saliva cortisol, hemogram biomarkers, and faecal myeloperoxidase (MPO) were measured. At the same days, skin lesions of all piglets were recorded on the front, middle, and rear body regions. Skin lesions were higher in the SS group than in the Ctrl group after the social challenge ($P<0.05$) but also increased between sampling days in the Ctrl piglets ($P<0.05$). No differences were observed in plasmatic and salivary cortisol ($P>0.05$) in any of the groups between sampling days. However, MPO, haemoglobin, red blood cells, leukocytes, basophils, lymphocytes, and monocytes counts decreased in both treatment groups ($P<0.05$) after the mixing. Thus, even if skin lesions increased more in the SS group relative to Ctrl, the magnitude of the stress response was similar between groups. We hypothesize that the Ctrl group was stressed by the aggressive behaviour occurring on adjacent pens. If this is true, it may suggest that piglets can be stressed by aggressive events occurring around them which may determine considering separation of groups in pig stress model studies.

Session 62 Theatre 4

Detecting animal contacts – a deep learning-based pig detection and tracking approach

M. Wutke[1], M. Gültas[2], A.O. Schmitt[1] and I. Traulsen[3]
[1]Georg-August University Göttingen, Breeding Informatics, Department of Animal Sciences, Margarethe von Wrangell-Weg 7, 37075, Germany, [2] South Westphalia University of Applied Sciences, Statistics and Data Science, Faculty of Agriculture, Lübecker Ring 2, 59494, Germany, [3]Georg-August University Göttingen, Livestock Systems, Department of Animal Sciences, Albrecht-Thaer-Weg 3, Germany; martin.wutke@uni-goettingen.de

The identification of social interactions is of fundamental importance for animal behavioural studies, addressing numerous problems like investigating the influence of social hierarchical structures or the drivers of agonistic behavioural disorders. In this work, we present a novel framework for the automated identification of social contacts in pigs. By applying a convolutional neural network (CNN) for the detection and localization of individual body parts, we are able to track the animals' movement trajectories over a period of time within a video. Based on the tracking and body part information, we identify social contacts in the form of head-head and head-tail contacts. Moreover, to enhance the applicability of our framework, we also used the individual animal IDs as well as the body part information to construct a network of social contacts as the final output, in which the intensity of the individual animal contacts is quantified and visualized. To evaluate our framework, we created two different test data sets to quantify the performance of our body part detection and tracking model. Consequently, by comparing the manually annotated body parts with the detected body parts from the CNN we achieved a Sensitivity, Precision, and F1-score of 94.2, 95.4 and 95.1% for the detection model. For the tracking model, we calculated the *Multiple Object Tracking Accuracy* (MOTA) and achieved a MOTA score of 94.4%. The findings of this study demonstrate the effectiveness of our keypoint-based tracking-by-detection strategy, which can be incorporate to address various problems like analysis of functional areas as part of an early warnings systems for the detection of agonistic behaviour. Moreover, our approach is not limited to pig detection and tracking, but can be applied to various animals to improve animal monitoring systems.

Session 62 Theatre 5

Test-retest reliability of selected welfare indicators for rearing piglets

J. Witt[1], J. Krieter[1], T. Wilder[1] and I. Czycholl[1,2]

[1]Institute of Animal Breeding and Husbandry, Olshausenstr. 40, 24098 Kiel, Germany, [2]Pig Improvement Company (PIC), 100 Bluegrass Cmmons Blvd. Ste 2200, Hendrsonville, TN 37075, USA; jwitt@tierzucht.uni-kiel.de

Objective tools for the assessment of animal welfare are needed. Welfare assessment schemes commonly recommend to use indicators that were developed for growing pigs also for rearing pigs, although the indicators have not been specifically tested for that age class. Therefore, the present study aimed at testing selected indicators from different self-monitoring protocols, e.g. the Welfare Quality® protocol for pigs, with regard to test–retest reliability (TRR), i.e. consistency over time in an on-farm study on rearing pigs. 20 different individual indicators (IND) were assessed weekly in the rearing period on 3 pig farms with a closed system in northern Germany by one trained observer between October 2020 to November 2021. 100 rearing piglets were randomly selected per batch and individually marked to assess them weekly. This procedure was repeated in 3 consecutive batches per farm, i.e. in total 900 rearing piglets were assessed. For evaluation of consistency over time, Spearman's rank correlation coefficient (RS), intraclass correlation coefficient (ICC) and limits of agreement (LoA) were calculated to compare the TRR between follow-up batches (different animals) and consecutive visits (same animals). In the comparison between the follow-up batches, only the IND 'Wounds on the body' and 'behaviour' were of acceptable (RS≥0.4; ICC≥0.4; LoA ε [0.1; 0.1]) to good (RS:≥0.7; ICC:≥0.7; LoA: ε [0.05; 0.05]) TRR. In the comparison of consecutive batches, 9 IND had a prevalence below 2.6%, so that no statement can be made about the TRR. 3 IND showed a good TRR, i.e. lameness (RS: 1.0, ICC: 0.8, LoA: -0.18 to 0.2), tail length (RS: 0.7, ICC: 0.84, LoA: -0.2 to 0.22) and hernias (RS: 0.8, ICC: 0.8, LoA: -0.2 to 0.18). 8 IND demonstrated acceptable (RS≥0.4; ICC≥0.4; LoA ε [0.1; 0.1]) TRR: wounds on the body, tail lesions, skin condition, tail posture, back posture, behaviour, claw injuries, ear lesions, These IND should be used in a special rearing period protocol. However, there is a need for further reliable IND to cover the four main principles of welfare. At the moment there is still a lack in the principles 'Good housing,' and 'Appropriate behaviour'.

Session 62 Theatre 6

How do horses express their stress: the effect of coping styles on subtle behavioural indicators?

A.-L. Maigrot, M. Roig-Pons, I. Bachmann and S. Briefer-Freymond

Agroscope, Swiss national stud farm (SNSF), Les longs prés, 1580 Avenches, Switzerland; anne-laure.maigrot@agoscope.admin.ch

Personality is defined as the tendency to express similar behaviours across situations and time. In horses, it is of main relevance as it influences the animal's relationship with humans as well as its ability to adapt to its living conditions. It also has an impact on the way the animal reacts to a stressful event, also called the coping style. Indeed, across different species, two main groups have been highlighted depending on the reaction of the animal in stressful situations: 'proactive' individuals, i.e. actively trying to avoid the stressor and to fly away; 'reactive' individuals, i.e. not showing any clearly visible behavioural response. In this exploratory study, we aim at finding new personality tests as well as more accurate stress indicators to better distinguish between 'proactive' and 'reactive' individuals and to find behavioural indicators specific to 'reactive' animals. To this aim, we tested 50 private horses with four well-established personality tests and three new tests developed at the SNSF in order to trigger different levels of stress. We recorded behavioural (head position, ears position, locomotion, postures and mimics) as well as physiological (heart rate variability) parameters. After selecting the data, we performed a principal component analyse (PCA) to extract factor in which behaviours were highly correlated to each other (correlation factor>0.5). This allowed us to highlight several groups of behaviour that had a great influence on the main dimensions of the PCA (eigenvalue>1). The head position, the locomotor activity, the ears position and the heart rate variability thus seem to be the most important indicators (P<0.32 for all). In addition, we performed an ascending hierarchical classification to classify our individuals in groups. Three groups were highlighted and highly correlated to those made by the experimenter analysing the behaviours (correlation factor=0.73). It thus seem that from our test we successfully could distinguish between 'proactive' and 'reactive' individuals. These results allow us to validate subtle indicators of stress in 'reactive' animals as well as the use of several tests to distinguish coping style in horses.

Session 62 Theatre 7

The effects of competition at the feeder on dominance in dairy cows

K. Sheng, B. Foris, J. Krahn, D.M. Weary and M.A.G. Von Keyserlingk
University of British Columbia, Animal Welfare Program, 2357 Main Mall, Vancouver, BC, V6T 1Z4, Canada; skysheng@mail.ubc.ca

In dairy cattle groups, access to resources can be affected by dominance relationships. In loose housing systems, cows compete to access resources such as feed, and dominance relationships may be upset by frequent regroupings. Dominance relationships in dairy cattle are typically assessed using video focusing on, for example, agonistic interactions at the feed bunk during times of high competition (e.g. after fresh feed delivery). Little is known about how the level of competition in the groups affects dominance. We hypothesised that under conditions of high competition (here operationalized as the percentage of filled feed bins occupied at the same time by cows in the group; CD), competitive replacements at the feeder are more reflective of animals' motivation to access fresh feed rather than their position in the social hierarchy. We investigated how the dominance changes in relation to different levels of CD. We monitored cows kept in a dynamic group of 48 with access to 48 lying stalls, 30 electronic feed bins and 5 water bins. Regrouping events took place every 16 days on average, such that we followed 159 lactating Holstein dairy cows over the 10-month study period. We used a validated algorithm to detect competitive replacements at the feed bins, and used these to calculate an Elo rating for each cow (as a measure of social dominance). We also recorded the corresponding CD at the feed bins which varied from 3% to 100%. A linear model revealed that as CD increased by 1% (reflective of a more competitive feeding environment), the variation among cows in Elo rating decreased by 0.52 (R^2=0.93). We observed little difference between the dominance score of individual cows when CD was high, suggesting that the natural hierarchy of the group might not be the primary regulator of resource access during these times. A breakdown in dominance hierarchy (indicated by the compressed Elo ratings) may reflect elevated social stress as cow compete for access to feed.

Session 63 Theatre 1

Effects of antioxidant supplementation on functions of mammary and immune cells in dairy cows

A. Corset[1,2], A. Boudon[2], P. Germon[3], A. Baldi[4] and M. Boutinaud[2]
[1]Biodevas laboratoires, 21 Rue des Chardons ZA de, L'Épine, 72460 Savigné-l'Évêque, France, [2]INRAE, Institut Agro Rennes Angers, UMR 1348 PEGASE, 16, le clos, 35590 Saint Gilles, France, [3]INRAE, UMR 1282 ISP, Centre de recherche Val de Loire, 37380 Nouzilly, France, [4]Univ Milan, Dept Vet Sci & Technol Food Safety, Via Festa del Perdono 7, 20122 Milano MI, Italy; angelique.corset@inrae.fr

In early lactation, dairy cows are highly susceptible to pathologies. This sensitivity may be due to dysfunctions of both inflammatory and antioxidant status of dairy cows linked to physiologic upheaval to produce milk. Dairy cows metabolisms change with increasing demands for energy and oxygen, which is a main cause of a decrease in antioxidant status through the production of reactive oxygen species (ROS). At the same time, excess ROS induces destabilisation of immune cells. An inflammatory state caused by oxidative stress is associated with a decrease in milk production. This review focuses on the knowledge linked to the effects of antioxidant supplementation on cellular functions of mammary tissue and the immune system in dairy cows. This nutritional strategy could reduce ROS production. Indeed, many studies performed *in vitro* and *in vivo* have proved that polyphenols (flavonoïds, resveratrol, lignans), vitamins (A, B, C, D, E) and minerals (selenium) show promising results to rebalance immune and antioxidant status at both systemic and local levels, especially in the mammary tissue. The supplementation of antioxidants can upregulate the Nrf2 signalling pathway and simultaneously downregulate NF-κB pathway, which is a major signalling pathway involved in the synthesis of several proinflammatory factors. The amplified Nrf2 cellular pathway increase the synthesis of antioxidant enzymes and reduce the stimulation of the NF-κB pathway. When NF-κB is inhibited by the antioxidants, the synthesis of pro-inflammatory molecules decrease. The regulation of these two cellular pathways can restore immune and antioxidant status. In addition, vitamins and minerals can participate in the capture of ROS by direct interaction. The rebalance of the immune and antioxidant status of mammary tissue results in a stabilisation or an increase in milk production and can reduce the somatic cell count in milk. A diet enriched with antioxidants can maintain the metabolic balance of dairy cows.

Session 63 Theatre 2

Scutellaria baicalensis plant upregulated antioxidant response genes of mammary cells in dairy cows

L. Nicolas[1], P. Roussel[1], P. Debournoux[1], A. Steen[2], F. Robert[2] and M. Boutinaud[1]
[1]INRAE, Institut Agro Rennes Angers, UMR 1348 PEGASE, 16, le Clos, 35590, France, [2]CCPA group DELTAVIT, Z.A. du Bois de Teillay, Quartier du Haut-Bois, 35150 Janzé, France; marion.boutinaud@inrae.fr

Antioxidant supplementation, especially with the plant *Scutellaria baicalensis* (SB), could limit the oxidative stress that is prone to appear at the beginning of lactation in the mammary tissue of dairy cows. Recent studies showed that SB extract could increase milk production and increase the survival of mammary epithelial cells (MEC) in bovine. These effects could be linked to a change in the antioxidant response of MEC. To study this potential effect, cows received a diet supplemented (n=8) or not (n=11) with 1 g/d of SB from calving until 150 d of lactation. Milk samples were collected at d30, d50, d130 and d150 after calving to purify the MEC from the milk by an immunomagnetic method in order to study by RT-PCR the mRNA level of genes involved in milk synthesis, cell death and the antioxidant response. Data were analysed using an ANOVA taking into account time (day), SB treatment and the interaction of both as fixed effects, and cow as a random effect. SB did not globally significantly affect the expression of genes involved in milk synthesis (LALBA, CNS3, and SLC2A1) or in cell death (BAX, BCL2). However, at d150, CSN3 mRNA level tended to be more expressed in milk purified MEC in SB treated cows than in control cows (P=0.07). SB treatment increased SOD1 mRNA levels compared with control treatment (+66%, P=0.05). SB did not affect the other genes involved in the antioxidant response (SOD2, TRxR1, GPX1, GPX3, CAT) excepted for a significant increase in GPX1 (+51%, P=0.03) at a specific time of milk sampling (d130). The gene expression of nuclear factor erythroid 2-related factor 2 (Nrf2) had a tendency to increase at d150 (P=0.06). This effect could trigger the previously observed reduced cell death in the milk purified MEC and the increase in GPX1 and SOD1 transcripts. In conclusion, SB induced positive effects on the expression of genes involved in the antioxidant response in milk purified MEC.

Session 63 Theatre 3

Can control of oxidative stress and inflammation lead to reduction of the use of antimicrobials?

C. Lauridsen
Aarhus University, Foulum, Department of Animal Science, Blichers Alle 20, 8830 Tjele, Denmark; charlotte.lauridsen@anis.au.dk

Oxidative stress and inflammatory reactions are part of normal defence mechanisms against pathogenic bacteria, and may be of special impact for the underlying pathophysiological mechanism during a normal life. Young animals such as newly-born pigs and pigs post weaning, as well as broiler chickens are having immature intestines making them very vulnerable towards invading microorganisms, and the typical reaction to an infection is localized inflammation of the gut, which appears during the immunological reaction to cope with the pathogens. Likewise, oxidative reactions appear during the immunological responses toward pathogenic bacteria. There is a lack of knowledge regarding the importance of these mediating reactions in relation to enteric infectious diseases and their pathogenicity of pigs and poultry. If these reactions get out of control, such reactions can have dramatic negative consequences on gut health and performance. Genetic factors including breeding towards high performance enhance risk of oxidative stress, and management factors such as weaning of piglets may enhance both oxidative stress and inflammatory reactions. In the present paper, we will look into especially dietary factors of the young animals, which can induce oxidative and inflammatory reactions in the gut, and factors, which potentially can treat these reactions. Handling of these reactions via dietary factors may improve the gut health and robustness, and thereby reduce the need of antibiotics to treat enteric infectious diseases.

Session 63 — Theatre 4

A comprehensive study of sow colostrum immune, antioxidant parameters and litter size correlations

P. Engler[1], D. Bussières[2], F. Guay[3], A. Demortreux[1] and M.E.A. Benarbia[1]
[1]Nor-Feed, 3 rue Amedeo Avogadro, 49070 Beaucouzé, France, [2]Groupe Cérès, 845, route Marie-Victorin, local, Lévis, QC, G7A 3S8, Canada, [3]Université Laval, Département des sciences animales, Québec, QC, G1V 0A6, Canada; paul.engler@norfeed.net

Colostrum is the first milk produced by mammals following the onset of birth and is of utmost importance for the survival, health and growth of the newborn. It is very rich in immune defences such as immunoglobulin G (IgG) and in antioxidants. Whilst this critical immunological aspect has been extensively studied, fewer studies have been conducted on the antioxidant composition of colostrum, especially in sows. The aim of the present work was to increase general knowledge on sow colostrum composition by studying the colostrum composition of sows of different parity levels in terms of immune and antioxidant parameters and evaluate potential correlations between these constitutive parameters and the litter size (LS) performances. The experiment included 4 groups of 19 sows (±2), established according to their parity ranks prior to insemination: [0-1], [2-3], [4] and [5+]. LS parameters were recorded for every sow, including proportion of total born alive (%BA), of stillborn (%SB), of mummies (%MM) and total born (TB). Samples of colostrum were manually collected, within 2-4 hours after the onset of farrowing. Brix value, IgG content, total antioxidant capacity (TAC), vitamin A, C and E content, glutathione peroxidase (GPx), total glutathione (GSH) and superoxide dismutase (SOD) were analysed. Whilst there was an overall effect of categorized parity the total LS ($P<0.10$), means did not significantly differ between groups. Several significant correlations between LS parameters and constitutive parameters of colostrum were found (%SB vs SOD, $P<0.05$; TB vs retinol, $P<0.001$), as well as correlations between constitutive parameters of the colostrum (IgG vs vitamin C, $P<0.001$; TAC vs SOD, $P<0.001$; SOD vs Vitamin E, $P<0.001$). Overall, these results showed the strong relationship between reproduction performances, colostrum immunity and colostrum antioxidant parameters. They help shine a fresh light on the understanding of oxidative stress management around parturition to ensure good performances and immune transmission to the piglets.

Session 63 — Poster 5

Oxidative stress of suckling calves in the immunological challenge with nutritional supplementation

M.S.V. Salles[1], F.J.F. Figueiroa[2], A. Saran Netto[2], V. Gomes[3], R.S. Marques[3], J.A.G. Silveira[4] and E.J. Facury Filho[4]
[1]Animal Science Institute, Ribeirão Preto, 14030-670, Brazil, [2]University of São Paulo, Pirassununga, 13635-900, Brazil, [3]University of São Paulo, São Paulo, 05508-270, Brazil, [4]University of Minas Gerais, Belo Horizonte, 31270-901, Brazil; marcia.saladini@gmail.com

The nutrition and health of young calves are important for the productive system because it affects the productive potential of cows' milk. The aim of this research was to evaluate the supplementation of selenium, iron, and vitamin E on the oxidative stress of calves challenged with Anaplasma marginale. Holstein male newborn calves (n=42) up to 60 days of age, in a randomized block design, were allocated in three treatments: C (control milk substitute); SeVitE (milk substitute supplemented with 0.3 mg organic selenium/kg + 50 IU vitamin E); SeVitEFe (milk substitute supplemented with 0.3 mg organic selenium/kg + 50 IU vitamin E + 100 mg Fe chelate/kg). The calves received 6 litres of substitute daily until 30 days of age and then 4 litres until weaning at 60 days, and received concentrate *ad libitum*. The animals were inoculated with Anaplasma marginale at 40 days of age. Blood samples were collected for glutathione peroxidase (GPx), glutathione (GSH), and total antioxidant status (TAS) analyses, just before inoculation and 20 days after the challenge. GPx plasma concentrations differed among treatments, but dependent on the collection time (interaction between treatment and collection, P=0.005, SEM=106.9). GPx values were higher for the SeVitEFe (1,578.74 U / g Hb) than for the SeVitE (1,160.73 U / g Hb) and control groups (968.18 U/g Hb), before inoculation. However, GPX values were similar between groups after the challenge (mean 1,262.60 U/g Hb). GSH plasma values did not change among treatments (mean=19.87 mg/dl, P=0.1352, SEM=1.56). The TAS was not affected by the treatments but tended to increase (P=0.088, SEM=0.01) in the calves' plasma after inoculation (0.79 and 0.88 mmol/l, before and after inoculation, respectively). A better antioxidant action against the pathogen was noted when the calves were supplemented with selenium, vitamin E, and iron. Financial support: FAPESP 2017/04165-5.

Development of precision feeding strategies for gestating sows

C. Gaillard and J.Y. Dourmad
PEGASE, INRAE, Institut Agro, Le Clos, 35590 Saint-Gilles, France; charlotte.gaillard@inrae.fr

In sows' conventional feeding (CF), diets composition is usually based on the average herd's nutrient requirements. Thus, sows can be under- or over- fed leading to extra feed costs and environmental losses. Nutritional models and new technology (sensors, automatons), bring opportunities to measure and integrate the individual variability into nutrient requirements estimations. The objective is therefore to go towards precision feeding (PF), combining on-farm data as input for a dynamic nutritional model with smart feeders to provide individual and daily-adjusted rations. A mechanistic model (InraPorc) was upgraded for gestating sows and applied to databases to calculate individual daily nutrient requirements. Herd historical data as well as the animal parity, body weight, backfat thickness and age at insemination were needed to predict some parameters required by the model (i.e. litter size and weight, sow's target body weight at the end of the gestation). There was a strong inter-and intra-individual variability of the nutrient requirements according to sows' characteristics, performance and day of gestation. Simulations showed that more sows had their requirements met with PF based on lysine supply than CF, especially for primiparous. With PF, protein intake, feed cost, nitrogen and phosphorus excretions were reduced by 25, 5, 17, and 15%, respectively. The results obtained during an on-farm trial confirmed those obtained by simulation. To improve the accuracy of the nutritional requirements estimations, new parameters could be added to the model like the individual physical daily activity. Indeed, the daily activity varies between and within sows, and impacts the energy requirements. An algorithm is therefore being developed to identify individual activities via video recordings. The results are promising as the neural networks are able to detect a sow lying, standing and eating with accuracies of 82, 73, and 87%, respectively, at the group level. The next step will be to be able to track and identify each individual to obtain its daily activity automatically. Until now, PF concerned only energy and protein supplies. For the future, other nutrient such as minerals and fiber could also be considered, but this requires improvement in the smart feeder design.

Session 64 Theatre 2

Targeted nutrition in gestating sows: opportunities to enhance sow performance and piglet health

P. Langendijk, M. Fleuren and G. Page
Trouw Nutrition, R&D, Stationsstraat 77, 3811 MH Amersfoort, the Netherlands; pieter.langendijk@trouwnutrition.com

In the four months between insemination and parturition, a pregnant sow and the litter she is carrying go through distinct developmental phases, although it is common to supply only one sow diet throughout gestation. This paper reviews how targeted nutrition can address the specific nutrient requirements in different gestational phases to improve sow performance and foetal development. During early gestation, there has been a lot of interest in how energy intake influences progesterone secretion and metabolism, and the consequences for embryo development and survival. In contrast to previous best practices, recent research indicates that a high feed allowance in early gestation can improve luteal function and progesterone secretion, and actually improve embryo survival. Within the embryonic phase, several developmental stages take place such as embryo migration, spacing, and implantation, that may be influenced by specific additives (i.e. neutraceuticals). A key example is arginine, known to improve placenta angiogenesis, and has repeatedly been shown to improve embryo survival. Others, such as some B vitamins, and some non-essential amino acids, may also have beneficial effects that remain largely unexplored. In the third trimester of gestation, most of the foetal weight gain occurs, and not surprisingly, most of the attempts to improve birth weight through nutrition have focused on this period. Extremely low feed intakes do limit foetal gain, however, within the range that is commercially practiced, increased feed allowance does not appear to benefit foetal gain and largely results in maternal gain. In contrast, placenta tissue mostly develops in mid-gestation, and recent research suggests that placenta development can be manipulated, with positive effects on birth weight. This type of targeted nutrition may benefit from specific neutraceuticals targeted at specific gestational phases rather than balancing macronutrients. Finally, mammary gland development occurs in late gestation, setting the stage for colostrum and milk production. This paper will review the role of targeted nutrition to support the above processes, and discuss opportunities for sow health and performance optimization, including carry-over effects to their progeny.

Effects of oligoelement supplements alone or associated with an hepatoprotector during late gestatio

M. Leblanc-Maridor[1], C. Brebion[1], C. Pirard[1], F. Maupertuis[2], C. Belloc[1] and A. Dubois[3]
[1]INRAE, Oniris, BIOEPAR, La Chantrerie, 44300 Nantes, France, [2]Chambre d'Agriculture des Pays de la Loire, 9 rue André Brouard, 49105 Angers, France, [3]Ferme expérimentale porcine des Trinottières, Trinottières, 49140 Montreuil-sur-Loir, France; leblanc.maridor@gmail.com

The aim of this study was to evaluate the effects of oligoelements supplementation with or without support of the liver function on the performances and health of sows and their piglets. A total of 84 Large White × Landrace multiparous sows from one farrow-to-finish farm were divided into three equivalent groups based on their parity and backfat thickness (BF). The CONTROL group received the pregnancy and lactating diets classically used in the farm. For the OLIGO and OLIGO+HEPATO groups, an oligoelement supplementation (B09MB2®+B22TEM®, Comptoirdesplantes.com, France) was added to the normal diet during 14 days before farrowing. Plants that support liver functioning and with anti-oxidant properties (Cynara cardunculus, Orthosiphon stamineus and Curcuma longa) were given to the OLIGO+HEPATO group for 7 days before farrowing (Carestim®, Carephyt, France). Data from sows (weight, BF, parity, breed and duration of lactation) and from their litter (weight at birth, at 24h, at weaning and every week until eight weeks old; mortality and health of piglets) were collected. Daily feed intake was recorded, and total feed intake was calculated. Body weight and feed intake of lactating sows that received both supplements did not differ from those of the CONTROL group. Nevertheless, 13 fat sows in the OLIGO group lost less BF after the lactation period. All groups had an equivalent mean litter size (17.4 piglets/sow) and a low preweaning mortality (11.8%). Piglet performances (weaning weight, average daily gain and gain:feed ratio) did not differ significantly. No diarrhoea or health problems were observed during the trial. For a homogeneous herd, this study underlines an absence of positive effects of oligoelement or hepatoprotector supplementations on sow or piglet performances. The positive effect observed on a few fat sows highlights the importance of proper veterinary diagnosis to target the specific sows for which supplementation could be beneficial.

A long time interval since the last meal impairs farrowing progress in sows

N. Quiniou
IFIP-Institut du Porc, BP 35104, 35651 Le Rheu cedex, France; nathalie.quiniou@ifip.asso.fr

Two trials were carried out on crossbred Large White × Landrace sows in a demonstration farm in summer 2020 (T1) and 2021 (T2) to characterize the birth intervals and farrowing duration associated to the time interval since the last meal. In T1, farrowing of sows from 5 batches were observed from 07:00 AM to 08:00 PM, corresponding to the working hours of the technical team, called 'the day' in contrast to 'the night' when no worker was in the farrowing unit. Sows were fed manually once (08:00 AM), twice (08:00 AM, 05:00 PM) or 3 times (08:00 AM, 12:30 PM, 5 PM) a day from the arrival in the farrowing unit to the farrowing day and the absence of feed in the trough was checked 15 min after the meal. In T2, two feed delivery times were automated through the use of electronic feeders. Video recordings were used to measure when the last meal initiated before farrowing occurred and the time of births. After farrowing, piglets were classified as born alive or stillborn piglets and weighed. Data obtained from sows that received farrowing assistance or had health problems (lameness, fever) were removed from the data set. Finally, data were collected on 37 sows from T1 and 63 from T2. The average parity (T1: 2.5, T2: 3.0), birth weight (T1: 1.33, T2: 1.37 kg), sow's body weight (T1: 285, T2: 286 kg) and backfat thickness (T1: 21.2, T2: 20.1 mm) were not different between T1 and T2 ($P>0.10$), but litter size was higher in T2 (18.7 vs 16.9 total born piglet (TB)/litter, $P=0.02$). Thereafter, sows were categorized depending on the time interval since the last meal: S (≤3 h), M (4-8 h) and L (≥9). Farrowing progression was compared until the 15th born piglets from litters with at least 15 TB (T1: n=27, T2: n=59). Compared to T1 sows (all observed during the day), most of L sows and 2/3 of M sows farrowed during the night in T2. In T1, farrowing progression seemed similar in S and M groups, but worse in L group. Similar results were obtained in T2 when data from sows that farrowed during the day or the night were considered separately: progression was similar in S and M groups during the day and slower in L compared to M during the night. These results support an increase in meal frequency before farrowing to avoid long interval between the last meal and the farrowing onset.

Poster presentation

P. Trevisi[1] and G. Bee[2]

[1]University of Bologna, Department of Agricultural and Food Sciences, Viale Fanin 46, Bologna, Italy, [2]Agroscope Posieux, Rte de la Tioleyre 4, 1725 Posieux, Switzerland; paolo.trevisi@unibo.it

In this time slot, we take time to present the posters submitted in this session.

Session 64 Theatre 6

Efficacy of phytase in gestating and lactating sows

P. Bikker, R.E. Van Genugten-Vos and G.P. Binnendijk

Wageningen University & Research, Wageningen Livestock Research, P.O. Box 338, 6700 AH Wageningen, the Netherlands; paul.bikker@wur.nl

The efficacy of microbial phytase to improve apparent total tract digestibility (ATTD) of phosphorous (P) in diet for growing pigs has been adequately demonstrated in numerous studies. Less studies have been published on the efficacy of phytase in reproductive sow diets and results vary substantially between studies. The present study was conducted with 4 treatments and 18 sows per treatment in two batches to quantify the effect of phytase in gestation and lactation. The treatments were: (1) a positive control (PC) with adequate calcium (Ca) and P content; (2) a negative control (NC) with 50% reduced digestible P; (3) NC + 500 FTU microbial phytase (OptiPhos® PLUS, a 6-phytase produced by a genetically modified strain of *Komagataella phaffii* (DSM 32854)) per kg; and (4) NC + 50,000 FTU per kg to determine potential adverse effects of an overdose. The Ca/P ratio was 1.25 in all diets. Diets were based on maize, oil seed meal and sugar beet pulp. Sows received gestation and lactation diets according to their treatment on individual basis during the entire gestation. Reproductive performance was registered, grab samples of faeces were collected during 5 consecutive days from day 98-102 in gestation and day 19-23 in lactation. Results were analysed using ANOVA with sow as experimental unit, dietary treatment as fixed factor and batch and parity as random factors. A non-inferiority test was conducted for the high phytase level. The ATTD of P was 19.4, 34.9 and 40.8% in gestation ($P<0.001$) and 28.9, 44.6, and 57.4% in lactation for treatments NC, 500 and 50,000 FTU/kg, respectively. The ATTD of Ca was enhanced by 4-5% units by inclusion of phytase, irrespective of its inclusion level. Reproductive performance was not significantly affected by dietary treatments, although the NC diet tended ($P=0.11$) to reduce the weaning weight of the piglets compared to the other treatments. In conclusion, use of phytase at 500 FTU/kg substantially improved ATTD of P and Ca, whereas a higher inclusion level of phytase can further increase digestibility of P without effect on digestibility of Ca. The non-inferiority test indicated that it is unlikely that an overdose of phytase hampers health and performance of sows and piglets.

Session 64 Theatre 7

The effects of feeding sows encapsulated SCFA and MCFA on IgG quality and subsequent piglet growth

M.G. Marchesi[1] and L. Jordaan[2]

[1]MGM Agri Consulting LTG, Britannia Chambers 26 George Street, WA10 1BZ, United Kingdom, [2]Devenish Nutrition LTD, BT13BG, Belfast, United Kingdom; mgm@mgmagriconsulting.com

This study investigated the effects of integrating gestation and lactation feed with encapsulated SCFA and MCFA on IgG quality and piglet growth. Encapsulated SCFA and MCFA were included into the treatment sow's late gestation feed for the last 56 days and continued in the lactation feed. Colostrum from the control (C) and treatment (T) sows were collected within three hours of farrowing and IgG levels were measured with a digital refractometer. These readings assigned the sows IgG quality into four categories namely, poor, borderline, adequate and very good. Approximately 200 piglets from each group of sows were individually tagged and weighed at weaning, and again after 33 days. A Chi-square test was used to analyse the IgG categories data. The piglet growth data was analysed by GLM ANOVA using SAS 9.4. Significant differences between treatment means were declared at $P \leq 0.05$ using Bonferroni's test. The T sows tended ($P=0.06$) to have superior IgG readings. The T sows piglets showed significantly better growth rate ($P<0.001$) after weaning compared to C. Growth rate sub-groups defined by the IgG categories also showed that the T piglets had significantly better growth ($P<0.05$) in each IgG category. The piglets were grouped according to their weaning weight in intervals of 1 kg (<6 to >9 kg). Here again, the T piglets significantly outgrew the C piglets in each sub-group. In fact, the lightest T piglets outgrew the heaviest C piglets. The use of encapsulated SCFA and MCFA is presumed to reduce the oxidative stress in the sows in late gestation, ensuring better and longer lasting IgG in the colostrum. Literature suggests that this type of treatment may also increase the villi height in the piglet intestine, known to be smaller in lighter pigs. The better the quality of IgG, the more capable the piglets are of overcoming the stress of weaning. This study showed that using SCFA and MCFA improved colostrum quality and subsequently significantly improved piglet performance. The use of a digital refractometer provides a reliable pen side method of assessing the sow's colostrum quality and can be used as a predictor of post weaning piglet growth rate.

Session 64 Theatre 8

Use of additives in the feeding of hyperprolific sows in the peripartum period – new approach

S. Ferreira[1,2], N. Guedes[3], M. Joaquim[3], A. Hamard[3] and D. Outor-Monteiro[1,2]

[1]Associate Laboratory for Animal and Veterinary Sciences (AL4AnimalS), Qta. de Prados, 5000, Vila real, Portugal, [2]CECAV, Veterinary and Animal Research Centre, Qta. de Prados, 5000, Vila Real, Portugal, [3]DIN-Groupe CCPA, Animal Nutrition, Zona Industrial da Catraia, Aptd 50, 3441-909 Sta. Comba Dão, Portugal; silviaferreira@utad.pt

Currently, there are several problems associated with the peripartum period in hyperprolific sows, such as hypocalcaemia, high neonatal mortality rates, intestinal constipation and others. The aim of this study was to evaluate the effects of a feed additive in diets for sows in the peripartum period (10 days before the expected date of farrowing to 4 days after farrowing) that includes increased levels of phytase to release more phosphorus, calcium chloride, Vitamin D and E, fast energy sources (dextrose) and slow energy sources (various types of fibre). The assay was performed using crossbred Topigs 70 line sows randomly distributed into three treatments: a control diet (C) corresponding to the lactation feed ($n=16$), a diet corresponding to the lactation feed plus 0.5 kg of the same feed (LAC) ($n=15$) and a third diet that included the lactation feed plus 0.5 kg of the test additive (ECLA) ($n=19$). The parameters evaluated were: dorsal fat thickness (DF) and lean deep muscle (LD) at farrowing, 14 days of lactation and at weaning; born alive piglets (BA), stillbirths (SB), total born piglets (TB), litter weight at farrowing (LWF), litter weight at weaning (LWW), neonatal mortality (NM), mortality at 12h (M12) and 48h (M48) postpartum. No significant differences were found for LD in any period, BA ($P=0.175$), SB ($P=0.257$), TB ($P=0.057$), NM ($P=0.417$), M12 ($P=0.189$) and M48 ($P=0.806$). For litter weight at weaning (LWW) there were significant differences ($P<0.005$), where treatment C has values of 88.5 kg (weaned piglets – WP=13.3), ECLA with 98.6 kg (WP=13.6) and LAC with 108.1 kg (WP=13.4). For LWW, the inclusion of the additive presented favourable results. However, we need to increase the number of animals in test and add some physiological or metabolic indicators in order to obtain a better evaluation of efficacy.
Acknowledgments: This work was supported by the project UI/BD/150836/2021 funded by the Portuguese Foundation for Science and Technology (FCT) and also by the DIN-CCPA.

Session 64 Poster 9

Dietary supplementation with *Pichia guilliermondii* yeast product improves sow and litter performance

E. Janvier[1], C. Oguey[2] and A. Samson[1]

[1]Neovia, Product Development and Application, Talhouët, 56250 Saint-Nolff, France, [2]ADM, Z. A. La Pièce 3, 1180 Rolle, Switzerland; arnaud.samson@adm.com

Sows' prolificacy increased tremendously over the past decades, thereby inducing lighter weight piglets at birth and increasing pre-weaning mortality. Therefore, the challenge is to improve litters' homogeneity and maximize piglets' survival while maintaining sows' body condition for further reproductive cycles. The objective of this study was to evaluate if a whole cell inactivated *Pichia guilliermondii* (Pg) yeast product could improve the body condition of the sows and the performance of their litters through a modulation of the immune system in sows and piglets. A total of 51 gilts and sows were randomly allotted to either a control diet (Con) or a diet supplemented with 0.1% of Pg yeast from breeding to weaning (21 d). Sow body condition and litter performance were analysed with ANOVA models considering the effects of diet, batch of sow, parity, backfat at breeding, and the interactions. Pre-weaning mortality and proportion of lightweight piglets were analysed with Chi-squared test. Average daily feed intake did not differ significantly between the two groups in gestation and lactation (P>0.10). Weight gain was significantly higher for the overall reproductive cycle for the Pg sows compared to the Con sows (P=0.05) and backfat loss was significantly lower (P=0.05). The number of piglets born alive was significantly higher for the Pg group compared to the Con group (15.2 and 14.6 respectively, P=0.02). The proportion of piglets weighing less than 0.8 kg at birth tended to be reduced in the Pg group compared to the Con group (P=0.07) and mortality during the suckling period was significantly reduced (P=0.03), resulting in more piglets being weaned from these sows. Finally, Pg supplementation of sows during gestation and lactation improves sows' body condition, which could favour future reproductive parameters, and optimize litter performance at birth and weaning.

Session 64 Poster 10

Effect of genes polymorphism on colostrum and milk composition and rearing performance of piglets

M. Szyndler-Nędza, A. Mucha, K. Ropka-Molik and K. Piórkowska

National Research Institute of Animal Production, Ul. Sarego 2, 31-047, Poland; aurelia.mucha@iz.edu.pl

It was confirmed that lactose content in colostrum collected within 1 h of farrowing significantly affected the body weight gains of the piglets during rearing. Was also reported, that sows producing low-lactose colostrum within 1 h of farrowing, compared to those producing colostrum high in lactose, reared significantly heavier piglets. These findings motivated a search for new polymorphisms in the genes related to lactogenesis, including lactose synthesis in colostrum and milk, which could be associated with the growth performance of the piglets. In our studies, polymorphisms in the genes were identified: of beta 1,4-galactosyltransferase-I (*B4GALT1*; ENSSSCT00000040755.2:c.*1924+1G>A), insulin receptor (*INSR*; ENSSSCT00000014815.4:c.*59T>C), 11-beta-hydroxysteroid dehydrogenase (*HSD11β2*; ENSSSCT00000003075.2:c.388C>T and ENSSSCP00000002996.2:p.Pro130Ser), progesterone receptor (*PGR*; ENSSSCT00000016339.3:c.229T>A). Subsequently, their effects on the colostrum and milk composition of sows and on the body weight gain of piglets were investigated. The study involved 55 litters of Polish Large White (PLW) sows and 57 litters of Polish Landrace (PL) sows, which gave birth to at least 11 piglets in the second reproductive cycle. Colostrum and milk were sampled during the lactation on d 1, 7, 14 and 21 from 112 sows of PLW and PL. The piglets' rearing performance was determined based on the number and body weight at 1, 7, 14 and 21 d of age. In the study five polymorphisms located in different genes were analysed. Among these polymorphisms under analysis, those identified in the *B4GALT1* genes had the most significant effect on colostrum and milk composition and the piglets' body weight changes. Sows of the $B4GALT1^{GG}$ genotype, when compared to those of $B4GALT1^{AA}$, produced colostrum and milk with a significantly lower lactose content and reared piglets with a substantially higher body weight (by 0.30 kg) at 21 d of age. The work financed from the National Research Institute of Animal Production (Task no. 01-11-02-11).

Session 64 Poster 11

Impact of a 3 or 4 week lactation on performances of sows and their progeny up to 69 days of age

N. Quiniou, D. Gaudré and I. Corrégé
IFIP – Institut du Porc, La Motte au Vicomte, 35360 Le Rheu, France; nathalie.quiniou@ifip.asso.fr

The lactation performances of Large White × Landrace sows over 3 and 4 wk were studied in 2 trials performed with 5 (trial 1, T1) or 2 (trial 2, T2) batches of 24 sows each. In T1, sows were all weaned after 4 wk. In T2, a group of sows was weaned after 3 wk, the other the same day but after 4 wk; growth performances of 252 weaned pig/group (called 3W or 4W pigs) were studied during 46 d. From 5 d after farrowing, sows were fed *ad libitum* with the same lactation diet. From 10 d of age to weaning, a standard creep feed was provided to 4W litters and a high-quality creep feed to 3W litters. After weaning, a 2-phase feeding program was used for 4W pigs, the phase 1 diet was used during 14 d. A 3-phase program was used for 3W pigs, as a starter diet replaced the phase 1 diet during the first 7 d. Body weight (BW, sows and piglets) and backfat thickness (BF, sows) were measured after 3 wk for all sows and 4 wk for those weaned at 4 wk. Feed intake (FI) of sows was measured (T1) or estimated (T2) using InraPorc model. Individual BW and FI per pen were measured 14, 21, 39 and 46 d after weaning (T2). After 4 wk of lactation, BF loss was 1 mm higher than after 3 wk in both trials (P<0.001). The BW loss was 12 kg higher (P<0.001; T1: -11 kg, 11.8 piglets weaned at 4 wk; T2: -12 kg, 12.5 piglets weaned), and FI of sows per weaned piglet was 5 kg higher with 16 vs 11 kg measured (P<0.001) and calculated in T1, and 14 vs 10 kg calculated in T2. Piglets weaned at 4 wk were 2.4 kg heavier (P<0.001; T1: 8.7 vs 6.1 kg; T2: 8.1 vs 6.0 kg). The 3W pigs caught up 15 d later with the BW of 4W pigs at weaning. At 69 d of age, both groups weighed 25.7 kg (P>0.10). The total intake per pig at 69 d was calculated as the sum of lactation feed/piglet weaned, corrected to account for the mortality rate after weaning (3.6 and 1.2% in 3W and 4W groups, respectively; P>0.10), starter, 1st and 2nd phase diets and the gestation diet required to recover sow body reserve after weaning. Weaning pigs at 3 wk in conditions associated to a low mortality rate after weaning reduces the net energy required (-27 MJ/pig at 69 d) and feed costs are reduced depending on the context of feed prices (cheap in 2018: -1.05 €; expensive in 2022: -0.58 €).

Session 64 Poster 12

Increasing feed allowance of lactating sows: what effects on sow and litter performances?

M. Girard[1], P. Stoll[1], G. Maïkoff[2] and G. Bee[1]
[1]Agroscope, Swine Research Unit, Route de la Tioleyre 4, 1725, Switzerland, [2]Agroscope, Research Contracts Animals, Route de la Tioleyre 4, 1725, Switzerland; marion.girard@agroscope.admin.ch

The improvement in sow productivity over the past decade raises the question of whether feeding recommendations in lactation are still appropriate for high prolific sows and their feed intake capacity. The present study aimed at testing the impact of increasing feed allowance in the lactation period on the sow performances and their litter. At farrowing, 90 sows, from 1st to 6th parity, were randomly assigned to one of two feed allowances in lactation: standard (ST) or high (HI). A single lactation diet was formulated to contain 14.1 MJ/kg digestible energy (DE), 186 g/kg crude protein, and 9.5 g/kg digestible lysine. The feed allowance of the ST sows was calculated to cover the energy and nutrient requirements of each sow based on the Swiss feeding recommendations. Sows of the HI group had *ad libitum* access to the feed. Litters were standardized to 12 piglets on average. The individual feed intake was recorded on a daily basis. Sow body weight and backfat thickness were measured at farrowing and then weekly until weaning. Piglets were weighed at birth and at weaning. Data were analysed with a mixed model considering the feeding system, the parity, and their first order interaction as fixed effects and the sow as random effect. As expected, the average daily DE and nutrient intakes were 20% greater (P<0.001) in the HI group, representing a 1.1 kg-greater average daily feed intake compared with the ST group. Neither the piglet weight at weaning (8.0 vs 7.7 kg, respectively) nor the average daily gain in the suckling period (230 vs 217 g/d, respectively) was affected by the feeding system. Nevertheless, when data are standardized by the lactation duration and the litter weight gain, HI sows lost less (P=0.04) weight compared with ST sows (5.9 vs 9.9 g/d per kg litter gain, respectively). Similarly, the daily backfat loss per litter weight gain was 3-fold lower (P=0.04) in the HI than in the ST sows. The similar litter performances suggested that milk production was unaffected by the increase in DE and nutrient intake, while the lower weight and backfat losses in the HI group reflect reduced mobilisation of body reserves.

In-depth analysis of supernumerary piglets management using new functionalities of PertMat

B. Badouard and S. Boulot
IFIP – Institut du Porc, La Motte au Vicomte, BP 35104, 35651 Le Rheu, France; brigitte.badouard@ifip.asso.fr

The expert tool PertMat developed by Ifip-Institut du Porc performs automatic pre-weaning mortality profiling and detection of main risk factors in pig farms. First results showed that large litter size was the major limiting factor in most of the cases and poor adjustment of liveborn piglets to teat numbers was suspected. Therefore, new functionalities have been implemented in PertMat to investigate the importance of supernumerary piglets, the efficiency of litter management practices and interactions between mortality risk factors. PertMat uses individual farm data collected by management software and stored in the French National Pig Management database (GTTT). It is supported by IFIP GT-DIRECT Web platform and no additional information is required. Farmers' results are compared to contemporary litters (#1,200 farms, over 700,000 litters each year). In a first step, the importance of supernumerary liveborn piglets is evaluated for each litter. Calculations take into account litter modifications and a functional teat number hypothesis. Depending on farms, registered litter management practices may include different solutions implemented at various rates: cross-fostering, nurse sows (1 or 2 step) and early weaning. In a given herd, PertMat analyses the impact of these different strategies on preweaning mortality and provides comparisons with other farms. This may help to deliver farm-specific comments about efficiency of supernumerary piglets management. Preweaning mortality is well known to have multifactorial inter-acting causes. Therefore first-step identification of single risk factors is completed by 9 to 20 cross analysis. This is applied to all calculated criteria: total born and live born mortality rates, still born and mummies rates. Analysis of the effect of litter size by parity, genetic, season or weekday provides more precise identification of limiting factors and action priorities. PertMat is a useful monitoring tool, but further evolutions may be required as more various practices have to be implemented to prevent the side effects of large litters.

Effects of sow supplementation with different fatty acids on the gut permeability of the litter

A. Heras-Molina[1], R. Escudero-Portugues[1], G. Gómez[1], H.D. Laviano[1], J. García-Casco[2], M. Muñoz[2], A. Bulnes[3], C. Óvilo[2], A. Rey[1] and C. López-Bote[1]
[1]UCM, Avda Puerta de Hierro s/n, 28040 Madrid, Spain, [2]INIA-CSIC, Avda. Padre Huidobro 7, 28040 Madrid, Spain, [3]CEU, Carrer Lluís Vives, 1, 46115 Valencia, Spain; anaherasm@ucm.es

Increase permeability of the gut barrier have been related to disease development. Moreover, maternal diet has been related to offspring's gut permeability alterations. Concretely, high fat diets can increase gut permeability, depending on their fatty acid composition. The aim of the study was to compare the effect of maternal diets with different fatty acid composition in the intestinal permeability of the piglets. Iberian sows (n=60) were randomly divided in 4 groups and were given a diet with different fatty acid composition: SFA/MUFA (lard), n6-PUFA (sunflower oil), n6 and n3-PUFA (calcium salt of linseed oil) and n6 and n3-PUFA (calcium salt of fish oil). The diet was maintained from day 85 of pregnancy to the end of lactation. Piglets were weaned at 28 days-old and6 days later, 14 male piglets from each group were selected, being then weighted and euthanized. Total intestines were collected and weighted, and a sample of duodenum was collected to perform permeability analyses by Evans Blue procedure. Statistical analyses were performed using R. Animals and their intestines had similar weights among groups. However, the different maternal fatty acid supplementation affected the intestinal permeability ($P<0.05$). Piglets from sows receiving fish-enriched diets showed significantly higher Evans blue concentration and, therefore, permeability ($P<0.05$) than the group from sows having a diet rich in SFA/MUFA (post-hoc test). After doing a post-hoc test, significant differences were found between piglets with a diet rich in SFA/MUFA and n6 and n3-PUFA from fish, with the latter showing higher Evans blue concentration and, therefore, permeability ($P<0.05$). Diets containing n3 have shown to increase intestinal permeability in pigs, possibly due to changes in the lipid raft composition and in the intracellular control of the endocytosis. However, it did not show effects on the piglets' weaning weight, in accordance with previous research.

Session 64 Poster 15

Use of recycled vegetable oil on the productive and reproductive performance of lactating sows

J.M. Uriarte, H.R. Guemez, J.A. Romo and J.M. Romo
Universidad Autonoma de Sinaloa, Blvd. Miguel Tamayo Espinoza de los Monteros, 2358, Tres Rios, 80020, Mexico; jumanul@uas.edu.mx

The objective of this investigation was to determine the influence of the use of recycled vegetable oil from restaurants in the productive and reproductive performance of sows in lactation. Twenty four lactating sows (Landrace × Yorkshire) were divided into three treatments with eight sows per treatment. On day 107 of gestation, the sows were moved to the mesh floor maternity cages in an environment regulated by the environment sows were moved into farrowing crates in an environmentally regulated (2.4×0.6 m) contained an area (2.4×0.5 m) for newborn pigs on each side, all diets were provided as dry powder, and the sows received free access to water throughout the experimental period After farrowing the sows were fasted for 12 hours, the daily feed ration gradually increased and the sows had *ad libitum* access to feed on the fourth day. The diets used were corn-soybean meal based, containing 0 (CONT), recycled vegetable oil 1.5% (RVO1.5), or recycled vegetable oil 2.0% (RVO2) for 30 days. The diets contained similar calculated levels of crude protein and metabolizable energy, and contained vitamins and minerals that exceeded National Research Council (1998) recommendations; sows were fed three times daily. On day 28, piglets were weaned and performances of lactating sows and nursery piglets were recorded. All data in this experiment were analysed in accordance with a completely randomized design. Results indicated that average daily feed intake (5.58, 5.55 and 5.49 kg for CONT, RVO1.5, and RVO2 respectively) of sows were not affected (P>0.05) by different dietary. There was no difference in average body weight of piglets on the day of birth, with 1.33, 1.36, and 1.35 kg, respectively (P>0.05). There was not difference in average body weight of piglets on the day 30, with 6.91, 6.75, and 7.05 kg, respectively (P>0.05) between treatments. The numbers of weaned piglets per sow (9.95, 9.80, and 9.80) were not affected by treatments. The days from weaning to service (P>0.05) did not show differences between the treatments (4 days on average). It concludes, that the substitution of virgin vegetable oil for recycled oil in the diet does not affect the productive and reproductive performance of lactating sows.

Session 64 Poster 16

Dietary PUFA manipulation through biofortified cow's milk improves gilt's offspring performance

L.G. Reis[1], T.H. Silva[1], G.M. Ravagnani[2], C.H.G. Martinez[2], M.S.V. Salles[3], A.F.C. Andrade[2], N.R.B. Cônsolo[1], S.M.M.K. Martins[2], A.M.C. Vidal[1] and A. Saran Netto[1]
[1]University of Sao Paulo/College of Animal Science and Food Engineering, Duque de Caxias Norte, 225, Pirassununga, SP, 13635900, Brazil, [2]University of Sao Paulo/College of Veterinary Medicine and Animal Science, Duque de Caxias Norte, 225, 13635900, Brazil, [3]APTA – Ribeirao Preto, Animal Science, Bandeirantes, 2419, 14030670, Brazil; saranetto@usp.br

Essential fatty acids (FA) must be consumed daily by humans. Considering that swine has been used as a model for applicability in humans, the aim of this study was to evaluate the impact of gilts supplementation with PUFA biofortified cow's milk on performance and FA profile of their offspring. Hybrid gilts (n=30; 34 days of old; 9.6±1.3 kg) were individually allocated into 3 treatment groups in a complete randomized design experiment, as follows: C = control, non-biofortified milk; ω-3 = milk from cows fed linseed oil; ω-6 = milk from cows fed soybean oil. The milk ω-6/ω-3 ratios were 7.86, 1.99 and 7.47 for control, ω-3, and ω-6 groups, respectively. The females received 200 ml (34 to 76 days), 300 ml (77 to 128 days), 400 ml (129 to 174 days), 500 ml (175 to 247 days), and 1 l (from oestrus synchronization up to the end of the lactation) per day. The piglets were weighed before the first suckling, at 7, 14, and 21 days. Blood samples were collected before the first suckling and 14 days by jugular vein puncture for serum FA analysis. All statistical analyses were performed using SAS. The treatments were analysed as orthogonal contrast, where contrast 1 was control vs ω-3 + ω-6, and contrast 2 was ω-3 vs ω-6. statistical significance was declared at P≤0.05. The serum EPA of ω-3 piglets was 69% higher than ω-6 piglets. Consequently, the ω-3 ARA/EPA ratio decreased compared to ω-6 group. Piglets from control group were born heavier compared to the other groups. The piglets from ω-3 and ω-6 groups, increased their BW from 0 to 21 days in 140 g than those in control group; however, at 21d of age the piglets from ω-6 group were heavier than those in ω-3 treatment. The gilts supplemented with PUFA biofortified cow's milk had altered serum FA profile, improving their offspring performance.

Chemical composition of colostrum and milk of sows depending on the subsequent lactation and feeding

K. Karpiesiuk[1], B. Jarocka[1], W. Kozera[1], Z. Antoszkiewicz[1], A. Okorski[1], A. Woźniakowska[1] and G. Żak[2]
[1]University of Warmia and Mazury in Olsztyn, ul. Oczapowskiego 5, 10-719 Olsztyn, Poland, [2]National Research Institute of Animal Production, ul. Sarego 2, 31-047 Kraków, Poland; grzegorz.zak@iz.edu.pl

The aim of this study was to determine the effect of nutrition and successive lactations on the content of selected chemical components and the fatty acid profile of sow colostrum and milk. The chemical composition of colostrum and milk samples collected from 108 PIC sows was evaluated. During the study, sows were allocated to two independent experiments. In Experiment I, sows were divided into three parity groups. In Experiment II, sows were divided into two groups based on the type of concentrate included in the diet. In both experiments, the content of selected chemical components was determined in sow colostrum and milk in view of the effect exerted by successive lactations (Experiment I, 54 animals) and nutrition (Experiment II, 54 animals). Colostrum and milk samples were obtained by hand milking on lactation days 1 and 10, respectively. The samples were subjected to chemical analyses to determine their content of dry matter, total protein (Kjeldahl method), crude fat (Soxhlet extraction) and crude ash, and the fatty acid profile. Fat was extracted by the Soxhlet method. Fatty acids were separated and determined by gas chromatography in a gas chromatograph (CP-3800, Varian, Walnut Creek, California, USA). Fatty acid methyl esters (FAMEs) were prepared according to the modified Peisker method (methanol:chloroform:concentrated sulfuric acid, 100:100:1, v/v). Milk collected from sows fed diet 2 was characterized by higher concentrations of monousaturated fatty acids (MUFAs), polyunsaturated fatty acids (PUFAs), unsaturated fatty acids (UFAs), hypocholesterolemic fatty acids (DFAs), n-3 and n-6 fatty acids, compared with milk collected from sows fed diet 1 that contained not only soybean oil but also hardened fish oil. Milk collected during the first and second lactation had higher fat content than milk collected during the third, fourth and subsequent lactations.

Dietary fish oil modifies blood fatty acids, oxylipins, and immune markers in sows and piglets

E. Llauradó-Calero, R. Lizardo, D. Torrallardona, E. Esteve-Garcia and N. Tous
IRTA, Animal Nutrition, Ctra Reus-El Morell, Km 3.8, 43120 Constantí, Spain; rosil.lizardo@irta.cat

Oxylipins are the major lipid mediators from polyunsaturated fatty acid (FA) metabolism in the body. Those from n-6 FA family are associated with a proinflammatory potential, while those derived from n-3 FAs tend to exert an anti-inflammatory activity. The aim of the current study was to include a fish oil rich in long chain n-3 FAs in sow diets and determine its influence on the blood FAs and oxylipins and the impact on the immune system of suckling piglets. Thirty-six sows were grouped by parity and body weight at insemination into 18 blocks. Whitin each block, they were randomly assigned to a control or an n-3 FA experimental diet. Feeds were distributed according to a feeding scale of 2.8 kg/d during gestation, and *ad libitum* during lactation. Blood samples were obtained from sows at day 108 of gestation, and at weaning, as well as from suckling piglets. Serum FAs were determined by gas chromatography, oxylipins in plasma by liquid chromatography-tandem mass spectrometry, and plasma immune markers by commercial ELISA kits. Results were analysed as a randomized block design using a Proc Glimmix procedure of SAS. Serum FA concentration, mainly eicosapentaenoic acid (EPA) and docosahexaenoic acid (DHA), and their plasma derived-oxygenated products were increased ($P<0.001$) both in suckling piglets and sows fed n-3 FA diet. Among these, oxylipins like 18-hydroxy-EPA, 13-hydroxy-DHA or 17-hydroxy-DHA, all related with inhibition of pro-inflammatory cytokines stand out in all sample types ($P<0.001$). On the other hand, immune markers like immunoglobulin M ($P<0.05$), interleukin-6 in sows ($P<0.05$ in gestation) and interleukin-1β in suckling piglets ($P<0.05$) were also increased by n-3 FA diet. To conclude, the inclusion of n-3 FAs from fish oil in sow's diets increases n-3 FA and their derived oxylipin concentrations modifying some plasma immune markers in sows and suckling piglets.

Session 65 — Theatre 1

Overview and discussion of posters #1

K. Nilsson
Swedish University of Agricultural Sciences, Department of Animal Breeding and Genetics, P.O. Box 7023, 75007, Sweden; katja.nilsson@slu.se

In this slot posters relating to the Portuguese Bísaro and other local breeds, will be shortly overviewed by the chair, with the opportunity for the audience to discuss with the authors.

Session 65 — Theatre 2

Genetic evaluation for productive traits in endangered Portuguese Malhado de Alcobaça pig

A. Vicente[1,2,3], J. Bastos[2], M. Silveira[4], I. Carolino[5] and N. Carolino[1,5,6]
[1]CIISA – Faculdade Medicina Veterinária -ULisboa / AL4AnimalS, Lisbon, 1300-477, Portugal, [2]FPAS – LGMA, Montijo, 2870-219, Portugal, [3]IPSantarém – Escola Superior Agrária, Santarém, 2001-904, Portugal, [4]Ruralbit, Rio Tinto, 4435-213, Portugal, [5]INIAV IP, Santarém, 2005-048, Portugal, [6]Escola Universitária Vasco da Gama, Coimbra, 3020-210, Portugal; apavicente@gmail.com

Malhado de Alcobaça (MA) is a swine breed from center west Portugal, recognized in 2003. Herdbook (LGMA) has information of >17,000 animals (1985-2022). It´s a very endangered breed with only 244 sows, 18 boars and 12 breeders. Within the scope of MA's genetic conservation plan, for genetic evaluation, genetic parameters and fixed effects were estimated for birth weight (PN), for adjusted weight at 30 d (P30) and at 90 d (P90). Pedigree records (n=16,883) and weighing records (nPN=9,198; nP30=5,799; nP90=2,429) from LGMA of 362 litters were compiled. Genetic parameters and fixed effects, genetic values and precisions were predicted for PN, P30 and P90, through BLUP, with a mixed model including fixed effects of breeder × year of birth, season of birth, sex, age of mother (linear / quadratic effect) and as random effects genetic value of animal, maternal genetic effect and permanent environmental effect of litter. Fixed effects of prolificacy and number of weaned piglets were included as covariates, respectively, in PN and P30/P90 analyses. Mean values for PN 1.32±0.28 kg, 7.16±1.6 kg P30 and 34.1±6.4 kg P90 were recorded. For PN, P30 and P90, respectively, a heritability for direct effects of 0.170±0.156, 0.145±0.161 and 0.293±0.175; a maternal heritability of 0.084±0.109, 0.124±0.173 and 0.250±0.186 and a genetic correlation between direct and maternal effects of -0.344, -0.524 and -0.174 was estimated. Permanent environmental effect of litter was 0.395±0.056 PN, 0.403±0.086 P30 and 0.129±0.047 P90. In fixed effects, a superiority was observed for males of +0.031 kg PN, +0.046 kg P30 and +0.201 P90. Breeder × year effect showed maximum differences of 0.93, 5.21 and 19.25 kg for PN, P30 and P90, respectively, with Spring being the best season of birth. Sow age at farrowing had a quadratic effect for all traits. For each increase of 1 piglet in prolificacy, PN, P30 and P90 were reduced by 4, 7 and 58 g, respectively. Ack: Proj CIISA UIDB/00276/2020.LA/P/0059/2020-AL4AnimalS.

Genome wide association (GWAS) for meat quality in D.O.P. 'Jamón de Teruel'

D. López-Carbonell, P. López-Buesa, H. Srihi, C. Burgos, J. Altarriba, M. Ramírez and L. Varona
Instituto Agroalimentario de Aragón (IA2), Universidad de Zaragoza, Departamento de Anatomía, Embriología y Genética Animal, C/ Miguel Servet 177, 50013 Zaragoza, Spain; 767339@unizar.es

'Jamón de Teruel' is a certificate of quality ('Denominación de Origen Protegida') whose objective is to produce high quality pig products in the Spanish province of Teruel. Pigs produced under 'Jamón de Teruel' are obtained from a maternal line (Large White × Landrace) and a paternal line (Duroc). The aim of this study is to find out regions of the genome associated with pig meat quality traits under a standard production environment. A total of 466 individuals, grown in a commercial farm, were measured post-sacrifice on Cold Carcass Weight (CCW), Backfat Thickness (BFT), Ham Weight (HW), pH 45 minutes approximately after sacrifice in *longissimus dorsi* (pH45), pH 24 hours or longer after sacrifice in 8 different points of *longissimus dorsi* (pH24). All phenotyped animals were genotyped with the *Illumina GPP Porcine HD Array* SNP device. Pedigree includes two generations, in which these animals are related with their sire (37 different sires) and dams (204 different dams). Once filtered raw genotypes, the number of SNP are 46,344. To begin with, a variance components estimation was performed using airemlf90. Results -trait mean ± standard deviation (heritability)- were: CCW: 104.85±8.13 (0.33); BFT: 22.55±4.35 (0.48); HW: 13.82±1.19 (0.29) pH45: 6.44±0.29 (0.08); and ph24: 5.64±0.18 (0.30). Single step GWAS methodology was applied using postgsf90. QTL regions were situated in chromosome 1 (HW and pH24), 4 (BFT), 6 (CCW and HW), 12 (pH45 and pH 24), 16 (BFT) and 17 (HW). Positional candidate genes within these QTL regions included ZFP30 (CCW and HW), WDR87 (CCW and HW), EFCAB3 (pH45), GJA5 (BFT), ACP6 (BFT), TBX18 (pH24), MPO (pH24), EPX (pH24), and others. Future research needs to be done to propose models that consider the paternal or maternal origin of SNP alleles.

Comparative genome-wide study on autochthonous Nero Siciliano pig breed

G. Chessari[1], S. Mastrangelo[2], S. Tumino[1], G. Senczuk[3], S. Chessa[4], B. Castiglioni[5], S. Bordonaro[1], D. Marletta[1] and A. Criscione[1]
[1]University of Catania, Via Valdisavoia, 95123 Catania, Italy, [2]University of Palermo, Viale delle Scienze, 90128 Palermo, Italy, [3]University of Molise, Via Francesco De Sanctis, 86100 Campobasso, Italy, [4]University of Torino, Largo Paolo Braccini, 10095 Grugliasco, Italy, [5]CNR, Inst Agr Biol & Biotechnol, Via Einstein, 26900 Lodi, Italy; giorgio.chessari@phd.unict.it

Local breeds represent a great cultural and genetic heritage fundamental for the exploitation of the territory. Nero Siciliano is an autochthonous pig breed reared in the north east of Sicily in semi-extensive or extensive systems, is well adapted to harsh environments, has great resistance to diseases, and fully enhances the food potential of the woods. In this study, using the PorcineSNP60v2 chip, we compared the genotyping data of Nero Siciliano (NS) with a sample of Italian wild boar (WB), four Italian local (IT) and four cosmopolitan (CM) breeds to investigate the genetic variability, population structure, autozygosity and signals of genomic differentiation. The NS showed a rate of genetic diversity comparable to CM breeds and higher than IT breeds. Multidimensional scaling, model-based clustering, and Neighbour network highlighted the NS proximity to WBIT, and an internal substructure probably due to family lines. The number of ROH identified in the NS is closer to the values observed in CM breeds and higher than that detected in IT breeds. Nero Siciliano reported more rich-homozygosity-islands (SSC8, SSC11, and SSC14) than the rest of breeds with the exception of Mora Romagnola and wild boar. Across breeds, SSC8 and SSC14 are the chromosomes most affected by ROH islands and overlapped QTLs related to immune system response, production, and reproduction. The Bayesian F_{st}-outlier approach detected the highest number of signals between NS and IT with six markers reported in association with production, reproduction and immune system response QTLs. The lowest number of outlier markers was observed between NS and WBIT showing signals involved in disease susceptibility, average daily gain, and reproduction. These results can help to better identify the genomic structure of Nero Siciliano and to plan breeding schemes in order to target the production system maintaining internal diversity.

Cinta Senese crossbreed: an excellent option for fresh meat production in typical Tuscan systems

R.E. Amarie[1], L. Casarosa[1], M. Tognocchi[1], S. Tinagli[1], A. Del Tongo[2], J. Goracci[2] and A. Serra[1]
[1]University of Pisa, Department of Agriculture, Food and Environment, via del Borghetto 80, 56124 Pisa, Italy, [22]Tenuta di Paganico Soc. Agr. SpA, Via della Stazione 10, 58045 Paganico, Italy; roxana.amarie@phd.unipi.it

Cinta Senese (CS) is a local pig breed of the Tuscany (Italy) often reared extensively and characterized by a slower growth rate and a predisposition to fat deposition, that makes CS suitable to meat processing; ham, salami, sausage from CS are very appreciated by the consumer due to the excellent organoleptic and nutritional characteristics. Nevertheless, the high-fat content discourages consumers from consuming fresh meat. This study aimed to assess the effectiveness of a CS crossbreed (I) (CS × (Large White × Duroc)) in an organic fresh meat production system. The research is included in a Rural development Regional project (reg. UE 1305/2013 – PSR 2014/2020 – FAR Maremma) and was carried out with 11 CS and 11 I. To evaluate the breed productivity the following parameters were considered: average daily gain (ADG), dressing yield, and carcass quality. The meat quality was assessed on Longissimus dorsi (LD) muscle by analysing physical parameters (colour, pH, and water holding capacity – WHC), proximate composition, fatty acids composition, and cholesterol content. Comparing to CS, I showed an higher ADG (376.2 vs 285.6 g/day; P=0.004) and a thinner back-fat thickness (3.14 vs 5.13, P<0.0001). No differences were relieved for weight, age, or slaughter yield. Nevertheless, the dressing percentage of CS was statistically lower for total lean (29.64% CS vs 48.71% I) and higher total fat content (63.67% CS vs 43.85% I). Finally, the crossbreed wasn't a significant variation factor concerning the physical parameters (colour, pH, and WHC), proximate composition, fatty acid composition and total cholesterol content. In conclusion, the investigated crossbreed produced higher-performing animals, with meat and carcasses leaner which, at the same time, preserved the excellent qualitative characteristics of the pure breed. These features make meat suitable for fresh consumption, offering a viable alternative to improve the productivity of this typical Tuscan production.

Phenotypic and genomic investigations on carcass and meat quality of a German local pig breed

A. Olschewsky[1], A. Kleinlein[1], D. Mörlein[2] and D. Hinrichs[1]
[1]University of Kassel, Animal Breeding Section, Nordbahnhofstraße 1a, 37213 Witzenhausen, Germany, [2]University of Göttingen, Department of Animal Sciences, Kellnerweg 6, 37077 Göttingen, Germany; olschewsky@uni-kassel.de

The Angler Saddleback pig is an endangered local breed from the north of Germany, which is characterized by increased carcass fatness as compared to modern genotypes. This results in a unique aroma, tenderness and good processing properties of the meat. However, these positive attributions to Angler Saddleback pig meat are not scientifically proven yet. Therefore, the objective of this study was to shed light on performance and meat quality traits of this breed as well as underlying genetic mechanisms. For this purpose, 73 Angler Saddleback pigs originating from six breeders were reared and fattened at an experimental farm in two consecutive trials for 214 to 239 days. Carcass weight, daily weight gain, lean meat percentage (LMP), intramuscular fat content (IMF), and fatty acid composition in backfat and IMF were assessed. Furthermore, all animals were genotyped with the Porcine SNP60 Beadchip (Illumina) and a genome-wide association study (GWAS) was conducted using GEMMA software. Average slaughter weight (117 kg), daily weight gain (527 g) and LMP (46%) indicate a lower performance than commercial breeds. However, with regard to LMP the extended fattening may have influenced the results. The higher carcass fatness is reflected in above average IMF (2.7%), increased proportions of saturated fatty acids (backfat=39%; IMF=33%) and reduced polyunsaturated fatty acids (backfat=13%; IMF=21%). Especially, IMF greatly varied between the animals (1% to 6%); this can be partly attributed to the influence of sex (P<0.01) and breeder (P=0.01). First results of GWAS analysis could not reveal new associations which might be due to the low sample size. In summary, this study provides for the first time a comprehensive overview of performance and meat quality of the Angler Saddleback pig. These results can be used when marketing meat and meat products of this endangered breed.

Productive traits and boar taint of early and late immunocastrated pigs of three Duroc crossbreeds

M. Font-I-Furnols[1], A. Brun[1], J. Soler[2], N. Panella-Riera[1], J. Reixach[3], N. Gomez[4], G. Mas[5] and M. Gispert[1]
[1]IRTA-Food Quality and Technology, Finca Camps i Armet, 17121 Monells, Spain, [2]IRTA, Infrastructure Platform, Veïnat de Sies s/n, 17121 Monells, Spain, [3]Selección Batallé S.A., Avda. Segadors s/n, 17421 Riudarenes, Spain, [4]GePork, Finca el Macià s/n, 08510 Masies de Roda, Spain, [5]UPB Genetic World S.L., Ctra. Berga, 13, 08670 Navàs, Spain; maria.font@irta.cat

A total of 144 pigs from 3 different Duroc commercial crossbreeds (A with 25% Duroc and B and C with 50% Duroc) were evaluated (n=48 each). For each crossbreed, 3 different treatments were considered, with 16 pigs each: ME: Entire male, T1: late immunocastrated (vaccinated with Improvac ® at 8 and 4 weeks before slaughter) and T3: early immunocastrated (vaccinated at 13 and 8 weeks before slaughter). Productive parameters were obtained individually for each pig. At slaughterplant, lean meat content of the carcass was objectively determined. Three trained panellist smelt fat after heating with a soldering iron (human nose method) and classified samples as no/slightly tainted and highly tainted. For crossbreed A, average daily gain (ADG) was higher for T1 than T2 and intermediate in EM, while no differences in feed conversion ratio (FCI) were found between T1 and T2, both being higher than EM, in agreement with the average daily feed intake (ADFI). For crossbreed B, no differences between treatments were found for ADG, ADFI nor FCR. Crossbreed C show no significant differences between treatments for ADG but T2 had higher ADFI and FCR than T1, EM being lower. Carcass weight was on average 99.0+7.8, 85.8+8.0 and 88.4+9.4 kg, for A, B and C crossbreds, respectively. Regarding carcass characteristics, immunocastrated from A and C were fatter than EM and there were no differences in those from crossbred B. No important differences in meat quality were detected. Only in crossbreed A, intramuscular fat was higher for immunocastrated than EM. All immunocastrated from all crossbreeds were boar taint free, while 7, 19 and 19% of EM from crossbreds A, B and C, respectively, were highly tainted. As a conclusion, the effect of immunocastration, early and late, in productive parameters, depends on the crossbreed and in all of them boar taint is reduced.

Genetic correlations between plasma metabolite levels and complete blood counts in healthy pigs

E. Dervishi[1], X. Bai[1], M.K. Dyck[1], J.C.S. Harding[2], F. Fortin[3], P.G. PigGen Canada[4], J. Cheng[5], J.C.M. Dekkers[5] and G.S. Plastow[1]
[1]University of Alberta, AFNS, 116 St and 85 Ave, T6G 2R3 Edmonton, Canada, [2]University of Saskatchewan Saskatoon, Large Animal Clinical Sciences, 52 Campus Drive Saskatoon Canada , S7N 5B4 Saskatchewan, Canada, [3]Centre de Developpement du Porc du Quebec inc. (CDPQ), 450-2590 Bd Laurier, G1V 4M6 Quebec City, QC, Canada, [4]PigGen Canada Research Consortium, Guelph, N1H 4G8 Ontario, Canada, [5]Iowa State University, Department of Animal Science, 1221 Kildee Hall, 50011 Ames IA, USA; dervishi@ualberta.ca

This study is part of a large research project investigating the underlying genetic mechanisms of disease resilience in grow-finisher pigs exposed to a natural polymicrobial disease challenge. The objective of this study was to estimate genetic and phenotypic correlations in young healthy nursery pigs prior to exposure between the concentration of 33 heritable metabolites in plasma and complete blood count (CBC) traits. Metabolites were quantified using nuclear magnetic resonance. The CBC traits included six white blood cell traits, seven red blood cell traits and two platelet traits. All animals were genotyped using a 650k Affymetrix Axiom Porcine Genotyping Array by Delta Genomics (Edmonton AB, Canada). Data from 968 healthy piglets at an average of 26 days of age was available. Genetic and phenotypic correlations were estimated using the BLUPF90 programs. Significance of correlations were determined using likelihood ratio tests with 1 degree of freedom. Results showed that phenotypic correlation estimates of plasma metabolites with blood traits were generally low. The highest phenotypic correlation was observed between hypoxanthine and haematocrit (0.2±0.04; P<0.05) and between oxoglutarate and haemoglobin (0.2±0.04; P<0.05). Some metabolites showed significant genetic correlation estimates, with the largest negative genetic correlation between L-alpha-aminobutyric acid and the width of the distribution of the size of red blood cells (-0.98±0.03; P<0.05). In addition, L-alpha-aminobutyric acid was genetically positively correlated with eosinophil concentration (0.75±0.51; P<0.05). These results suggest that some plasma metabolite phenotypes may be associated with the concentration of blood cells in healthy nursery pigs.

Tackling an old question: do Mangalitsa pigs of different colours really belong to different breeds?

S. Addo, L. Jung and D. Hinrichs
University of Kassel, Animal Breeding Department, Nordbahnhofstr. 1a, 37213 Witzenhausen, Germany; sowah.addo@uni-kassel.de

Mangalitsa pigs are broadly classified into three breeds based on coat colour variation. Due to their developmental history, over a decade old research called to question, the partitioning of these pigs into breeds. In this study we reinvestigated relatedness among the so-called breeds by detecting signatures of selection that may have played a role in breed development. Furthermore, we studied within- and between-breed diversity through the analysis of population structure, observed heterozygosity (Ho) and runs of homozygosity (ROH) patterns in our populations. Data consisted of 23 blond (BM), 30 Swallow-bellied (SM) and 24 Red (RM) Mangalitsa pigs genotyped with a customized version of the ProcineSNP60 v2 Genotyping Bead Chip. Using quality filtered data, we found low Ho estimates of 0.27, 0.28 and 0.29, and high genomic inbreeding (FROH) of 24.11%, 20.82% and 16.34% for BM, SM and RM, respectively. Detected ROH islands were not shared across breeds but located on chromosomes 7, 13 and 15 in RM; 11 and 17 in SM and only on chromosome 16 in BM. Furthermore, a fixation index (Fst) analysis revealed a number of selection signatures especially for the pairwise comparison of RM-SM and BM-SM. Average Fst estimates showed the closest relatedness between BM and RM (Fst=0.029), which is consistent with the population structure analysis in which some individuals of the two breeds clustered together. Our findings are not entirely conclusive but support the idea of rejecting the hypothesis that Mangalitsa individuals form just one unpartitioned population. A follow up study is investigating candidate genes proximal to the detected genome-wide significant variants.

Overview and discussion of posters #2

E.F. Knol
Topigs Norsvin Research Center, Schoenaker 6, 6641 SZ Beuningen, the Netherlands; egbert.knol@topigsnorsvin.com

In this slot the second set of posters will be presented either by the chair or in an open discussion between the authors and the audience.

Session 65 Poster 11

Can the inclusion of different olive oil cakes on diet affect carcass quality of Bísaro?

A. Leite[1], I. Ferreira[1], L. Vasconcelos[1], R. Dominguez[2], S. Rodrigues[1], D. Outor-Monteiro[3], V. Pinheiro[3], J.M. Lorenzo[2] and A. Teixeira[1]

[1]Centro de Investigação de Montanha (CIMO), Instituto Politécnico de Bragança, Campus de Santa Apolónia, 5300-253 Bragança, Portugal, [2]CTC, Parque Tecnológico de Galicia, San Cibrao das Viñas, 32900, Spain, [3]CECAV – UTAD, Quinta de Prados, 5000-801 Vila Real, Portugal; anaisabel.leite@ipb.pt

The present work aimed to evaluate the potential incorporation of olive by-products olive cake (crude olive cake, exhausted olive cake without and with olive oil and two-phase olive cake) in the diet of Bísaro pigs, a local breed reared in Trás-os-Montes region (northeast of Portugal) and to study its effect on the animal´s growth traits and carcass characteristics. The experiment was carried out on 40 Bísaro pigs selected from animals delivered for slaughter to Bragança-Portugal. Five different treatments with different olive cakes (T1 – basal diet; T2 – 10% crude olive oil; T3 – 10% olive cake two phases, T4 – 10% exhausted olive cake; and T5 – 10% exhausted olive cake + 1% olive oil) were. Body weight, pH (1 and 24 hours after slaughter) and carcass weight were similar in all treatments and no significant differences were observed. No significant differences were found between the treatments for the carcass measurements performed, except for the longissimus dorsi length at seventh rib ($P<0.05$) varying between 75.9 (T3) and 87.3 (T5) mm and fat depth measured at the last rib (P3 measurement) varying between 91.1 (T5) and 99.2 (T4) ($P<0.05$). The data provide the definition of a standard carcass for the breed and this ratio of body weight. Results indicate that this oil by-product can be used to feeding Bísaro pigs.

Session 65 Poster 12

Use of olive pomace in the Bísaro breed feeding – effect on processed meat products sensory quality

S.S.Q. Rodrigues, L. Vasconcelos, A. Leite, I. Ferreira, E. Pereira and A. Teixeira

CIMO, ESA Instituto Politécnico Bragança, Campus Sta Apolónia Apt 1172, 5301-855, Portugal; srodrigues@ipb.pt

This work aimed to evaluate the effect of feeding animals with olive pomace on the sensory characteristics of Bísaro pork transformed products: dry-cured loin and neck. Five treatments were studied considering the process of obtaining the olive pomace: pressed (PoPr), centrifuged (PoCf), and extracted (PoEx and PoExOO) olive pomace, compared with the control (Ct). Animals were fed with olive pomace for 2 weeks before slaughter. In 4 different times, 2 animals per treatment were slaughtered, meaning 10 animals each time, 40 animals in total. Dry-cured loins and necks were produced in the meat manufacturing industry Bísaro Salsicharia according to traditional practices. Twenty-one appearance, odour, texture, and taste attributes were evaluated by 8 members trained taste panel. All treatments were evaluated in duplicate in each of 3 sessions. A nonparametric ANOVA was performed for related samples, with pairwise comparisons by Friedman's test. Results showed no significant differences between treatments for all quantitative sensory attributes evaluated in the cured loins of Bísaro pork under study. In the cured neck, there was a significant influence of the treatment on the muscle/fat ratio. PoCf and PoEx had a significantly lower muscle/fat ratio, that is, more fat than muscle than Ct, and at the same time PoPr and PoExOO were not significantly different from one or the other. We can conclude that olive pomace can be used in pigs' diets with no significant influence on their processed meat products' sensory characteristics, adding value to an undervalued subproduct from olive oil production.

Session 65 Poster 13

Real-time ultrasonography: assessment of the subcutaneous fat layers in growing Bísaro pigs

S. Botelho-Fontela[1,2,3], G. Matos[3], A. Esteves[1,2,3], G. Paixão[1,2,3], R. Payan-Carreira[4], A. Teixeira[5] and S.R. Silva[1,2,3]
[1]AL4AnimalS, Quinta de Prados, 5000-801, Portugal, [2]CECAV, Quinta de Prados, 5000-801, Portugal, [3]UTAD, Quinta de Prados, 5000-801, Portugal, [4]CHRC, Rua do Cardeal Rei, 7000-849 Évora, Portugal, [5]CIMO, ESA/IPB, 5300-253 Bragança, Portugal; sbotelho@utad.pt

Fat content in pigs is a decisive factor in meat quality and carcass grading and a non-invasive *in vivo* assessment of subcutaneous fat (SF) is critical to understanding, among others, responses to castration procedures. This work used the real-time ultrasonography (RTU) for evaluating SF layers of entire male Bísaro pigs and in males submitted to three different castration treatments. The development of SF layers was monitored in 47 Bísaro male pigs from 41 to 53 weeks in four sessions [13 entire males–EM and 34 submitted to three different castration treatments: immunocastration at 9,13, 49 and 53 weeks old (IC1; n=8), immunocastration at 13, 17, 21, 49 and 53 weeks old (IC2; n=11) and surgically castrated (CC; n=15)]. The RTU images were captured on a laptop computer and then determined the total SF thickness and two SF layers: the outer (L1) and middle (L2). Analysis of variance was performed to compare the L1 and L2. The development of L1 and L2 relative to SF was determined with linear regressions of the log-transformed Huxley model (logY=*b*logX+log*a*; where *a* is a constant and *b* is the allometric coefficient. In general, for castrated animals, the L1 is thicker than L2 in all sessions (1.19 vs 0.61 cm, respectively), whereas this difference is smaller for the EM group (0.43 vs 0.59 cm, respectively). For the IC1 and EM groups, the development of the L1 relative to SF (b=1,187 and 1,332; b>1, P<0.05) was more rapid than that of the L2 (b=0.589 and 0.533; b<1, P<0.05); whereas for the CC and IC2 groups, the development of L1 and L2 relative to SF is isometric (b between 0.700 and 1.156; b=1, P<0.05). This study indicates that the RTU can be a helpful tool to monitor the SF layers variations in growing pigs. This project was funded by Icas-Bísaro Project (PDR 2020-101-031029), financed by EAFRD and Portuguese State under Ação 1.1 'Grupos Operacionais' – Medida 1. 'Inovação' do PDR 2020 – Programa de Desenvolvimento Rural do Continente, and project UIDB/CVT/00772/2020.

Session 65 Poster 14

Meat quality of Bísaro breed and terminal cross entire male pigs

R.P. Pinto[1], F. Mata[1], J.P. Araújo[1,2], J.L. Cerqueira[3] and M. Vaz-Velho[1]
[1]CISAS – Center for Research and Development in Agrifood Systems and Sustainability, Inst. Pol. Viana do Castelo, Rua da Escola Industrial e Comercial Nun'Alvares 34, 4900-347 V. Castelo, Portugal, [2]CIMO – Mountain Research Centre, Instituto Politécnico de Viana do Castelo, Portugal, Esc Superior Agrária, Refóios, 4990-706 P. Lima, Portugal, [3]CECAV – Veterinary and Animal Research Centre, Quinta de Prados, Apartado 1013 5000-801 Vila Real, 5000-801 Vila Real, Portugal; pedropi@esa.ipvc.pt

Entire male (EM) pig rearing can be advantageous over castration. Higher feed efficiency, and leaner carcasses with higher protein and unsaturated fat contents, can be achieved. To add, higher welfare standards are also fulfilled. However, EM pork may be tainted, compromising therefore its quality. The aim of this study was to investigate and compare pork quality from EM of the Portuguese autochthonous breed Bísaro (BI) and a terminal cross (Yorkshire × Landrace) × Pietrain (TC). Pigs were reared outdoors and slaughtered with a live weight and age of 121.8±24.0 kg and 223 days (BI) and 114.4±8.4 kg and 203 days (TC). Thawing loss (TL), pH, L*a*b* colour, intramuscular fat (IMF) and protein were determined in loin samples (M. longissimus dorsi). Skatole (SKA) and androsterone (AND) were analysed from backfat. Results showed no significant difference (P>0.05) in pH, a* and protein values. TC pigs had lower boar taint values: SKA (TC 8.7±4.7; BI 74.8±38.2, P<0.001); AND (TC 137.6±64.6; BI 269.6±159.9, P=0.043). BI had lower TL% (TC 7.7±1.3; BI 5.0±0.5, P<0.001), meat was darker (L*) (TC 54.7±3.3; BI 48.7±3.7, P<0.001), less yellow (b*) (TC 8.3±2.1; BI 7.0±1.2, P=0.015), and had a higher IMF% (TC 2.4±0.2; BI 3.8±0.5, P<0.001). BI is a local breed not subject to the intensive selection of global breeds used in terminal crosses, therefore boar tainting mitigation has been less effective in the former than the later. While selecting for quantitative traits (growth, reproductive performance), qualitative traits may have been neglected. Highly selected breeds' growth is early maturing, pushing sexual maturity forward. BI pork is at premium quality; however boar tainting is paramount in consumer's choice. Alternatives for taint control can be explored and tested in BI: selection for AND together with nutrition and management for SKA, may be effective.

Immunocastration to slaughter: effect on testis and boar taint compounds in adult male Bísaro pigs

S. Botelho-Fontela[1,2], R. Pereira Pinto[3], G. Paixão[1,2], M. Pires[1,2], M. Vaz Velho[3], R. Payan-Carreira[4] and A. Esteves[1,2]
[1]Associate Laboratory for Animal and Veterinary Sciences (AL4AnimalS), Quinta de Prados, 5000-801, Portugal, [2]CECAV – Veterinary and Animal Research Centre, Quinta de Prados, 5000-801, Portugal, [3]CISAS – Center for Research and Development in Agrifood Systems and Sustainability, Rua Escola Industrial e Comercial Nun'Álvares, 34, 4900-347 Viana do Castelo, Portugal, [4]CHRC, Departement of Veterinary Medicine, Rua do Cardeal Rei, 7000-849 Évora, Portugal; sbotelho@utad.pt

Boar taint is an unpleasant smell or taste mainly caused by androstenone (AND) and skatole (SKA). Immunocastration is the immunisation against gonadotropin-releasing hormone. It has been proved effective in arresting testicular function and controlling boar taint, which is achieved with two inoculations at least four weeks apart. Improvac®'s protocol recommends 4 to 5 weeks from second vaccination to slaughter for best results. This study aimed to assess the effect of long delays on boar taint compounds and morphologic and histologic traits of reproductive organs. Twenty-four adult male Bísaro pigs, housed in outdoor pens and fed with commercial maintenance diets, acorns and beets, were used. A group of entire males was used as a control group (EM; n=5); others were vaccinated with Improvac. Pigs were slaughtered 4 (IC1, n=5), 6 (IC2, n=5), 8 (IC3, n=5) and 10 (IC4, n=4) weeks after the second inoculation, aged between 8 to 13 months old. Reproductive organs were removed and measured, and testicular function was assessed histologically. Subcutaneous fat was sampled to determine AND and SKA levels. EM presented heavier and bigger testis than IC (P<0.01). AND levels were also higher in EM (P<0.01), proving the efficiency of immunocastration even in the adult phase. SKA levels were relatively low in all groups, with significant differences (P<0.01) found only between EM (the highest average) and IC1. This study was funded by Icas-Bísaro Project (PDR 2020-101-031029), financed by the European Agricultural Fund for Rural Development (EAFRD) and Portuguese State under Ação 1.1 'Grupos Operacionais' – Medida 1. 'Inovação' do PDR 2020 – Programa de Desenvolvimento Rural do Continente, and by project UIDB/CVT/00772/2020 – Portuguese Foundation for Science and Technology.

Low levels of androstenone and Skatole found in Bísaro Pigs reared in alternative systems

G. Paixão[1,2], S. Botelho-Fontela[1,2], R. Pinto[3], M. Pires[1,2], M. Velho[3], R. Payan-Carreira[1,2,4] and A. Esteves[1,2]
[1]AL4AnimalS – Associate Laboratory for Animal and Veterinary Sciences, Vila Real, 5000-801, Portugal, [2]CECAV – Veterinary and Animal Research Centre, University of Trás-os-Montes and Alto Douro, Quinta dos Prados, 5000-801 Vila Real, Portugal, [3]CISAS – Center for Research and Development in Agrifood Systems and Sustainability, Instituto Politécnico de Viana do Castelo, Viana do Castelo, 4900-347, Portugal, [4]CHRC – Departament of Veterinary Medicine, University of Évora, Évora, 7006-554, Portugal; sbotelho@utad.pt

Skatole (SKA) is produced by microbial degradation of the amino acid tryptophan and its metabolism modulated by gonadal steroids in the liver, whereas androstenone (AND) is a testicular steroid. The accumulation of these compounds in adipose tissue is primarily responsible for boar taint. Twenty-one (n=21) entire male pure-bred Bísaro pigs were used in this study. Animals were reared indoors until 4-5 months old and then moved to outdoor pens, as a free range system. At 13 months old animals were slaughtered, with an average live weight of 156.3±21.1 kg. They were fed commercial maintenance diets *ad libitum* supplemented with beets and acorns. Subcutaneous fat samples were collected from the neck region and then processed for quantitative analysis. AND and SKA concentrations varied significantly between individuals: SKA levels ranged from 0.00 to 0.14 μg/g, whereas AND ranged from 0.02 to 0.64 μg/g. The median AND level was low (0.14 μg/g), and only one sample was higher than 0.5 μg/g. For SKA, the average and median levels were also low, with 0.02 and 0.01 μg/g, respectively. The levels of AND and SKA obtained in this essay were markedly lower than levels achieved in similar studies using commercial crosses reared in industrial systems. This study was funded by Icas-Bísaro Project (PDR 2020-101-031029), financed by the European Agricultural Fund for Rural Development (EAFRD) and Portuguese State under Ação 1.1 'Grupos Operacionais' – Medida 1. 'Inovação' do PDR 2020 – Programa de Desenvolvimento Rural do Continente, and by project UIDB/CVT/00772/2020 – Portuguese Foundation for Science and Technology.

Modelling the growth in female and male piglets of Sarda breed

A. Cesarani, M.R. Mellino, A. Fenu, N.P.P. Macciotta and G. Battacone
University of Sassari, Dipartimento di Agraria, viale Italia 39, 07100 Sassari, Italy; acesarani@uniss.it

The knowledge about the growth behaviour of animals intended to meat production is of crucial interest in the livestock industry. Pig meat represents one of the most traditional products of Sardinia, the second biggest island of the Mediterranean Sea, where an autochthonous pig breed is raised mainly in marginal areas. Aim of this work was to study the growth in male and female piglets of Sarda pig breed during three periods: lactation, growing (i.e. post-weaning phase), or in the complete trial. A total of 31 piglets (14 males and 17 females) born from 5 sows were analysed. Each animal had 12±3 weight records from birth to about 9 months of age. Weights at birth, at weaning, at 6 and 9 months were analysed with a mixed model with sex as fixed and sow as random effect. Average daily gain (ADG) records during lactation, growing phase, or complete trial were analysed with a mixed model with age as covariate, sex as fixed effect, and piglet within sow as random effect. Males were heavier at birth (1.84 vs 1.65 kg, $P<0.001$), at weaning (9.62 vs 8.51 kg, $P<0.05$), at 6 months (67.9 vs 58.8 kg, $P<0.01$), and at 280 days of age (113.9 vs 89.6 kg, $P<0.001$). When considered, age was always highly significant ($P<0.001$). In all the three considered periods (i.e. lactation, growing, or complete trial), males showed larger ADG compared to females. The contribution of the sow decreased as the distance from the birth increased. The growth was modelled within period using three different models applied to the average weights for each day: linear, exponential, Gompertz. The best models (selected according to R^2 and BIC) were linear for the lactation phase, and Gompertz for the growing and complete periods. The results of this study showed that the Sarda breed can achieve discrete growth and body weight, allowing to valorise marginal areas.

Molecular approach to the genetic improvement of quality traits in dry-cured Iberian pork loin

P. Palma-Granados[1,2], M. Muñoz[1], M. Fernández-Barroso[1,2], C. Caraballo[1,2] and J. García-Casco[1,2]
[1]INIA-CSIC, Animal Breeding Department, Crta de la Coruña, km 7.5, 28040 Madrid, Spain, [2]Centro de I+D en Cerdo Ibérico, INIA-CSIC, Animal Breeding Department, Crta. EX101 km 4.7, 06300 Zafra, Spain; patricia.palma@inia.es

The main differential productive factor of Iberian pig regarding the conventional leaner breeds is the production of dry-cured products with a high organoleptic quality, which is the real reason for the success of its rearing in Spain. Therefore, a specific selection program for Iberian breed must be focused on meat quality and records collected from dry-cured samples should be considered. In this sense, water loss during the manufacturing process is one of the most relevant traits for the final quality of the cured products. In this work, association analyses between a panel of 32 SNPs designed for meat quality and the weight loss from the beginning to the end of ripening in dry-cured loins coming from 518 Iberian pigs from a breeding program has been performed. The model included as fixed effects: the livestock origin of the dams (3 levels) and the SNP effects taking values of 1 or -1 (homozygous) or 0 (heterozygous). The fresh loin weight was included as a covariate. The slaughter date that also allows adjusting the difference in loin drying time was included as a random effect (13 levels). In addition, genetic infinitesimal effects were included in the model by means of the kinship matrix. The association analyses were performed using Qxpak software. Mean value for weight loss was 40.1±6.1%. The estimated value of heritability (h^2) was 0.09. The only SNP significantly associated with water loss on dry-cured loin was CTSL_rs321623592A >T (a=-0.88±0.32; P=0.005). The observed high variability between samples in the measured quality trait may be due to the intrinsic differences between the animals and, above all, to the large number of environmental factors that could affect the final characteristics of this type of product. Although our results are not very promising, given that the value of dry-cured Iberian pork products is high and very relevant to the sector, it is worthy to search other SNPs mapped in candidate genes for dry-cured products quality and to assess other non-destructive traits easy and cheap to measure.

Carcass traits of Iberian × Duroc cross breed pigs according to age at the beginning of Montanera

A. Ortiz[1], D. Tejerina[1], S. García-Torres[1], P. Gaspar[2] and E. González[2]
[1]Centre of Scientific and Technological Research of Extremadura (CICYTEX-La Orden), Meat Quality Area, Av A5. Km 372, 06187 Guadajira, Badajoz, Spain, 06187, Spain, [2]Research Institute of Agricultural Resources (INURA), University of Extremadura, Department of Animal Production and Food Science, Avda. Adolfo Suarez, s/n, 06007 Badajoz, Spain, 06007, Spain; alberto.ortiz@juntaex.es

A more efficient management of the Iberian breed pigs raised under Montanera production system (extensive management at the final fattening phase in the dehesa where animal graze natural resources) could be attained using Iberian crossbred with Duroc pigs, because of better production parameters of these, and therefore reducing animal age at the beginning of Montanera. This study aimed to evaluate various ages of Iberian crossbred with Duroc pigs at the beginning of Montanera; 10, 12 and 14 months old, on carcass traits and primal cuts. For that, three animal batches of Iberian crossed with 50% Duroc pigs with average dates of birth successive and spaced 2 months from each other were used. During growing period, animal batches were fed with restrictions to start Montanera with similar body weight despite their different ages (10, 12 and 14 months old). After Montanera (63 days) animals were slaughtered and carcasses weight (including perirenal fat and kidneys), length (from the rear edge of the pubic symphysis to the front edge of the first rib) and subcutaneous backfat thickness (at last rib level) were measured. After quartering, ham length (from the front edge of the pubic symphysis to the hock joint) and perimeter (widest diameter), and weight of the ham, shoulder and loin were taken. The results showed an increase in carcass weight with age was observed, whilst carcass yield decreased. In terms of the primal cuts, only ham size was affected, with the lowest value being obtained by those from the youngest animals.

Effect of gender, castration, and diet on sensory characteristics of pork dry cured loins

S.S.Q. Rodrigues[1], L. Vasconcelos[1], E. Pereira[1], A. Leite[1], I. Ferreira[1], A. Teixeira[1], J. Alvarez[2] and I. Argemi-Armengol[2]
[1]CIMO, ESA, Instituto Politécnico Bragança, Campus Sta Apolónia Apt 1172, 5301-855, Portugal, [2]Universitat de Lleida, Rovira Roure 191, 25198 Lleida, Spain; srodrigues@ipb.pt

The sensory characteristics from pork dry-cured loins from immunocastrated females (F), surgically castrated males (CM), immunocastrated males (IM), fed with peas (P), or soybean meal (S) as the main dietary source of crude protein, were compared. The pigs were Duroc × Berkshire crossbreds slaughtered at 140 kg of body weight. Half loins were spiced and cured for 11 weeks (3 replicates per group). Twenty-two qualitative and quantitative appearance, odor, texture, and taste attributes were evaluated by a trained taste panel (n=8 people). All treatments were evaluated in duplicate in each of 3 sessions. Data were submitted to a non-parametric ANOVA, and pairwise comparisons were made using the Friedman test for related samples with SPSS. Results showed significant differences between fat colour from FS and IMS dry-cured loins. The highest differences were found in texture attributes, hardness, and juiciness. Gender, castration method, and feed influenced dry-cured loins hardness. CMS loins were significantly less hard than IMS and FS, and CMP. IMP loins were less hard than IMS. Juiciness was higher in IMS than IMP. About chewiness, pairwise comparisons indicated no significant differences between samples. IMP dry-cured loins were considered bitter than CMP. Only a small amount of sexual odour was detected by panellists, and no significant differences were found among the studied samples. Thus, immunocastration did not compromise the boar taint scoring and may be a good alternative to supply high quality meat products.

Session 65 Poster 21

Association of ghrelin gene polymorphisms with slaughter traits in pig

M. Tyra[1], K. Ropka-Molik[1], K. Piórkowska[1], M. Szyndler-Nędza[1], M. Małopolska[1], M. Babicz[2], A. Mucha[1], G. Żak[1] and R. Eckert[1]

[1]National Research Institute of animal Production, Sarego 2, 31-047 Kraków, Poland, [2]University of Life Sciences in Lublin, Akademicka 13, 20-950 Lublin, Poland; aurelia.mucha@iz.edu.pl

The GHRL gene was mapped on porcine SSC13 and was located close to QTL related to last rib backfat measurement. There was a significant homology between porcine SSC13 and human SSC3 on which the human GHRL gene is located. It has been hypothesized that mutations in the ghrelin gene in pigs may play a similar role as in humans and may be associated with obesity. The aim of study was to analyse regulatory regions and coding sequence of porcine GHRL gene as well as association analysis of selected SNPs with slaughter traits. The effect of c.-93A>G, 4428T>C and g.4486C>T polymorphisms at the ghrelin gene on slaughter performance were analysed in 346 gilts represented by three breeds (Polish Landrace, Duroc, Pietrain). Animals were fattened from 30 to 100 (±2.5) kg body weight. After slaughter, the carcasses were chilled for 24 hours (4 °C), weighted and the right half-carcasses were dissected and evaluated. A number of data were obtained including: meat weight in primary cuts, weight of ham, backfat thickness, carcass yield. It was demonstrated that carcasses composition traits were affected by c.-93A>G and g.4428T>C polymorphisms. The favourable results were obtained for pigs with the GG genotype at the c.-93A>G locus with were characterized by better carcass results than those with the AA genotype, e.g. higher ham weight and lower average backfat thickness. In pigs with the TT genotype at the g.4428T>C locus, we found lower mean backfat thickness than pigs with the CC genotype. Our results indicate that porcine GHRL polymorphism allows its use in breeding programs to improve selected carcasses characteristics. At the same time, the frequency of advantageous genotypes of these mutations is at a level guaranteeing the possibility of long-term use in breeding practice. These results emphasize the importance of further research on a larger population, with different sexes and breeds.

Session 65 Poster 22

Skatole and androstenone contents in backfat from different anatomical locations

N. Panella-Riera[1], R. Martín-Bernal[1], M. Egea[2], I. Peñaranda[2], M.B. Linares[2], M.B. Lopez[2], M.D. Garrido[2] and M. Font-I-Furnols[1]

[1]IRTA, Food Quality and Technology, Finca Camps i Armet, 17121, Spain, [2]UMU, Food Science and Technology, Faculty of Veterinary, University of Murcia, Campus Universitario Espinardo, 30100 Murcia, Spain; maria.font@irta.cat

Androstenone (AND) and skatole (SKA) are the main compounds responsible of boar taint. The measure of these compounds is usually carried out in subcutaneous fat of one anatomical region, and it is assumed that the levels are the same in all the carcass. There are some works that found differences in boar taint compounds by anatomical region, but results are not clear yet. The objective of this work was to evaluate the levels of AND and SKA on the backfat from 7 different locations of the pig carcass. For this purpose, carcasses from 16 Pietrain × (LargeWhite × Landrace) entire male pigs with 107.1+6.6 kg were used. Subcutaneous fat was taken from 7 anatomical positions: the caudal, central and cranial part of the loin, ham, belly, shoulder and tail. AND and SKA were analysed by stable isotope dilution analysis – headspace solid-phase microextraction – gas chromatography/mass spectrometry and results provided as concentration in liquid fat. Data analysis was carried out with SAS software. For the analysis of variance the logarithm of AND and SKA content was used. For the classification of the samples in low (L), medium (M) and high (H) levels, the thresholds used were 1.1 and 1.9 ppm liquid fat for AND and 0.16 and 0.32 ppm liquid fat for SKA. SKA levels were very low (0.06+0.04 ppm, between 0.02 and 0.20 ppm) and did not allow to have enough variability for the analysis. AND levels were high and with a big range of concentrations (2.08+1.76 ppm, between 0.07 and 9.53 ppm). Results showed not significant differences in the levels of AND and SKA by location. When the frequency of samples within location classified as L, M or H AND levels was studied, also no significant differences were observed. According to these results, in the conditions of this work, the levels of AND and SKA were not different among anatomical position.

Session 66 Theatre 1

D4Dairy – from data integration to decision support – lessons learned

C. Egger-Danner[1], K. Linke[1], B. Fuerst-Waltl[2], P. Klimek[3], O. Saukh[4], T. Wittek[5] and D4dairy-Consortium[1]
[1]ZuchtData, Dresdner Straße 89/B1/18, 1200, Austria, [2]University of Natural Ressources and Life Sciences, Gregor-Mendel-Str. 33, 1180 Vienna, Austria, [3]Medical University Vienna / Complexity Science Hub, Josefstaedter Str. 39, 1080 Vienna, Austria, [4]Graz University of Technology, Inffeldgasse 16/I, 8010 Graz, Austria, [5]University of Veterinary Medicine, Veterinaerplatz 1, 1210 Vienna, Austria; egger-danner@zuchtdata.at

To combine technological advances with a large number of novel data from different data sources, the project network D4Dairy comprising 13 scientific and 30 economic partners and 3 further collaboration partners within Europe was formed. This interdisciplinary network of expertise consists of researchers from different disciplines as well as company partners related to dairying along the value chain. The main aim of the project is to link and integrate different data sources, apply advanced technology and to generate data driven added value in terms of new tools and services for farmers and company partners with focus on prevention and early warning of diseases, advanced use in genetics, quality assurance measures to reduce the use of antimicrobials and to improve animal health and wellbeing. An example was developed to share data for research and joint use of data. This includes a data infrastructure for data processing and sharing with the respective legal background. Data from the Central Cattle Database including information on veterinary diagnoses and claw disorders as well as genetic and genomic information seved as a basis. This was expanded by data from technology providers, detailed information on housing, feeding and management aspects, as well as additional information related to health from about 300 farmers participating in different pilot studies within the project. The presentation will highlight challenges and achievements and share approaches and lessons learnt in delivering decision support tools for practical applications within the D4Dairy network of partners and data. This includes organizational as well as technical and scientific aspects.

Session 66 Theatre 2

Data integration in D4Dairy and new opportunities under the Data-Governance-Act and the Data Act

P. Majcen
Austrian Chamber of Agriculture, Schauflergasse 6, 1015 Vienna, Austria; p.majcen@lk-oe.at

The project D4Dairy shows how data integration can be handled in an optimal way. In this project, the cattle barn became a data room, where various animal and farm data were collected, merged and analysed resulting in new scientific results. Data protection, in particular personal reference, consents and further processing had to be clarified and it was recognized that the legal framework requires many resources in advance to check whether this is allowed at all. This raises the question of whether there is an easier way to do that. A large amount of data generated in the EU is archived or deleted immediately after they have been used once. In this way data protection requirements are easily fulfilled and data protection experts may approve that, but the potential of data cannot be explored if they are no longer available. Moreover, those who generate the data often do not have access to them, which currently leads to an imbalance on the data market. However, insufficiently exploiting the potential of data had justified reasons until now, like the different types of data and the fact, that member states do not handle their 'sovereign data management' in a uniform manner. The recently enacted Data Governance Act, as well as the proposal for a Data Act, are intended to revolutionize the European data space. The Data Governance Act provides a framework for the further use of data, more or less independent of its type or origin. Business to business data platforms shall enable data sharing. The Data Act shall among others clarify that both parties have access to all data collected by a machine. This leads to discussions on data ownership and should in any case strengthen the right to data portability. The contribution will cover the legal framework of the project D4Dairy with its issues and considerations related to data. The new governance framework and the proposal of the Data Act will be shown in detail. The declared aim is to give an answer to the question of whether the Data Governance Act and Data Act deliver what they promise, namely easier access, use and re-use of data with an open and broad scope of applications for scientific and commercial use.

Session 66 Theatre 3

Auxiliary traits for mastitis in Fleckvieh dairy cows from sensor based measurements

K. Schodl[1,2], B. Fuerst-Waltl[2], H. Schwarzenbacher[1], F. Steininger[1], C. Obmann[2], D4dairy-Consortium[1] and C. Egger-Danner[1]
[1]ZuchtData EDV-Dienstleistungen GmbH, Dresdner Str. 89, 1200 Vienna, Austria, [2]University of Natural Resources and Life Sciences, Vienna, Department of Sustainable Agricultural Systems, Gregor-Mendel-Str. 33, 1180 Vienna, Austria; schodl@zuchtdata.at

Mastitis still presents a widespread health problem in dairy herds with detrimental effects on animal welfare, productivity and economic revenues. However, heritabilities for mastitis diagnoses are commonly very low and thus auxiliary traits are of high interest. Sensor technology for measuring cow behaviour is increasingly used on dairy farms to improve herd management. Cows, which suffer from mastitis, may show differences in behavioural patterns, which may be detected by continuous sensor measurements. For breeding purposes, these continuous measurements additionally yield a high amount of data with the potential for large scale phenotyping. Thus, the aim of the present study was to develop potential auxiliary traits for mastitis based on variables derived from data from a sensor system for herd management, which measures activity and rumination time, and to estimate genetic parameters. In the course of the D4Dairy project, data from 2,708 Fleckvieh dairy cows on 33 farms equipped with Lely sensors and automatic milking systems (AMS) were collected between January 2020 and March 2021. Data comprised clinical diagnoses for udder health, bacteriological examinations of milk samples, somatic cell counts (SCC), daily milk yields from AMS, daily rumination time and activity from sensor data and cow and farm specific data from the Austrian central cattle database. Variables for genetic parameter estimation comprised clinical diagnoses, means and standard deviations of sensor variables, SCC, daily milk yields and milking intervals between -5 to 5 days after a mastitis diagnosis, during different lactation phases and during the whole lactation. Data are analysed by fitting bivariate linear animal models to estimate heritabilities and genetic correlations. Based on results of the genetic parameter estimations, suggestions for auxiliary traits for the complex udder health will be presented.

Session 66 Theatre 4

An integrative data-methodological approach to disease prevention in dairy cattle

C. Matzhold[1,2], E. Dervic[1,2], J. Lasser[2], C. Egger-Danner[3], F. Steininger[3] and P. Klimek[1,2]
[1]Medical University of Vienna, Spitalgasse 23, 1080 Vienna, Austria, [2]Complexity Science Hub Vienna, Josefstaedterstraße 39, 1080 Vienna, Austria, [3]3[4]ZuchtData EDV-Dienstleistungen GmbH, Dresdner Straße 89, 1200 Vienna, Austria; matzhold@csh.ac.at

Digitalization in livestock farming is generating an increasing amount of data, allowing for novel precision approaches to predict individual diseases. However, the full potential of such information has not yet been realised, as data often remains in separate silos, limiting its use to predict diseases. In this study, we combined 14 different data sources obtained as part of the D4Dairy project, allowing for comprehensive coverage of a cow's life cycle at any given point in time. The aggregated dataset includes information from approximately 457 farms with a total of 24,663 dairy cows. It includes information on a cow's health, husbandry, feeding-, housing- and milking-systems, as well as environmental conditions and genetic data such as breeding values. In addition, information on physiological parameters derived from sensory devices is available for a number of farms. We use statistical methods like logistic regression and machine learning techniques like XGboost to identify the best performing method for predicting an individual disease and identify its risk factors. As a result, an individual disease is predicted using a selection of features and the best performing method. Our preliminary results reveal that a disease is indeed best described as a product of a complex interplay between multiple characteristics of different domains of life. In terms of methodology, we find that the method's ability to unravel these disease traits and provide an accurate prediction depends on the available data. Less complex algorithm such as logistic regression, performs as well as or better than novel machine learning algorithms when the data set is large and complete enough. However, when data quality is compromised by missing or fewer observations, more advanced machine learning approaches such as XGBoost outperform simpler prediction algorithms. We believe that the obtained results of our data-driven analysis will be of interest both for future research and for practical decision-making processes.

Session 66 Theatre 5

Development of a decision support tool for targeted dry-off treatment of dairy-cows

W. Obritzhauser[1], C.L. Firth[1], K. Fuchs[2] and C. Egger-Danner[3]

[1]Institute of Food Safety, Food Technology & Veterinary Public Health, VetMedUni Vienna, Veterinärplatz 1, 1210 Vienna, Austria, [2]Data, Statistics and Risk Assessment, Austrian Agency for Health and Food Safety, Zinzendorfgasse 27/1, 8010 Graz, Austria, [3]ZuchtData EDV-Dienstleistungen GmbH, Dresdner Straße 89/B1/18, 1200 Vienna, Austria; egger-danner@zuchtdata.at

For decades, antibiotic dry cow therapy has been an integral part of farm management programmes to maintain a high level of udder health in dairy herds. Providing all cows in a herd with a long-acting antibiotic at the time of dry-off lowers tank milk somatic cell counts, increases the cure rate of chronic mastitis and reduces the number of new infections at the start of the subsequent lactation. However, the use of antibiotics also leads to an increased risk of selection for antibiotic resistance. Reducing the overall use of antibiotics is therefore a topic that is being discussed globally. Due to the increasing public health risk from antibiotic resistance, the use of antibiotics for dry cow therapy must also be viewed from a critical perspective. As part of the D4Dairy project, a protocol for the targeted drying-off of dairy cows was evaluated in a cohort study of 31 Austrian dairy farms. To determine the frequency of udder infections prior to dry-off, as well as the frequency of new infections, bacteriological milk cultures were carried out on all cows before dry-off and at the beginning of the subsequent lactation. Surveys were used to identify possible risk factors in dairy production systems and herd management that may have increased the incidence rate of mastitis. The use of antibiotic dry cow tubes was limited on 16 farms to evaluate whether the targeted use of such drugs can reduce the overall amount of antibiotics used on the farms, without causing a deterioration in the udder health of the herd. By linking the farm-specific and individual cow data collected, a decision support tool will be developed to help farmers make targeted and responsible decisions with respect to antibiotic dry cow therapy in their herd.

Session 66 Theatre 6

Exploring severity of clinical mastitis in four dimensions

Y. Song[1], M. Gote[1], L. D'anver[1], D. Meuwissen[1], I. Van Den Brulle[2], S. Piepers[2], S. De Vliegher[2], I. Adriaens[1,3] and B. Aernouts[1]

[1]KU Leuven, Biosystems department, Kleinhoefstraat 4, 2440 Geel, Belgium, 2440 Geel, Belgium, Belgium, [2]UGent, M-teamUGent, Salisburylaan 133, 9820 Merelbeke, Belgium, Belgium, [3]WUR, Animal Breeding and Genomics, 6700 AH Wageningen, 6700 AH Wageningen, the Netherlands; yifan.song@kuleuven.be

Clinical mastitis (CM), an inflammation of the udder usually caused by a bacterial infection, is a common and costly disease among dairy farms worldwide. Formally assessing CM severity in different dimensions can help to understand and quantify the impact of this disease. Severity can be assessed in the milk yield, immunological, clinical and pathogen dimension. Previous work mainly focussed on clinical severity alone. Our study aims to comprehensively describe the severity of CM in all four dimensions and analyse how they relate. Detailed mastitis data of 157 CM cases from 3 automatic milking system (AMS) farms were included in this study. The milk yield severity was defined based on relative quarter-level milk loss from day -7 to day 7 from detection, for the infected and uninfected quarters separately. The calculation of relative milk loss was based on the difference between realized milk yield and unexpected lactation curve. The immunological severity was defined as the maximum deviation of Somatic Cell Count (SCC) from a threshold (the average value of SCC (<500×1000 cells/ml) in this lactation), in which the contribution of the infected quarter-level yield was considered. Clinical level, ranging from 1 to 3, was calculated based on whether there are changes of milk, abnormalities of the udder and the general condition of the cows. The clinical severity was defined as the highest clinical level between day 0 and 7. The pathogen severity was defined based on the bacteriological culture result on day 0, in which we stratified between culture negative, and major or minor pathogen infected. The interconnection between these severities was measured with correlation and the results showed that each one positively correlated with the 3 other dimensions, and severity depends on parity and lactation stage. Further study is expected to link the severity with the probability of cure to explore the influence of CM on milk production and other physiological indicators for dairy cows in a short and long term.

Session 66 Theatre 7

Genetic parameters for mid-infrared-spectroscopy predicted mastitis and related phenotypes

L. Rienesl[1], B. Fuerst-Waltl[1], A. Koeck[2], C. Egger-Danner[2], N. Gengler[3], C. Grelet[4] and J. Soelkner[1]
[1]Institute of Livestock Sciences, University of Natural Resources and Life Sciences, Vienna, Gregor-Mendel-Straße 33, 1180 Vienna, Austria, [2]ZuchtData EDV-Dienstleistungen GmbH, Dresdner Straße 89/19, 1200 Vienna, Austria, [3]Gembloux Agro-Bio Tech, Université de Liège (ULg), Passage des Déportés 8, 5030 Gembloux, Belgium, [4]Walloon Agricultural Research Center (CRA-W), Chaussée de Namur 24, 5030 Gembloux, Belgium; lisa.rienesl@boku.ac.at

Genetic improvement of udder health in dairy cows is of high relevance as mastitis is one of the most frequent diseases. Since direct data on mastitis cases or diagnoses are often not available in large numbers, auxiliary traits, such as somatic cell count (SCC), are used for the genetic evaluation of udder health. In previous studies, models to predict clinical mastitis based on routinely collected mid-infrared (MIR) spectral data were developed. Those models can provide a probability of mastitis for each cow at every test-day, which is potentially useful as additional trait for an udder health index. The present study aimed to estimate the heritability of MIR-predicted-mastitis (MIRmastitis) and to estimate genetic correlations between MIRmastitis, SCC and clinical mastitis diagnosis (CM). Data were collected within the routine milk recording and health monitoring system of Austria from 2014 to 2020 and included records of ~59,000 Fleckvieh cows from ~2,600 farms. SCC was logarithmically transformed to somatic cell score (SCS). To estimate heritabilities, linear animal models were applied for all traits. Calving-age-class, calving year-season and days in milk were fitted as fixed effects; herd-test-day and animal permanent environment as random effects. Preliminary results show a heritability of 0.083 for MIRmastitis; the respective estimate was lower for CM (0.035) and higher for SCS (0.231). Results indicate a potential usability of MIRmastitis as auxiliary trait for genetic evaluation of udder health. Genetic correlations between traits, to be presented, will give more insight on the option of expanding the current udder health index (70% SCS, 30% CM) by MIRmastitis.

Session 66 Theatre 8

Ketosis and its auxiliary traits

B. Fuerst-Waltl[1], H. Schwarzenbacher[2], K. Schodl[1,2], A. Koeck[2], M. Suntinger[2], F. Steininger[2] and C. Egger-Danner[2]
[1]University of Natural Resources and Life Sciences, Vienna, Gregor-Mendel-Str. 33, 1180 Vienna, Austria, [2]ZuchtData EDV-Dienstleistungen GmbH, Dresdner Str. 89/B1/18, 1200 Vienna, Austria; birgit.fuerst-waltl@boku.ac.at

Clinical metabolic disorders usually have very low frequencies and low heritabilities. From a herd management perspective, an early detection of metabolic problems in the subclinical stage is important. Besides, also breeding may benefit from the additional consideration of subclinical cases or auxiliary traits from routine performance testing, herd management or automation technologies (e.g. automatic milking systems or sensors). Within the COMET-Project D4Dairy, the focus with regard to metabolic disorders is on the trait ketosis, both clinically and subclinically, and its potential auxiliary traits. Between January 2020 and March 2021, 99 pilot farms within the sub-project Genetics&Genomics not only provided validated health information but also recorded β-hydroxybutyrate (BHB) concentrations in the blood using the hand-held WellionVet BELUA device (Ketotest) 7 and 14 days postpartum. Additionally, the fat-protein ratio (FPR) and mid infrared (MIR) spectra were collected as part of the routine performance recording. Sensor information was also partly made available. Further information included data from a previous project, Efficient Cow, and another pilot study within the D4Dairy project, Milk-MIR-spectra, involving 49 farms with metabolic problems. In the latter, Ketotests were conducted in the course of the first and second routine milk recording after calving. While estimated heritabilities (h^2) for clinical ketosis were low ($\leq$0.01) throughout data sets, those for subclinical ketosis (based on Ketotest results) and the potential auxiliary traits were partly notably higher. Heritabilities for FPR ranged from about 0.10 to 0.20, those for Ketotests from about 0.05 to 0.26. The latter result was found for the farms with metabolic problems. Similarly, h2 values for MIR-predicted traits ranged from approx. 0.10 to 0.30. Higher h^2 and moderate to high genetic correlations between FPR, MIR-predicted traits and Ketotest but also to clinical ketosis indicate their usefulness as auxiliary traits. First genetic analyses on traits derived from sensor information have recently started.

On the image phenotyping for identification and development of dairy calves and heifers

C.D. Wallace[1,2], M.E. Wirch[2], R.R. Vicentini[3], U. Chauhan[1], R.V. Ventura[4] and Y.R. Montanholi[2]
[1]OneCup AI, 244-1231 Pacific Blvd, Vancouver, BC, V6Z 0E2, Canada, [2]Lakeland College, School of Agricultural Sciences, 5707 College Drive, Vermilion, AB, T9X 1K5, Canada, [3]Universidade Federal de Juiz de Fora, Núcleo de Estudos em Etologia e Bem-estar Animal, Rua José Lourenço Kelmer, s/n, Juiz de Fora, MG, 36036-900, Brazil, [4]Universidade de São Paulo, Faculdade de Medicina Veterinária e Zootecnia, Rua Duque de Caxias, 225, Pirassununga, SP, 13635-000, Brazil; courtwall017@gmail.com

The use of imaging-based artificial intelligence is growing in popularity to serve dairy farming. Image identification and development monitoring of replacement females, from birth to breeding, are under investigation to inform on imaging protocols and data handling to optimize tech-adoption. Continuous video footages of replacement Holstein calves were recorded using 11 cloud-connected surveillance cameras (Reolink, model RLC-511W-5MP) individually installed in 3 indoor maternity pens and 8 outdoor rearing pens. Animals were set into age ranges and were moved from pen to pen at 6 weeks periods. Cameras were fixed at 2 m height and angled to image the entire pen. Older heifers were oestrus monitored by accelerometers. Imaging collection has started on May 2021, a total of 117 replacement females were monitored to date. Animal identification is being devised through the development of convolutional neural network, and related techniques, to identify unique features based on colour pattern and whole-body biometrics to be validated by the ear tag. This model is to be further optimized by running the identification: (1) without the colour pattern; (2) only considering the head images; and (3) determining the minimum time elapsed between images to enable identification. Similarly, analysis to ultimately validate imaging phenotypes of oestrus are in course. Preliminary data support that the despite the higher accuracy with the most encompassing model, the optimization efforts with reduced models will yield alternatives not only for Holstein but also for non-spotted cattle to determine the ideal imaging protocol for visual identification. Moreover, the validation of imaging phenotypes for oestrus will bring complementary solutions to manage the reproductive performance in Holsteins.

Concentrate use efficiency in dairy cows: an investigation based on on-farm data

S.-J. Burn[1], B. Fuerst-Waltl[1], F. Steininger[2] and W. Zollitsch[1]
[1]University of Natural Resources and Life Sciences Vienna, Department of Sustainable Agricultural Systems, Institute of Livestock Sciences, Gregor Mendel-Str. 33, 1180 Vienna, Austria, [2]ZuchtData, Dresdner Str. 89/B1/18, 1200 Vienna, Austria; sarahjoeburn@gmail.com

Due to physiological, ecological and economical limitations on concentrate use, a high concentrate use efficiency is desirable. Feed consultants involved in the project 'D4Dairy' reported different concentrate use efficiencies between dairy farms. The aim of the present study was to investigate potential differences between cows and farms in concentrate use efficiency. Furthermore, the impact of selected factors was analysed. Performance data were analysed from 12 Austrian dairy farms with an automatic milking system and a total of 610 cows. Concentrate use efficiency was characterised by 4 indicators which were ratios between output (kg milk, kg ECM, MJ milk energy, g milk protein) and concentrate input (kg concentrate DM, MJ NEL from concentrate, g CP from concentrate, respectively). A distinct variability was recorded between cows within farms and within breeds. The difference of concentrate use efficiency between farms was significant for all 4 indicators (2.83-5.58 kg milk/kg concentrate DM; 2.88-5.72 kg ECM/kg concentrate DM; 1.16-2.49 MJ milk energy/MJ NEL from concentrate; 0.50-0.81 g milk protein/g CP from concentrate). Technique of feed provision had no significant effect on the concentrate use efficiency; the response to forage quality was inconclusive. Heavier cows and high dietary concentrate percentages lead to lower concentrate use efficiency. Milk yield had a moderate and positive effect on the concentrate use efficiency. Due to the relevance of the topic, it should be addressed in future studies, using appropriately large and differentiated data sets.

Milk progesterone sample rate and reactivity of farmers impacts reliability of insemination window

D. Meuwissen, K. Brosens, B. Aernouts and I. Adriaens
KU Leuven, Biosystems, Kasteelpark Arenberg 30, 3001 Heverlee, Belgium; dyan.meuwissen@kuleuven.be

Improving fertility of dairy cows on farm remains difficult, as it is a complex trait which depends on physiological, management and external factors. The availability of new sensor data to get a better grip on physiological status of the cows and to accurately predict the best insemination window based on milk progesterone (P4), offers new opportunities to analyse and understand fertility performance. Before external and physiological factors that impact conception success can be investigated, it is crucial that we have certainty on whether the insemination was correctly timed. This study uses an extensive dataset from 5 modern dairy farms with a Herd Navigator system, which farmers use to determine the optimal moment of insemination upon a drop in milk P4 preceding luteolysis. Despite that in many cases the estimation of the insemination moment goes well, inconsistencies in detection of luteolysis arise when sampling rate or milking frequency deviate. In this study, we analysed the factors influencing correct identification of the insemination window. To this end, we analysed different traits related to insemination and sampling frequency of the Herd Navigator and looked at their effect on success of conception across the 5 farms. We found that in farms with a better fertility performance, the time between true luteolysis and the insemination was more consistent, which was due to the reactivity of the farmer upon a luteolysis alert and the P4 sampling rate. It was found that the P4 sampling rate is influenced by the milking frequency, thereby causing a less reliable prediction of the insemination window for farms with an increased milking interval. Lastly, this study showed that conception success can be improved when the P4 sampling rate is accounted for. From this analysis, we can make more reliable selection of the correctly timed inseminations and thus where conception success depended on other factors. This is a first step towards improved understanding of fertility performance on farm, and as such, of a more sustainable dairy sector.

Development of a monitoring tool for metabolic imbalance issues based on milk recording data

S. Franceschini[1], C. Grelet[2], J. Leblois[3], N. Gengler[1] and H. Soyeurt[1]
[1]University of Liège – GxABT, TERRA Research and Teaching Centre, Passage des Déportés 2, 5030 Gembloux, Belgium, [2]Walloon Agricultural Research Center, Rue du Liroux 9, 5030 Gembloux, Belgium, [3]Walloon Breeders Association Group, Rue des Champs Elysées 4, 5590 Ciney, Belgium; sfranceschini@uliege.be

With increasing consumer concerns about animal health, monitoring tools are increasingly necessary. Often, they target only one or few biomarkers at the same time, making the detection of multifactorial health issue less powerful. This study aims to develop a multivariate estimator of animal health from a hierarchical clustering including 27 mid-infrared health-related predictions, milk yield, and somatic cells count. From 740,454 observations coming from first parity Holstein cows, a cluster related to the energy balance (EB) has been highlighted. It was then predicted using a partial least square discriminant analysis to estimate a probability for a cow to be in metabolic imbalance. Highly significant correlations between the probabilities and reference measures were found for biomarkers of EB issues: 0.70 for beta-hydroxybutyrate, 0.67 for non-esterified fatty acid and -0.60 for glucose. From these probabilities, a monitoring tool for EB has been developed to mark each observation into three categories: healthy, at-risk, and problematic. The predicted beta-hydroxybutyrate was 1.3 times higher in observations marked to be problematic than healthy ones and 2.0 times higher for acetone. As expected, the prevalences were related to the days in milk (DIM). Indeed, the prevalences for the first 20 DIM were of 34.87% for 'problematic', 18.78% for 'at-risk', and 46.36% for 'healthy'. The prevalences for the whole first lactation were 11.15% for 'problematic', 12.92% for 'at-risk', and 75.93% for 'healthy'. Those percentages are similar to those found in the literature using blood samples, confirming the interest of the developed tool. After validation with on-field information and extension to other lactations, those results are expected to allow better monitoring and breeding.

Session 66 Poster 13

Use of automatic milking systems data for early mastitis detection in Italian Holstein dairy cows

R. Moretti[1], E. Ponzo[1], M. Beretta[2], S. Chessa[1] and P. Sacchi[1]
[1]University of Turin, Department of Veterinary Sciences, Largo Paolo Braccini 2, 10095, Grugliasco, TO, Italy, [2]Agrilab, Regione Madonna Prati 318, 12044, Centallo, CN, Italy; riccardo.moretti@unito.it

Automatic Milking Systems (AMS) are introducing new perspectives in zootechnics, being one of the major data sources in the field of precision livestock farming applied to dairy cows. Specifically, AMS let a continuous monitoring of the herd through time, over a high number of parameters. Among these parameters, of paramount importance are those related to udder health and mastitis detection. Milk conductivity and somatic cells count are used as early indicators of mastitis. Different studies investigated the use of statistical models to improve the sensibility and sensitivity of AMS software's built-in alerts for breeders. However, results are so far less than optimal. In this study, alerts due to variations in milk conductivity (MC) and somatic cells count (SCC) were correlated with microbiological examination performed on milk samples collected the day after mastitis alerts from the AMS. A total of 270 observations from 165 Italian Holstein cows from two farms were analysed. Chi-squared test was used to test the relation between alerts and different bacterial classes (i.e. gram positive and negative) identified in samples by microbiological examination. A significant relationship was confirmed by the test ($P<0.001$). A logistic mixed regression model was subsequently fitted to investigate the effects of the abovementioned variables on both milk conductivity and somatic cell counts related alerts. The model showed that the presence of Gram-positive bacteria increased the odds of obtaining a SCC alarm ($P=0.046$), while the presence of Gram-negative bacteria increased the odds of obtaining a MC alarm ($P<0.001$). Furthermore, infection degree and changes in rumination time were also all correlated to bacterial classes and therefore influenced the probability to be classified as one of the two alerts. Additional investigations, increasing the number of farms and animals, are required to further validate the results obtained in this study.

Session 66 Poster 14

Compatibility of smart ear tags with twin pin fixing system in cows

N. Gobbo Oliveira Erünlü, J. Bérard and O. Wellnitz
Agroscope, Animal Production Systems and Animal Health, Route de la Tioleyre 4, 1725 Posieux, Switzerland; nicolle.gobbooliveiraeruenlue@agroscope.admin.ch

A new smart ear tag (AET) containing GPS, accelerometer, RFID and Bluetooth technologies was tested for compatibility in cows in a free stall barn in Switzerland. The AET is equipped with long lasting battery function via solar panel and uses a 'twin pin' fixing system. Maintaining the correct position of the ear tag is essential to ensure proper functioning of the AET. The right ears of 12 newborns (NB) were tagged with the AET, while the left ears were tagged simultaneously with standard single pin ear tags (SET). In 14 adolescents (AD) the existing SET in the right ear were removed and then tagged with the AET, while their left ear retained the SET. The existing SET hole could not be used, due to too large diameter. Ear growth, the state of the ear, general condition of the animal and cortisol in saliva was measured. Measurements were performed daily for the first week, weekly for the first month, and monthly for the second, third, and fourth month. All animals had exudate between day 7 and day 21 on AET. In NB similar responses were seen on the SET. Between month 2 and month 4, 42% of NB and 20% of AD showed exudate on the AET. In this period, 25% NB and 9% AD had exudate in the AET in all times, while 8% NB had exudate in both AET and SET, indicating an individual reaction to the ear tags. Ear length and width increased ($P<0.05$) in NB in the four month of examination without differences between ears with AET or SET. In the first week an increase in ear temperature was observed in all newly marked ears without difference between AET and SET ($P<0.05$). In the first 4 days, NB but not AD showed high saliva cortisol concentrations that decreased ($P<0.05$) until day 21 which is physiological in NB. 27% of all animals were once trapped in the fence with AET. Two animals tore the AET out of the ear. In conclusion, smart ear tags with long lasting battery function via solar panel that need a twin pin fixation in cows do not seem to induce systemic or local inflammations more frequently compared to standard ear tags but may have a higher risk of accidental injuries.

Session 66 Poster 15

Predicting pregnancy status of dairy cows with mid-infrared spectra: a tricky and limited approach

L. Rienesl[1], P. Pfeiffer[1], A. Koeck[2], C. Egger-Danner[2], N. Gengler[3], C. Grelet[4], L. Dale[5], A. Werner[5], F.-J. Auer[6], J. Leblois[7] and J. Sölkner[1]

[1]Institute of Livestock Sciences, University of Natural Resources and Life Sciences, Gregor-Mendel-Straße 33, 1180 Vienna, Austria, [2]ZuchtData EDV-Dienstleistungen GmbH, Dresdner Straße 89/19, 1120 Vienna, Austria, [3]Gembloux Agro-Bio Tech, Université de Liège, Passage des Déportés 8, 5030 Gembloux, Belgium, [4]Walloon Agricultural Research Center, Chaussée de Namur 24, 5030 Gembloux, Belgium, [5]LKV Baden-Wuerttemberg, Heinrich-Baumann Straße 1-3, 70190 Stuttgart, Germany, [6]LKV Austria GmbH, Dresdner Straße 89/19, 1120 Vienna, Austria, [7] Elevéo, Rue des Champs Elysées 4, 5590 Ciney, Belgium; soelkner@boku.ac.at

Diagnosis of pregnancy is an important component of successful reproduction management in dairy herds. Prediction of pregnancy status from mid-infrared (MIR) spectral data is of particular interest, as MIR spectra are routinely collected within milk recording schemes. The aim of this study was to investigate MIR-based prediction equations for pregnancy status and to assess potential limitations of the method. Data were collected within Austrian milk recording system from 2014 to 2020 and included test-day and reproductive data of ~40,000 cows. 212 selected MIR wavenumbers were used for modelling. Data (~400,000 records) were randomly split by farm into 50% calibration and 50% validation. Prediction models were developed using partial least square discriminant analysis. Applying a single prediction equation for all cows regardless lactation stage showed overall sensitivity of 0.86 and specificity of 0.84 in validation. A detailed investigation demonstrated that the model was not able to predict pregnant cases before days in milk (DIM) 120 (sensitivity<0.28) and open cases after DIM 120 (specificity<0.01). Pregnancy status is strongly linked to lactation stage; open in early and pregnant in later lactation. A model using DIM as predictor showed similar results (sensitivity=0.79; specificity=0.87) compared to the MIR-model. Prediction equations based on MIR-spectral data are mostly predicting lactation stage, not pregnancy status. Reducing this confounding by analyses within stages of lactation, sensitivities and specificities were in the range of 0.50-0.60, too low to be of practical use.

Session 67 Theatre 1

Rethinking Ivanov's legacy: artificial insemination as a technology

K. Rossiyanov

S.I.Vavilov Institute for the History of Science and Technology of the Russian Academy of Sciences, 14, Baltiiskaya ul., 125315, Russian Federation; rossiianov@yandex.ru

In his time, the Russian physiologist Il'ya Ivanovich Ivanov (1870-1932) was known for his pioneering work on the artificial insemination of farm animals. Analysing the historical background of his 1922 article that appeared in the 'Journal of Agricultural Science' and brought Ivanov international recognition, I emphasize the 'technological' dimension of his research on the use of artificial insemination in stud- and cattle-farming. Developing a new modification of the method, Ivanov created a tool that allowed, according to his words, the 'fullest utilization' of the sperm of 'particularly valuable males'. I argue that this approach had a profound and long term impact on the perception of the technique, shifting the emphasis from artificial insemination as a compensatory strategy in cases of infertility to understanding it as a technology, designed for the mass-improvement of stock. I also analyse other important aspects of his work, such as the use of artificial insemination in the experiments on hybridization between different species of animals, as well as his research on the preservation of sperm under low temperatures.

Asocial Darwinism or Darwinian socialism: artificial reproduction and socio-biological boundaries

A.B. Kojevnikov
University of British Columbia, History, 1873 East Mall, V6T 1Z1, Canada; a.nikov@ubc.ca

In the late 19th early 20th century, animal breeders' experiences provided an important source for the origin and development of Darwinism. Technological practices used in the artificial selection of domesticated species were generalized and extrapolated, in theory, also to the natural world, on the one hand, and to the human race on the other. New ideas and approaches to reproduction, insemination, and population control, likewise, travelled across the boundary between the animal and human species, in both directions. These transfers and analogies, also unavoidably, met with doubts regarding the limits of their applicability on either technological, economic, philosophical, or moral grounds. This paper will analyse and compare several such debates concerning group selection, artificial control of reproduction, cross-species hybridization, and eugenics. The positions and choices made by various participants often, if tacitly, depended on politically and culturally specific ways of drawing the distinction between the artificial and natural, and between biological and social.

Kaempferol modulates sperm cryocapacitation through changes in the levels of reactive oxygen species

E. Tvrdá, Š. Baňas, F. Benko, M. Ďuračka, M. Lenický and N. Lukáč
Institute of Applied Biology, Slovak University of Agriculture, Tr. A. Hlinku 2, 949 76 Nitra, Slovak Republic; evina.tvrda@gmail.com

An attractive strategy to overcome sperm cryocapacitation lies in the supplementation of biologically active substances exhibiting membrane-stabilizing, motility-promoting, and antioxidant effects. Out of these, kaempferol (KAE), a plant-derived flavonol has shown promise in the prevention of alterations to the sperm membranes, mitochondria, and DNA. Hence, the aim of this study was to assess the effects of selected KAE doses on the motility, capacitation status of cryopreserved bovine spermatozoa alongside its role in the production of major reactive oxygen species (ROS) classes, specifically superoxide (O_2^-), hydrogen peroxide (H_2O_2) and hydroxyl radical (•OH). Ejaculates from 12 breeding bulls were cryopreserved in a commercial extender containing 12.5 μM, 25 μM and 50 μM KAE or carrying no supplement. Sperm motility was evaluated with computer assisted semen analysis while the capacitation status was assessed with the chlortetracycline assay. Quantification of O_2^- was performed by the nitroblue tetrazolium test, H_2O_2 production was assessed with the Amplex Red assay while the concentration of •OH was quantified using the aminophenyl fluorescein reagent. Our results indicate that the presence of 12.5 and 25 μM KAE resulted in a higher sperm motility ($P<0.05$; $P<0.01$) and a concomitant decrease of prematurely capacitated spermatozoa ($P<0.05$; $P<0.01$). Exposure of cryopreserved spermatozoa to particularly 12.5 and 25 μM KAE lead to a significantly lower concentrations of all three major ROS types ($P<0.05$; $P<0.01$). We may suggest that KAE as an alternative cryosupplement may offer higher protection to premature capacitation, particularly by stabilizing the levels of ROS that may promote sperm cryodamage. This publication was supported by the Operational program Integrated Infrastructure within the project: Creation of nuclear herds of dairy cattle with requirement for high health status through the use of genomic selection, innovative biotechnological methods, and optimal management of breeding, NUKLEUS 313011V387, co-financed by the European Regional Development fund.

Session 67 Theatre 4

Cryoprotective ability of epicatechin on protein profile of cryopreserved bovine spermatozoa

F. Benko, Š. Baňas, M. Ďuračka, N. Lukáč and E. Tvrdá
Institute of Applied Biology, Faculty of Biotechnology and Food Sciences, SUA in Nitra, Tr. A. Hlinku 2, 94976 Nitra, Slovak Republic; xbenkof@uniag.sk

The goal of our research was to evaluate the cryoprotective ability of epicatechin (EPI) with respect to its impact on the expression of selected proteins (heat shock proteins 70 and 90; HSP70 and HSP90; pro-apoptotic Bax protein and anti-apoptotic Bcl-2 protein) in cryopreserved bovine spermatozoa. The ejaculates for the experiment were provided from 12 adult Holstein-Friesian bulls. Before cryopreservation, samples were distributed, and each fraction was enriched with different concentration of EPI (0, 25, 50 and 100 µmol/l) then cryopreserved and stored in liquid nitrogen at -196 °C. The motility of spermatozoa was assessed with CASA (Computer Assisted Semen Analysis) and the expression of the selected proteins was monitored by the Western blot technique. Based on the results, we observed that 50 and 100 µmol/l EPI statistically increased (P<0.0001) the post-thaw sperm motility against the control without EPI. In the case of HSP's, 100 µmol/l EPI statistically improved (P<0.01; P<0.0001) the expression of HSP90 and HSP70 against the control. Furthermore, 50 and 100 µmol/l EPI statistically decreased (P<0.0001) the expression of the Bax protein. No statistical change was observed in the expression of Bcl-2. In summary, we may confirm that 50 and 100 µmol/l EPI improved all selected parameters and exhibit cryoprotective properties against cryodamage.
Acknowledgement: This publication was supported by the Operational program Integrated Infrastructure within the project: Creation of nuclear herds of dairy cattle with a requirement for high health status through the use of genomic selection, innovative biotechnological methods, and optimal management of breeding, NUKLEUS 313011V387, co-financed by the European Regional Development Fund.

Session 67 Theatre 5

Novel nutritional approaches to enhance sustainability and productivity in cattle farming

S. Grossi, M. Dell'anno, L. Rossi and C.A. Sgoifo Rossi
Università degli Studi di Milano, Dipartimento di Medicina Veterinaria e Scienze Animali, Via dell'Università 6, 26900, Italy; silvia.grossi1994@libero.it

Three studies were set up to evaluate novel nutritional approaches to improve the sustainability of beef and dairy cattle farming. The effect of partial substitution of corn (CM) and soybean (SBM) meals with bakery former foodstuffs (BFF) and wheat wet distillers (WDGs) on the diet sustainability and cattle productivity was evaluated in Limousine heifers (n=408) (Control -CON, standard diet; Circular -CIR, diet with 1.5 kg of BFF and 1.5 kg of WDGs instead of 1.6 kg CM and 0.3 kg SBM). Greenhouse gases (GHG, kg CO_2 eq), water (H_2O, L) and land use (LU, m^2) for feed production, human-edible feeds consumption (HE, kg), production performances and apparent total tract digestibility (aTTD) were evaluated. CIR showed a reduction per kg of cold carcass weight (CCW) of 1.00 kg CO_2 eq of GHG, of 72.38 l of H_2O, of 1.20 m^2 of LU, and of 0.95 kg of HE (P<0.01), and sugar aTTD was improved (P<0.01). The effect of partial replacement of SBM with slow-release urea (SRU) on diet sustainability and productivity in Holstein cows (n=140; (1) Control; (2) Treatment, 0.22% d.m. SRU) was evaluated. Carbon footprint (CFP) of the diet, predicted methane (CH_4) emissions, production performances and aTTD were evaluated. Treatment increases production performances (P<0.01), aTTD of crude protein (P=0.012), NDF (P=0.039), and cellulose (P=0.033) and the environmental parameters (P<0.01). The effect of an essential oils, bioflavonoids and tannins blend was evaluated on CH_4, total gas and volatile fatty acids production *in vitro*, and on productivity of lactating Holstein cows (n=140; (1) Control; (2) Treatment, diet plus 10 g/head/d of the blend). Treatment reduced the *in vitro* total gas and CH_4 emissions (P<0.01 at 16, 20 and 24h), and acetic acid (P<0.01 at 16h and 24h) productions, while propionic acid was increased (P<0.01 at 16h and 24h). Treatment raised production performances (P<0.01), and aTTD of cellulose (P<0.01) and starch (P<0.01). Acting on diet composition, nutritional value and bioactive functionality can increase the sustainability and efficiency of cattle farming, in accordance to the circular economy principles.

Dietary energy source on Italian Holstein heifers feed efficiency

V. Fumo, F. Omodei Zorini, E. Alberti, M. Dell'anno, E. Zucca, G. Savoini and G. Invernizzi
University of Milan, Department of Veterinary Medicine and Animal Sciences, Via Dell'Universita', 6, 26900, Italy; guido.invernizzi@unimi.it

The objective of the trial was to study feed efficiency in Italian Holstein heifers fed diets characterized by different energy source. Sixteen heifers were divided in two homogenous groups and included in a crossover design. The trial lasted in total 10 weeks: after an adaptation of one week, a 4-weeks treatment period was applied followed by one week adaptation and a second 4-weeks treatment period. Diets were TMR isonitrogenous and isoenergetic composed by sorghum silage, hay, sunflower meal, and maize flour (Diet A) or hydrogenated fat (Diet B). Daily feed intake was measured by RIC system and live body weight, BCS, heart girth and wither height were recorded weekly. Once a week, ultrasound measures of subcutaneous, peritoneal, and retroperitoneal fat depot were performed. At the beginning and at the end of each experimental period, faeces, urine, rumen content and saliva pH were measured. Data was analysed by a MIXED repeated procedure of SAS and significance value was set at $P<0.05$. Feed intake was higher ($P<0.05$) when heifers were fed hydrogenated fat compared with corn-based diet. No differences were detected for LBW and BCS between dietary treatments. Heart girts and wither heights were not affected by dietary treatment but increased constantly during the trial. Faecal and urine pH values were lower when animals were fed corn-based diet compared with fat-included diet. Saliva and rumen content pH did not change between treatments. Mean average daily gain and feed conversion ratio were not affected by the dietary treatment ($P>0.05$). Ultrasound analysis detected significant higher values of subcutaneous fat when animals were treated with maize compared with fat-based diet ($P<0.05$). Furthermore, peritoneal, and retroperitoneal depots were higher ($P<0.05$) when animals were fed hydrogenated fat compared with corn-based diet. In conclusion, different depots of fat are directly related to dietary energy source without affecting efficiency measured as feed conversion ratio.

Fat supplement for dairy cow diets during early lactation – meta analysis

S. Lashkari[1], M.R. Weisbjerg[1], L. Foldager[1,2] and C.F. Børsting[1]
[1]Aarhus University, AU-Foulum, Department of Animal Science, Blichers allé 20, 8830 Tjele, Denmark, [2]Aarhus University, Bioinformatics Research Centre, Universitetsbyen 81, 8000 Aarhus C, Denmark; saman.l@anis.au.dk

The objective of the present study was to perform a meta-analysis to evaluate the effects of diet supplemented with different fat level on productive performance and plasma metabolites in early lactating dairy cows. The data set was formed from 16 peer-reviewed publications. The criteria for including experiments were that they were finalized no later than 100 days in milk. Fat sources were categorized in 3 sources as follows: Ca-soap of fatty acids (CaFA), saturated fatty acids (SFA), and polyunsaturated fatty acids (PUFA). For dry matter intake (DMI), there was an interaction between fat level and source ($P<0.01$), where each percentage-unit increase of fat level reduced DMI by 0.38 and 0.77 in CaFA and PUFA, respectively and increased by 0.55 kg/d in SFA. Corrected milk production (fat or energy corrected milk) increased by 0.45 kg/d for each percentage-unit increase in fat level ($P<0.01$). For milk fat percentage, there was an interaction between fat level and source ($P<0.01$), and each percentage-unit increase of fat level reduced milk fat percentage by 0.09, 0.21, and 0.05 in SFA, CaFA, and PUFA, respectively. Milk fat yield tended to be reduced by 0.02 kg/d per percentage-unit increase in fat level ($P=0.08$). For milk protein percentage, there was an interaction between fat level and source ($P<0.01$). Each percentage-unit increase of fat level reduced milk protein percentage by 0.07 and 0.02 in CaFA and PUFA, respectively and increased by 0.02 kg/d in SFA, whereas milk protein yield was not affected by fat level ($P=0.43$). Plasma non-esterified fatty acid (NEFA) concentration increased with 0.03 mmol/l per percentage-unit increase in fat level ($P<0.01$). Similarly, plasma β-hydroxybutyrate (BHB) concentration increased with 0.07 mmol/l per percentage-unit increase in fat level ($P<0.01$). In conclusion, increased fat supplementation increased corrected milk production despite the reduced DMI. Plasma NEFA and BHB concentration increased slightly by increasing fat level. However, adding dietary fat seems not to cause serious drawbacks for production performance and metabolism, despite of small increases in plasma NEFA and BHB concentration.

Effect of medium-chain fatty acids on productive and blood parameters in periparturient dairy cows

F. Gadeyne[1], M. Bustos[1], J. Gieling-Van Avezaath[1], L. Kroon[2] and A. Koopmans[2]
[1]Royal Agrifirm Group, Landgoedlaan 20, 7325 AW Apeldoorn, the Netherlands, [2]Schothorst Feed Research, Meerkoetenweg 26, 8218 NA Lelystad, the Netherlands; f.gadeyne@agrifirm.com

Periparturient cows undergo tremendous metabolic changes as they transition from pregnancy to lactation. In order to guarantee transition success and improve farmer's profitability, medium-chain fatty acids (MCFA) could be supplemented because of their impact on rumen fermentation and immunomodulating characteristics. A randomized block design was used to evaluate the effect of Aromabiotic® Cattle, a mixture of MCFA, on dry matter intake (DMI), milk yield (MY), milk composition and blood parameters around parturition. Thirty high-producing Holstein-Friesian dairy cows entered the study 28 days before expected calving and were followed until 70 days in lactation. Animals were assigned to 2 treatments (n=15): 1/ a control without MCFA (CON), or 2/ receiving 12.5 g/cow/day MCFA via the concentrate throughout both dry and lactation period (MCFA). Blood samples were taken at wk-1, 1 and 4 relative to calving. No differences were observed between treatments for DMI. In wk1, a tendency (P=0.098) for higher MY (+3.6 kg) was seen using MCFA and remained numerically 3 kg higher during the first 3 weeks after calving. Both before and after calving, a tendency for reduced lipopolysaccharide (LPS)-binding protein in blood was observed for MCFA vs CON (28 vs 111 ng/ml in wk-1, P=0.10, 37 vs 89 ng/ml in wk1, P=0.23, and 21 vs 77 ng/ml in wk4, P=0.08; resp.), indicating lower ruminal release of LPS. No significant differences were observed for non-esterified fatty acids (NEFA) or beta-hydroxy butyrate (BHBA) in blood, but values were numerically lower post-calving for MCFA vs CON (0.55 vs 0.79 mmol/l in wk1, P=0.22, and 0.27 vs 0.42 mmol/l in wk4, P=0.36, for NEFA; 0.67 vs 0.77 mmol/l in wk1, P=0.45, and 0.78 vs 0.93 mmol/l in wk4, P=0.37, for BHBA; resp.). The prevalence of clinical mastitis (1 vs 4) and repeated health issues (1 vs 5) was numerically lower for MCFA than CON. In conclusion, results suggested MCFA influences rumen health around parturition, thereby preventing an excessive immune response by LPS stimulation, which can mitigate early lactation negative energy balance and hence increase transition success.

Implementing selection index for nitrogen efficiency in Italian Holstein population

A. Fabris[1], G. Visentin[2], R. Finocchiaro[1], M. Marusi[1], F. Galluzzo[1], L. Mammi[2] and M. Cassandro[1,3]
[1]ANAFIBJ, via Bergamo 292, 26100 Cremona, Italy, [2]Alma Mater Studiorum, University of Bologna, Dipartimento di Scienze mediche veterinarie, via Zamboni 33, 40126 Bologna, Italy, [3]University of Padova, Dipartimento di Agronomia Animali Alimenti Risorse Naturali e Ambiente, Via VIII Febbraio 2, 35122 Padova, Italy; annafabris@anafi.it

ANAFIBJ has been working on improving environmental sustainability, as well as animal welfare and biodiversity. The aim of the present study was to develop a new selection index for Italian Holstein population, e.g. nitrogen efficiency, that will help farmers to breed more efficient animals. For this index, protein (P%) and urea (U%) content in milk were used. From them, a compound index P/U was derived as selection objective: P% and U% weights were derived following selection index theory, resulting in a combined indicator. Data ranged from 2017 to 2022 milk recording and included records between±3 DS from the mean and at least 4 records per cow; after editing, the dataset consisted of 20,824,821 P/U records on 1,908,431 cows belonging to 9,324 herds. Breeding values (EBV) were estimated using a repeatability animal model which included as fixed effects: days in milk (DIM – 10 classes of 30-d each, ranged from 5 to 305 d), parity (first, second and third lactation), age within parity (9 classes within parity), interaction between DIM and parity, and herd-test-day (328,654 classes) as contemporary group. The random effects were additive genetic animal, permanent environment and the residual. Phenotypic means for P% and U% were respectively 3.4%±0.4 and 0.024%±0.007 and heritabilities were 0.35 and 0.15, respectively. Genetic correlation between P/U genetic index and its phenotypic value is 0.42 and EBVs reliability averaged 0.94±0.05. As expected, bulls with higher EBVs had higher mean of P/U due to lower U%, while lower EBVs were associated with a decrease in P/U. On average, bulls with EBV equal to 100 have daughters' phenotypic mean equal to 141.46±10.42 P/U. ANAFIBJ will release this index to aid in the selection in minimizing nitrogen waste and maximizing milk protein output. Study supported by 'Latteco2 project, sottomisura 10.2 of the PSRN-Biodiversity 2020-2023'.

Mid-infrared spectroscopy for a rapid assessment of immunoglobulins G level in bovine colostrum

A. Costa, A. Goi, M. Penasa and M. De Marchi
University of Padova, Department of Agronomy, Food, Natural resources, Animals and Environment (DAFNAE), Viale dell'università 16, 35020 Legnaro, Italy; angela.costa@unipd.it

The concentration of immunoglobulins G (IgG, g/l) defines colostrum quality in cattle. By convention, colostrum with IgG<50 g/l is not recommended for calves feeding in the first h of life due to insufficient antibodies level. On average, 15% of cows produce colostrum of unacceptable quality in dairy farms, exposing the calf to greater risk of mortality and morbidity and impairing the future heifer's performance. In this study, first colostrum samples (521 Holstein cows) were collected between 2019 and 2020 in 9 farms within 6 h from calving. Each sample was aliquoted for IgG and protein content determination via gold standard, i.e. radial immunodiffusion and Kjeldahl, respectively, and for prediction of total protein content via mid-infrared spectroscopy (MIR) using the prediction model developed for mature milk. Before MIR analysis, colostrum samples were diluted in pure water (1:1) to reduce matrix density and avoid clogging issues. We demonstrated that MIR-predicted protein content was significantly correlated with both IgG (r=0.87) and protein content (r=0.97) measured with gold standard. Moreover, receiving operating characteristic analysis (ROC) showed that MIR-predicted protein content was able to accurately identify low- from high-quality colostrum samples regardless of the IgG threshold considered (50, 70, or 90 g/l). In parallel, we evaluated the discriminant ability of colostral refractive index (BRIX), whose performance were similar to those of MIR-predicted protein content. The area under the ROC curve was excellent, being 0.85 for MIR-predicted protein content and 0.83 for BRIX when IgG threshold was set at 50 g/l. The cut-off identified for MIR-predicted protein content was 13.08, 13.28, and 14.64% for IgG threshold at 50, 70, and 90 g/l, respectively. Findings do suggest that milk labs equipped with MIR devices may offer an indirect quality evaluation of bovine colostrum for screening purposes and to support farmers' decision making. Finally, our results are of interest for industries that use bovine colostrum as an ingredient.

Leukocytes and microRNA in colostrum following heat-treatment or freezing

T.L. Chandler, A. Newman, A.S. Sipka and S. Mann
Cornell University, Ithaca, NY, USA; tlc236@cornell.edu

Colostrum contains maternal leukocytes and microRNA (miRNA), potential regulators of immune function. These immunologically active components may be important for immune development in calves, but changes due to heat-treatment or freezing of colostrum have not been characterized in detail. Our objective was to compare miRNA abundance, leukocytes, and bacterial counts in colostrum after heat-treatment or freezing. Colostrum was harvested from cows (n=14) ≤8 h after calving, thoroughly mixed and immediately split into three aliquots. One aliquot each was either heat-treated (HT, 60 °C for 60 min), frozen (FR, -20 °C overnight), or cooled (COL, 4 °C). After HT, colostrum was rapidly cooled on ice and stored with COL at 4 °C overnight. The next day, colostrum was warmed to 40 °C before samples were processed further. Leukocytes were isolated by a series of centrifugation and wash steps. Cell viability was determined by trypan blue staining, and differential cell counts were performed using cytospin preparations. Micro-vesicles were isolated from whey by ultracentrifugation to isolate (on-column purification) and quantify miRNA using a fluorescent probe. Sample aliquots were submitted for bacterial counts (cfu) using standard diagnostic methods. Data were analysed by ANOVA (PROC MIXED, SAS v 9.4) with the fixed effect of treatment and random effect of cow. Cell viability averaged (SD) 60% (22%) in COL, but viable cells were not isolated from HT or FR. In COL, macrophages (52.8±6.5%) and neutrophils (30.0±6.5%) made up the greatest proportions of cells and lymphocytes (17.0±6.6%) were the lowest (P=0.002). Cell preservation in HT and FR cytospins was too poor to perform a reliable cell differential. Abundance of miRNA isolated from vesicles was decreased in HT (P=0.04) but increased in FR (P=0.02) compared to COL. Bacterial counts did not differ between COL and FR (LSM [95% CI] 5.0 [2.1 to 12.3] vs 5.3 [2.2 to 12.9] cfu/µl, P=0.77), but were decreased to 0 cfu in HT in 13 of 14 samples. We determined that heat-treating and freezing colostrum rendered colostral leukocytes unviable, altered the abundance of isolated miRNA, and that HT decreased bacterial contamination. The biological importance of observed changes in maternal leukocytes and miRNA abundance for the newborn calf warrant further investigation.

Performance of first parity Holstein, Danish Red and their crosses in an experimental setting

J.B. Clasen[1], H.M. Nielsen[1], L. Hein[2], K. Johansen[1], M. Vestergaard[3] and M. Kargo[1]
[1]Aarhus University, Center for Quantitative Genetics and Genomics, Blichers Allé, 8830 Tjele, Denmark, [2]SEGES, Agro Food Park, 8200 Aarhus N, Denmark, [3]Aarhus University, Animal Science, Blichers Allé, 8830 Tjele, Denmark; julie.clasen@qgg.au.dk

Crossbreeding between dairy cattle breeds is becoming popular in dairy production, calling for scientifically based comparison of breeds and their crosses for traits other than milk production. Crossbreeding experiments and breed comparisons conducted in controlled research environments are scarce. This study presents results from an on-station crossbreeding experiment in Denmark, including 16 first parity Holstein, 13 Danish Red and 16 F1 Danish Red × Holstein. The heifer calves were purchased at 4 weeks of age and was fed at either a low or medium feeding level as heifers. They began their first lactation between July and September 2021. Cows were housed in one group, fed the same total mixed ration (41% DM), and were otherwise handled equally. Data on daily milk yield, body weight, individual DM intake and pregnancy rate at first AI were analysed using a mixed model in SAS and compared between breeds and heifer feeding level. Preliminary results show that, in the period between 10 and 140 days in milk, daily DM intake was similar (P>0.05) between breed groups (17.2 to 18.0 kg DM/d), but cows fed at the low feeding level as heifers had a significantly higher (P<0.05) DM intake. There were significant differences (P<0.05) in milk yield between breed groups, but no effect of heifer feeding level. Crossbred cows produced 32.3 kg ECM per day, Holstein 29.7 kg, and Danish Red 26.2 kg. Danish Red cows tended to gain (1.5 kg/week) more than Holstein (0.5 kg/week) but significantly (P<0.05) more than crossbreds (-0.05 kg/week). Cows fed at a low feeding level as heifers gained significantly more than cows fed at a medium feeding level. The pregnancy rate at first AI was 66.7% for Danish Red, 53.8% for Holstein, and 57.1% for crossbreds. The conclusions so far are: (1) that given the same environment and feed intake, the two pure breeds may allocate energy intake differently for lactation and growth; and (2) the crossbred cows show a large and favourable heterosis for milk production.

Administration of eCG is not recommended in time-AI protocols for beef cattle in good body condition

A. Sanz[1], A. Noya[1], O. Escobedo†[1], M. Colazo[2], L. López De Armentia[1] and J.A. Rodríguez-Sánchez[3]
[1]CITA de Aragón – IA2 (Universidad de Zaragoza), Av Montañana 930, 50059 Zaragoza, Spain, [2]University of Alberta, T6G 2P5, Edmonton, Canada, [3]Terraiberica Desarrollos S.L., 44001, Teruel, Spain; asanz@aragon.es

The main objective of this study was to examine the effect of equine chorionic gonadotropin (eCG) administration on pregnancy per AI (P/AI) in Angus and Angus × Avileña cattle subjected to a 7-d ovulation synchronization protocol. On Day 0 (11:00), heifers (n=36, 460 kg, 3.7 BCS, 2.1 years old) and cows (n=256, 594 kg, 2.8 BCS, 4.8 years old) were randomly assigned to one of two intravaginal progesterone (P4) inserts (PRID 1.55 g, Ceva S.A., Spain or CIDR 1.38 g, Zoetis, Spain). On Day 7 (16:00), the P4 insert was removed, 500 µg of dinoprost (Enzaprost, Ceva S.A.) administered im and females were further subdivided to receive or not 500 IU of eCG im (Foligon, Merck Sharp & Dohme Animal Health, S.L., Spain). On Day 10 (11:00), all cattle received 100 µg of GnRH (Cystoreline, Ceva S.A.) and were inseminated by one technician with semen from three sires equally distributed among treatments. Ovarian ultrasonography was done in a subset of 47 cows at Day 0 and in all cattle 35 days after AI to determine cyclicity and pregnancy status, respectively. Data were analysed by Chi-square using FREQ and GLM procedures (SAS v 9.4). Three cows out of 47 cows were acyclic and remained non-pregnant, showing all of them a similar BCS (3.2). The overall P/AI was 51% (149/292). Cattle category, breed, P4 insert type or the inclusion of eCG did not affect P/AI (51 vs 52%, with and without eCG; P=0.890). Sire tended to affect P/AI (46, 45 and 60%; P=0.066). BCS did not differ between cattle that became pregnant or not. In summary, the inclusion of eCG did not affect P/AI, therefore, its use is not recommended in ovulation synchronization protocols for suckler cows and heifers in good body condition score, in order to reduce costs. Funded by Project FITE 2019 VACAFERTILTERUEL, Government of Aragón and FEDER.

Session 67 Poster 14

Challenges of the breeding systems of indigenous bovine populations in Algeria, Greece and Tunisia

D. Tsiokos[1], S. Boudalia[2], S. Ben Saïd[3], A. Mohamed-Brahmi[3], S. Smeti[4], A. Bousbia[2], Y. Gueroui[2], M. Anastasiadou[1] and G.K. Symeon[1]

[1]ELGO-DEMETER, Research Institute of Animal Science, Paralimni Giannitson, GR58100 Giannitsa, Greece, [2]8 May 1945 University, Guelma BP, 24000 Guelma, Algeria, [3]University of Jendouba, Boulifa, 7100 Le Kef, Tunisia, [4]INRA-Tunisia, Ariana, 2049 Tunis, Tunisia; gsymewn@yahoo.gr

The indigenous cattle populations represent an essential source for many rural communities in the Mediterranean countries. They combine unique qualities: valuable locally adapted genetic pool, substantial income to the local economies and added-value animal products. Nevertheless, their numbers are declining due to the preference of farmers towards foreign, more productive breeds. They also face continuous challenges such as fear of extinction, anarchic breeding schemes and harsh rearing conditions. The aim of this study was the identification of the socio-economic status of the breeders, the description of the breeding practices and the determination of the constraints and the proposition of solutions that could promote the sustainability of these breeding systems. The study involved data collection from different farms through the completion of a questionnaire designed for these purposes. A total number of 385 questionnaires was collected. In terms of socio-economic identification, the survey identified that the farmers are middle-aged, with high illiteracy rates and few possibilities of having a successor. The farm size is relatively small, especially in the African countries, and the breeding practices followed classify the production systems as extensive with low inputs. The major constraints faced are the increased feeding cost, the low productivity of the animals and the low selling prices of the products. On the other hand, higher selling prices and state funding are identified by the farmers as solutions to their problems, combined with products' advertisement and certification. The genetic improvement of the animals is moderately appreciated as a solution. Overall, the current study revealed that the indigenous cattle farming sector in Algeria, Tunisia and Greece faces several constraints but with the appropriate private and state interventions the flocks could be conserved and the production systems could be improved towards sustainability.

Session 68 Theatre 1

The adaptive capacities of small ruminants as a lever to design agroecological farming systems

F. Douhard[1], D. Hazard[1], R. Rupp[1], E. González-García[2], F. Stark[2] and A. Lurette[2]

[1]INRAE, UMR GENPHYSE, Castanet Tolosan, France, [2]INRAE, CIRAD, L'Institut Agro Montpellier SupAgro, Univ Montpellier, UMR SELMET, 34000 Montpellier, France; frederic.douhard@inrae.fr

As a sustainable food production may increasingly build on resilient, low-input farming systems, enhancing livestock reliance on natural processes is of growing interest. In this respect, small ruminants present various assets since they cover a broad diversity of breeds adapted to contrasting environments, including harsh and variable rangeland conditions where they evolved strong adaptive capacities. Yet, from an agroecological perspective, adaptive capacities remain poorly accounted for into current management and breeding strategies. Several traits reflecting aspects of resilience to undernutrition or to infections are amenable to selection and can thereby contribute to reduce the inputs needed for production or to an integrated health management. However, the consideration of adaptive capacities in breeding schemes generally targets conventional production systems and has only little scope for their re-design. Then, a more rigorous assessment of adaptive capacities would require to examine trade-offs among resilience traits, between resilience traits and production traits, during lifetime and across environments – which is virtually out of reach using classic experimental means. Finally, few studies have started to explore how improving individual adaptive capacities or managing their diversity within the herd can affect farm resilience, even though this aspect is key in agroecology. Given previous limitations, the use of systemic approaches integrating the mechanistic basis of adaptive capacities as well as their consequences on herd performances holds promise. Specifically, we illustrate at the animal level how resource allocation models could contribute to an integrative assessment of adaptive capacities, including their genetic trade-offs with production traits. In turn, the use of such individual models in herd simulation should allow identifying synergies between breeding and management strategies to promote herd resilience. Beyond, expanding this modelling at the farming system level may support the design of new agroecosystems, particularly those enhancing crop-livestock integration.

Session 68 Theatre 2

A model of energy allocation to predict adaptive capacities in meat sheep

F. Douhard[1], G. Dubreuil[1], S. Lobón[2], M. Joy[2], L. Puillet[3], N.C. Friggens[3] and A. Lurette[4]
[1]GenPhySE, Université de Toulouse, INRAE, ENVT, 31326 Castanet-Tolosan, France, France, [2]CITA, Avda. Montañana 930, 50059 Zaragoza, Spain, [3]UMR 0791 MoSAR, INRAE, AgroParisTech, Université Paris-Saclay, 75005 Paris, France, [4]INRAE, CIRAD, L'Institut Agro Montpellier SupAgro, Univ Montpellier, UMR SELMET, 34000 Montpellier, France; frederic.douhard@inrae.fr

The sustainability of Mediterranean livestock farming systems largely depends on sheep breeds capable of dealing with underfeeding. However, adaptive capacities may trade-off against production traits (e.g. litter size, lamb growth) due to the limited amount of energy to allocate between competing demands. How such trade-off can be underpinned by the changes in energy acquisition and allocation priorities during lifetime remains mostly unknown so far. In particular, different ewe body reserves dynamics (e.g. stable vs variable) over successive production cycles may reflect different adaptive strategies. Here, we propose an energy allocation model in meat sheep that predicts changes in body weight and body reserves over growth and reproduction in response to dietary energy availability. A critical assumption of our model is to consider an interdependency between energy acquisition and allocation. Specifically, we assumed a negative feedback from body reserves on the desired intake. We show that the strength of this feedback can differentiate adaptive strategies: a weak feedback is associated to a low ewe priority to maintain body reserves in favour of lamb growth ('risky strategy') whereas a high feedback associated with a high ewe priority to maintain body reserves penalizes lamb growth during underfeeding periods ('conservative strategy'). Finally, we assessed these various strategies by fitting our model to data from contrasting breeds from different environments: a prolific breed from an extensive pasture-based system (Romane, France), and two less prolific long-lived breeds from Spain, one from a semi-intensive system (Rasa Aragonesa), and the other from an extensive system in a mountain area (Churra Tensina). Our individual-based model shed lights on energy allocation mechanisms underlying adaptive capacities. It then provides a tool to explore how individual variability in feeding responses can be managed to improve farm resilience.

Session 68 Theatre 3

Dynamic interplay between reproduction, milk production and body reserves in Alpine goats

N. Gafsi[1,2], F. Bidan[1], B. Grimard[3], M. Legris[1], O. Martin[2] and L. Puillet[2]
[1]IDELE, MNE, 75595 Paris, France, [2]INRAE, AgroParisTech, UMR MoSAR, 75005 Paris, France, [3]ENVA, INRAE, UMR BREED, 78350 Jouy-en-Josas, France; nicolas.gafsi@inrae.fr

Variability in reproductive performance is a key aspect in dairy goat herds: it affects the distribution of physiological stages and therefore the way animals will respond to their environmental conditions (resource availability, thermal environment) both in terms of milk production (MP) and body reserves utilization. As a feedback, next reproductive cycle can be affected. Individual lifetime trajectory is thus dependent on the dynamic interplay between reproduction, MP and body reserves. Understanding this interplay is a crucial issue to optimize feeding and reproductive management strategies, under current farm environmental constraints but also under future constraints imposed by climate change impacts. The objective of this work was to evaluate the effects of MP and body reserves utilization on the success at artificial insemination (AI) in dairy goats. Routine data from an experimental station in South France (Le Pradel, French Livestock Institute) were used. The dataset included 574 Alpine goats (1,096 lactations from parity 1 to 9) over 25 years (1996 to 2021). The AI success of each goat was calculated using the interval between AI and next kidding (an interval of 160 days or less is considered as success at AI). A logistic regression model was used to analyse the relationships between AI success and individual factors (parity, stage of lactation at AI, failure at previous AI), thermal conditions (maximum temperature and THI around AI), MP (maximum MP during lactation, MP around AI, somatic cell count around AI, variation of MP 6 weeks before AI) and body reserves (body weight (BW) and body condition score (BCS) around AI, variation of BW and BCS 12 weeks before AI). Average AI success was 69% (±11%). AI success was significantly affected by MP around AI, lumbar BCS around AI, lumbar BCS dynamic before AI and maximum temperature. In contrast, classical factors of variation in AI success such as lactation rank, previous AI success or THI had no effect. These results will contribute to identify animals that best cope with environmental constraints and better manage animals at risk of reproductive failure.

Consequences of contrasting feed efficiency as lamb on later ewe performance

I. De Barbieri, G. Ferreira, Z. Ramos, E.A. Navajas and G. Ciappesoni
Instituto Nacional de Investigación Agropecuaria, Ruta 5 km 386, 45000, Uruguay; idebarbieri@inia.org.uy

The aim of this study was to evaluate the productive and reproductive performance of ewes with contrasting residual feed intake (RFI) measured as lamb in their first year of life (in the frame of H2020 project SMARTER n°772787). Two hundred and sixty-one Merino ewes from two cohorts (2018-2019), which belong to a Merino genetic information nucleus, were studied. Residual feed intake of each cohort was evaluated before the first shearing (9 to 13 months old). Then, as hoggets, they were first mated at 17 months of age, shorn prepartum and lambed in Spring (August-September). Ewe body condition score (CS) and body weight (BW) were measured at mating, shearing, prepartum and weaning and ewe wool fibre diameter and fleece weight at prepartum shearing. At weaning, number of lambs and lamb body weight weaned per ewe were recorded. Ewes were managed as a single flock, except at mating (4 weeks) and in accordance with the number of foetuses scanned (pregnancy rank, 0, 1, 2) from late pregnancy to weaning (4 months). The effect of lamb RFI group (25%low, 50%medium, 25%high) on ewe BW, CS, fleece weight, wool fibre diameter and weaned lambs (kg of weaned lambs/mated or lambed ewe) was analysed by a generalized linear model including RFI group, year, and pregnancy rank for those traits recorded after scanning as fixed effects The same model but with binomial distribution was utilised to evaluate the effect of RFI on ewe fertility and with poison distribution in the cases of prolificacy (0, 1, 2) and lambing percentage. Non-significant (P>0.05) differences were found in fertility, prolificacy, and lambing percentages among ewes of different RFI groups. Ewe BW and CS evaluated at different times, and fleece weight and wool fibre diameter production were not affected by RFI group (P>0.05). Finally, weaned lambs (kg of weaned lambs/mated or lambed ewe) showed no difference between RFI groups, ranging from 19.1 to 20.8 kg and from 28.4 to 30.6 kg for mated or lambed ewe, respectively. These preliminary results indicate that ewe performance in extensive grazing systems would not be negatively affected by producing with more efficient sheep (lower RFI). Larger databases and more years of evaluation are needed to confirm these first results.

Behavioural adaptation to rangeland and performances of lambs differing in experience and genetics

C. Ginane[1], N. Aigueperse[1], S. Parisot[2], C. Durand[2], A. Guittard[1], F. Tortereau[3], X. Boivin[1] and M.M. Mialon[1]
[1]INRAE, UMR1213, 63122 Saint-Genès-Champanelle, France, [2]INRAE, UE0321, 12250 Roquefort-sur-Soulzon, France, [3]INRAE, UMR1388, 31326 Auzeville Tolosane, France; cecile.ginane@inrae.fr

The use of rangelands for domestic livestock implies having resilient and efficient animals, able to adapt to environmental constraints of these generally low-producing environments. We studied the roles of early rearing mode and genetic lineage on this adaptation in lambs. Eighty Romane ewe lambs of divergent lines for feed efficiency (EFF-, EFF+) were reared from birth to 3 months, either indoors with artificial milking (AR) or in the rangeland with their mothers (MR). Then, they were grouped by 20 (separated for experience and mixed for genetic line) and placed in 4 experimental rangeland plots for 3.5 months. For the first 1.5 days after entry into the plots, the behavioural activity of 12 lambs per group (6/genetic line) was recorded by scan sampling on 3 sessions of 2.5h. The AR lambs spent 34% of the scans exploring the environment on morning 1 (vs 1% for MR lambs), but only 5% on morning 2. Rumination increased through sessions in AR and MR lambs and more strongly in EFF- than EFF+ lambs. Feeding represented 36% of scans in AR lambs vs 73% in MR ones on morning 1, similarly to morning 2. After a similar growth for AR and MR lambs from D0 to D90 (average daily gain: 240 g/d), the entry into the plots made ADG strongly decrease, particularly for AR lambs (D90-D120: 46 vs 124 g/d for MR ones). Beyond this age, lambs' growth was similar, maintaining the weight difference between AR and MR lambs. The genetic lineage had no effect excepted a trend for a greater ADG in EFF+ lambs between D90 and D200 (96 vs 86 g/d). Rearing lambs after weaning in rangeland conditions appears to be a major and a greater challenge for lambs initially reared indoors compared to those raised with their mothers in a similar environment. A reduced ability to find feed and shelter is probably involved. Nevertheless, the similar growth after one month in these conditons indicates a rapid improvement of adaptation in animals subjected to a totally new environment. This work has received funding from the EU Horizon 2020 research and innovation programme under grant agreement n°772787 (SMARTER).

Session 68 Theatre 6

Adaptation of Barki desert sheep to different feed shortage periods

F. Abou-Ammou, M. El-Shafie, A. Aboul-Naga, E. El-Wakel, A. Saber, T. Abdelsabour and T. Abdelkhalek
Animal Production Research Institute, Sheep and goat Department, Dokki, Cairo, 12611, Egypt; adelmaboulnaga@gmail.com

Barki desert sheep are facing periods of feed shortage (FS), which may extend to 3-4 months in the summer season. The trial was carried out at Borg Arab Farm, to determine effect of feed shortage periods (FS) on the performance of Barki sheep. The experimental work includes three trials: 1- short FS for one week(SFS), 2-medium FS for one month (MFS), and 3-long FS for 3 months (LFS). In the short feed shortage (SFS),10 control lactating ewes were fed individually on concentrate feed mixture and corn silage(C), where 20 ewes have 0 concentrate diet for 1 week (T). Short FS reduced significantly ($P \leq 0.01$) total milk yield (TMY) for 10 weeks of lactation (57.6 Vs 78.6 kg) and reduced total milk fat and protein by 24 and 32%, respectively($P \leq 0.05$). For the MFS trial, 10 lactating ewes were fed individually on concentrates and Alfalfa hay, while 20 treated ewes fed Alfalfa hay only for four weeks. Medium FS reduced TMY for 10 weeks of lactation in by 17%, milk fat yield by 22% and milk protein yield by 16.3%. In long FS trial, 45 ewe lambs of 2 months of age were fed concentrates and Alfalfa hay for 12 weeks. Treated group (30) fed for 12 weeks on 50% concentrates and 12 weeks on compensatory diet of 125% concentrates. The animals were return to normal feeding thereafter. LFS decrease daily gain of ewe lambs by 44.7 gm ($P \leq 0.05$) and increased during the compensatory period by 40 gm daily. Yearling weight of T group decreased insignificantly by 1.1 kg than control group. Differences in reproductive performance were insignificant except for weight at first oestrus which decreased by 1.9 kg ($P \leq 0.05$). Long FS are practiced further on lactated Barki ewes for 90 days; the trial is still in progress. The three trials confirm that desert Barki sheep are adapted for short-, medium- and long- periods feed shortage, and are showed compensatory performance thereafter.

Session 68 Theatre 7

Short-term adaptive capacity of dairy goats to a two-days nutritional challenge

M. Gindri, N. Friggens and L. Puillet
Université Paris Saclay, INRAE, AgroParisTech, 75005 Paris, France; marcelo.gindri@inrae.fr

Adaptive capacity, in the short or long term, in dairy goats, is an important trait for sustainable livestock production, given the increasingly variable environments due to climate change. Good adaptive capacity should favour productive lifespan and in turn long-term efficiency. The objective of this study was to evaluate milk yield (MY) and plasma glucose responses of first lactation goats to a two-day nutritional challenge. The goats were daughters of Alpine bucks divergently selected for longevity (LGV+ and LGV-; 190 days in lifespan difference). They were fed with one of two rearing diets (RD), differing in energy density, from weaning to mid-gestation. The experimental challenge started at 29±2.5 days in milk (DIM). At 33±2.5 DIM, all goats switched from a normal lactation diet into a straw-only diet for two days. Individual MY (kg/d) and plasma glucose (g/l) were recorded daily for 17 days (5 days before challenge, 2 days of challenge, and 10 days after the challenge). The time-series data was split into four-time variables representing the phases pre-challenge, rate of response to challenge, and linear and quadratic rates of recovery from challenge and used in a mixed model considering LGV and RD and the 4 time-variables as fixed effects. Animal was included as a random effect on all 4 of the time-variables. LGV affected only MY rate of recovery from challenge ($P \leq 0.046$) while RD did not affect any trait ($P \geq 0.17$). LGV- goats presented quicker and sharper recovery from challenge than goats from the longer longevity line. However, this LGV effect on MY linear rate of recovery from challenge was no longer significant when plasma glucose was used as a covariate ($P = 0.096$). These results suggest goats from the LGV- line drive glucose metabolism towards quick MY recovery after a two-day nutritional challenge in early lactation while LGV+ goats drive glucose metabolism towards other functions rather than quick MY recovery. Diet during early life seemed not to affect the goat's capacity of rebounding MY after short-term nutritional challenge in the following early lactation.

Session 68 Theatre 8

Performance of Scottish Blackface lambs post-weaning when offered a forage based diet

M.A. Dolan[1,2], T.M. Boland[1], N.A. Claffey[2] and F.P. Campion[2]

[1]University College Dublin, School of Agriculture and Food Science, Belfield, Dublin 4, Dublin 4, Ireland, [2]Teagasc, Animal & Bioscience Department, Mellows Campus, Athenry, Co. Galway, H65 R718, Ireland; mark.dolan@teagasc.ie

Hill and mountain farming systems in parts of Europe rely on the ability to sell lambs to specialized lamb finishers after weaning. These finishing systems require further development, particularly forage based finishing systems. The objective of this experiment was to examine the performance of Scottish Blackface (SB) and Texel cross Scottish blackface (TXSB) lambs when grazed *in-situ* on one of three brassica crops; forage rape (Brassica napus L. FR) hybrid brassica (Brassica napus L. HB) and kale (Brassica oleracea L. K). The experiment was a 3×2×2 factorial design blocked by live weight (LW) and balanced for breed and sex with 50 entire male and 25 castrate male lambs randomly allocated to one of the three brassica crops on two separate years. The mean crop yield across the two years for FR, HB and K was 6,933, 6,345 and 6,701 kg DM/ha, respectively. Lambs were acclimatized to the experimental diet for 10 days. Once lambs had acclimatized to the diet they were given *ad libitum* access to barley straw while grazing the brassica crops. Lamb LW was recorded at the start of the experiment and every fourteen days on all lambs up to slaughter or until supply of the crop was exhausted. Lambs grazing FR and HB had a higher average daily gain compared to K (142, 135 and 123 g/day±3.91; P<0.05). Lambs were drafted for slaughter at 40 kg LW and above with adequate fat cover. The percentage of lambs slaughtered from FR treatment was 49% followed by HB 44% and K 33% which was linked to crop yield. Mean pre-slaughter LW did not differ between treatment (44.1, 43.8, 43.8 kg±0.47; P>0.05). There was no difference in carcass weights between treatments (19.1, 18.7 and 18.3 kg±0.40; P>0.05) while forage type had no effect on kill out percentage (43.3, 42.4 and 41.9±0.81%; P>0.05) Results from this study show that lamb performance to slaughter is effected by the type of brassica crop offered but there was no difference post-slaughter performance between treatments. Further investigation is warranted into the effect of crop yield and lamb live weight at the start of the finishing period.

Session 68 Theatre 9

Effect of a methionine supplementation on the resilience of goats infected by *H. contortus*

L. Montout, H. Archimède, N. Minatchy, D. Feuillet, Y. Félicité and J.-C. Bambou

INRAE, GA (Animal Genetic), INRAE Antilles-Guyane, domaine de Duclos, 97170 Petit-Bourg, Guadeloupe, France; laura.montout@inrae.fr

Gastrointestinal nematode (GIN) infections are considered as one of the major constraints in small ruminant production at pasture. The negative impact of the drugs classically used to control these infections, on soil biodiversity coupled with concerns over the presence of residues in animal products led some states in the world to advocate a significant reduction of the use of chemical molecules in animal production. The nutritional manipulation of small ruminants has long been considered as a tool for the control of GIN infections. The aim of this study was to measure the impact of a rumen-protected methionine supplementation on the response of Creole goat kids experimentally infected with *H. contortus*. The animals were divided in three groups and infested by 10,000 larvae of *H. contortus*: a control group with no methionine supplementation (C: basal diet), a low methionine group (LM: basal diet + 4.5 g/head of rumen-protected methionine) and a high methionine group (HM: basal diet + 10 g/head rumen-protected of methionine). Parasitological, haematological and immunological data were collected each week during 42 day of experiment. The results showed that the level of parasitism, measured through the faecal eggs counts was significantly lower in the HM group. The first immunological and haematological parameters available by now, are correlated with the parasitological ones. Further analysis under course (RNAseq) will provide more information on the underlying mechanisms. In conclusion, the supplementation with rumen-protected methionine appear as a promising strategy for GIN control in goats.

Session 68 Theatre 10

Contribution of grazing areas to small ruminant production systems in the Mediterranean countries

S. Lobón[1], A. Aboul Naga[2], A. Mohamed-Brahmi[3], A. Tenza-Peral[1], E. Salah[2], M. Ameur[3], M. Joy[1], M. Nasraoui[4], T. Abdelsabour[2] and I. Casasús[1]

[1]CITA-IA2, Avd. Montañana 930, 50059, Spain, [2]APRI, Dokki, 12611 Cairo, Egypt, [3]Université de Jendouba, Boulifa, 7100, Tunisia, [4]INRAT, Ariana, 2049, Tunisia; slobon@cita-aragon.es

Climate change threatens livestock production systems because of its potential negative impacts on resources availability (e.g. pasture, supplements, water), which are expected to be important in Mediterranean countries. In order to foresee these potential impacts, it is necessary first to analyse the use of these resources in the different farming systems. We used a questionnaire to characterize land use, general flock management, feeding strategies and the products sold in small ruminant farming systems in Spain, Egypt and Tunisia (n=163 farms, 25 in Spain, 47 in Egypt, 90 in Tunisia). On average, Spanish farms had 216 ha utilised agricultural area (UAA) and 2,843 ha common lands, 228 livestock units (LU), 2 LU/ha UAA, 93 LU/work unit (WU), 1.55 prolificacy (n lambs/lambing) and 1.41 productivity (lambs sold/ewe/year). Egypt farms had 10 ha UAA and 51 ha common lands, 18 LU, 5 LU/ha UAA, 6 LU/WU, 1.34 prolificacy and 0.49 productivity. Tunisian farms had 82 ha UAA plus 86 ha common lands, 20 LU, 1 LU/ha UAA, 9 LU/WU, 1.17 prolificacy and 0.81 productivity. Flocks fed exclusively on grazed pastures during 8.1, 7.2 and 9.4 months in Spain, Egypt and Tunisia, respectively. The percentage of flocks that grazed on rangelands in each country was 81%, 31% and 92%, on forage crops it was 72, 81 and 64%, and on stubble 98, 55 and 68% from Spain, Egypt and Tunisia; respectively. During the rest of the year, flocks were supplemented with sources of energy (concentrate/cereals) and/or fibre (hay/straw/forages). Energy supplementation prevailed in Egypt (84%) and Tunisia (62%), followed by the combination of energy and fibre (13% and 36%), whereas in Spain the main type of supplementation was the combination of energy and fibre (52%) following by only energy (40%). Data obtained here will be fed into a computational model to identify strategies for strengthening the resilience and enhancing the efficiency of small ruminant farming systems in the Mediterranean basin.

Session 68 Theatre 11

Modelling as a tool to explore adaptation of Mediterranean sheep farming systems to climate change

A. Lurette[1], F. Douhard[2], L. Puillet[3], A. Madrid[4], M. Curtil[1,4] and F. Stark[1]

[1]UMR SELMET, Université de Montpellier, INRAE, CIRAD, L'Institut Agro – Montpellier, Montpellier, France, [2]UMR GenPhySE, Université de Toulouse, INRAE, ENVT, Castanet-Tolosan, France, [3]UMR MoSAR, Université Paris-Saclay, INRAE, AgroParisTech, Paris, France, [4]Idele, Service fourrages et pastoralisme, Castanet-Tolosan, France; amandine.lurette@inrae.fr

Mediterranean pastoral farming systems are increasingly subject to strong climatic constraints, which impact their access to grazing resources. To develop livestock farming systems adapted to climate change, combining resilient herds and an efficient use of various feed resources is central. Different combinations can be explored by modelling the impacts of climate change on feed resources and adaptation levers at the different levels of the farm organization (animal-herd-livestock system) however, this is methodologically challenging. This study aims at developing a simulation tool to represent, from animal to farm components, the multi-level implications of adaptation levers that can be mobilized by Mediterranean small ruminant farmers. These levers can be related to animal biology and/or management strategies. The simulation tool enables to evaluate relative and combined effects of levers on farm resilience and efficiency. It was developed with GAMA, an agent-based computer language, to allows the representation of each individual components (animals and areas) of the farming system. The simulator was calibrated on two contrasting pastoral sheep systems in the South of France: one grazing system uniquely based on rangelands and one complemented system with both rangelands and forage production. For these two contrasting situations, we tested the effects of three levers: (1) increasing the part of pastoral surfaces; (2) shifting the grazing periods; and (3) decreasing the flock size to better match with resources availability. The simulator was able to mimic the functioning of livestock farming systems and to evaluate for each situation the impact of adaptation levers on farm efficiency and resilience. Based on this prototype, other situations could be simulated according to climate change scenarios in the Mediterranean area and adaptation levers could be explored to address specifically these challenges at the farming system level.

Farmer views of best climate change adaptation strategies for sheep farming in the Mediterranean

D. Martín-Collado[1], S. Lobón[1], M. Joy[1], I. Casasús[1], A. Mohamed-Brahmi[2], Y. Yagoubi[2], F. Stark[3], A. Laurette[3], A. Abuoul Naga[4], E. Salah[4] and A. Tenza-Peral[1]
[1]CITA-IA2, Montañana, Zaragoza, Spain, [2]ESAK-INRAT, Boulifa le Kef, Ariana, Tunisia, [3]INRAE-UMR SELMET, L'Institut Agro, Montpellier, France, [4]APRI, Dokki, Cairo, Egypt; dmartin@cita-aragon.es

Climate Change (CC) impacts on agriculture will exacerbate in the next decades by a decrease of rainfall and increase of temperatures and extreme events. Pasture-based sheep farming systems in the Mediterranean basin are particularly vulnerable as they depend on local feed resources, which are directly damaged by CC. Research has been on livestock adaptation strategies, usually follows top-down approaches not accounting for the experiential knowledge of farmers, limiting its applicability at farm level. This work aimed to analyse farmers views about CC impact and the effect of periodic events of food shortage (FS) and heat stress (HS) on animal and farm performance, to determine the usual practices deployed to overcome these periodic events and the best herd and farm management strategies to adapt to CC. Data were collected through face-to-face surveys to 228 farmers in Egypt (n=47), France (34), Spain (45) and Tunisia (101), covering representative sheep farming systems in 5 different Köppen climatic regions. CC scenarios were developed for each region based on the IPCC projections. All farmers recognize that CC is real, however, most think that it is a natural and human-induced process or even just a natural process (Egypt). Perceived risk is lower in the most extreme arid and semi-arid regions and in irrigated farms. Anyhow FS and HS are generally perceived as increasingly important problems. Farms traditionally overcome periodic FS events mainly providing animals with food stuff produced on farm and kept for shortage periods, increasing purchased feed and/or modifying grazing schedules. 90, 75 and 50% of French, Tunisian, Spanish farmers respectively reported recent farm changes to adapt to CC. Modification of grazing, lambing and sowing periods (France and Spain), breed substitutions (Tunisia), improvements in farm buildings (all) were reported. Farmers identified three potential strategies named feed maximization, feed supplementation and herd downsizing, which feasibility varies across systems, countries and climatic regions.

Increasing resilience of small ruminants farming systems: 3 management strategies across countries

J. Quénon and V. Thénard
INRAE, AGIR, 24 chemin de Borde Rouge, Auzeville CS 52627, 31326 Castanet Tolosan Cedex, France; julien.quenon@inrae.fr

The sustainability of small ruminant livestock is one of the pillars of socio-economic sustainability and ecosystem services in many rural communities around the world. It is to help redefine these selection criteria that the European H2020 project SMARTER was established. The objective of this study was to analyse the genetic management practices of small ruminant breeders and the socio-technical elements that condition them. It is based on on-farm surveys conducted in five countries (France, Spain, Italy, Greece, Uruguay) among 208 breeders of 13 sheep and goat breeds. The information was collected in three sections. The first section dealt with general elements of structure and management of the system and the flock. The second section focused on genetic management practices: criteria for culling and replacement of females, selection criteria for males, use of EBV's and synthetic indexes, preferences for indexing new traits to increase the resilience of their system. The third section aimed to collect socio-technical information. We used a data abstraction method to standardize the representation of these data. A mixed data factor analysis (MDFA) followed by a hierarchical ascending classification allowed the characterization of three profiles of genetic management: (1) A profile of breeders (n=62) of small herds, with little knowledge or use of genetic selection tools (index, AI, performance testing). They do not feel that they need new traits to improve the sustainability of their system. (2) A profile of breeders (n=77) of multi-breed herds in very grazing systems. They are familiar with genetic tools: they currently use AI little and would like the indexes to include more traits related to health or robustness to make their system more resilient. (3) A profile of breeder-breeders (n=69) of large herds, with demanding culling practices. These breeders are satisfied with the indexes as they are to ensure the resilience of their system. These results are elements that can be used by selection organizations and companies to support their reflection on the evolution of selection objectives to increase the resilience of small ruminant breeding systems.

Session 68 Poster 14

Citrullination gene related detections by runs of homozigosity and resilience in sheep

L. Menegatto[1], L. Freitas[1], K. Costa[2], N. Stafuzza[2] and C. Paz[2]
[1]University of São Paulo, Department of Genetics, Ribeirão Preto Medical School, Bandeirantes Avenue, 3900, 14048900, Brazil, [2]Animal Science Institute, Carlos Tonani Higway, km 90, 14174000, Brazil; leonardomenegatto@gmail.com

Citrullination is a common process to occur by antibodies and many studies reported important genes like *PADI4*, which is particularly involved in citrullination catabolism to initiate extracellular neutrophil traps. We hypothesized that this process could present relationship with other genes besides only the resistance to hematophagous gastrointestinal nematodes, but also with resilience to these warms, considering that resilience trait is defined as a situation with high infection but without animal anaemia. We genotyped 497 Santa Inês sheep with Ovine SNP50 Genotyping BeadChip and we had clustered resilient animals with BLUPF90 by faecal egg count, corpuscular volume, eye colour chart, and total plasma protein parameters with repeated measurements over time, and pedigree information. After quality control with PreGSF90, we employed the runs homozigosity (ROHs) detection with PLINK and detectRUNS R package, and then we identified genes and its description by NCBI, considering shared results from *PADI3*, *PADI4* and *PADI6* genes with other important haplotypes. We detected 17 ROHs, with 2,861,053 b in average, identifying 14 regions which contains 319 genes, including a bitter taste detection (*TAS*) gene, whose process was associated with diversified diet, toxin neutralization, and immune system, beyond its description as wild gene introgressions which confer intoxication protection in European sheep. It was also identified thyroid oxidase gene (*TPO*), which action with innate system and neutrophils, macrophages, NK and dendritic cells; apolipoproteins (*APO*) genes, involved in cholesterol biosynthesis and its relationship with inflammation and innate immune system; and RNF186 gene, which is associated with intestinal permeability and inflammation. The results demonstrates that some haplotypes are potentially conserved since domestication or European past evolution, as an adaptation to new environments sheep colonization, and they are possible related with other resilient functions in a coordinated immunological response.

Session 68 Poster 15

Adaptive capacity of Mediterranean sheep over morphological traits at different physiological status

N. Atti[1], S. Lobón[2], A. Abounagua[3], D. Hazard[4], Y. Yagoubi[1], M. Joy[2], E. Salah[3], F. Douhard[4], S. Smeti[1] and E. Gonzalez-Garcia[4]
[1]INRAT, rue Hedi Karray Ariana, 2083, Tunisia, [2]CITA, Montañana, Zaragoza, 930, Spain, [3]APRI, Dokki, Giza, 72934, Egypt, [4]INRAE, SupAgro Montpellier, Place Pierre Viala, 34060, France; slobon@cita-aragon.es

Small ruminants represent a major livestock resource throughout the Mediterranean area. However, most of them are raised under low nutritional quantitative and qualitative inputs. In such conditions, the use of animal's body reserves (BR) is inevitable to cover their energy requirements. In this context, we studied, in the Adapt-Herd PRIMA Project, the productive performance of some Mediterranean sheep breeds from France, Egypt, Spain, and Tunisia. For this, and during two years, the body weight (BW) and body condition score (BCS) of ewes were recorded at different physiological stages (mating, middle and late pregnancy, lambing, during suckling and lamb's weaning). The intra and inter flock's variabilities in BW and BCS and related reproductive and productive parameters were recorded. A direct relationship between reproduction performance and ewes' BW, BCS was confirmed in Egyptian, French and Tunisian breeds. During pregnancy all flocks decreased the BCS, showing BR mobilization since mid-pregnancy. The decrease was more noticeable for Tunisian and Egyptian ewes, to face shortage nutrition during pregnancy which occurred in summer, and French Romane breed when ewes are reared on Mediterranean rangeland. All ewes mobilized their BR from late pregnancy to the end of suckling periods, to cover the lamb growth's requirements. However, when the lambs were weaned and/or their nutritional condition was improved, ewes were able to restore BRs, reaching higher BW and BCS. The lambs' BW at birth was affected by the prolificacy (single > twins > triple litters) for all breeds. During suckling, the growth rate was similar regardless the birth mode, which may be partially explained by high nutritional supply for the breeds with high prolificacy and the low prolificacy for other breeds. In conclusion, we confirm the essential role of BR mobilization and accretion in the expression of adaptive capacities in the current context of climate change in the Mediterranean region.

Genetic and nongenetic variation of fertility parameters in Comisana and Massese sheep

V. Floridia[1], G. Buonaiuto[2], L.M.E. Mammi[2], S. Grande[3], S. Biffani[4] and G. Visentin[2]
[1]University of Messina, DIMEVET, Messina, 98168 Messina, Italy, [2]University of Bologna, DIMEVET, Via Tolara di Sopra 50, 40064 Ozzano dell'Emilia (BO), Italy, [3]ASSONAPA, Via XXV Maggio 45, 00187 Roma, Italy, [4]Consiglio Nazionale delle Ricerche, IBBA, Via Bassini 15, 20133 Milano, Italy; giulio.visentin@unibo.it

Dairy sheep farming relies on the capacity to maximize milk yield and solids output concurrently with excellent fertility parameters. However, productive and reproductive performances are known to be antagonistically correlated from a genetic point of view, suggesting that selecting only for yield output can potentially decrease the genetic merit for fertility over time. Therefore, the aim of the present study was to identify sources of phenotypic variation and to estimate genetic parameters for fertility traits in two Italian dairy sheep. Data were collected between 2015 and 2020 in the ASSONAPA nucleus herd of Comisana and Massese breeds located in Asciano (Siena, Italy). The following phenotypes were investigated: the number of total born lambs, born-alive lambs, stillborn lambs, weaned lambs as well as stillbirth (stillborn lambs/total born), and survival (weaned lambs/total born). Data were analysed separately by breed and within breed by physiological status (primiparous and multiparous ewes). Fixed effects included in the linear mixed model were age and year at first lambing (only primiparous), lambing order and year of lambing (multiparous). Season of lambing fixed effect was included only for Massese (lamb) ewes. Additive genetic, permanent environment (for ewes), and the residual were fitted as random terms. Inbreeding coefficients were computed for each individual from pedigree data. Heritability estimates in lamb ewes ranged from 0.03 (Massese stillborn lambs) to 0.06 (Comisana born-alive lams). Heritability of ewe fertility traits was greater, varying from 0.07 (Massese weaned lambs) to 0.15 (Comisana total born lambs). Inbreeding coefficients were below 0.10 in almost all individuals for both breeds. Although the magnitude was not very large, genetic variation does exist for almost all fertility traits. These results suggest the inclusion of reproductive performance in an aggregate selection index for Comisana and Massese breeds.

Differential runs of homozigosity between resistant and resilient sheep in tropical regions

L. Menegatto[1], L. Freitas[1], K. Costa[2], N. Stafuzza[2] and C. Paz[2]
[1]University of São Paulo, Department of Genetics, Ribeirão Preto Medical School, Bandeirantes Avenue, 3900, 14048900, Brazil, [2]Animal Science Institute, Carlos Tonani Higway, km 90, 14174000, Brazil; leonardomenegatto@gmail.com

Infections by hematophagous gastrointestinal nematodes are a current obstacle to the sheep production, especially in tropical regions. Runs of homozigosity (ROHs) are an important method to detected conserved genomic regions, frequently enrichment with important genes. Our aim was to identify ROHs between resistant and resilient sheep and better understand the genetic patterns of these traits. We have clustered 497 Santa Inês sheep with BLUPF90 by faecal egg count, corpuscular volume, eye colour chart, total plasma protein with repeated measurements over time, and pedigree information in resistant, resilient and susceptible groups, which the first had low infection, while the second had high infection, but low anaemia. We have genotyped all this animals with Ovine SNP50 Genotyping BeadChip and we employed the ROHs detection with PLINK and detectRUNS R package after quality control with PreGSF90. Considering the differential haplotypes among groups, we identified genes and its ontologies by NCBI and Gorilla software, respectively. It was found 681 haplotypes, including 12 exclusive regions in resistant cluster and 14 exclusive regions in resilient cluster. We found 723 described coding-genes in differential regions, whose 233 were grouped to resistance and 319 to resilience clusters. The first group was enrichment to cytokines, leucin reach repet, and interferon associated genes, related to inflammation response and coagulation. The second group was enrichment to *TPO* and TH genes, which acts in phagocytosis function, and *MAST*, *TRIM*, and *RFN* genes, which are related to inflammation, permeability, immunoglobulin E and mucin production. Simultaneously, in both clusters it was identified 12 shared haplotypes, containing 171 genes, including inflammatory response associations, like interleukin and cyclin dependent kinase, besides globulin pathways and Zn fingers domain. In general, the resilience acts from broader biochemical functions, chronic response and homeostasis and erythrocyte association, while resistance presented functional enrichment to innate immunological system.

Session 68 — Poster 18

Genome-wide association study for gastrointestinal nematode parasite resistance in Santa Inês sheep

M.B. Mioto[1], N.B. Stafuzza[1], A.C. Freitas[1], R.L.D. Costa[2] and C.C.P. Paz[1,3]
[1]Animal Science Institute, Beef Cattle Research Center, Rod. Carlos Tonani, km94, 14174-000, Brazil, [2]Animal Science Institute, Diversified Animal Production Research Center, R. Heitor Penteado, 56, 13380-011, Brazil, [3]São Paulo University (USP), Av. Bandeirantes, 3900, 14040-900, Brazil; claudiacristinaparopaz@gmail.com

Gastrointestinal nematode infection is the most important cause of economic loss in sheep production. The inappropriate use of anthelmintics is increasing the resistance of gastrointestinal parasites, enforcing the search for sustainable alternatives of control. The aim of this study was to identify genomic regions associated with resistance to gastrointestinal endoparasites traits, such as, egg per gram of faeces (EPGlog), packed cell volume (PCV), degree of anaemia assessed by Famacha© score (FAM) and total plasma protein (TPP). The phenotypic data included 6,564 records collected from 1,725 Santa Inês animals. A total of 638 animals were genotyped with the Ovine SNP50 Genotyping BeadChip (Illumina, Inc.). Markers with unknown genomic position, located on sex chromosomes, with minor allele frequency <0.05, and call rate <90% were excluded. The variance components were estimated using a single animal trait model by Bayesian inference. For all studied traits, the fixed effects considered in the model were contemporary groups (farm, year and season of phenotype collection), sex and aged at the phenotype collection. The heritabilities estimated for EPGlog, PCV, FAM and TPP was 0.09, 0.27, 0.20, and 0.32 for EPGlog, PCV, FAM, and TPP, respectively. The single step GWAS identified 17, 19, 21 and 22 genomic regions explaining more than 1% of the additive genetic variance for EPGlog (57.9%), PCV (41.6%), FAM (32.9%) and TPP (61.1%), respectively. Those regions harbour important genes related with immune system functions, which could potentially benefit sheep breeding programs. Financial support: São Paulo Research Foundation (FAPESP) grant 2016/14522-7 and scholarship 2020/15760-4. C.C.P. Paz was the recipient of a productivity research fellowship from CNPq.

Session 68 — Poster 19

Resilience to underfeeding in dairy ewes diverging in feed efficiency: rumen fermentation

G. Hervás, P.G. Toral, A. Della Badia, A.G. Mendoza and P. Frutos
Instituto de Ganadería de Montaña (CSIC-Universidad de León), Finca Marzanas, 24346 Grulleros, Spain; a.dellabadia@csic.es

Sheep are traditionally well adapted to less-favoured environments. However, selection of high-performance breeds may have impaired their intrinsic resistance and resilience, which may compromise the response of dairy flocks to future challenges. Re-focusing breeding towards improved feed efficiency (FE) is expected, but no information seems to be available about the relationship between FE and resilience in dairy ewes. Because recent data suggest a key role of rumen function in FE in the ovine, a trial was conducted to compare ruminal fermentation patterns in lactating sheep phenotypically divergent for FE and subjected to a severe nutritional challenge. Daily intake and performance were recorded for 3 weeks in 40 Assaf ewes to estimate their FE (calculated as the difference between the actual and predicted intake estimated through net energy requirements for maintenance, production and weight change). Then, the feeding of the highest (H-FE) and the lowest (L-FE) feed efficiency ewes (n=10/group) was restricted only to straw for 3 days. Ruminal fluid was collected before, during and 10 days after the challenge, using a stomach tube, to examine fermentation characteristics. Concentrations of total volatile fatty acid (VFA) were lower in H-FE than L-FE, but no difference among groups were found for pH, ammonia concentration or molar proportions of VFA. As expected, the challenge increased rumen pH, and decreased ammonia and total VFA concentrations, but initial values were reached in the recovery period. Concerning molar proportions, propionate remained unaffected by undernourishment, butyrate was reduced, and acetate and minor VFA were increased. These proportions did not recover their initial value on day 10 after the challenge. In any event, no differences in temporal patterns of variation in ruminal fermentation were detected between H-FE and L-FE, which seems consistent with the similar impact of and recovery from the challenge in terms of milk production and composition. Overall, results suggest a similar resilience in sheep divergent for FE. Acknowledgements: project PID2020-113441RB-I00, MCIN/AEI; grant PRE2018-086174, MCIU/AEI/FSE, EU.

Session 69 Theatre 1

Chicken immune cell assay to model adaptive immune responses and its application in feed science

S. Kreuzer-Redmer[1], F. Larsberg[2], D. Hesse[2], G. Loh[3] and G.A. Brockmann[2]

[1]University of Veterinary Medicine Vienna, Nutrigenomics, Veterinärplatz 1, 1210 Wien, Austria, [2]Humboldt-Universität zu Berlin, Breeding Biology and Molecular Genetics, Invalidenstraße 42, 10115 Berlin, Germany, [3]Evonik Operations GmbH, Nutrition & Care, Kantstr. 2, 33790 Halle, Germany; susanne.kreuzer-redmer@vetmeduni.ac.at

Nowadays, there is growing concern about the resistance of pathogenic bacteria against antimicrobial growth promotors, residual antibiotics in meat products for human nutrition and public health risk from zoonotic pathogens. Recently, we established an *in vitro* chicken immune cell assay, which enables the investigation of direct interactions between immune cells and functional feed additives. The cell culture system was established by testing different anticoagulants for blood sampling, different PBMC isolation methods, additional L-glutamine, and different sera as cell culture supplementation. The cells were cultured in RPMI 1,640 medium at 41 °C and 5% CO_2. For flow cytometric analysis antibody sets for the differentiation of immune cell types were established. The response capacity of the system was validated using 10 µg/ml conA. Citrate tubes revealed the highest live cell count compared to EDTA ($P<0.05$) and heparin ($P<0.01$) tubes. Dextran-ficoll isolation decreased the unwanted relative thrombocyte count significantly ($P<0.01$). The addition of 10% chicken serum revealed the highest relative leukocyte count ($P<0.05$) and the lowest relative thrombocyte count compared to FCS ($P<0.05$). The cell viability did not differ. Additional L-glutamine had no effect. ConA activated CD4+ T-helper cells ($P<0.05$) and CD8+ cytotoxic T-cells ($P<0.1$) and serves as a positive control for the system. The established *in vitro* model with chicken PBMCs will help to reveal the mechanisms of immunmodulatory feed additives and improve animals' health. As one major application we tested the direct immunmodulatory effects of the spore forming bacteria *Bacillus amyloliquefaciens CECT 5940* and *Bacillus subtilis DSM 32315*, two licensed probiotics for chicken. We found that T-helper cells are stimulated by *B. amyloliquefaciens* and *B. sultilis. B. subtilis* could also activate cytotoxic T cells. Therefore, we were able provide evidence of direct immunmodulatory effects of these probiotics.

Session 69 Theatre 2

Dietary strategies to support disease resistance and resilience in swine and poultry

C.H.M. Smits

Trouw Nutrition Innovation, Stationsstraat 77, 3811 MH Amersfoort, the Netherlands; coen.smits@trouwnutrition.com

Antimicrobial resistance (AMR) is a threat for human and animal health care. AMR monitoring data from the Netherlands demonstrate that reduction in antibiotic use in swine and poultry coincides with reduced antimicrobial resistance, providing indirect evidence that reducing antibiotic use will contribute to lower the risk of AMR. Reducing antibiotic use in animal health care requires multidisciplinary and multi-stakeholder approach integrating genetics, health management, farm management practices and animal nutrition. Nutrition is one of the pillars for the strategy to prevent animal diseases and disorders. The strategy implies support of the gastrointestinal barriers in host defence and to reduce the risk of the presence of potentially harmful microbes. A high microbial diversity, stability has been associated with a healthy gut, contributing to robustness and resilience in challenge conditions. Promoting peristalsis and intestinal secretions can be achieved by including dietary fibres and by including coarsely milled ingredients. Lowering the protein content and other dietary measures to prevent imbalances in fermentation in the hindgut will in addition contribute to stability of the microbiota. Furthermore, selective well targeted use of combinations of feed additives can enforce animal's resistance and resilience. Organic acids Inhibit pathogens via reduction of the pH of the environment (water, crop, stomach) and via absorption of associated acids by bacteria, disturbing metabolism and proliferation. Synergies can be obtained in combination with other additives such as medium chain fatty acids, phytogenic compounds, prebiotic fibres and probiotics. Immune modulation is another route to follow besides steering the microbiota, but the desired direction is still complex. In general, anti-inflammatory action may be beneficial whereas, especially in early life, also immune-stimulatory effects may be desired to train the immune system and improve immune-competence. Dietary strategies to support disease resistance and resilience may not per se improve zootechnical performance. This will be dependent of the heath status. Regulation of claims of nutritional interventions in preventive health care must take this into consideration.

Session 69 Theatre 3

Nutrition and intestinal health in pigs – options and limits in practice

J. Zentek
Institute for Animal Nutrition, Freie Universität Berlin, Königin-Luise-Straße 49, Berlin, 14195, Germany; juergen.zentek@fu-berlin.de

The nutrition of pigs plays an essential role in maintaining intestinal health. In this context, many strategies have been developed, and it is obvious that the health of the digestive tract includes physiological homeostasis with resistance to infectious and non-infectious stressors. Nutrition supports optimal digestive capacity and efficiency, but also the barrier function, the immune system of the intestine and the orchestrated development tolerance and defence mechanisms with a functioning anti-inflammatory and antioxidant balance. The interaction of host and the microbiome of the digestive tract has a significant influence on digestive health. Optimal nutrition includes the selection of feed ingredients, balanced nutrient levels, the use of appropriate feed additives, many of which have positive zootechnical effects. In addition, feed technology, and optimum hygiene are mandatory to manage piglet health. Diet ingredients with their known nutritive and antinutritive factors play a major role for digestive health. Low protein concentrations with a balanced amino acid ratio reduce diarrhoea problem in practice. High protein diets or ingredients with low praecaecal digestibility can increase bacterial toxins, but also biogenic amines, which can lead to increased chloride secretion, for example. Among the macronutrients, the fibre component in feed has also received increased attention in recent years. While too much fibre tends to be perceived as antinutritive, a balanced use of fermentable or non-fermentable and fibre can have positive effects on animal health. Among feed additives, there are numerous approved additives that have a positive effect on animal health. These are the trace elements, acids and salts, enzymes, plant based products but especially the highly studied probiotics, now often combined with non-digestible carbohydrates in the sense of prebiotic/synbiotic concept. What is important for practice is that the use of feeding concepts and feed additives must be put on a rational basis. Further studies and developments are needed, including optimized diagnostics for a better understanding of microbiota-immune-host organism interactions.

Session 69 Theatre 4

Challenge and opportunities of the low protein diet for weaning pigs

P. Trevisi
University of Bologna, Department of Agro-Food Sciences and Technologies, Viale G. Fanin 44, 40127, Italy; paolo.trevisi@unibo.it

Nowadays, the Green Deal together with the Farm to Fork (F2F) strategy, provides the direction for the development of the EU agricultural sector for the next years. The main aim is to reduce/mitigate the environmental impact of the meat production to increase the sustainability and the resilience of the livestock production system. The pig production chain is a strategic asset for the EU but it is also imputed for a high environmental impact. Increasing the animal efficiency is the pre-requisite to sustain the transition trough the sustainable development of the pig sector. In this framework pig's health is essential and, especially, promoting the gut health is of utmost importance to reduce the production input and promote the efficiency of the breeding phase. As is well known, weaning transition is a critical phase, where the largest part of antibiotics is used to treat gut dysbiosis. Dietary strategies are part of the puzzle to contain the risk of bacterial diseases occurrence in this critical phase of the pig's life, as well as diet formulas based on the reduced amount of crude protein (CP) are widely adopted to mitigate the impact of the pig production and reduce the risk for pathogenic bacteria growth. Indeed, lowering the dietary CP can decrease the fermentation of undigested nitrogen compounds in the distal portion of the intestine and the release of their metabolites, reducing the risk of gut health impairment. Anyway, simply reducing the CP is not enough per se. The risk of amino acids unbalance is directly associated with rate of reduction of the CP, and this negatively impacts pig's efficiency and health. The aim of this talk is to progress on the discernment of the challenge and opportunities in pursuing this way to meet the F2F strategy while maintaining the competitiveness of the EU pig sector.

New dietary and management strategies that modulate the gut microbiota to optimize gut health

N. Everaert
KU Leuven, Department of Biosystems, Kasteelpark Arenberg 30, 3001 Heverlee, Belgium; nadia.everaert@kuleuven.be

The gut microbiota is a dynamic ecosystem in which the host and microorganisms compete but also cooperate in harvesting resources from feed. Manipulation of the animal-microbiota ecosystem has the potential to deliver benefits for health, performance and welfare of the animals. The diet is an important factor that modulates gut microbiota, as they metabolize undigested feed components into bioactive molecules and provide energy for the host. The short chain fatty acids, i.e. acetate, propionate, butyrate are major metabolites produced by gut bacteria from the degradation of complex carbohydrates, but also proteins. Besides butyrate, well-known to positively affect gut health, other special microbial factors may as well trigger increased immunity in the colonic mucosa, like γ-aminobutyric acid, a well-known inhibitory neurotransmitter that also has immunological functions. However, when proteolytic fermentation is dominant over saccharolytic fermentation, it is often linked to the growth of pathogens. Therefore, dietary and management strategies now aim to modulate the gut microbiota and optimize gut health. One can modulate the microbiota, affecting gut health positively, by changing the amount and type of cereals (or the co-products), rich in fibres, such as wheat and barley. Alternatively, dietary strategies can focus on the use of supplements, like prebiotic oligosaccharides such as milk oligosaccharides in the diet of piglets. Another strategy is the use of probiotics in early life, which can be done for piglets by maternal supplementation, or in artificial milk, or in chicks at the end of embryonic development by *in ovo* interventions. One concept gaining interest is to provide a simplified community to the animal, for example soon after hatch in chickens, or just after weaning to prevent post-weaning diarrhoea. Rearing systems (pasture, cage, litter), through the presence of different environment microbiota, as well shape the gut microbiota, affecting gut physiology. As management strategies, research looks at early social and environmental enrichment for piglets, accelerating the maturation of the gut microbiota, which may improve the health of the animals. During this talk, I will provide an overview of recent research outcome in this field.

Nutritional and health benefits of essential oils in dairy cows

L. Abdennebi-Najar[1,2], D.M. Pereira[3], S. Andrés[4], R.B. Pereira[3], R. Nehme[2], E. Vambergue[2], C. Gini[5], S. Even[6], M. Sabbah[1], H. Falleh[7], I. Hamrouni[7], R. Ksouri[7], M. Benjemaa[7], F.Z. Rahali[7], F.J. Giráldez[4], S. López[4], M.J. Ranilla[4], P. Cremonesi[8], S. Bouhallab[6] and F. Ceciliani[5]
[1]Sorbonne University-Centre de Recherche Saint-Antoine (CRSA), INSERM UMR_S_938, 75571 Paris cedex 12, France, [2]IDELE Institute, Quality and Health Department, 75595 Paris Cedex 12, France, [3]REQUIMTE/LAQV, Laboratory of Pharmacognosy, University of Porto, 4050-313, Portugal, [4]Instituto de Ganadería de Montaña (CSIC-Universidad de León), Finca Marzanas, 24346 León, Spain, [5]Università Degli Studi di Milano, Department of Veterinary and Animal Sciences, Via dell'Uni, 26900, Italy, [6]INRAE, Institut Agro, STLO, 35042 Rennes, France, [7]Laboratory of Aromatic and Medicinal Plants, Biotechnology Center of Borj-Cédria, BP 901, 2050 Hammam-Lif, Tunisia, [8]IBBA-CNR, Via Bassini, 15, 20133 Milan, Italy; latifa.najar@idele.fr

Scientific data supporting the efficacy of Essential oils (EOs) in livestock as anti-inflammatory, antibacterial and antioxidant molecules accumulate over time; however, the cumulative evidence is not always sufficient. Still, so far, the MILKQUA project is the first to evaluate, by a combined OMICS approach, the possible use of selected EOs in the nutrition sector and for the treatment of mastitis in dairy cows. Given the importance of the DOHaD concept (Developmental Origin of Health and Diseases), we emphasized EOs' impact, when included in calves' diet, on food efficiency and animal growth. We also tested EOs direct curative effects on inflamed mammary gland and isolated Blood Mononuclear Peripheral cells (PBMC). We also assessed the *in vitro* ruminal fermentation and change in the microbiome content in the presence of both natural and synthetic EOs compounds and decrypted, using several *in vitro* models, the EOs mechanism of action on the NfKB inflammatory pathway. We present in this communication the results of our interdisciplinary approach.

Genetic characterization of the indigenous Sanga cattle of Namibia

D.A. Januarie, E.D. Cason and F.W.C. Neser
University of Free State, Animal Sciences, P.O. Box 339, 9300, Bloemfontein, South Africa; deidre.januarie@gmail.com

Most of the Namibian indigenous Sanga cattle are found in the Northern Communal Areas (NCA) of Namibia, where movement of cloven-hoofed animals are controlled by the Veterinary Condon Fence (VCF). This study aims to determine the levels of genetic variability and inbreeding within the different ecotypes found north and south of the VCF. Hair samples from the Kavango- (n=124), Caprivi- (n=135), Ovambo- (n=269), Kunene- (n=118), Herero (n=17) Sanga ecotypes as well as Nguni (n=148) cattle were collected and analysed using a 150K Single Nucleotide Polymorphism (SNP) chip. Marker and individual quality control, and linkage disequilibrium (LD) filtering were performed on a total of 120,700 markers and 837 animals. Observed (Ho) and expected (He) heterozygosities, and inbreeding coefficients (*F*) were calculated. Low positive inbreeding coefficients (*F*), ranging from -0.0515 to 0.0113, were observed for all the ecotypes. Expected heterozygosity values for Kavango = 0.325±0.0001, Caprivi = 0.316±0.00006, Nguni = 0.322±0.0002, Kunene = 0.331±0.0004, Ovambo = 0.321±0.0001 and Herero = 0.346±0.00004 were observed. PCA scatter plots and admixture ($K=9$) showed four (4) cluster groups, despite some admixture being reflected. *Fst* coefficients between pairs of populations were calculated using PLINK (1.9) and low *Fst* coefficients were observed when compared to PCA plots. The low *Fst* coefficients showed a low level of differentiation and this could possibly be ascribed to the fact that all these Sanga ecotypes belong to one breed with common ancestory. The natural environment in which these ecotypes exist, clearly plays a role in the genetic diversity of the animals and the populations subjected to similar environments cluster closer together. Animals in the NCA are mostly subjected to natural selection in a harsh environment. This is because subsistence farming is practiced, and no clear breeding plans and objectives exist. However, more genetic exchange also occurs between these regions.

Using SEM models to increase of accuracy of QTL estimation in cattle

M. Jakimowicz[1], T. Suchocki[1,2], A. Zarnecki[2], M. Skarwecka[2] and J. Szyda[1,2]
[1]Wroclaw University of Environmental and Life Sciences, Department of Genetics, Kozuchowska 7, 51-631 Wroclaw, Poland, [2]National Research Institute of Animal Production, Krakowska 1, 32-083 Balice, Poland; tomasz.suchocki@upwr.edu.pl

The aim of this study was to investigate the relationships between fertility traits and to estimate their effects in Holstein – Frisian cattle, based on 5,200 Polish Holstein-Friesian cows. In the study phenotypes for conception rate of heifers, conception rate of cows and calving interval were used. To characterise relationships between traits, models that account for the recursiveness were used. First we used Bayesian networks (BN), which are direct acyclic graphs whose nodes represent variables (traits), and edges represent conditional dependencies (connections between traits). To learn the structure of BN, the hill climbing (HC) was used. HC is a local search algorithm that continuously moves in the direction of increasing value to find probabilities of the relationship between traits; then this probabilities were used as coefficients in modelling the dependencies between traits using structural equation modelling (SEM). In the end, the parameters estimated by SEM were used to carry out our SEM-based genome-wide association study.

Session 70 Theatre 3

PyMSQ: an efficient package for estimating Mendelian sampling quantities

A.A. Musa and N. Reinsch
Research Institute for Farm Animal Biology (FBN), Institute of Genetics and Biometry, Wilhelm-Stahl-Allee 2, 18196 Dummerstorf, Germany; musa@fbn-dummerstorf.de

Mendelian sampling variance (MSV) quantifies the expected genetic variability among full-sibs and, thus, the probability that a parent will produce offspring with extreme breeding values. Recently, a similarity matrix was introduced that was derived from the same data as MSV, with the diagonal elements representing a parent's similarity to itself, which equals its MSV. The off-diagonal elements of this matrix represent the similarity of parental haplotypes and the probability that parents will produce offspring with comparable genetic variability. While these quantities (similarity matrix and MSV) are extremely useful for selecting for short- and long-term genetic gains, as well as managing genetic diversity, their computational complexity precludes their use in breeding programs. Also, no software program exists that can compute the similarity matrix. Here, we present PyMSQ, a Python package that implements efficient algorithms for computing MSVs and similarity matrices of gametes produced by parents for a single trait using three primary data sources: marker (or allele substitution) effects, a genetic map, and phased genotypes. The package also calculates a variety of selection criteria, supports group-specific genetic maps, and extends all calculations to multiple traits, aggregate genotypes, and zygotes. MSVs and similarity matrices for milk traits were calculated for a large commercial German Holstein population of 71,163 animals to assess the package. Both MSVs and similarities varied greatly across traits, with the exhibition of low to perfect standardized similarities (0.1 to 1). Finally, we conducted a benchmark study to compare the time required to estimate MSV using PyMSQ and gamevar (a recent Fortran program). PyMSQ performed significantly faster in all scenarios, up to 240 times faster in some cases, depending on the number of markers and individuals. In summary, PyMSQ is an easy-to-use and efficient package for estimating Mendelian sampling quantities for large-scale breeding programs. The novel information on Mendelian sampling quantities will help breeders make optimal mating decisions for genetic gain and diversity management.

Session 70 Theatre 4

Genomic and phenotypic differentiation of the Tamarona line of the Spanish Retinta beef cattle breed

R.M. Morales[1], G. Anaya[1], S. Moreno-Jiménez[1], J.M. Jiménez[1,2], S. Demyda-Peyrás[3] and A. Molina[1]
[1]Cordoba University, Genetics, Campus Rabanales, NIV, 39⁰KM, 14014, Cordoba, Spain, [2]CEAG, The Council of Cadiz, Jerez de la Frontera, Spain, [3]Veterinary School, Animal Production, National University of La Plata, La Plata, Argentina; v22mocir@uco.es

Retinta breed was originally formed in the South of Spain by the union of two beef cattle populations from different ancestral trunks: the *Bos taurus Turdenatus*, origin of Retinta Extremeña subpopulation characterized by a dark red coat and the *Bos taurus Aquitanicus* origin of Rubia Andaluza subpopulation, which has a blond coat. After decades of crossbreeding, the degree of mixture of both populations is nowadays very high, being the dark red coat the predominant coloration. However, Rubia Andaluza animals are still bred in some herds, including a particular genetic line called Tamarona. This line has been selected for decades to obtain a larger morphotype (intended to be used for plowing) in closed breeding, maintaining the blond colour of the coat. In this study, we characterized the Tamarona line by performing a detailed characterization including genealogical, morphological, and genetic data with the aim to establish its degree of differentiation from the rest of the breed. The pedigree analysis showed a very reliable pedigree completeness of the Tamarona population. On the other hand, a reduced effective size (Ne=48), suggesting the onset of an incipient genetic bottleneck in the near future. The morphological characterization of 160 animals demonstrated a greater size and width of the horns and the predominance of the blond coat, in comparison with the average Retinta breed, but fitting within the official racial pattern. Finally, genomic data from 213 animals genotyped using the Axiom Bovine Genotyping v3 medium density SNP-array (including ≈67,000 SNPs per individual). The results revealed a certain degree of genetic diversity within the Tamarona population but a high degree of genetic differentiation with the rest of the Retinta breed, probably due to its selection targeting and reproductive isolation for more than a century during the formation of this line.

Session 70 Theatre 5

GWAS analyses confirms association of homogeneity with robustness

N. Formoso-Rafferty[1], J.P. Gutiérrez[2], I. Álvarez[3], F. Goyache[3] and I. Cervantes[2]
[1]E.T.S.I.A.A.B., Universidad Politécnica de Madrid, Producción Agraria, C/ Senda del Rey 18, 28040 Madrid, Spain, [2]Facultad de Veterinaria, Universidad Complutense de Madrid, Producción Animal, Avda. Puerta de Hierro s/n, 28040 Madrid, Spain, [3]SERIDA-Deva, Área de Genética y Reproducción Animal, SERIDA, 33394 Gijón, Asturias, Spain; nora.formosorafferty@upm.es

A divergent selection experiment for birth weight environmental variability in mice has been successfully performed during 27 generations. Selection for low variability (L-line) was beneficial for traits related to robustness such as birth weight homogeneity, litter size, survival and growth compared with the high variability line (H-line), showing that modifying the environmental variability of a trait has implications in animal robustness, and therefore in animal welfare. The objective of this study was to identify genomic regions associated to animal robustness in mice. 1,718 animals were genotyped by Affymetrix Mouse Diversity Genotyping Array. After quality control (MAF=0.05 and call rate=0.95) 173,546 SNPs were kept. Phenotypes used for the GWAS analysis were derived from birth weight residuals (BWR), being: the environmental variance within litter (VE), the mean BWR within litter (MBWR) and the individual BWR (IBWR). The BWR was the birth weight pre-corrected for sex, generation, parturition number and litter size effects, and MBWR and VE were obtained from BWR. GWAS analyses were performed via a single marker regression that fitted the additive and the dominance effects in the model, assigning the trait to the maternal genotype and also a model including the direct and maternal genetic effects were fitted for IBWR. Also, models fitted the first five eigenvectors of a genomic relationship matrix built ignoring the SNPs of the chromosome where the marker was located. For MBWR and VE the same effects as in the fixed model except sex were included. A threshold of 6.58 ($-\log_{10}(0.05)$/number SNPs) was established to check significance of the SNPs for either the additive or dominance effects. This preliminary analysis suggests that this experimental population could be informative on the genomic basis for robustness and birth weight homogeneity. But, it is necessary advances in statistical modelling to analyse a population so strongly structured.

Session 70 Theatre 6

Blood chimerism in bovine twins is common and affects the accuracy of genotyping

D. Lindtke[1], F. Seefried[2], C. Drögemüller[3] and M. Neuditschko[1]
[1]Agroscope, Rte de la Tioleyre 4, 1725 Posieux, Switzerland, [2]Qualitas AG, Chamerstrasse 56, 6300 Zug, Switzerland, [3]Institute of Genetics, University of Bern, 3001 Bern, Switzerland; dorothea.lindtke@agroscope.admin.ch

Multiple births in cattle are rare at less than 10%, whereof less than 10% of twins are monozygotic. In about 90% of multiple births, the formation of placental vascular anastosomes enables blood circulation between foetuses. The resulting exchange of bone marrow precursor cells leads to genetic chimerism in peripheral blood throughout lifetime. We report a case of blood chimerism between one pair of dizygotic twins born from artificial insemination of a Brown Swiss cow with mixed semen of an Angus and Simmental bull. Usage of mixed semen is common in Switzerland and intended to increase fertilization success. Low-pass whole-genome sequencing (WGS) of genomic DNA obtained from peripheral blood showed that 58 and 40% of blood was transferred from one sibling to the other during pregnancy. As expected due to chimerism, both twins have increased levels of genome-wide heterozygosity. In contrast, we found no evidence of chimerism in SNP microarray data generated from hair bulbs of the same two individuals. WGS of blood DNA from additional 281 Brown Swiss crosses revealed similarly increased levels of heterozygosity in 13 of 15 twins and further seven alleged singletons. Hence, our data suggests that blood chimerism is widespread in cattle and that single-born chimeras are common. Furthermore, our results strongly advise against the use of blood as a source of DNA for obtaining reliable genotype calls in cattle.

Genomic characterization of the Portuguese Mertolenga cattle

D. Gaspar[1,2], R. Gonzalez-Prendes[3], A. Usié[2,4], S. Guimarães[1], A.E. Pires[1], C. Bruno De Sousa[1], M. Makgahlela[5], J. Kantanen[6], R. Kugonza[7], N. Ghanem[8], R. Crooijmans[3] and C. Ginja[1]
[1]BIOPOIS CIBIO-InBIO, UPorto, Vairão, Portugal, [2]CEBAL, IPBeja, Beja, Portugal, [3]Wageningen University & Research, Animal Breeding and Genomics, Box 338, Wageningen, the Netherlands, [4]MED, Pólo da Mitra, Évora, Portugal, [5]Agricultural Research Council-Animal Production Institute, ARC Irene campus, Irene, South Africa, [6]Natural Resources Institute Finland, Jokioinen, Helsinki, Finland, [7]College of Agricultural and Environmental Sciences, Department of Agricultural Production, Box 7062, Kampala, Uganda, [8]Cairo University, Animal Production Department, El-Gammaa, Giza, Egypt; danibgaspar@gmail.com

Mertolenga is one of the largest native cattle breeds reared in Portugal under extensive conditions. Due to its high rusticity, this breed is well adapted to harsh environmental conditions and low input agricultural systems. Mertolenga exhibits remarkable coat colour diversity, i.e. red, roan and red spotted phenotypes, which may impart adaptive ability to cope with increasing exposure to heat stress. The purpose of this study was to: (1) estimate genomic diversity; (2) describe the genetic structure; (3) characterize runs of homozygosity (ROH) patterns underlying these coat colour phenotypes to infer inbreeding and adaptation signatures. Whole-genome resequencing data were obtained from DNA extracted from 29 blood samples. A total of 10,527,521 high-quality SNPs were used for downstream analysis. A Principal Component Analysis separated the different phenotypes in three well-differentiated clusters. These sub-groups had similar levels of mean genomic diversity ($0.28 \leq He \leq 0.32$) with negligible levels of inbreeding. ROH quantification with PLINK yielded 1,226 ROH segments comprising an average of 1.86 Gb across the genome in all individuals. On average, the highest number of ROH segments per animal (nROH=54) were identified in the roan sub-population, with the longest segment comprising 5.3 Mb. This was consistent with the highest F_{ROH} estimated in roan animals. We observed differences in ROH patterns among coat colour phenotypes which can inform on genes related to adaptation and help to improve management strategies in a changing climate.

Session 70 Theatre 8

Metabarcoding approaches to study microbial communities dynamics in milk and dairy products

L. Giagnoni[1,2], C. Spanu[2], A. Tondello[1], D. Saptarathi[1], A. Cecchinato[1], P. Stevanato[1], M. De Noni[3] and A. Squartini[1]
[1]University of Padova, DAFNAE, Viale dell'Università 16, 35020 Legnaro, Italy, [2]University of Sassari, Department of Veterinary Medicine, Via Vienna 2, 07100 Sassari, Italy, [3]Latteria Montello S.p.A., Via Fante d'Italia 26, 31040 Giavera del Montello, Italy; l.giagnoni@studenti.uniss.it

Thermoduric bacteria can survive during the pasteurization of raw milk, leading to a decrease but not to a complete elimination of the microbial load. During the refrigeration stages instead, microorganisms including psychrotrophic taxa, can grow and in turn affect cheese yield and sensory properties. The aim of this study was to evaluate the microbial communities of milk stored at different temperatures before being used to prepare fresh cheese. Cell numbers of bacteria and their taxonomic identity, with particular regard to thermoduric and psychrotrophic microorganisms, were analysed by culture-independent DNA-based metabarcoding approaches. The extraction and purification of the DNA was carried out using a Biosprint 96 platform. The quantity of the extracted DNA was analysed using a Qubit Fluorometer. The total bacterial content was assessed by RealTime quantitative PCR using universal eubacterial primers. For the sequencing, library preparation was performed on the V2-4-8 V3-6-7-9 regions of 16S rRNA gene using universal primers, the combination of the two primer pools allows for a more accurate sequence-based identification of a broad range of bacteria within a mixed population. Sequencing was performed using an ION S5 platform and dataset analysis was evaluated by a customized bioinformatics pipeline on the Silva database. The obtained results allowed to acquire a deeper knowledge of the microbial populations of fresh cheeses and their evolution over time in response to the starting milk conservation temperature. The data are of particular suitability to define and optimize cheese shelf-life, to improve its organoleptic properties, to support food business operators to ensure hygienic quality and compliance to food safety criteria.

Session 70 Theatre 9

Genetic heterogeneity of the POLLED locus in South African Bonsmara beef cattle

R. Grobler[1], C. Visser[2] and E. Van Marle-Köster[2]
[1]University of the Free State, Animal Science, Bloemfontein, 9301, Park West, South Africa, [2]University of Pretoria, Animal Science, Pretoria, 0002, Lynnwood, South Africa; groblerr1@ufs.ac.za

Due to the economic importance of polledness in modern cattle production systems, there has been global interest for the past four decades in characterizing the *POLLED* locus. Investigation and characterization of the polled phenotype is limited in indigenous African and Sanga cattle breeds, which are a hybrid between *Bos taurus* and *Bos indicus*. Population genetic analyses revealed that the genome of the Bonsmara breed is complex and have an admixture of European- and African taurine as well as indicine footprints. In this study, a genome wide association study (GWAS) was performed with the aim to identify genomic regions associated with the *POLLED* locus in South African Bonsmara beef cattle. A total of 74 Bonsmara animals were genotyped with the GGP150K Bovine SNP chip; 34 homozygous polled (P_CP_C) and 40 horned (pp) animals were selected as cases and controls, respectively. The genetic structure and potential within-breed population stratification of the sample group was assessed by principal component analysis (PCA) using GCTA software. To map the *POLLED* locus, a case-control association analyses was conducted between P_CP_C polled and pp horned animals using the --ASSOC function in PLINK. The results of the association analysis indicated a strong association signal above genome-wide significance (-log10 4.7 x10-7) on the proximal region of BTA1. The GWAS analysis also revealed seven genome-wide significant (-log10 4.7 x10-7) SNPs associated with polledness across BTA5, BTA15, BTA22 and BTA28. These results suggest a genetic heterogeneity of the *POLLED* locus and indicates that more than one gene might be responsible for the polled phenotype in South African Bonsmara beef cattle.

Session 70 Theatre 10

The genetic heritage of Colombian Creole sheep breeds

H.A. Revelo[1], V. Landi[2], D. Lopez-Alvarez[1], Y. Palacios[1] and L.A. Álvarez[1]
[1]Universidad Nacional de Colombia Sede Palmira, Ciencia Animal, Carretera 32, 12, 763531, Palmira, Colombia, [2]University of Bari Aldo Moro, Veterinary Medicine, via Marina Vecchia, 75, 70019, Italy; vincenzo.landi@uniba.it

Colombian Creole sheep are the result of the introduction of Iberian genetics types during the colonial period (1493-1500) and the subsequent introgression of other breeds, typically of African type. The population census corresponds to 16,291,201 animals distributed mainly in the region of 'La Guajira' (41.38%), 'Magdalena' (11.56%), 'Boyacá' (7.75%), 'Cesar' (7.69%), 'Córdoba' (6.42%), 'Santander' (3.33%), 'Sucre' (2.57%), 'Bolívar' (2.35%), 'Meta' (2.34%) and 'Cundinamarca '(2.20%). The rearing system is mainly linked to marginal area and to low economic input people. The objective of this work was to investigate the genetic diversity and structure of five Creole sheep populations and their genetic relationships with other breeds from several country. The samples were genotyped using Ovine SNPBeadChip (50K) for 228 animals proceeding from 5 breeds: Sudan (72), Ethiopian (49), Pelibuey (60), Criollo Wayuu (24) and Wolly Criollo (26). Dataset for comparison consisted in several data available in public repositories and included breeds from America, Europe and Africa. Analyses were performed using 42,701 SNPs obtained after quality control. A genetic diversity (He) of 0.396, 0.381 and 0.396 was found for Ethiopian, Sudan and Pelibuey, while values of 0.36 and 0.38 were found for Wayuu and Wolly creole The FST=0.06 (P<0.001), indicated an high genetic differentiation between breeds. Phylogenetic reconstruction analysis separated Caribbean (Ethiopian and Sudanese) sheep as well as Wolly Creole from Pelibuey. Additional analyses indicated that, Ethiopian and Sudan creole breeds, are genetically different. While for Wayuu and Wolly creole a close genetic relationship has been detected. At a global level, a clear separation from putative Iberian and African ancestor have been detected. These results support the need to assign breed status to these locally adapted populations, which have been conserved and used by producers since colonial times.

Session 70 Poster 11

Genetic characterization of Dexter cattle in Southern African

E.D. Cason[1], J.B. Van Wyk[1], M.D. Fair[1], P.D. Vermeulen[2] and F.W.C. Neser[1]
[1]University of the Free State, Animal Sciences, 205 Nelson Mandela Road, Park West, Bloemfontein, 9301, South Africa, [2]University of the Free State, Office of the Dean, Natural and Agricultural Sciences, 205 Nelson Mandela Road, Park West, Bloemfontein, 9301, South Africa; casoned@ufs.ac.za

The Dexter breed is a small, dual purpose, breed of cattle that originated in the Southwestern region of Ireland. The first registered Dexter's were imported into South Africa in 1917. Natural selection pressure should leave detectable signatures on the genome of these Africanised cattle. These signatures are often characterized by the fixation of genetic variants associated with genetic differentiation with other populations at particular loci. In the present study we investigate the population genetic structure and provide a first outline of potential selection signatures in Southern African Dexter cattle using single nucleotide polymorphism genotyping (SNP) data. Signatures of selection revealed genomic regions subject to natural and artificial selection that may provide background knowledge to understand the mechanisms that are involved in economic traits and adaptation to the South African environment. Principal component analysis (PCA) and genetic clustering emphasised the genetic distinctiveness of Dexter cattle relative to other South African cattle breeds. The study yields novel insights into the unique adaptive capacity in this population, emphasizing the need for the use of whole genome sequence data to gain a better understanding of the underlying molecular mechanisms.

Session 70 Poster 12

Identifying selection signatures based in Fst analysis for birth weight variability in mice

C. Ojeda-Marín[1], N. Formoso-Rafferty[2], I. Cervantes[1] and J.P. Gutiérrez[1]
[1]Universidad Complutense de Madrid, Dpto.Producción Animal, Avda Puerta del Hierro, s/n, 28040, Spain, [2]Universidad Politécnica de Madrid, Dpto. Producción Agraria, C/ Senda del Rey, 18, 28040, Spain; candelao@ucm.es

Wright's fixation index (F_{st}) can be used to detect selection signatures. It is based in the principle that in a set of neutral loci in 2 or more populations, differentiation is expected to be due to genetic drift affects in a similar manner. However, if divergent selection operates on one or several loci, markers that are involved show elevated Fst and serve as flags of genomic regions that may have been under selection. The aim of the study was to identify signatures of selection in a mice population that was divergently selected for birth weight during 26 generations to determine genomic regions that could explain the differences in performance between lines. A total of 842 high variability line, and 857 low variability line mice were genotyped using the Affymetrix Mouse Diversity Genotyping Array. All SNPs mapped on the sex chromosomes, SNPs with a genotyping rate lower than 97% and MAF lower than 0.05 were removed. A total of 173,642 SNPs were kept. F_{st} was used to estimate the difference between expected heterozygosity when there is assumed not population's subdivision (Ht) and expected average heterozygosity of the populations (Hs) divided by Ht. F_{st} was set in 1 to select candidate SNPs. Own code was used to estimated F_{st}. Genome-wide candidate regions for selection were established from each candidate SNP downstream regions of 100 Kb. Gene-annotation enrichment and functional annotation analyses were performed with BioMart Software (Ensembl Genes 99 database) and DAVID Bioinformatics Resources 6.8. Ps. A total of 999 SNPs presented F_{st}=1 at last generation, most of them were in the chr 1 (33%), chr 15 (25%) and chr 5 (11%). Gene-annotation enrichment analysis allowed significantly identify 4 functional term clusters and 18 genes related with immunity, fertility, embryo development, nervous system, homeostasis, cell function, heart function, tumour's appearance, and cancer confirming the important differences in robustness between lines already reported in this population. Further studies are required to apply other approaches for detection of signatures of selection.

Genome-wide association study revealed candidate genes for colour differentiation in Mangalitsa pigs

L. Jung, S. Addo and D. Hinrichs
University of Kassel, Animal Breeding, Nordbahnhofstr. 1a, 37213 Witzenhausen, Germany; lisa.jung@uni-kassel.de

The 19th century saw a systematic development of Mangalitsa pigs in Hungary relying on Serbian Sumadija and Hungarian local pigs as parent stock. Blond Mangalitsa (BM) arose, and was crossed either with Black Syrmian to develop Swallow-bellied (SM) or with Szalonta to develop Red Mangalitsa (RM) pigs. Different Mangalitsa pigs exhibit phenotypic variability particularly coat colour variation. The present study aimed at identifying genomic regions that may have impacted coat colour variation in these pigs. A total of 109 animals were genotyped with the ProcineSNP60 v2 Bead Chip. After quality control i.e. removal of both individuals and SNPs with less than 90% genotyping rate, and the removal of SNPs with minor allele frequency below 0.05, only 23,717 markers across 77 animals (23 BM, 24 RM and 30 SM) remained. We performed a multi-loci genome-wide association study (GWAS) using the Fixed and random model Circulating Probability Unification (FarmCPU) implemented in rMVP. Four different sets of GWAS were performed depending on the definition of phenotype. Firstly, the different breeds were coded as 1 (RM), 2 (BM) and 3 (SM) as labelled by colour. These codes were considered as phenotypes for the comparison of all three breeds in a single GWAS analysis. In each of the remaining three GWAS, one breed was compared against all other breeds. From the analyses, between 4 to 6 genome-wide significant variants were detected; the highest (DRGA0016741) being important for SM. Within 1 Mb distance on either side of these variants, a number of candidate genes (e.g. MAP1LC3A, EIF6, EDEM2, PROCR, ITCH, EIF6 and GSS) previously reported to play a role in melanogenesis were identified. We, however, propose a collaboration with other Mangalitsa projects and the joining of data that will enhance statistical power in confirming our results.

Genomic prediction in the Lesvos dairy sheep

A. Kominakis[1], I. Mastranestasis[2], V. Karagianni[3], M. Stavraki[3] and A.L. Hager-Theodorides[1]
[1]Agricultural University of Athens, Department of Animal Science, Iera Odos 75, 11855 Athens, Greece, [2]Breeder's Association of the Lesvos sheep, Anaxos Lesvos, 81109, Greece, [3]Ministry of Rural Development and Food, Centre of Animal Genetic Improvement of Athens, Verantzerou 46, 10438, Athens, Greece; a.hager@aua.gr

Aim of the present study was to estimate genomic breeding values (GEBVs) for total milk yield (MY) in the Lesvos dairy sheep. A total number of 520 Lesvos dairy ewes genotyped with the ovine 50k SNP chip array with records on total MY were used. Genomic prediction (GP) was carried out using the GBLUP method and genotypic data on 46.638 SNPs passing typical marker quality criteria (call rate >0.90, minor allele frequency >0.05 and Fisher's Hardy-Weinberg equilibrium test $P<10^{-4}$). A mixed linear model was applied that accounted for fixed effects (herd: 9 classes, year: 2 classes, lactation: 6 classes, lambing month: 4 classes, days in milk: linear covariate, population structure (first 5 PCs of genotypic data) and the random effects of animals via the Genomic Relationship Matrix (GRM). Predictions of the phenotypic values for the trait was also made via a 5-fold cross-validation (repeated 10 times) approach. GP accuracy was assessed via the Pearson correlations between EBVs and phenotypic values and between the predicted and the observed MY. Correlations between EBVs and observed MY were as high as 0.67 while the averaged (across the 5-folds) correlation between predicted and observed MY was 0.72 (range 0.67-0.79). These large correlations values are indicative of high accuracy of prediction. Current results appear promising for efficient genomic selection in this population, but they need to be verified in a larger sample.

Genotyping of PRNP gene using 7 codons in 17 goat populations

A. Canales[1,2], S. Nicar[1,2], J.A. Bouzada[3], J. Jordana[4], M. Amills[4], M.R. Fresno[5], A. Pons[6], M. Macri[1,2], J.V. Delgado[2] and M.A. Martínez[2]
[1]U. of Córdoba, Campus de Rabanales, 14071 Córdoba, Spain, [2]Animal Breeding Consulting, S.L., Córdoba Science and Technology Park Rabanales 21, 14071 Córdoba, Spain, [3]Laboratorio Central de Veterinaria, Ctra. M106, Km 1,4, 28110 Madrid, Spain, [4]CSIC-IRTA-UAB-UB, CRAG Building, 08193 Barcelona, Spain, [5]Instituto Canario de Investigaciones Agrarias, Finca 'Isamar', Ctra. de El Boquerón s/n, 38270 Tenerife, Spain, [6]Serveis Millora Agrària i Pesquera (SEMILLA), d'Eusebi Estada, 145, 00709 Palma, Spain; jbouzada@mapa.es

Scrapie is a fatal and neurodegenerative disease of sheep and goats that produces important economic losses for the farmers. It is also the earliest known member in the family of diseases classified as transmissible spongiform encephalopathies (TSE) or prion diseases. The susceptibility to this disease is influenced by the polymorphisms of the prion protein gene (PRNP), more specifically polymorphisms at codons 136, 154 and 171 of this gene in sheep. There is not clear evidence of a relationship between the polymorphisms of the PRNP gene and the resistance or susceptibility to Scrapie in goats, although the fact that sheep and goats share some polymorphism such as the R154H dimorphism could be of great interest as the association with Scrapie would be expected to be similar. The aim of this study is to report the polymorphisms of the PRNP gene in 17 goat populations (Florida, Payoya, Blanca de Rasquera, Blanca Andaluza, Blanca Celtiberica, Azpigorri, Palmera, Murciano-Granadina, Mallorquina, Majorera, Pitiüsa, Palmera, Tinerfeña, Malagueña, Saanen, Alpine and Tunisian goats). Three codons recommended by the EFSA (European Food Safety Authority), i.e. D146, S146 and K222, where analysed. Additionally, four codons were genotyped, i.e. I142, H143, R154 and R211. Primers to detect polymorphism in the 7 codons were designed: 142: (38T)GTGCCATGAGTAGGCCTCTTAT, 143: (24T)AAGTGCCATGAGTAGGCCTCTTATAC, 146: (22T)GGCCTCTTATACATTTTGGCA, 146.1: (16T)AGTAACGGTCCTCATAGTCAT (reverse complement), 154: (7T)TGACTATGAGGACCGTTACTATC, 211: (43T)GAAACTGACATCAAGATAATGGAGC and 222: GAGCAAATGTGCATCACCCAGTAC. This research could help to promote the adoption of selective breeding programs as a possible future strategy in caprine scrapie control, outbreak prevention and reduce economic losses to the farmer.

Genome-wide association study for mastitis resistance in Czech Holstein Cattle

K. Forejt[1], H. Vostra-Vydrova[2,3], N. Moravcikova[4], R. Kasarda[4], L. Zavadilova[2] and L. Vostry[3]
[1]Palacky University Olomouc, Faculty of Science, Department of Experimental Biology, Krizkovskeho 511, 779 00 Olomouc, Czech Republic, [2]Institut of Animal Science, Pratelstvi 815, 10400 Prague, Czech Republic, [3]Czech University of Life Science Prague, Kamycka 129, 16500 Prague, Czech Republic, [4]Slovak University of Agriculture in Nitra, Tr. A. Hlinku 2, 94976 Nitra, Slovak Republic; vostry@af.czu.cz

Clinical mastitis is an inflammatory disease of the mammary gland which largely impacts profitability and welfare in dairy farming. In the theoretical part of this thesis, a comprehensive overview of mastitis, its aetiology, epizootology and economic impacts has been compiled using international scientific resources. Furthermore, current findings regarding the potential link of certain genotypes to the resistance to mastitis and the appropriate bioinformatic methods to study it have been presented. In the experimental part of this thesis, DNA microarray data has been obtained from 1,258 Holstein cows via Illumina BovineSNP50 DNA Analysis BeadChip, examining 53,218 single-nucleotide polymorphisms (SNPs). Owners of these cows have provided binary phenotype data capturing the incidence of mastitis in each individual. After data pruning, controlling for genotype missingness, minor allele frequency and population stratification, the total of 51,557 SNPs from 1,042 animals has been analysed with the appropriate software using the general linear model (GLM). Two SNPs, BTA-121769-no-rs and BTB-00265951, have demonstrated statistically significant associations – both located on the chromosome BTA6. The detected SNPs have been further examined via genome mapping and discussed with relevant publications. They have been found to lie outside of transcribed regions but within the immediate vicinity of genes essential for the immune response, which further supports the case for their significance in the resistance to mastitis. he study was supported by the projects QK1910156, MZe-RO0718 and APVV-20-0161.

Session 70 Poster 17

Modification of imputation scheme for Polish Holstein-Friesian dairy cattle genomic evaluation

K. Żukowski[1] and T. Suchocki[2]

[1]National Research Institute of Animal Production, Department of Cattle Breeding, Krakowska 1, 32-083 Balice, Poland, [2]Wrocław University of Environmental and Life Sciences, Department of Animal Genetics, Kożuchowska 7, 51-631 Wroclaw, Poland; kacper.zukowski@iz.edu.pl

Presently, the genotype imputation was based on findhap.f90 software, and procedures implemented into the Polish Cattle Database System (CSNP) database scheme. Over twenty types of microarray were used in genotype imputation, including Illumina Bovine 50K chips, low-density chips like LD, GeneSeek and EuroG10K, and widely used in EurGenomics countries EuroGMD chip family. The study aimed to speed up and modify the current imputation procedures and CSNP database, including upgrading the genome map to ARS-UCD1.2 release to unnoticeable for dairy cattle genomic evaluation. The analysis was based on more than 130K cows and bulls deposited in (cSNP). The core of animals were 35K reference EuroGenomics bulls. The analysis included: (1) the currently used imputation model (findhap software and UMD3.1.1 genome release); (2) modified database procedures; and (3) final model with fimpute imputation software ARS-UCD1.2 with modified database procedures. All available and routinely used microarray types were used in the analyses. Altogether, the applied database modification procedures reduced the analysis time from days (1) to hours (2). The correlation coefficient between SNP effect for models (1) and (3) was 1, as well as the correlation between genomic breeding values for bulls and cows for models (2) and (3). Furthermore, the available procedures are more flexible in international projects such as SNPMACE.

Session 70 Poster 18

GENORIP project: a genomic approach to manage genomic variability in cattle dairy farms

C. Punturiero, F. Bernini, R. Milanesi, A. Bagnato and M.G. Strillacci

Università degli Studi di Milano, Department of Veterinary Medicine and Animal Sciences, Via dell'Università 6, 26900 Lodi, Italy; chiara.punturiero@unimi.it

For decades the Holstein dairy cattle populations have been undergoing a successful selection process to increase its performance. Nevertheless, it is well known that the strong directional selection determined loss of genetic variability and increase in inbreeding. In this study we investigated the existing genetic variability, the inbreeding and allele frequencies of monogenic traits in seven herds of Italian Friesian breed and provided insight to farmers on the value of genomic management of their herd. The project 'GENOmic tool for the management of REProduction in dairy cattle and for the control of inbreeding – GENORIP', aims to release a tool for the genomic management of herd reproduction. More than 3,500 females of Italian Friesian raised in 7 farms were sampled and genotyped with the GeneSeek Genomic Profiler (GGP) Bovine 100K. Principal component analysis (PCA) made it possible to graphically visualize how individuals and herds cluster according to their genotype. Herds appear similar but, within herd, cows show appreciable variability. Two genomic inbreeding coefficients were estimated and used: (1) based on observed and expected numbers of homozygous genotypes (FHOM) – within herd; (2) based on the Runs of Homozygosity (FROH). A total of 373,066 ROH were identified. The ROH distribution within herd, together with the genotype frequencies for disease, fertility and production mendelian traits, made it possible to identify genomic regions under different selection according to farmer selection strategies. This study released to each farmer the genomic make-up of their herd used jointly with the gEBV estimated by their national breeders association (ANAFIBJ) for herd reproductive management. GENORIP is genotyping 1,500 females in 50 additional farms to further explore genomic variation in the Italian Friesian and to show to a wider audience the possibility of adding value by using a precision genomic approach to herd management. Funded by EAFRD Rural Development Program 2014-2020, Management Autority Regione Lombardia – OP. 16.1.01 Project ID n. 201801062430 – 'Operational Group EIP AGRI' https://ec.europa.eu/eip/agriculture/en/eip-agri-projects/projects/operational-groups.

Session 70 Poster 19

Identification of SNPs affecting milk yield of Polish Holstein Friesian dairy cows

M. Kolenda[1], B. Sitkowska[1], D. Piwczyński[1], H. Önder[2], B. Kurnaz[2], U. Şen[3] and A. Dunisławska[1]
[1]Bydgoszcz University of Science and Technology, Department of Animal Biotechnology and Genetic, Kaliskiego 7, 85-796 Bydgoszcz, Poland, [2]Ondokuz Mayis University, Department of Animal Science, 55139 Samsun, Turkey, [3]Ondokuz Mayis University, Department of Agricultural Biotechnology, 55139 Samsun, Turkey; kolenda@pbs.edu.pl

The aim of the study was to determine the SNP included in microarrays associated with milk yield (MY) in full lactation in Polish Holstein Friesian cows. Genomic data were obtained during estimation of breeding value with the use customized SNP arrays. GenSel (https://de.cyverse.org/) was used to estimate breeding values for milk yield for 492 cows, a total of 13,481 SNPs were used. The lactation number was used as fixed factor and days in milk (DIM) used as random factor. During the research several SNPs were identified as those that impact MY (according to Bayes A method). These SNPs were identified on chromosome X, position 79,472,439 (UMD_3.1 reference genome) called SNP1 as well as on chromosome 27, position 32,513,708 (UMD_3.1), called SNP2. For SNP1 most frequent was BB genotype (57.7%) and the lest AA (5.7%,) while for SNP2 most frequent was AB (51.0%) and the least BB (15.0%). To examine the allele effects on MY covariance analysis were used. DIM and lactation number were used as covariate. To compare the means Bonferroni test was. Kendal's Tau correlation coefficient was used to evaluate the relationship among alleles. The highest estimated marginal mean for MY for SNP1 and SNP2 were obtained for BB genotype (13,802.38 and 14,232.09 kg, respectively), while the lowest for AA (13,111.59 and 13,148.45, respectively). Differences between genotypes for all SNPs were proven to be statistically significant. Further analysis of variance was performed for 305-days MY, and a post hoc Tukey test was used. Analysis showed for SNP1 (data for data for 671 cows) significant differences between MYs of cows with AB and BB genotype (11,699.04 kg and 12,114.31 kg, respectively) and for SNP2 (667 cows) AB and AA (12,081.85 kg vs 12,092.14 kg). The research shows that using microarrays data one can find new SNPs associated with impact on MY. This material has been supported by the Polish National Agency for Academic Exchange under Grant No. PPI/APM/2019/1/00003.

Session 70 Poster 20

Single step genomic prediction of milk yield in Polish Holstein Friesian dairy cattle

H. Önder[1], B. Sitkowska[2], M. Kolenda[2], D. Piwczyński[2], B. Kurnaz[1] and U. Şen[3]
[1]Ondokuz Mayis University, Department of Animal Science, Samsun, 55139, Turkey, [2]Bydgoszcz University of Science and Technology, Department of Animal Biotechnology and Genetic, Kaliskiego 7, 85-796 Bydgoszcz, Poland, [3]Ondokuz Mayis University, Department of Agricultural Biotechnology, Samsun, 55139, Turkey; sitkowskabeata@gmail.com

This study aimed to compare accuracy of Bayesian alphabet (A, B, C, Cπ) with single step genomic prediction of milk yield in Polish Holstein Friesian dairy cattle. Genomic data used in the study came form routine estimation of breeding value with the use customized SNP arrays that is done in Poland. Phenotypic data was obtained from Polish Federation of Cattle Breeders and Dairy Farmers. To estimate the genetic parameters such as heritability and genomic breeding values for milk yield on Polish Holstein Friesian population, 13,481 SNPs were used. 50,000 iteration were used and the first 5,000 iteration was burned-in. The lactation number (1, 2 and 3) was used as fixed factor and days in milk used as random factor. 231 of the SNPs were monomorphic and excluded from the analysis. All analysis were done using GenSel (https://de.cyverse.org/). The mean of 2pq was found as 0.376. The heritability values were estimated as 0.547, 0.5309, 0.543 and 0.545 for Bayes A, Bayes B, Bayes C and Bayes Cπ, respectively. The coefficient of determinations values were found to be 0.623, 0.619, 0.624 and 0.623 for same order of Bayesian alphabet. The π value was estimates as 0.529 for Bayes Cπ. The minimum computation time was 278 seconds for Bayes C and the maximum was 1,741 seconds for Bayes B. The results, according to coefficient of determination, show that the best model was Bayes C. Spearman Rank Correlation was made for animals ranked according to breeding value, and as a result, it was determined that the highest correlation was between Bayes A and Bayes Cπ (0.239) and the lowest correlation was between Bayes B and Bayes Cπ (0.083). This material has been supported by the Polish National Agency for Academic Exchange under Grant No. PPI/APM/2019/1/00003.

Session 70 Poster 21

Genome-wide association study for daily milk yield in Istrian dairy sheep

M. Špehar[1], J. Ramljak[2] and A. Kasap[2]
[1]Croatian Agency for Agriculture and Food, Svetošimunska cesta 25, 10000, Croatia, [2]University of Zagreb Faculty of Agriculture, Svetošimunska cesta 25, 10000, Croatia; marija.spehar@hapih.hr

The objective of this work was to identify genomic regions associated with daily milk yield in the Croatian native Istrian sheep breed traditionally selected for milk yield. A total of 719 animals (689 ewes and 30 rams) were genotyped using the Illumina OvineSNP50K BeadChip. Genotyped animals and SNPs were included in the analysis after quality control parameters for call rate per animal, call rate per SNP, and MAF set to 0.9, 0.9, and 0.05, respectively. Monomorphic markers and markers with unknown genome position or located on the sex chromosome were also removed. The final number of genotyped animals and markers was 693 and 43,793, respectively. The estimated breeding values obtained by single step genomic BLUP were used as phenotypes to detect significant associations between SNPs and milk yield across the genome. A GWAS analysis was performed with the rrBLUP package in R. Four models were considered: () without controlling for population structure (PS) or family relatedness; (2) controlling for PS effect; (3) controlling for relatedness; and (4) controlling for both PS and relatedness. The last one (4) was the most appropriate based on Q-Q plot analysis. After Bonferroni corrections, thresholds were set to $P<2.47\times10^{-6}$ and $P<4.94\times10^{-5}$ (suggestive level), corresponding to -log10(p) of 5.61 and 4.31, respectively. None of the markers showed significant effect on the former, while five were significant on the latter (suggestive) level (one on chromosomes 5, 6, 13 and two on chromosome 18). Further studies considering a larger sample size will be needed to improve the statistical power of the analysis.

Session 70 Poster 22

GWAS for clinical mastitis in Polish HF cows

P. Topolski[1], T. Suchocki[1,2] and A. Żarnecki[1]
[1]National Research Institute of Animal Production, Department of Cattle Genetics and Breeding, Krakowska 1, 32-083 Balice near Krakow, Poland, [2]Wroclaw University of Environmental and Life Sciences, Department of Genetics, Kozuchowska 7, 51-631 Wroclaw, Poland; piotr.topolski@iz.edu.pl

Mastitis is one of the most common reasons for the earlier culling of cows in dairy farming, associated with a significant economic implications and animal welfare. The low heritability of the trait causes that selection for resistance to mastitis is possible however, genetic response to traditional selection is small. Consequently, new genomic methods are being used to help a better understand the genetic variability underlying mastitis and present options for direct selection of such trait. In our research, we performed a genome-wide association study (GWAS) to identify genetic loci associated with clinical mastitis in Polish HF cows. Data for the study were drawn from the data base PLOWET used for veterinary-recording health traits in 4 experimental dairy farms of the National Research Institute of Animal Production and including a total of 1,499 cow genotypes; the incidence of clinical mastitis has been recorded in 712 cows. Each cow was genotyped or imputed to Illumina BovineSNP50K beadchip in version 2. As a final dataset we used 53,557 SNPs with MAF≥1% and minimal quality of genotyping ≥99%. The GWAS data were analysed using single SNP mixed model. We used FDR multiple testing correction. Finally, we found 6 statistical significant SNPs that may be associated with clinical mastitis located on 5 autosomes: 7, 8, 9, 13 and 29. The pathway analysis showed that part of these SNPs were located in or very near genes indirectly related to the immune system.

SNP-based genetic diversity in Pag sheep – preliminary results

J. Ramljak[1], M. Špehar[2], A. Kasap[1], B. Mioč[1], I. Širić[1], A. Ivanković[1], D. Barać[2], Z. Barać[3] and V. Držaić[1]
[1]University of Zagreb Faculty of Agriculture, Department of Animal Science and Technology, Svetošimunska 25, 10000 Zagreb, Croatia, [2]Croatian Agency for Agriculture and Food, Svetošimunska 25, 10000 Zagreb, Croatia, [3]Ministry of Agriculture, Ulica grada Vukovara 78, 10000 Zagreb, Croatia; jramljak@agr.hr

Sheep breeding has been present in Croatia for centuries, especially in coastal and island areas, where the harsh and unfavourable climate has affected the adaptability and survival of populations. Pag sheep is a local breed, bred exclusively in the restricted area, the Pag island. The economic importance for the inhabitants lies in the production of the well-known hard cheese and, to a lesser extent, lamb meat. In general, intensive production, limitation of the breeding area and control of gene flow lead to erosion of genetic variability and inbreeding. The aim of this research was to obtain preliminary results on the genetic structure of Pag sheep using SNP markers. A total of 100 randomly sampled animals from 10 flocks were genotyped using the Ovine SNP50 BeadChip. After quality control for call rates per SNP and animal and minor allele frequency, 45,224 autosomal SNPs and 100 individuals were available for analysis. Principal component analysis (PCA) was performed based on multidimensional scaling (MDS) to analyse population structure. Two clusters were separated by PCA (PC1, 3.79% and PC2, 3.73%) while the majority of individuals occupied an intermediate position. The population parameters determined in the studied sample were: observed heterozygosity (0.410), expected heterozygosity and Nei's genetic diversity (both 0.420) and inbreeding coefficient (-0.01), respectively. According to the results, the Pag sheep has preserved considerable genetic diversity with a low level of inbreeding. However, further studies with a larger data set and additional breeds are needed to confirm the results of this preliminary study.

Genetic characterization of Churra do Campo Portuguese sheep breed using a high-density SNP chip

M.F. Santos-Silva, N. Carolino, C. Rebelo De Andrade, C. Oliveira E Sousa, P. Jacob, I. Carolino and V. Landi
University Aldo Moro, Veterinary Medicine, SP.62 Casamassima km.3, 70010 Valenzano, Italy; fatima.santossilva@iniav.pt

Churra do Campo (CC), a local Portuguese sheep breed, has undergone a great population reduction, being almost disappeared at the end of last century. In order not to lose this important genetic resource a recovering program was established and at 2007 the Flock Book was created. Nowadays, 817 animals in 8 herds are recorded, the population still having an endangered extinction status that requires all possible support. Genetic characterization gives a picture of population's dynamics, status of genetic diversity, and inbreeding, crucial for the implementation of successful genomic programs, for selection and conservation of genetic resources. Modern genomic technology brought new insights for these studies and to investigate polymorphisms associate with unique population characteristics. This work, made in collaboration with CC Flock Book, used a high density 50K SNPs array, to assess genomic diversity of CC (48 individuals), and relationship with other sheep populations of different countries, with data available at HAPMAP consortium. After quality control, applying accepted published filters and merging data with public dataset, 42,347 loci remained for further analysis. Results revealed appreciable levels of genetic variability, with MAF medium value of 0.293, and 72% of marker> 0.20, observed heterozygozity of 0.37 (0.27 to 0.473), similar to most other population. Medium coefficient of inbreeding (F) of 0.05 was similar or lower than other populations compared, nevertheless individual heterogeneity observed must be considered in matching decisions. CC revealed some differentiation from other Portuguese breeds, but has a greater genetic proximity to these, than to breeds from other countries. This first study established with genomic characterization in CC breed offers novel insight for this population and its relationship with other sheep breed over the world.

Session 70 Poster 25

Analysis of the CNV inheritance in swine genome based on combined Illumina and Nanopore data

M. Frąszczak[1], M. Mielczarek[1,2], T. Strzała[1], B. Nowak[1] and J. Szyda[1,2]
[1]Wroclaw University of Environmental and Life Sciences, Biostatistics Group, Department of Genetics, Kożuchowska 7, 51-631 Wrocław, Poland, [2]National Research Institute of Animal Production, Krakowska 1, 32-083 Balice, Poland; magdalena.fraszczak@upwr.edu.pl

Sus scrofa is one of the most economically important livestock species worldwide and is considered as a model organism in human studies. This is because of its similarity in anatomy, physiology, metabolism, pathology, and pharmacology to human. Moreover, the porcine genome is three times closer than the mouse genome to that of the human. Copy number variations (CNVs) are defined as changes in the copy number of large genomic segments and can overlap multiple functional elements of the genome. CNVs can be inherited or arise *de novo*. A novel approach that combines short- and long-read sequencing, which has not yet been applied for swine genomes, can improve the detection of CNVs and provide better knowledge about the inheritance of CNVs in the domestic pig. In this study we aimed to determine the prevalence of CNVs formed *de novo* in the offspring genomes, which was possible thanks to the knowledge of the CNV variability of the parents' genomes. Additionally, we focused on the analysis of CNV inheritance in full siblings. In particular, the analysed dataset consisted of whole-genome DNA sequences of 12 pigs representing the Polish Large White breed, obtained using the Illumina HiSeq 2000 and Oxford Nanopore Technology. After variant pre-processing the total number of deletions per individual ranged between 216 and 548 and duplications between 237 and 537. The length of CNVs varied from 300 to 530,600 bp (average 3,939.44±10, 975.9) for deletions and from 900 to 346,200 bp (average 12,377.2±20, 154.3) for duplications. From 20% to 31%of deletions and from 15 to 25% of duplications were inherited in both siblings. While the number of deletions formed *de novo* varied from 233 to 333 and from 251 to 350 for duplications.

Session 70 Poster 26

Characterization of buffalo milk microbiota through NGS techniques

C. Ferrari[1], F. Luziatelli[2], L. Basiricò[1], M. Ruzzi[2] and U. Bernabucci[1]
[1]University of Tuscia, Department of Agriculture and Forests Sciences, via San Camillo De Lellis, snc, 01100, Italy, [2]University of Tuscia, Department for Innovation in Biological, Agrofood and Forest Systems, via San Camillo De Lellis, snc, 01100, Italy; bernab@unitus.it

The trial was carried out in three dairy buffalo farms located in central Italy. The three farms were representative of the production area. The diet was fed *ad libitum* as total mixed ration (TMR), distributed once a day in the morning using two-auger vertical mixer wagons. The buffaloes were milked two times a day in a herringbone milking parlour. The TMR of the three farms contained corn silage, hay, and concentrates; the difference among the diets was the energy and protein content. Also, the size of the three farms was different in terms of the number of lactating buffaloes: 190 farm A, 200 farm B, and 50 farm C. Bulk milk samples were collected in two different periods (December 2020 and January 2021), directly from the milk tank after accurate mixing, in 1000 ml plastic container adding Bronopol® as a preservative. Milk samples were kept at 4 °C and analysed within 24-h after collection. An aliquot was used for determining the content (%) of fat, protein, lactose, and solid not fat, pH and urea by I.R. spectrophotometry, and somatic cell count by fluorooptoelectronics (MilkoScanTM7 RM, FOSS, Denmark). Milk clotting parameters (RCT; k_{20}; a_{30}) were determined by lactodynamography (Mape System, Italy). Milk was then treated for DNA extraction and Next-Generation Sequencing (NSG) through the Illumina platforms and subsequent bioinformatics analysis was carried out. Chemical and metagenomic analyses were carried out to discriminate a possible triangulation between feed, chemical-rheological, and metagenomic parameters. Microbial metagenomic analysis provided interesting results in bacterial populations present in the samples indexable to the hygienic-health management of the farms. The milk metagenome analysis though NSG can be a valuable analytical technique for the management of dairy farms given the vastness of information obtainable. Acknowledgments: The authors thank the owners of the farms. This research was funded by LazioInnova, Gruppi di Ricerca2020, n. prot. A0375-2020- 36613.

Session 71 — Theatre 1

RNA-Seq reveals candidate microRNA (miRNA) linked to Cytochrome P450 pathway in dairy cows

A. Veshkini[1,2,3], H.M. Hammon[2], H. Sadri[3], B. Lazzari[4], V. Vendramin[5], H. Sauerwein[3] and F. Ceciliani[1]
[1]Department of Veterinary Medicine, University of Milan, 26900 Lodi, Italy, [2]Research Institute for Farm Animal Biology (FBN), 18196 Dummerstorf, Germany, [3]Institute of Animal Science, University of Bonn, 53115 Bonn, Germany, [4]IBBA-CNR, 20122, Milan, Italy, [5]IGA Technology Services, 33100, Udine, Italy; veshkini@fbn-dummerstorf.de

The endocrine, metabolic, and immunological changes during the transition into lactation have been characterized in dairy cows by univariate analyses, combinations therefrom, and, increasingly, also via multivariate OMICs technologies. Conversely, post-transcriptional regulation of gene expression is less studied in this context. We therefore aimed at comparing the miRNA-ome in plasma of transition dairy cows before and after calving. Total RNA was extracted from plasma collected at days -21 and +1 relative to calving from 32 Holstein dairy cows enrolled in a trial in which they received either saturated or polyunsaturated fatty acids via abomasal infusion. The sequences obtained by NovaSeq 6000 (Illumina, CA) were compared with *Bos taurus* miRbase database for identification of known miRNA using miRDeep2. Differentially expressed miRNA (DEM) were assessed using the GLM (generalized linear model) approach at a threshold of fold-change (FC) >1.5 and false discovery rate (FDR) <0.05 using edgeR package and searched against MiRWalk database. Out of 686 identified known miRNAs, 110 DEM were identified between days -21 and +1 relative to calving regardless of treatment effect. Based on mechanistic data from human and mice, 7 DEM were detected to target Cytochrome P450 (CYP) enzymes via their mRNA or by directly binding the promoter: miR-222, miR-20b, miR-126-5p, miR-101, miR-17-5p, and miR-195 were decreased on day +1 as compared to -21. miR-138 and a further DEM, miR-149-5p, for which prediction of targets via bioinformatics yielded an association with the CYP enzymes, increase from d-21 to d+1. The genes and proteins targeted by the specific DEM identified herein are directly or indirectly associated with the metabolism of xenobiotics by the CYP P450 pathway. Our data support the emerging importance of the CYP system in the transition from pregnancy to lactation in dairy cows.

Session 71 — Theatre 2

Plasma and milk metabolomics in primiparous and multiparous Holstein cows under heat stress

E. Jorge-Smeding[1], D. Rico[2], A. Ruíz-González[2], X. Wei[3], Y.H. Leung[1] and A. Kenez[1]
[1]City University of Hong Kong, Hong Kong, China, NA, China, P.R., [2]Centre de Recherche en Sciences Animales de Deschambault, Deschambault, QC, Canada, [3]Katholieke Universiteit Leuven, Leuven, Belgium; ejorgesmeding@gmail.com

Multiparous and primiparous dairy cows face different physiological conditions during lactation because of the growth requirements of primiparous cows. This parity-related difference might also affect their metabolic responses to heat stress (HS). Twelve Holstein cows (42.2±10.6 kg milk/d; 83±28 DIM; primiparous, P, n=6; multiparous, M, n=6) were enrolled to evaluate the effect of parity and HS on the plasma and milk metabolomic profiles. Treatments were heat stress (HS; THI=82), or pair feeding in thermo-neutrality (TN; THI=64) for 14 days in a Latin square design. Blood samples were collected on the last day of treatments and analysed by the Absolute IDQ p400 metabolomics kit (Biocrates, Innsbruck, Austria). ANOVA was used to evaluate treatment (TN, HS) and parity (P, M) effects and their interactions as fixed effect, and the cow as random effect. Raw-P values were adjusted by False discovery rate (FDR) correction. In the plasma, only α-amino adipic acid differed (FDR=0.043) according to parity as it was lower in P, and t4-hydroxyproline tended (FDR=0.057) to be greater in P than M. Thirteen metabolites differed (FDR<0.01) between TN and HS including glycine, phenylalanine, t4-hydroxyproline and 10 lipid species. No significant interaction effects were detected. In the milk, carnitine and 2 acyl-carnitines (C2:0, C5:0-OH) were higher (FDR≤0.025) in P than M cows, and 14 metabolites (glycine, α-amino adipic acid, taurine, 2 acyl-carnitines, 6 phosphatidyl-cholines, 2 sphingomyelins, and 1 triglyceride) differed (FDR≤0.04) between TN and HS, while no metabolite was affected by the interaction. Our results indicate that both the plasma and milk metabolome were affected by the parity, pointing to differences in AA and mitochondrial fatty acid oxidation for energy production.

Session 71 Theatre 3

Association of peripheral blood transcriptome and haemochromocytometric values in piglets

D. Luise, F. Correa, P. Bosi and P. Trevisi
University of Bologna, DISTAL, Bologna, 40127, Italy; paolo.trevis@unibo.it

The whole peripheral blood (PB) transcriptome can give synthetic information of the whole-body condition and health. This could be important for pigs in the period around weaning, given the relevance of their health and the call pressure from the various maturing tissues on this phase. Rapid haemochromocytometry, besides giving quantitative values on red and white series, could provide interesting association with some whole blood expressed genes. PB was obtained from 31 piglets, 16 on weaning day (26d of age) and 15 at 12d post-weaning. PB was analysed using mRNA-seq for transcriptome and automatic analyser CELL-DYN 3700R for haemochromocytometric values. The human globin gene was not depleted before sequencing. Single linear correlation between transcriptome and haemochromocytometric values were calculated for the whole sample and for each timepoint. Overall, correlations of transcriptome values with haemoglobin content, haematocrit values and mean corpuscular volumes were very poor. This could be due to the relative stability of these values in pigs. The presence of immature red blood cells, characterized by a higher content of transcript, rapidly decreased with age. Among the white blood cell contents, several interesting correlations were seen. In the whole set, the total lymphocyte numbers were positively associated with the expression of *CX3CR1* (r=0.708); *GZMA* (r=0.719); *GZMK* (r=0.711); *CD8A* (r=0.701) and *IKZF3* (r=0.641)(P<0.0001). Neutrophil numbers were correlated with *CDC42SE1* (r=0.708) *IKZF1* (r=0.596) and *MAP3K4* (r=0.593) (P<0.0001). These correlations were quite stable considering also the two timepoints separately. *GZMA* and *GZMK* encode for Granzyme A and K rsp that are serine esterase typical of cytotoxic T-lymphocyte; *CD8A* encodes in these cells for the main fraction that co-binds to type I major histocompatibility complex (MHC) molecules; *CX3CR1* for a receptor of a chemokine that governs the target adhesion of T-cells and monocytes. Results highlight the importance of lymphocyte counts to evidence the potential activity of cytotoxic T-lymphocytes. Concerning cells of erythroid origin, it could be interesting to connect counts of immature cells, such as reticulocytes, with the whole blood transcriptome of piglets of different ages.

Session 71 Theatre 4

Impact of ewes' GH2-Z genotypes on blood metabolites, mammary gland mRNA expression and milk traits

M.R. Marques[1], P. Mesquita[2,3], V. Pires[4], J.M.B. Ribeiro[1], R. Caldeira[4], A.P.L. Martins[5], C.C. Belo[1] and A.T. Belo[1]
[1]INIAV, UEISPSA, Quinta da Fonte Boa, 2005-048 Vale de Santarém, Portugal, [2]IPATIMUP, University of Porto, 4200-465 Porto, Portugal, [3]i3S, University of Porto, 4200-135 Porto, Portugal, [4]CIISA-FMV, Universidade de Lisboa, Avenida da Universidade Técnica, 1300-477 Lisboa, Portugal, [5]INIAV, UTI, Avenida da República, Quinta do Marquês, 2780-157 Oeiras, Portugal; rosario.marques@iniav.pt

Growth hormone (GH) is involved in mammary gland development, secretory cells maintenance, and productivity. The aim of the present research was to assess the impact of *GH2-Z* genotypes on blood metabolites, mammary gland mRNA expression, and milk yield in Serra da Estrela ewes. A total of 30 ewes were used for the study. The animals were divided based on their *GH2-Z* genotypes [AA (R9R/S63S), AB (R9C/S63S), and AE (R9R/S63G)]. Body condition score (BCS) and blood parameters: glucose, β-hydroxybutyrate (β-HBA), non-esterified fatty acids (NEFA), triglycerides, albumin, urea-N, GH, insulin growth factor-1 (IGF-1), and insulin were evaluated. Relative mRNA expression of GH receptor (GHR), signal transducer, and activator of transcription 5A (STAT5), IGF-1, and mucin 1 (Muc1) genes were assessed. Milk yield, composition [fat, protein, lactose, total solids (TS), and solids non-fat (SNF) were determined. Data were analysed using a SAS GLM procedure. Significant differences were observed in NEFA (P<0.05) and IGF-1 (P<0,05) with AB ewes presenting higher values compared to AE ewes (NEFA: 0.63±0.08 vs 0.34±0.08 mmol/l; IGF-1: 30.57±1.55 vs 23.37±2.08 ng/ml). Also, significant differences were detected in Urea-N with higher levels for AA in relation to AB ewes (31.73±1.17 vs 25.00±1.27 mg/dl; P<0.01). No differences were observed regarding the relative mRNA expression of the studied genes, and the milk yield and composition (P>0.05) with the exception of protein and SNF contents, both higher in milk from AB vs AE ewes (P<0.05). In conclusion, ewes' GH2-Z genotypes influenced NEFA, IGF-1, and urea-N circulatory levels in the blood, and protein and SNF content in milk. But, genotypes had no effect on gene expression in the mammary gland, nor in milk production levels in Serra da Estrela ewes. Funding: Project PTDC/CVT/112054/2009 funded by FCT.

Session 71 Theatre 5

The adverse effects of intrauterine growth restriction on body growth in pigs are mediated by KLB

F.X. Donadeu, S.O. Dan-Jumbo, Y. Cortes-Araya, M. Salavati, E. Clark, C. Stenhouse, C.J. Ashworth and C.L. Esteves
The Roslin Institute, The University of Edinburgh, Easter Bush, EH25 9RG, United Kingdom; xavier.donadeu@roslin.ed.ac.uk

Intrauterine growth restriction (IUGR) is a leading cause of neonatal morbidity and mortality in pigs as well as humans. Developmental adaptations in IUGR foetuses result in a permanent reduction in skeletal muscle mass and a tendency to accumulate body fat later in life, together with poor growth efficiency and reduced carcass value in affected animals, thus representing a significant source of lost productivity and economic loss for the pork industry. We performed a series of studies to investigate the molecular basis of altered skeletal muscle and adipose tissue development in naturally-occurring IUGR piglets. Through transcriptome analyses of foetal tissues we found that the Fibroblast Growth Factor 21 (FGF21) co-receptor, Beta-klotho (KLB), was expressed at distinctly higher levels in skeletal muscle from IUGR compared to normal weight (NW) littermates. Moreover, FGF21 concentrations in plasma were higher in IUGR foetuses. We then obtained progenitor cell populations from: (1) skeletal muscle; and (2) adipose tissue from IUGR and NW littermates, and determined *in vitro* the effects of KLB and FGF21 on the capacity of those cells to differentiate into several tissue lineages involved in body growth. We found that cells from IUGR littermates displayed: (1) reduced myogenesis; and (2) increased adipogenic and fibrogenic capacity simultaneous with a reduction in chondrogenesis and osteogenesis, compared to cells from NW littermates. Moreover, downregulation of KLB in progenitor cells using siRNAs promoted myogenesis and inhibited adipogenesis, whereas treatment with FGF21 had opposite and dose-dependent effects on those cells. In conclusion, our novel results identify FGF21 signalling through KLB as a potentially critical mechanism mediating the programming effects of IUGR on progenitor cell fate in developing pig tissues, and which result in preferential accumulation of fat at the expense of other tissues essential for normal growth, most notably skeletal muscle. Our data sheds new light into the pathogenesis of IUGR and provides new targets for preventing or ameliorating the negative consequences of IUGR on pig health and productivity.

Session 71 Theatre 6

Exploring fibre type, adipocyte size and gene expression relationship in two muscles in beef cattle

O. Urrutia, B. Soret, A. Arana, L. Alfonso and J.A. Mendizabal
Public University of Navarre, Campus Arrosadia, 31006, Spain; olaia.urrutia@unavarra.es

Meat quality is affected by intramuscular fat (IMF) content, a highly variable trait. Differences in fat accretion, muscle fibre type and metabolism might be related to anatomical regions and muscle functionalities, for instance glycolytic muscles have lower IMF level. The aim of this study was to analyse the relationship between muscle fibre contractile and metabolic characteristics, adipocyte size and gene expression in beef cattle. Muscles *Longissimus thoracis* (*LT*) that expressed MyHC isoforms I (16%), IIX (47%) and IIA (36%) and had unimodal adipocyte size distribution, and *Masseter* (*MS*), that only expressed MyHC I and had bimodal adipocyte population, from Pirenaica (n=16) and Friesian (n=16) young bulls aged 10-12 months old were used. Pearson correlations between adipocyte size, metabolic enzyme activities, fibre type and gene expression were calculated. In *LT*, MyHC I was positively correlated with isocitrate dehydrogenase (ICDH, 0.80, P=0.000) and citrate synthase (CS, 0.76, P=0.001) and negatively with lactate dehydrogenase (LDH, -0.52, P=0.04) and phosphofructokinase (-0.66, P=0.006) while MyHC IIX was negatively correlated with ICDH (-0.68, P=0.004) and CS (-0.71, P=0.002) but positively with LDH (0.59, P=0.017). Also in *LT*, adipocyte size was negatively correlated with *MYOD* gene (-0.498, P=0.05), suggesting its higher expression, the smaller adipocytes. IMF was negatively related with *SCD* (-0.74, P=0.001) and *HSL* (-0.58, P=0.020) in *LT* while in both muscles *FABP4* expression was positively related with *SCD* (*LT*: 0.85; *MS*: 0.64; P≤0.01), *ADIPOQ* (*LT*: 0.80; *MS*: 0.75; P≤0.001) and *leptin* (*LT*: 0.75; *MS*: 0.64; P≤0.01). In *MS*, adipocyte size and enzyme activities were not related but *leptin* positively correlated with small/large adipocytes ratio (0.58, P=0.020) and negatively with adipocyte size (-0.60, P=0.015). Some common features for *LT* and *MS* adipocytes such as the relation between *FAPB4* (fatty acid metabolism and transport into cells) with *ADIPOQ* (glucose and fatty acid metabolism) and *leptin* (energy homeostasis regulator) were found. There were differential traits as well, as the relation between adipocyte size and *leptin* in oxidative *MS* but with myogenic and lipolytic genes in more glycolytic *LT*.

Session 71 Theatre 7

Variability of colonic and ileal adult intestinal stem cell derived organoids in pigs

R.S.C. Rikkers[1], M.F.W. Te Pas[1], O. Madsen[2], S.K. Kar[1], D. Schokker[1], L. Kruijt[1], A.A.C. De Wit[1], L.M.G. Verschuren[3], S. Verstringe[4] and E.D. Ellen[1]

[1]Wageningen University & Research, Wageningen Livestock Research, Droevendaalsesteeg 1, 6700 AH Wageningen, the Netherlands, [2]Wageningen University & Research, Animal Breeding & Genomics, Droevendaalsesteeg 1, 6700 AH Wageningen, the Netherlands, [3]Topigs Norsvin Research Center B.V., Shoenaker 6, 6641 SZ Beuningen, the Netherlands, [4]Nutrition Sciences NV, Booiebos 5, 9031 Drongen, Belgium; roxann1.rikkers@wur.nl

Intestinal organoids of livestock animals are increasingly used as an *in vitro* research tool to investigate gut health and functioning and to unravel complex phenotypes. The transcriptomic resemblance of organoids with the original *in vivo* intestinal tissue and variability within cultured organoids has previously been studied for jejunum, but for colon and ileum such comparisons are still missing. Furthermore, for ileal organoids the sampling location might have an effect considering the presence or absence of Peyer's Patches (PP). The aims of this study were: (1) investigate similarities between tissue and organoid transcriptomic profiles at four different locations (ileum PP, ileum nonPP, mid colon and distal colon); and (2) investigate the variability within sampling site of these four locations. From six pigs, we harvested tissue samples from each location. These tissue samples were used to develop 3D organoids and we compared the transcriptomic RNA profiles of the original tissue and derived organoid samples. The results showed that tissue and organoid samples formed separate clusters based on their general gene expression profiles. This could be explained since tissue contain more diverse cell types compared to organoids. Average Pearson correlations of general gene expression profiles within ileum PP, ileum nonPP, mid colon and distal colon organoids were respectively, 0.97±0.03, 0.98±0.02, 0.94±0.06 and 0.97±0.02 suggesting low variability within location. Future work will focus on between and within variability of the four locations. Our results suggest that derived organoids resemble their original tissue quite well and at least for colonic organoids the exact sampling location has limited influence, and thus organoids could be a promising practical application to investigate colon function.

Session 71 Theatre 8

Using piecewise regression models to evaluate beef cow responses to nutritional challenges

A. Aliakbari, J. Pires, L. Barreto-Mendes, F. Blanc, I. Cassar-Malek, I. Ortigues-Marty and A. De La Torre

INRAE, UMR Herbivores, Université Clermont Auvergne, VetAgro Sup, Saint-Genès-Champanelle, 63122, France; amir.aliakbari@inrae.fr

Productive and metabolic adaptive responses to nutritional challenges are key components to study resilience and robustness regarding the challenges facing livestock production systems. Different methods have been proposed to evaluate responses to perturbations, however, there is no agreement on a reference method that accounts for the complex adaptive response of animals to FR. The objective of the present study was to assess the potential of piecewise models to quantify the responses of beef cows to nutritional challenges and describe between-cow variability. In total, 22 suckling Charolais cows were exposed to 4 nutritional challenges, each consisting of 4 days of feed restriction (FR1, FR2, FR3, and FR4; covered 50% of net energy requirements), followed by 3 days of *ad libitum* forage intake. Cows were allowed 14 days of *ad libitum* forage intake between the first challenge and the other three successive challenges. Data of daily milk yield (MY, kg/d) and plasma non-esterified fatty acid concentrations (NEFA, mM) were analysed using continuous piecewise regression models. The estimated individual regression coefficients for MY and NEFA from the models were used as indicators of adaptive responses. The coefficients were then grouped and compared using hierarchical clustering. The MY decreased in average -0.28, -0.58, -0.95, and -1.08 kg during FR1, FR2, FR3, and FR4, respectively, compared to the preceding *ad libitum* period. In contrast, NEFA increased 0.19, 0.16, 0.23, and 0.33 mM in FR1 to FR4, respectively, compared to the previous *ad libitum* period. Clusters of animals with greater MY change during FR presented smaller changes in plasma NEFA, compared to clusters of animals with smaller MY changes. These results showed that piecewise regression models are relevant to highlight differences between cows in the prioritisation of milk production underpinned by the use of body reserves. The continuous piecewise linear models allowed to reduce the complexity of beef cow responses to FR in the form of individual regression coefficients, which can be useful to define suitable proxies for resilience and robustness.

Effect of enhanced early life nutrition on pituitary miRNA and mRNA expression in the bull calf

K. Keogh[1], S. Coen[1], P. Lonergan[2], S. Fair[3] and D.A. Kenny[1]
[1]Teagasc, Animal and Bioscience Research Department, Grange, Dunsany, Co. Meath, Ireland, [2]University College Dublin, School of Agriculture and Food Science, Belfield, Dublin 4, Ireland, [3]University of Limerick, Laboratory of Animal Reproduction, Department of Biological Sciences, Limerick, Ireland; kate.a.keogh@teagasc.ie

Enhanced early life nutrition is known to advance reproductive development in the bull calf. This is mediated through earlier secretion of hypothalamic derived gonadotropin releasing hormone, leading to earlier release of the gonadotropins; follicle stimulating and luteinizing hormones from within the anterior pituitary gland. However, although the effect of enhanced nutrition towards earlier reproductive development is established, the precise molecular mechanisms regulating this effect remain to be elucidated fully. The objective of this study was to conduct an integrative analysis of global miRNA and mRNA expression profiles derived from the anterior pituitary of bull calves offered either a high or moderate plane of nutrition from 2 to 12 weeks of life. Holstein-Friesian bull calves (n=30; mean age: 17.5 days; mean bodyweight 48.8 kg), were assigned to either a high (H) or moderate (M) energy dietary treatment group. All calves were euthanised at 12 weeks of age, pituitary tissue harvested and global miRNAseq and mRNAseq analyses undertaken. Differential feeding up to 12 weeks of age resulted in greater growth rates in H calves (0.88 v 0.58 kg/day, P<0.001). Bioinformatic analyses revealed differential expression (Padj<0.1; fold change>1.5) of 5 miRNA and 37 mRNA between H and M groups. Target mRNA genes of 3 differentially expressed miRNA were also differentially expressed, displaying opposite direction of effect, indicating a direct relationship between the miRNA and corresponding mRNA. Of particular interest was miR-205, which was down-regulated in H calves with corresponding target mRNA genes (*CARTPT, PCSK1*) up-regulated in H calves. *CARTPT* is involved in appetite and energy balance regulation, whilst *PCSK1* is involved in the processing of hormones, including the gonadotropins. Results from this study indicate a role for miR-205 in mediating the interaction between enhanced metabolic status and reproductive development in bull calves. This study was funded by Science Foundation Ireland (16/IA/4474).

Session 71 Theatre 10

Comparative lipidomics profiling in porcine colostrum exosomes versus mature milk exosomes

R. Furioso Ferreira[1,2], M.H. Ghaffari[2], M. Audano[3], F. Ceciliani[4], D. Caruso[3], G. Savoini[4], A. Agazzi[4], V. Mrljak[1] and H. Sauerwein[2]
[1]University of Zagreb, Faculty of Veterinary Medicine, Heinzelova 55, 10000, Croatia, [2]University of Bonn, Institute of Animal Science, Physiology Unit, Katzenburgweg 7, 53115, Germany, [3]Università degli Studi di Milano, Department of Pharmacological and Biomolecular Sciences, Via Balzaretti 9, 20133 Milano, Italy, [4]Università degli Studi di Milano, Department of Veterinary Medicine, Via dell'Università 6, 26900 Lodi, Italy; rafaelaff.vet@gmail.com

Exosomes are membranous vesicles of endocytic origin, recently considered as major players in intercellular communication. Milk exosomes can mediate the regulation of the newborn's immune system and influence their growth and cellular development. Lipids are essential components of exosomal membranes, and the lipid components of exosomes have important implications on their function, acting not only in their structure, but in the exosome formation, uptake, and release to the extracellular environment. We aimed to assess differences in the lipidome of porcine colostrum exosomes versus milk exosomes. Milk samples were collected at day 0 (colostrum), 7, and 14 post-partum (mature milk). Exosomes were isolated by ultracentrifugation coupled with size exclusion chromatography and characterized by nanoparticle tracking analysis, transmission electron microscopy, and Western blotting for exosome marker. Lipids were extracted by the Folch method and the lipidome of isolated exosomes was determined following a liquid chromatography–quadrupole time-of-flight mass spectrometry approach. Data processing was carried out in MSDIAL with LipidBlast database and statistical analysis was carried out on the MetaboAnalyst 5.0 webtool. A total of 947 lipids from sixteen subclasses were identified in both colostrum and milk exosomes. When compared to colostrum exosomes, we identified 734 differentially abundant lipids in milk exosomes at day 7 and 779 differentially abundant lipids at day 14. The preliminary results unveil a distinct lipidomic profile in porcine milk exosomes in different lactating stages, with possible implications in their functional biology, including for strategies of uses of milk exosomes as vehicles for drug or additives delivery.

Session 71 Theatre 11

Pig saliva metabolome alterations under acute stress

L. Morgan[1,2], R.I.D. Birkler[2], S. Shaham-Niv[2], Y. Dong[2], L. Carmi[3], T. Wachsman[2], B. Yakobson[4], H. Cohen[5], J. Zohar[3], E. Gazit[2] and M. Bateson[1]

[1]Newcastle University, Newcastle, NE1 7RU, United Kingdom, [2]Tel Aviv University, Tel Aviv, 69978, Israel, [3]Chaim Sheba Medical Center, Ramat Gan, 52620, Israel, [4]Ministry of Agriculture and Rural Development, Rishon Lezion, 50250, Israel, [5]Ben-Gurion University of the Negev, Beer Sheva, 84105, Israel; liat.morgan@mail.huji.ac.il

Intensive farming, including several husbandry procedures, expose farm animals to stressful life, which might have both short and long term effects on their health and well-being. To understand the damaging effects of stress, improved objective physiological measures are needed. A single metabolite (small molecule) such as cortisol, which is in common use today, is insufficient for assessing individual health and welfare. We hypothesise that one solution to this problem is to use an untargeted metabolomics approach for measuring simultaneously high number of metabolites. 63 saliva samples were collected non-invasively, from 200 pigs at a group level. Saliva was collected from the same pigs twice: (1) In their familiar environment, as a control for homeostasis; (2) After 24 hours under known stressors for pigs; transport to the slaughterhouse, regrouping, and overnight at the new environment. Analyses were performed by ultrahigh-performance liquid-chromatography high-resolution mass spectrometry, in four different workflows, for broader metabolite coverage. Several thousand metabolites significantly altered under acute stress. Among others, amino acids, vitamins, hormones and phospholipids were significantly different (Adj $P<0.05$). While there was a 3.31 fold change difference in cortisol after acute stress, other metabolite features changed up to 333.33 fold change. Our results provide a fingerprint of the acute effect of stress and suggest candidate biomarkers with potential roles in impaired welfare. In addition, it demonstrates the potential of metabolomics in animal welfare and physiology research.

Session 71 Theatre 12

Large-scale analysis of chronic stress in dairy cows using hair cortisol and blood fructosamine

C. Grelet[1], J. Leblois[2], R. Reding[3], E.J.P. Strang[4], L. Dale[4], L. Manciaux[5], F.J. Auer[6], C. Egger-Danner[7], C. Happymoo[2], F. Dehareng[1] and H. Simon[1]

[1]Walloon Agricultural Research Center, Gembloux, 5030, Belgium, [2]Walloon Breeders Association Group, Ciney, 5590, Belgium, [3]Convis, Ettelbruck, 9010, Luxembourg, [4]LKV Baden Württemberg, Stuttgart, 70173, Germany, [5]Innoval, Noyal-sur-Vilaine, 35530, France, [6]LKV-Austria, Vienna, 86430, Austria, [7]ZuchtData, Vienna, 86430, Austria; h.simon@cra.wallonie.be

Stress in dairy herds can occur from multiples sources. When stress becomes chronic because of a long duration and inability of animals to adapt, it is likely to affect emotional state, health, immunity, fertility and milk production of cows. Therefore, it has a negative impact on welfare, economics and social acceptability of dairy farms. In a previous step of the HappyMoo project, two molecules were highlighted as chronic stress biomarkers: hair cortisol and blood fructosamine. However, those biomarkers have never been used for large scale monitoring in commercial farms and there is only few knowledge on the factors affecting them. The objective of this study is to evaluate the chronic stress at a large scale by measuring hair cortisol and blood fructosamine, and to get a better understanding of these biomarkers and their sources of variation. For this purpose, approximately 1,500 individual dairy cows were sampled for hair, blood and milk in Belgium, Luxembourg, Germany, France and Austria in more than a hundred of commercials farms. Herds were selected locally with the objective of gathering stressed and non-stressed herds. Hair samples were collected at the tail switch and analysed by ELISA for cortisol concentration, and blood samples analysed by ELISA for serum fructosamine concentration. Additional data were collected from milk recording organizations and through survey such as milk production, milk composition, parity, diet type, somatic cells, stocking density, floor quality, housing system, cleanliness, BCS, lameness, mastitis and any potential sources of stress. Regressions, ANOVA and multivariate analysis will aim to have a better understanding of stress in herds and highlight sources of variation affecting the hair cortisol and blood fructosamine. The samples were collected and are currently being analysed, and results will be presented and discussed during the conference.

Session 71 Poster 13

Fibre characterization of the Longissimus thoracis muscle in low marbling beef cattle during growth

B. Soret, J.A. Mendizabal, A. Arana, O. Urrutia and L. Alfonso
Public University of Navarre, Campus Arrosadia, 31006 Pamplona, Spain; soret@unavarra.es

Ruminant intramuscular adipose tissue is unique regarding some histological and metabolic aspects, such as its unimodal adipocyte size distribution and ability to use glucose as carbon source for de novo fatty acid synthesis. Muscle fibre and marbling determine beef quality and are tightly related: muscular environment can influence intramuscular adipocytes number and size and it is accepted that the expression of MyHC isoforms condition enzymatic and functional characteristics of muscle fibres. To analyse the relationship between those factors in beef cattle Longissimus thoracis muscle from 16 Pirenaica young bulls aged 6, 12 and 18 months old (M) fed with low (l) or high (h) energy ration from 12 to 18M and with low marbling tendency (2.5, 2.4, 3.3 and 3.6 marbling percentage respectively) were used. Activity of enzymes (µmol/min/g protein) of glycolytic (LDH and PFK) and oxidative metabolisms (ICDH and CS) and oxidative phosphorylation (COX) were determined. Relative amount of myosin heavy chain (MyHC) isoforms was quantified by SDS-PAGE. Diameter of adipocytes was measured after collagenase digestion by image analysis. Activity of PKF, CS and COX did not vary during growth; LDH activity tended to be lower at 12M and higher in 18Ml while ICDH followed the opposite pattern. MyHC IIa (fast oxi-glycolytic) and MyHC IIx (fast glycolytic) were the predominant type of fibre in all groups although the former did not change and the latter was predominant at 6M and decreased thereafter. On the contrary, slow-oxidative MyHC I isoform was lower in 6M group and increased with age ($P<0.05$). Adipocyte mean diameter was lower at 18Ml and higher in 6M and in 18Mh groups ($P<0.05$). No correlation between adipocyte size and enzyme activities or MyHC isoforms were found, although MyHC I correlated positively with MyHC IIa and negatively with MyHC IIx, LDH, COX and CS ($P<0.05$). In this work, no correlations of MyHC isoforms or enzyme activities with adipocyte diameter were found but a change toward more oxidative and less glycolytic isoforms with growth was clearly observed while the more divergent metabolism was found in 12M and 18M fed with the low energy ration, probably indicating an effect of muscle nutrient availability.

Session 71 Poster 14

The effect of PET microplastic on the number of VIP-immunoreactive neurons in the porcine duodenum

I. Gałęcka and J. Całka
Faculty of Veterinary Medicine, University of Warmia and Mazury in Olsztyn, Department of Clinical Physiology, Oczapowskiego 13, 10-719 Olsztyn, Poland; ismena.kordylewska@uwm.edu.pl

The increase in the production of various types of plastics poses a rising threat to health of humans and animals. Microplastic (MP) is defined as plastic particles with a diameter below 5 mm that are ubiquitous in the natural environment and in food. Polyethylene terephthalate (PET) is widely used for packaging of foods and beverages. The enteric nervous system (ENS) is characterized by high plasticity, which, under the influence of pathological factors or xenobiotics, changes the synthesis of neurotransmitters. Vasoactive intestinal polypeptide (VIP) is considered to be one of the most important substances involved in the intestinal regulatory processes. Vasoactive intestinal peptide is also one of the main neuroprotective factors within the ENS. The objective of this experiment was to evaluate the effect of high dose of PET microplastic administered orally in the enteric nervous system of the swine duodenum. This study was carried out on 10 immature gilts divided into 2 groups: C group – the animals were administered empty gelatine capsules; HD group – the animals were administrated PET MP in dose 1 g/animal/day. The experiment lasted 4 weeks. After this period, pigs were euthanized and fragments of duodenum were collected and fixed. The frozen sections were processed with the double immunofluorescent staining method using protein gene product 9.5 (PGP 9.5) as neuronal marker and against VIP as primary antibodies. Alexa Fluor 488 and 546 were used as secondary antibodies. Stained sections were examined under Zeiss Axio Imager.M2 fluorescence microscope. Analysis of the obtained results revealed changes in the number of VIP-positive neurons in the myenteric, outer and inner submucosal ganglia in the porcine duodenum as a response to microplastic supplementation. The conducted research suggests that PET microplastic presence in the gastrointestinal tract may affect the enteric nervous system by modulating VIP-immunoreactive neurons in the porcine duodenum. This study was supported by the National Science Centre, Poland – Preludium-19 grant no. 2020/37/N/NZ7/01383.

Session 71 — Poster 15

Bile acids in serum of dairy cows with high or normal body condition

L. Dicks[1], K. Schuh[2], C. Prehn[3], M.H. Ghaffari[1], H. Sadri[4], H. Sauerwein[1] and S. Häussler[1]

[1]University of Bonn, Institute of Animal Science, Katzenburgweg 7-9, 53115 Bonn, Germany, [2]University of Applied Sciences Bingen, Institute Feed Research GmbH, Berlinstr 109, 55411 Bingen am Rhein, Germany, [3]Helmholzzentrum München, Metabolomics & Proteomics Core, Ingolstädter Landstraße 1, 85764 Neuherberg, Germany, [4]University of Tabriz, Faculty of Veterinary Medicine, Shohadaye Ghavvas Blvd, 5166616471 Tabriz, Iran; ldicks@uni-bonn.de

Bile acids (BA) facilitate digestion and absorption of lipids and regulate cholesterol homeostasis, but they are also signalling molecules involved in lipid, glucose, and energy homeostasis. The aim of this study was to investigate the primary (p) and secondary (s) BA profiles in serum of dairy cows with high (HBCS) and normal body condition (NBCS). Fifteen wk ante partum (ap), multiparous Holstein cows were divided into either a HBCS (n=19) or a NBCS (n=19) group. For augmenting the difference, HBCS received a more energy-dense ration, i.e. 0.4 NE_L MJ/kg DM more than the NBCS cows, until dry-off. Thereafter, the same rations were fed to both groups during dry-off and lactation. One wk before calving, the groups differed by 0.7 BCS points and 1.1 cm backfat thickness. From 3 wk ap, DMI was documented weekly. BCS loss was greater during lactation in HBCS cows than in NBCS cows. Serum samples from wk -7, 1, 3, and 12 relative to parturition were assayed by LC-ESI-MS/MS with the Biocrates™ BA Kit (BIOCRATES Life Sciences AG, Austria), and 14 pBA and sBA were detected. Data were analysed with a linear mixed model and relationships were assessed by Spearman correlations (SPSS 28). The mean concentrations of pBA and sBA across all time-points were 1.3- and 1.2-fold higher in NBCS cows compared with HBCS cows. In both groups, pBA concentrations were higher (up to 3.4-fold) after calving compared with ap values. Regarding sBA in HBCS cows, concentrations increased 1.5-fold (wk 3) and 2.0-fold (wk 12) compared to ap values and 1.7-fold from ap until wk 3 in NBCS cows. DMI was greater in NBCS than in HBCS cows until calving and also during the lactation until wk 12 post partum. Although neither pBA nor sBA were correlated with DMI in either group, the higher BA values in NBCS cows may be due to the higher DMI, leading to a delayed response in serum BA.

Session 71 — Poster 16

PSE-like pork – looking for potential biomarkers and links with myopathy based on proteomic analysis

P. Suliga[1], S. Mebre-Abie[1], B. Egelandsdal[1], O. Alvseike[2], A. Johny[3], P. Kathiresan[1] and D. Münch[1]

[1]Norwegian University of Life Sciences, Faculty of Ecology and Natural Resource Management, Elizabeth Stephansens v. 15, 1430, Norway, [2]Norwegian Meat and Poultry Research Centre, Animalia, Lørenveien 38, 0585, Norway, [3]Norwegian Institute of Food, Fisheries and Aquaculture Research, Osloveien 1, 1430, Norway; pawel.suliga@nmbu.no

Both, for industry and research, there is a need for developing novel detection methods that can identify common pork quality defects, such as Pale Soft Exudative (PSE) meat. The objective of this study was to screen for unique biomarker proteins in PSE-like meat. To this end, we tested a group of randomly selected Musculus semimembranosus samples (n=84) for common quality parameters (pH_u, drip loss, lightness). In addition, bioimpedance measurements were included for accessing ultra-structural damage. Two groups of samples represented normal and poor ham quality based on a multivariate approach. Our LC-MS/MS analyses was performed on the sarcoplasmic fraction of muscle samples and revealed a total of 516 proteins. We identified 91 unique proteins for the group of poor-quality samples, and 164 unique proteins in normal-quality controls. This supports that a higher number of proteins being degraded in the poor-quality group. Protein enrichment analysis of the unique proteins in poor quality meat identified proteins of specific functional clusters related to cytoskeletal organization, muscle contraction and voltage-gated calcium channel signalling. In addition, several unique proteins in poor quality samples were previously associated with myopathies. Together, our study suggests novel protein candidates not only for quality defect detection but also for better understanding causal routes of quality deterioration and possible links with muscle diseases.

Session 71 Poster 17

Exosomes' isolation and proteomic profile of sheep mononuclear cells supernatant using two methods

M.G. Ciliberti, A. Della Malva, M. Di Corcia, A. Santillo, R. Marino, M. Albenzio and M. Caroprese
University of Foggia, Department of Agriculture, Food, Natural Resources, and Engineering (DAFNE), Via Napoli 25, 71122 Foggia, Italy; maria.ciliberti@unifg.it

Recent advances in exosomes (EV) studies suggested their crucial role as promising substrate in diagnostic human and animal studies. Different methods have been developed for the EV's isolation, among which ultracentrifugation-based, which is considered very time-consuming. Consequently, novel reagent-based methods are emerging for the isolation of intact EV. The aims of the present experiment were: (1) the isolation of EV from sheep peripheral blood mononuclear cells (PBMC) using two different methods: ultracentrifugation (UC) and reagent-based (REA) methods; and (2) the determination of total EV number and the related proteomic profile. A final concentration of 0.5×106 PBMC was cultured for 24 h at 37 °C and activated or not (CON) with mitogen (LPS). Supernatants were collected and subjected to UC: centrifugation at 300 g for 5 minutes, then the supernatant was subjected to microfiltration (0.22 µm), to avoid microvesicles contamination, followed by centrifugation at 17,000×*g* for 30 minutes, and a final ultracentrifugation at 35,000×*g* for 90 minutes at 4 °C for 2 times. The second method was based on a specific reagent (REA) for EV isolation characterized by a first centrifugation at 2,000×*g* for 30 minutes followed by overnight incubation of supernatants with REA at 4 °C, and the last centrifugation at 10,000×*g* for 1 h a 4 °C. The number of EV isolated was calculated using CD81 ELISA kit. Lysis of isolated EV was carried out using RIPA buffer and analysed by HPLC-MS/MS. Isolation methods significantly influenced the number of EVs (P<0.001); indeed, both EVs isolated with REA activated or not with mitogen resulted in a higher number than the EV number isolated with UC. The total protein number identified from the EV REA method was about 76 in CON and 74 in LPS treatment, among which 56 are in common. On the contrary, the number of identified proteins from EV isolated with UC was respectively 24 and 10 for EV isolated in LPS and CON treatment with only 4 proteins in common. These data demonstrated that the use of the EV isolation method based on a specific reagent resulted in increasing EV number and protein content.

Session 71 Poster 18

Effect of early weaning on the gene expression profiles in the skeletal muscle of Nelore beef calves

G.L. Pereira[1], R.A. Curi[1], L.A.L. Chardulo[1], W.A. Baldassini[1], O.R. Machado Neto[1], P. Moriel[2], A.P. Enara[3], G.L.B. Tinoco[3], G.H. Russo[3] and J.A. Torrecilhas[1]
[1]College of Medicine Veterinary and Animal Science, Unesp, Animal Breeding and Nutrition, Rua Prof. Doutor Walter Mauricio Correa, CEP 18618-168- Botucatu, SP, Brazil, [2]University of Florida (UF/IFAS), Animal Science, 2020 McCarty Hall D, P.O. Box 110270 Gainesville, FL 32611, USA, [3]College of Agricultural and Veterinary Sciences, Unesp, Animal Science, Via de Acesso Prof. Paulo Donato Castellane, CEP 14884-900- Jaboticabal, SP, Brazil; guilherme.luis@unesp.br

The Nellore breed has advantages over taurine breeds in terms of adaptation to warmer climates, which does not happen in meat quality traits. The early weaning has enabled a better nutritional intake of calves weaned from 90 days. This change in diet during the post-natal period of calves can drastically alter the metabolism, modifying the composition of muscle fibres and the metabolic dynamics of lipids. The aim of this study was to identify genes differentially expressed in the skeletal muscle tissue of Nellore calves submitted to different weaning protocols. For this, eight calves weaned at 120 days (Early Weaning – EW) and eight calves weaned at 205 days (Conventional Weaning – CW) were used. The EW calves were reallocated to a specific lot with a concentrate intake of 20 g of DM/kg LW (20% of CP; 75% TDN). At the end of the 205-day period, aliquots of the longissimus thoracis muscle were collected by biopsy and the total RNA of these aliquots were extracted and used for sequencing in NextSeq™ platform (Illumina®). Differential gene expression analysis, as well as functional enrichment KEGG pathways, were performed on *R* software using *edgeR*, GO.db and org.Bt.eg.db *packages*. It was identified 158 differentially expressed genes (FDR<0.05). Considering enriched metabolic pathways (P<0.05) which had up-regulated genes in EW we can emphasis Fatty acid biosynthesis, Biosynthesis of unsaturated fatty acids and Adipocytokine signalling pathways. Among pathways which had down-regulated genes in EW we can emphasis Fatty acid degradation and Fatty acid elongation pathways. PPAR signalling pathways had both up and down regulated genes in EW. In this sense, early weaning may exert some positive influence on the fatty acid carcass composition of Nellore young bulls.

Ghrelin expression in the skin of the sheep changes in relation to diets

F. Mercati[1], E. Palmioli[1], P. Scocco[2], S. Moscatelli[2], C. Dall'aglio[1], D. Marini[1], P. Anipchenko[1] and M. Maranesi[1]
[1]University of Perugia, Department of Veterinary Medicine, Via San Costanzo 4, 06126, Italy, [2]University of Camerino, School of Biosciences and Veterinary Medicine, Via Pontoni 5, 62032, Italy; elisa.palmioli@studenti.unipg.it

Ghrelin (GHRL) is a hormone involved in energy metabolism regulation. It induces appetite promoting the use of carbohydrates; inhibits lipid oxidation; stimulates gastric acid secretion and motility. GHRL is also an anti-inflammatory peptide localized in nonspecific immune organs such as the skin where it provides a protective role for innate immunity. This work aims to investigate the expression of the GHRL, as a molecule involved in energy metabolism, in the skin of the sheep fed differently to assess if the feeding can modulate the secretion of the molecule in the skin. Hence, GHRL was investigated by immunohistochemistry and Real time-PCR in the skin of 15 Comisana × Appenninica adult female sheep reared in a semi-natural pasture of the Italian Central Apennines. Samples were collected from the thoracic region at the maximum pasture flowering (MxF, 5 ewes) and at the maximum pasture dryness (MxD, 10 ewes). Five ewes of the second group were fed with 600 gr/die/head of barley and corn (1: 1) in addition to the fresh forage (Exp). Immunohistochemistry was performed on formalin-fixed and paraffin-embedded sections by using a polyclonal anti- GHRL antibody (Abcam Cambridge UK). The primer sequences used for the Real-time PCR were as follows: F: GGAACCTAAGAAGCCGTCAGG, R: ATTTCCAGCTCGTCCTCTGC, NCBI n. DQ152959.1 (*Ovis aries*). GHRL staining was mainly observed in a confined area of the anagen hair follicles at the level of the suprabulbar region. Positivity involved the inner layers of the outer root sheath and the inner root sheath. GHRL was also observed in the smooth muscle cells. A significant difference in GHRL expression (3.6-fold) was evidenced by Real-time PCR between M×F vs Exp group. No differences were evaluated between the other groups. Therefore, the dietary supplementation of animals seems to have a positive modulating effect for the skin GHRL transcript compared to animals fed in MxF. This is a preliminary report that introduces GHRL investigation in the sheep skin however, the influence of diet on this molecule needs further elucidation.

Session 71 Poster 20

morphological identification of the leptin and its receptor in the abomasum of the sheep

E.P. Palmioli[1], P.S. Scocco[2], C.D. Dall'aglio[1], K.D. Dobrzyn[3], M.B. Bellesi[2] and F.M. Mercati[1]
[1]University of Perugia, Department of Veterinary Medicine, Via San Costanzo 4, 06126 Perugia, Italy, [2]University of Camerino, School of Bioscience and Veterinary Medicine, Via Pontoni 5, 62032 Camerino, Italy, [3]University of Warmia and Mazury in Olsztyn, Department of Animal Anatomy and Physiology, Oczapowskiego St. 1A, 10-719 Olsztyn, Poland; elisa.palmioli@studenti.unipg.it

The growing summer drought stress is affecting the nutritional value of pasture, no longer sufficient to support the nutritional status of the animals. This study aimed to describe the localization and distribution of both leptin (Ob) and its receptor (Ob-R) in sheep abomasum to identify biological markers of nutritional status and therefore sheep welfare. The Ob, an adipokine mainly produced by adipose tissue, has been detected in the human and rat gastrointestinal tract, where it regulates the rate of gastric emptying. Furthermore, Ob regulates food intake by an anorexigenic action. Abomasum samples of 15 adult female sheep reared in a semi-natural pasture were used to identify Ob and Ob-R by immunohistochemistry. Formalin-fixed paraffin-embedded sections, microwaved in the citrate buffer (pH 6) for antigen retrieval, were incubated with mouse monoclonal anti-Ob and rabbit polyclonal anti-Ob-R primary antibodies. Horse anti-mouse and goat anti-rabbit biotin-conjugated secondary antibodies were used. Immunofluorescent double-label localization of the Ob system with different neuroendocrine hormones (synaptophysin, chromogranin and serotonin) was conducted to distinguish the gland cell types. Both Ob and Ob-R have been detected in the mucous layer of the abomasum. The positive cells were localized in the lower half of fundic glands, and they were labelled as chief cells based on their morphological characteristics. Double-label immunohistochemistry showed that the cells positive to serotonin did not stain with Ob and Ob-R, while the positive ones to synaptophysin and chromogranin partially colocalize with chief cells secreting Ob and Ob-R highlighting a different behaviour of the neuroendocrine cell populations. The abundant presence of Ob and Ob-R in the gastric glands suggests a role of the leptin system in the regulation of abomasum functions in sheep.

Session 72 Theatre 1

Application of a sustainability assessment tool on European pig farms

A.R. Ruckli[1], S.J. Hörtenhuber[1], P. Ferrari[2], J. Guy[3], J. Helmerichs[4], R. Hoste[5], C. Hubbard[3], N. Kasperczyk[6], C. Leeb[1], A. Malak-Rawlikowska[7], A. Valros[8] and S. Dippel[4]
[1]University of Natural Resources and Life Sciences, Gregor-Mendel-Str. 33, 1180 Vienna, Austria, [2]Centro Ricerche Produzioni Animali, C.R.P.A. S.p.A., 42121 Reggio Emilia, Italy, [3]Newcastle University, Kings Road, NE1 7RU Newcastle upon Tyne, United Kingdom, [4]Friedrich-Loeffler-Institut, Dörnbergstr. 25/27, 29223 Celle, Germany, [5]Wageningen University & Research, De Elst 1, P.O. Box 338, 6700 AH Wageningen, the Netherlands, [6]Justus-Liebig-Universität Gießen, Karl-Glöckner-Str. 21 C, 35394 Gießen, Germany, [7]Institute of Economics and Finance, Warsaw University of Life Sciences, 02-787 Warsaw, Poland, [8]University of Helsinki, P.O. Box 57, 00014 University of Helsinki, Finland; antonia.ruckli@boku.ac.at

Sustainability plays a crucial role in agriculture and improvements are urgently needed to secure food supply with limited natural resources. Therefore, within the project SusPigSys a tool was developed to assess sustainability of pig farms. The assessment tool covers the three dimensions of economy, environment and social wellbeing, as well as animal health and welfare as a new fourth dimension. The tool was developed in stages including the selection of indicators, their scaling and allocation to subthemes and themes as well as a weighting procedure. In total, 63 farms across seven countries (13 breeding, 27 breeding-to-finishing and 23 finishing farms) were analysed with this tool on a scale of 0 = poor to 100 = good. Farms performed best in the sustainability themes of Human-animal relationship (median: 75-78 depending on farm type), Water (76-80) and several Social themes (60-100), and worst in Animal Comfort (36-43), Biodiversity (39-42) and Economic resilience in breeding farms (23). Furthermore, some themes showed larger variability (e.g. Technical efficiency, Economic resilience, Fair trading practices) than others (e.g. Material and energy, Decent livelihood, Absence of injuries and disease). Variation in scores among farms shows room for improvement and highlights to farmers more sustainable farming practices.

Session 72 Theatre 2

How to include sustainability in breeding goals – an overview

O. Vangen[1], E. Gjerlaug-Enger[2] and K. Kolstad[3]
[1]Norwegian University of Life Sciences, Animal and aquacultural sciences, P.O. Box 5003, 1432 Aas, Norway, [2]Norsvin SA, Storhamargata 44, 2317 Hamar, Norway, [3]Norwegian University of Life Sciences, Faculty of Bioscience, P.O. Box 5003, 1432 Aas, Norway; odd.vangen@nmbu.no

Sustainable breeding goals has generally been defined as: (1) selection for many traits simultaneously to secure a balanced biology in the animals; (2) breeding in a long term perspective to secure a better balanced biology; (3) recording of traits in its natural production environment (to secure adaptation to the production environment); (4) account for biological limitations, and non-linear correlations between traits; (5) maintain a large enough effective population size to prevent inbreeding and development of genetic defects. How to apply such breeding goals in pigs? Fertility, health and product quality are included in many breeding programs in addition to the traditional production traits. However, selection for litter size without including survival of piglets is not sustainable. Survival of piglets has to be included in the breeding goal in order to obtain sustainability. Breeding for leanness is not optimal for mothering abilities or for product quality. Optimal leanness should be the breeding goal for body composition, especially in dam lines. Meat and fat quality parameters should be included in breeding goals, especially in sire lines. Health issues like leg weakness, shoulder sores and/or adaptation to outdoor environments need actions like recording of body conditions scores and conformation traits in the environment the pigs are kept under. For many of these traits, technologies like CT (computer tomography), NIR (near infrared reflectance) and other techniques are important tools to reach these goals. Number of traits in breeding goals have been largely expanded the last 20 years, in some maternal lines these more than 20 traits are included. As alternative production environments are increasing in pig breeding (like outdoor pig farming), recording of new traits is important to improve the sustainability in such production systems. Generally, breeding goals in pigs still have to be improved for further sustainability.

Contribution of pig farming to particulate emissions: analysis of emission factors and simulations

N. Guingand[1], M. Hassouna[2] and S. Lagadec[3]
[1]IFIP institut du Porc, La motte au vicomte, 35651 Le Rheu, France, [2]INRAE, UMR SAS, 65 rue de Saint Brieuc, 35000 Rennes, France, [3]CRAB, rue Le Lannou, 35042 Rennes, France; nadine.guingand@ifip.asso.fr

France is committed to reducing its emissions of fine particulate matter less than 2.5 microns in size (PM2.5) by 57% by 2030 compared to 2005 emissions. Compliance with the commitments is based on the annual inventory published by Citepa, which calculates the contribution of pig farming by multiplying the number of pigs by the emission factor (EF) published by the European Monitoring and Evaluation Programme (EMEP, 2019). First, we reviewed the evolution of EMEP's EF since 2006 as well as EF currently available in the literature, based on available information and most influential parameters. A database has been built including the main parameters influencing particules emissions and a total of 107 EF were collected on the three main animal categories (sows, piglets, pigs). EF show wide variability due to the diversity of housing characteristics and measurement methods used. Most of published EF concern fattening pigs but with few information about rearing conditions (e.g. type of floor, feed). Piglets and sows are poorly documented, as an example, no EF is available for piglets on strawdust litter. In parallel, EMEP's EF values have been drastically decreased since 2006. This evolution is explained by EMEP as the result of the increase of publications concerning PM. Nevertheless, since 2019, the distinction between emission factor for pigs bred on litter or slatted systems is stopped because of the lack of published EF concerning litter systems. Second, we compared five scenarios for PM emissions to the 2019 inventory. These scenarios were developed using EMEP's EF, but also average EF estimated with published EF, including French specific one. The results show the relevancy of establishing national EF including production specificities and enabling the calculation of representative inventories which can be the basis for realistic policies to reduce PM.

Is fattening immunocastrated pigs influencing the environmental impact of pork production?

I. Dittrich and J. Krieter
Kiel University, Institute of Animal Breeding and Husbandry, Olshausenstr. 40, 24098 Kiel, Germany; idittrich@tierzucht.uni-kiel.de

German pig production is challenged by the omission of castrating piglets surgically without anaesthesia as alternatives e.g. immunocastration are still questioned towards efficacy and rarely used. Despite this, immunocastrated pigs (IC) have a boar-like metabolism with reduced feed conversion ratio (FCR), increased average daily gain (ADG) and lean meat content (LM). These beneficial performance indicators also entail potential improvements of the environmental impact of pork as the consumed feed has a major impact on e.g. global warming potential (GWP). Thus, in this study a life cycle assessment (LCA; CML Baseline 3.06, SimaPro 9.2.0.1) was carried out to compare the GWP (kg CO_2 eq.), acidification potential (AP: g SO_2 eq.) and eutrophication potential (EP: g PO_4 eq.) of a simulated farrowing-to-finish farm (322 sows; 3,000 fattening places), fattening either surgical castrated (SC), IC or intact males (IM) with a standard feed. In a pig farm model, FCR (SC: 2.75 kg/kg, IC: 2.48 kg/kg, IM: 2.39 kg/kg), ADG (SC: 902 g/d, IC: 948 g/d, IM: 970 g/d) and LM (SC: 58.3%, IC: 59.6%, IM: 61.2%) were used to estimate pork production, feed consumption and fattened pigs per animal place and year. These were used as basis for LCA with the functional unit 1 kg pork at farm gate. In comparison to SC (GWP: 2.85), IC and IM produced 0.06 kg and 0.08 kg less GWP for each kg pork, respectively. Furthermore, IC and IM had lower AP and EP as SC (AP: 47.92 g, EP: 23.17 g), as AP was reduced by 0.43 g (IC) and 0.59 g (IM), respectively. EP was reduced by 0.56 g (IC) and 0.76 g (IM), respectively. Transferring these results to the current German pig production with 1% IC, an increase of IC to 20% and the simultaneous decrease of SC could reduce the environmental impact of domestic produced pork. As approximately 5.5 Million male pigs are fattened, the additional amount of IC would decrease the environmental impact of all male fattening pigs by at least 0.4% GWP, 0.2% AP and 0.4% EP. These decreases are equivalent to reduced GWP of 10,000 t CO_2 eq., AP of 100 t SO_2 eq. and EP of 800 t PO_4 eq. Summarising, promoting the fattening of IC as an alternative to SC has the potential to reduce environmental impact of pork production.

Poster summary – sustainable pig production

M. Aluwé
ILVO, Scheldeweg, 9090 Melle, Belgium; marijke.aluwe@ilvo.vlaanderen.be

In this time slot, we take time to present some posters submitted in this session.

Consumer views on animal welfare and organic and low-input farming: Results from a European survey

J.K. Niemi[1] and Ppilow Consortium[2]
[1]Natural Resources Institute Finland (Luke), Kampusranta 9, 60320, Finland, [2]PPILOW consortium, https://www.ppilow.eu/, France; jarkko.niemi@luke.fi

While low-input farming, such as free-range or organic production, is often considered having high animal welfare standards, several ways to enhance animal welfare in low-input production exist. To promote good farming practices, it is valuable to know how the general public responds to such practices. The aim of this study was to examine citizens' expectations and reactions to new approaches to organic and low-input pig and poultry production. A quantitative survey instrument was implemented in nine European countries (Finland, UK, France, Denmark, the Netherlands, Belgium, Germany, Italy, Romania) The data were representative of each country's adult population, included altogether 3,601 responses and were analysed statistically. Citizens viewed low-input organic and non-organic production more favourably than conventional indoor production. While close to one quarter of citizens were unwilling to pay a price premium for low-input products, about one third was willing to pay at least 20% premium in contingent evaluation. Hence, there is room for 'mid-market' products requiring a small price premium. Most pig and poultry practices suggested in the survey were considered desirable by the respondents. Practices such as adjusting the nutrition to ensure animal health, enhancing the opportunities to express natural behaviour, provision of enrichment and increasing the space allowance were found desirable. Letting animals to a pasture or outdoor yard was considered desirable more frequently in pig and egg than in broiler production. A substantial proportion of citizens did not have a clear view on which features of production they favoured (e.g. beak trimming, use of veterinary medicines). This suggests that there is a lack of knowledge among citizens and that they may have challenges in assessing complex production practices. More communication between farmers and citizens, and communication that conveys consistent messages through trusted sources of information, which differ by country, is needed. PPILOW project has received funding from the European Union's Horizon 2020 research and innovation programme under grant agreement No 816172.

Session 72 Theatre 7

Review on the effects of sustainable extensification of pig husbandry on pork quality

A. Ludwiczak[1], E. Sell-Kubiak[2] and Meat Quality Consortium[1,2]
[1]Poznan University of Life Sciences, Department of Animal Breeding and Product Quality Assessment, Słoneczna 1, 62-002 Suchy Las, Poland, [2]Poznan University of Life Sciences, Department of Genetics and Animal Breeding, Wołyńska 33, 60-637, Poland; ewa.sell-kubiak@puls.edu.pl

Pork quality is a complex trait with multiple factors affecting it throughout the production chain, including the husbandry practices. Extensive production provides more space, environmental enrichment and varied diet with foraging opportunities in comparison to intensive husbandry. Also native breeds are often only used in extensive production. Our goal was to provide a comprehensive review of the impact of extensification of husbandry practices on intrinsic pork quality parameters with the focus on: breed, diet, space and environmental enrichment. We have summarized the characteristics of pork from Italian, Polish and Spanish breeds. Native breeds are usually more robust and adapt more easily to changing conditions. Their meat has higher quality (e.g. nutrient, colour, tenderness) than meat from conventional breeds. Thus crosses of native breeds with Large White or Duroc are often used to improve their meat quality. Next to breed, diet is the most studied factor affecting pork quality. It influences the nutrient value of the meat and its flavour. Thus there are differences between meat from pigs fed only commercial diet and kept indoors and pigs with the access to the pasture and foraging opportunities. Pork from free-range pigs contains more unsaturated fatty acids compared with meat from traditionally reared pigs, which increases the risk of lipid oxidation. Thus a proper level of antioxidants in the feed is highly important. More space per pig and the environmental enrichment considered separately do not have a clear effect on intrinsic meat quality. However, in combination with diet, housing system and access to the pasture provide an opportunity for more animal-friendly and sustainable pig production. However, in most European countries pigs are kept indoors, on slatted floors, as these systems are considered more economically effective, less labour-intensive, more controllable considering biosecurity and zoonotic hazards when compared to indoors deep-bedded systems and free-range. This project was financed by the European Union grant no. 101000344.

Session 72 Theatre 8

Individual trajectories of pig farming in France: mechanisms, determinants and prospects

C. Roguet, B. Lécuyer and L. Le Clerc
IFIP-institut du porc, Economy, La Motte au Vicomte, 35651 Le Rheu, France; christine.roguet@ifip.asso.fr

Carried out in 2021 and supported by the French pig interprofession Inaporc, this study described how the structural concentration of pig farms operates at the individual scale and explains determinants of changes and their effects. It was based on analysis of individual trajectories from 2014-2020 of French pig sites in the comprehensive national identification database BDPORC and interviews with 8 technicians of cooperatives and 31 farmers in three contrasting territories of France. From 2014-2020, the number of sites that produced more than 300 pigs a year (2/3 of the total number of sites and 99,3% of the 23,2 millions slaughter pigs produced in 2020) declined linearly from 10,794 to 9,492, and their mean annual production increased from 3,116 to 3,618 pigs. More than 70% of this decrease was due to the drop in the number of sites with sows (farrowers and farrow-to-finishers) from around 5,200 to 4,200 (-18% in 6 years vs -6% for the number of sites without sows). The structural concentration is the result of four trajectories (T). More than 70% of sites remained active over the period without changing their productive orientation (T1). Their production increased by more than 1.2 million slaughter pigs, mainly due to the strong gain in sow productivity. Indeed, the sow herd in France decreased by 5.3% over the period (980,000 sows in 2020). The production gain of the sites of T1 trajectory compensated for the loss of production resulting from the negative balance between cessation (14% of sites, -1.6 M pigs, T2) and entry into activity (3% of sites, +0.3 M pigs, T3). Lastly, 8% of sites changed production orientation (T4) among which 70% were farrow-to-finishers who stopped farrowing to become post-weaners and/or finishers. This development raises the question of the supply of piglets. The changes were determined by the breeders preferences' for pig production, projects or collective work, or by the search for farm self-sufficiency, greater value for their products and streamlining of the work. In the future, the lack of project leaders and the cessation of farrowing could lead some cooperatives to invest more in farms to guarantee a supply of piglets to their members and maintain their production volumes.

Session 72 Theatre 9

Which economic consequences to expect for pig farms around an African swine fever outbreak?

A. Aubry and B. Duflot
IFIP – Institut du Porc, BP 35104, 35651 Le Rheu Cedex, France; alexia.aubry@ifip.asso.fr

African swine fever (ASF) represents a major threat to pig production, with serious economic losses. Due to ASF occurrence in wild boars in Belgium, France has drawn up a health emergency action plan in which a large-scale exercise was led in the department of Finistère, to anticipate actions to be taken in the event of a real crisis. The study concerned an ASF infection in a 460-sow farrow-to-finish farm, in a high pig density area. After confirmation of infection, the prefect ordered the slaughter of pigs, followed by complete cleaning and disinfection and sanitary emptying, mandatory before any herd repopulation. A 10 km surveillance area was also defined around the outbreak, with movements restriction for at least 35 days. This study concerned the economic impact for farmers in the outbreak proximity. The infected farm underwent loss of margin on operational costs from total slaughtering until first sale of pigs after repopulation. Calculations were made using farm average results and prices observed over the last three years. The estimated loss varied between €550,000 and €630,000, depending on the sanitary emptying duration. Adding costs of herd repopulation, valued at €415,000, the total loss for the farm reached €1 million, that is €2,000 per present sow. The cost of cleaning and disinfection was also considered, which could reach €300,000 if optimised, with pits cleaning. The slaughter pigs of the farms included in the surveillance area stayed longer than usual, generating feed costs and heavier carcasses, which remuneration was degraded. Euthanasia of one or more batches was sometimes necessary to maintain good breeding conditions. Regarding the economic context of July 2020, the average cost of movements restriction was estimated between €100 and €190 per present sow, depending on management constraints. Total loss for the exercise was estimated at €2.7 million, considering the 146 farms located in the area. This study underlines the major economic impact for farms located near the disease outbreak. While economic loss for the infected farm is partly covered by the administration, the costs linked to movements restriction are not. The exercise also studied other economic consequences, such as restrictions exports, that are not included in this article.

Session 72 Theatre 10

The PIGWEB TNA-program: transnational access to leading European pig research installations

C. De Cuyper[1], S. Millet[1], H.M. Hammon[2], C.C. Metges[2] and J. Van Milgen[3]
[1]ILVO, Scheldeweg 68, 9090, Belgium, [2]Institute of Nutritional Physiology 'Oskar Kellner', FBN, Wilhelm-Stahl-Allee 2, 18196 Dummerstorf, Germany, [3]PEGASE, INRAE, Institut Agro, INRA Saint Gilles, 16 Le Clos, 35590 Saint-Gilles, France; carolien.decuyper@ilvo.vlaanderen.be

PIGWEB is a European project that was launched on March 1, 2021. This five-year project gathers 16 partners from 10 different countries and fully embraces the European Green Deal, the ambition of the European Commission to make Europe the world's first climate-neutral continent by 2050. To identify levers that can be used to achieve the goals of the Green Deal, effective and convenient access to leading research infrastructures is crucial. Therefore, one of the major features of PIGWEB is the transnational access (TNA) program, allowing external academic and private researchers to access the partners' infrastructures through the submission of research proposals. A budget of about 1.5 million euro is reserved to provide free access to top-quality pig research installations across Europe in an easy and transparent way. In total, 28 experimental pig research installations are available. A two-stage process is used to select the TNA applications. For the first stage, a general proposal is expected, whereas in the second stage, a detailed proposal is submitted. First-stage proposals are reviewed based on their rationale, scientific quality, and valorisation strategy. The first of the three TNA calls, launched in September 2021, was a great success: 20 first-stage proposals were submitted. In general, the proposals were of very high quality and addressed different research topics offered through this call, with a strong interest in performance trials. Of the 20 first-stage proposals submitted, 11 have been selected to proceed to the second stage to submit a full proposal. The main selection criteria for the second stage are the scientific and ethical soundness of the proposal, together with the practical and financial feasibility. The second call will be launched in September 2022. First-stage proposals should be submitted before the end of December. More details on the installations offered, the eligibility and the procedures to be followed are available on the project website https://www.pigweb.eu/call-for-proposals.

Session 72 Theatre 11

Discussion: how can cooperation through a COST action support sustainable pig production

M. Aluwé
ILVO, Scheldeweg 68, 9090 Melle, Belgium; marijke.aluwe@ilvo.vlaanderen.be

Discussion session on the goals and activities of a COST-action on sustainable pig production.

Session 72 Poster 12

Derivation of multivariate indices of the environmental impact in extensive Iberian pig farms

E. Angon[1], J. García-Gudiño[2], I. Blanco-Penedo[3], M. Cantarero-Aparicio[1] and J. Perea[1]
[1]Universidad de Córdoba, Animal Production, Campus de Rabanales, 14071 Córdoba, Spain, [2]CICYTEX, Guadajira, 06187, Spain, [3]Universidad de Lleida, Animal Science, Av. de l'Alcalde Rovira Roure 191, 25198, Spain; eangon@uco.es

Traditionally, the production of Iberian pigs has been linked to the use of the natural resources of the dehesa ecosystem. Nowadays, the production of Iberian pigs has been intensified due to the increased demand for high-quality meat products. Thus, the production of Iberian pigs must be analysed not only by means of economic criteria but also by means of environmental criteria that allow the identification of strategies for more sustainable livestock production. Therefore, the objective was to determine the relationship between the economic value and the environmental impact of the fattening process of the Iberian pig through multivariate factor analysis (FA). This research was carried out in 36 traditional farms of Iberian pigs located in the Extremadura and Andalusian region (southwest of Spain). This approach was expected to generate new variables with possible technical meaning that could be used for management purposes. The FA revealed two common latent factors (F1 and F2), explaining 96.2% of the total variance. The first, most important first factor (80.1%) showed the positive relationship between Climate Change (CC), Acidification (AC), Eutrophication (EU), and Cumulative Energy Demand (CED); and the negative relationship between the economic value of the same variables. Farms with higher scores on this factor had a higher environmental impact and obtained a lower economic value from that impact. The second factor (F2) represented 16.2% of total variance and showed a negative relationship with Land Occupation (LO) and a positive relationship with the economic value of LO. Farms with higher scores on this factor need less land and more economic value from its use. The two extracted factors showed a well-defined technical meaning. They represent valuable indicators related to environmental aspects of Iberian pig farms that could be used for management purposes.

Session 72 Poster 13

Valorisation of effluents from pig production

A.C.G. Monteiro[1,2], V. Resende[3] and O. Moreira[3]

[1]IACA, Av. 5 de Outubro, 21, 2nd esq., 1050-047 Lisboa, Portugal, [2]FeedInov, Quinta da Fonte Boa, 2005-048 Vale de Santarém, Portugal, [3]INIAV, Estação Zootécnica Nacional, Quinta da Fonte Boa, 2005-048 Vale de Santarém, Portugal; ana.monteiro@feedinov.com

Recent agricultural and environmental policies are focused on decreasing the environmental impact of animal production, improving food systems, and decreasing the use of antibiotics and synthetic fertilizers. Effluents are an important part of the 'problem' and, as such, the promotion of an integrated approach to reduce and valorise the different nutrient flows generated within intensive animal production systems is a demand that cannot be ignored by this sector. Livestock production is concentrated in certain regions, some without enough area for land spreading valorisation of effluents. To be competitive and comply with legal requirements, the sector should promote circular economy, pursuing new alternatives for effluents management. The objectives of this project are: (1) optimization of effluents use as secondary raw materials, recovering energy and nutrients, improving farm nutrient balances and promoting sustainable management; (2) contribution to sustainable livestock intensification and landscape planning, to face climate change and resources scarcity; (3) a roadmap for effluents management, including technology portfolio, linked to farm and regional constraints. To meet the objectives, surveys were applied in the farms (pig, poultry and dairy). The present work will focus on pig production. The surveys were divided in: (1) general information; (2) feed management; (3) water and energy management; (4) animal housing; (5) effluents management. Farms had an average of 2,942 piglets, 1,580 growing pigs, 478 sows and 3 boars. Sows presented an average weight of 236 kg, 2,4 births/year and 15,8 piglets/litter. All animals were fed with compound feed, and piglets and growing pigs were fed *ad libitum*. Mean farm water consumption was 11,04 m^3/animal/year (drinking + washing). 50% of farms do solid separation of slurry by sieving with a rotating drum or sieves. Mean slurry production/farm was 8,387.7 tons/year (data from 3 farms). 58% of the effluents are used in the farm, and the rest are directly used in agriculture outside the farm.

Session 72 Poster 14

New reproduction index in selection of dam pig population

Z. Krupová, E. Krupa, E. Žáková, L. Zavadilová and E. Kašná

Institute of Animal Science, Přátelství 815, 10400, Czech Republic; krupova.zuzana@vuzv.cz

Reproduction ability of dam breeds belongs to the key parameters of the sustainable pig production. Litter size, farrowing interval and number of teats can be employed in breeding to synchronously enhance both the production and functional parameters of sows. Moreover, current production and economic parameters faced in pig sector should be regularly reflected in the selection process. Therefore, the reproduction index (RI) established for the Czech dam pig breeds 5 years ago was now updated to fit current conditions. Regarding the RI construction, four selection criteria, namely the number of piglets born in total (TNB), born alive (NBA) and weaned (NW) and farrowing interval of sows (FI) were employed to improve two breeding objectives (goals) i.e. NBA and FI. In the RI, each of the litter size trait is participating by 30% and FI by 10%. General principles for the selection index theory and for the selection response calculation were applied using the current genetic and economic parameters of dam breeds. Economic values of the goal traits expressed per sow and year (gained from the bio-economic model of the program package ECOWEIGHT) were 20.82 € per piglet born alive and -2.90 € per day of FI. Reliability of the breeding values estimated for selection criteria slightly increased by 7% over the last period and reached 51%, 49%, 47% and 19%, respectively. According to the current population parameters and RI construction the expected annual selection response is 0.150 piglet born alive and -0.023 day of FI. Optimising the index construction (to 23:11:50:16 ratio of traits) would result to more favourable genetic gain in both of breeding goals (+0.010 NBA and -0.013 of FI). Moreover, the reliability of animals' selection would be slightly improved from 51% to 56% and overall annual economic response in breeding objectives would be increased by 8% to 3.43 €. The results indicated that current reproduction index should be updated to reach a more favourable and thus sustainable response in the breeding goal traits. The study was supported by project MZE-RO0718-V003 and QK1910217 of the Czech Republic.

Session 72 — Poster 15

Barriers and levers for the development of organic pig farming

L. Montagne[1], J. Faure[1] and Origami Consortium[2]

[1]INRAE, Institut Agro, 35590, Saint Gilles, france, pegase, domaine de la prise, 35590 saint gilles, france, [2]inrae, https:// www6.inrae.fr/metabio/thematiques/gestion-des-ressources/origami, 147 Rue de l'Université, 75007 Paris, France; thomas.puech@inrae.fr

Despite an increasing market demand for pork organic products, the development of organic pig production in Europe remains slow compared to other animal and plant sectors. In France, pig production under organic farming specifications has developed recently (+83% of sows and +103% of slaughtered volume between 2015 and 2019) but represents only 1.7% of the total sow herd. The consortium Origami (ORganic for pIG FarMIng) funded by INRAe for 18 months (2021-22) aims at identifying 'multi-scale' brakes and levers linked to the development of organic pig farming to ultimately propose projects and actions to favour process optimization at each level. This network brings together research institute, technical institute and organic institute and is composed of about fifty attendees, with a diversity of area of expertise (biology, animal science, agronomy, food processing, economic and human sciences), levels of approach (from the animal to the sectors and territories) and geographical situations (metropolitan France and overseas). Interviews of stakeholders of the pork sector (organic, conventional or mixed; historical or new) were performed to update the challenges faced by the organic pig production. Brakes and levers identified are linked to technical (buildings, feed, health), economic (carcass value, price) and social (dialogue between stakeholders, communication with consumers) dimensions. Links between brakes and levers have been analysed regarding interactions between actors of the pig chain (production, slaughtering, processing, consumption, etc.). Using a participatory approach consortium participants were invited to express their point of view and expectations. Research problems and questions were then formalized. The shared culture and enlarged vision developed in this network will promote dialogue with multiple partners towards the emergence of interdisciplinary research projects.

Session 72 — Poster 16

Spanish consumers' attitudes towards fresh Iberian pork

A. Ortiz[1], C. Díaz-Caro[2], D. Tejerina[1], M. Escribano[3], E. Crespo[4] and P. Gaspar[3]

[1]Centre of Scientific and Technological Research of Extremadura (CICYTEX-La Orden), Meat Quality Area, Av A5. Km 372, 06187 Guadajira, Badajoz, Spain, [2]University of Extremadura, Department of Accounting and Finance, Avda. de la Universidad s/n. 10071 Cáceres, Spain, [3]School of Agricultural Engineering, University of Extremadura, Department of Animal Production and Food Science, Avda. Adolfo Suarez, s/n, 06007 Badajoz, Spain, [4]University of Extremadura, Department of Economics., Avda. Adolfo Suarez, s/n 06007 Badajoz, Spain; alberto.ortiz@juntaex.es

Iberian breed is a native breed of pig produced in the southwest of the Iberian Peninsula, which represents 6% of total pig production in Spain, providing products of great acceptance in national and international markets mainly due to the distinctive quality characteristics of its products and the extensive nature of its production system (located in agroforestry areas known as dehesa). Traditionally, the cured products of the Iberian pigs are the most accepted. However, the increase in consumption as fresh meat is recent, and especially of the tenderloin (*Illiopsoas et psoas minor*), the presa (*Serratus ventralis*), pluma (*Spinalis dorsi*) and secreto (*Latissimus dorsi*) commercial cuts. The consumption of these may not only responds to criteria of nutritional and sensory quality but also to ethical criteria associated with greater preservation of the environment, and therefore greater sustainability, ethical production and animal welfare of the production models. This study investigated the behaviour of consumers regarding four cuts of Iberian meat with greater presence in the market. For that, a sample of 1,501 consumers responded to an online survey about their consumption habits for these four cuts. Results showed differences in consumption frequencies among Iberian meat cuts studied, being tenderloin and secreto the most consumed. Additionally, sociodemographic characteristics and attitudes and lifestyles of consumers were determining factors in the consumption behaviour of the various Iberian commercial cuts. Therefore, these could be key aspects to take into account in designing future marketing strategies for these products, and could possible play an important role in the valorisation of the dehesa systems and their productions.

Session 72 Poster 17

Economic assessment of amino acid deficiency or feed restriction during the pig fattening period

A. Aubry and G. Daumas
IFIP – Institut du Porc, BP 35104, 35651 Le Rheu Cedex, France; alexia.aubry@ifip.asso.fr

The profitability of pig farms is influenced greatly by feed efficiency and carcass grading. The aim of this study was to assess the economic impact of two feeding strategies during the fattening period, compared to that of a control (T), which was *ad libitum* feeding with no amino acid deficiency. The first strategy was *ad libitum* feeding with a three-phase sequence limited in amino acids (CA). The second strategy was restricted feeding at 85% of the *ad libitum* without amino acid deficiency (RA). Each of the three feeding strategies was applied to 48 gilts and 48 barrows, crossbred between Pietrain sires and Large White × Landrace sows, reared in pens of 6 pigs. Technical performances were estimated using a general linear model by pen. Then, the feed conversion ratio (FCR) was standardized to the range of 30-120 kg. The lean meat content (LMC), assessed by the Image-Meater carcass classification method, was standardized to 120 kg. These criteria were used to assess the economic feeding margin of each strategy. Therefore, they were input in the 'Calculate' simulator, which was developed by IFIP in the French production conditions, available on the GT-DIRECT portal (https://gtdirect.ifip.asso.fr). Two extreme economic contexts were considered using two assumptions for fattening feed price (HIGH: 290 €/T and LOW: 152 €/T). FCR were 2.57, 2.75 and 2.74 kg/kg for T, CA and RA strategies respectively. LMC were 60.9, 60.4 and 62.4% for T, CA and RA strategies respectively. The CA margin was lower than that of T, by € 3.4 and € 5.6 per pig produced for the LOW and HIGH feed price contexts respectively. The predominant effect was that of feed price, which penalized feed cost due to an FCR higher by 0.18 kg/kg. The RA margin was also lower than that of T, by € 0.7 and € 2.8 per pig produced for the LOW and HIGH feed price contexts respectively. The higher FCR (+0.17 kg/kg) induced an increase in feed cost which was compensated only partially by the rise of output due to the better LMC (+1.5 percentage point). Feeding strategies allowing the FCR to be controlled are to be favoured, especially in a context of high feed prices, so as not to degrade the economic results of the farms. Meeting amino acid needs and mastering feed restriction remain essential.

Session 72 Poster 18

Immunocastration in heavy pig production: growth performance and carcass characteristics

G. Pesenti Rossi[1], M. Comin[1], M. Borciani[2], M. Caniatti[1], E. Dalla Costa[1], A. Gastaldo[2], M. Minero[1], A. Motta[2], F. Pilia[1] and S. Barbieri[1]
[1]Università degli Studi di Milano, Dipartimento di Medicina Veterinaria e Scienze Animali, via dell'Università 6, 26900 Lodi, Italy, [2]Fondazione C.R.P.A. Studi Ricerche, Viale Timavo 43/2, 42121 Reggio Emilia, Italy; gaia.pesenti@unimi.it

Immunocastration is an effective method to prevent boar taint, avoiding pain and stress due to surgical castration. Immunocastration maintains good productive performances, with faster growth rate and better feed conversion than barrows. Also, it is associated to heavier carcasses, higher percentage of lean meat and lower fat thickness. Few studies evaluated these aspects in heavy pig production: our aim is to compare growth performance and carcass characteristics in immunocastrated and surgically castrated pigs, raised for heavy pig production. 166 commercial-hybrid male pigs were randomly allocated to two treatment groups: Immunocastration (IC; n=83), pigs receiving 4 doses of Improvac® at 15, 22, 32, and 36 weeks of age; Surgical Castration (SC; n=83), pigs surgically castrated at 4 days of age. Animals were kept under the same feed and housing conditions, in compliance with Dir. 2008/120/EC. IC and SC pigs were slaughtered respectively at 40 and 41 weeks of age. Carcass classification was made accordingly to Decision 38/2014/EC using the Fat-O-Meter system. The average daily gain was 1,020 g in IC and 770 g in SC pigs. Despite the slightly shorter fattening period, IC pigs were significantly heavier (T-Test; P=0.007), with a mean weight of 180.99±14.54 kg, while SC pigs weighted 171.32±12.52 kg. Hot carcass weight also resulted significantly higher for immunocastrated pigs (T-Test; P=0.007): 150.54±12.48 kg for IC and 145.10±10.75 kg for SC. The lower mean fat and muscle thickness of IC (30.38±4.94 mm and 55.34±8.94 mm, respectively) resulted in a higher mean lean meat content (51,67%). Our results confirm that immunocastration is an interesting alternative to surgical castration in heavy pigs, as neither performance nor productive quality are negatively influenced. Further studies are required to evaluate sustainability in terms of animal welfare and economic impact in this production system.

Session 72 Poster 19

Effect of immunocastration on body lesions in heavy pigs: preliminary results

E. Dalla Costa[1], G. Pesenti Rossi[1], A. Motta[2], M. Borciani[2], A. Gastaldo[2], G. Berteselli[1], E. Canali[1], M. Minero[1] and S. Barbieri[1]

[1]Università degli Studi di Milano, Dipartimento di Medicina Veterinaria e Scienze Animali, Via dell'Università, 6, 26900 Lodi, Italy, [2]Fondazione CRPA Studi Ricerche di Reggio Emilia, Viale Timavo 43/2, 42121 Reggio Emilia, Italy; gaia.pesenti@unimi.it

Immunocastration is an interesting alternative to surgical castration in piglets. Studies have shown promising results in terms of production performance (e.g. maintaining adequate meat quality) also in heavy pig production; and improvement of animal welfare (e.g. preventing distress and pain caused by surgical castration). Aggressive and mounting behaviours, that often result in body lesions, seems to be reduced in light pig production; however, no studies have yet investigated animal welfare of heavy pigs subjected to immunocastration. This study aimed at evaluating the effect of immunocastration on welfare of heavy pigs by monitoring body lesions during growing and fattening period. Commercial-hybrid male pigs were randomly allocated to treatment groups: Immunocastration (IC; n=94), pigs receiving 4 doses of Improvac® at 15, 22, 32, and 36 weeks of age; Surgical Castration (SC; n=94), pigs surgically castrated at 4 days of age. IC and SC pigs received the same feeding regimen, they were housed in the same conditions and their management complied with Dir. 2008/120/EC. Before each Improvac administration, body lesions were recorded through direct observations and scored on a three-point scale (none, mild, severe). Independent T-test was used to determine differences between groups at each considered time point. Before the first administration of Improvac (15 weeks of age), IC pigs showed a significantly higher body lesion score (0.60±1.04) compared to SC (0.2±0.48) (P=0.001). The body lesion score remains higher in the other time points, but the difference between groups is not significant. A high level of agonistic behaviour before the suppression of testicular function suggests anticipating the vaccination protocol in relation to the onset of puberty and to increase the number of interventions in heavy pigs. Further research is needed to evaluate the sustainability of different timing of immunocastration, maintaining high level of animal welfare together with productive and economic benefits of the procedure.

Session 73 Theatre 1

Environmentally sustainable horse feeding and management

M. Saastamoinen

LUKE, Tietotie 2 C Jokioinen, 31600, Finland; markku.saastamoinen@luke.fi

Sustainable nutrition and management of horses are part of the socioeconomic effects of horse industry. Horse nutrition is based on forages and grazing, supplemented usually with grains and minerals. However, the diets are often supplemented – concerning mainly protein (nitrogen) and minerals – in amounts which exceed their requirements. This leads to increased faecal excretion of these nutrients. Concerning nitrogen, leaching from manure (dung and urine + beddings) is very potential, but also nitrous GH-gases are evaporated. In the case of minerals, phosphorus is the most harmful to the environment. In addition, some trace minerals can be toxic to plants, microorganisms and aquatic organisms. Leaching and evaporation can be happened from stables, manure stores and composts as well as from pastures and when applying manure. There is a positive relationship between the nutrient intake and faecal excretion. To reduce excretion, most important way is to balance the diets and improve the feed quality and, thus, availability of the nutrients. Fibre content and digestibility of the diet nutrients is necessary to consider. Supplementing forage diets with grains improves the digestibility of many nutrients. Also, regarding the methanogenesis in the gut of the horse, decrease of fibre (ADF) content may reduce methane production. Further, mitigation strategies have been studied in this context. Because horse manure may contain many important and valuable nutrients for plants, it is important to recover these preventing their leaching and evaporation. Pasturing of horses is beneficial to biodiversity if overgrazing is avoided. One aspect regarding the influence on the nature biodiversity, is to avoid spreading of weed seeds via dung when grazing and moving with horses. Horse industry has a potential to 'compensate' its possible harmful environmental effects by circulating nutrients, offering biomass for biogas production and carbon to soil, in the form of manure.

CAP'2ER® 'équins': an environmental footprint calculator tested on 39 French equine systems

A. Rzekęć[1], C. Vial[1,2] and S. Throude[3]

[1]IFCE, Jumenterie, 61310 Exmes, France, [2]INRAE, MoISA, Univ Montpellier, CIRAD, CIHEAM-IAMM, INRAE, Institut Agro, IRD, 2 Pl Pierre Viala, 34060 Montpellier, France, [3]IDELE, 149 rue de Bercy, 75012 Paris, France; agata.rzekec@ifce.fr

The agricultural sector is developing new tools to enhance knowledge about livestock activities consequences on climate change. The first environmental impact calculator for equine structures in France, CAP'2ER 'équins', has been created to help farmers better evaluate and manage environmental impacts of their activities. CAP'2ER has already existed for ruminant farming. It measures four impacts: climate change, air acidification, eutrophication and fossil energy consumption. Moreover, it quantifies three positive contributions: carbon storage, maintenance of biodiversity and nutritional performance. The tool requires 55 technical data collected on the farm and emission factors for each source of greenhouse gas pollution (housing, storage, spreading, grazing and inputs) found in the scientific literature and specific to the equine species. The CAP'2ER is based on algorithms based on IPCC (2019) for animal and farm emissions and the boundaries of the analysis were from cradle to farm gate, according to LCA analysis. In 2021, the tool was tested on a sample of 39 equine structures (30 riding centres and 9 breeding farms), as varied as possible. The estimated emissions were at 1,544 kg CO_2 eq/head on average (57% due to CO_2, 31% to CH_4 and 12% to N_2O). The calculated carbon storage was at 1,429 kg CO_2 eq/head on average. Thus, the sample offset its emissions by 93%, with variations depending on the production system. The main emission source is the purchase of cereal grains or processed feeds, fodder and bedding (38% of total emissions) by equestrian establishments, explained by their low food autonomy. A better optimisation of diets by limiting cereals and extending grazing time (thus decreasing the quantity of fodder and bedding), or even by increasing the surfaces in meadows would be some ways to decrease indirect CO_2 emissions. These first estimates depend on parameters that are not widely available in the literature and that require further studies. Research should develop consolidated emission factors that are more adapted to the various equine systems found in France and Europe.

Cattle and Esperia pony grazing sustains floristic diversity in thermo-Mediterranean garrigues

R. Primi[1], G. Filibeck[1], G. Salerno[2], M.M. Azzella[3], L. Cancellieri[1], C. Di Giovannantonio[4], A. Macciocchi[4], P.P. Danili[1] and B. Ronchi[1]

[1]Università degli Studi della Tuscia, Dip. di Scienze Agrarie e Forestali, Via S.C. de Lellis snc, 01100 Viterbo, Italy, [2]Università di Roma Tre, Dip. di Scienze, Viale G. Marconi 446, 00146 Roma, Italy, [3]Sapienza Università di Roma, Dip. di Pianificazione, Design e Tecnologia dell'Architettura, Via Gianturco 2, 00152 Roma, Italy, [4]Agenzia Regionale per lo sviluppo e l'innovazione dell'agricoltura nel Lazio, Via Rodolfo Lanciani 38, 00162 Roma, Italy; ronchi@unitus.it

Horses act as grazers or browsers according to feed availability, and several studies have shown their important role in maintaining biodiversity in agroecosystems. The objective of this study was to compare, in a mountain region of Central Italy, the effect on a Habitat of Community Interest (EU Directive 92/43 – Habitat Type 5330 – sub-type '32.23 Garigues dominated by *Ampelodesmos mauritanicus*') of: (1) successional grazing for 3 years of beef cattle and Pony di Esperia horses; (2) no grazing for more than 10 years; and (3) burning followed by 3 years of spontaneous evolution. In the grazed area the cattle had access from mid-April to the end of June of 2021, followed by the Esperia ponies which grazed throughout the month of July; the average livestock load was established at 0.4 LU/ha for bovine and 0.2 LU/ha for equines. In May 2021, structure and diversity of the grassland were assessed using 25 m^2 vegetation plots (n=17), randomly selected within the 3 treatments. Based on the abundance data, the vegetation types were identified and analysed through Cluster Analysis and PCA. Plant species richness (S) per plot (α-diversity) was compared through ANOVA. The α-diversity of the grazed plots (S=37±4) was significantly higher ($P<0.05$) than the burned (S=25±2) and non-grazed (S=24±3) ones. In the grazed area, all the plant species diagnostic of the habitat sub-type 32.23 were found. Thus, successional grazing of cattle and Esperia ponies had no negative influence on the conservation of the habitat. The driver that mainly determined richness was *A. mauritanicus* cover: mean value in grazed plots was 33%, vs 45% (burned areas) and 69% (abandoned areas), thus releasing competition and allowing a higher species density.

Session 73 — Theatre 4

Changes of pasture area in Molise region: a study over 30 years

A. Fatica and E. Salimei
Università del Molise, via de Sanctis 1, 86100 Campobasso, Italy; a.fatica@studenti.unimol.it

Molise region is characterized by pasture areas vulnerable although strategic from economic, social, environmental, and cultural standpoint. The present study is focused on Frosolone pasture areas declared UNESCO Intangible World Heritage site. A multidisciplinary investigation including grazing livestock population, botanical and productive characteristics of the turf, and climatic changes was carried out to study the evolution of the area over a 30-year period. Animal numbers in the selected area halved during the investigated period, as the average calculated adult bovine units (ABU) was 2,816 heads and 1,379 heads, respectively for 1992 and 2021, when equines declined from 288 to 94 heads, bovines from 2,126 to 1,121 heads, sheep from 2,327 to 1,006 heads, goat from 354 to 85 heads. The estimated ratio between ABU and pasture productivity, i.e. the average grazing index (GI) was 1.38 head/ha in 2021. Compared to literature data a steep decline was observed in GI (2.8 head/ha in 1992). Although the number of plant species decreased over the investigated period, from 42 (1992) to 15 (2021), Gramineae always resulted the most represented family. Notwithstanding the total average biomass production increased from 2.20 t DM/ha in 1992 to 5.00 t DM/ha in 2021, a progressive expansion of herbaceous and shrubby weed species has been observed as possible consequence of non-homogeneous pasture utilization. The observed veldt degradation and grazing pressure are consistent with the marginalization of inner Mediterranean lands in favour of a more productive intensification in flat areas. Considering the multiple benefits that mixed grazing provides to the local community, in terms of ecosystem services, including added nutritional value of traditional foods with territorial identity, it seems essential to re-evaluate the grazing management approach so that the nutrient requirements of grazing animals can be fulfilled in the respect of animal welfare and production, plant diversity and land protection.

Session 73 — Theatre 5

Long-term consequences of forage presentation on horses' welfare

M. Roig-Pons[1,2] and S. Briefer[1]
[1]Agroscope, Equine research group, Les Longs Prés, 1, 1580 Avenches, Switzerland, [2]Vetsuisse Faculty, VPHI, Animal Welfare Division, Langgastrasse 120, 3012 Bern, Switzerland; marie.roig-pons@agroscope.admin.ch

Slowfeeding dispensers have been developed to tackle one of the equine husbandry's dilemma: fulfilling the horses' behavioural and physiological needs in terms of ingestion without risking obesity. Slowfeeders' (SF) effects on horses remaining unknown, especially for long-term use, we aimed at exploring long-term associations between forage presentation and horses' welfare, using health and behavioural indicators. 356 horses from two cohorts (SF users, n=182; Control, n=174) were sampled, based on a preliminary survey about the target population (1,444 respondents; 4 strata identified: age, housing, training and shoeing). 344 additional horses will be assessed in Spring 2022. Health data comprised vibrissae and gums evaluation, pictures of the incisors and a musculoskeletal health assessment (MHA) performed by an osteopath. Horses' owners had to fill a survey about their horse's personality and reactivity to humans was assessed using two tests. Feeding management details (dispensers' characteristics, SF or not, period of use, etc.). were recorded. Inter- and intra-assessors reliability for the MHA protocol was verified ($r_{pearson}$>0.72 and Gwet's index = 0.91) and the sample size needed was calculated (n=700). Interim analysis on 2021 data showed that vibrissae length and use of slowfeeders are significantly associated: SF horses are 4.25 times more likely to display cropped vibrissae, compared with control horses (χ^2-test, P=1.55e^{-12}). Univariate analysis revealed that the use of SF is not associated with the total MHA score (t-test, P=0.37). Of all of our variables, only age (r=0.35, P=1.4e^{-10}, Pearson) and housing (one-way ANOVA, P=0.024) had a significant effect on the MHA score, which underlines the need for stratification. However, horses using dispensers higher than 60% of their height tended to display more osteopathic issues (P=0.10). Our preliminary results show that forage presentation has an impact on horses' vibrissae length but data from the second batch will be needed to draw conclusions on horses' personality and musculoskeletal health. By the time of the conference, we expect to compute associations between the overall feeding management and the horses' welfare.

Session 73 Theatre 6

A high starch vs a high fibre diet: a multidimensional approach to gut health in horse

F. Raspa[1], S. Chessa[1], I. Ferrocino[1], I. Vervuert[2], M.R. Corvaglia[1], L. Cocolin[1], R. Moretti[1], D. Bergero[1] and E. Valle[1]
[1]University of Turin, Largo Paolo Braccini 2, Grugliasco, 10095, Italy, [2]University of Leipzig, An den Tierkliniken 9, Leipzig, 04103, Germany; federica.raspa@unito.it

Feeding a starch-rich diet is still a common mistake in feeding practice for horses and it is usual in horses reared for meat production. Microbiota populations vary among the different gut compartments but knowledge about such diversity remain sparse according to the diet. The present study aimed to compare the effects of a high starch (HS) vs a high fibre (HF) diet on the microbiota in different horse gut compartments. Nineteen Bardigiano horses (12 ♀ and 7 ♂), 14.3±0.7 (mean ± sd) months of age, were randomly assigned to two group pens – HS (n=9) vs HF (n=10). After the fattening period, horses were slaughtered and samples of the gut chyme were aseptically collected from caecum (CAE), pelvic flexure (PF) and rectum (RE); and stored at −80 °C until DNA extraction and 16S rRNA amplicon target sequencing. Shannon index showed higher diversity (Kruskal Wallis test, P<0.01) in CAE and PF in horses fed the HF diet. β-diversity showed a clear separation of the microbial communities as a function of the diet (PERMANOVA; P<0.01). At the genus level, microbiota in CAE and PF were different between the two diets (ANOSIM, P<0.05), whereas no differences were found in RE. Particularly, in CAE Clostridiaceae, Ruminococcaceae and *Streptococcus* were mainly associated with HS; *Fibrobacter* with HF (FDR<0.05). In PF *Anaerovibrio*, *Clostridiaceae*, *Methanobrevibacter*, *Roseburia*, *Ruminococcus*, *Sarcinia* and Succinivibrionaceae were associated with HS; while *Desulfovibrio*, *Dorea*, *Oscillospira* and *Methanocorpusculum* with HF (FDR<0.05). In RE *Parabacteroides*, Succinivibrionaceae and *Treponema* were associated with HS; *Campylobacter*, Lachnispiraceae and Ruminococcaceae with HF (FDR<0.05). Diet highly influenced bacterial communities according to the different gut compartments. HS diet was associated with a reduction in bacterial diversity in CAE and PF. Highlighting the relationships between diet and gut microbiome, and their repercussions in microbiome-host metabolic and immune interactions are one of the main current challenges.

Session 73 Theatre 7

Use of protected and non-protected live yeast in the digestive health of horses

G. Pombo, C. Vasco, Y. Pereira, H. Mazzo, A. Araújo, R. Pereira, A. Silva, M. Duarte and A. Gobesso
USP, FMVZ/VNP, Rua Duque de Caxias, 13635-900, Brazil; gabriela.guarnieri@unifeb.edu.br

Investigation on the effect of supplementing protected live yeast as a pelleted concentrate in horses is warranted. This study aimed to evaluate the effect of supplementing protected and non-protected live yeast on nutrient apparent digestibility and digestive health parameters (faecal and gastric pH, short-chain fatty acids, lactate, and microbiome) in maintenance horses. Eight mature Arabian geldings (60±5 mo and 457±28 kg; mean ± SD) were randomly assigned to one of the dietary treatments in a replicated 4×4 Latin square with 21-d periods and a 2-wk washout between periods. Dietary treatments consisted of a basal diet (hay at 1%BW and pelleted concentrate at 0.75%BW) without yeast (Control), with 15 g non-protected live yeast/d (NPY), with 20 g protected live yeast/d (PY), or a combination of the NPY and PY daily doses (15 and 20 g/d, respectively). Treatments were compared by orthogonal contrasts, as follows: control vs yeast supplementation (C1), individual vs a combination of NPY and PY (C2), and NPY vs PY (C3). Yeast supplementation tended to increase digestibility of acid detergent fibre (C1; P=0.057; 34.8 vs 40.4%) and neutral detergent fibre (C1; P=0.064; 44.8 vs40.6%). Except for gastric acetate, there was no difference among yeast supplementation strategies (P>0.1). Gastric acetate was greater in yeast supplemented horses than control (C1; P=0.006; 5.3 vs 3.2 mmol/l), and greater when yeast was supplemented combined than individually (C2; P=0.044; 6.4 vs 4.8 mmol/l). Yeast supplementation strategy did not alter (P>0.1) faecal and gastric total and individual bacteria gene expression. However, *Fibrobacter succinogenes* modulated concentrations of faecal lactate (P=0.045; R^2=-0.800) and faecal propionate (P=0.043;R^2=-0.469), while Lactobacillus spp. modulated concentrations of faecal acetate (P=0.021; R^2=-0.529) and faecal propionate (P=0.009; R^2=0.236). The results indicate that supplementing protected yeast provides similar digestive health than non-protected yeast. However, yeast supplementation has potential to improve fibre digestibility and fermentative parameters regardless of the yeast type.

Session 73 — Theatre 8

Soil ingestion by grazing horses and heifers

C. Collas[1], L. Briot[2], G. Fleurance[2,3], D. Dozias[4], F. Launay[4], C. Feidt[1] and S. Jurjanz[1]
[1]URAFPA, Université de Lorraine-INRAE, Nancy, 54000, France, [2]DIR, IFCE, Exmes, 61310, France, [3]UMRH, INRAE-VetAgro Sup, Saint-Genès-Champanelle, 63122, France, [4]UEP, INRAE, Le Pin-au-Haras, 61310, France; claire.collas@univ-lorraine.fr

Grazing animals can ingest soil particles which can accumulate and damage the digestive tract, reducing diet digestibility and nutrient absorption and even causing sand colic in horses. Ingested soil also exposes animals to environmental contaminants as trace metals or organic pollutants. This can result in pathologies, non-compliance during anti-doping tests in horses or exceeding authorised thresholds in food of animal origin. In cattle, soil ingestion has never been studied in conditions of mixed grazing and in horses it is very little described. Our experimental design consisted in 3 groups: equine E (6 horses), cattle C (12 heifers), mixed M (3 horses and 6 heifers), grazing continuously at the same stocking rate between groups on 3 different plots from spring to autumn 2019 and 2020. Saddle horses (509 kg BW) and beef cattle (484 kg BW) were 2 and 1 years old respectively. The initial grazing area of 3 ha per group was extended to 6 ha from July. Each month, sward height and animal body weight (BW) were measured and daily soil ingestion was individually estimated (from grass, faeces and soil samples) using titanium as soil marker. Data were analysed by mixed models. Correlations of soil ingestions with rainfall, sward height and average daily gain were analysed. Both years showed significant period × animal species and period × grazing management interactions ($P<0.01$). In 2019, soil ingestion by C cattle (8%) was the highest, the lowest were for E and M horses (2.5%). In 2020, soil ingestions by cattle were higher than those by horses (5.1 vs 2.4%) whatever the grazing management. C cattle lose BW in the autumn 2019. This may result from the decrease in herbage availability which may have encouraged them to graze more closely to the ground compared to spring and summer (soil ingestion of 16.6%, Nov 2019). Soil ingestions were negatively correlated with sward heights for both years and positively correlated with rainfall in 2019 ($P<0.001$). This study of soil ingestion will be useful for risk assessment and recommendations for farmers to ensure animal health and food safety.

Session 73 — Theatre 9

Owner-reported use of feed supplements in Swiss riding horses

M.T. Dittmann[1,2], S.N. Latif[1] and M.A. Weishaupt[1]
[1]Vetsuisse Faculty, University of Zurich, Equine Department, Winterthurerstrasse 260, 8057 Zurich, Switzerland, [2]Research Institute of Organic Agriculture FiBL, Ackerstrasse 113, 5070 Frick, Switzerland; marie.dittmann@fibl.org

Recommended rations for domestic horses consist of a large proportion of grass-based roughage, small amounts of cereal based concentrates, salt and a mineral-vitamin-supplement. However, additional feed supplements (AFS) for different purposes have become increasingly popular among horse owners. This study's aim was to investigate, which AFS are fed to Swiss riding horses by their owners. Through an online survey, which was part of a study on the orthopaedic health of Swiss riding horses, information on the equines' diets was gathered. Among other inclusion criteria, owners could only take part, if their horse was ridden regularly and (based on their judgement) healthy. Of the 248 participants, 90% reported that their horse had pasture access at least 5 times a week during the season. Of all owners, 72% reported to know the amount of roughage and concentrates their horse was fed (9.7±3.7 kg and 2.0±1.5). Of all owners, 98% reported that their horse had access to salt, 74% reported the supplementation of a basic mineral-vitamin-feed. Surplus to these feeds, 61% of owners offered at least one AFS to their horse. The most common groups of AFS fed were supplements rich in magnesium (26%), Omega fatty acids (22%), vitamin E and/or selenium (16%), herbs (7%), mixed supplements for joint health (6%), hoof quality (5%), or gastrointestinal health (4%), general tonics (5%), plant-based oils (4%), or additional roughage or cereal based supplements (4%). Whether owners offered their horse AFS was not significantly associated with the age or sex of horse or owner, or the time the person had owned the horse. Compared to healthy horses, a higher proportion of horses with an owner-reported disease received AFS (56% vs 69%, $P=0.04$). Compared to horses owned by leisure riders, a higher proportion of horses owned by self-reported competition riders received AFS (52% vs 69%, $P=0.01$). The feeding of AFS was more common in Warmblood horses than in other breed categories (67% vs 51%, $P=0.01$). The results indicate, that many owners offer AFS with the intention to manage diseases or to support their horses' athletic performance.

Does protein concentrate supplement type effect mid-lactation milk production in grazing dairy cows?

Á. Murray[1,2], M. Dineen[2], T.J. Gilliland[1] and B. McCarthy[2]
[1]Queen's University Belfast, Institute of Global Food Security, Queen's University Belfast, Belfast, N. Ireland, BT9 5DL, United Kingdom, [2]Teagasc, Paddy O Keeffe Moorepark Teagasc, P61C996, Ireland; aine.murray@teagasc.ie

The objective of this study was to evaluate the effect of protein concentrate supplement type on milk production and composition in lactating dairy cows grazing mid-season perennial ryegrass (Lolium perenne L.; PRG) herbage. Twelve primiparous and 68 multiparous lactating dairy cows were blocked based on pre-study milk yield and parity and randomly assigned to 1 of 4 dietary treatments. The 4 dietary treatments were; a non-supplemented PRG control (PRG); PRG supplemented with 0.585 kg of dry matter (DM)/cow/day of a rumen escapable protein source (REP); PRG supplemented with 3 kg of DM/cow/day of a low protein concentrate supplement containing a rumen escapable protein source (RE-LP); and PRG supplemented with 3 kg of DM/cow/day of a low protein concentrate supplement containing a rumen degradable protein source (RD-LP). The study consisted of a 2-wk adaptation period and a 10-wk period of data collection. Weekly measurements recorded were milk yield, milk composition, body weight, and body condition score. Cows fed RE-LP and RE-DP diets had higher milk yield than cows fed the PRG and REP diets (P<0.001). Cows fed diets containing concentrate supplements (REP, RE-LP and RE-DP) had higher milk fat concentration and higher milk protein concentration when compared with cows fed the PRG control diet (P<0.05). Milk solids yield (kg fat + protein) was affected by diet (P<0.001), where RE-LP and RD-LP diets were highest (1.75 kg/d), REP was intermediate (1.63 kg/d) and the PRG diet was lowest (1.40 kg/d). The milk protein yield of cows fed RE-LP and RD-LP were highest (0.77 and 0.76 kg/d, respectively), REP was intermediate (0.71 kg/d), and PRG was lowest (0.63 kg/d). The results demonstrate that providing a low quantity of a rumen escapable protein source to cows consuming pasture-based diets can increase milk solids yield. The loss of response when compared to a rumen degradable protein source, at higher supplementation rates, warrants further investigation.

New insight on the transmission of passive immunity in foals

E. Portal[1], R. Brouillet[2], I. Gaudry[3], M. Hilaire[4], F. Barbe[5] and C. Villot[5]
[1]Institut National Universitaire Champollion, Lycée professionnel agricole La Cazotte, Moulin du Bousquet, 12 390 Rignac, France, [2]Centre de reproduction équine, Route de Bournac, 12400 Saint Affrique, France, [3]Clinique vétérinaire équine de Roqueville, 1454 chemin de Roqueville, 31450 Issus, France, [4]S.A.S. Terrya, Hip'podium, Moulin du Bousquet, 12 390 Rignac, France, [5]Lallemand SAS, 19 rue des Briquetiers, 31700 Blagnac, France; cvillot@lallemand.com

In this trial, 50 mares from 14 equestrian structures (16 control (CON) and 34 supplemented with a complex of micro-ingredients (probiotic, antioxidants), distributed 35 days before foaling (EXP)), were recruited to follow the course of foaling (T0) and assess the benefit of a nutritional solution provided to mares before T0. The colostrum was collected at T0 and at 3, 6 and 9h. Colostral IgG was analysed using a colotest. The foaling time and the placenta delivery time were also measured for each mare. The colotest data were analysed with 3 linear mixed repeated measures models with the mare as random factor and the group (CON/EXP), the physiological status (primiparous/multiparous), the access to grass (YES/NO), the time and interactions with time as fixed effects. Foaling time and placenta delivery time were analysed in the same way, without time. IgG decreased rapidly after T0 (0h: 88.2 g/l; 3h: 69.1 g/l; 6h: 28.1 g/l; 9h: 7.4 g/l), (P<0.001). The supplement increased IgG compared to CON at 0h (CON: 68.5; EXP: 98.5 g/l;+44%), 3h (CON: 52.1; EXP: 76.9 g/l;+48%), 6h (CON: 19.7; EXP: 31.8 g/l;+61%) and 9h (CON: 5.3; EXP: 8.5 g/l;+59%) (P<0.05). Higher IgG was observed in multiparous (49.8 g/l;+42%) than in primiparous mares (35.1 g/l) (P<0.1). Finally, grazing improved IgG at 0 h (YES: 106.9; NO: 73.6 g/l;+45%), 3h (YES: 87.9; NO: 53.3 g/l;+65%), 6h (YES: 36.1; NO: 21.4 g/l;+68%) and reduced it at 9h (YES: 7.1; NO: 7.7 g/l;-9%) (P<0.001). The supplement reduced foaling time (CON: 28.4; EXP: 21.3 min;-25%) and placenta delivery time (CON: 128.8; EXP: 25.7 min;-80%) (P<0.05). Understanding the factors influencing the colostrum quality allows promoting optimal conditions for the foal. This study brings new elements on the immune status of mares at the time of farrowing, contributing to the improvement of the immune status of the foal and to the reduction of a large part of neonatal losses.

Session 73 Poster 12

Case study on free faecal water in horses supplemented with live yeast probiotic *S. cerevisiae*

F. Barbe[1], D. Christensen[2], J. Dahl[3] and C. Villot[1]
[1]Lallemand Animal Nutrition, 19 rue des Briquetiers, 31702 Blagnac, France, [2]Lallemand Animal Nutrition, Sandfeldparken 7, 6933 Kibæk, Denmark, [3]Miljøfoder A/S, Rømersvej 3, 7430 Ikast, Denmark; cvillot@lallemand.com

Free faecal water (FFW) also named free faecal liquid (FFL) or faecal water syndrome (FWS) is proposed by Kienzle *et al.* (2016) as the condition in which horses produce normal faeces, but before, after or during defecation, faecal water runs out of the anus. It is often confused with diarrhoea, where the entire stool is loose or watery in the most severe cases. Usually, as no effect on general health and welfare is reported, FFW is not considered as a real pathology. However, horses with this symptom pollute their legs and tail, leading to secondary sanitary issues (skin irritation, flies biting, scratching, dermal infection, etc.). Different hypotheses have been raised to explain the occurrence of FFW: stress and anxiety leading to modification of the gastrointestinal tract environment (faster transit time due to increased gut motility, reduction of water absorption, increased intestinal permeability and changes in the microbiota balance with an increase of facultative anaerobic bacteria supporting the oxygen hypothesis, leading to dysbiosis and inflammation). As microbiota modulation could be a key factor in the management of this condition, the objective of this study was to investigate the effect of live yeast (LY) probiotic supplementation on FFW clinical signs. Four horses (2 Norwegian, 2 Warmblood, aged from 5 to 24 years) received LY probiotic supplementation *Saccharomyces cerevisiae* I-1077 (Levucell SC10 ME Titan) at 2 g/horse/day, providing 2×10^{10} cfu/horse/day during 4 weeks. Before supplementation and after 4 weeks, FFW was evaluated by visual inspection of the back (irritation, dirtiness) and horse digestibility was evaluated by fresh droppings double sieving on 2 grids, separating coarse and fine fractions. This trial confirms the beneficial effects of the live yeast tested for improved digestibility of the horse and adds valuable practical information regarding the potential of this live yeast to decrease FFW clinical signs: mainly back irritation and dirtiness. The improved digestibility in these horses could also improve water absorption, thereby reducing FFW condition.

Session 73 Poster 13

Suckling frequency during the first 24 hours of life of the newborn foal

L. Wimel[1], A. Destampes[2] and J. Auclair-Ronzaud[1]
[1]IFCE – Plateau Technique de Chamberet, 1 Impasse des Haras, 19370 Chamberet, France, [2]ENIL Saint Lô-Thère, Rue du Père Popielujko, 50000 Saint-Lô Thère, France; juliette.auclair-ronzaud@ifce.fr

Foal suckling frequency (SF) evolves during lactation while suckling duration (SD) appear to be steady. Less information, however, are available regarding the repartition of the suckling bouts along 24 hours. Some studies suggest that SF is reduced at night. The suckling behaviour of 24 foals was monitored in 2021 during the first 24 hours after birth. Foalings occurred between the 4th of April and the 20th of May and did not required any human assistance. Foals were housed in individual boxes with their dam. Water was provided *ad libitum* and hay, haylage and concentrates were provided twice a day to the mare, with adequate quantities for the energy intake to match mares needs. Time of the suckling and its duration were recorded, each foal being studied for two hours each two hours (i.e. 12 hours of continuous observation) using video recordings. 'Night' was considered to be from 10:00 PM to 7:00 AM and 62.5% of the foalings occurred during this period. Linear mixed models were computed for SF and SD with time after birth, birth period (day or night) and week of foaling as fixed effects and individual as random effect. When needed, Tukey post-hoc test was computed. Differences between day (7:00 AM to 10:00 PM) and night (10:00 PM to 7:00 AM) and between period with human activity (8:00 AM to 5:00 PM) and period without human activity (5:00 PM to 8:00 AM) were studied using Kruskal-Wallis test. Within the first hour of life, SF was significantly reduced ($P<0.05$) and, between 2 and 24 hours of life, SF became stable. Moreover, SF is increased ($P<0.05$) when foalings occurred at night. Regarding SD, however, an important inter-individual variation is observed with no significant differences between hours nor regarding other factors. Studying human activity, no difference appeared regarding neither SF nor SD. The same observation was made toward day vs night comparison even though a trend toward reduced SF at night was identified. In conclusion, foals seem to adopt from 24 hours of life a SD that can be found all along lactation and a SF that allow us to met their nutritional needs. A more complete study, however, is needed to be able to identify a difference between days and night suckling patterns.

Author index

B

D

H

O

P

S

T

U

V

W

X

Y

Z